STUDENT'S SOLUTIONS MANUAL

BEVERLY FUSFIELD

COLLEGE ALGEBRA & TRIGONOMETRY
SIXTH EDITION

and

PRECALCULUS
SIXTH EDITION

Margaret L. Lial
American River College

John Hornsby
University of New Orleans

David I. Schneider
University of Maryland

Callie J. Daniels
St. Charles Community College

PEARSON

Boston Columbus Indianapolis New York San Francisco
Amsterdam Cape Town Dubai London Madrid Milan Munich Paris Montreal Toronto
Delhi Mexico City Sao Paulo Sydney Hong Kong Seoul Singapore Taipei Tokyo

The author and publisher of this book have used their best efforts in preparing this book. These efforts include the development, research, and testing of the theories and programs to determine their effectiveness. The author and publisher make no warranty of any kind, expressed or implied, with regard to these programs or the documentation contained in this book. The author and publisher shall not be liable in any event for incidental or consequential damages in connection with, or arising out of, the furnishing, performance, or use of these programs.

Reproduced by Pearson from electronic files supplied by the author.

ISBN-13: 978-0-13-431434-1
ISBN-10: 0-13-431434-4

www.pearsonhighered.com

CONTENTS

7   Trigonometric Identities and Equations

8   Applications of Trigonometry

9   Systems and Matrices

Chapter R

REVIEW OF BASIC CONCEPTS

Section R.1   Sets

1. The elements of the set of natural number are {1, 2, 3, 4, ...}.

3. The set of all elements of the universal set U that do not belong to set A is the <u>complement</u> of set A.

5. The <u>union</u> of sets A and B is made up of all the elements belonging to set A or set B (or both).

7. Using set notation, the set $\{x|x$ is a natural number less than 6$\}$ is $\{1, 2, 3, 4, 5\}$.

9. $\{16, 18, 21, 50\} \cap \{15, 16, 17, 18\} = \{16, 18\}$

11. The set $\{4, 5, 6, ..., 15\}$ has a limited number of elements, so it is a finite set. Yes, 10 is an element of the set.

13. The set $\left\{1, \frac{1}{2}, \frac{1}{4}, \frac{1}{8}, ...\right\}$ has an unlimited number of elements, so it is an infinite set. No, 10 is not an element of the set.

15. The set $\{x \mid x$ is a natural number larger than 11$\}$, which can also be written as $\{11, 12, 13, 14, ...\}$, has an unlimited number of elements, so it is an infinite set. No, 10 is not an element of the set.

17. There are infinitely many fractions between 1 and 2, so $\{x \mid x$ is a fraction between 1 and 2$\}$ is an infinite set. No, 10 is not an element of the set.

19. The elements of the set $\{12, 13, 14, ..., 20\}$ are all the natural numbers from 12 to 20 inclusive. There are 9 elements in the set, $\{12, 13, 14, 15, 16, 17, 18, 19, 20\}$.

21. Each element of the set $\left\{1, \frac{1}{2}, \frac{1}{4}, ..., \frac{1}{32}\right\}$ after the first is found by multiplying the preceding number by $\frac{1}{2}$. There are 6 elements in the set, $\left\{1, \frac{1}{2}, \frac{1}{4}, \frac{1}{8}, \frac{1}{16}, \frac{1}{32}\right\}$.

23. To find the elements of the set $\{17, 22, 27, ..., 47\}$, start with 17 and add 5 to find the next number. There are 7 elements in the set, $\{17, 22, 27, 32, 37, 42, 47\}$.

25. When you list all elements in the set {all natural numbers greater than 8 and less than 15}, you obtain $\{9, 10, 11, 12, 13, 14\}$.

27. 6 is an element of the set $\{3, 4, 5, 6\}$, so we write $6 \in \{3, 4, 5, 6\}$.

29. 5 is not an element of $\{4, 6, 8, 10\}$, so we write $5 \notin \{4, 6, 8, 10\}$.

31. 0 is an element of $\{0, 2, 3, 4\}$, so we write $0 \in \{0, 2, 3, 4\}$.

33. $\{3\}$ is a subset of $\{2, 3, 4, 5\}$, not an element of $\{2, 3, 4, 5\}$, so we write $\{3\} \notin \{2, 3, 4, 5\}$.

35. $\{0\}$ is a subset of $\{0, 1, 2, 5\}$, not an element of $\{0, 1, 2, 5\}$, so we write $\{0\} \notin \{0, 1, 2, 5\}$.

37. 0 is not an element of $\varnothing$, because the empty set contains no elements. Thus, $0 \notin \varnothing$.

39. False. 3 is not one of the elements in $\{2, 5, 6, 8\}$.

41. True. 1 is one of the elements of $\{11, 5, 4, 3, 1\}$.

43. True. 9 is not one of the elements of $\{8, 5, 2, 1\}$.

45. True. Both sets contain exactly the same four elements.

47. False. These two sets are not equal because $\{5, 8, 9, 0\}$ contains the element 0, which is not an element of $\{5, 8, 9\}$.

49. True. 1 and 2 are the only natural numbers less than 3.

For Exercises 51−61, $A = \{2, 4, 6, 8, 10, 12\}$, $B = \{2, 4, 8, 10\}$, $C = \{4, 10, 12\}$, $D = \{2, 10\}$, and $U = \{2, 4, 6, 8, 10, 12, 14\}$.

51. True. This statement says "A is a subset of U." Because every element of A is also an element of U, the statement is true.

53. True. Because both elements of D, 2 and 10, are also elements of B, D is a subset of B.

55. False. Set A contains two elements, 6 and 12, that are not elements of B. Thus, A is not a subset of B.

57. True. The empty set is a subset of every set.

59. True. Because 4, 8, and 10 are all elements of B, $\{4, 8, 10\}$ is a subset of B.

61. False. Because B contains two elements, 4 and 8, that are not elements of D, B is not a subset of D.

63. Every element of $\{2, 4, 6\}$ is also an element of $\{2, 3, 4, 5, 6\}$, so $\{2, 4, 6\}$ is a subset of $\{2, 3, 4, 5, 6\}$.
We write $\{2, 4, 6\} \subseteq \{2, 3, 4, 5, 6\}$.

65. Because 0 is an element of $\{0, 1, 2\}$, but is not an element of $\{1, 2, 3, 4, 5\}$, $\{0, 1, 2\}$ is not a subset of $\{1, 2, 3, 4, 5\}$. We write $\{0, 1, 2\} \not\subseteq \{1, 2, 3, 4, 5\}$.

67. The empty set is a subset of every set, so $\varnothing \subseteq \{1, 4, 6, 8\}$.

69. True. 7 and 9 are the only elements belonging to both sets.

71. False. $\{1, 2, 7\} \cup \{1, 5, 9\} = \{1, 2, 5, 7, 9\}$, while $\{1, 2, 7\} \cap \{1, 5, 9\} = \{1\}$.

73. True. 2 is the only element belonging to both sets.

75. $\{3, 5, 9, 10\} \cap \varnothing = \{3, 5, 9, 10\}$
In order to belong to the intersection of two sets, an element must belong to both sets. Because the empty set contains no elements, $\{3, 5, 9, 10\} \cap \varnothing = \varnothing$, so the statement is false.

77. True. Because the two sets are equal, their union contains the same elements, namely 1, 2, and 4.

79. True.

For Exercises 81–109,
$U = \{0, 1, 2, 3, 4, 5, 6, 7, 8, 9, 10, 11, 12, 13\}$,
$M = \{0, 2, 4, 6, 8\}$, $N = \{1, 3, 5, 7, 9, 11, 13\}$,
$Q = \{0, 2, 4, 6, 8, 10, 12\}$, and $R = \{0, 1, 2, 3, 4\}$.

81. $M \cap R$
The only elements belonging to both M and R are 0, 2, and 4, so $M \cap R = \{0, 2, 4\}$.

83. $M \cup N$
The union of two sets contains all elements that belong to either set or to both sets.
$M \cup N = \{0, 1, 2, 3, 4, 5, 6, 7, 8, 9, 11, 13\}$

85. $M \cap N$
There are no elements which belong to both M and N, so $M \cap N = \varnothing$. M and N are disjoint sets.

87. $N \cup R = \{0, 1, 2, 3, 4, 5, 7, 9, 11, 13\}$

89. N'
The set N' is the complement of set N, which means the set of all elements in the universal set U that do not belong to N.
$N' = Q$ or $\{0, 2, 4, 6, 8, 10, 12\}$

91. $M' \cap Q$
First form M', the complement of M. M' contains all elements of U that are not elements of M. Thus, $M' = \{1, 3, 5, 7, 9, 10, 11, 12, 13\}$. Now form the intersection of M' and Q. Thus, we have $M' \cap Q = \{10, 12\}$.

93. $\varnothing \cap R$
Because the empty set contains no elements, there are no elements belonging to both $\varnothing$ and R. Thus, $\varnothing$ and R are disjoint sets, and $\varnothing \cap R = \varnothing$.

95. $N \cup \varnothing$
Because $\varnothing$ contains no elements, the only elements belonging to N or $\varnothing$ are the elements of N. Thus, $\varnothing$ and N are disjoint sets, and $N \cup \varnothing = N$ or $\{1, 3, 5, 7, 9, 11, 13\}$.

97. $(M \cap N) \cup R$
First, form the intersection of M and N. Because M and N have no common elements (they are disjoint), $M \cap N = \varnothing$. Thus, $(M \cap N) \cup R = \varnothing \cup R$. Now, because $\varnothing$ contains no elements, the only elements belonging to R or $\varnothing$ are the elements of R. Thus, $\varnothing$ and R are disjoint sets, and $\varnothing \cup R = R$ or $\{0, 1, 2, 3, 4\}$.

99. $(Q \cap M) \cup R$
First form the intersection of Q and M. We have $Q \cap M = \{0, 2, 4, 6, 8\} = M$. Now form the union of this set with R. We have $(Q \cap M) \cup R = M \cup R = \{0, 1, 2, 3, 4, 6, 8\}$.

101. $(M' \cup Q) \cap R$
First, find M', the complement of M. We have $M' = \{1, 3, 5, 7, 9, 10, 11, 12, 13\}$. Next, form the union of M' and Q. We have $M' \cup Q = \{0, 1, 2, 3, 4, 5, 6, 7, 8, 9, 10, 11, 12, 13\} = U$. Thus, we have $(M' \cup Q) \cap R = U \cap R = R$ or $\{0, 1, 2, 3, 4\}$.

103. $Q' \cap (N' \cap U)$

First, find Q', the complement of Q. We have $Q' = \{1, 3, 5, 7, 9, 11, 13\} = N$. Now find N', the complement of N. We have $N' = \{0, 2, 4, 6, 8, 10, 12\} = Q$. Next, form the intersection of N' and U. We have $N' \cap U = Q \cap U = Q$. Finally, we have $Q' \cap (N' \cap U) = Q' \cap Q = \varnothing$. Because the intersection of Q' and $(N' \cap U)$ is $\varnothing$, Q' and $(N' \cap U)$ are disjoint sets.

105. $\{x \mid x \in U, x \notin M\}$

This means all elements of U except those in set M. This gives the set $\{1, 3, 5, 7, 9, 10, 11, 12, 13\}$.

107. $\{x \mid x \in M \text{ and } x \in Q\}$

This means all elements that are members of both set M and set Q, or $M \cap Q$. Note that set M is a subset of set Q because all members of set M are included in set Q. Thus, the answer is set M or $\{0, 2, 4, 6, 8\}$.

109. $\{x \mid x \in M \text{ or } x \in Q\}$

This means all elements that are either members of set M or set Q, or $M \cup Q$. Note that set M is a subset of set Q because all members of set M are included in set Q. Thus, the answer is set Q or $\{0, 2, 4, 6, 8, 10, 12\}$.

Section R.2   Real Numbers and Their Properties

1. $\{0, 1, 2, 3, \ldots\}$ describes the set of <u>whole numbers</u>.

3. In the expression 6^3, 6 is the <u>base</u> and 3 is the <u>exponent</u>.

5. The distance on a number line from a number to 0 is the <u>absolute value</u> of that number.

7. $10^3 = 1000$

9. $|-4| = 4$

For Exercises 11–15,
$A = \left\{-6, -\frac{12}{4}, -\frac{5}{8}, -\sqrt{3}, 0, \frac{1}{4}, 1, 2\pi, 3, \sqrt{12}\right\}$.

11. 1 and 3 are natural numbers.

13. $-6, -\frac{12}{4}$ (or -3), 0, 1, and 3 are integers.

15. $-\sqrt{3}, 2\pi$ and $\sqrt{12}$ are irrational numbers.

17. $-2^4 = -(2 \cdot 2 \cdot 2 \cdot 2) = -16$

19. $(-2)^4 = (-2) \cdot (-2) \cdot (-2) \cdot (-2) = 16$

21. $(-3)^5 = (-3) \cdot (-3) \cdot (-3) \cdot (-3) \cdot (-3) = -243$

23. $-2 \cdot 3^4 = -2 \cdot (3 \cdot 3 \cdot 3 \cdot 3) = -2 \cdot 81 = -162$

25. $-2 \cdot 5 + 12 \div 3 = -10 + 12 \div 3$
$\qquad\qquad = -10 + 4 = -6$

27. $-4(9 - 8) + (-7)(2)^3 = -4(1) + (-7)(2)^3$
$\qquad\qquad = -4(1) + (-7) \cdot 8$
$\qquad\qquad = -4 + (-7) \cdot 8$
$\qquad\qquad = -4 + (-56) = -60$

29. $(4 - 2^3)(-2 + \sqrt{25}) = (4 - 8)(-2 + 5)$
$\qquad\qquad = (-4)(3) = -12$

31. $\left(-\frac{2}{9} - \frac{1}{4}\right) - \left[-\frac{5}{18} - \left(-\frac{1}{2}\right)\right]$
$= \left(-\frac{8}{36} - \frac{9}{36}\right) - \left(-\frac{5}{18} + \frac{9}{18}\right)$
$= \left(-\frac{17}{36}\right) - \left(\frac{4}{18}\right) = -\frac{17}{36} - \frac{8}{36} = -\frac{25}{36}$

33. $\dfrac{-8 + (-4)(-6) \div 12}{4 - (-3)} = \dfrac{-8 + 24 \div 12}{4 + 3}$
$\qquad\qquad = \dfrac{-8 + 2}{7}$
$\qquad\qquad = \dfrac{-6}{7} = -\dfrac{6}{7}$

For Exercises 35–47, $p = -4$, $q = 8$, and $r = -10$.

35. $-p^2 - 7q + r^2 = -(-4)^2 - 7 \cdot 8 + (-10)^2$
$\qquad\qquad = -(-4)^2 - 7 \cdot 8 + 100$
$\qquad\qquad = -16 - 7 \cdot 8 + 100$
$\qquad\qquad = -16 - 56 + 100$
$\qquad\qquad = -72 + 100 = 28$

37. $\dfrac{q + r}{q + p} = \dfrac{8 + (-10)}{8 + (-4)} = \dfrac{-2}{4} = -\dfrac{1}{2}$

39. $\dfrac{3q}{r} - \dfrac{5}{p} = \dfrac{3 \cdot 8}{-10} - \dfrac{5}{-4} = \dfrac{24}{-10} - \dfrac{5}{-4} = -\dfrac{12}{5} - \dfrac{5}{-4}$
$\qquad = -\dfrac{12}{5} + \dfrac{5}{4} = -\dfrac{48}{20} + \dfrac{25}{20} = -\dfrac{23}{20}$

41. $\dfrac{5r}{2p-3r} = \dfrac{5(-10)}{2(-4)-3(-10)} = \dfrac{-50}{-8+30}$

$\qquad = \dfrac{-50}{22} = -\dfrac{25}{11}$

43. $\dfrac{\dfrac{q}{2}-\dfrac{r}{3}}{\dfrac{3p}{4}+\dfrac{q}{8}} = \dfrac{\dfrac{8}{2}-\dfrac{-10}{3}}{\dfrac{3(-4)}{4}+\dfrac{8}{8}} = \dfrac{4+\dfrac{10}{3}}{-3+1} = \dfrac{\dfrac{22}{3}}{-2} = -\dfrac{11}{3}$

45. $\dfrac{-(p+2)^2-3r}{2-q} = \dfrac{-(-4+2)^2-3(-10)}{2-8}$

$\qquad = \dfrac{-(-2)^2-3(-10)}{-6}$

$\qquad = \dfrac{-4-3(-10)}{-6} = \dfrac{-4-(-30)}{-6}$

$\qquad = \dfrac{-4+30}{-6} = \dfrac{26}{-6} = -\dfrac{13}{3}$

47. $\dfrac{3p+3(4+p)^3}{r+8} = \dfrac{3(-4)+3(4+(-4))^3}{(-10)+8}$

$\qquad = \dfrac{-12+3(0)^3}{-2} = \dfrac{-12}{-2} = 6$

49. distributive   **51.** inverse

53. identity   **55.** commutative

57. associative   **59.** closure

61. No; in general $a-b \neq b-a$. Examples will vary, i.e. if $a = 15$ and $b = 0$, then $a - b = 15 - 0 = 15$, but $b - a = 0 - 15 = -15$.

63. $\dfrac{10}{11}(22z) = \left(\dfrac{10}{11} \cdot 22\right)z = 20z$

65. $(m+5)+6 = m+(5+6) = m+11$

67. $\dfrac{3}{8}\left(\dfrac{16}{9}y + \dfrac{32}{27}z - \dfrac{40}{9}\right)$

$\qquad = \dfrac{3}{8}\left(\dfrac{16}{9}y\right) + \dfrac{3}{8}\left(\dfrac{32}{27}z\right) - \dfrac{3}{8}\left(\dfrac{40}{9}\right)$

$\qquad = \left(\dfrac{3}{8}\cdot\dfrac{16}{9}\right)y + \left(\dfrac{3}{8}\cdot\dfrac{32}{27}\right)z - \dfrac{5}{3}$

$\qquad = \dfrac{2}{3}y + \dfrac{4}{9}z - \dfrac{5}{3}$

69. $8p - 14p = (8-14)p = -6p$

71. $-4(z-y) = -4z - (-4y) = -4z + 4y$

73. The process in your head should be the following:

$72 \cdot 17 + 28 \cdot 17 = (72+28)(17)$
$\qquad\qquad\qquad = (100)(17) = 1700$

75. The process in your head should be the following:

$123\dfrac{5}{8} \cdot 1\dfrac{1}{2} - 23\dfrac{5}{8} \cdot 1\dfrac{1}{2} = \left(123\dfrac{5}{8} - 23\dfrac{5}{8}\right)\left(1\dfrac{1}{2}\right)$

$\qquad\qquad = (100)\left(1\dfrac{1}{2}\right) = 150$

77. This statement is false because $|6-8| = |-2| = 2$ and $|6|-|8| = 6-8 = -2$. A corrected statement would be $|6-8| \neq |6|-|8|$ or $|6-8| = |8|-|6|$

79. This statement is true because $|-5|\cdot|6| = 5 \cdot 6 = 30$ and $|-5\cdot6| = |-30| = 30$.

81. This statement is false. For example if you let $a = 2$ and $b = 6$ then $|2-6| = |-4| = 4$ and $|a|-|b| = |2|-|6| = 2-6 = -4$. A corrected statement is $|a-b| = |b|-|a|$, if $b > a > 0$.

83. $|-10| = 10$

85. $-\left|\dfrac{4}{7}\right| = -\dfrac{4}{7}$

87. $-|-8| = -8$

89. $P_d = |P-120| = |116-120| = |-4| = 4$

The P_d value for a woman whose actual systolic blood pressure is 116 and whose normal value should be 120 is 4.

For Exercises 91–97, $x = -4$ and $y = 2$.

91. $|3x-2y| = |3(-4)-2(2)|$
$\qquad\qquad = |-12-4| = |-16| = 16$

93. $|-3x+4y| = |-3(-4)+4(2)|$
$\qquad\qquad = |12+8| = |20| = 20$

95. $\dfrac{2|y|-3|x|}{|xy|} = \dfrac{2|2|-3|-4|}{|-4(2)|}$

$\qquad = \dfrac{2(2)-3(4)}{|-8|} = \dfrac{4-12}{8} = \dfrac{-8}{8} = -1$

97. $\dfrac{|-8y+x|}{-|x|} = \dfrac{|-8(2)+(-4)|}{-|-4|} = \dfrac{|-16+(-4)|}{-|-4|} = \dfrac{|-20|}{-4} = \dfrac{20}{-4} = -5$

99. True because $|25| = 25$ and $|-25| = 25$.

101. False. $|5+(-13)| = |-8| = 8$ while $|5|+|-13| = 5+13 = 18$.

103. True because $|11|\cdot|-6| = 11\cdot 6 = 66$ and $|-66| = 66$.

105. $d(P,Q) = |-1-(-4)| = |-1+4| = |3| = 3$ or $d(P,Q) = |-4-(-1)| = |-4+1| = |-3| = 3$

107. $d(Q,R) = |8-(-1)| = |8+1| = |9| = 9$ or $d(Q,R) = |-1-8| = |-9| = 9$

109. $xy > 0$ if x and y have the same sign.

111. $\dfrac{x}{y} < 0$ if x and y have different signs.

113. Because x^3 has the same sign as x, $\dfrac{x^3}{y} > 0$ if x and y have the same sign.

115. Because $|-18-(+1)| = |-19| = 19$ and $|1-(-18)| = |19| = 19$, the number of strokes between their scores is 19.

117. $\text{BAC} = 48 \times 3.2 \times 0.075 \div 190 - 2 \times 0.015 \approx 0.031$

119. $\text{BAC} = 3 \times 20 \times 3.8 \times 0.075 \div 200 - 4 \times 0.015 \approx 0.026$
If the man's weight was greater, the BAC would be lower. For example, if the man weighed 250 pounds, his BAC would be $3 \times 20 \times 3.8 \times 0.075 \div 250 - 4 \times 0.015 = 0.0084$.

For exercises 121−131, we use the formula $R = \dfrac{\left(250\cdot\dfrac{C}{A}\right) + \left(1000\cdot\dfrac{T}{A}\right) + \left(12.5\cdot\dfrac{Y}{A}\right) + 6.25 - \left(1250\cdot\dfrac{I}{A}\right)}{3}$, where

$0 \le \frac{C}{A} \le 0.775$, $0 \le \frac{T}{A} \le 0.11875$, $0 \le \frac{Y}{A} \le 12.5$, and $0 \le \frac{I}{A} \le 0.095$.

121. First, find the value of the numerator.

$\left(250\cdot\dfrac{C}{A}\right) + \left(1000\cdot\dfrac{T}{A}\right) + \left(12.5\cdot\dfrac{Y}{A}\right) + 6.25 - \left(1250\cdot\dfrac{I}{A}\right)$

$= \left(250\cdot\dfrac{304}{435}\right) + \left(1000\cdot\dfrac{34}{435}\right) + \left(12.5\cdot\dfrac{3705}{435}\right) + 6.25 - \left(1250\cdot\dfrac{9}{435}\right) \approx 339.73$

$R \approx \dfrac{339.73}{3} \approx 113.2$

123. First, find the value of the numerator.

$\left(250\cdot\dfrac{C}{A}\right) + \left(1000\cdot\dfrac{T}{A}\right) + \left(12.5\cdot\dfrac{Y}{A}\right) + 6.25 - \left(1250\cdot\dfrac{I}{A}\right)$

$= \left(250\cdot\dfrac{408}{608}\right) + \left(1000\cdot\dfrac{32}{608}\right) + \left(12.5\cdot 6\dfrac{4952}{408}\right) + 6.25 - \left(1250\cdot\dfrac{9}{608}\right) \approx 309.95$

$R \approx \dfrac{309.95}{3} = 103.3$

125. First, find the value of the numerator.

$$\left(250 \cdot \frac{C}{A}\right) + \left(1000 \cdot \frac{T}{A}\right) + \left(12.5 \cdot \frac{Y}{A}\right) + 6.25 - \left(1250 \cdot \frac{I}{A}\right)$$

$$= \left(250 \cdot \frac{373}{582}\right) + \left(1000 \cdot \frac{33}{582}\right) + \left(12.5 \cdot \frac{4109}{582}\right) + 6.25 - \left(1250 \cdot \frac{9}{582}\right) \approx 292.10$$

$$R \approx \frac{292.10}{3} \approx 97.4$$

127. First, find the value of the numerator.

$$\left(250 \cdot \frac{C}{A}\right) + \left(1000 \cdot \frac{T}{A}\right) + \left(12.5 \cdot \frac{Y}{A}\right) + 6.25 - \left(1250 \cdot \frac{I}{A}\right)$$

$$= \left(250 \cdot \frac{380}{616}\right) + \left(1000 \cdot \frac{40}{616}\right) + \left(12.5 \cdot \frac{4761}{616}\right) + 6.25 - \left(1250 \cdot \frac{16}{616}\right) \approx 289.55$$

$$R \approx \frac{289.55}{3} \approx 96.5$$

129. First, find the value of the numerator.

$$\left(250 \cdot \frac{C}{A}\right) + \left(1000 \cdot \frac{T}{A}\right) + \left(12.5 \cdot \frac{Y}{A}\right) + 6.25 - \left(1250 \cdot \frac{I}{A}\right)$$

$$= \left(250 \cdot \frac{197}{312}\right) + \left(1000 \cdot \frac{17}{312}\right) + \left(12.5 \cdot \frac{2483}{312}\right) + 6.25 - \left(1250 \cdot \frac{8}{312}\right) \approx 286.02$$

$$R \approx \frac{286.02}{3} \approx 95.3$$

131. First, find the value of the numerator.

$$\left(250 \cdot \frac{C}{A}\right) + \left(1000 \cdot \frac{T}{A}\right) + \left(12.5 \cdot \frac{Y}{A}\right) + 6.25 - \left(1250 \cdot \frac{I}{A}\right)$$

$$= \left(250 \cdot \frac{415}{628}\right) + \left(1000 \cdot \frac{28}{628}\right) + \left(12.5 \cdot \frac{4694}{628}\right) + 6.25 - \left(1250 \cdot \frac{14}{628}\right) \approx 281.61$$

$$R \approx \frac{281.61}{3} \approx 93.9$$

133. First, find the value of the numerator.

$$\left(250 \cdot \frac{C}{A}\right) + \left(1000 \cdot \frac{T}{A}\right) + \left(12.5 \cdot \frac{Y}{A}\right) + 6.25 - \left(1250 \cdot \frac{I}{A}\right)$$

$$= \left(250 \cdot \frac{343}{502}\right) + \left(1000 \cdot \frac{45}{502}\right) + \left(12.5 \cdot \frac{4643}{502}\right) + 6.25 - \left(1250 \cdot \frac{6}{502}\right) \approx 367.38$$

$$R \approx \frac{367.38}{3} \approx 122.5$$

135. First, find the value of the numerator.

$$\left(250 \cdot \frac{C}{A}\right) + \left(1000 \cdot \frac{T}{A}\right) + \left(12.5 \cdot \frac{Y}{A}\right) + 6.25 - \left(1250 \cdot \frac{I}{A}\right)$$

$$= \left(250 \cdot \frac{324}{492}\right) + \left(1000 \cdot \frac{36}{492}\right) + \left(12.5 \cdot \frac{3900}{492}\right) + 6.25 - \left(1250 \cdot \frac{4}{492}\right) \approx 332.98$$

$$R \approx \frac{332.98}{3} \approx 110.0$$

Section R.3   Polynomials

1. The polynomial $2x^5 - x + 4$ is a trinomial of degree 5.

3. A polynomial containing exactly two terms is a binomial.

5. A convenient way to find the product of two binomials is to use the FOIL method.

7. True

9. False. $(x + y)^2 = x^2 + 2xy + y^2$

11. $\left(-4x^5\right)\left(4x^2\right) = (-4 \cdot 4)\left(x^5 x^2\right)$
$$= -16x^{5+2} = -16x^7$$

13. $n^6 \cdot n^4 \cdot n = n^{6+4+1} = n^{11}$

15. $9^3 \cdot 9^5 = 9^{3+5} = 9^8$

17. $\left(-3m^4\right)\left(6m^2\right)\left(-4m^5\right)$
$$= \left[(-3)(6)(-4)\right]\left(m^4 m^2 m^5\right)$$
$$= 72m^{4+2+5} = 72m^{11}$$

19. $\left(5x^2 y\right)\left(-3x^3 y^4\right) = \left[5(-3)\right]\left(x^2 x^3\right)\left(y y^4\right)$
$$= -15x^{2+3} y^{1+4} = -15x^5 y^5$$

21. $\left(\dfrac{1}{2} mn\right)\left(8m^2 n^2\right) = \left(\dfrac{1}{2} \cdot 8\right)\left(mm^2\right)\left(nn^2\right)$
$$= 4m^{1+2} n^{1+2} = 4m^3 n^3$$

23. $(2^2)^5 = 2^{2 \cdot 5} = 2^{10}$

25. $\left(-6x^2\right)^3 = (-6)^3 \left(x^2\right)^3 = (-6)^3 x^6 = -216x^6$

27. $-(4m^3 n^0)^2 = -\left[4^2 (m^3)^2 (n^0)^2\right]$
$$= -4^2 m^{3 \cdot 2} n^{0 \cdot 2}$$
$$= -4^2 m^6 n^0 = -\left(4^2\right) m^6 \cdot 1$$
$$= -4^2 m^6 = -16m^6$$

29. $\left(\dfrac{r^8}{s^2}\right)^3 = \dfrac{(r^8)^3}{(s^2)^3} = \dfrac{r^{8 \cdot 3}}{s^{2 \cdot 3}} = \dfrac{r^{24}}{s^6}$

31. $\left(\dfrac{-4m^2}{tp^2}\right)^4 = \dfrac{(-4)^4 (m^2)^4}{t^4 \left(p^2\right)^4} = \dfrac{(-4)^4 m^8}{t^4 p^8} = \dfrac{256m^8}{t^4 p^8}$

33. $-\left(\dfrac{x^3 y^5}{z}\right)^0 = -1$

35. **(a)** B; $6^0 = 1$ **(b)** C; $-6^0 = -1$

　　　(c) B; $(-6)^0 = 1$ **(d)** C; $-(-6)^0 = -1$

37. $-5x^{11}$ is a polynomial of degree 11. It is a monomial because it has one term.

39. $6x + 3x^4$ is a polynomial. It has degree 4 because the term $3x^4$ has the greatest degree. It is a binomial because there are two terms.

41. $-7z^5 - 2z^3 + 1$ is a polynomial. It has degree 5 because the term $-7z^5$ has the greatest degree. It is a trinomial because there are three terms.

43. $15a^2 b^3 + 12a^3 b^8 - 13b^5 + 12b^6$ is a polynomial. It has degree 11 because the term $12a^3 b^8$ has the greatest degree. It is not a monomial, binomial, or trinomial.

45. $\dfrac{3}{8} x^5 - \dfrac{1}{x^2} + 9$ is not a polynomial because one term has a variable in the denominator.

47. 5 is a polynomial of degree 0. It is a monomial.

49. $\left(5x^2 - 4x + 7\right) + \left(-4x^2 + 3x - 5\right)$
$$= \left[5 + (-4)\right] x^2 + (-4 + 3) x + \left[7 + (-5)\right]$$
$$= 1 \cdot x^2 + (-1) x + 2 = x^2 - x + 2$$

51. $2\left(12y^2 - 8y + 6\right) - 4\left(3y^2 - 4y + 2\right)$
$$= 2\left(12y^2\right) - 2(8y) + 2(6) - 4\left(3y^2\right)$$
$$\qquad\qquad\qquad - 4(-4y) - 4 \cdot 2$$
$$= 24y^2 - 16y + 12 - 12y^2 + 16y - 8$$
$$= 12y^2 + 4$$

53. $\left(6m^4 - 3m^2 + m\right) - \left(2m^3 + 5m^2 + 4m\right) + \left(m^2 - m\right)$
$$= 6m^4 - 3m^2 + m - 2m^3 - 5m^2 - 4m + m^2 - m$$
$$= 6m^4 - 2m^3 + (-3 - 5 + 1) m^2 + (1 - 4 - 1) m$$
$$= 6m^4 - 2m^3 + (-7) m^2 + (-4) m$$
$$= 6m^4 - 2m^3 - 7m^2 - 4m$$

55. $(4r - 1)(7r + 2) = 4r(7r) + 4r(2) - 1(7r) - 1(2)$
$$= 28r^2 + 8r - 7r - 2$$
$$= 28r^2 + r - 2$$

57. $x^2\left(3x-\dfrac{2}{3}\right)\left(5x+\dfrac{1}{3}\right)$

$= x^2\left[\left(3x-\dfrac{2}{3}\right)\left(5x+\dfrac{1}{3}\right)\right]$

$= x^2\left[(3x)(5x)+(3x)\left(\dfrac{1}{3}\right)-\dfrac{2}{3}(5x)-\dfrac{2}{3}\left(\dfrac{1}{3}\right)\right]$

$= x^2\left(15x^2+x-\dfrac{10}{3}x-\dfrac{2}{9}\right)$

$= x^2\left(15x^2+\dfrac{3}{3}x-\dfrac{10}{3}x-\dfrac{2}{9}\right)$

$= x^2\left(15x^2-\dfrac{7}{3}x-\dfrac{2}{9}\right)=15x^4-\dfrac{7}{3}x^3-\dfrac{2}{9}x^2$

59. $4x^2\left(3x^3+2x^2-5x+1\right)$

$= 4x^2\left(3x^3\right)+4x^2\left(2x^2\right)-4x^2\left(5x\right)+4x^2\cdot 1$

$= 12x^5+8x^4-20x^3+4x^2$

61. $(2z-1)\left(-z^2+3z-4\right)$

$= (2z-1)\left(-z^2\right)+(2z-1)(3z)-(2z-1)(4)$

$= 2z\left(-z^2\right)-1\left(-z^2\right)+2z(3z)-1(3z)$

$\quad -(2z)(4)-(-1)(4)$

$= -2z^3+z^2+6z^2-3z-8z-(-4)$

$= -2z^3+7z^2-11z+4$

We may also multiply vertically.

$$
\begin{array}{r}
-z^2+3z-4 \\
2z-1 \\
\hline
z^2-3z+4 \quad \leftarrow -1(-z^2+3z-4) \\
-2z^3+6z^2-8z \qquad \leftarrow 2z(-z^2+3z-4) \\
\hline
-2z^3+7z^2-11z+4
\end{array}
$$

63. $(m-n+k)(m+2n-3k)$

$= (m-n+k)(m)+(m-n+k)(2n)$

$\qquad -(m-n+k)(3k)$

$= m^2-mn+km+2mn-2n^2$

$\qquad +2kn-3km+3kn-3k^2$

$= m^2+mn-2n^2-2km+5kn-3k^2$

We may also multiply vertically.

$$
\begin{array}{r}
m-\ n+\ k \\
m+2n-3k \\
\hline
-3km+3kn-3k^2 \\
2mn-2n^2 \qquad\quad +2kn \\
m^2-mn \qquad\quad +km \\
\hline
m^2+mn-2n^2-2km+5kn-3k^2
\end{array}
$$

65. $(2x+3)(2x-3)\left(4x^2-9\right)$

Note that $(2x+3)(2x-3)$ is a special product, the difference of the sum and difference of squares.

$(2x+3)(2x-3)\left(4x^2-9\right)=\left(4x^2-9\right)\left(4x^2-9\right)$

Note that $\left(4x^2-9\right)\left(4x^2-9\right)$ is a special product, the square of a binomial.

$\left(4x^2-9\right)\left(4x^2-9\right)=16x^4-72x^2+81$

Thus,

$(2x+3)(2x-3)\left(4x^2-9\right)=16x^4-72x^2+81.$

67. $(x+1)(x+1)(x-1)(x-1)$

$= \left[(x+1)(x+1)\right]\left[(x-1)(x-1)\right]$

$= \left(x^2+2x+1\right)\left(x^2-2x+1\right)$

$= x^2\left(x^2-2x+1\right)+2x\left(x^2-2x+1\right)$

$\qquad +1\left(x^2-2x+1\right)$

$= x^4-2x^3+x^2+2x^3-4x^2+2x+x^2-2x+1$

$= x^4-2x^2+1$

Alternatively,

$(x+1)(x+1)(x-1)(x-1)$

$= (x+1)(x-1)(x+1)(x-1)$

$= \left(x^2-1\right)\left(x^2-1\right)=x^4-2x^2+1$

69. $(2m+3)(2m-3)=(2m)^2-3^2$

$\qquad\qquad\qquad\quad = 4m^2-9$

71. $(4x^2-5y)(4x^2+5y)=\left(4x^2\right)^2-(5y)^2$

$\qquad\qquad\qquad\qquad\quad = 16x^4-25y^2$

73. $(4m+2n)^2=(4m)^2+2(4m)(2n)+(2n)^2$

$\qquad\qquad\quad = 16m^2+16mn+4n^2$

75. $(5r-3t^2)^2=(5r)^2-2(5r)(3t^2)+(3t^2)^2$

$\qquad\qquad\quad = 25r^2-30rt^2+9t^4$

77. $\left[(2p-3)+q\right]^2$

$= (2p-3)^2+2(2p-3)(q)+q^2$

$= (2p)^2-2(2p)(3)+(3)^2+4pq-6q+q^2$

$= 4p^2-12p+9+4pq-6q+q^2$

79. $[(3q+5)-p][(3q+5)+p]$

$= (3q+5)^2-p^2$

$= [(3q)^2+2(3q)(5)+5^2]-p^2$

$= 9q^2+30q+25-p^2$

81. $[(3a+b)-1]^2 = (3a+b)^2 - 2(3a+b)(1) + 1^2$
$= (9a^2 + 6ab + b^2) - 2(3a+b) + 1$
$= 9a^2 + 6ab + b^2 - 6a - 2b + 1$

83. $(y+2)^3 = (y+2)^2(y+2)$
$= (y^2 + 4y + 4)(y+2)$
$= y^3 + 4y^2 + 4y + 2y^2 + 8y + 8$
$= y^3 + 6y^2 + 12y + 8$

85. $(q-2)^4 = (q-2)^2(q-2)^2$
$= (q^2 - 4q + 4)(q^2 - 4q + 4)$
$= q^4 - 4q^3 + 4q^2 - 4q^3 + 16q^2$
$\quad - 16q + 4q^2 - 16q + 16$
$= q^4 - 8q^3 + 24q^2 - 32q + 16$

87. $(p^3 - 4p^2 + p) - (3p^2 + 2p + 7)$
$= p^3 - 4p^2 + p - 3p^2 - 2p - 7$
$= p^3 - 7p^2 - p - 7$

89. $(7m + 2n)(7m - 2n) = (7m)^2 - (2n)^2$
$= 49m^2 - 4n^2$

91. $-3(4q^2 - 3q + 2) + 2(-q^2 + q - 4)$
$= -12q^2 + 9q - 6 - 2q^2 + 2q - 8$
$= -14q^2 + 11q - 14$

93. $p(4p - 6) + 2(3p - 8) = 4p^2 - 6p + 6p - 16$
$= 4p^2 - 16$

95. $-y(y^2 - 4) + 6y^2(2y - 3)$
$= -y^3 + 4y + 12y^3 - 18y^2$
$= 11y^3 - 18y^2 + 4y$

97.
$$
\begin{array}{r}
2x^5 + 7x^4 - 5x^2 + 7 \\
-2x^2 \overline{)-4x^7 - 14x^6 + 10x^4 - 14x^2} \\
\underline{-4x^7} \\
-14x^6 \\
\underline{-14x^6} \\
10x^4 \\
\underline{10x^4} \\
-14x^2 \\
\underline{-14x^2} \\
0
\end{array}
$$

$\dfrac{-4x^7 - 14x^6 + 10x^4 - 14x^2}{-2x^2}$
$\quad = 2x^5 + 7x^4 - 5x^2 + 7$

99.
$$
\begin{array}{r}
4x^2 + 5x + 10 \\
x - 2 \overline{)4x^3 - 3x^2 + 0x + 1} \\
\underline{4x^3 - 8x^2} \\
5x^2 + 0x \\
\underline{5x^2 - 10x} \\
10x + 1 \\
\underline{10x - 20} \\
21
\end{array}
$$

$\dfrac{4x^3 - 3x^2 + 1}{x - 2} = 4x^2 + 5x + 10 + \dfrac{21}{x - 2}$

101.
$$
\begin{array}{r}
2m^2 + m - 2 \\
3m + 2 \overline{)6m^3 + 7m^2 - 4m + 2} \\
\underline{6m^3 + 4m^2} \\
3m^2 - 4m \\
\underline{3m^2 + 2m} \\
-6m + 2 \\
\underline{-6m - 4} \\
6
\end{array}
$$

$\dfrac{6m^3 + 7m^2 - 4m + 2}{3m + 2} = 2m^2 + m - 2 + \dfrac{6}{3m + 2}$

103.
$$
\begin{array}{r}
x^2 \qquad + 2 \\
x^2 + 0x + 3 \overline{)x^4 + 0x^3 + 5x^2 + 5x + 27} \\
\underline{x^4 + 0x^3 + 3x^2} \\
2x^2 + 5x + 27 \\
\underline{2x^2 + 0x + 6} \\
5x + 21
\end{array}
$$

$\dfrac{x^4 + 5x^2 + 5x + 27}{x^2 + 3} = x^2 + 2 + \dfrac{5x + 21}{x^2 + 3}$

105. (a) The area of the largest square is
$s^2 = (x + y)^2$.

(b) The areas of the two squares are
x^2 and y^2. The area of each rectangle is
xy. Therefore, the area of the largest
square can be written as $x^2 + 2xy + y^2$.

(c) Answers will vary. Sample answer: The
expressions are equivalent because they
represent the same area.

(d) It reinforces the special product for
squaring a binomial:
$(x + y)^2 = x^2 + 2xy + y^2$.

107. (a) The volume is
$$V = \frac{1}{3}h(a^2 + ab + b^2)$$
$$= \frac{1}{3}(200)(314^2 + 314 \times 756 + 756^2)$$
$$\approx 60{,}501{,}000 \text{ ft}^3$$

(b) The shape becomes a rectangular box with a square base. Its volume is given by length × width × height or $b^2 h$.

(c) If we let $a = b$, then
$$V = \frac{1}{3}h(a^2 + ab + b^2) \text{ becomes}$$
$$V = \frac{1}{3}h(b^2 + bb + b^2), \text{ which simplifies}$$
to $V = hb^2$. Yes, the Egyptian formula gives the same result.

109. $x = 1950$
$$0.001259(1950)^2 - 5.039(1950) + 5044 \approx 5.3$$
The formula is 0.3 low.

111. $x = 1990$
$$0.001259(1990)^2 - 5.039(1990) + 5044 \approx 2.2$$
The formula is 0.2 low.

113. $\left(0.25^3\right)\left(400^3\right) = \left[(0.25)(400)\right]^3$
$$= 100^3 = 1{,}000{,}000$$

115. $\dfrac{4.2^5}{2.1^5} = \left(\dfrac{4.2}{2.1}\right)^5 = 2^5 = 32$

117. $99 \times 101 = (100 - 1)(100 + 1) = 100^2 - 1^2$
$$= 10{,}000 - 1 = 9999$$

119. $102^2 = (100 + 2)^2 = 100^2 + 2(100)(2) + 2^2$
$$= 10{,}000 + 400 + 4 = 10{,}404$$

Section R.4   Factoring Polynomials

1. The process of finding polynomials whose product equals a given polynomial is called <u>factoring</u>.

3. Factoring is the opposite of <u>multiplying</u>.

5. There is no factoring pattern for a <u>sum of squares</u> in the real number system. In particular, $x^2 + y^2$ does not factor as $(x + y)^2$, for real numbers x and y.

7. (a) B; $(2x - 3)(4x^2 + 6x + 9) = 8x^3 - 27$.

(b) C; $(2x + 3)(4x^2 - 6x + 9) = 8x^3 + 27$

(c) A; $(3 - 2x)(9 + 6x + 4x^2) = 27 - 8x^3$

9. B: $x^4 - 1 = (x^2 + 1)(x + 1)(x - 1)$. Choice A is not a complete factorization, because $x^2 - 1$ can be factored as $(x + 1)(x - 1)$. The other choices are not correct factorizations of $x^4 - 1$.

11. The greatest common factor is 12.
$$12m + 60 = 12(m) + 12(5) = 12(m + 5)$$

13. The greatest common factor is $8k$.
$$8k^3 + 24k = 8k(k^2) + 8k(3) = 8k(k^2 + 3)$$

15. The greatest common factor is xy.
$$xy - 5xy^2 = xy \cdot 1 - xy(5y) = xy(1 - 5y)$$

17. The greatest common factor is $-2p^2 q^4$.
$$-4p^3 q^4 - 2p^2 q^5$$
$$= \left(-2p^2 q^4\right)(2p) + \left(-2p^2 q^4\right)(q)$$
$$= -2p^2 q^4(2p + q)$$

19. The greatest common factor is $4k^2 m^3$.
$$4k^2 m^3 + 8k^4 m^3 - 12k^2 m^4$$
$$= \left(4k^2 m^3\right)(1) + \left(4k^2 m^3\right)(2k^2) - \left(4k^2 m^3\right)(3m)$$
$$= 4k^2 m^3\left(1 + 2k^2 - 3m\right)$$

21. The greatest common factor is $2(a + b)$
$$2(a + b) \ + \ 4m(a + b)$$
$$= \left[2(a + b)\right](1) + \left[2(a + b)\right](2m)$$
$$= \ 2(a + b)(1 \ + \ 2m)$$

23. $(5r - 6)(r + 3) - (2r - 1)(r + 3)$
$$= (r + 3)\left[(5r - 6) - (2r - 1)\right]$$
$$= (r + 3)\left[5r - 6 - 2r + 1\right]$$
$$= (r + 3)(3r - 5)$$

25. $2(m - 1) - 3(m - 1)^2 + 2(m - 1)^3$
$$= (m - 1)\left[2 - 3(m - 1) + 2(m - 1)^2\right]$$
$$= (m - 1)\left[2 - 3m + 3 + 2(m^2 - 2m + 1)\right]$$
$$= (m - 1)(2 - 3m + 3 + 2m^2 - 4m + 2)$$
$$= (m - 1)(2m^2 - 7m + 7)$$

27. The completely factored form of $4x^2y^5 - 8xy^3$ is $4xy^3\left(xy^2 - 2\right)$. The student did not factor out the greatest common factor, $4xy^3$.

29. $6st + 9t - 10s - 15 = (6st + 9t) - (10s + 15)$
$= 3t(2s + 3) - 5(2s + 3)$
$= (2s + 3)(3t - 5)$

31. $2m^4 + 6 - am^4 - 3a$
$= (2m^4 + 6) - (am^4 + 3a)$
$= 2(m^4 + 3) - a(m^4 + 3) = (m^4 + 3)(2 - a)$

33. $p^2q^2 - 10 - 2q^2 + 5p^2$
$= p^2q^2 - 2q^2 + 5p^2 - 10$
$= q^2(p^2 - 2) + 5(p^2 - 2)$
$= (p^2 - 2)(q^2 + 5)$

35. The positive factors of 6 could be 2 and 3, or 1 and 6. Because the middle term is negative, we know the factors of 4 must both be negative. As factors of 4, we could have -1 and -4, or -2 and -2. Try different combinations of these factors until the correct one is found.
$6a^2 - 11a + 4 = (2a - 1)(3a - 4)$

37. The positive factors of 3 are 1 and 3. Because the middle term is positive, we know the factors of 8 must both be positive. As factors of 8, we could have 1 and 8, or 2 and 4. Try different combinations of these factors until the correct one is found.
$3m^2 + 14m + 8 = (3m + 2)(m + 4)$

39. The positive factors of 15 are 1 and 15, or 3 and 5. Because the middle term is positive, we know the factors of 8 must both be positive. As factors of 8, we could have 1 and 8, or 2 and 4. Trying different combinations of these factors we find that $15p^2 + 24p + 8$ is prime.

41. Factor out the greatest common factor, $2a$:
$12a^3 + 10a^2 - 42a = 2a(6a^2 + 5a - 21)$. Now factor the trinomial by trial and error:
$6a^2 + 5a - 21 = (3a + 7)(2a - 3)$. Thus,
$12a^3 + 10a^2 - 42a = 2a(3a + 7)(2a - 3)$.

43. The positive factors of 6 could be 2 and 3, or 1 and 6. As factors of -6, we could have -1 and 6, -6 and 1, -2 and 3, or -3 and 2. Try different combinations of these factors until the correct one is found.
$6k^2 + 5kp - 6p^2 = (2k + 3p)(3k - 2p)$

45. The positive factors of 5 can only be 1 and 5. As factors of -6, we could have -1 and 6, -6 and 1, -2 and 3, or -3 and 2. Try different combinations of these factors until the correct one is found.
$5a^2 - 7ab - 6b^2 = (5a + 3b)(a - 2b)$

47. The positive factors of 12 could be 4 and 3, 2 and 6, or 1 and 12. The factors of -1 are 1 and -1. Try different combination of these factors until the correct one is found.
$12x^2 - xy - y^2 = (4x + y)(3x - y)$

49. Factor out the greatest common factor, $2a^2$:
$24a^4 + 10a^3b - 4a^2b^2 = 2a^2\left(12a^2 + 5ab - 2b^2\right)$
Now factor the trinomial by trial and error:
$12a^2 + 5ab - 2b^2 = (4a - b)(3a + 2b)$
Thus,
$24a^4 + 10a^3b - 4a^2b^2 = 2a^2\left(12a^2 + 5ab - 2b^2\right)$
$= 2a^2(4a - b)(3a + 2b)$

51. $9m^2 - 12m + 4 = (3m)^2 - 12m + 2^2$
$= (3m)^2 - 2(3m)(2) + 2^2$
$= (3m - 2)^2$

53. $32a^2 + 48ab + 18b^2$
$= 2(16a^2 + 24ab + 9b^2)$
$= 2[(4a)^2 + 24ab + (3b)^2]$
$= 2[(4a)^2 + 2(4a)(3b) + (3b)^2] = 2(4a + 3b)^2$

55. $4x^2y^2 + 28xy + 49 = (2xy)^2 + 28xy + 7^2$
$= (2xy)^2 + 2(2xy)(7) + 7^2$
$= (2xy + 7)^2$

57. $(a - 3b)^2 - 6(a - 3b) + 9$
$= (a - 3b)^2 - 6(a - 3b) + 3^2$
$= (a - 3b)^2 - 2(a - 3b)(3) + 3^2$
$= [(a - 3b) - 3]^2 = (a - 3b - 3)^2$

59. $9a^2 - 16 = (3a)^2 - 4^2$
$= (3a + 4)(3a - 4)$

61. $x^4 - 16 = (x^2 + 4)(x^2 - 4)$
$$= (x^2 + 4)(x + 2)(x - 2)$$

63. $25s^4 - 9t^2 = (5s^2)^2 - (3t)^2$
$$= (5s^2 + 3t)(5s^2 - 3t)$$

65. $(a + b)^2 - 16 = (a + b)^2 - 4^2$
$$= [(a + b) + 4][(a + b) - 4]$$
$$= (a + b + 4)(a + b - 4)$$

67. $p^4 - 625 = (p^2)^2 - 25^2 = (p^2 + 25)(p^2 - 25)$
$$= (p^2 + 25)(p^2 - 5^2)$$
$$= (p^2 + 25)(p + 5)(p - 5)$$

Note that $p^2 + 25$ is a prime factor.

69. $x^2 - 8x + 16 - y^2 = (x^2 - 8x + 16) - y^2$
$$= (x - 4)^2 - y^2$$
$$= (x - 4 + y)(x - 4 - y)$$

71. $y^2 - x^2 + 12x - 36 = y^2 - (x^2 - 12x + 36)$
$$= y^2 - (x - 6)^2$$
$$= (y + (x - 6))(y - (x - 6))$$
$$= (y + x - 6)(y - x + 6)$$

73. $8 - a^3 = 2^3 - a^3 = (2 - a)(2^2 + 2 \cdot a + a^2)$
$$= (2 - a)(4 + 2a + a^2)$$

75. $125x^3 - 27 = (5x)^3 - 3^3$
$$= (5x - 3)\left[(5x)^2 + 5x \cdot 3 + 3^2\right]$$
$$= (5x - 3)(25x^2 + 15x + 9)$$

77. $27y^9 + 125z^6$
$$= (3y^3)^3 + (5z^2)^3$$
$$= (3y^3 + 5z^2)\left[(3y^3)^2 - (3y^3)(5z^2) + (5z^2)^2\right]$$
$$= (3y^3 + 5z^2)(9y^6 - 15y^3z^2 + 25z^4)$$

79. $(r + 6)^3 - 216$
$$= (r + 6)^3 - 6^3$$
$$= \left[(r + 6) - 6\right]\left[(r + 6)^2 + (r + 6)(6) + 6^2\right]$$
$$= \left[(r + 6) - 6\right]\left[r^2 + 12r + 36 + (r + 6)(6) + 6^2\right]$$
$$= \left[r + 6 - 6\right]\left[r^2 + 12r + 36 + 6r + 36 + 36\right]$$
$$= r(r^2 + 18r + 108)$$

81. $27 - (m + 2n)^3$
$$= 3^3 - (m + 2n)^3$$
$$= \left[3 - (m + 2n)\right] \cdot \left[3^2 + (3)(m + 2n) + (m + 2n)^2\right]$$
$$= \left[3 - (m + 2n)\right] \cdot$$
$$\left[3^2 + (3)(m + 2n) + m^2 + 4mn + 4n^2\right]$$
$$= (3 - m - 2n)(9 + 3m + 6n + m^2 + 4mn + 4n^2)$$

83. Let $x = 3k - 1$. This substitution gives
$$7(3k - 1)^2 + 26(3k - 1) - 8 = 7x^2 + 26x - 8$$
$$= (7x - 2)(x + 4)$$
Replacing x with $3k - 1$ gives
$$7(3k - 1)^2 + 26(3k - 1) - 8$$
$$= [7(3k - 1) - 2][(3k - 1) + 4]$$
$$= (21k - 7 - 2)(3k - 1 + 4)$$
$$= (21k - 9)(3k + 3) = 3(7k - 3)(3)(k + 1)$$
$$= 9(7k - 3)(k + 1).$$

85. Let $x = a - 4$. This substitution gives
$$9(a - 4)^2 + 30(a - 4) + 25$$
$$= 9x^2 + 30x + 25$$
$$= (3x)^2 + 2(3x)(5) + 5^2 = (3x + 5)^2$$
Replacing x by $a - 4$ gives
$$9(a - 4)^2 + 30(a - 4) + 25$$
$$= [3(a - 4) + 5]^2$$
$$= (3a - 12 + 5)^2 = (3a - 7)^2$$

87. $(a + 1)^3 + 27$

Notice that this is the sum of cubes. Recall that $u^3 + v^3 = (u + v)(u^2 - uv + v^2)$.

$$(a + 1)^3 + 27$$
$$= (a + 1)^3 + 3^3$$
$$= \left[(a + 1) + 3\right]\left[(a + 1)^2 - 3(a + 1) + 3^2\right]$$
$$= (a + 4)(a^2 + 2a + 1 - 3a - 3 + 9)$$
$$= (a + 4)(a^2 - a + 7)$$

89. $(3x+4)^3 - 1$

Notice that this is the difference of cubes.

Recall that $u^3 - v^3 = (u-v)(u^2 + uv + v^2)$.

$$(3x+4)^3 - 1$$
$$= (3x+4)^3 - 1^2$$
$$= \left[(3x+4)-1\right]\left[(3x+4)^2 + (3x+4) + 1^2\right]$$
$$= (3x+3)(9x^2 + 24x + 16 + 3x + 4 + 1)$$
$$= 3(x+1)(9x^2 + 27x + 21)$$
$$= 3(x+1)3(3x^2 + 9x + 7)$$
$$= 9(x+1)(3x^2 + 9x + 7)$$

91. $m^4 - 3m^2 - 10$

Let $x = m^2$, so that $x^2 = (m^2)^2 = m^4$.

$x^2 - 3x - 10 = (x-5)(x+2)$.

Replacing x with m^2 gives

$m^4 - 3m^2 - 10 = (m^2 - 5)(m^2 + 2)$.

93. $12t^4 - t^2 - 35$

Let $w = t^2$, so $w^2 = (t^2)^2 = t^4$.

$12w^2 - w - 35 = (3w+5)(4w-7)$

Replacing w with t^2 gives

$12t^4 - t^2 - 35 = (3t^2 + 5)(4t^2 - 7)$.

95. $4b^2 + 4bc + c^2 - 16$
$$= (4b^2 + 4bc + c^2) - 16$$
$$= \left[(2b)^2 + 2(2b)(c) + c^2\right] - 16$$
$$= (2b+c)^2 - 4^2 = [(2b+c)+4][(2b+c)-4]$$
$$= (2b+c+4)(2b+c-4)$$

97. $x^2 + xy - 5x - 5y = (x^2 + xy) - (5x + 5y)$
$$= x(x+y) - 5(x+y)$$
$$= (x+y)(x-5)$$

99. $p^4(m-2n) + q(m-2n) = (m-2n)(p^4 + q)$

101. $4z^2 + 28z + 49 = (2z)^2 + 2(2z)(7) + 7^2$
$$= (2z+7)^2$$

103. $1000x^3 + 343y^3$
$$= (10x)^3 + (7y)^3$$
$$= (10x+7y)\left[(10x)^2 - (10x)(7y) + (7y)^2\right]$$
$$= (10x+7y)(100x^2 - 70xy + 49y^2)$$

105. $125m^6 - 216 = (5m^2)^3 - 6^3$
$$= (5m^2 - 6)\left[(5m^2)^2 + 5m^2(6) + 6^2\right]$$
$$= (5m^2 - 6)(25m^4 + 30m^2 + 36)$$

107. $64 + (3x+2)^3$
$$= 4^3 + (3x+2)^3$$
$$= \left[4 + (3x+2)\right]\left[4^2 - (4)(3x+2) + (3x+2)^2\right]$$
$$= \left[4 + (3x+2)\right]\begin{bmatrix}4^2 - (4)(3x+2)\\ +9x^2 + 12x + 4\end{bmatrix}$$
$$= (4 + 3x + 2)(16 - 12x - 8 + 9x^2 + 12x + 4)$$
$$= (3x+6)(9x^2 + 12)$$
$$= 3(x+2)(3)(3x^2 + 4) = 9(x+2)(3x^2 + 4)$$

109. $(x+y)^3 - (x-y)^3$

Let $a = x + y$ and let $b = x - y$. Then we have
$(x+y)^3 - (x-y)^3 = a^3 - b^3$
$$= (a-b)(a^2 + ab + b^2)$$

Replacing a and b gives
$(a-b)(a^2 - ab + b^2)$
$$= \left[(x+y) - (x-y)\right]$$
$$\cdot \left[(x+y)^2 + (x+y)(x-y) + (x-y)^2\right]$$
$$= 2y\left[x^2 + 2xy + y^2 + (x^2 - y^2)\right.$$
$$\left. + x^2 - 2xy + y^2\right]$$
$$= 2y(3x^2 + y^2)$$

111. $144z^2 + 121$

The sum of squares cannot be factored.

$144z^2 + 121$ is prime.

113. $(x+y)^2 - (x-y)^2$
$$= [(x+y) + (x-y)][(x+y) - (x-y)]$$
$$= (x+y+x-y)(x+y-x+y)$$
$$= (2x)(2y) = 4xy$$

115. Answers will vary. Sample answer: In general, a sum of squares is not factorable over the real number system. If there is a greatest common factor, as in $4x^2 + 16$, it may be factored out, as here, to obtain $4(x^2 + 4)$.

117. $49x^2 - \dfrac{1}{25} = \left(7x - \dfrac{1}{5}\right)\left(7x + \dfrac{1}{5}\right)$

119. $\dfrac{25}{9}x^4 - 9y^2 = \left(\dfrac{5}{3}x^2 + 3y\right)\left(\dfrac{5}{3}x^2 - 3y\right)$

121. $4z^2 + bz + 81 = (2z)^2 + bz + 9^2$ will be a

perfect trinomial if $bz = \pm 2(2z)(9) \Rightarrow$

$bz = \pm 36z \Rightarrow b = \pm 36$.

If $b = 36$, $4z^2 + 36z + 81 = (2z + 9)^2$.

If $b = -36$, $4z^2 - 36z + 81 = (2z - 9)^2$.

123. $100r^2 - 60r + c = (10r)^2 - \underbrace{2(10r)(3)}_{60r} + c$

will be a perfect trinomial if $c = 3^2 = 9$.

If $c = 9$, $100r^2 - 60r + 9 = (10r - 3)^2$.

125. $x^6 - 1 = (x^3)^2 - 1^2$

$= (x^3 + 1)(x^3 - 1)$ or $(x^3 - 1)(x^3 + 1)$

Use the patterns for the difference of cubes and sum of cubes to factor further. Because

$x^3 - 1 = (x - 1)(x^2 + x + 1)$ and

$x^3 + 1 = (x + 1)(x^2 - x + 1)$,

we obtain the factorization

$x^6 - 1 = (x^3 - 1)(x^3 + 1)$

$= (x - 1)(x^2 + x + 1)(x + 1)(x^2 - x + 1)$

127. From Exercise 125, we have

$x^6 - 1 = (x - 1)(x^2 + x + 1)(x + 1)(x^2 - x + 1)$.

From Exercise 126, we have

$x^6 - 1 = (x - 1)(x + 1)(x^4 + x^2 + 1)$.

Comparing these answers, we see that

$x^4 + x^2 + 1 = (x^2 - x + 1)(x^2 + x + 1)$.

129. The answer in Exercise 127 and the final line in Exercise 128 are the same.

Section R.5   Rational Expressions

1. The quotient of two polynomials in which the denominator is not equal to zero is a <u>rational expression</u>.

3. In the rational expression $\dfrac{x+1}{x-5}$, the domain cannot include the number <u>5</u>.

5. $\dfrac{2x}{5} \cdot \dfrac{10}{x^2} = \dfrac{2x}{5} \cdot \dfrac{2 \cdot 5}{x \cdot x} = \dfrac{2 \cdot 2 \cdot 5 \cdot x}{5 \cdot x \cdot x} = \dfrac{4}{x}$

7. $\dfrac{3}{x} + \dfrac{7}{x} = \dfrac{10}{x}$

9. $\dfrac{2x}{5} + \dfrac{x}{4} = \dfrac{4(2x)}{4(5)} + \dfrac{5x}{4(5)} = \dfrac{8x}{20} + \dfrac{5x}{20} = \dfrac{13x}{20}$

11. In the rational expression $\dfrac{x+3}{x-6}$, the solution to the equation $x - 6 = 0$ is excluded from the domain.

$x - 6 = 0 \Rightarrow x = 6$

The domain is $\{x \mid x \neq 6\}$.

13. In the rational expression $\dfrac{3x+7}{(4x+2)(x-1)}$, the solution to the equation $(4x + 2)(x - 1) = 0$ is excluded from the domain.

$(4x + 2)(x - 1) = 0$

$4x + 2 = 0$ or $x - 1 = 0$

$\quad 4x = -2$ $\qquad\qquad x = 1$

$\quad\, x = -\dfrac{1}{2}$

The domain is $\left\{x \;\middle|\; x \neq -\dfrac{1}{2},\, 1\right\}$.

15. In the rational expression $\dfrac{12}{x^2 + 5x + 6}$, the solution to the equation $x^2 + 5x + 6 = 0$ is excluded from the domain.

$x^2 + 5x + 6 = 0$

$(x + 3)(x + 2) = 0$

$x + 3 = 0$ or $x + 2 = 0$

$\quad x = -3$ $\qquad\qquad x = -2$

The domain is $\{x \mid x \neq -3, -2\}$.

17. In the rational expression $\dfrac{x^2 - 1}{x+1}$, the solution to the equation $x + 1$ is excluded from the domain: $x + 1 = 0 \Rightarrow x = -1$

The domain is $\{x \mid x \neq -1\}$.

19. In the rational expression $\dfrac{x^3 - 1}{x-1}$, the solution to the equation $x - 1$ is excluded from the domain: $x - 1 = 0 \Rightarrow x = 1$

The domain is $\{x \mid x \neq 1\}$.

21. $\dfrac{8x^2 + 16x}{4x^2} = \dfrac{8x(x+2)}{4x^2} = \dfrac{2 \cdot 4x(x+2)}{x \cdot 4x}$

$= \dfrac{2(x+2)}{x} = \dfrac{2x+4}{x}$ or $2 + \dfrac{4}{x}$

23. $\dfrac{3(3-t)}{(t+5)(t-3)} = \dfrac{3(3-t)(-1)}{(t+5)(t-3)(-1)} = \dfrac{-3}{t+5}$

25. $\dfrac{8k+16}{9k+18}=\dfrac{8(k+2)}{9(k+2)}=\dfrac{8}{9}$

27. $\dfrac{m^2-4m+4}{m^2+m-6}=\dfrac{(m-2)(m-2)}{(m-2)(m+3)}=\dfrac{m-2}{m+3}$

29. $\dfrac{8m^2+6m-9}{16m^2-9}=\dfrac{(2m+3)(4m-3)}{(4m+3)(4m-3)}=\dfrac{2m+3}{4m+3}$

31. $\dfrac{x^3+64}{x+4}=\dfrac{(x+4)\left(x^2-4x+16\right)}{x+4}$

$\qquad =x^2-4x+16$

33. $\dfrac{15p^3}{9p^2}\div\dfrac{6p}{10p^2}=\dfrac{15p^3}{9p^2}\cdot\dfrac{10p^2}{6p}=\dfrac{150p^5}{54p^3}$

$\qquad =\dfrac{25\cdot6p^5}{9\cdot6p^3}=\dfrac{25p^2}{9}$

35. $\dfrac{2k+8}{6}\div\dfrac{3k+12}{2}=\dfrac{2k+8}{6}\cdot\dfrac{2}{3k+12}$

$\qquad =\dfrac{2(k+4)(2)}{6(3)(k+4)}=\dfrac{2}{9}$

37. $\dfrac{x^2+x}{5}\cdot\dfrac{25}{xy+y}=\dfrac{x(x+1)}{5}\cdot\dfrac{25}{y(x+1)}$

$\qquad =\dfrac{25x(x+1)}{5y(x+1)}=\dfrac{5x}{y}$

39. $\dfrac{4a+12}{2a-10}\div\dfrac{a^2-9}{a^2-a-20}=\dfrac{4a+12}{2a-10}\cdot\dfrac{a^2-a-20}{a^2-9}$

$\qquad =\dfrac{4(a+3)}{2(a-5)}\cdot\dfrac{(a-5)(a+4)}{(a+3)(a-3)}$

$\qquad =\dfrac{2(a+4)}{a-3}=\dfrac{2a+8}{a-3}$

41. $\dfrac{p^2-p-12}{p^2-2p-15}\cdot\dfrac{p^2-9p+20}{p^2-8p+16}$

$\qquad =\dfrac{(p-4)(p+3)}{(p-5)(p+3)}\cdot\dfrac{(p-5)(p-4)}{(p-4)(p-4)}=1$

43. $\dfrac{m^2+3m+2}{m^2+5m+4}\div\dfrac{m^2+5m+6}{m^2+10m+24}$

$\qquad =\dfrac{m^2+3m+2}{m^2+5m+4}\cdot\dfrac{m^2+10m+24}{m^2+5m+6}$

$\qquad =\dfrac{(m+2)(m+1)}{(m+4)(m+1)}\cdot\dfrac{(m+6)(m+4)}{(m+3)(m+2)}=\dfrac{m+6}{m+3}$

45. $\dfrac{x^3+y^3}{x^3-y^3}\cdot\dfrac{x^2-y^2}{x^2+2xy+y^2}$

$\qquad =\dfrac{(x+y)(x^2-xy+y^2)}{(x-y)(x^2+xy+y^2)}\cdot\dfrac{(x+y)(x-y)}{(x+y)(x+y)}$

$\qquad =\dfrac{x^2-xy+y^2}{x^2+xy+y^2}$

47. $\dfrac{xz-xw+2yz-2yw}{z^2-w^2}\cdot\dfrac{4z+4w+xz+wx}{16-x^2}$

$\qquad =\dfrac{x(z-w)+2y(z-w)}{(z+w)(z-w)}\cdot\dfrac{4(z+w)+x(z+w)}{(4+x)(4-x)}$

$\qquad =\dfrac{(z-w)(x+2y)}{(z+w)(z-w)}\cdot\dfrac{(z+w)(4+x)}{(4+x)(4-x)}=\dfrac{x+2y}{4-x}$

49. Expressions (B) and (C) are both equivalent to -1, because the numerator and denominator are additive inverses.

B. $\dfrac{-x-4}{x+4}=\dfrac{-1(x+4)}{x+4}=-1$

C. $\dfrac{x-4}{4-x}=\dfrac{-1(4-x)}{4-x}=-1$

51. $\dfrac{3}{2k}+\dfrac{5}{3k}=\dfrac{3\cdot3}{2k\cdot3}+\dfrac{5\cdot2}{3k\cdot2}=\dfrac{9}{6k}+\dfrac{10}{6k}=\dfrac{19}{6k}$

53. $\dfrac{1}{6m}+\dfrac{2}{5m}+\dfrac{4}{m}=\dfrac{1\cdot5}{6m\cdot5}+\dfrac{2\cdot6}{5m\cdot6}+\dfrac{4\cdot30}{m\cdot30}$

$\qquad =\dfrac{5}{30m}+\dfrac{12}{30m}+\dfrac{120}{30m}=\dfrac{137}{30m}$

55. $\dfrac{1}{a}-\dfrac{b}{a^2}=\dfrac{1\cdot a}{a\cdot a}-\dfrac{b}{a^2}=\dfrac{a}{a^2}-\dfrac{b}{a^2}=\dfrac{a-b}{a^2}$

57. $\dfrac{5}{12x^2y}-\dfrac{11}{6xy}=\dfrac{5}{12x^2y}-\dfrac{11\cdot2x}{6xy\cdot2x}$

$\qquad =\dfrac{5}{12x^2y}-\dfrac{22x}{12x^2y}=\dfrac{5-22x}{12x^2y}$

59. $\dfrac{17y+3}{9y+7}-\dfrac{-10y-18}{9y+7}=\dfrac{(17y+3)-(-10y-18)}{9y+7}$

$\qquad =\dfrac{17y+3+10y+18}{9y+7}$

$\qquad =\dfrac{27y+21}{9y+7}=\dfrac{3(9y+7)}{9y+7}$

$\qquad =3$

61. $\dfrac{1}{x+z}+\dfrac{1}{x-z}=\dfrac{1\cdot(x-z)}{(x+z)(x-z)}+\dfrac{1\cdot(x+z)}{(x-z)(x+z)}=\dfrac{(x-z)+(x+z)}{(x-z)(x+z)}=\dfrac{x-z+x+z}{(x+z)(x-z)}=\dfrac{2x}{(x+z)(x-z)}$

63. Because $a-2=(-1)(2-a)$, we have

$$\dfrac{3}{a-2}-\dfrac{1}{2-a}=\dfrac{3}{a-2}-\dfrac{1(-1)}{(2-a)(-1)}=\dfrac{3}{a-2}-\dfrac{-1}{a-2}=\dfrac{3-(-1)}{a-2}=\dfrac{4}{a-2}$$

We may also use $2-a$ as the common denominator.

$$\dfrac{3}{a-2}-\dfrac{1}{2-a}=\dfrac{3(-1)}{(a-2)(-1)}-\dfrac{1}{2-a}=\dfrac{-3}{2-a}-\dfrac{1}{2-a}=\dfrac{-4}{2-a}$$

The two results, $\dfrac{4}{a-2}$ and $\dfrac{-4}{2-a}$, are equivalent rational expressions.

65. Because $2x-y=(-1)(y-2x)$,

$$\dfrac{x+y}{2x-y}-\dfrac{2x}{y-2x}=\dfrac{x+y}{2x-y}-\dfrac{2x(-1)}{(y-2x)(-1)}=\dfrac{x+y}{2x-y}-\dfrac{-2x}{2x-y}=\dfrac{x+y-(-2x)}{2x-y}=\dfrac{x+y+2x}{2x-y}=\dfrac{3x+y}{2x-y}$$

We may also use $y-2x$ as the common denominator. In this case, our result will be $\dfrac{-3x-y}{y-2x}$. The two results are equivalent rational expressions.

67. $\dfrac{4}{x+1}+\dfrac{1}{x^2-x+1}-\dfrac{12}{x^3+1}=\dfrac{4}{x+1}+\dfrac{1}{x^2-x+1}-\dfrac{12}{(x+1)(x^2-x+1)}$

$$=\dfrac{4(x^2-x+1)}{(x+1)(x^2-x+1)}+\dfrac{1(x+1)}{(x+1)(x^2-x+1)}-\dfrac{12}{(x+1)(x^2-x+1)}$$

$$=\dfrac{4(x^2-x+1)+(x+1)-12}{(x+1)(x^2-x+1)}=\dfrac{4x^2-4x+4+x+1-12}{(x+1)(x^2-x+1)}$$

$$=\dfrac{4x^2-3x-7}{(x+1)(x^2-x+1)}=\dfrac{(4x-7)(x+1)}{(x+1)(x^2-x+1)}=\dfrac{4x-7}{x^2-x+1}$$

69. $\dfrac{3x}{x^2+x-12}-\dfrac{x}{x^2-16}=\dfrac{3x}{(x-3)(x+4)}-\dfrac{x}{(x-4)(x+4)}=\dfrac{3x(x-4)}{(x-3)(x+4)(x-4)}-\dfrac{x(x-3)}{(x-4)(x+4)(x-3)}$

$$=\dfrac{3x(x-4)-x(x-3)}{(x-3)(x+4)(x-4)}=\dfrac{3x^2-12x-x^2+3x}{(x-3)(x+4)(x-4)}=\dfrac{2x^2-9x}{(x-3)(x+4)(x-4)}$$

71. $\dfrac{1+\frac{1}{x}}{1-\frac{1}{x}}=\dfrac{x\left(1+\frac{1}{x}\right)}{x\left(1-\frac{1}{x}\right)}=\dfrac{x\cdot1+x\left(\frac{1}{x}\right)}{x\cdot1-x\left(\frac{1}{x}\right)}=\dfrac{x+1}{x-1}$

73. $\dfrac{\frac{1}{x+1}-\frac{1}{x}}{\frac{1}{x}}=\dfrac{x(x+1)\left(\frac{1}{x+1}-\frac{1}{x}\right)}{x(x+1)\left(\frac{1}{x}\right)}=\dfrac{x(x+1)\left(\frac{1}{x+1}\right)-x(x+1)\left(\frac{1}{x}\right)}{x(x+1)\left(\frac{1}{x}\right)}=\dfrac{x-(x+1)}{x+1}=\dfrac{x-x-1}{x+1}=\dfrac{-1}{x+1}$

75. $\dfrac{1+\frac{1}{1-b}}{1-\frac{1}{1+b}}=\dfrac{(1-b)(1+b)\left(1+\frac{1}{1-b}\right)}{(1-b)(1+b)\left(1-\frac{1}{1+b}\right)}=\dfrac{(1-b)(1+b)(1)+(1-b)(1+b)\left(\frac{1}{1-b}\right)}{(1-b)(1+b)(1)-(1-b)(1+b)\left(\frac{1}{1+b}\right)}=\dfrac{(1-b)(1+b)+(1+b)}{(1-b)(1+b)-(1-b)}$

$$=\dfrac{(1+b)[(1-b)+1]}{(1-b)[(1+b)-1]}=\dfrac{(1+b)(2-b)}{(1-b)b}\quad\text{or}\quad\dfrac{(2-b)(1+b)}{b(1-b)}$$

77. $\dfrac{\frac{1}{a^3+b^3}}{\frac{1}{a^2+2ab+b^2}} = \dfrac{\frac{1}{(a+b)(a^2-ab+b^2)}}{\frac{1}{(a+b)^2}} = \dfrac{\frac{1}{(a+b)(a^2-ab+b^2)}(a+b)^2\left(a^2-ab+b^2\right)}{\frac{1}{(a+b)^2}(a+b)^2\left(a^2-ab+b^2\right)} = \dfrac{a+b}{a^2-ab+b^2}$

79. $\dfrac{m-\frac{1}{m^2-4}}{\frac{1}{m+2}} = \dfrac{m-\frac{1}{(m+2)(m-2)}}{\frac{1}{m+2}} = \dfrac{(m+2)(m-2)\left(m-\frac{1}{(m+2)(m-2)}\right)}{(m+2)(m-2)\left(\frac{1}{m+2}\right)}$

$= \dfrac{(m+2)(m-2)(m)-(m+2)(m-2)\left(\frac{1}{(m+2)(m-2)}\right)}{(m+2)(m-2)\left(\frac{1}{m+2}\right)}$

$= \dfrac{m(m+2)(m-2)-1}{m-2} = \dfrac{m(m^2-4)-1}{m-2} = \dfrac{m^3-4m-1}{m-2}$

81. $\dfrac{\frac{3}{p^2-16}+p}{\frac{1}{p-4}} = \dfrac{\frac{3}{(p+4)(p-4)}+p}{\frac{1}{p-4}} = \dfrac{(p+4)(p-4)\left[\frac{3}{(p+4)(p-4)}+p\right]}{(p+4)(p-4)\left[\frac{1}{p-4}\right]} = \dfrac{(p+4)(p-4)\left(\frac{3}{(p+4)(p-4)}\right)+(p+4)(p-4)p}{p+4}$

$= \dfrac{3+(p^2-16)p}{p+4} = \dfrac{3+p^3-16p}{p+4} = \dfrac{p^3-16p+3}{p+4}$

83. $\dfrac{\frac{y+3}{y}-\frac{4}{y-1}}{\frac{y}{y-1}+\frac{1}{y}} = \dfrac{y(y-1)\left(\frac{y+3}{y}-\frac{4}{y-1}\right)}{y(y-1)\left(\frac{y}{y-1}+\frac{1}{y}\right)} = \dfrac{y(y-1)\left(\frac{y+3}{y}\right)-y(y-1)\left(\frac{4}{y-1}\right)}{y(y-1)\left(\frac{y}{y-1}\right)+y(y-1)\left(\frac{1}{y}\right)} = \dfrac{(y-1)(y+3)-4y}{y^2+y-1}$

$= \dfrac{y^2+2y-3-4y}{y^2+y-1} = \dfrac{y^2-2y-3}{y^2+y-1}$

85. $\dfrac{\frac{1}{x+h}-\frac{1}{x}}{h} = \dfrac{x(x+h)\left(\frac{1}{x+h}-\frac{1}{x}\right)}{x(x+h)(h)} = \dfrac{x(x+h)\left(\frac{1}{x+h}\right)-x(x+h)\left(\frac{1}{x}\right)}{x(x+h)(h)} = \dfrac{x-(x+h)}{x(x+h)(h)} = \dfrac{x-x-h}{x(x+h)(h)}$

$= \dfrac{-h}{x(x+h)(h)} = \dfrac{-1}{x(x+h)}$

87. $\dfrac{\frac{1}{(x+h)^2+9}-\frac{1}{x^2+9}}{h} = \dfrac{\left[(x+h)^2+9\right]\left(x^2+9\right)\left[\frac{1}{(x+h)^2+9}-\frac{1}{x^2+9}\right]}{\left[(x+h)^2+9\right]\left(x^2+9\right)h}$

$= \dfrac{\left[(x+h)^2+9\right]\left(x^2+9\right)\left[\frac{1}{(x+h)^2+9}\right]-\left[(x+h)^2+9\right]\left(x^2+9\right)\left[\frac{1}{x^2+9}\right]}{\left[(x+h)^2+9\right]\left(x^2+9\right)h}$

$= \dfrac{\left(x^2+9\right)-\left[(x+h)^2+9\right]}{\left[(x+h)^2+9\right]\left(x^2+9\right)h} = \dfrac{x^2+9-\left[x^2+2xh+h^2+9\right]}{\left[(x+h)^2+9\right]\left(x^2+9\right)h} = \dfrac{x^2+9-x^2-2xh-h^2-9}{\left[(x+h)^2+9\right]\left(x^2+9\right)h}$

$= \dfrac{-2xh-h^2}{\left[(x+h)^2+9\right]\left(x^2+9\right)h} = \dfrac{h(-2x-h)}{\left[(x+h)^2+9\right]\left(x^2+9\right)h} = \dfrac{-2x-h}{\left[(x+h)^2+9\right]\left(x^2+9\right)}$

89. Altitude of 7000 feet, $x = 7$ (thousand)
The distance from the origin is
$$\frac{7 - 7}{0.639(7) + 1.75} = 0,$$
which represents 0 miles.

91. $y = \dfrac{6.7x}{100 - x}$; let $x = 75$ (75%)

$$y = \frac{6.7(75)}{100 - 75} = \frac{502.5}{25} = 20.1$$

The cost of removing 75% of the pollutant is 20.1 thousand dollars.

Section R.6   Rational Exponents

1. True

3. False. $\dfrac{a^6}{a^4} = a^2$

5. False. $(x + y)^{-1} = \dfrac{1}{x + y}$

7. (a) B; $4^{-2} = \dfrac{1}{4^2} = \dfrac{1}{16}$

    (b) D; $-4^{-2} = -\left(4^{-2}\right) = -\dfrac{1}{4^2} = -\dfrac{1}{16}$

    (c) B; $(-4)^{-2} = \dfrac{1}{(-4)^2} = \dfrac{1}{16}$

    (d) D; $-(-4)^{-2} = -\dfrac{1}{(-4)^2} = -\dfrac{1}{16}$

9. (a) E; $\left(\dfrac{4}{9}\right)^{3/2} = \left[\left(\dfrac{4}{9}\right)^{1/2}\right]^3 = \left(\dfrac{2}{3}\right)^3 = \dfrac{8}{27}$

    (b) G
$$\left(\frac{4}{9}\right)^{-3/2} = \left(\frac{9}{4}\right)^{3/2} = \left[\left(\frac{9}{4}\right)^{1/2}\right]^3$$
$$= \left(\frac{3}{2}\right)^3 = \frac{27}{8}$$

    (c) F
$$-\left(\frac{9}{4}\right)^{3/2} = -\left[\left(\frac{9}{4}\right)^{1/2}\right]^3 = -\left(\frac{3}{2}\right)^3 = -\frac{27}{8}$$

    (d) F
$$-\left(\frac{4}{9}\right)^{-3/2} = -\left(\frac{9}{4}\right)^{3/2} = -\left[\left(\frac{9}{4}\right)^{1/2}\right]^3$$
$$= -\left(\frac{3}{2}\right)^3 = -\frac{27}{8}$$

11. $(-4)^{-3} = \dfrac{1}{(-4)^3} = \dfrac{1}{-64} = -\dfrac{1}{64}$

13. $-(-5)^{-4} = -\dfrac{1}{(-5)^4} = -\dfrac{1}{625}$

15. $\left(\dfrac{1}{3}\right)^{-2} = \dfrac{1}{\left(\frac{1}{3}\right)^2} = \dfrac{1}{\frac{1}{9}} = 9$, or

$$\left(\frac{1}{3}\right)^{-2} = \frac{1}{\left(\frac{1}{3}\right)^2} = \left(\frac{3}{1}\right)^2 = 3^2 = 9$$

17. $(4x)^{-2} = \dfrac{1}{(4x)^2} = \dfrac{1}{16x^2}$

19. $4x^{-2} = 4 \cdot x^{-2} = 4 \cdot \dfrac{1}{x^2} = \dfrac{4}{x^2}$

21. $-a^{-3} = -\dfrac{1}{a^3}$

23. $\dfrac{4^8}{4^6} = 4^{8-6} = 4^2 = 16$

25. $\dfrac{x^{12}}{x^8} = x^{12-8} = x^4$

27. $\dfrac{r^7}{r^{10}} = r^{7-10} = r^{-3} = \dfrac{1}{r^3}$

29. $\dfrac{6^4}{6^{-2}} = 6^{4-(-2)} = 6^6$

Because 6^6 is a relatively large number, it is generally acceptable to leave it in exponential form instead of expanding it to 46,656.

31. $\dfrac{4r^{-3}}{6r^{-6}} = \dfrac{4}{6} \cdot \dfrac{r^{-3}}{r^{-6}} = \dfrac{2}{3} r^{-3-(-6)} = \dfrac{2}{3} r^3 = \dfrac{2r^3}{3}$

33. $\dfrac{16m^{-5}n^4}{12m^2n^{-3}} = \dfrac{16}{12} \cdot \dfrac{m^{-5}}{m^2} \cdot \dfrac{n^4}{n^{-3}} = \dfrac{4}{3} m^{-5-2} n^{4-(-3)}$

$$= \frac{4}{3} m^{-7} n^7 = \frac{4n^7}{3m^7}$$

35. $-4r^{-2}\left(r^4\right)^2 = -4r^{-2}r^8 = -4r^{-2+8} = -4r^6$

37. $\left(5a^{-1}\right)^4\left(a^2\right)^{-3} = 5^4a^{-4}a^{-6} = 5^4a^{-4+(-6)}$

$$= 5^4a^{-10} = \frac{5^4}{a^{10}} = \frac{625}{a^{10}}$$

39. $\dfrac{\left(p^{-2}\right)^0}{5p^{-4}} = \dfrac{p^0}{5p^{-4}} = \dfrac{1}{5}\cdot\dfrac{p^0}{p^{-4}}$

$$= \frac{1}{5}p^{0-(-4)} = \frac{1}{5}p^4 = \frac{p^4}{5}$$

41. $\dfrac{(3pq)q^2}{6p^2q^4} = \dfrac{3pq^{1+2}}{6p^2q^4} = \dfrac{3pq^3}{6p^2q^4}$

$$= \frac{3}{6}\cdot\frac{p}{p^2}\cdot\frac{q^3}{q^4} = \frac{1}{2}p^{1-2}q^{3-4}$$

$$= \frac{1}{2}p^{-1}q^{-1} = \frac{1}{2pq}$$

43. $\dfrac{4a^5\left(a^{-1}\right)^3}{\left(a^{-2}\right)^{-2}} = \dfrac{4a^5a^{-3}}{a^4} = \dfrac{4a^{5+(-3)}}{a^4}$

$$= \frac{4a^2}{a^4} = 4a^{2-4} = 4a^{-2} = \frac{4}{a^2}$$

45. $\dfrac{\left(5x\right)^{-2}\left(5x^3\right)^{-3}}{\left(5^{-2}x^{-3}\right)^3} = \dfrac{5^{-2}x^{-2}\cdot 5^{-3}x^{-9}}{5^{-6}x^{-9}} = \dfrac{5^{-5}x^{-11}}{5^{-6}x^{-9}}$

$$= 5^{-5-(-6)}x^{-11-(-9)} = 5x^{-2} = \frac{5}{x^2}$$

47. $169^{1/2} = 13$, because $13^2 = 169$.

49. $16^{1/4} = 2$, because $2^4 = 16$.

51. $\left(-\dfrac{64}{27}\right)^{1/3} = -\dfrac{4}{3}$, because $\left(-\dfrac{4}{3}\right)^3 = -\dfrac{64}{27}$.

53. $(-4)^{1/2}$ is not a real number, because no real number, when squared, will yield a negative quantity.

55. $8^{2/3} = \left(8^{1/3}\right)^2 = 2^2 = 4$

57. $100^{3/2} = \left(100^{1/2}\right)^3 = 10^3 = 1000$

59. $-81^{3/4} = -\left(81^{1/4}\right)^3 = -3^3 = -27$

61. $\left(\dfrac{27}{64}\right)^{-4/3} = \left(\dfrac{64}{27}\right)^{4/3} = \left[\left(\dfrac{64}{27}\right)^{1/3}\right]^4$

$$= \left(\frac{4}{3}\right)^4 = \frac{256}{81}$$

63. $3^{1/2}\cdot 3^{3/2} = 3^{1/2+3/2} = 3^{4/2} = 3^2 = 9$

65. $\dfrac{64^{5/3}}{64^{4/3}} = 64^{5/3-4/3} = 64^{1/3} = 4$

67. $y^{7/3}\cdot y^{-4/3} = y^{7/3+(-4/3)} = y^{3/3} = y^1 = y$

69. $\dfrac{k^{1/3}}{k^{2/3}\cdot k^{-1}} = \dfrac{k^{1/3}}{k^{2/3+(-1)}} = \dfrac{k^{1/3}}{k^{2/3+(-3/3)}}$

$$= \frac{k^{1/3}}{k^{-1/3}} = k^{1/3-(-1/3)} = k^{2/3}$$

71. $\dfrac{\left(x^{1/4}y^{2/5}\right)^{20}}{x^2} = \dfrac{x^5y^8}{x^2} = x^{5-2}y^8 = x^3y^8$

73. $\dfrac{\left(x^{2/3}\right)^2}{\left(x^2\right)^{7/3}} = \dfrac{x^{4/3}}{x^{14/3}} = x^{4/3-14/3} = x^{-10/3} = \dfrac{1}{x^{10/3}}$

75. $\left(\dfrac{16m^3}{n}\right)^{1/4}\left(\dfrac{9n^{-1}}{m^2}\right)^{1/2} = \dfrac{16^{1/4}m^{3/4}}{n^{1/4}}\cdot\dfrac{9^{1/2}n^{-1/2}}{m}$

$$= \frac{2m^{3/4}}{n^{1/4}}\cdot\frac{3n^{-1/2}}{m}$$

$$= (2\cdot 3)\frac{m^{3/4}}{m}\cdot\frac{n^{-1/2}}{n^{1/4}}$$

$$= 6m^{(3/4)-1}n^{(-1/2)-(1/4)}$$

$$= 6m^{(3/4)-(4/4)}n^{(-2/4)-(1/4)}$$

$$= 6m^{-1/4}n^{-3/4} = \frac{6}{m^{1/4}n^{3/4}}$$

77. $\dfrac{p^{1/5}p^{7/10}p^{1/2}}{\left(p^3\right)^{-1/5}} = \dfrac{p^{1/5+7/10+1/2}}{p^{-3/5}}$

$$= \frac{p^{2/10+7/10+5/10}}{p^{-6/10}} = \frac{p^{14/10}}{p^{-6/10}}$$

$$= p^{(14/10)-(-6/10)}$$

$$= p^{20/10} = p^2$$

79. (a) Let $w = 25$

$$t = 31,293w^{-1.5} = 31,293\cdot(25)^{-3/2}$$

$$= 31,293\cdot(25^{1/2})^{-3} = 31,293\cdot 5^{-3}$$

$$= \frac{31,293}{125} \approx 250 \text{ sec}$$

(b) To double the weight, replace w with $2w$ to get $\dfrac{31,293}{(2w)^{1.5}} = \dfrac{31,293}{2^{3/2}w^{1.5}} = \dfrac{1}{2^{3/2}}(t)$; so the holding time changes by a factor of

$$\dfrac{1}{2^{3/2}} \approx 0.3536.$$

81. $y^{5/8}(y^{3/8} - 10y^{11/8}) = y^{5/8}y^{3/8} - 10y^{5/8}y^{11/8}$
$$= y^{5/8+3/8} - 10y^{5/8+11/8}$$
$$= y - 10y^2$$

83. $-4k(k^{7/3} - 6k^{1/3}) = -4k^1 k^{7/3} + 24k^1 k^{1/3}$
$$= -4k^{10/3} + 24k^{4/3}$$

85. $(x + x^{1/2})(x - x^{1/2})$ has the form
$(a + b)(a - b) = a^2 - b^2$.
$(x + x^{1/2})(x - x^{1/2}) = x^2 - (x^{1/2})^2 = x^2 - x$

87. $\left(r^{1/2} - r^{-1/2}\right)^2$
$$= \left(r^{1/2}\right)^2 - 2\left(r^{1/2}\right)\left(r^{-1/2}\right) + \left(r^{-1/2}\right)^2$$
$$= r - 2r^{(1/2)+(-1/2)} + r^{-1} = r - 2r^0 + r^{-1}$$
$$= r - 2 \cdot 1 + r^{-1} = r - 2 + r^{-1} = r - 2 + \dfrac{1}{r}$$

89. Factor $4k^{-1} + k^{-2}$, using the common factor k^{-2}: $4k^{-1} + k^{-2} = k^{-2}(4k + 1)$

91. Factor $4t^{-2} + 8t^{-4}$, using the common factor $4t^{-4}$: $4t^{-2} + 8t^{-4} = 4t^{-4}(t^2 + 2)$

93. Factor $9z^{-1/2} + 2z^{1/2}$, using the common factor $z^{-1/2}$:
$$9z^{-1/2} + 2z^{1/2} = z^{-1/2}(9 + 2z)$$

95. Factor $p^{-3/4} - 2p^{-7/4}$, using the common factor $p^{-7/4}$:
$$p^{-3/4} - 2p^{-7/4} = p^{-7/4}(p^{4/4} - 2)$$
$$= p^{-7/4}(p - 2)$$

97. Factor $-4a^{-2/5} + 16a^{-7/5}$, using the common factor $4a^{-7/5}$:
$$-4a^{-2/5} + 16a^{-7/5} = 4a^{-7/5}(-a + 4)$$

99. Factor $(p+4)^{-3/2} + (p+4)^{-1/2} + (p+4)^{1/2}$, using the common factor $(p+4)^{-3/2}$.
$$(p+4)^{-3/2} + (p+4)^{-1/2} + (p+4)^{1/2}$$
$$= (p+4)^{-3/2} \cdot [1 + (p+4) + (p+4)^2]$$
$$= (p+4)^{-3/2} \cdot (1 + p + 4 + p^2 + 8p + 16)$$
$$= (p+4)^{-3/2}(p^2 + 9p + 21)$$

101. Factor
$$2(3x+1)^{-3/2} + 4(3x+1)^{-1/2} + 6(3x+1)^{1/2}$$
using the common factor $2(3x+1)^{-3/2}$:
$$2(3x+1)^{-3/2} + 4(3x+1)^{-1/2} + 6(3x+1)^{1/2}$$
$$= 2(3x+1)^{-3/2}\left[1 + 2(3x+1) + 3(3x+1)^2\right]$$
$$= 2(3x+1)^{-3/2}\left[1 + (6x+2) + 3(9x^2 + 6x + 1)\right]$$
$$= 2(3x+1)^{-3/2}\left[1 + 6x + 2 + 27x^2 + 18x + 3\right]$$
$$= 2(3x+1)^{-3/2}(27x^2 + 24x + 6)$$
$$= 2(3x+1)^{-3/2} \cdot 3(9x^2 + 8x + 2)$$
$$= 6(3x+1)^{-3/2}(9x^2 + 8x + 2)$$

103. Factor
$$4x(2x+3)^{-5/9} + 6x^2(2x+3)^{4/9} - 8x^3(2x+3)^{13/9}$$
using the common factor $2x(2x+3)^{-5/9}$:
$$4x(2x+3)^{-5/9} + 6x^2(2x+3)^{4/9} - 8x^3(2x+3)^{13/9}$$
$$= 2x(2x+3)^{-5/9}\left[2 + 3x(2x+3) - 4x^2(2x+3)^2\right]$$
$$= 2x(2x+3)^{-5/9}$$
$$\cdot \left[2 + (6x^2 + 9x) - 4x^2(4x^2 + 12x + 9)\right]$$
$$= 2x(2x+3)^{-5/9}$$
$$\cdot \left[2 + 6x^2 + 9x - 16x^4 - 48x^3 - 36x^2\right]$$
$$= 2x(2x+3)^{-5/9}\left[-16x^4 - 48x^3 - 30x^2 + 9x + 2\right]$$

105. $\dfrac{a^{-1} + b^{-1}}{(ab)^{-1}} = \dfrac{\frac{1}{a} + \frac{1}{b}}{\frac{1}{ab}} = \dfrac{\frac{1 \cdot b}{a \cdot b} + \frac{1 \cdot a}{b \cdot a}}{\frac{1}{ab}}$
$$= \dfrac{\frac{b+a}{ab}}{\frac{1}{ab}} = \dfrac{b+a}{ab} \cdot \dfrac{ab}{1} = b + a$$

107. $\dfrac{r^{-1}+q^{-1}}{r^{-1}-q^{-1}}\cdot\dfrac{r-q}{r+q}=\dfrac{\frac{1}{r}+\frac{1}{q}}{\frac{1}{r}-\frac{1}{q}}\cdot\dfrac{r-q}{r+q}$

$=\dfrac{rq\left(\frac{1}{r}+\frac{1}{q}\right)}{rq\left(\frac{1}{r}-\frac{1}{q}\right)}\cdot\dfrac{r-q}{r+q}$

$=\dfrac{q+r}{q-r}\cdot\dfrac{r-q}{r+q}=\dfrac{r-q}{q-r}$

$=\dfrac{r-q}{-1(r-q)}=-1$

109. $\dfrac{x-9y^{-1}}{\left(x-3y^{-1}\right)\left(x+3y^{-1}\right)}=\dfrac{x-\frac{9}{y}}{\left(x-\frac{3}{y}\right)\left(x+\frac{3}{y}\right)}$

$=\dfrac{x-\frac{9}{y}}{x^2-\frac{9}{y^2}}=\dfrac{y^2\left(x-\frac{9}{y}\right)}{y^2\left(x^2-\frac{9}{y^2}\right)}$

$=\dfrac{y^2x-9y}{y^2x^2-9}$ or $\dfrac{y(xy-9)}{x^2y^2-9}$

111. $\dfrac{(x^2+1)^4(2x)-x^2(4)(x^2+1)^3(2x)}{(x^2+1)^8}$

$=\dfrac{(2x)(x^2+1)^3\left[(x^2+1)-4x^2\right]}{(x^2+1)^8}$

$=\dfrac{(2x)\left[(x^2+1)-4x^2\right]}{(x^2+1)^5}=\dfrac{(2x)(1-3x^2)}{(x^2+1)^5}$

113. $\dfrac{4(x^2-1)^3+8x(x^2-1)^4}{16(x^2-1)^3}$

$=\dfrac{4(x^2-1)^3\left[1+2x(x^2-1)\right]}{16(x^2-1)^3}$

$=\dfrac{1+2x(x^2-1)}{4}=\dfrac{1+2x^3-2x}{4}$

115. $\dfrac{2(2x-3)^{1/3}-(x-1)(2x-3)^{-2/3}}{(2x-3)^{2/3}}$

$=\dfrac{(2x-3)^{1/3}\left[2-(x-1)(2x-3)^{-1}\right]}{(2x-3)^{2/3}}$

$=\dfrac{2-(x-1)(2x-3)^{-1}}{(2x-3)^{1/3}}=\dfrac{2-\frac{x-1}{2x-3}}{(2x-3)^{1/3}}$

$=\dfrac{2(2x-3)-(x-1)}{(2x-3)^{4/3}}=\dfrac{3x-5}{(2x-3)^{4/3}}$

117. Let x = length of side of cube. Then $3x$ = length of side of bigger cube (side tripled). x^3 is the volume of the cube, and $(3x)^3=3^3x^3=27x^3$ is the volume of the bigger cube. The volume will change by a factor of 27.

119. $0.2^{2/3}\cdot40^{2/3}=(0.2\cdot40)^{2/3}=(8^{1/3})^2=2^2=4$

121. $\dfrac{2^{2/3}}{2000^{2/3}}=\left(\dfrac{2}{2000}\right)^{2/3}$

$=\left(\dfrac{1}{1000}\right)^{2/3}\left[\left(\dfrac{1}{1000}\right)^{1/3}\right]^2$

$=\left(\dfrac{1}{10}\right)^2=\dfrac{1}{100}$

Section R.7   Radical Expressions

1. $\sqrt[3]{64}=64^{1/3}=4$

3. **(a)** F; $(-3x)^{1/3}=\sqrt[3]{-3x}$

(b) H; $(-3x)^{-1/3}=\dfrac{1}{(-3x)^{1/3}}=\dfrac{1}{\sqrt[3]{-3x}}$

(c) G; $(3x)^{1/3}=\sqrt[3]{3x}$

(d) C; $(3x)^{-1/3}=\dfrac{1}{\sqrt[3]{3x}}$

5. $\sqrt[5]{t^5}=t^{5/5}=t$

7. $\sqrt{50}=\sqrt{25\cdot2}=\sqrt{25}\sqrt{2}=5\sqrt{2}$

9. $3\sqrt{xy}-8\sqrt{xy}=-5\sqrt{xy}$

11. $\sqrt[3]{125}=125^{1/3}=5$

13. $\sqrt[4]{81}=81^{1/4}=3$

15. $\sqrt[3]{-125}=(-125)^{1/3}=-5$

17. $\sqrt[4]{-81}=(-81)^{1/4}$
This expression is not a real number.

19. $\sqrt[5]{32}=32^{1/5}=2$

21. $-\sqrt[5]{-32}=-(-32)^{1/5}=-(-2)=2$

23. $m^{2/3}=\sqrt[3]{m^2}$ or $\left(\sqrt[3]{m}\right)^2$

25. $(2m + p)^{2/3} = \sqrt[3]{(2m + p)^2}$ or $\left(\sqrt[3]{2m + p}\right)^2$

27. $\sqrt[5]{k^2} = k^{2/5}$

29. $-3\sqrt{5p^3} = -3(5p^3)^{1/2} = -3 \cdot 5^{1/2} p^{3/2}$

31. A

33. It is true for all $x \geq 0$.

35. $\sqrt[4]{x^4} = |x|$

37. $\sqrt{25k^4 m^2} = \sqrt{\left(5k^2 m\right)^2} = \left|5k^2 m\right| = 5k^2 |m|$

39. $\sqrt{(4x - y)^2} = |4x - y|$

41. $\sqrt[3]{81} = \sqrt[3]{27 \cdot 3} = \sqrt[3]{27} \cdot \sqrt[3]{3} = 3\sqrt[3]{3}$

43. $-\sqrt[4]{32} = -\sqrt[4]{16 \cdot 2} = -\sqrt[4]{16} \cdot \sqrt[4]{2} = -2\sqrt[4]{2}$

45. $\sqrt{14} \cdot \sqrt{3pqr} = \sqrt{14 \cdot 3pqr} = \sqrt{42pqr}$

47. $\sqrt[3]{7x} \cdot \sqrt[3]{2y} = \sqrt[3]{7x \cdot 2y} = \sqrt[3]{14xy}$

49. $-\sqrt{\dfrac{9}{25}} = -\dfrac{\sqrt{9}}{\sqrt{25}} = -\dfrac{3}{5}$

51. $-\sqrt[3]{\dfrac{5}{8}} = -\dfrac{\sqrt[3]{5}}{\sqrt[3]{8}} = -\dfrac{\sqrt[3]{5}}{2}$

53. $\sqrt[4]{\dfrac{m}{n^4}} = \dfrac{\sqrt[4]{m}}{\sqrt[4]{n^4}} = \dfrac{\sqrt[4]{m}}{n}$

55. $3\sqrt[5]{-3125} = 3\sqrt[5]{(-5)^5} = 3(-5) = -15$

57. $\sqrt[3]{16(-2)^4 (2)^8} = \sqrt[3]{2^4 \cdot (-2)^4 2^8} = \sqrt[3]{2^4 \cdot 2^4 \cdot 2^8}$
$= \sqrt[3]{2^{15} \cdot 2} = \sqrt[3]{2^{15}} \cdot \sqrt[3]{2}$
$= 2^5 \cdot \sqrt[3]{2} = 32\sqrt[3]{2}$

59. $\sqrt{8x^5 z^8} = \sqrt{2 \cdot 4 \cdot x^4 \cdot x \cdot z^8}$
$= \sqrt{4x^4 z^8} \cdot \sqrt{2x} = 2x^2 z^4 \sqrt{2x}$

61. $\sqrt[4]{x^4 + y^4}$ cannot be simplified further.

63. $\sqrt{\dfrac{2}{3x}} = \dfrac{\sqrt{2}}{\sqrt{3x}} = \dfrac{\sqrt{2}}{\sqrt{3x}} \cdot \dfrac{\sqrt{3x}}{\sqrt{3x}} = \dfrac{\sqrt{2 \cdot 3x}}{\sqrt{9x^2}} = \dfrac{\sqrt{6x}}{3x}$

65. $\sqrt{\dfrac{x^5 y^3}{z^2}} = \dfrac{\sqrt{x^5 y^3}}{\sqrt{z^2}} = \dfrac{\sqrt{x^4 xy^2 y}}{z}$
$= \dfrac{\sqrt{x^4 y^2} \cdot \sqrt{xy}}{z} = \dfrac{x^2 y\sqrt{xy}}{z}$

67. $\sqrt[3]{\dfrac{8}{x^4}} = \dfrac{\sqrt[3]{8}}{\sqrt[3]{x^4}} = \dfrac{2}{\sqrt[3]{x^3 \cdot x}} = \dfrac{2}{x\sqrt[3]{x}}$
$= \dfrac{2}{x\sqrt[3]{x}} \cdot \dfrac{\sqrt[3]{x^2}}{\sqrt[3]{x^2}} = \dfrac{2\sqrt[3]{x^2}}{x \cdot x} = \dfrac{2\sqrt[3]{x^2}}{x^2}$

69. $\sqrt[4]{\dfrac{g^3 h^5}{9r^6}} = \dfrac{\sqrt[4]{g^3 h^5}}{\sqrt[4]{9r^6}} = \dfrac{\sqrt[4]{h^4 \left(g^3 h\right)}}{\sqrt[4]{r^4 \left(9r^2\right)}} = \dfrac{h\sqrt[4]{g^3 h}}{r\sqrt[4]{9r^2}}$
$= \dfrac{h\sqrt[4]{g^3 h}}{r\sqrt[4]{3^2 r^2}} \cdot \dfrac{\sqrt[4]{3^2 r^2}}{\sqrt[4]{3^2 r^2}} = \dfrac{h\sqrt[4]{3^2 g^3 hr^2}}{r\sqrt[4]{3^4 r^4}}$
$= \dfrac{h\sqrt[4]{9g^3 hr^2}}{3r^2}$

71. $\sqrt[8]{3^4} = 3^{4/8} = 3^{1/2} = \sqrt{3}$

73. $\sqrt[3]{\sqrt{4}} = \sqrt[3]{4^{1/2}} = (4^{1/2})^{1/3} = 4^{1/6}$
$= (2^2)^{1/6} = 2^{2/6} = 2^{1/3} = \sqrt[3]{2}$

75. $\sqrt[4]{\sqrt[3]{2}} = \sqrt[4]{2^{1/3}} = (2^{1/3})^{1/4} = 2^{1/12} = \sqrt[12]{2}$

77. $8\sqrt{2x} - \sqrt{8x} + \sqrt{72x}$
$= 8\sqrt{2x} - \sqrt{4 \cdot 2x} + \sqrt{36 \cdot 2x}$
$= 8\sqrt{2x} - 2\sqrt{2x} + 6\sqrt{2x} = 12\sqrt{2x}$

79. $2\sqrt[3]{3} + 4\sqrt[3]{24} - \sqrt[3]{81}$
$= 2\sqrt[3]{3} + 4\sqrt[3]{8 \cdot 3} - \sqrt[3]{27 \cdot 3}$
$= 2\sqrt[3]{3} + 4(2)\sqrt[3]{3} - 3\sqrt[3]{3}$
$= 2\sqrt[3]{3} + 8\sqrt[3]{3} - 3\sqrt[3]{3} = 7\sqrt[3]{3}$

81. $\sqrt[4]{81x^6 y^3} - \sqrt[4]{16x^{10} y^3}$
$= \sqrt[4]{\left(81x^4\right) x^2 y^3} - \sqrt[4]{\left(16x^8\right) x^2 y^3}$
$= 3x\sqrt[4]{x^2 y^3} - 2x^2 \sqrt[4]{x^2 y^3}$

83. $5\sqrt{6} + 2\sqrt{10}$ cannot be simplified further

85. This product has the pattern
$(a + b)(a - b) = a^2 - b^2$, the difference of
squares.
$\left(\sqrt{2} + 3\right)\left(\sqrt{2} - 3\right) = \left(\sqrt{2}\right)^2 - 3^2 = 2 - 9 = -7$

87. This product has the pattern
$(a-b)(a^2+ab+b^2)=a^3-b^3$, the difference of cubes.

$$\left(\sqrt[3]{11}-1\right)\left(\sqrt[3]{11^2}+\sqrt[3]{11}+1\right)=\left(\sqrt[3]{11}\right)^3-1^3$$
$$=11-1=10$$

89. This product has the pattern
$(a+b)^2=a^2+2ab+b^2$, the square of a binomial.

$$\left(\sqrt{3}+\sqrt{8}\right)^2=\left(\sqrt{3}\right)^2+2\left(\sqrt{3}\right)\left(\sqrt{8}\right)+\left(\sqrt{8}\right)^2$$
$$=3+2\sqrt{24}+8=11+2\sqrt{4\cdot6}$$
$$=11+2(2)\sqrt{6}=11+4\sqrt{6}$$

91. This product can be found by using the FOIL method.

$$\left(3\sqrt{2}+\sqrt{3}\right)\left(2\sqrt{3}-\sqrt{2}\right)$$
$$=3\sqrt{2}\left(2\sqrt{3}\right)-3\sqrt{2}\left(\sqrt{2}\right)+\sqrt{3}\left(2\sqrt{3}\right)-\sqrt{3}\sqrt{2}$$
$$=6\sqrt{6}-3\cdot2+2\cdot3-\sqrt{6}=6\sqrt{6}-6+6-\sqrt{6}$$
$$=5\sqrt{6}$$

93.
$$\frac{\sqrt[3]{mn}\cdot\sqrt[3]{m^2}}{\sqrt[3]{n^2}}=\sqrt[3]{\frac{mnm^2}{n^2}}=\sqrt[3]{\frac{m^3}{n}}$$
$$=\frac{\sqrt[3]{m^3}}{\sqrt[3]{n}}\cdot\frac{\sqrt[3]{n^2}}{\sqrt[3]{n^2}}=\frac{m\sqrt[3]{n^2}}{n}$$

95.
$$\sqrt[3]{\frac{2}{x^6}}-\sqrt[3]{\frac{5}{x^9}}=\frac{\sqrt[3]{2}}{\sqrt[3]{x^6}}-\frac{\sqrt[3]{5}}{\sqrt[3]{x^9}}=\frac{\sqrt[3]{2}}{x^2}-\frac{\sqrt[3]{5}}{x^3}$$
$$=\frac{\sqrt[3]{2}\cdot x}{x^2\cdot x}-\frac{\sqrt[3]{5}}{x^3}=\frac{x\sqrt[3]{2}}{x^3}-\frac{\sqrt[3]{5}}{x^3}$$
$$=\frac{x\sqrt[3]{2}-\sqrt[3]{5}}{x^3}$$

97.
$$\frac{1}{\sqrt{2}}+\frac{3}{\sqrt{8}}+\frac{1}{\sqrt{32}}=\frac{1}{\sqrt{2}}+\frac{3}{\sqrt{4\cdot2}}+\frac{1}{\sqrt{16\cdot2}}$$
$$=\frac{1}{\sqrt{2}}+\frac{3}{2\sqrt{2}}+\frac{1}{4\sqrt{2}}$$
$$=\frac{1\cdot4}{4\sqrt{2}}+\frac{3\cdot2}{4\sqrt{2}}+\frac{1}{4\sqrt{2}}$$
$$=\frac{4}{4\sqrt{2}}+\frac{6}{4\sqrt{2}}+\frac{1}{4\sqrt{2}}$$
$$=\frac{11}{4\sqrt{2}}=\frac{11\sqrt{2}}{4\sqrt{2}\sqrt{2}}$$
$$=\frac{11\sqrt{2}}{4\cdot2}=\frac{11\sqrt{2}}{8}$$

99.
$$\frac{-4}{\sqrt[3]{3}}+\frac{1}{\sqrt[3]{24}}-\frac{2}{\sqrt[3]{81}}$$
$$=\frac{-4}{\sqrt[3]{3}}+\frac{1}{\sqrt[3]{8\cdot3}}-\frac{2}{\sqrt[3]{27\cdot3}}=\frac{-4}{\sqrt[3]{3}}+\frac{1}{2\sqrt[3]{3}}-\frac{2}{3\sqrt[3]{3}}$$
$$=\frac{-4\cdot6}{\sqrt[3]{3}\cdot6}+\frac{1\cdot3}{2\sqrt[3]{3}\cdot3}-\frac{2\cdot2}{3\sqrt[3]{3}\cdot2}$$
$$=\frac{-24}{6\sqrt[3]{3}}+\frac{3}{6\sqrt[3]{3}}-\frac{4}{6\sqrt[3]{3}}=\frac{-24+3-4}{6\sqrt[3]{3}}=\frac{-25}{6\sqrt[3]{3}}$$
$$=\frac{-25}{6\sqrt[3]{3}}\cdot\frac{\sqrt[3]{3^2}}{\sqrt[3]{3^2}}=\frac{-25\sqrt[3]{9}}{6\cdot3}=\frac{-25\sqrt[3]{9}}{18}$$

101.
$$\frac{\sqrt{3}}{\sqrt{5}+\sqrt{3}}=\frac{\sqrt{3}}{\sqrt{5}+\sqrt{3}}\cdot\frac{\sqrt{5}-\sqrt{3}}{\sqrt{5}-\sqrt{3}}$$
$$=\frac{\sqrt{3}\left(\sqrt{5}-\sqrt{3}\right)}{\left(\sqrt{5}\right)^2-\left(\sqrt{3}\right)^2}$$
$$=\frac{\sqrt{3}\sqrt{5}-\sqrt{3}\sqrt{3}}{5-3}=\frac{\sqrt{15}-3}{2}$$

103.
$$\frac{\sqrt{7}-1}{2\sqrt{7}+4\sqrt{2}}=\frac{\sqrt{7}-1}{2\sqrt{7}+4\sqrt{2}}\cdot\frac{2\sqrt{7}-4\sqrt{2}}{2\sqrt{7}-4\sqrt{2}}$$
$$=\frac{\left(\sqrt{7}-1\right)\left(2\sqrt{7}-4\sqrt{2}\right)}{\left(2\sqrt{7}+4\sqrt{2}\right)\left(2\sqrt{7}-4\sqrt{2}\right)}$$
$$=\frac{\sqrt{7}\cdot2\sqrt{7}-\sqrt{7}\cdot4\sqrt{2}-1\cdot2\sqrt{7}+1\cdot4\sqrt{2}}{\left(2\sqrt{7}\right)^2-\left(4\sqrt{2}\right)^2}$$
$$=\frac{2\cdot7-4\sqrt{14}-2\sqrt{7}+4\sqrt{2}}{4\cdot7-16\cdot2}$$
$$=\frac{14-4\sqrt{14}-2\sqrt{7}+4\sqrt{2}}{28-32}$$
$$=\frac{14-4\sqrt{14}-2\sqrt{7}+4\sqrt{2}}{-4}$$
$$=\frac{-2\left(-7+2\sqrt{14}+\sqrt{7}-2\sqrt{2}\right)}{-4}$$
$$=\frac{-7+2\sqrt{14}+\sqrt{7}-2\sqrt{2}}{2}$$

105.
$$\frac{p-4}{\sqrt{p}+2}=\frac{p-4}{\sqrt{p}+2}\cdot\frac{\sqrt{p}-2}{\sqrt{p}-2}=\frac{(p-4)\left(\sqrt{p}-2\right)}{\left(\sqrt{p}\right)^2-2^2}$$
$$=\frac{(p-4)\left(\sqrt{p}-2\right)}{p-4}=\sqrt{p}-2$$

107.
$$\frac{3m}{2+\sqrt{m+n}} = \frac{3m}{2+\sqrt{m+n}} \cdot \frac{2-\sqrt{m+n}}{2-\sqrt{m+n}}$$
$$= \frac{3m\left(2-\sqrt{m+n}\right)}{2^2 - \left(\sqrt{m+n}\right)^2}$$
$$= \frac{3m\left(2-\sqrt{m+n}\right)}{4-(m+n)}$$
$$= \frac{3m\left(2-\sqrt{m+n}\right)}{4-m-n}$$

109.
$$\frac{5\sqrt{x}}{2\sqrt{x}+\sqrt{y}} = \frac{5\sqrt{x}\left(2\sqrt{x}-\sqrt{y}\right)}{\left(2\sqrt{x}+\sqrt{y}\right)\left(2\sqrt{x}-\sqrt{y}\right)}$$
$$= \frac{5\sqrt{x}\left(2\sqrt{x}-\sqrt{y}\right)}{4x-y}$$

111. $S = 15.18\sqrt[9]{n} = 15.18\sqrt[9]{4} \approx 17.7$

The speed of the boat with a four-man crew is approx. 17.7 ft/sec

113. Windchill temperature
$$= 35.74 + 0.6215T - 35.75V^{0.16} + 0.4275TV^{0.16}$$
Windchill temperature
$$= 35.74 + 0.6215(10) - 35.75\left(30^{0.16}\right)$$
$$+ 0.4275(10)\left(30^{0.16}\right) \approx -12°$$

115. $\sqrt[4]{8} \cdot \sqrt[4]{2} = \sqrt[4]{8 \cdot 2} = \sqrt[4]{16} = 2$

117. $\dfrac{\sqrt[5]{320}}{\sqrt[5]{10}} = \sqrt[5]{\dfrac{320}{10}} = \sqrt[5]{32} = 2$

119. $\dfrac{\sqrt[3]{15}}{\sqrt[3]{5}} \cdot \sqrt[3]{9} = \sqrt[3]{3} \cdot \sqrt[3]{9} = \sqrt[3]{27} = 3$

121. $\dfrac{355}{113} = 3.1415929\ldots$ and $\pi \approx 3.1415926\ldots$, so it gives six decimal places of accuracy.

123. $\dfrac{3927}{1250} = 3.1416$ and $\pi \approx 3.1415926\ldots$, so it first differs in the fourth decimal place.

Chapter R   Review Exercises

1. The elements of the set $\{6, 8, 10, \ldots, 20\}$ are the even numbers from 6 to 20 inclusive. The elements in the set are $\{6, 8, 10, 12, 14, 16, 18, 20\}$.

3. True. The set of negative integers $= \{\ldots, -4, -3, -2, -1\}$, while the set of whole numbers $= \{0, 1, 2, 3, \ldots\}$. The two sets do not intersect, and so they are disjoint.

5. True

7. False. The two sets are not equal because they do not have the same elements.

9. True   **11.** True

13. $A' = \{2, 6, 9, 10\}$ **15.** $B \cap E = \varnothing$

17. $D \cap \varnothing = \varnothing$

19. $(C \cap D) \cup B = \{1, 3\} \cup B = \{1, 2, 3, 4, 6, 8\}$

21. $\varnothing' = \{1, 2, 3, 4, 5, 6, 7, 8, 9, 10\}$ or U

23. $-12, -6, -\sqrt{4}$ (or -2), 0, and 6 are integers.

25. $\dfrac{4\pi}{5}$ is an irrational number and a real number.

27. 0 is a whole number, an integer, a rational number, and a real number.

29. Answers will vary. Sample answer: The reciprocal of a product is the product of the reciprocals.

31. Answers will vary. Sample answer: A product raised to a power is the product of the factors raised to the power.

33. Answers will vary. Sample answer: A quotient raised to a power is the quotient of the numerator raised to the power and the denominator raised to the power.

35. commutative   **37.** associative

39. identity

41. The year 2012 corresponds to $x = 10$.
$0.0112 \cdot 10^2 + 0.4663 \cdot 10 + 1.513 = 7.296$
7.296 million strudents took at least one online college course in 2012.

43. $(-4-1)(-3-5) - 2^3 = (-5)(-8) - 8$
$$= 40 - 8 = 32$$

45. $\left(-\dfrac{5}{9} - \dfrac{2}{3}\right) - \dfrac{5}{6} = \left(-\dfrac{5}{9} - \dfrac{6}{9}\right) - \dfrac{5}{6}$
$$= \dfrac{-11}{9} - \dfrac{5}{6} = -\dfrac{22}{18} - \dfrac{15}{18} = -\dfrac{37}{18}$$

47. $\dfrac{6(-4)-3^2(-2)^3}{-5[-2-(-6)]} = \dfrac{6(-4)-9(-8)}{-5[-2+6]}$

$\quad = \dfrac{6(-4)-9(-8)}{-5(4)}$

$\quad = \dfrac{-24-(-72)}{-20} = \dfrac{-24+72}{-20}$

$\quad = \dfrac{48}{-20} = -\dfrac{12}{5}$

49. Let $a=-1$, $b=-2$, $c=4$.

$\quad -c(2a-5b) = -4[2(-1)-5(-2)]$

$\qquad = -4\left[-2-(-10)\right] = -4(-2+10)$

$\qquad = -4(8) = -32$

51. Let $a=-1$, $b=-2$, $c=4$.

$\quad \dfrac{9a+2b}{a+b+c} = \dfrac{9(-1)+2(-2)}{-1+(-2)+4}$

$\qquad = \dfrac{-9+(-4)}{-3+4} = \dfrac{-13}{1} = -13$

53. $(3q^3-9q^2+6)+(4q^3-8q+3)$

$\quad = 3q^3-9q^2+6+4q^3-8q+3$

$\quad = 7q^3-9q^2-8q+9$

55. $(8y-7)(2y^2+7y-3)$

$\quad = 8y(2y^2+7y-3)-7(2y^2+7y-3)$

$\quad = 16y^3+56y^2-24y-14y^2-49y+21$

$\quad = 16y^3+42y^2-73y+21$

57. $(3k-5m)^2 = (3k)^2-2(3k)(5m)+(5m)^2$

$\qquad = 9k^2-30km+25m^2$

59. $\dfrac{30m^3-9m^2+22m+5}{5m+1}$

$$
\begin{array}{r}
6m^2 - 3m + 5 \\
5m+1\overline{)30m^3-9m^2+22m+5}\quad \text{s} \\
\underline{30m^3+6m^2} \\
-15m^2+22m \\
\underline{-15m^2-3m} \\
25m+5 \\
\underline{25m+5} \\
0
\end{array}
$$

Thus,

$\dfrac{30m^3-9m^2+22m+5}{5m+1} = 6m^2-3m+5.$

61. $\dfrac{3b^3-8b^2+12b-30}{b^2+4}$

Insert the missing term in the divisor with a
0 coefficient.

$$
\begin{array}{r}
3b - 8 \\
b^2+0b+4\overline{)3b^3-8b^2+12b-30} \\
\underline{3b^3+0b^2+12b} \\
-8b^2+0b-30 \\
\underline{-8b^2+0b-32} \\
2
\end{array}
$$

Thus, $\dfrac{3b^3-8b^2+12b-30}{b^2+4} = 3b-8+\dfrac{2}{b^2+4}$.

63. $3(z-4)^2+9(z-4)^3 = 3(z-4)^2[1+3(z-4)]$

$\qquad = 3(z-4)^2(1+3z-12)$

$\qquad = 3(z-4)^2(3z-11)$

65. $z^2-6zk-16k^2 = (z-8k)(z+2k)$

67. $48a^8-12a^7b-90a^6b^2$

$\quad = 6a^6(8a^2-2ab-15b^2)$

$\quad = 6a^6(4a+5b)(2a-3b)$

69. $49m^8-9n^2 = (7m^4)^2-(3n)^2$

$\qquad = (7m^4+3n)(7m^4-3n)$

71. $6(3r-1)^2+(3r-1)-35$

Let $x=3r-1$. With this substitution,

$6(3r-1)^2+(3r-1)-35$ becomes

$6x^2+x-35$.

Factor the trinomial by trial and error.

$6x^2+x-35 = (3x-7)(2x+5)$

Replacing x with $3r-1$ gives

$[3(3r-1)-7][2(3r-1)+5]$

$= (9r-3-7)(6r-2+5) = (9r-10)(6r+3)$

$= 3(9r-10)(2r+1).$

73. $xy+2x-y-2 = (xy+2x)-(y+2)$

$\qquad = x(y+2)-(y+2)$

$\qquad = (y+2)(x-1)$

75. $(3x-4)^2+(x-5)(2)(3x-4)(3)$

$\quad = (3x-4)[(3x-4)+(x-5)(2)(3)]$

$\quad = (3x-4)[3x-4+6x-30]$

$\quad = (3x-4)(9x-34)$

77. $\dfrac{k^2+k}{8k^3}\cdot\dfrac{4}{k^2-1} = \dfrac{k(k+1)(4)}{8k^3(k+1)(k-1)}$

$\qquad = \dfrac{4k}{8k^3(k-1)} = \dfrac{1}{2k^2(k-1)}$

79. $\dfrac{x^2+x-2}{x^2+5x+6} \div \dfrac{x^2+3x-4}{x^2+4x+3}$

$= \dfrac{x^2+x-2}{x^2+5x+6} \cdot \dfrac{x^2+4x+3}{x^2+3x-4}$

$= \dfrac{(x+2)(x-1)}{(x+3)(x+2)} \cdot \dfrac{(x+3)(x+1)}{(x+4)(x-1)} = \dfrac{x+1}{x+4}$

81. $\dfrac{p^2-36q^2}{p^2-12pq+36q^2} \cdot \dfrac{p^2-5pq-6q^2}{p^2+2pq+q^2}$

$= \dfrac{(p+6q)(p-6q)}{(p-6q)^2} \cdot \dfrac{(p-6q)(p+q)}{(p+q)^2}$

$= \dfrac{p+6q}{p+q}$

83. $\dfrac{m}{4-m} + \dfrac{3m}{m-4} = \dfrac{m(-1)}{(4-m)(-1)} + \dfrac{3m}{m-4}$

$= \dfrac{-m}{m-4} + \dfrac{3m}{m-4} = \dfrac{2m}{m-4}$

We may also use $4-m$ as the common denominator. In this case, the result will be

$\dfrac{-2m}{4-m}$. The two results are equivalent rational expressions.

85. $\dfrac{p^{-1}+q^{-1}}{1-(pq)^{-1}} = \dfrac{\frac{1}{p}+\frac{1}{q}}{1-\frac{1}{pq}} = \dfrac{pq\left(\frac{1}{p}+\frac{1}{q}\right)}{pq\left(1-\frac{1}{pq}\right)}$

$= \dfrac{pq\left(\frac{1}{p}\right)+pq\left(\frac{1}{q}\right)}{pq(1)-pq\left(\frac{1}{pq}\right)} = \dfrac{q+p}{pq-1}$

87. $\left(-\dfrac{5}{4}\right)^{-2} = \left(-\dfrac{4}{5}\right)^2 = \dfrac{16}{25}$

89. $(5z^3)(-2z^5) = -10z^{3+5} = -10z^8$

91. $(-6p^5w^4m^{12})^0 = 1$ Definition of a^0

93. $\dfrac{-8y^7p^{-2}}{y^{-4}p^{-3}} = -8y^{7-(-4)}p^{(-2)-(-3)} = -8y^{11}p$

95. $\dfrac{(p+q)^4(p+q)^{-3}}{(p+q)^6} = (p+q)^{4+(-3)-6}$

$= (p+q)^{-5} = \dfrac{1}{(p+q)^5}$

97. $(7r^{1/2})(2r^{3/4})(-r^{1/6}) = -14r^{1/2+3/4+1/6}$

$= -14r^{17/12}$

99. $\dfrac{y^{5/3}\cdot y^{-2}}{y^{-5/6}} = \dfrac{y^{5/3+(-2)}}{y^{-5/6}} = \dfrac{y^{5/3+(-6/3)}}{y^{-5/6}} = \dfrac{y^{-1/3}}{y^{-5/6}}$

$= y^{-1/3-(-5/6)} = y^{-2/6+5/6}$

$= y^{3/6} = y^{1/2}$

101. $\dfrac{\left(p^{15}q^{12}\right)^{-4/3}}{\left(p^{24}q^{16}\right)^{-3/4}} = \dfrac{p^{15(-4/3)}q^{12(-4/3)}}{p^{24(-3/4)}q^{16(-3/4)}} = \dfrac{p^{-20}q^{-16}}{p^{-18}q^{-12}}$

$= p^{-20-(-18)}q^{-16-(-12)}$

$= p^{-2}q^{-4} = \dfrac{1}{p^2q^4}$

103. $\sqrt{200} = \sqrt{100\cdot 2} = \sqrt{100}\cdot\sqrt{2} = 10\sqrt{2}$

105. $\sqrt[4]{1250} = \sqrt[4]{625\cdot 2} = \sqrt[4]{625}\cdot\sqrt[4]{2} = 5\sqrt[4]{2}$

107. $-\sqrt[3]{\dfrac{2}{5p^2}} = -\dfrac{\sqrt[3]{2}}{\sqrt[3]{5p^2}} = -\dfrac{\sqrt[3]{2}}{\sqrt[3]{5p^2}}\cdot\dfrac{\sqrt[3]{25p}}{\sqrt[3]{25p}}$

$= -\dfrac{\sqrt[3]{2\cdot 25p}}{\sqrt[3]{125p^3}} = -\dfrac{\sqrt[3]{50p}}{5p}$

109. $\sqrt[4]{\sqrt[3]{m}} = \left(\sqrt[3]{m}\right)^{1/4} = (m^{1/3})^{1/4} = m^{1/3\cdot 1/4}$

$= m^{1/12} = \sqrt[12]{m}$

111. This product has the pattern

$(a+b)(a^2-ab+b^2) = a^3+b^3$, the sum of two cubes.

$\left(\sqrt[3]{2}+4\right)\left(\sqrt[3]{2^2}-4\sqrt[3]{2}+16\right)$

$= \left(\sqrt[3]{2}+4\right)\left[\left(\sqrt[3]{2}\right)^2 - \sqrt[3]{2}\cdot 4 + 4^2\right]$

$= \left(\sqrt[3]{2}\right)^3 + 4^3 = 2+64 = 66$

113. $\sqrt{18m^3} - 3m\sqrt{32m} + 5\sqrt{m^3}$

$= \sqrt{9m^2\cdot 2m} - 3m\sqrt{16\cdot 2m} + 5\sqrt{m^2m}$

$= 3m\sqrt{2m} - 4\cdot 3m\sqrt{2m} + 5m\sqrt{m}$

$= 3m\sqrt{2m} - 12m\sqrt{2m} + 5m\sqrt{m}$

$= -9m\sqrt{2m} + 5m\sqrt{m}$ or $m\left(-9\sqrt{2m}+5\sqrt{m}\right)$

115. $\dfrac{6}{3-\sqrt{2}} = \dfrac{6}{3-\sqrt{2}}\cdot\dfrac{3+\sqrt{2}}{3+\sqrt{2}}$

$= \dfrac{6\left(3+\sqrt{2}\right)}{9-2} = \dfrac{6\left(3+\sqrt{2}\right)}{7}$

117. $x(x^2+5) = x\cdot x^2 + x\cdot 5 = x^3 + 5x$

119. $(m^2)^3 = m^{2\cdot 3} = m^6$

121. $\dfrac{\left(\frac{a}{b}\right)}{2} = \left(\dfrac{a}{b}\right) \cdot \left(\dfrac{1}{2}\right) = \dfrac{a}{2b}$

123. $\dfrac{1}{(-2)^3} = \dfrac{1}{-8} = (-2)^{-3}$ or

$\dfrac{1}{(-2)^3} = \dfrac{1}{-8} = -\dfrac{1}{8} = -2^{-3}$

125. $\left(\dfrac{8}{7} + \dfrac{a}{b}\right)^{-1} = \left(\dfrac{8b}{7b} + \dfrac{7a}{7b}\right)^{-1} = \left(\dfrac{8b + 7a}{7b}\right)^{-1}$

$= \dfrac{7b}{8b + 7a}$

Chapter R   Test

1. False. $B' = \{2, 4, 6, 7, 8\}$

2. True

3. False. $(B \cap C) \cup D = \{1\} \cup D = \{1, 4\}$

4. True

5. **(a)** $-13, -\dfrac{12}{4}$ (or -3), 0, and $\sqrt{49}$ (or 7) are integers.

(b) $-13, -\dfrac{12}{4}$ (or -3), 0, $\dfrac{3}{5}$, 5.9 (or $\frac{59}{10}$), and $\sqrt{49}$ (or 7) are rational numbers.

(c) All numbers in the set are real numbers.

6. Let $x = -2$, $y = -4$, $z = 5$.

$\left|\dfrac{x^2 + 2yz}{3(x+z)}\right| = \left|\dfrac{(-2)^2 + 2(-4)(5)}{3(-2+5)}\right| = \left|\dfrac{4 + (-40)}{3(3)}\right|$

$= \left|\dfrac{-36}{9}\right| = |-4| = 4$

7. **(a)** associative property

(b) commutative property

(c) distributive property

(d) inverse property

8. $A = 650$, $C = 446$, $Y = 5162$, $T = 39$, $I = 12$
First, find the value of the numerator.

$\left(250 \cdot \dfrac{C}{A}\right) + \left(1000 \cdot \dfrac{T}{A}\right) + \left(12.5 \cdot \dfrac{Y}{A}\right)$

$+ 6.25 - \left(1250 \cdot \dfrac{I}{A}\right)$

$= \left(250 \cdot \dfrac{446}{650}\right) + \left(1000 \cdot \dfrac{39}{650}\right) + \left(12.5 \cdot \dfrac{5162}{650}\right)$

$+ 6.25 - \left(1250 \cdot \dfrac{12}{650}\right)$

≈ 313.98

$R \approx \dfrac{313.98}{3} \approx 104.7$

Drew Brees' quarterback rating was about 104.7.

9. $(x^2 - 3x + 2) - (x - 4x^2) + 3x(2x + 1)$

$= x^2 - 3x + 2 - x + 4x^2 + 6x^2 + 3x$

$= 11x^2 - x + 2$

10. $(6r - 5)^2 = (6r)^2 - 2(6r)(5) + 5^2$

$= 36r^2 - 60r + 25$

11. $(t + 2)(3t^2 - t + 4)$

$\begin{array}{r} 3t^2 - t + 4 \\ t + 2 \\ \hline 6t^2 - 2t + 8 \\ 3t^3 - t^2 + 4t \\ \hline 3t^3 + 5t^2 + 2t + 8 \end{array}$

12. $\dfrac{2x^3 - 11x^2 + 28}{x - 5}$

$\begin{array}{r} 2x^2 - x - 5 \\ x - 5 \overline{)2x^3 - 11x^2 + 0x + 28} \\ \underline{2x^3 - 10x^2} \\ -x^2 + 0x \\ \underline{-x^2 + 5x} \\ -5x + 28 \\ \underline{-5x + 25} \\ 3 \end{array}$

Thus,

$\dfrac{2x^3 - 11x^2 + 28}{x - 5} = 2x^2 - x - 5 + \dfrac{3}{x - 5}$.

13. For the year 2005, $x = 6$.
Adjusted poverty threshold

$\approx 2.719x^2 + 196.1x + 8718$

$= 2.719 \cdot 6^2 + 196.1 \cdot 6 + 8718$

$= 97.884 + 1176.6 + 8718 = 9992.484 \approx \$9992$

14. For the year 2012, $x = 13$
Adjusted poverty threshold
$$\approx 2.719x^2 + 196.1x + 8718$$
$$= 2.719 \cdot 13^2 + 196.1 \cdot 13 + 8718$$
$$= 459.511 + 2549.3 + 8718$$
$$= 11,726.811 \approx \$11,727$$

15. $6x^2 - 17x + 7 = (3x - 7)(2x - 1)$

16. $x^4 - 16 = (x^2)^2 - 4^2$
$$= (x^2 + 4)(x^2 - 4)$$
$$= (x^2 + 4)(x^2 - 2^2)$$
$$= (x^2 + 4)(x + 2)(x - 2)$$

17. $24m^3 - 14m^2 - 24m = 2m(12m^2 - 7m - 12)$
$$= 2m(4m + 3)(3m - 4)$$

18. $x^3y^2 - 9x^3 - 8y^2 + 72$
$$= (x^3y^2 - 9x^3) - (8y^2 - 72)$$
$$= x^3(y^2 - 9) - 8(y^2 - 9) = (x^3 - 8)(y^2 - 9)$$
$$= (x^3 - 2^3)(y^2 - 3^2)$$
$$= (x - 2)(x^2 + 2x + 4)(y + 3)(y - 3)$$

19. $(a - b)^2 + 2(a - b) = (a - b)(a - b + 2)$

20. $1 - 27x^6 = (1 - 3x^2)(1 + 3x^2 + 9x^4)$

21. $\dfrac{5x^2 - 9x - 2}{30x^3 + 6x^2} \div \dfrac{x^4 - 3x^2 - 4}{2x^8 + 6x^7 + 4x^6}$
$$= \dfrac{5x^2 - 9x - 2}{30x^3 + 6x^2} \cdot \dfrac{2x^8 + 6x^7 + 4x^6}{x^4 - 3x^2 - 4}$$
$$= \dfrac{(5x + 1)(x - 2)}{6x^2(5x + 1)} \cdot \dfrac{2x^6(x^2 + 3x + 2)}{(x^2 - 4)(x^2 + 1)}$$
$$= \dfrac{(5x + 1)(x - 2)}{6x^2(5x + 1)} \cdot \dfrac{2x^6(x + 2)(x + 1)}{(x + 2)(x - 2)(x^2 + 1)}$$
$$= \dfrac{2x^6(x + 1)}{6x^2(x^2 + 1)} = \dfrac{x^4(x + 1)}{3(x^2 + 1)}$$

22. $\dfrac{x}{x^2 + 3x + 2} + \dfrac{2x}{2x^2 - x - 3}$
$$= \dfrac{x}{(x + 2)(x + 1)} + \dfrac{2x}{(2x - 3)(x + 1)}$$
The least common denominator is
$(x + 2)(x + 1)(2x - 3)$.

$$= \dfrac{x(2x - 3)}{(x + 2)(x + 1)(2x - 3)} + \dfrac{2x(x + 2)}{(2x - 3)(x + 1)(x + 2)}$$
$$= \dfrac{2x^2 - 3x}{(x + 2)(x + 1)(2x - 3)} + \dfrac{2x^2 + 4x}{(x + 2)(x + 1)(2x - 3)}$$
$$= \dfrac{2x^2 - 3x + 2x^2 + 4x}{(x + 2)(x + 1)(2x - 3)} = \dfrac{4x^2 + x}{(x + 2)(x + 1)(2x - 3)}$$
$$= \dfrac{x(4x + 1)}{(x + 2)(x + 1)(2x - 3)}$$

23. $\dfrac{a + b}{2a - 3} - \dfrac{a - b}{3 - 2a} = \dfrac{a + b}{2a - 3} - \dfrac{(a - b)(-1)}{(3 - 2a)(-1)}$
$$= \dfrac{a + b}{2a - 3} + \dfrac{a - b}{2a - 3} = \dfrac{2a}{2a - 3}$$
If $3 - 2a$ is used as the common denominator,
the result will be $\dfrac{-2a}{3 - 2a}$. The rational
expressions $\dfrac{2a}{2a - 3}$ and $\dfrac{-2a}{3 - 2a}$ are
equivalent.

24. $\dfrac{y - 2}{y - \frac{4}{y}} = \dfrac{y(y - 2)}{y\left(y - \frac{4}{y}\right)} = \dfrac{y^2 - 2y}{y^2 - 4}$
$$= \dfrac{y(y - 2)}{(y + 2)(y - 2)} = \dfrac{y}{y + 2}$$

25. $\sqrt{18x^5y^8} = \sqrt{(9x^4y^8)(2x)}$
$$= \sqrt{9x^4y^8} \cdot \sqrt{2x} = 3x^2y^4\sqrt{2x}$$

26. $\sqrt{32x} + \sqrt{2x} - \sqrt{18x}$
$$= \sqrt{16 \cdot 2x} + \sqrt{2x} - \sqrt{9 \cdot 2x}$$
$$= 4\sqrt{2x} + \sqrt{2x} - 3\sqrt{2x} = 2\sqrt{2x}$$

27. $(\sqrt{x} - \sqrt{y})(\sqrt{x} + \sqrt{y}) = (\sqrt{x})^2 - (\sqrt{y})^2$
$$= x - y$$

28. $\dfrac{14}{\sqrt{11} - \sqrt{7}} = \dfrac{14}{\sqrt{11} - \sqrt{7}} \cdot \dfrac{\sqrt{11} + \sqrt{7}}{\sqrt{11} + \sqrt{7}}$
$$= \dfrac{14(\sqrt{11} + \sqrt{7})}{11 - 7} = \dfrac{14(\sqrt{11} + \sqrt{7})}{4}$$
$$= \dfrac{7(\sqrt{11} + \sqrt{7})}{2}$$

29. $\left(\dfrac{x^{-2}y^{-1/3}}{x^{-5/3}y^{-2/3}}\right)^3 = \dfrac{x^{-6}y^{-1}}{x^{-5}y^{-2}}$
$$= x^{-6-(-5)}y^{-1-(-2)} = x^{-1}y = \dfrac{y}{x}$$

30. $\left(-\dfrac{64}{27}\right)^{-2/3} = \left(-\dfrac{27}{64}\right)^{2/3} = \left[\left(-\dfrac{27}{64}\right)^{1/3}\right]^2$

$\qquad = \left(-\dfrac{3}{4}\right)^2 = \dfrac{9}{16}$

31. False. For all real numbers x, $\sqrt{x^2} = |x|$.

32. Let $L = 3.5$.

$t = 2\pi\sqrt{\dfrac{L}{32}} = 2\pi\sqrt{\dfrac{3.5}{32}} \approx 2.1$

The period of a pendulum 3.5 ft long is approximately 2.1 seconds.

Chapter 1

EQUATIONS AND INEQUALITIES

Section 1.1   Linear Equations

1. An <u>equation</u> is a statement that two expressions are equal.

3. A linear equation is a <u>first-degree equation</u> because the greatest degree of the variable is 1.

5. A <u>contradiction</u> is an equation that has no solution.

7. True. The left side can be written as
$$5(x-8) = 5[x+(-8)] = 5x + 5(-8)$$
$$= 5x + (-40) = 5x - 40,$$
which is the same as the right side. Therefore, the statement is an identity.

9. False. Solving the literal equation $A = \frac{1}{2}bh$ for h gives
$$A = \frac{1}{2}bh$$
$$2A = bh$$
$$\frac{2A}{b} = h$$

11. $5x + 4 = 3x - 4$
$2x + 4 = -4$
$2x = -8 \Rightarrow x = -4$
Solution set: $\{-4\}$

13. $6(3x-1) = 8 - (10x - 14)$
$18x - 6 = 8 - 10x + 14$
$18x - 6 = 22 - 10x$
$28x - 6 = 22$
$28x = 28 \Rightarrow x = 1$
Solution set: $\{1\}$

15. $\frac{5}{6}x - 2x + \frac{4}{3} = \frac{5}{3}$
$6 \cdot \left[\frac{5}{6}x - 2x + \frac{4}{3} \right] = 6 \cdot \frac{5}{3}$
$5x - 12x + 8 = 10$
$-7x + 8 = 10$
$-7x = 2 \Rightarrow x = -\frac{2}{7}$
Solution set: $\left\{ -\frac{2}{7} \right\}$

17. $3x + 5 - 5(x+1) = 6x + 7$
$3x + 5 - 5x - 5 = 6x + 7$
$-2x = 6x + 7$
$-8x = 7 \Rightarrow x = \frac{7}{-8} = -\frac{7}{8}$
Solution set: $\left\{ -\frac{7}{8} \right\}$

19. $2[x - (4+2x) + 3] = 2x + 2$
$2(x - 4 - 2x + 3) = 2x + 2$
$2(-x - 1) = 2x + 2$
$-2x - 2 = 2x + 2$
$-2 = 4x + 2$
$-4 = 4x$
$-1 = x$
Solution set: $\{-1\}$

21. $\frac{1}{14}(3x-2) = \frac{x+10}{10}$
$70 \cdot \left[\frac{1}{14}(3x-2) \right] = 70 \cdot \left[\frac{x+10}{10} \right]$
$5(3x-2) = 7(x+10)$
$15x - 10 = 7x + 70$
$8x - 10 = 70$
$8x = 80 \Rightarrow x = 10$
Solution set: $\{10\}$

23. $0.2x - 0.5 = 0.1x + 7$
$10(0.2x - 0.5) = 10(0.1x + 7)$
$2x - 5 = x + 70$
$x - 5 = 70$
$x = 75$
Solution set: $\{75\}$

25. $-4(2x-6) + 8x = 5x + 24 + x$
$-8x + 24 + 8x = 6x + 24$
$24 = 6x + 24$
$0 = 6x \Rightarrow 0 = x$
Solution set: $\{0\}$

27. $0.5x + \dfrac{4}{3}x = x + 10$

$\dfrac{1}{2}x + \dfrac{4}{3}x = x + 10$

$6\left(\dfrac{1}{2}x + \dfrac{4}{3}x\right) = 6(x + 10)$

$3x + 8x = 6x + 60$

$11x = 6x + 60$

$5x = 60 \Rightarrow x = 12$

Solution set: $\{12\}$

29. $0.08x + 0.06(x + 12) = 7.72$

$100\big[0.08x + 0.06(x + 12)\big] = 100 \cdot 7.72$

$8x + 6(x + 12) = 772$

$8x + 6x + 72 = 772$

$14x + 72 = 772$

$14x = 700 \Rightarrow x = 50$

Solution set: $\{50\}$

31. $4(2x + 7) = 2x + 22 + 3(2x + 2)$

$8x + 28 = 2x + 22 + 6x + 6$

$8x + 28 = 8x + 28$

$28 = 28 \Rightarrow 0 = 0$

identity; $\{$all real numbers$\}$

33. $2(x - 8) = 3x - 16$

$2x - 16 = 3x - 16$

$-16 = x - 16 \Rightarrow 0 = x$

conditional equation; $\{0\}$

35. $4(x + 7) = 2(x + 12) + 2(x + 1)$

$4x + 28 = 2x + 24 + 2x + 2$

$4x + 28 = 4x + 26$

$28 = 26$

contradiction; $\varnothing$

37. $0.3(x + 2) - 0.5(x + 2) = -0.2x - 0.4$

$10\big[0.3(x + 2) - 0.5(x + 2)\big] = 10\big[-0.2x - 0.4\big]$

$3(x + 2) - 5(x + 2) = -2x - 4$

$3x + 6 - 5x - 10 = -2x - 4$

$-2x - 4 = -2x - 4$

$0 = 0$

identity; $\{$all real numbers$\}$

39. $V = lwh$

$\dfrac{V}{wh} = \dfrac{lwh}{wh}$

$l = \dfrac{V}{wh}$

41. $P = a + b + c$

$P - a - b = c$

$c = P - a - b$

43. $\mathcal{A} = \dfrac{1}{2}h(B + b)$

$2\mathcal{A} = 2\left[\dfrac{1}{2}h(B + b)\right]$

$2\mathcal{A} = h(B + b)$

$2\mathcal{A} = Bh + bh$

$2\mathcal{A} - bh = Bh$

$\dfrac{2\mathcal{A} - bh}{h} = \dfrac{Bh}{h}$

$B = \dfrac{2\mathcal{A} - bh}{h} = \dfrac{2\mathcal{A}}{h} - b$

45. $S = 2\pi rh + 2\pi r^2$

$S - 2\pi r^2 = 2\pi rh$

$\dfrac{S - 2\pi r^2}{2\pi r} = \dfrac{2\pi rh}{2\pi r}$

$h = \dfrac{S - 2\pi r^2}{2\pi r} = \dfrac{S}{2\pi r} - r$

47. $S = 2lw + 2wh + 2hl$

$S - 2lw = 2wh + 2hl$

$S - 2lw = (2w + 2l)h$

$\dfrac{S - 2lw}{2w + 2l} = \dfrac{(2w + 2l)h}{2w + 2l}$

$h = \dfrac{S - 2lw}{2w + 2l}$

49. $2(x - a) + b = 3x + a$

$2x - 2a + b = 3x + a$

$-3a + b = x$

$x = -3a + b$

51. $ax + b = 3(x - a)$

$ax + b = 3x - 3a$

$3a + b = 3x - ax$

$3a + b = (3 - a)x$

$x = \dfrac{3a + b}{3 - a}$

53. $\dfrac{x}{a - 1} = ax + 3$

$(a - 1)\left[\dfrac{x}{a - 1}\right] = (a - 1)(ax + 3)$

$x = a^2x + 3a - ax - 3$

$3 - 3a = a^2x - ax - x$

$3 - 3a = (a^2 - a - 1)x$

$x = \dfrac{3 - 3a}{a^2 - a - 1}$

55.
$$a^2 x + 3x = 2a^2$$
$$\left(a^2 + 3\right)x = 2a^2$$
$$x = \frac{2a^2}{a^2 + 3}$$

57.
$$3x = (2x-1)(m+4)$$
$$3x = 2xm + 8x - m - 4$$
$$m + 4 = 2xm + 5x$$
$$m + 4 = (2m+5)x$$
$$\frac{m+4}{2m+5} = x$$
$$x = \frac{m+4}{2m+5}$$

59. (a) Here, $r = 0.04$, $P = 3150$, and
$$t = \frac{6}{12} = \frac{1}{2} \text{ (year)}.$$
$$I = Prt = 3150(0.04)\left(\frac{1}{2}\right) = \$63$$
The interest is \$63.

(b) The amount Miguel must pay Julio at the end of the six months is
$$\$3150 + \$63 = \$3213.$$

61.
$$F = \frac{9}{5}C + 32$$
$$F = \frac{9}{5} \cdot 20 + 32 = 36 + 32 = 68$$
Therefore, $20°C = 68°F$.

63.
$$C = \frac{5}{9}(F - 32)$$
$$C = \frac{5}{9}(50 - 32) = \frac{5}{9} \cdot 18 = 10$$
Therefore, $50°F = 10°C$.

65.
$$C = \frac{5}{9}(F - 32)$$
$$C = \frac{5}{9}(100 - 32) = \frac{5}{9} \cdot 68 \approx 37.8$$
Therefore, $100°F \approx 37.8°C$.

67.
$$C = \frac{5}{9}(F - 32)$$
$$C = \frac{5}{9}(867 - 32) = \frac{5}{9} \cdot 835 \approx 463.9$$
Therefore, $865°F \approx 463.9°C$.

69.
$$C = \frac{5}{9}(F - 32)$$
$$C = \frac{5}{9}(113 - 32) = \frac{5}{9} \cdot (81) = 45$$
Therefore, $113°F = 45°C$.

Section 1.2 Applications and Modeling with Linear Equations

1. Distance = rate × time, so time = distance ÷ rate. Divide 400 miles by 50 mph to obtain 8 hours.

3. 2% is 0.02, so multiply \$500 by 0.02 and by 4 yrs to get interest of \$40

5. 75% is $\frac{3}{4}$, so multiply 120 L by $\frac{3}{4}$, to get 90 L acid.

7. Concentration A, 24%, cannot possibly be the concentration of the mixture because it is less than both the original concentrations.

9. D

11. In the formula $P = 2l + 2w$, let
$P = 294$ and $w = 57$.
$$294 = 2l + 2 \cdot 57$$
$$294 = 2l + 114$$
$$180 = 2l \Rightarrow 90 = l$$
The length is 90 cm.

13. Let x = the length of the shortest side. Then $2x$ = the length of each of the longer sides.
The perimeter of a triangle is the sum of the measures of the three sides.
$$x + 2x + 2x = 30 \Rightarrow 5x = 30 \Rightarrow x = 6$$
The length of the shortest side is 6 cm.

15. Let x = the length of the shortest side. Then $2x - 200$ = the length of the longest side and the length of the middle side is
$(2x - 200) - 200 = 2x - 400$.
The perimeter of a triangle is the sum of the measures of the three sides.
$$x + (2x - 200) + (2x - 400) = 2400$$
$$x + 2x - 200 + 2x - 400 = 2400$$
$$5x - 600 = 2400$$
$$5x = 3000 \Rightarrow x = 600$$
The length of the shortest side is 600 ft. The middle side is $2 \cdot 600 - 400 = 1200 - 400$
$= 800$ ft. The longest side is $2 \cdot 600 - 200$
$= 1200 - 200 = 1000$ ft.

17. Let h = the height of box.
Use the formula for the surface area of a rectangular box.
$$S = 2lw + 2wh + 2hl$$
$$496 = 2 \cdot 18 \cdot 8 + 2 \cdot 8h + 2h \cdot 18$$
$$496 = 288 + 16h + 36h$$
$$496 = 288 + 52h$$
$$208 = 52h \Rightarrow 4 = h$$
The height of the box is 4 ft.

19. Let x = the time (in hours) spent on the way to the business appointment.

	r	t	d
Morning	50	x	$50x$
Afternoon	40	$x + \frac{1}{4}$	$40\left(x + \frac{1}{4}\right)$

The distance on the way to the business appointment is the same as the return trip, so
$$50x = 40\left(x + \tfrac{1}{4}\right)$$
$$50x = 40x + 10$$
$$10x = 10 \Rightarrow x = 1$$
Because she drove 1 hr, her distance traveled would be $50 \cdot 1 = 50$ mi.

21. Let x = David's speed (in mph) on bike. Then $x + 4.5$ = David's speed (in mph) driving.

	r	t	d
Car	$x + 4.5$	$20\text{ min} = \frac{1}{3}$ hr	$\frac{1}{3}(x + 4.5)$
Bike	x	$45\text{ min} = \frac{3}{4}$ hr	$\frac{3}{4}x$

The distance by bike and car are the same, so
$$\tfrac{1}{3}(x + 4.5) = \tfrac{3}{4}x$$
$$12\left[\tfrac{1}{3}(x + 4.5)\right] = 12\left[\tfrac{3}{4}x\right]$$
$$4(x + 4.5) = 9x$$
$$4x + 18 = 9x$$
$$18 = 5x \Rightarrow \frac{18}{5} = x$$

Because his rate is $\frac{18}{5}$ (or 3.6) mph, David travels $\frac{3}{4}\left(\frac{18}{5}\right) = \frac{27}{10} = 2.7$ mi to work.

23. Let x = time (in hours) it takes for Russ and Janet to be 1.5 mi apart.

	r	t	d
Mary	7	x	$7x$
Janet	5	x	$5x$

Because Mary's rate is faster than Janet's, she travels farther than Janet in the same amount of time. To have the difference between Mary and Janet to be 1.5 mi, solve the following equation.
$$7x - 5x = 1.5 \Rightarrow 2x = 1.5 \Rightarrow x = 0.75$$
It will take 0.75 hr = 45 min for Mary and Janet to be 1.5 mi apart.

25. We need to determine how many meters are in 26 miles.
$$26\text{ mi} \cdot \frac{5,280\text{ ft}}{1\text{ mi}} \cdot \frac{1\text{ m}}{3.281\text{ ft}} \approx 41,840.9\text{ m}$$
Usain Bolt's rate in the 100-m dash would be
$$r = \frac{d}{t} = \frac{100}{9.69} \text{ meters per second.}$$
Thus, the time it would take for Usain to run the 26-mi marathon would be
$$t = \frac{d}{r} = \frac{41,840.9}{\frac{100}{9.69}} = 41,840.9 \cdot \frac{9.69}{100} \approx 4054\text{ sec.}$$
Because there are 60 seconds in one minute and $60 \cdot 60 = 3,600$ seconds in one hour,
$$4054\text{ sec} = 1 \cdot 3600 + 7 \cdot 60 + 34\text{ sec}$$
or 1 hr, 7 min, 34 sec. This is about $\frac{1}{2}$ the world record time.

27. Let x = speed (in km/hr) of Callie's boat. When Callie is traveling upstream, the current slows her down, so we subtract the speed of the current from the speed of the boat. When she is traveling downstream, the current speeds her up, so we add the speed of the current to the speed of the boat.

	r	t	d
Upstream	$x - 5$	$20\text{ min} = \frac{1}{3}$ hr	$\frac{1}{3}(x - 5)$
Downstream	$x + 5$	$15\text{ min} = \frac{1}{4}$ hr	$\frac{1}{4}(x + 5)$

Because the distance upstream and downstream are the same, we must solve the following equation.
$$\tfrac{1}{3}(x - 5) = \tfrac{1}{4}(x + 5)$$
$$12\left[\tfrac{1}{3}(x - 5)\right] = 12\left[\tfrac{1}{4}(x + 5)\right]$$
$$4(x - 5) = 3(x + 5)$$
$$4x - 20 = 3x + 15$$
$$x - 20 = 15 \Rightarrow x = 35$$
The speed of Callie's boat is 35 km per hour.

29. Let x = the amount of 5% acid solution (in gallons).

Strength	Gallons of Solution	Gallons of Pure Acid
5%	x	$0.05x$
10%	5	$0.10 \cdot 5 = 0.5$
7%	$x + 5$	$0.07(x + 5)$

The number of gallons of pure acid in the 5% solution plus the number of gallons of pure acid in the 10% solution must equal the number of gallons of pure acid in the 7% solution.

$$0.05x + 0.5 = 0.07(x + 5)$$
$$0.05x + 0.5 = 0.07x + 0.35$$
$$0.5 = 0.02x + 0.35$$
$$0.15 = 0.02x$$
$$\frac{0.15}{0.02} = x \Rightarrow x = 7.5 = 7\frac{1}{2} \text{ gal}$$

$7\frac{1}{2}$ gallons of the 5% solution should be added.

31. Let x = the amount of 100% alcohol solution (in liters).

Strength	Liters of Solution	Liters of Pure Alcohol
100%	x	$1x = x$
10%	7	$0.10 \cdot 7 = 0.7$
30%	$x + 7$	$0.30(x + 7)$

The number of liters of pure alcohol in the 100% solution plus the number of liters of pure alcohol in the 10% solution must equal the number of liters of pure alcohol in the 30% solution.

$$x + 0.7 = 0.30(x + 7)$$
$$x + 0.7 = 0.30x + 2.1$$
$$0.7x + 0.7 = 2.1$$
$$0.7x = 1.4$$
$$x = \frac{1.4}{0.7} = \frac{14}{7} = 2 \text{ L}$$

2 L of the 100% solution should be added.

33. Let x = the amount of water (in mL).

Strength	Milliliters of Solution	Milliliters of Salt
6%	8	$0.06(8) = 0.48$
0%	x	$0(x) = 0$
4%	$8 + x$	$0.04(8 + x)$

The number of milliliters of salt in the 6% solution plus the number of milliliters of salt in the water (0% solution) must equal the number of milliliters in the 4% solution.

$$0.48 + 0 = 0.04(8 + x)$$
$$0.48 = 0.32 + 0.04x$$
$$0.16 = 0.04x$$
$$\frac{0.16}{0.04} = x \Rightarrow x = \frac{16}{4} = 4 \text{ mL}$$

To reduce the saline concentration to 4%, 4 mL of water should be added.

35. Let x = amount of the short-term note. Then $240,000 - x$ = amount of the long-term note.

Note Amount	Interest Rate	Time (years)	Interest Paid
x	2%	1	$x(0.02)(1)$
$240,000 - x$	2.5%	1	$(240,000 - x)(0.025)(1)$
			5500

The amount of interest from the 2% note plus the amount of interest from the 2.5% note must equal the total amount of interest.

$$0.02x + 0.025(240,000 - x) = 5500$$
$$0.02x + 6000 - 0.025x = 5500$$
$$-0.005x + 6000 = 5500$$
$$-0.005x = -500$$
$$x = 100,000$$

The amount of the short-term note is $100,000 and the amount of the long-term note is $240,000 - \$100,000 = \$140,000$.

37. Let x = amount invested at 2.5%. Then $2x$ = amount invested at 3%.

Amount in Account	Interest Rate	Interest
x	2.5%	$0.025x$
$2x$	3%	$0.03(2x) = 0.06x$
		850

The amount of interest from the 2.5% account plus the amount of interest from the 3% account must equal the total amount of interest.

$$0.025x + 0.03(2x) = 850$$
$$0.025x + 0.06x = 850$$
$$0.085x = 850 \Rightarrow x = \$10,000$$

Janet deposited $10,000 at 2.5% and $2(\$10,000) = \$20,000$ at 3%.

39. 30% of $200,000 is $60,000, so after paying her income tax, Linda had $140,000 left to invest. Let x = amount invested at 1.5%. Then $140,000 - x$ = amount invested at 4%.

Amount Invested	Interest Rate	Interest
x	1.5%	$0.015x$
$140,000 - x$	4%	$0.04(140,000 - x)$
$140,000$		4350

$$0.015x + 0.04(140,000 - x) = 4350$$
$$0.015x + 5600 - 0.04x = 4350$$
$$-0.025x + 5600 = 4350$$
$$-0.025x = -1250$$
$$x = \$50,000$$

Linda invested $50,000 at 1.5% and $140,000 - \$50,000 = \$90,000$ at 4%.

41. **(a)** $y = 100 - 0.02x$
$y = 100 - 0.02(2400) = 100 - 48 = 52$
The annual cost is $52.

(b) $50 = 100 - 0.02x$
$-50 = -0.02x$
$2500 = x$
The annual cost of membership will be $50 if the club purchases are $2500.

(c) $0 = 100 - 0.02x$
$-100 = -0.02x$
$5000 = x$
The annual cost of membership will be $0 if the club purchases are $5000.

43. **(a)** Let x = the number of hours. Then $F = 100(140)x = 14,000x$.

(b) Because $33\ \mu g/ft^3$ causes irritation, the room would need $33 \cdot 800 = 26,400\ \mu g$ to cause irritation.
$$F = 14,000x$$
$$26,400 = 14,000x$$
$$\frac{26,400}{14,000} = x$$
$$x \approx 1.9\ \text{hr}$$
It will take about 1.9 hours for concentrations to reach $33\ \mu g/ft^3$.

45. **(a)** In 2018, $x = 4$.
$y = 0.3143x + 21.95$
$y = 0.3143 \cdot 4 + 21.95 = 23.2072$
The projected enrollment for fall 2018 is approximately 23.2 million.

(b) $y = 0.3143x + 21.95$
$24 = 0.3143x + 21.95$
$2.05 = 0.3143x$
$$x = \frac{2.05}{0.3143} \approx 6.52$$
$2014 + 6 = 2020$
Enrollment is projected to reach 24 million during the year 2020.

(c) They are quite close.

(d) The year 2000 is represented by $x = -14$.
$y = 0.3143x + 21.95$
$y = 0.3143(-14) + 21.95 \approx 17.5$
According to the model, the enrollment was approximately 17.5 million

(e) Answers will vary. Sample answer: When using the model for predictions, it is best to stay within the scope of the sample data.

Section 1.3   Complex Numbers

1. By definition, $i = \sqrt{-1}$, and therefore, $i^2 = \underline{-1}$.

3. The numbers $6 + 5i$ and $6 - 5i$, which differ only in the sign of their imaginary parts, are <u>complex conjugates</u>.

5. To find the quotient of two complex numbers in standard form, multiply both the numerator and the denominator by the complex conjugate of the <u>denominator</u>.

7. True. $\sqrt{-4} \cdot \sqrt{-9} = 2i \cdot 3i = 6i^2 = -6$

9. False.
$$(-2 + 7i) - (10 - 6i) = -2 + 7i - 10 + 6i$$
$$= -12 + 13i$$

11. -4 is real and complex.

13. $13i$ is complex, pure imaginary, and nonreal complex.

15. $5 + i$ is complex and nonreal complex.

17. π is real and complex.

19. $\sqrt{-25} = 5i$ is complex, pure imaginary, and nonreal complex.

21. $\sqrt{-25} = i\sqrt{25} = 5i$

23. $\sqrt{-10} = i\sqrt{10}$

25. $\sqrt{-288} = i\sqrt{288} = i\sqrt{144 \cdot 2} = 12i\sqrt{2}$

27. $-\sqrt{-18} = -i\sqrt{18} = -i\sqrt{9 \cdot 2} = -3i\sqrt{2}$

29. $\sqrt{-13} \cdot \sqrt{-13} = i\sqrt{13} \cdot i\sqrt{13}$
$$= i^2 \left(\sqrt{13}\right)^2 = -1 \cdot 13 = -13$$

31. $\sqrt{-3} \cdot \sqrt{-8} = i\sqrt{3} \cdot i\sqrt{8} = i^2\sqrt{3 \cdot 8}$
$$= -1 \cdot \sqrt{24} = -\sqrt{4 \cdot 6} = -2\sqrt{6}$$

33. $\dfrac{\sqrt{-30}}{\sqrt{-10}} = \dfrac{i\sqrt{30}}{i\sqrt{10}} = \sqrt{\dfrac{30}{10}} = \sqrt{3}$

35. $\dfrac{\sqrt{-24}}{\sqrt{8}} = \dfrac{i\sqrt{24}}{\sqrt{8}} = i\sqrt{\dfrac{24}{8}} = i\sqrt{3}$

37. $\dfrac{\sqrt{-10}}{\sqrt{-40}} = \dfrac{i\sqrt{10}}{i\sqrt{40}} = \sqrt{\dfrac{10}{40}} = \sqrt{\dfrac{1}{4}} = \dfrac{1}{2}$

39. $\dfrac{\sqrt{-6} \cdot \sqrt{-2}}{\sqrt{3}} = \dfrac{i\sqrt{6} \cdot i\sqrt{2}}{\sqrt{3}} = i^2\sqrt{\dfrac{6 \cdot 2}{3}}$
$$= -1 \cdot \sqrt{\dfrac{12}{3}} = -\sqrt{4} = -2$$

41. $\dfrac{-6 - \sqrt{-24}}{2} = \dfrac{-6 - \sqrt{-4 \cdot 6}}{2} = \dfrac{-6 - 2i\sqrt{6}}{2}$
$$= \dfrac{2\left(-3 - i\sqrt{6}\right)}{2} = -3 - i\sqrt{6}$$

43. $\dfrac{10 + \sqrt{-200}}{5} = \dfrac{10 + \sqrt{-100 \cdot 2}}{5}$
$$= \dfrac{10 + 10i\sqrt{2}}{5} = \dfrac{5\left(2 + 2i\sqrt{2}\right)}{5}$$
$$= 2 + 2i\sqrt{2}$$

45. $\dfrac{-3 + \sqrt{-18}}{24} = \dfrac{-3 + \sqrt{-9 \cdot 2}}{24} = \dfrac{-3 + 3i\sqrt{2}}{24}$
$$= \dfrac{3\left(-1 + i\sqrt{2}\right)}{24} = \dfrac{-1 + i\sqrt{2}}{8}$$
$$= -\dfrac{1}{8} + \dfrac{\sqrt{2}}{8}i$$

47. $(3 + 2i) + (9 - 3i) = (3 + 9) + \left[2 + (-3)\right]i$
$$= 12 + (-1)i = 12 - i$$

49. $(-2 + 4i) - (-4 + 4i)$
$$= \left[-2 - (-4)\right] + (4 - 4)i$$
$$= 2 + 0i = 2$$

51. $(2 - 5i) - (3 + 4i) - (-2 + i)$
$$= \left[2 - 3 - (-2)\right] + (-5 - 4 - 1)i$$
$$= 1 + (-10)i = 1 - 10i$$

53. $-i\sqrt{2} - 2 - \left(6 - 4i\sqrt{2}\right) - \left(5 - i\sqrt{2}\right)$
$$= (-2 - 6 - 5) + \left[-\sqrt{2} - \left(-4\sqrt{2}\right) - \left(-\sqrt{2}\right)\right]i$$
$$= -13 + 4i\sqrt{2}$$

55. $(2 + i)(3 - 2i)$
$$= 2(3) + 2(-2i) + i(3) + i(-2i)$$
$$= 6 - 4i + 3i - 2i^2 = 6 - i - 2(-1)$$
$$= 6 - i + 2 = 8 - i$$

57. $(2 + 4i)(-1 + 3i)$
$$= 2(-1) + 2(3i) + 4i(-1) + 4i(3i)$$
$$= -2 + 6i - 4i + 12i^2 = -2 + 2i + 12(-1)$$
$$= -2 + 2i - 12 = -14 + 2i$$

59. $(3 - 2i)^2 = 3^2 - 2(3)(2i) + (2i)^2$
$$= 9 - 12i - 4 = 5 - 12i$$

61. $(3 + i)(3 - i) = 3^2 - i^2 = 9 - (-1) = 10$

63. $(-2 - 3i)(-2 + 3i) = (-2)^2 - (3i)^2 = 4 - 9i^2$
$$= 4 - 9(-1) = 13$$

65. $\left(\sqrt{6} + i\right)\left(\sqrt{6} - i\right) = \left(\sqrt{6}\right)^2 - i^2$
$$= 6 - (-1) = 6 + 1 = 7$$

67. $i(3 - 4i)(3 + 4i) = i\left[(3 - 4i)(3 + 4i)\right]$
$$= i\left[3^2 - (4i)^2\right]$$
$$= i\left[9 - 16i^2\right]$$
$$= i\left[9 - 16(-1)\right]$$
$$= i(9 + 16) = 25i$$

69. $3i(2 - i)^2 = 3i\left(2^2 - 2(2i) + i^2\right)$
$$= 3i(4 - 4i - 1) = 3i(3 - 4i)$$
$$= 9i - 12i^2 = 9i - 12(-1)$$
$$= 12 + 9i$$

71. $(2 + i)(2 - i)(4 + 3i) = \left[(2 + i)(2 - i)\right](4 + 3i)$
$$= \left[2^2 - i^2\right](4 + 3i)$$
$$= \left[4 - (-1)\right](4 + 3i)$$
$$= 5(4 + 3i) = 20 + 15i$$

73. $\dfrac{6+2i}{1+2i} = \dfrac{(6+2i)(1-2i)}{(1+2i)(1-2i)}$

$\phantom{\dfrac{6+2i}{1+2i}}= \dfrac{6-12i+2i-4i^2}{1^2-(2i)^2} = \dfrac{6-10i-4(-1)}{1-4i^2}$

$\phantom{\dfrac{6+2i}{1+2i}}= \dfrac{6-10i+4}{1-4(-1)} = \dfrac{10-10i}{1+4} = \dfrac{10-10i}{5}$

$\phantom{\dfrac{6+2i}{1+2i}}= \dfrac{10}{5} - \dfrac{10}{5}i = 2-2i$

75. $\dfrac{2-i}{2+i} = \dfrac{(2-i)(2-i)}{(2+i)(2-i)} = \dfrac{2^2-2(2i)+i^2}{2^2-i^2}$

$\phantom{\dfrac{2-i}{2+i}}= \dfrac{4-4i+(-1)}{4-(-1)} = \dfrac{3-4i}{5} = \dfrac{3}{5} - \dfrac{4}{5}i$

77. $\dfrac{1-3i}{1+i} = \dfrac{(1-3i)(1-i)}{(1+i)(1-i)} = \dfrac{1-i-3i+3i^2}{1^2-i^2}$

$\phantom{\dfrac{1-3i}{1+i}}= \dfrac{1-4i+3(-1)}{1-(-1)} = \dfrac{1-4i-3}{2}$

$\phantom{\dfrac{1-3i}{1+i}}= \dfrac{-2-4i}{2} = \dfrac{-2}{2} - \dfrac{4}{2}i = -1-2i$

79. $\dfrac{-5}{i} = \dfrac{-5(-i)}{i(-i)} = \dfrac{5i}{-i^2}$

$\phantom{\dfrac{-5}{i}}= \dfrac{5i}{-(-1)} = \dfrac{5i}{1} = 5i \text{ or } 0+5i$

81. $\dfrac{8}{-i} = \dfrac{8 \cdot i}{-i \cdot i} = \dfrac{8i}{-i^2}$

$\phantom{\dfrac{8}{-i}}= \dfrac{8i}{-(-1)} = \dfrac{8i}{1} = 8i \text{ or } 0+8i$

83. $\dfrac{2}{3i} = \dfrac{2(-3i)}{3i \cdot (-3i)} = \dfrac{-6i}{-9i^2} = \dfrac{-6i}{-9(-1)}$

$\phantom{\dfrac{2}{3i}}= \dfrac{-6i}{9} = -\dfrac{2}{3}i \text{ or } 0-\dfrac{2}{3}i$

Note: In the above solution, we multiplied the numerator and denominator by the complex conjugate of $3i$, namely $-3i$. Because there is a reduction in the end, the same results can be achieved by multiplying the numerator and denominator by $-i$.

85. $I = 5+7i, Z = 6+4i$

$E = IZ$

$E = (5+7i)(6+4i)$

$= 5(6) + 5(4i) + 7i(6) + 7i(4i)$

$= 30+20i+42i+28i^2$

$= 30+62i-28$

$= 2+62i$

87. $I = 10+4i, \ E = 88+128i$

$E = IZ$

$88+128i = (10+4i)Z$

$Z = \dfrac{88+128i}{10+4i}$

$Z = \dfrac{(88+128i)(10-4i)}{(10+4i)(10-4i)}$

$Z = \dfrac{88(10)+88(-4i)+128i(10)+128i(-4i)}{10^2-(4i)^2}$

$Z = \dfrac{880-352i+1280i-512i^2}{100-(-16)}$

$Z = \dfrac{880-352i+1280i+512}{116}$

$Z = \dfrac{1392+928i}{116} = 12+8i$

89. $i^{25} = i^{24} \cdot i = \left(i^4\right)^6 \cdot i = 1^6 \cdot i = i$

91. $i^{22} = i^{20} \cdot i^2 = \left(i^4\right)^5 \cdot (-1) = 1^5 \cdot (-1) = -1$

93. $i^{23} = i^{20} \cdot i^3 = \left(i^4\right)^5 \cdot i^3 = 1^5 \cdot (-i) = -i$

95. $i^{32} = \left(i^4\right)^8 = 1^8 = 1$

97. $i^{-13} = i^{-16} \cdot i^3 = \left(i^4\right)^{-4} \cdot i^3 = 1^{-4} \cdot (-i) = -i$

99. $\dfrac{1}{i^{-11}} = i^{11} = i^8 \cdot i^3 = \left(i^4\right)^2 \cdot i^3 = 1^2 \cdot (-i) = -i$

101. We need to show that $\left(\dfrac{\sqrt{2}}{2} + \dfrac{\sqrt{2}}{2}i\right)^2 = i.$

$\left(\dfrac{\sqrt{2}}{2} + \dfrac{\sqrt{2}}{2}i\right)^2$

$= \left(\dfrac{\sqrt{2}}{2}\right)^2 + 2 \cdot \dfrac{\sqrt{2}}{2} \cdot \dfrac{\sqrt{2}}{2}i + \left(\dfrac{\sqrt{2}}{2}i\right)^2$

$= \dfrac{2}{4} + 2 \cdot \dfrac{2}{4}i + \dfrac{2}{4}i^2 = \dfrac{1}{2} + i + \dfrac{1}{2}i^2$

$= \dfrac{1}{2} + i + \dfrac{1}{2}(-1) = \dfrac{1}{2} + i - \dfrac{1}{2} = i$

Thus, $\dfrac{\sqrt{2}}{2} + \dfrac{\sqrt{2}}{2}i$ is a square root of i.

103. We need to show that $\left(\dfrac{\sqrt{3}}{2}+\dfrac{1}{2}i\right)^3 = i$.

$$\left(\dfrac{\sqrt{3}}{2}+\dfrac{1}{2}i\right)^3$$

$$=\left(\dfrac{\sqrt{3}}{2}+\dfrac{1}{2}i\right)\left(\dfrac{\sqrt{3}}{2}+\dfrac{1}{2}i\right)^2$$

$$=\left(\dfrac{\sqrt{3}}{2}+\dfrac{1}{2}i\right)\left[\left(\dfrac{\sqrt{3}}{2}\right)^2 + 2\cdot\dfrac{\sqrt{3}}{2}\cdot\dfrac{1}{2}i + \left(\dfrac{1}{2}i\right)^2\right]$$

$$=\left(\dfrac{\sqrt{3}}{2}+\dfrac{1}{2}i\right)\left[\dfrac{3}{4}+\dfrac{\sqrt{3}}{2}i+\dfrac{1}{4}i^2\right]$$

$$=\left(\dfrac{\sqrt{3}}{2}+\dfrac{1}{2}i\right)\left[\dfrac{3}{4}+\dfrac{\sqrt{3}}{2}i+\dfrac{1}{4}(-1)\right]$$

$$=\left(\dfrac{\sqrt{3}}{2}+\dfrac{1}{2}i\right)\left[\dfrac{3}{4}+\dfrac{\sqrt{3}}{2}i-\dfrac{1}{4}\right]$$

$$=\left(\dfrac{\sqrt{3}}{2}+\dfrac{1}{2}i\right)\left(\dfrac{2}{4}+\dfrac{\sqrt{3}}{2}i\right)$$

$$=\left(\dfrac{\sqrt{3}}{2}+\dfrac{1}{2}i\right)\left(\dfrac{1}{2}+\dfrac{\sqrt{3}}{2}i\right)$$

$$=\dfrac{\sqrt{3}}{2}\cdot\dfrac{1}{2}+\dfrac{\sqrt{3}}{2}\cdot\dfrac{\sqrt{3}}{2}i+\dfrac{1}{2}\cdot\dfrac{1}{2}i+\dfrac{1}{2}\dfrac{\sqrt{3}}{2}i^2$$

$$=\dfrac{\sqrt{3}}{4}+\dfrac{3}{4}i+\dfrac{1}{4}i+\dfrac{\sqrt{3}}{4}i^2$$

$$=\dfrac{\sqrt{3}}{4}+\dfrac{4}{4}i+\dfrac{\sqrt{3}}{4}(-1)=\dfrac{\sqrt{3}}{4}+i+\left(-\dfrac{\sqrt{3}}{4}\right)=i$$

105. If $-2+i$ is a solution of the equation, then substituting that value for x makes a true statement.

$$x^2+4x+5=0$$

$$(-2+i)^2+4(-2+i)+5=0$$

$$4-4i+i^2-8+4i+5=0$$

$$4-4i-1-8+4i+5=0$$

$$0=0$$

107. If $-3+4i$ is a solution of the equation, then substituting that value for x makes a true statement.

$$x^2+6x+25=0$$

$$(-3+4i)^2+6(-3+4i)+25=0$$

$$9-24i+16i^2-18+24i+25=0$$

$$9-24i-16-18+24i+25=0$$

$$0=0$$

Section 1.4 Quadratic Equations

1. G; $x^2=25\Rightarrow x=\pm\sqrt{25}=\pm5$

3. C; $x^2+5=0\Rightarrow x^2=-5\Rightarrow x=\pm\sqrt{-5}=\pm i\sqrt{5}$

5. H; $x^2=-20\Rightarrow x=\pm\sqrt{-20}=\pm2i\sqrt{5}$

7. D; $x-5=0\Rightarrow x=5$

9. D is the only one set up for direct use of the zero-factor property.

$$(3x-1)(x-7)=0$$

$$3x-1=0 \quad\text{or}\quad x-7=0$$

$$x=\tfrac{1}{3}\quad\text{or}\qquad x=7$$

Solution set: $\left\{\tfrac{1}{3},7\right\}$

11. C is the only one that does not require Step 1 of the method of completing the square.

$$x^2+x=12 \qquad\text{Note:}$$

$$x^2+x+\tfrac{1}{4}=12+\tfrac{1}{4}\qquad\left[\tfrac{1}{2}\cdot1\right]^2=\left(\tfrac{1}{2}\right)^2=\tfrac{1}{4}$$

$$\left(x+\tfrac{1}{2}\right)^2=\tfrac{49}{4}$$

$$x+\tfrac{1}{2}=\pm\sqrt{\tfrac{49}{4}}$$

$$x+\tfrac{1}{2}=\pm\tfrac{7}{2}\Rightarrow x=-\tfrac{1}{2}\pm\tfrac{7}{2}$$

$$-\tfrac{1}{2}-\tfrac{7}{2}=\tfrac{-8}{2}=-4 \text{ and } -\tfrac{1}{2}+\tfrac{7}{2}=\tfrac{6}{2}=3$$

Solution set: $\{-4,3\}$

13. $$x^2-5x+6=0$$
$$(x-2)(x-3)=0$$
$$x-2=0\Rightarrow x=2 \text{ or } x-3=0\Rightarrow x=3$$
Solution set: $\{2,3\}$

15. $$5x^2-3x-2=0$$
$$(5x+2)(x-1)=0$$
$$5x+2=0\Rightarrow x=-\tfrac{2}{5} \text{ or } x-1=0\Rightarrow x=1$$
Solution set: $\left\{-\tfrac{2}{5},1\right\}$

17. $$-4x^2+x=-3$$
$$0=4x^2-x-3$$
$$0=(4x+3)(x-1)$$
$$4x+3=0\Rightarrow x=-\tfrac{3}{4} \text{ or } x-1=0\Rightarrow x=1$$
Solution set: $\left\{-\tfrac{3}{4},1\right\}$

19. $$x^2-100=0$$
$$(x+10)(x-10)=0$$
$$x+10=0\Rightarrow x=-10 \text{ or } x-10=0\Rightarrow x=10$$
Solution set: $\{-10, 10\}$

21. $4x^2 - 4x + 1 = 0$

$\quad (2x - 1)^2 = 0$

$\quad\quad 2x - 1 = 0 \Rightarrow 2x = 1 \Rightarrow x = \frac{1}{2}$

Solution set: $\left\{ \frac{1}{2} \right\}$

23. $25x^2 + 30x + 9 = 0$

$\quad (5x + 3)^2 = 0$

$\quad\quad 5x + 3 = 0 \Rightarrow 5x = -3 \Rightarrow x = -\frac{3}{5}$

Solution set: $\left\{ -\frac{3}{5} \right\}$

25. $x^2 = 16$

$\quad x = \pm\sqrt{16} = \pm 4$

Solution set: $\{ \pm 4 \}$

27. $27 - x^2 = 0$

$\quad 27 = x^2$

$\quad\quad x = \pm\sqrt{27} = \pm 3\sqrt{3}$

Solution set: $\left\{ \pm 3\sqrt{3} \right\}$

29. $x^2 = -81$

$\quad x = \pm\sqrt{-81} = \pm 9i$

Solution set: $\{ \pm 9i \}$

31. $(3x - 1)^2 = 12$

$\quad 3x - 1 = \pm\sqrt{12}$

$\quad\quad 3x = 1 \pm 2\sqrt{3} \Rightarrow x = \frac{1 \pm 2\sqrt{3}}{3}$

Solution set: $\left\{ \frac{1 \pm 2\sqrt{3}}{3} \right\}$

33. $(x + 5)^2 = -3$

$\quad x + 5 = \pm\sqrt{-3}$

$\quad x + 5 = \pm i\sqrt{3}$

$\quad\quad x = -5 \pm i\sqrt{3}$

Solution set: $\left\{ -5 \pm i\sqrt{3} \right\}$

35. $(5x - 3)^2 = -3$

$\quad 5x - 3 = \pm\sqrt{-3}$

$\quad 5x - 3 = \pm i\sqrt{3}$

$\quad 5x = 3 \pm i\sqrt{3}$

$\quad\quad x = \frac{3 \pm i\sqrt{3}}{5} = \frac{3}{5} \pm \frac{\sqrt{3}}{5}i$

Solution set: $\left\{ \frac{3}{5} \pm \frac{\sqrt{3}}{5}i \right\}$

37. $x^2 - 4x + 3 = 0$

$\quad x^2 - 4x + 4 = -3 + 4 \quad$ Note: $\left[\frac{1}{2} \cdot (-4) \right]^2 = (-2)^2 = 4$

$\quad\quad (x - 2)^2 = 1$

$\quad\quad\quad x - 2 = \pm\sqrt{1}$

$\quad\quad\quad x - 2 = \pm 1$

$\quad\quad\quad\quad x = 2 \pm 1$

$\quad 2 - 1 = 1$ and $2 + 1 = 3$

Solution set: $\{ 1, 3 \}$

39. $2x^2 - x - 28 = 0$

$\quad x^2 - \frac{1}{2}x - 14 = 0 \quad$ Multiply by $\frac{1}{2}$.

$\quad x^2 - \frac{1}{2}x + \frac{1}{16} = 14 + \frac{1}{16}$

$\quad\quad$ Note: $\left[\frac{1}{2} \cdot \left(-\frac{1}{2} \right) \right]^2 = \left(-\frac{1}{4} \right)^2 = \frac{1}{16}$

$\quad\quad \left(x - \frac{1}{4} \right)^2 = \frac{225}{16}$

$\quad\quad\quad x - \frac{1}{4} = \pm\sqrt{\frac{225}{16}}$

$\quad\quad\quad x - \frac{1}{4} = \pm\frac{15}{4}$

$\quad\quad\quad\quad x = \frac{1}{4} \pm \frac{15}{4}$

$\quad \frac{1}{4} - \frac{15}{4} = \frac{-14}{4} = -\frac{7}{2}$ and $\frac{1}{4} + \frac{15}{4} = \frac{16}{4} = 4$

Solution set: $\left\{ -\frac{7}{2}, 4 \right\}$

41. $x^2 - 2x - 2 = 0$

$\quad x^2 - 2x + 1 = 2 + 1$

$\quad\quad$ Note: $\left[\frac{1}{2} \cdot (-2) \right]^2 = (-1)^2 = 1$

$\quad\quad (x - 1)^2 = 3$

$\quad\quad\quad x - 1 = \pm\sqrt{3}$

$\quad\quad\quad\quad x = 1 \pm \sqrt{3}$

Solution set: $\left\{ 1 \pm \sqrt{3} \right\}$

43. $\quad 2x^2 + x = 10$

$\quad\quad x^2 + \frac{1}{2}x = 5$

$\quad x^2 + \frac{1}{2}x + \frac{1}{16} = 5 + \frac{1}{16} \quad$ Note: $\left[\frac{1}{2} \cdot \frac{1}{2} \right]^2 = \left(\frac{1}{4} \right)^2 = \frac{1}{16}$

$\quad\quad \left(x + \frac{1}{4} \right)^2 = \frac{81}{16}$

$\quad\quad\quad x + \frac{1}{4} = \pm\sqrt{\frac{81}{16}}$

$\quad\quad\quad x + \frac{1}{4} = \pm\frac{9}{4}$

$\quad\quad\quad\quad x = -\frac{1}{4} \pm \frac{9}{4}$

$\quad -\frac{1}{4} - \frac{9}{4} = \frac{-10}{4} = -\frac{5}{2}$ and $-\frac{1}{4} + \frac{9}{4} = \frac{8}{4} = 2$

Solution set: $\left\{ -\frac{5}{2}, 2 \right\}$

45. $-2x^2 + 4x + 3 = 0$

$x^2 - 2x - \frac{3}{2} = 0$

$x^2 - 2x + 1 = \frac{3}{2} + 1$ Note: $\left[\frac{1}{2} \cdot (-2)\right]^2 = (-1)^2 = 1$

$(x - 1)^2 = \frac{5}{2}$

$x - 1 = \pm\sqrt{\frac{5}{2}} = \pm\frac{\sqrt{10}}{2}$

$x = 1 \pm \frac{\sqrt{10}}{2} = \frac{2 \pm \sqrt{10}}{2}$

Solution set: $\left\{\frac{2 \pm \sqrt{10}}{2}\right\}$

47. $-4x^2 + 8x = 7$

$x^2 - 2x = -\frac{7}{4}$

$x^2 - 2x + 1 = -\frac{7}{4} + 1$ Note: $\left[\frac{1}{2} \cdot (-2)\right]^2 = (-1)^2 = 1$

$(x - 1)^2 = \frac{-3}{4}$

$x - 1 = \pm\sqrt{\frac{-3}{4}} = \pm\frac{i\sqrt{3}}{2}$

$x = 1 \pm \frac{\sqrt{3}}{2} i$

Solution set: $\left\{1 \pm \frac{\sqrt{3}}{2} i\right\}$

49. Francisco is incorrect because $c = 0$ and the quadratic formula, $x = \frac{-b \pm \sqrt{b^2 - 4ac}}{2a}$, can be evaluated with $a = 1$, $b = -8$, and $c = 0$.

51. $x^2 - x - 1 = 0$

Let $a = 1$, $b = -1$, and $c = -1$.

$x = \frac{-b \pm \sqrt{b^2 - 4ac}}{2a}$

$= \frac{-(-1) \pm \sqrt{(-1)^2 - 4(1)(-1)}}{2(1)}$

$= \frac{1 \pm \sqrt{1 + 4}}{2} = \frac{1 \pm \sqrt{5}}{2}$

Solution set: $\left\{\frac{1 \pm \sqrt{5}}{2}\right\}$

53. $x^2 - 6x = -7$

$x^2 - 6x + 7 = 0$

Let $a = 1$, $b = -6$, and $c = 7$.

$x = \frac{-b \pm \sqrt{b^2 - 4ac}}{2a}$

$= \frac{-(-6) \pm \sqrt{(-6)^2 - 4(1)(7)}}{2(1)} = \frac{6 \pm \sqrt{36 - 28}}{2}$

$= \frac{6 \pm \sqrt{8}}{2} = \frac{6 \pm 2\sqrt{2}}{2} = 3 \pm \sqrt{2}$

Solution set: $\left\{3 \pm \sqrt{2}\right\}$

55. $x^2 = 2x - 5$

$x^2 - 2x + 5 = 0$

Let $a = 1$, $b = -2$, and $c = 5$.

$x = \frac{-b \pm \sqrt{b^2 - 4ac}}{2a}$

$= \frac{-(-2) \pm \sqrt{(-2)^2 - 4(1)(5)}}{2(1)}$

$= \frac{2 \pm \sqrt{4 - 20}}{2} = \frac{2 \pm \sqrt{-16}}{2} = \frac{2 \pm 4i}{2} = 1 \pm 2i$

Solution set: $\left\{1 \pm 2i\right\}$

57. $-4x^2 = -12x + 11$

$0 = 4x^2 - 12x + 11$

Let $a = 4$, $b = -12$, and $c = 11$.

$x = \frac{-b \pm \sqrt{b^2 - 4ac}}{2a}$

$= \frac{-(-12) \pm \sqrt{(-12)^2 - 4(4)(11)}}{2(4)}$

$= \frac{12 \pm \sqrt{144 - 176}}{8} = \frac{12 \pm \sqrt{-32}}{8}$

$= \frac{12 \pm 4i\sqrt{2}}{8} = \frac{12}{8} \pm \frac{4\sqrt{2}}{8} i = \frac{3}{2} \pm \frac{\sqrt{2}}{2} i$

Solution set: $\left\{\frac{3}{2} \pm \frac{\sqrt{2}}{2} i\right\}$

59. $\frac{1}{2}x^2 + \frac{1}{4}x - 3 = 0$

$4\left(\frac{1}{2}x^2 + \frac{1}{4}x - 3\right) = 4 \cdot 0$

$2x^2 + x - 12 = 0$

Let $a = 2$, $b = 1$, and $c = -12$.

$x = \dfrac{-b \pm \sqrt{b^2 - 4ac}}{2a} = \dfrac{-1 \pm \sqrt{1^2 - 4(2)(-12)}}{2(2)}$

$= \dfrac{-1 \pm \sqrt{1 + 96}}{4} = \dfrac{-1 \pm \sqrt{97}}{4}$

Solution set: $\left\{\frac{-1 \pm \sqrt{97}}{4}\right\}$

61. $0.2x^2 + 0.4x - 0.3 = 0$

$10\left(0.2x^2 + 0.4x - 0.3\right) = 10 \cdot 0$

$2x^2 + 4x - 3 = 0$

Let $a = 2$, $b = 4$, and $c = -3$.

$x = \dfrac{-b \pm \sqrt{b^2 - 4ac}}{2a}$

$= \dfrac{-4 \pm \sqrt{4^2 - 4(2)(-3)}}{2(2)} = \dfrac{-4 \pm \sqrt{16 + 24}}{4}$

$= \dfrac{-4 \pm \sqrt{40}}{4} = \dfrac{-4 \pm 2\sqrt{10}}{4} = \dfrac{-2 \pm \sqrt{10}}{2}$

Solution set: $\left\{\frac{-2 \pm \sqrt{10}}{2}\right\}$

63. $(4x - 1)(x + 2) = 4x$

$4x^2 + 7x - 2 = 4x \Rightarrow 4x^2 + 3x - 2 = 0$

Let $a = 4$, $b = 3$, and $c = -2$.

$x = \dfrac{-b \pm \sqrt{b^2 - 4ac}}{2a} = \dfrac{-3 \pm \sqrt{3^2 - 4(4)(-2)}}{2(4)}$

$= \dfrac{-3 \pm \sqrt{9 + 32}}{8} = \dfrac{-3 \pm \sqrt{41}}{8}$

Solution set: $\left\{\frac{-3 \pm \sqrt{41}}{8}\right\}$

65. $(x - 9)(x - 1) = -16$

$x^2 - 10x + 9 = -16$

$x^2 - 10x + 25 = 0$

Let $a = 1$, $b = -10$, and $c = 25$.

$x = \dfrac{-(-10) \pm \sqrt{(-10)^2 - (4)(1)(25)}}{2(1)}$

$= \dfrac{10 \pm \sqrt{100 - 100}}{2} = \dfrac{10 \pm 0}{2} = 5$

Solution set: $\{5\}$

67. $x^3 - 8 = 0$

$x^3 - 2^3 = 0$

$(x - 2)(x^2 + 2x + 4) = 0$

$x - 2 = 0 \Rightarrow x = 2$ or

$x^2 + 2x + 4 = 0$

$a = 1$, $b = 2$, and $c = 4$

$x = \dfrac{-b \pm \sqrt{b^2 - 4ac}}{2a}$

$= \dfrac{-2 \pm \sqrt{2^2 - 4(1)(4)}}{2(1)} = \dfrac{-2 \pm \sqrt{4 - 16}}{2}$

$= \dfrac{-2 \pm \sqrt{-12}}{2} = \dfrac{-2 \pm 2i\sqrt{3}}{2} = -1 \pm \sqrt{3}i$

Solution set: $\left\{2, -1 \pm \sqrt{3}i\right\}$

69. $x^3 + 27 = 0$

$x^3 + 3^3 = 0$

$(x + 3)(x^2 - 3x + 9) = 0$

$x + 3 = 0 \Rightarrow x = -3$ or

$x^2 - 3x + 9 = 0$

$a = 1$, $b = -3$, and $c = 9$

$x = \dfrac{-b \pm \sqrt{b^2 - 4ac}}{2a}$

$= \dfrac{-(-3) \pm \sqrt{(-3)^2 - 4(1)(9)}}{2(1)} = \dfrac{3 \pm \sqrt{9 - 36}}{2}$

$= \dfrac{3 \pm \sqrt{-27}}{2} = \dfrac{3 \pm 3i\sqrt{3}}{2} = \dfrac{3}{2} \pm \dfrac{3\sqrt{3}}{2}i$

Solution set: $\left\{-3, \frac{3}{2} \pm \frac{3\sqrt{3}}{2}i\right\}$

71. $s = \dfrac{1}{2}gt^2$

$2s = 2\left[\dfrac{1}{2}gt^2\right] \Rightarrow 2s = gt^2 \Rightarrow \dfrac{2s}{g} = \dfrac{gt^2}{g} \Rightarrow$

$t^2 = \dfrac{2s}{g} \Rightarrow t = \pm\sqrt{\dfrac{2s}{g}} = \dfrac{\pm\sqrt{2s}}{\sqrt{g}} \cdot \dfrac{\sqrt{g}}{\sqrt{g}} = \dfrac{\pm\sqrt{2sg}}{g}$

73. $F = \dfrac{kMv^2}{r}$

$rF = r\left[\dfrac{kMv^2}{r}\right] \Rightarrow Fr = kMv^2 \Rightarrow$

$\dfrac{Fr}{kM} = \dfrac{kMv^2}{kM} \Rightarrow v^2 = \dfrac{Fr}{kM} \Rightarrow v = \pm\sqrt{\dfrac{Fr}{kM}} \Rightarrow$

$v = \dfrac{\pm\sqrt{Fr}}{\sqrt{kM}} \cdot \dfrac{\sqrt{kM}}{\sqrt{kM}} = \dfrac{\pm\sqrt{FrkM}}{kM}$

75.
$$r = r_0 + \frac{1}{2}at^2$$
$$r - r_0 = \frac{1}{2}at^2$$
$$2(r - r_0) = at^2$$
$$\frac{2(r - r_0)}{a} = t^2$$
$$\pm\sqrt{\frac{2(r - r_0)}{a}} = \frac{\pm\sqrt{2a(r - r_0)}}{a} = t$$

77. $h = -16t^2 + v_0 t + s_0$
$$16t^2 - v_0 t + h - s_0 = 0$$
$$16t^2 - v_0 t + (h - s_0) = 0 \qquad a = 16, b = -v_0,$$
$$c = h - s_0$$

$$t = \frac{-b \pm \sqrt{b^2 - 4ac}}{2a}$$
$$= \frac{-(-v_0) \pm \sqrt{(-v_0)^2 - 4(16)(h - s_0)}}{2(16)}$$
$$= \frac{v_0 \pm \sqrt{v_0^2 - 64(h - s_0)}}{32}$$
$$= \frac{v_0 \pm \sqrt{v_0^2 - 64h + 64s_0}}{32}$$

79.
$$4x^2 - 2xy + 3y^2 = 2$$
$$4x^2 - 2xy + 3y^2 - 2 = 0$$

(a) Solve for x in terms of y.
$$4x^2 - (2y)x + (3y^2 - 2) = 0$$
$$a = 4, b = -2y, \text{ and } c = 3y^2 - 2$$
$$x = \frac{-b \pm \sqrt{b^2 - 4ac}}{2a}$$
$$= \frac{-(-2y) \pm \sqrt{(-2y)^2 - 4(4)(3y^2 - 2)}}{2(4)}$$
$$= \frac{2y \pm \sqrt{4y^2 - 16(3y^2 - 2)}}{8}$$
$$= \frac{2y \pm \sqrt{4y^2 - 48y^2 + 32}}{8}$$
$$= \frac{2y \pm \sqrt{32 - 44y^2}}{8} = \frac{2y \pm \sqrt{4(8 - 11y^2)}}{8}$$
$$= \frac{2y \pm 2\sqrt{8 - 11y^2}}{8} = \frac{y \pm \sqrt{8 - 11y^2}}{4}$$

(b) Solve for y in terms of x.
$$3y^2 - (2x)y + (4x^2 - 2) = 0$$
$$a = 3, b = -2x, \text{ and } c = 4x^2 - 2$$
$$y = \frac{-b \pm \sqrt{b^2 - 4ac}}{2a}$$
$$= \frac{-(-2x) \pm \sqrt{(-2x)^2 - 4(3)(4x^2 - 2)}}{2(3)}$$
$$= \frac{2x \pm \sqrt{4x^2 - 12(4x^2 - 2)}}{6}$$
$$= \frac{2x \pm \sqrt{4x^2 - 48x^2 + 24}}{6}$$
$$= \frac{2x \pm \sqrt{24 - 44x^2}}{6} = \frac{2x \pm \sqrt{4(6 - 11x^2)}}{6}$$
$$= \frac{2x \pm 2\sqrt{6 - 11x^2}}{6} = \frac{x \pm \sqrt{6 - 11x^2}}{3}$$

81.
$$2x^2 + 4xy - 3y^2 = 2$$
$$2x^2 + 4xy - 3y^2 - 2 = 0$$

a. Solve for x in terms of y.
$$2x^2 + (4y)x + (-3y^2 - 2) = 0$$
$$a = 2, b = 4y, \text{ and } c = -3y^2 - 2$$
$$x = \frac{-b \pm \sqrt{b^2 - 4ac}}{2a}$$
$$= \frac{-(4y) \pm \sqrt{(4y)^2 - 4(2)(-3y^2 - 2)}}{2(2)}$$
$$= \frac{-4y \pm \sqrt{16y^2 - 8(-3y^2 - 2)}}{4}$$
$$= \frac{-4y \pm \sqrt{16y^2 + 24y^2 + 16}}{4}$$
$$= \frac{-4y \pm \sqrt{40y^2 + 16}}{4}$$
$$= \frac{-4y \pm \sqrt{4(10y^2 + 4)}}{4}$$
$$= \frac{-4y \pm 2\sqrt{10y^2 + 4}}{4} = \frac{-2y \pm \sqrt{10y^2 + 4}}{2}$$

b. Solve for y in terms of x.

$$-3y^2 + (4x)y + (2x^2 - 2) = 0$$

$$a = -3, \ b = 4x, \ c = 2x^2 - 2$$

$$x = \frac{-b \pm \sqrt{b^2 - 4ac}}{2a}$$

$$= \frac{-(4x) \pm \sqrt{(4x)^2 - 4(-3)(2x^2 - 2)}}{2(-3)}$$

$$= \frac{-4x \pm \sqrt{16x^2 + 12(2x^2 - 2)}}{-6}$$

$$= \frac{-4x \pm \sqrt{16x^2 + 24x^2 - 24}}{-6}$$

$$= \frac{-4x \pm \sqrt{40x^2 - 24}}{-6}$$

$$= \frac{-4x \pm \sqrt{4(10x^2 - 6)}}{-6}$$

$$= \frac{-4x \pm 2\sqrt{10x^2 - 6}}{-6}$$

$$= \frac{2x \pm \sqrt{10x^2 - 6}}{3}$$

83. $x^2 - 8x + 16 = 0$

$a = 1, b = -8,$ and $c = 16$

$b^2 - 4ac = (-8)^2 - 4(1)(16) = 64 - 64 = 0$

One rational solution (a double solution)

85. $3x^2 + 5x + 2 = 0$

$a = 3, b = 5,$ and $c = 2$

$b^2 - 4ac = 5^2 - 4(3)(2) = 25 - 24 = 1 = 1^2$

Two distinct rational solutions

87. $4x^2 = -6x + 3$

$4x^2 + 6x - 3 = 0$

$a = 4, b = 6,$ and $c = -3$

$b^2 - 4ac = 6^2 - 4(4)(-3) = 36 + 48 = 84$

Two distinct irrational solutions

89. $9x^2 + 11x + 4 = 0$

$a = 9, b = 11,$ and $c = 4$

$b^2 - 4ac = 11^2 - 4(9)(4) = 121 - 144 = -23$

Two distinct nonreal complex solutions

91. $8x^2 - 72 = 0$

$a = 8, b = 0,$ and $c = -72$

$b^2 - 4ac = 0^2 - 4(8)(-72) = 2304 = 48^2$

Two distinct rational solutions

93. No, it is not possible for the solution set of a quadratic equation with integer coefficients to consist of a single irrational number. Additional responses will vary.

In exercises 95–97, there are other possible answers.

95. $x = 4$ or $x = 5$

$x - 4 = 0$ or $x - 5 = 0$

$$(x - 4)(x - 5) = 0$$

$$x^2 - 5x - 4x + 20 = 0$$

$$x^2 - 9x + 20 = 0$$

$a = 1, b = -9,$ and $c = 20$

97. $x = 1 + \sqrt{2}$ or $x = 1 - \sqrt{2}$

$x - (1 + \sqrt{2}) = 0$ or $x - (1 - \sqrt{2}) = 0$

$$\left[x - (1 + \sqrt{2})\right]\left[x - (1 - \sqrt{2})\right] = 0$$

$$x^2 - x(1 - \sqrt{2}) - x(1 + \sqrt{2})$$
$$+ (1 + \sqrt{2})(1 - \sqrt{2}) = 0$$

$$x^2 - x + x\sqrt{2} - x - x\sqrt{2} + \left[1^2 - (\sqrt{2})^2\right] = 0$$

$$x^2 - 2x + (1 - 2) = 0$$

$$x^2 - 2x - 1 = 0$$

$a = 1, b = -2,$ and $c = -1$

Chapter 1 Quiz
(Sections 1.1–1.4)

1. $3(x - 5) + 2 = 1 - (4 + 2x)$

$3x - 15 + 2 = 1 - 4 - 2x$

$3x - 13 = -3 - 2x$

$5x - 13 = -3$

$\qquad 5x = 10 \Rightarrow x = 2$

Solution set $\{2\}$

3. $ay + 2x = y + 5x$

$ay - 3x = y$

$\quad -3x = y - ay = y(1 - a)$

$\quad 3x = y(a - 1)$

$\dfrac{3x}{a - 1} = y$

5. Substitute 2008 for x in the equation:

$y = 0.128(2008) - 250.43 \approx 6.59$

So, the model predicts that the minimum hourly wage for 2008 was $6.59. The model predicts a wage that is $0.04 greater than the actual wage.

7. $\dfrac{7-2i}{2+4i} = \dfrac{7-2i}{2+4i} \cdot \dfrac{2-4i}{2-4i} = \dfrac{14-28i-4i+(-8)}{4-(-16)}$

$= \dfrac{6-32i}{20} = \dfrac{6}{20} - \dfrac{32}{20}i = \dfrac{3}{10} - \dfrac{8}{5}i$

9. $x^2 - 29 = 0 \Rightarrow x^2 = 29 \Rightarrow x = \pm\sqrt{29}$

Solution set: $\{\pm 29\}$

Section 1.5 Applications and Modeling with Quadratic Equations

1. A. The length of the parking area is $2x + 200$, while the width is x, so the area is $(2x + 200)x$. Set the area equal to 40,000 to obtain $x(2x + 200) = 40{,}000$.

3. D. Use the Pythagorean theorem with $a = x$, $b = 2x - 2$, and $c = x + 4$.

$x^2 + (2x-2)^2 = (x+4)^2$

5. A. Let $x =$ the width, so $x + 5 =$ the length. If 2 in. are cut from each corner, then the width of the box is $x - 4$ and the length of the box is $x + 5 - 4$ or $x + 1$. The height of the box is 2.

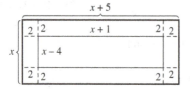

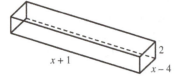

Then, the volume of the box is represented by $V = (x+1)(x-4)2 = 64$.

7. B. We are seeking the height given the time $t = 2$ seconds. Therefore, the correct equation is $s = -16(2)^2 + 45(2)$.

9. Let $x =$ the first integer. Then $x + 1 =$ the next consecutive integer.

$x(x+1) = 56 \Rightarrow x^2 + x = 56$

$x^2 + x - 56 = 0 \Rightarrow (x+8)(x-7) = 0$

$x + 8 = 0 \Rightarrow x = -8$ or $x - 7 = 0 \Rightarrow x = 7$

If $x = -8$, then $x + 1 = -7$. If $x = 7$, then $x + 1 = 8$. So the two integers are -8 and -7, or 7 and 8.

11. Let $x =$ the first even integer. Then $x + 2 =$ the next consecutive even integer.

$x(x+2) = 168 \Rightarrow x^2 + 2x = 168$

$x^2 + 2x - 168 = 0 \Rightarrow (x+14)(x-12) = 0$

$x + 14 = 0 \Rightarrow x = -14$ or

$x - 12 = 0 \Rightarrow x = 12$

If $x = -14$, then $x + 2 = -12$. If $x = 12$, then $x + 2 = 14$. so, the two even integers are -14 and -12, or 12 and 14.

13. Let $x =$ the first odd integer. Then $x + 2 =$ the next consecutive odd integer.

$x(x+2) = 63 \Rightarrow x^2 + 2x = 63 \Rightarrow$

$x^2 + 2x - 63 = 0 \Rightarrow (x+9)(x-7) = 0$

$x + 9 = 0 \Rightarrow x = -9$ or

$x - 7 = 0 \Rightarrow x = 7$

If $x = -9$, then $x + 2 = -7$. If $x = 7$, then $x + 2 = 9$. so, the two odd integers are -9 and -7, or 7 and 9.

15. Let $x =$ the first odd integer. Then $x + 2 =$ the next consecutive odd integer.

$x^2 + (x+2)^2 = 202$

$x^2 + x^2 + 4x + 4 = 202$

$2x^2 + 4x + 4 = 202 \Rightarrow 2x^2 + 4x - 198 = 0$

$2(x^2 + 2x - 99) = 0 \Rightarrow x^2 + 2x - 99 = 0$

$(x+11)(x-9) = 0$

$x + 11 = 0 \Rightarrow x = -11$ or

$x - 9 = 0 \Rightarrow x = 9$

If $x = -11$, then $x + 2 = -9$. If $x = 9$, then $x + 2 = 11$. So the two integers are -11 and -9, or 9 and 11.

17. Let $x =$ the first even integer. Then $x + 2 =$ the next consecutive even integer.

$(x+2)^2 - x^2 = 84$

$x^2 + 4x + 4 - x^2 = 84 \Rightarrow 4x + 4 = 84 \Rightarrow$

$4x = 80 \Rightarrow x = 20$

If $x = 20$, then $x + 2 = 22$. The two integers are 20 and 22.

19. Let $x =$ the length of one leg, $x + 2 =$ the length of the other leg, and $x + 4 =$ the length of the hypotenuse. (Remember that the hypotenuse is the longest side in a right triangle.) The Pythagorean theorem gives

$x^2 + (x+2)^2 = (x+4)^2$

$x^2 + x^2 + 4x + 4 = x^2 + 8x + 16$

$x^2 - 4x - 12 = 0 \Rightarrow (x-6)(x+2) = 0$

$x - 6 = 0 \Rightarrow x = 6$ or

$x + 2 = 0 \Rightarrow x = -2$

(continued on next page)

(*continued*)

Length cannot be negative, so reject that solution. If $x = 6$, then $x + 2 = 8$ and $x + 4 = 10$. The sides of the right triangle are 6, 8, and 10.

21. Let x = the length of the side of the smaller square. Then $x + 3$ = the length of the side of the larger square.

$$(x+3)^2 + x^2 = 149$$
$$x^2 + 6x + 9 + x^2 = 149 \Rightarrow 2x^2 + 6x - 140 = 0$$
$$x^2 + 3x - 70 = 0 \Rightarrow (x-7)(x+10) = 0$$
$$x - 7 = 0 \Rightarrow x = 7 \text{ or}$$
$$x + 10 = 0 \Rightarrow x = -10$$

Length cannot be negative, so reject that solution. If $x = 7$, then $x + 3 = 10$. The length of the side of smaller square is 7 in., and the length of the side of the larger square is 10 in.

23. Use the figure and equation A from Exercise 1.

$$x(2x + 200) = 40,000$$
$$2x^2 + 200x = 40,000$$
$$2x^2 + 200x - 40,000 = 0$$
$$x^2 + 100x - 20,000 = 0$$
$$(x - 100)(x + 200) = 0$$
$$x = 100 \text{ or } x = -200$$

The negative solution is not meaningful. If $x = 100$, then $2x + 200 = 400$. The dimensions of the lot are 100 yd by 400 yd.

25. Let x = the width of the strip of floor around the rug.

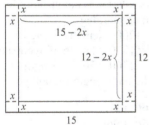

The dimensions of the carpet are $15 - 2x$ by $12 - 2x$. Because $A = lw$, the equation for the carpet area is $(15 - 2x)(12 - 2x) = 108$.
Put this equation in standard form and solve by factoring.

$$(15 - 2x)(12 - 2x) = 108$$
$$180 - 30x - 24x + 4x^2 = 108$$
$$180 - 54x + 4x^2 = 108$$
$$4x^2 - 54x + 72 = 0$$
$$2x^2 - 27x + 36 = 0$$
$$(2x - 3)(x - 12) = 0$$

$$2x - 3 = 0 \Rightarrow x = \frac{3}{2}$$
$$x - 12 = 0 \Rightarrow x = 12$$

The solutions of the quadratic equation are $\frac{3}{2}$ and 12. We eliminate 12 as meaningless in this problem. If $x = \frac{3}{2}$, then $15 - 2x = 12$ and $12 - 2x = 9$. The dimensions of the carpet are 9 ft by 12 ft.

27. Let x = the width of the metal. The dimensions of the base of the box are $x - 4$ by $x + 6$.

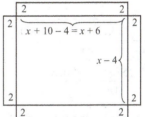

Because the formula for the volume of a box is $V = lwh$, we have

$$(x+6)(x-4)(2) = 832$$
$$(x+6)(x-4) = 416$$
$$x^2 - 4x + 6x - 24 = 416$$
$$x^2 + 2x - 24 = 416$$
$$x^2 + 2x - 440 = 0 \Rightarrow (x+22)(x-20) = 0$$
$$x + 22 = 0 \Rightarrow x = -22 \text{ or}$$
$$x - 20 = 0 \Rightarrow x = 20$$

The negative solution is not meaningful. If $x = 20$, then $x + 10 = 30$. The dimensions of the sheet of metal are 20 in by 30 in.

29. Let h = height and r = radius.
Surface area = $2\pi rh + 2\pi r^2$
$$8\pi = 2\pi r(3) + 2\pi r^2$$
$$8\pi = 6\pi r + 2\pi r^2$$
$$0 = 2\pi r^2 + 6\pi r - 8\pi$$
$$0 = 2\pi(r^2 + 3r - 4) \Rightarrow 0 = (r+4)(r-1)$$
$$r + 4 = 0 \Rightarrow r = -4 \text{ or } r - 1 = 0 \Rightarrow r = 1$$

The r represents the radius of a cylinder, so −4 is not reasonable. The radius of the circular top is 1 ft.

31. Let x = length of side of square. Area = x^2 and perimeter = $4x$
$$x^2 = 4x \Rightarrow x^2 - 4x = 0 \Rightarrow x(x - 4) = 0 \Rightarrow$$
$$x = 0 \text{ or } x = 4$$

We reject 0 because x must be greater than 0. The side of the square measures 4 units.

33. Let h = height and r = radius.

Area of side = $2\pi rh$ and Area of circle = πr^2
Surface area = area of side + area of top + area of bottom

Surface area = $2\pi rh + \pi r^2 + \pi r^2 = 2\pi rh + 2\pi r^2$

$371 = 2\pi r(12) + 2\pi r^2$

$371 = 24\pi r + 2\pi r^2$

$0 = 2\pi r^2 + 24\pi r - 371$

$a = 2\pi$, $b = 24\pi$, and $c = -371$

$r = \dfrac{-b \pm \sqrt{b^2 - 4ac}}{2a}r$

$= \dfrac{-24\pi \pm \sqrt{(24\pi)^2 - 4(2\pi)(-371)}}{2(2\pi)}$

$= \dfrac{-24\pi \pm \sqrt{576\pi^2 + 2968\pi}}{4\pi}$

$r \approx -15.75$ or $r \approx 3.75$

The negative solution is not meaningful. The radius of the circular top is approximately 3.75 cm.

35. Let h = the height of the dock.
Then $2h + 3$ = the length of the rope from the boat to the top of the dock.
Apply the Pythagorean theorem to the triangle shown in the text.

$h^2 + 12^2 = (2h + 3)^2$

$h^2 + 144 = (2h)^2 + 2(6h) + 3^2$

$h^2 + 144 = 4h^2 + 12h + 9$

$0 = 3h^2 + 12h - 135$

$0 = h^2 + 4h - 45 \Rightarrow 0 = (h + 9)(h - 5)$

$h + 9 = 0 \Rightarrow h = -9$ or $h - 5 = 0 \Rightarrow h = 5$

The negative solution is not meaningful. The height of the dock is 5 ft.

37. Let r = radius of circle and x = length of side of square. The radius is $\frac{1}{2}$ the length of the side of the square. Area = x^2

$800 = x^2 \Rightarrow x = \sqrt{800} = 20\sqrt{2} \Rightarrow$

$r = 10\sqrt{2}$

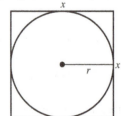

The radius is $10\sqrt{2}$ feet.

39. Let x = length of ladder
Distance from building to ladder = $8 + 2 = 10$.
Distance from ground to window = 13
Apply the Pythagorean theorem.

$a^2 + b^2 = c^2$

$10^2 + 13^2 = x^2 \Rightarrow 100 + 169 = x^2 \Rightarrow$

$269 = x^2 \Rightarrow \pm\sqrt{269} = x$

$x \approx -16.4$ or $x \approx 16.4$

The negative solution is not meaningful. The worker will need a 16.4-ft ladder.

41. Let x = length of short leg, $x + 700$ = length of long leg, and $x + 700 + 100$ or $x + 800$ = length of hypotenuse.

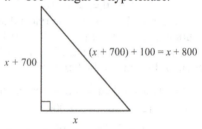

Apply the Pythagorean theorem.

$c^2 = a^2 + b^2$

$(x + 800)^2 = x^2 + (x + 700)^2$

$x^2 + 1600x + 640{,}000$

$\qquad = x^2 + x^2 + 1400x + 490{,}000$

$0 = x^2 - 200x - 150{,}000$

$0 = (x + 300)(x - 500)$

$x + 300 = 0 \Rightarrow x = -300$ or

$x - 500 = 0 \Rightarrow x = 500$

The negative solution is not meaningful.
500 = length of short leg
$500 + 700 = 1200$ = length of long leg
$1200 + 100 = 1300$ = length of hypotenuse
$500 + 1200 + 1300 = 3000$ = length of walkway. The total length is 3000 yd.

43. (a) $s = -16t^2 + v_0 t$

$s = -16t^2 + 96t$

$80 = -16t^2 + 96t$

$16t^2 - 96t + 80 = 0$

$a = 16$, $b = -96$ and $c = 80$

$t = \dfrac{-b \pm \sqrt{b^2 - 4ac}}{2a}$

$= \dfrac{-(-96) \pm \sqrt{(-96)^2 - 4(16)(80)}}{2(16)}$

(continued on next page)

(*continued*)

$$t = \frac{96 \pm \sqrt{9216 - 5120}}{32}$$

$$= \frac{96 \pm \sqrt{4096}}{32} = \frac{96 \pm 64}{32}$$

$$t = \frac{96 - 64}{32} = 1 \text{ or } t = \frac{96 + 64}{32} = 5$$

The projectile will reach 80 ft at 1 sec and 5 sec.

(b) $s = -16t^2 + 96t$
$0 = -16t^2 + 96t$
$0 = -16t(t - 6)$
$-16t = 0 \Rightarrow t = 0$ or $t - 6 = 0 \Rightarrow t = 6$
The projectile will return to the ground after 6 sec.

45. **(a)** $s = -16t^2 + v_0 t$
$s = -16t^2 + 32t$
$80 = -16t^2 + 32t$
$16t^2 - 32t + 80 = 0$
$t^2 - 2t + 5 = 0$
$a = 1, b = -2 \text{ and } c = 5$

$$t = \frac{-b \pm \sqrt{b^2 - 4ac}}{2a}$$

$$= \frac{-(-2) \pm \sqrt{(-2)^2 - 4(1)(5)}}{2(1)}$$

$$= \frac{2 \pm \sqrt{4 - 20}}{2} = \frac{2 \pm \sqrt{-16}}{2} = \frac{2 \pm 4i}{2}$$

$$= 1 \pm 2i$$

The projectile will not reach 80 ft.

(b) $s = -16t^2 + 32t$
$0 = -16t^2 + 32t \Rightarrow 0 = -16t(t - 2) \Rightarrow$
$-16t = 0 \Rightarrow t = 0 \text{ or } t - 2 = 0 \Rightarrow t = 2$
The projectile will return to the ground after 2 sec.

47.   The height of the ball is given by
$s = -2.7t^2 + 30t + 6.5$.

(a)   When the ball is 12 ft above the moon's surface, $s = 12$. Set $s = 12$ and solve for t.

$12 = -2.7t^2 + 30t + 6.5$

$2.7t^2 - 30t + 5.5 = 0$

Use the quadratic formula with $a = 2.7$, $b = -30$, and $c = 5.5$.

$$t = \frac{30 \pm \sqrt{900 - 4(2.7)(5.5)}}{2(2.7)} = \frac{30 \pm \sqrt{840.6}}{5.4}$$

$$\frac{30 + \sqrt{840.6}}{5.4} \approx 10.92 \text{ or } \frac{30 - \sqrt{840.6}}{5.4} \approx 0.19$$

Therefore, the ball reaches 12 ft first after 0.19 sec (on the way up), then again after 10.92 sec (on the way down).

(b)   When the ball returns to the surface, $s = 0$.

$0 = -2.7t^2 + 30t + 6.5$

Use the quadratic formula with $a = -2.7$, $b = 30$, and $c = 6.5$.

$$t = \frac{-30 \pm \sqrt{900 - 4(-2.7)(6.5)}}{2(-2.7)}$$

$$= \frac{-30 \pm \sqrt{970.2}}{-5.4}$$

$$\frac{-30 + \sqrt{970.2}}{-5.4} \approx -0.21 \text{ or }$$

$$\frac{-30 - \sqrt{970.2}}{-5.4} \approx 11.32$$

The negative solution is not meaningful. Therefore, the ball hits the moon's surface after 11.32 sec.

49. **(a)**   The year 2007 corresponds to $x = 13$.

$y = 0.2313x^2 + 2.600x + 35.17$

$y = 0.2313(13)^2 + 2.600(13) + 35.17$

≈ 108.0597

In 2007, the NFL salary cap was approximately $108.1 million.

(b)   We must solve for x when $y = 90$.

$90 = 0.2313x^2 + 2.600x + 35.17$

$0 = 0.2313x^2 + 2.600x - 54.83$

Use the quadratic formula with $a = 0.2313$, $b = 2.600$, and $c = -54.83$.

$$x = \frac{-b \pm \sqrt{b^2 - 4ac}}{2a}$$

$$x = \frac{-2.600 \pm \sqrt{2.600^2 - 4(0.2313)(-54.83)}}{2(0.2313)}$$

$$= \frac{-2.600 \pm \sqrt{57.4887}}{0.4626}$$

$$\approx 10.8, \ -22.0$$

The negative solution is not meaningful. Therefore, the salary cap reached 90 million dollars during 2004.

51. (a) Let $x = 50$.
$$T = 0.00787(50)^2 - 1.528(50) + 75.89 \approx 19.2$$
The exposure time when $x = 50$ ppm is approximately 19.2 hr.

(b) Let $T = 3$ and solve for x.
$$3 = 0.00787x^2 - 1.528x + 75.89$$
$$0.00787x^2 - 1.528x + 72.89 = 0$$
Use the quadratic formula with $a = 0.00787$, $b = -1.528$, and $c = 72.89$.
$$x = \frac{-(-1.528) \pm \sqrt{(-1.528)^2 - 4(0.00787)(72.89)}}{2(0.00787)}$$
$$= \frac{1.528 \pm \sqrt{2.334784 - 2.2945772}}{0.01574}$$
$$= \frac{1.528 \pm \sqrt{0.0402068}}{0.01574}$$
$$\frac{1.528 + \sqrt{0.0402068}}{0.01574} \approx 109.8 \text{ or}$$
$$\frac{1.528 - \sqrt{0.0402068}}{0.01574} \approx 84.3$$
We reject the potential solution 109.8 because it is not in the interval [50, 100]. So, 84.3 ppm carbon monoxide concentration is necessary for a person to reach the 4% to 6% CoHb level in 3 hr.

53. (a) 2014 is represented by $x = 6$. Substitute $x = 6$ into the equation to find y:
$$y = 0.0429x^2 - 9.73x + 606$$
$$y = 0.0429(6)^2 - 9.73(6) + 606$$
$$\approx 549.2 \text{ million tons}$$
In 2014, emissions were about 549.2 million tons.

(b) Let $y = 500$ and solve for x.
$$500 = 0.0429x^2 - 9.73x + 606$$
$$0.0429x^2 - 9.73x + 106 = 0$$
Use the quadratic formula with $a = 0.0429$, $b = -9.73$, and $c = 106$.
$$x = \frac{-(-9.73) \pm \sqrt{(-9.73)^2 - 4(0.0429)(106)}}{2(0.0429)}$$
$$= \frac{9.73 \pm \sqrt{76.4833}}{0.0858} \approx 11.5 \text{ or } 215.3$$
The model predicts that the emissions will reach 500 million tons about 11.5 years after 2008, which is during 2019.

55. The year 2010 is represented by $x = 3$.
$$y = 710.55x^2 + 1333.7x + 32399$$
$$y = 710.55(3)^2 + 1333.7(3) + 32399$$
$$\approx 42,795$$
In 2010, the revenue from Internet publishing and broadcasting was about $42,795 million.

57. For each $20 increase in rent over $300, one unit will remain vacant. Therefore, for x $20 increases, x units will remain vacant. Therefore, the number of rented units will be $80 - x$.

59. $300 + 20x$ is the rent for each apartment, and $80 - x$ is the number of apartments that will be rented at that cost. The revenue generated will then be the product of $80 - x$ and $300 + 20x$, so the correct expression is
$$R = (80 - x)(300 + 20x)$$
$$= 24,000 + 1600x - 300x - 20x^2$$
$$= 24,000 + 1300x - 20x^2.$$

61. Let $x =$ number of passengers in excess of 75. Then $225 - 5x =$ the cost per passenger (in dollars) and $75 + x =$ the number of passengers.
(Cost per passenger)(Number of passengers) = Revenue
$$(225 - 5x)(75 + x) = 16,000$$
$$16,875 + 225x - 375x - 5x^2 = 16,000$$
$$16,875 - 150x - 5x^2 = 16,000$$
$$0 = 5x^2 + 150x - 875$$
$$0 = x^2 + 30x - 175 \Rightarrow 0 = (x + 35)(x - 5)$$
$$x + 35 = 0 \Rightarrow x = -35 \text{ or } x - 5 = 0 \Rightarrow x = 5$$
The negative solution is not meaningful. Because there are 5 passengers in excess of 75, the total number of passengers is 80.

63. Let $x =$ number of weeks the manager should wait. Then $100 + 5x =$ number of pounds and $0.40 - 0.02x =$ cost per pound
(Cost per pound)(Number of pounds) = Revenue
$$(0.40 - 0.02x)(100 + 5x) = 38.40$$
$$40 + 2x - 2x - 0.1x^2 = 38.40$$
$$40 - 0.1x^2 = 38.40$$
$$-0.1x^2 = -1.6$$
$$-10(-0.1x^2) = -10(-1.6)$$
$$x^2 = 16 \Rightarrow x = \pm 4$$
The negative solution is not meaningful. The farmer should wait 4 weeks to get an average revenue of $38.40 per tree.

Section 1.6 Other Types of Equations and Applications

1. A <u>rational equation</u> is an equation that has a rational expression for one or more terms.

3. If a job can be completed in 4 hours, then the rate of work is <u>1/4</u> of the job per hour.

5. An equation such as $x^{3/2} = 8$ is an equation with a <u>rational exponent</u> because it contains a variable raised to an exponent that is a rational number.

7. D   **9.** E

11. $\dfrac{5}{2x+3} - \dfrac{1}{x-6} = 0$

$2x+3 \neq 0 \Rightarrow x \neq -\frac{3}{2}$ and $x-6 \neq 0 \Rightarrow x \neq 6$.

13. $\dfrac{3}{x-2} + \dfrac{1}{x+1} = \dfrac{3}{x^2-x-2}$

or $\dfrac{3}{x-2} + \dfrac{1}{x+1} = \dfrac{3}{(x-2)(x+1)}$

$x-2 \neq 0 \Rightarrow x \neq 2$ and $x+1 \neq 0 \Rightarrow x \neq -1$

15. $\dfrac{1}{4x} - \dfrac{2}{x} = 3$

$4x \neq 0 \Rightarrow x \neq 0$

17. $\dfrac{2x+5}{2} - \dfrac{3x}{x-2} = x$

The least common denominator is $2(x-2)$, which is equal to 0 if $x = 2$. Therefore, 2 cannot possibly be a solution of this equation.

$2(x-2)\left[\dfrac{2x+5}{2} - \dfrac{3x}{x-2}\right] = 2(x-2)(x)$

$(x-2)(2x+5) - 2(3x) = 2x(x-2)$

$2x^2 + 5x - 4x - 10 - 6x = 2x^2 - 4x$

$-5x - 10 = -4x \Rightarrow -10 = x$

The restriction $x \neq 2$ does not affect the result. Therefore, the solution set is $\{-10\}$.

19. $\dfrac{x}{x-3} = \dfrac{3}{x-3} + 3$

The least common denominator is $x - 3$, which is equal to 0 if $x = 3$. Therefore, 3 cannot possibly be a solution of this equation.

$(x-3)\left(\dfrac{x}{x-3}\right) = (x-3)\left[\dfrac{3}{x-3} + 3\right]$

$x = 3 + 3(x-3)$

$x = 3 + 3x - 9$

$x = 3x - 6 \Rightarrow -2x = -6 \Rightarrow x = 3$

The only possible solution is 3. However, the variable is restricted to real numbers except 3. Therefore, the solution set is: $\varnothing$.

21. $\dfrac{-2}{x-3} + \dfrac{3}{x+3} = \dfrac{-12}{x^2-9}$ or

$\dfrac{-2}{x-3} + \dfrac{3}{x+3} = \dfrac{-12}{(x+3)(x-3)}$

The least common denominator is $(x+3)(x-3)$, which is equal to 0 if $x = -3$ or $x = 3$. Therefore, -3 and 3 cannot possibly be solutions of this equation.

$(x+3)(x-3)\left[\dfrac{-2}{x-3} + \dfrac{3}{x+3}\right]$

$= (x+3)(x-3)\left(\dfrac{-12}{(x+3)(x-3)}\right)$

$-2(x+3) + 3(x-3) = -12$

$-2x - 6 + 3x - 9 = -12$

$-15 + x = -12 \Rightarrow x = 3$

The only possible solution is 3. However, the variable is restricted to real numbers except -3 and 3. Therefore, the solution set is: $\varnothing$.

23. $\dfrac{4}{x^2+x-6} - \dfrac{1}{x^2-4} = \dfrac{2}{x^2+5x+6}$ or

$\dfrac{4}{(x+3)(x-2)} - \dfrac{1}{(x+2)(x-2)} = \dfrac{2}{(x+2)(x+3)}$

The least common denominator is $(x+3)(x-2)(x+2)$, which is equal to 0 if $x = -3$ or $x = 2$ or $x = -2$. Therefore, -3 and 2 and -2 cannot possibly be solutions of this equation.

$(x+3)(x-2)(x+2)$

$\cdot\left[\dfrac{4}{(x+3)(x-2)} - \dfrac{1}{(x+2)(x-2)}\right]$

$= (x+3)(x-2)(x+2)\left(\dfrac{2}{(x+2)(x+3)}\right)$

$4(x+2) - 1(x+3) = 2(x-2)$

$4x + 8 - x - 3 = 2x - 4$

$3x + 5 = 2x - 4 \Rightarrow x + 5 = -4 \Rightarrow x = -9$

The restrictions $x \neq -3$, $x \neq 2$, and $x \neq -2$ do not affect the result. Therefore, the solution set is $\{-9\}$.

25. $\dfrac{2x+1}{x-2}+\dfrac{3}{x}=\dfrac{-6}{x^2-2x}$ or

$$\dfrac{2x+1}{x-2}+\dfrac{3}{x}=\dfrac{-6}{x(x-2)}$$

Multiply each term in the equation by the least common denominator, $x(x-2)$, assuming $x \neq 0, 2$.

$$x(x-2)\left[\dfrac{2x+1}{x-2}+\dfrac{3}{x}\right]=x(x-2)\left(\dfrac{-6}{x(x-2)}\right)$$

$$x(2x+1)+3(x-2)=-6$$
$$2x^2+x+3x-6=-6$$
$$2x^2+4x-6=-6$$
$$2x^2+4x=0$$
$$2x(x+2)=0$$

$2x=0 \Rightarrow x=0$ or $x+2=0 \Rightarrow x=-2$

Because of the restriction $x \neq 0$, the only valid solution is -2. The solution set is $\{-2\}$.

27. $\dfrac{x}{x-1}-\dfrac{1}{x+1}=\dfrac{2}{x^2-1}$ or

$$\dfrac{x}{x-1}-\dfrac{1}{x+1}=\dfrac{2}{(x+1)(x-1)}$$

Multiply each term in the equation by the least common denominator, $(x+1)(x-1)$, assuming $x \neq \pm 1$.

$$(x+1)(x-1)\left[\dfrac{x}{x-1}-\dfrac{1}{x+1}\right]$$
$$=(x+1)(x-1)\left(\dfrac{2}{(x+1)(x-1)}\right)$$

$$x(x+1)-(x-1)=2$$
$$x^2+x-x+1=2$$
$$x^2+1=2$$
$$x^2-1=0$$
$$(x+1)(x-1)=0$$

$x+1=0 \Rightarrow x=-1$ or
$x-1=0 \Rightarrow x=1$

Because of the restriction $x \neq \pm 1$, the solution set is $\varnothing$.

29. $\dfrac{5}{x^2}-\dfrac{43}{x}=18$

Multiply each term in the equation by the least common denominator, x^2, assuming $x \neq 0$.

$$x^2\left[\dfrac{5}{x^2}-\dfrac{43}{x}\right]=x^2(18)$$

$$5-43x=18x^2$$
$$0=18x^2+43x-5$$
$$0=(2x+5)(9x-1)$$

$2x+5=0 \Rightarrow x=-\tfrac{5}{2}$ or

$9x-1=0 \Rightarrow x=\tfrac{1}{9}$

The restriction $x \neq 0$ does not affect the result.

Therefore, the solution set is $\left\{-\tfrac{5}{2},\tfrac{1}{9}\right\}$.

31. $2=\dfrac{3}{2x-1}+\dfrac{-1}{(2x-1)^2}$

Multiply each term in the equation by the least common denominator, $(2x-1)^2$, assuming $x \neq \tfrac{1}{2}$.

$$(2x-1)^2(2)=(2x-1)^2\left[\dfrac{3}{2x-1}+\dfrac{-1}{(2x-1)^2}\right]$$

$$2(4x^2-4x+1)=3(2x-1)-1$$
$$8x^2-8x+2=6x-3-1$$
$$8x^2-8x+2=6x-4$$
$$8x^2-14x+6=0$$
$$2(4x^2-7x+3)=0$$
$$2(4x-3)(x-1)=0$$

$4x-3=0 \Rightarrow x=\tfrac{3}{4}$ or
$x-1=0 \Rightarrow x=1$

The restriction $x \neq \tfrac{1}{2}$ does not affect the result.

Therefore the solution set is $\left\{\tfrac{3}{4},1\right\}$.

33. $\dfrac{2x-5}{x}=\dfrac{x-2}{3}$

Multiply each term in the equation by the least common denominator, $3x$, assuming $x \neq 0$.

$$3x\left(\dfrac{2x-5}{x}\right)=3x\left(\dfrac{x-2}{3}\right)$$

$$3(2x-5)=x(x-2) \Rightarrow 6x-15=x^2-2x \Rightarrow$$
$$0=x^2-8x+15=(x-3)(x-5)$$

$x-3=0 \Rightarrow x=3$ or $x-5=0 \Rightarrow x=5$

The restriction $x \neq 0$ does not affect the result. Therefore, the solution set is $\{3, 5\}$.

35. $\dfrac{2x}{x-2} = 5 + \dfrac{4x^2}{x-2}$

Multiply each term in the equation by the least common denominator, $x-2$, assuming $x \neq 2$.

$$(x-2)\left(\dfrac{2x}{x-2}\right) = (x-2)\left[5 + \dfrac{4x^2}{x-2}\right]$$
$$2x = 5(x-2) + 4x^2$$
$$2x = 5x - 10 + 4x^2$$
$$0 = 4x^2 + 3x - 10$$
$$0 = (x+2)(4x-5)$$

$x + 2 = 0 \Rightarrow x = -2$ or $4x - 5 = 0 \Rightarrow x = \frac{5}{4}$

The restriction $x \neq 2$ does not affect the result.

Therefore the solution set is $\left\{-2, \frac{5}{4}\right\}$.

37. Let x = the amount of time (in hours) it takes Joe and Sam to paint the house.

	r	t	Part of the Job Accomplished
Joe	$\frac{1}{3}$	x	$\frac{1}{3}x$
Sam	$\frac{1}{5}$	x	$\frac{1}{5}x$

Because Joe and Sam must accomplish 1 job (painting a house), we must solve the following equation.

$$\tfrac{1}{3}x + \tfrac{1}{5}x = 1$$
$$15\left[\tfrac{1}{3}x + \tfrac{1}{5}x\right] = 15 \cdot 1$$
$$5x + 3x = 15 \Rightarrow 8x = 15 \Rightarrow x = \dfrac{15}{8} = 1\tfrac{7}{8}$$

It takes Joe and Sam $1\frac{7}{8}$ hr working together to paint the house.

39. Let x = the amount of time (in hours) it takes plant A to produce the pollutant. Then $2x$ = the amount of time (in hours) it takes plant B to produce the pollutant.

	Rate	Time	Part of the Job Accomplished
Pollution from A	$\frac{1}{x}$	26	$26\left(\frac{1}{x}\right)$
Pollution from B	$\frac{1}{2x}$	26	$26\left(\frac{1}{2x}\right)$

Because plant A and B accomplish 1 job (producing the pollutant), we must solve the following equation.

$$26\left(\tfrac{1}{x}\right) + 26\left(\tfrac{1}{2x}\right) = 1$$
$$\tfrac{26}{x} + \tfrac{13}{x} = 1$$
$$x\left[\tfrac{39}{x}\right] = x \cdot 1 \Rightarrow 39 = x$$

Plant B will take $2 \cdot 39 = 78$ hr to produce the pollutant.

41. Let x = the amount of time (in hours) to fill the pool with both pipes open.

	Rate	Time	Part of the Job Accomplished
Inlet pipe	$\frac{1}{5}$	x	$\frac{1}{5}x$
Outlet pipe	$\frac{1}{8}$	x	$\frac{1}{8}x$

Filling the pool is 1 whole job, but because the outlet pipe empties the pool, its contribution should be subtracted from the contribution of the inlet pipe.

$$\tfrac{1}{5}x - \tfrac{1}{8}x = 1$$
$$40\left[\tfrac{1}{5}x - \tfrac{1}{8}x\right] = 40 \cdot 1 \Rightarrow 8x - 5x = 40 \Rightarrow$$
$$3x = 40 \Rightarrow x = \dfrac{40}{3} = 13\tfrac{1}{3}\ \text{hr}$$

It took $13\frac{1}{3}$ hr to fill the pool.

43. Let x = the amount of time (in minutes) to fill the sink with both pipes open.

	Rate	Time	Part of the Job Accomplished
Tap	$\frac{1}{5}$	x	$\frac{1}{5}x$
Drain	$\frac{1}{10}$	x	$\frac{1}{10}x$

Filling the sink is 1 whole job, but because the sink is draining, its contribution should be subtracted from the contribution of the taps.

$$\tfrac{1}{5}x - \tfrac{1}{10}x = 1$$
$$10\left[\tfrac{1}{5}x - \tfrac{1}{10}x\right] = 10 \cdot 1 \Rightarrow 2x - x = 10 \Rightarrow x = 10$$

It will take 10 minutes to fill the sink if Mark forgets to put in the stopper.

45. $x - \sqrt{2x+3} = 0$

$$x = \sqrt{2x+3} \Rightarrow x^2 = \left(\sqrt{2x+3}\right)^2$$

$$x^2 = 2x + 3 \Rightarrow x^2 - 2x - 3 = 0 \Rightarrow$$

$$(x+1)(x-3) = 0 \Rightarrow x = -1 \text{ or } x = 3$$

Check $x = -1$.

$$x - \sqrt{2x+3} = 0$$

$$-1 - \sqrt{2(-1)+3} \overset{?}{=} 0$$

$$-1 - \sqrt{-2+3} = 0$$

$$-1 - \sqrt{1} = 0 \Rightarrow -1 - 1 = 0 \Rightarrow -2 = 0$$

This is a false statement. -1 is not a solution.

Check $x = 3$.

$$x - \sqrt{2x+3} = 0$$

$$3 - \sqrt{2(3)+3} \overset{?}{=} 0$$

$$3 - \sqrt{6+3} = 0$$

$$3 - \sqrt{9} = 0 \Rightarrow 3 - 3 = 0 \Rightarrow 0 = 0$$

This is a true statement. 3 is a solution.

Solution set: $\{3\}$

47. $\sqrt{3x+7} = 3x + 5$

$$\left(\sqrt{3x+7}\right)^2 = (3x+5)^2$$

$$3x + 7 = 9x^2 + 30x + 25$$

$$0 = 9x^2 + 27x + 18$$

$$0 = 9\left(x^2 + 3x + 2\right) = 9(x+2)(x+1)$$

$x = -2$ or $x = -1$

Check $x = -2$.

$$\sqrt{3x+7} = 3x + 5$$

$$\sqrt{3(-2)+7} \overset{?}{=} 3(-2)+5$$

$$\sqrt{-6+7} = -6+5$$

$$\sqrt{1} = -1 \Rightarrow 1 = -1$$

This is a false statement. -2 is not a solution.

Check $x = -1$

$$\sqrt{3x+7} = 3x + 5$$

$$\sqrt{3(-1)+7} \overset{?}{=} 3(-1)+5$$

$$\sqrt{-3+7} = -3+5$$

$$\sqrt{4} = 2 \Rightarrow 2 = 2$$

This is a true statement. -1 is a solution.

Solution set: $\{-1\}$

49. $\sqrt{4x+5} - 6 = 2x - 11$

$$\sqrt{4x+5} = 2x - 5$$

$$\left(\sqrt{4x+5}\right)^2 = (2x-5)^2$$

$$4x + 5 = 4x^2 - 20x + 25$$

$$0 = 4x^2 - 24x + 20$$

$$0 = 4\left(x^2 - 6x + 5\right) = 4(x-1)(x-5)$$

$$x = 1 \text{ or } x = 5$$

Check $x = 1$.

$$\sqrt{4x+5} - 6 = 2x - 11$$

$$\sqrt{4(1)+5} - 6 \overset{?}{=} 2(1) - 11$$

$$\sqrt{4+5} - 6 = 2 - 11$$

$$\sqrt{9} - 6 = -9$$

$$3 - 6 = -9 \Rightarrow -3 = -9$$

This is a false statement. 1 is not a solution.

Check $x = 5$.

$$\sqrt{4x+5} - 6 = 2x - 11$$

$$\sqrt{4(5)+5} - 6 \overset{?}{=} 2(5) - 11$$

$$\sqrt{20+5} - 6 = 10 - 11$$

$$\sqrt{25} - 6 = -1 \Rightarrow 5 - 6 = -1 \Rightarrow -1 = -1$$

This is a true statement. 5 is a solution.

Solution set: $\{5\}$

51. $\sqrt{4x} - x + 3 = 0$

$$\sqrt{4x} = x - 3$$

$$\left(\sqrt{4x}\right)^2 = (x-3)^2$$

$$4x = x^2 - 6x + 9$$

$$0 = x^2 - 10x + 9 = (x-1)(x-9)$$

$$x = 1 \text{ or } x = 9$$

Check $x = 1$.

$$\sqrt{4x} - x + 3 = 0$$

$$\sqrt{4(1)} - 1 + 3 \overset{?}{=} 0$$

$$\sqrt{4} - 1 + 3 = 0$$

$$2 - 1 + 3 = 0 \Rightarrow 4 = 0$$

This is a false statement. 1 is not a solution.

Check $x = 9$.

$$\sqrt{4x} - x + 3 = 0$$

$$\sqrt{4(9)} - 9 + 3 \overset{?}{=} 0$$

$$\sqrt{36} - 9 + 3 = 0$$

$$6 - 9 + 3 = 0 \Rightarrow 0 = 0$$

This is a true statement. 9 is a solution.

Solution set: $\{9\}$

53. $\sqrt{x} - \sqrt{x-5} = 1$

$\sqrt{x} = 1 + \sqrt{x-5} \Rightarrow \left(\sqrt{x}\right)^2 = \left(1 + \sqrt{x-5}\right)^2$

$x = 1 + 2\sqrt{x-5} + (x-5)$

$x = x + 2\sqrt{x-5} - 4 \Rightarrow 4 = 2\sqrt{x-5}$

$2 = \sqrt{x-5} \Rightarrow 2^2 = \left(\sqrt{x-5}\right)^2$

$4 = x - 5 \Rightarrow 9 = x$

Check $x = 9$.

$\sqrt{x} - \sqrt{x-5} = 1$

$\sqrt{9} - \sqrt{9-5} \overset{?}{=} 1$

$3 - \sqrt{4} = 1 \Rightarrow 3 - 2 = 1 \Rightarrow 1 = 1$

This is a true statement.

Solution set is: $\{9\}$

55. $\sqrt{x+7} + 3 = \sqrt{x-4}$

$\left(\sqrt{x+7} + 3\right)^2 = \left(\sqrt{x-4}\right)^2$

$(x+7) + 6\sqrt{x+7} + 9 = x - 4$

$x + 6\sqrt{x+7} + 16 = x - 4 \Rightarrow 6\sqrt{x+7} = -20$

$3\sqrt{x+7} = -10 \Rightarrow \left(3\sqrt{x+7}\right)^2 = (-10)^2$

$9(x+7) = 100 \Rightarrow 9x + 63 = 100$

$9x = 37 \Rightarrow x = \dfrac{37}{9}$

Check $x = \frac{37}{9}$.

$\sqrt{x+7} + 3 = \sqrt{x-4}$

$\sqrt{\frac{37}{9} + 7} + 3 \overset{?}{=} \sqrt{\frac{37}{9} - 4}$

$\sqrt{\frac{37}{9} + \frac{63}{9}} + 3 = \sqrt{\frac{37}{9} - \frac{36}{9}}$

$\sqrt{\frac{100}{9}} + 3 = \sqrt{\frac{1}{9}}$

$\frac{10}{3} + 3 = \frac{1}{3} \Rightarrow \frac{10}{3} + \frac{9}{3} = \frac{1}{3} \Rightarrow \frac{19}{3} = \frac{1}{3}$

This is a false statement.

Solution set: $\varnothing$

57. $\sqrt{2x+5} - \sqrt{x+2} = 1$

$\sqrt{2x+5} = \sqrt{x+2} + 1$

$\left(\sqrt{2x+5}\right)^2 = \left(\sqrt{x+2} + 1\right)^2$

$2x + 5 = (x+2) + 2\sqrt{x+2} + 1$

$2x + 5 = x + 3 + 2\sqrt{x+2}$

$x + 2 = 2\sqrt{x+2}$

$(x+2)^2 = \left(2\sqrt{x+2}\right)^2$

$x^2 + 4x + 4 = 4(x+2)$

$x^2 + 4x + 4 = 4x + 8$

$0 = x^2 - 4$

$0 = (x+2)(x-2) \Rightarrow x = \pm 2$

Check $x = 2$.

$\sqrt{2x+5} - \sqrt{x+2} = 1$

$\sqrt{2(2)+5} - \sqrt{2+2} \overset{?}{=} 1$

$\sqrt{4+5} - \sqrt{4} \overset{?}{=} 1$

$\sqrt{9} - \sqrt{4} \overset{?}{=} 1$

$3 - 2 \overset{?}{=} 1 \Rightarrow 1 = 1$

This is a true statement. 2 is a solution.

Check $x = -2$.

$\sqrt{2x+5} - \sqrt{x+2} = 1$

$\sqrt{2(-2)+5} - \sqrt{-2+2} \overset{?}{=} 1$

$\sqrt{-4+5} - \sqrt{0} \overset{?}{=} 1$

$\sqrt{1} - \sqrt{0} \overset{?}{=} 1$

$1 - 0 \overset{?}{=} 1 \Rightarrow 1 = 1$

This is a true statement. -2 is a solution.

Solution set: $\{\pm 2\}$

59.

$\sqrt{3x} = \sqrt{5x+1} - 1$

$\left(\sqrt{3x}\right)^2 = \left(\sqrt{5x+1} - 1\right)^2$

$3x = (5x+1) - 2\sqrt{5x+1} + 1$

$3x = 5x + 2 - 2\sqrt{5x+1}$

$2\sqrt{5x+1} = 2 + 2x \Rightarrow \sqrt{5x+1} = 1 + x$

$\left(\sqrt{5x+1}\right)^2 = (1+x)^2 \Rightarrow 5x+1 = 1 + 2x + x^2$

$0 = x^2 - 3x \Rightarrow 0 = x(x-3) \Rightarrow$

$x = 0 \text{ or } x = 3$

Check $x = 0$.

$\sqrt{3x} = \sqrt{5x+1} - 1$

$\sqrt{3(0)} \overset{?}{=} \sqrt{5(0)+1} - 1$

$\sqrt{0} = \sqrt{0+1} - 1 \Rightarrow 0 = \sqrt{1} - 1$

$0 = 1 - 1 \Rightarrow 0 = 0$

This is a true statement. 0 is a solution.

Check $x = 3$.

$\sqrt{3x} = \sqrt{5x+1} - 1$

$\sqrt{3(3)} \overset{?}{=} \sqrt{5(3)+1} - 1$

$\sqrt{9} = \sqrt{15+1} - 1 \Rightarrow 3 = \sqrt{16} - 1$

$3 = 4 - 1 \Rightarrow 3 = 3$

This is a true statement. 3 is a solution.

Solution set: $\{0, 3\}$

61.
$$\sqrt{x+2} = 1 - \sqrt{3x+7}$$
$$\left(\sqrt{x+2}\right)^2 = (1-\sqrt{3x+7})^2$$
$$x+2 = 1 - 2\sqrt{3x+7} + (3x+7)$$
$$x+2 = 3x+8 - 2\sqrt{3x+7}$$
$$2\sqrt{3x+7} = 2x+6$$
$$2\sqrt{3x+7} = 2(x+3)$$
$$\sqrt{3x+7} = x+3 \Rightarrow \left(\sqrt{3x+7}\right)^2 = (x+3)^2$$
$$3x+7 = x^2 + 6x + 9 \Rightarrow 0 = x^2 + 3x + 2$$
$$0 = (x+2)(x+1)$$
$$x = -2 \text{ or } x = -1$$

Check $x = -2$.
$$\sqrt{x+2} = 1 - \sqrt{3x+7}$$
$$\sqrt{-2+2} \overset{?}{=} 1 - \sqrt{3(-2)+7}$$
$$\sqrt{0} = 1 - \sqrt{-6+7}$$
$$0 = 1 - \sqrt{1}$$
$$0 = 1 - 1 \Rightarrow 0 = 0$$
This is a true statement. -2 is a solution.

Check $x = -1$.
$$\sqrt{x+2} = 1 - \sqrt{3x+7}$$
$$\sqrt{-1+2} \overset{?}{=} 1 - \sqrt{3(-1)+7}$$
$$\sqrt{1} = 1 - \sqrt{-3+7}$$
$$1 = 1 - \sqrt{4}$$
$$1 = 1 - 2 \Rightarrow 1 = -1$$
This is a false statement.
-1 is not a solution.
Solution set: $\{-2\}$

63.
$$\sqrt{2\sqrt{7x+2}} = \sqrt{3x+2}$$
$$\left(\sqrt{2\sqrt{7x+2}}\right)^2 = \left(\sqrt{3x+2}\right)^2$$
$$2\sqrt{7x+2} = 3x+2$$
$$\left(2\sqrt{7x+2}\right)^2 = (3x+2)^2$$
$$4(7x+2) = 9x^2 + 12x + 4$$
$$28x + 8 = 9x^2 + 12x + 4$$
$$0 = 9x^2 - 16x - 4$$
$$0 = (9x+2)(x-2)$$
$$x = -\tfrac{2}{9} \text{ or } x = 2$$

Check $x = -\tfrac{2}{9}$.
$$\sqrt{2\sqrt{7x+2}} = \sqrt{3x+2}$$
$$\sqrt{2\sqrt{7\left(-\tfrac{2}{9}\right)+2}} \overset{?}{=} \sqrt{3\left(-\tfrac{2}{9}\right)+2}$$
$$\sqrt{2\sqrt{-\tfrac{14}{9}+2}} = \sqrt{-\tfrac{2}{3}+2}$$
$$\sqrt{2\sqrt{-\tfrac{14}{9}+\tfrac{18}{9}}} = \sqrt{-\tfrac{2}{3}+\tfrac{6}{3}}$$
$$\sqrt{2\sqrt{\tfrac{4}{9}}} = \sqrt{\tfrac{4}{3}} \Rightarrow \sqrt{2\left(\tfrac{2}{3}\right)} = \tfrac{\sqrt{4}}{\sqrt{3}}$$

$$\sqrt{\tfrac{4}{3}} = \tfrac{2}{\sqrt{3}} \cdot \tfrac{\sqrt{3}}{\sqrt{3}} \Rightarrow \tfrac{\sqrt{4}}{\sqrt{3}} = \tfrac{2\sqrt{3}}{3}$$
$$\tfrac{2}{\sqrt{3}} \cdot \tfrac{\sqrt{3}}{\sqrt{3}} = \tfrac{2\sqrt{3}}{3} \Rightarrow \tfrac{2\sqrt{3}}{3} = \tfrac{2\sqrt{3}}{3}$$
This is a true statement.
$-\tfrac{2}{9}$ is a solution.

Check $x = 2$.
$$\sqrt{2\sqrt{7x+2}} = \sqrt{3x+2}$$
$$\sqrt{2\sqrt{7(2)+2}} \overset{?}{=} \sqrt{3(2)+2}$$
$$\sqrt{2\sqrt{14+2}} = \sqrt{6+2}$$
$$\sqrt{2\sqrt{16}} = \sqrt{8}$$
$$\sqrt{2(4)} = 2\sqrt{2}$$
$$\sqrt{8} = 2\sqrt{2} \Rightarrow 2\sqrt{2} = 2\sqrt{2}$$
This is a true statement. 2 is a solution.
Solution set: $\left\{-\tfrac{2}{9}, 2\right\}$

65.
$$3 - \sqrt{x} = \sqrt{2\sqrt{x}-3}$$
$$\left(3-\sqrt{x}\right)^2 = \left(\sqrt{2\sqrt{x}-3}\right)^2$$
$$9 - 6\sqrt{x} + x = 2\sqrt{x} - 3$$
$$12 + x = 8\sqrt{x}$$
$$(12+x)^2 = \left(8\sqrt{x}\right)^2$$
$$144 + 24x + x^2 = 64x$$
$$x^2 - 40x + 144 = 0$$
$$(x-36)(x-4) = 0 \Rightarrow x = 36 \text{ or } x = 4$$

Check $x = 36$.
$$3 - \sqrt{x} = \sqrt{2\sqrt{x}-3}$$
$$3 - \sqrt{36} \overset{?}{=} \sqrt{2\sqrt{36}-3}$$
$$3 - 6 = \sqrt{2(6)-3}$$
$$-3 = \sqrt{12-3} \Rightarrow -3 = \sqrt{9} \Rightarrow -3 = 3$$
This is a false statement. 36 is not a solution.

Check $x = 4$.
$$3 - \sqrt{x} = \sqrt{2\sqrt{x}-3}$$
$$3 - \sqrt{4} \overset{?}{=} \sqrt{2\sqrt{4}-3}$$
$$3 - 2 = \sqrt{2(2)-3}$$
$$1 = \sqrt{4-3} \Rightarrow 1 = \sqrt{1} \Rightarrow 1 = 1$$
This is a true statement. 4 is a solution.
Solution set: $\{4\}$

67. $\sqrt[3]{4x+3} = \sqrt[3]{2x-1}$

$\left(\sqrt[3]{4x+3}\right)^3 = \left(\sqrt[3]{2x-1}\right)^3$

$4x+3 = 2x-1 \Rightarrow 2x = -4 \Rightarrow x = -2$

Check $x = -2$.

$\sqrt[3]{4(-2)+3} = \sqrt[3]{2(-2)-1}$

$\sqrt[3]{-5} \overset{?}{=} \sqrt[3]{-5} \Rightarrow -\sqrt[3]{5} = -\sqrt[3]{5}$

This is a true statement. -2 is a solution.

Solution set: $\{-2\}$

69. $\sqrt[3]{5x^2-6x+2} - \sqrt[3]{x} = 0$

$\sqrt[3]{5x^2-6x+2} = \sqrt[3]{x}$

$\left(\sqrt[3]{5x^2-6x+2}\right)^3 = \left(\sqrt[3]{x}\right)^3$

$5x^2 - 6x + 2 = x$

$5x^2 - 7x + 2 = 0$

$(5x-2)(x-1) = 0 \Rightarrow x = \frac{2}{5}$ or $x = 1$

Check $x = \frac{2}{5}$.

$\sqrt[3]{5x^2-6x+2} - \sqrt[3]{x} = 0$

$\sqrt[3]{5\left(\frac{2}{5}\right)^2 - 6\left(\frac{2}{5}\right)+2} - \sqrt[3]{\frac{2}{5}} \overset{?}{=} 0$

$\sqrt[3]{5\left(\frac{4}{25}\right) - \frac{12}{5}+2} - \sqrt[3]{\frac{2}{5}} = 0$

$\sqrt[3]{\frac{4}{5} - \frac{12}{5} + \frac{10}{5}} - \sqrt[3]{\frac{2}{5}} = 0$

$\sqrt[3]{\frac{2}{5}} - \sqrt[3]{\frac{2}{5}} = 0 \Rightarrow 0 = 0$

This is a true statement. $\frac{2}{5}$ is a solution.

Check $x = 1$.

$\sqrt[3]{5x^2-6x+2} - \sqrt[3]{x} = 0$

$\sqrt[3]{5(1)^2 - 6(1)+2} - \sqrt[3]{1} \overset{?}{=} 0$

$\sqrt[3]{5(1) - 6 + 2} - 1 = 0$

$\sqrt[3]{5 - 6 + 2} - 1 = 0$

$\sqrt[3]{1} - 1 = 0 \Rightarrow 1 - 1 = 0 \Rightarrow 0 = 0$

This is a true statement. 1 is a solution.

Solution set: $\left\{\frac{2}{5}, 1\right\}$

71. $\sqrt[4]{x-15} = 2 \Rightarrow \left(\sqrt[4]{x-15}\right)^4 = 2^4 \Rightarrow$

$x - 15 = 16 \Rightarrow x = 31$

Check $x = 31$.

$\sqrt[4]{x-15} = 2 \Rightarrow \sqrt[4]{31-15} \overset{?}{=} 2$

$\sqrt[4]{16} = 2 \Rightarrow 2 = 2$

This is a true statement.

Solution set: $\{31\}$

73. $\sqrt[4]{x^2+2x} = \sqrt[4]{3} \Rightarrow \left(\sqrt[4]{x^2+2x}\right)^4 = \left(\sqrt[4]{3}\right)^4$

$x^2 + 2x = 3 \Rightarrow x^2 + 2x - 3 = 0$

$(x+3)(x-1) = 0 \Rightarrow x = -3$ or $x = 1$

Check $x = -3$.

$\sqrt[4]{x^2+2x} = \sqrt[4]{3}$

$\sqrt[4]{(-3)^2 + 2(-3)} \overset{?}{=} \sqrt[4]{3}$

$\sqrt[4]{9-6} = \sqrt[4]{3} \Rightarrow \sqrt[4]{3} = \sqrt[4]{3}$

This is a true statement. -3 is a solution.

Check $x = 1$.

$\sqrt[4]{x^2+2x} = \sqrt[4]{3}$

$\sqrt[4]{1^2 + 2(1)} \overset{?}{=} \sqrt[4]{3}$

$\sqrt[4]{1+2} = \sqrt[4]{3} \Rightarrow \sqrt[4]{3} = \sqrt[4]{3}$

This is a true statement. 1 is a solution.

Solution set: $\{-3, 1\}$

75. $x^{3/2} = 125$

$\left(x^{3/2}\right)^{2/3} = 125^{2/3}$

$x = 5^2 = 25$

Check $x = 25$.

$x^{3/2} = 125$

$25^{3/2} \overset{?}{=} 125$

$125 = 125$

This is a true statement. 25 is a solution.

Solution set: $\{25\}$

77. $(x^2+24x)^{1/4} = 3 \Rightarrow \left[\left(x^2+24x\right)^{1/4}\right]^4 = 3^4 \Rightarrow$

$x^2 + 24x = 81 \Rightarrow x^2 + 24x - 81 = 0 \Rightarrow$

$(x+27)(x-3) = 0 \Rightarrow x + 27 = 0 \Rightarrow x = -27$ or

$x - 3 = 0 \Rightarrow x = 3$

Check $x = -27$.

$(x^2+24x)^{1/4} = 3$

$\left[(-27)^2 + 24(-27)\right]^{1/4} \overset{?}{=} 3$

$(729 - 648)^{1/4} = 3$

$81^{1/4} = 3 \Rightarrow 3 = 3$

This is a true statement. -27 is a solution.

Check $x = 3$.

$(x^2+24x)^{1/4} = 3$

$\left[3^2 + 24(3)\right]^{1/4} \overset{?}{=} 3$

$(9 + 72)^{1/4} = 3 \Rightarrow 81^{1/4} = 3 \Rightarrow 3 = 3$

This is a true statement. 3 is a solution.

Solution set: $\{-27, 3\}$

79.
$$(x-3)^{2/5} = 4$$
$$\left[(x-3)^{2/5}\right]^{5/2} = \pm 4^{5/2}$$
$$x - 3 = \pm 32$$
$$x - 3 = 32 \Rightarrow x = 35$$
$$x - 3 = -32 \Rightarrow x = -29$$

Check $x = 35$.
$$(x-3)^{2/5} = 4$$
$$(35-3)^{2/5} \overset{?}{=} 4$$
$$32^{2/5} = 4 \Rightarrow 4 = 4$$

This is a true statement. 35 is a solution.
Check $x = -29$.
$$(x-3)^{2/5} = 4$$
$$(-29-3)^{2/5} \overset{?}{=} 4$$
$$(-32)^{2/5} = 4$$
$$(-2)^2 = 4 \Rightarrow 4 = 4$$

This is a true statement. -29 is a solution.
Solution set: $\{-29, 35\}$

81.
$$(2x+5)^{1/3} - (6x-1)^{1/3} = 0$$
$$(2x+5)^{1/3} = (6x-1)^{1/3}$$
$$\left[(2x+5)^{1/3}\right]^3 = \left[(6x-1)^{1/3}\right]^3$$
$$2x + 5 = 6x - 1$$
$$-4x = -6$$
$$x = \frac{3}{2}$$

Check $x = \frac{3}{2}$.
$$(2x+5)^{1/3} - (6x-1)^{1/3} = 0$$
$$\left(2 \cdot \frac{3}{2} + 5\right)^{1/3} - \left(6 \cdot \frac{3}{2} - 1\right)^{1/3} \overset{?}{=} 0$$
$$8^{1/3} - 8^{1/3} = 0 \Rightarrow 0 = 0$$

This is a true statement. $\frac{3}{2}$ is a solution.

Solution set: $\left\{\frac{3}{2}\right\}$

83.
$$(2x-1)^{2/3} = x^{1/3}$$
$$[(2x-1)^{2/3}]^3 = (x^{1/3})^3$$
$$(2x-1)^2 = x \Rightarrow 4x^2 - 4x + 1 = x$$
$$4x^2 - 5x + 1 = 0 \Rightarrow (4x-1)(x-1) = 0 \Rightarrow$$
$$x = \frac{1}{4} \text{ or } x = 1$$

Check $x = \frac{1}{4}$.
$$(2x-1)^{2/3} = x^{1/3} \Rightarrow \left[2\left(\frac{1}{4}\right) - 1\right]^{2/3} \overset{?}{=} \left(\frac{1}{4}\right)^{1/3}$$
$$\left[\frac{1}{2} - 1\right]^{2/3} = \frac{1}{\sqrt[3]{4}} \cdot \frac{\sqrt[3]{2}}{\sqrt[3]{2}} \Rightarrow \left[-\frac{1}{2}\right]^{2/3} = \frac{\sqrt[3]{2}}{2} \Rightarrow$$
$$\left[\left(-\frac{1}{2}\right)^2\right]^{1/3} = \frac{\sqrt[3]{2}}{2} \Rightarrow \left(\frac{1}{4}\right)^{1/3} = \frac{\sqrt[3]{2}}{2}$$
$$\frac{1}{\sqrt[3]{4}} \cdot \frac{\sqrt[3]{2}}{\sqrt[3]{2}} = \frac{\sqrt[3]{2}}{2} \Rightarrow \frac{\sqrt[3]{2}}{2} = \frac{\sqrt[3]{2}}{2}$$

This is a true statement. $\frac{1}{4}$ is a solution.

Check $x = 1$.
$$(2x-1)^{2/3} = x^{1/3} \Rightarrow \left[2(1) - 1\right]^{2/3} \overset{?}{=} (1)^{1/3}$$
$$[2-1]^{2/3} = 1 \Rightarrow 1^{2/3} = 1 \Rightarrow 1 = 1$$

This is a true statement. 1 is a solution.
Solution set: $\left\{\frac{1}{4}, 1\right\}$

85.
$$x^{2/3} = 2x^{1/3} \Rightarrow \left(x^{2/3}\right)^3 = \left(2x^{1/3}\right)^3 \Rightarrow$$
$$x^2 = 8x \Rightarrow x^2 - 8x = 0 \Rightarrow x(x-8) = 0 \Rightarrow$$
$$x = 0 \text{ or } x = 8$$

Check $x = 0$.
$$x^{2/3} = 2x^{1/3}$$
$$0^{2/3} \overset{?}{=} 2\left(0^{1/3}\right) \Rightarrow 0 = 2 \cdot 0 \Rightarrow 0 = 0$$

This is a true statement. 0 is a solution.
Check $x = 8$.
$$x^{2/3} = 2x^{1/3} \Rightarrow 8^{2/3} \overset{?}{=} 2\left(8^{1/3}\right)$$
$$\left(8^2\right)^{1/3} = 2 \cdot 2 \Rightarrow 64^{1/3} = 4 \Rightarrow 4 = 4$$

This is a true statement. 8 is a solution.
Solution set: $\{0, 8\}$

87. $2x^4 - 7x^2 + 5 = 0$
Let $u = x^2$; then $u^2 = x^4$. With this substitution, the equation becomes
$2u^2 - 7u + 5 = 0$.
$$2u^2 - 7u + 5 = 0 \Rightarrow (u-1)(2u-5) = 0 \Rightarrow$$
$$u = 1 \text{ or } u = \frac{5}{2}$$

To find x, replace u with x^2.
$$x^2 = 1 \Rightarrow x = \pm 1 \text{ or}$$
$$x^2 = \frac{5}{2} \Rightarrow x = \pm\sqrt{\frac{5}{2}} = \pm\frac{\sqrt{5}}{\sqrt{2}} \cdot \frac{\sqrt{2}}{\sqrt{2}} = \pm\frac{\sqrt{10}}{2}$$

Solution set: $\left\{\pm 1, \pm\frac{\sqrt{10}}{2}\right\}$

89. $x^4 + 2x^2 - 15 = 0$

Let $u = x^2$; then $u^2 = x^4$.

$u^2 + 2u - 15 = 0 \Rightarrow (u - 3)(u + 5) = 0$.

$\qquad u = 3 \text{ or } u = -5$

To find x, replace u with x^2.

$x^2 = 3 \Rightarrow x = \pm\sqrt{3}$ or

$x^2 = -5 \Rightarrow x = \pm\sqrt{-5} = \pm i\sqrt{5}$

Solution set: $\left\{\pm\sqrt{3}, \pm i\sqrt{5}\right\}$

91. $(x-1)^{2/3} + (x-1)^{1/3} - 12 = 0$

Let $u = (x-1)^{1/3}$. Then

$u^2 = \left[(x-1)^{1/3}\right]^2 = (x-1)^{2/3}$.

$u^2 + u - 12 = 0 \Rightarrow (u+4)(u-3) = 0 \Rightarrow$

$u = -4 \text{ or } u = 3$

To find x, replace u with $(x-1)^{1/3}$.

$(x-1)^{1/3} = -4 \Rightarrow \left[(x-1)^{1/3}\right]^3 = (-4)^3 \Rightarrow$

$x - 1 = -64 \Rightarrow x = -63$ or

$(x-1)^{1/3} = 3 \Rightarrow \left[(x-1)^{1/3}\right]^3 = 3^3 \Rightarrow$

$x - 1 = 27 \Rightarrow x = 28$

Check $x = -63$.

$(x-1)^{2/3} + (x-1)^{1/3} - 12 = 0$

$(-63-1)^{2/3} + (-63-1)^{1/3} - 12 \overset{?}{=} 0$

$(-64)^{2/3} + (-64)^{1/3} - 12 = 0$

$\left[(-64)^{1/3}\right]^2 - 4 - 12 = 0$

$(-4)^2 - 4 - 12 = 0$

$16 - 4 - 12 = 0 \Rightarrow 0 = 0$

This is a true statement. -63 is a solution.

Check $x = 28$.

$(x-1)^{2/3} + (x-1)^{1/3} - 12 = 0$

$(28-1)^{2/3} + (28-1)^{1/3} - 12 \overset{?}{=} 0$

$27^{2/3} + 27^{1/3} - 12 = 0$

$\left[27^{1/3}\right]^2 + 3 - 12 = 0$

$3^2 + 3 - 12 = 0$

$9 + 3 - 12 = 0 \Rightarrow 0 = 0$

This is a true statement. 28 is a solution.

Solution set: $\{-63, 28\}$

93. $(x+1)^{2/5} - 3(x+1)^{1/5} + 2 = 0$

Let $u = (x+1)^{1/5}$. Then

$u^2 = \left[(x+1)^{1/5}\right]^2 = (x+1)^{2/5}$.

$u^2 - 3u + 2 = 0 \Rightarrow (u-1)(u-2) = 0 \Rightarrow$

$u = 1 \text{ or } u = 2$

To find x, replace u with $(x+1)^{1/5}$.

$(x+1)^{1/5} = 1 \Rightarrow \left[(x+1)^{1/5}\right]^5 = 1^5 \Rightarrow$ or

$x + 1 = 1 \Rightarrow x = 0$

$(x+1)^{1/5} = 2 \Rightarrow \left[(x+1)^{1/5}\right]^5 = 2^5 \Rightarrow$

$x + 1 = 32 \Rightarrow x = 31$

Check $x = 0$.

$(x+1)^{2/5} - 3(x+1)^{1/5} + 2 = 0$

$(0+1)^{2/5} - 3(0+1)^{1/5} + 2 \overset{?}{=} 0$

$1^{2/5} - 3(1)^{1/5} + 2 = 0$

$1 - 3(1) + 2 = 0 \Rightarrow 0 = 0$

This is a true statement. 0 is a solution.

Check $x = 31$.

$(x+1)^{2/5} - 3(x+1)^{1/5} + 2 = 0$

$(31+1)^{2/5} - 3(31+1)^{1/5} + 2 \overset{?}{=} 0$

$32^{2/5} - 3(32)^{1/5} + 2 = 0$

$\left[(32)^{1/5}\right]^2 - 3(2) + 2 = 0$

$2^2 - 6 + 2 = 0$

$4 - 6 + 2 = 0 \Rightarrow 0 = 0$

This is a true statement. 31 is a solution.

Solution set: $\{0, 31\}$

95. $4(x+1)^4 - 13(x+1)^2 = -9$

$4(x+1)^4 - 13(x+1)^2 + 9 = 0$

Let $u = (x+1)^2$. Then $u^2 = (x+1)^4$.

$4u^2 - 13u + 9 = 0$

$(4u - 9)(u - 1) = 0 \Rightarrow u = \frac{9}{4} \text{ or } u = 1$

To find x, replace u with $(x+1)^2$.

$(x+1)^2 = \frac{9}{4} \Rightarrow x + 1 = \pm\frac{3}{2} \Rightarrow$

$x = -1 \pm \frac{3}{2} \Rightarrow x = -\frac{5}{2} \text{ or } x = \frac{1}{2}$

$(x+1)^2 = 1 \Rightarrow x + 1 = \pm 1 \Rightarrow$

$x = -1 \pm 1 \Rightarrow x = -2 \text{ or } x = 0$

Be sure to check all possible solutions in the original equation.

Solution set: $\left\{-\frac{5}{2}, -2, 0, \frac{1}{2}\right\}$

97. $6(x+2)^4 - 11(x+2)^2 = -4$

$6(x+2)^4 - 11(x+2)^2 + 4 = 0$

Let $u = (x+2)^2$; then $u^2 = (x+2)^4$.

$6u^2 - 11u + 4 = 0 \Rightarrow (3u-4)(2u-1) = 0 \Rightarrow$

$u = \frac{4}{3}$ or $u = \frac{1}{2}$

To find x, replace u with $(x+2)^2$.

$(x+2)^2 = \frac{4}{3} \Rightarrow x+2 = \pm\sqrt{\frac{4}{3}} = \pm\frac{2\sqrt{3}}{3}$ or

$\quad x = -2 \pm \frac{2\sqrt{3}}{3} = -\frac{6}{3} \pm \frac{2\sqrt{3}}{3} = \frac{-6\pm2\sqrt{3}}{3}$

$(x+2)^2 = \frac{1}{2} \Rightarrow x+2 = \pm\sqrt{\frac{1}{2}} = \pm\frac{\sqrt{2}}{2}$

$\quad x = -2 \pm \frac{\sqrt{2}}{2} = -\frac{4}{2} \pm \frac{\sqrt{2}}{2} = \frac{-4\pm\sqrt{2}}{2}$

Solution set: $\left\{ \frac{-6\pm2\sqrt{3}}{3}, \frac{-4\pm\sqrt{2}}{2} \right\}$

99. $10x^{-2} + 33x^{-1} - 7 = 0$

Let $u = x^{-1}$; then $u^2 = x^{-2}$.

$10u^2 + 33u - 7 = 0 \Rightarrow (2u+7)(5u-1) = 0$

$u = -\frac{7}{2}$ or $u = \frac{1}{5}$

To find x, replace u with x^{-1}.

$x^{-1} = -\frac{7}{2} \Rightarrow x = -\frac{2}{7}$ or $x^{-1} = \frac{1}{5} \Rightarrow x = 5$

Solution set: $\left\{ -\frac{2}{7}, 5 \right\}$

101. $x^{-2/3} + x^{-1/3} - 6 = 0$

Let $u = x^{-1/3}$; then $u^2 = \left(x^{-1/3}\right)^2 = x^{-2/3}$.

$u^2 + u - 6 = 0 \Rightarrow (u+3)(u-2) = 0$

$u = -3$ or $u = 2$

To find x, replace u with $x^{-1/3}$.

$x^{-1/3} = -3 \Rightarrow \left(x^{-1/3}\right)^{-3} = (-3)^{-3} \Rightarrow$ or

$\quad x = \frac{1}{(-3)^3} \Rightarrow x = -\frac{1}{27}$

$x^{-1/3} = 2 \Rightarrow \left(x^{-1/3}\right)^{-3} = 2^{-3} \Rightarrow$

$\quad x = \frac{1}{2^3} \Rightarrow x = \frac{1}{8}$

Check $x = -\frac{1}{27}$.

$x^{-2/3} + x^{-1/3} - 6 = 0$

$\left(-\frac{1}{27}\right)^{-2/3} + \left(-\frac{1}{27}\right)^{-1/3} - 6 \overset{?}{=} 0$

$(-27)^{2/3} + (-27)^{1/3} - 6 = 0$

$\left[(-27)^{1/3}\right]^2 - 3 - 6 = 0$

$(-3)^2 - 3 - 6 = 0$

$9 - 3 - 6 = 0 \Rightarrow 0 = 0$

This is a true statement.

Check $x = \frac{1}{8}$.

$x^{-2/3} + x^{-1/3} - 6 = 0$

$\left(\frac{1}{8}\right)^{-2/3} + \left(\frac{1}{8}\right)^{-1/3} - 6 \overset{?}{=} 0$

$8^{2/3} + 8^{1/3} - 6 = 0$

$\left(8^{1/3}\right)^2 + 2 - 6 = 0$

$2^2 + 2 - 6 = 0$

$4 + 2 - 6 = 0 \Rightarrow 0 = 0$

This is a true statement.

Solution set: $\left\{ -\frac{1}{27}, \frac{1}{8} \right\}$

103. $16x^{-4} - 65x^{-2} + 4 = 0$

Let $u = x^{-2}$; then $u^2 = x^{-4}$. Solve the resulting equation by factoring:

$16u^2 - 65u + 4 = 0 \Rightarrow (u-4)(16u-1) = 0 \Rightarrow$

$\quad u = 4$ or $u = \frac{1}{16}$

Find x by replacing u with x^{-2}:

$x^{-2} = 4 \Rightarrow x^2 = \frac{1}{4} \Rightarrow x = \pm\frac{1}{2}$

$x^{-2} = \frac{1}{16} \Rightarrow x^2 = 16 \Rightarrow x = \pm4$

Check $x = \frac{1}{2}$

$16\left(\frac{1}{2}\right)^{-4} - 65\left(\frac{1}{2}\right)^{-2} + 4 = 0$

$16(2)^4 - 65(2)^2 + 4 \overset{?}{=} 0$

$16(16) - 65(4) + 4 = 0$

$256 - 260 + 4 = 0$

$0 = 0$

This is a true statement, so $\frac{1}{2}$ is a solution.

Check $x = -\frac{1}{2}$

$16\left(-\frac{1}{2}\right)^{-4} - 65\left(-\frac{1}{2}\right)^{-2} + 4 = 0$

$16(-2)^4 - 65(-2)^2 + 4 \overset{?}{=} 0$

$16(16) - 65(4) + 4 = 0$

$256 - 260 + 4 = 0$

$0 = 0$

This is a true statement, so $-\frac{1}{2}$ is a solution.

Check $x = 4$

$16(4)^{-4} - 65(4)^{-2} + 4 = 0$

$16\left(\frac{1}{4}\right)^4 - 65\left(\frac{1}{4}\right)^2 + 4 \overset{?}{=} 0$

$16\left(\frac{1}{256}\right) - 65\left(\frac{1}{16}\right) + 4 = 0$

$\frac{1}{16} - \frac{65}{16} + 4 = 0$

$0 = 0$

This is a true statement, so 4 is a solution.

(*continued on next page*)

(*continued*)

Check $x = -4$

$$16(-4)^{-4} - 65(-4)^{-2} + 4 = 0$$
$$16\left(-\tfrac{1}{4}\right)^4 - 65\left(-\tfrac{1}{4}\right)^2 + 4 \overset{?}{=} 0$$
$$16\left(\tfrac{1}{256}\right) - 65\left(\tfrac{1}{16}\right) + 4 = 0$$
$$\tfrac{1}{16} - \tfrac{65}{16} + 4 = 0$$
$$0 = 0$$

This is a true statement, so -4 is a solution.

Solution set: $\left\{\pm\tfrac{1}{2}, \pm 4\right\}$

105. $d = k\sqrt{h}$ for h

$$\frac{d}{k} = \sqrt{h} \Rightarrow \frac{d^2}{k^2} = h$$

So, $h = \dfrac{d^2}{k^2}$.

107. $m^{3/4} + n^{3/4} = 1$ for m

$$m^{3/4} = 1 - n^{3/4}$$

Raise both sides to the $\tfrac{4}{3}$ power.

$$\left(m^{3/4}\right)^{4/3} = \left(1 - n^{3/4}\right)^{4/3}$$
$$m = \left(1 - n^{3/4}\right)^{4/3}$$

109.

$$\frac{E}{e} = \frac{R+r}{r} \text{ for } e$$
$$er\left(\frac{E}{e}\right) = er\left(\frac{R+r}{r}\right)$$

Multiply both sides by er.

$$Er = eR + er$$
$$Er = e(R + r)$$
$$\frac{Er}{R+r} = e$$

So, $e = \dfrac{Er}{R+r}$.

111. $x - \sqrt{x} - 12 = 0$

Let $u = \sqrt{x}$; then $u^2 = x$. Solve the resulting equation by factoring.

$$u^2 - u - 12 = 0 \Rightarrow (u - 4)(u + 3) = 0$$
$$u = 4 \text{ or } u = -3$$

To find x, replace u with $\sqrt{x}$.

$$\sqrt{x} = 4 \Rightarrow \left(\sqrt{x}\right)^2 = 4^2 \Rightarrow x = 16 \text{ or}$$
$$\sqrt{x} = -3 \Rightarrow \left(\sqrt{x}\right)^2 = (-3)^2 \Rightarrow x = 9$$

But $\sqrt{9} \neq -3$

So when $u = -3$, there is no solution for x.

Solution set: $\{16\}$

113. Answers will vary.

Summary Exercises on Solving Equations

1. $4x - 3 = 2x + 3 \Rightarrow 2x - 3 = 3 \Rightarrow$
$2x = 6 \Rightarrow x = 3$
Solution set: $\{3\}$

3. $x(x + 6) = 9 \Rightarrow x^2 + 6x = 9 \Rightarrow x^2 + 6x - 9 = 0$

Solve by completing the square.

$$x^2 + 6x + 9 = 9 + 9$$
$$\text{Note: } \left[\tfrac{1}{2} \cdot 6\right]^2 = 3^2 = 9$$
$$(x + 3)^2 = 18 \Rightarrow x + 3 = \pm\sqrt{18} \Rightarrow$$
$$x + 3 = \pm 3\sqrt{2} \Rightarrow x = -3 \pm 3\sqrt{2}$$

Solve by the quadratic formula.

Let $a = 1, b = 6,$ and $c = -9$.

$$x = \frac{-b \pm \sqrt{b^2 - 4ac}}{2a}$$
$$= \frac{-6 \pm \sqrt{6^2 - 4(1)(-9)}}{2(1)}$$
$$= \frac{-6 \pm \sqrt{36 + 36}}{2} = \frac{-6 \pm \sqrt{72}}{2}$$
$$= \frac{-6 \pm 6\sqrt{2}}{2} = -3 \pm 3\sqrt{2}$$

Solution set: $\left\{-3 \pm 3\sqrt{2}\right\}$

5.

$$\sqrt{x+2} + 5 = \sqrt{x+15}$$
$$\left(\sqrt{x+2} + 5\right)^2 = \left(\sqrt{x+15}\right)^2$$
$$(x+2) + 10\sqrt{x+2} + 25 = x + 15 \Rightarrow$$
$$x + 27 + 10\sqrt{x+2} = x + 15 \Rightarrow$$
$$27 + 10\sqrt{x+2} = 15 \Rightarrow$$
$$10\sqrt{x+2} = -12 \Rightarrow 5\sqrt{x+2} = -6$$
$$\left(5\sqrt{x+2}\right)^2 = (-6)^2 \Rightarrow$$
$$25(x+2) = 36 \Rightarrow 25x + 50 = 36$$
$$25x = -14 \Rightarrow x = -\frac{14}{25}$$

Check $x = -\frac{14}{25}$.

$$\sqrt{x+2} + 5 = \sqrt{x+15}$$
$$\sqrt{-\tfrac{14}{25} + 2} + 5 \overset{?}{=} \sqrt{-\tfrac{14}{25} + 15}$$
$$\sqrt{-\tfrac{14}{25} + \tfrac{50}{25}} + 5 = \sqrt{-\tfrac{14}{25} + \tfrac{375}{25}}$$
$$\sqrt{\tfrac{36}{25}} + 5 = \sqrt{\tfrac{361}{25}} \Rightarrow \tfrac{6}{5} + 5 = \tfrac{19}{5} \Rightarrow \tfrac{31}{5} = \tfrac{19}{5}$$

This is a false statement. Solution set: $\varnothing$

7. $\dfrac{3x+4}{3} - \dfrac{2x}{x-3} = x$

The least common denominator is $3(x-3)$, which is equal to 0 if $x = 3$. Therefore, 3 cannot possibly be a solution of this equation.

$$3(x-3)\left[\dfrac{3x+4}{3} - \dfrac{2x}{x-3}\right] = 3(x-3)(x)$$
$$(x-3)(3x+4) - 3(2x) = 3x(x-3)$$
$$3x^2 + 4x - 9x - 12 - 6x = 3x^2 - 9x$$
$$3x^2 - 11x - 12 = 3x^2 - 9x$$
$$-11x - 12 = -9x$$
$$-12 = 2x \Rightarrow -6 = x$$

The restriction $x \ne 3$ does not affect the result. Therefore, the solution set is $\{-6\}$.

9. $5 - \dfrac{2}{x} + \dfrac{1}{x^2} = 0$

The least common denominator is x^2, which is equal to 0 if $x = 0$. Therefore, 0 cannot possibly be a solution of this equation.

$$x^2\left[5 - \dfrac{2}{x} + \dfrac{1}{x^2}\right] = x^2(0) \Rightarrow 5x^2 - 2x + 1 = 0$$

Solve by completing the square.
$$x^2 - \tfrac{2}{5}x + \tfrac{1}{5} = 0 \qquad \text{Multiply by } \tfrac{1}{5}.$$
$$x^2 - \tfrac{2}{5}x + \tfrac{1}{25} = -\tfrac{1}{5} + \tfrac{1}{25}$$
$$\text{Note:}\left[\tfrac{1}{2}\cdot\left(-\tfrac{2}{5}\right)\right]^2 = \left(-\tfrac{1}{5}\right)^2 = \tfrac{1}{25}$$
$$\left(x - \tfrac{1}{5}\right)^2 = -\tfrac{5}{25} + \tfrac{1}{25} = \tfrac{-4}{25}$$
$$x - \tfrac{1}{5} = \pm\sqrt{\tfrac{-4}{25}}$$
$$x - \tfrac{1}{5} = \pm\tfrac{2}{5}i \Rightarrow x = \tfrac{1}{5} \pm \tfrac{2}{5}i$$

Solve by the quadratic formula.
Let $a = 5$, $b = -2$, and $c = 1$.

$$x = \dfrac{-b \pm \sqrt{b^2 - 4ac}}{2a}$$
$$= \dfrac{-(-2) \pm \sqrt{(-2)^2 - 4(5)(1)}}{2(5)}$$
$$= \dfrac{2 \pm \sqrt{4 - 20}}{10} = \dfrac{2 \pm \sqrt{-16}}{10}$$
$$= \dfrac{2 \pm 4i}{10} = \dfrac{2}{10} \pm \dfrac{4}{10}i = \dfrac{1}{5} \pm \dfrac{2}{5}i$$

The restriction $x \ne 0$ does not affect the result. Therefore, the solution set is $\left\{\tfrac{1}{5} \pm \tfrac{2}{5}i\right\}$.

11. $x^{-2/5} - 2x^{-1/5} - 15 = 0$

Let $u = x^{-1/5}$; then $u^2 = \left(x^{-1/5}\right)^2 = x^{-2/5}$.

$$u^2 - 2u - 15 = 0 \Rightarrow (u+3)(u-5) = 0 \Rightarrow$$
$$u = -3 \text{ or } u = 5$$

To find x, replace u with $x^{-1/5}$.

$$x^{-1/5} = -3 \Rightarrow \left(x^{-1/5}\right)^{-5} = (-3)^{-5}$$
$$x = \dfrac{1}{(-3)^5} \Rightarrow x = -\dfrac{1}{243} \qquad \text{or}$$

$$x^{-1/5} = 5 \Rightarrow \left(x^{-1/5}\right)^{-5} = 5^{-5} \Rightarrow x = \dfrac{1}{5^5} \Rightarrow$$
$$x = \dfrac{1}{3125}$$

Check $x = -\dfrac{1}{243}$.

$$x^{-2/5} - 2x^{-1/5} - 15 = 0$$
$$\left(-\tfrac{1}{243}\right)^{-2/5} - 2\left(-\tfrac{1}{243}\right)^{-1/5} - 15 = 0 \text{ ?}$$
$$(-243)^{2/5} - 2(-243)^{1/5} - 15 = 0$$
$$\left[(-243)^{1/5}\right]^2 - 2(-3) - 15 = 0$$
$$(-3)^2 + 6 - 15 = 0 \Rightarrow 9 + 6 - 15 = 0 \Rightarrow 0 = 0$$

This is a true statement. $-\dfrac{1}{243}$ is a solution.

Check $x = \dfrac{1}{3125}$.

$$x^{-2/5} - 2x^{-1/5} - 15 = 0$$
$$\left(\tfrac{1}{3125}\right)^{-2/5} - 2\left(\tfrac{1}{3125}\right)^{-1/5} - 15 \overset{?}{=} 0$$
$$(3125)^{2/5} - 2(3125)^{1/5} - 15 = 0$$
$$\left[(3125)^{1/5}\right]^2 - 2(5) - 15 = 0$$
$$5^2 - 10 - 15 = 0$$
$$25 - 10 - 15 = 0 \Rightarrow 0 = 0$$

This is a true statement. $\dfrac{1}{3125}$ is a solution.

Solution set: $\left\{-\dfrac{1}{243}, \dfrac{1}{3125}\right\}$

13. $x^4 - 3x^2 - 4 = 0$

Let $u = x^2$; then $u^2 = x^4$.
$$u^2 - 3u - 4 = 0 \Rightarrow (u+1)(u-4) = 0 \Rightarrow$$
$$u = -1 \text{ or } u = 4$$

To find x, replace x with x^2.
$$x^2 = -1 \Rightarrow x = \pm\sqrt{-1} = \pm i \text{ or}$$
$$x^2 = 4 \Rightarrow x = \pm\sqrt{4} = \pm 2$$

Solution set: $\{\pm i, \pm 2\}$

15. $\sqrt[3]{2x+1} = \sqrt[3]{9} \Rightarrow \left(\sqrt[3]{2x+1}\right)^3 = \left(\sqrt[3]{9}\right)^3$

$2x+1 = 9 \Rightarrow 2x = 8 \Rightarrow x = 4$

Check $x = 4$.

$\sqrt[3]{2x+1} = \sqrt[3]{9} \Rightarrow \sqrt[3]{2(4)+1} \overset{?}{=} \sqrt[3]{9}$

$\sqrt[3]{8+1} = \sqrt[3]{9} \Rightarrow \sqrt[3]{9} = \sqrt[3]{9}$

This is a true statement.
Solution set: $\{4\}$

17. $3\left[2x-(6-2x)+1\right] = 5x$

$3(2x-6+2x+1) = 5x$

$3(4x-5) = 5x$

$12x-15 = 5x \Rightarrow -15 = -7x \Rightarrow$

$\dfrac{-15}{-7} = x \Rightarrow x = \dfrac{15}{7}$

Solution set: $\left\{\frac{15}{7}\right\}$

19. $(14-2x)^{2/3} = 4$

$\left[(14-2x)^{2/3}\right]^3 = 4^3$

$(14-2x)^2 = 64$

$196 - 56x + 4x^2 = 64$

$4x^2 - 56x + 132 = 0$

$4\left(x^2 - 14x + 33\right) = 0$

$4(x-3)(x-11) = 0 \Rightarrow x = 3$ or $x = 11$

Check $x = 3$.

$(14-2x)^{2/3} = 4$

$\left[14-2(3)\right]^{2/3} \overset{?}{=} 4$

$(14-6)^{2/3} = 4 \Rightarrow 8^{2/3} = 4$

$\left(8^{1/3}\right)^2 = 4 \Rightarrow 2^2 = 4 \Rightarrow 4 = 4$

This is a true statement.
Check $x = 11$.

$(14-2x)^{2/3} = 4$

$\left[14-2(11)\right]^{2/3} \overset{?}{=} 4$

$(14-22)^{2/3} = 4 \Rightarrow (-8)^{2/3} = 4$

$\left[(-8)^{1/3}\right]^2 = 4 \Rightarrow (-2)^2 = 4 \Rightarrow 4 = 4$

This is a true statement.
Solution set: $\{3, 11\}$

21. $\dfrac{3}{x-3} = \dfrac{3}{x-3}$

The least common denominator is $(x-3)$

which is equal to 0 if $x = 3$. Therefore, 3
cannot possibly be a solution of this equation.
Solution set: $\{x \mid x \neq 3\}$.

Section 1.7   Inequalities

1. F. The inequality $x < -6$ includes all real
numbers less than -6 not including -6. The
correct interval notation is $(-\infty, -6)$.

3. A. The inequality $-2 < x \leq 6$ includes all real
numbers from -2 to 6, not including -2, but
including 6. The correct interval notation is
$(-2, 6]$.

5. I. The inequality $x \geq -6$ includes all real
numbers greater than or equal to -6, so it
includes -6. The correct interval notation is
$[-6, \infty)$.

7. B. The interval shown on the number line
includes all real numbers between -2 and 6,
including -2, but not including 6. The correct
interval notation is $[-2, 6)$.

9. E. The interval shown on the number line
includes all real numbers less than -3, not
including -3, and greater than 3, not including
3. The correct interval notation is
$(-\infty, -3) \cup (3, \infty)$,

11. Answers will vary. Sample answer: A square
bracket is used to show that a number is part
of the solution set, while a parenthesis is used
to indicate that a number is not part of the
solution set.

13. $-2x+8 \leq 16 \Rightarrow -2x+8-8 \leq 16-8 \Rightarrow$

$-2x \leq 8 \Rightarrow \dfrac{-2x}{-2} \geq \dfrac{8}{-2} \Rightarrow x \geq -4$

Solution set: $[-4, \infty)$

15. $-2x-2 \leq 1+x$

$-2x-2+2 \leq 1+x+2$

$-2x \leq x+3 \Rightarrow -2x-x \leq 3 \Rightarrow$

$-3x \leq 3 \Rightarrow \dfrac{-3x}{-3} \geq \dfrac{3}{-3} \Rightarrow x \geq -1$

Solution set: $[-1, \infty)$

17. $3(x+5)+1 \geq 5+3x$

$3x+15+1 \geq 5+3x \Rightarrow 16 \geq 5$

The inequality is true when x is any real
number.
Solution set: $(-\infty, \infty)$

19. $8x - 3x + 2 < 2(x + 7)$
$$5x + 2 < 2x + 14$$
$$5x + 2 - 2x < 2x + 14 - 2x$$
$$3x + 2 < 14$$
$$3x + 2 - 2 < 14 - 2$$
$$3x < 12$$
$$\frac{3x}{3} < \frac{12}{3}$$
$$x < 4$$
Solution set: $(-\infty, 4)$

21. $\dfrac{4x + 7}{-3} \le 2x + 5$
$$(-3)\left(\frac{4x + 7}{-3}\right) \ge (-3)(2x + 5)$$
$$4x + 7 \ge -6x - 15$$
$$4x + 7 + 6x \ge -6x - 15 + 6x$$
$$10x + 7 \ge -15$$
$$10x + 7 - 7 \ge -15 - 7$$
$$10x \ge -22$$
$$\frac{10x}{10} \ge \frac{-22}{10}$$
$$x \ge -\frac{11}{5}$$
Solution set: $\left[-\frac{11}{5}, \infty\right)$

23. $\dfrac{1}{3}x + \dfrac{2}{5}x - \dfrac{1}{2}(x + 3) \le \dfrac{1}{10}$
$$30\left[\frac{1}{3}x + \frac{2}{5}x - \frac{1}{2}(x + 3)\right] \le 30\left[\frac{1}{10}\right]$$
$$10x + 12x - 15(x + 3) \le 3$$
$$10x + 12x - 15x - 45 \le 3$$
$$7x - 45 \le 3$$
$$7x - 45 + 45 \le 3 + 45$$
$$7x \le 48$$
$$\frac{7x}{7} \le \frac{48}{7}$$
$$x \le \frac{48}{7}$$
Solution set: $\left(-\infty, \frac{48}{7}\right]$

25. $C = 50x + 5000$; $R = 60x$
The product will at least break even when $R \ge C$. Set $R \ge C$ and solve for x.
$$60x \ge 50x + 5000 \Rightarrow 10x \ge 5000 \Rightarrow x \ge 500$$
The break-even point is at $x = 500$.
This product will at least break even if the number of units of picture frames produced is in the interval $[500, \infty)$.

27. $C = 105x + 900$; $R = 85x$
The product will at least break even when $R \ge C$. Set $R \ge C$ and solve for x.
$$85x \ge 105x + 900 \Rightarrow -20x \ge 900 \Rightarrow x \le -45$$
The product will never break even.

29. $-5 < 5 + 2x < 11$
$$-5 - 5 < 5 + 2x - 5 < 11 - 5$$
$$-10 < 2x < 6$$
$$\frac{-10}{2} < \frac{2x}{2} < \frac{6}{2}$$
$$-5 < x < 3$$
Solution set: $(-5, 3)$

31. $10 \le 2x + 4 \le 16$
$$10 - 4 \le 2x + 4 - 4 \le 16 - 4$$
$$6 \le 2x \le 12$$
$$\frac{6}{2} \le \frac{2x}{2} \le \frac{12}{2}$$
$$3 \le x \le 6$$
Solution set: [3, 6]

33. $-11 > -3x + 1 > -17$
$$-11 - 1 > -3x + 1 - 1 > -17 - 1$$
$$-12 > -3x > -18$$
$$\frac{-12}{-3} < \frac{-3x}{-3} < \frac{-18}{-3}$$
$$4 < x < 6$$
Solution set: (4, 6)

35. $-4 \le \dfrac{x + 1}{2} \le 5$
$$2(-4) \le 2\left(\frac{x + 1}{2}\right) \le 2(5)$$
$$-8 \le x + 1 \le 10$$
$$-8 - 1 \le x + 1 - 1 \le 10 - 1$$
$$-9 \le x \le 9$$
Solution set: [−9, 9]

37. $-3 \le \dfrac{3x - 4}{-5} < 4$
$$(-5)(-3) \ge (-5)\left(\frac{3x - 4}{-5}\right) > (-5)(4)$$
$$15 \ge 3x - 4 > -20$$
$$15 + 4 \ge 3x - 4 + 4 > -20 + 4$$
$$19 \ge 3x > -16$$
$$-\frac{16}{3} < x \le \frac{19}{3}$$
Solution set: $\left(-\frac{16}{3}, \frac{19}{3}\right]$

39. $x^2 - x - 6 > 0$

Step 1: Find the values of x that satisfy

$x^2 - x - 6 = 0.$

$x^2 - x - 6 = 0 \Rightarrow (x+2)(x-3) = 0$

$x + 2 = 0 \Rightarrow x = -2$ or $x - 3 = 0 \Rightarrow x = 3$

Step 2: The two numbers divide a number line into three regions.

Interval A   Interval B   Interval C
$(-\infty, -2)$ $(-2, 3)$ $(3, \infty)$

$-2 \quad 0 \quad 3$

Step 3: Choose a test value to see if it satisfies the inequality, $x^2 - x - 6 > 0$.

Interval	Test Value	Is $x^2 - x - 6 > 0$ True or False?
A: $(-\infty, -2)$	-3	$(-3)^2 - (-3) - 6 \overset{?}{>} 0$ $6 > 0$ True
B: $(-2, 3)$	0	$0^2 - 0 - 6 \overset{?}{>} 0$ $-6 > 0$ False
C: $(3, \infty)$	4	$4^2 - 4 - 6 \overset{?}{>} 0$ $6 > 0$ True

Solution set: $(-\infty, -2) \cup (3, \infty)$

41. $2x^2 - 9x \le 18$

Step 1: Find the values of x that satisfy the corresponding equation.

$2x^2 - 9x = 18$

$2x^2 - 9x - 18 = 0$

$(2x + 3)(x - 6) = 0$

$2x + 3 = 0 \Rightarrow x = -\frac{3}{2}$ or $x - 6 = 0 \Rightarrow x = 6$

Step 2: The two numbers divide a number line into three regions.

Interval A   Interval B   Interval C
$(-\infty, -\frac{3}{2})$ $(-\frac{3}{2}, 6)$ $(6, \infty)$

$-\frac{3}{2} \quad 0 \qquad 6$

Step 3: Choose a test value to see if it satisfies the inequality, $2x^2 - 9x \le 18$

Interval	Test Value	Is $2x^2 - 9x \le 18$ True or False?
A: $\left(-\infty, -\frac{3}{2}\right)$	-2	$2(-2)^2 - 9(-2) \overset{?}{\le} 18$ $26 \le 18$ False
B: $\left(-\frac{3}{2}, 6\right)$	0	$2(0)^2 - 9(0) \overset{?}{\le} 18$ $0 \le 18$ True
C: $(6, \infty)$	7	$2(7)^2 - 9(7) \overset{?}{\le} 18$ $35 \le 18$ False

Solution set: $\left[-\frac{3}{2}, 6\right]$

43. $-x^2 - 4x - 6 \le -3$

Step 1: Find the values of x that satisfy the corresponding equation.

$-x^2 - 4x - 6 = -3$

$x^2 + 4x + 3 = 0$

$(x + 3)(x + 1) = 0$

$x + 3 = 0 \Rightarrow x = -3$ or $x + 1 = 0 \Rightarrow x = -1$

Step 2: The two numbers divide a number line into three regions.

Interval A   Interval B   Interval C
$(-\infty, -3)$ $(-3, -1)$ $(-1, \infty)$

$-3 \quad -1 \quad 0$

Step 3: Choose a test value to see if it satisfies the inequality, $-x^2 - 4x - 6 \le -3$

Interval	Test Value	Is $-x^2 - 4x - 6 \le -3$ True or False?
A: $(-\infty, -3)$	-4	$-(-4)^2 - 4(-4) - 6 \overset{?}{\le} -3$ $-6 \le -3$ True
B: $(-3, -1)$	-2	$-(-2)^2 - 4(-2) - 6 \overset{?}{\le} -3$ $-2 \le -3$ False
C: $(-1, \infty)$	0	$-(0)^2 - 4(0) - 6 \overset{?}{\le} -3$ $-6 \le -3$ True

Solution set: $(-\infty, -3] \cup [-1, \infty)$

45. $x(x-1) \le 6 \Rightarrow x^2 - x \le 6 \Rightarrow x^2 - x - 6 \le 0$

Step 1: Find the values of x that satisfy $x^2 - x - 6 = 0$.

$$x^2 - x - 6 = 0$$
$$(x+2)(x-3) = 0$$

$$x + 2 = 0 \Rightarrow x = -2 \quad \text{or} \quad x - 3 = 0 \Rightarrow x = 3$$

Step 2: The two numbers divide a number line into three regions.

Interval A | Interval B | Interval C
$(-\infty, -2)$ | $(-2, 3)$ | $(3, \infty)$

$-2 \quad 0 \quad 3$

Step 3: Choose a test value to see if it satisfies the inequality, $x(x-1) \le 6$.

Interval	Test Value	Is $x(x-1) \le 6$ True or False?
A: $(-\infty, -2)$	-3	$-3(-3-1) \overset{?}{\le} 6$ $12 \le 6$ False
B: $(-2, 3)$	0	$0(0-1) \overset{?}{\le} 6$ $0 \le 6$ True
C: $(3, \infty)$	4	$4(4-1) \overset{?}{\le} 6$ $12 \le 6$ False

Solution set: $[-2, 3]$

47. $x^2 \le 9$

Step 1: Find the values of x that satisfy $x^2 \le 9$

$$x^2 = 9$$
$$x^2 - 9 = 0$$
$$(x+3)(x-3) = 0$$

$$x + 3 = 0 \Rightarrow x = -3 \quad \text{or} \quad x - 3 = 0 \Rightarrow x = 3$$

Step 2: The two numbers divide a number line into three regions.

Interval A | Interval B | Interval C
$(-\infty, -3)$ | $(-3, 3)$ | $(3, \infty)$

$-3 \quad 0 \quad 3$

Step 3: Choose a test value to see if it satisfies the inequality, $x^2 \le 9$.

Interval	Test Value	Is $x^2 \le 9$ True or False?
A: $(-\infty, -3)$	-4	$(-4)^2 \overset{?}{\le} 9$ $16 \le 9$ False
B: $(-3, 3)$	0	$(0)^2 \overset{?}{\le} 9$ $0 \le 9$ True
C: $(3, \infty)$	4	$(4)^2 \overset{?}{\le} 9$ $16 \le 9$ False

Solution set: $[-3, 3]$

49. $x^2 + 5x + 7 < 0$

Step 1: Find the values of x that satisfy $x^2 + 5x + 7 = 0$.

Use the quadratic formula to solve the equation.

Let $a = 1$, $b = 5$, and $c = 7$.

$$x = \frac{-b \pm \sqrt{b^2 - 4ac}}{2a} = \frac{-5 \pm \sqrt{5^2 - 4(1)(7)}}{2(1)}$$

$$= \frac{-5 \pm \sqrt{25 - 28}}{2} = \frac{-5 \pm \sqrt{-3}}{2}$$

$$= \frac{-5 \pm i\sqrt{3}}{2} = \frac{-5}{2} \pm \frac{\sqrt{3}}{2}i$$

Step 2: The number line is one region, $(-\infty, \infty)$.

Interval A
$(-\infty, \infty)$

0

Step 3: Because there are no real values of x that satisfy $x^2 + 5x + 7 = 0$, $x^2 + 5x + 7$ is either always positive or always negative. By substituting an arbitrary value such as $x = 0$, we see that $x^2 + 5x + 7$ will be positive and thus the solution set is $\varnothing$.

Interval	Test Value	Is $x^2 + 5x - 2 < 0$ True or False?
A: $(-\infty, \infty)$	0	$0^2 + 5(0) + 7 \overset{?}{<} 0$ $7 < 0$ False

Solution set: $\varnothing$

51. $x^2 - 2x \le 1 \Rightarrow x^2 - 2x - 1 \le 0$

Step 1: Find the values of x that satisfy $x^2 - 2x - 1 = 0$.

Use the quadratic formula to solve the equation.

Let $a = 1$, $b = -2$, and $c = -1$.

$$x = \frac{-b \pm \sqrt{b^2 - 4ac}}{2a}$$

$$= \frac{-(-2) \pm \sqrt{(-2)^2 - 4(1)(-1)}}{2(1)}$$

$$= \frac{2 \pm \sqrt{4+4}}{2} = \frac{2 \pm \sqrt{8}}{2} = \frac{2 \pm 2\sqrt{2}}{2} = 1 \pm \sqrt{2}$$

$1 - \sqrt{2} \approx -0.4$ or $1 + \sqrt{2} \approx 2.4$

Step 2: The two numbers divide a number line into three regions.

Step 3: Choose a test value to see if it satisfies the inequality, $x^2 - 2x \le 1$.

Interval	Test Value	Is $x^2 - 2x \le 1$ True or False?
A: $\left(-\infty, 1 - \sqrt{2}\right)$	-1	$(-1)^2 - 2(-1) \overset{?}{\le} 1$ $3 \le 1$ False
B: $\left(1 - \sqrt{2}, 1 + \sqrt{2}\right)$	0	$0^2 - 2(0) \overset{?}{\le} 1 \Rightarrow 0 \le 1$ True
C: $\left(1 + \sqrt{2}, \infty\right)$	3	$3^2 - 2(3) \overset{?}{\le} 1 \Rightarrow 3 \le 1$ False

Solution set: $\left[1 - \sqrt{2}, 1 + \sqrt{2}\right]$

53. A. $(x+3)^2$ is equal to zero when $x = -3$. For any other real number, $(x+3)^2$ is positive.

$(x+3)^2 \ge 0$ has solution set $(-\infty, \infty)$.

55. $\dfrac{x-3}{x+5} \le 0$

Because one side of the inequality is already 0, we start with Step 2.

Step 2: Determine the values that will cause either the numerator or denominator to equal 0.

$x - 3 = 0 \Rightarrow x = 3$ or $x + 5 = 0 \Rightarrow x = -5$

The values -5 and 3 divide the number line into three regions. Use an open circle on -5 because it makes the denominator equal 0.

Step 3: Choose a test value to see if it satisfies the inequality, $\dfrac{x-3}{x+5} \le 0$.

Interval	Test Value	Is $\frac{x-3}{x+5} \le 0$ True or False?
A: $(-\infty, -5)$	-6	$\frac{-6-3}{-6+5} \overset{?}{\le} 0$ $9 \le 0$ False
B: $(-5, 3)$	0	$\frac{0-3}{0+5} \overset{?}{\le} 0$ $-\frac{3}{5} \le 0$ True
C: $(3, \infty)$	4	$\frac{4-3}{4+5} \overset{?}{\le} 0$ $\frac{1}{9} \le 0$ False

Interval B satisfies the inequality. The endpoint -5 is not included because it makes the denominator 0.

Solution set: $(-5, 3]$

57. $\dfrac{1-x}{x+2} < -1$

Step 1: Rewrite the inequality so that 0 is on one side and there is a single fraction on the other side.

$$\frac{1-x}{x+2} < -1 \Rightarrow \frac{x-1}{x+2} > 1$$

$$\frac{x-1}{x+2} - 1 > 0 \Rightarrow \frac{x-1}{x+2} - \frac{x+2}{x+2} > 0$$

$$\frac{x-1-(x+2)}{x+2} > 0 \Rightarrow \frac{x-1-x-2}{x+2} > 0$$

$$\frac{-3}{x+2} > 0$$

Step 2: Because the numerator is a constant, determine the values that will cause denominator to equal 0.

$x + 2 = 0 \Rightarrow x = -2$

The value -2 divides the number line into two regions.

Step 3: Choose a test value to see if it satisfies the inequality, $\dfrac{1-x}{x+2} < -1$

(continued on next page)

(*continued*)

Interval	Test Value	Is $\frac{1-x}{x+2}<-1$ True or False?
A: $(-\infty,-2)$	-3	$\frac{-1-(-3)}{-3+2}\overset{?}{<}-1$ $-2<-1$ True
B: $(-2,\infty)$	-1	$\frac{1-(-1)}{-1+2}\overset{?}{<}1$ $2<-1$ False

Solution set: $(-\infty,-2)$

59. $\dfrac{3}{x-6}\le 2$

Step 1: Rewrite the inequality so that 0 is on one side and there is a single fraction on the other side.

$$\frac{3}{x-6}-2\le 0 \Rightarrow \frac{3}{x-6}-\frac{2(x-6)}{x-6}\le 0$$

$$\frac{3-2(x-6)}{x-6}\le 0 \Rightarrow \frac{3-2x+12}{x-6}\le 0$$

$$\frac{15-2x}{x-6}\le 0$$

Step 2: Determine the values that will cause either the numerator or denominator to equal 0.

$15-2x=0 \Rightarrow x=\frac{15}{2}$ or $x-6=0 \Rightarrow x=6$

The values 6 and $\frac{15}{2}$ divide the number line into three regions. Use an open circle on 6 because it makes the denominator equal 0.

Interval A   Interval B   Interval C
$(-\infty,6)$ $\left(6,\frac{15}{2}\right)$ $\left(\frac{15}{2},\infty\right)$

       6   7 $\frac{15}{2}$

Step 3: Choose a test value to see if it satisfies the inequality, $\dfrac{3}{x-6}\le 2$.

Interval	Test Value	Is $\frac{3}{x-6}\le 2$ True or False?
A: $(-\infty,6)$	0	$\frac{3}{0-6}\overset{?}{\le}2$ $-\frac{1}{2}\le 2$ True
B: $\left(6,\frac{15}{2}\right)$	7	$\frac{3}{7-6}\overset{?}{\le}2$ $3\le 2$ False
C: $\left(\frac{15}{2},\infty\right)$	8	$\frac{3}{8-6}\overset{?}{\le}2$ $\frac{3}{2}\le 2$ True

Intervals A and C satisfiy the inequality. The endpoint 6 is not included because it makes the denominator 0.

Solution set: $\left(-\infty,6\right)\cup\left[\frac{15}{2},\infty\right)$

61. $\dfrac{-4}{1-x}<5$

Step 1: Rewrite the inequality to compare a single fraction to 0:

$$\frac{-4}{1-x}<5 \Rightarrow \frac{-4}{1-x}-5<0$$

$$\frac{-4}{1-x}-\frac{5(1-x)}{1-x}<0 \Rightarrow \frac{-4-5(1-x)}{1-x}<0$$

$$\frac{-4-5+5x}{1-x}<0 \Rightarrow \frac{-9+5x}{1-x}<0$$

Step 2: Determine the values that will cause either the numerator or denominator to equal 0.

$-9+5x=0 \Rightarrow x=\frac{9}{5}$ or $1-x=0 \Rightarrow x=1$

The values 1 and $\frac{9}{5}$ divide the number line into three regions.

Interval A   Interval B   Interval C
$(-\infty,1)$ $\left(1,\frac{9}{5}\right)$ $\left(\frac{9}{5},\infty\right)$

   0   1 $\frac{9}{5}$ 2

Step 3: Choose a test value to see if it satisfies the inequality, $\dfrac{-4}{1-x}<5$

Interval	Test Value	Is $\frac{-4}{1-x}<5$ True or False?
A: $(-\infty,1)$	0	$\frac{-4}{1-0}\overset{?}{<}5$ $-4<5$ True
B: $\left(1,\frac{9}{5}\right)$	$\frac{6}{5}$	$\frac{-4}{1-\frac{6}{5}}\overset{?}{<}5$ $20<5$ False
C: $\left(\frac{9}{5},\infty\right)$	2	$\frac{-4}{1-2}\overset{?}{<}5$ $4<5$ True

Solution set: $\left(-\infty,1\right)\cup\left(\frac{9}{5},\infty\right)$

63. $\dfrac{10}{3+2x} \le 5$

Step 1: Rewrite the inequality so that 0 is on one side and there is a single fraction on the other side.

$$\dfrac{10}{3+2x} - 5 \le 0 \Rightarrow \dfrac{10}{3+2x} - \dfrac{5(3+2x)}{3+2x} \le 0$$

$$\dfrac{10 - 5(3+2x)}{3+2x} \le 0 \Rightarrow \dfrac{10 - 15 - 10x}{3+2x} \le 0$$

$$\dfrac{-10x - 5}{3+2x} \le 0$$

Step 2: Determine the values that will cause either the numerator or denominator to equal 0.

$-10x - 5 = 0 \Rightarrow x = -\frac{1}{2}$ or $3 + 2x = 0 \Rightarrow x = -\frac{3}{2}$

The values $-\frac{3}{2}$ and $-\frac{1}{2}$ divide the number line into three regions. Use an open circle on $-\frac{3}{2}$ because it makes the denominator equal 0.

Interval A $(-\infty, -\frac{3}{2})$ Interval B $(-\frac{3}{2}, -\frac{1}{2})$ Interval C $(-\frac{1}{2}, \infty)$

$-2 \quad -\frac{3}{2} \quad -1 \quad -\frac{1}{2} \quad 0$

Step 3: Choose a test value to see if it satisfies the inequality, $\dfrac{10}{3+2x} \le 5$.

Interval	Test Value	Is $\frac{10}{3+2x} \le 5$ True or False?
A: $\left(-\infty, -\frac{3}{2}\right)$	-2	$\frac{10}{3+2(-2)} \overset{?}{\le} 5$ $-10 \le 5$ True
B: $\left(-\frac{3}{2}, -\frac{1}{2}\right)$	-1	$\frac{10}{3+2(-1)} \overset{?}{\le} 5$ $10 \le 5$ False
C: $\left(-\frac{1}{2}, \infty\right)$	0	$\frac{10}{3+2(0)} \overset{?}{\le} 5$ $\frac{10}{3} \le 5$ True

Intervals A and C satisfy the inequality. The endpoint $-\frac{3}{2}$ is not included because it makes the denominator 0.

Solution set: $\left(-\infty, -\frac{3}{2}\right) \cup \left[-\frac{1}{2}, \infty\right)$

65. $\dfrac{7}{x+2} \ge \dfrac{1}{x+2}$

Step 1: Rewrite the inequality so that 0 is on one side and there is a single fraction on the other side.

$$\dfrac{7}{x+2} - \dfrac{1}{x+2} \ge 0 \Rightarrow \dfrac{6}{x+2} \ge 0$$

Step 2: Because the numerator is a constant, determine the value that will cause the denominator to equal 0.

$x + 2 = 0 \Rightarrow x = -2$

The value -2 divides the number line into two regions. Use an open circle on -2 because it makes the denominator equal 0.

Interval A $(-\infty, -2)$ Interval B $(-2, \infty)$

$-2 \quad 0$

Step 3: Choose a test value to see if it satisfies the inequality, $\dfrac{7}{x+2} \ge \dfrac{1}{x+2}$.

Interval	Test Value	Is $\frac{7}{x+2} \ge \frac{1}{x+2}$ True or False?
A: $(-\infty, -2)$	-3	$\frac{7}{-3+2} \overset{?}{\ge} \frac{1}{-3+2}$ $-7 \ge -1$ False
B: $(-2, \infty)$	0	$\frac{7}{0+2} \overset{?}{\ge} \frac{1}{0+2}$ $\frac{7}{2} \ge \frac{1}{2}$ True

Interval B satisfies the inequality. The endpoint -2 is not included because it makes the denominator 0.

Solution set: $(-2, \infty)$

67. $\dfrac{3}{2x-1} > \dfrac{-4}{x}$

Step 1: Rewrite the inequality so that 0 is on one side and there is a single fraction on the other side.

$$\dfrac{3}{2x-1} + \dfrac{4}{x} > 0$$

$$\dfrac{3x}{x(2x-1)} + \dfrac{4(2x-1)}{x(2x-1)} > 0$$

$$\dfrac{3x + 4(2x-1)}{x(2x-1)} > 0$$

$$\dfrac{3x + 8x - 4}{x(2x-1)} > 0 \Rightarrow \dfrac{11x - 4}{x(2x-1)} > 0$$

Step 2: Determine the values that will cause either the numerator or denominator to equal 0.

$11x - 4 = 0 \Rightarrow x = \frac{4}{11}$ or $x = 0$ or

$2x - 1 = 0 \Rightarrow x = \frac{1}{2}$

The values 0, $\frac{4}{11}$, and $\frac{1}{2}$ divide the number line into four regions.

Interval A $(-\infty, 0)$ Interval B $(0, \frac{4}{11})$ Interval C $(\frac{4}{11}, \frac{1}{2})$ Interval D $(\frac{1}{2}, \infty)$

$0 \quad \frac{4}{11} \quad \frac{1}{2}$

(continued on next page)

(continued)

Step 3: Choose a test value to see if it satisfies

the inequality, $\dfrac{3}{2x-1} > \dfrac{-4}{x}$.

Interval	Test Value	Is $\frac{3}{2x-1} > \frac{-4}{x}$ True or False?
A: $(-\infty, 0)$	-1	$\dfrac{3}{2(-1)-1} \overset{?}{>} \dfrac{-4}{-1}$ $-1 > 4$ False
B: $\left(0, \frac{4}{11}\right)$	$\frac{1}{11}$	$\dfrac{3}{2\left(\frac{1}{11}\right)-1} \overset{?}{>} \dfrac{-4}{\frac{1}{11}}$ or $-\dfrac{11}{3} \overset{?}{>} -44$ $-3\frac{2}{3} > -44$ True
C: $\left(\frac{4}{11}, \frac{1}{2}\right)$	$\frac{9}{22}$	$\dfrac{3}{2\left(\frac{9}{22}\right)-1} \overset{?}{>} \dfrac{-4}{\frac{9}{22}}$ or $-\dfrac{33}{2} \overset{?}{>} -\dfrac{88}{9}$ $-16\frac{1}{2} > -9\frac{7}{9}$ False
D: $\left(\frac{1}{2}, \infty\right)$	1	$\dfrac{3}{2(1)-1} \overset{?}{>} \dfrac{-4}{1}$ $3 > -4$ True

Solution set: $\left(0, \frac{4}{11}\right) \cup \left(\frac{1}{2}, \infty\right)$

69. $\dfrac{4}{2-x} \ge \dfrac{3}{1-x}$

Step 1: Rewrite the inequality so that 0 is on one side and there is a single fraction on the other side.

$$\frac{4}{2-x} \ge \frac{3}{1-x}$$
$$\frac{4}{2-x} - \frac{3}{1-x} \ge 0$$
$$\frac{4(1-x)}{(2-x)(1-x)} - \frac{3(2-x)}{(1-x)(2-x)} \ge 0$$
$$\frac{4(1-x) - 3(2-x)}{(x-2)(1-x)} \ge 0$$
$$\frac{4 - 4x - 6 + 3x}{(2-x)(1-x)} \ge 0$$
$$\frac{-2 - x}{(2-x)(1-x)} \ge 0$$

Step 2: Determine the values that will cause either the numerator or denominator to equal 0.
$-2 - x = 0 \Rightarrow x = -2$ or $2 - x = 0 \Rightarrow x = 2$ or $1 - x = 0 \Rightarrow x = 1$

The values -2, 1, and 2 divide the number line into four regions. Use an open circle on 1 and 2 because they make the denominator equal 0.

Interval A $(-\infty,-2)$	Interval B $(-2, 1)$	Interval C $(1, 2)$	Interval D $(2, \infty)$

$$\xleftarrow{\quad\bullet\quad\quad+\quad\quad\circ\quad\quad\circ\quad}\rightarrow$$
$$\quad -2 \qquad 0 \qquad 1 \qquad 2$$

Step 3: Choose a test value to see if it satisfies

the inequality, $\dfrac{4}{2-x} \ge \dfrac{3}{1-x}$

Interval	Test Value	Is $\frac{4}{2-x} \ge \frac{3}{1-x}$ True or False?
A: $(-\infty, -2)$	-3	$\dfrac{4}{-2-(-3)} \overset{?}{\ge} \dfrac{3}{1-(-3)}$ $4 \ge \frac{3}{4}$ True
B: $(-2, 1)$	0	$\dfrac{4}{2-0} \overset{?}{\ge} \dfrac{3}{1-0}$ $2 \ge 3$ False
C: $(1, 2)$	1.5	$\dfrac{4}{2-1.5} \overset{?}{\ge} \dfrac{3}{1-1.5}$ $8 \ge -6$ True
D: $(2, \infty)$	3	$\dfrac{4}{2-3} \overset{?}{\ge} \dfrac{3}{1-3}$ $-4 \ge -\frac{3}{2}$ False

Intervals A and C satisfy the inequality. The endpoints 1 and 2 are not included because they make the denominator 0.
Solution set: $(-\infty, -2] \cup (1, 2)$

71. $\dfrac{x+3}{x-5} \le 1$

Step 1: Rewrite the inequality so that 0 is on one side and there is a single fraction on the other side.

$$\frac{x+3}{x-5} - 1 \le 0 \Rightarrow \frac{x+3}{x-5} - \frac{x-5}{x-5} \le 0$$
$$\frac{x+3-(x-5)}{x-5} \le 0 \Rightarrow \frac{x+3-x+5}{x-5} \le 0$$
$$\frac{8}{x-5} \le 0$$

Step 2: Because the numerator is a constant, determine the value that will cause the denominator to equal 0.
$x - 5 = 0 \Rightarrow x = 5$

The value 5 divides the number line into two regions. Use an open circle on 5 because it makes the denominator equal 0.

Interval A $(-\infty, 5)$	Interval B $(5, \infty)$

$$\xleftarrow{\quad+\quad+\quad+\quad+\quad\circ\quad+\quad+\quad}\rightarrow$$
$$\qquad\quad 0 \qquad\qquad 5$$

Step 3: Choose a test value to see if it satisfies

the inequality, $\dfrac{x+3}{x-5} \le 1$.

(continued on next page)

(continued)

Interval	Test Value	Is $\frac{x+3}{x-5} \leq 1$ True or False?
A: $(-\infty, 5)$	0	$\frac{0+3}{0-5} \overset{?}{\leq} 1$ $-\frac{3}{5} \leq 1$ True
B: $(5, \infty)$	6	$\frac{6+3}{6-5} \overset{?}{\leq} 1$ $9 \leq 1$ False

Interval A satisfies the inequality. The endpoint 5 is not included because it makes the denominator 0.

Solution set: $(-\infty, 5)$

73. $\dfrac{2x-3}{x^2+1} \geq 0$

Because one side of the inequality is already 0, we start with Step 2.

Step 2: Determine the values that will cause either the numerator or denominator to equal 0.

$2x - 3 = 0$ or $x^2 + 1 = 0$

$x = \frac{3}{2}$ has no real solutions

$\frac{3}{2}$ divides the number line into two intervals.

Interval A $(-\infty, \frac{3}{2})$ Interval B $(\frac{3}{2}, \infty)$

$0\quad 1\quad \frac{3}{2}\quad 2$

Step 3: Choose a test value to see if it satisfies the inequality, $\dfrac{2x-3}{x^2+1} \geq 0$.

Interval	Test Value	Is $\frac{2x-3}{x^2+1} \geq 0$ True or False?
A: $\left(-\infty, \frac{3}{2}\right)$	0	$\frac{2(0)-3}{0^2+1} \overset{?}{\geq} 0$ $-3 \geq 0$ False
B: $\left(\frac{3}{2}, \infty\right)$	2	$\frac{2(2)-3}{2^2+1} \overset{?}{\geq} 0$ $\frac{1}{5} \geq 0$ True

Solution set: $\left[\frac{3}{2}, \infty\right)$

75. $\dfrac{(5-3x)^2}{(2x-5)^3} > 0$

Because one side of the inequality is already 0, we start with Step 2.

Step 2: Determine the values that will cause either the numerator or denominator to equal 0.

$5 - 3x = 0 \Rightarrow x = \frac{5}{3}$ or $2x - 5 = 0 \Rightarrow x = \frac{5}{2}$

The values $\frac{5}{3}$ and $\frac{5}{2}$ divide the number line into three intervals.

Interval A $(-\infty, \frac{5}{3})$ Interval B $(\frac{5}{3}, \frac{5}{2})$ Interval C $(\frac{5}{2}, \infty)$

$1\qquad\quad 2\qquad\quad 3$

Step 3: Choose a test value to see if it satisfies the inequality, $\dfrac{(5-3x)^2}{(2x-5)^3} > 0$.

Interval	Test Value	Is $\frac{(5-3x)^2}{(2x-5)^3} > 0$ True or False?
A: $\left(-\infty, \frac{5}{3}\right)$	0	$\frac{(5-3\cdot0)^2}{(2\cdot0-5)^3} \overset{?}{>} 0$ $-\frac{1}{5} > 0$ False
B: $\left(\frac{5}{3}, \frac{5}{2}\right)$	2	$\frac{(5-3\cdot2)^2}{(2\cdot2-5)^3} \overset{?}{>} 0$ $-1 > 0$ False
C: $\left(\frac{5}{2}, \infty\right)$	3	$\frac{(5-3\cdot3)^2}{(2\cdot3-5)^3} \overset{?}{>} 0$ $16 > 0$ True

Solution set: $\left(\frac{5}{2}, \infty\right)$

77. $\dfrac{(2x-3)(3x+8)}{(x-6)^3} \geq 0$

Because one side of the inequality is already 0, we start with Step 2.

Step 2: Determine the values that will cause either the numerator or denominator to equal 0.

$2x - 3 = 0$ or $3x + 8 = 0$ or $x - 6 = 0$

$x = \frac{3}{2}$ or $x = -\frac{8}{3}$ or $x = 6$

The values $-\frac{8}{3}$, $\frac{3}{2}$, and 6 divide the number line into four intervals. Use an open circle on 6 because it makes the denominator equal 0.

Interval A $(-\infty, -\frac{8}{3})$ Interval B $(-\frac{8}{3}, \frac{3}{2})$ Interval C $(\frac{3}{2}, 6)$ Interval D $(6, \infty)$

$-\frac{8}{3}\qquad 0\quad \frac{3}{2}\qquad\quad 6$

Step 3: Choose a test value to see if it satisfies the inequality, $\dfrac{(2x-3)(3x+8)}{(x-6)^3} \geq 0$.

(continued on next page)

(continued)

Interval	Test Value	Is $\frac{(2x-3)(3x+8)}{(x-6)^3} \geq 0$ True or False?
A: $\left(-\infty, -\frac{8}{3}\right)$	-3	$\frac{[2(-3)-3][3(-3)+8]}{(-3-6)^3} \overset{?}{\geq} 0$ $-\frac{1}{81} \geq 0$ False
B: $\left(-\frac{8}{3}, \frac{3}{2}\right)$	0	$\frac{(2\cdot0-3)(3\cdot0+8)}{(0-6)^3} \overset{?}{\geq} 0$ $\frac{1}{9} \geq 0$ True
C: $\left(\frac{3}{2}, 6\right)$	2	$\frac{(2\cdot2-3)(3\cdot2+8)}{(2-6)^3} \overset{?}{\geq} 0$ $-\frac{7}{32} \geq 0$ False
D: $(6, \infty)$	7	$\frac{(2\cdot7-3)(3\cdot7+8)}{(7-6)^3} \overset{?}{\geq} 0$ $319 \geq 0$ True

Solution set: $\left[-\frac{8}{3}, \frac{3}{2}\right] \cup (6, \infty)$

79. (a) Let $R = 7.6$ and then solve for x.
$$0.2844x + 5.535 > 7.6$$
$$0.2844x > 2.065 \Rightarrow x > 7.3$$
The model predicts that the receipts exceeded $7.6 billion about 7.3 years after 1993, which was in 2000.

(b) Let $R = 10$ and then solve for x.
$$0.2844x + 5.535 > 10$$
$$0.2844x > 4.465 \Rightarrow x > 15.7$$
The model predicts that the receipts exceeded $10 billion about 15.7 years after 1993, which was in 2008.

81. $-16t^2 + 220t \geq 624$
Step 1: Find the values of t that satisfy
$-16t^2 + 220t = 624$.
$$-16t^2 + 220t = 624 \Rightarrow 0 = 16t^2 - 220t + 624$$
$$0 = 4t^2 - 55t + 156$$
$$0 = (t-4)(4t-39)$$
$t - 4 = 0 \Rightarrow t = 4$ or $4t - 39 = 0 \Rightarrow t = \frac{39}{4} = 9.75$
Step 2: The two numbers divide a number line into three regions, where $t \geq 0$.

Interval A [0,4) Interval B (4,9.75) Interval C (9.75,∞)

0   4   9.75

Step 3: Choose a test value to see if it satisfies the inequality, $-16t^2 + 220t \geq 624$.

Interval	Test Value	Is $-16t^2 + 220t \geq 624$ True or False?
A: $(0, 4)$	1	$-16\cdot1^2 + 220\cdot1 \overset{?}{\geq} 624$ $204 \geq 624$ False
B: $(4, 9.75)$	5	$-16\cdot5^2 + 220\cdot5 \overset{?}{\geq} 624$ $700 \geq 624$ True
C: $(9.75, \infty)$	10	$-16\cdot10^2 + 220\cdot10 \overset{?}{\geq} 624$ $600 \geq 624$ False

The projectile will be at least 624 feet above ground between 4 sec and 9.75 sec (inclusive).

83. $-16t^2 + 44t + 4 \geq 32$
Step 1: Find the values of t that satisfy
$-16t^2 + 44t + 4 = 32$.
$$-16t^2 + 44t + 4 = 32$$
$$-16t^2 + 44t - 28 = 0$$
$$4t^2 - 11t + 7 = 0$$
$$(4t - 7)(t - 1) = 0$$
$4t - 7 = 0 \Rightarrow t = \frac{7}{4} = 1.75$ or $t - 1 = 0 \Rightarrow t = 1$
Step 2: The two numbers divide a number line into three regions, where $t \geq 0$.

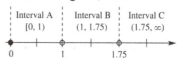

Interval A [0, 1) Interval B (1, 1.75) Interval C (1.75, ∞)

0   1   1.75

Step 3: Choose a test value to see if it satisfies the inequality, $-16t^2 + 44t + 4 \geq 32$

Interval	Test Value	Is $-16t^2 + 44t + 4 \geq 32$ True or False?
A: $(-\infty, 1)$	0	$-16\cdot0^2 + 44\cdot0 + 4 \overset{?}{\geq} 32$ $4 \geq 32$ False
B: $(1, 1.75)$	1.5	$-16\cdot1.5^2 + 44\cdot1.5 + 4 \overset{?}{\geq} 32$ $34 \geq 32$ True
C: $(1.75, \infty)$	2	$-16\cdot2^2 + 44\cdot2 + 4 \overset{?}{\geq} 32$ $28 \geq 32$ False

The baseball will be at least 32 feet above ground between 1 sec and 1.75 sec (inclusive).

85. $2t^2 - 5t - 12 < 0$

Step 1: Find the values of t that satisfy

$2t^2 - 5t - 12 = 0$.

$2t^2 - 5t - 12 = 0 \Rightarrow (2t + 3)(t - 4) = 0$

$2t + 3 = 0 \Rightarrow t = -\frac{3}{2} = -1.5$ or

$t - 4 = 0 \Rightarrow t = 4$

Step 2: The two numbers divide a number line into three regions.

Interval A	Interval B	Interval C
$(-\infty, -1.5)$	$(-1.5, 4)$	$(4, \infty)$

```
    +-----+-----o-----+-----o----->
    0         -1.5         4
```

Step 3: Choose a test value to see if it satisfies the inequality, $2t^2 - 5t - 12 < 0$.

Interval	Test Value	Is $2t^2 - 5t - 12 < 0$ True or False?
A: $(-\infty, -1.5)$	-2	$2(-2)^2 - 5(-2) - 12 \overset{?}{<} 0$ $6 < 0$ False
B: $(-1.5, 4)$	0	$2 \cdot 0^2 - 5 \cdot 0 - 12 \overset{?}{<} 0$ $-12 < 0$ True
C: $(4, \infty)$	5	$2 \cdot 5^2 - 5 \cdot 5 - 12 \overset{?}{<} 0$ $13 < 0$ False

The velocity will be negative between -1.5 sec and 4 sec.

87. $(3x - 4)(x + 2)(x + 6) = 0$

Set each factor to zero and solve.

$3x - 4 = 0 \Rightarrow x = \frac{4}{3}$ or $x + 2 = 0 \Rightarrow x = -2$ or

$x + 6 = 0 \Rightarrow x = -6$

Solution set: $\left\{ \frac{4}{3}, -2, -6 \right\}$

89.

Interval	Test Value	Is $(3x - 4)(x + 2)(x + 6) \leq 0$ True or False?
A: $(-\infty, -6)$	-10	$[3(-10) - 4][-10 + 2]$ $\cdot [-10 + 6] \overset{?}{\leq} 0$ $-1088 \leq 0$ True
B: $(-6, -2)$	-4	$[3(-4) - 4][-4 + 2]$ $\cdot [-4 + 6] \overset{?}{\leq} 0$ $64 \leq 0$ False

Interval	Test Value	Is $(3x - 4)(x + 2)(x + 6) \leq 0$ True or False?
C: $\left(-2, \frac{4}{3} \right)$	0	$[3(0) - 4][0 + 2][0 + 6] \overset{?}{\leq} 0$ $-48 \leq 0$ True
D: $\left(\frac{4}{3}, \infty \right)$	4	$[3(4) - 4][4 + 2][4 + 6] \overset{?}{\leq} 0$ $480 \leq 0$ False

91. $(2x - 3)(x + 2)(x - 3) \geq 0$

Step 1: Solve $(2x - 3)(x + 2)(x - 3) = 0$.

Set each factor to zero and solve.

$2x - 3 = 0 \Rightarrow x = \frac{3}{2}$ or $x + 2 = 0 \Rightarrow x = -2$ or

$x - 3 = 0 \Rightarrow x = 3$

Solution set: $\left\{ -2, \frac{3}{2}, 3 \right\}$

Step 2: Plot the solutions $-2, \frac{3}{2}$, and 3 on a number line.

```
    +--+--+--+--+--+--+--+--+-->
                -2    0   3/2  3
```

Step 3: Choose a test value to see if it satisfies the inequality, $(2x - 3)(x + 2)(x - 3) \geq 0$.

Interval	Test Value	Is $(2x - 3)(x + 2)(x - 3) \geq 0$ True or False?
A: $(-\infty, -2)$	-3	$[2(-3) - 3][-3 + 2]$ $\cdot [-3 - 3] \overset{?}{\geq} 0$ $-54 \geq 0$ False
B: $\left(-2, \frac{3}{2} \right)$	0	$[2(0) - 3][0 + 2]$ $\cdot [0 - 3] \overset{?}{\geq} 0$ $18 \geq 0$ True
C: $\left(\frac{3}{2}, 3 \right)$	2	$[2(2) - 3][2 + 2]$ $\cdot [2 - 3] \overset{?}{\geq} 0$ $-4 \geq 0$ False
D: $(3, \infty)$	4	$[2(4) - 3][4 + 2]$ $\cdot [4 - 3] \overset{?}{\geq} 0$ $30 \geq 0$ True

Solution set: $\left[-2, \frac{3}{2} \right] \cup [3, \infty)$

93. $4x - x^3 \geq 0$

Step 1: Solve $4x - x^3 = 0$.

$$4x - x^3 = 0 \Rightarrow x\left(4 - x^2\right) = 0 \Rightarrow$$
$$x\left(2 + x\right)\left(2 - x\right) = 0$$

Set each factor to zero and solve.

$x = 0$ or $2 + x = 0 \Rightarrow x = -2$ or
$2 - x = 0 \Rightarrow x = 2$

Solution set: $\{-2, 0, 2\}$

Step 2: The values -2, 0, and 2 divide the number line into four intervals.

Step 3: Choose a test value to see if it satisfies the inequality, $4x - x^3 \geq 0$.

Interval	Test Value	Is $4x - x^3 \geq 0$ True or False?
A: $(-\infty, -2)$	-3	$4(-3) - (-3)^3 \overset{?}{\geq} 0$ $15 \geq 0$ True
B: $(-2, 0)$	-1	$4(-1) - (-1)^3 \overset{?}{\geq} 0$ $-3 \geq 0$ False
C: $(0, 2)$	1	$4(1) - 1^3 \overset{?}{\geq} 0$ $3 \geq 0$ True
D: $(2, \infty)$	3	$4(3) - 3^3 \overset{?}{\geq} 0$ $-15 \geq 0$ False

Solution set: $(-\infty, -2] \cup [0, 2]$

95. $(x + 1)^2 (x - 3) < 0$

Step 1: Solve $(x + 1)^2 (x - 3) = 0$.

Set each distinct factor to zero and solve.
$x + 1 = 0 \Rightarrow x = -1$ or $x - 3 = 0 \Rightarrow x = 3$

Solution set: $\{-1, 3\}$

Step 2: The values -1 and 3 divide the number line into three intervals.

Step 3: Choose a test value to see if it satisfies the inequality, $(x + 1)^2 (x - 3) < 0$.

Interval	Test Value	Is $(x + 1)^2 (x - 3) < 0$ True or False?
A: $(-\infty, -1)$	-2	$(-2 + 1)^2 (-2 - 3) \overset{?}{<} 0$ $-5 < 0$ True
B: $(-1, 3)$	0	$(0 + 1)^2 (0 - 3) \overset{?}{<} 0$ $-3 < 0$ True
C: $(3, \infty)$	4	$(4 + 1)^2 (4 - 3) \overset{?}{<} 0$ $25 < 0$ False

Solution set: $(-\infty, -1) \cup (-1, 3)$

97. $x^3 + 4x^2 - 9x \geq 36$

Step 1: Solve $x^3 + 4x^2 - 9x = 36$

$$x^3 + 4x^2 - 9x = 36$$
$$x^3 + 4x^2 - 9x - 36 = 0$$
$$x^2 (x + 4) - 9(x + 4) = 0$$
$$(x + 4)(x^2 - 9) = 0$$
$$(x + 4)(x + 3)(x - 3) = 0$$

Set each factor to zero and solve.
$x + 4 = 0 \Rightarrow x = -4$ or $x + 3 = 0 \Rightarrow x = -3$ or
$x - 3 = 0 \Rightarrow x = 3$

Solution set: $\{-4, -3, 3\}$

Step 2: The values -4, -3, and 3 divide the number line into four intervals.

Step 3: Choose a test value to see if it satisfies the inequality, $x^3 + 4x^2 - 9x \geq 36$

Interval	Test Value	Is $x^3 + 4x^2 - 9x \geq 36$ True or False?
A: $(-\infty, -4)$	-5	$(-5)^3 + 4(-5)^2 - 9(-5) \overset{?}{\geq} 36$ $20 \geq 36$ False
B: $(-4, -3)$	-3.5	$(-3.5)^3 + 4(-3.5)^2$ $-9(-3.5) \overset{?}{\geq} 36$ $37.625 \geq 36$ True
C: $(-3, 3)$	0	$0^3 + 4(0)^2 - 9(0) \overset{?}{\geq} 36$ $0 \geq 36$ False
D: $(3, \infty)$	4	$4^3 + 4(4)^2 - 9(4) \overset{?}{\geq} 36$ $92 \geq 36$ True

Solution set: $[-4, -3] \cup [3, \infty)$

99. $x^2(x+4)^2 \geq 0$

Step 1: Solve $x^2(x+4)^2 = 0$.

Set each distinct factor to zero and solve.

$x = 0$ or $x + 4 = 0 \Rightarrow x = -4$

Solution set: $\{-4, 0\}$

Step 2: The values -4 and 0 divide the number line into three intervals.

Step 3: Choose a test value to see if it satisfies the inequality, $x^2(x+4)^2 \geq 0$.

Interval	Test Value	Is $x^2(x+4)^2 \geq 0$ True or False?
A: $(-\infty, -4)$	-5	$(-5)^2(-5+4)^2 \overset{?}{\geq} 0$ $25 \geq 0$ True
B: $(-4, 0)$	-1	$(-1)^2(-1+4)^2 \overset{?}{\geq} 0$ $9 \geq 0$ True
C: $(0, \infty)$	1	$1^2(1+4)^2 \overset{?}{\geq} 0$ $25 \geq 0$ True

Solution set: $(-\infty, \infty)$

Section 1.8   Absolute Value Equations and Inequalities

1. F. The solution set includes any value of x whose absolute value is 7; thus $x = 7$ or $x = -7$ are both solutions.

3. D. The solution set is all real numbers because the absolute value of any real number is always greater than -7. The graph shows the entire number line.

5. G. The solution set includes any value of x whose absolute value is less than 7; thus x must be between -7 and 7, not including -7 or 7.

7. C. The solution set includes any value of x whose absolute value is less than or equal to 7; thus x must be between -7 and 7, including -7 and 7.

9. $|3x - 1| = 2$

$3x - 1 = 2 \Rightarrow 3x = 3 \Rightarrow x = 1$ or

$3x - 1 = -2 \Rightarrow 3x = -1 \Rightarrow x = -\frac{1}{3}$

Solution set: $\left\{-\frac{1}{3}, 1\right\}$

11. $|5 - 3x| = 3$

$5 - 3x = 3 \Rightarrow 2 = 3x \Rightarrow \frac{2}{3} = x$ or

$5 - 3x = -3 \Rightarrow 8 = 3x \Rightarrow \frac{8}{3} = x$

Solution set: $\left\{\frac{2}{3}, \frac{8}{3}\right\}$

13. $\left|\dfrac{x-4}{2}\right| = 5$

$\dfrac{x-4}{2} = 5 \Rightarrow x - 4 = 10 \Rightarrow x = 14$ or

$\dfrac{x-4}{2} = -5 \Rightarrow x - 4 = -10 \Rightarrow x = -6$

Solution set: $\{-6, 14\}$

15. $\left|\dfrac{5}{x-3}\right| = 10$

$\dfrac{5}{x-3} = 10 \Rightarrow 5 = 10(x-3) \Rightarrow 5 = 10x - 30 \Rightarrow$

$35 = 10x \Rightarrow x = \frac{35}{10} = \frac{7}{2}$ or

$\dfrac{5}{x-3} = -10 \Rightarrow 5 = -10(x-3) \Rightarrow$

$5 = -10x + 30 \Rightarrow -25 = -10x \Rightarrow x = \frac{-25}{-10} = \frac{5}{2}$

Solution set: $\left\{\frac{5}{2}, \frac{7}{2}\right\}$

17. $\left|\dfrac{6x+1}{x-1}\right| = 3$

$\dfrac{6x+1}{x-1} = 3 \Rightarrow 6x + 1 = 3(x-1) \Rightarrow$

$6x + 1 = 3x - 3 \Rightarrow 3x = -4 \Rightarrow x = -\frac{4}{3}$ or

$\dfrac{6x+1}{x-1} = -3 \Rightarrow 6x + 1 = -3(x-1) \Rightarrow$

$6x + 1 = -3x + 3 \Rightarrow 9x = 2 \Rightarrow x = \frac{2}{9}$

Solution set: $\left\{-\frac{4}{3}, \frac{2}{9}\right\}$

19. $|2x - 3| = |5x + 4|$

$2x - 3 = 5x + 4 \Rightarrow -7 = 3x \Rightarrow -\frac{7}{3} = x$ or

$2x - 3 = -(5x + 4) \Rightarrow 2x - 3 = -5x - 4 \Rightarrow$

$7x = -1 \Rightarrow x = \frac{-1}{7} = -\frac{1}{7}$

Solution set: $\left\{-\frac{7}{3}, -\frac{1}{7}\right\}$

21. $|4 - 3x| = |2 - 3x|$

$4 - 3x = 2 - 3x \Rightarrow 4 = 2$ False   or

$4 - 3x = -(2 - 3x) \Rightarrow 4 - 3x = -2 + 3x \Rightarrow$

$6 = 6x \Rightarrow 1 = x$

Solution set: $\{1\}$

23. $|5x - 2| = |2 - 5x|$

$5x - 2 = 2 - 5x \Rightarrow 10x = 4 \Rightarrow x = \frac{4}{10} = \frac{2}{5}$ or

$5x - 2 = -(2 - 5x) \Rightarrow 5x - 2 = -2 + 5x \Rightarrow$
$\qquad 0 = 0$ True

Solution set: $(-\infty, \infty)$

25. Answers will vary. Sample answer: If x is positive, then $-5x$ will be negative. Because the outcome of an absolute value can never be negative, a positive value of x is not possible.

27. $|2x + 5| < 3$

$-3 < 2x + 5 < 3$
$-8 < 2x < -2$
$-4 < x < -1$

Solution set: $(-4, -1)$

29. $|2x + 5| \geq 3$

$2x + 5 \leq -3 \Rightarrow 2x \leq -8 \Rightarrow x \leq -4$ or
$2x + 5 \geq 3 \Rightarrow 2x \geq -2 \Rightarrow x \geq -1$
Solution set: $(-\infty, -4] \cup [-1, \infty)$

31. $\left|\frac{1}{2} - x\right| < 2$

$-2 < \frac{1}{2} - x < 2$
$2(-2) < 2\left(\frac{1}{2} - x\right) < 2(2)$
$-4 < 1 - 2x < 4$
$-5 < -2x < 3$
$\frac{5}{2} > x > -\frac{3}{2}$

Solution set: $\left(-\frac{3}{2}, \frac{5}{2}\right)$

33. $4|x - 3| > 12 \Rightarrow |x - 3| > 3$

$x - 3 < -3 \Rightarrow x < 0$ or $x - 3 > 3 \Rightarrow x > 6$
Solution set: $(-\infty, 0) \cup (6, \infty)$

35. $|5 - 3x| > 7$

$5 - 3x < -7 \Rightarrow -3x < -12 \Rightarrow x > 4$ or
$5 - 3x > 7 \Rightarrow -3x > 2 \Rightarrow x < -\frac{2}{3}$

Solution set: $\left(-\infty, -\frac{2}{3}\right) \cup (4, \infty)$

37. $|5 - 3x| \leq 7$

$-7 \leq 5 - 3x \leq 7$
$-12 \leq -3x \leq 2$
$4 \geq x \geq -\frac{2}{3}$
$-\frac{2}{3} \leq x \leq 4$

Solution set: $\left[-\frac{2}{3}, 4\right]$

39. $\left|\frac{2}{3}x + \frac{1}{2}\right| \leq \frac{1}{6}$

$-\frac{1}{6} \leq \frac{2}{3}x + \frac{1}{2} \leq \frac{1}{6}$
$6\left(-\frac{1}{6}\right) \leq 6\left(\frac{2}{3}x + \frac{1}{2}\right) \leq 6\left(\frac{1}{6}\right)$
$-1 \leq 4x + 3 \leq 1$
$-4 \leq 4x \leq -2$
$-1 \leq x \leq -\frac{1}{2}$

Solution set: $\left[-1, -\frac{1}{2}\right]$

41. $|0.01x + 1| < 0.01$

$-0.01 < 0.01x + 1 < 0.01$
$-1 < x + 100 < 1$
$-101 < x < -99$
Solution set: $(-101, -99)$

43. $|4x + 3| - 2 = -1 \Rightarrow |4x + 3| = 1$

$4x + 3 = 1 \Rightarrow 4x = -2 \Rightarrow x = \frac{-2}{4} = -\frac{1}{2}$ or
$4x + 3 = -1 \Rightarrow 4x = -4 \Rightarrow x = -1$
Solution set: $\left\{-1, -\frac{1}{2}\right\}$

45. $|6 - 2x| + 1 = 3 \Rightarrow |6 - 2x| = 2$

$6 - 2x = 2 \Rightarrow -2x = -4 \Rightarrow x = 2$ or
$6 - 2x = -2 \Rightarrow -2x = -8 \Rightarrow x = 4$
Solution set: $\{2, 4\}$

47. $|3x + 1| - 1 < 2 \Rightarrow |3x + 1| < 3$

$-3 < 3x + 1 < 3$
$-4 < 3x < 2$
$-\frac{4}{3} < x < \frac{2}{3}$

Solution set: $\left(-\frac{4}{3}, \frac{2}{3}\right)$

49. $\left|5x + \frac{1}{2}\right| - 2 < 5 \Rightarrow \left|5x + \frac{1}{2}\right| < 7$

$-7 < 5x + \frac{1}{2} < 7$
$2(-7) < 2\left(5x + \frac{1}{2}\right) < 2(7)$
$-14 < 10x + 1 < 14$
$-15 < 10x < 13$
$-\frac{15}{10} < x < \frac{13}{10} \Rightarrow -\frac{3}{2} < x < \frac{13}{10}$

Solution set: $\left(-\frac{3}{2}, \frac{13}{10}\right)$

51. $|10 - 4x| + 1 \geq 5 \Rightarrow |10 - 4x| \geq 4$

$10 - 4x \leq -4 \Rightarrow -4x \leq -14 \Rightarrow x \geq \frac{-14}{-4} \Rightarrow x \geq \frac{7}{2}$
or
$10 - 4x \geq 4 \Rightarrow -4x \geq -6 \Rightarrow x \leq \frac{-6}{-4} \Rightarrow x \leq \frac{3}{2}$

Solution set: $\left(-\infty, \frac{3}{2}\right] \cup \left[\frac{7}{2}, \infty\right)$

53. $|3x - 7| + 1 < -2 \Rightarrow |3x - 7| < -3$

An absolute value cannot be negative.
Solution set: $\varnothing$

55. Because the absolute value of a number is always nonnegative, the inequality $|10 - 4x| \geq -4$ is always true. The solution set is $(-\infty, \infty)$.

57. There is no number whose absolute value is less than any negative number. The solution set of $|6 - 3x| < -11$ is $\varnothing$.

59. The absolute value of a number will be 0 if that number is 0. Therefore $|8x + 5| = 0$ is equivalent to $8x + 5 = 0$, which has solution set $\left\{-\frac{5}{8}\right\}$.

61. Any number less than zero will be negative. There is no number whose absolute value is a negative number. The solution set of $|4.3x + 9.8| < 0$ is $\varnothing$.

63. Because the absolute value of a number is always nonnegative, $|2x + 1| < 0$ is never true, so $|2x + 1| \leq 0$ is only true when $|2x + 1| = 0$.

$|2x + 1| = 0 \Rightarrow 2x + 1 = 0 \Rightarrow 2x = -1 \Rightarrow x = -\frac{1}{2}$

Solution set: $\left\{-\frac{1}{2}\right\}$

65. $|3x + 2| > 0$ will be false only when $3x + 2 = 0$, which occurs when $x = -\frac{2}{3}$. So the solution set for $|3x + 2| > 0$ is $\left(-\infty, -\frac{2}{3}\right) \cup \left(-\frac{2}{3}, \infty\right)$.

67. $|p - q| = 2$, which is equivalent to $|q - p| = 2$, indicates that the distance between p and q is 2 units.

69. "m is no more than 2 units from 7" means that m is 2 units or less from 7. Thus the distance between m and 7 is less than or equal to 2, or $|m - 7| \leq 2$.

71. "p is within 0.0001 unit of 9" means that p is less than 0.0001 unit from 9. Thus the distance between p and 9 is less than 0.0001, or $|p - 9| < 0.0001$.

73. "r is no less than 1 unit from 29" means that r is 1 unit or more from 29. Thus the distance between r and 29 is greater than or equal to 1, or $|r - 29| \geq 1$.

75. Because we want y to be within 0.002 unit of 6, we have $|y - 6| < 0.002$ or

$|5x + 1 - 6| < 0.002$.

$\begin{aligned} |5x - 5| &< 0.002 \\ -0.002 < 5x - 5 &< 0.002 \\ 4.998 < 5x &< 5.002 \\ 0.9996 < x &< 1.0004 \end{aligned}$

Values of x in the interval $(0.9996, 1.0004)$ would satisfy the condition.

77. $|x - 8.2| \leq 1.5$

$\begin{aligned} -1.5 \leq x - 8.2 &\leq 1.5 \\ 6.7 \leq x &\leq 9.7 \end{aligned}$

The range of weights, in pounds, is $[6.7, 9.7]$.

79. 780 is 50 more than 730 and 680 is 50 less than 730, so all of the temperatures in the acceptable range are within 50° of 730°. That is $|F - 730| \leq 50$.

81. $|R_L - 26.75| \leq 1.42$

$\begin{aligned} -1.42 \leq R_L - 26.75 &\leq 1.42 \\ 25.33 \leq R_L &\leq 28.17 \end{aligned}$

$|R_E - 38.75| \leq 2.17$

$\begin{aligned} -2.17 \leq R_E - 38.75 &\leq 2.17 \\ 36.58 \leq R_E &\leq 40.92 \end{aligned}$

83. 6 and the opposite of 6, namely –6

85. $x^2 - x = -6 \Rightarrow x^2 - x + 6 = 0$

The quadratic formula, $x = \dfrac{-b \pm \sqrt{b^2 - 4ac}}{2a}$, can be evaluated with $a = 1$, $b = -1$, and $c = 6$.

$\begin{aligned} x &= \frac{-b \pm \sqrt{b^2 - 4ac}}{2a} \\ &= \frac{-(-1) \pm \sqrt{(-1)^2 - 4(1)(6)}}{2(1)} = \frac{1 \pm \sqrt{1 - 24}}{2} \\ &= \frac{1 \pm \sqrt{-23}}{2} = \frac{1 \pm i\sqrt{23}}{2} = \frac{1}{2} \pm \frac{\sqrt{23}}{2}i \end{aligned}$

Solution set: $\left\{\frac{1}{2} \pm \frac{\sqrt{23}}{2}i\right\}$

87. $\left|3x^2 + x\right| = 14 \Rightarrow 3x^2 + x = 14$ or $3x^2 + x = -14$

$$3x^2 + x = 14$$
$$3x^2 + x - 14 = 0$$
$$(3x+7)(x-2) = 0$$
$$3x + 7 = 0 \Rightarrow x = -\tfrac{7}{3}$$
$$x - 2 = 0 \Rightarrow x = 2$$

$3x^2 + x = -14 \Rightarrow 3x^2 + x + 14 = 0$

We must use the quadratic formula with $a = 3$, $b = 1$, and $c = 14$.

$$x = \frac{-b \pm \sqrt{b^2 - 4ac}}{2a}$$
$$x = \frac{-1 \pm \sqrt{1^2 - 4 \cdot 3 \cdot 14}}{2 \cdot 3}$$
$$= \frac{-1 \pm \sqrt{-167}}{6}$$
$$= \frac{-1 \pm i\sqrt{167}}{6}$$
$$= -\frac{1}{6} \pm \frac{i\sqrt{167}}{6}$$

Solution set: $\left\{ -\tfrac{7}{3},\ 2,\ -\tfrac{1}{6} \pm \tfrac{i\sqrt{167}}{6} \right\}$

89. $\left|4x^2 - 23x - 6\right| = 0$

Because 0 and the opposite of 0 represent the same value, only one equation needs to be solved.

$$4x^2 - 23x - 6 = 0 \Rightarrow (4x+1)(x-6) = 0$$
$$4x + 1 = 0 \quad \text{or} \quad x - 6 = 0$$
$$x = -\tfrac{1}{4} \quad \text{or} \quad x = 6$$

Solution set: $\left\{ -\tfrac{1}{4}, 6 \right\}$

91. $\left|x^2 + 1\right| - \left|2x\right| = 0$

$$\left|x^2 + 1\right| - \left|2x\right| = 0 \Rightarrow \left|x^2 + 1\right| = \left|2x\right|$$
$$x^2 + 1 = 2x \Rightarrow x^2 - 2x + 1 = 0$$
$$(x-1)^2 = 0 \Rightarrow x - 1 = 0 \Rightarrow x = 1 \quad \text{or}$$
$$x^2 + 1 = -2x$$
$$x^2 + 2x + 1 = 0$$
$$(x+1)^2 = 0$$
$$x + 1 = 0 \Rightarrow x = -1$$

Solution set: $\{-1, 1\}$

93. Any number less than zero will be negative. There is no number whose absolute value is a negative number. The solution set of $\left|x^4 + 2x^2 + 1\right| < 0$ is $\varnothing$.

95. $\left|\dfrac{x-4}{3x+1}\right| \geq 0$

This inequality will be true, except where $\frac{x-4}{3x+1}$ is undefined. This occurs when

$$3x + 1 = 0, \text{ or } x = -\tfrac{1}{3}.$$

Solution set: $\left(-\infty, -\tfrac{1}{3}\right) \cup \left(-\tfrac{1}{3}, \infty\right)$

Chapter 1 Review Exercises

1. $2x + 8 = 3x + 2 \Rightarrow 8 = x + 2 \Rightarrow 6 = x$
Solution set: $\{6\}$

3. $5x - 2(x+4) = 3(2x+1)$
$$5x - 2x - 8 = 6x + 3 \Rightarrow 3x - 8 = 6x + 3 \Rightarrow$$
$$-8 = 3x + 3 \Rightarrow -11 = 3x \Rightarrow -\frac{11}{3} = x$$

Solution set: $\left\{ -\tfrac{11}{3} \right\}$

5. $A = \dfrac{24f}{B(p+1)}$ for f (approximate annual interest rate)

$$B(p+1)A = B(p+1)\left(\frac{24f}{B(p+1)} \right)$$
$$AB(p+1) = 24f$$
$$\frac{AB(p+1)}{24} = f$$
$$f = \frac{AB(p+1)}{24}$$

7. A and B cannot be equations used to find the number of pennies in a jar. The number of pennies must be a whole number.

A. $5x + 3 = 11 \Rightarrow 5x = 8 \Rightarrow x = \tfrac{8}{5}$

B. $12x + 6 = -4 \Rightarrow 12x = -10 \Rightarrow$
$$x = -\tfrac{10}{12} = -\tfrac{5}{6}$$

C. $100x = 50(x+3)$
$$100x = 50x + 150 \Rightarrow 50x = 150 \Rightarrow x = 3$$

D. $6(x+4) = x + 24$
$$6x + 24 = x + 24$$
$$5x + 24 = 24 \Rightarrow 5x = 0 \Rightarrow x = 0$$

9. Let x = the original length of the square (in inches). Because the perimeter of a square is 4 times the length of one side, we have $4(x-4)=\frac{1}{2}(4x)+10.$ Solve this equation for x to determine the length of each side of the original square.
$4x-16=2x+10$
$2x-16=10 \Rightarrow 2x=26 \Rightarrow x=13$
The original square is 13 in. on each side.

11. Let x = the amount of 100% alcohol solution (in liters).

Strength	Liters of Solution	Liters of Pure Alcohol
100%	x	$1x=x$
10%	12	$0.10 \cdot 12 = 1.2$
30%	$x+12$	$0.30(x+12)$

The number of liters of pure alcohol in the 100% solution plus the number of liters of pure alcohol in the 10% solution must equal the number of liters of pure alcohol in the 30% solution.
$$x+1.2=0.30(x+12)$$
$$x+1.2=0.30x+3.6$$
$$0.7x+1.2=3.6$$
$$0.7x=2.4$$
$$x=\frac{2.4}{.7}=\frac{24}{7}=3\tfrac{3}{7} \text{ L}$$

$3\tfrac{3}{7}$ L of the 100% solution should be added.

13. Let x = time (in hours) the mother spent driving to meet plane.
Because Mary Lynn has been in the plane for 15 minutes, and 15 minutes is $\frac{1}{4}$ hr, she has been traveling by plane for $x+\frac{1}{4}$ hr.

	d	r	t
Mary Lynn by plane	420		$x+\frac{1}{4}$
Mother by car	20	40	x

The time driven by Mary Lynn's mother can be found by $20=40x \Rightarrow x=\frac{1}{2}$ hr. Mary Lynn, therefore, flew for
$\frac{1}{2}+\frac{1}{4}=\frac{2}{4}+\frac{1}{4}=\frac{3}{4}$ hr. The rate of Mary Lynn's plane can be found by
$$r=\frac{d}{t}=\frac{420}{\frac{3}{4}}=420\cdot\frac{4}{3}=560 \text{ km per hour.}$$

15. **(a)** In one year, the maximum amount of lead ingested would be
$$0.05\,\frac{\text{mg}}{\text{liter}}\cdot 2\,\frac{\text{liters}}{\text{day}}\cdot 365.25\,\frac{\text{days}}{\text{year}}$$
$$=36.525\,\frac{\text{mg}}{\text{year}}.$$
The maximum amount A of lead (in milligrams) ingested in x years would be $A=36.525x.$

(b) If $x=72$, then $A=36.525(72)=2629.8$ mg. The EPA maximum lead intake from water over a lifetime is 2629.8 mg.

17. **(a)** Using 1956 for $x=0$, then for 1990, $x=34.$
$y=0.1132x+0.4609$
$y=0.1132\cdot 34+0.4609 \approx 4.31$
According to the model, the minimum wage in 1990 was \$4.31. This is \$0.51 more than the actual value of \$3.80.

(b) Let $y=\$5.85$ and then solve for x.
$y=0.1132x+0.4609$
$5.85=0.1132x+0.4609$
$5.3891=0.1132x$
$47.6 \approx x$
The model predicts the minimum wage to be \$5.85 about 47.6 years after 1956, which is mid-2003. This is close to the minimum wage changing to \$5.85 in 2007.

19. $(6-i)+(7-2i)=(6+7)+\big[-1+(-2)\big]i$
$=13+(-3)i=13-3i$

21. $15i-(3+2i)-11=(-3-11)+(15-2)i$
$=-14+13i$

23. $(5-i)(3+4i)=5(3)+5(4i)-i(3)-i(4i)$
$=15+20i-3i-4i^2$
$=15+17i-4(-1)$
$=15+17i+4=19+17i$

25. $(5-11i)(5+11i)=5^2-(11i)^2$ Product of the sum and difference of two terms
$=25-121i^2$
$=25-121(-1)$
$=25+121=146$

27. $-5i(3-i)^2 = -5i\left[3^2 - 2(3)(i) + i^2\right]$

$\qquad = -5i\left[9 - 6i + (-1)\right]$

$\qquad = -5i(8 - 6i) = -40i + 30i^2$

$\qquad = -40i + 30(-1) = -40i + (-30)$

$\qquad = -30 - 40i$

29. $\dfrac{-12-i}{-2-5i} = \dfrac{(-12-i)(-2+5i)}{(-2-5i)(-2+5i)}$

$\qquad = \dfrac{24 - 60i + 2i - 5i^2}{(-2)^2 - (5i)^2} = \dfrac{24 - 58i - 5(-1)}{4 - 25i^2}$

$\qquad = \dfrac{24 - 58i + 5}{4 - 25(-1)} = \dfrac{29 - 58i}{4 + 25} = \dfrac{29 - 58i}{29}$

$\qquad = \dfrac{29}{29} - \dfrac{58}{29}i = 1 - 2i$

31. $i^{11} = i^8 \cdot i^3 = 1 \cdot (-i) = -i$

33. $i^{1001} = i^{1000} \cdot i = \left(i^4\right)^{250} \cdot i = 1^{250} \cdot i = i$

35. $i^{-27} = i^{-28} \cdot i = \left(i^4\right)^{-7} \cdot i = 1^{-7} \cdot i = i$

37. $(x+7)^2 = 5 \Rightarrow x + 7 = \pm\sqrt{5} \Rightarrow x = -7 \pm \sqrt{5}$

Solution set: $\{-7 \pm \sqrt{5}\}$

39. $2x^2 + x - 15 = 0$

$(x+3)(2x-5) = 0$

$x + 3 = 0 \Rightarrow x = -3$ or $2x - 5 = 0 \Rightarrow x = \frac{5}{2}$

Solution set: $\left\{-3, \frac{5}{2}\right\}$

41. $-2x^2 + 11x = -21 \Rightarrow -2x^2 + 11x + 21 = 0$

$2x^2 - 11x - 21 = 0 \Rightarrow (2x+3)(x-7) = 0$

$2x + 3 = 0 \Rightarrow x = -\frac{3}{2}$ or $x - 7 = 0 \Rightarrow x = 7$

Solution set: $\left\{-\frac{3}{2}, 7\right\}$

43. $(2x+1)(x-4) = x \Rightarrow 2x^2 - 8x + x - 4 = x \Rightarrow$

$2x^2 - 7x - 4 = x \Rightarrow 2x^2 - 8x - 4 = 0 \Rightarrow$

$x^2 - 4x - 2 = 0$

Solve by completing the square.

$x^2 - 4x - 2 = 0$

$x^2 - 4x + 4 = 2 + 4$

$\qquad$ Note: $\left[\frac{1}{2} \cdot (-4)\right]^2 = (-2)^2 = 4$

$(x-2)^2 = 6 \Rightarrow x - 2 = \pm\sqrt{6} \Rightarrow x = 2 \pm \sqrt{6}$

Solve by the quadratic formula.

Let $a = 1$, $b = -4$, and $c = -2$.

$x = \dfrac{-b \pm \sqrt{b^2 - 4ac}}{2a}$

$\quad = \dfrac{-(-4) \pm \sqrt{(-4)^2 - 4(1)(-2)}}{2(1)} = \dfrac{4 \pm \sqrt{16 + 8}}{2}$

$\quad = \dfrac{4 \pm \sqrt{24}}{2} = \dfrac{4 \pm 2\sqrt{6}}{2} = 2 \pm \sqrt{6}$

Solution set: $\left\{2 \pm \sqrt{6}\right\}$

45. $x^2 - \sqrt{5}x - 1 = 0$

Using the quadratic formula would be the most direct approach.

$a = 1$, $b = -\sqrt{5}$, and $c = -1$.

$x = \dfrac{-b \pm \sqrt{b^2 - 4ac}}{2a}$

$\quad = \dfrac{-\left(-\sqrt{5}\right) \pm \sqrt{(-\sqrt{5})^2 - 4 \cdot 1 \cdot (-1)}}{2 \cdot 1}$

$\quad = \dfrac{\sqrt{5} \pm \sqrt{5+4}}{2} = \dfrac{\sqrt{5} \pm \sqrt{9}}{2} = \dfrac{\sqrt{5} \pm 3}{2}$

Solution set: $\left\{\frac{\sqrt{5} \pm 3}{2}\right\}$

47. **D.** The equation $(7x + 4)^2 = 11$ has two real, distinct solutions because the positive number 11 has a positive square root and a negative square root.

49. $-6x^2 + 2x = -3 \Rightarrow -6x^2 + 2x + 3 = 0$

$a = -6$, $b = 2$, and $c = 3$

$b^2 - 4ac = 2^2 - 4(-6)(3) = 4 + 72 = 76$

The equation has two distinct irrational solutions because the discriminant is positive but not a perfect square.

51. $-8x^2 + 10x = 7 \Rightarrow 0 = 8x^2 - 10x + 7$

$a = 8$, $b = -10$, and $c = 7$

$b^2 - 4ac = (-10)^2 - 4(8)(7) = 100 - 224 = -124$

The equation has two distinct nonreal complex solutions because the discriminant is negative.

53. $x(9x + 6) = -1 \Rightarrow 9x^2 + 6x = -1 \Rightarrow$

$9x^2 + 6x + 1 = 0$

$a = 9$, $b = 6$, and $c = 1$

$b^2 - 4ac = 6^2 - 4(9)(1) = 36 - 36 = 0$

The equation has one rational solution (a double solution) because the discriminant is equal to zero.

55. The projectile will be 750 ft above the ground whenever $-16t^2 + 220t = 750$.

Solve this equation for t.
$$-16t^2 + 220t = 750$$
$$-16t^2 + 220t - 750 = 0$$
$$16t^2 - 220t + 750 = 0$$
$$8t^2 - 110t + 375 = 0$$
$$(4t - 25)(2t - 15) = 0$$

$4t - 25 = 0$ or $2t - 15 = 0$
$t = \frac{25}{4} = 6.25$ or $t = \frac{15}{2} = 7.5$

The projectile will be 750 ft high at 6.25 sec and at 7.5 sec.

57. Let $x =$ width of border.

Apply the formula $A = LW$ to both the outside and inside rectangles.

Inside area= Outside area − Border area
$$(12 - 2x)(10 - 2x) = 12 \cdot 10 - 21$$

$$120 - 24x - 20x + 4x^2 = 120 - 21$$
$$120 - 44x + 4x^2 = 99$$
$$4x^2 - 44x + 120 = 99$$
$$4x^2 - 44x + 21 = 0$$
$$(2x - 21)(2x - 1) = 0$$

$2x = 21 \Rightarrow x = \frac{21}{2} = 10\frac{1}{2}$ or
$2x - 1 = 0 \Rightarrow x = \frac{1}{2}$

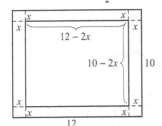

The border width cannot be $10\frac{1}{2}$ because this exceeds the width of the outside rectangle, so reject this solution. The width of the border is $\frac{1}{2}$ ft.

59. In 2009, $x = 19$.
$$y = 1.016x^2 + 12.49x + 197.8$$
$$y = 1.016(19)^2 + 12.49(19) + 197.8 \approx 801.9$$

According to the model, in 2009 the U.S. government spent approximately $801.9 billion on medical care.

61. $4x^4 + 3x^2 - 1 = 0$

Let $u = x^2$; then $u^2 = x^4$.

With this substitution, the equation becomes
$4u^2 + 3u - 1 = 0$.

Solve this equation by factoring.
$$(u + 1)(4u - 1) = 0$$

$u + 1 \Rightarrow u = -1$ or $4u - 1 = 0 \Rightarrow u = \frac{1}{4}$

To find x, replace u with x^2.

$x^2 = -1 \Rightarrow x = \pm\sqrt{-1} \Rightarrow x = \pm i$ or
$x^2 = \frac{1}{4} \Rightarrow x = \pm\sqrt{\frac{1}{4}} \Rightarrow x = \pm\frac{1}{2}$

Solution set: $\left\{\pm i, \pm\frac{1}{2}\right\}$

63. $\dfrac{2}{x} - \dfrac{4}{3x} = 8 + \dfrac{3}{x}$

$$3x\left(\frac{2}{x} - \frac{4}{3x}\right) = 3x\left(8 + \frac{3}{x}\right)$$
$$6 - 4 = 24x + 9$$
$$2 = 24x + 9$$
$$-7 = 24x \Rightarrow -\frac{7}{24} = x$$

Solution set: $\left\{-\frac{7}{24}\right\}$

65. $\dfrac{10}{4x - 4} = \dfrac{1}{1 - x} \Rightarrow \dfrac{10}{4(x - 1)} = \dfrac{1}{1 - x} \Rightarrow$

$$\frac{10}{4(x - 1)} = \frac{(-1) \cdot 1}{(-1)(1 - x)} \Rightarrow \frac{10}{4(x - 1)} = \frac{-1}{x - 1}$$

Multiply each term in the equation by the least common denominator, $4(x - 1)$, assuming $x \neq 1$.

$$4(x - 1)\left[\frac{10}{4(x - 1)}\right] = 4(x - 1)\left(\frac{-1}{x - 1}\right)$$
$$10 = -4 \Rightarrow 14 = 0$$

This is a false statement, the solution set is $\varnothing$.

Alternate solution:

$$\frac{10}{4x - 4} = \frac{1}{1 - x} \text{ or } \frac{10}{4(x - 1)} = \frac{1}{1 - x}$$

Multiply each term in the equation by the least common denominator, $4(x - 1)(1 - x)$, assuming $x \neq 1$.

(continued on next page)

(continued)

$$4(x-1)(1-x)\left[\frac{10}{4(x-1)}\right]$$
$$= 4(x-1)(1-x)\left(\frac{1}{1-x}\right)$$
$$10(1-x) = 4(x-1)$$
$$10 - 10x = 4x - 4$$
$$10 = 14x - 4$$
$$14 = 14x$$
$$1 = x$$

Because of the restriction $x \neq 1$, the solution set is $\varnothing$.

67. $(x-4)^{2/5} = 9$

$$\left[(x-4)^{2/5}\right]^{5/2} = \pm 9^{5/2} = \pm \left(9^{1/2}\right)^5$$
$$x - 4 = (\pm 3)^5$$

$x - 4 = (-3)^5$ or $x - 4 = 3^5$
$x - 4 = -243$ $\qquad x - 4 = 243$
$x = -239$ $\qquad x = 247$

Solution set: $\{-239, 247\}$

69. $(x-2)^{2/3} = x^{1/3}$

$$\left[(x-2)^{2/3}\right]^3 = \left(x^{1/3}\right)^3$$
$$(x-2)^2 = x$$
$$x^2 - 4x + 4 = x$$
$$x^2 - 5x + 4 = 0$$
$$(x-4)(x-1) = 0$$
$x - 4 = 0 \Rightarrow x = 4$ or $x - 1 = 0 \Rightarrow x = 1$

Check $x = 1$.
$$(x-2)^{2/3} = x^{1/3} \Rightarrow$$
$$(1-2)^{2/3} = 1^{1/3} \Rightarrow (-1)^{2/3} = 1$$
$$\left[(-1)^{1/3}\right]^2 = 1 \Rightarrow (-1)^2 = 1 \Rightarrow 1 = 1$$

This is a true statement. 1 is a solution.
Check $x = 4$.
$$(x-2)^{2/3} = x^{1/3} \Rightarrow$$
$$(4-2)^{2/3} = 4^{1/3} \Rightarrow (2)^{2/3} = 4^{1/3}$$
$$\left[2^2\right]^{1/3} = 4^{1/3} \Rightarrow 4^{1/3} = 4^{1/3}$$

This is a true statement. 4 is a solution.
Solution set: $\{1, 4\}$

71. $\sqrt{x+2} - x = 2 \Rightarrow \sqrt{x+2} = 2 + x$

$$\left(\sqrt{x+2}\right)^2 = (2+x)^2 \Rightarrow x + 2 = 4 + 4x + x^2$$
$$0 = x^2 + 3x + 2 \Rightarrow 0 = (x+2)(x+1)$$
$x + 2 = 0 \Rightarrow x = -2$ or $x + 1 = 0 \Rightarrow x = -1$

Check $x = -2$.
$$\sqrt{x+2} = 2 + x$$
$$\sqrt{-2+2} = 2 + (-2)$$
$$\sqrt{0} = 0 \Rightarrow 0 = 0$$

This is a true statement. -2 is a solution.
Check $x = -1$.
$$\sqrt{x+2} = 2 + x$$
$$\sqrt{-1+2} = 2 + (-1)?$$
$$\sqrt{1} = 1 \Rightarrow 1 = 1$$

This is a true statement. -1 is a solution.
Solution set: $\{-2, -1\}$

73. $\sqrt{4x-2} = \sqrt{3x+1}$

$$\left(\sqrt{4x-2}\right)^2 = \left(\sqrt{3x+1}\right)^2$$
$$4x - 2 = 3x + 1$$
$$x - 2 = 1$$
$$x = 3$$

Check $x = 3$.
$$\sqrt{4x-2} = \sqrt{3x+1}$$
$$\sqrt{4(3)-2} = \sqrt{3(3)+1}$$
$$\sqrt{12-2} = \sqrt{9+1}$$
$$\sqrt{10} = \sqrt{10}$$

This is a true statement.
Solution set: $\{3\}$

75. $\sqrt{x+3} - \sqrt{3x+10} = 1$

$$\sqrt{x+3} = 1 + \sqrt{3x+10}$$
$$\left(\sqrt{x+3}\right)^2 = \left(1 + \sqrt{3x+10}\right)^2$$
$$x + 3 = 1 + 2\sqrt{3x+10} + (3x+10)$$
$$x + 3 = 3x + 11 + 2\sqrt{3x+10}$$
$$-2x - 8 = 2\sqrt{3x+10}$$
$$x + 4 = -\sqrt{3x+10}$$
$$(x+4)^2 = \left(-\sqrt{3x+10}\right)^2$$
$$x^2 + 8x + 16 = 3x + 10$$
$$x^2 + 5x + 6 = 0 \Rightarrow (x+2)(x+3) = 0$$
$x + 3 = 0 \Rightarrow x = -3$ or $x + 2 = 0 \Rightarrow x = -2$

Check $x = -3$.
$$\sqrt{x+3} - \sqrt{3x+10} = 1$$
$$\sqrt{-3+3} - \sqrt{3(-3)+10} = 1$$
$$\sqrt{0} - \sqrt{-9+10} = 1$$
$$0 - \sqrt{1} = 1$$
$$0 - 1 = 1 \Rightarrow -1 = 1$$

This is a false statement. -3 is not a solution.
Check $x = -2$.
$$\sqrt{x+3} - \sqrt{3x+10} = 1$$
$$\sqrt{-2+3} - \sqrt{3(-2)+10} = 1$$
$$\sqrt{1} - \sqrt{-6+10} = 1$$
$$1 - \sqrt{4} = 1$$
$$1 - 2 = 1 \Rightarrow -1 = 1$$

This is a false statement. -2 is not a solution.
Because neither of the proposed solutions satisfies the original equation, the equation has no solution. Solution set: $\varnothing$

77. $\sqrt[3]{6x+2} - \sqrt[3]{4x} = 0$
$$\sqrt[3]{6x+2} = \sqrt[3]{4x}$$
$$\left(\sqrt[3]{6x+2}\right)^3 = \left(\sqrt[3]{4x}\right)^3$$
$$6x + 2 = 4x \Rightarrow 2 = -2x \Rightarrow -1 = x$$

Check $x = -1$.
$$\sqrt[3]{6x+2} - \sqrt[3]{4x} = 0$$
$$\sqrt[3]{6(-1)+2} - \sqrt[3]{4(-1)} = 0$$
$$\sqrt[3]{-6+2} - \sqrt[3]{-4} = 0$$
$$\sqrt[3]{-4} - \left(-\sqrt[3]{4}\right) = 0$$
$$-\sqrt[3]{4} + \sqrt[3]{4} = 0 \Rightarrow 0 = 0$$

This is a true statement.
Solution set: $\{-1\}$

79. $\dfrac{x}{x+2} + \dfrac{1}{x} + 3 = \dfrac{2}{x^2+2x} \Rightarrow$
$$\dfrac{x}{x+2} + \dfrac{1}{x} + 3 = \dfrac{2}{x(x+2)}$$

Multiply each term in the equation by the least common denominator, $x(x+2)$, assuming $x \neq 0, -2$.
$$x(x+2)\left[\dfrac{x}{x+2} + \dfrac{1}{x} + 3\right] = x(x+2)\left(\dfrac{2}{x(x+2)}\right)$$
$$x^2 + (x+2) + 3x(x+2) = 2$$
$$x^2 + x + 2 + 3x^2 + 6x = 2$$
$$4x^2 + 7x + 2 = 2$$
$$4x^2 + 7x = 0$$
$$x(4x+7) = 0$$

$x = 0$ or $4x + 7 = 0 \Rightarrow x = -\frac{7}{4}$
Because of the restriction $x \neq 0$, the only valid solution is $-\frac{7}{4}$. The solution set is $\left\{-\frac{7}{4}\right\}$.

81. $(2x+3)^{2/3} + (2x+3)^{1/3} - 6 = 0$

Let $u = (2x+3)^{1/3}$. Then
$$u^2 = [(2x+3)^{1/3}]^2 = (2x+3)^{2/3}.$$
With this substitution, the equation becomes $u^2 + u - 6 = 0$. Solve by factoring.
$$(u+3)(u-2) = 0$$
$$u + 3 = 0 \Rightarrow u = -3 \quad \text{or} \quad u - 2 = 0 \Rightarrow u = 2$$
To find x, replace u with $(2x+3)^{1/3}$.
$$(2x+3)^{1/3} = -3 \Rightarrow \left[(2x+3)^{1/3}\right]^3 = (-3)^3 \Rightarrow$$
$$2x + 3 = -27 \Rightarrow 2x = -30 \Rightarrow x = -15 \quad \text{or}$$
$$(2x+3)^{1/3} = 2 \Rightarrow \left[(2x+3)^{1/3}\right]^3 = 2^3 \Rightarrow$$
$$2x + 3 = 8 \Rightarrow 2x = 5 \Rightarrow x = \tfrac{5}{2}$$

Check $x = -15$.
$$(2x+3)^{2/3} + (2x+3)^{1/3} = 6$$
$$[2(-15)+3]^{2/3} + [2(-15)+3]^{1/3} = 6$$
$$(-30+3)^{2/3} + (-30+3)^{1/3} = 6$$
$$(-27)^{2/3} + (-27)^{1/3} = 6$$
$$\left[(-27)^{1/3}\right]^2 + (-3) = 6$$
$$(-3)^2 - 3 = 6$$
$$9 - 3 = 6 \Rightarrow 6 = 6$$

This is a true statement. -15 is a solution.
Check $x = \frac{5}{2}$.
$$(2x+3)^{2/3} + (2x+3)^{1/3} = 6$$
$$\left[2\left(\tfrac{5}{2}\right)+3\right]^{2/3} + \left[2\left(\tfrac{5}{2}\right)+3\right]^{1/3} = 6$$
$$(5+3)^{2/3} + (5+3)^{1/3} = 6$$
$$8^{2/3} + 8^{1/3} = 6$$
$$\left[8^{1/3}\right]^2 + 2 = 6$$
$$(2)^2 + 2 = 6$$
$$4 + 2 = 6 \Rightarrow 6 = 6$$

This is a true statement. $\frac{5}{2}$ is a solution.
Solution set: $\left\{-15, \tfrac{5}{2}\right\}$

83. $-9x + 3 < 4x + 10$
$$-13x < 7$$
$$x > -\tfrac{7}{13}$$

Solution set: $\left(-\tfrac{7}{13}, \infty\right)$

85. $-5x - 4 \geq 3(2x - 5)$
$-5x - 4 \geq 6x - 15$
$-11x - 4 \geq -15$
$-11x \geq -11$
$x \leq 1$
Solution set: $(-\infty, 1]$

87. $5 \leq 2x - 3 \leq 7$
$8 \leq 2x \leq 10$
$4 \leq x \leq 5$
Solution set: $[4, 5]$

89. $x^2 + 3x - 4 \leq 0$
Step 1: Find the values of x that satisfy
$x^2 + 3x - 4 = 0$.
$x^2 + 3x - 4 = 0$
$(x + 4)(x - 1) = 0$
$x + 4 = 0 \Rightarrow x = -4$ or $x - 1 = 0 \Rightarrow x = 1$
Step 2: The two numbers divide a number line into three regions.

Step 3: Choose a test value to see if it satisfies the inequality, $x^2 + 3x - 4 \leq 0$.

Interval	Test Value	Is $x^2 + 3x - 4 \leq 0$ True or False?
A: $(-\infty, -4)$	-5	$(-5)^2 + 3(-5) - 4 \overset{?}{\leq} 0$ $6 \leq 0$ False
B: $(-4, 1)$	0	$0^2 + 3(0) - 4 \overset{?}{\leq} 0$ $-4 \leq 0$ True
C: $(1, \infty)$	2	$2^2 + 3(2) - 4 \overset{?}{\leq} 0$ $6 \leq 0$ False

Solution set: $[-4, 1]$

91. $6x^2 - 11x < 10$
Step 1: Find the values of x that satisfy
$6x^2 - 11x = 10$.
$6x^2 - 11x = 10$
$6x^2 - 11x - 10 = 0$
$(3x + 2)(2x - 5) = 0$
$3x + 2 = 0 \Rightarrow x = -\frac{2}{3}$ or $2x - 5 = 0 \Rightarrow x = \frac{5}{2}$
Step 2: The two numbers divide a number line into three regions.

Step 3: Choose a test value to see if it satisfies the inequality, $6x^2 - 11x < 10$

Interval	Test Value	Is $6x^2 - 11x < 10$ True or False?
A: $\left(-\infty, -\frac{2}{3}\right)$	-1	$6(-1)^2 - 11(-1) \overset{?}{<} 10$ $17 < 10$ False
B: $\left(-\frac{2}{3}, \frac{5}{2}\right)$	0	$6 \cdot 0^2 - 11 \cdot 0 - 10 \overset{?}{<} 10$ $-10 < 10$ True
C: $\left(\frac{5}{2}, \infty\right)$	3	$6 \cdot 3^2 - 11 \cdot 3 - 10 \overset{?}{<} 10$ $11 < 10$ False

Solution set: $\left(-\frac{2}{3}, \frac{5}{2}\right)$

93. $x^3 - 16x \leq 0$
Step 1: Solve $x^3 - 16x = 0$.
$x^3 - 16x = 0 \Rightarrow x(x^2 - 16) = 0 \Rightarrow$
$x(x + 4)(x - 4) = 0$
Set each factor to zero and solve.
$x = 0$ or $x + 4 = 0 \Rightarrow x = -4$ or
$x - 4 = 0 \Rightarrow x = 4$
Step 2: The values -4, 0, and, 4 divide the number line into four intervals.

Step 3: Choose a test value to see if it satisfies the inequality, $x^3 - 16x \leq 0$.

Interval	Test Value	Is $x^3 - 16x \leq 0$ True or False?
A: $(-\infty, -4)$	-5	$(-5)^3 - 16(-5) \overset{?}{\leq} 0$ $-45 \leq 0$ True
B: $(-4, 0)$	-1	$(-1)^3 - 16(-1) \overset{?}{\leq} 0$ $15 \leq 0$ False

(continued on next page)

(*continued*)

Interval	Test Value	Is $x^3 - 16x \le 0$ True or False?
C: $(0, 4)$	1	$1^3 - 16 \cdot 1 \overset{?}{\le} 0$ $-15 \le 0$ True
D: $(4, \infty)$	5	$5^3 - 16 \cdot 5 \overset{?}{\le} 0$ $45 \le 0$ False

Solution set: $(-\infty, -4] \cup [0, 4]$

95. $\dfrac{3x+6}{x-5} > 0$

Because one side of the inequality is already 0, we start with Step 2.

Step 2: Determine the values that will cause either the numerator or denominator to equal 0.

$3x + 6 = 0 \Rightarrow x = -2$ or $x - 5 = 0 \Rightarrow x = 5$

The values -2 and 5 to divide the number line into three regions.

Step 3: Choose a test value to see if it satisfies the inequality, $\dfrac{3x+6}{x-5} > 0$.

Interval	Test Value	Is $\frac{3x+6}{x-5} > 0$ True or False?
A: $(-\infty, -2)$	-3	$\frac{3(-3)+6}{-3-5} \overset{?}{>} 0$ $\frac{3}{8} > 0$ True
B: $(-2, 5)$	0	$\frac{3(0)+6}{0-5} \overset{?}{>} 0$ $-\frac{6}{5} > 0$ False
C: $(5, \infty)$	6	$\frac{3(6)+6}{6-5} \overset{?}{>} 0$ $24 > 0$ True

Solution set: $(-\infty, -2) \cup (5, \infty)$

97. $\dfrac{3x-2}{x} - 4 > 0$

Step 1: Rewrite the inequality so that 0 is on one side and there is a single fraction on the other side.

$\dfrac{3x-2}{x} - 4 > 0 \Rightarrow \dfrac{3x-2}{x} - \dfrac{4x}{x} > 0 \Rightarrow$

$\dfrac{3x-2-4x}{x} > 0 \Rightarrow \dfrac{-x-2}{x} > 0$

Step 2: Determine the values that will cause either the numerator or denominator to equal 0.

$-x - 2 = 0 \Rightarrow x = -2$ or $x = 0$

The values -2 and 0 divide the number line into three regions.

Step 3: Choose a test value to see if it satisfies the inequality, $\dfrac{3x-2}{x} - 4 > 0$.

Interval	Test Value	Is $\frac{3x-2}{x} - 4 > 0$ True or False?	
A: $(-\infty, -2)$	-3	$\frac{3(-3)-2}{-3} - 4 \overset{?}{>} 0$ $-\frac{1}{3} > 0$ False	
B: $(-2, 0)$	-1	$\frac{3(-1)-2}{-1} - 4 \overset{?}{>} 0$ $1 > 0$ True	
C: $(0, \infty)$	1	$\frac{3 \cdot 1 - 2}{1} - 4 \overset{?}{>} 0$ $-3 > 0$ False	

Solution set: $(-2, 0)$

99. $\dfrac{3}{x-1} \le \dfrac{5}{x+3}$

Step 1: Rewrite the inequality so that 0 is on one side and there is a single fraction on the other side.

$$\dfrac{3}{x-1} - \dfrac{5}{x+3} \le 0$$

$$\dfrac{3(x+3)}{(x-1)(x+3)} - \dfrac{5(x-1)}{(x+3)(x-1)} \le 0$$

$$\dfrac{3(x+3) - 5(x-1)}{(x-1)(x+3)} \le 0$$

$$\dfrac{3x+9 - 5x+5}{(x-1)(x+3)} \le 0$$

$$\dfrac{-2x+14}{(x-1)(x+3)} \le 0$$

Step 2: Determine the values that will cause either the numerator or denominator to equal 0.

$-2x + 14 = 0 \Rightarrow x = 7$ or $x - 1 = 0 \Rightarrow x = 1$ or $x + 3 = 0 \Rightarrow x = -3$

The values -3, 1 and 7 divide the number line into four regions. Use an open circle on -3 and 1 because they make the denominator equal 0.

(*continued on next page*)

(continued)

Interval A Interval B Interval C Interval D
$(-\infty, -3)$ $(-3, 1)$ $(1, 7)$ $(7, \infty)$

Step 3: Choose a test value to see if it satisfies

the inequality, $\dfrac{3}{x-1} \le \dfrac{5}{x+3}$.

Interval	Test Value	Is $\frac{3}{x-1} \le \frac{5}{x+3}$ True or False?
A: $(-\infty, -3)$	-4	$\frac{3}{-4-1} \overset{?}{\le} \frac{5}{-4+3}$ $-\frac{3}{5} \le -5$ False
B: $(-3, 1)$	0	$\frac{3}{0-1} \overset{?}{\le} \frac{5}{0+3}$ $-3 \le \frac{5}{3}$ True
C: $(1, 7)$	2	$\frac{3}{2-1} \overset{?}{\le} \frac{5}{2+3}$ $3 \le 1$ False
D: $(7, \infty)$	8	$\frac{3}{8-1} \overset{?}{\le} \frac{5}{8+3}$ $\frac{3}{7} \overset{?}{\le} \frac{5}{11}$ $\frac{33}{77} \le \frac{35}{77}$ True

Intervals B and D satisfy the inequality. The endpoints -3 and 1 are not included because they make the denominator 0.
Solution set: $(-3, 1) \cup [7, \infty)$

101. (a) Let x = the ozone concentration after the Purafil air filter is used.

$x = 140 - 0.43(140) = 79.8$

The ozone concentration after the Purafil air filter is used is 79.8 ppb.

(b) Let x = the maximum initial concentration of ozone.

$x - 0.43x \le 50 \Rightarrow 0.57x \le 50$

$x \le 87.7$ (approximately)

The filter will reduce ozone concentrations that don't exceed 87.7 ppb.

103. $s = -16t^2 + 320$

(a) When $s = 0$, the projectile will be at ground level.

$0 = -16t^2 + 320t \Rightarrow 16t^2 - 320t = 0 \Rightarrow$

$t^2 - 20t = 0 \Rightarrow t(t - 20) = 0 \Rightarrow$

$t = 0$ or $t = 20$

The projectile will return to the ground after 20 sec.

(b) Solve $s > 576$ for t.

$320t - 16t^2 > 576$

$0 > 16t^2 - 320t + 576$

$0 > t^2 - 20t + 36$

Step 1: Find the values of t that satisfy $t^2 - 20t + 36 = 0$.

$t^2 - 20t + 36 = 0 \Rightarrow (t - 2)(t - 18) = 0 \Rightarrow$

$t - 2 = 0 \Rightarrow t = 2$ or $t - 18 = 0 \Rightarrow t = 18$

Step 2: The two numbers divide a number line into three regions.

Interval A Interval B Interval C
$(0, 2)$ $(2, 18)$ $(18, \infty)$

Step 3: Choose a test value to see if it satisfies the inequality,

$320t - 16t^2 > 576$.

Interval	Test Value	Is $320t - 16t^2 > 576$ True or False?
A: $(0, 2)$	1	$320(1) - 16(1)^2 \overset{?}{>} 576$ $304 > 576$ False
B: $(2, 18)$	3	$320(3) - 16(3)^2 \overset{?}{>} 576$ $816 > 576$ True
C: $(18, \infty)$	20	$320(20) - 16(20)^2 \overset{?}{>} 576$ $0 > 576$ False

The projectile will be more than 576 ft above the ground between 2 and 18 sec.

105. Answers will vary. 3 cannot be in the solution set because when 3 is substituted into $\frac{14x+9}{x-3}$, division by zero occurs.

107. $|x + 4| = 7$

$x + 4 = 7 \Rightarrow x = 3$ or $x + 4 = -7 \Rightarrow x = -11$

Solution set: $\{-11, 3\}$

109. $\left|\dfrac{7}{2 - 3x}\right| - 9 = 0 \Rightarrow \left|\dfrac{7}{2 - 3x}\right| = 9$

$\dfrac{7}{2 - 3x} = 9 \Rightarrow 7 = 9(2 - 3x) \Rightarrow 7 = 18 - 27x \Rightarrow$

$-11 = -27x \Rightarrow \frac{-11}{-27} = x \Rightarrow x = \frac{11}{27}$ or

$\dfrac{7}{2 - 3x} = -9 \Rightarrow 7 = -9(2 - 3x) \Rightarrow$

$7 = -18 + 27x \Rightarrow 25 = 27x \Rightarrow \frac{25}{27} = x \Rightarrow x = \frac{25}{27}$

Solution set: $\left\{\frac{11}{27}, \frac{25}{27}\right\}$

111. $|5x-1| = |2x+3|$

$5x-1 = 2x+3 \Rightarrow 3x-1 = 3 \Rightarrow 3x = 4 \Rightarrow x = \frac{4}{3}$

or

$5x-1 = -(2x+3) \Rightarrow 5x-1 = -2x-3 \Rightarrow$

$7x-1 = -3 \Rightarrow 7x = -2 \Rightarrow x = -\frac{2}{7}$

Solution set: $\left\{-\frac{2}{7}, \frac{4}{3}\right\}$

113. $|2x+9| \le 3$

$-3 \le 2x+9 \le 3$

$-12 \le 2x \le -6$

$-6 \le x \le -3$

Solution set: $[-6, -3]$

115. $|7x-3| > 4$

$7x-3 < -4 \Rightarrow 7x < -1 \Rightarrow x < -\frac{1}{7}$ or

$7x-3 > 4 \Rightarrow 7x > 7 \Rightarrow x > 1$

Solution set: $\left(-\infty, -\frac{1}{7}\right) \cup (1, \infty)$

117. $|3x+7| - 5 < 5 \Rightarrow |3x+7| < 10$

$-10 < 3x+7 < 10$

$-17 < 3x < 3$

$-\frac{17}{3} < x < 1$

Solution set: $\left(-\frac{17}{3}, 1\right)$

119. Because the absolute value of a number is always nonnegative, the inequality $|4x-12| \ge -3$ is always true. The solution set is $(-\infty, \infty)$.

121. Because the absolute value of a number is always nonnegative, $|x^2+4x| < 0$ is never true, so $|x^2+4x| \le 0$ is only true when $|x^2+4x| = 0$.

$|x^2+4x| = 0 \Rightarrow x^2+4x = 0 \Rightarrow x(x+4) = 0$

$x = 0$ or $x+4 = 0$

$x = 0$ or $x = -4$

Solution set: $\{-4, 0\}$

123. "k is 12 units from 6" means that the distance between k and 6 is 12 units, or $|k-6| = 12$ or $|6-k| = 12$.

125. "t is no less than 0.01 unit from 5" means that t is 0.01 unit or more from 5. Thus, the distance between t and 5 is greater than or equal to 0.01, or $|t-5| \ge 0.01$ or $|5-t| \ge 0.01$.

Chapter 1   Test

1. $3(x-4) - 5(x+2) = 2 - (x+24)$

$3x-12 - 5x-10 = 2 - x - 24$

$-2x-22 = -x-22$

$-22 = x-22$

$0 = x$

Solution set: $\{0\}$

2. $\frac{2}{3}x + \frac{1}{2}(x-4) = x - 4$

$6\left[\frac{2}{3}x + \frac{1}{2}(x-4)\right] = 6(x-4)$

$4x + 3(x-4) = 6x - 24$

$4x + 3x - 12 = 6x - 24$

$7x - 12 = 6x - 24$

$x - 12 = -24$

$x = -12$

Solution set: $\{-12\}$

3. $6x^2 - 11x - 7 = 0$

$(2x+1)(3x-7) = 0$

$2x+1 = 0 \Rightarrow x = -\frac{1}{2}$ or $3x-7 = 0 \Rightarrow x = \frac{7}{3}$

Solution set: $\left\{-\frac{1}{2}, \frac{7}{3}\right\}$

4. $(3x+1)^2 = 8$

$3x+1 = \pm\sqrt{8} = \pm 2\sqrt{2}$

$3x = -1 \pm 2\sqrt{2} \Rightarrow x = \frac{-1 \pm 2\sqrt{2}}{3}$

Solution set: $\left\{\frac{-1 \pm 2\sqrt{2}}{3}\right\}$

5. $3x^2 + 2x = -2$

Solve by completing the square.

$3x^2 + 2x = -2$

$3x^2 + 2x + 2 = 0$

$x^2 + \frac{2}{3}x + \frac{2}{3} = 0 \Rightarrow x^2 + \frac{2}{3}x + \frac{1}{9} = -\frac{2}{3} + \frac{1}{9}$

Note: $\left[\frac{1}{2} \cdot \frac{2}{3}\right]^2 = \left(\frac{1}{3}\right)^2 = \frac{1}{9}$

$\left(x + \frac{1}{3}\right)^2 = -\frac{5}{9} \Rightarrow x + \frac{1}{3} = \pm\sqrt{-\frac{5}{9}} \Rightarrow$

$x + \frac{1}{3} = \pm\frac{\sqrt{5}}{3}i \Rightarrow x = -\frac{1}{3} \pm \frac{\sqrt{5}}{3}i$

(continued on next page)

(continued)

Solve by the quadratic formula.
Let $a = 3$, $b = 2$, and $c = 2$.

$$x = \frac{-b \pm \sqrt{b^2 - 4ac}}{2a}$$

$$= \frac{-2 \pm \sqrt{2^2 - 4(3)(2)}}{2(3)} = \frac{-2 \pm \sqrt{4 - 24}}{6}$$

$$= \frac{-2 \pm \sqrt{-20}}{6} = \frac{-2 \pm 2i\sqrt{5}}{6}$$

$$= -\frac{2}{6} \pm \frac{2\sqrt{5}}{6} i = -\frac{1}{3} \pm \frac{\sqrt{5}}{3} i$$

Solution set: $\left\{ -\frac{1}{3} \pm \frac{\sqrt{5}}{3} i \right\}$

6.
$$\frac{12}{x^2 - 9} = \frac{2}{x - 3} - \frac{3}{x + 3}$$

$$\frac{12}{(x+3)(x-3)} + \frac{3}{x+3} = \frac{2}{x-3}$$

Multiply each term in the equation by the least common denominator, $(x+3)(x-3)$ assuming $x \neq -3, 3$.

$$(x+3)(x-3)\left[\frac{12}{(x+3)(x-3)} + \frac{3}{x+3} \right]$$

$$= (x+3)(x-3)\left(\frac{2}{x-3} \right)$$

$$12 + 3(x-3) = 2(x+3)$$
$$12 + 3x - 9 = 2x + 6$$
$$3x + 3 = 2x + 6$$
$$x + 3 = 6 \Rightarrow x = 3$$

The only possible solution is 3. However, the variable is restricted to real numbers except –3 and 3. Therefore, the solution set is $\varnothing$.

7. $\dfrac{4x}{x-2} + \dfrac{3}{x} = \dfrac{-6}{x^2 - 2x}$ or $\dfrac{4x}{x-2} + \dfrac{3}{x} = \dfrac{-6}{x(x-2)}$

Multiply each term in the equation by the least common denominator, $x(x - 2)$ assuming $x \neq 0, 2$.

$$x(x-2)\left[\frac{4x}{x-2} + \frac{3}{x} \right] = x(x-2)\left(\frac{-6}{x(x-2)} \right)$$

$$4x^2 + 3(x-2) = -6 \Rightarrow 4x^2 + 3x - 6 = -6$$
$$4x^2 + 3x = 0 \Rightarrow x(4x+3) = 0$$

$x = 0$ or $4x + 3 = 0 \Rightarrow x = -\frac{3}{4}$

Because of the restriction $x \neq 0$, the only valid solution is $-\frac{3}{4}$. The solution set is $\left\{ -\frac{3}{4} \right\}$.

8. $\sqrt{3x + 4} + 5 = 2x + 1 \Rightarrow \sqrt{3x + 4} = 2x - 4$

$$\left(\sqrt{3x+4} \right)^2 = (2x - 4)^2$$

$$3x + 4 = 4x^2 - 16x + 16$$
$$0 = 4x^2 - 19x + 12$$
$$0 = (4x - 3)(x - 4)$$

$4x - 3 = 0 \Rightarrow x = \frac{3}{4}$ or $x - 4 = 0 \Rightarrow x = 4$

Check $x = \frac{3}{4}$.

$$\sqrt{3x + 4} + 5 = 2x + 1$$
$$\sqrt{3\left(\frac{3}{4}\right) + 4} + 5 = 2\left(\frac{3}{4}\right) + 1$$
$$\sqrt{\frac{9}{4} + 4} + 5 = \frac{5}{2} \Rightarrow \sqrt{\frac{25}{4}} + 5 = \frac{5}{2}$$
$$\frac{5}{2} + 5 = \frac{5}{2} \Rightarrow \frac{15}{2} = \frac{5}{2}$$

This is a false statement. $\frac{3}{4}$ is a not solution.

Check $x = 4$.

$$\sqrt{3x + 4} + 5 = 2x + 1$$
$$\sqrt{3(4) + 4} + 5 = 2(4) + 1$$
$$\sqrt{12 + 4} + 5 = 8 + 1$$
$$\sqrt{16} + 5 = 8 + 1$$
$$4 + 5 = 9 \Rightarrow 9 = 9$$

This is a true statement. 4 is a solution.
Solution set: {4}

9. $\sqrt{-2x + 3} + \sqrt{x + 3} = 3$

$$\sqrt{-2x + 3} = 3 - \sqrt{x + 3}$$
$$\left(\sqrt{-2x+3} \right)^2 = \left(3 - \sqrt{x+3} \right)^2$$
$$-2x + 3 = 9 - 6\sqrt{x+3} + (x+3)$$
$$-2x + 3 = 12 + x - 6\sqrt{x+3}$$
$$-3x - 9 = -6\sqrt{x+3}$$
$$x + 3 = 2\sqrt{x+3}$$
$$(x+3)^2 = \left(2\sqrt{x+3} \right)^2$$
$$x^2 + 6x + 9 = 4(x+3)$$
$$x^2 + 6x + 9 = 4x + 12$$
$$x^2 + 2x - 3 = 0 \Rightarrow (x+3)(x-1) = 0$$

$x + 3 = 0 \Rightarrow x = -3$ or $x - 1 = 0 \Rightarrow x = 1$

Check $x = -3$.

$$\sqrt{-2x + 3} + \sqrt{x + 3} = 3$$
$$\sqrt{-2(-3) + 3} + \sqrt{-3 + 3} = 3$$
$$\sqrt{6 + 3} + \sqrt{0} = 3$$
$$\sqrt{9} + 0 = 3$$
$$3 + 0 = 3 \Rightarrow 3 = 3$$

This is a true statement. –3 is a solution.

Check $x = 1$.

$$\sqrt{-2x+3} + \sqrt{x+3} = 3$$
$$\sqrt{-2(1)+3} + \sqrt{1+3} = 3$$
$$\sqrt{-2+3} + \sqrt{4} = 3$$
$$\sqrt{1} + 2 = 3$$
$$1 + 2 = 3 \Rightarrow 3 = 3$$

This is a true statement. 1 is a solution.
Solution set: $\{-3, 1\}$

10. $\sqrt[3]{3x-8} = \sqrt[3]{9x+4}$

$$\left(\sqrt[3]{3x-8}\right)^3 = \left(\sqrt[3]{9x+4}\right)^3$$
$$3x - 8 = 9x + 4 \Rightarrow -8 = 6x + 4 \Rightarrow$$
$$-12 = 6x \Rightarrow -2 = x$$

Check $x = -2$.

$$\sqrt[3]{3x-8} = \sqrt[3]{9x+4}$$
$$\sqrt[3]{3(-2)-8} = \sqrt[3]{9(-2)+4}$$
$$\sqrt[3]{-6-8} = \sqrt[3]{-18+4}$$
$$\sqrt[3]{-14} = \sqrt[3]{-14} \Rightarrow -\sqrt[3]{14} = -\sqrt[3]{14}$$

This is a true statement.
Solution set: $\{-2\}$

11. $x^4 - 17x^2 + 16 = 0$

Let $u = x^2$; then $u^2 = x^4$.
With this substitution, the equation becomes
$u^2 - 17u + 16 = 0$.
Solve this equation by factoring.
$(u-1)(u-16) = 0$
$u - 1 = 0 \Rightarrow u = 1$ or $u - 16 = 0 \Rightarrow u = 16$
To find x, replace u with x^2.
$x^2 = 1 \Rightarrow x = \pm\sqrt{1} \Rightarrow x = \pm 1$ or
$x^2 = 16 \Rightarrow x = \pm\sqrt{16} \Rightarrow x = \pm 4$
Solution set: $\{\pm 1, \pm 4\}$

12. $(x+3)^{2/3} + (x+3)^{1/3} - 6 = 0$

Let $u = (x+3)^{1/3}$. Then
$$u^2 = \left[(x+3)^{1/3}\right]^2 = (x+3)^{2/3}.$$
$u^2 + u - 6 = 0 \Rightarrow (u+3)(u-2) = 0$
$u + 3 = 0 \Rightarrow u = -3$ or $u - 2 = 0 \Rightarrow u = 2$

To find x, replace u with $(x+3)^{1/3}$.

$$(x+3)^{1/3} = -3 \Rightarrow \left[(x+3)^{1/3}\right]^3 = (-3)^3 \Rightarrow$$
$x + 3 = -27 \Rightarrow x = -30$ or
$$(x+3)^{1/3} = 2 \Rightarrow \left[(x+3)^{1/3}\right]^3 = 2^3 \Rightarrow$$
$x + 3 = 8 \Rightarrow x = 5$

Check $x = -30$.

$$(x+3)^{2/3} + (x+3)^{1/3} - 6 = 0$$
$$(-30+3)^{2/3} + (-30+3)^{1/3} - 6 = 0$$
$$(-27)^{2/3} + (-27)^{1/3} - 6 = 0$$
$$\left[(-27)^{1/3}\right]^2 + (-3) - 6 = 0$$
$$(-3)^2 - 3 - 6 = 0$$
$$9 - 3 - 6 = 0 \Rightarrow 0 = 0$$

This is a true statement. -30 is a solution.
Check $x = 5$.

$$(x+3)^{2/3} + (x+3)^{1/3} - 6 = 0$$
$$(5+3)^{2/3} + (5+3)^{1/3} - 6 = 0$$
$$8^{2/3} + 8^{1/3} - 6 = 0$$
$$\left[8^{1/3}\right]^2 + 2 - 6 = 0$$
$$2^2 + 2 - 6 = 0$$
$$4 + 2 - 6 = 0 \Rightarrow 0 = 0$$

This is a true statement. 5 is a solution.
Solution set: $\{-30, 5\}$

13. $|4x+3| = 7$

$4x + 3 = 7 \Rightarrow 4x = 4 \Rightarrow x = 1$ or
$4x + 3 = -7 \Rightarrow 4x = -10 \Rightarrow x = -\frac{10}{4} = -\frac{5}{2}$

Solution set: $\left\{-\frac{5}{2}, 1\right\}$

14. $|2x+1| = |5-x|$

$2x + 1 = 5 - x \Rightarrow 3x + 1 = 5 \Rightarrow 3x = 4 \Rightarrow x = \frac{4}{3}$
or
$2x + 1 = -(5-x) \Rightarrow 2x + 1 = -5 + x \Rightarrow x = -6$

Solution set: $\left\{-6, \frac{4}{3}\right\}$

15.
$$S = 2HW + 2LW + 2LH$$
$$S - 2LH = 2HW + 2LW$$
$$S - 2LH = W(2H + 2L)$$
$$\frac{S - 2LH}{2H + 2L} = W$$
$$W = \frac{S - 2LH}{2H + 2L}$$

16. **(a)** $(9 - 3i) - (4 + 5i) = (9 - 4) + (-3 - 5)i$
$$= 5 - 8i$$

(b) $(4 + 3i)(-5 + 3i) = -20 + 12i - 15i + 9i^2$
$$= -20 - 3i + 9(-1)$$
$$= -20 - 3i - 9 = -29 - 3i$$

(c) $(8+3i)^2 = 8^2 + 2(8)(3i) + (3i)^2$

$\qquad = 64 + 48i + 9i^2$

$\qquad = 64 + 48i + 9(-1)$

$\qquad = 64 + 48i - 9 = 55 + 48i$

(d) $\dfrac{3+19i}{1+3i} = \dfrac{(3+19i)(1-3i)}{(1+3i)(1-3i)}$

$\qquad = \dfrac{3 - 9i + 19i - 57i^2}{1 - (3i)^2}$

$\qquad = \dfrac{3 + 10i - 57(-1)}{1 - 9i^2} = \dfrac{3 + 10i + 57}{1 - 9(-1)}$

$\qquad = \dfrac{60 + 10i}{1 + 9} = \dfrac{60 + 10i}{10} = 6 + i$

17. (a) $i^{42} = i^{40} \cdot i^2 = \left(i^4\right)^{10} \cdot (-1) = 1^{10} \cdot (-1) = -1$

(b) $i^{-31} = i^{-32} \cdot i = \left(i^4\right)^{-8} \cdot i = 1^{-8} \cdot i = i$

(c) $\dfrac{1}{i^{19}} = i^{-19} = i^{-20} \cdot i = \left(i^4\right)^{-5} \cdot i = 1^{-5} \cdot i = i$

18. (a) Minimum:

$1120\dfrac{\text{gal}}{\text{min}} \cdot 60 \dfrac{\text{min}}{\text{hr}} \cdot 12 \dfrac{\text{hr}}{\text{day}} = 806,400 \dfrac{\text{gal}}{\text{day}}$

The equation that will calculate the minimum amount of water pumped after x days would be $A = 806,400x$.

(b) $A = 806,400x$ when $x = 30$ would be

$A = 806,400(30) = 24,192,000$ gal.

(c) Because there would be $806,400 \dfrac{\text{gal}}{\text{day}}$

minimum and each pool requires 20,000 gal, there would be a minimum of

$\dfrac{806,400}{20,000} = 40.32$ pools that could be

filled each day. The equation that will calculate the minimum number of pools that could be filled after x days would be $P = 40.32x$. Approximately 40 pools could be filled each day.

(d) Solve $P = 40.32x$ where $P = 1000$.

$1000 = 40.32x \Rightarrow x = \dfrac{1000}{40.32} \approx 24.8$ days.

A minimum of 1000 pools could be filled in 25 days.

19. Let $w = $ width of rectangle. Then

$\qquad 2w - 20 = $ length of rectangle.

Use the formula for the perimeter of a rectangle.

$\qquad P = 2l + 2w$

$\qquad 620 = 2(2w - 20) + 2w$

$\qquad 620 = 4w - 40 + 2w$

$\qquad 620 = 6w - 40 \Rightarrow 660 = 6w \Rightarrow 110 = w$

The width is 110 m and the length is

$2(110) - 20 = 220 - 20 = 200$ m.

20. Let $x = $ amount of cashews (in pounds). Then $35 - x = $ amount of walnuts (in pounds).

	Cost per Pound	Amount of Nuts	Total Cost
Cashews	7.00	x	$7.00x$
Walnuts	5.50	$35 - x$	$5.50(35 - x)$
Mixture	6.50	35	$35 \cdot 6.50$

Solve the following equation.

$\qquad 7.00x + 5.50(35 - x) = 35 \cdot 6.50$

$\qquad 7x + 192.5 - 5.5x = 227.5$

$\qquad 1.5x + 192.5 = 227.5$

$\qquad 1.5x = 35$

$\qquad x = \dfrac{35}{1.5} = \dfrac{350}{15} = \dfrac{70}{3} = 23\frac{1}{3}$

The fruit and nut stand owner should mix

$23\frac{1}{3}$ lbs of cashews with $35 - 23\frac{1}{3} = 11\frac{2}{3}$ lbs

of walnuts.

21. Let $x = $ average speed upriver.

Then $x + 5 = $ average speed on return trip.

	r	t	d
Upriver	x	1.2	$1.2x$
Downriver	$x+5$	0.9	$0.9(x+5)$

Because the distance upriver and downriver are the same, we solve the following.

$\qquad 1.2x = 0.9(x+5)$

$\qquad 1.2x = 0.9x + 4.5 \Rightarrow 0.3x = 4.5 \Rightarrow x = 15$

The average speed of the boat upriver is 15 mph.

22. $y = -0.461x + 6.32$

(a) The year 2014 is represented by $x = 10$.

$\qquad y = -0.461(10) + 6.32 = 1.71$

According to the model, in 2014 about 1.7% of college freshmen smoked.

(b) $\qquad 4.9 = -0.461x + 6.32$

$\qquad -1.42 = -0.461x \Rightarrow 3.1 \approx x$

According to the model, 4.9% of freshman smoked about 3.1 years after 2004 or in 2007.

23. $s = -16t^2 + 96t$

  (a) Let $s = 80$ and solve for t.

  $80 = -16t^2 + 96t \Rightarrow 16t^2 - 96t + 80 = 0$

  $t^2 - 6t + 5 = 0$

  $(t - 1)(t - 5) = 0$

  $t - 1 = 0 \Rightarrow t = 1$ or $t - 5 = 0 \Rightarrow t = 5$

  The projectile will reach a height of 80 ft at 1 sec and 5 sec.

  (b) Let $s = 0$ and solve for t.

  $0 = -16t^2 + 96t$

  $0 = -16t(t - 6)$

  $t = 0$ or $t - 6 = 0 \Rightarrow t = 6$

  The projectile will return to the ground at 6 sec.

24. $-2(x - 1) - 12 < 2(x + 1)$

  $-2x + 2 - 12 < 2x + 2$

  $-2x - 10 < 2x + 2$

  $-4x - 10 < 2$

  $-4x < 12$

  $x > -3$

  Solution set: $(-3, \infty)$

25. $-3 \leq \dfrac{1}{2}x + 2 \leq 3$

  $2(-3) \leq 2\left(\dfrac{1}{2}x + 2\right) \leq 2(3)$

  $-6 \leq x + 4 \leq 6$

  $-10 \leq x \leq 2$

  Solution set: $[-10, 2]$

26. $2x^2 - x \geq 3$

  Step 1: Find the values of x that satisfy $2x^2 - x = 3$.

  $2x^2 - x = 3$

  $2x^2 - x - 3 = 0$

  $(x + 1)(2x - 3) = 0$

  $x + 1 = 0 \Rightarrow x = -1$ or $2x - 3 = 0 \Rightarrow x = \dfrac{3}{2}$

  Step 2: The two numbers divide a number line into three regions.

Interval A	Interval B	Interval C
$(-\infty, -1)$	$(-1, \frac{3}{2})$	$(\frac{3}{2}, \infty)$

  $-1 \quad 0 \quad 1 \quad \frac{3}{2} \quad 2$

  Step 3: Choose a test value to see if it satisfies the inequality, $2x^2 - x \geq 3$

Interval	Test Value	Is $2x^2 - x \geq 3$ True or False?
A: $(-\infty, -1)$	-2	$2(-2)^2 - (-2) \overset{?}{\geq} 3$ $\quad 10 \geq 3$ True
B: $\left(-1, \frac{3}{2}\right)$	0	$2 \cdot 0^2 - 0 \overset{?}{\geq} 3$ $\quad 0 \geq 3$ False
C: $\left(\frac{3}{2}, \infty\right)$	2	$2 \cdot 2^2 - 2 \overset{?}{\geq} 3$ $\quad 6 \geq 3$ True

  Solution set: $(-\infty, -1] \cup \left[\frac{3}{2}, \infty\right)$

27. $\dfrac{x + 1}{x - 3} < 5$

  Step 1: Rewrite the inequality so that 0 is on one side and there is a single fraction on the other side.

  $\dfrac{x + 1}{x - 3} < 5 \Rightarrow \dfrac{x + 1}{x - 3} - 5 < 0$

  $\dfrac{x + 1}{x - 3} - \dfrac{5(x - 3)}{x - 3} < 0 \Rightarrow \dfrac{x + 1 - 5(x - 3)}{x - 3} < 0$

  $\dfrac{x + 1 - 5x + 15}{x - 3} < 0 \Rightarrow \dfrac{-4x + 16}{x - 3} < 0$

  Step 2: Determine the values that will cause either the numerator or denominator to equal 0.

  $-4x + 16 = 0 \Rightarrow x = 4$ or $x - 3 = 0 \Rightarrow x = 3$

  The values 3 and 4 divide the number line into three regions.

Interval A	Interval B	Interval C
$(-\infty, 3)$	$(3, 4)$	$(4, \infty)$

  $2 \quad 3 \quad 4$

  Step 3: Choose a test value to see if it satisfies the inequality, $\frac{x+1}{x-3} < 5$.

Interval	Test Value	Is $\frac{x+1}{x-3} < 5$ True or False?
A: $(-\infty, 3)$	0	$\dfrac{0+1}{0-3} \overset{?}{<} 5$ $\quad -\frac{1}{3} < 5$ True
B: $(3, 4)$	3.5	$\dfrac{3.5+1}{3.5-3} \overset{?}{<} 5$ $\quad 9 < 5$ False
C: $(4, \infty)$	5	$\dfrac{5+1}{5-3} \overset{?}{<} 5$ $\quad 3 < 5$ True

  Solution set: $(-\infty, 3) \cup (4, \infty)$

28. $|2x - 5| < 9$

$-9 < 2x - 5 < 9$

$-4 < 2x < 14$

$-2 < x < 7$

Solution set: $(-2, 7)$

29. $|2x + 1| - 11 \geq 0 \Rightarrow |2x + 1| \geq 11$

$\quad 2x + 1 \leq -11 \quad$ or $\quad 2x + 1 \geq 11$

$\qquad 2x \leq -12 \qquad\qquad 2x \geq 10$

$\qquad x \leq -6 \quad$ or $\qquad x \geq 5$

Solution set: $(-\infty, -6] \cup [5, \infty)$

30. $|3x + 7| \leq 0 \Rightarrow 3x + 7 \leq 0 \Rightarrow x \leq -\frac{7}{3}$

However, if $x < -\frac{7}{3}$, then $3x + 7 < 0$, and $|3x + 7|$ is not defined. Thus, the solution set of $|3x + 7| \leq 0$ is $\left\{-\frac{7}{3}\right\}$.

Chapter 2

GRAPHS AND FUNCTIONS

Section 2.1 Rectangular Coordinates and Graphs

1. The point $(-1, 3)$ lies in quadrant <u>II</u> in the rectangular coordinate system.

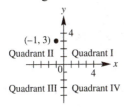

3. Any point that lies on the x-axis has y-coordinate equal to <u>0</u>.

5. The x-intercept of the graph of $2x + 5y = 10$ is <u>(5, 0)</u>. Find the x-intercept by letting $y = 0$ and solving for x.
$$2x + 5(0) = 10 \Rightarrow 2x = 10 \Rightarrow x = 5$$

7. True

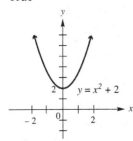

9. False. The midpoint of the segment joining $(0, 0)$ and $(4, 4)$ is
$$\left(\frac{4+0}{2}, \frac{4+0}{2}\right) = \left(\frac{4}{2}, \frac{4}{2}\right) = (2, 2).$$

11. Any three of the following:
$(2, -5), (-1, 7), (3, -9), (5, -17), (6, -21)$

13. Any three of the following: $(1999, 35)$, $(2001, 29), (2003, 22), (2005, 23), (2007, 20)$, $(2009, 20)$

15. $P(-5, -6), Q(7, -1)$

    (a) $d(P, Q) = \sqrt{[7-(-5)]^2 + [-1-(-6)]^2}$
    $$= \sqrt{12^2 + 5^2} = \sqrt{169} = 13$$

    (b) The midpoint M of the segment joining points P and Q has coordinates
    $$\left(\frac{-5+7}{2}, \frac{-6+(-1)}{2}\right) = \left(\frac{2}{2}, -\frac{7}{2}\right)$$
    $$= \left(1, -\frac{7}{2}\right).$$

17. $P(8, 2), Q(3, 5)$

    (a) $d(P, Q) = \sqrt{(3-8)^2 + (5-2)^2}$
    $$= \sqrt{(-5)^2 + 3^2}$$
    $$= \sqrt{25 + 9} = \sqrt{34}$$

    (b) The midpoint M of the segment joining points P and Q has coordinates
    $$\left(\frac{8+3}{2}, \frac{2+5}{2}\right) = \left(\frac{11}{2}, \frac{7}{2}\right).$$

19. $P(-6, -5), Q(6, 10)$

    (a) $d(P, Q) = \sqrt{[6-(-6)]^2 + [10-(-5)]^2}$
    $$= \sqrt{12^2 + 15^2} = \sqrt{144 + 225}$$
    $$= \sqrt{369} = 3\sqrt{41}$$

    (b) The midpoint M of the segment joining points P and Q has coordinates
    $$\left(\frac{-6+6}{2}, \frac{-5+10}{2}\right) = \left(\frac{0}{2}, \frac{5}{2}\right) = \left(0, \frac{5}{2}\right).$$

21. $P(3\sqrt{2}, 4\sqrt{5}), Q(\sqrt{2}, -\sqrt{5})$

    (a) $d(P, Q)$
    $$= \sqrt{(\sqrt{2} - 3\sqrt{2})^2 + (-\sqrt{5} - 4\sqrt{5})^2}$$
    $$= \sqrt{(-2\sqrt{2})^2 + (-5\sqrt{5})^2}$$
    $$= \sqrt{8 + 125} = \sqrt{133}$$

    (b) The midpoint M of the segment joining points P and Q has coordinates
    $$\left(\frac{3\sqrt{2} + \sqrt{2}}{2}, \frac{4\sqrt{5} + (-\sqrt{5})}{2}\right)$$
    $$= \left(\frac{4\sqrt{2}}{2}, \frac{3\sqrt{5}}{2}\right) = \left(2\sqrt{2}, \frac{3\sqrt{5}}{2}\right).$$

23. Label the points $A(-6, -4)$, $B(0, -2)$, and $C(-10, 8)$. Use the distance formula to find the length of each side of the triangle.

$$d(A, B) = \sqrt{[0-(-6)]^2 + [-2-(-4)]^2}$$
$$= \sqrt{6^2 + 2^2} = \sqrt{36+4} = \sqrt{40}$$

$$d(B, C) = \sqrt{(-10-0)^2 + [8-(-2)]^2}$$
$$= \sqrt{(-10)^2 + 10^2} = \sqrt{100+100}$$
$$= \sqrt{200}$$

$$d(A, C) = \sqrt{[-10-(-6)]^2 + [8-(-4)]^2}$$
$$= \sqrt{(-4)^2 + 12^2} = \sqrt{16+144} = \sqrt{160}$$

Because $\left(\sqrt{40}\right)^2 + \left(\sqrt{160}\right)^2 = \left(\sqrt{200}\right)^2$, triangle ABC is a right triangle.

25. Label the points $A(-4, 1)$, $B(1, 4)$, and $C(-6, -1)$.

$$d(A, B) = \sqrt{[1-(-4)]^2 + (4-1)^2}$$
$$= \sqrt{5^2 + 3^2} = \sqrt{25+9} = \sqrt{34}$$

$$d(B, C) = \sqrt{(-6-1)^2 + (-1-4)^2}$$
$$= \sqrt{(-7)^2 + (-5)^2} = \sqrt{49+25} = \sqrt{74}$$

$$d(A, C) = \sqrt{[-6-(-4)]^2 + (-1-1)^2}$$
$$= \sqrt{(-2)^2 + (-2)^2} = \sqrt{4+4} = \sqrt{8}$$

Because $(\sqrt{8})^2 + (\sqrt{34})^2 \neq (\sqrt{74})^2$ because $8 + 34 = 42 \neq 74$, triangle ABC is not a right triangle.

27. Label the points $A(-4, 3)$, $B(2, 5)$, and $C(-1, -6)$.

$$d(A, B) = \sqrt{[2-(-4)]^2 + (5-3)^2}$$
$$= \sqrt{6^2 + 2^2} = \sqrt{36+4} = \sqrt{40}$$

$$d(B, C) = \sqrt{(-1-2)^2 + (-6-5)^2}$$
$$= \sqrt{(-3)^2 + (-11)^2}$$
$$= \sqrt{9+121} = \sqrt{130}$$

$$d(A, C) = \sqrt{[-1-(-4)]^2 + (-6-3)^2}$$
$$= \sqrt{3^2 + (-9)^2} = \sqrt{9+81} = \sqrt{90}$$

Because $\left(\sqrt{40}\right)^2 + \left(\sqrt{90}\right)^2 = \left(\sqrt{130}\right)^2$, triangle ABC is a right triangle.

29. Label the given points $A(0, -7)$, $B(-3, 5)$, and $C(2, -15)$. Find the distance between each pair of points.

$$d(A, B) = \sqrt{(-3-0)^2 + [5-(-7)]^2}$$
$$= \sqrt{(-3)^2 + 12^2} = \sqrt{9+144}$$
$$= \sqrt{153} = 3\sqrt{17}$$

$$d(B, C) = \sqrt{[2-(-3)]^2 + (-15-5)^2}$$
$$= \sqrt{5^2 + (-20)^2} = \sqrt{25+400}$$
$$= \sqrt{425} = 5\sqrt{17}$$

$$d(A, C) = \sqrt{(2-0)^2 + [-15-(-7)]^2}$$
$$= \sqrt{2^2 + (-8)^2} = \sqrt{68} = 2\sqrt{17}$$

Because $d(A, B) + d(A, C) = d(B, C)$ or $3\sqrt{17} + 2\sqrt{17} = 5\sqrt{17}$, the points are collinear.

31. Label the points $A(0, 9)$, $B(-3, -7)$, and $C(2, 19)$.

$$d(A, B) = \sqrt{(-3-0)^2 + (-7-9)^2}$$
$$= \sqrt{(-3)^2 + (-16)^2} = \sqrt{9+256}$$
$$= \sqrt{265} \approx 16.279$$

$$d(B, C) = \sqrt{[2-(-3)]^2 + [19-(-7)]^2}$$
$$= \sqrt{5^2 + 26^2} = \sqrt{25+676}$$
$$= \sqrt{701} \approx 26.476$$

$$d(A, C) = \sqrt{(2-0)^2 + (19-9)^2}$$
$$= \sqrt{2^2 + 10^2} = \sqrt{4+100}$$
$$= \sqrt{104} \approx 10.198$$

Because $d(A, B) + d(A, C) \neq d(B, C)$

or $\quad \sqrt{265} + \sqrt{104} \neq \sqrt{701}$
$$16.279 + 10.198 \neq 26.476,$$
$$26.477 \neq 26.476,$$

the three given points are not collinear. (Note, however, that these points are very close to lying on a straight line and may appear to lie on a straight line when graphed.)

33. Label the points $A(-7, 4)$, $B(6, -2)$, and $C(-1, 1)$.

$$d(A, B) = \sqrt{\left[6 - (-7)\right]^2 + (-2 - 4)^2}$$
$$= \sqrt{13^2 + (-6)^2} = \sqrt{169 + 36}$$
$$= \sqrt{205} \approx 14.3178$$

$$d(B, C) = \sqrt{(-1 - 6)^2 + \left[1 - (-2)\right]^2}$$
$$= \sqrt{(-7)^2 + 3^2} = \sqrt{49 + 9}$$
$$= \sqrt{58} \approx 7.6158$$

$$d(A, C) = \sqrt{\left[-1 - (-7)\right]^2 + (1 - 4)^2}$$
$$= \sqrt{6^2 + (-3)^2} = \sqrt{36 + 9}$$
$$= \sqrt{45} \approx 6.7082$$

Because $d(B, C) + d(A, C) \neq d(A, B)$ or
$$\sqrt{58} + \sqrt{45} \neq \sqrt{205}$$
$$7.6158 + 6.7082 \neq 14.3178$$
$$14.3240 \neq 14.3178,$$

the three given points are not collinear. (Note, however, that these points are very close to lying on a straight line and may appear to lie on a straight line when graphed.)

35. Midpoint $(5, 8)$, endpoint $(13, 10)$

$$\frac{13 + x}{2} = 5 \quad \text{and} \quad \frac{10 + y}{2} = 8$$
$$13 + x = 10 \quad \text{and} \quad 10 + y = 16$$
$$x = -3 \quad \text{and} \quad y = 6.$$

The other endpoint has coordinates $(-3, 6)$.

37. Midpoint $(12, 6)$, endpoint $(19, 16)$

$$\frac{19 + x}{2} = 12 \quad \text{and} \quad \frac{16 + y}{2} = 6$$
$$19 + x = 24 \quad \text{and} \quad 16 + y = 12$$
$$x = 5 \quad \text{and} \quad y = -4.$$

The other endpoint has coordinates $(5, -4)$.

39. Midpoint (a, b), endpoint (p, q)

$$\frac{p + x}{2} = a \quad \text{and} \quad \frac{q + y}{2} = b$$
$$p + x = 2a \quad \text{and} \quad q + y = 2b$$
$$x = 2a - p \quad \text{and} \quad y = 2b - q$$

The other endpoint has coordinates $(2a - p, 2b - q)$.

41. The endpoints of the segment are $(1990, 21.3)$ and $(2012, 30.1)$.

$$M = \left(\frac{1990 + 2012}{2}, \frac{21.3 + 30.9}{2}\right)$$
$$= (2001, 26.1)$$

The estimate is 26.1%. This is very close to the actual figure of 26.2%.

43. The points to use are $(2011, 23021)$ and $(2013, 23834)$. Their midpoint is

$$\left(\frac{2011 + 2013}{2}, \frac{23{,}021 + 23{,}834}{2}\right)$$
$$= (2012, 23427.5).$$

In 2012, the poverty level cutoff was approximately $23,428.

45. The midpoint M has coordinates

$$\left(\frac{x_1 + x_2}{2}, \frac{y_1 + y_2}{2}\right).$$

$$d(P, M)$$
$$= \sqrt{\left(\frac{x_1 + x_2}{2} - x_1\right)^2 + \left(\frac{y_1 + y_2}{2} - y_1\right)^2}$$
$$= \sqrt{\left(\frac{x_1 + x_2}{2} - \frac{2x_1}{2}\right)^2 + \left(\frac{y_1 + y_2}{2} - \frac{2y_1}{2}\right)^2}$$
$$= \sqrt{\left(\frac{x_2 - x_1}{2}\right)^2 + \left(\frac{y_2 - y_1}{2}\right)^2}$$
$$= \sqrt{\frac{(x_2 - x_1)^2}{4} + \frac{(y_2 - y_1)^2}{4}}$$
$$= \sqrt{\frac{(x_2 - x_1)^2 + (y_2 - y_1)^2}{4}}$$
$$= \tfrac{1}{2}\sqrt{(x_2 - x_1)^2 + (y_2 - y_1)^2}$$

$$d(M, Q)$$
$$= \sqrt{\left(x_2 - \frac{x_1 + x_2}{2}\right)^2 + \left(y_2 - \frac{y_1 + y_2}{2}\right)^2}$$
$$= \sqrt{\left(\frac{2x_2}{2} - \frac{x_1 + x_2}{2}\right)^2 + \left(\frac{2y_2}{2} - \frac{y_1 + y_2}{2}\right)^2}$$
$$= \sqrt{\left(\frac{x_2 - x_1}{2}\right)^2 + \left(\frac{y_2 - y_1}{2}\right)^2}$$
$$= \sqrt{\frac{(x_2 - x_1)^2}{4} + \frac{(y_2 - y_1)^2}{4}}$$
$$= \sqrt{\frac{(x_2 - x_1)^2 + (y_2 - y_1)^2}{4}}$$
$$= \tfrac{1}{2}\sqrt{(x_2 - x_1)^2 + (y_2 - y_1)^2}$$

(continued on next page)

(*continued*)

$$d(P,Q) = \sqrt{(x_2 - x_1)^2 + (y_2 - y_1)^2}$$

Because $\frac{1}{2}\sqrt{(x_2 - x_1)^2 + (y_2 - y_1)^2}$

$$+ \frac{1}{2}\sqrt{(x_2 - x_1)^2 + (y_2 - y_1)^2}$$

$$= \sqrt{(x_2 - x_1)^2 + (y_2 - y_1)^2},$$

this shows $d(P,M) + d(M,Q) = d(P,Q)$ and
$d(P,M) = d(M,Q).$

In exercises 47–57, other ordered pairs are possible.

47. (a)

x	y	
0	−2	y-intercept:
		$x = 0 \Rightarrow$
		$y = \frac{1}{2}(0) - 2 = -2$
4	0	x-intercept:
		$y = 0 \Rightarrow$
		$0 = \frac{1}{2}x - 2 \Rightarrow$
		$2 = \frac{1}{2}x \Rightarrow 4 = x$
2	−1	additional point

(b)

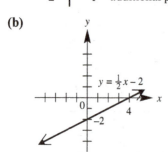

49. (a)

x	y	
0	$\frac{5}{3}$	y-intercept:
		$x = 0 \Rightarrow$
		$2(0) + 3y = 5 \Rightarrow$
		$3y = 5 \Rightarrow y = \frac{5}{3}$
$\frac{5}{2}$	0	x-intercept:
		$y = 0 \Rightarrow$
		$2x + 3(0) = 5 \Rightarrow$
		$2x = 5 \Rightarrow x = \frac{5}{2}$
4	−1	additional point

(b)

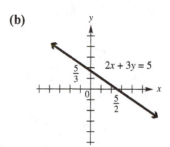
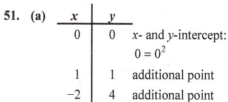

51. (a)

x	y	
0	0	x- and y-intercept:
		$0 = 0^2$
1	1	additional point
−2	4	additional point

(b)

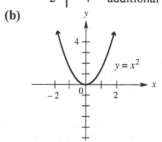

53. (a)

x	y	
3	0	x-intercept:
		$y = 0 \Rightarrow$
		$0 = \sqrt{x - 3} \Rightarrow$
		$0 = x - 3 \Rightarrow 3 = x$
4	1	additional point
7	2	additional point

no y-intercept:

$$x = 0 \Rightarrow y = \sqrt{0 - 3} \Rightarrow y = \sqrt{-3}$$

(b)

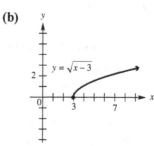

55. (a)

x	y	
0	2	y-intercept:
		$x = 0 \Rightarrow$
		$y = |0 - 2| \Rightarrow$
		$y = |-2| \Rightarrow y = 2$
2	0	x-intercept:
		$y = 0 \Rightarrow$
		$0 = |x - 2| \Rightarrow$
		$0 = x - 2 \Rightarrow 2 = x$
-2	4	additional point
4	2	additional point

(b)

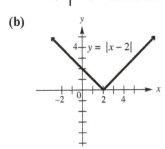

57. (a)

x	y	
0	0	x- and y-intercept:
		$0 = 0^3$
-1	-1	additional point
2	8	additional point

(b)

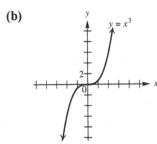

59. Points on the x-axis have y-coordinates equal to 0. The point on the x-axis will have the same x-coordinate as point (4, 3). Therefore, the line will intersect the x-axis at (4, 0).

61. Because (a, b) is in the second quadrant, a is negative and b is positive. Therefore, $(a, -b)$ will have a negative x–coordinate and a negative y-coordinate and will lie in quadrant III.
$(-a, b)$ will have a positive x-coordinate and a positive y-coordinate and will lie in quadrant I.
$(-a, -b)$ will have a positive x-coordinate and a negative y-coordinate and will lie in quadrant IV.
(b, a) will have a positive x-coordinate and a negative y-coordinate and will lie in quadrant IV.

63. To determine which points form sides of the quadrilateral (as opposed to diagonals), plot the points.

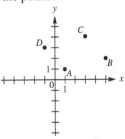

Use the distance formula to find the length of each side.

$$d(A, B) = \sqrt{(5 - 1)^2 + (2 - 1)^2}$$
$$= \sqrt{4^2 + 1^2} = \sqrt{16 + 1} = \sqrt{17}$$

$$d(B, C) = \sqrt{(3 - 5)^2 + (4 - 2)^2}$$
$$= \sqrt{(-2)^2 + 2^2} = \sqrt{4 + 4} = \sqrt{8}$$

$$d(C, D) = \sqrt{(-1 - 3)^2 + (3 - 4)^2}$$
$$= \sqrt{(-4)^2 + (-1)^2}$$
$$= \sqrt{16 + 1} = \sqrt{17}$$

$$d(D, A) = \sqrt{[1 - (-1)]^2 + (1 - 3)^2}$$
$$= \sqrt{2^2 + (-2)^2} = \sqrt{4 + 4} = \sqrt{8}$$

Because $d(A, B) = d(C, D)$ and $d(B, C) = d(D, A)$, the points are the vertices of a parallelogram. Because $d(A, B) \neq d(B, C)$, the points are not the vertices of a rhombus.

Section 2.2   Circles

1. The circle with equation $x^2 + y^2 = 49$ has center with coordinates (0, 0) and radius equal to 7.

3. The graph of $(x - 4)^2 + (y + 7)^2 = 9$ has center with coordinates (4, –7).

5. This circle has center (3, 2) and radius 5. This is graph B.

7. This circle has center (–3, 2) and radius 5. This is graph D.

9. The graph of $x^2 + y^2 = 0$ has center (0, 0) and radius 0. This is the point (0, 0). Therefore, there is one point on the graph.

11. (a) Center (0, 0), radius 6
$$\sqrt{(x - 0)^2 + (y - 0)^2} = 6$$
$$(x - 0)^2 + (y - 0)^2 = 6^2 \Rightarrow x^2 + y^2 = 36$$

(b)

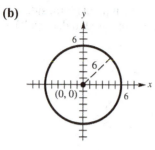

$$x^2 + y^2 = 36$$

13. (a) Center (2, 0), radius 6

$$\sqrt{(x-2)^2 + (y-0)^2} = 6$$
$$(x-2)^2 + (y-0)^2 = 6^2$$
$$(x-2)^2 + y^2 = 36$$

(b)

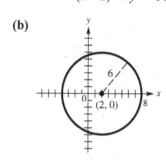

$$(x-2)^2 + y^2 = 36$$

15. (a) Center (0, 4), radius 4

$$\sqrt{(x-0)^2 + (y-4)^2} = 4$$
$$x^2 + (y-4)^2 = 16$$

(b)

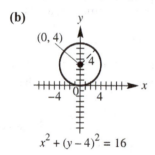

$$x^2 + (y-4)^2 = 16$$

17. (a) Center (–2, 5), radius 4

$$\sqrt{\left[x-(-2)\right]^2 + (y-5)^2} = 4$$
$$[x-(-2)]^2 + (y-5)^2 = 4^2$$
$$(x+2)^2 + (y-5)^2 = 16$$

(b)

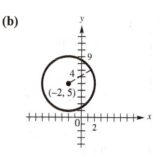

$$(x+2)^2 + (y-5)^2 = 16$$

19. (a) Center (5, –4), radius 7

$$\sqrt{(x-5)^2 + \left[y-(-4)\right]^2} = 7$$
$$(x-5)^2 + [y-(-4)]^2 = 7^2$$
$$(x-5)^2 + (y+4)^2 = 49$$

(b)

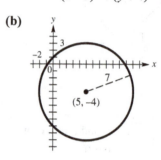

$$(x-5)^2 + (y+4)^2 = 49$$

21. (a) Center $\left(\sqrt{2}, \sqrt{2}\right)$, radius $\sqrt{2}$

$$\sqrt{\left(x-\sqrt{2}\right)^2 + \left(y-\sqrt{2}\right)^2} = \sqrt{2}$$
$$\left(x-\sqrt{2}\right)^2 + \left(y-\sqrt{2}\right)^2 = 2$$

(b)

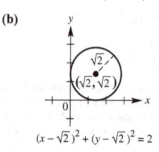

$$\left(x-\sqrt{2}\right)^2 + \left(y-\sqrt{2}\right)^2 = 2$$

23. (a) The center of the circle is located at the midpoint of the diameter determined by the points (1, 1) and (5, 1). Using the midpoint formula, we have

$$C = \left(\frac{1+5}{2}, \frac{1+1}{2}\right) = (3,1).$$ The radius is

one-half the length of the diameter:

$$r = \frac{1}{2}\sqrt{(5-1)^2 + (1-1)^2} = 2$$

The equation of the circle is

$$(x-3)^2 + (y-1)^2 = 4$$

(b) Expand $(x-3)^2 + (y-1)^2 = 4$ to find the equation of the circle in general form:

$$(x-3)^2 + (y-1)^2 = 4$$
$$x^2 - 6x + 9 + y^2 - 2y + 1 = 4$$
$$x^2 + y^2 - 6x - 2y + 6 = 0$$

25. (a) The center of the circle is located at the midpoint of the diameter determined by the points (–2, 4) and (–2, 0). Using the midpoint formula, we have

$$C = \left(\frac{-2 + (-2)}{2}, \frac{4+0}{2}\right) = (-2, 2).$$

The radius is one-half the length of the diameter:

$$r = \frac{1}{2}\sqrt{[-2 - (-2)]^2 + (4-0)^2} = 2$$

The equation of the circle is

$$(x+2)^2 + (y-2)^2 = 4$$

(b) Expand $(x+2)^2 + (y-2)^2 = 4$ to find the equation of the circle in general form:

$$(x+2)^2 + (y-2)^2 = 4$$
$$x^2 + 4x + 4 + y^2 - 4y + 4 = 4$$
$$x^2 + y^2 + 4x - 4y + 4 = 0$$

27. $x^2 + y^2 + 6x + 8y + 9 = 0$

Complete the square on x and y separately.

$$\left(x^2 + 6x\right) + \left(y^2 + 8y\right) = -9$$
$$\left(x^2 + 6x + 9\right) + \left(y^2 + 8y + 16\right) = -9 + 9 + 16$$
$$(x+3)^2 + (y+4)^2 = 16$$

Yes, it is a circle. The circle has its center at (–3, –4) and radius 4.

29. $x^2 + y^2 - 4x + 12y = -4$

Complete the square on x and y separately.

$$\left(x^2 - 4x\right) + \left(y^2 + 12y\right) = -4$$
$$\left(x^2 - 4x + 4\right) + \left(y^2 + 12y + 36\right) = -4 + 4 + 36$$
$$(x-2)^2 + (y+6)^2 = 36$$

Yes, it is a circle. The circle has its center at (2, –6) and radius 6.

31. $4x^2 + 4y^2 + 4x - 16y - 19 = 0$

Complete the square on x and y separately.

$$4\left(x^2 + x\right) + 4\left(y^2 - 4y\right) = 19$$
$$4\left(x^2 + x + \tfrac{1}{4}\right) + 4\left(y^2 - 4y + 4\right) =$$
$$19 + 4\left(\tfrac{1}{4}\right) + 4(4)$$

$$4\left(x + \tfrac{1}{2}\right)^2 + 4(y-2)^2 = 36$$
$$\left(x + \tfrac{1}{2}\right)^2 + (y-2)^2 = 9$$

Yes, it is a circle with center $\left(-\tfrac{1}{2}, 2\right)$ and radius 3.

33. $x^2 + y^2 + 2x - 6y + 14 = 0$

Complete the square on x and y separately.

$$\left(x^2 + 2x\right) + \left(y^2 - 6y\right) = -14$$
$$\left(x^2 + 2x + 1\right) + \left(y^2 - 6y + 9\right) = -14 + 1 + 9$$
$$(x+1)^2 + (y-3)^2 = -4$$

The graph is nonexistent.

35. $x^2 + y^2 - 6x - 6y + 18 = 0$

Complete the square on x and y separately.

$$\left(x^2 - 6x\right) + \left(y^2 - 6y\right) = -18$$
$$\left(x^2 - 6x + 9\right) + \left(y^2 - 6y + 9\right) = -18 + 9 + 9$$
$$(x-3)^2 + (y-3)^2 = 0$$

The graph is the point (3, 3).

37. $9x^2 + 9y^2 - 6x + 6y - 23 = 0$

Complete the square on x and y separately.

$$\left(9x^2 - 6x\right) + \left(9y^2 + 6y\right) = 23$$
$$9\left(x^2 - \tfrac{2}{3}x\right) + 9\left(y^2 + \tfrac{2}{3}y\right) = 23$$
$$\left(x^2 - \tfrac{2}{3}x + \tfrac{1}{9}\right) + \left(y^2 + \tfrac{2}{3}y + \tfrac{1}{9}\right) = \tfrac{23}{9} + \tfrac{1}{9} + \tfrac{1}{9}$$
$$\left(x - \tfrac{1}{3}\right)^2 + \left(y + \tfrac{1}{3}\right)^2 = \tfrac{25}{9} = \left(\tfrac{5}{3}\right)^2$$

Yes, it is a circle with center $\left(\tfrac{1}{3}, -\tfrac{1}{3}\right)$ and radius $\tfrac{5}{3}$.

39. The equations of the three circles are

$$(x-7)^2 + (y-4)^2 = 25,$$
$$(x+9)^2 + (y+4)^2 = 169, \text{ and}$$
$$(x+3)^2 + (y-9)^2 = 100.$$ From the graph of the three circles, it appears that the epicenter is located at (3, 1).

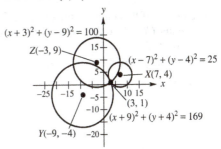

(continued on next page)

(*continued*)

Check algebraically:

$(x-7)^2 + (y-4)^2 = 25$

$(3-7)^2 + (1-4)^2 = 25$

$4^2 + 3^2 = 25 \Rightarrow 25 = 25$

$(x+9)^2 + (y+4)^2 = 169$

$(3+9)^2 + (1+4)^2 = 169$

$12^2 + 5^2 = 169 \Rightarrow 169 = 169$

$(x+3)^2 + (y-9)^2 = 100$

$(3+3)^2 + (1-9)^2 = 100$

$6^2 + (-8)^2 = 100 \Rightarrow 100 = 100$

(3, 1) satisfies all three equations, so the epicenter is at (3, 1).

41. From the graph of the three circles, it appears that the epicenter is located at $(-2, -2)$.

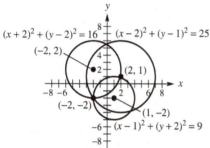

Check algebraically:

$(x-2)^2 + (y-1)^2 = 25$

$(-2-2)^2 + (-2-1)^2 = 25$

$(-4)^2 + (-3)^2 = 25$

$25 = 25$

$(x+2)^2 + (y-2)^2 = 16$

$(-2+2)^2 + (-2-2)^2 = 16$

$0^2 + (-4)^2 = 16$

$16 = 16$

$(x-1)^2 + (y+2)^2 = 9$

$(-2-1)^2 + (-2+2)^2 = 9$

$(-3)^2 + 0^2 = 9$

$9 = 9$

$(-2, -2)$ satisfies all three equations, so the epicenter is at $(-2, -2)$.

43. The radius of this circle is the distance from the center $C(3, 2)$ to the *x*-axis. This distance is 2, so $r = 2$.

$(x-3)^2 + (y-2)^2 = 2^2 \Rightarrow$

$(x-3)^2 + (y-2)^2 = 4$

45. Label the points $P(x, y)$ and $Q(1, 3)$.

If $d(P, Q) = 4$, $\sqrt{(1-x)^2 + (3-y)^2} = 4 \Rightarrow$

$(1-x)^2 + (3-y)^2 = 16$.

If $x = y$, then we can either substitute *x* for *y* or *y* for *x*. Substituting *x* for *y* we solve the following:

$(1-x)^2 + (3-x)^2 = 16$

$1 - 2x + x^2 + 9 - 6x + x^2 = 16$

$2x^2 - 8x + 10 = 16$

$2x^2 - 8x - 6 = 0$

$x^2 - 4x - 3 = 0$

To solve this equation, we can use the quadratic formula with $a = 1$, $b = -4$, and $c = -3$.

$x = \dfrac{-(-4) \pm \sqrt{(-4)^2 - 4(1)(-3)}}{2(1)}$

$= \dfrac{4 \pm \sqrt{16 + 12}}{2} = \dfrac{4 \pm \sqrt{28}}{2}$

$= \dfrac{4 \pm 2\sqrt{7}}{2} = 2 \pm \sqrt{7}$

Because $x = y$, the points are

$\left(2 + \sqrt{7}, 2 + \sqrt{7}\right)$ and $\left(2 - \sqrt{7}, 2 - \sqrt{7}\right)$.

47. Let $P(x, y)$ be a point whose distance from $A(1, 0)$ is $\sqrt{10}$ and whose distance from $B(5, 4)$ is $\sqrt{10}$. $d(P, A) = \sqrt{10}$, so

$\sqrt{(1-x)^2 + (0-y)^2} = \sqrt{10} \Rightarrow$

$(1-x)^2 + y^2 = 10$. $d(P, B) = \sqrt{10}$, so

$\sqrt{(5-x)^2 + (4-y)^2} = \sqrt{10} \Rightarrow$

$(5-x)^2 + (4-y)^2 = 10$. Thus,

$(1-x)^2 + y^2 = (5-x)^2 + (4-y)^2$

$1 - 2x + x^2 + y^2 =$

$\qquad 25 - 10x + x^2 + 16 - 8y + y^2$

$1 - 2x = 41 - 10x - 8y$

$8y = 40 - 8x$

$y = 5 - x$

Substitute $5 - x$ for *y* in the equation $(1-x)^2 + y^2 = 10$ and solve for *x*.

$(1-x)^2 + (5-x)^2 = 10 \Rightarrow$

$1 - 2x + x^2 + 25 - 10x + x^2 = 10$

$2x^2 - 12x + 26 = 10 \Rightarrow 2x^2 - 12x + 16 = 0$

$x^2 - 6x + 8 = 0 \Rightarrow (x-2)(x-4) = 0 \Rightarrow$

$x - 2 = 0 \quad \text{or} \quad x - 4 = 0$

$x = 2 \quad \text{or} \quad x = 4$

(*continued on next page*)

(*continued*)

To find the corresponding values of y use the equation $y = 5 - x$. If $x = 2$, then $y = 5 - 2 = 3$. If $x = 4$, then $y = 5 - 4 = 1$. The points satisfying the conditions are $(2, 3)$ and $(4, 1)$.

49. Label the points $A(3, y)$ and $B(-2, 9)$. If $d(A, B) = 12$, then

$$\sqrt{(-2-3)^2 + (9-y)^2} = 12$$
$$\sqrt{(-5)^2 + (9-y)^2} = 12$$
$$(-5)^2 + (9-y)^2 = 12^2$$
$$25 + 81 - 18y + y^2 = 144$$
$$y^2 - 18y - 38 = 0$$

Solve this equation by using the quadratic formula with $a = 1$, $b = -18$, and $c = -38$:

$$y = \frac{-(-18) \pm \sqrt{(-18)^2 - 4(1)(-38)}}{2(1)}$$
$$= \frac{18 \pm \sqrt{324 + 152}}{2(1)} = \frac{18 \pm \sqrt{476}}{2}$$
$$= \frac{18 \pm \sqrt{4(119)}}{2} = \frac{18 \pm 2\sqrt{119}}{2} = 9 \pm \sqrt{119}$$

The values of y are $9 + \sqrt{119}$ and $9 - \sqrt{119}$.

51. Let $P(x, y)$ be the point on the circle whose distance from the origin is the shortest. Complete the square on x and y separately to write the equation in center-radius form:

$$x^2 - 16x + y^2 - 14y + 88 = 0$$
$$x^2 - 16x + 64 + y^2 - 14y + 49 =$$
$$-88 + 64 + 49$$
$$(x - 8)^2 + (y - 7)^2 = 25$$

So, the center is $(8, 7)$ and the radius is 5.

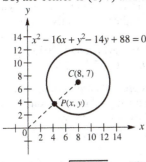

$d(C, O) = \sqrt{8^2 + 7^2} = \sqrt{113}$. Because the length of the radius is 5, $d(P, O) = \sqrt{113} - 5$.

53. The midpoint M has coordinates

$$\left(\frac{-1+5}{2}, \frac{3+(-9)}{2} \right) = \left(\frac{4}{2}, \frac{-6}{2} \right) = (2, -3).$$

55. Use points $C(2, -3)$ and $Q(5, -9)$.

$$d(C, Q) = \sqrt{(5-2)^2 + [-9-(-3)]^2}$$
$$= \sqrt{3^2 + (-6)^2} = \sqrt{9 + 36}$$
$$= \sqrt{45} = 3\sqrt{5}$$

The radius is $3\sqrt{5}$.

57. The center-radius form for this circle is

$$(x - 2)^2 + (y + 3)^2 = (3\sqrt{5})^2 \Rightarrow$$
$$(x - 2)^2 + (y + 3)^2 = 45.$$

59. Label the endpoints of the diameter $P(-1, 2)$ and $Q(11, 7)$. The midpoint M of the segment joining P and Q has coordinates

$$\left(\frac{-1+11}{2}, \frac{2+7}{2} \right) = \left(5, \frac{9}{2} \right).$$

The center is $C\left(5, \frac{9}{2}\right)$. To find the radius, we can use points $C\left(5, \frac{9}{2}\right)$ and $P(-1, 2)$.

$$d(C, P) = \sqrt{[5-(-1)]^2 + \left(\frac{9}{2}-2\right)^2}$$
$$= \sqrt{6^2 + \left(\frac{5}{2}\right)^2} = \sqrt{\frac{169}{4}} = \frac{13}{2}$$

We could also use points $C\left(5, \frac{9}{2}\right)$ and $Q(11, 7)$.

$$d(C, Q) = \sqrt{(5-11)^2 + \left(\frac{9}{2}-7\right)^2}$$
$$= \sqrt{(-6)^2 + \left(-\frac{5}{2}\right)^2} = \sqrt{\frac{169}{4}} = \frac{13}{2}$$

Using the points P and Q to find the length of the diameter, we have

$$d(P, Q) = \sqrt{(-1-11)^2 + (2-7)^2}$$
$$= \sqrt{(-12)^2 + (-5)^2}$$
$$= \sqrt{169} = 13$$

$$\frac{1}{2} d(P, Q) = \frac{1}{2}(13) = \frac{13}{2}$$

The center-radius form of the equation of the circle is

$$\left(x - 5\right)^2 + \left(y - \frac{9}{2}\right)^2 = \left(\frac{13}{2}\right)^2$$
$$\left(x - 5\right)^2 + \left(y - \frac{9}{2}\right)^2 = \frac{169}{4}$$

61. Label the endpoints of the diameter $P(1, 4)$ and $Q(5, 1)$. The midpoint M of the segment joining P and Q has coordinates

$$\left(\frac{1+5}{2}, \frac{4+1}{2} \right) = \left(3, \frac{5}{2} \right).$$

The center is $C\left(3, \frac{5}{2} \right)$.

The length of the diameter PQ is

$$\sqrt{(1-5)^2 + (4-1)^2} = \sqrt{(-4)^2 + 3^2} = \sqrt{25} = 5.$$

The length of the radius is $\frac{1}{2}(5) = \frac{5}{2}$.

The center-radius form of the equation of the circle is

$$(x-3)^2 + \left(y - \frac{5}{2} \right)^2 = \left(\frac{5}{2} \right)^2$$
$$(x-3)^2 + \left(y - \frac{5}{2} \right)^2 = \frac{25}{4}$$

Section 2.3 Functions

1. The domain of the relation $\{(3,5), (4,9), (10,13)\}$ is $\{3, 4, 10\}$.

3. The equation $y = 4x - 6$ defines a function with independent variable $\underline{x}$ and dependent variable $\underline{y}$.

5. For the function $f(x) = -4x + 2$,

$$f(-2) = -4(-2) + 2 = 8 + 2 = \underline{10}.$$

7. The function in Exercise 6, $g(x) = \sqrt{x}$, has domain $\underline{[0, \infty)}$.

For exercises 9 and 10, use this graph.

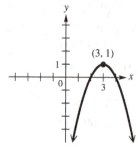

9. The largest open interval over which the function graphed here increases is $\underline{(-\infty, 3)}$.

11. The relation is a function because for each different x-value there is exactly one y-value. This correspondence can be shown as follows.

$\{5, 3, 4, 7\}$ x-values

$\{1, 2, 9, 8\}$ y-values

13. Two ordered pairs, namely (2, 4) and (2, 6), have the same x-value paired with different y-values, so the relation is not a function.

15. The relation is a function because for each different x-value there is exactly one y-value. This correspondence can be shown as follows.

$\{-3, 4, -2\}$ x-values

$\{1, 7\}$ y-values

17. The relation is a function because for each different x-value there is exactly one y-value. This correspondence can be shown as follows.

$\{3, 7, 10\}$ x-values

$\{-4\}$ y-values

19. Two sets of ordered pairs, namely (1, 1) and (1, −1) as well as (2, 4) and (2, −4), have the same x-value paired with different y-values, so the relation is not a function.
domain: $\{0, 1, 2\}$; range: $\{-4, -1, 0, 1, 4\}$

21. The relation is a function because for each different x-value there is exactly one y-value.
domain: $\{2, 3, 5, 11, 17\}$; range: $\{1, 7, 20\}$

23. The relation is a function because for each different x-value there is exactly one y-value. This correspondence can be shown as follows.

$\{0, -1, -2\}$ x-values

$\{0, \ 1, \ 2\}$ y-values
Domain: $\{0, -1, -2\}$; range: $\{0, 1, 2\}$

25. The relation is a function because for each different year, there is exactly one number for visitors.
domain: $\{2010, 2011, 2012, 2013\}$
range: $\{64.9, 63.0, 65.1, 63.5\}$

27. This graph represents a function. If you pass a vertical line through the graph, one x-value corresponds to only one y-value.
domain: $(-\infty, \infty)$; range: $(-\infty, \infty)$

29. This graph does not represent a function. If you pass a vertical line through the graph, there are places where one value of x corresponds to two values of y.
domain: $[3, \infty)$; range: $(-\infty, \infty)$

31. This graph represents a function. If you pass a vertical line through the graph, one x-value corresponds to only one y-value.

domain: $(-\infty, \infty)$; range: $(-\infty, \infty)$

33. $y = x^2$ represents a function because y is always found by squaring x. Thus, each value of x corresponds to just one value of y. x can be any real number. Because the square of any real number is not negative, the range would be zero or greater.

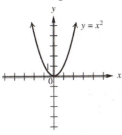

domain: $(-\infty, \infty)$; range: $[0, \infty)$

35. The ordered pairs $(1, 1)$ and $(1, -1)$ both satisfy $x = y^6$. This equation does not represent a function. Because x is equal to the sixth power of y, the values of x are nonnegative. Any real number can be raised to the sixth power, so the range of the relation is all real numbers.

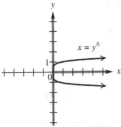

domain: $[0, \infty)$ range: $(-\infty, \infty)$

37. $y = 2x - 5$ represents a function because y is found by multiplying x by 2 and subtracting 5. Each value of x corresponds to just one value of y. x can be any real number, so the domain is all real numbers. Because y is twice x, less 5, y also may be any real number, and so the range is also all real numbers.

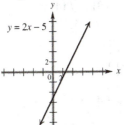

domain: $(-\infty, \infty)$; range: $(-\infty, \infty)$

39. By definition, y is a function of x if every value of x leads to exactly one value of y. Substituting a particular value of x, say 1, into $x + y < 3$ corresponds to many values of y. The ordered pairs $(1, -2)$, $(1, 1)$, $(1, 0)$, $(1, -1)$, and so on, all satisfy the inequality. Note that the points on the graphed line do not satisfy the inequality and only indicate the boundary of the solution set. This does not represent a function. Any number can be used for x or for y, so the domain and range of this relation are both all real numbers.

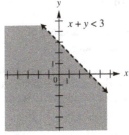

domain: $(-\infty, \infty)$; range: $(-\infty, \infty)$

41. For any choice of x in the domain of $y = \sqrt{x}$, there is exactly one corresponding value of y, so this equation defines a function. Because the quantity under the square root cannot be negative, we have $x \geq 0$. Because the radical is nonnegative, the range is also zero or greater.

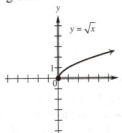

domain: $[0, \infty)$; range: $[0, \infty)$

43. Because $xy = 2$ can be rewritten as $y = \frac{2}{x}$, we can see that y can be found by dividing x into 2. This process produces one value of y for each value of x in the domain, so this equation is a function. The domain includes all real numbers except those that make the denominator equal to zero, namely $x = 0$. Values of y can be negative or positive, but never zero. Therefore, the range will be all real numbers except zero.

(continued on next page)

(*continued*)

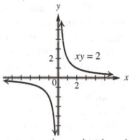

domain: $(-\infty, 0) \cup (0, \infty)$;

range: $(-\infty, 0) \cup (0, \infty)$

45. For any choice of x in the domain of $y = \sqrt{4x+1}$ there is exactly one corresponding value of y, so this equation defines a function. Because the quantity under the square root cannot be negative, we have $4x+1 \geq 0 \Rightarrow 4x \geq -1 \Rightarrow x \geq -\frac{1}{4}$. Because the radical is nonnegative, the range is also zero or greater.

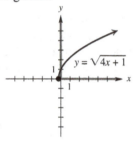

domain: $\left[-\frac{1}{4}, \infty\right)$; range: $[0, \infty)$

47. Given any value in the domain of $y = \frac{2}{x-3}$, we find y by subtracting 3, then dividing into 2. This process produces one value of y for each value of x in the domain, so this equation is a function. The domain includes all real numbers except those that make the denominator equal to zero, namely $x = 3$. Values of y can be negative or positive, but never zero. Therefore, the range will be all real numbers except zero.

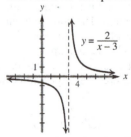

domain: $(-\infty, 3) \cup (3, \infty)$;

range: $(-\infty, 0) \cup (0, \infty)$

49. B. The notation $f(3)$ means the value of the dependent variable when the independent variable is 3.

51. $f(x) = -3x + 4$
$f(0) = -3 \cdot 0 + 4 = 0 + 4 = 4$

53. $g(x) = -x^2 + 4x + 1$
$g(-2) = -(-2)^2 + 4(-2) + 1$
$\quad\quad = -4 + (-8) + 1 = -11$

55. $f(x) = -3x + 4$
$f\left(\frac{1}{3}\right) = -3\left(\frac{1}{3}\right) + 4 = -1 + 4 = 3$

57. $g(x) = -x^2 + 4x + 1$
$g\left(\frac{1}{2}\right) = -\left(\frac{1}{2}\right)^2 + 4\left(\frac{1}{2}\right) + 1$
$\quad\quad = -\frac{1}{4} + 2 + 1 = \frac{11}{4}$

59. $f(x) = -3x + 4$
$f(p) = -3p + 4$

61. $f(x) = -3x + 4$
$f(-x) = -3(-x) + 4 = 3x + 4$

63. $f(x) = -3x + 4$
$f(x+2) = -3(x+2) + 4$
$\quad\quad = -3x - 6 + 4 = -3x - 2$

65. $f(x) = -3x + 4$
$f(2m-3) = -3(2m-3) + 4$
$\quad\quad = -6m + 9 + 4 = -6m + 13$

67. (a) $f(2) = 2$ (b) $f(-1) = 3$

69. (a) $f(2) = 15$ (b) $f(-1) = 10$

71. (a) $f(2) = 3$ (b) $f(-1) = -3$

73. (a) $f(-2) = 0$ (b) $f(0) = 4$
(c) $f(1) = 2$ (d) $f(4) = 4$

75. (a) $f(-2) = -3$ (b) $f(0) = -2$
(c) $f(1) = 0$ (d) $f(4) = 2$

77. (a) $x + 3y = 12$
$3y = -x + 12$
$y = \dfrac{-x + 12}{3}$
$y = -\frac{1}{3}x + 4 \Rightarrow f(x) = -\frac{1}{3}x + 4$

(b) $f(3) = -\frac{1}{3}(3) + 4 = -1 + 4 = 3$

79. (a) $y + 2x^2 = 3 - x$
$$y = -2x^2 - x + 3$$
$$f(x) = -2x^2 - x + 3$$

(b) $f(3) = -2(3)^2 - 3 + 3$
$$= -2 \cdot 9 - 3 + 3 = -18$$

81. (a) $4x - 3y = 8$
$$4x = 3y + 8$$
$$4x - 8 = 3y$$
$$\frac{4x - 8}{3} = y$$
$$y = \tfrac{4}{3}x - \tfrac{8}{3} \Rightarrow f(x) = \tfrac{4}{3}x - \tfrac{8}{3}$$

(b) $f(3) = \tfrac{4}{3}(3) - \tfrac{8}{3} = \tfrac{12}{3} - \tfrac{8}{3} = \tfrac{4}{3}$

83. $f(3) = 4$

85. $f(3)$ is the y-component of the coordinate, which is -4.

87. (a) $(-2, 0)$ **(b)** $(-\infty, -2)$

(c) $(0, \infty)$

89. (a) $(-\infty, -2); (2, \infty)$

(b) $(-2, -2)$ **(c)** none

91. (a) $(-1, 0); (1, \infty)$

(b) $(-\infty, -1); (0, 1)$

(c) none

93. (a) Yes, it is the graph of a function.

(b) $[0, 24]$

(c) When $t = 8$, $y = 1200$ from the graph. At 8 A.M., approximately 1200 megawatts is being used.

(d) The most electricity was used at 17 hr or 5 P.M. The least electricity was used at 4 A.M.

(e) $f(12) \approx 1900$

At 12 noon, electricity use is about 1900 megawatts.

(f) increasing from 4 A.M. to 5 P.M.; decreasing from midnight to 4 A.M. and from 5 P.M. to midnight

95. (a) At $t = 12$ and $t = 20$, $y = 55$ from the graph. Therefore, after about 12 noon until about 8 P.M. the temperature was over 55°.

(b) At $t = 6$ and $t = 22$, $y = 40$ from the graph. Therefore, until about 6 A.M. and after 10 P.M. the temperature was below 40°.

(c) The temperature at noon in Bratenahl, Ohio was 55°. Because the temperature in Greenville is 7° higher, we are looking for the time at which Bratenahl, Ohio was at or above 48°. This occurred at approximately 10 A.M and 8:30 P.M.

(d) The temperature is just below 40° from midnight to 6 A.M., when it begins to rise until it reaches a maximum of just below 65° at 4 P.M. It then begins to fall util it reaches just under 40° again at midnight.

Section 2.4 Linear Functions

1. B; $f(x) = 3x + 6$ is a linear function with y-intercept $(0, 6)$.

3. C; $f(x) = -8$ is a constant function.

5. A; $f(x) = 5x$ is a linear function whose graph passes through the origin, $(0, 0)$.
$f(0) = 5(0) = 0$.

7. $m = -3$ matches graph C because the line falls rapidly as x increases.

9. $m = 3$ matches graph D because the line rises rapidly as x increases.

11. $f(x) = x - 4$
Use the intercepts.
$f(0) = 0 - 4 = -4$: y-intercept
$0 = x - 4 \Rightarrow x = 4$: x-intercept
Graph the line through $(0, -4)$ and $(4, 0)$.

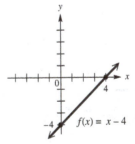

The domain and range are both $(-\infty, \infty)$.

13. $f(x) = \frac{1}{2}x - 6$

Use the intercepts.

$f(0) = \frac{1}{2}(0) - 6 = -6$: y-intercept

$0 = \frac{1}{2}x - 6 \Rightarrow 6 = \frac{1}{2}x \Rightarrow x = 12$: x-intercept

Graph the line through (0, −6) and (12, 0).

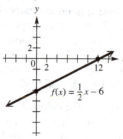

The domain and range are both $(-\infty, \infty)$.

15. $f(x) = 3x$

The x-intercept and the y-intercept are both zero. This gives us only one point, (0, 0). If $x = 1$, $y = 3(1) = 3$. Another point is (1, 3). Graph the line through (0, 0) and (1, 3).

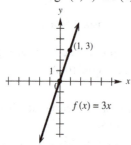

The domain and range are both $(-\infty, \infty)$.

17. $f(x) = -4$ is a constant function.

The graph of $f(x) = -4$ is a horizontal line with a y-intercept of −4.

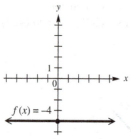

domain: $(-\infty, \infty)$; range: {−4}

19. $f(x) = 0$ is a constant function whose graph is the x-axis.

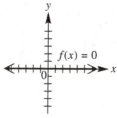

domain: $(-\infty, \infty)$; range: {0}

21. $-4x + 3y = 12$

Use the intercepts.

$-4(0) + 3y = 12 \Rightarrow 3y = 12 \Rightarrow$
$y = 4$: y-intercept

$-4x + 3(0) = 12 \Rightarrow -4x = 12 \Rightarrow$
$x = -3$: x-intercept

Graph the line through (0, 4) and (−3, 0).

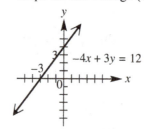

The domain and range are both $(-\infty, \infty)$.

23. $3y - 4x = 0$

Use the intercepts.

$3y - 4(0) = 0 \Rightarrow 3y = 0 \Rightarrow y = 0$: y-intercept

$3(0) - 4x = 0 \Rightarrow -4x = 0 \Rightarrow x = 0$: x-intercept

The graph has just one intercept. Choose an additional value, say 3, for x.

$3y - 4(3) = 0 \Rightarrow 3y - 12 = 0 \Rightarrow$
$3y = 12 \Rightarrow y = 4$

Graph the line through (0, 0) and (3, 4):

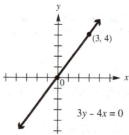

The domain and range are both $(-\infty, \infty)$.

25. $x = 3$ is a vertical line, intersecting the x-axis at (3, 0).

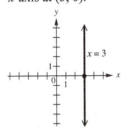

domain: {3}; range: $(-\infty, \infty)$

27. $2x + 4 = 0 \Rightarrow 2x = -4 \Rightarrow x = -2$ is a vertical line intersecting the x-axis at (–2, 0).

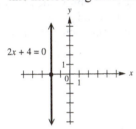

domain: {–2}; range: $(-\infty, \infty)$

29. $-x + 5 = 0 \Rightarrow x = 5$ is a vertical line intersecting the x-axis at (5, 0).

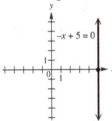

domain: {5}; range: $(-\infty, \infty)$

31. $y = 5$ is a horizontal line with y-intercept 5. Choice A resembles this.

33. $x = 5$ is a vertical line with x-intercept 5. Choice D resembles this.

35.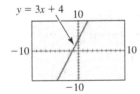

37.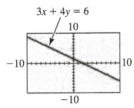

39. The rise is 2.5 feet while the run is 10 feet so the slope is $\frac{2.5}{10} = 0.25 = 25\% = \frac{1}{4}$. So A = 0.25, C = $\frac{2.5}{10}$, D = 25%, and E = $\frac{1}{4}$ are all expressions of the slope.

41. Through (2, –1) and (–3, –3)
Let $x_1 = 2$, $y_1 = -1$, $x_2 = -3$, and $y_2 = -3$.
Then rise = $\Delta y = -3 - (-1) = -2$ and run = $\Delta x = -3 - 2 = -5$.
The slope is $m = \dfrac{\text{rise}}{\text{run}} = \dfrac{\Delta y}{\Delta x} = \dfrac{-2}{-5} = \dfrac{2}{5}$.

43. Through (5, 8) and (3, 12)
Let $x_1 = 5$, $y_1 = 8$, $x_2 = 3$, and $y_2 = 12$.
Then rise = $\Delta y = 12 - 8 = 4$ and run = $\Delta x = 3 - 5 = -2$.
The slope is $m = \dfrac{\text{rise}}{\text{run}} = \dfrac{\Delta y}{\Delta x} = \dfrac{4}{-2} = -2$.

45. Through (5, 9) and (–2, 9)
$m = \dfrac{\Delta y}{\Delta x} = \dfrac{y_2 - y_1}{x_2 - x_1} = \dfrac{9 - 9}{-2 - 5} = \dfrac{0}{-7} = 0$

47. Horizontal, through (5, 1)
The slope of every horizontal line is zero, so $m = 0$.

49. Vertical, through (4, –7)
The slope of every vertical line is undefined; m is undefined.

51. (a) $y = 3x + 5$
Find two ordered pairs that are solutions to the equation.
If $x = 0$, then $y = 3(0) + 5 \Rightarrow y = 5$.
If $x = -1$, then
$y = 3(-1) + 5 \Rightarrow y = -3 + 5 \Rightarrow y = 2$.
Thus two ordered pairs are (0, 5) and (–1, 2)
$m = \dfrac{\text{rise}}{\text{run}} = \dfrac{y_2 - y_1}{x_2 - x_1} = \dfrac{2 - 5}{-1 - 0} = \dfrac{-3}{-1} = 3$.

(b)

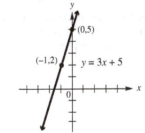

53. $2y = -3x$

Find two ordered pairs that are solutions to the equation. If $x = 0$, then $2y = 0 \Rightarrow y = 0$.

If $y = -3$, then $2(-3) = -3x \Rightarrow -6 = -3x \Rightarrow$ $x = 2$. Thus two ordered pairs are $(0, 0)$ and $(2, -3)$.

$$m = \frac{\text{rise}}{\text{run}} = \frac{y_2 - y_1}{x_2 - x_1} = \frac{-3 - 0}{2 - 0} = -\frac{3}{2}.$$

(b)

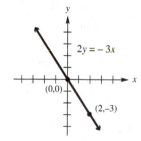

55. $5x - 2y = 10$

Find two ordered pairs that are solutions to the equation. If $x = 0$, then $5(0) - 2y = 10 \Rightarrow$ $y = -5$. If $y = 0$, then $5x - 2(0) = 10 \Rightarrow$ $5x = 10 \Rightarrow x = 2$.

Thus two ordered pairs are $(0, -5)$ and $(2, 0)$.

$$m = \frac{\text{rise}}{\text{run}} = \frac{y_2 - y_1}{x_2 - x_1} = \frac{0 - (-5)}{2 - 0} = \frac{5}{2}.$$

(b)

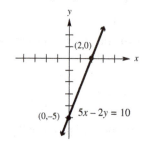

57. Through $(-1, 3)$, $m = \frac{3}{2}$

First locate the point $(-1, 3)$. Because the slope is $\frac{3}{2}$, a change of 2 units horizontally (2 units to the right) produces a change of 3 units vertically (3 units up). This gives a second point, $(1, 6)$, which can be used to complete the graph.

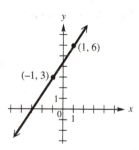

59. Through $(3, -4)$, $m = -\frac{1}{3}$. First locate the point $(3, -4)$. Because the slope is $-\frac{1}{3}$, a change of 3 units horizontally (3 units to the right) produces a change of -1 unit vertically (1 unit down). This gives a second point, $(6, -5)$, which can be used to complete the graph.

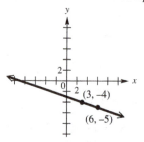

61. Through $\left(-\frac{1}{2}, 4\right)$, $m = 0$.

The graph is the horizontal line through $\left(-\frac{1}{2}, 4\right)$.

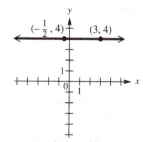

63. Through $\left(-\frac{5}{2}, 3\right)$, undefined slope. The slope is undefined, so the line is vertical, intersecting the x-axis at $\left(-\frac{5}{2}, 0\right)$.

65. The average rate of change is

$$m = \frac{f(b) - f(a)}{b - a}$$

$$\frac{20 - 4}{0 - 4} = \frac{-16}{4} = -\$4 \text{ (thousand) per year. The}$$

value of the machine is decreasing \$4000 each year during these years.

67. The graph is a horizontal line, so the average rate of change (slope) is 0. The percent of pay raise is not changing—it is 3% each year.

69. $m = \dfrac{f(b) - f(a)}{b - a} = \dfrac{2562 - 5085}{2012 - 1980} = \dfrac{-2523}{32}$

    $= -78.8$ thousand per year

The number of high school dropouts decreased by an average of 78.8 thousand per year from 1980 to 2012.

71. **(a)** The slope of –0.0167 indicates that the average rate of change of the winning time for the 5000 m run is 0.0167 min less. It is negative because the times are generally decreasing as time progresses.

(b) The Olympics were not held during World Wars I (1914–1919) and II (1939–1945).

(c) $y = -0.0167(2000) + 46.45 = 13.05 \text{ min}$

The model predicts a winning time of 13.05 minutes. The times differ by $13.35 - 13.05 = 0.30$ min.

73. $\dfrac{f(2013) - f(2008)}{2013 - 2008} = \dfrac{335,652 - 270,334}{2013 - 2008}$

$$= \frac{65,318}{5} = 13,063.6$$

The average annual rate of change from 2008 through 2013 is about 13,064 thousand.

75. **(a)** $m = \dfrac{f(b) - f(a)}{b - a} = \dfrac{56.3 - 138}{2013 - 2003}$

    $= \dfrac{-81.7}{10} = -8.17$

The average rate of change was –8.17 thousand mobile homes per year.

(b) The negative slope means that the number of mobile homes decreased by an average of 8.17 thousand each year from 2003 to 2013.

77. **(a)** $C(x) = 10x + 500$

(b) $R(x) = 35x$

(c) $\begin{aligned} P(x) &= R(x) - C(x) \\ &= 35x - (10x + 500) \\ &= 35x - 10x - 500 = 25x - 500 \end{aligned}$

(d) $\begin{aligned} C(x) &= R(x) \\ 10x + 500 &= 35x \\ 500 &= 25x \\ 20 &= x \end{aligned}$

20 units; do not produce

79. **(a)** $C(x) = 400x + 1650$

(b) $R(x) = 305x$

(c) $\begin{aligned} P(x) &= R(x) - C(x) \\ &= 305x - (400x + 1650) \\ &= 305x - 400x - 1650 \\ &= -95x - 1650 \end{aligned}$

(d) $\begin{aligned} C(x) &= R(x) \\ 400x + 1650 &= 305x \\ 95x + 1650 &= 0 \\ 95x &= -1650 \\ x &\approx -17.37 \text{ units} \end{aligned}$

This result indicates a negative "break-even point," but the number of units produced must be a positive number. A calculator graph of the lines

$y_1 = C(x) = 400x + 1650$ and

$y_2 = R(x) = 305x$ in the window

[0, 70] × [0, 20000] or solving the inequality $305x < 400x + 1650$ will show that $R(x) < C(x)$ for all positive values of x (in fact whenever x is greater than –17.4). Do not produce the product because it is impossible to make a profit.

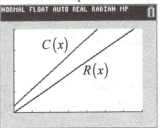

81. $C(x) = R(x) \Rightarrow 200x + 1000 = 240x \Rightarrow$
$1000 = 40x \Rightarrow 25 = x$
The break-even point is 25 units.
$C(25) = 200(25) + 1000 = \$6000$ which is the same as $R(25) = 240(25) = \$6000$

83. The first two points are $A(0, -6)$ and $B(1, -3)$.
$$m = \frac{-3 - (-6)}{1 - 0} = \frac{3}{1} = 3$$

85. If we use any two points on a line to find its slope, we find that the slope is <u>the same</u> in all cases.

87. The second and fourth points are $B(1, -3)$ and $D(3, 3)$.
$$d(B, D) = \sqrt{(3-1)^2 + [3 - (-3)]^2}$$
$$= \sqrt{2^2 + 6^2} = \sqrt{4 + 36}$$
$$= \sqrt{40} = 2\sqrt{10}$$

89. $\sqrt{10} + 2\sqrt{10} = 3\sqrt{10}$; The sum is $3\sqrt{10}$, which is equal to the answer in Exercise 88.

91. The midpoint of the segment joining $A(0, -6)$ and $G(6, 12)$ has coordinates $\left(\frac{0+6}{2}, \frac{-6+12}{2}\right) = \left(\frac{6}{2}, \frac{6}{2}\right) = (3,3)$. The midpoint is $M(3, 3)$, which is the same as the middle entry in the table.

Chapter 2 Quiz
(Sections 2.1–2.4)

1. $d(A, B) = \sqrt{(x_2 - x_1)^2 + (y_2 - y_1)^2}$
$$= \sqrt{(-8 - (-4))^2 + (-3 - 2)^2}$$
$$= \sqrt{(-4)^2 + (-5)^2} = \sqrt{16 + 25} = \sqrt{41}$$

3.

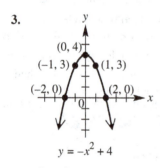

$$y = -x^2 + 4$$

5. $x^2 + y^2 - 4x + 8y + 3 = 0$

Complete the square on x and y separately.
$$(x^2 - 4x + 4) + (y^2 + 8y + 16) = -3 + 4 + 16 \Rightarrow$$
$$(x - 2)^2 + (y + 4)^2 = 17$$

The radius is $\sqrt{17}$ and the midpoint of the circle is $(2, -4)$.

7. Domain: $(-\infty, \infty)$; range: $[0, \infty)$

9. **(a)** $m = \frac{11 - 5}{5 - 1} = \frac{6}{4} = \frac{3}{2}$

(b) $m = \frac{4 - 4}{-1 - (-7)} = \frac{0}{6} = 0$

(c) $m = \frac{-4 - 12}{6 - 6} = \frac{-16}{0} \Rightarrow$ the slope is undefined.

Section 2.5 Equations of Lines and Linear Models

1. The graph of the line $y - 3 = 4(x - 8)$ has slope <u>4</u> and passes through the point (8, <u>3</u>).

3. The vertical line through the point (–4, 8) has equation <u>$x = -4$</u>.

5. Any line parallel to the graph of $6x + 7y = 0$ must have slope <u>$-\frac{6}{7}$</u>. Note that
$$6x + 7y = 0 \Rightarrow 7y = -6x \Rightarrow y = -\frac{6}{7}x$$

7. $y = \frac{1}{4}x + 2$ is graphed in D.

The slope is $\frac{1}{4}$ and the y-intercept is (0, 2).

9. $y - (-1) = \frac{3}{2}(x - 1)$ is graphed in C. The slope is $\frac{3}{2}$ and a point on the graph is (1, –1).

11. Through (1, 3), $m = -2$.
Write the equation in point-slope form.
$$y - y_1 = m(x - x_1) \Rightarrow y - 3 = -2(x - 1)$$
Then, change to standard form.
$$y - 3 = -2x + 2 \Rightarrow 2x + y = 5$$

13. Through (–5, 4), $m = -\frac{3}{2}$
Write the equation in point-slope form.
$$y - 4 = -\frac{3}{2}\left[x - (-5)\right]$$
Change to standard form.
$$2(y - 4) = -3(x + 5)$$
$$2y - 8 = -3x - 15$$
$$3x + 2y = -7$$

15. Through (–8, 4), undefined slope
Because undefined slope indicates a vertical line, the equation will have the form $x = a$. The equation of the line is $x = -8$.

17. Through (5, –8), $m = 0$
This is a horizontal line through (5, –8), so the equation is $y = -8$.

19. Through $(-1, 3)$ and $(3, 4)$
First find m.
$$m = \frac{4-3}{3-(-1)} = \frac{1}{4}$$
Use either point and the point-slope form.
$$y - 4 = \tfrac{1}{4}(x - 3)$$
$$4y - 16 = x - 3$$
$$-x + 4y = 13$$
$$x - 4y = -13$$

21. x-intercept $(3, 0)$, y-intercept $(0, -2)$
The line passes through $(3, 0)$ and $(0, -2)$. Use these points to find m.
$$m = \frac{-2-0}{0-3} = \frac{2}{3}$$
Using slope-intercept form we have
$$y = \tfrac{2}{3}x - 2.$$

23. Vertical, through $(-6, 4)$
The equation of a vertical line has an equation of the form $x = a$. Because the line passes through $(-6, 4)$, the equation is $x = -6$. (Because the slope of a vertical line is undefined, this equation cannot be written in slope-intercept form.)

25. Horizontal, through $(-7, 4)$
The equation of a horizontal line has an equation of the form $y = b$. Because the line passes through $(-7, 4)$, the equation is $y = 4$.

27. $m = 5$, $b = 15$
Using slope-intercept form, we have
$$y = 5x + 15.$$

29. Through $(-2, 5)$, slope $= -4$
$$y - 5 = -4(x - (-2))$$
$$y - 5 = -4(x + 2)$$
$$y - 5 = -4x - 8$$
$$y = -4x - 3$$

31. slope 0, y-intercept $\left(0, \tfrac{3}{2}\right)$
These represent $m = 0$ and $b = \tfrac{3}{2}$.
Using slope-intercept form, we have
$$y = (0)x + \tfrac{3}{2} \Rightarrow y = \tfrac{3}{2}.$$

33. The line $x + 2 = 0$ has x-intercept <u>(–2, 0)</u>. It <u>does not</u> have a y-intercept. The slope of his line is <u>undefined</u>. The line $4y = 2$ has y-intercept $\left(0, \tfrac{1}{2}\right)$. It <u>does not</u> have an x-intercept. The slope of this line is <u>0</u>.

35. $y = 3x - 1$
This equation is in the slope-intercept form, $y = mx + b$.
slope: 3;
y-intercept: $(0, -1)$

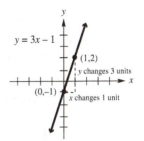

37. $4x - y = 7$
Solve for y to write the equation in slope-intercept form.
$$-y = -4x + 7 \Rightarrow y = 4x - 7$$
slope: 4; y-intercept: $(0, -7)$

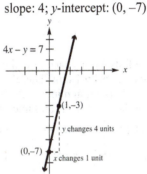

39. $4y = -3x \Rightarrow y = -\tfrac{3}{4}x$ or $y = -\tfrac{3}{4}x + 0$
slope: $-\tfrac{3}{4}$; y-intercept $(0, 0)$

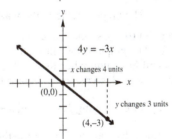

41. $x + 2y = -4$
Solve the equation for y to write the equation in slope-intercept form.
$$2y = -x - 4 \Rightarrow y = -\tfrac{1}{2}x - 2$$
slope: $-\tfrac{1}{2}$; y-intercept: $(0, -2)$

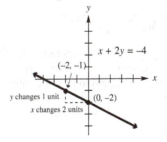

43. $y - \frac{3}{2}x - 1 = 0$

Solve the equation for y to write the equation in slope-intercept form.

$y - \frac{3}{2}x - 1 = 0 \Rightarrow y = \frac{3}{2}x + 1$

slope: $\frac{3}{2}$; y-intercept: (0, 1)

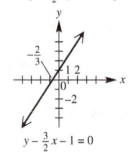

$y - \frac{3}{2}x - 1 = 0$

45. (a) The line falls 2 units each time the x value increases by 1 unit. Therefore the slope is −2. The graph intersects the y-axis at the point (0, 1) and intersects the x-axis at $\left(\frac{1}{2}, 0\right)$, so the y-intercept is (0, 1) and the x-intercept is $\left(\frac{1}{2}, 0\right)$.

(b) An equation defining f is $y = -2x + 1$.

47. (a) The line falls 1 unit each time the x value increases by 3 units. Therefore the slope is $-\frac{1}{3}$. The graph intersects the y-axis at the point (0, 2), so the y-intercept is (0, 2). The graph passes through (3, 1) and will fall 1 unit when the x value increases by 3, so the x-intercept is (6, 0).

(b) An equation defining f is $y = -\frac{1}{3}x + 2$.

49. (a) The line falls 200 units each time the x value increases by 1 unit. Therefore the slope is −200. The graph intersects the y-axis at the point (0, 300) and intersects the x-axis at $\left(\frac{3}{2}, 0\right)$, so the y-intercept is (0, 300) and the x-intercept is $\left(\frac{3}{2}, 0\right)$.

(b) An equation defining f is $y = -200x + 300$.

51. (a) through (−1, 4), parallel to $x + 3y = 5$
Find the slope of the line $x + 3y = 5$ by writing this equation in slope-intercept form.

$x + 3y = 5 \Rightarrow 3y = -x + 5 \Rightarrow$
$y = -\frac{1}{3}x + \frac{5}{3}$

The slope is $-\frac{1}{3}$.

Because the lines are parallel, $-\frac{1}{3}$ is also the slope of the line whose equation is to be found. Substitute $m = -\frac{1}{3}$, $x_1 = -1$, and $y_1 = 4$ into the point-slope form.

$y - y_1 = m(x - x_1)$
$y - 4 = -\frac{1}{3}[x - (-1)]$
$y - 4 = -\frac{1}{3}(x + 1)$
$3y - 12 = -x - 1 \Rightarrow x + 3y = 11$

(b) Solve for y.
$3y = -x + 11 \Rightarrow y = -\frac{1}{3}x + \frac{11}{3}$

53. (a) through (1, 6), perpendicular to $3x + 5y = 1$
Find the slope of the line $3x + 5y = 1$ by writing this equation in slope-intercept form.

$3x + 5y = 1 \Rightarrow 5y = -3x + 1 \Rightarrow$
$y = -\frac{3}{5}x + \frac{1}{5}$

This line has a slope of $-\frac{3}{5}$. The slope of any line perpendicular to this line is $\frac{5}{3}$, because $-\frac{3}{5}\left(\frac{5}{3}\right) = -1$. Substitute $m = \frac{5}{3}$, $x_1 = 1$, and $y_1 = 6$ into the point-slope form.

$y - 6 = \frac{5}{3}(x - 1)$
$3(y - 6) = 5(x - 1)$
$3y - 18 = 5x - 5$
$-13 = 5x - 3y$ or $5x - 3y = -13$

(b) Solve for y.
$3y = 5x + 13 \Rightarrow y = \frac{5}{3}x + \frac{13}{3}$

55. (a) through (4, 1), parallel to $y = -5$
Because $y = -5$ is a horizontal line, any line parallel to this line will be horizontal and have an equation of the form $y = b$. Because the line passes through (4, 1), the equation is $y = 1$.

(b) The slope-intercept form is $y = 1$.

57. (a) through (−5, 6), perpendicular to $x = -2$.
Because $x = -2$ is a vertical line, any line perpendicular to this line will be horizontal and have an equation of the form $y = b$. Because the line passes through (−5, 6), the equation is $y = 6$.

(b) The slope-intercept form is $y = 6$.

59. (a) Find the slope of the line $3y + 2x = 6$.

$3y + 2x = 6 \Rightarrow 3y = -2x + 6 \Rightarrow$

$y = -\frac{2}{3}x + 2$

Thus, $m = -\frac{2}{3}$. A line parallel to

$3y + 2x = 6$ also has slope $-\frac{2}{3}$.

Solve for k using the slope formula.

$$\frac{2 - (-1)}{k - 4} = -\frac{2}{3}$$

$$\frac{3}{k - 4} = -\frac{2}{3}$$

$$3(k - 4)\left(\frac{3}{k - 4}\right) = 3(k - 4)\left(-\frac{2}{3}\right)$$

$$9 = -2(k - 4)$$

$$9 = -2k + 8$$

$$2k = -1 \Rightarrow k = -\frac{1}{2}$$

(b) Find the slope of the line $2y - 5x = 1$.

$2y - 5x = 1 \Rightarrow 2y = 5x + 1 \Rightarrow$

$y = \frac{5}{2}x + \frac{1}{2}$

Thus, $m = \frac{5}{2}$. A line perpendicular to $2y$

$- 5x = 1$ will have slope $-\frac{2}{5}$, because

$\frac{5}{2}\left(-\frac{2}{5}\right) = -1$.

Solve this equation for k.

$$\frac{3}{k - 4} = -\frac{2}{5}$$

$$5(k - 4)\left(\frac{3}{k - 4}\right) = 5(k - 4)\left(-\frac{2}{5}\right)$$

$$15 = -2(k - 4)$$

$$15 = -2k + 8$$

$$2k = -7 \Rightarrow k = -\frac{7}{2}$$

61. (a) First find the slope using the points
$(0, 6312)$ and $(3, 7703)$.

$$m = \frac{7703 - 6312}{3 - 0} = \frac{1391}{3} \approx 463.67$$

The y-intercept is $(0, 6312)$, so the
equation of the line is
$y = 463.67x + 6312$.

(b) The value $x = 4$ corresponds to the year
2013.

$y = 463.67(4) + 6312 = 8166.68$

The model predicts that average tuition
and fees were $8166.68 in 2013. This is
$96.68 more than the actual amount.

63. (a) First find the slope using the points
$(0, 22036)$ and $(4, 24525)$.

$$m = \frac{24525 - 22036}{4 - 0} = \frac{2489}{4} = 622.25$$

The y-intercept is $(0, 22036)$, so the
equation of the line is
$y = 622.25x + 22{,}036$.

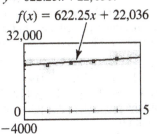

$f(x) = 622.25x + 22{,}036$

The slope of the line indicates that the
average tuition increase is about $622 per
year from 2009 through 2013.

(b) The year 2012 corresponds to $x = 3$.

$y = 622.25(3) + 22{,}036 = 23{,}902.75$

According to the model, average tuition
and fees were $23,903 in 2012. This is
$443 more than the actual amount
$23,460.

(c) Using the linear regression feature, the
equation of the line of best fit is
$y = 653x + 21{,}634$.

65. (a) The ordered pairs are $(0, 32)$ and
$(100, 212)$.

The slope is $m = \frac{212 - 32}{100 - 0} = \frac{180}{100} = \frac{9}{5}$.

Use $(x_1, y_1) = (0, 32)$ and $m = \frac{9}{5}$ in the

point-slope form.

$y - y_1 = m(x - x_1)$

$y - 32 = \frac{9}{5}(x - 0)$

$y - 32 = \frac{9}{5}x$

$y = \frac{9}{5}x + 32 \Rightarrow F = \frac{9}{5}C + 32$

(b) $F = \frac{9}{5}C + 32$

$5F = 9(C + 32)$

$5F = 9C + 160 \Rightarrow 9C = 5F - 160 \Rightarrow$

$9C = 5(F - 32) \Rightarrow C = \frac{5}{9}(F - 32)$

(c) $F = C \Rightarrow F = \frac{5}{9}(F - 32) \Rightarrow$
$9F = 5(F - 32) \Rightarrow 9F = 5F - 160 \Rightarrow$
$4F = -160 \Rightarrow F = -40$
$F = C$ when F is $-40°$.

67. (a) Because we want to find C as a function of I, use the points (12026, 10089) and (14167, 11484), where the first component represents the independent variable, I. First find the slope of the line.
$$m = \frac{11484 - 10089}{14167 - 12026} = \frac{1395}{2141} \approx 0.6516$$
Now use either point, say (12026, 10089), and the point-slope form to find the equation.
$$C - 10089 = 0.6516(I - 12026)$$
$$C - 10089 \approx 0.6516I - 7836$$
$$C \approx 0.6516I + 2253$$

(b) Because the slope is 0.6516, the marginal propensity to consume is 0.6516.

69. Write the equation as an equivalent equation with 0 on one side: $2x + 7 - x = 4x - 2 \Rightarrow$
$2x + 7 - x - 4x + 2 = 0$. Now graph
$y = 2x + 7 - x - 4x + 2$ in the window
$[-5, 5] \times [-5, 5]$ to find the x-intercept:

Solution set: $\{3\}$

71. Write the equation as an equivalent equation with 0 on one side: $3(2x + 1) - 2(x - 2) = 5 \Rightarrow$
$3(2x + 1) - 2(x - 2) - 5 = 0$. Now graph
$y = 3(2x + 1) - 2(x - 2) - 5$ in the window
$[-5, 5] \times [-5, 5]$ to find the x-intercept:

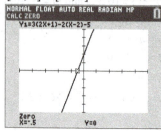

Solution set: $\left\{-\frac{1}{2}\right\}$ or $\{-0.5\}$

73. (a) $-2(x - 5) = -x - 2$
$-2x + 10 = -x - 2$
$10 = x - 2$
$12 = x$
Solution set: $\{12\}$

(b) Answers will vary. Sample answer: The solution does not appear in the standard viewing window x-interval $[10, -10]$. The minimum and maximum values must include 12.

75. $A(-1, 4)$, $B(-2, -1)$, $C(1, 14)$

For A and B, $m = \dfrac{-1 - 4}{-2 - (-1)} = \dfrac{-5}{-1} = 5$

For B and C, $m = \dfrac{14 - (-1)}{1 - (-2)} = \dfrac{15}{3} = 5$

For A and C, $m = \dfrac{14 - 4}{1 - (-1)} = \dfrac{10}{2} = 5$

Since all three slopes are the same, the points are collinear.

77. $A(-1, -3)$, $B(-5, 12)$, $C(1, -11)$

For A and B, $m = \dfrac{12 - (-3)}{-5 - (-1)} = -\dfrac{15}{4}$

For B and C, $m = \dfrac{-11 - 12}{1 - (-5)} = -\dfrac{23}{6}$

For A and C, $m = \dfrac{-11 - (-3)}{1 - (-1)} = -\dfrac{8}{2} = -4$

Since all three slopes are not the same, the points are not collinear.

79. $d(O, P) = \sqrt{(x_1 - 0)^2 + (m_1 x_1 - 0)^2}$
$\qquad = \sqrt{x_1^2 + m_1^2 x_1^2}$

81. $d(P, Q) = \sqrt{(x_2 - x_1)^2 + (m_2 x_2 - m_1 x_1)^2}$

83. $-2m_1 m_2 x_1 x_2 - 2x_1 x_2 = 0$
$\quad -2x_1 x_2(m_1 m_2 + 1) = 0$

85. If two nonvertical lines are perpendicular, then the product of the slopes of these lines is -1.

Summary Exercises on Graphs, Circles, Functions, and Equations

1. $P(3, 5)$, $Q(2, -3)$

(a) $d(P, Q) = \sqrt{(2 - 3)^2 + (-3 - 5)^2}$
$\qquad = \sqrt{(-1)^2 + (-8)^2}$
$\qquad = \sqrt{1 + 64} = \sqrt{65}$

(b) The midpoint M of the segment joining points P and Q has coordinates

$$\left(\frac{3+2}{2}, \frac{5+(-3)}{2}\right) = \left(\frac{5}{2}, \frac{2}{2}\right) = \left(\frac{5}{2}, 1\right).$$

(c) First find m: $m = \dfrac{-3-5}{2-3} = \dfrac{-8}{-1} = 8$

Use either point and the point-slope form.

$$y - 5 = 8(x - 3)$$

Change to slope-intercept form.

$$y - 5 = 8x - 24 \Rightarrow y = 8x - 19$$

3. $P(-2, 2), Q(3, 2)$

(a) $d(P, Q) = \sqrt{[3-(-2)]^2 + (2-2)^2}$
$$= \sqrt{5^2 + 0^2} = \sqrt{25+0} = \sqrt{25} = 5$$

(b) The midpoint M of the segment joining points P and Q has coordinates

$$\left(\frac{-2+3}{2}, \frac{2+2}{2}\right) = \left(\frac{1}{2}, \frac{4}{2}\right) = \left(\frac{1}{2}, 2\right).$$

(c) First find m: $m = \dfrac{2-2}{3-(-2)} = \dfrac{0}{5} = 0$

All lines that have a slope of 0 are horizontal lines. The equation of a horizontal line has an equation of the form $y = b$. Because the line passes through (3, 2), the equation is $y = 2$.

5. $P(5, -1), Q(5, 1)$

(a) $d(P, Q) = \sqrt{(5-5)^2 + [1-(-1)]^2}$
$$= \sqrt{0^2 + 2^2} = \sqrt{0+4} = \sqrt{4} = 2$$

(b) The midpoint M of the segment joining points P and Q has coordinates

$$\left(\frac{5+5}{2}, \frac{-1+1}{2}\right) = \left(\frac{10}{2}, \frac{0}{2}\right) = (5, 0).$$

(c) First find m.

$$m = \frac{1-(-1)}{5-5} = \frac{2}{0} = \text{undefined}$$

All lines that have an undefined slope are vertical lines. The equation of a vertical line has an equation of the form $x = a$. The line passes through (5, 1), so the equation is $x = 5$. (Because the slope of a vertical line is undefined, this equation cannot be written in slope-intercept form.)

7. $P\left(2\sqrt{3}, 3\sqrt{5}\right), Q\left(6\sqrt{3}, 3\sqrt{5}\right)$

(a) $d(P, Q) = \sqrt{\left(6\sqrt{3} - 2\sqrt{3}\right)^2 + \left(3\sqrt{5} - 3\sqrt{5}\right)^2}$
$$= \sqrt{\left(4\sqrt{3}\right)^2 + 0^2} = \sqrt{48} = 4\sqrt{3}$$

(b) The midpoint M of the segment joining points P and Q has coordinates

$$\left(\frac{2\sqrt{3} + 6\sqrt{3}}{2}, \frac{3\sqrt{5} + 3\sqrt{5}}{2}\right)$$
$$= \left(\frac{8\sqrt{3}}{2}, \frac{6\sqrt{5}}{2}\right) = \left(4\sqrt{3}, 3\sqrt{5}\right).$$

(c) First find m: $m = \dfrac{3\sqrt{5} - 3\sqrt{5}}{6\sqrt{3} - 2\sqrt{3}} = \dfrac{0}{4\sqrt{3}} = 0$

All lines that have a slope of 0 are horizontal lines. The equation of a horizontal line has an equation of the form $y = b$. Because the line passes through $\left(2\sqrt{3}, 3\sqrt{5}\right)$, the equation is

$$y = 3\sqrt{5}.$$

9. Through $(-2, 1)$ and $(4, -1)$

First find m: $m = \dfrac{-1-1}{4-(-2)} = \dfrac{-2}{6} = -\dfrac{1}{3}$

Use either point and the point-slope form.

$$y - (-1) = -\tfrac{1}{3}(x - 4)$$

Change to slope-intercept form.

$$3(y+1) = -(x-4) \Rightarrow 3y + 3 = -x + 4 \Rightarrow$$
$$3y = -x + 1 \Rightarrow y = -\tfrac{1}{3}x + \tfrac{1}{3}$$

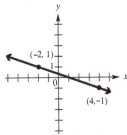

11. the circle with center $(2, -1)$ and radius 3

$$(x-2)^2 + [y-(-1)]^2 = 3^2$$
$$(x-2)^2 + (y+1)^2 = 9$$

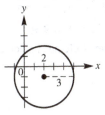

13. the line through $(3, -5)$ with slope $-\frac{5}{6}$

Write the equation in point-slope form.

$$y - (-5) = -\frac{5}{6}(x - 3)$$

Change to standard form.

$$6(y + 5) = -5(x - 3) \Rightarrow 6y + 30 = -5x + 15$$

$$6y = -5x - 15 \Rightarrow y = -\frac{5}{6}x - \frac{15}{6}$$

$$y = -\frac{5}{6}x - \frac{5}{2}$$

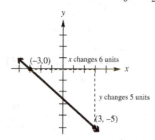

15. a line through $(-3, 2)$ and parallel to the line $2x + 3y = 6$

First, find the slope of the line $2x + 3y = 6$ by writing this equation in slope-intercept form.

$$2x + 3y = 6 \Rightarrow 3y = -2x + 6 \Rightarrow y = -\frac{2}{3}x + 2$$

The slope is $-\frac{2}{3}$. Because the lines are parallel, $-\frac{2}{3}$ is also the slope of the line whose equation is to be found. Substitute $m = -\frac{2}{3}$, $x_1 = -3$, and $y_1 = 2$ into the point-slope form.

$$y - y_1 = m(x - x_1) \Rightarrow y - 2 = -\frac{2}{3}\left[x - (-3)\right] \Rightarrow$$

$$3(y - 2) = -2(x + 3) \Rightarrow 3y - 6 = -2x - 6 \Rightarrow$$

$$3y = -2x \Rightarrow y = -\frac{2}{3}x$$

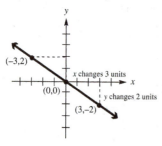

17. $x^2 - 4x + y^2 + 2y = 4$

Complete the square on x and y separately.

$$\left(x^2 - 4x\right) + \left(y^2 + 2y\right) = 4$$

$$\left(x^2 - 4x + 4\right) + \left(y^2 + 2y + 1\right) = 4 + 4 + 1$$

$$\left(x - 2\right)^2 + \left(y + 1\right)^2 = 9$$

Yes, it is a circle. The circle has its center at $(2, -1)$ and radius 3.

19. $x^2 - 12x + y^2 + 20 = 0$

Complete the square on x and y separately.

$$\left(x^2 - 12x\right) + y^2 = -20$$

$$\left(x^2 - 12x + 36\right) + y^2 = -20 + 36$$

$$\left(x - 6\right)^2 + y^2 = 16$$

Yes, it is a circle. The circle has its center at $(6, 0)$ and radius 4.

21. $x^2 - 2x + y^2 + 10 = 0$

Complete the square on x and y separately.

$$\left(x^2 - 2x\right) + y^2 = -10$$

$$\left(x^2 - 2x + 1\right) + y^2 = -10 + 1$$

$$\left(x - 1\right)^2 + y^2 = -9$$

No, it is not a circle.

23. The equation of the circle is

$$(x - 4)^2 + (y - 5)^2 = 4^2.$$

Let $y = 2$ and solve for x:

$$(x - 4)^2 + (2 - 5)^2 = 4^2 \Rightarrow$$

$$(x - 4)^2 + (-3)^2 = 4^2 \Rightarrow (x - 4)^2 = 7 \Rightarrow$$

$$x - 4 = \pm\sqrt{7} \Rightarrow x = 4 \pm \sqrt{7}$$

The points of intersection are $\left(4 + \sqrt{7}, 2\right)$ and $\left(4 - \sqrt{7}, 2\right)$

25. (a) The equation can be rewritten as

$$-4y = -x - 6 \Rightarrow y = \frac{1}{4}x + \frac{6}{4} \Rightarrow y = \frac{1}{4}x + \frac{3}{2}.$$

x can be any real number, so the domain is all real numbers and the range is also all real numbers.

domain: $(-\infty, \infty)$; range: $(-\infty, \infty)$

(b) Each value of x corresponds to just one value of y. $x - 4y = -6$ represents a function.

$$y = \frac{1}{4}x + \frac{3}{2} \Rightarrow f(x) = \frac{1}{4}x + \frac{3}{2}$$

$$f(-2) = \frac{1}{4}(-2) + \frac{3}{2} = -\frac{1}{2} + \frac{3}{2} = \frac{2}{2} = 1$$

27. (a) $(x+2)^2 + y^2 = 25$ is a circle centered at $(-2, 0)$ with a radius of 5. The domain will start 5 units to the left of –2 and end 5 units to the right of –2. The domain will be $[-2-5, 2+5] = [-7, 3]$. The range will start 5 units below 0 and end 5 units above 0. The range will be $[0-5, 0+5]$ $= [-5, 5]$.

(b) Because $(-2, 5)$ and $(-2, -5)$ both satisfy the relation, $(x+2)^2 + y^2 = 25$ does not represent a function.

Section 2.6   Graphs of Basic Functions

1. The equation $f(x) = x^2$ matches graph E.

The domain is $(-\infty, \infty)$.

3. The equation $f(x) = x^3$ matches graph A.

The range is $(-\infty, \infty)$.

5. Graph F is the graph of the identity function. Its equation is $f(x) = x$.

7. The equation $f(x) = \sqrt[3]{x}$ matches graph H.

No, there is no interval over which the function is decreasing.

9. The graph in B is discontinuous at many points. Assuming the graph continues, the range would be $\{..., -3, -2, -1, 0, 1, 2, 3, ...\}$.

11. The function is continuous over the entire domain of real numbers $(-\infty, \infty)$.

13. The function is continuous over the interval $[0, \infty)$.

15. The function has a point of discontinuity at $(3, 1)$. It is continuous over the interval $(-\infty, 3)$ and the interval $(3, \infty)$.

17. $f(x) = \begin{cases} 2x & \text{if } x \leq -1 \\ x-1 & \text{if } x > -1 \end{cases}$

(a) $f(-5) = 2(-5) = -10$

(b) $f(-1) = 2(-1) = -2$

(c) $f(0) = 0 - 1 = -1$

(d) $f(3) = 3 - 1 = 2$

19. $f(x) = \begin{cases} 2+x & \text{if } x < -4 \\ -x & \text{if } -4 \leq x \leq 2 \\ 3x & \text{if } x > 2 \end{cases}$

(a) $f(-5) = 2 + (-5) = -3$

(b) $f(-1) = -(-1) = 1$

(c) $f(0) = -0 = 0$

(d) $f(3) = 3 \cdot 3 = 9$

21. $f(x) = \begin{cases} x-1 & \text{if } x \leq 3 \\ 2 & \text{if } x > 3 \end{cases}$

Draw the graph of $y = x - 1$ to the left of $x = 3$, including the endpoint at $x = 3$. Draw the graph of $y = 2$ to the right of $x = 3$, and note that the endpoint at $x = 3$ coincides with the endpoint of the other ray.

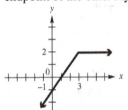

23. $f(x) = \begin{cases} 4-x & \text{if } x < 2 \\ 1+2x & \text{if } x \geq 2 \end{cases}$

Draw the graph of $y = 4 - x$ to the left of $x = 2$, but do not include the endpoint. Draw the graph of $y = 1 + 2x$ to the right of $x = 2$, including the endpoint.

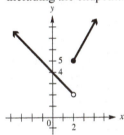

25. $f(x) = \begin{cases} -3 & \text{if } x \leq 1 \\ -1 & \text{if } x > 1 \end{cases}$

Graph the line $y = -3$ to the left of $x = 1$, including the endpoint. Draw $y = -1$ to the right of $x = 1$, but do not include the endpoint.

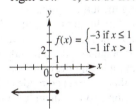

27. $f(x) = \begin{cases} 2+x & \text{if } x < -4 \\ -x & \text{if } -4 \le x \le 5 \\ 3x & \text{if } x > 5 \end{cases}$

Draw the graph of $y = 2 + x$ to the left of –4, but do not include the endpoint at $x = 4$. Draw the graph of $y = -x$ between –4 and 5, including both endpoints. Draw the graph of $y = 3x$ to the right of 5, but do not include the endpoint at $x = 5$.

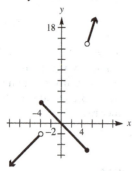

29. $f(x) = \begin{cases} -\frac{1}{2}x^2 + 2 & \text{if } x \le 2 \\ \frac{1}{2}x & \text{if } x > 2 \end{cases}$

Graph the curve $y = -\frac{1}{2}x^2 + 2$ to the left of $x = 2$, including the endpoint at (2, 0). Graph the line $y = \frac{1}{2}x$ to the right of $x = 2$, but do not include the endpoint at (2, 1). Notice that the endpoints of the pieces do not coincide.

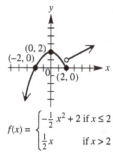

$f(x) = \begin{cases} -\frac{1}{2}x^2 + 2 & \text{if } x \le 2 \\ \frac{1}{2}x & \text{if } x > 2 \end{cases}$

31. $f(x) = \begin{cases} 2x & \text{if } -5 \le x < -1 \\ -2 & \text{if } -1 \le x < 0 \\ x^2 - 2 & \text{if } 0 \le x \le 2 \end{cases}$

Graph the line $y = 2x$ between $x = -5$ and $x = -1$, including the left endpoint at (–5, –10), but not including the right endpoint at (–1, –2). Graph the line $y = -2$ between $x = -1$ and $x = 0$, including the left endpoint at (–1, –2) and not including the right endpoint at (0, –2). Note that (–1, –2) coincides with the first two sections, so it is included. Graph the curve $y = x^2 - 2$ from $x = 0$ to $x = 2$, including the endpoints at (0, –2) and (2, 2).

Note that (0, –2) coincides with the second two sections, so it is included. The graph ends at $x = -5$ and $x = 2$.

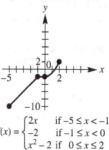

$f(x) = \begin{cases} 2x & \text{if } -5 \le x < -1 \\ -2 & \text{if } -1 \le x < 0 \\ x^2 - 2 & \text{if } 0 \le x \le 2 \end{cases}$

33. $f(x) = \begin{cases} x^3 + 3 & \text{if } -2 \le x \le 0 \\ x + 3 & \text{if } 0 < x < 1 \\ 4 + x - x^2 & \text{if } 1 \le x \le 3 \end{cases}$

Graph the curve $y = x^3 + 3$ between $x = -2$ and $x = 0$, including the endpoints at (–2, –5) and (0, 3). Graph the line $y = x + 3$ between $x = 0$ and $x = 1$, but do not include the endpoints at (0, 3) and (1, 4). Graph the curve $y = 4 + x - x^2$ from $x = 1$ to $x = 3$, including the endpoints at (1, 4) and (3, –2). The graph ends at $x = -2$ and $x = 3$.

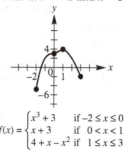

$f(x) = \begin{cases} x^3 + 3 & \text{if } -2 \le x \le 0 \\ x + 3 & \text{if } 0 < x < 1 \\ 4 + x - x^2 & \text{if } 1 \le x \le 3 \end{cases}$

35. The solid circle on the graph shows that the endpoint (0, –1) is part of the graph, while the open circle shows that the endpoint (0, 1) is not part of the graph. The graph is made up of parts of two horizontal lines. The function which fits this graph is

$f(x) = \begin{cases} -1 & \text{if } x \le 0 \\ 1 & \text{if } x > 0. \end{cases}$

domain: $(-\infty, \infty)$; range: $\{-1, 1\}$

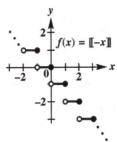

37. The graph is made up of parts of two horizontal lines. The solid circle shows that the endpoint (0, 2) of the one on the left belongs to the graph, while the open circle shows that the endpoint (0, −1) of the one on the right does not belong to the graph. The function that fits this graph is

$$f(x) = \begin{cases} 2 \text{ if } x \le 0 \\ -1 \text{ if } x > 1. \end{cases}$$

domain: $(-\infty, 0] \cup (1, \infty)$; range: $\{-1, 2\}$

39. For $x \le 0$, that piece of the graph goes through the points $(-1, -1)$ and $(0, 0)$. The slope is 1, so the equation of this piece is $y = x$. For $x > 0$, that piece of the graph is a horizontal line passing through $(2, 2)$, so its equation is $y = 2$. We can write the function as

$$f(x) = \begin{cases} x \text{ if } x \le 0 \\ 2 \text{ if } x > 0. \end{cases}$$

domain: $(-\infty, \infty)$ range: $(-\infty, 0] \cup \{2\}$

41. For $x < 1$, that piece of the graph is a curve passes through $(-8, -2)$, $(-1, -1)$ and $(1, 1)$, so the equation of this piece is $y = \sqrt[3]{x}$. The right piece of the graph passes through $(1, 2)$ and $(2, 3)$. $m = \dfrac{2-3}{1-2} = 1$, and the equation of the line is $y - 2 = x - 1 \Rightarrow y = x + 1$. We can write the function as $f(x) = \begin{cases} \sqrt[3]{x} \text{ if } x < 1 \\ x + 1 \text{ if } x \ge 1 \end{cases}$

domain: $(-\infty, \infty)$ range: $(-\infty, 1) \cup [2, \infty)$

43. $f(x) = [\![-x]\!]$

Plot points.

x	$-x$	$f(x) = [\![-x]\!]$
−2	2	2
−1.5	1.5	1
−1	1	1
−0.5	0.5	0
0	0	0
0.5	−0.5	−1
1	−1	−1
1.5	−1.5	−2
2	−2	−2

More generally, to get $y = 0$, we need
$0 \le -x < 1 \Rightarrow 0 \ge x > -1 \Rightarrow -1 < x \le 0$.
To get $y = 1$, we need $1 \le -x < 2 \Rightarrow$
$-1 \ge x > -2 \Rightarrow -2 < x \le -1$.
Follow this pattern to graph the step function.

domain: $(-\infty, \infty)$; range: $\{..., -2, -1, 0, 1, 2, ...\}$

45. $f(x) = [\![2x]\!]$

To get $y = 0$, we need $0 \le 2x < 1 \Rightarrow 0 \le x < \frac{1}{2}$.

To get $y = 1$, we need $1 \le 2x < 2 \Rightarrow \frac{1}{2} \le x < 1$.

To get $y = 2$, we need $2 \le 2x < 3 \Rightarrow 1 \le x < \frac{3}{2}$.

Follow this pattern to graph the step function.

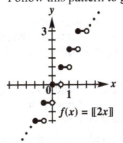

domain: $(-\infty, \infty)$; range: $\{..., -2, -1, 0, 1, 2, ...\}$

47. The cost of mailing a letter that weighs more than 1 ounce and less than 2 ounces is the same as the cost of a 2-ounce letter, and the cost of mailing a letter that weighs more than 2 ounces and less than 3 ounces is the same as the cost of a 3-ounce letter, etc.

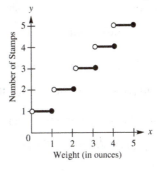

49.

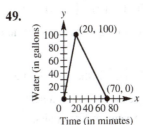

51. (a) For $0 \le x \le 8$, $m = \dfrac{49.8 - 34.2}{8 - 0} = 1.95$,

so $y = 1.95x + 34.2$. For $8 < x \le 13$,

$m = \dfrac{52.2 - 49.8}{13 - 8} = 0.48$, so the equation

is $y - 52.2 = 0.48(x - 13) \Rightarrow$

$y = 0.48x + 45.96$

(b) $f(x) = \begin{cases} 1.95x + 34.2 & \text{if } 0 \le x \le 8 \\ 0.48x + 45.96 & \text{if } 8 < x \le 13 \end{cases}$

53. (a) The initial amount is 50,000 gallons. The final amount is 30,000 gallons.

(b) The amount of water in the pool remained constant during the first and fourth days.

(c) $f(2) \approx 45,000$; $f(4) = 40,000$

(d) The slope of the segment between (1, 50000) and (3, 40000) is −5000, so the water was being drained at 5000 gallons per day.

55. (a) There is no charge for additional length, so we use the greatest integer function. The cost is based on multiples of two feet, so $f(x) = 0.8 \left[\!\left[\frac{x}{2} \right]\!\right]$ if $6 \le x \le 18$.

(b) $f(8.5) = 0.8 \left[\!\left[\frac{8.5}{2} \right]\!\right] = 0.8(4) = \$3.20$

$f(15.2) = 0.8 \left[\!\left[\frac{15.2}{2} \right]\!\right] = 0.8(7) = \$5.60$

Section 2.7   Graphing Techniques

1. To graph the function $f(x) = x^2 - 3$, shift the graph of $y = x^2$ down <u>3</u> units.

3. The graph of $f(x) = (x + 4)^2$ is obtained by shifting the graph of $y = x^2$ to the <u>left</u> 4 units.

5. The graph of $f(x) = -\sqrt{x}$ is a reflection of the graph of $f(x) = \sqrt{x}$ across the <u>x</u>-axis.

7. To obtain the graph of $f(x) = (x + 2)^3 - 3$, shift the graph of $y = x^3$ <u>2</u> units to the left and <u>3</u> units down.

9. The graph of $f(x) = |-x|$ is the same as the graph of $y = |x|$ because reflecting it across the <u>y</u>-axis yields the same ordered pairs.

11. (a) B; $y = (x - 7)^2$ is a shift of $y = x^2$, 7 units to the right.

(b) D; $y = x^2 - 7$ is a shift of $y = x^2$, 7 units downward.

(c) E; $y = 7x^2$ is a vertical stretch of $y = x^2$, by a factor of 7.

(d) A; $y = (x + 7)^2$ is a shift of $y = x^2$, 7 units to the left.

(e) C; $y = x^2 + 7$ is a shift of $y = x^2$, 7 units upward.

13. (a) B; $y = x^2 + 2$ is a shift of $y = x^2$, 2 units upward.

(b) A; $y = x^2 - 2$ is a shift of $y = x^2$, 2 units downward.

(c) G; $y = (x + 2)^2$ is a shift of $y = x^2$, 2 units to the left.

(d) C; $y = (x - 2)^2$ is a shift of $y = x^2$, 2 units to the right.

(e) F; $y = 2x^2$ is a vertical stretch of $y = x^2$, by a factor of 2.

(f) D; $y = -x^2$ is a reflection of $y = x^2$, across the x-axis.

(g) H; $y = (x - 2)^2 + 1$ is a shift of $y = x^2$, 2 units to the right and 1 unit upward.

(h) E; $y = (x + 2)^2 + 1$ is a shift of $y = x^2$, 2 units to the left and 1 unit upward.

(i) I; $y = (x + 2)^2 - 1$ is a shift of $y = x^2$, 2 units to the left and 1 unit down.

15. (a) F; $y = |x - 2|$ is a shift of $y = |x|$ 2 units to the right.

(b) C; $y = |x| - 2$ is a shift of $y = |x|$ 2 units downward.

(c) H; $y = |x| + 2$ is a shift of $y = |x|$ 2 units upward.

(d) D; $y = 2|x|$ is a vertical stretch of $y = |x|$ by a factor of 2.

(e) G; $y = -|x|$ is a reflection of
$y = |x|$ across the x-axis.

(f) A; $y = |-x|$ is a reflection of $y = |x|$
across the y-axis.

(g) E; $y = -2|x|$ is a reflection of $y = 2|x|$
across the x-axis. $y = 2|x|$ is a vertical
stretch of $y = |x|$ by a factor of 2.

(h) I; $y = |x-2|+2$ is a shift of $y = |x|$ 2
units to the right and 2 units upward.

(i) B; $y = |x+2|-2$ is a shift of $y = |x|$ 2
units to the left and 2 units downward.

17. $f(x) = 3|x|$

x	$h(x) = |x|$	$f(x) = 3|x|$
-2	2	6
-1	1	3
0	0	0
1	1	3
2	2	6

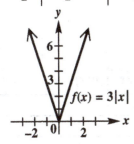

19. $f(x) = \frac{2}{3}|x|$

x	$h(x) = |x|$	$f(x) = \frac{2}{3}|x|$
-3	3	2
-2	2	$\frac{4}{3}$
-1	1	$\frac{2}{3}$
0	0	0
1	1	$\frac{2}{3}$
2	2	$\frac{4}{3}$
3	3	2

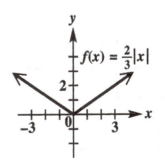

21. $f(x) = 2x^2$

x	$h(x) = x^2$	$f(x) = 2x^2$
-2	4	8
-1	1	2
0	0	0
1	1	2
2	4	8

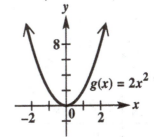

23. $f(x) = \frac{1}{2}x^2$

x	$h(x) = x^2$	$f(x) = \frac{1}{2}x^2$
-2	4	2
-1	1	$\frac{1}{2}$
0	0	0
1	1	$\frac{1}{2}$
2	4	2

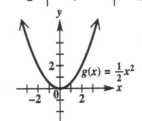

25. $f(x) = -\frac{1}{2}x^2$

x	$h(x) = x^2$	$f(x) = -\frac{1}{2}x^2$
-3	9	$-\frac{9}{2}$
-2	4	-2
-1	1	$-\frac{1}{2}$
0	0	0
1	1	$-\frac{1}{2}$
2	4	-2
3	9	$-\frac{9}{2}$

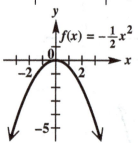

27. $f(x) = -3|x|$

x	$h(x) = |x|$	$f(x) = -3|x|$
-2	2	-6
-1	1	-3
0	0	0
1	1	-3
2	2	-6

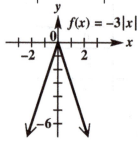

29. $h(x) = \left|-\frac{1}{2}x\right|$

x	$f(x) = |x|$	$h(x) = \left|-\frac{1}{2}x\right|$ $= \left|-\frac{1}{2}\right|\|x\| = \frac{1}{2}|x|$
-4	4	2
-3	3	$\frac{3}{2}$
-2	2	1
-1	1	$\frac{1}{2}$
0	0	0
1	1	$\frac{1}{2}$
2	2	1
3	3	$\frac{3}{2}$
4	4	2

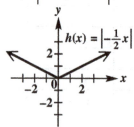

31. $h(x) = \sqrt{4x}$

x	$f(x) = \sqrt{x}$	$h(x) = \sqrt{4x} = 2\sqrt{x}$
0	0	0
1	1	2
2	$\sqrt{2}$	$2\sqrt{2}$
3	$\sqrt{3}$	$2\sqrt{3}$
4	2	4

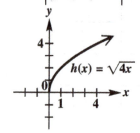

33. $f(x) = -\sqrt{-x}$

x	$h(x) = \sqrt{-x}$	$f(x) = -\sqrt{-x}$
-4	2	-2
-3	$\sqrt{3}$	$-\sqrt{3}$
-2	$\sqrt{2}$	$-\sqrt{2}$
-1	1	-1
0	0	0

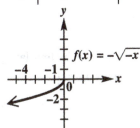

35. (a) $y = f(x+4)$ is a horizontal translation of f, 4 units to the left. The point that corresponds to (8, 12) on this translated function would be $(8-4, 12) = (4, 12)$.

(b) $y = f(x) + 4$ is a vertical translation of f, 4 units up. The point that corresponds to (8, 12) on this translated function would be $(8, 12+4) = (8, 16)$.

37. (a) $y = f(4x)$ is a horizontal shrinking of f, by a factor of 4. The point that corresponds to (8, 12) on this translated function is $\left(8 \cdot \frac{1}{4}, 12\right) = (2, 12)$.

(b) $y = f\left(\frac{1}{4}x\right)$ is a horizontal stretching of f, by a factor of 4. The point that corresponds to (8, 12) on this translated function is $(8 \cdot 4, 12) = (32, 12)$.

39. (a) The point that is symmetric to (5, –3) with respect to the x-axis is (5, 3).

(b) The point that is symmetric to (5, –3) with respect to the y-axis is (–5, –3).

(c) The point that is symmetric to (5, –3) with respect to the origin is (–5, 3).

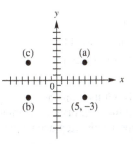

41. (a) The point that is symmetric to (–4, –2) with respect to the x-axis is (–4, 2).

(b) The point that is symmetric to (–4, –2) with respect to the y-axis is (4, –2).

(c) The point that is symmetric to (–4, –2) with respect to the origin is (4, 2).

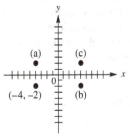

43. The graph of $y = |x - 2|$ is symmetric with respect to the line $x = 2$.

45. $y = x^2 + 5$

Replace x with $-x$ to obtain $y = (-x)^2 + 5 = x^2 + 5$. The result is the same as the original equation, so the graph is symmetric with respect to the y-axis. Because y is a function of x, the graph cannot be symmetric with respect to the x-axis. Replace x with $-x$ and y with $-y$ to obtain

$-y = (-x)^2 + 2 \Rightarrow -y = x^2 + 2 \Rightarrow y = -x^2 - 2.$

The result is not the same as the original equation, so the graph is not symmetric with respect to the origin. Therefore, the graph is symmetric with respect to the y-axis only.

47. $x^2 + y^2 = 12$

Replace x with $-x$ to obtain

$(-x)^2 + y^2 = 12 \Rightarrow x^2 + y^2 = 12$.

The result is the same as the original equation, so the graph is symmetric with respect to the y-axis. Replace y with $-y$ to obtain

$x^2 + (-y)^2 = 12 \Rightarrow x^2 + y^2 = 12$

The result is the same as the original equation, so the graph is symmetric with respect to the x-axis. Because the graph is symmetric with respect to the x-axis and y-axis, it is also symmetric with respect to the origin.

49. $y = -4x^3 + x$

Replace x with $-x$ to obtain

$y = -4(-x)^3 + (-x) \Rightarrow y = -4(-x^3) - x \Rightarrow$
$y = 4x^3 - x.$

The result is not the same as the original equation, so the graph is not symmetric with respect to the y-axis. Replace y with $-y$ to

obtain $-y = -4x^3 + x \Rightarrow y = 4x^3 - x.$

The result is not the same as the original equation, so the graph is not symmetric with respect to the x-axis. Replace x with $-x$ and y with $-y$ to obtain

$-y = -4(-x)^3 + (-x) \Rightarrow -y = -4(-x^3) - x \Rightarrow$
$-y = 4x^3 - x \Rightarrow y = -4x^3 + x.$

The result is the same as the original equation, so the graph is symmetric with respect to the origin. Therefore, the graph is symmetric with respect to the origin only.

51. $y = x^2 - x + 8$

Replace x with $-x$ to obtain

$y = (-x)^2 - (-x) + 8 \Rightarrow y = x^2 + x + 8.$

The result is not the same as the original equation, so the graph is not symmetric with respect to the y-axis. Because y is a function of x, the graph cannot be symmetric with respect to the x-axis. Replace x with $-x$ and y with $-y$ to obtain $-y = (-x)^2 - (-x) + 8 \Rightarrow$

$-y = x^2 + x + 8 \Rightarrow y = -x^2 - x - 8.$

The result is not the same as the original equation, so the graph is not symmetric with respect to the origin. Therefore, the graph has none of the listed symmetries.

53. $f(x) = -x^3 + 2x$

$f(-x) = -(-x)^3 + 2(-x)$
$\quad = x^3 - 2x = -(-x^3 + 2x) = -f(x)$

The function is odd.

55. $f(x) = 0.5x^4 - 2x^2 + 6$

$f(-x) = 0.5(-x)^4 - 2(-x)^2 + 6$
$\quad = 0.5x^4 - 2x^2 + 6 = f(x)$

The function is even.

57. $f(x) = x^3 - x + 9$

$f(x) = (-x)^3 - (-x) + 9$
$\quad = -x^3 + x + 9 = -(x^3 - x - 9) \neq -f(x)$

The function is neither.

59. $f(x) = x^2 - 1$

This graph may be obtained by translating the graph of $y = x^2$ 1 unit downward.

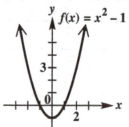

61. $f(x) = x^2 + 2$

This graph may be obtained by translating the graph of $y = x^2$ 2 units upward.

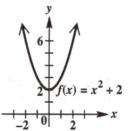

63. $g(x) = (x - 4)^2$

This graph may be obtained by translating the graph of $y = x^2$ 4 units to the right.

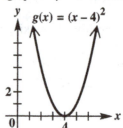

65. $g(x) = (x + 2)^2$

This graph may be obtained by translating the graph of $y = x^2$ 2 units to the left.

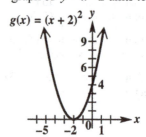

67. $g(x) = |x| - 1$

The graph is obtained by translating the graph of $y = |x|$ 1 unit downward.

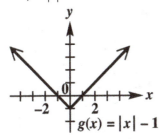

69. $h(x) = -(x + 1)^3$

This graph may be obtained by translating the graph of $y = x^3$ 1 unit to the left. It is then reflected across the x-axis.

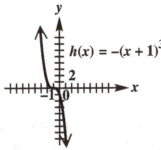

71. $h(x) = 2x^2 - 1$

This graph may be obtained by translating the graph of $y = x^2$ 1 unit down. It is then stretched vertically by a factor of 2.

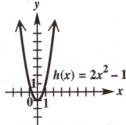

73. $f(x) = 2(x - 2)^2 - 4$

This graph may be obtained by translating the graph of $y = x^2$ 2 units to the right and 4 units down. It is then stretched vertically by a factor of 2.

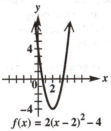

75. $f(x) = \sqrt{x + 2}$

This graph may be obtained by translating the graph of $y = \sqrt{x}$ two units to the left.

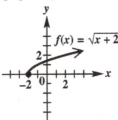

77. $f(x) = -\sqrt{x}$

This graph may be obtained by reflecting the graph of $y = \sqrt{x}$ across the x-axis.

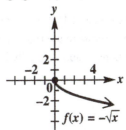

79. $f(x) = 2\sqrt{x} + 1$

This graph may be obtained by stretching the graph of $y = \sqrt{x}$ vertically by a factor of two and then translating the resulting graph one unit up.

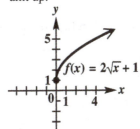

81. $g(x) = \frac{1}{2}x^3 - 4$

This graph may be obtained by stretching the graph of $y = x^3$ vertically by a factor of $\frac{1}{2}$, then shifting the resulting graph down four units.

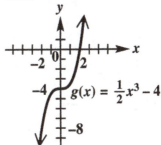

83. $g(x) = (x+3)^3$

This graph may be obtained by shifting the graph of $y = x^3$ three units left.

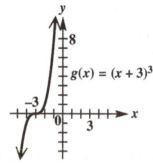

85. $f(x) = \frac{2}{3}(x-2)^2$

This graph may be obtained by translating the graph of $y = x^2$ two units to the right, then stretching the resulting graph vertically by a factor of $\frac{2}{3}$.

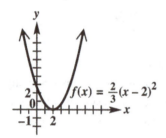

87. (a) $y = g(-x)$

The graph of $g(x)$ is reflected across the y-axis.

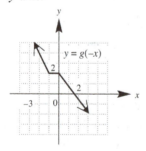

(b) $y = g(x-2)$

The graph of $g(x)$ is translated to the right 2 units.

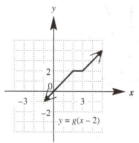

(c) $y = -g(x)$

The graph of $g(x)$ is reflected across the x-axis.

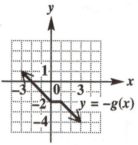

(d) $y = -g(x) + 2$

The graph of $g(x)$ is reflected across the x-axis and translated 2 units up.

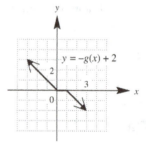

89. It is the graph of $f(x) = |x|$ translated 1 unit to the left, reflected across the x-axis, and translated 3 units up. The equation is $y = -|x+1| + 3$.

91. It is the graph of $f(x) = \sqrt{x}$ translated one unit right and then three units down. The equation is $y = \sqrt{x-1} - 3$.

93. It is the graph of $g(x) = \sqrt{x}$ translated 4 units to the left, stretched vertically by a factor of 2, and translated four units down. The equation is $y = 2\sqrt{x+4} - 4$.

95. Because $f(3) = 6$, the point $(3, 6)$ is on the graph. Because the graph is symmetric with respect to the origin, the point $(-3, -6)$ is on the graph. Therefore, $f(-3) = -6$.

97. Because $f(3) = 6$, the point $(3, 6)$ is on the graph. The graph is symmetric with respect to the line $x = 6$ and the point $(3, 6)$ is 3 units to the left of the line $x = 6$, so the image point of $(3, 6)$, 3 units to the right of the line $x = 6$ is $(9, 6)$. Therefore, $f(9) = 6$.

99. Because $(3, 6)$ is on the graph, $(-3, -6)$ must also be on the graph. Therefore, $f(-3) = -6$.

101. $f(x) = 2x + 5$

Translate the graph of $f(x)$ up 2 units to obtain the graph of
$$t(x) = (2x + 5) + 2 = 2x + 7.$$
Now translate the graph of $t(x) = 2x + 7$ left 3 units to obtain the graph of
$$g(x) = 2(x + 3) + 7 = 2x + 6 + 7 = 2x + 13.$$
(Note that if the original graph is first translated to the left 3 units and then up 2 units, the final result will be the same.)

103. (a) Because $f(-x) = f(x)$, the graph is symmetric with respect to the *y*-axis.

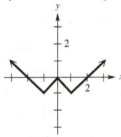

(b) Because $f(-x) = -f(x)$, the graph is symmetric with respect to the origin.

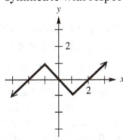

Chapter 2 Quiz
(Sections 2.5–2.7)

1. (a) First, find the slope: $m = \dfrac{9 - 5}{-1 - (-3)} = 2$

Choose either point, say, $(-3, 5)$, to find the equation of the line:
$$y - 5 = 2(x - (-3)) \Rightarrow y = 2(x + 3) + 5 \Rightarrow$$
$$y = 2x + 11.$$

(b) To find the *x*-intercept, let $y = 0$ and solve for *x*: $0 = 2x + 11 \Rightarrow x = -\frac{11}{2}$. The

x-intercept is $\left(-\frac{11}{2}, 0\right)$.

3. (a) $x = -8$ **(b)** $y = 5$

5. $f(x) = 0.40[\![x]\!] + 0.75$
$f(5.5) = 0.40[\![5.5]\!] + 0.75$
$\quad = 0.40(5) + 0.75 = 2.75$
A 5.5-minute call costs $2.75.

7. $f(x) = -x^3 + 1$

Reflect the graph of $f(x) = x^3$ across the *x*-axis, and then translate the resulting graph one unit up.

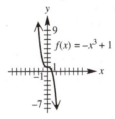

9. This is the graph of $g(x) = \sqrt{x}$, translated four units to the left, reflected across the *x*-axis, and then translated two units down. The equation is $y = -\sqrt{x + 4} - 2$.

Section 2.8 Function Operations and Composition

In exercises 1–10, $f(x) = x + 1$ and $g(x) = x^2$.

1. $(f + g)(2) = f(2) + g(2)$
$$= (2 + 1) + 2^2 = 7$$

3. $(fg)(2) = f(2) \cdot g(2)$
$$= (2 + 1) \cdot 2^2 = 12$$

5. $(f \circ g)(2) = f(g(2)) = f(2^2) = 2^2 + 1 = 5$

7. *f* is defined for all real numbers, so its domain is $(-\infty, \infty)$.

9. $f + g$ is defined for all real numbers, so its domain is $(-\infty, \infty)$.

In Exercises 11–17, $f(x) = x^2 + 3$ and $g(x) = -2x + 6$.

11. $(f + g)(3) = f(3) + g(3)$
$$= \left[(3)^2 + 3\right] + \left[-2(3) + 6\right]$$
$$= 12 + 0 = 12$$

13. $(f - g)(-1) = f(-1) - g(-1)$
$$= [(-1)^2 + 3] - [-2(-1) + 6]$$
$$= 4 - 8 = -4$$

15. $(fg)(4) = f(4) \cdot g(4)$
$$= [4^2 + 3] \cdot [-2(4) + 6]$$
$$= 19 \cdot (-2) = -38$$

17. $\left(\dfrac{f}{g}\right)(-1) = \dfrac{f(-1)}{g(-1)} = \dfrac{(-1)^2 + 3}{-2(-1) + 6} = \dfrac{4}{8} = \dfrac{1}{2}$

19. $f(x) = 3x + 4,\ g(x) = 2x - 5$

$(f + g)(x) = f(x) + g(x)$
$= (3x + 4) + (2x - 5) = 5x - 1$

$(f - g)(x) = f(x) - g(x)$
$= (3x + 4) - (2x - 5) = x + 9$

$(fg)(x) = f(x) \cdot g(x) = (3x + 4)(2x - 5)$
$= 6x^2 - 15x + 8x - 20$
$= 6x^2 - 7x - 20$

$\left(\dfrac{f}{g}\right)(x) = \dfrac{f(x)}{g(x)} = \dfrac{3x + 4}{2x - 5}$

The domains of both f and g are the set of all real numbers, so the domains of $f + g$, $f - g$, and fg are all $(-\infty, \infty)$. The domain of $\dfrac{f}{g}$ is the set of all real numbers for which $g(x) \neq 0$. This is the set of all real numbers except $\frac{5}{2}$, which is written in interval notation as $\left(-\infty, \frac{5}{2}\right) \cup \left(\frac{5}{2}, \infty\right)$.

21. $f(x) = 2x^2 - 3x,\ g(x) = x^2 - x + 3$

$(f + g)(x) = f(x) + g(x)$
$= (2x^2 - 3x) + (x^2 - x + 3)$
$= 3x^2 - 4x + 3$

$(f - g)(x) = f(x) - g(x)$
$= (2x^2 - 3x) - (x^2 - x + 3)$
$= 2x^2 - 3x - x^2 + x - 3$
$= x^2 - 2x - 3$

$(fg)(x) = f(x) \cdot g(x)$
$= (2x^2 - 3x)(x^2 - x + 3)$
$= 2x^4 - 2x^3 + 6x^2 - 3x^3 + 3x^2 - 9x$
$= 2x^4 - 5x^3 + 9x^2 - 9x$

$\left(\dfrac{f}{g}\right)(x) = \dfrac{f(x)}{g(x)} = \dfrac{2x^2 - 3x}{x^2 - x + 3}$

The domains of both f and g are the set of all real numbers, so the domains of $f + g$, $f - g$, and fg are all $(-\infty, \infty)$. The domain of $\dfrac{f}{g}$ is the set of all real numbers for which $g(x) \neq 0$. If $x^2 - x + 3 = 0$, then by the quadratic formula $x = \frac{1 \pm i\sqrt{11}}{2}$. The equation has no real solutions. There are no real numbers which make the denominator zero. Thus, the domain of $\dfrac{f}{g}$ is also $(-\infty, \infty)$.

23. $f(x) = \sqrt{4x - 1},\ g(x) = \dfrac{1}{x}$

$(f + g)(x) = f(x) + g(x) = \sqrt{4x - 1} + \dfrac{1}{x}$

$(f - g)(x) = f(x) - g(x) = \sqrt{4x - 1} - \dfrac{1}{x}$

$(fg)(x) = f(x) \cdot g(x)$
$= \sqrt{4x - 1}\left(\dfrac{1}{x}\right) = \dfrac{\sqrt{4x - 1}}{x}$

$\left(\dfrac{f}{g}\right)(x) = \dfrac{f(x)}{g(x)} = \dfrac{\sqrt{4x - 1}}{\frac{1}{x}} = x\sqrt{4x - 1}$

Because $4x - 1 \geq 0 \Rightarrow 4x \geq 1 \Rightarrow x \geq \frac{1}{4}$, the domain of f is $\left[\frac{1}{4}, \infty\right)$. The domain of g is $(-\infty, 0) \cup (0, \infty)$. Considering the intersection of the domains of f and g, the domains of $f + g$, $f - g$, and fg are all $\left[\frac{1}{4}, \infty\right)$. Because $\frac{1}{x} \neq 0$ for any value of x, the domain of $\dfrac{f}{g}$ is also $\left[\frac{1}{4}, \infty\right)$.

25. $M(2008) \approx 280$ and $F(2008) \approx 470$, thus
$T(2008) = M(2008) + F(2008)$
$= 280 + 470 = 750$ (thousand).

27. Looking at the graphs of the functions, the slopes of the line segments for the period 2008–2012 are much steeper than the slopes of the corresponding line segments for the period 2004–2008. Thus, the number of associate's degrees increased more rapidly during the period 2008–2012.

29. $(T - S)(2000) = T(2000) - S(2000)$
$= 19 - 13 = 6$
It represents the dollars in billions spent for general science in 2000.

31. Spending for space and other technologies spending decreased in the years 1995–2000 and 2010–2015.

33. **(a)** $(f + g)(2) = f(2) + g(2)$
$= 4 + (-2) = 2$

(b) $(f - g)(1) = f(1) - g(1) = 1 - (-3) = 4$

(c) $(fg)(0) = f(0) \cdot g(0) = 0(-4) = 0$

(d) $\left(\dfrac{f}{g}\right)(1) = \dfrac{f(1)}{g(1)} = \dfrac{1}{-3} = -\dfrac{1}{3}$

35. (a) $(f+g)(-1) = f(-1) + g(-1) = 0 + 3 = 3$

(b) $(f-g)(-2) = f(-2) - g(-2)$
$$= -1 - 4 = -5$$

(c) $(fg)(0) = f(0) \cdot g(0) = 1 \cdot 2 = 2$

(d) $\left(\dfrac{f}{g}\right)(2) = \dfrac{f(2)}{g(2)} = \dfrac{3}{0} = \text{undefined}$

37. (a) $(f+g)(2) = f(2) + g(2) = 7 + (-2) = 5$

(b) $(f-g)(4) = f(4) - g(4) = 10 - 5 = 5$

(c) $(fg)(-2) = f(-2) \cdot g(-2) = 0 \cdot 6 = 0$

(d) $\left(\dfrac{f}{g}\right)(0) = \dfrac{f(0)}{g(0)} = \dfrac{5}{0} = \text{undefined}$

39.

x	$f(x)$	$g(x)$	$(f+g)(x)$	$(f-g)(x)$	$(fg)(x)$	$\left(\dfrac{f}{g}\right)(x)$
-2	0	6	$0+6=6$	$0-6=-6$	$0 \cdot 6 = 0$	$\frac{0}{6} = 0$
0	5	0	$5+0=5$	$5-0=5$	$5 \cdot 0 = 0$	$\frac{5}{0} = \text{undefined}$
2	7	-2	$7+(-2)=5$	$7-(-2)=9$	$7(-2)=-14$	$\frac{7}{-2} = -3.5$
4	10	5	$10+5=15$	$10-5=5$	$10 \cdot 5 = 50$	$\frac{10}{5} = 2$

41. Answers may vary. Sample answer: Both the slope formula and the difference quotient represent the ratio of the vertical change to the horizontal change. The slope formula is stated for a line while the difference quotient is stated for a function f.

43. $f(x) = 2 - x$

(a) $f(x+h) = 2 - (x+h) = 2 - x - h$

(b) $f(x+h) - f(x) = (2-x-h) - (2-x)$
$$= 2 - x - h - 2 + x = -h$$

(c) $\dfrac{f(x+h) - f(x)}{h} = \dfrac{-h}{h} = -1$

45. $f(x) = 6x + 2$

(a) $f(x+h) = 6(x+h) + 2 = 6x + 6h + 2$

(b) $f(x+h) - f(x)$
$$= (6x+6h+2) - (6x+2)$$
$$= 6x + 6h + 2 - 6x - 2 = 6h$$

(c) $\dfrac{f(x+h) - f(x)}{h} = \dfrac{6h}{h} = 6$

47. $f(x) = -2x + 5$

(a) $f(x+h) = -2(x+h) + 5$
$$= -2x - 2h + 5$$

(b) $f(x+h) - f(x)$
$$= (-2x - 2h + 5) - (-2x + 5)$$
$$= -2x - 2h + 5 + 2x - 5 = -2h$$

(c) $\dfrac{f(x+h) - f(x)}{h} = \dfrac{-2h}{h} = -2$

49. $f(x) = \dfrac{1}{x}$

(a) $f(x+h) = \dfrac{1}{x+h}$

(b) $f(x+h) - f(x)$
$$= \dfrac{1}{x+h} - \dfrac{1}{x} = \dfrac{x - (x+h)}{x(x+h)}$$
$$= \dfrac{-h}{x(x+h)}$$

(c) $\dfrac{f(x+h) - f(x)}{h} = \dfrac{\frac{-h}{x(x+h)}}{h} = \dfrac{-h}{hx(x+h)}$
$$= -\dfrac{1}{x(x+h)}$$

51. $f(x) = x^2$

(a) $f(x+h) = (x+h)^2 = x^2 + 2xh + h^2$

(b) $f(x+h) - f(x) = x^2 + 2xh + h^2 - x^2$
$$= 2xh + h^2$$

(c) $\dfrac{f(x+h) - f(x)}{h} = \dfrac{2xh + h^2}{h}$
$$= \dfrac{h(2x+h)}{h}$$
$$= 2x + h$$

53. $f(x) = 1 - x^2$

    (a) $f(x+h) = 1 - (x+h)^2$
$$= 1 - (x^2 + 2xh + h^2)$$
$$= 1 - x^2 - 2xh - h^2$$

    (b) $f(x+h) - f(x)$
$$= (1 - x^2 - 2xh - h^2) - (1 - x^2)$$
$$= 1 - x^2 - 2xh - h^2 - 1 + x^2$$
$$= -2xh - h^2$$

    (c) $\dfrac{f(x+h) - f(x)}{h} = \dfrac{-2xh - h^2}{h}$
$$= \dfrac{h(-2x - h)}{h}$$
$$= -2x - h$$

55. $f(x) = x^2 + 3x + 1$

    (a) $f(x+h) = (x+h)^2 + 3(x+h) + 1$
$$= x^2 + 2xh + h^2 + 3x + 3h + 1$$

    (b) $f(x+h) - f(x)$
$$= \left(x^2 + 2xh + h^2 + 3x + 3h + 1\right)$$
$$- \left(x^2 + 3x + 1\right)$$
$$= x^2 + 2xh + h^2 + 3x + 3h + 1 - x^2 - 3x - 1$$
$$= 2xh + h^2 + 3h$$

    (c) $\dfrac{f(x+h) - f(x)}{h} = \dfrac{2xh + h^2 + 3h}{h}$
$$= \dfrac{h(2x + h + 3)}{h}$$
$$= 2x + h + 3$$

57. $g(x) = -x + 3 \Rightarrow g(4) = -4 + 3 = -1$
$$(f \circ g)(4) = f\left[g(4)\right] = f(-1)$$
$$= 2(-1) - 3 = -2 - 3 = -5$$

59. $g(x) = -x + 3 \Rightarrow g(-2) = -(-2) + 3 = 5$
$$(f \circ g)(-2) = f\left[g(-2)\right] = f(5)$$
$$= 2(5) - 3 = 10 - 3 = 7$$

61. $f(x) = 2x - 3 \Rightarrow f(0) = 2(0) - 3 = 0 - 3 = -3$
$$(g \circ f)(0) = g\left[f(0)\right] = g(-3)$$
$$= -(-3) + 3 = 3 + 3 = 6$$

63. $f(x) = 2x - 3 \Rightarrow f(2) = 2(2) - 3 = 4 - 3 = 1$
$$(f \circ f)(2) = f\left[f(2)\right] = f(1) = 2(1) - 3 = -1$$

65. $(f \circ g)(2) = f[g(2)] = f(3) = 1$

67. $(g \circ f)(3) = g[f(3)] = g(1) = 9$

69. $(f \circ f)(4) = f[f(4)] = f(3) = 1$

71. $(f \circ g)(1) = f[g(1)] = f(9)$
However, $f(9)$ cannot be determined from the table given.

73. **(a)** $(f \circ g)(x) = f(g(x)) = f(5x + 7)$
$$= -6(5x + 7) + 9$$
$$= -30x - 42 + 9 = -30x - 33$$
The domain and range of both f and g are $(-\infty, \infty)$, so the domain of $f \circ g$ is $(-\infty, \infty)$.

    (b) $(g \circ f)(x) = g(f(x)) = g(-6x + 9)$
$$= 5(-6x + 9) + 7$$
$$= -30x + 45 + 7 = -30x + 52$$
The domain of $g \circ f$ is $(-\infty, \infty)$.

75. **(a)** $(f \circ g)(x) = f(g(x)) = f(x + 3) = \sqrt{x + 3}$
The domain and range of g are $(-\infty, \infty)$, however, the domain and range of f are $[0, \infty)$. So, $x + 3 \geq 0 \Rightarrow x \geq -3$. Therefore, the domain of $f \circ g$ is $[-3, \infty)$.

    (b) $(g \circ f)(x) = g(f(x)) = g\left(\sqrt{x}\right) = \sqrt{x} + 3$
The domain and range of g are $(-\infty, \infty)$, however, the domain and range of f are $[0, \infty)$. Therefore, the domain of $g \circ f$ is $[0, \infty)$.

77. **(a)** $(f \circ g)(x) = f(g(x)) = f(x^2 + 3x - 1)$
$$= (x^2 + 3x - 1)^3$$
The domain and range of f and g are $(-\infty, \infty)$, so the domain of $f \circ g$ is $(-\infty, \infty)$.

    (b) $(g \circ f)(x) = g(f(x)) = g(x^3)$
$$= \left(x^3\right)^2 + 3\left(x^3\right) - 1$$
$$= x^6 + 3x^3 - 1$$
The domain and range of f and g are $(-\infty, \infty)$, so the domain of $g \circ f$ is $(-\infty, \infty)$.

79. (a) $(f \circ g)(x) = f(g(x)) = f(3x) = \sqrt{3x-1}$

The domain and range of g are $(-\infty, \infty)$, however, the domain of f is $[1, \infty)$, while the range of f is $[0, \infty)$. So,

$3x - 1 \geq 0 \Rightarrow x \geq \frac{1}{3}$. Therefore, the

domain of $f \circ g$ is $\left[\frac{1}{3}, \infty\right)$.

(b) $(g \circ f)(x) = g(f(x)) = g\left(\sqrt{x-1}\right)$

$= 3\sqrt{x-1}$

The domain and range of g are $(-\infty, \infty)$, however, the range of f is $[0, \infty)$. So $x - 1 \geq 0 \Rightarrow x \geq 1$. Therefore, the domain of $g \circ f$ is $[1, \infty)$.

81. (a) $(f \circ g)(x) = f(g(x)) = f(x+1) = \frac{2}{x+1}$

The domain and range of g are $(-\infty, \infty)$, however, the domain of f is $(-\infty, 0) \cup (0, \infty)$. So, $x + 1 \neq 0 \Rightarrow x \neq -1$. Therefore, the domain of $f \circ g$ is $(-\infty, -1) \cup (-1, \infty)$.

(b) $(g \circ f)(x) = g(f(x)) = g\left(\frac{2}{x}\right) = \frac{2}{x} + 1$

The domain and range of f is $(-\infty, 0) \cup (0, \infty)$, however, the domain and range of g are $(-\infty, \infty)$. So $x \neq 0$. Therefore, the domain of $g \circ f$ is $(-\infty, 0) \cup (0, \infty)$.

83. (a) $(f \circ g)(x) = f(g(x)) = f\left(-\frac{1}{x}\right) = \sqrt{-\frac{1}{x} + 2}$

The domain and range of g are $(-\infty, 0) \cup (0, \infty)$, however, the domain of f is $[-2, \infty)$. So, $-\frac{1}{x} + 2 \geq 0 \Rightarrow$

$x < 0$ or $x \geq \frac{1}{2}$ (using test intervals). Therefore, the domain of $f \circ g$ is

$(-\infty, 0) \cup \left[\frac{1}{2}, \infty\right)$.

(b) $(g \circ f)(x) = g(f(x)) = g\left(\sqrt{x+2}\right) = -\frac{1}{\sqrt{x+2}}$

The domain of f is $[-2, \infty)$ and its range is $[0, \infty)$. The domain and range of g are $(-\infty, 0) \cup (0, \infty)$. So $x + 2 > 0 \Rightarrow x > -2$. Therefore, the domain of $g \circ f$ is $(-2, \infty)$.

85. (a) $(f \circ g)(x) = f(g(x)) = f\left(\frac{1}{x+5}\right) = \sqrt{\frac{1}{x+5}}$

The domain of g is $(-\infty, -5) \cup (-5, \infty)$, and the range of g is $(-\infty, 0) \cup (0, \infty)$. The domain of f is $[0, \infty)$. Therefore, the domain of $f \circ g$ is $(-5, \infty)$.

(b) $(g \circ f)(x) = g(f(x)) = g\left(\sqrt{x}\right) = \frac{1}{\sqrt{x}+5}$

The domain and range of f is $[0, \infty)$. The domain of g is $(-\infty, -5) \cup (-5, \infty)$. Therefore, the domain of $g \circ f$ is $[0, \infty)$.

87. (a) $(f \circ g)(x) = f(g(x)) = f\left(\frac{1}{x}\right) = \frac{1}{1/x - 2} = \frac{x}{1-2x}$

The domain and range of g are $(-\infty, 0) \cup (0, \infty)$. The domain of f is $(-\infty, -2) \cup (-2, \infty)$, and the range of f is $(-\infty, 0) \cup (0, \infty)$. So, $\frac{x}{1-2x} < 0 \Rightarrow x < 0$ or

$0 < x < \frac{1}{2}$ or $x > \frac{1}{2}$ (using test intervals).

Thus, $x \neq 0$ and $x \neq \frac{1}{2}$. Therefore, the domain of $f \circ g$ is

$(-\infty, 0) \cup \left(0, \frac{1}{2}\right) \cup \left(\frac{1}{2}, \infty\right)$.

(b) $(g \circ f)(x) = g(f(x)) = g\left(\frac{1}{x-2}\right) = \frac{1}{1/(x-2)}$

$= x - 2$

The domain and range of g are $(-\infty, 0) \cup (0, \infty)$. The domain of f is $(-\infty, 2) \cup (2, \infty)$, and the range of f is $(-\infty, 0) \cup (0, \infty)$. Therefore, the domain of $g \circ f$ is $(-\infty, 2) \cup (2, \infty)$.

89. $g[f(2)] = g(1) = 2$ and $g[f(3)] = g(2) = 5$
Since $g[f(1)] = 7$ and $f(1) = 3$, $g(3) = 7$.

x	$f(x)$	$g(x)$	$g[f(x)]$
1	3	2	7
2	1	5	2
3	2	7	5

91. Answers will vary. In general, composition of functions is not commutative. Sample answer:

$(f \circ g)(x) = f(2x-3) = 3(2x-3) - 2$

$= 6x - 9 - 2 = 6x - 11$

$(g \circ f)(x) = g(3x-2) = 2(3x-2) - 3$

$= 6x - 4 - 3 = 6x - 7$

Thus, $(f \circ g)(x)$ is not equivalent to

$(g \circ f)(x)$.

93. $(f \circ g)(x) = f[g(x)] = 4\left[\frac{1}{4}(x-2)\right] + 2$

$\qquad = \left(4 \cdot \frac{1}{4}\right)(x-2) + 2$

$\qquad = (x-2) + 2 = x - 2 + 2 = x$

$(g \circ f)(x) = g[f(x)] = \frac{1}{4}[(4x+2) - 2]$

$\qquad = \frac{1}{4}(4x + 2 - 2) = \frac{1}{4}(4x) = x$

95. $(f \circ g)(x) = f[g(x)] = \sqrt[3]{5\left(\frac{1}{5}x^3 - \frac{4}{5}\right) + 4}$

$\qquad = \sqrt[3]{x^3 - 4 + 4} = \sqrt[3]{x^3} = x$

$(g \circ f)(x) = g[f(x)] = \frac{1}{5}\left(\sqrt[3]{5x + 4}\right)^3 - \frac{4}{5}$

$\qquad = \frac{1}{5}(5x + 4) - \frac{4}{5} = \frac{5x}{5} + \frac{4}{5} - \frac{4}{5}$

$\qquad = \frac{5x}{5} = x$

In Exercises 97–101, we give only one of many possible answers.

97. $h(x) = (6x - 2)^2$

Let $g(x) = 6x - 2$ and $f(x) = x^2$.

$(f \circ g)(x) = f(6x - 2) = (6x - 2)^2 = h(x)$

99. $h(x) = \sqrt{x^2 - 1}$

Let $g(x) = x^2 - 1$ and $f(x) = \sqrt{x}$.

$(f \circ g)(x) = f(x^2 - 1) = \sqrt{x^2 - 1} = h(x)$.

101. $h(x) = \sqrt{6x} + 12$

Let $g(x) = 6x$ and $f(x) = \sqrt{x} + 12$.

$(f \circ g)(x) = f(6x) = \sqrt{6x} + 12 = h(x)$

103. $f(x) = 12x, \ g(x) = 5280x$

$(f \circ g)(x) = f[g(x)] = f(5280x)$

$\qquad = 12(5280x) = 63{,}360x$

The function $f \circ g$ computes the number of inches in x miles.

105. $\mathcal{A}(x) = \frac{\sqrt{3}}{4}x^2$

(a) $\mathcal{A}(2x) = \frac{\sqrt{3}}{4}(2x)^2 = \frac{\sqrt{3}}{4}(4x^2) = \sqrt{3}x^2$

(b) $\mathcal{A}(16) = \mathcal{A}(2 \cdot 8) = \sqrt{3}(8)^2$

$\qquad = 64\sqrt{3}$ square units

107. (a) $r(t) = 4t$ and $\mathcal{A}(r) = \pi r^2$

$(\mathcal{A} \circ r)(t) = \mathcal{A}[r(t)]$

$\qquad = \mathcal{A}(4t) = \pi(4t)^2 = 16\pi t^2$

(b) $(\mathcal{A} \circ r)(t)$ defines the area of the leak in terms of the time t, in minutes.

(c) $\mathcal{A}(3) = 16\pi(3)^2 = 144\pi$ ft^2

109. Let $x =$ the number of people less than 100 people that attend.

(a) x people fewer than 100 attend, so $100 - x$ people do attend $N(x) = 100 - x$

(b) The cost per person starts at \$20 and increases by \$5 for each of the x people that do not attend. The total increase is \$5x, and the cost per person increases to \$20 + \$5x. Thus, $G(x) = 20 + 5x$.

(c) $C(x) = N(x) \cdot G(x) = (100 - x)(20 + 5x)$

(d) If 80 people attend, $x = 100 - 80 = 20$.

$C(20) = (100 - 20)[20 + 5(20)]$

$\qquad = (80)(20 + 100)$

$\qquad = (80)(120) = \$9600$

111. (a) $g(x) = \frac{1}{2}x$

(b) $f(x) = x + 1$

(c) $(f \circ g)(x) = f(g(x)) = f\left(\frac{1}{2}x\right) = \frac{1}{2}x + 1$

(d) $(f \circ g)(60) = \frac{1}{2}(60) + 1 = \$31$

Chapter 2   Review Exercises

1. $P(3, -1), Q(-4, 5)$

$d(P, Q) = \sqrt{(-4 - 3)^2 + [5 - (-1)]^2}$

$\qquad = \sqrt{(-7)^2 + 6^2} = \sqrt{49 + 36} = \sqrt{85}$

Midpoint:

$\left(\frac{3 + (-4)}{2}, \frac{-1 + 5}{2}\right) = \left(\frac{-1}{2}, \frac{4}{2}\right) = \left(-\frac{1}{2}, 2\right)$

3. $A(-6, 3), B(-6, 8)$

$d(A, B) = \sqrt{[-6 - (-6)]^2 + (8 - 3)^2}$

$\qquad = \sqrt{0 + 5^2} = \sqrt{25} = 5$

Midpoint:

$\left(\frac{-6 + (-6)}{2}, \frac{3 + 8}{2}\right) = \left(\frac{-12}{2}, \frac{11}{2}\right) = \left(-6, \frac{11}{2}\right)$

5. Let B have coordinates (x, y). Using the midpoint formula, we have

$$(8, 2) = \left(\frac{-6+x}{2}, \frac{10+y}{2}\right) \Rightarrow$$

$$\begin{array}{c|c} \dfrac{-6+x}{2} = 8 & \dfrac{10+y}{2} = 2 \\ -6 + x = 16 & 10 + y = 4 \\ x = 22 & y = -6 \end{array}$$

The coordinates of B are $(22, -6)$.

7. Center $(-2, 3)$, radius 15

$$(x-h)^2 + (y-k)^2 = r^2$$
$$[x-(-2)]^2 + (y-3)^2 = 15^2$$
$$(x+2)^2 + (y-3)^2 = 225$$

9. Center $(-8, 1)$, passing through $(0, 16)$
The radius is the distance from the center to any point on the circle. The distance between $(-8, 1)$ and $(0, 16)$ is

$$r = \sqrt{(0-(-8))^2 + (16-1)^2} = \sqrt{8^2 + 15^2}$$
$$= \sqrt{64 + 225} = \sqrt{289} = 17.$$

The equation of the circle is

$$[x-(-8)]^2 + (y-1)^2 = 17^2$$
$$(x+8)^2 + (y-1)^2 = 289$$

11. The center of the circle is $(0, 0)$. Use the distance formula to find the radius:

$$r^2 = (3-0)^2 + (5-0)^2 = 9 + 25 = 34$$

The equation is $x^2 + y^2 = 34$.

13. The center of the circle is $(0, 3)$. Use the distance formula to find the radius:

$$r^2 = (-2-0)^2 + (6-3)^2 = 4 + 9 = 13$$

The equation is $x^2 + (y-3)^2 = 13$.

15. $x^2 - 4x + y^2 + 6y + 12 = 0$

Complete the square on x and y to put the equation in center-radius form.

$$(x^2 - 4x) + (y^2 + 6y) = -12$$
$$(x^2 - 4x + 4) + (y^2 + 6y + 9) = -12 + 4 + 9$$
$$(x-2)^2 + (y+3)^2 = 1$$

The circle has center $(2, -3)$ and radius 1.

17.
$$2x^2 + 14x + 2y^2 + 6y + 2 = 0$$
$$x^2 + 7x + y^2 + 3y + 1 = 0$$
$$(x^2 + 7x) + (y^2 + 3y) = -1$$
$$\left(x^2 + 7x + \tfrac{49}{4}\right) + \left(y^2 + 3y + \tfrac{9}{4}\right) = -1 + \tfrac{49}{4} + \tfrac{9}{4}$$
$$\left(x + \tfrac{7}{2}\right)^2 + \left(y + \tfrac{3}{2}\right)^2 = -\tfrac{4}{4} + \tfrac{49}{4} + \tfrac{9}{4}$$
$$\left(x + \tfrac{7}{2}\right)^2 + \left(y + \tfrac{3}{2}\right)^2 = \tfrac{54}{4}$$

The circle has center $\left(-\tfrac{7}{2}, -\tfrac{3}{2}\right)$ and radius

$$\sqrt{\tfrac{54}{4}} = \tfrac{\sqrt{54}}{\sqrt{4}} = \tfrac{\sqrt{9 \cdot 6}}{\sqrt{4}} = \tfrac{3\sqrt{6}}{2}.$$

19. This is not the graph of a function because a vertical line can intersect it in two points. domain: $[-6, 6]$; range: $[-6, 6]$

21. This is not the graph of a function because a vertical line can intersect it in two points. domain: $(-\infty, \infty)$; range: $(-\infty, -1] \cup [1, \infty)$

23. This is not the graph of a function because a vertical line can intersect it in two points. domain: $[0, \infty)$; range: $(-\infty, \infty)$

25. $y = 6 - x^2$

Each value of x corresponds to exactly one value of y, so this equation defines a function.

27. The equation $y = \pm\sqrt{x-2}$ does not define y as a function of x. For some values of x, there will be more than one value of y. For example, ordered pairs $(3, 1)$ and $(3, -1)$ satisfy the relation.

29. In the function $f(x) = -4 + |x|$, we may use any real number for x. The domain is $(-\infty, \infty)$.

31. $f(x) = \sqrt{6-3x}$

In the function $y = \sqrt{6-3x}$, we must have $6 - 3x \geq 0$.
$$6 - 3x \geq 0 \Rightarrow 6 \geq 3x \Rightarrow 2 \geq x \Rightarrow x \leq 2$$
Thus, the domain is $(-\infty, 2]$.

In exercises 33–35, $f(x) = -2x^2 + 3x - 6$.

33. $f(3) = -2 \cdot 3^2 + 3 \cdot 3 - 6$
$$= -2 \cdot 9 + 3 \cdot 3 - 6$$
$$= -18 + 9 - 6 = -15$$

35. $f(0) = -2(0)^2 + 3(0) - 6 = -6$

37. $2x - 5y = 5 \Rightarrow -5y = -2x + 5 \Rightarrow y = \frac{2}{5}x - 1$

The graph is the line with slope $\frac{2}{5}$ and

y-intercept $(0, -)1$. It may also be graphed using intercepts. To do this, locate the x-intercept: $y = 0$

$2x - 5(0) = 5 \Rightarrow 2x = 5 \Rightarrow x = \frac{5}{2}$

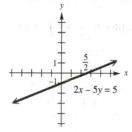

39. $2x + 5y = 20 \Rightarrow 5y = -2x + 20 \Rightarrow y = -\frac{2}{5}x + 4$

The graph is the line with slope of $-\frac{2}{5}$ and

y-intercept $(0, 4)$. It may also be graphed using intercepts. To do this, locate the x-intercept: x-intercept: $y = 0$

$2x + 5(0) = 20 \Rightarrow 2x = 20 \Rightarrow x = 10$

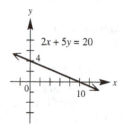

41. $f(x) = x$

The graph is the line with slope 1 and y-intercept $(0, 0)$, which means that it passes through the origin. Use another point such as $(1, 1)$ to complete the graph.

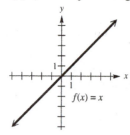

43. $x = -5$

The graph is the vertical line through $(-5, 0)$.

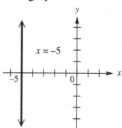

45. $y + 2 = 0 \Rightarrow y = -2$

The graph is the horizontal line through $(0, -2)$.

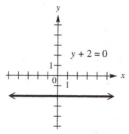

47. Line through $(0, 5)$, $m = -\frac{2}{3}$

Note that $m = -\frac{2}{3} = \frac{-2}{3}$.

Begin by locating the point $(0, 5)$. Because the slope is $\frac{-2}{3}$, a change of 3 units horizontally (3 units to the right) produces a change of -2 units vertically (2 units down). This gives a second point, $(3, 3)$, which can be used to complete the graph.

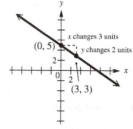

49. through $(2, -2)$ and $(3, -4)$

$m = \dfrac{y_2 - y_1}{x_2 - x_1} = \dfrac{-4 - (-2)}{3 - 2} = \dfrac{-2}{1} = -2$

51. through $(0, -7)$ and $(3, -7)$

$m = \dfrac{-7 - (-7)}{3 - 0} = \dfrac{0}{3} = 0$

53. $11x + 2y = 3$

Solve for y to put the equation in slope-intercept form.

$2y = -11x + 3 \Rightarrow y = -\frac{11}{2}x + \frac{3}{2}$

Thus, the slope is $-\frac{11}{2}$.

55. $x - 2 = 0 \Rightarrow x = 2$
The graph is a vertical line, through $(2, 0)$. The slope is undefined.

57. Initially, the car is at home. After traveling for 30 mph for 1 hr, the car is 30 mi away from home. During the second hour the car travels 20 mph until it is 50 mi away. During the third hour the car travels toward home at 30 mph until it is 20 mi away. During the fourth hour the car travels away from home at 40 mph until it is 60 mi away from home. During the last hour, the car travels 60 mi at 60 mph until it arrived home.

59. (a) We need to first find the slope of a line that passes between points $(0, 30.7)$ and $(12, 82.9)$

$$m = \frac{y_2 - y_1}{x_2 - x_1} = \frac{82.9 - 30.7}{12 - 0} = \frac{52.2}{12} = 4.35$$

Now use the point-intercept form with $b = 30.7$ and $m = 4.35$.
$y = 4.35x + 30.7$
The slope, 4.35, indicates that the number of e-filing taxpayers increased by 4.35% each year from 2001 to 2013.

(b) For 2009, we evaluate the function for $x = 8$. $y = 4.35(8) + 30.7 = 65.5$
65.5% of the tax returns are predicted to have been filed electronically.

61. (a) through $(3, -5)$ with slope -2
Use the point-slope form.
$$y - y_1 = m(x - x_1)$$
$$y - (-5) = -2(x - 3)$$
$$y + 5 = -2(x - 3)$$
$$y + 5 = -2x + 6$$
$$y = -2x + 1$$

(b) Standard form: $y = -2x + 1 \Rightarrow 2x + y = 1$

63. (a) through $(2, -1)$ parallel to $3x - y = 1$
Find the slope of $3x - y = 1$.
$3x - y = 1 \Rightarrow -y = -3x + 1 \Rightarrow y = 3x - 1$
The slope of this line is 3. Because parallel lines have the same slope, 3 is also the slope of the line whose equation is to be found. Now use the point-slope form with $(x_1, y_1) = (2, -1)$ and $m = 3$.
$$y - y_1 = m(x - x_1)$$
$$y - (-1) = 3(x - 2)$$
$$y + 1 = 3x - 6 \Rightarrow y = 3x - 7$$

(b) Standard form:
$y = 3x - 7 \Rightarrow -3x + y = -7 \Rightarrow 3x - y = 7$

65. (a) through $(2, -10)$, perpendicular to a line with an undefined slope
A line with an undefined slope is a vertical line. Any line perpendicular to a vertical line is a horizontal line, with an equation of the form $y = b$. The line passes through $(2, -10)$, so the equation of the line is $y = -10$.

(b) Standard form: $y = -10$

67. (a) through $(-7, 4)$, perpendicular to $y = 8$
The line $y = 8$ is a horizontal line, so any line perpendicular to it will be a vertical line. Because x has the same value at all points on the line, the equation is $x = -7$. It is not possible to write this in slope-intercept form.

(b) Standard form: $x = -7$

69. $f(x) = |x| - 3$

The graph is the same as that of $y = |x|$, except that it is translated 3 units downward.

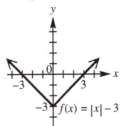

71. $f(x) = -(x + 1)^2 + 3$

The graph of $f(x) = -(x + 1)^2 + 3$ is a translation of the graph of $y = x^2$ to the left 1 unit, reflected over the x-axis and translated up 3 units.

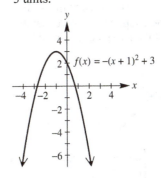

73. $f(x) = [\![x - 3]\!]$

To get $y = 0$, we need $0 \le x - 3 < 1 \Rightarrow$
$3 \le x < 4$. To get $y = 1$, we
need $1 \le x - 3 < 2 \Rightarrow 4 \le x < 5$.
Follow this pattern to graph the step function.

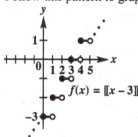

75. $f(x) = \begin{cases} -4x + 2 & \text{if } x \le 1 \\ 3x - 5 & \text{if } x > 1 \end{cases}$

Draw the graph of $y = -4x + 2$ to the left of
$x = 1$, including the endpoint at $x = 1$. Draw
the graph of $y = 3x - 5$ to the right of $x = 1$,
but do not include the endpoint at $x = 1$.
Observe that the endpoints of the two pieces
coincide.

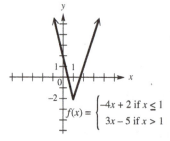

77. $f(x) = \begin{cases} |x| & \text{if } x < 3 \\ 6 - x & \text{if } x \ge 3 \end{cases}$

Draw the graph of $y = |x|$ to the left of $x = 3$,
but do not include the endpoint. Draw the
graph of $y = 6 - x$ to the right of $x = 3$,
including the endpoint. Observe that the
endpoints of the two pieces coincide.

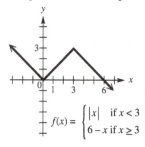

79. True. The graph of an even function is
symmetric with respect to the y-axis.

81. False. For example, $f(x) = x^2$ is even and
$(2, 4)$ is on the graph but $(2, -4)$ is not.

83. True. The constant function $f(x) = 0$ is both
even and odd. Because $f(-x) = 0 = f(x)$,
the function is even. Also
$f(-x) = 0 = -0 = -f(x)$, so the function is
odd.

85. $x + y^2 = 10$

Replace x with $-x$ to obtain $(-x) + y^2 = 10$.
The result is not the same as the original
equation, so the graph is not symmetric with
respect to the y-axis. Replace y with $-y$ to
obtain $x + (-y)^2 = 10 \Rightarrow x + y^2 = 10$. The
result is the same as the original equation, so
the graph is symmetric with respect to the
x-axis. Replace x with $-x$ and y with $-y$ to
obtain $(-x) + (-y)^2 = 10 \Rightarrow (-x) + y^2 = 10$.
The result is not the same as the original
equation, so the graph is not symmetric with
respect to the origin. The graph is symmetric
with respect to the x-axis only.

87. $x^2 = y^3$

Replace x with $-x$ to obtain
$(-x)^2 = y^3 \Rightarrow x^2 = y^3$. The result is the same
as the original equation, so the graph is
symmetric with respect to the y-axis. Replace
y with $-y$ to obtain $x^2 = (-y)^3 \Rightarrow x^2 = -y^3$.
The result is not the same as the original
equation, so the graph is not symmetric with
respect to the x-axis. Replace x with $-x$ and y
with $-y$ to obtain $(-x)^2 = (-y)^3 \Rightarrow x^2 = -y^3$.
The result is not the same as the original
equation, so the graph is not symmetric with
respect to the origin. Therefore, the graph is
symmetric with respect to the y-axis only.

89. $6x + y = 4$

Replace x with $-x$ to obtain $6(-x) + y = 4 \Rightarrow$
$-6x + y = 4$. The result is not the same as the
original equation, so the graph is not
symmetric with respect to the y-axis. Replace
y with $-y$ to obtain
$6x + (-y) = 4 \Rightarrow 6x - y = 4$. The result is not
the same as the original equation, so the graph
is not symmetric with respect to the x-axis.
Replace x with $-x$ and y with $-y$ to obtain
$6(-x) + (-y) = 4 \Rightarrow -6x - y = 4$. This
equation is not equivalent to the original one,
so the graph is not symmetric with respect to
the origin. Therefore, the graph has none of
the listed symmetries.

91. $y = 1$

This is the graph of a horizontal line through $(0, 1)$. It is symmetric with respect to the y-axis, but not symmetric with respect to the x-axis and the origin.

93. $x^2 - y^2 = 0$

Replace x with $-x$ to obtain

$(-x)^2 - y^2 = 0 \Rightarrow x^2 - y^2 = 0$. The result is the same as the original equation, so the graph is symmetric with respect to the y-axis.

Replace y with $-y$ to obtain

$x^2 - (-y)^2 = 0 \Rightarrow x^2 - y^2 = 0$. The result is the same as the original equation, so the graph is symmetric with respect to the x-axis.

Because the graph is symmetric with respect to the x-axis and with respect to the y-axis, it must also by symmetric with respect to the origin.

95. To obtain the graph of $g(x) = -|x|$, reflect the graph of $f(x) = |x|$ across the x-axis.

97. To obtain the graph of $k(x) = 2|x - 4|$, translate the graph of $f(x) = |x|$ to the right 4 units and stretch vertically by a factor of 2.

99. If the graph of $f(x) = 3x - 4$ is reflected about the y-axis, we obtain a graph whose equation is $y = f(-x) = 3(-x) - 4 = -3x - 4$.

101. (a) To graph $y = f(x) + 3$, translate the graph of $y = f(x)$, 3 units up.

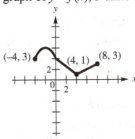

(b) To graph $y = f(x - 2)$, translate the graph of $y = f(x)$, 2 units to the right.

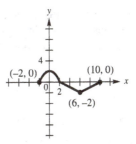

(c) To graph $y = f(x + 3) - 2$, translate the graph of $y = f(x)$, 3 units to the left and 2 units down.

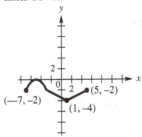

(d) To graph $y = |f(x)|$, keep the graph of $y = f(x)$ as it is where $y \geq 0$ and reflect the graph about the x-axis where $y < 0$.

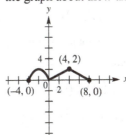

For Exercises 103–109, $f(x) = 3x^2 - 4$ and $g(x) = x^2 - 3x - 4$.

103. $(fg)(x) = f(x) \cdot g(x)$
$= (3x^2 - 4)(x^2 - 3x - 4)$
$= 3x^4 - 9x^3 - 12x^2 - 4x^2 + 12x + 16$
$= 3x^4 - 9x^3 - 16x^2 + 12x + 16$

105. $(f + g)(-4) = f(-4) + g(-4)$
$= [3(-4)^2 - 4] + [(-4)^2 - 3(-4) - 4]$
$= [3(16) - 4] + [16 - 3(-4) - 4]$
$= [48 - 4] + [16 + 12 - 4]$
$= 44 + 24 = 68$

107. $\left(\dfrac{f}{g}\right)(3) = \dfrac{f(3)}{g(3)} = \dfrac{3 \cdot 3^2 - 4}{3^2 - 3 \cdot 3 - 4} = \dfrac{3 \cdot 9 - 4}{9 - 3 \cdot 3 - 4}$
$= \dfrac{27 - 4}{9 - 9 - 4} = \dfrac{23}{-4} = -\dfrac{23}{4}$

109. The domain of $(fg)(x)$ is the intersection of the domain of $f(x)$ and the domain of $g(x)$. Both have domain $(-\infty, \infty)$, so the domain of $(fg)(x)$ is $(-\infty, \infty)$.

111. $f(x) = 2x + 9$

$f(x + h) = 2(x + h) + 9 = 2x + 2h + 9$

$f(x + h) - f(x) = (2x + 2h + 9) - (2x + 9)$
$= 2x + 2h + 9 - 2x - 9 = 2h$

Thus, $\dfrac{f(x + h) - f(x)}{h} = \dfrac{2h}{h} = 2.$

For Exercises 113–117,

$f(x) = \sqrt{x - 2}$ and $g(x) = x^2$.

113. $(g \circ f)(x) = g[f(x)] = g\left(\sqrt{x - 2}\right)$

$= \left(\sqrt{x - 2}\right)^2 = x - 2$

115. $f(x) = \sqrt{x - 2}$, so $f(3) = \sqrt{3 - 2} = \sqrt{1} = 1$.
Therefore,

$(g \circ f)(3) = g\left[f(3)\right] = g(1) = 1^2 = 1.$

117. $(g \circ f)(-1) = g\left(f(-1)\right) = g\left(\sqrt{-1 - 2}\right) = g\left(\sqrt{-3}\right)$

Because $\sqrt{-3}$ is not a real number, $(g \circ f)(-1)$ is not defined.

119. $(f + g)(1) = f(1) + g(1) = 7 + 1 = 8$

121. $(fg)(-1) = f(-1) \cdot g(-1) = 3(-2) = -6$

123. $(g \circ f)(-2) = g[f(-2)] = g(1) = 2$

125. $(f \circ g)(2) = f[g(2)] = f(2) = 1$

127. Let x = number of yards.
$f(x) = 36x$, where $f(x)$ is the number of inches.
$g(x) = 1760x$, where $g(x)$ is the number of yards. Then
$(g \circ f)(x) = g\left[f(x)\right] = 1760(36x) = 63{,}360x.$
There are $63{,}360x$ inches in x miles

128. Use the definition for the perimeter of a rectangle.
P = length + width + length + width
$P(x) = 2x + x + 2x + x = 6x$
This is a linear function.

129. If $V(r) = \frac{4}{3}\pi r^3$ and if the radius is increased by 3 inches, then the amount of volume gained is given by
$V_g(r) = V(r + 3) - V(r) = \frac{4}{3}\pi(r + 3)^3 - \frac{4}{3}\pi r^3.$

Chapter 2   Test

1. (a) The domain of $f(x) = \sqrt{x} + 3$ occurs when $x \geq 0$. In interval notation, this correlates to the interval in D, $[0, \infty)$.

(b) The range of $f(x) = \sqrt{x - 3}$ is all real numbers greater than or equal to 0. In interval notation, this correlates to the interval in D, $[0, \infty)$.

(c) The domain of $f(x) = x^2 - 3$ is all real numbers. In interval notation, this correlates to the interval in C, $(-\infty, \infty)$.

(d) The range of $f(x) = x^2 + 3$ is all real numbers greater than or equal to 3. In interval notation, this correlates to the interval in B, $[3, \infty)$.

(e) The domain of $f(x) = \sqrt[3]{x - 3}$ is all real numbers. In interval notation, this correlates to the interval in C, $(-\infty, \infty)$.

(f) The range of $f(x) = \sqrt[3]{x} + 3$ is all real numbers. In interval notation, this correlates to the interval in C, $(-\infty, \infty)$.

(g) The domain of $f(x) = |x| - 3$ is all real numbers. In interval notation, this correlates to the interval in C, $(-\infty, \infty)$.

(h) The range of $f(x) = |x + 3|$ is all real numbers greater than or equal to 0. In interval notation, this correlates to the interval in D, $[0, \infty)$.

(i) The domain of $x = y^2$ is $x \geq 0$ because when you square any value of y, the outcome will be nonnegative. In interval notation, this correlates to the interval in D, $[0, \infty)$.

(j) The range of $x = y^2$ is all real numbers. In interval notation, this correlates to the interval in C, $(-\infty, \infty)$.

2. Consider the points $(-2, 1)$ and $(3, 4)$.

$m = \dfrac{4 - 1}{3 - (-2)} = \dfrac{3}{5}$

3. We label the points $A(-2,1)$ and $B(3,4)$.

$$d(A, B) = \sqrt{[3-(-2)]^2 + (4-1)^2}$$
$$= \sqrt{5^2 + 3^2} = \sqrt{25+9} = \sqrt{34}$$

4. The midpoint has coordinates
$$\left(\frac{-2+3}{2}, \frac{1+4}{2}\right) = \left(\frac{1}{2}, \frac{5}{2}\right).$$

5. Use the point-slope form with
$(x_1, y_1) = (-2,1)$ and $m = \frac{3}{5}$.

$$y - y_1 = m(x - x_1)$$
$$y - 1 = \frac{3}{5}[x - (-2)]$$
$$y - 1 = \frac{3}{5}(x+2) \Rightarrow 5(y-1) = 3(x+2) \Rightarrow$$
$$5y - 5 = 3x + 6 \Rightarrow 5y = 3x + 11 \Rightarrow$$
$$-3x + 5y = 11 \Rightarrow 3x - 5y = -11$$

6. Solve $3x - 5y = -11$ for y.
$$3x - 5y = -11$$
$$-5y = -3x - 11$$
$$y = \frac{3}{5}x + \frac{11}{5}$$
Therefore, the linear function is
$$f(x) = \frac{3}{5}x + \frac{11}{5}.$$

7. (a) The center is at $(0, 0)$ and the radius is 2, so the equation of the circle is
$$x^2 + y^2 = 4.$$

(b) The center is at $(1, 4)$ and the radius is 1, so the equation of the circle is
$$(x-1)^2 + (y-4)^2 = 1$$

8. $x^2 + y^2 + 4x - 10y + 13 = 0$

Complete the square on x and y to write the equation in standard form:
$$x^2 + y^2 + 4x - 10y + 13 = 0$$
$$\left(x^2 + 4x + \quad\right) + \left(y^2 - 10y + \quad\right) = -13$$
$$\left(x^2 + 4x + 4\right) + \left(y^2 - 10y + 25\right) = -13 + 4 + 25$$
$$(x+2)^2 + (y-5)^2 = 16$$
The circle has center $(-2, 5)$ and radius 4.

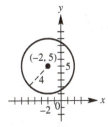

$x^2 + y^2 + 4x - 10y + 13 = 0$

9. (a) This is not the graph of a function because some vertical lines intersect it in more than one point. The domain of the relation is $[0, 4]$. The range is $[-4, 4]$.

(b) This is the graph of a function because no vertical line intersects the graph in more than one point. The domain of the function is $(-\infty, -1) \cup (-1, \infty)$. The range is $(-\infty, 0) \cup (0, \infty)$. As x is getting larger on the intervals $(-\infty, -1)$ and $(-1, \infty)$, the value of y is decreasing, so the function is decreasing on these intervals. (The function is never increasing or constant.)

10. Point A has coordinates $(5, -3)$.

(a) The equation of a vertical line through A is $x = 5$.

(b) The equation of a horizontal line through A is $y = -3$.

11. The slope of the graph of $y = -3x + 2$ is -3.

(a) A line parallel to the graph of $y = -3x + 2$ has a slope of -3.
Use the point-slope form with
$(x_1, y_1) = (2, 3)$ and $m = -3$.
$$y - y_1 = m(x - x_1)$$
$$y - 3 = -3(x - 2)$$
$$y - 3 = -3x + 6 \Rightarrow y = -3x + 9$$

(b) A line perpendicular to the graph of $y = -3x + 2$ has a slope of $\frac{1}{3}$ because
$$-3\left(\tfrac{1}{3}\right) = -1.$$
$$y - 3 = \tfrac{1}{3}(x - 2)$$
$$3(y-3) = x - 2 \Rightarrow 3y - 9 = x - 2 \Rightarrow$$
$$3y = x + 7 \Rightarrow y = \tfrac{1}{3}x + \tfrac{7}{3}$$

12. (a) $(2, \infty)$ **(b)** $(0, 2)$

(c) $(-\infty, 0)$ **(d)** $(-\infty, \infty)$

(e) $(-\infty, \infty)$ **(f)** $[-1, \infty)$

13. To graph $f(x) = |x-2| - 1$, we translate the graph of $y = |x|$, 2 units to the right and 1 unit down.

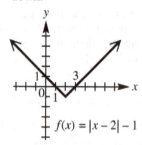

$$f(x) = |x-2| - 1$$

14. $f(x) = [\![x+1]\!]$

To get $y = 0$, we need $0 \le x+1 < 1 \Rightarrow -1 \le x < 0$. To get $y = 1$, we need $1 \le x+1 < 2 \Rightarrow 0 \le x < 1$. Follow this pattern to graph the step function.

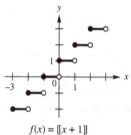

$$f(x) = [\![x+1]\!]$$

15. $f(x) = \begin{cases} 3 & \text{if } x < -2 \\ 2 - \frac{1}{2}x & \text{if } x \ge -2 \end{cases}$

For values of x with $x < -2$, we graph the horizontal line $y = 3$. For values of x with $x \ge -2$, we graph the line with a slope of $-\frac{1}{2}$ and a y-intercept of $(0, 2)$. Two points on this line are $(-2, 3)$ and $(0, 2)$.

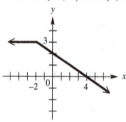

$$f(x) = \begin{cases} 3 & \text{if } x < -2 \\ 2 - \frac{1}{2}x & \text{if } x \ge -2 \end{cases}$$

16. (a) Shift $f(x)$, 2 units vertically upward.

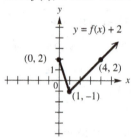

(b) Shift $f(x)$, 2 units horizontally to the left.

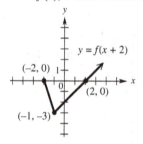

(c) Reflect $f(x)$, across the x-axis.

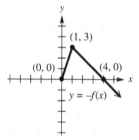

(d) Reflect $f(x)$, across the y-axis.

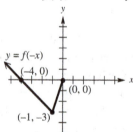

(e) Stretch $f(x)$, vertically by a factor of 2.

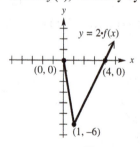

17. Starting with $y = \sqrt{x}$, we shift it to the left 2 units and stretch it vertically by a factor of 2. The graph is then reflected over the x-axis and then shifted down 3 units.

18. $3x^2 - 2y^2 = 3$

(a) Replace y with $-y$ to obtain
$3x^2 - 2(-y)^2 = 3 \Rightarrow 3x^2 - 2y^2 = 3$.
The result is the same as the original equation, so the graph is symmetric with respect to the x-axis.

(b) Replace x with $-x$ to obtain
$3(-x)^2 - 2y^2 = 3 \Rightarrow 3x^2 - 2y^2 = 3$.
The result is the same as the original equation, so the graph is symmetric with respect to the y-axis.

(c) The graph is symmetric with respect to the x-axis and with respect to the y-axis, so it must also be symmetric with respect to the origin.

19. $f(x) = 2x^2 - 3x + 2$, $g(x) = -2x + 1$

(a) $(f - g)(x) = f(x) - g(x)$
$= \left(2x^2 - 3x + 2\right) - \left(-2x + 1\right)$
$= 2x^2 - 3x + 2 + 2x - 1$
$= 2x^2 - x + 1$

(b) $\left(\dfrac{f}{g}\right)(x) = \dfrac{f(x)}{g(x)} = \dfrac{2x^2 - 3x + 2}{-2x + 1}$

(c) We must determine which values solve the equation $-2x + 1 = 0$.
$-2x + 1 = 0 \Rightarrow -2x = -1 \Rightarrow x = \frac{1}{2}$
Thus, $\frac{1}{2}$ is excluded from the domain, and the domain is $\left(-\infty, \frac{1}{2}\right) \cup \left(\frac{1}{2}, \infty\right)$.

(d) $f(x) = 2x^2 - 3x + 2$
$f(x + h) = 2(x + h)^2 - 3(x + h) + 2$
$= 2\left(x^2 + 2xh + h^2\right) - 3x - 3h + 2$
$= 2x^2 + 4xh + 2h^2 - 3x - 3h + 2$
$f(x + h) - f(x)$
$= (2x^2 + 4xh + 2h^2 - 3x - 3h + 2)$
$\qquad - (2x^2 - 3x + 2)$
$= 2x^2 + 4xh + 2h^2 - 3x$
$\qquad - 3h + 2 - 2x^2 + 3x - 2$
$= 4xh + 2h^2 - 3h$
$\dfrac{f(x + h) - f(x)}{h} = \dfrac{4xh + 2h^2 - 3h}{h}$
$= \dfrac{h(4x + 2h - 3)}{h}$
$= 4x + 2h - 3$

(e) $(f + g)(1) = f(1) + g(1)$
$= (2 \cdot 1^2 - 3 \cdot 1 + 2) + (-2 \cdot 1 + 1)$
$= (2 \cdot 1 - 3 \cdot 1 + 2) + (-2 \cdot 1 + 1)$
$= (2 - 3 + 2) + (-2 + 1)$
$= 1 + (-1) = 0$

(f) $(fg)(2) = f(2) \cdot g(2)$
$= (2 \cdot 2^2 - 3 \cdot 2 + 2) \cdot (-2 \cdot 2 + 1)$
$= (2 \cdot 4 - 3 \cdot 2 + 2) \cdot (-2 \cdot 2 + 1)$
$= (8 - 6 + 2) \cdot (-4 + 1)$
$= 4(-3) = -12$

(g) $g(x) = -2x + 1 \Rightarrow g(0) = -2(0) + 1$
$= 0 + 1 = 1$. Therefore,
$(f \circ g)(0) = f\left[g(0)\right]$
$= f(1) = 2 \cdot 1^2 - 3 \cdot 1 + 2$
$= 2 \cdot 1 - 3 \cdot 1 + 2$
$= 2 - 3 + 2 = 1$

For exercises 20 and 21, $f(x) = \sqrt{x + 1}$ and $g(x) = 2x - 7$.

20. $(f \circ g) = f(g(x)) = f(2x - 7)$
$= \sqrt{(2x - 7) + 1} = \sqrt{2x - 6}$
The domain and range of g are $(-\infty, \infty)$, while the domain of f is $[0, \infty)$. We need to find the values of x which fit the domain of f:
$2x - 6 \geq 0 \Rightarrow x \geq 3$. So, the domain of $f \circ g$ is $[3, \infty)$.

21. $(g \circ f) = g(f(x)) = g\left(\sqrt{x + 1}\right)$
$= 2\sqrt{x + 1} - 7$
The domain and range of g are $(-\infty, \infty)$, while the domain of f is $[0, \infty)$. We need to find the values of x which fit the domain of f:
$x + 1 \geq 0 \Rightarrow x \geq -1$. So, the domain of $g \circ f$ is $[-1, \infty)$.

22. (a) $C(x) = 3300 + 4.50x$

(b) $R(x) = 10.50x$

(c) $P(x) = R(x) - C(x)$
$= 10.50x - (3300 + 4.50x)$
$= 6.00x - 3300$

(d) $P(x) > 0$
$6.00x - 3300 > 0$
$6.00x > 3300$
$x > 550$
She must produce and sell 551 items before she earns a profit.

Chapter 3

POLYNOMIAL AND RATIONAL FUNCTIONS

Section 3.1   Quadratic Functions and Models

1. A polynomial function with leading term $3x^5$ has degree 5.

3. The highest point on the graph of a parabola that opens down is the vertex of the parabola.

5. The vertex of the graph of $f(x) = x^2 + 2x + 4$ has x-coordinate –1.

7. C. $y = (x+4)^2 + 2$ has a vertex $(-4, 2)$ and opens up because $a = 1$.

9. D. $y = -(x+4)^2 + 2$ has vertex $(-4, 2)$ and opens down because $a = -1$.

11. $f(x) = (x+3)^2 - 4$

   (a) domain: $(-\infty, \infty)$; range: $[-4, \infty)$

   (b) vertex: $(h, k) = (-3, -4)$

   (c) axis: $x = -3$

   (d) To find the y-intercept, let $x = 0$.
$$y = (0+3)^2 - 4 = 3^2 - 4 = 9 - 4 = 5$$
y-intercept: $(0, 5)$

   (e) To find the x-intercepts, let $f(x) = 0$.
$$0 = (x+3)^2 - 4$$
$$(x+3)^2 = 4$$
$$x + 3 = \pm\sqrt{4} = \pm 2$$
$$x = -3 \pm 2$$
$$x = -3 - 2 = -5 \text{ or } x = -3 + 2 = -1$$
x-intercepts: $(-5, 0)$ and $(-1, 0)$

13. $f(x) = -2(x+3)^2 + 2$

   (a) domain: $(-\infty, \infty)$; range: $(-\infty, 2]$

   (b) vertex: $(h, k) = (-3, 2)$

   (c) axis: $x = -3$

   (d) To find the y-intercept, let $x = 0$.
$$y = -2(0+3)^2 + 2 = -2 \cdot 3^2 + 2$$
$$= -2 \cdot 9 + 2 = -18 + 2 = -16$$
y-intercept: $(0, -16)$

   (e) To find the x-intercepts, let $f(x) = 0$.
$$0 = -2(x+3)^2 + 2$$
$$(x+3)^2 = 1$$
$$x + 3 = \pm\sqrt{1} = \pm 1$$
$$x = -3 \pm 1$$
$$x = -3 - 1 = -4 \text{ or } x = -3 + 1 = -2$$
x-intercepts: $(-4, 0)$ and $(-2, 0)$

15. $f(x) = (x-4)^2 - 3$

Because a > 0, the parabola opens upward. The vertex is at $(4, -3)$. The correct graph, therefore, is B.

17. $f(x) = (x+4)^2 - 3$

Because a > 0, the parabola opens upward. The vertex is at $(-4, -3)$. The correct graph, therefore, is D.

19. For parts (a), (b), and (c), see the following graph.

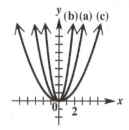

   (d) If the absolute value of the coefficient is greater than 1, it causes the graph to be stretched vertically, so it is narrower. If the absolute value of the coefficient is between 0 and 1, it causes the graph to shrink vertically, so it is wider.

21. For parts (a), (b), and (c), see the following graph.

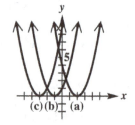

(e) The graph of $y = (x - h)^2$ is translated h units to the right if h is positive and $|h|$ units to the left if h is negative.

23. $f(x) = (x - 2)^2$

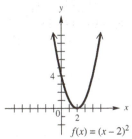

$f(x) = (x - 2)^2$

This equation is of the form $y = (x - h)^2$, with $h = 2$. The graph opens upward and has the same shape as that of $y = x^2$. It is a horizontal translation of the graph of $y = x^2$, 2 units to the right. Two points on the graph are $(1, 1)$ and $(3, 1)$.

(a) The vertex is $(2, 0)$.

(b) The axis is the vertical line $x = 2$.

(c) The domain is $(-\infty, \infty)$.

(d) The smallest value of y is 0 and the graph opens upward, so the range is $[0, \infty)$.

(e) The function is increasing on $(2, \infty)$.

(f) The function is decreasing on $(-\infty, 2)$.

25. $f(x) = (x + 3)^2 - 4 = [x - (-3)]^2 + (-4)$

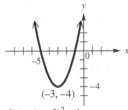

$f(x) = (x + 3)^2 - 4$

This equation is of the form $y = (x - h)^2 + k$, with $h = -3$ and $k = -4$. The graph opens upward and has the same shape as $y = x^2$. It is a translation of $y = x^2$, 3 units to the left and 4 units down. Two points on the graph are $(-4, -3)$ and $(-2, -3)$.

(a) The vertex is $(-3, -4)$.

(b) The axis is the vertical line $x = -3$.

(c) The domain is $(-\infty, \infty)$.

(d) The smallest value of y is -4 and the graph opens upward, so the range is $[-4, \infty)$.

(e) The function is increasing on $(-3, \infty)$.

(f) The function is decreasing on $(-\infty, -3)$.

27. $f(x) = -\frac{1}{2}(x + 1)^2 - 3 = -\frac{1}{2}[x - (-1)]^2 + (-3)$

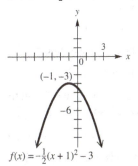

$f(x) = -\frac{1}{2}(x + 1)^2 - 3$

This equation is of the form $y = a(x - h)^2 + k$, with $h = -1$, $k = -3$, $a = -\frac{1}{2}$. The graph opens downward and is wider than $y = x^2$. It is a translation of the graph $y = -\frac{1}{2}x^2$, 1 unit to the left and 3 units down. Additional points on the graph are $\left(-2, -3\frac{1}{2}\right)$ and $\left(0, -3\frac{1}{2}\right)$.

(a) The vertex is $(-1, -3)$.

(b) The axis is the vertical line $x = -1$.

(c) The domain is $(-\infty, \infty)$.

(d) The largest value of y is -3 and the graph opens downward, so the range is $(-\infty, -3]$.

(e) The function is increasing on $(-\infty, -1)$.

(f) The function is decreasing on $(-1, \infty)$.

29. $f(x) = x^2 - 2x + 3$

Rewrite by completing the square on x.

$f(x) = x^2 - 2x + 3 = (x^2 - 2x + 1 - 1) + 3$

$\qquad$ Note: $\left[\frac{1}{2}(-2)\right]^2 = (-1)^2 = 1$

$\qquad = (x^2 - 2x + 1) - 1 + 3 = (x - 1)^2 + 2$

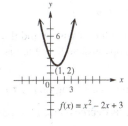

$f(x) = x^2 - 2x + 3$

(continued on next page)

(*continued*)

This equation is of the form $y = (x - h)^2 + k$, with $h = 1$ and $k = 2$. The graph opens upward and has the same shape as $y = x^2$. It is a translation of the graph $y = x^2$, 1 unit to the right and 2 units up. Two points on the graph are (0, 3) and (2, 3).

(a) The vertex is (1, 2).

(b) The axis is the vertical line $x = 1$.

(c) The domain is $(-\infty, \infty)$.

(d) The smallest value of y is 2 and the graph opens upward, so the range is $[2, \infty)$.

(e) The function is increasing on $(1, \infty)$.

(f) The function is decreasing on $(-\infty, 1)$.

31. $f(x) = x^2 - 10x + 21$

Rewrite by completing the square on x.

$f(x) = x^2 - 10x + 21 = (x^2 - 10x + 25 - 25) + 21$

$\qquad\qquad$ Note: $\left[\frac{1}{2}(-10)\right]^2 = (-5)^2 = 25$

$\qquad = (x^2 - 10x + 25) - 25 + 21 = (x - 5)^2 - 4$

$\qquad = (x - 5)^2 + (-4)$

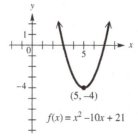

$f(x) = x^2 - 10x + 21$

This equation is of the form $y = (x - h)^2 + k$, with $h = 5$ and $k = -4$. The graph opens upward and has the same shape as $y = x^2$. It is a translation of the graph $y = x^2$, 5 units to the right and 4 units down. Two points on the graph are (4, -3) and (6, -3).

(a) The vertex is (5, -4).

(b) The axis is the vertical line $x = 5$.

(c) The domain is $(-\infty, \infty)$.

(d) The smallest value of y is -4 and the graph opens upward, so the range is $[-4, \infty)$.

(e) The function is increasing on $(5, \infty)$.

(f) The function is decreasing on $(-\infty, 5)$.

33. $f(x) = -2x^2 - 12x - 16$

Rewrite by completing the square on x.

$f(x) = -2x^2 - 12x - 16 = -2(x^2 + 6x) - 16$

$\qquad = -2(x^2 + 6x + 9 - 9) - 16$

$\qquad\qquad$ Note: $\left[\frac{1}{2}(6)\right]^2 = 3^2 = 9$

$\qquad = -2(x^2 + 6x + 9) + 18 - 16$

$\qquad = -2(x + 3)^2 + 2 = -2[x - (-3)]^2 + 2$

$f(x) = -2x^2 - 12x - 16$

This equation is of the form $y = a(x - h)^2 + k$, with $h = -3$, $k = 2$, and $a = -2$. The graph opens downward and is narrower than $y = x^2$.

It is a translation of the graph $y = -2x^2$, 3 units to the left and 2 units up. Additional points on the graph are (-4, 0) and (-2, 0).

(a) The vertex is (-3, 2).

(b) The axis is the vertical line $x = -3$.

(c) The domain is $(-\infty, \infty)$.

(d) The largest value of y is 2 and the graph opens downward, so the range is $(-\infty, 2]$.

(e) The function is increasing on $(-\infty, -3)$.

(f) The function is decreasing on $(-3, \infty)$.

35. $f(x) = -\frac{1}{2}x^2 - 3x - \frac{1}{2}$

Rewrite by completing the square on x.

$f(x) = -\frac{1}{2}x^2 - 3x - \frac{1}{2}$

$\qquad = -\frac{1}{2}(x^2 + 6x + 1 + 8) - \frac{1}{2}(-8)$

$\qquad\qquad$ Note: $\left[\frac{1}{2}(6)\right]^2 = 3^2 = 9$

$\qquad = -\frac{1}{2}(x^2 + 6x + 9) + 4 = -(x + 3)^2 + 4$

$\qquad = -\frac{1}{2}[x - (-3)]^2 + 4$

This equation is of the form $y = a(x - h)^2 + k$, with $h = -3$, $k = 4$, and $a = -1$. The graph opens downward and is wider than $y = x^2$. It is a translation of the graph $y = -x^2$, 3 units to the left and 4 units up. Two points on the graph are (-1, 2) and (1, -4).

(*continued on next page*)

(*continued*)

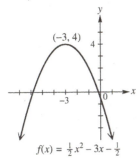

$$f(x) = \tfrac{1}{2}x^2 - 3x - \tfrac{1}{2}$$

(a) The vertex is $(-3, 4)$.

(b) The axis is the vertical line $x = -3$.

(c) The domain is $(-\infty, \infty)$.

(d) The largest value of y is 4 and the graph opens downward, so the range is $(-\infty, 4]$.

(e) The function is increasing on $(-\infty, -3)$.

(f) The function is decreasing on $(-3, \infty)$.

37. The minimum value of $f(x)$ is $f(-3) = 3$.

39. There are no real solutions to the equation $f(x) = 1$ because the value of $f(x)$ is never less than 3.

41. $a < 0$, $b^2 - 4ac = 0$

The correct choice is E. $a < 0$ indicates that the parabola will open downward, while $b^2 - 4ac = 0$ indicates that the graph will have exactly one x-intercept.

43. $a < 0$, $b^2 - 4ac < 0$

The correct choice is D. $a < 0$ indicates that the parabola will open downward, while $b^2 - 4ac < 0$ indicates that the graph will have no x-intercepts.

45. $a > 0$, $b^2 - 4ac > 0$

The correct choice is C. $a > 0$ indicates that the parabola will open upward, while $b^2 - 4ac > 0$ indicates that the graph will have two x-intercepts.

47. The vertex of the parabola in the figure is $(2, -1)$ and the y-intercept is 0. The equation takes the form $f(x) = a(x - 2)^2 - 1$. When

$x = 0$, $f(x) = 0$, so $0 = a(0 - 2)^2 - 1 \Rightarrow$

$0 = 4a - 1 \Rightarrow a = \tfrac{1}{4}$. The equation is

$f(x) = \tfrac{1}{4}(x - 2)^2 - 1$.

This function may also be written as

$$f(x) = \tfrac{1}{4}(x - 2)^2 - 1$$
$$= \tfrac{1}{4}\left(x^2 - 4x + 4\right) - 1 = \tfrac{1}{4}x^2 - x.$$

49. The vertex of the parabola in the figure is $(1, 4)$ and the y-intercept is 2. The equation takes the form $f(x) = a(x - 1)^2 + 4$. When

$x = 0$, $f(x) = 2$, so $2 = a(0 - 1)^2 + 4 \Rightarrow$

$2 = a + 4 \Rightarrow a = -2$. The equation is

$f(x) = -2(x - 1)^2 + 4$. This function may also

be written as $f(x) = -2(x^2 - 2x + 1) + 4$

$= -2x^2 + 4x - 2 + 4 = -2x^2 + 4x + 2$.

51. Linear; the points lie in a pattern that suggests a positive slope.

53. Quadratic; the points lie in a pattern suggesting a parabola opening upward, so $a > 0$.

55. Quadratic; the points lie in a pattern that suggests a parabola opening downward, so $a < 0$.

57. (a) $v_0 = 200$, $s_0 = 50$, and

$s(t) = -16t^2 + v_0 t + s_0$, so

$s(t)$ or $f(t) = -16t^2 + 200t + 50$.

(b) Algebraic Solution:

Find the coordinates of the vertex of the parabola. Using the vertex formula with $a = -16$ and $b = 200$.

$$x = -\frac{b}{2a} = -\frac{200}{2(-16)} = 6.25 \text{ and}$$

$$y = -16(6.25)^2 + 200(6.25) + 50 = 675.$$

The vertex is $(6.25, 675)$. Because $a < 0$, this is the maximum point.

Graphing Calculator Solution:

Graph $y_1 = -16x^2 + 200x + 50$ in the window $[-1, 15] \times [0, 800]$.

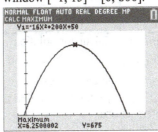

Thus, the number of seconds to reach maximum height is 6.25 seconds. The maximum height is 675 ft.

(c) Algebraic Solution:

To find the time interval in which the rocket will be more than 300 ft above ground level, solve the inequality

$-16t^2 + 200t + 50 > 300.$

$-16t^2 + 200t + 50 > 300 \Rightarrow$

$-16t^2 + 200t - 250 > 0 \Rightarrow$

$-8t^2 + 100t - 125 > 0$

Solve the corresponding equation

$-8t^2 + 100t - 125 = 0.$

Use the quadratic formula with $a = -8$, $b = 100$, and $c = -125$.

$$t = \frac{-100 \pm \sqrt{100^2 - 4(-8)(-125)}}{2(-8)}$$

$$= \frac{-100 \pm \sqrt{10,000 - 4000}}{-16}$$

$$= \frac{-100 \pm \sqrt{6000}}{-16}$$

$$t = \frac{-100 + \sqrt{6000}}{-16} \approx 1.41 \text{ or}$$

$$t = \frac{-100 - \sqrt{6000}}{-16} \approx 11.09$$

The values 1.41 and 11.09 divide the number line into three intervals: $(-\infty, 1.41)$, $(1.41, 11.09)$, and $(11.09, \infty)$. Use a test point in each interval to determine where the inequality is satisfied.

Interval	Test Value	Is $-16t^2 + 200t + 50 > 300$ True or False?
$(-\infty, 1.41)$	0	$-16 \cdot 0^2 + 200 \cdot 0 + 50 \overset{?}{>} 300$ $50 > 300$ False
$(1.41, 11.09)$	2	$-16 \cdot 2^2 + 200 \cdot 2 + 50 \overset{?}{>} 300$ $386 > 300$ True
$(11.09, \infty)$	12	$-16 \cdot 12^2 + 200 \cdot 12 + 50 \overset{?}{>} 300$ $146 > 300$ False

Graphing Calculator Solution:

Graph $y_1 = -16x^2 + 200x + 50$ and $y_2 = 300$ in the window $[-1, 15] \times [0, 800]$. Then find the intersections.

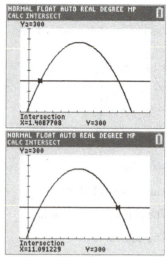

The rocket will be more than 300 ft above the ground between 1.41 sec and 11.09 sec.

(d) Algebraic Solution:

To find the number of seconds for the toy rocket to hit the ground, let $f(t) = 0$ and solve for t: $-16t^2 + 200t + 50 = 0$

Use the quadratic formula with $a = -16$, $b = 200$, and $c = 50$.

$$t = \frac{-200 \pm \sqrt{200^2 - 4(-16)(50)}}{2(-16)}$$

$$= \frac{-200 \pm \sqrt{40,000 + 3200}}{-32}$$

$$= \frac{-200 \pm \sqrt{43,200}}{-32} \approx -0.25 \text{ or } 12.75$$

We reject the negative solution.

Graphing Calculator Solution:

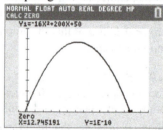

It will take approximately 12.75 seconds for the toy rocket to hit the ground.

59. (a) The length of the other side is $640 - 2x$.

(b) In order for the two lengths to be positive, $0 < x < 320$.

(c) $\mathcal{A}(x) = x(640 - 2x) = 640x - 2x^2$

$= -2x^2 + 640x$

(d) Algebraic Solution:
Solve the inequality
$30,000 < -2x^2 + 640x < 40,000.$ Treat
this as two inequalities,
$-2x^2 + 640x > 30,000$ and
$-2x^2 + 640x < 40,000.$
For $-2x^2 + 640x > 30,000,$ we solve the
corresponding equation
$-2x^2 + 640x = 30,000.$
This equation is equivalent to
$-2x^2 + 640x - 30,000 = 0$ or
$x^2 - 320x + 15,000 = 0.$
Use the quadratic formula with $a = 1,$
$b = -320,$ and $c = 15,000.$

$$x = \frac{-(-320) \pm \sqrt{(-320)^2 - 4(1)(15,000)}}{2(1)}$$
$$= \frac{320 \pm \sqrt{102,400 - 60,000}}{2}$$
$$= \frac{320 \pm \sqrt{42,400}}{2}$$
$$x = \frac{320 - \sqrt{42,400}}{2} \approx 57.04 \text{ or}$$
$$x = \frac{320 + \sqrt{42,400}}{2} \approx 262.96$$

The values 57.04 and 262.96 divide the
number line into three intervals:
$(-\infty, 57.04),$ $(57.04, 262.96),$ and
$(262.96, \infty).$ Because $x > 0,$ the first
interval becomes $(0, 57.04).$ Use a test
point in each interval to determine where
the inequality is satisfied.

Interval	Test Value	Is $-2x^2 + 640x > 30,000$ True or False?
$(0, 57.04)$	1	$-2 \cdot 1^2 + 640 \cdot 1 \overset{?}{>} 30,000$ $638 > 30,000$ False
$(57.04, 262.96)$	60	$-2 \cdot 60^2 + 640 \cdot 60 \overset{?}{>} 30,000$ $31,200 > 30,000$ True
$(262.96, \infty)$	300	$-2 \cdot 300^2 + 640 \cdot 300 \overset{?}{>} 30,000$ $12,000 > 30,000$ False

Thus, the first inequality is satisfied when
the measure of x is in the interval
$(57.04, 262.96).$

For $-2x^2 + 640x < 40,000,$ we solve the
corresponding equation
$-2x^2 + 640x = 40,000.$
This equation is equivalent to
$-2x^2 + 640x - 40,000 = 0$ or
$x^2 - 320x + 20,000 = 0$
Use the quadratic formula with $a = 1,$
$b = -320,$ and $c = 20,000.$

$$x = \frac{-(-320) \pm \sqrt{(-320)^2 - 4(1)(20,000)}}{2(1)}$$
$$= \frac{320 \pm \sqrt{102,400 - 80,000}}{2}$$
$$= \frac{320 \pm \sqrt{22,400}}{2}$$
$$x = \frac{320 - \sqrt{22,400}}{2} \approx 85.17 \text{ or}$$
$$x = \frac{320 + \sqrt{22,400}}{2} \approx 234.83$$

The values 85.17 and 234.83 divide the
number line into three intervals:
$(-\infty, 85.17),$ $(85.17, 234.83),$ and
$(234.83, \infty).$ Use a test point in each
interval to determine where the inequality
is satisfied.

Interval	Test Value	Is $-2x^2 + 640x < 40,000$ True or False?
$(-\infty, 85.17)$	0	$-2 \cdot 0^2 + 640 \cdot 0 \overset{?}{<} 40,000$ $0 < 40,000$ True
$(85.17, 234.83)$	90	$-2 \cdot 90^2 + 640 \cdot 90 \overset{?}{<} 40,000$ $41,400 < 40,000$ False
$(234.83, \infty)$	300	$-2 \cdot 300^2 + 640 \cdot 300 \overset{?}{<} 40,000$ $12,000 < 40,000$ True

Thus, the second inequality is satisfied
when the measure of x is in the interval
$(-\infty, 85.17)$ or $(234.83, \infty).$ We must
now seek the intersection of the intervals
$(57.04, 262.96),$ and $(-\infty, 85.17)$ or
$(234.83, \infty).$ If we use the real number
line as an aid, we can see that the solution
will be between 57.04 ft and 85.17 ft or
234.83 ft and 262.96 ft.

(continued on next page)

(continued)

57.04 85.17   234.83 262.96

Graphing Calculator Solution:

Graph $y_1 = -2x^2 + 640x$, $y_2 = 30000$,

and $y_3 = 40000$ in the window

$[0, 320] \times [0, 60000]$. Then find the

intersections.

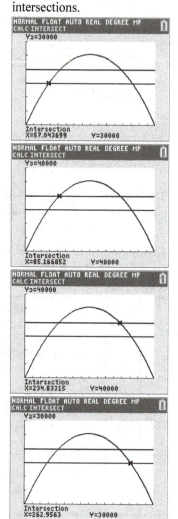

We can see from the graphs that the
quadratic function lies between the lines
when x is between 57.04 ft and 85.17 ft
or 234.83 ft and 262.96 ft.

(e) Algebraic Solution:
Find the coordinates of the vertex of the
parabola, $\mathcal{A}(x) = -2x^2 + 640x$. Use the
vertex formula with $a = -2$ and $b = 640$.

We have $x = -\dfrac{b}{2a} = -\dfrac{640}{2(-2)} = 160$ and

$y = -2(160)^2 + 640(160) = 51,200.$

This is the maximum point because $a < 0$.
Graphing Calculator Solution:

Graph $y_1 = -2x^2 + 640x$ in the window
$[0, 320] \times [0, 60000]$. Then find the
maximum.

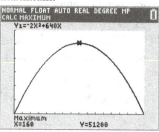

Thus, the length of the two parallel sides
would be 160 ft and the third side would
be $640 - 2(160) = 620 - 320 = 320$ ft.

The maximum area would be 51,200 ft^2.

61. (a) The length of the original piece of
cardboard would be $2x$.

(b) The length of the rectangular box would
be $2x - 4$ and the width would be $x - 4$,
where $x > 4$.

(c) $V(x) = (2x - 4)(x - 4)(2)$

$\qquad = (2x^2 - 12x + 16)(2)$

$\qquad = 4x^2 - 24x + 32$

(d) Algebraic Solution:
Solve the equation $4x^2 - 24x + 32 = 320$.

$4x^2 - 24x + 32 = 320$

$4x^2 - 24x - 288 = 0$

$x^2 - 6x - 72 = 0$

$(x + 6)(x - 12) = 0$

$x + 6 = 0 \Rightarrow x = -6$ or

$x - 12 = 0 \Rightarrow x = 12$

We discard the negative solution.

Graphing Calculator Solution:

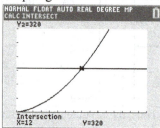

If $x = 12$, then the dimensions of the
bottom of the box will be $12 - 4 = 8$ in.
by $2(12) - 4 = 24 - 4 = 20$ in.

(e) Algebraic Solution:
Solve the inequality

$400 < 4x^2 - 24x + 32 < 500$. Treat this as

two inequalities, $4x^2 - 24x + 32 > 400$

and $4x^2 - 24x + 32 < 500$.

For $4x^2 - 24x + 32 > 400$, we solve the

corresponding equation

$4x^2 - 24x + 32 = 400$.

This equation is equivalent to

$4x^2 - 24x - 368 = 0$ or $x^2 - 6x - 92 = 0$

Use the quadratic formula with $a = 1$,

$b = -6$, and $c = -92$.

$$x = \frac{-(-6) \pm \sqrt{(-6)^2 - 4(1)(-92)}}{2(1)}$$

$$= \frac{6 \pm \sqrt{36 + 368}}{2} = \frac{6 \pm \sqrt{404}}{2}$$

$$= \frac{6 \pm 2\sqrt{101}}{2} = 3 \pm \sqrt{101}$$

$x = 3 - \sqrt{101} \approx -7.05$ or

$x = 3 + \sqrt{101} \approx 13.05$

Because $x > 4$, we need only check the

intervals: $(4, 13.0)$ and $(13.0, \infty)$. Use a

test point in each interval to determine

where the inequality is satisfied.

Interval	Test Value	Is $4x^2 - 24x + 32 > 400$ True or False?
$(4, 13.05)$	5	$4 \cdot 5^2 - 24 \cdot 5 + 32 \overset{?}{>} 400$ $12 > 400$ False
$(13.05, \infty)$	14	$4 \cdot 14^2 - 24 \cdot 14 + 32 \overset{?}{>} 400$ $480 > 400$ True

Thus, the first inequality is satisfied when

the length, x, is in the interval $(13.05, \infty)$.

For $4x^2 - 24x + 32 < 500$, we solve the

corresponding equation

$4x^2 - 24x + 32 = 500$. This equation is

equivalent to $4x^2 - 24x - 468 = 0$ or

$x^2 - 6x - 117 = 0$. Use the quadratic

formula with $a = 1$, $b = -6$, and

$c = -117$.

$$x = \frac{-(-6) \pm \sqrt{(-6)^2 - 4(1)(-117)}}{2(1)}$$

$$= \frac{6 \pm \sqrt{36 + 468}}{2} = \frac{6 \pm \sqrt{504}}{2}$$

$$= \frac{6 \pm 2\sqrt{126}}{2} = 3 \pm \sqrt{126}$$

$x = 3 - \sqrt{126} \approx -8.22$ or

$x = 3 + \sqrt{126} \approx 14.22$

Because $x > 4$, we need only check the

intervals: $(4, 14.22)$ and $(14.22, \infty)$. Use

a test point in each interval to determine

where the inequality is satisfied.

Interval	Test Value	Is $4x^2 - 24x + 32 < 500$ True or False?
$(4, 14.22)$	5	$4 \cdot 5^2 - 24 \cdot 5 + 32 \overset{?}{<} 500$ $12 < 500$ True
$(14.22, \infty)$	15	$4 \cdot 15^2 - 24 \cdot 15 + 32 \overset{?}{<} 500$ $572 < 500$ False

Thus, the second inequality is satisfied

when the length, x, is in the interval

$(4, 14.22)$. We must now seek the

intersection of the intervals $(13.05, \infty)$

and $(4, 14.22)$. This intersection is

$(13.05, 14.22)$.

Graphing Calculator Solution:

Graph $y_1 = 4x^2 - 24x + 32$, $y_2 = 400$,

and $y_3 = 500$ in the window $[4, 20] \times$

$[0, 600]$ and find the intersections.

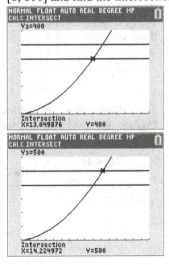

(*continued on next page*)

(continued)

We can see from the graphs that the quadratic function lies between the lines when x is between 13.05 in and 14.22 in. Thus, the volume will be between 400 in.³ and 500 in.³, when the original width is between 13.05 in. and 14.22 in.

63. $y = \dfrac{-16x^2}{0.434v^2} + 1.15x + 8$

(a) Let $y = 10$ and $x = 15$.

$$10 = \frac{-16(15)^2}{0.434v^2} + 1.15(15) + 8$$

$$10 = \frac{-16(225)}{0.434v^2} + 1.15(15) + 8$$

$$10 = \frac{-3600}{0.434v^2} + 25.25$$

$$\frac{3600}{0.434v^2} = 15.25$$

$$3600 = 6.6185v^2 \Rightarrow \frac{3600}{6.6185} = v^2$$

$$\pm\sqrt{\frac{3600}{6.6185}} = v \Rightarrow v \approx \pm 23.32$$

Because v represents a velocity, only the positive square root is meaningful. The basketball should have an initial velocity of 23.32 ft per sec.

(b) $y = \dfrac{-16x^2}{0.434(23.32)^2} + 1.15x + 8$

Algebraic Solution:

The parabola opens downward, so the vertex is the maximum point. Use the vertex formula to find the x-coordinate of the vertex.

$$x = -\frac{b}{2a} = -\frac{1.15}{2\left(\dfrac{-16}{0.434(23.32)^2}\right)} \approx 8.482$$

To find the y-coordinate of the vertex, evaluate the quadratic function when $x = 8.482$.

$$y = \frac{-16(8.482)^2}{0.434(23.32)^2} + 1.15(8.482) + 8$$

$$\approx 12.88$$

Graphing Calculator Solution:

Graph $y_1 = \dfrac{-16x^2}{0.434(23.32)^2} + 1.15x + 8$ in the window $[0, 20] \times [0, 20]$.

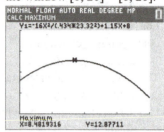

The y-coordinate of the vertex represents the maximum height of the basketball, which is approximately 12.88 ft.

65. Let $x = $ one number. Then $20 - x = $ the other number. Now find the maximum of $x(20 - x)$ by finding the vertex of the function:

$$f(x) = x(20 - x) = 20x - x^2 = -(x^2 - 20x)$$

$$= -(x^2 - 20x + 100) + 100 \quad \text{complete the square}$$

$$= -(x - 10)^2 + 100$$

The maximum occurs at the vertex. So the two numbers are 10 and $20 - 10 = 10$. Confirm graphically:

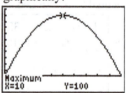

67. To find the number of seconds the object will take to reach its maximum height, find the vertex of the function:

$$s(t) = -16t^2 + 32t = -16(t^2 - 2t)$$

$$= -16(t^2 - 2t + 1) + 16 \quad \text{complete the square}$$

$$= -16(t - 1)^2 + 16$$

The object will reach its maximum height at $t = 1$ second. The maximum height is the value of the function at the vertex, 16 feet. Confirm graphically:

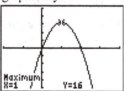

69. $f(x) = 0.0095x^2 - 0.0076x + 2.660$

The year 2018 is represented by $x = 16$

$f(16) = 0.0095(16)^2 - 0.0076(16) + 2.660$
$= 4.9704$

The model predicts that a pound of chocolate chip cookies will cost $4.97 is 2018.

71. $f(x) = 0.7714x^2 - 3.693x + 297.9$

We seek the x-coordinate of the vertex.
Algebraic Solution:
The parabola opens downward, so the vertex is the maximum point. Use the vertex formula to find the x-coordinate of the vertex.

$$x = -\frac{b}{2a} = -\frac{-3.693}{2(0.7714)} \approx 2.4$$

Graphing Calculator Solution:
Graph $y_1 = 0.7714x^2 - 3.693x + 297.9$ in the window $[0, 10] \times [0, 400]$.

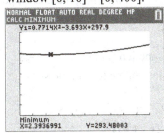

Based on the model, spending on shoes and clothing reached a maximum about two years after 2000, during the year 2002.

73. (a) Plot the 5 points given.

(b) Use the quadratic regression function on the graphing calculator to find the equation. The equation is

$f(x) = -0.2857x^2 + 1.503x + 19.13$.

(c) Plotting the points together with $f(x)$, we see that f models the data almost exactly.

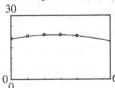

(d) $x = 5$ corresponds to the year 2013.
Algebraic Solution:

$f(5) = -0.2857(5)^2 + 1.503(5) + 19.13$
≈ 19.5

Graphing Calculator Solution:

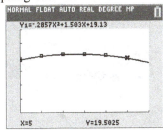

According to the model, in the year 2013, the total fall enrollments were approximately 19.5 million.

75. (a) Plot the 9 points given.

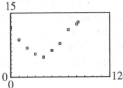

(b) Let the point $(40, 4.7)$, be the vertex.
Then, $f(x) = a(x - 40)^2 + 4.7$.

Next let the point $(20, 6.9)$ lie on the graph of the function and solve for a.

$f(20) = a(20 - 40)^2 + 4.7 = 6.9$
$a(-20)^2 + 4.7 = 6.9$
$400a + 4.7 = 6.9$

$$400a = 2.2 \Rightarrow a = \frac{2.2}{400} = 0.0055$$

Thus, $f(x) = 0.0055(x - 40)^2 + 4.7$.

(c) Plotting the points together with $f(x)$, we see that there is a relatively good fit.

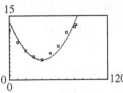

(d) The quadratic regression curve is

$g(x) = 0.0041x^2 - 0.3130x + 11.43$

(e) The year 2019 corresponds to $x = 89$.

$f(89) = 0.0055(89 - 40)^2 + 4.7 \approx 17.9$

$g(89) = 0.0041(89)^2 - 0.3130(89) + 11.43$
≈ 16.0

(*continued on next page*)

(continued)

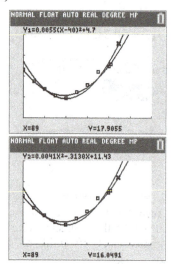

Using f, about 17.9% of the U.S. population will be foreign-born in 2019. Using g, about 16.0% of the U.S. population will be foreign-born in 2019. The unrounded regression equation gives 16.2%.

77. $y = x^2 - 10x + c$

An x-intercept occurs where $y = 0$, or $0 = x^2 - 10x + c$. There will be exactly one x-intercept if this equation has exactly one solution, or if the discriminant is zero.
$$b^2 - 4ac = 0 \Rightarrow (-10)^2 - 4(1)c = 0 \Rightarrow$$
$$100 - 4c = 0 \Rightarrow 100 = 4c \Rightarrow c = 25$$

79. x-intercepts $(2, 0)$ and $(5, 0)$, and y-intercept $(0, 5)$

The x-intercepts are $(2, 0)$ and $(5, 0)$, so f has linear factors of $x - 2$ and $x - 5$.
$$f(x) = a(x - 2)(x - 5) = a(x^2 - 7x + 10)$$
The y-intercept is $(0, 5)$, so $f(0) = 5$.
$$f(x) = a(0^2 - 7(0) + 10) = 5 \Rightarrow$$
$$10a = 5 \Rightarrow a = \tfrac{1}{2}$$
The required quadratic function is
$$f(x) = \tfrac{1}{2}(x^2 - 7x + 10) = \tfrac{1}{2}x^2 - \tfrac{7}{2}x + 5.$$

81. Use the distance formula,
$$d(P, R) = \sqrt{(x_1 - x_2)^2 + (y_1 - y_2)^2}, \text{ to find the}$$
distance between the points $P(x, 2x)$ and $R(1, 7)$, where P is any point on the line $y = 2x$.

$$d(P, R) = \sqrt{(x - 1)^2 + (2x - 7)^2}$$
$$= \sqrt{(x^2 - 2x + 1) + (4x^2 - 28x + 49)}$$
$$= \sqrt{5x^2 - 30x + 50}$$

Consider the equation $y = 5x^2 - 30x + 50$. This is the equation of a parabola opening upward, so the expression $5x^2 - 30x + 50$ has a minimum value, which is the y-value of the vertex. Complete the square to find the vertex.
$$y = 5x^2 - 30x + 50 = 5(x^2 - 6x) + 50$$
$$= 5(x^2 - 6x + 9 - 9) + 50$$
$$= 5(x^2 - 6x + 9) - 45 + 50 = 5(x - 3)^2 + 5$$
The vertex is $(3, 5)$, so the minimum value of $5x^2 - 30x + 50$ is 5 when $x = 3$. Thus, the minimum value of $\sqrt{5x^2 - 30x + 50}$ is $\sqrt{5}$ when $x = 3$. The point on the line $y = 2x$ for which $x = 3$ is $(3, 6)$. Thus, the closest point on the line $y = 2x$ to the point $(1, 7)$ is the point $(3, 6)$.

83. Graph the function $f(x) = x^2 + 2x - 8$. Complete the square to find the vertex, $(-1, -9)$.
$$y = x^2 + 2x - 8 = (x^2 + 2x) - 8$$
$$= (x^2 + 2x + 1 - 1) - 8$$
$$= (x^2 + 2x + 1) - 1 - 8 = (x + 1)^2 - 9$$
Use a table of values to find points on the graph.

x	y
-5	7
-4	0
-3	-5
-2	-8
-1	-9
0	-8
1	-5
2	0
3	7

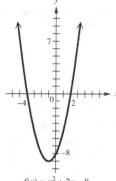

$f(x) = x^2 + 2x - 8$

85. The solution set of the inequality $x^2 + 2x - 8 < 0$ consists of all x-values for which the graph of f lies below the x-axis. By examining the graph of the function $f(x) = x^2 + 2x - 8$ from Exercise 83, we see that the graph lies below the x-axis when x is between the -4 and 2. Thus, the solution set is the open interval $(-4, 2)$.

87. $f(x) = x^2 - x - 6$

Complete the square to find the vertex..

$$y = x^2 - x - 6 = (x^2 - x) - 6$$
$$= \left(x^2 - x + \tfrac{1}{4} - \tfrac{1}{4}\right) - 6$$
$$= \left(x^2 - x + \tfrac{1}{4}\right) - \tfrac{1}{4} - 6 = \left(x - \tfrac{1}{2}\right)^2 - \tfrac{25}{4}$$

The vertex is $\left(\tfrac{1}{2}, -\tfrac{25}{4}\right)$. Now use a table of values to find points on the graph.

x	y
−4	14
−3	6
−2	0
−1	−4
0	−6
1	−6
2	−4
3	0
4	6

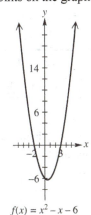

$f(x) = x^2 - x - 6$

The zeros of the function are at $x = -2$ and $x = 3$. The value of the function is less than 0 between these values. Thus, the solution set of $x^2 - x - 6 < 0$ is $(-2, 3)$.

89. $2x^2 - 9x \geq 18 \Rightarrow f(x) = 2x^2 - 9x - 18$

Complete the square to find the vertex.

$$y = 2x^2 - 9x - 18$$
$$y = 2\left(x^2 - \tfrac{9}{2}x\right) - 18 = 2\left(x^2 - \tfrac{9}{2}x + \tfrac{81}{16} - \tfrac{81}{16}\right) - 18$$
$$= 2\left(x^2 - \tfrac{9}{2}x + \tfrac{81}{16}\right) - 2\left(\tfrac{81}{16}\right) - 18$$
$$= 2\left(x - \tfrac{9}{4}\right)^2 - \tfrac{81}{8} - 18 = 2\left(x - \tfrac{9}{4}\right)^2 - \tfrac{225}{8}$$

The vertex is $\left(\tfrac{9}{4}, -\tfrac{225}{8}\right)$. Now use a table of values to find points on the graph.

x	y
−2	8
−1.5	0
−1	−7
0	−18
1	−25
2	−28
3	−27
4	−22
5	−13
6	0
7	17

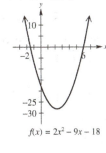

$f(x) = 2x^2 - 9x - 18$

The zeros of the function are at $x = -1.5$ and $x = 6$. The value of the function is less than 0 between these values. Thus, the solution set of $2x^2 - 9x \geq 18$ or $2x^2 - 9x - 18 \geq 0$ is $(-\infty, -1.5] \cup [6, \infty)$.

91. $f(x) = -x^2 + 4x + 1$

Complete the square to find the vertex.

$$y = -x^2 + 4x + 1$$
$$y = -\left(x^2 - 4x\right) + 1 = -\left(x^2 - 4x + 4 - 4\right) + 1$$
$$= -\left(x^2 - 4x + 4\right) + 4 + 1$$
$$= -(x - 2)^2 + 5$$

The vertex is $(2, 5)$ and the graph opens downward. Now use a table of values to find points on the graph.

x	y
−2	−11
−1	−4
0	1
1	4
2	5
3	4
4	1
5	−4
6	−11

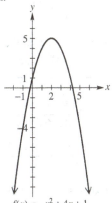

$f(x) = -x^2 + 4x + 1$

Use the quadratic formula to determine the zeros of the function.

$$x = \frac{-b \pm \sqrt{b^2 - 4ac}}{2a} = \frac{-4 \pm \sqrt{4^2 - 4(-1)(1)}}{2(-1)}$$
$$= \frac{-4 \pm \sqrt{20}}{-2} = \frac{-4 \pm 2\sqrt{5}}{-2} = 2 \pm \sqrt{5}$$

The value of the function is greater than or equal to 0 between these values. Thus, the solution set of $-x^2 + 4x + 1 \geq 0$ is $\left[2 - \sqrt{5}, \, 2 + \sqrt{5}\right]$.

Section 3.2   Synthetic Division

1. In arithmetic, the result of the division

$$5\overline{)19}$$
$$\begin{array}{r} 3 \\ \underline{15} \\ 4 \end{array}$$

can be written $19 = 5 \cdot \underline{3} + \underline{4}$.

3. To perform the division in Exercise 2 using synthetic division, we begin by writing the following.

$$1\overline{)1 \quad 2 \quad 3}$$

5. To perform the division $x - 3\overline{)x^3 + 6x^2 + 2x}$

using synthetic division, we begin by writing the following.

$$3\overline{)1 \quad 6 \quad 2 \quad 0}$$

7. Because $x + 1$ is in the form $x + k$, $k = -1$.

$$
\begin{array}{r}
-1\overline{)1 \quad 3 \quad 11 \quad 9} \\
\underline{-1 \; -2 \; -9} \\
1 \quad 2 \quad 9 \quad 0
\end{array}
$$

$$\frac{x^3 + 3x^2 + 11x + 9}{x + 1} = x^2 + 2x + 9$$

9. Express $x + 1$ in the form $x - k$ by writing it as $x - (-1)$. Thus $k = -1$.

$$
\begin{array}{r}
-1\overline{)5 \quad 5 \quad 2 \quad -1 \quad -3} \\
\underline{-5 \quad 0 \; -2 \quad 3} \\
5 \quad 0 \quad 2 \; -3 \quad 0
\end{array}
$$

$$\frac{5x^4 + 5x^3 + 2x^2 - x - 3}{x + 1} = 5x^3 + 2x - 3$$

11. Express $x + 4$ in the form $x - k$ by writing it as $x - (-4)$. Thus $k = -4$.

$$
\begin{array}{r}
-4\overline{)1 \quad 4 \quad 2 \quad 9 \quad 4} \\
\underline{-4 \quad 0 \; -8 \; -4} \\
1 \quad 0 \quad 2 \quad 1 \quad 0
\end{array}
$$

$$\frac{x^4 + 4x^3 + 2x^2 + 9x + 4}{x + 4} = x^3 + 2x + 1$$

13. Express $x + 2$ in the form $x - k$ by writing it as $x - (-2)$. Thus $k = -2$.

$$
\begin{array}{r}
-2\overline{)1 \quad 3 \quad 2 \quad 2 \quad 3 \quad 1} \\
\underline{-2 \; -2 \quad 0 \; -4 \quad 2} \\
1 \quad 1 \quad 0 \quad 2 \; -1 \quad 3
\end{array}
$$

$$\frac{x^5 + 3x^4 + 2x^3 + 2x^2 + 3x + 1}{x + 2}$$
$$= x^4 + x^3 + 2x - 1 + \frac{3}{x + 2}$$

15. Because $x - 2$ is in the form $x - k$, $k = 2$.

$$
\begin{array}{r}
2\overline{)-9 \quad 8 \; -7 \quad 2} \\
\underline{-18 \; -20 \; -54} \\
-9 \; -10 \; -27 \; -52
\end{array}
$$

$$\frac{-9x^3 + 8x^2 - 7x + 2}{x - 2} = -9x^2 - 10x - 27 + \frac{-52}{x - 2}$$

17. Because $x - \frac{1}{3}$ is in the form $x - k$, $k = \frac{1}{3}$.

$$
\begin{array}{r}
\frac{1}{3}\overline{)\frac{1}{3} \; -\frac{2}{9} \quad \frac{2}{27} \; -\frac{1}{81}} \\
\underline{\frac{1}{9} \; -\frac{1}{27} \quad \frac{1}{81}} \\
\frac{1}{3} \; -\frac{1}{9} \quad \frac{1}{27} \quad 0
\end{array}
$$

$$\frac{\frac{1}{3}x^3 - \frac{2}{9}x^2 + \frac{2}{27}x - \frac{1}{81}}{x - \frac{1}{3}} = \frac{1}{3}x^2 - \frac{1}{9}x + \frac{1}{27}$$

19. Because $x - 2$ is in the form $x - k$, $k = 2$. The constant term is missing, so include a 0.

$$
\begin{array}{r}
2\overline{)1 \; -3 \; -4 \quad 12 \quad 0} \\
\underline{2 \; -2 \; -12 \quad 0} \\
1 \; -1 \; -6 \quad 0 \quad 0
\end{array}
$$

$$\frac{x^4 - 3x^3 - 4x^2 + 12x}{x - 2} = x^3 - x^2 - 6x$$

21. Because $x - 1$ is in the form $x - k$, $k = 1$. The x^2- and x-terms are missing, so include 0's as their coefficients.

$$
\begin{array}{r}
1\overline{)1 \quad 0 \quad 0 \; -1} \\
\underline{1 \quad 1 \quad 1} \\
1 \quad 1 \quad 1 \quad 0
\end{array}
$$

$$\frac{x^3 - 1}{x - 1} = x^2 + x + 1$$

23. Express $x + 1$ in the form $x - k$ by writing it as $x - (-1)$. Thus $k = -1$. Because the x^4, x^3, x^2 and x – terms are missing, include 0's as their coefficients.

$$
\begin{array}{r}
-1\overline{)1 \quad 0 \quad 0 \quad 0 \quad 0 \quad 1} \\
\underline{-1 \quad 1 \; -1 \quad 1 \; -1} \\
1 \; -1 \quad 1 \; -1 \quad 1 \quad 0
\end{array}
$$

$$\frac{x^5 + 1}{x + 1} = x^4 - x^3 + x^2 - x + 1$$

25. $f(x) = 2x^3 + x^2 + x - 8;\ k = -1$

Use synthetic division to write the polynomial in the form $f(x) = (x - k)q(x) + r.$

$$
\begin{array}{r|rrr}
-1 & 2 & 1 & 1 & -8 \\
   &   & -2 & 1 & -2 \\
\hline
   & 2 & -1 & 2 & -10
\end{array}
$$

$f(x) = \left[x - (-1)\right](2x^2 - x + 2) - 10$

$\qquad = (x + 1)(2x^2 - x + 2) - 10$

27. $f(x) = x^3 + 4x^2 + 5x + 2;\ k = -2$

$$
\begin{array}{r|rrr}
-2 & 1 & 4 & 5 & 2 \\
   &   & -2 & -4 & -2 \\
\hline
   & 1 & 2 & 1 & 0
\end{array}
$$

$f(x) = \left[x - (-2)\right](x^2 + 2x + 1) + 0$

$\qquad = (x + 2)(x^2 + 2x + 1) + 0$

29. $f(x) = 4x^4 - 3x^3 - 20x^2 - x;\ k = 3$

$$
\begin{array}{r|rrrrr}
3 & 4 & -3 & -20 & -1 & 0 \\
  &   & 12 & 27 & 21 & 60 \\
\hline
  & 4 & 9 & 7 & 20 & 60
\end{array}
$$

$f(x) = (x - 3)(4x^3 + 9x^2 + 7x + 20) + 60$

31. $f(x) = 3x^4 + 4x^3 - 10x^2 + 15;\ k = -1$

$$
\begin{array}{r|rrrrr}
-1 & 3 & 4 & -10 & 0 & 15 \\
   &   & -3 & -1 & 11 & -11 \\
\hline
   & 3 & 1 & -11 & 11 & 4
\end{array}
$$

$f(x) = \left[x - (-1)\right](3x^3 + x^2 - 11x + 11) + 4$

$\qquad = (x + 1)(3x^3 + x^2 - 11x + 11) + 4$

33. $f(x) = x^2 + 5x + 6;\ k = -2$

$$
\begin{array}{r|rrr}
-2 & 1 & 5 & 6 \\
   &   & -2 & -6 \\
\hline
   & 1 & 3 & 0
\end{array}
$$

$f(-2) = 0$

35. $f(x) = 2x^2 - 3x - 3;\ k = 2$

$$
\begin{array}{r|rrr}
2 & 2 & -3 & -3 \\
  &   & 4 & 2 \\
\hline
  & 2 & 1 & -1
\end{array}
$$

$f(2) = -1$

37. $f(x) = x^3 - 4x^2 + 2x + 1;\ k = -1$

$$
\begin{array}{r|rrr}
-1 & 1 & -4 & 2 & 1 \\
   &   & -1 & 5 & -7 \\
\hline
   & 1 & -5 & 7 & -6
\end{array}
$$

$f(-1) = -6$

39. $f(x) = x^2 - 5x + 1;\ k = 2 + i$

$$
\begin{array}{r|rr}
2+i & 1 & -5 & 1 \\
    &   & 2+i & -7-i \\
\hline
    & 1 & -3+i & -6-i
\end{array}
$$

$f(2 + i) = -6 - i$

41. $f(x) = x^2 + 4;\ k = 2i$

$$
\begin{array}{r|rr}
2i & 1 & 0 & 4 \\
   &   & 2i & -4 \\
\hline
   & 1 & 2i & 0
\end{array}
$$

$f(2i) = 0$

43. $f(x) = 2x^5 - 10x^3 - 19x^2 - 50;\ k = 3$

$$
\begin{array}{r|rrrrrr}
3 & 2 & 0 & -10 & -19 & 0 & -50 \\
  &   & 6 & 18 & 24 & 15 & 45 \\
\hline
  & 2 & 6 & 8 & 5 & 15 & -5
\end{array}
$$

$f(3) = -5$

45. $f(x) = 6x^4 + x^3 - 8x^2 + 5x + 6;\ k = \frac{1}{2}$

$$
\begin{array}{r|rrrr}
\frac{1}{2} & 6 & 1 & -8 & 5 & 6 \\
            &   & 3 & 2 & -3 & 1 \\
\hline
            & 6 & 4 & -6 & 2 & 7
\end{array}
$$

$f\left(\frac{1}{2}\right) = 7$

47. To determine if $k = 2$ is a zero of
$f(x) = x^2 + 2x - 8,$ divide synthetically.

$$
\begin{array}{r|rr}
2 & 1 & 2 & -8 \\
  &   & 2 & 8 \\
\hline
  & 1 & 4 & 0
\end{array}
$$

Yes, 2 is a zero of $f(x)$ because $f(2) = 0.$

49. To determine if $k = 2$ is a zero of
$f(x) = x^3 - 3x^2 + 4x - 4,$ divide synthetically.

$$
\begin{array}{r|rrr}
2 & 1 & -3 & 4 & -4 \\
  &   & 2 & -2 & 4 \\
\hline
  & 1 & -1 & 2 & 0
\end{array}
$$

Yes, 2 is a zero of $f(x)$ because $f(2) = 0.$

51. To determine if $k = 1$ is a zero of
$f(x) = 2x^3 - 6x^2 - 9x + 4$, divide
synthetically.

$$
\begin{array}{r|rrrr}
1 & 2 & -6 & -9 & 4 \\
  &   & 2 & -4 & -13 \\
\hline
  & 2 & -4 & -13 & -9
\end{array}
$$

No, 1 is not a zero of $f(x)$ because
$f(1) = -9$.

53. To determine if $k = 0$ is a zero of
$f(x) = x^3 + 7x^2 + 10x$, divide synthetically.

$$
\begin{array}{r|rrrr}
0 & 1 & 7 & 10 & 0 \\
  &   & 0 & 0 & 0 \\
\hline
  & 1 & 7 & 10 & 0
\end{array}
$$

Yes, 0 is a zero of $f(x)$ because $f(0) = 0$.

55. To determine if $k = \frac{2}{5}$ is a zero of
$f(x) = 5x^4 + 2x^3 - x + 3$, divide synthetically.

$$
\begin{array}{r|rrrrr}
\frac{2}{5} & 5 & 2 & 0 & -1 & 3 \\
  &   & 2 & \frac{8}{5} & \frac{16}{25} & -\frac{18}{125} \\
\hline
  & 5 & 4 & \frac{8}{5} & -\frac{9}{25} & \frac{357}{125}
\end{array}
$$

No, $\frac{2}{5}$ is not a zero of $f(x)$ because
$f\left(\frac{2}{5}\right) = \frac{357}{125}$.

57. To determine if $k = 1 - i$ is a zero of
$f(x) = x^2 - 2x + 2$, divide synthetically.

$$
\begin{array}{r|rrr}
1-i & 1 & -2 & 2 \\
  &   & 1-i & -2 \\
\hline
  & 1 & -1-i & 0
\end{array}
$$

Yes, $1 - i$ is a zero of $f(x)$ because
$f(1 - i) = 0$.

59. To determine if $k = 2 + i$ is a zero of
$f(x) = x^2 + 3x + 4$, divide synthetically.

$$
\begin{array}{r|rrr}
2+i & 1 & 3 & 4 \\
  &   & 2+i & 9+7i \\
\hline
  & 1 & 5+i & 13+7i
\end{array}
$$

No, $2 + i$ is not a zero of $f(x)$ because
$f(2 + i) = 13 + 7i$.

61. To determine if $k = -\frac{3}{2}$ is a zero of
$f(x) = 4x^4 + x^2 + 17x + 3$, divide
synthetically. The x^3 term is missing, so
insert a 0.

$$
\begin{array}{r|rrrrr}
-\frac{3}{2} & 4 & 0 & 1 & 17 & 3 \\
  &   & -6 & 9 & -15 & -3 \\
\hline
  & 4 & -6 & 10 & 2 & 0
\end{array}
$$

Yes, $-\frac{3}{2}$ is a zero of $f(x)$ because
$f\left(-\frac{3}{2}\right) = 0$.

63. To determine if $k = 1 + i$ is a zero of
$f(x) = x^3 + 3x^2 - x + 1$, divide synthetically.

$$
\begin{array}{r|rrrr}
1+i & 1 & 3 & -1 & 1 \\
  &   & 1+i & 3+5i & -3+7i \\
\hline
  & 1 & 4+i & 2+5i & -2+7i
\end{array}
$$

No, $1 + i$ is not a zero of $f(x)$ because
$f(1 + i) = -2 + 7i$.

65.
$$
\begin{array}{r|rrrr}
-2 & 1 & -2 & -1 & 2 \\
  &   & -2 & 8 & -14 \\
\hline
  & 1 & -4 & 7 & -12
\end{array}
$$
$f(-2) = -12$
The coordinates of the corresponding point
are $(-2, -12)$.

67.
$$
\begin{array}{r|rrrr}
-\frac{1}{2} & 1 & -2 & -1 & 2 \\
  &   & -\frac{1}{2} & \frac{5}{4} & -\frac{1}{8} \\
\hline
  & 1 & -\frac{5}{2} & \frac{1}{4} & \frac{15}{8}
\end{array}
$$
$f\left(-\frac{1}{2}\right) = \frac{15}{8}$
The coordinates of the corresponding point
are $\left(-\frac{1}{2}, \frac{15}{8}\right)$.

69.
$$
\begin{array}{r|rrrr}
1 & 1 & -2 & -1 & 2 \\
  &   & 1 & -1 & -2 \\
\hline
  & 1 & -1 & -2 & 0
\end{array}
$$
$f(1) = 0$
The coordinates of the corresponding point
are $(1, 0)$.

71.
$$
\begin{array}{r|rrrr}
2 & 1 & -2 & -1 & 2 \\
  &   & 2 & 0 & -2 \\
\hline
  & 1 & 0 & -1 & 0
\end{array}
$$
$f(2) = 0$
The coordinates of the corresponding point
are $(2, 0)$.

73.

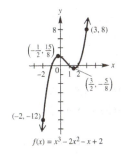

$f(x) = x^3 - 2x^2 - x + 2$

Section 3.3 Zeros of Polynomial Functions

1. True. Because $x - 1$ is a factor of
   $f(x) = x^6 - x^4 + 2x^2 - 2$, we are assured that
   $f(1) = 0$. This statement is justified by the
   factor theorem.

3. False. For $f(x) = (x + 2)^4 (x - 3)$, the number
   2 is a zero of multiplicity 4. To find the zero,
   set the factor equal to 0: $x + 2 = 0 \Rightarrow x = -2$
   2 is not a zero of the function; therefore, the
   statement is false. (It would be true to say that
   -2 is a zero of multiplicity 4.)

5. True. A polynomial function having degree 6
   and only real coefficients may have no real
   zeros.

7. False. If $z = 7 - 6i$, then $\overline{z} = 7 + 6i$.

9. $x^3 - 5x^2 + 3x + 1;\ x - 1$

   Let $f(x) = x^3 - 5x^2 + 3x + 1$. By the factor
   theorem, $x - 1$ will be a factor of $f(x)$ if and
   only if $f(1) = 0$. Use synthetic division and
   the remainder theorem.

   $$\begin{array}{r|rrrr} 1 & 1 & -5 & 3 & 1 \\ & & 1 & -4 & -1 \\ \hline & 1 & -4 & -1 & 0 \end{array}$$

   Because $f(1) = 0$, $x - 1$ is a factor of $f(x)$.

11. $2x^4 + 5x^3 - 8x^2 + 3x + 13;\ x + 1$

    Let $f(x) = 2x^4 + 5x^3 - 8x^2 + 3x + 13$. By the
    factor theorem, $x + 1$ will be a factor of $f(x)$
    if and only if $f(-1) = 0$. Use synthetic
    division and the remainder theorem.

    $$\begin{array}{r|rrrrr} -1 & 2 & 5 & -8 & 3 & 13 \\ & & -2 & -3 & 11 & -14 \\ \hline & 2 & 3 & -11 & 14 & -1 \end{array}$$

Because the remainder is -1, $f(-1) = -1$, so
$x + 1$ is not a factor of $f(x)$.

13. $-x^3 + 3x - 2;\ x + 2$

    Let $f(x) = -x^3 + 3x - 2$. By the factor
    theorem, $x + 2$ will be a factor of $f(x)$ if and
    only if $f(-2) = 0$. Use synthetic division and
    the remainder theorem.

    $$\begin{array}{r|rrrr} -2 & -1 & 0 & 3 & -2 \\ & & 2 & -4 & 2 \\ \hline & -1 & 2 & -1 & 0 \end{array}$$

    Because $f(-2) = 0$, $x + 2$ is a factor of $f(x)$.

15. $4x^2 + 2x + 54;\ x - 4$

    Let $f(x) = 4x^2 + 2x + 54$. By the factor
    theorem, $x - 4$ will be a factor of $f(x)$ if and
    only if $f(4) = 0$. Use synthetic division and
    the remainder theorem.

    $$\begin{array}{r|rrr} 4 & 4 & 2 & 54 \\ & & 16 & 72 \\ \hline & 4 & 18 & 126 \end{array}$$

    Because the remainder is 126, $f(4) = 126$, so
    $x - 4$ is not a factor of $f(x)$.

17. $x^3 + 2x^2 + 3;\ x - 1$

    Let $f(x) = x^3 + 2x^2 + 1$. By the factor
    theorem, $x - 1$ will be a factor of $f(x)$ if and
    only if $f(1) = 0$. Use synthetic division and
    the remainder theorem.

    $$\begin{array}{r|rrrr} 1 & 1 & 2 & 0 & 3 \\ & & 1 & 3 & 3 \\ \hline & 1 & 3 & 3 & 6 \end{array}$$

    Because $f(1) = 6$, $x - 1$ is not a factor of
    $f(x)$.

19. $2x^4 + 5x^3 - 2x^2 + 5x + 6;\ x + 3$

    Let $f(x) = 2x^4 + 5x^3 - 2x^2 + 5x + 6$. By the
    factor theorem, $x + 3$ will be a factor of $f(x)$ if
    and only if $f(-3) = 0$. Use synthetic division
    and the remainder theorem.

    $$\begin{array}{r|rrrrr} -3 & 2 & 5 & -2 & 5 & 6 \\ & & -6 & 3 & -3 & -6 \\ \hline & 2 & -1 & 1 & 2 & 0 \end{array}$$

    Because $f(-3) = 0$, $x + 3$ is a factor of $f(x)$.

21. $f(x) = 2x^3 - 3x^2 - 17x + 30$; $k = 2$

Because 2 is a zero of $f(x)$, $x - 2$ is a factor.

Divide $f(x)$ by $x - 2$.

$$2\overline{)2 \ -3 \ -17 \ \ \ 30}$$
$$ \ \ \ \ 4 \ \ \ \ 2 \ -30$$
$$\overline{2 \ \ \ 1 \ -15 \ \ \ \ 0}$$

Thus,

$$f(x) = (x - 2)(2x^2 + x - 15)$$
$$= (x - 2)(2x - 5)(x + 3).$$

23. $f(x) = 6x^3 + 13x^2 - 14x + 3$; $k = -3$

Because -3 is a zero of $f(x)$, $x + 3$ is a factor.

Divide $f(x)$ by $x + 3$.

$$-3\overline{)6 \ \ \ 13 \ -14 \ \ \ 3}$$
$$ \ -18 \ \ \ 15 \ -3$$
$$\overline{6 \ -5 \ \ \ \ 1 \ \ \ \ 0}$$

Thus,

$$f(x) = (x + 3)(6x^2 - 5x + 1)$$
$$= (x + 3)(3x - 1)(2x - 1)$$

25. $f(x) = 6x^3 + 25x^2 + 3x - 4$; $k = -4$

Because -4 is a zero of $f(x)$, $x + 4$ is a factor. Divide $f(x)$ by $x + 4$.

$$-4\overline{)6 \ \ \ 25 \ \ \ \ 3 \ -4}$$
$$ \ -24 \ -4 \ \ \ \ 4$$
$$\overline{6 \ \ \ \ 1 \ -1 \ \ \ \ 0}$$

Thus,

$$f(x) = (x + 4)(6x^2 + x - 1)$$
$$= (x + 4)(3x - 1)(2x + 1)$$

27. $f(x) = x^3 + (7 - 3i)x^2 + (12 - 21i)x - 36i$

$k = 3i$

Because $3i$ is a zero of $f(x)$, $x - 3i$ is a factor. Divide $f(x)$ by $x - 3i$.

$$3i\overline{)1 \ \ \ 7 - 3i \ \ 12 - 21i \ \ -36i}$$
$$ \ \ \ \ 3i \ \ \ \ \ \ \ 21i \ \ \ \ \ \ 36i$$
$$\overline{1 \ \ \ 7 \ \ \ \ \ \ \ 12 \ \ \ \ \ \ \ \ \ \ 0}$$

Thus,

$$f(x) = (x - 3i)(x^2 + 7x + 12)$$
$$= (x - 3i)(x + 4)(x + 3)$$

29. $f(x) = 2x^3 + (3 - 2i)x^2 + (-8 - 5i)x + (3 + 3i)$

$k = 1 + i$

Because $1 + i$ is a zero of $f(x)$, $x - (1 + i)$ is a factor. Divide $f(x)$ by $x - (1 + i)$.

$$1 + i\overline{)2 \ \ \ 3 - 2i \ \ -8 - 5i \ \ \ 3 + 3i}$$
$$ \ 2 + 2i \ \ \ 5 + 5i \ -3 - 3i$$
$$\overline{2 \ \ \ 5 \ \ \ \ \ \ \ -3 \ \ \ \ \ \ \ \ \ \ 0}$$

Thus,

$$f(x) = \left[x - (1 + i)\right](2x^2 + 5x - 3)$$
$$= \left[x - (1 + i)\right](2x - 1)(x + 3)$$

31. $f(x) = x^4 + 2x^3 - 7x^2 - 20x - 12$; $k = -2$ (multiplicity 2)

Because -2 is a zero of $f(x)$, $x + 2$ is a factor. Divide $f(x)$ by $x + 2$.

$$-2\overline{)1 \ \ \ 2 \ -7 \ -20 \ -12}$$
$$ \ -2 \ \ \ \ 0 \ \ \ 14 \ \ \ 12$$
$$\overline{1 \ \ \ 0 \ -7 \ \ -6 \ \ \ \ 0}$$

Thus,

$$f(x) = x^4 + 2x^3 - 7x^2 - 20x - 12$$
$$= (x + 2)(x^3 - 7x - 6)$$

Because -2 has multiplicity 2, divide the quotient polynomial by $x + 2$.

$$-2\overline{)1 \ \ \ 0 \ -7 \ -6}$$
$$ \ -2 \ \ \ 4 \ \ \ 6$$
$$\overline{1 \ -2 \ -3 \ \ \ 0}$$

Thus,

$$f(x) = (x + 2)(x^3 - 7x - 6)$$
$$= (x + 2)^2(x^2 - 2x - 3)$$
$$= (x + 2)^2(x + 1)(x - 3)$$

33. $f(x) = x^3 - x^2 - 4x - 6$; 3

Because 3 is a zero, first divide $f(x)$ by $x - 3$.

$$3\overline{)1 \ -1 \ -4 \ -6}$$
$$ \ \ \ \ 3 \ \ \ 6 \ \ \ 6$$
$$\overline{1 \ \ \ 2 \ \ \ 2 \ \ \ 0}$$

This gives $f(x) = (x - 3)(x^2 + 2x + 2)$.

Because $x^2 + 2x + 2$ cannot be factored, use the quadratic formula with $a = 1$, $b = 2$, and $c = 2$ to find the remaining two zeros.

$$x = \frac{-2 \pm \sqrt{4 - 4(1)(2)}}{2(1)} = \frac{-2 \pm \sqrt{4 - 8}}{2}$$
$$= \frac{-2 \pm \sqrt{-4}}{2} = \frac{-2 \pm 2i}{2} = -1 \pm i$$

The remaining zeros are $-1 \pm i$.

35. $f(x) = x^3 - 7x^2 + 17x - 15;\ 2 - i$

Because $2 - i$ is a zero, first divide $f(x)$ by $x - (2 - i)$.

$$
\begin{array}{r}
2-i\,\overline{)\,1 \quad -7 \qquad 17 \qquad -15} \\
\underline{\quad 2-i \quad -11+3i \quad 15} \\
1 \quad -5-i \quad 6+3i \qquad 0
\end{array}
$$

By the conjugate zeros theorem, $2 + i$ is also a zero, so divide the quotient polynomial from the first synthetic division by $x - (2 + i)$.

$$
\begin{array}{r}
2+i\,\overline{)\,1 \quad -5-i \qquad 6+3i} \\
\underline{\quad 2+i \quad -6-3i} \\
1 \quad -3 \qquad\quad 0
\end{array}
$$

This gives
$$f(x) = [x - (2 - i)][x - (2 + i)](x - 3).$$
The remaining zeros are $2 + i$ and 3.

37. $f(x) = x^4 + 5x^2 + 4;\ -i$

Because $-i$ is a zero, first divide $f(x)$ by $x + i$.

$$
\begin{array}{r}
-i\,\overline{)\,1 \quad 0 \quad 5 \quad 0 \quad 4} \\
\underline{\quad -i \quad -1 \quad -4i \quad -4} \\
1 \quad -i \quad 4 \quad -4i \quad 0
\end{array}
$$

By the conjugate zeros theorem, i is also a zero, so divide the quotient polynomial from the first synthetic division by $x - i$.

$$
\begin{array}{r}
i\,\overline{)\,1 \quad -i \quad 4 \quad -4i} \\
\underline{\quad i \quad 0 \quad 4i} \\
1 \quad 0 \quad 4 \quad 0
\end{array}
$$

The remaining zeros will be zeros of the new quotient polynomial, $x^2 + 4$. Find the remaining zeros by using the square root property.
$$x^2 + 4 = 0 \Rightarrow x^2 = -4 \Rightarrow x = \pm\sqrt{-4} \Rightarrow$$
$$x = \pm 2i$$
The other zeros are i and $\pm 2i$.

39. (a) $f(x) = x^3 - 2x^2 - 13x - 10$

p must be a factor of $a_0 = -10$ and q must be a factor of $a_3 = 1$. Thus, p can be $\pm 1, \pm 2, \pm 5, \pm 10$ and q can be ± 1. The possible rational zeros, $\frac{p}{q}$, are $\pm 1, \pm 2,$ $\pm 5, \pm 10$.

(b) The remainder theorem shows that -1 is a zero.

$$
\begin{array}{r}
-1\,\overline{)\,1 \quad -2 \quad -13 \quad -10} \\
\underline{\quad -1 \quad\ 3 \quad\ 10} \\
1 \quad -3 \quad -10 \quad\ 0
\end{array}
$$

The new quotient polynomial will be $x^2 - 3x - 10$.
$$x^2 - 3x - 10 = 0$$
$$(x + 2)(x - 5) = 0$$
$$x + 2 = 0 \Rightarrow x = -2 \text{ or } x - 5 = 0 \Rightarrow x = 5$$
The rational zeros are $-1, -2,$ and 5.

(c) Because the three zeros are $-1, -2,$ and 5, the factors are $x + 1, x + 2,$ and $x - 5$.
$$f(x) = (x + 1)(x + 2)(x - 5)$$

41. (a) $f(x) = x^3 + 6x^2 - x - 30$

p must be a factor of $a_0 = -30$ and q must be a factor of $a_3 = 1$. Thus, p can be $\pm 1, \pm 2, \pm 3, \pm 5, \pm 6, \pm 10, \pm 15, \pm 30$ and q can be ± 1. The possible zeros, $\frac{p}{q}$, are $\pm 1,$ $\pm 2, \pm 3, \pm 5, \pm 6, \pm 10, \pm 15, \pm 30$.

(b) The remainder theorem shows that -5 is a zero.

$$
\begin{array}{r}
-5\,\overline{)\,1 \quad 6 \quad -1 \quad -30} \\
\underline{\quad -5 \quad -5 \quad\ 30} \\
1 \quad 1 \quad -6 \quad\ 0
\end{array}
$$

The new quotient polynomial will be $x^2 + x - 6$.
$$x^2 + x - 6 = 0$$
$$(x + 3)(x - 2) = 0$$
$$x + 3 = 0 \Rightarrow x = -3 \text{ or } x - 2 = 0 \Rightarrow x = 2$$
The rational zeros are $-5, -3,$ and 2.

(c) The three zeros are $-5, -3,$ and 2, so the factors are $x + 5, x + 3,$ and $x - 2$.
$$f(x) = (x + 5)(x + 3)(x - 2)$$

43. (a) $f(x) = 6x^3 + 17x^2 - 31x - 12$

p must be a factor of $a_0 = -12$ and q must be a factor of $a_3 = 6$. Thus, p can be $\pm 1,$ $\pm 2, \pm 3, \pm 4, \pm 6, \pm 12$ and q can be $\pm 1, \pm 2,$ $\pm 3, \pm 6$. The possible zeros, $\frac{p}{q}$, are $\pm 1,$

$\pm 2, \pm 3, \pm 4, \pm 6, \pm 12, \pm\frac{1}{2}, \pm\frac{3}{2}, \pm\frac{1}{3},$

$\pm\frac{2}{3}, \pm\frac{4}{3}, \pm\frac{1}{6}.$

(b) The remainder theorem shows that -4 is a zero.

$$-4)\overline{\begin{array}{rrrr} 6 & 17 & -31 & -12 \\ & -24 & 28 & 12 \\ \hline 6 & -7 & -3 & 0 \end{array}}$$

The new quotient polynomial is
$6x^2 - 7x - 3$.
$6x^2 - 7x - 3 = 0 \Rightarrow (3x + 1)(2x - 3) = 0$
$3x + 1 = 0 \Rightarrow x = -\frac{1}{3}$ or
$2x - 3 = 0 \Rightarrow x = \frac{3}{2}$
The rational zeros are $-4, -\frac{1}{3}$, and $\frac{3}{2}$.

(c) The three zeros are $-4, -\frac{1}{3}$, and $\frac{3}{2}$, so the factors are $x + 4$, $3x + 1$, and $2x - 3$.
$f(x) = (x + 4)(3x + 1)(2x - 3)$

45. (a) $f(x) = 24x^3 + 40x^2 - 2x - 12$
p must be a factor of $a_0 = -12$ and q must be a factor of $a_3 = 24$. Thus, p can be ± 1, $\pm 2, \pm 3, \pm 4, \pm 6, \pm 12$ and q can be $\pm 1, \pm 2$, $\pm 3, \pm 4, \pm 6, \pm 8, \pm 12, \pm 24$. The possible zeros, $\frac{p}{q}$, are $\pm 1, \pm 2, \pm 3, \pm 4, \pm 6, \pm 12$,
$\pm \frac{1}{2}, \pm \frac{3}{2}, \pm \frac{1}{3}, \pm \frac{2}{3}, \pm \frac{4}{3}, \pm \frac{1}{4}$,
$\pm \frac{3}{4}, \pm \frac{1}{6}, \pm \frac{1}{8}, \pm \frac{3}{8}, \pm \frac{1}{12}, \pm \frac{1}{24}$.

(b) The remainder theorem shows that $-\frac{3}{2}$ is a zero.

$$-\tfrac{3}{2})\overline{\begin{array}{rrrr} 24 & 40 & -2 & -12 \\ & -36 & -6 & 12 \\ \hline 24 & 4 & -8 & 0 \end{array}}$$

The new quotient polynomial is
$24x^2 + 4x - 8$.
$24x^2 + 4x - 8 = 0 \Rightarrow 4(6x^2 + x - 2) = 0$
$\Rightarrow 4(3x + 2)(2x - 1) = 0$
$3x + 2 = 0 \Rightarrow x = -\frac{2}{3}$ or
$2x - 1 = 0 \Rightarrow x = \frac{1}{2}$
The rational zeros are $-\frac{3}{2}, -\frac{2}{3}$, and $\frac{1}{2}$.

(c) The three rational zeros are $-\frac{3}{2}, -\frac{2}{3}$, and $\frac{1}{2}$, so the factors are $3x + 2$, $2x + 3$, and $2x - 1$.
$f(x) = 2(2x + 3)(3x + 2)(2x - 1)$
Note: Because $-\frac{3}{2}$ is a zero,
$$\begin{aligned} f(x) &= 24x^3 + 40x^2 - 2x - 12 \\ &= \left[x - \left(-\tfrac{3}{2}\right)\right]\left(24x^2 + 4x - 8\right) \\ &= \left(x + \tfrac{3}{2}\right)\left[4(6x^2 + x - 2)\right] \\ &= 4\left(x + \tfrac{3}{2}\right)(3x + 2)(2x - 1) \\ &= 2\left[2\left(x + \tfrac{3}{2}\right)\right](3x + 2)(2x - 1) \\ &= 2(2x + 3)(3x + 2)(2x - 1) \end{aligned}$$

47. $f(x) = (x - 2)^3 \left(x^2 - 7\right)$
To find the zeros, let $f(x) = 0$. Set each factor equal to zero and solve for x.
$(x - 2)^3 \left(x^2 - 7\right) = 0$
$(x - 2)^3 = 0$ or $x^2 - 7 = 0$
$\quad x - 2 = 0 \qquad\qquad x^2 = 7$
$\quad\quad x = 2 \qquad\qquad\quad x = \pm\sqrt{7}$
The zeros are 2 (multiplicity 3) and $\pm\sqrt{7}$.

49. $f(x) = 3x(x - 2)(x + 3)(x^2 - 1)$
To find the zeros, let $f(x) = 0$.
Set each factor equal to zero and solve for x.
$3x = 0 \Rightarrow x = 0$
$x - 2 = 0 \Rightarrow x = 2$
$x + 3 = 0 \Rightarrow x = -3$
$x^2 - 1 = 0 \Rightarrow x^2 = 1 \Rightarrow x = \pm 1$
The zeros are 0, 2, –3, 1, and –1.

51. $f(x) = (x^2 + x - 2)^5 (x - 1 + \sqrt{3})^2$
To find the zeros, let $f(x) = 0$.
Set each factor equal to zero and solve for x.
$(x^2 + x - 2)^5 = 0 \Rightarrow x^2 + x - 2 = 0 \Rightarrow$
$(x + 2)(x - 1) = 0 \Rightarrow$
$x = -2$, multiplicity 5 or $x = 1$, multiplicity 5
$(x - 1 + \sqrt{3})^2 = 0 \Rightarrow x - 1 + \sqrt{3} = 0 \Rightarrow$
$x = 1 - \sqrt{3}$, multiplicity 2
The zeros are –2 (multiplicity 5), 1 (multiplicity 5) and $1 - \sqrt{3}$ (multiplicity 2).

53. Zeros of -3, 1, and 4, $f(2) = 30$

These three zeros give

$x - (-3) = x + 3$, $x - 1$, and $x - 4$ as factors of

$f(x)$. Because $f(x)$ is to have degree 3,

these are the only possible factors by the

number of zeros theorem. Therefore, $f(x)$

has the form $f(x) = a(x + 3)(x - 1)(x - 4)$ for

some real number a. To find a, use the fact

that $f(2) = 30$.

$f(2) = a(2 + 3)(2 - 1)(2 - 4) = 30 \Rightarrow$

$a(5)(1)(-2) = 30 \Rightarrow -10a = 30 \Rightarrow$

$a = -3$

Thus,

$f(x) = -3(x + 3)(x - 1)(x - 4)$

$= -3(x^2 + 2x - 3)(x - 4)$

$= -3(x^3 - 2x^2 - 11x + 12)$

$= -3x^3 + 6x^2 + 33x - 36$

55. Zeros of -2, 1, and 0; $f(-1) = -1$

These three zeros give

$x - (-2) = x + 2$, $x - 1$, and $x - 0 = x$ as factors

of $f(x)$. Because $f(x)$ is to have degree 3,

these are the only possible factors by the

number of zeros theorem. Therefore, $f(x)$

has the form $f(x) = a(x + 2)(x - 1)x$ for some

real number a. To find a, use the fact that

$f(-1) = -1$.

$f(-1) = a(-1 + 2)(-1 - 1)(-1) = -1 \Rightarrow$

$a(1)(-2)(-1) = -1 \Rightarrow 2a = -1 \Rightarrow a = -\frac{1}{2}$

Thus,

$f(x) = -\frac{1}{2}(x + 2)(x - 1)x$

$= -\frac{1}{2}(x^2 + x - 2)x = -\frac{1}{2}x^3 - \frac{1}{2}x^2 + x$

57. Zero of -3 having multiplicity 3; $f(3) = 36$

These three zeros give

$x - (-3) = x + 3$, $x - (-3) = x + 3$, and

$x - (-3) = x + 3$ as factors of $f(x)$. Because

$f(x)$ is to have degree 3, these are the only

possible factors by the number of zeros

theorem. Therefore, $f(x)$ has the form

$f(x) = a(x + 3)(x + 3)(x + 3) = a(x + 3)^3$ for

some real number a. To find a, use the fact

that $f(3) = 36$.

$f(3) = a(3 + 3)^3 = 36 \Rightarrow a(6)^3 = 36 \Rightarrow$

$216a = 36 \Rightarrow a = \frac{1}{6}$

Thus,

$f(x) = \frac{1}{6}(x + 3)^3 = \frac{1}{6}(x + 3)^2(x + 3)$

$= \frac{1}{6}(x^2 + 6x + 9)(x + 3)$

$= \frac{1}{6}(x^3 + 9x^2 + 27x + 27)$

$= \frac{1}{6}x^3 + \frac{3}{2}x^2 + \frac{9}{2}x + \frac{9}{2}$

59. Zero of 0 and zero of 1 having multiplicity 2;

$f(2) = 10$

These three zeros give x, $x - 1$, and $x - 1$ as

factors of $f(x)$. Because $f(x)$ is to have

degree 3, these are the only possible factors by

the number of zeros theorem. Therefore, $f(x)$

has the form

$f(x) = ax(x - 1)(x - 1) = ax(x - 1)^2$ for some

real number a. To find a, use the fact that

$f(2) = 10$.

$f(2) = a(2)(2 - 1)^2 = 10 \Rightarrow 2a = 10 \Rightarrow a = 5$

Thus,

$f(x) = 5x(x - 1)^2 = 5x(x^2 - 2x + 1)$

$= 5x^3 - 10x^2 + 5x$

In Exercises 61−77, we must find a polynomial of

least degree having only real coefficients and the

given zeros. For each of these exercises, other

answers may be possible.

61. $5 + i$ and $5 - i$

$f(x) = [x - (5 + i)][x - (5 - i)]$

$= (x - 5 - i)(x - 5 + i)$

$= [(x - 5) - i][(x - 5) + i]$

$= (x - 5)^2 - i^2$

$= (x^2 - 10x + 25) - i^2$

$= x^2 - 10x + 25 + 1 = x^2 - 10x + 26$

63. 0, i, and $1 + i$

By the conjugate zeros theorem, $-i$ and $1 - i$

must also be zeros.

$f(x) = x(x - i)(x + i)[x - (1 + i)][x - (1 - i)]$

$= x(x^2 - i^2)(x - 1 - i)(x - 1 + i)$

$= x(x^2 + 1)(x - 1 - i)(x - 1 + i)$

$= x(x^2 + 1)[(x - 1) - i][(x - 1) + i]$

$= x(x^2 + 1)[(x - 1)^2 - i^2]$

$= x(x^2 + 1)[(x^2 - 2x + 1) - i^2]$

(continued on next page)

(*continued*)

$$f(x) = x(x^2+1)(x^2-2x+1+1) = (x^3+x)(x^2-2x+2)$$
$$= x^5 - 2x^4 + 3x^3 - 2x^2 + 2x$$

65. $1+\sqrt{2}$, $1-\sqrt{2}$, and 1

$$f(x) = \left[x-\left(1+\sqrt{2}\right)\right]\left[x-\left(1-\sqrt{2}\right)\right](x-1) = \left(x-1-\sqrt{2}\right)\left(x-1+\sqrt{2}\right)(x-1)$$
$$= \left[(x-1)-\sqrt{2}\right]\left[(x-1)+\sqrt{2}\right](x-1) = \left[(x-1)^2 - \left(\sqrt{2}\right)^2\right](x-1)$$
$$= \left(x^2-2x+1-2\right)(x-1) = \left(x^2-2x-1\right)(x-1) = x^3 - 3x^2 + x + 1$$

67. $2-i$, 3, and -1

By the conjugate zeros theorem, $2+i$ must also be a zero.

$$f(x) = \left[x-(2+i)\right]\left[x-(2-i)\right](x-3)(x+1) = \left[(x-2-i)(x-2+i)\right]\left[(x-3)(x+1)\right]$$
$$= \left[(x-2)-i\right]\left[(x-2)+i\right]\left(x^2-2x-3\right) = \left[(x-2)^2 - i^2\right]\left(x^2-2x-3\right)$$
$$= \left[\left(x^2-4x+4\right)-i^2\right]\left(x^2-2x-3\right) = \left(x^2-4x+4+1\right)\left(x^2-2x-3\right)$$
$$= \left(x^2-4x+5\right)\left(x^2-2x-3\right) = x^4 - 6x^3 + 10x^2 + 2x - 15$$

69. 2 and $3+i$

By the conjugate zeros theorem, $3-i$ must also be a zero.

$$f(x) = (x-2)\left[x-(3+i)\right]\left[x-(3-i)\right] = (x-2)\left[(x-3)-i\right]\left[(x-3)+i\right] = (x-2)\left[(x-3)^2 - i^2\right]$$
$$= (x-2)\left(x^2-6x+9+1\right) = (x-2)\left(x^2-6x+10\right) = x^3 - 8x^2 + 22x - 20$$

71. $1-\sqrt{2}$, $1+\sqrt{2}$, and $1-i$

By the conjugate zeros theorem, $1+i$ must also be a zero.

$$f(x) = \left[x-\left(1-\sqrt{2}\right)\right]\left[x-\left(1+\sqrt{2}\right)\right]\cdot\left[x-(1-i)\right]\left[x-(1+i)\right]$$
$$= \left(x-1+\sqrt{2}\right)\left(x-1-\sqrt{2}\right)\cdot(x-1+i)(x-1-i) = \left[(x-1)+\sqrt{2}\right]\left[(x-1)-\sqrt{2}\right]\cdot\left[(x-1)+i\right]\left[(x-1)-i\right]$$
$$= \left[(x-1)^2-\left(\sqrt{2}\right)^2\right]\left[(x-1)^2-i^2\right] = \left(x^2-2x+1-2\right)\left[\left(x^2-2x+1\right)-i^2\right] = \left(x^2-2x-1\right)\left(x^2-2x+1+1\right)$$
$$= \left(x^2-2x-1\right)\left(x^2-2x+2\right) = x^4 - 4x^3 + 5x^2 - 2x - 2$$

73. $2-i$ and $6-3i$

By the conjugate zeros theorem, $2+i$ and $6+3i$ must also be zeros.

$$f(x) = \left[x-(2-i)\right]\left[x-(2+i)\right]\cdot\left[x-(6-3i)\right]\left[x-(6+3i)\right] = \left[(x-2+i)(x-2-i)\right]\cdot\left[(x-6+3i)(x-6-3i)\right]$$
$$= \left[(x-2)+i\right]\left[(x-2)-i\right]\cdot\left[(x-6)+3i\right]\left[(x-6)-3i\right] = \left[(x-2)^2 - i^2\right]\left[(x-6)^2 - (3i)^2\right]$$
$$= \left[\left(x^2-4x+4\right)-i^2\right]\cdot\left[\left(x^2-12x+36\right)-9i^2\right] = \left(x^2-4x+4+1\right)\left(x^2-12x+36+9\right)$$
$$= \left(x^2-4x+5\right)\left(x^2-12x+45\right) = x^4 - 16x^3 + 98x^2 - 240x + 225$$

75. $4, 1 - 2i,$ and $3 + 4i$

By the conjugate zeros theorem, $1 + 2i$ and $3 - 4i$ must also be zeros.

$$f(x) = (x-4)\big[x-(1-2i)\big]\big[x-(1+2i)\big]\cdot\big[x-(3+4i)\big]\big[x-(3-4i)\big]$$
$$= (x-4)(x-1+2i)(x-1-2i)\cdot(x-3-4i)(x-3+4i)$$
$$= (x-4)\big[(x-1)+2i\big]\big[(x-1)-2i\big]\cdot\big[(x-3)-4i\big]\big[(x-3)+4i\big]$$
$$= (x-4)\big[(x-1)^2 - (2i)^2\big]\big[(x-3)^2 - (4i)^2\big] = (x-4)\big(x^2 - 2x+1-4i^2\big)\cdot\big(x^2 - 6x+9-16i^2\big)$$
$$= (x-4)\big(x^2 - 2x+1+4\big)\big(x^2 - 6x+9+16\big) = (x-4)\big(x^2 - 2x+5\big)\big(x^2 - 6x+25\big)$$
$$= (x-4)\big(x^4 - 8x^3 + 42x^2 - 80x + 125\big) = x^5 - 12x^4 + 74x^3 - 248x^2 + 445x - 500$$

77. $1 + 2i, 2$ (multiplicity 2).

By the conjugate zeros theorem, $1 - 2i$ must also be a zero.

$$f(x) = (x-2)^2\big[x-(1+2i)\big]\big[x-(1-2i)\big] = \big(x^2 - 4x+4\big)\big[(x-1)-2i\big]\big[(x-1)+2i\big]$$
$$= \big(x^2 - 4x+4\big)\big[(x-1)^2 - (2i)^2\big] = \big(x^2 - 4x+4\big)\big(x^2 - 2x+1-4i^2\big)$$
$$= \big(x^2 - 4x+4\big)\big(x^2 - 2x+1+4\big) = \big(x^2 - 4x+4\big)\big(x^2 - 2x+5\big) = x^4 - 6x^3 + 17x^2 - 28x + 20$$

79. $f(x) = 2x^3 - 4x^2 + 2x + 7$

$f(x) = \underset{1}{2x^3 - \underset{2}{4x^2}} + 2x + 7$ has 2 variations in sign. f has either 2 or $2 - 2 = 0$ positive real zeros.

$f(-x) = -2x^3 - 4x^2 \underset{1}{-2x + 7}$ has 1 variation in sign. f has 1 negative real zero. Because the degree of the function is 3, the remaining zeros are nonreal complex. Thus, the different possibilities are

Positive	Negative	Nonreal Complex
2	1	0
0	1	2

81. $f(x) = 4x^3 - x^2 + 2x - 7$

$f(x) = \underset{1}{4x^3 - \underset{2}{x^2} + \underset{3}{2x - 7}}$ has 3 variations in sign. f has either 3 or $3 - 2 = 1$ positive real zeros.

$f(-x) = -4x^3 - x^2 - 2x - 7$ has no variations in sign, so f has no negative real zeros. Because the degree of the function is 3, the remaining zeros are nonreal complex. Thus, the different possibilities are

Positive	Negative	Nonreal Complex
3	0	0
1	0	2

83. $f(x) = 5x^4 + 3x^2 + 2x - 9$

$f(x) = 5x^4 + 3x^2 \underset{1}{+ 2x - 9}$ has 1 variation in sign. f has 1 positive real zero.

$f(-x) = 5x^4 + \underset{1}{3x^2 - 2x - 9}$ has 1 variation in sign. f has 1 negative real zero. Because the degree of the function is 4, the remaining zeros are nonreal complex. Thus, the different possibilities are

Positive	Negative	Nonreal Complex
1	1	2

85. $f(x) = -8x^4 + 3x^3 - 6x^2 + 5x - 7$

$f(x) = \underset{1}{-8x^4 + \underset{2}{3x^3} - \underset{3}{6x^2} + \underset{4}{5x - 7}}$ has 4 variation in sign. f has 4 or $4 - 2 = 2$ or $4 - 4 = 0$ positive real zeros.

$f(-x) = -8x^4 - 3x^3 - 6x^2 - 5x - 7$ has no variations in sign, so f has no negative real zeros. Because the degree of the function is 4, the remaining zeros are nonreal complex. Thus, the different possibilities are

Positive	Negative	Nonreal Complex
4	0	0
2	0	2
0	0	4

87. $f(x) = x^5 + 3x^4 - x^3 + 2x + 3$

$f(x) = x^5 \underbrace{+3x^4 - x^3}_{1} \underbrace{+ 2x + 3}_{2}$ has 2 variations

in sign. f has 2 or $2 - 2 = 0$ positive real zeros.

$f(-x) = \underbrace{-x^5 + 3x^4}_{1} \underbrace{+ x^3 - 2x}_{2} \underbrace{+ 3}_{3}$ has 3

variations in sign. f has 3 or $3 - 2 = 1$ negative real zeros. Because the degree of the function is 5, the remaining zeros are nonreal complex. Thus, the different possibilities are

Positive	Negative	Nonreal Complex
2	3	0
2	1	2
0	3	2
0	1	4

89. $f(x) = 2x^5 - 7x^3 + 6x + 8$

$f(x) = \underbrace{2x^5 - 7x^3}_{1} \underbrace{+ 6x + 8}_{2}$ has two variations

in sign, so f has 2 or $2 - 2 = 0$ positive real zeros.

$f(-x) = \underbrace{-2x^5 + 7x^3}_{1} \underbrace{- 6x}_{2} \underbrace{+ 8}_{3}$ has three

variations in sign, so f has 3 or $3 - 2 = 1$ negative real zeros. Because the degree of the function is 5, the remaining zeros are nonreal complex. Thus, the different possibilities are

Positive	Negative	Nonreal Complex
2	3	0
2	1	2
0	3	2
0	1	4

91. $f(x) = 5x^6 - 6x^5 + 7x^3 - 4x^2 + x + 2$

$f(x) = \underbrace{5x^6 - 6x^5}_{1} \underbrace{+ 7x^3 - 4x^2}_{3} \underbrace{+ x + 2}_{4}$ has four

variations in sign, so f has 4 or $4 - 2 = 2$ or $4 - 4 = 0$ positive real zeros.

$f(-x) = 5x^6 \underbrace{+6x^5 - 7x^3}_{1} - 4x^2 \underbrace{-x + 2}_{2}$ has two

variations in sign, so f has 2 or $2 - 2 = 0$ negative real zeros. Because the degree of the function is 6, the remaining zeros are nonreal complex.

Positive	Negative	Nonreal Complex
4	2	0
4	0	2
2	2	2

Positive	Negative	Nonreal Complex
2	0	4
0	2	4
0	0	6

93. $f(x) = 7x^5 + 6x^4 + 2x^3 + 9x^2 + x + 5$

$f(x) = 7x^5 + 6x^4 + 2x^3 + 9x^2 + x + 5$ has no variations in sign, so f has no positive real zeros.

$f(-x) = \underbrace{-7x^5 + 6x^4}_{1} \underbrace{- 2x^3 + 9x^2}_{2} \underbrace{- x}_{4} \underbrace{+ 5}_{5}$ has

five variations in sign, so f has 5 or $5 - 2 = 3$ or $5 - 4 = 1$ negative real zeros. Because the degree of the function is 5, the remaining zeros are nonreal complex. Thus, the different possibilities are

Positive	Negative	Nonreal Complex
0	5	0
0	3	2
0	1	4

95. $f(x) = x^4 + 2x^3 - 3x^2 + 24x - 180$

p must be a factor of $a_0 = -180$ and q must be a factor of $a_4 = 1$. Thus, p can be $\pm 1, \pm 2, \pm 3,$ $\pm 4, \pm 5, \pm 6, \pm 9, \pm 10, \pm 12, \pm 15, \pm 18, \pm 20,$ $\pm 30, \pm 36, \pm 45, \pm 60, \pm 90,$ and ± 180. q can be ± 1. The possible rational zeros, $\frac{p}{q}$, are $\pm 1,$ $\pm 2, \pm 3, \pm 4, \pm 5, \pm 6, \pm 9, \pm 10, \pm 12, \pm 15, \pm 18,$ $\pm 20, \pm 30, \pm 36, \pm 45, \pm 60, \pm 90,$ and ± 180. Using the remainder theorem and synthetic division, we find that one zero is $x = -5$.

$$\begin{array}{r|rrrr} -5 & 1 & 2 & -3 & 24 & -180 \\ & & -5 & 15 & -60 & 180 \\ \hline & 1 & -3 & 12 & -36 & 0 \end{array}$$

Setting the quotient $x^3 - 3x^2 + 12x - 36$ equal to zero and factoring by grouping, we have:

$0 = x^3 - 3x^2 + 12x - 36$
$= x^2(x - 3) + 12(x - 3)$
$= (x^2 + 12)(x - 3) \Rightarrow$

$x^2 + 12 = 0 \Rightarrow x^2 = -12 \Rightarrow x = \sqrt{-12} \Rightarrow$
$x = \pm 2i\sqrt{3}$ or $x - 3 = 0 \Rightarrow x = 3$

Thus, the zeros of $f(x)$ are $\left\{-5, 3, \pm 2i\sqrt{3}\right\}$.

97. $f(x) = x^4 + x^3 - 9x^2 + 11x - 4$

p must be a factor of $a_0 = -4$ and q must be a factor of $a_4 = 1$. Thus, p can be $\pm 1, \pm 2$, and ± 4. q can be ± 1. The possible rational zeros, $\frac{p}{q}$, are $\pm 1, \pm 2$, and ± 4. Using the remainder theorem and synthetic division, we find that one zero is $x = -4$.

$$
\begin{array}{r|rrrr}
-4 & 1 & 1 & -9 & 11 & -4 \\
   &   & -4 & 12 & -12 & 4 \\
\hline
   & 1 & -3 & 3 & -1 & 0
\end{array}
$$

Set the quotient $x^3 - 3x^2 + 3x - 1$ equal to zero and then factor. Note that $x^3 - 3x^2 + 3x - 1$ is a perfect cube.

$x^3 - 3x^2 + 3x - 1 = 0$

$(x-1)^3 = 0 \Rightarrow x = 1$ (with multiplicity 3)

Thus, the zeros of $f(x)$ are $\{-4, 1, 1, 1\}$.

99. $f(x) = 2x^5 + 11x^4 + 16x^3 + 15x^2 + 36x$

$a_0 = 0$, so 0 is a zero of the function.

Factoring, we have

$2x^5 + 11x^4 + 16x^3 + 15x^2 + 36x$

$= x(2x^4 + 11x^3 + 16x^2 + 15x + 36)$

To find a zero of the quotient, $2x^4 + 11x^3 + 16x^2 + 15x + 36$, p must be a factor of $a_0 = 36$ and q must be a factor of $a_4 = 2$.

Thus, p can be $\pm 1, \pm 2, \pm 3, \pm 4, \pm 6, \pm 9, \pm 12, \pm 18$, or ± 36. q can be ± 1 or ± 2. The possible rational zeros, $\frac{p}{q}$, are

$\pm \frac{1}{2}, \pm \frac{3}{2}, \pm \frac{9}{2}, \pm 1, \pm 2, \pm 3, \pm 4, \pm 6, \pm 8, \pm 9, \pm 12$ ± 18, or ± 36. Using the remainder theorem and synthetic division, we find that one zero is $x = -3$.

$$
\begin{array}{r|rrrrr}
-3 & 2 & 11 & 16 & 15 & 36 \\
   &   & -6 & -15 & -3 & -36 \\
\hline
   & 2 & 5 & 1 & 12 & 0
\end{array}
$$

Now find a zero of the quotient $2x^3 + 5x^2 + x + 12$. p must be a factor of $a_0 = 12$ and q must be a factor of $a_3 = 2$.

Thus, p can be $\pm 1, \pm 2, \pm 3, \pm 4, \pm 6$, or ± 12. q can be ± 1 or ± 2. The possible rational zeros, $\frac{p}{q}$, are $\pm \frac{1}{2}, \pm \frac{3}{2}, \pm 1, \pm 2, \pm 3, \pm 4, \pm 6$, or ± 12.

Using the remainder theorem and synthetic division, we find that one zero is $x = -3$

$$
\begin{array}{r|rrrr}
-3 & 2 & 5 & 1 & 12 \\
   &   & -6 & 3 & -12 \\
\hline
   & 2 & -1 & 4 & 0
\end{array}
$$

Setting the quotient $2x^2 - x + 4$ equal to zero and using the quadratic equation with $a = 2$, $b = -1$, and $c = 4$ to solve this, we have:

$$x = \frac{-(-1) \pm \sqrt{(-1)^2 - 4(2)(4)}}{2(2)}$$

$$= \frac{1 \pm \sqrt{1 - 32}}{4} = \frac{1 \pm i\sqrt{31}}{4}$$

Thus, the zeros of $f(x)$ are

$$\left\{ -3, -3, 0, \frac{1 \pm i\sqrt{31}}{4} \right\}.$$

101. $f(x) = x^5 - 6x^4 + 14x^3 - 20x^2 + 24x - 16$

p must be a factor of $a_0 = -16$ and q must be a factor of $a_5 = 1$. Thus, p can be $\pm 1, \pm 2, \pm 4, \pm 8$, or ± 16. q can be ± 1. The possible rational zeros, $\frac{p}{q}$, are $\pm 1, \pm 2, \pm 4, \pm 8$, or ± 16. Using the remainder theorem and synthetic division, we find that one zero is $x = 2$

$$
\begin{array}{r|rrrrrr}
2 & 1 & -6 & 14 & -20 & 24 & -16 \\
  &   & 2 & -8 & 12 & -16 & 16 \\
\hline
  & 1 & -4 & 6 & -8 & 8 & 0
\end{array}
$$

To find a zero of the quotient, $x^4 - 4x^3 + 6x^2 - 8x + 8$, p must be a factor of $a_0 = 8$ and q must be a factor of $a_4 = 1$. Thus, p can be $\pm 1, \pm 2, \pm 4$, or ± 8. q can be ± 1. The possible rational zeros, $\frac{p}{q}$, are $\pm 1, \pm 2, \pm 4$, or ± 8. Using the remainder theorem and synthetic division, we find that one zero is $x = 2$.

$$
\begin{array}{r|rrrrr}
2 & 1 & -4 & 6 & -8 & 8 \\
  &   & 2 & -4 & 4 & -8 \\
\hline
  & 1 & -2 & 2 & -4 & 0
\end{array}
$$

Now find a zero of the quotient $x^3 - 2x^2 + 2x - 4$. p must be a factor of $a_0 = -4$ and q must be a factor of $a_3 = 1$. Thus, p can be $\pm 1, \pm 2$, or ± 4. q can be ± 1. The possible rational zeros, $\frac{p}{q}$, are $\pm 1, \pm 2$, or ± 4.

Using the remainder theorem and synthetic division, we find that one zero is $x = 2$.

(continued on next page)

(*continued*)

$$2\overline{)\,1 \quad -2 \quad 2 \quad -4\,}$$
$$ \underline{\quad\quad 2 \quad 0 \quad 4\quad}$$
$$ 1 \quad 0 \quad 2 \quad 0$$

Setting the quotient $x^2 + 2$ equal to zero, we have: $x^2 + 2 = 0 \Rightarrow x^2 = -2 \Rightarrow x = \pm i\sqrt{2}$

Thus, the zeros of $f(x)$ are $\left\{2, 2, 2, \pm i\sqrt{2}\right\}$.

103. $f(x) = 2x^4 - x^3 + 7x^2 - 4x - 4$

p must be a factor of $a_0 = -4$ and q must be a factor of $a_4 = 2$. Thus, p can be $\pm 1, \pm 2,$ and ± 4. q can be ± 1 or ± 2. The possible rational zeros, $\frac{p}{q}$, are $\pm\frac{1}{2}, \pm 1, \pm 2,$ and ± 4. Using the remainder theorem and synthetic division, we find that one zero is $x = 1$.

$$1\overline{)\,2 \quad -1 \quad 7 \quad -4 \quad -4\,}$$
$$ \underline{\quad\quad 2 \quad 1 \quad 8 \quad 4\quad}$$
$$ 2 \quad 1 \quad 8 \quad 4 \quad 0$$

Setting the quotient $2x^3 + x^2 + 8x + 4$ equal to zero and factoring by grouping, we have:

$$2x^3 + x^2 + 8x + 4 = 0$$
$$x^2(2x + 1) + 4(2x + 1) = 0$$
$$(x^2 + 4)(2x + 1) = 0 \Rightarrow$$
$$x^2 + 4 = 0 \Rightarrow x^2 = -4 \Rightarrow x = \pm 2i \text{ or}$$
$$2x + 1 = 0 \Rightarrow x = -\tfrac{1}{2}$$

Thus, the zeros of $f(x)$ are $\left\{-\tfrac{1}{2}, 1, \pm 2i\right\}$.

105. $f(x) = 5x^3 - 9x^2 + 28x + 6$

p must be a factor of $a_0 = 6$ and q must be a factor of $a_3 = 5$. Thus, p can be $\pm 1, \pm 2, \pm 3,$ or ± 6. q can be ± 1 or ± 5. The possible rational zeros, $\frac{p}{q}$, are $\pm 1, \pm 2, \pm 3, \pm 6, \pm\frac{1}{5}, \pm\frac{2}{5}, \pm\frac{3}{5},$ or $\pm\frac{6}{5}$. Using the remainder theorem and synthetic division, we find that one zero is $x = -\tfrac{1}{5}$

$$-\tfrac{1}{5}\overline{)\,5 \quad -9 \quad 28 \quad 6\,}$$
$$\phantom{-\tfrac{1}{5})5\ } \underline{\quad\quad -1 \quad 2 \quad -6\quad}$$
$$\phantom{-\tfrac{1}{5})\,} 5 \quad -10 \quad 30 \quad 0$$

Setting the quotient
$$5x^2 - 10x + 30 = 5(x^2 - 2x + 6) \text{ equal to zero}$$
and using the quadratic equation with $a = 1$, $b = -2$, and $c = 6$ to solve this, we have:

$$x = \frac{-(-2) \pm \sqrt{(-2)^2 - 4(1)(6)}}{2(1)} = \frac{2 \pm \sqrt{4 - 24}}{2}$$
$$= \frac{2 \pm \sqrt{-20}}{2} = \frac{2 \pm 2i\sqrt{5}}{2} = 1 \pm i\sqrt{5}$$

Thus, the zeros of $f(x)$ are $\left\{-\tfrac{1}{5}, 1 \pm i\sqrt{5}\right\}$.

107. $f(x) = x^4 + 29x^2 + 100$

Letting $u = x^2$, we have
$x^4 + 29x^2 + 100 = u^2 + 29u + 100$. Set the function equal to zero, and solve for u using the quadratic formula with $a = 1$, $b = 29$ and $c = 100$.

$$u = \frac{-29 \pm \sqrt{29^2 - 4(1)(100)}}{2(1)}$$
$$= \frac{-29 \pm \sqrt{441}}{2} = \frac{-29 \pm 21}{2} = -25 \text{ or } u = -4$$

$u = -25 \Rightarrow x^2 = -25 \Rightarrow x = \pm 5i$
$u = -4 \Rightarrow x^2 = -4 \Rightarrow x = \pm 2i$

Thus, the zeros of $f(x)$ are $\left\{\pm 2i, \pm 5i\right\}$.

109. $f(x) = x^4 + 2x^2 + 1$

Setting $f(x)$ equal to 0, then factoring to solve for x, we have

$$x^4 + 2x^2 + 1 = 0$$
$$(x^2 + 1)^2 = 0$$
$$x^2 + 1 = 0 \Rightarrow x^2 = -1 \Rightarrow$$
$$x = \pm i \text{ (with multiplicity 2)}$$

Thus, the zeros of $f(x)$ are $\left\{\pm i, \pm i\right\}$.

111. $f(x) = x^4 - 6x^3 + 7x^2$

Setting $f(x)$ equal to zero and solving for x, we have

$$x^4 - 6x^3 + 7x^2 = 0 \Rightarrow x^2(x^2 - 6x + 7) = 0 \Rightarrow$$
$$x^2 = 0 \Rightarrow$$
$$x = 0 \text{ (with multiplicity 2) or}$$
$$x^2 - 6x + 7 = 0 \Rightarrow$$
$$x = \frac{-(-6) \pm \sqrt{(-6)^2 - 4(1)(7)}}{2(1)}$$
$$= \frac{6 \pm \sqrt{36 - 28}}{2} = \frac{6 \pm \sqrt{8}}{2} = 3 \pm \sqrt{2}$$

Thus, the zeros of $f(x)$ are $\left\{0, 0, 3 \pm \sqrt{2}\right\}$.

113. $f(x) = x^4 - 8x^3 + 29x^2 - 66x + 72$

p must be a factor of $a_0 = 72$ and q must be a factor of $a_4 = 1$. Thus, p can be $\pm1, \pm2, \pm3,$ $\pm4, \pm6, \pm8, \pm9, \pm12, \pm18, \pm24, \pm36,$ or ±72.

q can be ±1. The possible rational zeros, $\frac{p}{q}$, are $\pm1, \pm2, \pm3, \pm4, \pm6, \pm8, \pm9, \pm12, \pm18, \pm24,$ $\pm36,$ or ±72.

Using the remainder theorem and synthetic division, we find that one zero is $x = 3$.

$$3\overline{)1 \quad -8 \quad 29 \quad -66 \quad 72}$$
$$\underline{ \quad 3 \quad -15 \quad 42 \quad -72}$$
$$1 \quad -5 \quad 14 \quad -24 \quad 0$$

Now find a zero of the quotient $x^3 - 5x^2 + 14x - 24$. p must be a factor of $a_0 = -24$ and q must be a factor of $a_3 = 1$. Thus, p can be $\pm1, \pm2, \pm3, \pm4, \pm6, \pm8, \pm12,$ or ±24, while q can be ±1.

Using the remainder theorem and synthetic division, we find that one zero is $x = 3$.

$$3\overline{)1 \quad -5 \quad 14 \quad -24}$$
$$\underline{ \quad 3 \quad -6 \quad 24}$$
$$1 \quad -2 \quad 8 \quad 0$$

Setting the quotient $x^2 - 2x + 8$ equal to zero and solving for x using the quadratic formula with $a = 1$, $b = -2$, and $c = 8$, we have

$$x = \frac{-(-2) \pm \sqrt{(-2)^2 - 4(1)(8)}}{2(1)}$$

$$= \frac{2 \pm \sqrt{4 - 32}}{2} = \frac{2 \pm \sqrt{-28}}{2} = 1 \pm i\sqrt{7}$$

Thus, the zeros of $f(x)$ are $\left\{3, 3, 1 \pm i\sqrt{7}\right\}$.

115. $f(x) = x^6 - 9x^4 - 16x^2 + 144$

Let $u = x^2$. Then $x^6 - 9x^4 - 16x^2 + 144 \Rightarrow$ $u^3 - 9u^2 - 16u + 144$. Set $f(u)$ equal to 0, then factor by grouping to solve for u:

$$u^3 - 9u^2 - 16u + 144 = 0$$
$$u^2(u - 9) - 16(u - 9) = 0$$
$$(u^2 - 16)(u - 9) = 0 \Rightarrow$$

$u^2 - 16 = 0 \Rightarrow u^2 = 16 \Rightarrow u = \pm4$ or
$u - 9 = 0 \Rightarrow u = 9$

$u = 4 \Rightarrow x^2 = 4 \Rightarrow x = \pm2$
$u = -4 \Rightarrow x^2 = -4 \Rightarrow x = \pm2i$
$u = 9 \Rightarrow x^2 = 9 \Rightarrow x = \pm3$

Thus, the zeros of $f(x)$ are $\left\{\pm2, \pm3, \pm2i\right\}$.

For exercises 117–121, let $c = a + bi$ and $d = m + ni$.

117. $\overline{c + d} = \overline{(a + bi) + (m + ni)}$

$$= \overline{(a + m) + (b + n)i} = (a + m) - (b + n)i$$
$$= a + m - bi - ni = (a - bi) + (m - ni)$$
$$= \bar{c} + \bar{d}$$

119. If a is a real number, then a is of the form $a + 0 \cdot i$. $\bar{a} = \overline{a + 0 \cdot i} = a - 0i = a$. Therefore, the statement is true.

121. $x = \sqrt[3]{\frac{n}{2} + \sqrt{\left(\frac{n}{2}\right)^2 + \left(\frac{m}{3}\right)^3}} - \sqrt[3]{\frac{-n}{2} + \sqrt{\left(\frac{n}{2}\right)^2 + \left(\frac{m}{3}\right)^3}}$

can be used to solve a cubic equation $x^3 + mx = n$.

For $x^3 + 9x = 26$, $m = 9$ and $n = 26$. Substitute

$$x = \sqrt[3]{\frac{26}{2} + \sqrt{\left(\frac{26}{2}\right)^2 + \left(\frac{9}{3}\right)^3}}$$
$$- \sqrt[3]{-\frac{26}{2} + \sqrt{\left(\frac{26}{2}\right)^2 + \left(\frac{9}{3}\right)^3}}$$
$$= \sqrt[3]{13 + \sqrt{13^2 + 3^3}} - \sqrt[3]{-13 + \sqrt{13^2 + 3^3}}$$
$$= \sqrt[3]{13 + \sqrt{169 + 27}} - \sqrt[3]{-13 + \sqrt{169 + 27}}$$
$$= \sqrt[3]{13 + 14} - \sqrt[3]{-13 + 14}$$
$$= \sqrt[3]{27} - \sqrt[3]{1} = 3 - 1 = 2$$

Section 3.4 Polynomial Functions: Graphs, Applications, and Models

1. $y = x^3 - 3x^2 - 6x + 8$

The range of an odd-degree polynomial is $(-\infty, \infty)$. The y-intercept of the graph is 8. The graph fitting these criteria is A.

3. Graph C crosses the x-axis at one point, so the function has one real zero.

5. A polynomial of degree 3 can have at most 2 turning points. Graphs B and D have more than 2 turning points, so they cannot be graphs of cubic polynomial functions.

7. Graph B touches the x-axis at -5, so the function has 2 real zeros of -5. The two other real zeros are where the graph crosses the x-axis, at 0 and 3.

$$f(x) = x^4 + 7x^3 - 5x^2 - 75x = x(x + 5)^2(x - 3)$$

9. $f(x) = 2x^4$ is in the form $f(x) = ax^n$.

$|a| = 2 > 1$, so the graph is narrower than $f(x) = x^4$. It includes the points $(-2, 32)$, $(-1, 2)$, $(0, 0)$, $(1, 2)$, and $(2, 32)$. Connect these points with a smooth curve.

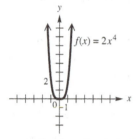

(a) The function is increasing on $(0, \infty)$.

(b) The function is decreasing on $(-\infty, 0)$.

11. $f(x) = -\frac{2}{3}x^5$ is in the form $f(x) = ax^n$.

$|a| = \frac{2}{3} < 1$, so the graph is broader than that of $f(x) = x^5$. Because $a = -\frac{2}{3}$ is a negative, the graph is the reflection of $f(x) = \frac{2}{3}x^5$ about the x-axis. It includes the points $\left(-2, \frac{64}{3}\right)$, $\left(-1, \frac{2}{3}\right)$, $(0, 0)$, $\left(1, -\frac{2}{3}\right)$, and $\left(2, -\frac{64}{3}\right)$. Connect these points with a smooth curve.

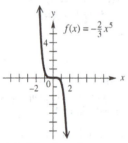

(a) The function never increases.

(b) The function is decreasing on $(-\infty, \infty)$.

13. $f(x) = \frac{1}{2}x^3 + 1$ is in the form $f(x) = ax^n + k$, with $|a| = \frac{1}{2} < 0$ and $k = 1$. The graph of $f(x) = \frac{1}{2}x^3 + 1$ looks like $y = x^3$ but is broader and is translated 1 unit up. The graph includes the points $(-2, -3)$, $\left(-1, \frac{1}{2}\right)$, $(0, 1)$, $\left(1, \frac{3}{2}\right)$, and $(2, 5)$.

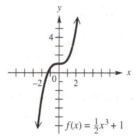

(a) The function is increasing on $(-\infty, \infty)$.

(b) The function never decreases.

15. $f(x) = -(x+1)^3 + 1 = -\left[x - (-1)\right]^3 + 1$
The graph can be obtained by reflecting the graph of $f(x) = x^3$ about the x-axis and then translating it 1 unit to the left and 1 unit up.

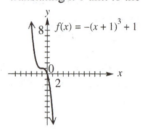

(a) The function never increases.

(b) The function is decreasing on $(-\infty, \infty)$.

17. $f(x) = (x-1)^4 + 2$ This graph has the same shape as $y = x^4$, but is translated 1 unit to the right and 2 units up.

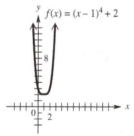

(a) The function is increasing on $(1, \infty)$.

(b) The function is decreasing on $(-\infty, 1)$.

19. $f(x) = \frac{1}{2}(x-2)^2 + 4$ is in the form $f(x) = a(x-h)^n + k$, with $|a| = \frac{1}{2} < 1$.

The graph is broader than that of $f(x) = x^2$. Because $h = 2$, the graph has been translated 2 units to the right. Also, because $k = 4$, the graph has been translated 4 units up.

(continued on next page)

(*continued*)

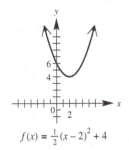

$$f(x) = \tfrac{1}{2}(x-2)^2 + 4$$

(a) The function is increasing on $(2, \infty)$.

(b) The function is decreasing on $(-\infty, 2)$.

21.   **23.**

25.   **27.**

29. $f(x) = x^3 + 5x^2 + 2x - 8$

Step 1: p must be a factor of $a_0 = -8$ and q must be a factor of $a_3 = 1$. Thus, p can be ± 1, $\pm 2, \pm 4, \pm 8$ and q can be ± 1. The possible rational zeros, $\frac{p}{q}$, are $\pm 1, \pm 2, \pm 4,$ and ± 8. The remainder theorem shows that -4 is a zero.

$$-4\overline{)\begin{array}{rrrr} 1 & 5 & 2 & -8 \\ & -4 & -4 & 8 \\ \hline 1 & 1 & -2 & 0 \end{array}}$$

The new quotient polynomial is $x^2 + x - 2$.
$$x^2 + x - 2 = 0$$
$$(x+2)(x-1) = 0$$
$$x + 2 = 0 \Rightarrow x = -2 \quad \text{or} \quad x - 1 = 0 \Rightarrow x = 1$$
The rational zeros are $-4, -2,$ and 1. The three zeros are $-4, -2,$ and 1, so the factors are $x + 4, x + 2,$ and $x - 1$ and thus
$$f(x) = (x+4)(x+2)(x-1).$$

Step 2: $f(0) = -8$, so plot $(0, -8)$.

Step 3: The x-intercepts divide the x-axis into four intervals:

Interval	Test Point	Value of $f(x)$	Sign of $f(x)$	Graph Above or Below x-Axis
$(-\infty, -4)$	-5	-18	Negative	Below
$(-4, -2)$	-3	4	Positive	Above

Interval	Test Point	Value of $f(x)$	Sign of $f(x)$	Graph Above or Below x-Axis
$(-2, 1)$	0	-8	Negative	Below
$(1, \infty)$	2	24	Positive	Above

Plot the x-intercepts, y-intercept and test points (the y-intercept is one of the test points) with a smooth curve to get the graph.

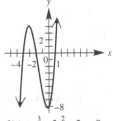

$$f(x) = x^3 + 5x^2 + 2x - 8$$

31. $f(x) = 2x(x-3)(x+2)$

Step 1: Set each factor equal to 0 and solve the resulting equations to find the zeros of the function.
$$2x = 0 \Rightarrow x = 0 \text{ or } x - 3 = 0 \Rightarrow x = 3 \text{ or}$$
$$x + 2 = 0 \Rightarrow x = -2$$
The three zeros, $-2, 0,$ and 3, divide the x-axis into four regions. Test a point in each region to find the sign of $f(x)$ in that region.

Step 2: $f(0) = 0$, so plot $(0, 0)$.

Step 3: The x-intercepts divide the x-axis into four intervals.

Interval	Test Point	Value of $f(x)$	Sign of $f(x)$	Graph Above or Below x-Axis
$(-\infty, -2)$	-3	-36	Negative	Below
$(-2, 0)$	-1	8	Positive	Above
$(0, 3)$	1	-12	Negative	Below
$(3, \infty)$	4	48	Positive	Above

Plot the x-intercepts, y-intercept (which is also an x-intercept in this exercise), and test points with a smooth curve to get the graph.

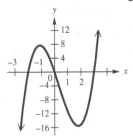

$$f(x) = 2x(x-3)(x+2)$$

33. $f(x) = x^2(x-2)(x+3)^2$

Step 1: Set each factor equal to 0 and solve the resulting equations to find the zeros of the function.

$x^2 = 0 \Rightarrow x = 0$ or $x - 2 = 0 \Rightarrow x = 2$ or

$(x+3)^2 = 0 \Rightarrow x + 3 = 0 \Rightarrow x = -3$

The zeros are –3, 0, and 2; divide the x-axis into four regions. Test a point in each region to find the sign of $f(x)$ in that region.

Step 2: $f(0) = 0$, so plot $(0,0)$.

Step 3: The x-intercepts divide the x-axis into four intervals.

Interval	Test Point	Value of $f(x)$	Sign of $f(x)$	Graph Above or Below x-Axis
$(-\infty, -3)$	–4	–96	Negative	Below
$(-3, 0)$	–1	–12	Negative	Below
$(0, 2)$	1	–16	Negative	Below
$(2, \infty)$	3	324	Positive	Above

Plot the x-intercepts, y-intercept (which is also an x-intercept in this exercise), and test points with a smooth curve to get the graph.

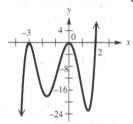

$f(x) = x^2(x-2)(x+3)^2$

35. $f(x) = (3x-1)(x+2)^2$

Step 1: Set each factor equal to 0 and solve the resulting equations to find the zeros of the function.

$3x - 1 = 0 \Rightarrow x = \frac{1}{3}$ or $x + 2 = 0 \Rightarrow x = -2$

The zeros are –2 and $\frac{1}{3}$, which divide the x-axis into three regions. Test a point in each region to find the sign of $f(x)$ in that region.

Step 2: $f(0) = -4$, so plot $(0, -4)$.

Step 3: The x-intercepts divide the x-axis into three intervals.

Interval	Test Point	Value of $f(x)$	Sign of $f(x)$	Graph Above or Below x-Axis
$(-\infty, -2)$	–3	–10	Negative	Below
$\left(-2, \frac{1}{3}\right)$	0	–4	Negative	Below
$\left(\frac{1}{3}, \infty\right)$	1	18	Positive	Above

Plot the x-intercepts, y-intercept, and test points (the y-intercept is one of the test points) with a smooth curve to get the graph.

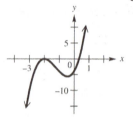

$f(x) = (3x-1)(x+2)^2$

37. $f(x) = x^3 + 5x^2 - x - 5$

Step 1: Find the zeros of the function by factoring by grouping.

$f(x) = x^3 + 5x^2 - x - 5 = x^2(x+5) - 1(x+5)$
$= (x+5)(x^2-1) = (x+5)(x+1)(x-1)$

$x + 5 = 0 \Rightarrow x = -5$ or $x + 1 = 0 \Rightarrow x = -1$ or $x - 1 = 0 \Rightarrow x = 1$

The zeros are –5, –1, and 1, which divide the x-axis into four regions. Test a point in each region to find the sign of $f(x)$ in that region.

Step 2: $f(0) = -5$, so plot $(0, -5)$.

Step 3: The x-intercepts divide the x-axis into four intervals.

Interval	Test Point	Value of $f(x)$	Sign of $f(x)$	Graph Above or Below x-Axis
$(-\infty, -5)$	–6	–35	Negative	Below
$(-5, -1)$	–2	9	Positive	Above
$(-1, 1)$	0	–5	Negative	Below
$(1, \infty)$	2	21	Positive	Above

Plot the x-intercepts, y-intercept, and test points (the y-intercept is one of the test points) with a smooth curve to get the graph.

(continued on next page)

(continued)

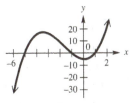

$f(x) = x^3 + 5x^2 - x - 5$

39. $f(x) = x^3 - x^2 - 2x$

Step 1: Find the zeros of the function by factoring out the common factor, x, and then factoring the resulting quadratic factor. Set each factor equal to 0 and solve the resulting equations.

$$f(x) = x^3 - x^2 - 2x$$
$$= x(x^2 - x - 2) = x(x+1)(x-2)$$

$$x = 0 \quad \text{or} \quad x+1 = 0 \quad \text{or} \quad x - 2 = 0$$
$$x = -1 \qquad\qquad x = 2$$

The zeros are -1, 0, and 2, which divide the x-axis into four regions. Test a point in each region to find the sign of $f(x)$ in that region.

Step 2: $f(0) = 0$, so plot $(0, 0)$.

Step 3: The x-intercepts divide the x-axis into four intervals.

Interval	Test Point	Value of $f(x)$	Sign of $f(x)$	Graph Above or Below x-Axis
$(-\infty, -1)$	-2	-8	Negative	Below
$(-1, 0)$	$-\frac{1}{2}$	$\frac{5}{8}$	Positive	Above
$(0, 2)$	1	-2	Negative	Below
$(2, \infty)$	3	12	Positive	Above

Plot the x-intercepts, y-intercept (which is also an x-intercept in this exercise), and test points with a smooth curve to get the graph.

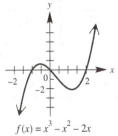

$f(x) = x^3 - x^2 - 2x$

41. $f(x) = 2x^3(x^2 - 4)(x - 1)$

Step 1: Find the zeros of the function by factoring the difference of two squares in the polynomial. Set each factor equal to 0 and solve the resulting equations.

$$f(x) = 2x^3(x^2 - 4)(x - 1)$$
$$= 2x^3(x + 2)(x - 2)(x - 1)$$

$$2x^3 = 0 \Rightarrow x = 0 \text{ or } x + 2 = 0 \Rightarrow x = -2 \text{ or}$$
$$x - 2 = 0 \Rightarrow x = 2 \text{ or } x - 1 = 0 \Rightarrow x = 1$$

The zeros are -2, 0, 1, and 2, which divide the x-axis into five regions. Test a point in each region to find the sign of $f(x)$ in that region.

Step 2: $f(0) = 0$, so plot $(0, 0)$.

Step 3: The x-intercepts divide the x-axis into four intervals.

Interval	Test Point	Value of $f(x)$	Sign of $f(x)$	Graph Above or Below x-Axis
$(-\infty, -2)$	-3	1080	Positive	Above
$(-2, 0)$	-1	-12	Negative	Below
$(0, 1)$	$\frac{1}{2}$	$\frac{15}{32} \approx 0.5$	Positive	Above
$(1, 2)$	$\frac{3}{2}$	$-\frac{189}{32} \approx -5.9$	Negative	Below
$(2, \infty)$	3	540	Positive	Above

Plot the x-intercepts, y-intercept (which is also an x-intercept in this exercise), and test points with a smooth curve to get the graph.

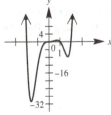

$f(x) = 2x^3(x^2 - 4)(x - 1)$

43. $f(x) = 2x^3 - 5x^2 - x + 6$

Step 1: Find the zeros of the function.
p must be a factor of $a_0 = 6$ and q must be a factor of $a_3 = 2$. Thus, p can be ± 1, ± 2, ± 3, or ± 6 and q can be ± 1 or ± 2. The possible rational zeros, $\frac{p}{q}$, are $\pm\frac{1}{2}$, ± 1, $\pm\frac{3}{2}$, ± 2, ± 3, or ± 6. The remainder theorem shows that 2 is a zero.

(continued on next page)

(*continued*)

$$2\overline{)\begin{array}{cccc} 2 & -5 & -1 & 6 \\ & 4 & -2 & -6 \\ \hline 2 & -1 & -3 & 0 \end{array}}$$

The new quotient polynomial is $2x^2 - x - 3$.

$2x^2 - x - 3 = 0$

$(2x - 3)(x + 1) = 0$

$2x - 3 = 0 \Rightarrow x = \frac{3}{2}$ or $x + 1 = 0 \Rightarrow x = -1$

The rational zeros are -1, $\frac{3}{2}$, and 2. The three

zeros are -1, $\frac{3}{2}$, and 2, so the factors are

$x + 1$, $x - \frac{3}{2}$, and $x - 2$ and thus

$f(x) = 2(x + 1)\left(x - \frac{3}{2}\right)(x - 2)$ or

$f(x) = (x + 1)(2x - 3)(x - 2)$.

Step 2: $f(0) = 6$, so plot $(0, 6)$.

Step 3: The x-intercepts divide the x-axis into four intervals.

Interval	Test Point	Value of $f(x)$	Sign of $f(x)$	Graph Above or Below x-Axis
$(-\infty, -1)$	-2	-28	Negative	Below
$\left(-1, \frac{3}{2}\right)$	1	2	Positive	Above
$\left(\frac{3}{2}, 2\right)$	$\frac{7}{4}$	$-\frac{11}{32}$	Negative	Below
$(2, \infty)$	3	12	Positive	Above

Plot the x-intercepts, y-intercept, and test points with a smooth curve to get the graph.

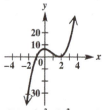

$f(x) = 2x^3 - 5x^2 - x + 6$

45. $f(x) = 3x^4 - 7x^3 - 6x^2 + 12x + 8$

Step 1: Find the zeros of the function.
p must be a factor of $a_0 = 8$ and q must be a factor of $a_4 = 3$. Thus, p can be $\pm 1, \pm 2, \pm 4,$ or ± 8, and q can be ± 1 or ± 3. The possible rational zeros, $\frac{p}{q}$, are $\pm\frac{1}{3}, \pm\frac{2}{3}, \pm 1, \pm\frac{4}{3}, \pm 2,$

$\pm\frac{8}{3}, \pm 4,$ or ± 8. The remainder theorem shows that 2 is a zero.

$$2\overline{)\begin{array}{ccccc} 3 & -7 & -6 & 12 & 8 \\ & 6 & -2 & -16 & -8 \\ \hline 3 & -1 & -8 & -4 & 0 \end{array}}$$

The new quotient polynomial is

$3x^3 - x^2 - 8x - 4$. Find the zeros of this

function. p must be a factor of $a_0 = -4$ and q

must be a factor of $a_3 = 3$. Thus, p can be ± 1,

± 2, or ± 4, and q can be ± 1 or ± 3. The possible

rational zeros, $\frac{p}{q}$, are $\pm\frac{1}{3}, \pm\frac{2}{3}, \pm 1, \pm\frac{4}{3}, \pm 2,$ or

± 4. The remainder theorem shows that -1 is a zero.

$$-1\overline{)\begin{array}{cccc} 3 & -1 & -8 & -4 \\ & -3 & 4 & 4 \\ \hline 3 & -4 & -4 & 0 \end{array}}$$

The new quotient polynomial is $3x^2 - 4x - 4$.

$3x^2 - 4x - 4 = 0$

$(3x + 2)(x - 2) = 0$

$3x + 2 = 0 \Rightarrow x = -\frac{2}{3}$ or $x - 2 = 0 \Rightarrow x = 2$

The rational zeros are -1, $-\frac{2}{3}$, 2, and 2. The

four zeros are -1, $-\frac{2}{3}$, 2, and 2, so the factors

are $x + 1$, $x + \frac{2}{3}$, $x - 2$, and $x - 2$, and thus

$f(x) = 3(x + 1)\left(x + \frac{2}{3}\right)(x - 2)^2$ or

$f(x) = (x + 1)(3x + 2)(x - 2)^2$.

Step 2: $f(0) = 8$, so plot $(0, 8)$.

Step 3: The x-intercepts divide the x-axis into four intervals.

Interval	Test Point	Value of $f(x)$	Sign of $f(x)$	Graph Above or Below x-Axis
$(-\infty, -1)$	-2	64	Positive	Above
$\left(-1, -\frac{2}{3}\right)$	$-\frac{5}{6}$	$-\frac{289}{432}$	Negative	Below
$\left(-\frac{2}{3}, 2\right)$	1	10	Positive	Above
$(2, \infty)$	3	44	Positive	Above

Plot the x-intercepts, y-intercept, and test points with a smooth curve to get the graph.

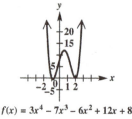

$f(x) = 3x^4 - 7x^3 - 6x^2 + 12x + 8$

47. $f(x) = 2x^2 - 7x + 4$; 2 and 3

Use synthetic division to find $f(2)$ and $f(3)$.

$$\begin{array}{r} 2\overline{)\ 2\ -7\ \ \ 4} \\ \underline{\ \ \ \ \ \ 4\ -6} \\ 2\ -3\ -2 \end{array} \qquad \begin{array}{r} 3\overline{)\ 2\ -7\ \ \ 4} \\ \underline{\ \ \ \ \ \ 6\ -3} \\ 2\ -1\ \ \ 1 \end{array}$$

Because $f(2) = -2$ is negative and $f(3) = 1$ is positive, there is a zero between 2 and 3.

49. $f(x) = 2x^3 - 5x^2 - 5x + 7$; 0 and 1

Because $f(0) = 7$ can easily be determined, use synthetic division only to find $f(1)$.

$$\begin{array}{r} 1\overline{)\ 2\ -5\ -5\ \ \ 7} \\ \underline{\ \ \ \ \ \ 2\ -3\ -8} \\ 2\ -3\ -8\ -1 \end{array}$$

Because $f(0) = 7$ is positive and $f(1) = -1$ is negative, there is a zero between 0 and 1.

51. $f(x) = 2x^4 - 4x^2 + 4x - 8$; 1 and 2

Use synthetic division to find $f(1)$ and $f(2)$.

$$\begin{array}{r} 1\overline{)\ 2\ \ \ 0\ -4\ \ \ 4\ -8} \\ \underline{\ \ \ \ \ \ 2\ \ \ 2\ -2\ \ \ 2} \\ 2\ \ \ 2\ -2\ \ \ 2\ -6 \end{array} \quad \begin{array}{r} 2\overline{)\ 2\ \ \ 0\ -4\ \ \ 4\ -8} \\ \underline{\ \ \ \ \ \ 4\ \ \ 8\ \ \ 8\ \ 24} \\ 2\ \ \ 4\ \ \ 4\ \ 12\ \ 16 \end{array}$$

Because $f(1) = -6$ is negative and $f(2) = 16$ is positive, there is a zero between 1 and 2.

53. $f(x) = x^4 + x^3 - 6x^2 - 20x - 16$; 3.2 and 3.3

Use synthetic division to find $f(3.2)$ and $f(3.3)$.

$$\begin{array}{r} 3.2\overline{)\ 1\ \ \ \ 1\ \ \ \ \ \ -6\ \ \ \ \ \ \ -20\ \ \ \ \ \ \ \ -16} \\ \underline{\ \ \ \ \ \ \ \ \ 3.2\ \ \ 13.44\ \ \ 23.808\ \ \ 12.1856} \\ 1\ \ \ \ 4.2\ \ \ \ 7.44\ \ \ \ \ 3.808\ \ \ -3.8144 \end{array}$$

$$\begin{array}{r} 3.3\overline{)\ 1\ \ \ \ 1\ \ \ \ \ \ -6\ \ \ \ \ \ \ -20\ \ \ \ \ \ \ \ -16} \\ \underline{\ \ \ \ \ \ \ \ \ 3.3\ \ \ 14.19\ \ \ 27.027\ \ \ 23.1891} \\ 1\ \ \ \ 4.3\ \ \ \ 8.19\ \ \ \ \ 7.027\ \ \ \ \ \ 7.1891 \end{array}$$

Because $f(3.2) = -3.8144$ is negative and $f(3.3) = 7.1891$ is positive, there is a zero between 3.2 and 3.3.

55. $f(x) = x^4 - 4x^3 - 20x^2 + 32x + 12$; -1 and 0

Because $f(0) = 12$ can easily be determined, use synthetic division only to find $f(-1)$.

$$\begin{array}{r} -1\overline{)\ 1\ \ \ -4\ -20\ \ \ 32\ \ \ 12} \\ \underline{\ \ \ \ \ \ \ \ \ -1\ \ \ \ \ 5\ \ \ 15\ -47} \\ 1\ \ \ -5\ -15\ \ \ 47\ -35 \end{array}$$

Because $f(-1) = -35$ is negative and $f(0) = 12$ is positive, there is a zero between -1 and 0.

57. $f(x) = x^4 - x^3 + 3x^2 - 8x + 8$; no real zero greater than 2.

Because $f(x)$ has real coefficients and the leading coefficient, 1, is positive, use the boundedness theorem. Divide $f(x)$ synthetically by $x - 2$. Because $2 > 0$ and all numbers in the last row are nonnegative, $f(x)$ has no zero greater than 2.

$$\begin{array}{r} 2\overline{)\ 1\ -1\ \ \ 3\ -8\ \ \ 8} \\ \underline{\ \ \ \ \ \ \ 2\ \ \ 2\ \ 10\ \ \ 4} \\ 1\ \ \ 1\ \ \ 5\ \ \ 2\ \ 12 \end{array}$$

59. $f(x) = x^4 + x^3 - x^2 + 3$; no real zero less than -2.

Because $f(x)$ has real coefficients and the leading coefficient is positive, use the boundedness theorem. Divide $f(x)$ synthetically by $x + 2 = x - (-2)$. Because $-2 < 0$ and the numbers in the last row alternate in sign, $f(x)$ has no zero less than -2.

$$\begin{array}{r} -2\overline{)\ 1\ \ \ 1\ -1\ \ \ 0\ \ \ 3} \\ \underline{\ \ \ \ \ \ \ \ \ -2\ \ \ 2\ -2\ \ \ 4} \\ 1\ -1\ \ \ 1\ -2\ \ \ 7 \end{array}$$

61. $f(x) = 3x^4 + 2x^3 - 4x^2 + x - 1$; no real zero greater than 1

Because $f(x)$ has real coefficients and the leading coefficient, 3, is positive, use the boundedness theorem. Divide $f(x)$ synthetically by $x - 1$. Because $1 > 0$ and all numbers in the last row are nonnegative, $f(x)$ has no zero greater than 1.

$$\begin{array}{r} 1\overline{)\ 3\ \ \ 2\ -4\ \ \ 1\ -1} \\ \underline{\ \ \ \ \ \ \ 3\ \ \ 5\ \ \ 1\ \ \ 2} \\ 3\ \ \ 5\ \ \ 1\ \ \ 2\ \ \ 1 \end{array}$$

63. $f(x) = x^5 - 3x^3 + x + 2$; no real zero greater than 2.

Because $f(x)$ has real coefficients and the leading coefficient, 1, is positive, use the boundedness theorem. Divide $f(x)$ synthetically by $x - 2$. Because $2 > 0$ and all numbers in the last row are nonnegative, $f(x)$ has no zero greater than 2.

$$\begin{array}{r} 2\overline{)\ 1\ \ \ 0\ -3\ \ \ 0\ \ \ 1\ \ \ 2} \\ \underline{\ \ \ \ \ \ \ 2\ \ \ 4\ \ \ 2\ \ \ 4\ \ 10} \\ 1\ \ \ 2\ \ \ 1\ \ \ 2\ \ \ 5\ \ 12 \end{array}$$

65. The graph shows that the zeros are –6, 2, and 5. The polynomial function has the form $f(x) = a(x+6)(x-2)(x-5)$. Because $(0, 30)$ is on the graph, $f(0) = 30$.
$$f(0) = a(0+6)(0-2)(0-5) \Rightarrow 30 = 60a \Rightarrow$$
$$\tfrac{1}{2} = a$$
A cubic polynomial that has the graph shown is $f(x) = \tfrac{1}{2}(x+6)(x-2)(x-5)$ or
$$f(x) = \tfrac{1}{2}x^3 - \tfrac{1}{2}x^2 - 16x + 30.$$

67. The graph shows that the zeros are -1 and 1. There is a turning point at both points. Because the graph crosses and is tangent to the x-axis at both $x = -1$ and $x = 1$, these are zeros with odd multiplicity. This has to be 3 in order to find a polynomial of least degree. The polynomial function has the form $f(x) = a(x-1)^3(x+1)^3$. Because $(0, -1)$ is on the graph, $f(0) = -1$.
$$f(0) = -1 = a(0-1)^3(0+1)^3 \Rightarrow a = 1$$
So the function is
$$f(x) = (x-1)^3(x+1)^3$$
$$= (x^3 - 3x^2 + 3x - 1)(x^3 + 3x^2 + 3x + 1)$$
$$= x^6 - 3x^4 + 3x^2 - 1$$

69. The graph shows that the zeros are -3 and 3. The graph is tangent to the x-axis at both $x = -3$ and $x = 3$, so these are zeros with even multiplicity. This has to be 2 in order to find a polynomial of least degree. The polynomial function has the form $f(x) = a(x-3)^2(x+3)^2$. Because $(0, 81)$ is on the graph, $f(0) = 81$.
$$f(0) = 81 = a(0-3)^2(0+3)^2 \Rightarrow 81 = 81a \Rightarrow$$
$$a = 1.$$
So the function is
$$f(x) = (x-3)^2(x+3)^2$$
$$= (x^2 - 6x + 9)(x^2 + 6x + 9)$$
$$= x^4 - 18x^2 + 81$$

71.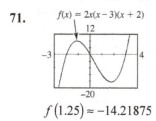
$f(x) = 2x(x-3)(x+2)$
$$f(1.25) \approx -14.21875$$

73.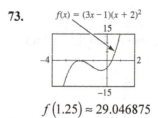
$f(x) = (3x-1)(x+2)^2$
$$f(1.25) \approx 29.046875$$

75. $f(x) = 2x^2 - 7x + 4$; 2 and 3
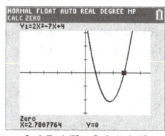
$[-4.7, 4.7] \times [-3.1, 3.1]$
The real zero between 2 and 3 is approximately 2.7807764.

77. $f(x) = 2x^4 - 4x^2 + 4x - 8$; 1 and 2
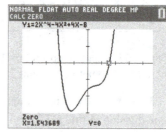
$[-4.7, 4.7] \times [-15, 10]$
The real zero between 1 and 2 is approximately 1.543689.

79. $f(x) = x^3 + 3x^2 - 2x - 6$

The highest degree term is x^3, so the graph will have end behavior similar to the graph of $f(x) = x^3$, which is downward at the left and upward at the right. There is at least one real zero because the polynomial is of odd degree. There are at most three real zeros because the polynomial is third-degree. A graphing calculator can be used to approximate each zero. The function is graphed in the window $[-4, 4] \times [-10, 5]$.

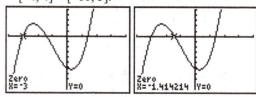

(continued on next page)

(*continued*)

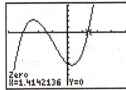

The graphs show that the zeros are approximately –3.0, –1.4 and 1.4.

81. $f(x) = -2x^4 - x^2 + x + 5$

The highest degree term is $-2x^4$ so the graph will have the same end behavior as the graph of $f(x) = -x^4$, which is downward at both the left and the right. Because $f(0) = 5 > 0$, the end behavior and the intermediate value theorem tell us that there must be at least one zero on each side of the y-axis, that is, at least one negative and one positive zero. A graphing calculator can be used to approximate each zero. The function is graphed in the window $[-4, 4] \times [-10, 10]$.

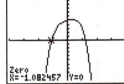

The graphs show that the only zeros are approximately –1.1 and 1.2

83. $f(x) = 2x^3 - 5x^2 - x + 1; [-1, 0]$

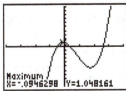

$[-4, 4] \times [-10, 10]$

The turning point is (–0.09, 1.05).

85. $f(x) = 2x^3 - 5x^2 - x + 1; [1.4, 2]$

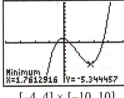

$[-4, 4] \times [-10, 10]$

The turning point is (1.76, –5.34).

87. $f(x) = x^3 + 4x^2 - 8x - 8; [-3.8, -3]$

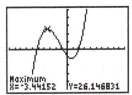

$[-10, 10] \times [-50, 50]$

The turning point is (–3.44, 26.15).

89. Answers will vary.

91. **(a)** In order for the length, width and height of the box to be positive quantities, $0 < x < 6$.

(b) The width of the rectangular base of the box would be $12 - 2x$. The length would be $18 - 2x$ and the height would be x. The volume would be the product of these three measures; therefore,

$$V(x) = x(18 - 2x)(12 - 2x)$$
$$= 4x^3 - 60x^2 + 216x$$

(c) When x is approximately 2.35 in. the maximum volume will be approximately 228.16 in.^{3}. Graph $y_1 = 4x^3 - 60x^2 + 216$ in the window $[0, 6] \times [0, 300]$.

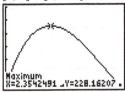

(d) When x is between 0.42 in. (approximately) and 5 in., the volume will be greater than 80 in.^{2}. Graph $y_1 = 4x^3 - 60x^2 + 216$ and $y_2 = 80$ in the window $[0, 6] \times [0, 300]$. Then find the intersections.

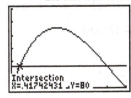 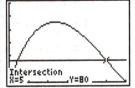

93. **(a)** length of the leg = $x - 1$. The domain is $x > 1$ or $(1, \infty)$.

(b) By the Pythagorean theorem,
$$a^2 + b^2 = c^2 \Rightarrow a^2 + (x-1)^2 = x^2 \Rightarrow$$
$$a^2 = x^2 - (x-1)^2 \Rightarrow a = \sqrt{x^2 - (x-1)^2}$$
Thus, the length of the other leg is
$$\sqrt{x^2 - (x-1)^2}.$$

(c) $\mathcal{A} = \frac{1}{2}bh \Rightarrow 84 = \frac{1}{2}(x-1)\sqrt{x^2 - (x-1)^2}$

Multiply by 2: $168 = (x-1)\sqrt{x^2 - (x-1)^2}$
Square both sides.
$$28,224 = (x-1)^2 \left[x^2 - (x-1)^2 \right]$$
$$28,224 = (x^2 - 2x + 1) \cdot \left[x^2 - (x^2 - 2x + 1) \right]$$
$$28,224 = (x^2 - 2x + 1)(2x - 1)$$
$$28,224 = 2x^3 - 5x^2 + 4x - 1$$
$$2x^3 - 5x^2 + 4x - 28,225 = 0$$

(d) Graph $y_1 = 2x^3 - 5x^2 + 4x - 28225$ in the window $[0, 30] \times [-28225, 26685]$. Then find the zero. We obtain $x = 25$. If $x = 25$, $x - 1 = 24$, and
$$\sqrt{x^2 - (x-1)^2} = \sqrt{625 - 576} = 7.$$ The hypotenuse is 25 in.; the legs are 24 in. and 7 in.

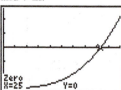

95. Use the following volume formulas:
$$V_{\text{cylinder}} = \pi r^2 h \text{ and}$$
$$V_{\text{hemisphere}} = \frac{1}{2} V_{\text{sphere}} = \frac{1}{2} \left(\frac{4}{3} \pi r^3 \right) = \frac{2}{3} \pi r^3$$
$$\pi r^2 h + 2 \left(\frac{2}{3} \pi r^3 \right) = \text{Total volume of tank } (V)$$
Let $V = 144\pi$, $h = 12$, and $r = x$.
$$\pi x^2 (12) + \frac{4}{3} \pi x^3 = 144\pi$$
$$\frac{4}{3} x^3 + 12x^2 = 144$$
$$\frac{4}{3} x^3 + 12x^2 - 144 = 0$$
$$4x^3 + 36x^2 - 432 = 0$$
$$x^3 + 9x^2 - 108 = 0$$

The end behavior of the graph of $f(x) = x^3 + 9x^2 - 108$, together with the negative y-intercept, tells us that this cubic polynomial must have one positive zero. (We are not interested in negative zeros because x represents the radius.)

Algebraic Solution:

Use synthetic division or a graphing calculator to locate this zero.

$$\begin{array}{r|rrrr} 3 & 1 & 9 & 0 & -108 \\ & & 3 & 36 & 108 \\ \hline & 1 & 12 & 36 & 0 \end{array}$$

Note that the bottom line of the synthetic division above implies the polynomial function $g(x) = x^2 + 12x + 36 = (x + 6)^2$, which has the zero -6 (multiplicity 2).

Graphing Calculator Solution:

Graph $y_1 = x^3 + 9x^2 - 108$ in the window $[-10, 10] \times [-150, 30]$ and then find the zero.

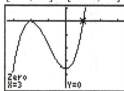

A radius of 3 feet would cause the volume of the tank to be 144π ft^3.

97. $f(x) = \frac{\pi}{3} x^3 - 5\pi x^2 + \frac{500\pi d}{3}$

Use a graphing calculator for the three parts of this exercise.

(a) When $d = 0.8$ we have
$$f(x) = \frac{\pi}{3} x^3 - 5\pi x^2 + \frac{500\pi (0.8)}{3}.$$
Graph this function in the window $[0, 10] \times [-100, 500]$. Then find the zero.

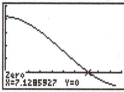

The approximate depth is 7.13 cm. The ball floats partly above the surface.

(b) When $d = 2.7$ we have

$$f(x) = \tfrac{\pi}{3}x^3 - 5\pi x^2 + \tfrac{500\pi(2.7)}{3}.$$

Graph this function in the window $[0, 10] \times [-100, 1500]$.

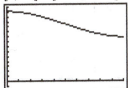

There is no x-intercept in this window. The sphere is more dense than water and sinks below the surface.

(c) When $d = 1$ we have

$$f(x) = \tfrac{\pi}{3}x^3 - 5\pi x^2 + \tfrac{500\pi}{3}.$$

Graph this function in the window $[0, 10] \times [-100, 600]$. Then find the zero.

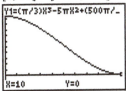

By tracing on the curve we see that the approximate depth is 10 cm. The balloon is submerged with its top even with the surface.

99. (a)

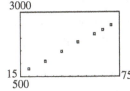

(b) The best-fitting linear function is $y = 33.93x + 113.4$.

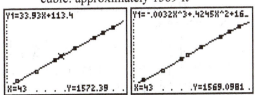

(c) The best-fitting cubic function is

$$y = -0.0032x^3 + 0.4245x^2 + 16.64x + 323.1.$$

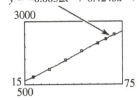

(d) linear: approximately 1572 ft
cubic: approximately 1569 ft

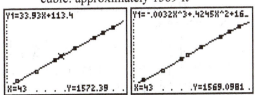

(e) The cubic function is a slightly better fit because only one data point is not on the curve.

101. The function in B, $g(x) = (x - 2004) + 27.1$

provides the best model, even though it's not a close fit.

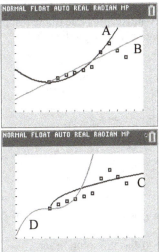

103. $f(x) = x^3 - 3x^2 - 6x + 8 = (x-4)(x-1)(x+2)$

Step 1: Set each factor equal to 0 and solve the resulting equations to find the zeros of the function.

$x - 4 = 0 \Rightarrow x = 4$ or $x - 1 = 0 \Rightarrow x = 1$ or $x + 2 = 0 \Rightarrow x = -2$

The zeros are -2, 1, and 4, which divide the x-axis into four regions. Test a point in each region to find the sign of $f(x)$ in that region.

Step 2: $f(0) = 8$, so plot $(0, 8)$.

(*continued on next page*)

(*continued*)

Step 3: The *x*-intercepts divide the *x*-axis into four intervals.

Interval	Test Point	Value of $f(x)$	Sign of $f(x)$	Graph Above or Below *x*-Axis
$(-\infty, -2)$	-3	-28	Negative	Below
$(-2, 1)$	0	8	Positive	Above
$(1, 4)$	2	-8	Negative	Below
$(4, \infty)$	5	28	Positive	Above

Plot the *x*-intercepts, *y*-intercept, and test points (the *y*-intercept is one of the test points) with a smooth curve to get the graph.

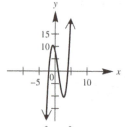

$$f(x) = x^3 - 3x^2 - 6x + 8$$
$$= (x - 4)(x - 1)(x + 2)$$

(a) $f(x) = 0$ has the solution set $\{-2, 1, 4\}$.

(b) $f(x) < 0$ has the solution set
$$(-\infty, -2) \cup (1, 4).$$

(c) $f(x) > 0$ has the solution set
$$(-2, 1) \cup (4, \infty).$$

105. $f(x) = 2x^4 - 9x^3 - 5x^2 + 57x - 45$
$$= (x - 3)^2 (2x + 5)(x - 1)$$

Step 1: Set each factor equal to 0 and solve the resulting equations to find the zeros of the function.
$$(x - 3)^2 = 0 \Rightarrow x = 3 \text{ or } 2x + 5 = 0 \Rightarrow x = -\tfrac{5}{2}$$
$$\text{or } x - 1 = 0 \Rightarrow x = 1$$

The zeros are $-\tfrac{5}{2} = -2.5$, 1, and 3, which divide the *x*-axis into four regions. Test a point in each region to find the sign of $f(x)$ in that region.

Step 2: $f(0) = -45$, so plot $(0, -45)$.

Step 3: The *x*-intercepts divide the *x*-axis into four intervals.

Interval	Test Point	Value of $f(x)$	Sign of $f(x)$	Graph Above or Below *x*-Axis
$(-\infty, -2.5)$	-3	144	Positive	Above
$(-2.5, 1)$	0	-45	Negative	Below
$(1, 3)$	2	9	Positive	Above
$(3, \infty)$	4	39	Positive	Above

Plot the *x*-intercepts, *y*-intercept, and test points (the *y*-intercept is one of the test points) with a smooth curve to get the graph.

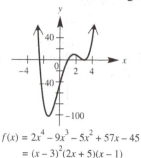

$$f(x) = 2x^4 - 9x^3 - 5x^2 + 57x - 45$$
$$= (x - 3)^2 (2x + 5)(x - 1)$$

(a) $f(x) = 0$ has the solution set $\{-2.5, 1, 3$ (multiplicity 2)$\}$.

(b) $f(x) < 0$ has the solution set $(-2.5, 1)$.

(c) $f(x) > 0$ has the solution set
$$(-\infty, -2.5) \cup (1, 3) \cup (3, \infty).$$

107. $f(x) = -x^4 - 4x^3 + 3x^2 + 18x$
$$= x(2 - x)(x + 3)^2$$

Step 1: Set each factor equal to 0 and solve the resulting equations to find the zeros of the function.
$$x = 0 \text{ or } 2 - x = 0 \Rightarrow 2 = x \text{ or}$$
$$(x + 3)^2 = 0 \Rightarrow x = -3$$

The zeros are 0, 2, and -3, which divide the *x*-axis into four regions. Test a point in each region to find the sign of $f(x)$ in that region.

Step 2: $f(0) = 0$, so plot $(0, 0)$.

(*continued on next page*)

(*continued*)

Step 3: The *x*-intercepts divide the *x*-axis into four intervals.

Interval	Test Point	Value of $f(x)$	Sign of $f(x)$	Graph Above or Below *x*-Axis
$(-\infty, -3)$	-4	-24	Negative	Below
$(-3, 0)$	-1	-12	Negative	Below
$(0, 2)$	1	16	Positive	Above
$(2, \infty)$	3	-108	Negative	Below

Plot the *x*-intercepts, *y*-intercept (which is also an *x*-intercept in this exercise), and test points with a smooth curve to get the graph.

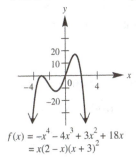

$$f(x) = -x^4 - 4x^3 + 3x^2 + 18x$$
$$= x(2 - x)(x + 3)^2$$

(a) $f(x) = 0$ has the solution set $\{-3 \text{ (multiplicity 2)}, 0, 2\}$.

(b) $f(x) \geq 0$ has the solution set $\{-3\} \cup [0, 2]$.

(c) $f(x) \leq 0$ has the solution set $(-\infty, 0] \cup [2, \infty)$.

Summary Exercises on Polynomial Functions, Zeros, and Graphs

1. $f(x) = 6x^3 - 41x^2 + 26x + 24$

(a) $f(x) = \underbrace{6x^3 - 41x^2}_{1 \quad 2} + 26x + 24$ has two variations in sign, so *f* has 2 or $2 - 2 = 0$ positive real zeros.

$f(-x) = -6x^3 - 41x^2 \underbrace{- 26x}_{1} + 24$ has one variation in sign, so *f* has 1 negative real zero. Because the degree of the function is 3, the remaining zeros are nonreal complex.

Positive	Negative	Nonreal Complex
2	1	0
0	1	2

(b) *p* must be a factor of $a_0 = 24$ and *q* must be a factor of $a_3 = 6$. Thus, *p* can be ± 1, ± 2, ± 3, ± 4, ± 6, ± 8, ± 12, or ± 24, and *q* can be ± 1, ± 2, ± 3, or ± 6. The possible rational zeros, $\frac{p}{q}$, are ± 1, ± 2, ± 3, ± 4, ± 6, ± 8, ± 12, ± 24, $\pm \frac{1}{2}$, $\pm \frac{3}{2}$, $\pm \frac{1}{3}$, $\pm \frac{2}{3}$, $\pm \frac{4}{3}$, $\pm \frac{8}{3}$, $\pm \frac{1}{6}$.

(c) The remainder theorem shows that 6 is a zero.

$$6\overline{)\begin{array}{cccc} 6 & -41 & 26 & 24 \\ & 36 & -30 & -24 \\ \hline 6 & -5 & -4 & 0 \end{array}}$$

Setting the new quotient polynomial $6x^2 - 5x - 4$ equal to 0 and factoring, we have

$$6x^2 - 5x - 4 = 0$$
$$(2x + 1)(3x - 4) = 0$$

$2x + 1 = 0$	$3x - 4 = 0$
$x = -\frac{1}{2}$	$x = \frac{4}{3}$

Thus, the rational zeros of $f(x)$ are $\left\{ 6, -\frac{1}{2}, \frac{4}{3} \right\}$.

(d) We identified three rational zeros, so there are no nonreal complex zeros.

3. $f(x) = 3x^4 - 5x^3 + 14x^2 - 20x + 8$

(a) $f(x) = \underbrace{3x^4}_{1} \underbrace{- 5x^3}_{2} \underbrace{+ 14x^2}_{3} \underbrace{- 20x}_{4} + 8$ has four variations in sign, so *f* has 4 or $4 - 2 = 2$ or $4 - 4 = 0$ positive real zeros.

$f(-x) = 3x^4 + 5x^3 + 14x^2 + 20x + 8$ has no variations in sign, so *f* has no negative real zeros. The degree of the function is 4, so the remaining zeros are nonreal complex.

Positive	Negative	Nonreal Complex
4	0	0
2	0	2
0	0	4

(b) p must be a factor of $a_0 = 8$ and q must be a factor of $a_3 = 3$. Thus, p can be ± 1, ± 2, ± 4, or ± 8, and q can be ± 1 or ± 3. The possible rational zeros, $\frac{p}{q}$, are ± 1, ± 2,

± 4, ± 8, $\pm \frac{1}{3}$, $\pm \frac{2}{3}$, $\pm \frac{4}{3}$, $\pm \frac{8}{3}$.

(c) The remainder theorem shows that 1 is a zero.

$$
\begin{array}{r|rrrrr}
1) & 3 & -5 & 14 & -20 & 8 \\
  &   & 3 & -2 & 12 & -8 \\
\hline
  & 3 & -2 & 12 & -8 & 0
\end{array}
$$

Setting the new quotient polynomial $3x^3 - 2x^2 + 12x - 8$ equal to 0 and factoring by grouping, we have

$$3x^3 - 2x^2 + 12x - 8 = 0$$
$$x^2(3x - 2) + 4(3x - 2) = 0$$
$$(3x - 2)(x^2 + 4) = 0$$

$$
\begin{array}{c|c}
3x - 2 = 0 & x^2 + 4 = 0 \\
x = \frac{2}{3} & x^2 = -4 \\
 & x = \pm 2i
\end{array}
$$

Thus, the rational zeros of $f(x)$ are 1 and $\frac{2}{3}$.

(d) The nonreal complex zeros are $\pm 2i$.

5. $f(x) = 6x^4 - 5x^3 - 11x^2 + 10x - 2$

(a) $f(x) = \underbrace{6x^4}_{} - \underbrace{5x^3}_{1} \underbrace{- 11x^2}_{2} \underbrace{+ 10x}_{3} - 2$ has

three variations in sign, so f has 3 or $3 - 2 = 1$ positive real zeros.

$f(-x) = 6x^4 + 5x^3 + 11x^2 \underbrace{+ 10x - 2}_{1}$ has

one variation in sign, so f has 1 negative real zero. The degree of the function is 4, so the remaining zeros are nonreal complex.

Positive	Negative	Nonreal Complex
3	1	0
1	1	2

(b) p must be a factor of $a_0 = -2$ and q must be a factor of $a_4 = 6$. Thus, p can be ± 1 or ± 2, and q can be ± 1, ± 2, ± 3, or ± 6. The possible rational zeros, $\frac{p}{q}$, are ± 1, ± 2,

$\pm \frac{1}{2}$, $\pm \frac{1}{3}$, $\pm \frac{2}{3}$, $\pm \frac{1}{6}$.

(c) The remainder theorem shows that $\frac{1}{3}$ is a zero.

$$
\begin{array}{r|rrrrr}
\frac{1}{3}) & 6 & -5 & -11 & 10 & -2 \\
 &   & 2 & -1 & -4 & 2 \\
\hline
 & 6 & -3 & -12 & 6 & 0
\end{array}
$$

Setting the new quotient polynomial $6x^3 - 3x^2 - 12x + 6$ equal to 0 and factoring by grouping, we have

$$6x^3 - 3x^2 - 12x + 6 = 0$$
$$3x^2(2x - 1) - 6(2x - 1) = 0$$
$$(3x^2 - 6)(2x - 1) = 0$$

$$
\begin{array}{c|c}
3x^2 - 6 = 0 & 2x - 1 = 0 \\
3x^2 = 6 & x = \frac{1}{2} \\
x^2 = 2 & \\
x = \pm\sqrt{2} &
\end{array}
$$

Thus, the rational zeros of $f(x)$ are $\frac{1}{3}$ and $\frac{1}{2}$.

(d) The real complex zeros are $\pm\sqrt{2}$.

7. $f(x) = x^5 - 6x^4 + 16x^3 - 24x^2 + 16x$

(a) $f(x) = \underbrace{x^5}_{} \underbrace{- 6x^4}_{1} \underbrace{+ 16x^3}_{2} \underbrace{- 24x^2}_{3} \underbrace{+ 16x}_{4}$ has

four variations in sign, so f has 4 or $4 - 2 = 2$ or $4 - 4 = 0$ positive real zeros.

$f(-x) = -x^5 - 6x^4 - 16x^3 - 24x^2 - 16x$

has no variations in sign, so f has no negative real zeros. The degree of the function is 5, so the remaining zeros are nonreal complex. However, nonreal complex roots occur in pairs, so the fifth zero is not positive, negative, or nonreal complex.

Positive	Negative	Nonreal Complex
4	0	0
2	0	2
0	0	4

If we factor out x, we have

$$f(x) = x^5 - 6x^4 + 16x^3 - 24x^2 + 16x$$
$$= x(x^4 - 6x^3 + 16x^2 - 24x + 16)$$

If we set $f(x) = 0$, then $x = 0$. Thus 0 is the remaining zero.

(b) One factor is x, so we must find the zeros of the other factor, $x^4 - 6x^3 + 16x^2 - 24x + 16$. p must be a factor of $a_0 = 16$ and q must be a factor of $a_4 = 1$. Thus, p can be ± 1, ± 2, ± 4, ± 8, or ± 16, and q can be ± 1. The possible rational zeros $\frac{p}{q}$, of this factor, are ± 1, ± 2, ± 4, ± 8, or ± 16. The possible rational zeros of the original function $f(x)$ are 0, ± 1, ± 2, ± 4, ± 8, or ± 16.

(c) The remainder theorem shows that 2 is a zero.

$$2\overline{)\,1 \quad -6 \quad 16 \quad -24 \quad 16}$$
$$ \quad 2 \quad -8 \quad 16 \quad -16$$
$$\overline{\,1 \quad -4 \quad 8 \quad -8 \quad 0}$$

The new quotient polynomial is $x^3 - 4x^2 + 8x - 8$. For this polynomial, p must be a factor of $a_0 = -8$ and q must be a factor of $a_3 = 1$. Thus, p can be ± 1, ± 2, ± 4, or ± 8, and q can be ± 1 The possible rational zeros $\frac{p}{q}$, are ± 1, ± 2, ± 4, or ± 8. The remainder theorem shows that 2 is a zero.

$$2\overline{)\,1 \quad -4 \quad 8 \quad -8}$$
$$ \quad 2 \quad -4 \quad 8$$
$$\overline{\,1 \quad -2 \quad 4 \quad 0}$$

Now use the quadratic formula to find the zeros of the quotient polynomial $x^2 - 2x + 4$.

$$x = \frac{-(-2) \pm \sqrt{(-2)^2 - 4(1)(4)}}{2(1)}$$
$$= \frac{2 \pm \sqrt{-12}}{2} = \frac{2 \pm 2i\sqrt{3}}{2} = 1 \pm i\sqrt{3}$$

Thus, the rational zeros are 0 and 2 (multiplicity 2).

(d) The nonreal complex zeros are $1 \pm i\sqrt{3}$.

9. $f(x) = 8x^4 + 8x^3 - x - 1$

(a) $f(x) = 8x^4 \underbrace{+ 8x^3 - x}_{1} - 1$ has one variation in sign, so f has 1 positive real zero.

$f(-x) = \underbrace{8x^4 - 8x^3}_{1} \underbrace{+ 8x}_{2} \underbrace{- 1}_{3}$ has three variations in sign, so f has 3 or $3 - 1 = 0$ negative real zeros.

The degree of the function is 4, so the remaining zeros are nonreal complex.

Positive	Negative	Nonreal Complex
1	3	0
1	1	2

(b) p must be a factor of $a_0 = -1$ and q must be a factor of $a_4 = 8$. Thus, p can be ± 1, and q can be ± 1, ± 2, ± 4, or ± 8. The possible rational zeros, $\frac{p}{q}$, are ± 1, $\pm \frac{1}{2}$, $\pm \frac{1}{4}$, $\pm \frac{1}{8}$.

(c) We can factor $f(x)$, so set $f(x) = 0$, then solve to find the rational zeros.
$$8x^4 + 8x^3 - x - 1 = 0$$
$$8x^3(x+1) - (x+1) = 0$$
$$(8x^3 - 1)(x+1) = 0$$
$$(2x-1)(4x^2 + 2x + 1)(x+1) = 0$$

This gives $2x - 1 = 0 \Rightarrow x = \frac{1}{2}$ and $x + 1 = 0 \Rightarrow x = -1$. Thus, the rational zeros are $\frac{1}{2}$ and -1.

(d) To find the zeros of the factor $4x^2 + 2x + 1$, use the quadratic formula.
$$x = \frac{-2 \pm \sqrt{2^2 - 4(4)(1)}}{2(4)} = \frac{-2 \pm \sqrt{-12}}{8}$$
$$= \frac{-2 \pm 2i\sqrt{3}}{8} = -\frac{1}{4} \pm \frac{\sqrt{3}}{4}i$$

The real complex zeros are $-\frac{1}{4} \pm \frac{\sqrt{3}}{4}i$.

11. $f(x) = x^4 + 3x^3 - 3x^2 - 11x - 6$

(a) $f(x) = x^4 \underbrace{+3x^3 - 3x^2 - 11x - 6}_{1}$ has 1 variation in sign. f has 1 positive real zero.

$f(-x) = \underbrace{x^4 - 3x^3}_{1} \underbrace{-3x^2 + 11x}_{2} \underbrace{- 6}_{3}$ has 3 variations in sign. f has 3 or $3 - 2 = 1$ negative real zeros. Thus, the different possibilities are

Positive	Negative	Nonreal Complex
1	3	0
1	1	2

(b) p must be a factor of $a_0 = -6$ and q must be a factor of $a_4 = 1$. Thus, p can be ± 1, $\pm 2, \pm 3, \pm 6$ and q can be ± 1. The possible zeros, $\frac{p}{q}$, are $\pm 1, \pm 2, \pm 3, \pm 6$.

(c) The remainder theorem shows that 2 is zero.

$$2)\overline{\begin{array}{rrrrr} 1 & 3 & -3 & -11 & -6 \\ & 2 & 10 & 14 & 6 \\ \hline 1 & 5 & 7 & 3 & 0 \end{array}}$$

The new quotient polynomial is $x^3 + 5x^2 + 7x + 3$. Because f has 1 positive real zero, try negative values. The remainder theorem shows that -3 is zero.

$$-3)\overline{\begin{array}{rrrr} 1 & 5 & 7 & 3 \\ & -3 & -6 & -3 \\ \hline 1 & 2 & 1 & 0 \end{array}}$$

The new quotient polynomial is $x^2 + 2x + 1$. Factor this polynomial and set equal to zero to find the remaining zeros.

$$x^2 + 2x + 1 = 0 \Rightarrow (x+1)^2 = 0 \Rightarrow$$
$$x + 1 = 0 \Rightarrow x = -1$$

The rational zeros are $-3, -1$ (multiplicity 2), and 2.

(d) All zeros have been found, and they are all rational.

(e) All zeros have been found, and they are all real.

(f) The x-intercepts are $(-3, 0)$, $(-1, 0)$, and $(2, 0)$.

(g) $f(0) = -6$, so the y-intercept is $(0, -6)$.

(h)

$$4)\overline{\begin{array}{rrrrr} 1 & 3 & -3 & -11 & -6 \\ & 4 & 28 & 100 & 356 \\ \hline 1 & 7 & 25 & 89 & 350 \end{array}}$$

The corresponding point on the graph is $(4, 350)$.

(i)

(j) The x-intercepts divide the x-axis into four intervals:

Interval	Test Point	Value of $f(x)$	Sign of $f(x)$	Graph Above or Below x-Axis
$(-\infty, -3)$	-4	54	Positive	Above
$(-3, -1)$	-2	-4	Negative	Below
$(-1, 2)$	0	-6	Negative	Below
$(2, \infty)$	3	96	Positive	Above

Plot the x-intercepts, y-intercept and test points with a smooth curve to get the graph.

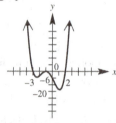

$$f(x) = x^4 + 3x^3 - 3x^2 - 11x - 6$$

13. $f(x) = 2x^5 - 10x^4 + x^3 - 5x^2 - x + 5$

(a) $f(x) = \underset{1}{2x^5} - \underset{2}{10x^4} + \underset{3}{x^3} - 5x^2 \underset{4}{-x} + 5$ has 4 variations in sign. f has 4 or $4 - 2 = 2$ or $2 - 2 = 0$ positive real zeros.

$$f(-x) = -2x^5 - 10x^4 - x^3 \underset{1}{-5x^2} + x + 5$$

has 1 variation in sign. f has 1 negative real zero.

Thus, the different possibilities are

Positive	Negative	Nonreal Complex
4	1	0
2	1	2
0	1	4

(b) p must be a factor of $a_0 = 5$ and q must be a factor of $a_5 = 2$. Thus, p can be ± 1, ± 5 and q can be $\pm 1, \pm 2$. The possible zeros, $\frac{p}{q}$, are $\pm 1, \pm 5, \pm \frac{1}{2}, \pm \frac{5}{2}$.

(c) The remainder theorem shows that 5 is zero.

$$5)\overline{\begin{array}{rrrrrr} 2 & -10 & 1 & -5 & -1 & 5 \\ & 10 & 0 & 5 & 0 & -5 \\ \hline 2 & 0 & 1 & 0 & -1 & 0 \end{array}}$$

(continued on next page)

(continued)

The new quotient polynomial will be $2x^4 + x^2 - 1$. At this point, if we try to determine which of the possible rational zero will result in a remainder of zero, we will find that none of them work. We have found the only rational zero of our polynomial function. The rational zero is 5.

(d) The new quotient polynomial is $2x^4 + x^2 - 1$, so we can factor the polynomial and set equal to zero to find the remaining zeros.

$$2x^4 + x^2 - 1 = 0 \Rightarrow (x^2 + 1)(2x^2 - 1) = 0$$

Because $x^2 + 1 = 0$ does not yield any real solutions, we will examine $2x^2 - 1 = 0$ first.

$$2x^2 - 1 = 0 \Rightarrow 2x^2 = 1 \Rightarrow x^2 = \tfrac{1}{2} \Rightarrow$$
$$x = \pm\sqrt{\tfrac{1}{2}} = \pm\tfrac{\sqrt{2}}{2}$$

The other real zeros are $-\tfrac{\sqrt{2}}{2}$ and $\tfrac{\sqrt{2}}{2}$.

(e) Examining $x^2 + 1 = 0$, we have

$$x^2 = -1 \Rightarrow x = \pm\sqrt{-1} = \pm i.$$ Thus, the complex zeros are $-i$ and i.

(f) The x-intercepts are $\left(-\tfrac{\sqrt{2}}{2}, 0\right)$, $\left(\tfrac{\sqrt{2}}{2}, 0\right)$, and $(5, 0)$.

(g) $f(0) = 5$, so the y-intercept is $(0, 5)$.

(h)

$$\begin{array}{r|rrrrr} 4) & 2 & -10 & 1 & -5 & -1 & 5 \\ & & 8 & -8 & -28 & -132 & -532 \\ \hline & 2 & -2 & -7 & -33 & -133 & -527 \end{array}$$

The corresponding point on the graph is $(4, -527)$.

(i)

(j) The x-intercepts divide the x-axis into four intervals:

Interval	Test Point	Value of $f(x)$	Sign of $f(x)$	Graph Above or Below x-Axis
$\left(-\infty, -\tfrac{\sqrt{2}}{2}\right)$	-1	-12	Negative	Below
$\left(-\tfrac{\sqrt{2}}{2}, \tfrac{\sqrt{2}}{2}\right)$	0	5	Positive	Above
$\left(\tfrac{\sqrt{2}}{2}, 5\right)$	1	-8	Negative	Below
$(5, \infty)$	6	2627	Positive	Above

Plot the x-intercepts, y-intercept and test points (the y-intercept is also a test point) with a smooth curve to get the graph.

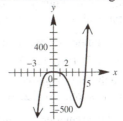

$$f(x) = 2x^5 - 10x^4 + x^3 - 5x^2 - x + 5$$

15. $f(x) = -2x^4 - x^3 + x + 2$

(a) $f(x) = -2x^4 \underset{1}{-x^3} + x + 2$ has 1 variation in sign. f has 1 positive real zero.

$f(x) = \underset{1}{-2x^4} + \underset{2}{x^3} - \underset{3}{x} + 2$ has 3 variations in sign. f has 3 or $3 - 2 = 1$ negative real zeros. Thus, the different possibilities are

Positive	Negative	Nonreal Complex
1	3	0
1	1	2

(b) p must be a factor of $a_0 = 2$ and q must be a factor of $a_4 = -2$. Thus, p can be ± 1, ± 2 and q can be ± 1, ± 2. The possible zeros, $\tfrac{p}{q}$, are ± 1, ± 2, $\pm\tfrac{1}{2}$.

(c) The remainder theorem shows that -1 is zero.

$$\begin{array}{r|rrrrr} -1) & -2 & -1 & 0 & 1 & 2 \\ & & 2 & -1 & 1 & -2 \\ \hline & -2 & 1 & -1 & 2 & 0 \end{array}$$

(continued on next page)

(*continued*)

The new quotient polynomial will be $-2x^3 + x^2 - x + 2$. The signs are alternating in the bottom row of the synthetic division, so we know that -1 is a lower bound. The remainder theorem shows that 1 is zero.

$$
\begin{array}{r|rrrr}
1) & -2 & 1 & -1 & 2 \\
  &   & -2 & -1 & -2 \\
\hline
  & -2 & -1 & -2 & 0
\end{array}
$$

The new quotient polynomial is $-2x^2 - x - 2 = -\left(2x^2 + x + 2\right)$. At this point, if we try to determine which of the possible rational zeros will result in a remainder of zero, we will find that none of them work. We have found the only rational zeros of our polynomial function. The rational zeros are -1 and 1.

(d) If we examine the discriminant of $2x^2 + x + 2$, $1^2 - 4(2)(2) = 1 - 16 = -15$, we see that it is negative, which implies there are no other real zeros.

(e) To find the remaining complex zeros, we must solve $2x^2 + x + 2 = 0$. To solve this equation, we use the quadratic formula with $a = 2$, $b = 1$, and $c = 2$.

$$
x = \frac{-1 \pm \sqrt{1^2 - 4(2)(2)}}{2(2)} = \frac{-1 \pm \sqrt{1 - 16}}{4}
$$

$$
= \frac{-1 \pm \sqrt{-15}}{4} = \frac{-1 \pm i\sqrt{15}}{4}
$$

Thus, the other complex zeros are $-\frac{1}{4} + \frac{\sqrt{15}}{4}i$ and $-\frac{1}{4} - \frac{\sqrt{15}}{4}i$

(f) The x-intercepts are $(-1, 0)$ and $(1, 0)$.

(g) $f(0) = 2$, so the y-intercept is $(0, 2)$.

(h)
$$
\begin{array}{r|rrrrr}
4) & -2 & -1 & 0 & 1 & 2 \\
  &   & -8 & -36 & -144 & -572 \\
\hline
  & -2 & -9 & -36 & -143 & -570
\end{array}
$$

The corresponding point on the graph is $(4, -570)$.

(i)

(j) The x-intercepts divide the x-axis into three intervals:

Interval	Test Point	Value of $f(x)$	Sign of $f(x)$	Graph Above or Below x-Axis
$(-\infty, -1)$	-2	-24	Negative	Below
$(-1, 1)$	0	2	Positive	Above
$(1, \infty)$	2	-36	Negative	Below

Plot the x-intercepts, y-intercept and test points (the y-intercept is also a test point) with a smooth curve to get the graph.

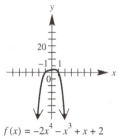

$f(x) = -2x^4 - x^3 + x + 2$

17. $f(x) = 3x^4 - 14x^2 - 5$

(a) $f(x) = 3x^4 \underbrace{- 14x^2 - 5}_{1}$ has 1 variation in sign. f has 1 positive real zero.

$f(-x) = 3x^4 \underbrace{- 14x^2 - 5}_{1}$ has 1 variation in sign. f has 1 negative real zeros. Thus, the different possibilities are

Positive	Negative	Nonreal Complex
1	1	2

(b) p must be a factor of $a_0 = -5$ and q must be a factor of $a_4 = 3$. Thus, p can be ± 1, ± 5 and q can be $\pm 1, \pm 3$. The possible zeros, $\frac{p}{q}$, are ± 1, ± 5, $\pm \frac{1}{3}$, $\pm \frac{5}{3}$.

(c) If we try to determine which of the possible rational zeros will result in a remainder of zero, we will find that none of them work. There are no rational zeros.

(d) We can factor the polynomial, $3x^4 - 14x^2 - 5$, and set equal to zero to the remaining zeros.

$$
3x^4 - 14x^2 - 5 = 0
$$

$$
\left(3x^2 + 1\right)\left(x^2 - 5\right) = 0
$$

(*continued on next page*)

(continued)

Because $3x^2 + 1 = 0$ does not yield any real solutions, we will examine $x^2 - 5 = 0$ first.

$$x^2 - 5 = 0 \Rightarrow x^2 = 5 \Rightarrow x = \pm\sqrt{5}$$

The real zeros are $-\sqrt{5}$ and $\sqrt{5}$.

(e) Examining $3x^2 + 1 = 0$, we have

$$3x^2 = -1 \Rightarrow x^2 = -\frac{1}{3} \Rightarrow x = \pm\sqrt{-\frac{1}{3}} = \pm i\frac{\sqrt{3}}{3}.$$

Thus, the complex zeros are $-\frac{\sqrt{3}}{3}i$ and $\frac{\sqrt{3}}{3}i$.

(f) The x-intercepts are $\left(-\sqrt{5}, 0\right)$ and $\left(\sqrt{5}, 0\right)$.

(g) $f(0) = -5$, so the y-intercept is $(0, -5)$.

(h)
$$
\begin{array}{r}
4)\overline{3 \quad 0 \quad -14 \quad 0 \quad -5} \\
\underline{12 \quad 48 \quad 136 \quad 544} \\
3 \quad 12 \quad 34 \quad 136 \quad 539
\end{array}
$$

The corresponding point on the graph is $(4, 539)$.

(i)

(j) The x-intercepts divide the x-axis into three intervals:

Interval	Test Point	Value of $f(x)$	Sign of $f(x)$	Graph Above or Below x-Axis
$\left(-\infty, -\sqrt{5}\right)$	-3	112	Positive	Above
$\left(-\sqrt{5}, \sqrt{5}\right)$	0	-5	Negative	Below
$\left(\sqrt{5}, \infty\right)$	3	112	Positive	Above

Plot the x-intercepts, y-intercept and test points (the y-intercept is also a test point) with a smooth curve to get the graph.

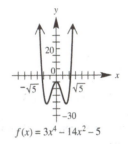

$$f(x) = 3x^4 - 14x^2 - 5$$

19. $f(x) = -3x^4 + 22x^3 - 55x^2 + 52x - 12$

(a) $f(x) = \underset{1}{-3x^4} + \underset{2}{22x^3} - \underset{3}{55x^2} + \underset{4}{52x} - 12$

has 4 variations in sign. f has 4 or $4 - 2 = 2$ or $2 - 2 = 0$ positive real zeros.

$$f(-x) = -3x^4 - 22x^3 - 55x^2 - 52x - 12$$

has 0 variations in sign. f has 0 negative real zeros. Thus, the different possibilities are

Positive	Negative	Nonreal Complex
4	0	0
2	0	2
0	0	4

(b) p must be a factor of $a_0 = -12$ and q must be a factor of $a_4 = -3$. Thus, p can be $\pm1, \pm2, \pm3, \pm4, \pm6, \pm12$ and q can be $\pm1, \pm3$. The possible zeros, $\frac{p}{q}$, are

$$\pm1, \ \pm2, \ \pm3, \ \pm4, \ \pm6, \ \pm12, \ \pm\tfrac{1}{3},$$
$$\pm\tfrac{2}{3}, \ \pm\tfrac{4}{3}.$$

(c) The remainder theorem shows that 2 is a zero.

$$
\begin{array}{r}
2)\overline{-3 \quad 22 \ -55 \quad 52 \ -12} \\
\underline{-6 \quad 32 \ -46 \quad 12} \\
-3 \quad 16 \ -23 \quad 6 \quad 0
\end{array}
$$

The new quotient polynomial will be $-3x^3 + 16x^2 - 23x + 6$. The remainder theorem shows that 3 is zero.

$$
\begin{array}{r}
3)\overline{-3 \quad 16 \ -23 \quad 6} \\
\underline{-9 \quad 21 \ -6} \\
-3 \quad 7 \ -2 \quad 0
\end{array}
$$

The new quotient polynomial is $-3x^2 + 7x - 2 = -\left(3x^2 - 7x + 2\right)$. Factor $3x^2 - 7x + 2$ and set equal to zero to find the remaining zeros.

(continued on next page)

(continued)

$$3x^2 - 7x + 2 = 0 \Rightarrow (3x - 1)(x - 2) = 0$$

$$3x - 1 = 0 \Rightarrow x = \tfrac{1}{3} \text{ or } x - 2 = 0 \Rightarrow x = 2$$

The rational zeros are

$\tfrac{1}{3}$, 2 (multiplicity 2), and 3.

(d) All zeros have been found, and they are all rational.

(e) All zeros have been found, and they are all real.

(f) The x-intercepts are $\left(\tfrac{1}{3}, 0\right), (2, 0)$, and $(3, 0)$.

(g) $f(0) = -12$, so the y-intercept is $(0, -12)$.

(h)
$$4\overline{)-3 \quad 22 \ -55 \quad 52 \ -12}$$
$$\underline{\quad\quad -12 \quad 40 \ -60 \ -32}$$
$$-3 \quad 10 \ -15 \ -8 \ -44$$

The corresponding point on the graph is $(4, -44)$.

(i)

(j) The x-intercepts divide the x-axis into four intervals:

Interval	Test Point	Value of $f(x)$	Sign of $f(x)$	Graph Above or Below x-Axis
$\left(-\infty, \tfrac{1}{3}\right)$	0	-12	Negative	Below
$\left(\tfrac{1}{3}, 2\right)$	1	4	Positive	Above
$(2, 3)$	$\tfrac{5}{2}$	$\tfrac{13}{16}$	Positive	Above
$(3, \infty)$	4	-44	Negative	Below

Plot the x-intercepts, y-intercept and test points (the y-intercept is also a test point) with a smooth curve to get the graph.

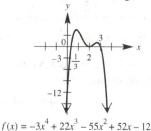

$$f(x) = -3x^4 + 22x^3 - 55x^2 + 52x - 12$$

Section 3.5   Rational Functions: Graphs, Applications, and Models

1. $f(x) = \dfrac{1}{x}$

Domain: $(-\infty, 0) \cup (0, \infty)$

Range: $(-\infty, 0) \cup (0, \infty)$

3. $f(x) = \dfrac{1}{x}$

Increasing: nowhere

Decreasing: $(-\infty, 0) \cup (0, \infty)$

Constant: nowhere

5. $y = \dfrac{1}{x - 3} + 2$

Vertical asymptote: $x = 3$

Horizontal asymptote: $y = 2$

7. Because $f(-x) = \dfrac{1}{(-x)^2} = \dfrac{1}{x^2} = f(x)$, the function is an even function. It exhibits symmetry with respect to y-axis.

9. Graphs A, B, and C have a domain of $(-\infty, 3) \cup (3, \infty)$.

11. Graph A has a range of $(-\infty, 0) \cup (0, \infty)$.

13. Graph A has a single solution to the equation $f(x) = 3$.

15. Graphs A, C, and D have the x-axis as a horizontal asymptote.

17. $f(x) = \dfrac{2}{x}$

To obtain the graph of $f(x) = \tfrac{2}{x}$, stretch the graph of $y = \tfrac{1}{x}$ vertically by a factor of 2. Just as with the graph of $y = \tfrac{1}{x}$, $y = 0$ is the horizontal asymptote and $x = 0$ is the vertical asymptote.

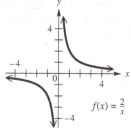

$$f(x) = \tfrac{2}{x}$$

(a) Domain: $(-\infty, 0) \cup (0, \infty)$

(b) Range: $(-\infty, 0) \cup (0, \infty)$

(c) The function is never increasing.

(d) The function decreases over $(-\infty, 0)$ and $(0, \infty)$.

19. $f(x) = \dfrac{1}{x+2}$

To obtain the graph of $f(x) = \frac{1}{x+2}$ shift the graph of $y = \frac{1}{x}$ to the left 2 units. Just as with $y = \frac{1}{x}$, $y = 0$ is the horizontal asymptote, but this graph has $x = -2$ as its vertical asymptote (this affects the domain).

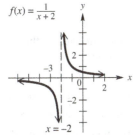

(a) Domain: $(-\infty, -2) \cup (-2, \infty)$

(b) Range: $(-\infty, 0) \cup (0, \infty)$

(c) The function is never increasing.

(d) The function decreases over $(-\infty, -2)$ and $(-2, \infty)$.

21. $f(x) = \dfrac{1}{x} + 1$

To obtain the graph of $f(x) = \frac{1}{x} + 1$, translate the graph of $y = \frac{1}{x}$ 1 unit up. Just as with $y = \frac{1}{x}$, $x = 0$ is the vertical asymptote, but this graph has $y = 1$ as its horizontal asymptote (this affects the range).

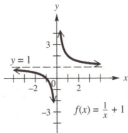

a. Domain: $(-\infty, 0) \cup (0, \infty)$

b. Range: $(-\infty, 1) \cup (1, \infty)$

(c) The function is never increasing.

(d) The function decreases over $(-\infty, 0)$ and $(0, \infty)$.

23. $f(x) = -\dfrac{2}{x^2}$

To obtain the graph of $f(x) = -\frac{2}{x^2}$, stretch the graph of $y = \frac{1}{x^2}$ by a factor of 2, and then reflect the graph will be reflected across the x-axis.

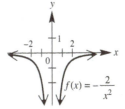

(a) Domain: $(-\infty, 0) \cup (0, \infty)$

(b) Range: $(-\infty, 0)$

(c) The function increases over $(0, \infty)$.

(d) The function decreases over $(-\infty, 0)$.

25. $f(x) = \dfrac{1}{(x-3)^2}$

To obtain the graph of $f(x) = \frac{1}{(x-3)^2}$, shift the graph of $y = \frac{1}{x^2}$ 3 units to the right. Just as with $y = \frac{1}{x^2}$, $y = 0$ is the horizontal asymptote, but this graph has $x = 3$ as its vertical asymptote.

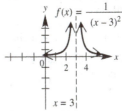

(a) Domain: $(-\infty, 3) \cup (3, \infty)$

(b) Range: $(0, \infty)$

(c) The function increases over $(-\infty, 3)$.

(d) The function decreases over $(3, \infty)$.

27. $f(x) = \dfrac{-1}{(x+2)^2} - 3$

To obtain the graph of $f(x) = \dfrac{-1}{(x+2)^2} - 3$, shift the graph of $y = \dfrac{1}{x^2}$ 2 units to the left, reflect the graph across the x-axis, and then shfit the graph 3 units down. Unlike the graph of $y = \dfrac{1}{x^2}$, the vertical asymptote of $f(x) = \dfrac{-1}{(x+2)^2} - 3$ is $x = -2$ and the horizontal asymptote is $y = -3$.

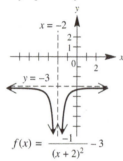

$f(x) = \dfrac{-1}{(x+2)^2} - 3$

(a) Domain: $(-\infty, -2) \cup (-2, \infty)$

(b) Range: $(-\infty, -3)$

(c) The function increases over $(-2, \infty)$.

(d) The function decreases over $(-\infty, -2)$.

29. D. The graph of $f(x) = \dfrac{x+7}{x+1}$ has the vertical asymptote at $x = -1$.

31. G. The graph of $f(x) = \dfrac{1}{x+4}$ has the x-axis as its horizontal asymptote, and the y-axis is not its vertical asymptote. The line $x = -4$ is its vertical asymptote.

33. E. The graph of $f(x) = \dfrac{x^2-16}{x+4}$ has a "hole" in its graph located at $(-4, -8)$ because
$f(x) = \dfrac{x^2-16}{x+4} = \dfrac{(x+4)(x-4)}{x+4} = x - 4,\ x \neq -4$.

35. F. The graph of $f(x) = \dfrac{x^2+3x+4}{x-5}$ has an oblique asymptote because
$f(x) = x + 8 + \dfrac{44}{x-5}$.

37. $f(x) = \dfrac{3}{x-5}$. To find the vertical asymptote, set the denominator equal to zero.
$x - 5 = 0 \Rightarrow x = 5$
The equation of the vertical asymptote is $x = 5$. To find the horizontal asymptote, divide each term by the largest power of x in the expression.
$$f(x) = \dfrac{\frac{3}{x}}{\frac{x}{x} - \frac{5}{x}} = \dfrac{\frac{3}{x}}{1 - \frac{5}{x}}$$
As $|x| \to \infty$, $\frac{1}{x}$ approaches 0, thus $f(x)$ approaches $\frac{0}{1-0} = \frac{0}{1} = 0$. The line $y = 0$ (that is, the x-axis) is the horizontal asymptote.

39. $f(x) = \dfrac{4 - 3x}{2x + 1}$

To find the vertical asymptote, set the denominator equal to zero.
$2x + 1 = 0 \Rightarrow x = -\dfrac{1}{2}$
The equation of the vertical asymptote is $x = -\frac{1}{2}$. To find the horizontal asymptote, divide each term by the largest power of x in the expression.
$$f(x) = \dfrac{\frac{4}{x} - \frac{3x}{x}}{\frac{2x}{x} + \frac{1}{x}} = \dfrac{\frac{4}{x} - 3}{2 + \frac{1}{x}}$$
As $|x| \to \infty$, $\frac{1}{x}$ approaches 0, so $f(x)$ approaches $\frac{0-3}{2+0} = -\frac{3}{2}$. The equation of the horizontal asymptote is $y = -\frac{3}{2}$.

41. $f(x) = \dfrac{x^2 - 1}{x + 3}$

The vertical asymptote is $x = -3$, found by solving $x + 3 = 0$. The numerator is of degree exactly one more than the denominator, so there is no horizontal asymptote, but there may be an oblique asymptote. To find it, divide the numerator by the denominator.

$$-3 \overline{)\begin{array}{rrr} 1 & 0 & -1 \\ & -3 & 9 \\ \hline 1 & -3 & 8 \end{array}}$$

Thus, $f(x) = \dfrac{x^2-1}{x+3} = x - 3 + \dfrac{8}{x+3}$.

For very large values of $|x|$, $\frac{8}{x+3}$ is close to 0, and the graph approaches the line $y = x - 3$.

43. $f(x) = \dfrac{x^2 - 2x - 3}{2x^2 - x - 10}$

To find the vertical asymptotes, set the denominator equal to zero and solve.

$2x^2 - x - 10 = 0 \Rightarrow (x+2)(2x-5) = 0 \Rightarrow$

$x + 2 = 0 \Rightarrow x = -2$ or $2x - 5 = 0 \Rightarrow x = \frac{5}{2}$

Thus, the vertical asymptotes are $x = -2$ and $x = \frac{5}{2}$.

To determine the horizontal asymptote, divide the numerator and denominator by x^2.

$f(x) = \dfrac{\frac{x^2}{x^2} - \frac{2x}{x^2} - \frac{3}{x^2}}{\frac{2x^2}{x^2} - \frac{x}{x^2} - \frac{10}{x^2}} = \dfrac{1 - \frac{2}{x} - \frac{3}{x^2}}{2 - \frac{1}{x} - \frac{10}{x^2}}$

As $|x| \to \infty$, $\frac{1}{x}$ and $\frac{1}{x^2}$ approach 0, so $f(x)$ approaches $\frac{1-0-0}{2-0-0} = \frac{1}{2}$. Thus, the equation of the horizontal asymptote is $y = \frac{1}{2}$.

45. $f(x) = \dfrac{x^2 + 1}{x^2 + 9}$

To find the vertical asymptotes, set the denominator equal to zero and solve.

$x^2 + 9 = 0 \Rightarrow x^2 = -9 \Rightarrow x = \pm\sqrt{-9} = \pm 3i$

Thus, there are no vertical asymptotes.

To determine the horizontal asymptote, divide the numerator and denominator by x^2.

$f(x) = \dfrac{\frac{x^2}{x^2} + \frac{1}{x^2}}{\frac{x^2}{x^2} + \frac{9}{x^2}} = \dfrac{1 + \frac{1}{x^2}}{1 + \frac{9}{x^2}}$

As $|x| \to \infty$, $\frac{1}{x^2}$ approaches 0, so $f(x)$ approaches $\frac{1+0}{1+0} = \frac{1}{1} = 1$. Thus, the equation of the horizontal asymptote is $y = 1$.

47. (a) Translating $y = \frac{1}{x}$ three units to the right gives $y = \frac{1}{x-3}$. Translating $y = \frac{1}{x-3}$ two units up yields

$y = \frac{1}{x-3} + 2 = \frac{1}{x-3} + \frac{2x-6}{x-3} = \frac{2x-5}{x-3}$. So

$f(x) = \frac{2x-5}{x-3}$.

(b) $f(x) = \frac{2x-5}{x-3}$ has a zero when $2x - 5 = 0$

or $x = \frac{5}{2}$.

(c) $f(x) = \frac{2x-5}{x-3}$ has a horizontal asymptote at $y = 2$ and a vertical asymptote at $x = 3$.

49. (a) $f(x) = x + 1 + \frac{x^2-x}{x^4+1}$ has an oblique asymptote of $y = x + 1$.

(b) In order to determine when the function intersects the oblique asymptote, we must determine the values that make $\frac{x^2-x}{x^4+1} = 0$.

Because $\frac{x^2-x}{x^4+1} = \frac{x(x-1)}{x^4+1}$, and $x(x-1) = 0$ when

$x = 0$ or $x = 1$, the function crosses its asymptote at $x = 0$ and $x = 1$.

(c) For large values of x, $\frac{x^2-x}{x^4+1} > 0$. Thus as $x \to \infty$, the function approaches its asymptote from above.

51. Function A, because the denominator can never be equal to 0.

53. From the graph, the vertical asymptote is $x = 2$, the horizontal asymptote is $y = 4$, and there is no oblique asymptote. The function is defined for all real numbers x such that $x \neq 2$, therefore the domain is $(-\infty, 2) \cup (2, \infty)$.

55. From the graph, the vertical asymptotes are $x = -2$ and $x = 2$, the horizontal asymptote is $y = -4$, and there is no oblique asymptote. The function is defined for all real numbers x such that $x \neq \pm 2$, therefore the domain is $(-\infty, -2) \cup (-2, 2) \cup (2, \infty)$.

57. From the graph, there is no vertical asymptote, the horizontal asymptote is $y = 0$, and there is no oblique asymptote. The function is defined for all x; therefore, the domain is $(-\infty, \infty)$.

59. From the graph, the vertical asymptote is $x = -1$, there is no horizontal asymptote, and the oblique asymptote passes through the points $(0, -1)$ and $(1, 0)$. Thus, the equation of the oblique asymptote is $y = x - 1$. The function is defined for all $x \neq -1$, therefore the domain is $(-\infty, -1) \cup (-1, \infty)$.

For exercises 61–99, follow the 7 steps outlined on page 367 in the text to sketch the graph. The solutions for exercises 61 and 62 show all steps. The solutions for exercises 63–100 are abbreviated.

61. $f(x) = \dfrac{x+1}{x-4}$

Step 1: Find the vertical asymptote by setting the denominator equal to 0: $x - 4 = 0 \Rightarrow x = 4$

Step 2: Find the horizontal asymptote. The numerator and denominator have the same degree, so the horizontal asymptote has equation $y = \dfrac{a_n}{b_n} = \dfrac{1}{1} = 1$. There is no oblique asymptote.

Step 3: Find the y-intercept:

$f(0) = \dfrac{0+1}{0-4} = -\dfrac{1}{4}$

Step 4: Find the x-intercept:

$\dfrac{x+1}{x-4} = 0 \Rightarrow x + 1 = 0 \Rightarrow x = -1$

Step 5: Determine whether the graph will intersect the horizontal asymptote:

$f(x) = \dfrac{x+1}{x-4} = 1 \Rightarrow x + 1 = x - 4 \Rightarrow 0 = -5 \Rightarrow$

the graph does not intersect its horizontal asymptote.

Step 6: Plot a point in each of the intervals $(-\infty, -1), (-1, 4)$, and $(4, \infty)$.

Interval	Test Point	Value of $f(x)$	Sign of $f(x)$	Graph Above or Below x-Axis
$(-\infty, -1)$	-5	$\frac{4}{9}$	Positive	Above
$(-1, 4)$	1	$-\frac{2}{3}$	Negative	Below
$(4, \infty)$	5	6	Positive	Above

Step 7: Plot the intercepts and test points to sketch the graph.

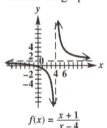

$f(x) = \dfrac{x+1}{x-4}$

63. $f(x) = \dfrac{x+2}{x-3}$

vertical asymptote: $x = 3$

horizontal asymptote: $y = 1$

y-intercept: $-\dfrac{2}{3}$; x-intercept: -2

$f(x)$ does not intersect the horizontal asymptote

Interval	Test Point	Value of $f(x)$	Sign of $f(x)$	Graph Above or Below x-Axis
$(-\infty, -2)$	-5	$\frac{3}{8}$	Positive	Above
$(-2, 3)$	1	$-\frac{3}{2}$	Negative	Below
$(3, \infty)$	5	$\frac{7}{2}$	Positive	Above

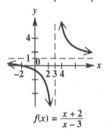

$f(x) = \dfrac{x+2}{x-3}$

65. $f(x) = \dfrac{4-2x}{8-x}$

vertical asymptote: $x = 8$

horizontal asymptote: $y = 2$

y-intercept: $\dfrac{1}{2}$; x-intercept: 2

$f(x)$ does not intersect the horizontal asymptote

Interval	Test Point	Value of $f(x)$	Sign of $f(x)$	Graph Above or Below x-Axis
$(-\infty, 2)$	-5	$\frac{14}{13}$	Positive	Above
$(2, 8)$	4	-1	Negative	Below
$(8, \infty)$	10	8	Positive	Above

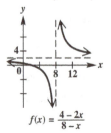

$f(x) = \dfrac{4-2x}{8-x}$

67. $f(x) = \dfrac{3x}{x^2 - x - 2} = \dfrac{3x}{(x-2)(x+1)}$

vertical asymptotes: $x = -1$, $x = 2$
horizontal asymptote: $y = 0$
y-intercept: 0; x-intercept: 0
$f(x)$ intersects the horizontal asymptote

Interval	Test Point	Value of $f(x)$	Sign of $f(x)$	Graph Above or Below x-Axis
$(-\infty, -1)$	-5	$-\frac{15}{28}$	Negative	Below
$(-1, 0)$	$-\frac{1}{2}$	$\frac{6}{5}$	Positive	Above
$(0, 2)$	1	$-\frac{3}{2}$	Negative	Below
$(2, \infty)$	5	$\frac{5}{6}$	Positive	Above

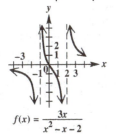

$f(x) = \dfrac{3x}{x^2 - x - 2}$

69. $f(x) = \dfrac{5x}{x^2 - 1}$

vertical asymptotes: $x = -1$, $x = 1$
horizontal asymptote: $y = 0$
y-intercept: 0; x-intercept: 0
$f(x)$ intersects the horizontal asymptote

Interval	Test Point	Value of $f(x)$	Sign of $f(x)$	Graph Above or Below x-Axis
$(-\infty, -1)$	-2	$-\frac{10}{3}$	Negative	Below
$(-1, 0)$	$-\frac{1}{2}$	$\frac{10}{3}$	Positive	Above
$(0, 1)$	$\frac{1}{2}$	$-\frac{10}{3}$	Negative	Below
$(1, \infty)$	2	$\frac{10}{3}$	Positive	Above

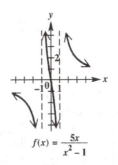

$f(x) = \dfrac{5x}{x^2 - 1}$

71. $f(x) = \dfrac{(x+6)(x-2)}{(x+3)(x-4)}$

vertical asymptotes: $x = -3$, $x = 4$
horizontal asymptote: $y = 1$
y-intercept: 1; x-intercepts: -6, 2
$f(x)$ intersects the horizontal asymptote

Interval	Test Point	Value of $f(x)$	Sign of $f(x)$	Graph Above or Below x-Axis
$(-\infty, -6)$	-8	$\frac{1}{3}$	Positive	Above
$(-6, -3)$	-4	$-\frac{3}{2}$	Negative	Below
$(-3, 2)$	1	$\frac{7}{12}$	Positive	Above
$(2, 4)$	3	$-\frac{3}{2}$	Negative	Below
$(4, \infty)$	6	$\frac{8}{3}$	Positive	Above

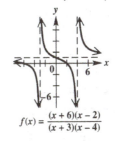

$f(x) = \dfrac{(x+6)(x-2)}{(x+3)(x-4)}$

73. $f(x) = \dfrac{3x^2 + 3x - 6}{x^2 - x - 12} = \dfrac{3(x+2)(x-1)}{(x-4)(x+3)}$

vertical asymptotes: $x = -3$, $x = 4$
horizontal asymptote: $y = 3$
y-intercept: $\frac{1}{2}$; x-intercepts: -2, 1
$f(x)$ intersects the horizontal asymptote

Interval	Test Point	Value of $f(x)$	Sign of $f(x)$	Graph Above or Below x-Axis
$(-\infty, -3)$	-4	$\frac{15}{4}$	Positive	Above
$(-3, -2)$	$-\frac{5}{2}$	$-\frac{21}{13}$	Negative	Below
$(-2, 1)$	-1	$\frac{3}{5}$	Positive	Above
$(1, 4)$	2	$-\frac{6}{5}$	Negative	Below
$(4, \infty)$	6	$\frac{20}{3}$	Positive	Above

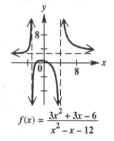

$f(x) = \dfrac{3x^2 + 3x - 6}{x^2 - x - 12}$

75. $f(x) = \dfrac{9x^2 - 1}{x^2 - 4} = \dfrac{(3x+1)(3x-1)}{(x+2)(x-2)}$

vertical asymptotes: $x = -2$, $x = 2$
horizontal asymptote: $y = 9$
y-intercept: $\frac{1}{4}$; x-intercepts: $-\frac{1}{3}, \frac{1}{3}$

$f(x)$ does not intersect the horizontal asymptote

Interval	Test Point	Value of $f(x)$	Sign of $f(x)$	Graph Above or Below x-Axis
$(-\infty, -2)$	-3	16	Positive	Above
$\left(-2, -\frac{1}{3}\right)$	-1	$-\frac{8}{3}$	Negative	Below
$\left(-\frac{1}{3}, \frac{1}{3}\right)$	$\frac{1}{4}$	$\frac{1}{9}$	Positive	Above
$\left(\frac{1}{3}, 2\right)$	1	$-\frac{8}{3}$	Negative	Below

Interval	Test Point	Value of $f(x)$	Sign of $f(x)$	Graph Above or Below x-Axis
$(2, \infty)$	3	16	Positive	Above

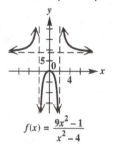

$f(x) = \dfrac{9x^2 - 1}{x^2 - 4}$

77. $f(x) = \dfrac{(x-3)(x+1)}{(x-1)^2}$

vertical asymptote: $x = 1$
horizontal asymptote: $y = 1$
y-intercept: -3; x-intercepts: -1, 3
$f(x)$ intersects the horizontal asymptote

Interval	Test Point	Value of $f(x)$	Sign of $f(x)$	Graph Above or Below x-Axis
$(-\infty, -1)$	-2	$\frac{5}{9}$	Positive	Above
$(-1, 1)$	0	-3	Negative	Below
$(1, 3)$	2	-3	Negative	Below
$(3, \infty)$	4	$\frac{5}{9}$	Positive	Above

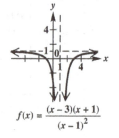

$f(x) = \dfrac{(x-3)(x+1)}{(x-1)^2}$

79. $f(x) = \dfrac{x}{x^2 - 9}$

vertical asymptotes: $x = -3$, $x = 3$
horizontal asymptote: $y = 0$
y-intercept: 0; x-intercept: 0
$f(x)$ intersects the horizontal asymptote

Interval	Test Point	Value of $f(x)$	Sign of $f(x)$	Graph Above or Below x-Axis
$(-\infty, -3)$	-4	$-\frac{4}{7}$	Negative	Below
$(-3, 0)$	-1	$\frac{1}{8}$	Positive	Above
$(0, 3)$	1	$-\frac{1}{8}$	Negative	Below
$(3, \infty)$	4	$\frac{4}{7}$	Positive	Above

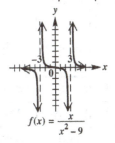

$f(x) = \dfrac{x}{x^2 - 9}$

81. $f(x) = \dfrac{1}{x^2 + 1}$

vertical asymptote: none
horizontal asymptote: $y = 0$
y-intercept: 1; x-intercept: none
$f(x)$ does not intersect the horizontal asymptote

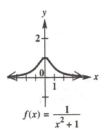

$f(x) = \dfrac{1}{x^2 + 1}$

83. $f(x) = \dfrac{(x + 4)^2}{(x - 1)(x + 5)}$

vertical asymptotes: $x = -5$, $x = 1$
horizontal asymptote: $y = 1$
y-intercept: $-\frac{16}{5}$; x-intercept: -4
$f(x)$ intersects the horizontal asymptote

Interval	Test Point	Value of $f(x)$	Sign of $f(x)$	Graph Above or Below x-Axis
$(-\infty, -5)$	-6	$\frac{4}{7}$	Positive	Above
$(-5, -4)$	$-\frac{9}{2}$	$-\frac{1}{11}$	Negative	Below
$(-4, 1)$	-2	$-\frac{4}{9}$	Negative	Below
$(1, \infty)$	2	$\frac{36}{7}$	Positive	Above

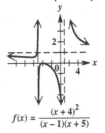

$f(x) = \dfrac{(x + 4)^2}{(x - 1)(x + 5)}$

85. $f(x) = \dfrac{20 + 6x - 2x^2}{8 + 6x - 2x^2} = \dfrac{(x - 5)(x + 2)}{(x - 4)(x + 1)}$

vertical asymptotes: $x = -1$, $x = 4$
horizontal asymptote: $y = 1$
y-intercept: $\frac{5}{2}$; x-intercepts: -2, 5
$f(x)$ does not intersect the horizontal asymptote

Interval	Test Point	Value of $f(x)$	Sign of $f(x)$	Graph Above or Below x-Axis
$(-\infty, -2)$	-4	$\frac{3}{4}$	Positive	Above
$(-2, -1)$	$-\frac{3}{2}$	$-\frac{13}{11}$	Negative	Below
$(-1, 4)$	2	2	Positive	Above
$(4, 5)$	$\frac{9}{2}$	$-\frac{13}{11}$	Negative	Below
$(5, \infty)$	6	$\frac{4}{7}$	Positive	Above

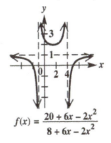

$f(x) = \dfrac{20 + 6x - 2x^2}{8 + 6x - 2x^2}$

87. $f(x) = \dfrac{x^2 + 1}{x + 3}$

vertical asymptote: $x = -3$

oblique asymptote: $y = x - 3$

y-intercept: $\frac{1}{3}$; x-intercepts: none

$f(x)$ does not intersect the oblique asymptote

Interval	Test Point	Value of $f(x)$	Sign of $f(x)$	Graph Above or Below x-Axis
$(-\infty, -3)$	-4	-17	Negative	Below
$(-3, \infty)$	4	$\frac{17}{7}$	Positive	Above

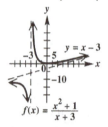

89. $f(x) = \dfrac{x^2 + 2x}{2x - 1} = \dfrac{x(x + 2)}{2x - 1}$

vertical asymptote: $x = \frac{1}{2}$

oblique asymptote: $y = \frac{x}{2} + \frac{5}{4}$

y-intercept: 0; x-intercepts: -2, 0

$f(x)$ does not intersect the oblique asymptote

Interval	Test Point	Value of $f(x)$	Sign of $f(x)$	Graph Above or Below x-Axis
$(-\infty, -2)$	-3	$-\frac{3}{7}$	Negative	Below
$(-2, 0)$	-1	$\frac{1}{3}$	Positive	Above
$\left(0, \frac{1}{2}\right)$	$\frac{1}{4}$	$-\frac{9}{8}$	Negative	Below
$\left(\frac{1}{2}, \infty\right)$	2	$\frac{8}{3}$	Positive	Above

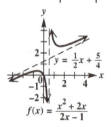

91. $f(x) = \dfrac{x^2 - 9}{x + 3} = \dfrac{(x + 3)(x - 3)}{x + 3}$

The function degenerates into the line

$f(x) = x - 3, x \neq -3$

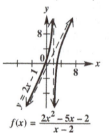

93. $f(x) = \dfrac{2x^2 - 5x - 2}{x - 2}$

vertical asymptote: $x = 2$

oblique asymptote: $y = 2x - 1$

y-intercept: 1; x-intercepts: $\approx -0.4, \approx 2.9$

$f(x)$ does not intersect the oblique asymptote

Interval	Test Point	Value of $f(x)$	Sign of $f(x)$	Graph Above or Below x-Axis
$(-\infty, -0.4)$	-2	-4	Negative	Below
$(-0.4, 2)$	1	5	Positive	Above
$(2, 2.9)$	2.5	-4	Negative	Below
$(2.9, \infty)$	4	5	Positive	Above

95. $f(x) = \dfrac{x^2 - 1}{x^2 - 4x + 3} = \dfrac{(x + 1)(x - 1)}{(x - 3)(x - 1)}$

$= \dfrac{x + 1}{x - 3}, x \neq 1$

vertical asymptote: $x = 3$

horizontal asymptote: $y = 1$

y-intercept: $-\frac{1}{3}$; x-intercept: -1

$f(x)$ is not defined for $x = 1$, so there is a hole in the graph.

$f(x)$ does not intersect the horizontal asymptote

(*continued on next page*)

(continued)

Interval	Test Point	Value of $f(x)$	Sign of $f(x)$	Graph Above or Below x-Axis
$(-\infty, -1)$	-4	$\frac{3}{7}$	Positive	Above
$(-1, 1)$	$\frac{1}{2}$	$-\frac{3}{5}$	Negative	Below
$(1, 3)$	2	-3	Negative	Below
$(3, \infty)$	4	5	Positive	Above

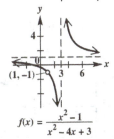

$$f(x) = \frac{x^2 - 1}{x^2 - 4x + 3}$$

97. $f(x) = \dfrac{(x^2 - 9)(2 + x)}{(x^2 - 4)(3 + x)} = \dfrac{(x + 3)(x - 3)(x + 2)}{(x + 2)(x - 2)(x + 3)}$

$\qquad = \dfrac{x - 3}{x - 2}, \ x \neq -3, -2$

vertical asymptote: $x = 2$
horizontal asymptote: $y = 1$
y-intercept: $\frac{3}{2}$; x-intercept: 3
$f(x)$ is not defined for $x = -3$ and $x = -2$, so there are two holes in the graph.
$f(x)$ does not intersect the horizontal asymptote

Interval	Test Point	Value of $f(x)$	Sign of $f(x)$	Graph Above or Below x-Axis
$(-\infty, -3)$	-4	$\frac{7}{6}$	Positive	Above
$(-3, -2)$	$-\frac{5}{2}$	$\frac{11}{9}$	Positive	Above
$(-2, 2)$	1	2	Positive	Above
$(2, 3)$	$\frac{5}{2}$	-1	Negative	Below
$(3, \infty)$	4	$\frac{1}{2}$	Positive	Above

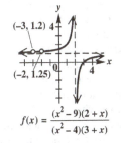

$$f(x) = \frac{(x^2 - 9)(2 + x)}{(x^2 - 4)(3 + x)}$$

99. $f(x) = \dfrac{x^4 - 20x^2 + 64}{x^4 - 10x^2 + 9}$

$\qquad = \dfrac{(x + 4)(x - 4)(x + 2)(x - 2)}{(x + 3)(x - 3)(x + 1)(x - 1)}$

vertical asymptotes: $x = -3$, $x = -1$, $x = 1$, $x = 3$
horizontal asymptote: $y = 1$
y-intercept: $\frac{64}{9}$; x-intercepts: $-4, -2, 2, 4$
$f(x)$ intersects the horizontal asymptote

Interval	Test Point	Value of $f(x)$	Sign of $f(x)$	Graph Above or Below x-Axis
$(-\infty, -4)$	-5	0.49	Positive	Above
$(-4, -3)$	-3.5	-0.85	Negative	Below
$(-3, -2)$	-2.5	1.52	Positive	Above
$(-2, -1)$	-1.5	-2.85	Negative	Below
$(-1, 1)$	0.5	9	Positive	Above
$(1, 2)$	1.5	-2.85	Negative	Below
$(2, 3)$	2.5	1.52	Positive	Above
$(3, 4)$	3.5	-0.85	Negative	Below
$(4, \infty)$	5	0.49	Positive	Above

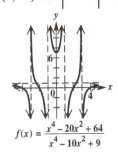

$$f(x) = \frac{x^4 - 20x^2 + 64}{x^4 - 10x^2 + 9}$$

101. The graph has a vertical asymptote, $x = 2$, so the factor $x - 2$ is in the denominator of the function. The horizontal asymptote is $y = 1$, so the numerator and denominator have the same degree. There is a "hole" in the graph at $x = -2$, so the factor $x + 2$ is in the denominator and numerator also. The x-intercept is 3, so that when $f(x) = 0$, $x = 3$. This condition exists if $x - 3$ is a factor of the numerator. Putting these conditions together, we have a possible function

$$f(x) = \frac{(x-3)(x+2)}{(x-2)(x+2)} \text{ or } f(x) = \frac{x^2 - x - 6}{x^2 - 4}.$$

103. The graph has vertical asymptotes at $x = 4$ and $x = 0$, so $x - 4$ and x are factors in the denominator of the function. The only x-intercept is 2, so that when $f(x) = 0$, $x = 2$. This condition exists if $x - 2$ is a factor of the numerator. The graph has a horizontal asymptote $y = 0$, so the degree of the denominator is larger than the degree of the numerator. Putting these conditions together, we have a possible function

$$f(x) = \frac{x-2}{x(x-4)} \text{ or } f(x) = \frac{x-2}{x^2 - 4x}.$$

105. The graph has a vertical asymptote $x = 1$ so $x - 1$ is a factor in the denominator of the function. The x-intercepts are 0 and 2, so that when $f(x) = 0$, $x = 0$ or $x = 2$. Such conditions exist if x and $x - 2$ are factors of the numerator. The horizontal asymptote is $y = -1$, so the numerator and denominator have the same degree. The numerator will have degree 2, so the denominator must also have degree 2. Also, from the horizontal asymptote, we have $y = -1 = \frac{a_n}{b_n}$. Putting these conditions together, we have a possible function

$$f(x) = \frac{-x(x-2)}{(x-1)^2} \text{ or } f(x) = \frac{-x^2 + 2x}{x^2 - 2x + 1}.$$

107. Several answers are possible. One answer is

$$f(x) = \frac{(x-3)(x+1)}{(x-1)^2}.$$

109. $f(x) = \dfrac{x+1}{x-4}$

The function is graphed in the window $[-7.4, 11.4] \times [-6.2, 6.2]$.

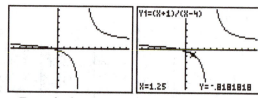

From the last screen we can see that $f(1.25) = -0.8\overline{1}$. If you want the fractional equivalent, you can go back to your homescreen and obtain Y_1 from the VARS menu. Thus $f(1.25) = -\frac{9}{11}$.

111. $f(x) = \dfrac{x^2 + 2x}{2x - 1}$

The function is graphed in the window $[-4.7, 4.7] \times [-3.1, 6.2]$.

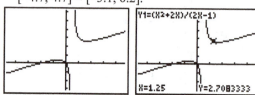

From the last screen we can see that $f(1.25) = 2.708\overline{3}$. If you want the fractional equivalent, you can go back to your homescreen and obtain Y_1 from the VARS menu. Thus $f(1.25) = \frac{65}{24}$.

113. **(a)** $T(r) = \dfrac{2r - k}{2r^2 - 2kr}; k = 25 \Rightarrow$

$$T(r) = \frac{2r - 25}{2r^2 - 2(25)r} = \frac{2r - 25}{2r^2 - 50r}$$

Graph $y_1 = \frac{2x-25}{2x^2-50x}$ and $y_2 = 0.5$ (because 30 sec = 0.5 min) on the same screen in the window $[25, 30] \times [-2, 2]$.

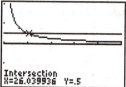

Using the "intersect" option in the CALC menu, we find that the graphs intersect at $x \approx 26$, which represents $r \approx 26$ in the given function. Therefore, there must be an average admittance rate of 26 vehicles per minute.

(b) $\frac{26}{5.3} \approx 4.9$ or 5 parking attendants must be on duty to keep the wait less than 30 seconds.

115. (a) Graph $y_1 = \dfrac{8710x^2 - 69,400x + 470,000}{1.08x^2 - 324x + 82,200}$

and $y_2 = 300$ in the window $[20, 70] \times [-50, 650]$, then find the intersection.

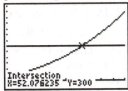

The graphs intersect when $x \approx 52$ miles per hour.

(b) Use the TABLE feature of the calculator to find the values.

x	$d(x)$	x	$d(x)$
20	**34**	50	**273**
25	**56**	55	**340**
30	**85**	60	**415**
35	**121**	65	**499**
40	**164**	70	**591**
45	**215**		

(c) By comparing values in the table generated in part b, it appears that when the speed is doubled, the breaking distance is more than doubled.

(d) If the breaking distance doubled whenever the speed doubled, then there would be a linear relationship between the speed and the breaking distance.

117. $R(x) = \dfrac{80x - 8000}{x - 110}$

(a) $R(55) = \dfrac{80(55) - 8000}{55 - 110}$
$\approx \$65.5$ tens of millions

(b) $R(60) = \dfrac{80(60) - 8000}{60 - 110}$
$= \$64$ tens of millions

(c) $R(70) = \dfrac{80(70) - 8000}{70 - 110}$
$= \$60$ tens of millions

(d) $R(90) = \dfrac{80(90) - 8000}{90 - 110}$
$= \$40$ tens of millions

(e) $R(100) = \dfrac{80(100) - 8000}{100 - 110} = \$0$

Exercises 119–127 refer to the function
$f(x) = \dfrac{x^4 - 3x^3 - 21x^2 + 43x + 60}{x^4 - 6x^3 + x^2 + 24x - 20}$.

119. The degree of the numerator equals the degree of the denominator, so the graph has a horizontal asymptote at $y = \frac{1}{1} = 1$.

121. (a) Use synthetic division where
$g(x) = x^4 - 6x^3 + x^2 + 24x - 20$
and $k = 1$.

$$\begin{array}{r|rrrrr} 1 & 1 & -6 & 1 & 24 & -20 \\ & & 1 & -5 & -4 & 20 \\ \hline & 1 & -5 & -4 & 20 & 0 \end{array}$$

Now use synthetic division where
$h(x) = x^3 - 5x^2 - 4x + 20$ and $k = 2$.

$$\begin{array}{r|rrrr} 2 & 1 & -5 & -4 & 20 \\ & & 2 & -6 & -20 \\ \hline & 1 & -3 & -10 & 0 \end{array}$$

The resulting polynomial quotient is $x^2 - 3x - 10$, which factors to $(x + 2)(x - 5)$. Thus, the complete factorization of the denominator would be $(x - 1)(x - 2)(x + 2)(x - 5)$.

(b) $f(x) = \dfrac{(x+4)(x+1)(x-3)(x-5)}{(x-1)(x-2)(x+2)(x-5)}$

123. Although $x - 5$ is a factor of the numerator, it will not yield an x-intercept because it is also a factor of the denominator. The other three factors of the numerator, namely $(x + 4)$, $(x + 1)$, and $(x - 3)$ will yield x-intercepts of $(-4, 0), (-1, 0)$, and $(3, 0)$ because any of these will yield $f(x) = 0$.

125. Although $x - 5$ is a factor of the denominator, it will not yield a vertical asymptote because it is also a factor of the numerator. The other three factors of the denominator (namely $x - 1, x - 2$, and $x + 2$) will yield vertical asymptotes of $x = 1, x = 2$, and $x = -2$ because any of these will yield a denominator of zero.

127. The vertical asymptotes are $x = 1$, $x = 2$, and $x = -2$ and the x-intercepts occur at -4, -1, and 3. We should also consider the "hole" in the graph which occurs at $x = 5$ This divides the real number line into eight intervals.

Interval	Test Point	Value of $f(x)$	Sign of $f(x)$	Graph Above or Below x-Axis
$(-\infty, -4)$	-5	$\frac{16}{63}$	Positive	Above
$(-4, -2)$	-3	$-\frac{3}{5}$	Negative	Below
$(-2, -1)$	$-\frac{3}{2}$	$\frac{9}{7}$	Positive	Above
$(-1, 1)$	0	-3	Negative	Below
$(1, 2)$	$\frac{3}{2}$	$\frac{165}{7}$	Positive	Above
$(2, 3)$	$\frac{5}{2}$	$-\frac{91}{27}$	Negative	Below
$(3, 5)$	4	$\frac{10}{9}$	Positive	Above
$(5, \infty)$	6	$\frac{21}{16}$	Positive	Above

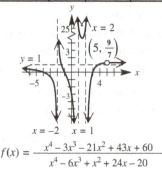

$$f(x) = \frac{x^4 - 3x^3 - 21x^2 + 43x + 60}{x^4 - 6x^3 + x^2 + 24x - 20}$$

Chapter 3 Quiz
(Sections 3.1–3.5)

1. (a)

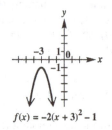

$f(x) = -2(x + 3)^2 - 1$

The equation is of the form $y = a(x - h)^2 + k$, with $h = -3$, $k = -1$, and $a = -2$. The graph opens downward and is narrower than $y = x^2$. It is a horizontal translation of the graph of $y = -2x^2$, 3 units to the left and 1 unit down. The vertex is $(-3, -1)$. The axis is $x = -3$. The domain is $(-\infty, \infty)$. The largest value of y is -1 and the graph opens downward, so the range is $(-\infty, -1]$. The function is increasing on $(-\infty, -3)$ and decreasing on $(-3, \infty)$.

(b)

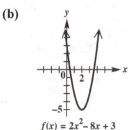

$f(x) = 2x^2 - 8x + 3$

Rewrite the function by completing the square on x.

$$f(x) = 2x^2 - 8x + 3$$
$$= 2(x^2 - 4x + 4) + (3 - 8)$$
$$\text{Note: } 2\left[\tfrac{1}{2}(-4)\right]^2 = 2(-2)^2 = 8$$
$$= 2(x - 2)^2 - 5$$

This equation is of the form $y = a(x - h)^2 + k$, with $h = 2$, $k = -5$, and $a = 2$. The graph opens upward and is narrower than $y = x^2$. It is a horizontal translation of the graph of $y = 2x^2$, 2 units to the right and 5 units down. The vertex is $(2, -5)$. The axis is $x = 2$. The domain is $(-\infty, \infty)$. The smallest value of y is -5 and the graph opens upward, so the range is $[-5, \infty)$. The function is increasing on $(2, \infty)$ and decreasing on $(-\infty, 2)$.

3. $f(x) = 2x^4 + x^3 - 3x + 4;\ k = 2$

Use synthetic division to write the polynomial in the form $f(x) = (x-k)q(x) + r$. The x^2-term is missing, so include 0 as its coefficient.

$$
\begin{array}{r|rrrrr}
2 & 2 & 1 & 0 & -3 & 4 \\
  &   & 4 & 10 & 20 & 34 \\
\hline
  & 2 & 5 & 10 & 17 & 38
\end{array}
$$

$f(x) = (x-2)(2x^3 + 5x^2 + 10x + 17) + 38$

The remainder is 38, so $k = 2$ is not a zero of the function. $k(2) = 38$.

5. Because $3 - i$ is a factor, $3 + i$ is also a factor. So

$f(x)$
$= \big[(x - (3 - i)\big]\big[(x - (3 + i)\big]\big[(x - (-2)\big](x - 3)$
$= (x - 3 + i)(x - 3 - i)(x + 2)(x - 3)$
$= (x^2 - 6x + 1)(x^2 - x - 6)$
$= x^4 - 7x^3 + 10x^2 + 26x - 60$

7. $f(x) = 2x^4 - 9x^3 - 5x^2 + 57x - 45$

Step 1: p must be a factor of $a_0 = -45$ and q must be a factor of $a_4 = 2$. Thus, p can be $\pm 1, \pm 3, \pm 5, \pm 9, \pm 15,$ or ± 45, and q can be ± 1 or ± 2. The possible rational zeros, $\frac{p}{q}$, are

$\pm\frac{1}{2}, \pm 1,\ \pm\frac{3}{2}, \pm\frac{5}{2}, \pm 3, \pm\frac{9}{2}, \pm 5, \pm\frac{15}{2}, \pm 9,\ \pm 15,$
$\pm\frac{45}{2},$ or ± 45. The remainder theorem shows that 1 is a zero.

$$
\begin{array}{r|rrrrr}
1 & 2 & -9 & -5 & 57 & -45 \\
  &   & 2 & -7 & -12 & 45 \\
\hline
  & 2 & -7 & -12 & 45 & 0
\end{array}
$$

The new quotient polynomial is
$2x^3 - 7x^2 - 12x + 45$.

p must be a factor of $a_0 = 45$ and q must be a factor of $a_3 = 2$. Thus, p can be $\pm 1, \pm 3, \pm 5,$ $\pm 9, \pm 15,$ or ± 45, and q can be ± 1 or ± 2. The possible rational zeros, $\frac{p}{q}$, are $\pm\frac{1}{2}, \pm 1,$

$\pm\frac{3}{2}, \pm\frac{5}{2}, \pm 3, \pm\frac{9}{2}, \pm 5, \pm\frac{15}{2},\ \pm 9, \pm 15, \pm\frac{45}{2},$ or ± 45. The remainder theorem shows that 3 is a zero.

$$
\begin{array}{r|rrrr}
3 & 2 & -7 & -12 & 45 \\
  &   & 6 & -3 & -45 \\
\hline
  & 2 & -1 & -15 & 0
\end{array}
$$

The new quotient polynomial is
$2x^2 - x - 15 = 0.$
$(2x + 5)(x - 3) = 0$

$2x + 5 = 0 \Rightarrow x = -\frac{5}{2}$ or $x - 3 = 0 \Rightarrow x = 3$

The rational zeros are $-\frac{5}{2}$, 1, and 3. Note that 3 is a zero of multiplicity 2, so the graph is tangent to the x-axis at $x = 3$. The factors are $2x + 5,\ x - 1,$ and $(x - 3)^2$ and thus
$f(x) = (2x + 5)(x - 1)(x - 3)^2.$

Step 2: $f(0) = -45$, so plot $(0, -45)$.

Step 3: The x-intercepts divide the x-axis into four intervals:

Interval	Test Point	Value of $f(x)$	Sign of $f(x)$	Graph Above or Below x-Axis
$\left(-\infty, -\frac{5}{2}\right)$	-3	144	Positive	Above
$\left(-\frac{5}{2}, 1\right)$	-1	-96	Negative	Below
$(1, 3)$	2	9	Positive	Above
$(3, \infty)$	4	39	Positive	Above

Plot the x-intercepts, y-intercept and test points (the y-intercept is one of the test points) with a smooth curve to get the graph.

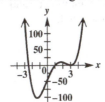

$f(x) = 2x^4 - 9x^3 - 5x^2 + 57x - 45$
$\quad = (x - 3)^2(2x + 5)(x - 1)$

9. $f(x) = \dfrac{3x + 1}{x^2 + 7x + 10} = \dfrac{3x + 1}{(x + 5)(x + 2)}$

Step 1: The graph has vertical asymptotes where the denominator equals zero, that is, at $x = -5$ and $x = -2$.

Step 2: The degree of the numerator is less than the degree of the denominator, so the graph has a horizontal asymptote at $y = 0$ (the x-axis).

Step 3: The y-intercept is
$$f(0) = \frac{3(0) + 1}{0^2 + 7(0) + 10} = \frac{1}{10}$$

(continued on next page)

(continued)

Step 4: Find any *x*-intercepts by solving $f(x) = 0$:

$$\frac{3x+1}{x^2+7x+10} = \frac{3x+1}{(x+2)(x+5)} = 0 \Rightarrow$$

$$3x+1 = 0 \Rightarrow x = -\tfrac{1}{3}$$

Step 5: The graph will intersect the horizontal asymptote, $y = 0$, when $f(x) = 0$. From step 4, that is at $x = -\tfrac{1}{3}$.

Step 6: Plot a point in each of the intervals determined by the *x*-intercept and the vertical asymptotes. There are four intervals,

$(-\infty, -5), (-5, -2), \left(-2, -\tfrac{1}{3}\right)$, and $\left(-\tfrac{1}{3}, \infty\right)$.

Interval	Test Point	Value of $f(x)$	Sign of $f(x)$	Graph Above or Below x-Axis
$(-\infty, -5)$	-6	-4.25	Negative	Below
$(-5, -2)$	-4	5.5	Positive	Above
$\left(-2, -\tfrac{1}{3}\right)$	-1	-0.5	Negative	Below
$\left(-\tfrac{1}{3}, \infty\right)$	3	0.25	Positive	Above

Use the asymptotes, intercepts, and these points to sketch the graph.

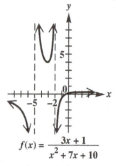

$$f(x) = \frac{3x+1}{x^2+7x+10}$$

Summary Exercises on Solving Equations and Inequalities

1. (a) $f(x) > 0$ for $\left\{x \mid x \in (-1, 1) \cup (1, \infty)\right\}$

   (b) $f(x) \le 0$ for $\left\{x \mid x \in (-\infty, -1] \cup \{1\}\right\}$

3. (a) $f(x) = 0$ for $x = 0$

   (b) $f(x) > 0$ for $\left\{x \mid x \in (-3, 0) \cup (3, \infty)\right\}$

5. $\dfrac{5x+8}{-2} = 2x - 10$

   $5x + 8 = -4x + 20$

   $9x = 12 \Rightarrow x = \dfrac{12}{9} = \dfrac{4}{3}$

   Solution set: $\left\{\tfrac{4}{3}\right\}$

7. $(x-5)^{-4} - 13(x-5)^{-2} = -36$

   Let $u = (x-5)^{-2}$. Then we have

   $$u^2 - 13u = -36$$
   $$u^2 - 13u + 36 = 0$$
   $$(u-4)(u-9) = 0$$
   $$u = 4,\ 9$$

   Now substitute.
   If $u = 4$, then

   $$4 = (x-5)^{-2}$$
   $$\frac{1}{4} = (x-5)^2$$
   $$\pm\sqrt{\frac{1}{4}} = x - 5$$
   $$\pm\frac{1}{2} = x - 5 \Rightarrow x = 5 \pm \frac{1}{2} = \frac{11}{2} \text{ or } \frac{9}{2}$$

   If $u = 9$, then

   $$9 = (x-5)^{-2}$$
   $$\frac{1}{9} = (x-5)^2$$
   $$\pm\sqrt{\frac{1}{9}} = x - 5$$
   $$\pm\frac{1}{3} = x - 5 \Rightarrow x = 5 \pm \frac{1}{3} = \frac{16}{3} \text{ or } \frac{14}{3}$$

   Solution set: $\left\{\tfrac{9}{2}, \tfrac{11}{2}, \tfrac{14}{3}, \tfrac{16}{3}\right\}$

9. $\sqrt{2x-5} - \sqrt{x-3} = 1$

   $$\sqrt{2x-5} = 1 + \sqrt{x-3}$$
   $$\left(\sqrt{2x-5}\right)^2 = \left(1 + \sqrt{x-3}\right)^2$$
   $$2x - 5 = 1 + 2\sqrt{x-3} + x - 3$$
   $$x - 3 = 2\sqrt{x-3}$$
   $$(x-3)^2 = \left(2\sqrt{x-3}\right)^2$$
   $$x^2 - 6x + 9 = 4x - 12$$
   $$x^2 - 10x + 21 = 0$$
   $$(x-3)(x-7) = 0 \Rightarrow x = 3,\ x = 7$$

   Be sure to check each proposed solution because extraneous solutions may have been introduced when both sides of the equation were squared.
   Solution set: $\{3, 7\}$

11. $x^{2/3} + \dfrac{1}{2} = \dfrac{3}{4}$

$$x^{2/3} = \dfrac{1}{4}$$

$$\left(x^{2/3}\right)^{3/2} = \pm\left(\dfrac{1}{4}\right)^{3/2}$$

$$x = \left(\pm\dfrac{1}{2}\right)^3 = \pm\dfrac{1}{8}$$

Be sure to check each proposed solution because extraneous solutions may have been introduced when both sides of the equation were raised to the power.

Solution set: $\left\{\pm\frac{1}{8}\right\}$

13. $25x^2 - 20x + 4 > 0$

The expression is an inequality. We will graph the associated function

$f(x) = 25x^2 - 20x + 4,$ which is a parabola.

$h = -\dfrac{b}{2a} = -\dfrac{-20}{2(25)} = \dfrac{2}{5}$ and

$k = f(h) = f\left(\dfrac{2}{5}\right) = 25\left(\dfrac{2}{5}\right)^2 - 20\left(\dfrac{2}{5}\right) + 4 = 0.$

Thus, the vertex is $\left(\frac{2}{5}, 0\right)$. The parabola opens upward and is narrower than the graph of $y = x^2$. The vertex is the x-intercept and the parabola opens upward, so all values of the function except the value for $x = \frac{2}{5}$ are greater than 0. The solution set of the inequality is $\left(-\infty, \frac{2}{5}\right) \cup \left(\frac{2}{5}, \infty\right)$.

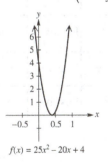

$f(x) = 25x^2 - 20x + 4$

For exercises 15–21, follow the three steps outlined on page 344 or the seven steps outlined on page 367 (as appropriate) to sketch the graph. The solutions shown here may be abbreviated.

15. $x^4 - 2x^3 - 3x^2 + 4x + 4 = 0$

The expression is an equation.

Step 1: Find the zeros of the function

$f(x) = x^4 - 2x^3 - 3x^2 + 4x + 4.$ p must be a factor of $a_0 = 4$ and q must be a factor of $a_4 = 1.$

Thus, p can be $\pm 1, \pm 2,$ or $\pm 4,$ and q can be $\pm 1.$ The possible rational zeros, $\frac{p}{q}$, are $\pm 1, \pm 2,$ or $\pm 4.$ The remainder theorem shows that 2 is a zero.

$$
\begin{array}{r|rrrrr}
2) & 1 & -2 & -3 & 4 & 4 \\
   &   & 2 & 0 & -6 & -4 \\
\hline
   & 1 & 0 & -3 & -2 & 0
\end{array}
$$

The new quotient polynomial is $x^3 - 3x - 2.$ We must find the zeros of this function. p must be a factor of $a_0 = -2$ and q must be a factor of $a_3 = 1.$ Thus, p can be ± 1 or $\pm 2,$ and q can be $\pm 1.$ The possible rational zeros, $\frac{p}{q}$, are ± 1 or $\pm 2.$ The remainder theorem shows that -1 is a zero.

$$
\begin{array}{r|rrrr}
-1) & 1 & 0 & -3 & -2 \\
    &   & -1 & 1 & 2 \\
\hline
    & 1 & -1 & -2 & 0
\end{array}
$$

The new quotient polynomial is $x^2 - x - 2.$

$x^2 - x - 2 = 0 \Rightarrow (x+1)(x-2) = 0 \Rightarrow$

$x = -1, 2$

The rational zeros are -1 (multiplicity 2) and 2 (multiplicity 2). The solution set of the equation is $\{-1, 2\}.$ We now complete the graphing process.

Step 2: $f(0) = 4,$ so the y-intercept is $(0, 4).$

Step 3: The x-intercepts divide the x-axis into three intervals.

Interval	Test Point	Value of $f(x)$	Sign of $f(x)$	Graph Above or Below x-Axis
$(-\infty, -1)$	-2	16	Positive	Above
$(-1, 2)$	1	4	Positive	Above
$(2, \infty)$	3	16	Positive	Above

Plot the x-intercepts, y-intercept, and test points with a smooth curve to get the graph.

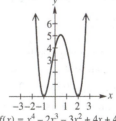

$f(x) = x^4 - 2x^3 - 3x^2 + 4x + 4$

Examining the graph, we see that the solution set of the original inequality is $\{-1, 2\}.$

17. $-x^4 - x^3 + 12x^2 = 0$

The expression is an equation.

Step 1: Find the zeros of the function

$f(x) = -x^4 - x^3 + 12x^2 = 0$.

$-x^4 - x^3 + 12x^2 = 0$

$-x^2(x^2 + x - 12) = 0$

$x^2(x + 4)(x - 3) = 0 \Rightarrow x = 0, -4, 3$

The solution set of the equation is $\{0, -4, 3\}$.
We now complete the graphing process.

Step 2: $f(0) = 0$, so the y-intercept is $(0, 0)$.

Step 3: The x-intercepts divide the x-axis into four intervals.

Interval	Test Point	Value of $f(x)$	Sign of $f(x)$	Graph Above or Below x-Axis
$(-\infty, -4)$	-5	-200	Negative	Below
$(-4, 0)$	-1	12	Positive	Above
$(0, 3)$	1	10	Positive	Above
$(3, \infty)$	4	-128	Negative	Below

Plot the x-intercepts, y-intercept, and test points with a smooth curve to get the graph.

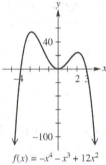

$f(x) = -x^4 - x^3 + 12x^2$

Examining the graph, we see that the solution set of the original inequality is $\{-4, 0, 3\}$.

19. $\dfrac{2x^2 - 13x + 15}{x^2 - 3x} \geq 0$

The expression is an inequality. We will graph

$f(x) = \dfrac{2x^2 - 13x + 15}{x^2 - 3x}$

Step 1: Find the vertical asymptotes by setting the denominator equal to 0.

$x^2 - 3x = 0 \Rightarrow x(x - 3) = 0 \Rightarrow x = 0, x = 3$

The vertical asymptotes are $x = 0$ (the y-axis) and $x = 3$.

Step 2: Find the horizontal asymptote. The numerator and denominator have the same degree, so the horizontal asymptote has equation $y = \dfrac{a_n}{b_n} = \dfrac{2}{1} = 2$. There is no oblique asymptote.

Step 3: There is no the y-intercept because the y-axis is one of the vertical asymptotes.

Step 4: Find the x-intercept:

$\dfrac{2x^2 - 13x + 15}{x^2 - 3x} = 0 \Rightarrow 2x^2 - 13x + 15 = 0 \Rightarrow$

$(2x - 3)(x - 5) = 0 \Rightarrow x = \frac{3}{2}, 5$

The x-intercepts are $\left(\frac{3}{2}, 0\right)$ and $(5, 0)$.

Step 5: Determine whether the graph will intersect the horizontal asymptote:

$\dfrac{2x^2 - 13x + 15}{x^2 - 3x} = 2 \Rightarrow$

$2x^2 - 13x + 15 = 2x^2 - 6x \Rightarrow 15 = 7x \Rightarrow$

$x = \frac{15}{7}$

$f(x)$ intersects the horizontal asymptote

Step 6: Plot a point in each of the intervals $(-\infty, 0), \left(0, \frac{3}{2}\right), \left(\frac{3}{2}, 3\right), (3, 5)$, and $(5, \infty)$.

Interval	Test Point	Value of $f(x)$	Sign of $f(x)$	Graph Above or Below x-Axis
$(-\infty, 0)$	-1	$\frac{15}{2}$	Positive	Above
$\left(0, \frac{3}{2}\right)$	1	-2	Negative	Below
$\left(\frac{3}{2}, 3\right)$	2	$\frac{3}{2}$	Positive	Above
$(3, 5)$	4	$-\frac{5}{4}$	Negative	Below
$(5, \infty)$	6	$\frac{1}{2}$	Positive	Above

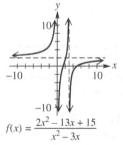

$f(x) = \dfrac{2x^2 - 13x + 15}{x^2 - 3x}$

By examining the graph, we find that the solution set for the original inequality is $(-\infty, 0) \cup \left[\frac{3}{2}, 3\right) \cup [5, \infty)$.

21. $\dfrac{x-1}{(x-3)^2} = 0$

The expression is an equation. We will graph

$$f(x) = \dfrac{x-1}{(x-3)^2}$$

Step 1: Find the vertical asymptotes by setting the denominator equal to 0.

$$(x-3)^2 = 0 \Rightarrow x - 3 = 0 \Rightarrow x = 3$$

The vertical asymptote is $x = 3$.

Step 2: The numerator has lesser degree than the denominator, so the x-axis is the horizontal asymptote. There is no oblique asymptote.

Step 3: Find the y-intercept:

$$f(0) = \dfrac{0-1}{(0-3)^2} = -\dfrac{1}{9}$$

Step 4: Find the x-intercept:

$$\dfrac{x-1}{(x-3)^2} = 0 \Rightarrow x - 1 = 0 \Rightarrow x = 1$$

The x-intercept is $(1, 0)$.

Step 5: The graph will intersect the horizontal asymptote.

Step 6: Plot a point in each of the intervals $(-\infty, 1)$, $(1, 3)$, and $(3, \infty)$.

Interval	Test Point	Value of $f(x)$	Sign of $f(x)$	Graph Above or Below x-Axis
$(-\infty, 1)$	-2	$-\dfrac{3}{25}$	Negative	Below
$(1, 3)$	2	1	Positive	Above
$(3, \infty)$	4	3	Positive	Above

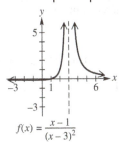

$$f(x) = \dfrac{x-1}{(x-3)^2}$$

By examining the graph, we find that the solution set for the original equation is $\{1\}$.

Section 3.6 Variation

1. For $k > 0$, if y varies directly as x, then when x increases, y <u>increases</u>, and when x decreases, y <u>decreases</u>.

3. When $x = 10$, $y = 60$ because y varies directly as x. Because x is doubled, so is y.

5. 6

7. *Step 1*: $y = kx$

Step 2: Substitute $x = 4$ and $y = 20$ to find k.

$$20 = k(4) \Rightarrow k = 5$$

Step 3: $y = 5x$

Step 4: Now find y when $x = -6$.

$$y = 5(-6) = -30$$

9. *Step 1*: $m = kxy$

Step 2: Substitute $m = 10$, $x = 2$, and $y = 14$ to find k.

$$10 = k(2)(14) \Rightarrow 10 = 28k \Rightarrow k = \dfrac{5}{14}$$

Step 3: $m = \dfrac{5}{14}xy$

Step 4: Now find m when $x = 21$ and $y = 8$.

$$m = \dfrac{5}{14}(21)(8) = 60$$

11. *Step 1*: $y = \dfrac{k}{x}$

Step 2: Substitute $x = 3$ and $y = 10$ to find k.

$$10 = \dfrac{k}{3} \Rightarrow k = 30$$

Step 3: $y = \dfrac{30}{x}$

Step 4: Now find y when $x = 20$.

$$y = \dfrac{30}{20} = \dfrac{3}{2}$$

13. *Step 1*: $r = \dfrac{km^2}{s}$

Step 2: Substitute $r = 12$, $m = 6$, and $s = 4$ to find k.

$$12 = \dfrac{k \cdot 6^2}{4} \Rightarrow 12 = \dfrac{36k}{4} \Rightarrow 12 = 9k \Rightarrow k = \dfrac{4}{3}$$

Step 3: $r = \dfrac{4}{3} \cdot \dfrac{m^2}{s} = \dfrac{4m^2}{3s}$

Step 4: Now find r when $m = 6$ and $s = 20$.

$$r = \dfrac{4(6^2)}{3(20)} = \dfrac{4(36)}{60} = \dfrac{144}{60} = \dfrac{12}{5}$$

15. *Step 1:* $a = \dfrac{kmn^2}{y^3}$

Step 2: Substitute $a = 9$, $m = 4$, $n = 9$, and $y = 3$ to find k.

$9 = \dfrac{k(4)(9^2)}{3^3} \Rightarrow 9 = \dfrac{324k}{27} \Rightarrow 9 = 12k \Rightarrow$

$k = \dfrac{3}{4}$

Step 3: $a = \dfrac{3}{4} \cdot \dfrac{mn^2}{y^3} = \dfrac{3mn^2}{4y^3}$

Step 4: Now find a when $m = 6$, $n = 2$, and $y = 5$.

$a = \dfrac{3(6)(2^2)}{4(5^3)} = \dfrac{3(6)(4)}{4(125)} = \dfrac{18}{125}$

17. y varies directly as x, $y = kx$, is a straight-line model. It matches graph C.

19. y varies directly as the second power of x, $y = kx^2$, matches graph A.

21. $C = 2\pi r$, where C is the circumference of a circle of radius r
The circumference of a circle varies directly as (or is proportional to) its radius.

23. $r = \dfrac{d}{t}$, where r is the speed when traveling d miles in t hours
The average speed varies directly as (or is proportional to) the distance traveled and inversely as the time.

25. $s = kx^3$, where s is the strength of a muscle of length x
The strength of a muscle varies directly as (or is proportional to) the cube of its length.

27. *Step 1:* Let C be the circumference of the circle (in inches), r is the radius of the circle (in inches). $C = kr$
Step 2: Substitute $C = 43.96$ and $r = 7$ to find k.
$43.96 = k(7) \Rightarrow k = 6.28$

Step 3: $C = 6.28r$
Step 4: Now find C when $r = 11$.
$C = 6.28(11) = 69.08$

A radius of 11 in. yields a circumference of 69.08 in.

29. *Step 1:* Let R be the resistance (in ohms); t is the temperature (in degrees Kelvin, K). $R = kt$
Step 2: Substitute $R = 646$ and $t = 190$ to find k: $646 = k(190) \Rightarrow k = \dfrac{646}{190} = \dfrac{17}{5}$

Step 3: $R = \dfrac{17}{5}t$

Step 4: Now find R when $t = 250$.
$R = \dfrac{17}{5}(250) = 850$

A resistance of 850 ohms occurs at 250K.

31. *Step 1:* Let d be the distance a person can see (in km), h is the height above Earth (in meters). $d = k\sqrt{h}$
Step 2: Substitute $d = 18$ and $h = 144$ to find k.
$18 = k\sqrt{144} \Rightarrow 18 = 12k \Rightarrow k = \dfrac{18}{12} = \dfrac{3}{2}$

Step 3: $d = \dfrac{3}{2}\sqrt{h}$

Step 4: Now find d when $h = 64$.
$d = \dfrac{3}{2}\sqrt{64} = \dfrac{3}{2}(8) = 12$

A person can see 12 km from a point 64 m above earth.

33. *Step 1:* Let d be the distance the spring stretches (in inches), f is the force applied (in lbs). $d = kf$
Step 2: Substitute $d = 8$ and $f = 15$ to find k.
$8 = k(15) \Rightarrow k = \dfrac{8}{15}$

Step 3: $d = \dfrac{8}{15}f$

Step 4: Now find d when $f = 30$.
$d = \dfrac{8}{15}(30) = 16$

The spring will stretch 16 in.

35. *Step 1:* Let v be the speed of the pulley (in rpm). D is the diameter of the pulley (in inches). $v = \dfrac{k}{D}$
Step 2: Substitute $v = 150$ and $D = 3$ to find k.

$150 = \dfrac{k}{3} \Rightarrow k = 450$

Step 3: $v = \dfrac{450}{d}$

Step 4: Now find v when $D = 5$: $v = \dfrac{450}{5} = 90$

The 5-in. diameter pulley will have a speed of 90 revolutions per minute.

37. *Step 1*: Let R be the resistance (in ohms), D is the diameter (in inches). $R = \dfrac{k}{D^2}$

Step 2: Substitute $D = 0.01$ and $R = 0.4$ to find k.

$0.4 = \dfrac{k}{0.01^2} \Rightarrow 0.4 = \dfrac{k}{0.0001} \Rightarrow k = 0.00004$

Step 3: $R = \dfrac{0.00004}{D^2}$

Step 4: Now find R when $D = 0.03$

$R = \dfrac{0.00004}{0.03^2} = \dfrac{0.00004}{0.0009} \approx 0.0444$

The resistance is approximately 0.0444 ohm.

39. *Step 1*: Let i be the interest (in dollars). p is the principal (in dollars). t is the time (in years). $i = kpt$

Step 2: Substitute $i = 70$, $p = 1000$, and $t = 2$ to find k.

$70 = k(1000)(2) \Rightarrow 70 = 2000k$

$k = \dfrac{70}{2000} = 0.035$ (or 3.5%)

Step 3: $i = 0.035pt$

Step 4: Now find i when $p = 5000$ and $t = 5$.

$i = 0.035(5000)(5) = 875$

The amount of interest is \$875.

41. *Step 1*: Let F be the force of the wind (in lbs), A is the area (in ft^2), v is the velocity of the wind (in mph). $F = kAv^2$

Step 2: Substitute $F = 50$, $A = \frac{1}{2}$, and $v = 40$ to find k.

$50 = k\left(\frac{1}{2}\right)(40^2) = k\left(\frac{1}{2}\right)(1600)$

$50 = 800k \Rightarrow k = \dfrac{50}{800} = \dfrac{1}{16}$

Step 3: $F = \frac{1}{16}Av^2$

Step 4: Now find F when $v = 80$, and $A = 2$.

$F = \frac{1}{16}(2)(80^2) = \frac{1}{16}(2)(6400) = 800$

The force would be 800 pounds.

43. *Step 1*: Let L be the load (in metric tons), D is the diameter (in meters), h is the height (in meters). $L = \dfrac{kD^4}{h^2}$

Step 2: Substitute $h = 9$, $D = 1$, and $L = 8$ to find k.

$8 = \dfrac{k(1^4)}{9^2} \Rightarrow k = 648$

Step 3: $L = \dfrac{648D^4}{h^2}$

Step 4: Now find L when $D = \frac{2}{3}$ and $h = 12$.

$L = \dfrac{648\left(\frac{2}{3}\right)^4}{12^2} = \dfrac{648\left(\frac{16}{81}\right)}{144} = \dfrac{128}{144} = \dfrac{8}{9}$

A column 12 m high and $\frac{2}{3}$ m in diameter will support $\frac{8}{9}$ metric ton.

45. *Step 1*: Let p be the period of pendulum (in sec), l is the length of pendulum (in cm), a is the acceleration due to gravity $\left(\text{in } \frac{\text{cm}}{\text{sec}^2}\right)$.

$p = \dfrac{k\sqrt{l}}{\sqrt{a}}$

Step 2: Substitute $p = 6\pi$, $l = 289$, and $a = 980$ to find k.

$6\pi = \dfrac{k\sqrt{289}}{\sqrt{980}} \Rightarrow 6\pi = \dfrac{17k}{14\sqrt{5}} \Rightarrow$

$84\sqrt{5}\pi = 17k \Rightarrow k = \dfrac{84\sqrt{5}\pi}{17}$

Step 3: $p = \dfrac{84\sqrt{5}\pi}{17} \cdot \dfrac{\sqrt{l}}{\sqrt{a}} = \dfrac{84\sqrt{5}\pi\sqrt{l}}{17\sqrt{a}}$

Step 4: Now find p when $l = 121$ and $a = 980$.

$p = \dfrac{84\pi\sqrt{5}\sqrt{121}}{17\sqrt{980}} = \dfrac{84\pi\sqrt{5}(11)}{17(14\sqrt{5})} = \dfrac{924\pi\sqrt{5}}{238\sqrt{5}} = \dfrac{66\pi}{17}$

The period is $\frac{66\pi}{17}$ sec.

47. *Step 1*: Let B be the BMI, w is the weight (in lbs), h is the height (in inches). $B = \dfrac{kw}{h^2}$

Step 2: Substitute $w = 177$, $B = 24$, and $h = 72$ (6 feet) to find k.

$24 = \dfrac{k(177)}{72^2} \Rightarrow 124,416 = 177k \Rightarrow$

$k = \dfrac{124,416}{177} = \dfrac{41,472}{59}$

Step 3: $B = \dfrac{41,472}{59} \cdot \dfrac{w}{h^2} = \dfrac{41,472w}{59h^2}$

Step 4: Now find B when $w = 130$ and $h = 66$.

$B = \dfrac{41,472(130)}{59(66^2)} = \dfrac{5,391,360}{59(4356)} = \dfrac{5,391,360}{257,004} \approx 20.98$

The BMI would be approximately 21.

49. *Step 1:* Let R be the radiation, t is the temperature (in degrees kelvin, K). $R = kt^4$:
Step 2: Substitute $R = 213.73$ and $t = 293$ to find k.

$$213.73 = k\left(293^4\right) \Rightarrow k = \frac{213.73}{293^4} \approx 2.9 \times 10^{-8}$$

Step 3: $R = (2.9 \times 10^{-8})t^4$

Step 4: Now find R when $t = 335$.

$$R = \left(2.9 \times 10^{-8}\right) \cdot 335^4 \approx 365.24$$

The radiation of heat would be 365.24.

51. *Step 1:* Let p be the person's pelidisi, w is the person's weight (in g), h is the person's sitting height (in cm). $p = \dfrac{k\sqrt[3]{w}}{h}$

Step 2: Substitute $w = 48820$, $h = 78.7$ and $p = 100$ to find k.

$$100 = \frac{k\sqrt[3]{48,820}}{78.7} \Rightarrow 7870 = \sqrt[3]{48,820}k \Rightarrow$$
$$k = \frac{7870}{\sqrt[3]{48,820}} \approx 215.33$$

Step 3: $p = \dfrac{215.33\sqrt[3]{w}}{h}$

Step 4: Now find p when $w = 54,430$ and $h = 88.9$

$$p = \frac{215.33\sqrt[3]{54,430}}{88.9} \approx 92$$

This person's pelidisi is 92. The individual is undernourished because his pelidisi is below 100.

53. $y = \dfrac{k}{x}$

If x is doubled, then $y_1 = \dfrac{k}{2x} = \dfrac{1}{2} \cdot \dfrac{k}{x} = \dfrac{1}{2}y.$
Thus, y is half as large as it was before.

55. $y = kx$

If x is replaced by $\frac{1}{3}x$, then we have

$y_1 = k\left(\frac{1}{3}x\right) = \frac{1}{3} \cdot kx = \frac{1}{3}y.$ Thus, y is one-third as large as it was before.

57. $p = \dfrac{kr^3}{t^2}$

If r is replaced by $\frac{1}{2}r$ and t is replaced by $2t$, then we have the following.

$$p_1 = \frac{k\left(\frac{1}{2}r\right)^3}{(2t)^2} = \frac{k\left(\frac{1}{8}r^3\right)}{4t^2} = \frac{kr^3}{32t^2} = \frac{1}{32} \cdot \frac{kr^3}{t^2}$$

Thus, so p is $\frac{1}{32}$ as large as it was before.

Chapter 3   Review Exercises

1.

$f(x) = 3(x+4)^2 - 5$

$(-4, -5)$

Because

$$f(x) = 3(x+4)^2 - 5 = 3\left[x - (-4)\right]^2 + (-5),$$

the function has the form
$f(x) = a(x-h)^2 + k$ with
$a = 3$, $h = -4$, and $k = -5$. The graph is a parabola that opens upward with vertex $(h, k) = (-4, -5)$. The axis is
$x = h \Rightarrow x = -4$. To find the x-intercepts, let $f(x) = 0$.

$$3(x+4)^2 - 5 = 0 \Rightarrow 3(x+4)^2 = 5 \Rightarrow$$
$$(x+4)^2 = \frac{5}{3} \Rightarrow x + 4 = \pm\sqrt{\frac{5}{3}} \Rightarrow$$
$$x = -4 \pm \sqrt{\frac{5}{3}} = -4 \pm \frac{\sqrt{15}}{3} = \frac{-12 \pm \sqrt{15}}{3}$$

Thus, the x-intercepts are $\left(\frac{-12 \pm \sqrt{15}}{3}, 0\right).$

To find the y-intercept, let $x = 0$.

$$f(0) = 3(0+4)^2 - 5 = 3(16) - 5 = 48 - 5 = 43$$

The y-intercept is $(0, 43)$.
The domain is $(-\infty, \infty)$. The lowest point on the graph is $(-4, -5)$, so the range is $[-5, \infty)$. The function is increasing on $(-4, \infty)$ and decreasing on $(-\infty, -4)$.

3.

$f(x) = -3x^2 - 12x - 1$

$(-2, 11)$

Complete the square so the function has the form $f(x) = a(x-h)^2 + k$.

$$f(x) = -3x^2 - 12x - 1 = -3\left(x^2 + 4x\right) - 1$$
$$= -3\left(x^2 + 4x + 4 - 4\right) - 1$$

(continued on next page)

(continued)

$$f(x) = -3(x+2)^2 + 12 - 1$$
$$= -3\left[x - (-2)\right]^2 + 11$$

This parabola is in now in standard form with $a = -3$, $h = -2$, and $k = 11$. It opens downward with vertex $(h, k) = (-2, 11)$. The axis is $x = h \Rightarrow x = -2$. To find the x-intercepts, let $f(x) = 0$.

$$-3(x+2)^2 + 11 = 0 \Rightarrow -3(x+2)^2 = -11$$
$$(x+2)^2 = \tfrac{11}{3} \Rightarrow x + 2 = \pm\sqrt{\tfrac{11}{3}}$$
$$x = -2 \pm \sqrt{\tfrac{11}{3}} = -2 \pm \tfrac{\sqrt{33}}{3}$$
$$= \tfrac{-6 \pm \sqrt{33}}{3}$$

Thus, the x-intercepts are $\left(\tfrac{-6 \pm \sqrt{33}}{3}, 0\right)$.

To find the y-intercept, let $x = 0$.

$$f(0) = -3(0^2) - 12(0) - 1 = -1$$

The y-intercept is $(0, -1)$.
The domain is $(-\infty, \infty)$. Because the highest point on the graph is $(-2, 11)$, the range is $(-\infty, 11]$. The function is increasing on $(-\infty, -2)$ and decreasing on $(-2, \infty)$.

5. $f(x) = a(x-h)^2 + k; \ a > 0$
   The graph is a parabola that opens upward. The coordinates of the lowest point of the graph are represented by the vertex, (h, k).

7. For the graph to have one or more x-intercepts, $f(x) = 0$ must have real number solutions.

   $$a(x-h)^2 + k = 0 \Rightarrow a(x-h)^2 = -k$$
   $$(x-h)^2 = \tfrac{-k}{a} \Rightarrow x - h = \pm\sqrt{\tfrac{-k}{a}}$$
   $$x = h \pm \sqrt{\tfrac{-k}{a}}$$

   These solutions are real only if $\tfrac{-k}{a} \geq 0$.
   Because $a > 0$, this condition is equivalent to $-k \geq 0$ or $k \leq 0$. The x-intercepts for these conditions are given by $\left(h \pm \sqrt{\tfrac{-k}{a}}, 0\right)$.

9. Let $x =$ the width of the rectangular region.
   $180 - 2x =$ the length of the region.

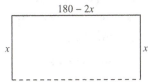

Because $\mathcal{A} = lw$, we have
$$\mathcal{A}(x) = (180 - 2x)x = -2x^2 + 180x.$$

The graph of $A(x)$ is a parabola that opens downward, so the maximum area is the y-coordinate of the vertex.

$$\mathcal{A}(x) = -2x^2 + 180x = -2(x^2 - 90x)$$
$$= -2(x^2 - 90x + 2025) + 4050$$
$$= -2(x - 45)^2 + 4050$$

The x-coordinate of the vertex, $x = 45$, is the width that gives the maximum area. The length will be $L = 180 - 2(45) = 90$. Thus, the dimensions of the region are 90 m $\times$ 45 m.

11. Because $V(x)$ has different equations on two different intervals, it is a piecewise-defined function.

$$V(x) = \begin{cases} 2x^2 - 32x + 150 & \text{if } 1 \leq x < 8 \\ 31x - 226 & \text{if } 8 \leq x \leq 12 \end{cases}$$

Note: Because $2 \cdot 8^2 - 32 \cdot 8 + 150 = 22$ and $31 \cdot 8 - 226 = 22$, the point $(8, 22)$ can be found using either rule, and the graph will have no breaks.

(a) In January, $x = 1$.
$$V(x) = 2x^2 - 32x + 150$$
$$V(1) = 2(1)^2 - 32(1) + 150 = 120$$

(b) In May, $x = 5$.
$$V(x) = 2x^2 - 32x + 150$$
$$V(5) = 2(5)^2 - 32(5) + 150 = 40$$

(c) In August, $x = 8$.
$$V(x) = 31x - 226$$
$$V(8) = 31(8) - 226 = 22$$

(d) In October, $x = 10$.
$$V(x) = 31x - 226$$
$$V(10) = 31(10) - 226 = 84$$

(e) In December, $x = 12$.
$$V(x) = 31x - 226$$
$$V(12) = 31(12) - 226 = 146$$

(f)

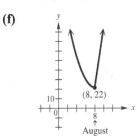

(continued on next page)

(continued)

The graph shows a minimum point at $(8, 22)$. Thus, in August ($x = 8$), the fewest volunteers are available.

13. $f(x) = -2.64x^2 + 5.47x + 3.54$

The discriminant is $b^2 - 4ac$ in the standard quadratic equation $y = ax^2 + bx + c$. Here we have $a = -2.64$, $b = 5.47$, and $c = 3.54$. Thus, $b^2 - 4ac = (5.47)^2 - 4(-2.64)(3.54) = 67.3033$. Because the discriminant is 67.3033, a positive number, there are two x-intercepts.

15. (a) $f(x) > 0$ on the open interval $(-0.52, 2.59)$.

(b) $f(x) < 0$ on $(-\infty, -0.52) \cup (2.59, \infty)$.

17. $\dfrac{x^3 + x^2 - 11x - 10}{x - 3}$

Because $x - 3$ is in the form $x - k$, $k = 3$.

$$
\begin{array}{r|rrrr}
3 & 1 & 1 & -11 & -10 \\
  &   & 3 & 12 & 3 \\
\hline
  & 1 & 4 & 1   & -7
\end{array}
$$

$$\frac{x^3 + x^2 - 11x - 10}{x - 3} = x^2 + 4x + 1 + \frac{-7}{x - 3}$$

19. $\dfrac{2x^3 - x + 6}{x + 4}$

Express $x + 4$ in the form $x - k$ by writing it as $x - (-4)$. Thus we have $k = -4$. The x^2-term is missing, so include a 0 as its coefficient.

$$
\begin{array}{r|rrrr}
-4 & 2 & 0 & -1 & 6 \\
   &   & -8 & 32 & -124 \\
\hline
   & 2 & -8 & 31 & -118
\end{array}
$$

$$\frac{2x^3 - x + 6}{x + 4} = 2x^2 - 8x + 31 + \frac{-118}{x + 4}$$

21. $f(x) = 5x^3 - 3x^2 + 2x - 6$; $k = 2$

Use synthetic division to write the polynomial in the form $f(x) = (x - k)q(x) + r$.

$$
\begin{array}{r|rrrr}
2 & 5 & -3 & 2 & -6 \\
  &   & 10 & 14 & 32 \\
\hline
  & 5 & 7 & 16 & 26
\end{array}
$$

$$f(x) = (x - 2)(5x^2 + 7x + 16) + 26$$

23. $f(x) = -x^3 + 5x^2 - 7x + 1$; find $f(2)$.

$$
\begin{array}{r|rrrr}
2 & -1 & 5 & -7 & 1 \\
  &   & -2 & 6 & -2 \\
\hline
  & -1 & 3 & -1 & -1
\end{array}
$$

The synthetic division shows that $f(2) = -1$.

25. $f(x) = 5x^4 - 12x^2 + 2x - 8$; find $f(2)$.

$$
\begin{array}{r|rrrrr}
2 & 5 & 0 & -12 & 2 & -8 \\
  &   & 10 & 20 & 16 & 36 \\
\hline
  & 5 & 10 & 8 & 18 & 28
\end{array}
$$

The synthetic division shows that $f(2) = 28$.

27. $f(x) = x^3 + 2x^2 + 3x + 2$; $k = -1$

$$
\begin{array}{r|rrrr}
-1 & 1 & 2 & 3 & 2 \\
   &   & -1 & -1 & -2 \\
\hline
   & 1 & 1 & 2 & 0
\end{array}
$$

Because $f(-1) = 0$, we have that -1 is a zero of the function.

29. By the conjugate zeros theorem, $7 - 2i$ is also a zero.

In Exercises 31 through 33, other answers are possible.

31. Zeros: $-1, 4, 7$
$$
\begin{aligned}
f(x) &= \left[x - (-1)\right](x - 4)(x - 7) \\
&= (x + 1)(x - 4)(x - 7) \\
&= (x + 1)\left[(x - 4)(x - 7)\right] \\
&= (x + 1)\left(x^2 - 11x + 28\right) \\
&= x^3 - 10x^2 + 17x + 28
\end{aligned}
$$

33. Zeros: $\sqrt{3}, -\sqrt{3}, 2, 3$.
$$
\begin{aligned}
f(x) &= \left(x - \sqrt{3}\right)\left(x + \sqrt{3}\right)(x - 2)(x - 3) \\
&= \left[\left(x - \sqrt{3}\right)\left(x + \sqrt{3}\right)\right]\left[(x - 2)(x - 3)\right] \\
&= \left(x^2 - 3\right)\left(x^2 - 5x + 6\right) \\
&= x^4 - 5x^3 + 3x^2 + 15x - 18
\end{aligned}
$$

35. Zeros: $2, 4, -i$
By the conjugate zeros theorem, if $-i$ is a zero of a polynomial function, then i is also a zero of the function.
$$
\begin{aligned}
f(x) &= (x - 2)(x - 4)(x - i)(x + i) \\
&= \left[(x - 2)(x - 4)\right]\left[(x - i)(x + i)\right] \\
&= \left(x^2 - 6x + 8\right)\left(x^2 + 1\right) \\
&= x^4 - 6x^3 + 9x^2 - 6x + 8
\end{aligned}
$$

37. $f(x) = 2x^3 - 9x^2 - 6x + 5$

p must be a factor of $a_0 = 5$ and q must be a factor of $a_3 = 2$. Thus, p can be $\pm 1, \pm 5,$ and q can be $\pm 1, \pm 2$. The possible zeros, $\frac{p}{q}$, are $\pm 1, \pm 5, \pm \frac{1}{2}, \pm \frac{5}{2}$. The remainder theorem shows that $\frac{1}{2}$ is a zero.

$$\frac{1}{2}\overline{)2 \quad -9 \quad -6 \quad 5}$$
$$\phantom{\frac{1}{2})}\underline{ \quad 1 \quad -4 \quad -5}$$
$$\phantom{\frac{1}{2})}2 \quad -8 \quad -10 \quad 0$$

The new quotient polynomial is therefore $2x^2 - 8x - 10.$

$2x^2 - 8x - 10 = 0$
$x^2 - 4x - 5 = 0$
$(x+1)(x-5) = 0$
$x + 1 = 0 \Rightarrow x = -1$ or $x - 5 = 0 \Rightarrow x = 5$

The rational zeros are $\frac{1}{2}, -1,$ and $5.$

39. $f(x) = 3x^3 - 8x^2 + x + 2$

(a) To show there is a zero between -1 and 0, we need to find $f(-1)$ and $f(0)$.

Find $f(-1)$ by synthetic division and $f(0)$ by evaluating the function.

$$-1\overline{)3 \quad -8 \quad 1 \quad 2}$$
$$\underline{ \quad -3 \quad 11 \quad -12}$$
$$3 \quad -11 \quad 12 \quad -10$$

$f(0) = 3(0^3) - 8(0^2) + (0) + 2 = 2$

Because $f(-1) = -10 < 0$ and $f(0) = 2 > 0$, there must be a zero between -1 and 0.

(b) To show there is a zero between 2 and 3, we need to find $f(2)$ and $f(3)$. Find these by synthetic division.

$$2\overline{)3 \quad -8 \quad 1 \quad 2}$$
$$\underline{ \quad 6 \quad -4 \quad -6}$$
$$3 \quad -2 \quad -3 \quad -4$$

$$3\overline{)3 \quad -8 \quad 1 \quad 2}$$
$$\underline{ \quad 9 \quad 3 \quad 12}$$
$$3 \quad 1 \quad 4 \quad 14$$

Because $f(2) = -4 < 0$ and $f(3) = 14 > 0$, there must be a zero between 2 and 3.

(c) Using a graphing calculator, we find that the zero between 2 and 3 is about 2.414.

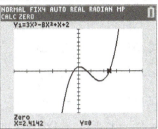

41. $f(x) = 6x^4 + 13x^3 - 11x^2 - 3x + 5$

(a) To show that there is no zero greater than 1, we must show that 1 is an upper bound.

$$1\overline{)6 \quad 13 \quad -11 \quad -3 \quad 5}$$
$$\underline{ \quad 6 \quad 19 \quad 8 \quad 5}$$
$$6 \quad 19 \quad 8 \quad 5 \quad 10$$

Because the 1 is positive and the bottom row in the synthetic division has all positive numbers, there is no zero greater than 1.

(b) To show that there is no zero less than -3, we must show that -3 is a lower bound.

$$-3\overline{)6 \quad 13 \quad -11 \quad -3 \quad 5}$$
$$\underline{ \quad -18 \quad 15 \quad -12 \quad 45}$$
$$6 \quad -5 \quad 4 \quad -15 \quad 50$$

Because the -3 is negative and the bottom row of the synthetic division alternates in sign, there is no zero less than -3.

43. To determine if $x + 1$ is a factor of $f(x) = x^3 + 2x^2 + 3x + 2$, we can find $f(-1)$ by synthetic division

$$-1\overline{)1 \quad 2 \quad 3 \quad 2}$$
$$\underline{ \quad -1 \quad -1 \quad -2}$$
$$1 \quad 1 \quad 2 \quad 0$$

Because $f(-1) = 0$, $x + 1$ is a factor of $f(x)$.

45. Because $f(x) = a\left[x - (-2)\right](x-1)(x-4)$
$= a(x+2)(x-1)(x-4),$ we can solve for a.
$16 = a(2+2)(2-1)(2-4)$
$16 = -8a \Rightarrow a = -2$
The polynomial function is $f(x) = -2(x+2)(x-1)(x-4)$. It can be expanded as follows.

(continued on next page)

(*continued*)

$$f(x) = -2(x+2)(x-1)(x-4)$$
$$= -2\left[(x+2)(x-1)\right](x-4)$$
$$= -2\left(x^2 + x - 2\right)(x-4)$$
$$= -2\left(x^3 - 3x^2 - 6x + 8\right)$$
$$= -2x^3 + 6x^2 + 12x - 16$$

47. $f(x) = 2x^4 - x^3 + 7x^2 - 4x - 4$; 1 and $-2i$ are zeros.

Use synthetic division where $k = 1$.

$$\begin{array}{r|rrrrr} 1) & 2 & -1 & 7 & -4 & -4 \\ & & 2 & 1 & 8 & 4 \\ \hline & 2 & 1 & 8 & 4 & 0 \end{array}$$

Synthetically divide $-2i$ into the quotient polynomial.

$$\begin{array}{r|rrrr} -2i) & 2 & 1 & 8 & 4 \\ & & -4i & -8-2i & -4 \\ \hline & 2 & 1-4i & -2i & 0 \end{array}$$

Because $-2i$ is a zero, $2i$ is also a zero. Synthetically divide $2i$ into the quotient polynomial.

$$\begin{array}{r|rrr} 2i) & 2 & 1-4i & -2i \\ & & 4i & 2i \\ \hline & 2 & 1 & 0 \end{array}$$

Because the quotient polynomial represented is $2x + 1$, we can set it equal to zero to find the remaining zero.

$$2x + 1 = 0 \Rightarrow x = -\frac{1}{2}$$

Thus, all the zeros are 1, $-\frac{1}{2}$ and $\pm 2i$.

49. To determine the value of k such that when the polynomial $x^3 - 3x^2 + kx - 4$ is divided by $x - 2$, the remainder is 5, we perform synthetic division.

$$\begin{array}{r|rrrr} 2) & 1 & -3 & k & -4 \\ & & 2 & -2 & 2k-4 \\ \hline & 1 & -1 & k-2 & 2k-8 \end{array}$$

To ensure that 5 is the remainder set $2k - 8$ equal to 5 and solve.

$$2k - 8 = 5 \Rightarrow 2k = 13 \Rightarrow k = \frac{13}{2}$$

Thus, the value of s is $\frac{13}{2}$.

51. Any polynomial that can be factored as $a(x-b)^3$, where a and b are real numbers, will be a cubic polynomial function having exactly one real zero. One example is $f(x) = 2(x-1)^3$.

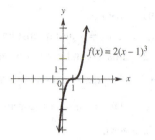

53. The polynomial has a leading term of $10x^7$.

(a) The domain is $(-\infty, \infty)$.

(b) The range is $(-\infty, \infty)$.

(c) As $x \to \infty$, $f(x) \to \infty$ and as $x \to -\infty$, $f(x) \to -\infty$.

(d) There are at most 7 zeros.

(e) There are at most 6 turning points.

55. $f(x) = (x-2)^2(x+3)$

Step 1: Set each factor equal to 0 and solve the resulting equations to find the zeros of the function.

$$(x-2)^2 = 0 \Rightarrow x = 2 \text{ or } x+3 = 0 \Rightarrow x = -3$$

The zeros are -3 and 2, which divide the x-axis into three regions. Test a point in each region to find the sign of $f(x)$ in that region.

Step 2: $f(0) = (0-2)^2(0+3) = 4 \cdot 3 = 12$, so plot $(0, 12)$.

Step 3: The x-intercepts divide the x-axis into three intervals.

Interval	Test Point	Value of $f(x)$	Sign of $f(x)$	Graph Above or Below x-Axis
$(-\infty, -3)$	-4	-36	Negative	Below
$(-3, 2)$	1	4	Positive	Above
$(2, \infty)$	3	6	Positive	Above

Plot the x-intercepts, y-intercept, and test points with a smooth curve to get the graph.

(*continued on next page*)

(continued)

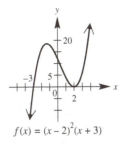

$$f(x) = (x-2)^2(x+3)$$

57. $f(x) = 2x^3 + x^2 - x$

Step 1: Factor the polynomial function and set each factor equal to 0 and solve the resulting equations to find the zeros of the function.

$$2x^3 + x^2 - x = x(2x^2 + x - 1)$$
$$= x(x+1)(2x-1)$$

$x = 0$ or $x + 1 = 0 \Rightarrow x = -1$ or
$2x - 1 = 0 \Rightarrow x = \frac{1}{2}$

The zeros are -1, 0, and $\frac{1}{2}$, which divide the x-axis into four regions. Test a point in each region to find the sign of $f(x)$ in that region.

Step 2: $f(0) = 0$, so plot $(0, 0)$.

Step 3: The x-intercepts divide the x-axis into four intervals.

Interval	Test Point	Value of $f(x)$	Sign of $f(x)$	Graph Above or Below x-Axis
$(-\infty, -1)$	-2	-10	Negative	Below
$(-1, 0)$	$-\frac{1}{2}$	$\frac{1}{2}$	Positive	Above
$\left(0, \frac{1}{2}\right)$	$\frac{1}{4}$	$-\frac{5}{32}$	Negative	Below
$\left(\frac{1}{2}, \infty\right)$	1	2	Positive	Above

Plot the x-intercepts, y-intercept (the y-intercept is also an x-intercept), and test points with a smooth curve to get the graph.

$$f(x) = 2x^3 + x^2 - x$$

59. $f(x) = x^4 + x^3 - 3x^2 - 4x - 4$

Step 1: The first step is to find the zeros of the polynomial function. p must be a factor of $a_0 = -4$ and q must be a factor of $a_4 = 1$. Thus, p can be ± 1, ± 2, or ± 4, and q can be ± 1. The possible zeros, $\frac{p}{q}$, are ± 1, ± 2, or ± 4. The remainder theorem shows that 2 is a zero.

$$
\begin{array}{r}
2\,\overline{)\,1 \quad 1 \quad -3 \quad -4 \quad -4} \\
\underline{\quad\;\; 2 \quad\;\; 6 \quad\;\; 6 \quad\;\; 4} \\
1 \quad 3 \quad\;\; 3 \quad\;\; 2 \quad\;\; 0
\end{array}
$$

The new quotient polynomial therefore is $x^3 + 3x^2 + 3x + 2$.

p must be a factor of $a_0 = 2$ and q must be a factor of $a_3 = 1$. Thus, p can be ± 1, or ± 2, and q can be ± 1. The possible zeros, $\frac{p}{q}$, are ± 1, or ± 2. The remainder theorem shows that -2 is a zero.

$$
\begin{array}{r}
-2\,\overline{)\,1 \quad 3 \quad\;\; 3 \quad\;\; 2} \\
\underline{\quad\;\; -2 \;\; -2 \;\; -2} \\
1 \quad 1 \quad\;\; 1 \quad\;\; 0
\end{array}
$$

The new quotient polynomial therefore is $x^2 + x + 1$. Setting this equal to 0 and using the quadratic formula with $a = 1$, $b = 1$, and $c = 1$, we have

$$x = \frac{-1 \pm \sqrt{1^2 - 4(1)(1)}}{2(1)} = \frac{-1 \pm \sqrt{-3}}{2}$$
$$= -\frac{1}{2} \pm \frac{i\sqrt{3}}{2}$$

The rational zeros are -2 and 2, which divide the x-axis into three regions.

Step 2: $f(0) = -4$, so plot $(0, -4)$.

Step 3: The x-intercepts divide the x-axis into three intervals.

Interval	Test Point	Value of $f(x)$	Sign of $f(x)$	Graph Above or Below x-Axis
$(-\infty, -2)$	-3	35	Positive	Above
$(-2, 2)$	-1	-3	Negative	Below
$(2, \infty)$	3	65	Positive	Above

Plot the x-intercepts, y-intercept, and test points with a smooth curve to get the graph.

(continued on next page)

(*continued*)

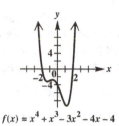

$f(x) = x^4 + x^3 - 3x^2 - 4x - 4$

61. $f(x) = (x-2)^2(x-5)$ is a cubic polynomial with positive y-values for $x > 5$, so it matches graph C.

63. $f(x) = (x-2)^2(x-5)^2$ is a quartic polynomial that is always greater than or equal to 0, so it matches graph E.

65. $f(x) = -(x-2)(x-5)$ is a quadratic polynomial that opens down, so it matches graph B.

67. Using the calculator, we find that the real zeros are 7.6533119, 1, and –0.6533119.

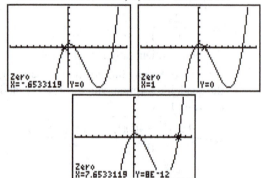

69. (a)

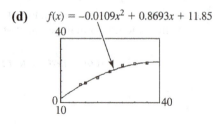

(b) Find the quadratic regression.
$$f(x) = -0.0109x^2 + 0.8693x + 11.85$$

(c) Find the cubic regression.
$$f(x) = -0.00087x^3 + 0.0456x^2 - 0.2191x + 17.83$$

(d) $f(x) = -0.0109x^2 + 0.8693x + 11.85$

$f(x) = -0.00087x^3 + 0.00456x^2 - 0.2191x + 17.83$

(e) Both functions approximate the data well. The quadratic function is probably better for prediction, because it is unlikely that the percent of out-of-pocket spending would decrease after 2025 (as the cubic function shows) unless changes were made in Medicare law.

71. $V_{\text{box}} = LWH$

If we let $L = x + 11$, $W = 3x$, and $H = x$, we have the volume of the rectangular box is $V(x) = (x+11)(3x)x$. The volume is 720 in.^{3}, so we need to solve the equation
$$720 = (x+11)(3x)x \Rightarrow 720 = 3x^3 + 33x^2$$
$$0 = 3x^3 + 33x^2 - 720 \Rightarrow x^3 + 11x^2 - 240 = 0$$

Algebraic Solution:

p must be a factor of $a_0 = -240$ and q must be a factor of $a_3 = 1$. Including the $\pm$, there are 40 factors of p and 2 factors of q. There are 80 possible zeros, $\frac{p}{q}$. Because this list is so long, we will simply show that 4 is a zero by the remainder theorem.

$$
\begin{array}{r|rrrr}
4 & 1 & 11 & 0 & -240 \\
  &   & 4 & 60 & 240 \\
\hline
  & 1 & 15 & 60 & 0 \\
\end{array}
$$

The new quotient polynomial is $x^2 + 15x + 60$. If we examine the discriminant, $b^2 - 4ac$, we will see that the quotient polynomial cannot be factored further over the reals
$$b^2 - 4ac = 15^2 - 4(1)(60)$$
$$= 225 - 240 = -15 < 0$$

The discriminant is negative, so 4 is our only real solution to the equation
$x^3 + 11x^2 - 240 = 0$.

(*continued on next page*)

(*continued*)

Graphing Calculator Solution:

Graph $y_1 = x^3 + 11x^2 - 240$ in the window $[-15, 10] \times [-250, 50]$.

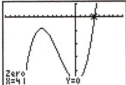

The graph shows that 4 is the only real zero. Because the only real solution is 4, the dimensions of the rectangular box are 4 in. by $3(4) = 12$ in. by $4 + 11 = 15$ in. or 12 in. $\times$ 4 in. $\times$ 15 in.

73. $f(x) = \dfrac{4}{x - 1}$

Step 1: The graph has a vertical asymptote where $x - 1 = 0 \Rightarrow x = 1$

Step 2: The degree of the numerator (which is considered degree zero) is less than the degree of the denominator, so the graph has a horizontal asymptote at $y = 0$ (the x-axis).

Step 3: The y-intercept is

$f(0) = \frac{4}{0-1} = \frac{4}{-1} = -4.$

Step 4: Any x-intercepts are found by solving

$f(x) = 0: \dfrac{4}{x - 1} = 0 \Rightarrow 4 = 0$

This is a false statement, so there is no x-intercept.

Step 5: The graph will intersect the horizontal asymptote when $\frac{4}{x-1} = 0 \Rightarrow 4 = 0$

This is a false statement, so the graph does not intersect the horizontal asymptote.

Step 6: Because the vertical asymptote is $x = 1$ and there is no x-intercept, we must determine values in two intervals.

Interval	Test Point	Value of $f(x)$	Sign of $f(x)$	Graph Above or Below x-Axis
$(-\infty, 1)$	-2	$-\frac{4}{3}$	Negative	Below
$(1, \infty)$	2	4	Positive	Above

Step 7: Use the asymptotes, intercepts, and these points to sketch the graph.

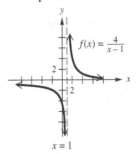

75. $f(x) = \dfrac{6x}{x^2 + x - 2} = \dfrac{6x}{(x + 2)(x - 1)}$

Step 1: The graph has a vertical asymptote where $(x + 2)(x - 1) = 0$, that is, when $x = -2$ and $x = 1$.

Step 2: Because the degree of the numerator is less than the degree of the denominator, the graph has a horizontal asymptote at $y = 0$ (the x-axis).

Step 3: The y-intercept is

$f(0) = \frac{6(0)}{0^2 + 0 - 2} = \frac{0}{0 + 0 - 2} = \frac{0}{-2} = 0.$

Step 4: Any x-intercepts are found by solving $f(x) = 0$.

$\dfrac{6x}{x^2 + x - 2} = 0 \Rightarrow 6x = 0 \Rightarrow x = 0$

The only x-intercept is 0.

Step 5: The graph will intersect the horizontal asymptote when $\dfrac{6x}{x^2 + x - 2} = 0.$

$\frac{6x}{x^2 + x - 2} = 0 \Rightarrow 6x = 0 \Rightarrow x = 0$

Step 6: The vertical asymptotes are $x = -2$ and $x = 1$ and the x-intercept occurs at 0, so we must determine values in four intervals.

Interval	Test Point	Value of $f(x)$	Sign of $f(x)$	Graph Above or Below x-Axis
$(-\infty, -2)$	-3	$-\frac{9}{2}$	Negative	Below
$(-2, 0)$	-1	3	Positive	Above
$(0, 1)$	$\frac{1}{2}$	$-\frac{12}{5}$	Negative	Below
$(1, \infty)$	2	3	Positive	Above

(*continued on next page*)

(continued)

Step 7: Use the asymptotes, intercepts, and these points to sketch the graph.

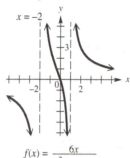

$$f(x) = \frac{6x}{x^2 + x - 2}$$

77. $f(x) = \dfrac{x^2 + 4}{x + 2}$

Step 1: The graph has a vertical asymptote where $x + 2 = 0 \Rightarrow x = -2$.

Step 2: Because the degree of the numerator is one more than the degree of the denominator, the graph has an oblique asymptote. Divide $x^2 + 4$ by $x + 2$.

$$
\begin{array}{r}
-2\overline{)\begin{array}{ccc} 1 & 0 & 4 \end{array}} \\
\underline{\begin{array}{cc} -2 & 4 \end{array}} \\
\begin{array}{ccc} 1 & -2 & 8 \end{array}
\end{array}
$$

$$f(x) = \frac{x^2 + 4}{x + 2} = x - 2 + \frac{8}{x + 3}$$

The oblique asymptote is the line $y = x - 2$.

Step 3: The y-intercept is

$$f(0) = \frac{0^2 + 4}{0 + 2} = \frac{0 + 4}{2} = \frac{4}{2} = 2.$$

Step 4: Any x-intercepts are found by solving $f(x) = 0$.

$$\frac{x^2 + 4}{x + 2} = 0 \Rightarrow x^2 + 4 = 0 \Rightarrow x = \pm 2i$$

There are no x-intercepts.

Step 5: The graph will intersect the oblique asymptote when $\frac{x^2 + 4}{x + 2} = x - 2$.

$$\frac{x^2 + 4}{x + 2} = x - 2 \Rightarrow x^2 + 4 = (x + 2)(x - 2)$$
$$x^2 + 4 = x^2 - 4 \Rightarrow 4 = -4$$

This is a false statement, so the graph does not intersect the oblique asymptote.

Step 6: Because the vertical asymptote is $x = -2$ and there are no x-intercepts, we must determine values in two intervals.

Interval	Test Point	Value of $f(x)$	Sign of $f(x)$	Graph Above or Below x-Axis
$(-\infty, -2)$	-3	-13	Negative	Below
$(-2, \infty)$	3	$\frac{13}{5}$	Positive	Above

Step 7: Use the asymptotes, y-intercept, and these points to sketch the graph.

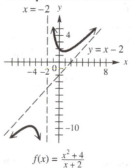

$$f(x) = \frac{x^2 + 4}{x + 2}$$

79. $f(x) = \dfrac{-2}{x^2 + 1}$

Step 1: The graph has a vertical asymptote where $x^2 + 1 = 0$. Because we have $x^2 + 1 = 0 \Rightarrow x = \pm i$, there are no vertical asymptotes.

Step 2: Because the degree of the numerator (which is considered degree zero) is less than the degree of the denominator, the graph has a horizontal asymptote at $y = 0$ (the x-axis).

Step 3: The y-intercept is

$$f(0) = \frac{-2}{0^2 + 1} = \frac{-2}{0 + 1} = \frac{-2}{1} = -2.$$

Step 4: Any x-intercepts are found by solving $f(x) = 0$.

$$\frac{-2}{x^2 + 1} = 0 \Rightarrow -2 = 0$$

This is a false statement, so there is no x-intercept.

Step 5: The graph will intersect the horizontal asymptote when $\frac{-2}{x^2 + 1} = 0$.

$$\frac{-2}{x^2 + 1} = 0 \Rightarrow -2 = 0$$

This is a false statement, so the graph does not intersect the horizontal asymptote.

Step 6: Because there are no vertical asymptotes and no x-intercept, we can see from the y-intercept that the function is always negative and, therefore, below the x-axis. Because of the limited information we have, we should explore symmetry and calculate a few more points on the graph.

(continued on next page)

(*continued*)

x	y
-3	$-\frac{1}{5}$
$\frac{1}{2}$	$-\frac{8}{5}$
1	-1

Because $f(-x) = \frac{-2}{(-x)^2+1} = \frac{-2}{x^2+1} = f(x)$, the graph is symmetric about the *y*-axis.

Step 7: Use the asymptote, *y*-intercept, these points, and symmetry to sketch the graph.

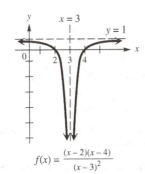

$f(x) = \frac{-2}{x^2+1}$

81. (a)

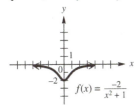

$f(x) = \frac{(x-2)(x-4)}{(x-3)^2}$

(b) The graph has a vertical asymptote $x = 3$, so $x - 3$ is in the denominator of the function. The *x*-intercepts are $(2, 0)$ and $(4, 0)$, so that when $f(x) = 0$, $x = 2$ or $x = 4$. This would exist if $x - 2$ and $x - 4$ were factors of the numerator. The horizontal asymptote is $y = 1$, so the numerator and denominator have the same degree. The numerator will have degree 2, so the denominator will also have degree 2. Putting these conditions together, we get a possible function $f(x) = \frac{(x-2)(x-4)}{(x-3)^2}$.

83. The graph has a vertical asymptote $x = 1$, so $x - 1$ is in the denominator of the function. The *x*-intercept is 2, so that when $f(x) = 0$, $x = 2..$ This would exist if $x - 2$ was a factor of the numerator. The horizontal asymptote is $y = -3$, so the numerator and denominator have the same degree. They both have degree 1. Also, from the horizontal asymptote, we have $y = -3 = \frac{a_n}{b_n} = \frac{-3}{1}$.

Putting these conditions together, we have a possible function $f(x) = \frac{-3(x-2)}{x-1}$ or $f(x) = \frac{-3x+6}{x-1}$.

85. $C(x) = \frac{6.7x}{100-x}$

(a) Graph $y_1 = \frac{6.7x}{100-x}$ in the window $[0, 100] \times [0, 100]$.

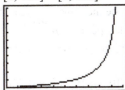

(b) To find the cost to remove 95% of the pollutant, find $C(95)$.

$$C(95) = \frac{6.7(95)}{100-95} = 127.3$$

This can also be found using the graphing calculator.

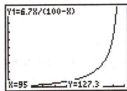

It would cost $127.3 thousand to remove 95% of the pollutant.

87. *Step 1*: $x = ky$

Step 2: Substitute $x = 20$ and $y = 14$ to find k. $20 = k(14) \Rightarrow k = \frac{20}{14} = \frac{10}{7}$

Step 3: $x = \frac{10}{7}y$

Step 4: Now find *y* when $x = 50$.

$50 = \frac{10}{7}y \Rightarrow 350 = 10y \Rightarrow y = 35$

89. *Step 1*: $t = \frac{k}{s}$

Step 2: Substitute $t = 3$ and $s = 5$ to find k.

$3 = \frac{k}{5} \Rightarrow k = 15$

Step 3: $t = \frac{15}{s}$

Step 4: Now find *s* when $t = 20$.

$20 = \frac{15}{s} \Rightarrow 20s = 15 \Rightarrow s = \frac{15}{20} = \frac{3}{4}$

91. *Step 1:* $f = kg^2h$

Step 2: Substitute $f = 50$, $g = 5$, and $h = 4$ to find k.

$$50 = k\left(5^2\right)(4) \Rightarrow 50 = 100k \Rightarrow k = \frac{50}{100} = \frac{1}{2}$$

Step 3: $f = \frac{1}{2}g^2h$

Step 4: Now find f when $g = 3$ and $h = 6$.

$$f = \frac{1}{2}\left(3^2\right)(6) = 27$$

93. *Step 1:* Let p be the power, v is the wind velocity (in km per hr). $p = kv^3$

Step 2: Substitute $p = 10{,}000$ and $v = 10$ to find k.

$$10{,}000 = k \cdot 10^3 \Rightarrow 10{,}000 = 1000k \Rightarrow k = 10$$

Step 3: $p = 10v^3$

Step 4: Now find p when $v = 15$.

$$p = 10 \cdot 15^3 = 33{,}750$$

33,750 units of power are produced.

Chapter 3   Test

1. $f(x) = -2x^2 + 6x - 3$

Complete the square so the function has the form $f(x) = a(x - h)^2 + k$.

$$f(x) = -2x^2 + 6x - 3 = -2\left(x^2 - 3x\right) - 3$$
$$= -2\left(x^2 - 3x + \frac{9}{4} - \frac{9}{4}\right) - 3$$

Note: $\left[\frac{1}{2}(-3)\right]^2 = \frac{9}{4}$

$$= -2\left(x - \frac{3}{2}\right)^2 + \frac{9}{2} - 3 = -2\left(x - \frac{3}{2}\right)^2 + \frac{3}{2}$$

This parabola is in now in standard form with $a = -2$, $h = \frac{3}{2}$, and $k = \frac{3}{2}$. It opens downward with vertex $(h, k) = \left(\frac{3}{2}, \frac{3}{2}\right)$. The axis is $x = h \Rightarrow x = \frac{3}{2}$. To find the x-intercepts, let $f(x) = 0$.

$$-2\left(x - \frac{3}{2}\right)^2 + \frac{3}{2} = 0 \Rightarrow -2\left(x - \frac{3}{2}\right)^2 = -\frac{3}{2}$$
$$\left(x - \frac{3}{2}\right)^2 = \frac{3}{4} \Rightarrow x - \frac{3}{2} = \pm\sqrt{\frac{3}{4}}$$
$$x = \frac{3}{2} \pm \sqrt{\frac{3}{4}} = \frac{3}{2} \pm \frac{\sqrt{3}}{2} = \frac{3 \pm \sqrt{3}}{2}$$

Thus the x-intercepts are $\left(\frac{3 \pm \sqrt{3}}{2}, 0\right)$.

Note: If you use the quadratic formula to solve $-2x^2 + 6x - 3 = 0$ where $a = -2$, $b = 6$, and $c = -3$, you will arrive at the equivalent answer for the x-intercepts.

To find the y-intercept, let $x = 0$.

$$f(0) = -2\left(0^2\right) + 6(0) - 3 = -3$$

The y-intercept is $(0, -3)$. The domain is $(-\infty, \infty)$. The highest point on the graph is $\left(\frac{3}{2}, \frac{3}{2}\right)$, so the range is $\left(-\infty, \frac{3}{2}\right]$. The function is increasing on $\left(-\infty, \frac{3}{2}\right)$ and decreasing on $\left(\frac{3}{2}, \infty\right)$.

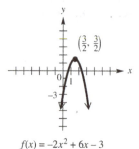

$f(x) = -2x^2 + 6x - 3$

2. $s(t) = -16t^2 + 88t + 48$

(a) The rocket reaches its maximum height at the vertex of the function.

$$h = -\frac{b}{2a} = -\frac{88}{2(-16)} = 2.75$$

The rocket reaches its maximum height at 2.75 seconds.

(b) The maximum height is given by $s(2.75)$.

$$s(2.75) = -16(2.75)^2 + 88(2.75) + 48$$
$$= 169$$

The maximum height is 169 ft.

(c) We must solve $-16t^2 + 88t + 48 > 100$.

$$-16t^2 + 88t + 48 = 100$$
$$-16t^2 + 88t - 52 = 0$$
$$t = \frac{-88 \pm \sqrt{88^2 - 4(-16)(-52)}}{2(-16)} \approx 0.7, \, 4.8$$

Because the vertex lies between these two t-values, we know that the rocket will be more than 100 ft above ground level between 0.7 sec and 4.8 sec.

We can verify this graphically by graphing $y_1 = -16x^2 + 88x + 48$ and $y_2 = 100$ in the window $[0, 6] \times [-20, 180]$ and then finding the intersections.

(continued on next page)

(*continued*)

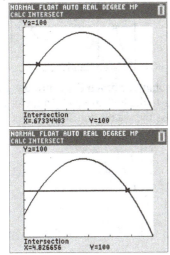

(d) The rocket returns to the ground when $s = 0$.

$$-16t^2 + 88t + 48 = 0$$

$$t = \frac{-88 \pm \sqrt{88^2 - 4(-16)(48)}}{2(-16)} = -0.5, \; 6$$

The negative solution has no meaning, so we disregard it. The rocket returns to ground after 6 seconds.

3. $\dfrac{3x^3 + 4x^2 - 9x + 6}{x + 2}$

Express $x + 2$ in the form $x - k$ by writing it as $x - (-2)$. Thus $k = -2$.

$$\begin{array}{r}
-2 \overline{)3 \quad 4 \; -9 \quad 6} \\
\underline{-6 \quad 4 \; 10} \\
3 \; -2 \; -5 \; 16
\end{array}$$

Thus,

$$\frac{3x^3 + 4x^2 - 9x + 6}{x + 2} = 3x^2 - 2x - 5 + \frac{16}{x + 2}$$

4. $\dfrac{2x^3 - 11x^2 + 25}{x - 5}$

Because $x - 5$ is in the form $x - k$, $k = 5$.

$$\begin{array}{r}
5 \overline{)2 \; -11 \quad 0 \quad 25} \\
\underline{10 \; -5 \; -25} \\
2 \; -1 \; -5 \quad 0
\end{array}$$

$$\frac{2x^3 - 11x^2 + 28}{x - 5} = 2x^2 - x - 5$$

5. $f(x) = 2x^3 - 9x^2 + 4x + 8; \; k = 5$

$$\begin{array}{r}
5 \overline{)2 \; -9 \quad 4 \quad 8} \\
\underline{10 \quad 5 \; 45} \\
2 \quad 1 \; 9 \; 53
\end{array}$$

Thus, $f(5) = 53$.

6. $6x^4 - 11x^3 - 35x^2 + 34x + 24; \; x - 3$

Let $f(x) = 6x^4 - 11x^3 - 35x^2 + 34x + 24$. By the factor theorem, $x - 3$ will be a factor of $f(x)$ only if $f(3) = 0$.

$$\begin{array}{r}
3 \overline{)6 \; -11 \; -35 \quad 34 \quad 24} \\
\underline{18 \quad 21 \; -42 \; -24} \\
6 \quad 7 \; -14 \; -8 \quad 0
\end{array}$$

Because $f(3) = 0$, $x - 3$ is a factor of $f(x)$. The other factor is $6x^3 + 7x^2 - 14x - 8$.

7. $f(x) = x^3 + 8x^2 + 25x + 26; \; -2$

Because -2 is a zero, first divide $f(x)$ by $x + 2$.

$$\begin{array}{r}
-2 \overline{)1 \quad 8 \quad 25 \quad 26} \\
\underline{-2 \; -12 \; -26} \\
1 \quad 6 \quad 13 \quad 0
\end{array}$$

This gives $f(x) = (x + 2)(x^2 + 6x + 13)$.

Because $x^2 + 6x + 13$ cannot be factored, use the quadratic formula with $a = 1$, $b = 6$, and $c = 13$ to find the remaining two zeros.

$$x = \frac{-6 \pm \sqrt{6^2 - 4(1)(13)}}{2(1)} = \frac{-6 \pm \sqrt{36 - 52}}{2}$$

$$= \frac{-6 \pm \sqrt{-16}}{2} = \frac{-6 \pm 4i}{2} = -3 \pm 2i$$

The zeros are -2, $-3 - 2i$, and $-3 + 2i$.

8. Zeros of -1, 2 and i; $f(3) = 80$

By the conjugate zeros theorem, $-i$ is also a zero. The polynomial has the following form.

$$f(x) = a(x + 1)(x - 2)(x - i)(x + i)$$

Use the condition $f(3) = 80$ to find a.

$$80 = a(3 + 1)(3 - 2)(3 - i)(3 + i)$$

$$80 = a(4)(1)\left[(3 - i)(3 + i)\right]$$

$$80 = a\left[(4)(1)\right]\left[3^2 - i^2\right]$$

$$80 = a(4)\left[9 - (-1)\right]$$

$$80 = a(4)(10) \Rightarrow 80 = 40a \Rightarrow a = 2$$

(*continued on next page*)

(*continued*)

Thus we have the following.

$$f(x) = 2(x+1)(x-2)(x-i)(x+i)$$
$$= 2\big[(x+1)(x-2)\big]\big[(x-i)(x+i)\big]$$
$$= 2(x^2 - x - 2)(x^2 + 1)$$
$$= 2(x^4 - x^3 - x^2 - x - 2)$$
$$= 2x^4 - 2x^3 - 2x^2 - 2x - 4$$

9. Because $f(x) = x^4 + 8x^2 + 12$

$= (x^2 + 6)(x^2 + 2)$, the zeros are $\pm i\sqrt{6}$ and

$\pm i\sqrt{2}$. Moreover, $f(x) > 0$ for all x; thus, the graph never crosses or touches the x-axis, so $f(x)$ has no real zeros.

10. (a) $f(x) = x^3 - 5x^2 + 2x + 7$; find $f(1)$ and

$f(2)$ synthetically.

```
1)1 −5   2   7
     1 −4  −2
  1 −4  −2   5
```

```
2)1 −5   2   7
     2  −6  −8
  1 −3  −4  −1
```

By the intermediate value theorem, because $f(1) = 5 > 0$ and $f(2) = -1 < 0$, there must be at least one real zero between 1 and 2.

(b) $f(x) = x^3 - 5x^2 + 2x + 7$ has 2 variations

in sign. f has 2 or $2 - 2 = 0$ positive real zeros.

$f(-x) = -x^3 - 5x^2 - 2x + 7$ has 1

variation in sign. f has 1 negative real zero. Thus, the different possibilities are

Positive	Negative	Nonreal Complex
2	1	0
0	1	2

(c) Using the calculator, we find that the real zeros are 4.0937635, 1.8370381, and −0.9308016. The function is graphed in the window $[-2, 5] \times [-15, 15]$.

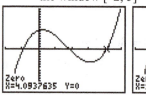
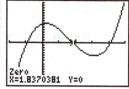

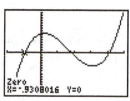

11. To obtain the graph of g, shift the graph of f 5 units to the left, stretch by a factor of 2, reflect across the x-axis, and shift 3 units up.

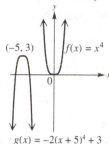

12. $f(x) = -x^7 + x - 4$

Because $f(x)$ is of odd degree and the sign of a_n is negative, the end behavior is ⌐. The correct graph is C.

13. $f(x) = x^3 - 5x^2 + 3x + 9$

Step 1: The first step is to find the zeros of the polynomial function. p must be a factor of $a_0 = 9$ and q must be a factor of $a_3 = 1$. Thus, p can be $\pm 1, \pm 3, \pm 9$ and q can be ± 1.

The possible zeros, $\frac{p}{q}$, are $\pm 1, \pm 3, \pm 9$. The remainder theorem shows that 3 is a zero.

```
3)1  −5   3   9
      3  −6  −9
  1  −2  −3   0
```

The new quotient polynomial is $x^2 - 2x - 3$.

$x^2 - 2x - 3 = 0 \Rightarrow (x+1)(x-3) = 0$

$x + 1 = 0 \Rightarrow x = -1$ or $x - 3 = 0 \Rightarrow x = 3$

The rational zeros are -1 and 3 (multiplicity 2) which divide the x-axis into three regions. Test a point in each region to find the sign of $f(x)$ in that region.

Step 2: $f(0) = 9$, so plot $(0, 9)$.

(*continued on next page*)

(continued)

Step 3: The x-intercepts divide the x-axis into three intervals.

Interval	Test Point	Value of $f(x)$	Sign of $f(x)$	Graph Above or Below x-Axis
$(-\infty, -1)$	-2	-25	Negative	Below
$(-1, 3)$	0	9	Positive	Above
$(3, \infty)$	4	5	Positive	Above

Plot the x-intercepts, y-intercept, and test points (the y-intercept is one of the test points) with a smooth curve to get the graph.

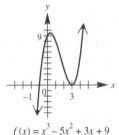

$f(x) = x^3 - 5x^2 + 3x + 9$

14. $f(x) = 2x^2 (x-2)^2$

Step 1: Set each factor equal to 0 and solve the resulting equations to find the zeros of the function.

$2x^2 = 0 \Rightarrow x = 0$ or $(x-2)^2 = 0 \Rightarrow x = 2$

The zeros are 0 (multiplicity 2) and 2 (multiplicity 2), which divide the x-axis into three regions. Test a point in each region to find the sign of $f(x)$ in that region.

Step 2: $f(0) = 0$, so plot $(0, 0)$.

Step 3: The x-intercepts divide the x-axis into three intervals.

Interval	Test Point	Value of $f(x)$	Sign of $f(x)$	Graph Above or Below x-Axis
$(-\infty, 0)$	-1	18	Positive	Above
$(0, 2)$	1	2	Positive	Above
$(2, \infty)$	3	18	Positive	Above

Plot the x-intercepts, y-intercept (the y-intercept is an x-intercept), and test points with a smooth curve to get the graph.

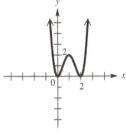

$f(x) = 2x^2 (x-2)^2$

15. $f(x) = -x^3 - 4x^2 + 11x + 30$

Step 1: The first step is to find the zeros of the polynomial function. p must be a factor of $a_0 = 30$ and q must be a factor of $a_3 = -1$. Thus, p can be $\pm 1, \pm 2, \pm 3, \pm 5, \pm 6, \pm 10, \pm 15, \pm 30$ and q can be ± 1. The possible zeros, $\frac{p}{q}$, are $\pm 1, \pm 2, \pm 3, \pm 5, \pm 6, \pm 10, \pm 15, \pm 30$.

The remainder theorem shows that 3 is a zero.

$$3 \overline{)\begin{array}{rrrr} -1 & -4 & 11 & 30 \\ & -3 & -21 & -30 \\ \hline -1 & -7 & -10 & 0 \end{array}}$$

The new quotient polynomial is $-x^2 - 7x - 10$.

$-x^2 - 7x - 10 = 0 \Rightarrow x^2 + 7x + 10 = 0$
$(x+5)(x+2) = 0$
$x + 5 = 0 \Rightarrow x = -5$ or $x + 2 = 0 \Rightarrow x = -2$

The rational zeros are $-5, -2,$ and $3,$ which divide the x-axis into four regions. Test a point in each region to find the sign of $f(x)$ in that region.

Step 2: $f(0) = 30$, so plot $(0, 30)$.

Step 3: The x-intercepts divide the x-axis into four intervals.

Interval	Test Point	Value of $f(x)$	Sign of $f(x)$	Graph Above or Below x-Axis
$(-\infty, -5)$	-6	36	Positive	Above
$(-5, -2)$	-3	-12	Negative	Below
$(-2, 3)$	0	30	Positive	Above
$(3, \infty)$	4	-54	Negative	Below

Plot the x-intercepts, y-intercept, and test points (the y-intercept is one of the test points) with a smooth curve to get the graph.

(continued on next page)

(*continued*)

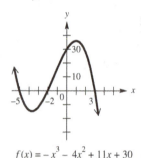

$$f(x) = -x^3 - 4x^2 + 11x + 30$$

16. The zeros are –3 and 2. The graph of *f* touches the x-axis at 2, so the zero 2 has multiplicity 2.

Thus, $f(x) = a(x-2)^2(x+3)$. Also, the point $(0, 24)$ is on the graph, so we have

$$f(0) = 24.$$

$$24 = a(0-2)^2(0+3) \Rightarrow 24 = a(-2)^2(3)$$
$$24 = a(4)(3) \Rightarrow 24 = 12a \Rightarrow a = 2$$

The polynomial function is the following.

$$\begin{aligned} f(x) &= 2(x-2)^2(x+3) \\ &= 2(x^2 - 4x + 4)(x+3) \\ &= 2(x^3 - x^2 - 8x + 12) \\ &= 2x^3 - 2x^2 - 16x + 24 \end{aligned}$$

17. $f(t) = 1.06t^3 - 24.6t^2 + 180t$

   (a) $f(2) = 1.06(2^3) - 24.6(2)^2 + 180(2)$
   $$= 270.08$$
   This can also be found using the graphing calculator.

   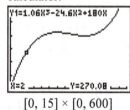

   $[0, 15] \times [0, 600]$

   (b) From the graph we see that the amount of change is increasing from $t = 0$ to $t = 5.9$ and from $t = 9.5$ to $t = 15$ and decreasing from $t = 5.9$ to $t = 9.5$.

   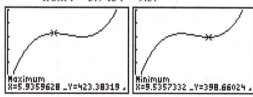

18. $f(x) = \dfrac{3x-1}{x-2}$

   Step 1: The graph has a vertical asymptote where $x - 2 = 0$, that is, when $x = 2$.

   Step 2: The degree of the numerator equals the degree of the denominator, so the graph has a horizontal asymptote at $y = \frac{3}{1} = 3$.

   Step 3: The y-intercept is

   $$f(0) = \frac{3(0)-1}{0-2} = \frac{0-1}{-2} = \frac{-1}{-2} = \frac{1}{2}.$$

   Step 4: Any x-intercepts are found by solving $f(x) = 0$.

   $$\frac{3x-1}{x-2} = 0 \Rightarrow 3x - 1 = 0 \Rightarrow 3x = 1 \Rightarrow x = \frac{1}{3}$$

   The only x-intercept is $\frac{1}{3}$.

   Step 5: The graph will intersect the horizontal asymptote when $\frac{3x-1}{x-2} = 3$.

   $$\frac{3x-1}{x-2} = 3 \Rightarrow 3x - 1 = 3(x-2) \Rightarrow$$
   $$3x - 1 = 3x - 6 \Rightarrow -1 = -6$$

   This is a false statement, so the graph does not intersect the horizontal asymptote.

   Step 6: Because the vertical asymptote is $x = 2$ and the x-intercept occurs at $\frac{1}{3}$, we must determine values in three intervals.

Interval	Test Point	Value of $f(x)$	Sign of $f(x)$	Graph Above or Below x-Axis
$\left(-\infty, \frac{1}{3}\right)$	0	$\frac{1}{2}$	Positive	Above
$\left(\frac{1}{3}, 2\right)$	1	–2	Negative	Below
$(2, \infty)$	3	8	Positive	Above

   Step 7: Use the asymptotes, intercepts, and test points (the y-intercept is one of the test points) to sketch the graph.

   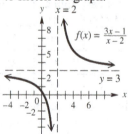

19. $f(x) = \frac{x^2-1}{x^2-9} = \frac{(x+1)(x-1)}{(x+3)(x-3)}$

Step 1: The graph has a vertical asymptote where $(x+3)(x-3) = 0$, that is, when

$x = -3$ and $x = 3$.

Step 2: Because the degree of the numerator equals the degree of the denominator, the graph has a horizontal asymptote at $y = \frac{1}{1} = 1$.

Step 3: The y-intercept is

$f(0) = \frac{0^2-1}{0^2-9} = \frac{-1}{-9} = \frac{1}{9}$.

Step 4: Any x-intercepts are found by solving $f(x) = 0$.

$\frac{(x+1)(x-1)}{(x+3)(x-3)} = 0 \Rightarrow (x+1)(x-1) = 0$

$x = -1$ or $x = 1$

The x-intercepts are -1 and 1.

Step 5: The graph will intersect the horizontal asymptote when $\frac{x^2-1}{x^2-9} = 1$.

$\frac{x^2-1}{x^2-9} = 1 \Rightarrow x^2 - 1 = x^2 - 9 \Rightarrow -1 = -9$

This is a false statement, so the graph does not intersect the horizontal asymptote.

Step 6: Because the vertical asymptotes are $x = -3$ and $x = 3$, and the x-intercepts occur at -1 and 1, we must determine values in five intervals.

Interval	Test Point	Value of $f(x)$	Sign of $f(x)$	Graph Above or Below x-Axis
$(-\infty, -3)$	-4	$\frac{15}{7}$	Positive	Above
$(-3, -1)$	-2	$-\frac{3}{5}$	Negative	Below
$(-1, 1)$	0	$\frac{1}{9}$	Positive	Above
$(1, 3)$	2	$-\frac{3}{5}$	Negative	Below
$(3, \infty)$	4	$\frac{15}{7}$	Positive	Above

Step 7: Use the asymptotes, intercepts, and test points (the y-intercept is one of the test points) to sketch the graph. Note: This function is an even function, thus the graph is symmetric with respect to the y-axis.

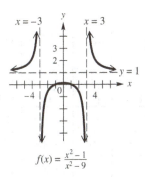

$f(x) = \frac{x^2-1}{x^2-9}$

20. $f(x) = \frac{2x^2 + x - 6}{x - 1}$

(a) Because the degree of the numerator is one more than the degree of the denominator, the graph has an oblique asymptote. Divide $2x^2 + x - 6$ by $x - 1$.

$$\begin{array}{r} 1)\overline{\begin{array}{rrr} 2 & 1 & -6 \\ & 2 & 3 \\ \hline 2 & 3 & -3 \end{array}} \end{array}$$

$f(x) = \frac{2x^2+x-6}{x-1} = 2x + 3 - \frac{3}{x-1}$

The oblique asymptote is the line $y = 2x + 3$.

(b) To find the x-intercepts, let $f(x) = 0$.

$\frac{2x^2+x-6}{x-1} = 0 \Rightarrow 2x^2 + x - 6 = 0$

$(x+2)(2x-3) = 0$

$x = -2$ or $x = \frac{3}{2}$

The x-intercepts are $(-2, 0)$ and $\left(\frac{3}{2}, 0\right)$.

(c) $f(0) = \frac{2(0^2)+0-6}{0-1} = \frac{2(0)+0-6}{-1}$

$= \frac{0+0-6}{-1} = \frac{-6}{-1} = 6$

The y-intercept is $(0, 6)$.

(d) To find the vertical asymptote, set the denominator equal to zero and solve for x.

$x - 1 = 0 \Rightarrow x = 1$

The equation of the vertical asymptote is $x = 1$.

(e) Use the information from (a)–(d) and a few additional points to graph the function. Because the vertical asymptote is $x = 1$ and the x-intercepts occur at -2 and $\frac{3}{2}$, we must determine values in four intervals.

(continued on next page)

(*continued*)

Interval	Test Point	Value of f (x)	Sign of $f(x)$	Graph Above or Below x-Axis
$(-\infty, -2)$	-3	$-\frac{9}{4}$	Negative	Below
$(-2, 1)$	0	6	Positive	Above
$\left(1, \frac{3}{2}\right)$	$\frac{5}{4}$	$-\frac{13}{2}$	Negative	Below
$\left(\frac{3}{2}, \infty\right)$	2	4	Positive	Above

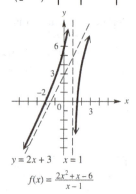

$y = 2x + 3$ $x = 1$

$$f(x) = \frac{2x^2 + x - 6}{x - 1}$$

21. *Step 1:* $y = k\sqrt{x}$

Step 2: Substitute $y = 12$ and $x = 4$ to find k.

$12 = k\sqrt{4} \Rightarrow 12 = 2k \Rightarrow k = 6$

Step 3: $y = 6\sqrt{x}$

Step 4: Now find y when $x = 100$.

$y = 6\sqrt{100} = 6 \cdot 10 = 60$

22. *Step 1:* Let w be the weight of the object (in kg), d is the distance from the center of Earth (in km). $w = \dfrac{k}{d^2}$

Step 2: Substitute $w = 90$ and $d = 6400$ to find k.

$$90 = \frac{k}{6400^2} \Rightarrow k = 3,686,400,000$$

Step 3: $w = \dfrac{3,686,400,000}{d^2}$

Step 4: Now find w when $d = 800 + 6400 = 7200$.

$$w = \frac{3,686,400,000}{7200^2} = \frac{3,686,400,000}{51,840,000} = \frac{640}{9} \approx 71.1$$

The man weighs $\frac{640}{9}$ kg or approximately 71.1 kg.

Chapter 4

INVERSE, EXPONENTIAL, AND LOGARITHMIC FUNCTIONS

Section 4.1   Inverse Functions

1. Yes, it is one-to-one, because every number in the list of registered passenger cars is used only once.

3. In order for a function to have an inverse, it must be <u>one-to-one</u>.

5. The domain of f is equal to the <u>range</u> of f^{-1}, and the range of f is equal to the <u>domain</u> of f^{-1}.

7. If $f(x) = x^3$, then $f^{-1}(x) = \underline{\sqrt[3]{x}}$.

9. If a function f has an inverse and $f(-3) = 6$, then $f^{-1}(6) = \underline{-3}$.

11. This is a one-to-one function because every horizontal line intersects the graph in no more than one point.

13. This function is not one-to-one because there are infinitely many horizontal lines that intersect the graph in two points.

15. This function is one-to-one because every horizontal line will intersect the graph in exactly one point.

17. $y = 2x - 8$

    Using the definition of a one-to-one function, we have $f(a) = f(b) \Rightarrow 2a - 8 = 2b - 8 \Rightarrow 2a = 2b \Rightarrow a = b$. So the function is one-to-one.

19. $y = \sqrt{36 - x^2}$

    If $x = 6$, $y = \sqrt{36 - 6^2} = \sqrt{36 - 36} = \sqrt{0} = 0$.
    If $x = -6$,
    $y = \sqrt{36 - (-6)^2} = \sqrt{36 - 36} = \sqrt{0} = 0$.

    Two different values of x lead to the same value of y, so the function is not one-to-one.

21. $y = 2x^3 - 1$

    Looking at this function in the window $[-5, 5]$ by $[-5, 5]$, we can see that it appears that any horizontal line passed through the function will intersect the graph in at most one place.

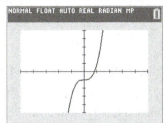

Another way of showing that a function is one-to-one is to assume that you have two equal y-values $\left(f(a) = f(b)\right)$ and show that they must have come from the same x-value $(a = b)$.

$f(a) = f(b) \Rightarrow 2a^3 - 1 = 2b^3 - 1 \Rightarrow$
$2a^3 = 2b^3 \Rightarrow a^3 = b^3 \Rightarrow \sqrt[3]{a^3} = \sqrt[3]{b^3} \Rightarrow a = b$

So, the function is one-to-one.

23. $y = -\dfrac{1}{x + 2}$

    Looking at this function in the window $[-7, 3]$ by $[-5, 5]$, we can see that it appears that any horizontal line passed through the function will intersect the graph in at most one place.

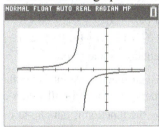

We could also show that $f(a) = f(a)$ implies $a = b$.

$f(a) = f(b) \Rightarrow -\frac{1}{a+2} = -\frac{1}{b+2} \Rightarrow$
$b + 2 = a + 2 \Rightarrow b = a$

So, the function is one-to-one.

25. $y = 2(x + 1)^2 - 6$

    Looking at this function in the window $[-10, 10]$ by $[-10, 10]$, we can see that it appears that any horizontal line passed through the function will intersect the graph in two places, except a horizontal line through the vertex.

    (continued on next page)

(*continued*)

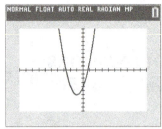

For example, $f(0) = 2(0+1)^2 - 6 = -4$ and

$f(-2) = 2(-2+1)^2 - 6 = -4$

So, the function is not one-to-one.

27. $y = \sqrt[3]{x+1} - 3$

Looking at this function in the window [–10, 10] by [–10, 10], we can see that it appears that any horizontal line passed through the function will intersect the graph in at most one place.

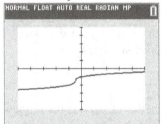

We could also show that $f(a) = f(a)$ implies $a = b$.

$f(a) = f(b) \Rightarrow \sqrt[3]{a+1} - 3 = \sqrt[3]{b+1} - 3 \Rightarrow$
$\sqrt[3]{a+1} = \sqrt[3]{b+1} \Rightarrow a+1 = b+1 \Rightarrow a = b$

So, the function is one-to-one.

29. For a constant function defined over the set of real numbers, $f(x) = c$ for all values of c.
Therefore, the function is not one-to-one.

31. The inverse operation of tying your shoelaces would be untying your shoelaces, because untying "undoes" tying.

33. The inverse operation of entering a room would be leaving a room, because leaving "undoes" entering.

35. The inverse operation of screwing in a light bulb would be unscrewing the light bulb.

37. For each point (x, y) for the first function, there is a point (y, x) for the second function, so $f(x)$ and $g(x)$ are inverses of each other.

39. The point (3, 5) is on $f(x)$, but the point (5, 3) is not on $g(x)$ (there is another example), so the functions are not inverses of each other.

41. $f(x) = 2x + 4, g(x) = \frac{1}{2}x - 2$

$(f \circ g)(x) = 2\left(\frac{1}{2}x - 2\right) + 4 = x - 4 + 4 = x$

$(g \circ f)(x) = \frac{1}{2}(2x + 4) - 2 = x + 2 - 2 = x$

Because $(f \circ g)(x) = x$ and $(g \circ f)(x) = x$, these functions are inverses.

43. $f(x) = -3x + 12, g(x) = -\frac{1}{3}x - 12$

$(f \circ g)(x) = -3\left(-\frac{1}{3}x - 12\right) + 12$
$= x + 36 + 12 = x + 48$

Because $(f \circ g)(x) \neq x$, the functions are not inverses. It is not necessary to check $(g \circ f)(x)$.

45. $f(x) = \frac{x+1}{x-2}, g(x) = \frac{2x+1}{x-1}$

$(f \circ g)(x) = \frac{\frac{2x+1}{x-1} + 1}{\frac{2x+1}{x-1} - 2} = \frac{\frac{2x+1+x-1}{x-1}}{\frac{2x+1-2(x-1)}{x-1}} = \frac{3x}{3} = x$

$(g \circ f)(x) = \frac{2\left(\frac{x+1}{x-2}\right) + 1}{\frac{x+1}{x-2} - 1} = \frac{\frac{2x+2}{x-2} + 1}{\frac{x+1-(x-2)}{x-2}}$

$= \frac{\frac{2x+2+x-2}{x-2}}{\frac{x+1-(x-2)}{x-2}} = \frac{3x}{3} = x$

Because $(f \circ g)(x) = x$ and $(g \circ f)(x) = x$, these functions are inverses.

47. $f(x) = \frac{2}{x+6}, g(x) = \frac{6x+2}{x}$

$(f \circ g)(x) = \frac{2}{\frac{6x+2}{x} + 6} = \frac{2}{\frac{6x+2+6x}{x}} = \frac{2}{1} \cdot \frac{x}{12x+2}$

$= \frac{2x}{12x+2} = \frac{x}{6x+1} \neq x$

Because $(f \circ g)(x) \neq x$, the functions are not inverses. It is not necessary to check $(g \circ f)(x)$.

49. $f(x) = x^2 + 3$, domain $[0, \infty)$;

$g(x) = \sqrt{x-3}$, domain $[3, \infty)$

$(f \circ g)(x) = f\left(\sqrt{x-3}\right) = \left(\sqrt{x-3}\right)^2 + 3 = x$

$(g \circ f)(x) = g\left(x^2 + 3\right) = \sqrt{x^2 + 3 - 3}$

$\qquad = \sqrt{x^2} = |x| = x$ for $[0, \infty)$

Because $(f \circ g)(x) = x$ and $(g \circ f)(x) = x$, these functions are inverses.

51. Each y-value corresponds to only one x-value, so this function is one-to one and has an inverse. The inverse is: $\{(6, -3), (1, 2), (8, 5)\}$.

53. The y-value -3 corresponds to two different x-values, so this function is not one-to-one.

55. These functions are inverses because their graphs are symmetric with respect to the line $y = x$.

57. These functions are not inverses because their graphs are not symmetric with respect to the line $y = x$.

59. $f(x) = 3x - 4$

The function is one-to-one.

(a) *Step 1*: Replace $f(x)$ with y and then interchange x and y: $x = 3y - 4$

Step 2: Solve for y.

$x = 3y - 4 \Rightarrow x + 4 = 3y \Rightarrow \frac{x+4}{3} = y \Rightarrow$

$y = \frac{x+4}{3} = \frac{1}{3}x + \frac{4}{3}$

Step 3: Replace y with $f^{-1}(x)$.

$f^{-1}(x) = \frac{1}{3}x + \frac{4}{3}$

(b) The graph of the original function, $f(x) = 3x - 4$, is a line with slope 3 and y-intercept -4. Because $f^{-1}(x) = \frac{1}{3}x + \frac{4}{3}$, the graph of the inverse function is a line with slope $\frac{1}{3}$ and y-intercept $\frac{4}{3}$.

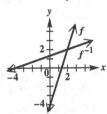

(c) For both $f(x)$ and $f^{-1}(x)$, the domain is $(-\infty, \infty)$ and the range is $(-\infty, \infty)$.

61. $f(x) = -4x + 3$

This function is one-to-one.

(a) *Step 1*: Replace $f(x)$ with y and interchange x and y.

$y = -4x + 3$

$x = -4y + 3$

Step 2: Solve for y.

$x = -4y + 3 \Rightarrow x - 3 = -4y \Rightarrow \frac{x-3}{-4} = y \Rightarrow$

$y = \frac{x-3}{-4} = -\frac{1}{4}x + \frac{3}{4}$

Step 3: Replace y with $f^{-1}(x)$.

$f^{-1}(x) = -\frac{1}{4}x + \frac{3}{4}$

(b) The graph of the original function, $f(x) = -4x + 3$, is a line with slope -4 and y-intercept 3. Because $f^{-1}(x) = -\frac{1}{4}x + \frac{3}{4}$, the graph of the inverse function is a line with slope $-\frac{1}{4}$ and y-intercept $\frac{3}{4}$.

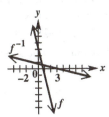

(c) For both $f(x)$ and $f^{-1}(x)$, the domain is $(-\infty, \infty)$ and the range is $(-\infty, \infty)$.

63. $f(x) = x^3 + 1$

This function is one-to-one.

(a) *Step 1*: Replace $f(x)$ with y and interchange x and y.

$y = x^3 + 1 \Rightarrow x = y^3 + 1$

Step 2: Solve for y.

$x = y^3 + 1 \Rightarrow x - 1 = y^3 \Rightarrow$

$\sqrt[3]{x-1} = y \Rightarrow y = \sqrt[3]{x-1}$

Step 3: Replace y with $f^{-1}(x)$.

$f^{-1}(x) = \sqrt[3]{x-1}$

(b) Tables of ordered pairs will be helpful in drawing the graphs of these functions.

x	$f(x)$	x	$f^{-1}(x)$
-2	-7	-7	-2
-1	0	0	-1
0	1	1	0
1	2	2	1
2	9	9	2

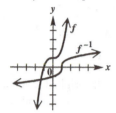

(c) For both $f(x)$ and $f^{-1}(x)$, the domain is $(-\infty, \infty)$ and the range is $(-\infty, \infty)$.

65. $f(x) = x^2 + 8$

This is not a one-to-one function because two different x-values can correspond to the same y-value ($2^2 + 8 = 12$ and $(-2)^2 + 8 = 12$, for example), so this function is not one-to-one. Thus, the function has no inverse function.

67. $f(x) = \dfrac{1}{x}$, $x \neq 0$

The function is one-to-one.

(a) *Step 1*: Replace $f(x)$ with y and interchange x and y.
$$y = \frac{1}{x} \Rightarrow x = \frac{1}{y}$$
Step 2: Solve for y.
$$x = \frac{1}{y} \Rightarrow xy = 1 \Rightarrow y = \frac{1}{x}$$
Step 3: Replace y with $f^{-1}(x)$.
$$f^{-1}(x) = \frac{1}{x}, \ x \neq 0 = f(x).$$

(b) Tables of ordered pairs will be helpful in drawing the graph of this function $\left(\text{in this case, } f(x) = f^{-1}(x)\right)$.

x	$f(x) = f^{-1}(x)$
-2	$-\frac{1}{2}$
-1	-1
$-\frac{1}{2}$	-2
$\frac{1}{2}$	2
1	1
2	$\frac{1}{2}$

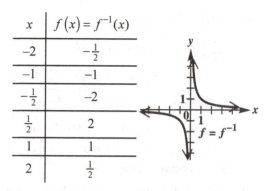

(c) For both $f(x)$ and $f^{-1}(x)$, the domain and range are both $(-\infty, 0) \cup (0, \infty)$.

69. $f(x) = \dfrac{1}{x - 3}$, $x \neq 3$

This function is one-to-one.

(a) *Step 1*: Replace $f(x)$ with y and interchange x and y.
$$y = \frac{1}{x - 3} \Rightarrow x = \frac{1}{y - 3}$$
Step 2: Solve for y.
$$x = \frac{1}{y - 3} \Rightarrow x(y - 3) = 1 \Rightarrow xy - 3x = 1 \Rightarrow$$
$$xy = 1 + 3x \Rightarrow y = \frac{1 + 3x}{x}$$
Step 3: Replace y with $f^{-1}(x)$.
$$f^{-1}(x) = \frac{1 + 3x}{x}, \ x \neq 0$$

(b) To graph $f(x) = \frac{1}{x-3}$, we can determine that there are no x-intercepts. The y-intercept is $f(0) = \frac{1}{0-3} = -\frac{1}{3}$. There is a vertical aymptote when the denominator is zero, $x - 3 = 0$, which implies $x = 3$ is the vertical asymptote. Also, the horizontal asymptote is $y = 0$ because the degree of the numerator is less than the denominator. Examining the following intervals, we have test points which will be helpful in drawing the graph of f as well as f^{-1}.

Interval	Test Point	Value of $f(x)$	Sign of $f(x)$	Graph Above or Below x-Axis
$(-\infty, 3)$	2	-1	Negative	Below
$(3, \infty)$	4	1	Positive	Above

(*continued on next page*)

(continued)

Plot the vertical asymptote, y-intercept, and test points with a smooth curve to get the graph of f.

To graph $f^{-1}(x) = \frac{1+3x}{x}$, we can determine that the x-intercept occurs when $1 + 3x = 0 \Rightarrow x = -\frac{1}{3}$. There is no y-intercept because 0 is not in the domain f^{-1}. There is a vertical asymptote when the denominator is zero, namely $x = 0$. Also, the degree of the numerator is the same as the denominator, so the horizontal asymptote is $y = \frac{3}{1} = 3$.

Using this information along with the points $(-1, 2)$ and $(1, 4)$, we can sketch f^{-1}.

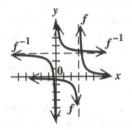

(c) Domain of f = range of
$f^{-1} = (-\infty, 3) \cup (3, \infty)$

Domain of f^{-1} = range of
$f = (-\infty, 0) \cup (0, \infty)$

71. $f(x) = \frac{x+1}{x-3}, \ x \neq 3$

This function is one-to-one.

(a) *Step 1*: Replace $f(x)$ with y and interchange x and y.

$$y = \frac{x+1}{x-3} \Rightarrow x = \frac{y+1}{y-3}$$

Step 2: Solve for y.

$$x = \frac{y+1}{y-3} \Rightarrow x(y-3) = y+1 \Rightarrow$$
$$xy - 3x = y + 1 \Rightarrow xy - y = 3x + 1 \Rightarrow$$
$$y(x-1) = 3x + 1 \Rightarrow y = \frac{3x+1}{x-1}$$

Step 3: Replace y with $f^{-1}(x)$.

$$f^{-1}(x) = \frac{3x+1}{x-1}, \ x \neq 1$$

(b) To graph $f(x) = \frac{x+1}{x-3}$, find the x-intercept:

$\frac{x+1}{x-3} = 0 \Rightarrow x + 1 = 0 \Rightarrow x = -1$.

The y-intercept is $f(0) = \frac{0+1}{0-3} = -\frac{1}{3}$.

There is a vertical aymptote when the denominator is zero, $x - 3 = 0 \Rightarrow x = 3$ is the vertical asymptote. The degree of the numerator equals the degree of the denominator, so the horizontal asymptote is $y = \frac{1}{1} = 1$. Examining the following intervals, we have test points which will be helpful in drawing the graph of f as well as f^{-1}.

Interval	Test Point	Value of $f(x)$	Sign of $f(x)$	Graph Above or Below x-Axis
$(-\infty, -1)$	-3	$\frac{1}{3}$	Positive	Above
$(-1, 3)$	$\frac{1}{2}$	$-\frac{3}{5}$	Negative	Below
$(3, \infty)$	4	5	Positive	Above

Plot the vertical asymptote, y-intercept, and test points with a smooth curve to get the graph of f.

To graph $f^{-1}(x) = \frac{3x+1}{x-1}$ we can determine that the x-intercept occurs when $3x + 1 = 0 \Rightarrow x = -\frac{1}{3}$. The y-intercept is $f(0) = \frac{3(0)+1}{0-1} = -1$. There is a vertical aymptote when the denominator is zero, $x - 1 = 0 \Rightarrow x = 1$. Also, because the degree of the numerator is the same as the denominator, the horizontal asymptote is $y = \frac{3}{1} = 3$. Examining the following intervals, we have test points which will be helpful in drawing the graph of f as well as f^{-1}.

Interval	Test Point	Value of $f(x)$	Sign of $f(x)$	Graph Above or Below x-Axis
$\left(-\infty, -\frac{1}{3}\right)$	-2	$\frac{5}{3}$	Positive	Above
$\left(-\frac{1}{3}, 1\right)$	$\frac{1}{2}$	-5	Negative	Below
$(1, \infty)$	2	7	Positive	Above

(continued on next page)

(*continued*)

Plot the vertical asymptote, y-intercept, and test points with a smooth curve to get the graph of f^{-1}.

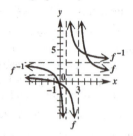

(c) Domain of f = range of

$$f^{-1} = (-\infty, 3) \cup (3, \infty);$$

Domain of f^{-1} = range of

$$f = (-\infty, 1) \cup (1, \infty)$$

73. $f(x) = \dfrac{2x+6}{x-3}$, $x \neq 3$

This function is one-to-one.

(a) *Step 1*: Replace $f(x)$ with y and interchange x and y.

$$y = \frac{2x+6}{x-3} \Rightarrow x = \frac{2y+6}{y-3}$$

Step 2: Solve for y.

$$x = \frac{2y+6}{y-3} \Rightarrow x(y-3) = 2y+6 \Rightarrow$$

$$xy - 3x = 2y + 6 \Rightarrow xy - 2y = 3x + 6 \Rightarrow$$

$$y(x-2) = 3x+6 \Rightarrow y = \frac{3x+6}{x-2}$$

Step 3: Replace y with $f^{-1}(x)$.

$$f^{-1}(x) = \frac{3x+6}{x-2}, \ x \neq 2.$$

(b) To graph $f(x) = \frac{2x+6}{x-3}$, find the x-intercept:

$$\frac{2x+6}{x-3} = 0 \Rightarrow 2x+6 = 0 \Rightarrow 2x = -6 \Rightarrow$$
$$x = -3$$

The y-intercept is $f(0) = \frac{2(0)+6}{0-3} = -2$.

There is a vertical aymptote when the denominator is zero, $x - 3 = 0 \Rightarrow x = 3$ is the vertical asymptote. The degree of the numerator equals the degree of the denominator, so the horizontal asymptote is $y = \frac{2}{1} = 2$. Examining the following intervals, we have test points which will be helpful in drawing the graph of f as well as f^{-1}.

Interval	Test Point	Value of $f(x)$	Sign of $f(x)$	Graph Above or Below x-Axis
$(-\infty, -3)$	-9	1	Positive	Above
$(-3, 3)$	-1	-1	Negative	Below
$(3, \infty)$	6	6	Positive	Above

Plot the vertical asymptote, y-intercept, and test points with a smooth curve to get the graph of f.

To graph $f^{-1}(x) = \frac{3x+6}{x-2}$, find the x-intercept:

$$\frac{3x+6}{x-2} = 0 \Rightarrow 3x+6 = 0 \Rightarrow 3x = -6 \Rightarrow$$
$$x = -2$$

The y-intercept is $f(0) = \frac{3(0)+6}{0-2} = -3$.

There is a vertical aymptote when the denominator is zero, $x - 2 = 0 \Rightarrow x = 2$ is the vertical asymptote. The degree of the numerator equals the degree of the denominator, so the horizontal asymptote is $y = \frac{3}{1} = 3$.

Examining the following intervals, we have test points which will be helpful in drawing the graph of f as well as f^{-1}.

Interval	Test Point	Value of $f(x)$	Sign of $f(x)$	Graph Above or Below x-Axis
$(-\infty, -2)$	-4	1	Positive	Above
$(-2, 2)$	-1	-1	Negative	Below
$(2, \infty)$	3	15	Positive	Above

Plot the vertical asymptote, y-intercept, and test points with a smooth curve to get the graph of f.

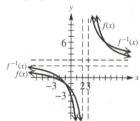

(c) Domain of f = range of

$$f^{-1} = (-\infty, 3) \cup (3, \infty);$$

Domain of f^{-1} = range of

$$f = (-\infty, 2) \cup (2, \infty)$$

75. $f(x) = \sqrt{x+6}$

This function is one-to-one.

(a) *Step 1*: Replace $f(x)$ with y and interchange x and y.

$y = \sqrt{x+6} \Rightarrow x = \sqrt{y+6}$

Step 2: Solve for y. In this problem we must consider that the range of f will be the domain of f^{-1}.

$x = \sqrt{y+6}$

$x^2 = \left(\sqrt{y+6}\right)^2$, for $x \geq 0$

$x^2 = y+6$, for $x \geq 0$

$x^2 - 6 = y$, for $x \geq 0$

Step 3: Replace y with $f^{-1}(x)$.

$f^{-1}(x) = x^2 - 6$, for $x \geq 0$

(b) Tables of ordered pairs will be helpful in drawing the graphs of these functions.

x	$f(x)$	x	$f^{-1}(x)$
–6	0	0	–6
–5	1	1	–5
–2	2	2	–2
3	3	3	3

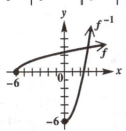

(c) Domain of f = range of f^{-1} = $[-6, \infty)$;

Range of f = domain of f^{-1} = $[0, \infty)$

77. Draw the mirror image of the original graph across the line $y = x$.

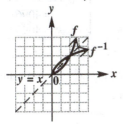

79. Carefully draw the mirror image of the original graph across the line $y = x$.

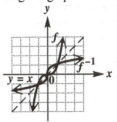

81. Draw the mirror image of the original graph across the line $y = x$.

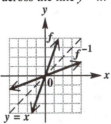

83. To find $f^{-1}(4)$, find the point with y-coordinate equal to 4. That point is (4, 4). The graph of f^{-1} contains (4, 4). Hence

$f^{-1}(4) = 4$.

85. To find $f^{-1}(0)$, find the point with y-coordinate equal to 0. That point is $(2, 0)$. The graph of f^{-1} contains $(0, 2)$.

Hence $f^{-1}(0) = 2$.

87. To find $f^{-1}(-3)$, find the point with y-coordinate equal to –3. That point is $(-2, -3)$. The graph of f^{-1} contains $(-3, -2)$. Hence $f^{-1}(-3) = -2$.

89. $f^{-1}(1000)$ represents the number of dollars required to build 1000 cars.

91. If a line has slope a, the slope of its reflection in the line $y = x$ will be reciprocal of a, which is $\frac{1}{a}$.

93. $f(x) = 6x^3 + 11x^2 - 6$

The graph is shown in the window $[-3, 2]$ by $[-10, 10]$. The horizontal line test shows that this function is not one-to-one.

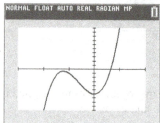

95. $f(x) = \dfrac{x-5}{x+3}$, $x \neq -3$

The graph is shown in the window $[-8, 8]$ by $[-6, 8]$. The horizontal line test shows that this function is one-to-one.

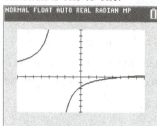

Find the equation of f^{-1}.

Step 1: Replace $f(x) = \frac{x-5}{x+3}$ with y and interchange x and y.

$y = \frac{x-5}{x+3} \Rightarrow x = \frac{y-5}{y+3}$

Step 2: Solve for y.

$x = \frac{y-5}{y+3} \Rightarrow x(y+3) = y-5 \Rightarrow$

$xy + 3x = y - 5 \Rightarrow xy - y = -5 - 3x \Rightarrow$

$y(x-1) = -5 - 3x \Rightarrow y = \frac{-5-3x}{x-1}$

Step 3: Replace y with $f^{-1}(x)$.

$f^{-1}(x) = \frac{-5-3x}{x-1}$, $x \neq 1$

97. Given $f(x) = 3x - 2$, find $f^{-1}(x)$.

Step 1: Replace $f(x) = 3x - 2$ with y and interchange x and y.

$y = 3x - 2 \Rightarrow x = 3y - 2$

Step 2: Solve for y.

$x = 3y - 2 \Rightarrow x + 2 = 3y \Rightarrow \frac{x+2}{3} = y$

Step 3: Replace y with $f^{-1}(x)$.

$f^{-1}(x) = \frac{x+2}{3}$

37; $f^{-1}(37) = \frac{37+2}{3}$ $= \frac{39}{3} = 13$; M	25; $f^{-1}(25) = \frac{25+2}{3}$ $= \frac{27}{3} = 9$; I
19; $f^{-1}(19) = \frac{19+2}{3}$ $= \frac{21}{3} = 7$; G	61; $f^{-1}(61) = \frac{61+2}{3}$ $= \frac{63}{3} = 21$; U
13; $f^{-1}(13) = \frac{13+2}{3}$ $= \frac{15}{3} = 5$; E	34; $f^{-1}(34) = \frac{34+2}{3}$ $= \frac{36}{3} = 12$; L
22; $f^{-1}(22) = \frac{22+2}{3}$ $= \frac{24}{3} = 8$; H	1; $f^{-1}(1) = \frac{1+2}{3}$ $= \frac{3}{3} = 1$; A
55; $f^{-1}(55) = \frac{55+2}{3}$ $= \frac{57}{3} = 19$; S	1; $f^{-1}(1) = \frac{1+2}{3}$ $= \frac{3}{3} = 1$; A
52; $f^{-1}(52) = \frac{52+2}{3}$ $= \frac{54}{3} = 18$; R	52; $f^{-1}(52) = \frac{52+2}{3}$ $= \frac{54}{3} = 18$; R
25; $f^{-1}(25) = \frac{25+2}{3}$ $= \frac{27}{3} = 9$; I	64; $f^{-1}(64) = \frac{64+2}{3}$ $= \frac{66}{3} = 22$; V
13; $f^{-1}(13) = \frac{13+2}{3}$ $= \frac{15}{3} = 5$; E	10; $f^{-1}(10) = \frac{10+2}{3}$ $= \frac{12}{3} = 4$; D

The message is MIGUEL HAS ARRIVED.

99. Given $f(x) = x^3 - 1$, we have the following.

S = 19; $f(19) = 19^3 - 1$ $= 6859 - 1$ $= 6858$	E = 5; $f(5) = 5^3$ $= 125 - 1 = 124$
N = 14; $f(14) = 14^3 - 1$ $= 2744 - 1$ $= 2743$	D = 4; $f(4) = 4^3 - 1$ $= 64 - 1 = 63$
H = 8; $f(8) = 8^3 - 1$ $= 512 - 1 = 511$	E = 5; $f(5) = 5^3 - 1$ $= 125 - 1 = 124$
L = 12; $f(12) = 12^3 - 1$ $= 1728 - 1$ $= 1727$	P = 16; $f(16) = 16^3 - 1$ $= 4096 - 1$ $= 4095$

Given $f(x) = x^3 - 1$, find $f^{-1}(x)$.

Step 1: Replace $f(x) = x^3 - 1$ with y and interchange x and y.

$y = x^3 - 1 \Rightarrow x = y^3 - 1$

Step 2: Solve for y.

$x = y^3 - 1 \Rightarrow x + 1 = y^3 \Rightarrow \sqrt[3]{x+1} = y$

Step 3: Replace y with $f^{-1}(x)$.

$f^{-1}(x) = \sqrt[3]{x+1}$

Section 4.2   Exponential Functions

1. If $f(x) = 4^x$, then $f(2) = \underline{16}$ and $f(-2) = \underline{\frac{1}{16}}$.

3. If $0 < a < 1$, then the graph of $f(x) = a^x$ <u>falls</u> from left to right.

5. The graph of $f(x) = 8^x$ passes through the points $\left(-1, \underline{\frac{1}{8}}\right)$, $(0, \underline{1})$, and $(1, \underline{8})$.

7. $\left(\dfrac{1}{4}\right)^x = 64$

   $\left(4^{-1}\right)^x = 64$

   $4^{-x} = 4^3$

   $-x = 3 \Rightarrow x = -3$

   Solution set: $\{-3\}$

9. $A = 2000\left(1 + \dfrac{0.03}{4}\right)^{8(4)} \approx 2540.22$

For exercises 11–25, $f(x) = 3^x$ and $g(x) = \left(\frac{1}{4}\right)^x$.

11. $f(2) = 3^2 = 9$

13. $f(-2) = 3^{-2} = \dfrac{1}{3^2} = \dfrac{1}{9}$

15. $g(2) = \left(\dfrac{1}{4}\right)^2 = \dfrac{1}{16}$

17. $g(-2) = \left(\dfrac{1}{4}\right)^{-2} = 4^2 = 16$

19. $f\left(\dfrac{3}{2}\right) = \left(\sqrt{3}\right)^3 = \sqrt{27} = \sqrt{9 \cdot 3} = 3\sqrt{3}$

21. $g\left(\dfrac{3}{2}\right) = \left(\dfrac{1}{4}\right)^{3/2} = \dfrac{1}{\left(\sqrt{4}\right)^3} = \dfrac{1}{2^3} = \dfrac{1}{8}$

23. $f(2.34) = 3^{2.34} \approx 13.076$

25. $g(-1.68) = \left(\dfrac{1}{4}\right)^{-1.68} \approx 10.267$

27. The y-intercept of $f(x) = 3^x$ is 1, and the x-axis is a horizontal asymptote. Make a table of values.

x	$f(x)$
-2	$\frac{1}{9} \approx 0.1$
-1	$\frac{1}{3} \approx 0.3$
$-\frac{1}{2}$	≈ 0.6
0	1
$\frac{1}{2}$	≈ 1.7
1	3
2	9

Plot these points and draw a smooth curve through them. This is an increasing function. The domain is $(-\infty, \infty)$ and the range is $(0, \infty)$ and is one-to-one.

29. The y-intercept of $f(x) = \left(\frac{1}{3}\right)^x$ is 1, and the x-axis is a horizontal asymptote. Make a table of values.

x	$f(x)$
-2	9
-1	3
$-\frac{1}{2}$	≈ 1.7
0	1
$\frac{1}{2}$	≈ 0.6
1	$\frac{1}{3} \approx 0.3$
2	$\frac{1}{9} \approx 0.1$

Plot these points and draw a smooth curve through them. This is a decreasing function. The domain is $(-\infty, \infty)$ and the range is $(0, \infty)$ and is one-to-one. Note: Because

$f(x) = \left(\frac{1}{3}\right)^x = \left(3^{-1}\right)^x = 3^{-x}$, the graph of

$f(x) = \left(\frac{1}{3}\right)^x$ is the reflection of the graph of

$f(x) = 3^x$ (Exercise 27) about the y-axis.

31. The y-intercept of $f(x) = \left(\frac{3}{2}\right)^x$ is 1, and the x-axis is a horizontal asymptote. Make a table of values.

x	$f(x)$
-2	≈ 0.4
-1	≈ 0.7
$-\frac{1}{2}$	≈ 0.8
0	1
$\frac{1}{2}$	≈ 1.2
1	1.5
2	2.25

Plot these points and draw a smooth curve through them. This is an increasing function. The domain is $(-\infty, \infty)$ and the range is $(0, \infty)$ and is one-to-one.

33. The y-intercept of $f(x) = \left(\frac{1}{10}\right)^{-x}$ is 1, and the x-axis is a horizontal asymptote. Make a table of values.

x	$f(x)$
-2	0.01
-1	0.1
$-\frac{1}{2}$	≈ 0.3
0	1
$\frac{1}{2}$	≈ 3.2
1	10
2	100

Plot these points and draw a smooth curve through them. This is an increasing function. The domain is $(-\infty, \infty)$ and the range is $(0, \infty)$ and is one-to-one.

35. The y-intercept of $f(x) = 4^{-x}$ is 1, and the x-axis is a horizontal asymptote. Make a table of values.

x	$f(x)$
-2	16
-1	4
$-\frac{1}{2}$	2
0	1
$\frac{1}{2}$	0.5
1	0.25
2	0.0625

Plot these points and draw a smooth curve through them. This is a decreasing function. The domain is $(-\infty, \infty)$ and the range is $(0, \infty)$ and is one-to-one. Note: The graph of $f(x) = 4^{-x}$ is the reflection of the graph of $f(x) = 4^x$ (Exercise 28) about the y-axis.

37. The y-intercept of $f(x) = 2^{|x|}$ is 1. Make a table of values.

x	$f(x)$
-2	4
-1	2
$-\frac{1}{2}$	≈ 1.4
0	1
$\frac{1}{2}$	≈ 1.4
1	2
2	4

Plot these points and draw a smooth curve through them. The domain is $(-\infty, \infty)$ and the range is $[1, \infty)$ and is not one-to-one. Note: For $x < 0$, $|x| = -x$, so the graph is the same as that of $f(x) = 2^{-x}$. For $x \geq 0$, we have $|x| = x$, so the graph is the same as that of $f(x) = 2^x$. Because $|-x| = |x|$, the graph is symmetric with respect to the y-axis.

For Exercises 39–49, refer to the following graph of $f(x) = 2^x$.

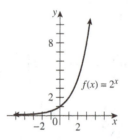

39. The graph of $f(x) = 2^x + 1$ is obtained by translating the graph of $f(x) = 2^x$ up one unit.

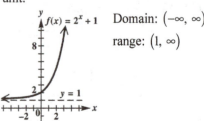

Domain: $(-\infty, \infty)$
range: $(1, \infty)$

41. Because $f(x) = 2^{x+1} = 2^{x-(-1)}$, the graph is
obtained by translating the graph of
$f(x) = 2^x$ to the left one unit.

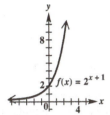

Domain: $(-\infty, \infty)$; range: $(0, \infty)$

43. The graph of $f(x) = -2^{x+2}$ is obtained by
translating the graph of $f(x) = 2^x$ to the left 2
units and then reflecting the graph across the
x-axis.

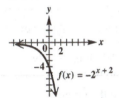

Domain: $(-\infty, \infty)$; range: $(-\infty, 0)$

45. The graph of $f(x) = 2^{-x}$ is obtained by
reflecting the graph across the y-axis.

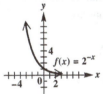

Domain: $(-\infty, \infty)$; range: $(0, \infty)$

47. The graph of $f(x) = 2^{x-1} + 2$ is obtained by
translating the graph of $f(x) = 2^x$ to the right
one unit and up two units.

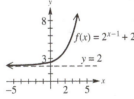

Domain: $(-\infty, \infty)$; range: $(2, \infty)$

49. The graph of $f(x) = 2^{x+2} - 4$ is obtained by
translating the graph of $f(x) = 2^x$ to the left
two units and down four units.

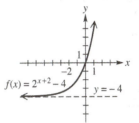

Domain: $(-\infty, \infty)$; range: $(-4, \infty)$

For Exercises 51–61, refer to the following graph of
$f(x) = \left(\frac{1}{3}\right)^x$.

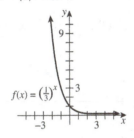

51. The graph of $f(x) = \left(\frac{1}{3}\right)^x - 2$ obtained by
translating the graph of $f(x) = \left(\frac{1}{3}\right)^x$ down 2
units.

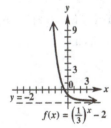

Domain: $(-\infty, \infty)$; range: $(-2, \infty)$

53. Because $f(x) = \left(\frac{1}{3}\right)^{x+2} = \left(\frac{1}{3}\right)^{x-(-2)}$, the graph
is obtained by translating the graph of
$f(x) = \left(\frac{1}{3}\right)^x$ 2 units to the left.

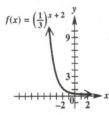

Domain: $(-\infty, \infty)$; range: $(0, \infty)$

55. The graph of $f(x) = \left(\frac{1}{3}\right)^{-x+1}$ is obtained by translating the graph of $f(x) = \left(\frac{1}{3}\right)^{x}$ left one unit and then reflecting the resulting graph across the y-axis.

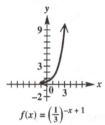

$f(x) = \left(\frac{1}{3}\right)^{-x+1}$

Domain: $(-\infty, \infty)$; range: $(0, \infty)$

57. The graph of $f(x) = \left(\frac{1}{3}\right)^{-x}$ is obtained by reflecting the graph of $f(x) = \left(\frac{1}{3}\right)^{x}$ across the y-axis.

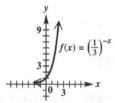

$f(x) = \left(\frac{1}{3}\right)^{-x}$

Domain: $(-\infty, \infty)$; range: $(0, \infty)$

59. The graph of $f(x) = \left(\frac{1}{3}\right)^{x-2} + 2$ is obtained by translating the graph of $f(x) = \left(\frac{1}{3}\right)^{x}$ two units to the right and two units up.

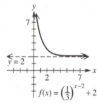

$y = 2$

$f(x) = \left(\frac{1}{3}\right)^{x-2} + 2$

Domain: $(-\infty, \infty)$; range: $(2, \infty)$

61. The graph of $f(x) = \left(\frac{1}{3}\right)^{x+2} - 1$ is obtained by translating the graph of $f(x) = \left(\frac{1}{3}\right)^{x}$ two units to the left and one unit down.

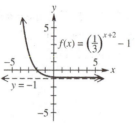

$f(x) = \left(\frac{1}{3}\right)^{x+2} - 1$

$y = -1$

Domain: $(-\infty, \infty)$; range: $(-1, \infty)$

63. The horizontal asymptote is $y = -2$, so the graph has been shifted down two units. The general form of the equation is $f(x) = a^{x+b} - 2$. The base is either 2 or 3, so try $a = 3$. Then substitute the coordinates of a point in the equation and solve for b:

$7 = 3^{2+b} - 2 \Rightarrow 9 = 3^{2+b} \Rightarrow 3^2 = 3^{2+b} \Rightarrow$
$2 = 2 + b \Rightarrow 0 = b$

So, the equation is $f(x) = 3^x - 2$. Verify that the coordinates of other two points given satisfy the equation.

Alternate solution: Working backward and shifting the graph up two units to transform the given graph into the graph of $y = 3^x$, it goes through the points (2, 9), (1, 3), and (0, 1), which is the y-intercept. $9 = 3^2$, so $a = 3$, and the equation is $f(x) = 3^x - 2$.

Verify by checking that the coordinates of the points satisfy the equation.

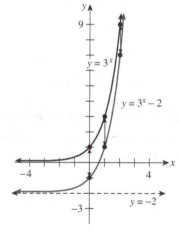

$y = 3^x$

$y = 3^x - 2$

$y = -2$

65. The horizontal asymptote is $y = -1$, so the graph has been shifted down one unit. The general form of the equation is

$f(x) = a^{x+b} - 1$. The base is either 2 or 3, so try $a = 2$. Then substitute the coordinates of a point in the equation and solve for b:

$1 = 2^{-2+b} - 1 \Rightarrow 2 = 2^{-2+b} \Rightarrow 2^1 = 2^{-2+b} \Rightarrow$
$1 = -2 + b \Rightarrow 3 = b$

So, the equation is $f(x) = 2^{x+3} - 1$. Verify that the coordinates of other two points given satisfy the equation.

Alternate solution: Working backward and shifting the graph up one unit and right three units to transform the given graph into the graph of $y = 2^x$, it goes through the points $(3, 8)$, $(1, 2)$, and $(0, 1)$, which is the y-intercept. $8 = 2^3$, so $a = 2$, and the equation is $f(x) = 2^{x+3} - 1$. Verify by checking that the coordinates of the points satisfy the equation.

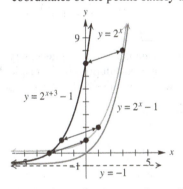

67. The horizontal asymptote is $y = 3$, so the graph has been shifted up three units. The graph has also been reflected across the x-axis. The general form of the equation is

$f(x) = -a^{x+b} + 3$. The base is either 2 or 3, so try $a = 2$. Then substitute the coordinates of a point in the equation and solve for b:

$-1 = -2^{0+b} + 3 \Rightarrow -4 = -2^b \Rightarrow 4 = 2^b \Rightarrow$
$2^2 = 2^b \Rightarrow 2 = b$

So, the equation is $f(x) = -2^{x+2} + 3$. Verify that the coordinates of other two points given satisfy the equation.

Alternate solution: Working backward and shifting the graph down three units and right two units to transform the given graph into the graph of $y = -\left(2^x\right)$, it goes through the points $(0, -1)$, $(1, -2)$, and $(2, -4)$.

The y-intercept is $(0, -1)$. $-2 = -\left(2^1\right)$, so $a = 2$, and the equation is $f(x) = -2^{x+2} + 3$.

Verify by checking that the coordinates of the points satisfy the equation.

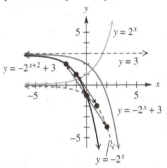

69. The horizontal asymptote is $y = 1$, so the graph has been shifted up one unit. The graph has also been reflected across the y-axis. So the general form of the equation is

$f(x) = a^{-x+b} + 1$. The base is either 2 or 3, so try $a = 3$. Then substitute the coordinates of a point in the equation and solve for b:

$4 = 3^{-(-1)+b} + 1 \Rightarrow 3 = 3^{1+b} \Rightarrow 3^1 = 3^{1+b} \Rightarrow$
$1 = 1 + b \Rightarrow 0 = b$

So, the equation is $f(x) = 3^{-x} + 1$. Verify by checking that the coordinates of the other two points satisfy the equation.

Alternate solution: Working backward and shifting the graph down one unit to transform the given graph into the graph of $y = 3^{-x}$, it goes through the points $(-1, 3)$, $(0, 1)$, and $\left(1, \frac{1}{3}\right)$. The y-intercept is $(0, 1)$. $3 = 3^{-(-1)}$, so $a = 3$, and the equation is $f(x) = 3^{-x} + 1$.

Verify by checking that the coordinates of the points satisfy the equation.

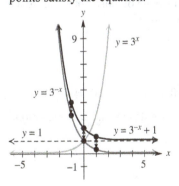

71. $4^x = 2 \Rightarrow \left(2^2\right)^x = 2^1 \Rightarrow 2^{2x} = 2^1 \Rightarrow$
$2x = 1 \Rightarrow x = \frac{1}{2}$
Solution set: $\left\{\frac{1}{2}\right\}$

73. $\left(\frac{5}{2}\right)^x = \frac{4}{25} \Rightarrow \left(\frac{2}{5}\right)^{-x} = \left(\frac{2}{5}\right)^2 \Rightarrow -x = 2 \Rightarrow$
$x = -2$
Solution set: $\{-2\}$

75. $2^{3-2x} = 8 \Rightarrow 2^{3-2x} = 2^3 \Rightarrow 3 - 2x = 3 \Rightarrow$
$-2x = 0 \Rightarrow x = 0$
Solution set: $\{0\}$

77. $e^{4x-1} = \left(e^2\right)^x \Rightarrow e^{4x-1} = e^{2x} \Rightarrow 4x - 1 = 2x \Rightarrow$
$-1 = -2x \Rightarrow \frac{1}{2} = x$
Solution set: $\left\{\frac{1}{2}\right\}$

79. $27^{4x} = 9^{x+1} \Rightarrow \left(3^3\right)^{4x} = \left(3^2\right)^{x+1} \Rightarrow$
$3^{3(4x)} = 3^{2(x+1)} \Rightarrow 3^{12x} = 3^{2x+2} \Rightarrow$
$12x = 2x + 2 \Rightarrow 10x = 2 \Rightarrow x = \frac{1}{5}$
Solution set: $\left\{\frac{1}{5}\right\}$

81. $4^{x-2} = 2^{3x+3} \Rightarrow \left(2^2\right)^{x-2} = 2^{3x+3} \Rightarrow$
$2^{2(x-2)} = 2^{3x+3} \Rightarrow 2^{2x-4} = 2^{3x+3} \Rightarrow$
$2x - 4 = 3x + 3 \Rightarrow -4 = x + 3 \Rightarrow -7 = x$
Solution set: $\{-7\}$

83. $x^{2/3} = 4 \Rightarrow \left(x^{2/3}\right)^{3/2} = \pm 4^{3/2} \Rightarrow$
$x = \left(\pm\sqrt{4}\right)^3 \Rightarrow x = \left(\pm 2\right)^3 \Rightarrow x = \pm 8$

Recall from Chapter 1 that it is necessary to check all proposed solutions in the original equation when you raise both sides to a power.
Check $x = -8$.

$4 = x^{2/3}$

$4 \overset{?}{=} (-8)^{2/3}$

$4 = \left(\sqrt[3]{-8}\right)^2 \Rightarrow 4 = (-2)^2 \Rightarrow 4 = 4$

This is a true statement. -8 is a solution.
Check $r = 8$.

$4 = x^{2/3}$

$4 \overset{?}{=} 8^{2/3}$

$4 = \left(\sqrt[3]{8}\right)^2 \Rightarrow 4 = 2^2 \Rightarrow 4 = 4$

This is a true statement. 8 is a solution.
Solution set: $\{-8, 8\}$

85. $x^{5/2} = 32 \Rightarrow \left(x^{5/2}\right)^{2/5} = 32^{2/5} \Rightarrow x = 32^{2/5} \Rightarrow$
$x = \left(\sqrt[5]{32}\right)^2 \Rightarrow x = 2^2 \Rightarrow x = 4$

Recall from Chapter 1 that it is necessary to check all proposed solutions in the original equation when you raise both sides to a power.
Check $x = 4$.

$x^{5/2} = 32$

$4^{5/2} \overset{?}{=} 32$

$\left(\sqrt{4}\right)^5 = 32 \Rightarrow 2^5 = 32 \Rightarrow 32 = 32$

This is a true statement.
Solution set: $\{4\}$

87. $x^{-6} = \frac{1}{64} \Rightarrow \frac{1}{x^6} = \frac{1}{64} \Rightarrow x^6 = 64 \Rightarrow$
$x^6 = (\pm 2)^6 \Rightarrow x = \pm 2$
Solution set: $\{-2, 2\}$

89. $x^{5/3} = -243 \Rightarrow \left(x^{5/3}\right)^{3/5} = (-243)^{3/5} \Rightarrow$
$x = (-3)^3 = -27$

Recall from Chapter 1 that it is necessary to check all proposed solutions in the original equation when you raise both sides to a power.
Check $x = -27$.

$x^{5/3} = -243$

$(-27)^{5/3} \overset{?}{=} -243$

$(-3)^5 = -243$

This is a true statement. -27 is a solution.
Solution set: $\{-27\}$

91. $\left(\frac{1}{e}\right)^{-x} = \left(\frac{1}{e^2}\right)^{x+1} \Rightarrow \left(e^{-1}\right)^{-x} = \left(e^{-2}\right)^{x+1} \Rightarrow$
$e^x = e^{-2(x+1)} \Rightarrow e^x = e^{-2x-2} \Rightarrow$
$x = -2x - 2 \Rightarrow 3x = -2 \Rightarrow x = -\frac{2}{3}$
Solution set: $\left\{-\frac{2}{3}\right\}$

93. $\left(\sqrt{2}\right)^{x+4} = 4^x \Rightarrow \left(2^{1/2}\right)^{x+4} = \left(2^2\right)^x \Rightarrow$
$2^{(1/2)(x+4)} = 2^{2x} \Rightarrow 2^{(1/2)x+2} = 2^{2x} \Rightarrow$
$\frac{1}{2}x + 2 = 2x \Rightarrow 2 = \frac{3}{2}x \Rightarrow \frac{2}{3}\cdot 2 = x \Rightarrow x = \frac{4}{3}$
Solution set: $\left\{\frac{4}{3}\right\}$

95. $\dfrac{1}{27}=x^{-3}\Rightarrow 3^{-3}=x^{-3}\Rightarrow x=3$

Alternate solution:

$\dfrac{1}{27}=x^{-3}\Rightarrow \dfrac{1}{27}=\dfrac{1}{x^3}\Rightarrow 27=x^3\Rightarrow$

$x=\sqrt[3]{27}=3$

Solution set: $\{3\}$

97. (a) Use the compound interest formula to find the future value, $A=P\left(1+\frac{r}{n}\right)^{tn}$, given $n=2$, $P=8906.54$, $r=0.03$, and $t=9$.

$A=P\left(1+\frac{r}{m}\right)^{tm}=(8906.54)\left(1+\frac{0.03}{2}\right)^{9(2)}$

$\quad=(8906.54)(1+0.015)^{18}\approx 11,643.88$

Rounding to the nearest cent, the future value is $11,643.88. The amount of interest would be
$11,643.88 − $8906.54 = $2737.34.

(b) Use the continuous compounding interest formula to find the future value, $A=Pe^{rt}$, given $P=8906.54$, $r=0.03$, and $t=9$.

$A=Pe^{rt}=8906.54e^{0.03(9)}=8906.54e^{0.27}$

$\quad\approx 11,667.25$

Rounding to the nearest cent, the future value is $11,667.25. The amount of interest would be
$11,667.25 − $8906.54 = $2760.71.

99. Use the compound interest formula to find the present amount, $A=P\left(1+\frac{r}{n}\right)^{tn}$, given $n=4$, $A=25,000$, $r=0.032$, and $t=\frac{11}{4}$.

$A=P\left(1+\frac{r}{n}\right)^{tn}$

$25,000=P\left(1+\frac{0.032}{4}\right)^{(11/4)(4)}$

$25,000=P(1.008)^{11}$

$P=\dfrac{25,000}{(1.008)^{11}}\approx \$22,902.04$

Rounding to the nearest cent, the present value is $22,902.04.

101. Use the compound interest formula to find the present value, $A=P\left(1+\frac{r}{n}\right)^{tn}$, given $n=4$, $A=5000$, $r=0.035$, and $t=10$.

$A=P\left(1+\frac{r}{n}\right)^{tn}$

$5,000=P\left(1+\frac{0.035}{4}\right)^{10(4)}=P(1.00875)^{40}$

$P=\dfrac{5,000}{(1.00875)^{40}}\approx \$3528.808535$

Rounding to the nearest cent, the present value is $3528.81.

103. Use the compound interest formula to find the interest rate, $A=P\left(1+\frac{r}{n}\right)^{tn}$, given $n=4$, $A=1500$, $P=1200$, and $t=9$.

$A=P\left(1+\frac{r}{n}\right)^{tn}$

$1500=1200\left(1+\frac{r}{4}\right)^{9(4)}$

$1500=1200\left(1+\frac{r}{4}\right)^{36}$

$1.25=\left(1+\frac{r}{4}\right)^{36}$

$(1.25)^{1/36}=1+\frac{r}{4}\Rightarrow (1.25)^{1/36}-1=\frac{r}{4}\Rightarrow$

$4\left[(1.25)^{1/36}-1\right]=r\Rightarrow r\approx 0.025$

The interest rate, to the nearest tenth, is 2.5%.

105. For each bank we need to calculate $\left(1+\frac{r}{n}\right)^n$.

The base, $1+\frac{r}{n}$, is greater than 1, so we need only compare the three values calculated to determine which bank will yield the least amount of interest. It is understood that the amount of time, t, and the principal, P, are the same for all three banks.

Bank A: Calculate $\left(1+\frac{r}{n}\right)^n$ where $n=1$ and $r=0.064$.

$\left(1+\frac{0.064}{1}\right)^1=(1+0.064)^1=(1.064)^1=1.064$

Bank B: Calculate $\left(1+\frac{r}{n}\right)^n$ where $n=12$ and $r=0.063$.

$\left(1+\frac{0.063}{12}\right)^{12}=(1+0.00525)^{12}=(1.00525)^{12}$

$\quad\approx 1.064851339$

Bank C: Calculate $\left(1+\frac{r}{n}\right)^n$ where $n=4$ and $r=0.0635$.

$\left(1+\frac{0.0635}{4}\right)^4=(1+0.015875)^4=(1.015875)^4$

$\quad\approx 1.06502816$

Bank A will charge you the least amount of interest, even though it has the highest stated rate.

107. (a)

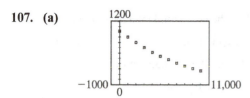

(b) From the graph above, we can see that the data are not linear but exponentially decreasing.

(c)

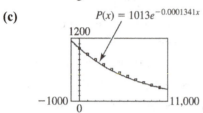

$P(x) = 1013e^{-0.0001341x}$

(d)
$$P(x) = 1013e^{-0.0001341x}$$
$$P(1500) = 1013e^{-0.0001341(1500)}$$
$$\approx 1013(0.817790) \approx 828$$
$$P(11,000) = 1013e^{-0.0001341(11,000)}$$
$$\approx 1013(0.228756) \approx 232$$

When the altitude is 1500 m, the function P gives a pressure of 828 mb, which is less than the actual value of 846 mb. When the altitude is 11,000 m, the function P gives a pressure of 232 mb, which is more than the actual value of 227 mb.

109. (a) Evaluate $f(x) = 50,000(1 + 0.06)^x$ where $n = 4$.
$$f(x) = 50,000(1 + 0.06)^4$$
$$= 50,000(1.06)^4 \approx 63,123.848$$
Total population after 4 years is about 63,000.

(b) Evaluate $f(x) = 30,000(1 + 0.12)^x$ where $n = 3$.
$$f(x) = 30,000(1 + 0.12)^3$$
$$= 30,000(1.12)^3 = 42,147.84$$
There would be about 42,000 deer after 3 years.

(c) Evaluate $f(x) = 45,000(1 + 0.08)^x$ where $n = 5$.
$$f(x) = 45,000(1 + 0.08)^5$$
$$= 45,000(1.08)^5 \approx 66,119.76346$$

There would be about 66,000 deer after 5 years. Thus, we can expect about $66,000 - 45,000 = 21,000$ additional deer after 5 years.

111. $5e^{3x} = 75$

Graph $y_1 = 5e^{3x}$ and $y_2 = 75$ in the window $[-2, 2] \times [0, 100]$. Then find the intersection of the two graphs.

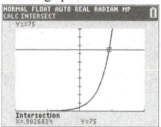

Solution set: $\{0.9\}$

113. $3x + 2 = 4^x$

Graph $y_1 = 3x + 2$ and $y_2 = 4^x$ in the window $[-2, 2] \times [-2, 8]$. Then find the intersection of the two graphs. We can see from the following screens that there are two solutions.

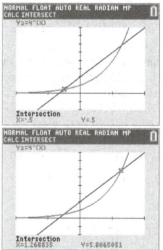

Solution set: $\{-0.5, 1.3\}$

115. Answers will vary. Sample answer: The variable is located in the base of a power function but in the exponent of an exponential function.

117. If the graph of the exponential function $f(x) = a^x$ contains the point $(3, 8)$, we have $a^3 = 8$. This implies $a = \sqrt[3]{8} = 2$. Thus, the equation which satisfies the given condition is $f(x) = 2^x$.

119. If the graph of the exponential function $f(x) = a^x$ contains the point $(-3, 64)$, we have $a^{-3} = 64$. This implies $a^3 = \frac{1}{64} \Rightarrow a = \sqrt[3]{\frac{1}{64}} = \frac{1}{4}$. Thus, the equation which satisfies the given condition is $f(x) = \left(\frac{1}{4}\right)^x$.

121. $f(t) = 3^{2t+3} = 3^{2t} \cdot 3^3 = 27 \cdot \left(3^2\right)^t = 27 \cdot 9^t$

123. $f(t) = \left(\frac{1}{3}\right)^{1-2t} = \left(\frac{1}{3}\right)^1 \left(\frac{1}{3}\right)^{-2t}$
$= \left(\frac{1}{3}\right)(3)^{2t} = \left(\frac{1}{3}\right)\left(3^2\right)^t = \left(\frac{1}{3}\right)9^t$

125. $e^1 \approx 1 + 1 + \frac{1^2}{2 \cdot 1} + \frac{1^3}{3 \cdot 2 \cdot 1} + \frac{1^4}{4 \cdot 3 \cdot 2 \cdot 1} + \frac{1^5}{5 \cdot 4 \cdot 3 \cdot 2 \cdot 1}$
$= 1 + 1 + \frac{1}{2} + \frac{1}{6} + \frac{1}{24} + \frac{1}{120} \approx 2.717$

127. Yes; $f(x) = a^x$ is a one-to-one function. Therefore, an inverse function exists for f.

129. Because $f(x) = a^x$ has an inverse, we find it as follows: $y = a^x \Rightarrow x = a^y$

131. If $a = e$, the equation for $f^{-1}(x)$ will be given by $x = e^y$.

Section 4.3   Logarithmic Functions

1. (a) C; $\log_2 16 = 4$ because $2^4 = 16$.

(b) A; $\log_3 1 = 0$ because $3^0 = 1$.

(c) E; $\log_{10} 0.1 = -1$ because $10^{-1} = 0.1$.

(d) B; $\log_2 \sqrt{2} = \frac{1}{2}$ because $2^{1/2} = \sqrt{2}$.

(e) F; $\log_e \left(\frac{1}{e^2}\right) = -2$ because $e^{-2} = \frac{1}{e^2}$.

(f) D; $\log_{1/2} 8 = -3$ because $\left(\frac{1}{2}\right)^{-3} = 8$.

3. $\log_2 8 = 3$ is equivalent to $2^3 = 8$.

5. $\log_x \frac{16}{81} = 2 \Rightarrow x^2 = \frac{16}{81} \Rightarrow x = \frac{4}{9}$
Solution set: $\left\{\frac{4}{9}\right\}$

7. $f(x) = \log_5 x$
We can write the exponential form of $f(x) = y = \log_5 x$ as $x = 5^y$ in order to find ordered pairs that satisfy the equation. It is easier to choose values for y and find the corresponding values of x. Make a table of values.

x	$y = \log_5 x$
$\frac{1}{25} = 0.04$	-2
$\frac{1}{5} = 0.2$	-1
1	0
5	1
25	2

The graph can also be found by reflecting the graph of $f(x) = 5^x$ about the line $y = x$. The graph has the y-axis as a vertical asymptote. The domain is $(0, \infty)$ and the range is $(-\infty, \infty)$.

9. $\log_{10} \frac{2x}{7} = \log_{10}(2x) - \log_{10} 7$
$= \log_{10} 2 + \log_{10} x - \log_{10} 7$

11. $3^4 = 81$ is equivalent to $\log_3 81 = 4$.

13. $\left(\frac{2}{3}\right)^{-3} = \frac{27}{8}$ is equivalent to $\log_{2/3} \frac{27}{8} = -3$.

15. $\log_6 36 = 2$ is equivalent to $6^2 = 36$.

17. $\log_{\sqrt{3}} 81 = 8$ is equivalent to $\left(\sqrt{3}\right)^8 = 81$.

19. $x = \log_5 \frac{1}{625} \Rightarrow 5^x = \frac{1}{625} \Rightarrow 5^x = \frac{1}{5^4} \Rightarrow$
$5^x = 5^{-4} \Rightarrow x = -4$
Solution set: $\{-4\}$

21. $\log_x \frac{1}{32} = 5 \Rightarrow x^5 = \frac{1}{32} \Rightarrow x^5 = \frac{1}{2^5} = \left(\frac{1}{2}\right)^5 \Rightarrow$
$x = \frac{1}{2}$
Solution set: $\left\{\frac{1}{2}\right\}$

23. $x = \log_8 \sqrt[4]{8} \Rightarrow 8^x = \sqrt[4]{8} \Rightarrow 8^x = 8^{1/4} \Rightarrow x = \frac{1}{4}$
Solution set: $\left\{\frac{1}{4}\right\}$

25. $x = 3^{\log_3 8}$

Writing as a logarithmic equation, we have
$\log_3 8 = \log_3 x \Rightarrow x = 8$

Using the Theorem of Inverses on page 440, we can directly state that $x = 8$.

Solution set: $\{8\}$

27. $x = 2^{\log_2 9}$

Writing as a logarithmic equation, we have
$\log_2 9 = \log_2 x \Rightarrow x = 9$

Using the Theorem of Inverses on page 440, we can directly state that $x = 9$.

Solution set: $\{9\}$

29. $\log_x 25 = -2 \Rightarrow x^{-2} = 25 \Rightarrow x^{-2} = 5^2 \Rightarrow$

$\left(x^{-2}\right)^{-1/2} = \left(5^2\right)^{-1/2} \Rightarrow x = 5^{-1}$

Do not include a $\pm$ since the base, x, cannot be negative.

$x = \frac{1}{5}$

Solution set: $\left\{\frac{1}{5}\right\}$

31. $\log_4 x = 3 \Rightarrow 4^3 = x \Rightarrow 64 = x$

Solution set: $\{64\}$

33. $x = \log_4 \sqrt[3]{16} \Rightarrow 4^x = \sqrt[3]{16} \Rightarrow 4^x = (16)^{1/3} \Rightarrow$

$4^x = \left(4^2\right)^{1/3} \Rightarrow 4^x = 4^{2/3} \Rightarrow x = \frac{2}{3}$

Solution set: $\left\{\frac{2}{3}\right\}$

35. $\log_9 x = \frac{5}{2} \Rightarrow 9^{5/2} = x \Rightarrow (3)^5 = 243 = x$

Note that we do not include $\sqrt{9} = -3$ because logarithms are not defined for negative numbers.

Solution set: $\{243\}$

37. $\log_{1/2}(x+3) = -4 \Rightarrow x + 3 = \left(\frac{1}{2}\right)^{-4} \Rightarrow$

$x + 3 = 2^4 \Rightarrow x + 3 = 16 \Rightarrow x = 13$

Solution set: $\{13\}$

39. $\log_{(x+3)} 6 = 1 \Rightarrow (x+3)^1 = 6 \Rightarrow x + 3 = 6 \Rightarrow$

$x = 3$

Solution set: $\{3\}$

41. $3x - 15 = \log_x 1 \ (x > 0, \ x \neq 1)$

Note that $\log_x 1 = 0$ because $x^0 = 1$ for any number x.

Thus,

$3x - 15 = \log_x 1 \Rightarrow 3x - 15 = 0 \Rightarrow 3x = 15 \Rightarrow$

$x = 5$

Solution set: $\{5\}$

For Exercises 43–45, refer to the following graph of $f(x) = \log_2 x$.

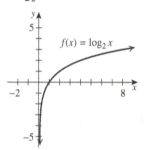

43. The graph of $f(x) = (\log_2 x) + 3$ is obtained by translating the graph of $f(x) = \log_2 x$ up 3 units.

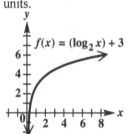

Domain: $(0, \infty)$; range: $(-\infty, \infty)$

45. To find the graph of $f(x) = \left|\log_2(x+3)\right|$, translate the graph of $f(x) = \log_2 x$ to the left 3 units to obtain the graph of $\log_2(x+3)$. (See exercise 44.) For the portion of the graph where $f(x) \geq 0$, that is, where $x \geq -2$, use the same graph as in exercise 44. For the portion of the graph in exercise 44 where $f(x) < 0$, $-3 < x < -2$, reflect the graph about the x-axis. In this way, each negative value of $f(x)$ on the graph in exercise 44 is replaced by its opposite, which is positive. The graph has a vertical asymptote at $x = -3$.

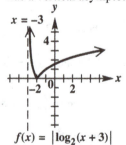

$f(x) = \left|\log_2(x+3)\right|$

Domain: $(-3, \infty)$; range: $[0, \infty)$

For Exercise 47, refer to the following graph of $f(x) = \log_{1/2} x$.

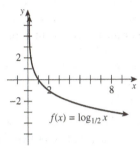

47. The graph of $f(x) = \log_{1/2}(x-2)$ is obtained by translating the graph of $f(x) = \log_{1/2} x$ to the right 2 units. The graph has a vertical asymptote at $x = 2$.

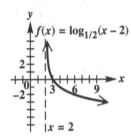

Domain: $(2, \infty)$; range: $(-\infty, \infty)$

49. Because $f(x) = \log_2 x$ has a vertical asymptote, the y-axis (the line $x = 0$), x-intercept of 1, and is increasing, the correct choice is the graph in E.

51. Because $f(x) = \log_2 \frac{1}{x} = \log_2 x^{-1} = -\log_2 x$, it has a vertical asymptote, which is the y-axis (the line $x = 0$), has an x-intercept 1, and is the reflection of $f(x) = \log_2 x$ across the x-axis, it is decreasing and the correct choice is the graph in B.

53. Because $f(x) = \log_2(x-1)$ represents the horizontal shift of $f(x) = \log_2 x$ to the right 1 unit, the function has a vertical asymptote (the line $x = 1$), has an x-intercept when $x - 1 = 1 \Rightarrow x = 2$, and is increasing, the correct choice is the graph in F.

55. $f(x) = \log_5 x$

We can write the exponential form of $f(x) = y = \log_5 x$ as $x = 5^y$ in order to find ordered pairs that satisfy the equation. It is easier to choose values for y and find the corresponding values of x. Make a table of values.

x	$y = \log_5 x$
$\frac{1}{25} = 0.04$	-2
$\frac{1}{5} = 0.2$	-1
1	0
5	1
25	2

The graph can also be found by reflecting the graph of $f(x) = 5^x$ about the line $y = x$. The graph has the y-axis as a vertical asymptote.

57. $f(x) = \log_5(x+1)$

We can write the exponential form of $f(x) = y = \log_5(x+1)$ as

$x + 1 = 5^y \Rightarrow x = 5^y - 1$ in order to find ordered pairs that satisfy the equation. It is easier to choose values for y and find the corresponding values of x. Make a table of values.

$x = 5^y - 1$	$y = \log_5(x+1)$
-0.96	-2
-0.8	-1
0	0
4	1
24	2

The vertical asymptote is $x = -1$.

59. $f(x) = \log_{1/2}(1-x)$

We can write the exponential form of

$f(x) = y = \log_{1/2}(1-x)$ as $1 - x = \left(\frac{1}{2}\right)^y \Rightarrow$

$x = 1 - \left(\frac{1}{2}\right)^y$ in order to find ordered pairs that satisfy the equation. It is easier to choose values for y and find the corresponding values of x. Make a table of values.

x	$y = \log_{1/2}(1-x)$
-3	-2
-1	-1
0	0
$\frac{1}{2} = 0.5$	1
$\frac{3}{4} = 0.75$	2

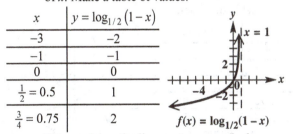

The graph has the line $x = 1$ as a vertical asymptote.

61. $f(x) = \log_3(x-1) + 2$

We can write the exponential form of
$f(x) = y = \log_3(x-1) + 2$ as

$y - 2 = \log_3(x-1) \Rightarrow x - 1 = 3^{y-2} \Rightarrow$

$x = 3^{y-2} + 1$ in order to find ordered pairs that
satisfy the equation. It is easier to choose
values for y and find the corresponding values
of x. Make a table of values.

$x = 3^{y-2} + 1$	$y = \log_3(x-1) + 2$
$\frac{82}{81} \approx 1.01$	-2
$\frac{28}{27} \approx 1.04$	-1
$\frac{10}{9} \approx 1.11$	0
$\frac{4}{3} \approx 1.33$	1
2	2
4	3
10	4

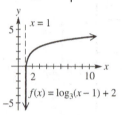

The vertical asymptote is $x = 1$.

63. $f(x) = \log_{1/2}(x+3) - 2$

We can write the exponential form of
$f(x) = y = \log_{1/2}(x+3) - 2$ as

$y + 2 = \log_{1/2}(x+3) \Rightarrow \left(\frac{1}{2}\right)^{y+2} = x + 3 \Rightarrow$

$x = \left(\frac{1}{2}\right)^{y+2} - 3$

in order to find ordered pairs that satisfy the
equation. It is easier to choose values for y and
find the corresponding values of x. Make a
table of values.

x	$y = \log_{1/2}(x+3) - 2$
5	-5
1	-4
-1	-3
-2.75	0
-2.88	1

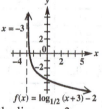

The vertical asymptote is the line $x = -3$.

65. The vertical asymptote is $x = -1$, so the graph
has been shifted left one unit. So the general
form of the equation is $f(x) = \log_a(x+1) + k$.

The base is either 2 or 3, so try $a = 2$. Then
substitute the coordinates of a point in the
equation and solve for k:

$-2 = \log_2(1+1) + k \Rightarrow -2 - k = \log_2 2 \Rightarrow$

$2^{-2-k} = 2 \Rightarrow 2^{-2-k} = 2^1 \Rightarrow$

$-2 - k = 1 \Rightarrow -3 = k$

So, the equation is $f(x) = \log_2(x+1) - 3$.

Verify that the coordinates of other two points
given satisfy the equation.

Alternate solution: Working backward and
shifting the graph up three units and right one
unit to transform the given graph into the
graph of $y = \log_2 x$, it goes through the
points (1, 0), which is the x-intercept, (2, 1),
and (8, 3). $3 = \log_2 8$, so $a = 2$, and the
equation is $f(x) = \log_2(x+1) - 3$. Verify by
checking that the coordinates of the points
shown on the graph satisfy the equation.

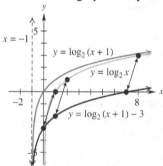

67. The graph has been reflected across the y-axis,
so the general form of the equation is
$f(x) = \log_a(-x - b) + k$. The vertical
asymptote of the graph is $x = 3$, so the graph
has been shifted right three units, and $b = -3$.
So the general form of the equation is
$f(x) = \log_a(-x+3) + k$. The base is either 2
or 3, so try $a = 2$. Then substitute the
coordinates of a point in the equation and
solve for k:

$-1 = \log_2(-1+3) + k \Rightarrow -1 - k = \log_2 2 \Rightarrow$

$2^{-1-k} = 2 \Rightarrow 2^{-1-k} = 2^1 \Rightarrow$

$-1 - k = 1 \Rightarrow -2 = k$

So, the equation is $f(x) = \log_2(-x+3) - 2$.

Verify that the coordinates of other two points
given satisfy the equation.

(*continued on next page*)

(*continued*)

Alternate solution: Working backward and shifting the graph up two units and left three units to transform the given graph into the graph of $y = \log_2(-x)$, it goes through the points $(-1, 0)$, which is the x-intercept, $(-2, 1)$, and $(-4, 2)$. $2 = \log_2[-(-4)]$, so $a = 2$, and the equation is $f(x) = \log_2(-x+3) - 2$. Verify by checking that the coordinates of the points shown on the graph satisfy the equation.

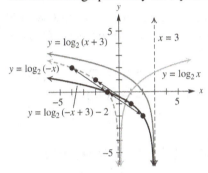

69. The graph has been reflected across the x-axis, so the general form of the equation is $f(x) = -\log_a(x - b) + k$. The vertical asymptote of the original graph is $x = 1$, so the graph has been shifted right one unit and $b = 1$. So the general form of the equation is $f(x) = -\log_a(x - 1) + k$. The base is either 2 or 3, so try $a = 3$. Then substitute the coordinates of a point in the equation and solve for k:

$-1 = -\log_3(4 - 1) + k \Rightarrow$

$-(-1 - k) = \log_3 3 \Rightarrow 3^{1+k} = 3 \Rightarrow$

$3^{1+k} = 3^1 \Rightarrow 1 + k = 1 \Rightarrow k = 0$

So, the equation is $f(x) = -\log_3(x - 1)$.

Verify that the coordinates of other two points given satisfy the equation.

Alternate solution: Working backward and shifting the graph left one unit, it goes through the points $(1, 0)$, which is the x-intercept, $\left(\frac{1}{3}, 1\right)$, and $(3, -1)$. $-1 = -\log_3 3$, so $a = 3$, and the equation is $f(x) = -\log_3(x - 1)$. Verify by checking that the coordinates of the points shown on the graph satisfy the equation.

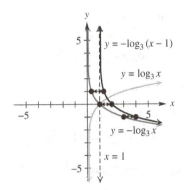

71. $\log_2 \frac{6x}{y} = \log_2 6x - \log_2 y$
$\qquad = \log_2 6 + \log_2 x - \log_2 y$

73. $\log_5 \frac{5\sqrt{7}}{3} = \log_5 5\sqrt{7} - \log_5 3$
$\qquad = \log_5 5 + \log_5 \sqrt{7} - \log_5 3$
$\qquad = 1 + \log_5 7^{1/2} - \log_5 3$
$\qquad = 1 + \frac{1}{2}\log_5 7 - \log_5 3$

75. $\log_4(2x + 5y)$

This is a sum, so none of the logarithm properties apply and this expression cannot be simplified.

77. $\log_2 \sqrt{\frac{5r^3}{z^5}} = \log_2 \left(\frac{5r^3}{z^5}\right)^{1/2} = \frac{1}{2}\log_2 \frac{5r^3}{z^5}$
$\qquad = \frac{1}{2}\left(\log_2 5r^3 - \log_2 z^5\right)$
$\qquad = \frac{1}{2}\left(\log_2 5 + \log_2 r^3 - \log_2 z^5\right)$
$\qquad = \frac{1}{2}\left(\log_2 5 + 3\log_2 r - 5\log_2 z\right)$

79. $\log_2 \frac{ab}{cd} = \log_2(ab) - \log_2(cd)$
$\qquad = \log_2 a + \log_2 b - \left(\log_2 c + \log_2 d\right)$
$\qquad = \log_2 a + \log_2 b - \log_2 c - \log_2 d$

81. $\log_3 \frac{\sqrt{x}\sqrt[3]{y}}{w^2 \sqrt{z}} = \log_3 \left(x^{1/2}y^{1/3}\right) - \log_3 \left(w^2 z^{1/2}\right)$
$\qquad = \log_3 x^{1/2} + \log_3 y^{1/3}$
$\qquad\qquad - \left(\log_3 w^2 + \log_3 z^{1/2}\right)$
$\qquad = \frac{1}{2}\log_3 x + \frac{1}{3}\log_3 y$
$\qquad\qquad - \left(2\log_3 w + \frac{1}{2}\log_3 z\right)$
$\qquad = \frac{1}{2}\log_3 x + \frac{1}{3}\log_3 y$
$\qquad\qquad - 2\log_3 w - \frac{1}{2}\log_3 z$

83. $\log_a x + \log_a y - \log_a m = \log_a xy - \log_a m$
$\qquad\qquad\qquad = \log_a \frac{xy}{m}$

85. $\log_a m - \log_a n - \log_a t$

$= \log_a m - \left(\log_a n + \log_a t\right)$

$= \log_a m - \log_a (nt) = \log_a \frac{m}{nt}$

87. $\frac{1}{3}\log_b x^4 y^5 - \frac{3}{4}\log_b x^2 y$

$= \log_b \left(x^4 y^5\right)^{1/3} - \log_b \left(x^2 y\right)^{3/4}$

$= \log_b \left(x^{4/3} y^{5/3}\right) - \log_b \left(x^{3/2} y^{3/4}\right)$

$= \log_b \dfrac{x^{4/3} y^{5/3}}{x^{3/2} y^{3/4}} = \log_b \left(x^{4/3 - 3/2} y^{5/3 - 3/4}\right)$

$= \log_b \left(x^{-1/6} y^{11/12}\right)$

89. $2\log_a (z+1) + \log_a (3z+2)$

$= \log_a (z+1)^2 + \log_a (3z+2)$

$= \log_a \left[(z+1)^2 (3z+2)\right]$

91. $-\frac{2}{3}\log_5 5m^2 + \frac{1}{2}\log_5 25m^2$

$= \log_5 \left(5m^2\right)^{-2/3} + \log_5 \left(25m^2\right)^{1/2}$

$= \log_5 \left[\left(5m^2\right)^{-2/3} \cdot \left(25m^2\right)^{1/2}\right]$

$= \log_5 \left(5^{-2/3} m^{-4/3} \cdot 5m\right)$

$= \log_5 \left(5^{-2/3} \cdot 5^1 \cdot m^{-4/3} \cdot m^1\right)$

$= \log_5 \left(5^{1/3} \cdot m^{-1/3}\right) = \log_5 \dfrac{5^{1/3}}{m^{1/3}} = \log_5 \sqrt[3]{\dfrac{5}{m}}$

93. $\log_{10} 6 = \log_{10} (2 \cdot 3) = \log_{10} 2 + \log_{10} 3$

$= 0.3010 + 0.4771 = 0.7781$

95. $\log_{10} \frac{3}{2} = \log_{10} 3 - \log_{10} 2$

$= 0.4771 - 0.3010 = 0.1761$

97. $\log_{10} \frac{9}{4} = \log_{10} 9 - \log_{10} 4 = \log_{10} 3^2 - \log_{10} 2^2$

$= 2\log_{10} 3 - 2\log_{10} 2$

$= 2(0.4771) - 2(0.3010)$

$= 0.9542 - 0.6020 = 0.3522$

99. $\log_{10} \sqrt{30} = \log_{10} 30^{1/2} = \frac{1}{2}\log_{10} 30$

$= \frac{1}{2}\log_{10} (10 \cdot 3)$

$= \frac{1}{2}\left(\log_{10} 10 + \log_{10} 3\right)$

$= \frac{1}{2}(1 + 0.4771) = \frac{1}{2}(1.4771)$

≈ 0.7386

101. (a) If the x-values are representing years, 3 months is $\frac{3}{12} = \frac{1}{4} = 0.25$ yr and 6 months is $\frac{6}{12} = \frac{1}{2} = 0.5$ yr.

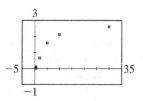

(b) A logarithmic function will model the data best.

103. $f(x) = \log_a x$ and $f(3) = 2$

$2 = \log_a 3 \Rightarrow a^2 = 3 \Rightarrow \left(a^2\right)^{1/2} = 3^{1/2} \Rightarrow a = \sqrt{3}$

(There is no $\pm$ because a must be positive and not equal to 1.) We now have $f(x) = \log_{\sqrt{3}} x$.

(a) $f\left(\frac{1}{9}\right) = \log_{\sqrt{3}} \frac{1}{9} \Rightarrow y = \log_{\sqrt{3}} \frac{1}{9} \Rightarrow$

$\left(\sqrt{3}\right)^y = \frac{1}{9} \Rightarrow \left(3^{1/2}\right)^y = \frac{1}{3^2} \Rightarrow$

$3^{y/2} = 3^{-2} \Rightarrow \frac{y}{2} = -2 \Rightarrow y = -4$

(b) $f(27) = \log_{\sqrt{3}} 27 \Rightarrow y = \log_{\sqrt{3}} 27 \Rightarrow$

$\left(\sqrt{3}\right)^y = 27 \Rightarrow \left(3^{1/2}\right)^y = 3^3 \Rightarrow$

$3^{y/2} = 3^3 \Rightarrow \frac{y}{2} = 3 \Rightarrow y = 6$

(c) $f(9) = \log_{\sqrt{3}} 9 \Rightarrow y = \log_{\sqrt{3}} 9 \Rightarrow$

$\left(\sqrt{3}\right)^y = 9 \Rightarrow \left(3^{1/2}\right)^y = 3^2 \Rightarrow$

$3^{y/2} = 3^2 \Rightarrow \frac{y}{2} = 2 \Rightarrow y = 4$

(d) $f\left(\frac{\sqrt{3}}{3}\right) = \log_{\sqrt{3}} \left(\frac{\sqrt{3}}{3}\right) \Rightarrow y = \log_{\sqrt{3}} \left(\frac{\sqrt{3}}{3}\right) \Rightarrow$

$\left(\sqrt{3}\right)^y = \frac{\sqrt{3}}{3} \Rightarrow \left(3^{1/2}\right)^y = \frac{3^{1/2}}{3^1} \Rightarrow$

$\left(3^{1/2}\right)^y = 3^{1/2-1} = 3^{-1/2} \Rightarrow$

$3^{y/2} = 3^{-1/2} \Rightarrow \frac{y}{2} = -\frac{1}{2} \Rightarrow y = -1$

105. In the formula $A = P\left(1+\frac{r}{n}\right)^{tn}$ we substitute $A = 2P$ because we want the present value to be doubled in the future. Thus, we need to solve for t in the equation $2P = P\left(1+\frac{r}{n}\right)^{tn}$.

$2P = P\left(1+\frac{r}{n}\right)^{tn} \Rightarrow 2 = \left(1+\frac{r}{n}\right)^{tn}$

$2^{\frac{1}{tn}} = \left[\left(1+\frac{r}{n}\right)^{tn}\right]^{\frac{1}{tn}} \Rightarrow 2^{\frac{1}{tn}} = \left(1+\frac{r}{n}\right)$

$\log_2 \left(1+\frac{r}{n}\right) = \frac{1}{tn} \Rightarrow tn = \frac{1}{\log_2 \left(1+\frac{r}{n}\right)}$

$t = \frac{1}{n\log_2 \left(1+\frac{r}{n}\right)} \Rightarrow t = \frac{1}{\log_2 \left(1+\frac{r}{n}\right)^n}$

107. $\log_{10} x = x - 2$

Graph $y_1 = \log x$ and $y_2 = x - 2$ in the window $[-1, 3] \times [-5, 5]$. The x-coordinates of the intersection points will be the solutions to the given equation. There are two points of intersection, hence there are two possible solutions.

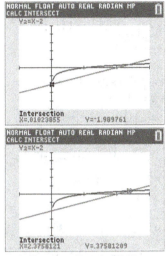

Solution set: $\{0.01, 2.38\}$

109. Prove that $\log_a \frac{x}{y} = \log_a x - \log_a y$.

Let $m = \log_a x$ and $n = \log_a y$. Changing to exponential form we have $a^m = x$ and $a^n = y$. Because $\frac{x}{y} = \frac{a^m}{a^n}$ we have $\frac{x}{y} = a^{m-n}$. Changing to logarithmic form, we have $\log_a \frac{x}{y} = m - n$. Substituting for m and n we have $\log_a \frac{x}{y} = \log_a x - \log_a y$.

Summary Exercises on Inverse, Exponential, and Logarithmic Functions

1. $f(x) = 3x - 4,\ g(x) = \frac{1}{3}x + \frac{4}{3}$

$(f \circ g)(x) = f\left[g(x)\right] = 3\left(\frac{1}{3}x + \frac{4}{3}\right) - 4$
$= x + 4 - 4 = x$
$(g \circ f)(x) = g\left[f(x)\right] = \frac{1}{3}(3x - 4) + \frac{4}{3}$
$= \frac{3x}{3} = x$

Because $(f \circ g)(x) = x$ and $(g \circ f)(x) = x$, these functions are inverses.

3. $f(x) = 1 + \log_2 x,\ g(x) = 2^{x-1}$

$(f \circ g)(x) = f\left[g(x)\right] = f\left(2^{x-1}\right) = 1 + \log_2 2^{x-1}$
$= 1 + (x - 1)\log_2 2 = 1 + x - 1 = x$

$(g \circ f)(x) = g\left[f(x)\right] = g\left(1 + \log_2 x\right)$
$= 2^{(1 + \log_2 x - 1)} = 2^{\log_2 x} = x$

Because $(f \circ g)(x) = x$ and $(g \circ f)(x) = x$, these functions are inverses.

5. Any horizontal line passing through the graph of this function will touch the graph in at most one place, so the function is one-to-one. A sketch of the graph of the inverse function is as follows.

7. A horizontal line passing through the graph of this function will touch the graph in more than one place, so the function is not one-to-one.

9. Because $f(x) = \log_3 (x + 2)$ has a vertical asymptote when $x + 2 = 0 \Rightarrow x = -2$, x-intercept when $x + 2 = 1 \Rightarrow x = -1$, and is increasing, the correct choice is the graph in B.

11. Because $f(x) = \log_2 (5 - x)$ has a vertical asymptote when $5 - x = 0 \Rightarrow x = 5$, x-intercept when $5 - x = 1 \Rightarrow x = 4$. The correct choice is the graph in C.

13. The functions in Exercises 9 and 12 are inverses. The functions in Exercises 10 and 11 are inverses.

15. $f(x) = 3x - 6$

This function is one-to-one.

Step 1: Replace $f(x)$ with y and interchange x and y. $y = 3x - 6 \Rightarrow x = 3y - 6$

Step 2: Solve for y.
$x = 3y - 6 \Rightarrow x + 6 = 3y \Rightarrow \frac{x+6}{3} = y$

Step 3: Replace y with $f^{-1}(x)$:
$f^{-1}(x) = \frac{x+6}{3} = \frac{1}{3}x + 2$

For both $f(x)$ and $f^{-1}(x)$, the domain is $(-\infty, \infty)$ and the range is $(-\infty, \infty)$.

17. $f(x) = 3x^2$

If $x = 1$, $f(1) = 3(1)^2 = 3(1) = 3$. Also if

$x = -1$, $f(-1) = 3(-1)^2 = 3(1) = 3$.

Two different values of x lead to the same value of y, so the function is not one-to-one.

19. $f(x) = \sqrt[3]{5 - x^4}$

If $x = 1$, $f(1) = \sqrt[3]{5 - 1^4} = \sqrt[3]{5 - 1} = \sqrt[3]{4}$. Also if

$x = -1$, $f(-1) = \sqrt[3]{5 - (-1)^4} = \sqrt[3]{5 - 1} = \sqrt[3]{4}$.

Two different values of x lead to the same value of y, the function is not one-to-one.

21. $\left(\frac{1}{10}\right)^{-3} = 1000$ is equivalent to

$\log_{1/10} 1000 = -3$.

23. $\left(\sqrt{3}\right)^4 = 9$ is equivalent to $\log_{\sqrt{3}} 9 = 4$.

25. $2^x = 32$ is equivalent to $\log_2 32 = x$.

27. $3x = 7^{\log_7 6} \Rightarrow 3x = 6 \Rightarrow x = 2$

Solution set: $\{2\}$

29. $x = \log_6 \frac{1}{216} \Rightarrow 6^x = \frac{1}{216} \Rightarrow 6^x = 6^{-3} \Rightarrow$

$x = -3$

Solution set: $\{-3\}$

31. $\log_{10} 0.01 = x$

$\quad 10^x = 0.01 \Rightarrow 10^x = 10^{-2} \Rightarrow x = -2$

Solution set: $\{-2\}$

33. $\log_x 1 = 0 \Rightarrow x^0 = 1$

This is a true statement for all real numbers greater than 0, excluding 1.

Solution set: $(0,1) \cup (1, \infty)$

35. $\log_x \sqrt[3]{5} = \frac{1}{3} \Rightarrow x^{1/3} = \sqrt[3]{5} \Rightarrow x^{1/3} = 5^{1/3} \Rightarrow$

$\left(x^{1/3}\right)^3 = \left(5^{1/3}\right)^3 \Rightarrow x = 5$

Recall from Chapter 1 that it is necessary to check all proposed solutions in the original equation when you raise both sides to a power.

Check $x = 5$.

$\overline{}$

$\log_x \sqrt[3]{5} = \frac{1}{3}$

$\log_5 \sqrt[3]{5} \overset{?}{=} \frac{1}{3}$

$\log_5 5^{1/3} = \frac{1}{3} \Rightarrow \frac{1}{3}\log_5 5 = \frac{1}{3} \Rightarrow \frac{1}{3} \cdot 1 = \frac{1}{3} \Rightarrow \frac{1}{3} = \frac{1}{3}$

This is a true statement.

Solution set: $\{5\}$

37. $\log_{10}\left(\log_2 2^{10}\right) = x \Rightarrow$

$\log_{10}\left(10 \log_2 2\right) = x \Rightarrow \log_{10}\left(10 \cdot 1\right) = x \Rightarrow$

$\log_{10} 10 = x \Rightarrow x = 1$

Solution set: $\{1\}$

38. $x = \log_{4/5}\frac{25}{16} \Rightarrow \left(\frac{4}{5}\right)^x = \frac{25}{16} \Rightarrow \left(\frac{4}{5}\right)^x = \left(\frac{16}{25}\right)^{-1} \Rightarrow$

$\left(\frac{4}{5}\right)^x = \left[\left(\frac{4}{5}\right)^2\right]^{-1} \Rightarrow \left(\frac{4}{5}\right)^x = \left(\frac{4}{5}\right)^{-2} \Rightarrow x = -2$

Solution set: $\{-2\}$

39. $2x - 1 = \log_6 6^x \Rightarrow 2x - 1 = x \Rightarrow$

$\quad -1 = -x \Rightarrow 1 = x$

Solution set: $\{1\}$

41. $2^x = \log_2 16 \Rightarrow 2^{\left(2^x\right)} = 16 \Rightarrow 2^{\left(2^x\right)} = 2^4 \Rightarrow$

$2^x = 4 \Rightarrow 2^x = 2^2 \Rightarrow x = 2$

Solution set: $\{2\}$

43. $\left(\frac{1}{3}\right)^{x+1} = 9^x \Rightarrow \left(3^{-1}\right)^{x+1} = \left(3^2\right)^x \Rightarrow$

$3^{(-1)(x+1)} = 3^{2x} \Rightarrow 3^{-x-1} = 3^{2x} \Rightarrow$

$-x - 1 = 2x \Rightarrow x = -\frac{1}{3}$

Solution set: $\left\{-\frac{1}{3}\right\}$

Section 4.4 Evaluating Logarithms and the Change-of-Base Theorem

1. For $f(x) = a^x$, where $a > 0$, the function is *increasing* over its entire domain.

3. If $f(x) = 5^x$, the rule for $f^{-1}(x)$ is

$f^{-1}(x) = \log_5 x$.

5. A base e logarithm is called a <u>natural</u> logarithm, while a base 10 logarithm is called a <u>common</u> logarithm.

7. $\log_2 0$ is undefined because there is no power of 2 that yields a result of 0. In other words, the equation $2^x = 0$ has no solution.

9. $\log 8 = 0.90308999$

11. $\log 10^{12} = 12$

13. $\log 0.1 = \log 10^{-1} = -1$

15. $\log 63 \approx 1.7993$

17. $\log 0.0022 \approx -2.6576$

19. $\log(387 \times 23) \approx 3.9494$

21. $\log\left(\dfrac{518}{342}\right) \approx 0.1803$

23. $\log 387 + \log 23 \approx 3.9494$

25. $\log 518 - \log 342 \approx 0.1803$

27. $\log(387 \times 23) = \log 387 + \log 23$ by the product property of logarithms. The logarithm of the product of two numbers is equal to the sum of the logarithms of the numbers.

29. Grapefruit, $\left[H_3O^+\right] = 6.3 \times 10^{-4}$

$$\begin{aligned} pH &= -\log\left[H_3O^+\right] = -\log\left(6.3 \times 10^{-4}\right) \\ &= -\left(\log 6.3 + \log 10^{-4}\right) = -(0.7793 - 4) \\ &= -0.7993 + 4 \approx 3.2 \end{aligned}$$

The pH of grapefruit is 3.2.

31. Crackers, $\left[H_3O^+\right] = 3.9 \times 10^{-9}$

$$\begin{aligned} pH &= -\log\left[H_3O^+\right] = -\log\left(3.9 \times 10^{-9}\right) \\ &= -\left(\log 3.9 + \log 10^{-9}\right) = -(0.59106 - 9) \\ &= -(-8.409) \approx 8.4 \end{aligned}$$

The pH of crackers is 8.4.

33. Soda pop, pH = 2.7

$$\begin{aligned} pH &= -\log\left[H_3O^+\right] \\ 2.7 &= -\log\left[H_3O^+\right] \\ -2.7 &= \log\left[H_3O^+\right] \Rightarrow \left[H_3O^+\right] = 10^{-2.7} \end{aligned}$$

Evaluating $10^{-2.7}$ with a calculator gives $\left[H_3O^+\right] \approx 2.0 \times 10^{-3}$

35. Beer, pH = 4.8

$$\begin{aligned} pH &= -\log\left[H_3O^+\right] \\ 4.8 &= -\log\left[H_3O^+\right] \\ -4.8 &= \log\left[H_3O^+\right] \Rightarrow \left[H_3O^+\right] = 10^{-4.8} \end{aligned}$$

Evaluating $10^{-4.8}$ with a calculator gives $\left[H_3O^+\right] \approx 1.6 \times 10^{-5}$

37. Wetland, $\left[H_3O^+\right] = 2.49 \times 10^{-5}$

$$\begin{aligned} pH &= -\log\left[H_3O^+\right] = -\log\left(2.49 \times 10^{-5}\right) \\ &= -\left(\log 2.49 + \log 10^{-5}\right) \\ &= -\log 2.49 - (-5) = -\log 2.49 + 5 \end{aligned}$$

$pH \approx 4.6$

The pH is between 4.0 and 6.0, so it is a poor fen.

39. Wetland, $\left[H_3O^+\right] = 2.49 \times 10^{-2}$

$$\begin{aligned} pH &= -\log\left[H_3O^+\right] = -\log\left(2.49 \times 10^{-2}\right) \\ &= -\left(\log 2.49 + \log 10^{-2}\right) \\ &= -\log 2.49 - (-2) = -\log 2.49 + 2 \end{aligned}$$

$pH \approx 1.6$

The pH is less than 3.0, so it is a bog.

41. Wetland, $\left[H_3O^+\right] = 2.49 \times 10^{-7}$

$$\begin{aligned} pH &= -\log\left[H_3O^+\right] = -\log\left(2.49 \times 10^{-7}\right) \\ &= -\left(\log 2.49 + \log 10^{-7}\right) \\ &= -\log 2.49 - (-7) = -\log 2.49 + 7 \end{aligned}$$

$pH \approx 6.6$

The pH is greater than 6.0, so it is a rich fen.

43. **(a)** $\log 398.4 \approx 2.60031933$

(b) $\log 39.84 \approx 1.60031933$

(c) $\log 3.984 \approx 0.6003193298$

(d) The whole number parts will vary, but the decimal parts will be the same.

45. $\ln e^{1.6} = 1.6$

47. $\ln\left(\dfrac{1}{e^2}\right) = \ln e^{-2} = -2$

49. $\ln \sqrt{e} = \ln e^{1/2} = \dfrac{1}{2}\ln e = \dfrac{1}{2}$

51. $\ln 28 \approx 3.3322$

53. $\ln 0.00013 \approx -8.9480$

55. $\ln(27 \times 943) \approx 10.1449$

57. $\ln\left(\dfrac{98}{13}\right) \approx 2.0200$

59. $\ln 27 + \ln 943 \approx 10.1449$

61. $\ln 98 - \ln 13 \approx 2.0200$

63. $d = 10 \log \frac{I}{I_0}$, where d is the decibel rating.

    (a) $d = 10 \log \frac{100 I_0}{I_0}$

        $= 10 \log_{10} 100 = 10(2) = 20$

    (b) $d = 10 \log \frac{1000 I_0}{I_0}$

        $= 10 \log_{10} 1000 = 10(3) = 30$

    (c) $d = 10 \log \frac{100,000 I_0}{I_0}$

        $= 10 \log_{10} 100,000 = 10(5) = 50$

    (d) $d = 10 \log \frac{1,000,000 I_0}{I_0}$

        $= 10 \log_{10} 1,000,000 = 10(6) = 60$

    (e) $I = 2 I_0$

        $d = 10 \log \frac{2 I_0}{I_0} = 10 \log 2 \approx 3.0103$

    The described rating is increased by about 3 decimals.

65. $r = \log_{10} \frac{I}{I_0}$, where r is the Richter scale rating of an earthquake.

    (a) $r = \log_{10} \frac{1000 I_0}{I_0} = \log_{10} 1000 = 3$

    (b) $r = \log_{10} \frac{1,000,000 I_0}{I_0} = \log_{10} 1,000,000 = 6$

    (c) $r = \log_{10} \frac{100,000,000 I_0}{I_0}$

        $= \log_{10} 100,000,000 = 8$

67. From exercise 65, the magnitude of an earthquake, measured on the Richter scale, is $r = \log_{10} \frac{I}{I_0}$, where I is the amplitude registered on a seismograph 100 km from the epicenter of the earthquake, and I_0 is the amplitude of an earthquake of a certain small size. So, $8.8 = \log_{10} \frac{I}{I_0} \Rightarrow \frac{I}{I_0} = 10^{8.8} \Rightarrow$

$I = 10^{8.8} I_0 \approx 631,000,000 I_0$

69. The year 2016 is represented by $x = 66$.

$f(66) = -273 + 90.6 \ln 66 \approx 106.6$

According to the model, about 106.6 thousand bachelor's degrees in psychology will be earned in the year 2016. To estimate the number of degrees in years beyond 2012, we must assume that the model continues to be logarithmic.

71. If $a = 0.36$, then

$S(n) = a \ln\left(1 + \frac{n}{a}\right) = 0.36 \ln\left(1 + \frac{n}{0.36}\right)$.

    (a) $S(100) = 0.36 \ln\left(1 + \frac{100}{0.36}\right) \approx 2.0269 \approx 2$

    (b) $S(200) = 0.36 \ln\left(1 + \frac{200}{0.36}\right) \approx 2.2758 \approx 2$

    (c) $S(150) = 0.36 \ln\left(1 + \frac{150}{0.36}\right) \approx 2.1725 \approx 2$

    (d) $S(10) = 0.36 \ln\left(1 + \frac{10}{0.36}\right) \approx 1.2095 \approx 1$

73. The index of diversity H for 2 species is given by $H = -\left[P_1 \log_2 P_1 + P_2 \log_2 P_2\right]$. When $P_1 = \frac{50}{100} = 0.5$ and $P_2 = \frac{50}{100} = 0.5$ we have $H = -\left[0.5 \log_2 0.5 + 0.5 \log_2 0.5\right]$. Because $\log_2 0.5 = \log_2 \frac{1}{2} = \log_2 2^{-1} = -1$, we have $H = -\left[0.5(-1) + (-1)\right] = -(-1) = 1$. Thus, the index of diversity is 1.

75. $T(k) = 1.03k \ln \frac{C}{C_0}$

Because $10 \le k \le 16$ and $\frac{C}{C_0} = 2$, the range for $T = 1.03k \ln \frac{C}{C_0}$ will be between $T(10)$ and $T(16)$. Because

$T(10) = 1.03(10) \ln 2 \approx 7.1$ and

$T(16) = 1.03(16) \ln 2 \approx 11.4$, the predicted increased global temperature due to the greenhouse effect from a doubling of the carbon dioxide in the atmosphere is between 7°F and 11°F.

77. $t = \left(1.26 \times 10^9\right) \frac{\ln[1 + 8.33(0.103)]}{\ln 2} \approx 1.126 \times 10^9$

The rock sample is approximately 1.13 billion years old.

For Exercises 79–89, the solutions will be evaluated at the intermediate steps to four decimal places. However, the final answers are obtained without rounding the intermediate steps.

79. $\log_2 5 = \frac{\ln 5}{\ln 2} \approx \frac{1.6094}{0.6931} \approx 2.3219$ or

$\log_2 5 = \frac{\log 5}{\log 2} \approx \frac{0.6990}{0.3010} \approx 2.3219$

81. $\log_8 0.59 = \frac{\log 0.59}{\log 8} \approx \frac{-0.2291}{0.9031} \approx -0.2537$ or

$\log_8 0.59 = \frac{\ln 0.59}{\ln 8} \approx \frac{-0.5276}{2.0794} \approx -0.2537$

83. $\log_{1/2} 3 = \dfrac{\log 3}{\log\left(\frac{1}{2}\right)} \approx \dfrac{0.4771}{-0.3010} \approx -1.5850$ or

$\log_{1/2} 3 = \dfrac{\ln 3}{\ln\left(\frac{1}{2}\right)} \approx \dfrac{1.0986}{-0.6931} \approx -1.5850$

85. $\log_\pi e = \dfrac{\ln e}{\ln \pi} \approx \dfrac{1}{1.1447} \approx 0.8736$ or

$\log_\pi e = \dfrac{\log e}{\log \pi} \approx \dfrac{0.4343}{0.4971} \approx 0.8736$

87. Because $\sqrt{13} = 13^{1/2}$, we have

$\log_{\sqrt{13}} 12 = \dfrac{\ln 12}{\ln \sqrt{13}} = \dfrac{\ln 12}{\frac{1}{2}\ln 13} \approx \dfrac{2.4849}{1.2825} \approx 1.9376$

or

$\log_{\sqrt{13}} 12 = \dfrac{\log 12}{\log \sqrt{13}} = \dfrac{\log 12}{\frac{1}{2}\log 13} \approx \dfrac{1.0792}{0.5570}$

≈ 1.9376

89. $\log_{0.32} 5 = \dfrac{\log 5}{\log 0.32} \approx \dfrac{0.6990}{-0.4949} \approx -1.4125$

or $\log_{0.32} 5 = \dfrac{\ln 5}{\ln 0.32} \approx \dfrac{1.6094}{-1.1394} \approx -1.4125$

91. $\ln\left(b^4 \sqrt{a}\right) = \ln\left(b^4 a^{1/2}\right) = \ln b^4 + \ln a^{1/2}$
$= 4\ln b + \frac{1}{2}\ln a = 4v + \frac{1}{2}u$

93. $\ln\sqrt{\dfrac{a^3}{b^5}} = \ln\left(\dfrac{a^3}{b^5}\right)^{1/2} = \ln\left(\dfrac{a^{3/2}}{b^{5/2}}\right)$

$= \ln a^{3/2} - \ln b^{5/2} = \frac{3}{2}\ln a - \frac{5}{2}\ln b$

$= \frac{3}{2}u - \frac{5}{2}v$

95. $g(x) = e^x$

(a) $g(\ln 4) = e^{\ln 4} = 4$

(b) $g\left[\ln\left(5^2\right)\right] = e^{\ln 5^2} = 5^2$ or 25

(c) $g\left[\ln\left(\dfrac{1}{e}\right)\right] = e^{\ln(1/e)} = \dfrac{1}{e}$

97. $f(x) = \ln x$

(a) $f\left(e^6\right) = \ln e^6 = 6$

(b) $f\left(e^{\ln 3}\right) = \ln e^{\ln 3} = \ln 3$

(c) $f\left(e^{2\ln 3}\right) = \ln e^{2\ln 3} = 2\ln 3$ or $\ln 9$.

99. $2\ln 3x = \ln(3x)^2 = \ln\left(3^2 x^2\right) = \ln 9x^2$

It is equivalent to D.

101. $f(x) = \ln|x|$

The domain of f is all real numbers except 0: $(-\infty, 0)\cup(0, \infty)$ and the range is $(-\infty, \infty)$.

Because $f(-x) = \ln|-x| = \ln|x| = f(x)$, this is an even function and symmetric with respect to the y-axis.

103. $f(x) = \ln\left(e^2 x\right) = \ln e^2 + \ln x = 2 + \ln x$

$f(x) = \ln\left(e^2 x\right)$ is a vertical shift of the graph of $g(x) = \ln x$, 2 units up.

105. $f(x) = \ln\dfrac{x}{e^2} = \ln x - 2\ln e = \ln x - 2$

$f(x) = \ln\dfrac{x}{e^2}$ is a vertical shift of the graph of $g(x) = \ln x$, 2 units down.

Chapter 4 Quiz
(Sections 4.1−4.4)

1. *Step 1*: Replace $f(x)$ with y and interchange x and y: $y = \sqrt[3]{3x - 6} \Rightarrow x = \sqrt[3]{3y - 6}$

Step 2: Solve for y.

$x = \sqrt[3]{3y - 6} \Rightarrow x^3 = 3y - 6 \Rightarrow \dfrac{x^3 + 6}{3} = y$

Step 3: Replace y with $f^{-1}(x)$.

$f^{-1}(x) = \dfrac{x^3 + 6}{3}$

3.

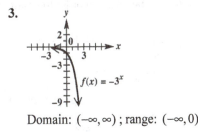

$f(x) = -3^x$

Domain: $(-\infty, \infty)$; range: $(-\infty, 0)$

5. Use the compound interest formula to find the future value (the amount in the account).

$$A = P\left(1+\frac{r}{n}\right)^{tn}, \ P = 15,000, \ r = 0.027, \ t = 8$$

(a) $n = 1$

$$A = 15,000\left(1+\frac{0.027}{1}\right)^{8\cdot1}$$
$$= (15,000)(1.027)^8 \approx \$18,563.28$$

There will be about $18,563.28 after one year.

(b) $n = 4$

$$A = 15,000\left(1+\frac{0.027}{4}\right)^{8\cdot4}$$
$$= (15,000)(1.00675)^{32} \approx \$18,603.03$$

There will be about $18,603.03 after one year.

(c) $n = 12$

$$A = 15,000\left(1+\frac{0.027}{12}\right)^{8\cdot12}$$
$$= (15,000)(1.00225)^{96} \approx \$18,612.02$$

There will be about $19,838.86 after one year.

(d) $n = 365$

$$A = 15,000\left(1+\frac{0.027}{365}\right)^{8\cdot365} \approx \$18,616.39$$

There will be about $18,616.39 after one year.

7. The expression $\log_6 25$ represents the exponent to which 6 must be raised in order to obtain 25.

9. $\log_3 \dfrac{\sqrt{x}\cdot y}{pq^4} = \log_3\left(\sqrt{x}\cdot y\right) - \log_3\left(pq^4\right)$
$$= \log_3\left(x^{1/2}\right) + \log_3 y$$
$$\quad - \left(\log_3 p + \log_3 q^4\right)$$
$$= \tfrac{1}{2}\log_3 x + \log_3 y$$
$$\quad - \log_3 p - 4\log_3 q$$

11. $\log_3 40 = \dfrac{\log 40}{\log 3} \approx \dfrac{1.6021}{0.4771} \approx 3.3578$

We could also have used the natural logarithm.
$$\log_3 40 = \dfrac{\ln 40}{\ln 3} \approx \dfrac{3.689}{1.0986} \approx 3.3578$$

Section 4.5 Exponential and Logarithmic Equations

1. B. By taking the common logarithm on each side, we have
$$10^x = 150$$
$$\log 10^x = \log 150$$
$$x\log 10 = \log 150$$
$$x = \log 150$$

3. E. By changing the equation to exponential form, we have
$$\log_4\left(x^2 - 10\right) = 2$$
$$x^2 - 10 = 4^2$$
$$x^2 - 10 = 16$$
Solve the equation using the steps for solving a quadratic equation.

5. D. By letting $u = e^x$ and writing the equation in quadratic form, we have
$$2e^{2x} - 5e^x - 3 = 0$$
$$2u^2 - 5u - 3 = 0$$
Solve the equation using the steps for solving a quadratic equation. Then substitute $e^x = u$ in each solution and solve for x.

7. Because x is the exponent to which 7 must be raised in order to obtain 19, the solution is
$$\log_7 19 \text{ or } \dfrac{\log 19}{\log 7} \text{ or } \dfrac{\ln 19}{\ln 7}.$$

9. Because x is the exponent to which $\frac{1}{2}$ must be raised in order to obtain 12, the solution is
$$\log_{1/2} 12 \text{ or } \dfrac{\log 12}{\log\left(\frac{1}{2}\right)} \text{ or } \dfrac{\ln 12}{\ln\left(\frac{1}{2}\right)}.$$

11. $3^x = 7$
$$\ln 3^x = \ln 7$$
$$x\ln 3 = \ln 7$$
$$x = \dfrac{\ln 7}{\ln 3} \approx 1.771$$
Solution set: $\{1.771\}$

13. $\left(\dfrac{1}{2}\right)^x = 5 \Rightarrow \ln\left(\dfrac{1}{2}\right)^x = \ln 5 \Rightarrow x\ln\dfrac{1}{2} = \ln 5$
$$x = \dfrac{\ln 5}{\ln\frac{1}{2}} \approx -2.322$$
Solution set: $\{-2.322\}$

15. $0.8^x = 4 \Rightarrow \ln\left(0.8^x\right) = \ln 4 \Rightarrow x\ln 0.8 = \ln 4 \Rightarrow$

$x = \dfrac{\ln 4}{\ln 0.8} \approx -6.213$

Solution set: $\{-6.213\}$

17. $\qquad 4^{x-1} = 3^{2x} \Rightarrow \ln\left(4^{x-1}\right) = \ln\left(3^{2x}\right) \Rightarrow$

$(x-1)\ln 4 = 2x\ln 3$

$x\ln 4 - \ln 4 = 2x\ln 3$

$x\ln 4 - 2x\ln 3 = \ln 4 \Rightarrow x(\ln 4 - 2\ln 3) = \ln 4 \Rightarrow$

$x = \dfrac{\ln 4}{\ln 4 - 2\ln 3} \approx -1.710$

Solution set: $\{-1.710\}$

19. $\qquad 6^{x+1} = 4^{2x-1} \Rightarrow \ln\left(6^{x+1}\right) = \ln\left(4^{2x-1}\right)$

$(x+1)\ln 6 = (2x-1)\ln 4$

$x\ln 6 + \ln 6 = 2x\ln 4 - \ln 4$

$\ln 6 + \ln 4 = 2x\ln 4 - x\ln 6$

$\ln 6 + \ln 4 = x(2\ln 4 - \ln 6)$

$x = \dfrac{\ln 6 + \ln 4}{2\ln 4 - \ln 6} \approx 3.240$

Solution set: $\{3.240\}$

21. $e^{x^2} = 100 \Rightarrow \ln\left(e^{x^2}\right) = \ln 100$

$x^2 = \ln 100 \Rightarrow x = \pm\sqrt{\ln 100} = \pm 2.146$

Solution set: $\left\{\pm 2.146\right\}$

23. $e^{3x-7} \cdot e^{-2x} = 4e \Rightarrow e^{x-7} = 4e$

$\ln\left(e^{x-7}\right) = \ln(4e) \Rightarrow (x-7)\ln e = \ln 4 + \ln e$

$x - 7 = \ln 4 + 1 \Rightarrow x = \ln 4 + 8 \approx 9.386$

Solution set: $\{9.386\}$

25. $\left(\frac{1}{3}\right)^x = -3$ has no solution because $\frac{1}{3}$ raised to any power is positive.

Solution set: $\varnothing$

27. $0.05(1.15)^x = 5 \Rightarrow 1.15^x = \dfrac{5}{0.05} = 100 \Rightarrow$

$\log\left(1.15^x\right) = \log 100 \Rightarrow x\log 1.15 = \log 100 \Rightarrow$

$x = \dfrac{\log 100}{\log 1.15} = \dfrac{2}{\log 1.15} \approx 32.950$

Solution set: $\{32.950\}$

29. $3(2)^{x-2} + 1 = 100 \Rightarrow 3(2)^{x-2} = 99 \Rightarrow$

$2^{x-2} = 33 \Rightarrow \ln\left(2^{x-2}\right) = \ln 33 \Rightarrow$

$(x-2)\ln 2 = \ln 33 \Rightarrow x = \dfrac{\ln 33}{\ln 2} + 2 \approx 7.044$

Solution set: $\{7.044\}$

31. $2(1.05)^x + 3 = 10$

$2(1.05)^x = 7$

$1.05^x = 3.5$

$\ln\left(1.05^x\right) = \ln 3.5$

$x\ln 1.05 = \ln 3.5$

$x = \dfrac{\ln 3.5}{\ln 1.05} \approx 25.677$

Solution set: $\{25.677\}$

33. $\qquad 5(1.015)^{x-1980} = 8$

$1.015^{(x-1980)} = 1.6$

$\ln\left(1.015^{(x-1980)}\right) = \ln 1.6$

$(x-1980)\ln 1.015 = \ln 1.6$

$x = \dfrac{\ln 1.6}{\ln 1.015} + 1980$

≈ 2011.568

Solution set: $\{2011.568\}$

35. $e^{2x} - 6e^x + 8 = 0$

Let $u = e^x$.

$u^2 - 6u + 8 = 0$

$(u-4)(u-2) = 0$

$u - 4 = 0$	$u - 2 = 0$
$u = 4$	$u = 2$
$e^x = 4$	$e^x = 2$
$\ln e^x = \ln 4$	$\ln e^x = \ln 2$
$x\ln e = \ln 4$	$x\ln e = \ln 2$
$x = \ln 4$	$x = \ln 2$

Solution set: $\{\ln 2,\ \ln 4\}$

37. $2e^{2x} + e^x = 6 \Rightarrow 2e^{2x} + e^x - 6 = 0$

Let $u = e^x$.

$2u^2 + u - 6 = 0$

$(2u-3)(u+2) = 0$

$2u - 3 = 0$	$u + 2 = 0$
$2u = 3$	$u = -2$
$u = \frac{3}{2}$	$e^x = -2$
$e^x = \frac{3}{2}$	Disregard this value
$\ln e^x = \ln\frac{3}{2}$	because e^x is always
$x\ln e = \ln\frac{3}{2}$	positive.
$x = \ln\frac{3}{2}$	

Solution set: $\left\{\ln\frac{3}{2}\right\}$

39. $5^{2x} + 3(5^x) = 28 \Rightarrow 5^{2x} + 3(5^x) - 28 = 0$

Let $u = 5^x$.

$u^2 + 3u - 28 = 0$

$(u - 4)(u + 7) = 0$

$\begin{array}{l|l} u - 4 = 0 & u + 7 = 0 \\ u = 4 & u = -7 \\ 5^x = 4 & 5^x = -7 \\ \log 5^x = \log 4 & \text{Disregard this} \\ x \log 5 = \log 4 & \text{value because } 5^x \\ x = \dfrac{\log 4}{\log 5} = \log_5 4 & \text{is always positive.} \end{array}$

Solution set: $\{\log_5 4\}$

41. $5 \ln x = 10 \Rightarrow \ln x = 2 \Rightarrow e^2 = x$

Solution set: $\{e^2\}$

43. $\ln(4x) = 1.5$

$4x = e^{1.5}$

$x = \dfrac{e^{1.5}}{4}$

Solution set: $\left\{\dfrac{e^{1.5}}{4}\right\}$

45. $\log(2 - x) = 0.5$

$2 - x = 10^{0.5}$

$2 - x = \sqrt{10} \Rightarrow x = 2 - \sqrt{10}$

Solution set: $\left\{2 - \sqrt{10}\right\}$

47. $\log_6(2x + 4) = 2$

$2x + 4 = 6^2$

$2x + 4 = 36$

$2x = 32 \Rightarrow x = 16$

Solution set: $\{16\}$

49. $\log_4\left(x^3 + 37\right) = 3 \Rightarrow x^3 + 37 = 4^3 \Rightarrow$

$x^3 = 27 \Rightarrow x = 3$

Solution set: $\{3\}$

51. $\ln x + \ln x^2 = 3$

$\ln\left(x \cdot x^2\right) = 3$

$\ln x^3 = 3$

$x^3 = e^3 \Rightarrow x = e$

Solution set: $\{e\}$

53. $\log_3\left[(x + 5)(x - 3)\right] = 2$

$(x + 5)(x - 3) = 3^2$

$x^2 + 2x - 15 = 9$

$x^2 + 2x - 24 = 0$

$(x + 6)(x - 4) = 0$

$\begin{array}{l|l} x + 6 = 0 & x - 4 = 0 \\ x = -6 & x = 4 \end{array}$

Solution set: $\{-6, 4\}$

55. $\log_2\left[(2x + 8)(x + 4)\right] = 5$

$(2x + 8)(x + 4) = 2^5$

$2x^2 + 16x + 32 = 32$

$2x^2 + 16x = 0$

$2x(x + 8) = 0$

$\begin{array}{l|l} 2x = 0 & x + 8 = 0 \\ x = 0 & x = -8 \end{array}$

Solution set: $\{-8, 0\}$

57. $\log x + \log(x + 15) = 2$

$\log\left[x(x + 15)\right] = 2$

$x^2 + 15x = 10^2 = 100$

$x^2 + 15x - 100 = 0$

$(x - 5)(x + 20) = 0$

$\begin{array}{l|l} x - 5 = 0 & x + 20 = 0 \\ x = 5 & x = -20 \end{array}$

Because $\log(-20)$ is undefined, we reject this proposed solution.

Solution set: $\{5\}$

59. $\log(x + 25) = \log(x + 10) + \log 4$

$\log(x + 25) = \log\left[4(x + 10)\right]$

$\log(x + 25) = \log(4x + 40)$

$x + 25 = 4x + 40$

$-3x = 15 \Rightarrow x = -5$

Solution set: $\{-5\}$

61. $\log(x - 10) - \log(x - 6) = \log 2$

$\log\left(\dfrac{x - 10}{x - 6}\right) = \log 2$

$\dfrac{x - 10}{x - 6} = 2$

$x - 10 = 2(x - 6)$

$x - 10 = 2x - 12$

$x = 2$

When the proposed solution is substituted for x in $\log(x - 10)$, a negative argument results. This is not allowed, so we reject this proposed solution.

Solution set: $\varnothing$

63. $\ln(7-x)+\ln(1-x)=\ln(25-x)$
$$\ln\left[(7-x)(1-x)\right]=\ln(25-x)$$
$$7-8x+x^2=25-x$$
$$x^2-7x-18=0$$
$$(x-9)(x+2)=0 \Rightarrow x=9, \ x=-2$$

If 9 is substituted for x in $\ln(7-x)$, the argument becomes -2. This is not allowed, so reject this proposed solution.
Solution set: $\{-2\}$

65. $\log_8(x+2)+\log_8(x+4)=\log_8 8$
$$\log_8\left[(x+2)(x+4)\right]=\log_8 8$$
$$x^2+6x+8=8$$
$$x^2+6x=0$$
$$x(x+6)=0 \Rightarrow x=0, \ -6$$

If -6 is substituted for x in $\log_8(x+2)$, the argument becomes -4. This is not allowed, so we reject this proposed solution.
Solution set: $\{0\}$

67. $\log_2(x^2-100)-\log_2(x+10)=1$
$$\log_2\left(\frac{x^2-100}{x+10}\right)=1$$
$$\frac{x^2-100}{x+10}=2^1=2$$
$$x^2-100=2x+20$$
$$x^2-2x-120=0$$
$$(x+10)(x-12)=0$$
$$x=-10, \ 12$$

If -10 is substituted for x in $\log_2(x+10)$, the argument becomes 0. This is not allowed, so we reject this proposed solution.
Solution set: $\{12\}$

69. $\log x+\log(x-21)=\log 100 \Rightarrow$
$$\log\left[x(x-21)\right]=2 \Rightarrow \log\left(x^2-21x\right)=2 \Rightarrow$$
$$x^2-21x=10^2 \Rightarrow x^2-21x-100=0 \Rightarrow$$
$$(x-25)(x+4)=0 \Rightarrow x=25 \text{ or } x=-4$$

The negative solution $(x=-4)$ is not in the domain of $\log x$, so it must be discarded.
Solution set: $\{25\}$

71. $$\log(9x+5)=3+\log(x+2)$$
$$\log(9x+5)-\log(x+2)=3$$
$$\log\frac{9x+5}{x+2}=3$$
$$\frac{9x+5}{x+2}=10^3=1000$$
$$9x+5=1000x+2000$$
$$-991x=1995$$
$$x=-\frac{1995}{991}\approx-2.013$$

If $-\frac{1995}{991}$ is substituted for x in $\log_2(x+2)$, the argument is negative. This is not allowed, so we reject this proposed solution.
Solution set: $\varnothing$

73. $\ln(4x-2)-\ln 4=-\ln(x-2)$
$$\ln\frac{4x-2}{4}=-\ln(x-2) \Rightarrow \frac{4x-2}{4}=\frac{1}{x-2}$$
$$(4x-2)(x-2)=4 \Rightarrow 4x^2-10x+4=4$$
$$4x^2-10x=0 \Rightarrow 2x(2x-5)=0 \Rightarrow$$
$$2x=0 \Rightarrow x=0 \text{ or } 2x-5=0 \Rightarrow x=\frac{5}{2}=2.5$$

Because $x=0$ is not in the domain of $\ln(x-2)$, it must be discarded.
Solution set: $\{2.5\}$

75. $\log_5(x+2)+\log_5(x-2)=1$
$$\log_5\left[(x+2)(x-2)\right]=1 \Rightarrow x^2-4=5^1$$
$$x^2-9=0$$
$$(x-3)(x+3)=0 \Rightarrow x=\pm 3$$

-3 is not in the domain, so reject it.
Solution set: $\{3\}$

77. $\log_2(2x-3)+\log_2(x+1)=1$
$$\log_2\left[(2x-3)(x+1)\right]=1$$
$$2x^2-x-3=2^1$$
$$2x^2-x-5=0$$
$$x=\frac{-(-1)\pm\sqrt{(-1)^2-4(2)(-5)}}{2(2)}=\frac{1\pm\sqrt{41}}{4}$$

Because the negative solution $\left(x=\frac{1-\sqrt{41}}{4}\right)$ is not in the domain of $\log(x+1)$, it must be discarded.
Solution set: $\left\{\frac{1+\sqrt{41}}{4}\right\}$

79. $\ln e^x-2\ln e=\ln e^4 \Rightarrow x-2=4 \Rightarrow x=6$
Solution set: $\{6\}$

81. $\log_2(\log_2 x)=1 \Rightarrow \log_2 x=2^1 \Rightarrow$
$$\log_2 x=2 \Rightarrow x=2^2 \Rightarrow x=4$$
Solution set: $\{4\}$

83. $\log x^2 = (\log x)^2 \Rightarrow 2\log x = (\log x)^2 \Rightarrow$

$(\log x)^2 - 2\log x = 0 \Rightarrow \log x(\log x - 2) = 0$

$\log_{10} x = 0$ or $\log_{10} x - 2 = 0$

$\quad x = 10^0 \qquad\qquad \log_{10} x = 2$

$\quad x = 1 \qquad\qquad\quad x = 10^2 = 100$

Solution set: $\{1, 100\}$

85. Answers will vary. Sample answer: Do not immediately reject a negative answer when solving equations involving logarithms. Examine what happens to an answer when substituted back into the original equation. An answer (whether negative, positive, or zero) that causes a nonpositive value in the argument of the logarithm must be rejected, regardless of its sign.

87. $p = a + \dfrac{k}{\ln x}$, for x

$p - a = \dfrac{k}{\ln x} \Rightarrow (\ln x)(p - a) = k \Rightarrow$

$\ln x = \dfrac{k}{p - a} \Rightarrow x = e^{k/(p-a)}$

89. $T = T_0 + (T_1 - T_0)10^{-kt}$, for t

$T - T_0 = (T_1 - T_0)10^{-kt}$

$\dfrac{T - T_0}{T_1 - T_0} = 10^{-kt}$

$\log_{10}\left(\dfrac{T - T_0}{T_1 - T_0}\right) = -kt$

$\dfrac{\log\left(\frac{T-T_0}{T_1-T_0}\right)}{-k} = \dfrac{-kt}{-k}$

$\quad t = -\dfrac{1}{k}\log\left(\dfrac{T - T_0}{T_1 - T_0}\right)$

91. $I = \dfrac{E}{R}\left(1 - e^{-Rt/2}\right)$, for t

$RI = R\left[\dfrac{E}{R}\left(1 - e^{-Rt/2}\right)\right]$

$RI = E\left(1 - e^{-Rt/2}\right)$

$\dfrac{RI}{E} = 1 - e^{-Rt/2}$

$\dfrac{RI}{E} - 1 = -e^{-Rt/2}$

$1 - \dfrac{RI}{E} = e^{-Rt/2}$

$\ln\left(1 - \dfrac{RI}{E}\right) = \ln e^{-Rt/2}$

$-\dfrac{Rt}{2} = \ln\left(1 - \dfrac{RI}{E}\right)$

$-\dfrac{2}{R}\left(-\dfrac{Rt}{2}\right) = -\dfrac{2}{R}\ln\left(1 - \dfrac{RI}{E}\right)$

$t = -\dfrac{2}{R}\ln\left(1 - \dfrac{RI}{E}\right)$

93. $y = A + B\left(1 - e^{-Cx}\right)$, for x

$y - A = B\left(1 - e^{-Cx}\right)$

$\dfrac{y - A}{B} = 1 - e^{-Cx}$

$\dfrac{y - A}{B} - 1 = \dfrac{y - A - B}{B} = -e^{-Cx}$

$\dfrac{A + B - y}{B} = e^{-Cx}$

$\ln\left(\dfrac{A + B - y}{B}\right) = -Cx$

$\dfrac{\ln\left(\dfrac{A + B - y}{B}\right)}{-C} = x$

95. $\log A = \log B - C\log x$, for A

$\log A = \log B - C\log x$

$\log A = \log\dfrac{B}{x^C}$

$A = \dfrac{B}{x^C}$

97. $A = P\left(1 + \dfrac{r}{n}\right)^{tn}$, for t

$\dfrac{A}{P} = \left(1 + \dfrac{r}{n}\right)^{tn}$

$\log\left(\dfrac{A}{P}\right) = nt\log\left(1 + \dfrac{r}{n}\right)$

$\dfrac{\log\left(\frac{A}{P}\right)}{n\log\left(1 + \dfrac{r}{n}\right)} = t$

99. $A = P\left(1 + \dfrac{r}{n}\right)^{tn}$

To solve for A, substitute $P = 10{,}000$, $r = 0.03$, $n = 4$, and $t = 5$.

$A = 10{,}000\left(1 + \dfrac{0.03}{4}\right)^{(4)(5)}$

$A = 10{,}000\left(1.0075\right)^{20} \approx 11{,}611.84$

There will be $11,611.84 in the account.

101. $A = P\left(1+\dfrac{r}{n}\right)^{tn}$

To solve for t, substitute $A = 30,000$, $P = 27,000$, $r = 0.04$, and $n = 4$.

$$30,000 = 27,000\left(1+\dfrac{0.04}{4}\right)^{t(4)}$$

$$\dfrac{30,000}{27,000} = (1+0.01)^{4t} \Rightarrow \dfrac{10}{9} = 1.01^{4t} \Rightarrow$$

$$\ln\dfrac{10}{9} = \ln\left(1.01^{4t}\right) \Rightarrow \ln\dfrac{10}{9} = 4t\ln 1.01 \Rightarrow$$

$$t = \dfrac{\ln\frac{10}{9}}{4\ln 1.01} \approx 2.6$$

To the nearest tenth of a year, Tom will be ready to buy a truck in 2.6 yr.

103. $A = P\left(1+\dfrac{r}{n}\right)^{tn}$

To solve for r, substitute $A = 2500$, $P = 2000$, $t = 8.5$, and $n = 2$.

$$2500 = 2000\left(1+\dfrac{r}{2}\right)^{(8.5)(2)} \Rightarrow$$

$$1.25 = \left(1+\dfrac{r}{2}\right)^{17} \Rightarrow \sqrt[17]{1.25} = 1+\dfrac{r}{2}$$

$$\sqrt[17]{1.25} - 1 = \dfrac{r}{2} \Rightarrow r = 2\left(\sqrt[17]{1.25} - 1\right) \approx 0.0264$$

The interest rate is about 2.64%.

105. **(a)** $f(3000) = 86.3\ln 3000 - 680 \approx 10.9$

At 3000 ft, about 10.9% of the moisture falls as snow.

(b) $f(4000) = 86.3\ln 4000 - 680 \approx 35.8$

At 4000 ft, about 35.8% of the moisture falls as snow.

(c) $f(7000) = 86.3\ln 7000 - 680 \approx 84.1$

At 7000 ft, about 84.1% of the moisture falls as snow.

107. Double the 2006 value is $2(12,837) = 25,674$.

$$f(x) = 13,017(1.05)^x$$

$$25,674 = 13,017(1.05)^x \Rightarrow \dfrac{25,674}{13,017} = 1.05^x \Rightarrow$$

$$\ln\dfrac{25,674}{13,017} = \ln 1.05^x \Rightarrow \ln\dfrac{25,674}{13,017} = x\ln 1.05$$

$$x = \dfrac{\ln\frac{25,674}{13,017}}{\ln 1.05} \approx 13.9$$

The cost of a year's tuition, room and board, and fees at a public university will be double the cost in 2006 about 13.9 years later or in 2019.

109. $f(x) = \dfrac{67.21}{1+1.081e^{-x/24.71}}$

(a) The year 2014 is represented by $x = 64$.

$$f(64) = \dfrac{67.21}{1+1.081e^{-64/24.71}} \approx 62.2$$

In 2014, about 62% of U.S. women were in the civilian labor force.

(b) $\dfrac{67.21}{1+1.081e^{-x/24.71}} = 55$

$$67.21 = 55\left(1+1.081e^{-x/24.71}\right)$$

$$67.21 = 55 + 59.455e^{-x/24.71}$$

$$12.21 = 59.455e^{-x/24.71}$$

$$\dfrac{12.21}{59.455} = e^{-x/24.71}$$

$$\ln\left(\dfrac{12.21}{59.455}\right) = \ln\left(e^{-x/24.71}\right)$$

$$\ln\left(\dfrac{12.21}{59.455}\right) = -\dfrac{x}{24.71}$$

$$-24.71\ln\left(\dfrac{12.21}{59.455}\right) \approx 39.1 = x$$

55% of U.S. women were in the civilian labor force about 39.1 years after 1950 or in 1989.

111. **(a)** Change this equation to exponential form, then isolate P.

$$\ln(1-P) = -0.0034 - 0.0053x$$

$$1-P = e^{-0.0034-0.0053x}$$

$$P(x) = 1 - e^{-0.0034-0.0053x}$$

(b) $P(x) = 1 - e^{-0.0034 - 0.0053x}$

From the graph one can see that initially there is a rapid reduction of carbon dioxide emissions. However, after a while there is little benefit in raising taxes further.

(c) $P(60) = 1 - e^{-0.00340-0.0053(60)}$

$$\approx 0.275 \text{ or } 27.5\%$$

The reduction in carbon emissions from a tax of $60 per ton of carbon is 27.5%.

(d) We must determine x when $P = 0.5$.

$$0.5 = 1 - e^{-0.0034 - 0.0053x}$$
$$0.5 - 1 = -e^{-0.0034 - 0.0053x}$$
$$-0.5 = -e^{-0.0034 - 0.0053x}$$
$$0.5 = e^{-0.0034 - 0.0053x}$$
$$\ln 0.5 = -0.0034 - 0.0053x$$
$$\ln 0.5 + 0.0034 = -0.0053x$$
$$x = \frac{\ln 0.5 + 0.0034}{-0.0053} \approx 130.14$$

The value $T = \$130.14$ will give a 50% reduction in carbon emissions.

113. *Step 1*: Replace $f(x)$ with y and interchange x and y.

$$y = e^{x-5} \Rightarrow x = e^{y-5}$$

Step 2: Solve for y.

$$x = e^{y-5} \Rightarrow \ln x = \ln e^{y-5} \Rightarrow$$
$$\ln x = y - 5 \Rightarrow \ln x + 5 = y$$

Step 3: Replace y with $f^{-1}(x)$.

$$f^{-1}(x) = \ln x + 5$$

Domain: $(0, \infty)$; range: $(-\infty, \infty)$

115. *Step 1*: Replace $f(x)$ with y and interchange x and y.

$$y = e^{x+1} - 4 \Rightarrow x = e^{y+1} - 4$$

Step 2: Solve for y.

$$x = e^{y+1} - 4 \Rightarrow x + 4 = e^{y+1}$$
$$\ln(x+4) = y + 1 \Rightarrow \ln(x+4) - 1 = y$$

Step 3: Replace y with $f^{-1}(x)$.

$$f^{-1}(x) = \ln(x+4) - 1$$

Domain: $(-4, \infty)$; range: $(-\infty, \infty)$

117. *Step 1*: Replace $f(x)$ with y and interchange x and y.

$$y = 2\ln 3x \Rightarrow x = 2\ln 3y$$

Step 2: Solve for y.

$$x = 2\ln 3y \Rightarrow \frac{x}{2} = \ln 3y \Rightarrow 3y = e^{x/2}$$
$$y = \tfrac{1}{3}e^{x/2}$$

Step 3: Replace y with $f^{-1}(x)$.

$$f^{-1}(x) = \tfrac{1}{3}e^{x/2}$$

Domain: $(-\infty, \infty)$; range: $(0, \infty)$

119. $e^x + \ln x = 5$

Graph $y_1 = e^x + \ln x$ and $y_2 = 5$ in the window $[-3, 3] \times [-1, 6]$, then find the intersection.

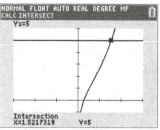

The two graphs intersect at approximately $(1.52, 5)$. The x-coordinate of this point is the solution of the equation.
Solution set: $\{1.52\}$

121. $2e^x + 1 = 3e^{-x}$

Graph $y_1 = 2e^x + 1$ and $y_2 = 3e^{-x}$ in the window $[-3, 3] \times [-1, 6]$ and then find the intersection of the two graphs.

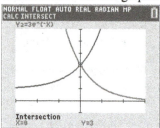

The two curves intersect at the point $(0, 3)$. The x-coordinate of this point is the solution of the equation. Solution set: $\{0\}$

123. $\log x = x^2 - 8x + 14$

Graph $y_1 = \log x$ and $y_2 = x^2 - 8x + 14$ in the window $[-1, 7] \times [-3, 3]$ and then find the intersections of the two graphs.

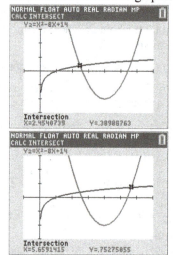

The two graphs intersect at approximately $(2.45, 0.39)$ and $(5.66, 0.75)$. The x-coordinates of these points represent the solution of the equation.
Solution set: $\{2.45, 5.66\}$

125. In the last line, $2 < 1$, the sign should have flipped because the value of $\log \frac{1}{3} \approx -0.4771$ is negative and whenever you divide (or multiply) both sides of an inequality by a negative value, the sign flips.

Section 4.6   Applications and Models of Exponential Growth and Decay

1. B. The equation $2e^{0.02x} = 6$ represents an initial amount of 2 and a final amount 6 (triple 2).

3. C. The equation $y = 2e^{0.02(3)}$ represents the amount after 3 yr.

5. B. The equation $y = 200e^{-0.0241(10)}$ represents the amount os Strontium-90 that will remain after 10 years.

7. C. Use the equation $\frac{1}{2}y_0 = y_0 e^{-0.0241t}$ to find the half-life of Strontium-90.

For exercises 9–13, use the function for exponential growth or decay, $y = y_0 e^{kt}$.

9. $20 = 60e^{3k} \Rightarrow \frac{1}{3} = e^{3k} \Rightarrow \ln \frac{1}{3} = 3k \Rightarrow$
$k = \frac{1}{3}\ln\frac{1}{3}$

11. Because the half-life is 100 days, after 100 days there are 5 mg remaining.
$5 = 10e^{100k} \Rightarrow \frac{1}{2} = e^{100k} \Rightarrow \ln\frac{1}{2} = 100k \Rightarrow$
$k = \frac{1}{100}\ln\frac{1}{2}$

13. $0.6 = 2.4e^{2k} \Rightarrow \frac{1}{4} = e^{2k} \Rightarrow \ln\frac{1}{4} = 2k \Rightarrow$
$k = \frac{1}{2}\ln\frac{1}{4}$

15. $A(t) = 500e^{-0.032t}$

   (a) $t = 4 \Rightarrow A(4) = 500e^{-0.032(4)} \approx 440$
      After 4 years, about 440 g remain.

   (b) $t = 8 \Rightarrow A(8) = 500e^{-0.032(8)} \approx 387$
      After 8 years, about 387 g remain.

   (c) $t = 20 \Rightarrow A(20) = 500e^{-0.032(20)} \approx 264$
      After 20 years, about 264 g remain.

   (d) Find t when $A(t) = 250$.
$$250 = 500e^{-0.032t}$$
$$0.5 = e^{-0.032t}$$
$$\ln 0.5 = \ln e^{-0.032t} \Rightarrow \ln 0.5 = -0.032t$$
$$t = \frac{\ln 0.5}{-0.032} \approx 21.66$$
      The half-life is about 22 years.

17. $A(t) = A_0 e^{-0.00043t}$

Find t when $A(t) = \frac{1}{2}A_0$.

$$\frac{1}{2}A_0 = A_0 e^{-0.00043t} \Rightarrow \frac{1}{2} = e^{-0.00043t} \Rightarrow$$
$$\ln\frac{1}{2} = \ln e^{-0.00043t} \Rightarrow \ln\frac{1}{2} = -0.00043t \Rightarrow$$
$$t = \frac{\ln\frac{1}{2}}{-0.00043} \approx 1611.97$$
The half-life is about 1600 years.

19. First use the given values to find k.
$$A_0 = 12,\ t = 4,\ A(4) = 6$$
$$A(t) = A_0 e^{kt} \Rightarrow 6 = 12e^{4k} \Rightarrow 0.5 = e^{4k} \Rightarrow$$
$$\ln 0.5 = 4k \Rightarrow k = \frac{\ln 0.5}{4} \approx -0.173$$
The exponential decay equation is
$$A(t) = 12e^{-0.173t}$$
To find the amount present after 7 years, let $t = 7$.
$$A(7) = 12e^{-0.173(7)} \approx 3.57$$
After 7 years, about 3.6 g of the substance will be present.

21. $A(t) = A_0 e^{-0.087t}$

Find t when $A(t) = 0.25A_0$.

$$0.25A_0 = A_0 e^{-0.087t}$$
$$0.25 = e^{-0.087t}$$
$$\ln 0.25 = \ln e^{-0.087t} \Rightarrow \ln 0.25 = -0.087t$$
$$t = \frac{\ln 0.25}{-0.087} \approx 15.93$$
It will take about 16 days to decay to 25% of the initial amount.

23. From Example 5, we have the amount of radiocarbon present after t years is given by $y = y_0 e^{-0.0001216t}$, where y_0 is the initial amount present. Letting $y = \frac{1}{3} y_0$, solve for t.

$$\frac{1}{3} y_0 = y_0 e^{-0.0001216t} \Rightarrow \frac{1}{3} = e^{-0.0001216t} \Rightarrow$$

$$\ln \frac{1}{3} = \ln e^{-0.0001216t} \Rightarrow \ln \frac{1}{3} = -0.0001216t \Rightarrow$$

$$\frac{1}{-0.0001216} \ln \frac{1}{3} \approx 9034.64 = t$$

The Egyptian died about 9000 yr ago.

25. Because $y = y_0 e^{-0.0001216t}$, where y_0 is the initial amount present. If we let $y = 0.15 y_0$, we can solve for t.

$$0.15 y_0 = y_0 e^{-0.0001216t}$$

$$0.15 = e^{-0.0001216t} \Rightarrow \ln 0.15 = \ln e^{-0.0001216t} \Rightarrow$$

$$\ln 0.15 = -0.0001216t \Rightarrow$$

$$\frac{1}{-0.0001216} \ln 0.15 \approx 15{,}601.32 = t$$

The paintings are about 15,600 yr old.

27. From Example 6, we have the temperature, $f(t)$, of a body at time t after being introduced into an environment having constant temperature T_0 is $f(t) = T_0 + Ce^{-kt}$, where C and k are constants.

From the given information, when $t = 0$, $T_0 = 0$, and the temperature of the water is $f(0) = 100$.

$$100 = 0 + Ce^{-0k} \Rightarrow 100 = C$$

Thus, we have $f(t) = 0 + 100e^{-kt} = 100e^{-kt}$.

Also, when $t = \frac{24}{60} = \frac{2}{5}$ hr, $f\left(\frac{2}{5}\right) = 50$. Using this information, we have

$$50 = 100e^{-(2/5)k} \Rightarrow \frac{1}{2} = e^{-(2/5)k}$$

$$\ln \frac{1}{2} = \ln e^{-(2/5)k} \Rightarrow \ln \frac{1}{2} = -\frac{2}{5} k$$

$$\ln 1 - \ln 2 = -\frac{2}{5} k \Rightarrow \ln 2 = \frac{2}{5} k$$

$$k = \frac{5}{2} \ln 2 \approx 1.733$$

Thus, the model is $f(t) = 100e^{-1.733t}$.

To find the temperature after $\frac{96}{60} = \frac{8}{5}$ hrs, we find $f\left(\frac{8}{5}\right)$. Because

$$f\left(\frac{8}{5}\right) = 100e^{-1.733(8/5)} \approx 6.25,$$ the temperature after 96 minutes is about $6.25°C$.

Note: We could have used the exact value of k to perform the calculations.

$$f\left(\frac{8}{5}\right) = 100e^{-\left(\frac{5}{2} \ln 2\right)\left(\frac{8}{5}\right)} = 100e^{-4 \ln 2}$$

$$= 100e^{\ln \frac{1}{16}} = 100 \cdot \frac{1}{16} = 6.25$$

29. Given $P = 60{,}000$ and $t = 5$, substitute $r = 0.03$ and $n = 4$ into the compound interest formula, $A = P\left(1 + \frac{r}{n}\right)^{tn}$. We have

$$A = 60{,}000\left(1 + \frac{0.03}{4}\right)^{5(4)}$$

$$= 60{,}000(1.0075)^{20} \approx 69{,}671.05$$

The interest from this investment would be $69{,}671.05 - $60{,}000 = $9{,}671.05$.

Given $P = 60{,}000$ and $t = 5$, substitute $r = 0.0275$ into the continuous compounding formula, $A = Pe^{rt}$.

$$A = 60{,}000e^{0.0275(5)} \approx 68{,}844.10$$

The interest from this investment would be $68{,}844.10 - $60{,}000 = $8{,}844.10$.

(a) The investment that offers 3% compounded quarterly will earn more interest than the investment that offers 2.75% compounded continuously.

(b) Note that $9{,}671.05 - $8{,}844.10 = 826.95. The investment that offers 3% compounded quarterly will earn $826.95 more in interest.

31. $A = Pe^{rt}$, $A = 2P$,

$$2P = Pe^{0.025t}$$

$$2 = e^{0.025t}$$

$$\ln 2 = 0.025t$$

$$27.73 \approx t$$

The doubling time is about 27.73 yr if interest is compounded continuously.

33. $A = Pe^{rt}$

$$3P = Pe^{0.03t} \Rightarrow 3 = e^{0.03t} \Rightarrow \ln 3 = \ln e^{0.03t}$$

$$\ln 3 = 0.03t \Rightarrow t = \frac{\ln 3}{0.03} \approx 36.62$$

It will take about 36.62 years for the investment to triple.

35. (a) 1969 is represented by $t = 4$

$$M(4) = 434e^{-0.08(4)} \approx 315$$

There were 315 continuously serving members in 1969.

(b) 1973 is represented by $t = 8$

$M(8) = 434e^{-0.08(8)} \approx 229$

There were 229 continuously serving members in 1973.

(c) 1979 is represented by $t = 14$

$M(14) = 434e^{-0.08(14)} \approx 142$

There were 142 continuously serving members in 1973.

37. (a) A point associated with the graph of $O(x) = P_0 a^{x-2000}$ is (2000, 1). Because

$P(2000) = P_0 a^{2000-2000} = 1,$ we have

$P_0 a^0 = 1 \Rightarrow P_0 = 1.$ Thus, we have

$P(x) = a^{x-2000}.$ The point (2025, 1.4) is projected to be on the graph, so we have the following.

$P(2025) = a^{2025-2000} = 1.4$

$a^{25} = 1.4 \Rightarrow a = \sqrt[25]{1.4} \approx 1.01355$

(b) From part (a) we have

$P(x) = (1.01355)^{x-2000}.$ To find the population projected for 2020, we must find $P(2020).$ Thus

$P(2020) = (1.01355)^{2020-2000} = (1.01355)^{20}$
≈ 1.3 billion is the estimated population in 2020.

(c) We must solve $(1.01355)^{x-2000} = 1.5$ for x.

$(1.01355)^{x-2000} = 1.5$

$\ln(1.01355)^{x-2000} = \ln 1.5$

$(x - 2000)\ln(1.01355) = \ln 1.5$

$x - 2000 = \dfrac{\ln 1.5}{\ln 1.01355}$

$2000 + \dfrac{\ln 1.5}{\ln 1.01355} \approx 2030.13 = x$

In 2030, it is projected that the population will reach 1.5 billion.

39. $f(x) = 7446e^{0.0305x}$

(a) $t = 4$ represents 2012.

$f(4) = 7446e^{0.0305(4)} \approx 8412$

In 2012, out-of-pocket spending on health care was about $8412 per household.

(b) Find x when $f(x) = 7915$.

$7446e^{0.0305x} = 7915 \Rightarrow e^{0.0305x} = \dfrac{7915}{7446} \Rightarrow$

$0.0305x = \ln\left(\dfrac{7915}{7446}\right) \Rightarrow x = \dfrac{\ln \frac{7915}{7446}}{0.0305} \approx 2.0$

Out-of-pocket spending on health care reached $7915 per household during 2010 (2 years after 2008).

41. $H = 12,744e^{0.0264t}$

(a) For 2005, $t = 5$.

$H = 12,744e^{0.0264(5)} \approx \$14,542$

(b) For 2009, $t = 9$.

$H = 12,744e^{0.0264(9)} \approx \$16,162$

(c) For 2012, $t = 12$.

$H = 12,744e^{0.0264(12)} \approx \$17,494$

43. (a) We must find values for y_0 and k in the formula $y = y_0 e^{kx}$. From the graph, $y_0 = 14,225$ and $y_{10} = 19,915$. Using these values, we have

$19,915 = 14,225e^{10k}$

$\dfrac{19,915}{14,225} = e^{10k}$

$\ln\left(\dfrac{19,915}{14,225}\right) = 10k$

$k = \dfrac{\ln \frac{19,915}{14,225}}{10} \approx 0.034$

Therefore, the function is $y = 14,225e^{0.034x}.$

(b) $21,500 = 14,225e^{0.034x}$

$\dfrac{21,500}{14,225} = e^{0.034x}$

$\ln \dfrac{21,500}{14,225} = 0.034x$

$x = \dfrac{\ln \frac{21,500}{14,225}}{0.034} \approx 12.1$

Projected enrollment will be 21,500 about 12 years after the school year 2013–14, or in the school year 2025–26.

45. $f(t) = 15,000e^{-0.05t}$

(a) At the beginning of the epidemic, $t = 0$.

$f(0) = 15,000e^{-0.05(0)} = 15,000$

At the beginning of the epidemic, 15,000 people were susceptible.

(b) After 10 days, $t = 10$.

$$f(0) = 15,000e^{-0.05(10)} \approx 9098$$

After 10 days, approximately 9098 people were susceptible.

(c) After 3 weeks, $t = 21$.

$$f(21) = 15,000e^{-0.05(21)} \approx 5249$$

After three weeks, approximately 5249 people were susceptible.

47. $f(t) = 500e^{0.1t}$

(a) $f(2) = 500e^{0.1(2)} \approx 611$

At two days, the bacteria count is approximately 611 million.

(b) $f(4) = 500e^{0.1(4)} \approx 746$

At four days, the bacteria count is approximately 746 million.

(c) $f(7) = 500e^{0.1(7)} \approx 1007$

At one week (seven days), the bacteria count is approximately 1007 million.

49. $f(t) = 200(0.90)^t$

Find t when $f(t) = 50$.

$$50 = 200(0.90)^t \Rightarrow 0.25 = (0.90)^t \Rightarrow$$

$$\ln 0.25 = \ln\left[(0.90)^t\right] \Rightarrow \ln 0.25 = t \ln 0.90 \Rightarrow$$

$$t = \frac{\ln 0.25}{\ln 0.90} \Rightarrow t = \frac{\ln 0.25}{\ln 0.90} \approx 13.2$$

The dose will reach a level of 50 mg in about 13.2 hr.

51. $A(t) = 100e^{0.0264t}$

We want to find the year in which the CPI will be 225.

$$225 = 100e^{0.0264t} \Rightarrow 2.25 = e^{0.0264t} \Rightarrow$$

$$\ln 2.25 = \ln e^{0.0264t} \Rightarrow \ln 2.25 = 0.0264t \Rightarrow$$

$$t = \frac{\ln 2.25}{0.0264} \approx 30.7$$

Costs will be 125% higher than in 1990 in 2020, 30 years after 1990.

53. $S(t) = 50,000e^{-0.1t}$

Find t when $S(t) = 25,000$.

$$25,000 = 50,000e^{-0.1t} \Rightarrow 0.5 = e^{-0.1t} \Rightarrow$$

$$\ln 0.5 = \ln e^{-0.1t} \Rightarrow \ln 0.5 = -0.1t \Rightarrow$$

$$t = \frac{\ln 0.5}{-0.1} \approx 6.9$$

It will take about 6.9 yr for sales to fall to half the initial sales.

55. Use the formula for continuous compounding with $r = 0.06$.

$$A = Pe^{rt} \Rightarrow 2P = Pe^{0.06t} \Rightarrow 2 = e^{0.06t} \Rightarrow$$

$$\ln 2 = \ln e^{0.06t} \Rightarrow \ln 2 = 0.06t \Rightarrow$$

$$t = \frac{\ln 2}{0.06} \approx 11.6$$

It will take about 11.6 yr before twice as much electricity is needed.

57. $f(x) = \dfrac{0.9}{1 + 271e^{-0.122x}}$

(a) $f(25) = \dfrac{0.9}{1 + 271e^{-0.122(25)}} \approx 0.065$

$f(65) = \dfrac{0.9}{1 + 271e^{-0.122(65)}} \approx 0.820$

Among people age 25, 6.5% have some CHD, while among people age 65, 82% have some CHD.

(b)

$$0.50 = \frac{0.9}{1 + 271e^{-0.122x}}$$

$$0.50\left(1 + 271e^{-0.122x}\right) = 0.9$$

$$0.5 + 135.5e^{-0.122x} = 0.9$$

$$135.5e^{-0.122x} = 0.4$$

$$e^{-0.122x} = \frac{0.4}{135.5}$$

$$\ln e^{-0.122x} = \ln \frac{0.4}{135.5}$$

$$-0.122x = \ln \frac{0.4}{135.5}$$

$$x = \frac{\ln \frac{0.4}{135.5}}{-0.122} \approx 47.75$$

At about 48, the likelihood of coronary heart disease is 50%.

Summary Exercises on Functions: Domains and Defining Equations

1. $f(x) = 3x - 6$

Domain: $(-\infty, \infty)$

3. $f(x) = |x + 4|$

Domain: $(-\infty, \infty)$

5. $f(x) = \dfrac{-2}{x^2 + 7}$

The domain is the set of all real numbers such that $x^2 + 7 \neq 0 \Rightarrow$ there is no real solution.

Domain: $(-\infty, \infty)$

7. $f(x) = \dfrac{x^2 + 7}{x^2 - 9}$

The domain is the set of all real numbers such that $x^2 - 9 \neq 0 \Rightarrow x \neq -3$ or $x \neq 3$
Domain: $(-\infty, -3) \cup (-3, 3) \cup (3, \infty)$

9. $f(x) = \log_5 \left(16 - x^2\right)$

The domain is the set of all real numbers such that $16 - x^2 > 0 \Rightarrow 16 > x^2 \Rightarrow 4 > x$ and $-4 < x$.
Domain: $(-4, 4)$

11. $f(x) = \sqrt{x^2 - 7x - 8}$

The domain is the set of all real numbers such that $x^2 - 7x - 8 \geq 0$. Solve the equation to find the test intervals: $x^2 - 7x - 8 = 0 \Rightarrow$
$(x - 8)(x + 1) = 0 \Rightarrow x = 8$ or $x = -1$

Interval	Test Point	Value of $x^2 - 7x - 8$	Sign of $x^2 - 7x - 8$
$(-\infty, -1)$	-2	10	Positive
$(-1, 8)$	0	-8	Negative
$(8, \infty)$	10	22	Positive

Domain: $(-\infty, -1] \cup [8, \infty)$

13. $f(x) = \dfrac{1}{2x^2 - x + 7}$

The domain is the set of all real numbers such that $2x^2 - x + 7 \neq 0$
$x = \dfrac{-(-1) \pm \sqrt{(-1)^2 - 4(2)(7)}}{2(2)} = \dfrac{1 \pm \sqrt{-55}}{4} \Rightarrow$ there are
no real solutions. Domain: $(-\infty, \infty)$

15. $f(x) = \sqrt{x^3 - 1}$

The domain is the set of all real numbers such that $x^3 - 1 \geq 0 \Rightarrow x^3 \geq 1 \Rightarrow x \geq 1$
Domain: $[1, \infty)$

17. $f(x) = e^{x^2 + x + 4}$

The domain is the set of values such that $x^2 + x + 4$ is real. Domain: $(-\infty, \infty)$

19. $f(x) = \sqrt{\dfrac{-1}{x^3 - 1}}$

The domain is the set of all real numbers such that $\sqrt{\dfrac{-1}{x^3 - 1}}$ is defined or $\dfrac{-1}{x^3 - 1} \geq 0$. $\dfrac{-1}{x^3 - 1}$ is
defined for $x^3 - 1 \neq 0 \Rightarrow x^3 \neq 1 \Rightarrow x \neq 1$

Interval	Test Point	Value of $\dfrac{-1}{x^3 - 1}$	Sign of $\dfrac{-1}{x^3 - 1}$
$(-\infty, 1)$	0	1	Positive
$(1, \infty)$	2	$-\dfrac{1}{7}$	Negative

Domain: $(-\infty, 1)$

21. $f(x) = \ln\left(x^2 + 1\right)$

The domain is the set of all real numbers such that $x^2 + 1 > 0$. Domain: $(-\infty, \infty)$

23. $f(x) = \log\left(\dfrac{x + 2}{x - 3}\right)^2$

Because $\left(\dfrac{x+2}{x-3}\right)^2 \geq 0$ for all real numbers, the domain of $f(x)$ is the set of all real numbers such that $\dfrac{x+2}{x-3} \neq 0$. $\dfrac{x+2}{x-3} = 0$ when $x = -2$, and
$\dfrac{x+2}{x-3}$ is undefined when $x = 3$.
Domain: $(-\infty, -2) \cup (-2, 3) \cup (3, \infty)$

25. $f(x) = e^{|1/x|}$

The domain is the set of all real numbers such that $\dfrac{1}{x}$ is defined, or $x \neq 0$
Domain: $(-\infty, 0) \cup (0, \infty)$

27. $f(x) = x^{100} - x^{50} + x^2 + 5$
Domain: $(-\infty, \infty)$

29. $f(x) = \sqrt[4]{16 - x^4}$

The domain is the set of all real numbers such that $16 - x^4 \geq 0$. Solve the equation to find the test intervals: $16 - x^4 = 0 \Rightarrow$
$(2 - x)(2 + x)(4 + x^2) = 0 \Rightarrow x = 2$ or $x = -2$

Interval	Test Point	Value of $16 - x^4$	Sign of $16 - x^4$
$(-\infty, -2)$	-3	-65	Negative
$(-2, 2)$	0	16	Positive
$(2, \infty)$	3	-65	Negative

Domain: $[-2, 2]$

31. $f(x) = \sqrt{\dfrac{x^2 - 2x - 63}{x^2 + x - 12}}$

The domain is the set of real numbers such that $\dfrac{x^2-2x-63}{x^2+x-12} \geq 0$. $\dfrac{x^2-2x-63}{x^2+x-12}$ is not defined for

$x^2 + x - 12 = 0 \Rightarrow (x+4)(x-3) = 0 \Rightarrow$

$x \neq -4$ or $x \neq 3$. Solve $\dfrac{x^2-2x-63}{x^2+x-12} = 0$ to find

the test intervals: $\dfrac{x^2-2x-63}{x^2+x-12} = 0 \Rightarrow$

$x^2 - 2x - 63 = 0 \Rightarrow (x-9)(x+7) = 0 \Rightarrow x = 9$ or $x = -7$.

Interval	Test Point	Value of $\dfrac{x^2-2x-63}{x^2+x-12}$	Sign
$(-\infty, -7)$	-10	$\dfrac{19}{26}$	Positive
$(-7, -4)$	-5	$-\dfrac{7}{2}$	Negative
$(-4, 3)$	0	$\dfrac{21}{4}$	Positive
$(3, 9)$	5	$-\dfrac{8}{3}$	Negative
$(9, \infty)$	10	$\dfrac{17}{98}$	Positive

Domain: $(-\infty, -7] \cup (-4, 3) \cup [9, \infty)$

33. $f(x) = \left|\sqrt{5-x}\right|$

The domain is the set of real numbers such that $5 - x \geq 0 \Rightarrow 5 \geq x$

Domain: $(-\infty, 5]$

35. $f(x) = \log\left|\dfrac{1}{4-x}\right|$

The domain is the set of real numbers such that $\left|\dfrac{1}{4-x}\right| > 0 \Rightarrow \dfrac{1}{4-x} > 0 \Rightarrow 4 - x > 0 \Rightarrow 4 > x$

or $-\dfrac{1}{4-x} < 0 \Rightarrow -4 + x < 0 \Rightarrow x < 4$

Domain: $(-\infty, 4) \cup (4, \infty)$

37. $f(x) = 6^{\sqrt{x^2-25}}$

The domain is the set of real numbers such that $\sqrt{x^2 - 25}$ is a real number.

$x^2 - 25 \geq 0 \Rightarrow x^2 \geq 25 \Rightarrow x \geq 5$ or $x \leq -5$

Domain: $(-\infty, -5] \cup [5, \infty)$

39. $f(x) = \ln\left(\dfrac{-3}{(x+2)(x-6)}\right)$

The domain is the set of real numbers such that $\dfrac{-3}{(x+2)(x-6)} > 0$ and $(x+2)(x-6) \neq 0 \Rightarrow$

$x \neq -2$ or $x \neq 6$.

Interval	Test Point	Value of $\dfrac{-3}{(x+2)(x-6)}$	Sign of $\dfrac{-3}{(x+2)(x-6)}$
$(-\infty, -2)$	-3	$-\dfrac{1}{3}$	Negative
$(-2, 6)$	0	$\dfrac{1}{4}$	Positive
$(6, \infty)$	7	$-\dfrac{1}{3}$	Negative

Domain: $(-2, 6)$

41. Choice A can be written as a function of x.
$3x + 2y = 6 \Rightarrow y = f(x) = -\dfrac{3}{2}x + 3$

43. Choice C can be written as a function of x.
$x^3 + y^3 = 5 \Rightarrow y = f(x) = \sqrt[3]{5 - x^3}$

45. Choice A can be written as a function of x.
$x = \dfrac{2-y}{y+3} \Rightarrow y = f(x) = \dfrac{2-3x}{x+1}$

47. Choice D can be written as a function of x.
$2x = \dfrac{1}{y^3} \Rightarrow y = f(x) = \sqrt[3]{\dfrac{1}{2x}}$

49. Choice C can be written as a function of x.
$\dfrac{x}{4} - \dfrac{y}{9} = 0 \Rightarrow y = f(x) = \dfrac{9x}{4}$

Chapter 4   Review Exercises

1. This is not a one-to-one function because a horizontal line can intersect the graph in more than one point.

3. This is a one-to-one function because every horizontal line intersects the graph in no more than one point.

5. $y = (x+3)^2$

If $x = -2$, $y = (-2+3)^2 = 1^2 = 1$.

If $x = -4$, $y = (-4+3)^2 = (-1)^2 = 1$.

Two different values of x lead to the same value of y, so the function is not one-to-one.

7. $f(x) = x^3 - 3$

   This function is one-to-one.

   Step 1: Replace $f(x)$ with y and interchange

   x and y. $y = x^3 - 3 \Rightarrow x = y^3 - 3$

   Step 2: Solve for y.

   $x = y^3 - 3 \Rightarrow x + 3 = y^3 \Rightarrow y = \sqrt[3]{x+3}$

   Step 3: Replace y with $f^{-1}(x)$.

   $f^{-1}(x) = \sqrt[3]{x+3}$

9. $f^{-1}(\$50,000)$ represents the number of years after 2004 required for the investment to reach $50,000.

11. To have an inverse, a function must be a <u>one-to-one</u> function.

13. $y = \log_{0.3} x$

    The point $(1, 0)$ is on the graph of every function of the form $y = \log_a x$, so the correct choice must be either B or C. The base is $a = 0.3$ and $0 < 0.3 < 1$, so $y = \log_{0.3} x$ is a decreasing function, and so the correct choice must be B.

15. $y = \ln x = \log_e x$

    The point $(1, 0)$ is on the graph of every function of the form $y = \log_a x$, so the correct choice must be either B or C. The base is $a = e$ and $e > 1$, so $y = \ln x$ is an increasing function, and so the correct choice must be C.

17. $2^5 = 32$ is written in logarithmic form as $\log_2 32 = 5$.

19. $\left(\frac{3}{4}\right)^{-1} = \frac{4}{3}$ is written in logarithmic form as

    $\log_{3/4} \frac{4}{3} = -1$.

21. $\log 1000 = 3$ is written in exponential form as

    $10^3 = 1000$.

23. $\ln \sqrt{e} = \frac{1}{2}$ is written in exponential form as

    $e^{1/2} = \sqrt{e}$.

25. Let $f(x) = a^x$ be the required function. Then

    $f(-4) = \frac{1}{16} \Rightarrow a^{-4} = \frac{1}{16} \Rightarrow a^{-4} = 2^{-4} \Rightarrow a = 2$.

    The base is 2.

27. $\log_3 \dfrac{mn}{5r} = \log_3 mn - \log_3 5r$

    $\quad = \log_3 m + \log_3 n - \left(\log_3 5 + \log_3 r\right)$

    $\quad = \log_3 m + \log_3 n - \log_3 5 - \log_3 r$

29. $\log 0.0411 \approx -1.3862$

31. $\ln 144,000 \approx 11.8776$

33. To find $\log_{2/3} \frac{5}{8}$, use the change-of-base theorem. We have

    $\log_{2/3} \dfrac{5}{8} = \dfrac{\log \frac{5}{8}}{\log \frac{2}{3}} = \dfrac{\ln \frac{5}{8}}{\ln \frac{2}{3}} \approx 1.1592$.

35. $16^{x+4} = 8^{3x-2} \Rightarrow \left(2^4\right)^{x+4} = \left(2^3\right)^{3x-2} \Rightarrow$

    $2^{4x+16} = 2^{9x-6} \Rightarrow 4x + 16 = 9x - 6 \Rightarrow$

    $22 = 5x \Rightarrow \frac{22}{5} = x$

    Solution set: $\left\{\frac{22}{5}\right\}$

37. $3^{2x-5} = 13 \Rightarrow \ln 3^{2x-5} = \ln 13 \Rightarrow$

    $(2x-5)\ln 3 = \ln 13 \Rightarrow 2x - 5 = \dfrac{\ln 13}{\ln 3} \Rightarrow$

    $2x = 5 + \dfrac{\ln 13}{\ln 3} \Rightarrow x = \dfrac{1}{2}\left(5 + \dfrac{\ln 13}{\ln 3}\right) \approx 3.667$

    or

    $3^{2x-5} = 13 \Rightarrow \ln 3^{2x-5} = \ln 13 \Rightarrow$

    $(2x-5)\ln 3 = \ln 13 \Rightarrow 2x \ln 3 - 5 \ln 3 = \ln 13 \Rightarrow$

    $x \ln 3^2 - \ln 3^5 = \ln 13 \Rightarrow$

    $x \ln 9 - \ln 243 = \ln 13 \Rightarrow$

    $x \ln 9 = \ln 13 + \ln 243 \Rightarrow x \ln 9 = \ln 3159 \Rightarrow$

    $x = \dfrac{\ln 3159}{\ln 9} \approx 3.667$

    Solution set: $\{3.667\}$

39. $\quad 6^{x+3} = 4^x \Rightarrow \ln 6^{x+3} = \ln 4^x \Rightarrow$

    $\quad (x+3)\ln 6 = x \ln 4$

    $\quad x \ln 6 + 3 \ln 6 = x \ln 4$

    $\quad x \ln 6 - x \ln 4 = -3 \ln 6$

    $\quad x(\ln 6 - \ln 4) = -3 \ln 6 \Rightarrow x\left(\ln \frac{6}{4}\right) = -\ln 6^3 \Rightarrow$

    $\quad\quad x\left(\ln \frac{3}{2}\right) = -\ln 216$

    $\quad\quad\quad x = \dfrac{-\ln 216}{\ln \frac{3}{2}} \approx -13.257$

    Solution set: $\{-13.257\}$

41. $e^{2-x} = 12 \Rightarrow \ln e^{2-x} = \ln 12 \Rightarrow$

    $2 - x = \ln 12 \Rightarrow -x = -2 + \ln 12 \Rightarrow$

    $\quad x = 2 - \ln 12 \approx -0.485$

    Solution set: $\{-0.485\}$

43. $10e^{3x-7} = 5 \Rightarrow e^{3x-7} = \frac{1}{2} \Rightarrow$

$\ln e^{3x-7} = \ln\frac{1}{2} \Rightarrow 3x - 7 = \ln\frac{1}{2} \Rightarrow$

$\quad 3x = \ln\frac{1}{2} + 7 \Rightarrow x = \frac{1}{3}\left(\ln\frac{1}{2} + 7\right) \approx 2.102$

Solution set: {2.102}

45. $\qquad 6^{x-3} = 3^{4x+1}$

$\ln 6^{x-3} = \ln 3^{4x+1}$

$(x-3)\ln 6 = (4x+1)\ln 3$

$x\ln 6 - 3\ln 6 = 4x\ln 3 + \ln 3$

$x\ln 6 - \ln 6^3 = x\ln 3^4 + \ln 3$

$x\ln 6 - \ln 216 = x\ln 81 + \ln 3$

$x\ln 6 - x\ln 81 = \ln 3 + \ln 216$

$x(\ln 6 - \ln 81) = \ln 3 + \ln 216$

$$x = \frac{\ln 3 + \ln 216}{\ln 6 - \ln 81} = \frac{\ln(3 \cdot 216)}{\ln\frac{6}{81}}$$

$$= \frac{\ln 648}{\ln\frac{2}{27}} \, x \approx -2.487$$

Solution set: {–2.487}

47. $e^{6x} \cdot e^x = e^{21} \Rightarrow e^{7x} = e^{21} \Rightarrow 7x = 21 \Rightarrow x = 3$

Solution set: {3}

49. $2e^{2x} - 5e^x - 3 = 0$

Let $u = e^x$.

$2u^2 - 5u - 3 = 0$

$(u-3)(2u+1) = 0$

$\begin{array}{c|c} u - 3 = 0 & 2u + 1 = 0 \\ u = 3 & 2u = -1 \\ e^x = 3 & u = -\dfrac{1}{2} \\ \ln e^x = \ln 3 & \\ x = \ln 3 & e^x = -\dfrac{1}{2} \end{array}$

Disregard the proposed solution $-\frac{1}{2}$ because

e^x is always positive.

Solution set: {ln 3}

51. $4(1.06)^x + 2 = 8 \Rightarrow 4(1.06)^x = 6 \Rightarrow$

$(1.06)^x = 1.5 \Rightarrow \ln(1.06^x) = \ln 1.5 \Rightarrow$

$x\ln 1.06 = \ln 1.5 \Rightarrow x = \dfrac{\ln 1.5}{\ln 1.06} \approx 6.959$

53. $3\ln x = 13 \Rightarrow \ln x = \frac{13}{3} \Rightarrow x = e^{13/3}$

Solution set: $\left\{e^{13/3}\right\}$

55. $\log(2x+7) = 0.25 \Rightarrow 2x + 7 = 10^{0.25} = \sqrt[4]{10} \Rightarrow$

$x = \frac{\sqrt[4]{10} - 7}{2}$

Solution set: $\left\{\frac{\sqrt[4]{10}-7}{2}\right\}$

57. $\log_2(x^3 + 5) = 5 \Rightarrow x^3 + 5 = 2^5 \Rightarrow x^3 = 27 \Rightarrow$

$x = 3$

Solution set: {3}

59. $\log_4\left[(3x+1)(x-4)\right] = 2 \Rightarrow$

$(3x+1)(x-4) = 4^2 \Rightarrow 3x^2 - 11x - 4 = 16 \Rightarrow$

$3x^2 - 11x - 20 = 0 \Rightarrow (3x+4)(x-5) = 0 \Rightarrow$

$x = -\frac{4}{3} \text{ or } x = 5$

Solution set: $\left\{-\frac{4}{3}, 5\right\}$

61. $\log x + \log(13 - 3x) = 1 \Rightarrow$

$\log[x(13-3x)] = 1 \Rightarrow 13x - 3x^2 = 10^1 \Rightarrow$

$13x - 3x^2 = 10 \Rightarrow 3x^2 - 13x + 10 = 0 \Rightarrow$

$(3x-10)(x-1) = 0 \Rightarrow x = \frac{10}{3} \text{ or } x = 1$

Solution set: $\left\{1, \frac{10}{3}\right\}$

63. $\ln(6x) - \ln(x+1) = \ln 4$

$\ln\dfrac{6x}{x+1} = \ln 4 \Rightarrow \dfrac{6x}{x+1} = 4$

$6x = 4(x+1)$

$6x = 4x + 4$

$2x = 4 \Rightarrow x = 2$

Solution set: {2}

65. $\ln\left[\ln\left(e^{-x}\right)\right] = \ln 3$

$\ln(-x) = \ln 3 \Rightarrow -x = 3 \Rightarrow x = -3$

Solution set: {–3}

67. $\dfrac{d}{10} = \log\left(\dfrac{I}{I_0}\right) \Rightarrow 10^{d/10} = \dfrac{I}{I_0} \Rightarrow$

$I_0\left(10^{d/10}\right) = I \Rightarrow I_0 = \dfrac{I}{10^{d/10}}$

69. Graph $y_1 = e^x$ and $y_2 = 4 - \ln x$ in the window $[-3, 5] \times [-2, 6]$. Then find the intersection of the two graphs.

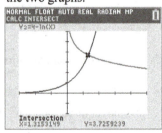

Solution set: {1.315}

71. (a) $6 = \log_{10}\dfrac{I}{I_0} \Rightarrow \dfrac{I}{I_0} = 10^6 \Rightarrow$

$I = 10^6 I_0 = 1{,}000{,}000 I_0$

(b) $8.2 = \log_{10} \dfrac{I}{I_0} \Rightarrow \dfrac{I}{I_0} = 10^{8.2} \Rightarrow$

$I = 10^{8.2} I_0 \approx 158{,}500{,}000 I_0$

(c) Consider the ratio of the magnitudes.

$\dfrac{158{,}500{,}000 I_0}{1{,}000{,}000 I_0} = 158.5$

The earthquake with a measure of 8.2 was 158.5 times as great as the earthquake that measured 6.0.

73. Substitute $A = 4700$, $P = 3500$, $t = 10$, $n = 1$ into the formula $A = P\left(1 + \frac{r}{n}\right)^{tn}$.

$4700 = 3500\left(1 + \dfrac{r}{1}\right)^{10(1)} \Rightarrow \dfrac{47}{35} = (1+r)^{10} \Rightarrow$

$\left(\dfrac{47}{35}\right)^{1/10} = 1 + r \Rightarrow \left(\dfrac{47}{35}\right)^{1/10} - 1 = r \Rightarrow$

$r \approx 0.030$

The annual interest rate, to the nearest tenth, is 3.0%.

75. First, substitute $P = 10{,}000$, $r = 0.03$, $t = 12$, and $n = 1$ into the formula $A = P\left(1 + \frac{r}{n}\right)^{tn}$.

$A = 10{,}000\left(1 + \dfrac{0.03}{1}\right)^{12(1)} = 10{,}000(1.03)^{12}$

$\approx 14{,}257.61$

After the first 12 years, there will be about \$14,257.61 in the account. To finish off the 21-year period, substitute $P = 14{,}257.61$, $r = 0.04$, $t = 9$, and $n = 2$ into the formula $A = P\left(1 + \frac{r}{n}\right)^{tn}$.

$A = 14{,}257.61\left(1 + \dfrac{0.04}{2}\right)^{9(2)}$

$= 14{,}257.61(1.02)^{18} \approx 20{,}363.38$

At the end of the 21-year period, about \$20,363.38 will be in the account.

77. To find t, substitute $a = 2$, $P = 1$, and $r = 0.04$ into $A = Pe^{rt}$ and solve.

$2 = 1 \cdot e^{0.04t} \Rightarrow 2 = e^{0.04t} \Rightarrow \ln 2 = \ln e^{0.04t} \Rightarrow$

$\ln 2 = 0.04t \Rightarrow t = \dfrac{\ln 2}{0.04} \approx 17.3$

It would take about 17.3 yr.

79. Half the 2010 total payroll is 72.7 Using the function $f(x) = 146.02e^{-0.112x}$, we solve for x when $f(x) = 72.7$.

$146.02e^{-0.112x} = 72.7$

$e^{-0.112x} = \dfrac{72.7}{146.02}$

$\ln e^{-0.112x} = \ln \dfrac{72.7}{146.02}$

$-0.112x = \ln \dfrac{72.7}{146.02}$

$x = \dfrac{\ln \frac{72.7}{146.02}}{-0.112} \approx 6.2$

According to the model, the total payroll will be half the 2010 value about 6 years after 2010, or in 2016.

81. (a) We must determine how much money will be in the account after 35 years. The money is being compounded continuously, so we use $A = Pe^{rt}$.

$A = 5000e^{0.04 \cdot 35} \approx \$20{,}076$

After paying 25% of this for taxes, there is $20{,}076 \cdot 0.75 = \$15{,}207$ remaining.

(b) $A = 3750e^{0.03 \cdot 35} \approx \$10{,}716.19$
You will have about \$10,716.

(c) You will have $\$20{,}076 - \$10{,}716 = \$4491$ more money with the IRA.

(d) $A = 3750e^{0.04 \cdot 35} \approx \$15{,}207$
The two balances are the same.

Chapter 4   Test

1. (a) $f(x) = \sqrt[3]{2x - 7}$
Because it is a cube root, $2x - 7$ may be any real number and the domain is $(-\infty, \infty)$
The cube root of any real number is also any real number, so the range is $(-\infty, \infty)$

(b) $f(x) = \sqrt[3]{2x - 7}$
The graph is a stretched translation of $y = \sqrt[3]{x}$, which passes the horizontal line test and, thus, is a one-to-one function.

(c) *Step 1:* Replace $f(x)$ with y and interchange x and y.

$$y = \sqrt[3]{2x - 7} \Rightarrow x = \sqrt[3]{2y - 7}$$

Step 2: Solve for y.

$$x = \sqrt[3]{2y - 7} \Rightarrow x^3 = \left(\sqrt[3]{2y - 7}\right)^3 \Rightarrow$$

$$x^3 = 2y - 7 \Rightarrow x^3 + 7 = 2y \Rightarrow \frac{x^3 + 7}{2} = y$$

Step 3: Replace y with $f^{-1}(x)$.

$$f^{-1}(x) = \frac{x^3 + 7}{2}$$

(d) The domain and range of f are $(-\infty, \infty)$, so the domain and range of f^{-1} are also $(-\infty, \infty)$.

(e)

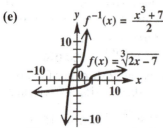

These graphs are reflections of each other across the line $y = x$.

2. (a) $y = \log_{1/3} x$

The point $(1, 0)$ is on the graph of every function of the form $y = \log_a x$, so the correct choice must be either B or C. The base is $a = \frac{1}{3}$ and $0 < \frac{1}{3} < 1$, so $y = \log_{1/3} x$ is a decreasing function, and so the correct choice must be B.

(b) $y = e^x$

The point $(0, 1)$ is on the graph because $e^0 = 1$, so the correct choice must be either A or D. The base is e and $e > 1$, so $y = e^x$ is an increasing function, and so the correct choice must be A.

(c) $y = \ln x$ or $y = \log_e x$

The point $(1, 0)$ is on the graph of every function of the form $y = \log_a x$, so the correct choice must be B or C. The base is $a = e$ and $e > 1$, so $y = \ln x$ is an increasing function, and the correct choice must be C.

(d) $y = \left(\frac{1}{3}\right)^x$

The point $(0, 1)$ is on the graph because $\left(\frac{1}{3}\right)^0 = 1$, so the correct choice must be either A or D. The base is $\frac{1}{3}$ and $0 < \frac{1}{3} < 1$, so $y = \left(\frac{1}{3}\right)^x$ is a decreasing function, and so the correct choice must be D.

3. $\left(\frac{1}{8}\right)^{2x-3} = 16^{x+1} \Rightarrow \left(2^{-3}\right)^{2x-3} = \left(2^4\right)^{x+1} \Rightarrow$

$2^{-3(2x-3)} = 2^{4(x+1)} \Rightarrow 2^{-6x+9} = 2^{4x+4} \Rightarrow$

$-6x + 9 = 4x + 4 \Rightarrow -10x + 9 = 4 \Rightarrow$

$-10x = -5 \Rightarrow x = \frac{1}{2}$

Solution set: $\left\{\frac{1}{2}\right\}$

4. (a) $4^{3/2} = 8$ is written in logarithmic form as $\log_4 8 = \frac{3}{2}$.

(b) $\log_8 4 = \frac{2}{3}$ is written in exponential form as $8^{2/3} = 4$.

5.

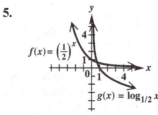

They are inverses of each other.

6. $\log_7 \frac{x^2 \sqrt[4]{y}}{z^3} = \log_7 x^2 + \log_7 \sqrt[4]{y} - \log_7 z^3$

$$= \log_7 x^2 + \log_7 y^{1/4} - \log_7 z^3$$

$$= 2\log_7 x + \frac{1}{4}\log_7 y - 3\log_7 z$$

7. $\log 2388 \approx 3.3780$

8. $\ln 2388 \approx 7.7782$

9. $\log_9 13 = \frac{\log 13}{\log 9} = \frac{\ln 13}{\ln 9} \approx 1.1674$

10. $x^{2/3} = 25 \Rightarrow \left(x^{2/3}\right)^{3/2} = \pm 25^{3/2} \Rightarrow$

$x = (\pm 5)^3 = \pm 125$

Solution set: $\{\pm 125\}$

11. $12^x = 1 \Rightarrow x\log 12 = \log 1 \Rightarrow x\log 12 = 0 \Rightarrow$

$x = 0$

Solution set: $\{0\}$

12. $9^x = 4 \Rightarrow x \log 9 = \log 4 \Rightarrow x = \dfrac{\log 4}{\log 9} \approx 0.631$

Solution set: $\{0.631\}$

13. $16^{2x+1} = 8^{3x} \Rightarrow 2^{4(2x+1)} = 2^{3(3x)} \Rightarrow$
$4(2x+1) = 3(3x) = 8x + 4 = 9x \Rightarrow x = 4$

Solution set: $\{4\}$

14.
$$2^{x+1} = 3^{x-4}$$
$$(x+1)\log 2 = (x-4)\log 3$$
$$x\log 2 + \log 2 = x\log 3 - 4\log 3$$
$$x\log 2 - x\log 3 = -\log 2 - 4\log 3$$
$$x(\log 2 - \log 3) = -\log 2 - 4\log 3$$
$$x = \frac{-\log 2 - 4\log 3}{\log 2 - \log 3} \approx 12.548$$

Solution set: $\{12.548\}$

15.
$$e^{0.4x} = 4^{x-2}$$
$$0.4x = (x-2)\ln 4$$
$$0.4x = x\ln 4 - 2\ln 4$$
$$0.4x - x\ln 4 = -2\ln 4$$
$$x(0.4 - \ln 4) = -2\ln 4$$
$$x = \frac{-2\ln 4}{0.4 - \ln 4} \approx 2.811$$

Solution set: $\{2.811\}$

16. $2e^{2x} - 5e^x + 3 = 0$

Let $u = e^x$.
$$2u^2 - 5u + 3 = 0$$
$$(u-1)(2u-3) = 0$$

$u - 1 = 0$	$2u - 3 = 0$
$u = 1$	$u = \frac{3}{2}$
$e^x = 1$	$e^x = \frac{3}{2}$
$\ln e^x = \ln 1$	$\ln e^x = \ln \frac{3}{2} \Rightarrow x = \ln \frac{3}{2}$
$x = 0$	

Solution set: $\left\{0, \ln \frac{3}{2}\right\}$ or $\{0, 0.405\}$

17. $\log_x \dfrac{9}{16} = 2 \Rightarrow x^2 = \dfrac{9}{16} \Rightarrow x = \pm \dfrac{3}{4}$

It is not possible for a logarithm to have a negative base, so we reject the negative solution. Solution set: $\left\{\frac{3}{4}\right\}$

18. $\log_2\left[(x-4)(x-2)\right] = 3$
$$(x-4)(x-2) = 2^3$$
$$x^2 - 6x + 8 = 8$$
$$x^2 - 6x = 0$$
$$x(x-6) = 0 \Rightarrow x = 0 \text{ or } x = 6$$

Solution set: $\{0, 6\}$

19. $\log_2 x + \log_2(x+2) = 3 \Rightarrow \log_2[x(x+2)] = 3$
$$x^2 + 2x = 2^3 \Rightarrow x^2 + 2x - 8 = 0$$
$$(x+4)(x-2) = 0 \Rightarrow x = -4 \text{ or } x = 2$$

The negative solution is not in the domain, so it must be discarded.
Solution set: $\{2\}$

20. $\ln x - 4\ln 3 = \ln\left(\frac{1}{5}x\right)$

$\ln x - \ln 3^4 = \ln \dfrac{x}{5}$

$\ln x - \ln 81 = \ln \dfrac{x}{5} \Rightarrow \ln \dfrac{x}{81} = \ln \dfrac{x}{5}$

$\dfrac{x}{81} = \dfrac{x}{5} \Rightarrow \dfrac{5}{81} = 1 \Rightarrow$ there is no

solution.
Solution set: $\varnothing$

21. $\log_3(x+1) - \log_3(x-3) = 2$

$\log_3 \dfrac{x+1}{x-3} = 2$

$\dfrac{x+1}{x-3} = 3^2 \Rightarrow \dfrac{x+1}{x-3} = 9$

$x + 1 = 9(x-3)$

$x + 1 = 9x - 27$

$-8x = -28 \Rightarrow x = \frac{28}{8} = \frac{7}{2}$

Solution set: $\left\{\frac{7}{2}\right\}$

22. Answers will vary. Sample answer:
$\log_5 27$ is the exponent to which 5 must be raised in order to obtain 27. To approximate $\log_5 27$ on your calculator, use the change-of-base formula: $\log_5 27 = \frac{\log 27}{\log 5} = \frac{\ln 27}{\ln 5} \approx 2.048$.

23. $v(t) = 176(1 - e^{-0.18t})$

Find the time t at which $v(t) = 147$.
$$147 = 176(1 - e^{-0.18t}) \Rightarrow \frac{147}{176} = 1 - e^{-0.18t} \Rightarrow$$
$$-e^{-0.18t} = \frac{147}{176} - 1 \Rightarrow -e^{-0.18t} = -\frac{29}{176}$$
$$e^{-0.18t} = \frac{29}{176} \Rightarrow \ln e^{-0.18t} = \ln \frac{29}{176} \Rightarrow$$
$$-0.18t = \ln \frac{29}{176} \Rightarrow t = \frac{\ln \frac{29}{176}}{-0.18} \approx 10.02$$

It will take the skydiver about 10 sec to attain the speed of 147 ft per sec (100 mph).

24. (a) Substitute $P = 5000$, $A = 18,000$, $r = 0.03$, and $n = 12$ into the formula

$$A = P\left(1 + \frac{r}{n}\right)^{tn}.$$

$$18,000 = 5000\left(1 + \frac{0.03}{12}\right)^{t(12)}$$

$$3.6 = (1.0025)^{12t}$$

$$\ln 3.6 = \ln (1.0025)^{12t}$$

$$\ln 3.6 = 12t \ln (1.0025)$$

$$t = \frac{\ln 3.6}{12 \ln (1.0025)} \approx 42.8$$

It will take about 42.8 years.

(b) Substitute $P = 5000$, $A = 18,000$, and $r = 0.03$, and into the formula $A = Pe^{rt}$.

$$18,000 = 5000e^{0.03t} \Rightarrow 3.6 = e^{0.03t} \Rightarrow$$

$$\ln 3.6 = \ln e^{0.03t} \Rightarrow \ln 3.6 = 0.03t \Rightarrow$$

$$t = \frac{\ln 3.6}{0.03} \approx 42.7$$

It will take about 42.7 years.

25. Substitute $A = 3P$ and $r = 0.028$ into the continuous compounding formula $A = Pe^{rt}$, then solve for t:

$$A = Pe^{rt}$$

$$3P = Pe^{0.028t} \Rightarrow 3 = e^{0.028t} \Rightarrow$$

$$\ln 3 = \ln e^{0.028t} \Rightarrow \ln 3 = 0.028t \Rightarrow$$

$$\frac{\ln 3}{0.028} = t \Rightarrow t \approx 39.2$$

It will take about 39.2 years for any amount of money to triple at 2.8% annual interest.

26. $A(t) = 600e^{-0.05t}$

(a) $A(12) = 600e^{-0.05(12)} = 600e^{-0.6} \approx 329.3$

The amount of radioactive material present after 12 days is about 329.3 g.

(b) Because

$$A(0) = 600e^{-0.05(0)} = 600e^0 = 600 \text{ g is}$$

the amount initially present, we seek to find t when $A(t) = \frac{1}{2}(600) = 300$ g is present.

$$300 = 600e^{-0.05t} \Rightarrow 0.5 = e^{-0.05t} \Rightarrow$$

$$\ln 0.5 = \ln e^{-0.05t} \Rightarrow \ln 0.5 = -0.05t \Rightarrow$$

$$t = \frac{\ln 0.5}{-0.05} \approx 13.9$$

The half-life of the material is about 13.9 days.

Chapter 5

TRIGONOMETRIC FUNCTIONS

Section 5.1   Angles

1. One degree, written $1°$, represents $\frac{1}{360}$ of a complete rotation.

3. If the measure of an angle is $x°$, its supplement can be expressed as $\underline{180° - x°}$.

5. The measure of an angle that is its own supplement is $\underline{90°}$.

7. One second, written $1''$, is $\frac{1}{60}$ of a minute.

9. $55.25°$ written in degrees and minutes is $\underline{55°15'}$.

11. $30°$
    (a) $90° - 30° = 60°$
    (b) $180° - 30° = 150°$

13. $45°$
    (a) $90° - 45° = 45°$
    (b) $180° - 45° = 135°$

15. $54°$
    (a) $90° - 54° = 36°$
    (b) $180° - 54° = 126°$

17. $1°$
    (a) $90° - 1° = 89°$
    (b) $180° - 1° = 179°$

19. $14°20'$
    (a) $90° - 14°20' = 89°60' - 14°20' = 75°40'$
    (b) $180° - 14°20' = 179°60' - 14°20'$
        $= 165°40'$

21. $20°10'30''$
    (a) $90° - 20°10'30'' = 89°59'60'' - 20°10'30''$
        $= 69°49'30''$

    (b) $180° - 20°10'30''$
        $= 179°59'60'' - 20°10'30''$
        $= 159°49'30''$

23. The two angles form a straight angle.
    $7x + 11x = 180 \Rightarrow 18x = 180 \Rightarrow x = 10$
    The measures of the two angles are
    $(7x)° = \left[7(10)\right]° = 70°$ and
    $(11x)° = \left[11(10)\right]° = 110°$.

25. The two angles form a right angle.
    $4x + 2x = 90 \Rightarrow 6x = 90 \Rightarrow x = 15$
    The two angles have measures of
    $(4x)° = \left[4(15)\right]° = 60°$ and
    $(2x)° = \left[2(15)\right]° = 30°$.

27. The two angles form a straight angle.
    $(-4x) + (-14x) = 180 \Rightarrow -18x = 180 \Rightarrow$
    $x = -10$
    The measures of the two angles are
    $(-4x)° = \left[-4(-10)\right]° = 40°$ and
    $(-14x)° = \left[-14(-10)\right]° = 140°$.

29. The sum of the measures of two supplementary angles is $180°$.
    $(10x + 7) + (7x + 3) = 180$
    $17x + 10 = 180$
    $17x = 170 \Rightarrow x = 10$
    The measures of the two angles are
    $(10x + 7)° = \left[10(10) + 7\right]° = (100 + 7)° = 107°$
    and $(7x + 3)° = \left[7(10) + 3\right]° = (70 + 3)° = 73°$.

31. The sum of the measures of two complementary angles is $90°$.
    $(9x + 6) + 3x = 90 \Rightarrow 12x + 6 = 90 \Rightarrow$
    $12x = 84 \Rightarrow x = 7$
    The measures of the two angles are
    $(9x + 6)° = \left[9(7) + 6\right]° = (63 + 6)° = 69°$ and
    $(3x)° = \left[3(7)\right]° = 21°$.

33. $\dfrac{25 \text{ minutes}}{60 \text{ minutes}} = \dfrac{x}{360°}$
    $x = \dfrac{25}{60}(360) = 25(6) = 150°$

35. At 15 minutes after the hour, the minute hand is $\frac{1}{4}$ the way around, so the hour hand is $\frac{1}{4}$ of the way between the 3 and 4. Thus, the hour hand is located 16.25 minutes past 12. The minute hand is 15 minutes after the 12. The smaller angle formed by the hands of the clock can be found by solving the proportion

$$\frac{(16.25-15)\text{ minutes}}{60\text{ minutes}} = \frac{x}{360°}.$$

$$\frac{(16.25-15)\text{ minutes}}{60\text{ minutes}} = \frac{x}{360°} \Rightarrow \frac{1.25}{60} = \frac{x}{360} \Rightarrow$$

$$x = \frac{1.25}{60}(360) = 1.25(6) = 7.5° = 7°30'$$

37. At 20 minutes after the hour, the minute hand is $\frac{1}{3}$ the way around, so the hour hand is $\frac{1}{3}$ of the way between the 8 and 9. Thus, the hour hand is located $41\frac{2}{3}$ minutes past 12. The minute hand is 20 minutes after the 12. The smaller angle formed by the hands of the clock can be found by solving the proportion

$$\frac{\left(41\frac{2}{3}-20\right)\text{ minutes}}{60\text{ minutes}} = \frac{x}{360°}.$$

$$\frac{\left(41\frac{2}{3}-20\right)\text{ minutes}}{60\text{ minutes}} = \frac{x}{360°} \Rightarrow \frac{21\frac{2}{3}}{60} = \frac{x}{360} \Rightarrow$$

$$x = \frac{21\frac{2}{3}}{60}(360) = \left(21\frac{2}{3}\right)(6) = 130°$$

39.
$$\begin{array}{r} 62°\ 18' \\ +21°\ 41' \\ \hline 83°\ 59' \end{array}$$

41. $97°42' + 81°37' = 178°79' = 179°19'$

43. $47°\ 29' - 71°18' = -\left(71°18' - 47°\ 29'\right)$
$$= -\left(70°\ 78' - 47°\ 29'\right)$$
$$= -23°\ 49'$$

45. $90° - 51°\ 28' = 89°\ 60' - 51°\ 28' = 38°32'$

47. $180° - 119°\ 26' = 179°\ 60' - 119°\ 26' = 60°\ 34'$

49. $90° - 72°\ 58'\ 11'' = 89°\ 59'\ 60'' - 72°\ 58'\ 11''$
$$= 17°01'\ 49''$$

51. $26°20' + 18°17' - 14°10' = 44°37' - 14°10'$
$$= 30°27'$$

53. $35°30' = 35° + \frac{30}{60}° = 35° + 0.5° = 35.5°$

55. $112°15' = 112° + \frac{15}{60}° = 112° + 0.25° = 112.25°$

57. $-60°12' = -\left(60° + \frac{12}{60}°\right) = -\left(60° + 0.2°\right)$
$$= -60.2°$$

59. $20°54'36'' = 20° + \frac{54}{60}° + \frac{36}{3600}°$
$$= 20° + 0.900° + 0.01° = 20.91°$$

61. $91°35'54'' = 91° + \frac{35}{60}° + \frac{54}{3600}°$
$$\approx 91° + 0.5833° + 0.0150°$$
$$\approx 91.598°$$

63. $274°18'59'' = 274° + \frac{18}{60}° + \frac{59}{3600}°$
$$\approx 274° + 0.3000° + 0.0164°$$
$$\approx 274.316°$$

65. $39.25° = 39° + 0.25° = 39° + 0.25(60')$
$$= 39° + 15' + 0'' = 39°15'00''$$

67. $126.76° = 126° + 0.76° = 126° + 0.76(60')$
$$= 126° + 45.6' = 126° + 45' + 0.6'$$
$$= 126° + 45' + 0.6(60'')$$
$$= 126° + 45' + 36'' = 126°45'36''$$

69. $-18.515° = -\left(18° + 0.515°\right)$
$$= -\left(18° + 0.515(60')\right)$$
$$= -\left(18° + 30.9'\right) = -\left(18° + 30' + 0.9'\right)$$
$$= -\left(18° + 30' + 0.9(60'')\right)$$
$$= -\left(18° + 30' + 54''\right) = -18°30'54''$$

71. $31.4296° = 31° + 0.4296° = 31° + 0.4296(60')$
$$= 31° + 25.776' = 31° + 25' + 0.776'$$
$$= 31° + 25' + 0.776(60'')$$
$$= 31°25'46.56'' \approx 31°25'47''$$

73. $89.9004° = 89° + 0.9004° = 89° + 0.9004(60')$
$$= 89° + 54.024' = 89° + 54' + 0.024'$$
$$= 89° + 54' + 0.024(60'')$$
$$= 89°54'1.44'' \approx 89°54'01''$$

75. $178.5994° = 178° + 0.5994°$
$$= 178° + 0.5994(60')$$
$$= 178° + 35.964'$$
$$= 178° + 35' + 0.964'$$
$$= 178° + 35' + 0.964(60'')$$
$$= 178°35'57.84'' \approx 178°35'58''$$

77. $32°$ is coterminal with $360° + 32° = 392°$.

79. $26°30'$ is coterminal with
$360° + 26°30' = 386°30'$.

81. $-40°$ is coterminal with $360° + (-40°) = 320°$

83. $-125°30'$ is coterminal with $360° + (-125°30') =$
$359°60' - 125°30' = 234°30'$.

85. 361° is coterminal with 361° − 360° = 1°.

87. −361° is coterminal with
−361° + 2(360°) = 359°.

89. 539° is coterminal with 539° − 360° = 179°.

91. 850° is coterminal with
850° − 2(360°) = 850° − 720° = 130°.

93. 5280° is coterminal with
5280° − 14 · 360° = 5280° − 5040° = 240°.

95. −5280° is coterminal with
−5280° + 15 · 360° = −5280° + 5400° = 120°.

In exercises 97–100, answers may vary.

97. 90° is coterminal with
$$90° + 360° = 450°$$
$$90° + 2(360°) = 810°$$
$$90° − 360° = −270°$$
$$90° − 2(360°) = −630°$$

99. 0° is coterminal with
$$0° + 360° = 360°$$
$$0° + 2(360°) = 720°$$
$$0° − 360° = −360°$$
$$0° − 2(360°) = −720°$$

101. 30°
A coterminal angle can be obtained by adding an integer multiple of 360°.
$$30° + n · 360°$$

103. 135°
A coterminal angle can be obtained by adding an integer multiple of 360°.
$$135° + n · 360°$$

105. −90°
A coterminal angle can be obtained by adding an integer multiple of 360°.
$$−90° + n · 360°$$

107. 0°
A coterminal angle can be obtained by adding integer multiple of 360°.
$$0° + n · 360° = n · 360°$$

109. The answers to Exercises 107 and 108 give the same set of angles because 0° is coterminal with 360°.

For Exercises 111–121, angles other than those given are possible.

111.

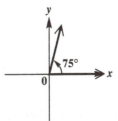

**435°; −285°;
quadrant I**
75° is coterminal with 75° + 360° = 435° and 75° − 360° = −285°. These angles are in quadrant I.

113.

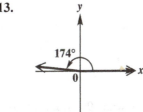

**534°; −186°;
quadrant II**
174° is coterminal with 174° + 360° = 534° and 174° − 360° = −186°. These angles are in quadrant II.

115.

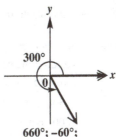

**660°; −60°;
quadrant IV**
300° is coterminal with 300° + 360° = 660° and 300° − 360° = −60°. These angles are in quadrant IV.

117.

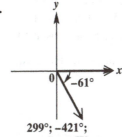

**299°; −421°;
quadrant IV**
−61° is coterminal with −61° + 360° = 299° and −61° − 360° = −421°. These angles are in quadrant IV.

119.

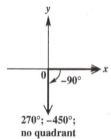

450°; –270°;
no quadrant

90° is coterminal with $90° + 360° = 450°$ and $90° - 360° = -270°$. These angles are not in a quadrant.

121.

450°; –270° (y axis diagram)

270°; –450°;
no quadrant

–90° is coterminal with $-90° + 360° = 270°$ and $-90° - 360° = -450°$. These angles are not in a quadrant.

123. 45 revolutions per min $= \frac{45}{60}$ revolution per sec
$= \frac{3}{4}$ revolution per sec

A turntable will make $\frac{3}{4}$ revolution in 1 sec.

125. 600 rotations per min $= \frac{600}{60}$ rotations per sec
$= 10$ rotations per sec
$= 5$ rotations per $\frac{1}{2}$ sec
$= 5(360°)$ per $\frac{1}{2}$ sec
$= 1800°$ per $\frac{1}{2}$ sec

A point on the edge of the tire will move 1800° in $\frac{1}{2}$ sec.

127. 75° per min = $75°(60)$ per hr = 4500° per hr
$= \frac{4500°}{360°}$ rotations per hr
$= 12.5$ rotations per hr

The pulley makes 12.5 rotations in 1 hour.

129. Earth rotates 360° in 24 hr. 360° is equal to $360(60') = 21,600'$.

$$\frac{24\,\text{hr}}{21,600'} = \frac{x}{1'} \Rightarrow$$

$$x = \frac{24}{21,600}\,\text{hr} = \frac{24}{21,600}(60\,\text{min}) = \frac{1}{15}\,\text{min}$$

$$= \frac{1}{15}(60\,\text{sec}) = 4\,\text{sec}$$

It should take the motor 4 sec to rotate the telescope through an angle of 1 min.

Section 5.2   Trigonometric Functions

1. $r = \sqrt{(-3)^2 + (-3)^2} = \sqrt{18} = 3\sqrt{2}$

3. $\cos\theta = \frac{x}{r} = \frac{-3}{3\sqrt{2}} = -\frac{1}{\sqrt{2}} = -\frac{\sqrt{2}}{2}$

5. $\sin\theta = \frac{1}{2}$ is possible because the range of $\sin\theta$ is $[-1, 1]$. Furthermore, when $\sin\theta = \frac{1}{2}$, $\csc\theta = \frac{1}{\frac{1}{2}} = 2$. Thus, $\sin\theta = \frac{1}{2}$ and $\csc\theta = 2$ is possible.

7. $\sin\theta > 0$, $\csc\theta < 0$ is impossible because $\csc\theta = \frac{1}{\sin\theta}$, so both functions must have the same sign.

9. $\cot\theta = -1.5$ is possible because the range of $\cot\theta$ is $(-\infty, \infty)$.

11.

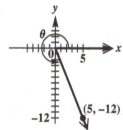

$x = 5$, $y = -12$, and
$$r = \sqrt{x^2 + y^2} = \sqrt{(-12)^2 + 5^2}$$
$$= \sqrt{144 + 25} = \sqrt{169} = 13$$

$\sin\theta = \frac{y}{r} = \frac{-12}{13} = -\frac{12}{13}$

$\cos\theta = \frac{x}{r} = \frac{5}{13}$

$\tan\theta = \frac{y}{x} = \frac{-12}{5} = -\frac{12}{5}$

$\cot\theta = \frac{x}{y} = \frac{5}{-12} = -\frac{5}{12}$

$\sec\theta = \frac{r}{x} = \frac{13}{5}$

$\csc\theta = \frac{r}{y} = \frac{13}{-12} = -\frac{13}{12}$

13.

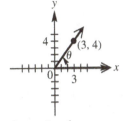

$x = 3$, $y = 4$ and

$$r = \sqrt{x^2 + y^2} = \sqrt{3^2 + 4^2} = \sqrt{25} = 5$$

$$\sin\theta = \frac{y}{r} = \frac{4}{5}; \quad \cos\theta = \frac{x}{r} = \frac{3}{5}$$

$$\tan\theta = \frac{y}{x} = \frac{4}{3}; \quad \cot\theta = \frac{x}{y} = \frac{3}{4}$$

$$\sec\theta = \frac{r}{x} = \frac{5}{3}; \quad \csc\theta = \frac{r}{y} = \frac{5}{4}$$

15.

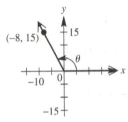

$x = -8$, $y = 15$, and

$$r = \sqrt{x^2 + y^2} = \sqrt{(-8)^2 + 15^2}$$
$$= \sqrt{64 + 225} = \sqrt{289} = 17$$

$$\sin\theta = \frac{y}{r} = \frac{15}{17}$$

$$\cos\theta = \frac{x}{r} = \frac{-8}{17} = -\frac{8}{17}$$

$$\tan\theta = \frac{y}{x} = \frac{15}{-8} = -\frac{15}{8}$$

$$\cot\theta = \frac{x}{y} = \frac{-8}{15} = -\frac{8}{15}$$

$$\sec\theta = \frac{r}{x} = \frac{17}{-8} = -\frac{17}{8}$$

$$\csc\theta = \frac{r}{y} = \frac{17}{15}$$

17.

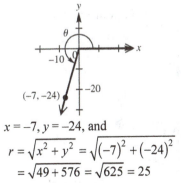

$x = -7$, $y = -24$, and

$$r = \sqrt{x^2 + y^2} = \sqrt{(-7)^2 + (-24)^2}$$
$$= \sqrt{49 + 576} = \sqrt{625} = 25$$

$$\sin\theta = \frac{y}{r} = \frac{-24}{25} = -\frac{24}{25}$$

$$\cos\theta = \frac{x}{r} = \frac{-7}{25} = -\frac{7}{25}$$

$$\tan\theta = \frac{y}{x} = \frac{-24}{-7} = \frac{24}{7}$$

$$\cot\theta = \frac{x}{y} = \frac{-7}{-24} = \frac{7}{24}$$

$$\sec\theta = \frac{r}{x} = \frac{25}{-7} = -\frac{25}{7}$$

$$\csc\theta = \frac{r}{y} = \frac{25}{-24} = -\frac{25}{24}$$

19.

$x = 0$, $y = 2$, and

$$r = \sqrt{x^2 + y^2} = \sqrt{0^2 + 2^2} = \sqrt{0 + 4} = \sqrt{4} = 2$$

$$\sin\theta = \frac{y}{r} = \frac{2}{2} = 1$$

$$\cos\theta = \frac{x}{r} = \frac{0}{2} = 0$$

$$\tan\theta = \frac{y}{x} = \frac{2}{0} \text{ undefined}$$

$$\cot\theta = \frac{x}{y} = \frac{0}{2} = 0$$

$$\sec\theta = \frac{r}{x} = \frac{2}{0} \text{ undefined}$$

$$\csc\theta = \frac{r}{y} = \frac{2}{2} = 1$$

21.

$x = 4$, $y = 0$, and

$$r = \sqrt{x^2 + y^2} = \sqrt{(-4)^2 + 0^2}$$
$$= \sqrt{16 + 0} = \sqrt{16} = 4$$

$$\sin\theta = \frac{y}{r} = \frac{0}{4} = 0$$

$$\cos\theta = \frac{x}{r} = \frac{-4}{4} = -1$$

(*continued on next page*)

(*continued*)

$$\tan\theta = \frac{y}{x} = \frac{0}{-4} = 0$$

$$\cot\theta = \frac{x}{y} = \frac{-4}{0} \quad \text{undefined}$$

$$\sec\theta = \frac{r}{x} = \frac{4}{-4} = -1$$

$$\csc\theta = \frac{r}{y} = \frac{4}{0} \quad \text{undefined}$$

23.

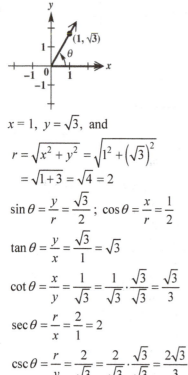

$x = 1,\ y = \sqrt{3},$ and

$$r = \sqrt{x^2 + y^2} = \sqrt{1^2 + \left(\sqrt{3}\right)^2}$$
$$= \sqrt{1 + 3} = \sqrt{4} = 2$$

$$\sin\theta = \frac{y}{r} = \frac{\sqrt{3}}{2};\ \cos\theta = \frac{x}{r} = \frac{1}{2}$$

$$\tan\theta = \frac{y}{x} = \frac{\sqrt{3}}{1} = \sqrt{3}$$

$$\cot\theta = \frac{x}{y} = \frac{1}{\sqrt{3}} = \frac{1}{\sqrt{3}} \cdot \frac{\sqrt{3}}{\sqrt{3}} = \frac{\sqrt{3}}{3}$$

$$\sec\theta = \frac{r}{x} = \frac{2}{1} = 2$$

$$\csc\theta = \frac{r}{y} = \frac{2}{\sqrt{3}} = \frac{2}{\sqrt{3}} \cdot \frac{\sqrt{3}}{\sqrt{3}} = \frac{2\sqrt{3}}{3}$$

25.

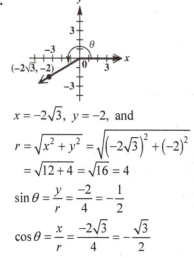

$x = -2\sqrt{3},\ y = -2,$ and

$$r = \sqrt{x^2 + y^2} = \sqrt{\left(-2\sqrt{3}\right)^2 + (-2)^2}$$
$$= \sqrt{12 + 4} = \sqrt{16} = 4$$

$$\sin\theta = \frac{y}{r} = \frac{-2}{4} = -\frac{1}{2}$$

$$\cos\theta = \frac{x}{r} = \frac{-2\sqrt{3}}{4} = -\frac{\sqrt{3}}{2}$$

$$\tan\theta = \frac{y}{x} = \frac{-2}{-2\sqrt{3}} = \frac{1}{\sqrt{3}} = \frac{\sqrt{3}}{3}$$

$$\cot\theta = \frac{x}{y} = \frac{-2\sqrt{3}}{-2} = \sqrt{3}$$

$$\sec\theta = \frac{r}{x} = \frac{4}{-2\sqrt{3}} = -\frac{2}{\sqrt{3}} = -\frac{2\sqrt{3}}{3}$$

$$\csc\theta = \frac{r}{y} = \frac{4}{-2} = -2$$

In Exercises 27–41, $r = \sqrt{x^2 + y^2}$, which is positive.

27. In quadrant II, x is negative, so $\dfrac{x}{r}$ is negative.

29. In quadrant IV, x is positive and y is negative, so $\dfrac{y}{x}$ is negative.

31. In quadrant II, y is positive, so $\dfrac{y}{r}$ is positive.

33. In quadrant IV, x is positive, so $\dfrac{x}{r}$ is positive.

35. In quadrant II, x is negative and y is positive, so $\dfrac{x}{y}$ is negative.

37. In quadrant III, x is negative and y is negative, so $\dfrac{y}{x}$ is positive.

39. In quadrant III, x is negative, so $\dfrac{r}{x}$ is negative.

41. In quadrant I, x is positive and y is positive, so $\dfrac{x}{y}$ is positive.

43. Because $x \geq 0$, the graph of the line $2x + y = 0$ is shown to the right of the y-axis. A point on this line is $(1, -2)$ because $2(1) + (-2) = 0$. The corresponding value of r is
$$r = \sqrt{1^2 + (-2)^2} = \sqrt{1 + 4} = \sqrt{5}.$$

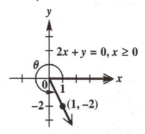

(*continued on next page*)

(*continued*)

$$\sin\theta = \frac{y}{r} = \frac{-2}{\sqrt{5}} = -\frac{2}{\sqrt{5}} \cdot \frac{\sqrt{5}}{\sqrt{5}} = -\frac{2\sqrt{5}}{5}$$

$$\cos\theta = \frac{x}{r} = \frac{1}{\sqrt{5}} = \frac{1}{\sqrt{5}} \cdot \frac{\sqrt{5}}{\sqrt{5}} = \frac{\sqrt{5}}{5}$$

$$\tan\theta = \frac{y}{x} = \frac{-2}{1} = -2$$

$$\cot\theta = \frac{x}{y} = \frac{1}{-2} = -\frac{1}{2}$$

$$\sec\theta = \frac{r}{x} = \frac{\sqrt{5}}{1} = \sqrt{5}$$

$$\csc\theta = \frac{r}{y} = \frac{\sqrt{5}}{-2} = -\frac{\sqrt{5}}{2}$$

45. Because $x \le 0$, the graph of the line $-4x + 7y = 0$ is shown to the left of the y-axis. A point on this line is $(-7, -4)$ because $-4(-7) + 7(-4) = 0$. The corresponding value of r is $r = \sqrt{(-7)^2 + (-4)^2} = \sqrt{49 + 16} = \sqrt{65}$.

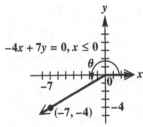

$$\sin\theta = \frac{y}{r} = \frac{-4}{\sqrt{65}} = -\frac{4}{\sqrt{65}} \cdot \frac{\sqrt{65}}{\sqrt{65}} = -\frac{4\sqrt{65}}{65}$$

$$\cos\theta = \frac{x}{r} = \frac{-7}{\sqrt{65}} = -\frac{7}{\sqrt{65}} \cdot \frac{\sqrt{65}}{\sqrt{65}} = -\frac{7\sqrt{65}}{65}$$

$$\tan\theta = \frac{y}{x} = \frac{-4}{-7} = \frac{4}{7}; \quad \cot\theta = \frac{x}{y} = \frac{-7}{-4} = \frac{7}{4}$$

$$\sec\theta = \frac{r}{x} = \frac{\sqrt{65}}{-7} = -\frac{\sqrt{65}}{7}$$

$$\csc\theta = \frac{r}{y} = \frac{\sqrt{65}}{-4} = -\frac{\sqrt{65}}{4}$$

47. Because $x \ge 0$, the graph of the line $x + y = 0$ is shown to the right of the y-axis. A point on this line is $(2, -2)$ because $2 + (-2) = 0$. The corresponding value of r is $r = \sqrt{2^2 + (-2)^2} = \sqrt{4 + 4} = \sqrt{8} = 2\sqrt{2}$.

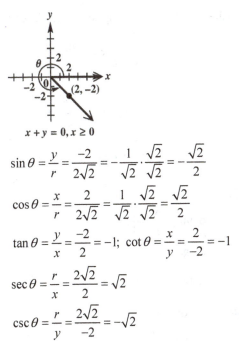

$x + y = 0, x \ge 0$

$$\sin\theta = \frac{y}{r} = \frac{-2}{2\sqrt{2}} = -\frac{1}{\sqrt{2}} \cdot \frac{\sqrt{2}}{\sqrt{2}} = -\frac{\sqrt{2}}{2}$$

$$\cos\theta = \frac{x}{r} = \frac{2}{2\sqrt{2}} = \frac{1}{\sqrt{2}} \cdot \frac{\sqrt{2}}{\sqrt{2}} = \frac{\sqrt{2}}{2}$$

$$\tan\theta = \frac{y}{x} = \frac{-2}{2} = -1; \quad \cot\theta = \frac{x}{y} = \frac{2}{-2} = -1$$

$$\sec\theta = \frac{r}{x} = \frac{2\sqrt{2}}{2} = \sqrt{2}$$

$$\csc\theta = \frac{r}{y} = \frac{2\sqrt{2}}{-2} = -\sqrt{2}$$

49. Because $x \le 0$, the graph of the line $-\sqrt{3}x + y = 0$ is shown to the left of the y-axis. A point on this line is $\left(-1, -\sqrt{3}\right)$ because $-\sqrt{3}(-1) - \sqrt{3} = \sqrt{3} - \sqrt{3} = 0$

The corresponding value of r is

$$r = \sqrt{(-1)^2 + \left(-\sqrt{3}\right)^2} = \sqrt{1 + 3} = \sqrt{4} = 2.$$

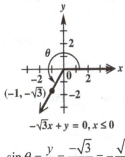

$-\sqrt{3}x + y = 0, x \le 0$

$$\sin\theta = \frac{y}{r} = \frac{-\sqrt{3}}{2} = -\frac{\sqrt{3}}{2}$$

$$\cos\theta = \frac{x}{r} = \frac{-1}{2} = -\frac{1}{2}$$

$$\tan\theta = \frac{y}{x} = \frac{-\sqrt{3}}{-1} = \sqrt{3}$$

$$\cot\theta = \frac{x}{y} = \frac{-1}{-\sqrt{3}} = \frac{1}{\sqrt{3}} \cdot \frac{\sqrt{3}}{\sqrt{3}} = \frac{\sqrt{3}}{3}$$

$$\sec\theta = \frac{r}{x} = \frac{2}{-1} = -2$$

$$\csc\theta = \frac{r}{y} = \frac{2}{-\sqrt{3}} = -\frac{2}{\sqrt{3}} \cdot \frac{\sqrt{3}}{\sqrt{3}} = -\frac{2\sqrt{3}}{3}$$

Use the figure below to help solve exercises 51–79.

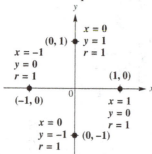

51. $\cos 90°$

$$\cos 90° = \frac{x}{r} = \frac{0}{1} = 0$$

53. $\tan 180°$

$$\tan 180° = \frac{y}{x} = \frac{0}{-1} = 0$$

55. $\sec 180°$

$$\sec 180° = \frac{r}{x} = \frac{1}{-1} = -1$$

57. $\sin(-270°)$

The quadrantal angle $\theta = -270°$ is coterminal with $-270° + 360° = 90°$.

$$\sin(-270°) = \sin 90 = \frac{y}{r} = \frac{1}{1} = 1$$

59. $\cot 540°$

The quadrantal angle $\theta = 540°$ is coterminal with $540° - 360° = 180°$.

$$\cot 540° = \cot 180° = \frac{x}{y} = \frac{-1}{0} \text{ undefined}$$

61. $\csc(-450°)$

The quadrantal angle $\theta = -450°$ is coterminal with $720° - 450° = 270°$.

$$\csc(-450°) = \csc 270° = \frac{r}{y} = \frac{1}{-1} = -1$$

63. $\sin 1800°$

The quadrantal angle $\theta = 1800°$ is coterminal with $1800° - 5(360°) = 1800° - 1800° = 0°$

$$\sin 1800° = \sin 0° = \frac{y}{r} = \frac{0}{1} = 0$$

65. $\csc 1800°$

The quadrantal angle $\theta = 1800°$ is coterminal with $1800° - 5(360°) = 1800° - 1800° = 0°$

$$\csc 1800° = \csc 0° = \frac{r}{y} = \frac{1}{0} \text{ undefined}$$

67. $\sec 1800°$

The quadrantal angle $\theta = 1800°$ is coterminal with $1800° - 5(360°) = 1800° - 1800° = 0°$

$$\sec 1800° = \sec 0° = \frac{r}{x} = \frac{1}{1} = 1$$

69. $\cos(-900°)$

The quadrantal angle $\theta = -900°$ is coterminal with $-900° + 3(360°) = -900° + 1080° = 180°$.

$$\cos(-900°) = \cos 180° = \frac{x}{r} = \frac{-1}{1} = -1$$

71. $\cos 90° + 3\sin 270°$

$$\cos 90° = \frac{x}{r} = \frac{0}{1} = 0$$

$$\sin 270° = \frac{y}{r} = \frac{-1}{1} = -1$$

$$\cos 90° + 3\sin 270° = 0 + 3(-1) = -3$$

73. $3\sec 180° - 5\tan 360°$

$$\sec 180° = \frac{r}{x} = \frac{1}{-1} = -1 \text{ and}$$

$$\tan 360° = \tan 0° = \frac{y}{x} = \frac{0}{1} = 0$$

$$3\sec 180° - 5\tan 360° = 3(-1) - 5(0)$$
$$= -3 - 0 = -3$$

75. $\tan 360° + 4\sin 180° + 5\cos^2 180°$

$$\tan 360° = \tan 0° = \frac{y}{x} = \frac{0}{1} = 0,$$

$$\sin 180° = \frac{y}{r} = \frac{0}{1} = 0, \text{ and}$$

$$\cos 180° = \frac{x}{r} = \frac{-1}{1} = -1$$

$$\tan 360° + 4\sin 180° + 5\cos^2 180°$$
$$= 0 + 4(0) + 5(-1)^2 = 0 + 0 + 5(1) = 5$$

77. $-2\sin^4 0° + 3\tan^2 0°$

$$\sin 0° = \frac{y}{r} = \frac{0}{1} = 0; \tan 0° = \frac{y}{x} = \frac{0}{1} = 0$$

$$-2\sin^4 0° + 3\tan^2 0° = -2(0)^4 + 3(0)^2 = 0$$

79. $\sin^2\left(-90°\right)+\cos^2\left(-90°\right)$

$\sin\left(-90°\right)=\sin\left(-90°+360°\right)$

$\qquad=\sin 270°=\dfrac{y}{r}=\dfrac{-1}{1}=-1$

$\cos\left(-90°\right)=\cos\left(-90°+360°\right)$

$\qquad=\cos 270°=\dfrac{x}{r}=\dfrac{0}{1}=0$

$\sin^2\left(-90°\right)+\cos^2\left(-90°\right)=\left(-1\right)^2+0^2=1$

81. $\cos\left[\left(2n+1\right)\cdot 90°\right]$

This angle is a quadrantal angle whose terminal side lies on either the positive part of the y-axis or the negative part of the y-axis. Any point on these terminal sides would have the form $\left(0,k\right),$ where k is any real number, $k\neq 0.$

$\cos\left[\left(2n+1\right)\cdot 90°\right]=\dfrac{x}{r}=\dfrac{0}{\sqrt{0^2+k^2}}=0$

83. $\tan\left[n\cdot 180°\right]$

The angle is a quadrantal angle whose terminal side lies on either the positive part of the x-axis or the negative part of the x-axis. Any point on these terminal sides would have the form $(k,\,0),$ where k is any real number, $k\neq 0.$

$\tan\left[n\cdot 180°\right]=\dfrac{y}{x}=\dfrac{0}{k}=0$

85. $\tan\left[\left(2n+1\right)\cdot 90°\right]$

This angle is a quadrantal angle whose terminal side lies on either the positive part of the y-axis or the negative part of the y-axis. Any point on these terminal sides would have the form $\left(0,k\right),$ where k is any real number, $k\neq 0.$

$\tan\left[\left(2n+1\right)\cdot 90°\right]=\dfrac{y}{x}=\dfrac{k}{0}$ undefined

87. $\sec\theta=\dfrac{1}{\cos\theta}=\dfrac{1}{\frac{2}{3}}=\dfrac{3}{2}$

89. $\csc\theta=\dfrac{1}{\sin\theta}=\dfrac{1}{-\frac{3}{7}}=-\dfrac{7}{3}$

91. $\cot\theta=\dfrac{1}{\tan\theta}=\dfrac{1}{5}$

93. $\sin\theta=\dfrac{1}{\csc\theta}=\dfrac{1}{\frac{\sqrt{8}}{2}}=\dfrac{2}{\sqrt{8}}=\dfrac{2}{2\sqrt{2}}\cdot\dfrac{\sqrt{2}}{\sqrt{2}}=\dfrac{\sqrt{2}}{2}$

95. $\sin\theta=\dfrac{1}{\csc\theta}=\dfrac{1}{1.25}=0.8$

97. A $74°$ angle in standard position lies in quadrant I, so all its trigonometric functions are positive.

99. A $218°$ angle in standard position lies in quadrant III, so its sine, cosine, secant, and cosecant are negative, while its tangent and cotangent are positive.

101. A $178°$ angle in standard position lies in quadrant II, so its sine and cosecant are positive, while its cosine, secant, tangent and cotangent are negative.

103. A $-80°$ angle in standard position lies in quadrant IV, so its cosine and secant are positive, while its sine, cosecant, tangent and cotangent are negative.

105. An $855°$ angle in standard position is coterminal with a $135°$ angle, and thus, lies in quadrant II. So its sine and cosecant are positive, while its cosine, secant, tangent and cotangent are negative.

107. A $-345°$ angle in standard position lies in quadrant I, so all its trigonometric functions are positive.

109. $\sin\theta>0,$ so $\csc\theta$ is also greater than 0. The functions are greater than 0 (positive) in quadrants I and II.

111. $\cos\theta>0$ in quadrants I and IV, while $\sin\theta>0$ in quadrants I and II. Both conditions are met only in quadrant I.

113. $\tan\theta<0$ in quadrants II and IV, while $\cos\theta<0$ in quadrants II and III. Both conditions are met only in quadrant II.

115. $\sec\theta>0$ in quadrants I and IV, while $\csc\theta>0$ in quadrants I and II. Both conditions are met only in quadrant I.

117. $\sec\theta<0$ in quadrants II and III, while $\csc\theta<0$ in quadrants III and IV. Both conditions are met only in quadrant III.

119. $\sin\theta<0,$ so $\csc\theta$ is also less than 0. The functions are less than 0 (negative) in quadrants III and IV.

121. Impossible because the range of $\sin\theta$ is $[-1,\,1].$

123. Possible because the range of $\cos\theta$ is $[-1,\,1].$

125. Possible because the range of $\tan\theta$ is $(-\infty, \infty)$.

127. Impossible because the range of $\sec\theta$ is $(-\infty, -1] \cup [1, \infty)$.

129. Possible because the range of $\csc\theta$ is $(-\infty, -1] \cup [1, \infty)$.

131. If $\sin\theta = \dfrac{3}{5}$, then $y = 3$ and $r = 5$. So

$$r^2 = x^2 + y^2 \Rightarrow 5^2 = x^2 + 3^2 \Rightarrow 25 = x^2 + 9 \Rightarrow$$
$$16 = x^2 \Rightarrow \pm 4 = x. \ \theta \text{ is in quadrant II, so}$$
$$x = -4. \text{ Therefore, } \cos\theta = -\dfrac{4}{5}.$$

Alternatively, use the identity

$$\sin^2\theta + \cos^2\theta = 1: \left(\dfrac{3}{5}\right)^2 + \cos^2\theta = 1 \Rightarrow$$

$$\dfrac{9}{25} + \cos^2\theta = 1 \Rightarrow \cos^2\theta = \dfrac{16}{25} \Rightarrow \cos\theta = \pm\dfrac{4}{5}$$

θ is in quadrant II, so $\cos\theta$ is negative.

Thus, $\cos\theta = -\dfrac{4}{5}$.

133. If $\cot\theta = -\dfrac{1}{2}$ and θ is in quadrant IV, then $x = 1$ and $y = -2$. So

$$r^2 = x^2 + y^2 \Rightarrow r^2 = 1^2 + (-2)^2 \Rightarrow$$
$$r^2 = 1 + 4 \Rightarrow r^2 = 5 \Rightarrow r = \sqrt{5}$$

Therefore, $\csc\theta = \dfrac{r}{y} = -\dfrac{\sqrt{5}}{2}$. Alternatively,

use the identity $1 + \cot^2\theta = \csc^2\theta$:

$$1 + \left(-\dfrac{1}{2}\right)^2 = \csc^2\theta \Rightarrow 1 + \dfrac{1}{4} = \csc^2\theta \Rightarrow$$

$$\dfrac{5}{4} = \csc^2\theta \Rightarrow \pm\dfrac{\sqrt{5}}{2} = \csc\theta. \ \theta \text{ is in quadrant}$$

IV, so $\csc\theta$ is negative. Thus, $\csc\theta = -\dfrac{\sqrt{5}}{2}$

135. If $\sin\theta = \dfrac{1}{2}$, then $y = 1$ and $r = 2$. So

$$r^2 = x^2 + y^2 \Rightarrow 2^2 = x^2 + 1^2 \Rightarrow 4 = x^2 + 1 \Rightarrow$$
$$3 = x^2 \Rightarrow \pm\sqrt{3} = x$$

θ is in quadrant II, so $x = -\sqrt{3}$. Therefore,

$$\tan\theta = -\dfrac{1}{\sqrt{3}} = -\dfrac{1}{\sqrt{3}} \cdot \dfrac{\sqrt{3}}{\sqrt{3}} = -\dfrac{\sqrt{3}}{3}.$$

Alternatively, use the identity $\sin^2\theta + \cos^2\theta = 1$ to obtain

$$\left(\dfrac{1}{2}\right)^2 + \cos^2\theta = 1 \Rightarrow \cos^2\theta = \dfrac{3}{4} \Rightarrow$$

$$\cos\theta = \pm\dfrac{\sqrt{3}}{2}.$$

θ is in quadrant II, so $\cos\theta = -\dfrac{\sqrt{3}}{2}$. Then,

$$\tan\theta = \dfrac{\sin\theta}{\cos\theta} = \dfrac{\frac{1}{2}}{-\frac{\sqrt{3}}{2}} = -\dfrac{1}{\sqrt{3}}$$

$$= -\dfrac{1}{\sqrt{3}} \cdot \dfrac{\sqrt{3}}{\sqrt{3}} = -\dfrac{\sqrt{3}}{3}$$

137. Using the identity $1 + \cot^2\theta = \csc^2\theta$ gives

$$1 + \cot^2\theta = (-1.45)^2 \Rightarrow \cot^2\theta = 1.1025 \Rightarrow$$
$$\cot\theta = \pm 1.05$$

Because θ is in quadrant III, $\cot\theta = 1.05$.

For Exercises 139–147, remember that r is always positive.

139. $\tan\theta = -\dfrac{15}{8} = \dfrac{15}{-8}$, with θ in quadrant II

$\tan\theta = \dfrac{y}{x}$ and θ is in quadrant II, so let

$y = 15$ and $x = -8$.

$$x^2 + y^2 = r^2 \Rightarrow (-8)^2 + 15^2 = r^2 \Rightarrow$$
$$64 + 225 = r^2 \Rightarrow 289 = r^2 \Rightarrow r = 17$$

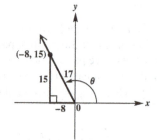

$$\sin\theta = \dfrac{y}{r} = \dfrac{15}{17}; \ \cos\theta = \dfrac{x}{r} = \dfrac{-8}{17} = -\dfrac{8}{17}$$

$$\tan\theta = \dfrac{y}{x} = \dfrac{15}{-8} = -\dfrac{15}{8}$$

$$\cot\theta = \dfrac{x}{y} = \dfrac{-8}{15} = -\dfrac{8}{15}$$

$$\sec\theta = \dfrac{r}{x} = \dfrac{17}{-8} = -\dfrac{17}{8}; \ \csc\theta = \dfrac{r}{y} = \dfrac{17}{15}$$

141. $\sin\theta = \dfrac{\sqrt{5}}{7}$, with θ in quadrant I

$\sin\theta = \dfrac{y}{r}$ and θ in quadrant I, so let

$y = \sqrt{5}, r = 7$.

$x^2 + y^2 = r^2 \Rightarrow x^2 + \left(\sqrt{5}\right)^2 = 7^2 \Rightarrow$

$x^2 + 5 = 49 \Rightarrow x^2 = 44 \Rightarrow x = \pm\sqrt{44} \Rightarrow$

$x = \pm 2\sqrt{11}$

θ is in quadrant I, so $x = 2\sqrt{11}$.

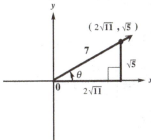

Drawing not to scale

$\sin\theta = \dfrac{y}{r} = \dfrac{\sqrt{5}}{7}$

$\cos\theta = \dfrac{x}{r} = \dfrac{2\sqrt{11}}{7}$

$\tan\theta = \dfrac{y}{x} = \dfrac{\sqrt{5}}{2\sqrt{11}} = \dfrac{\sqrt{5}}{2\sqrt{11}} \cdot \dfrac{\sqrt{11}}{\sqrt{11}} = \dfrac{\sqrt{55}}{22}$

$\cot\theta = \dfrac{x}{y} = \dfrac{2\sqrt{11}}{\sqrt{5}} = \dfrac{2\sqrt{11}}{\sqrt{5}} \cdot \dfrac{\sqrt{5}}{\sqrt{5}} = \dfrac{2\sqrt{55}}{5}$

$\sec\theta = \dfrac{r}{x} = \dfrac{7}{2\sqrt{11}} = \dfrac{7}{2\sqrt{11}} \cdot \dfrac{\sqrt{11}}{\sqrt{11}} = \dfrac{7\sqrt{11}}{22}$

$\csc\theta = \dfrac{r}{y} = \dfrac{7}{\sqrt{5}} = \dfrac{7}{\sqrt{5}} \cdot \dfrac{\sqrt{5}}{\sqrt{5}} = \dfrac{7\sqrt{5}}{5}$

143. $\cot\theta = \dfrac{\sqrt{3}}{8}$, with θ in quadrant I

$\cot\theta = \dfrac{x}{y}$ and θ is in quadrant I, so let

$x = \sqrt{3}$ and $y = 8$.

$x^2 + y^2 = r^2 \Rightarrow \left(\sqrt{3}\right)^2 + (8)^2 = r^2 \Rightarrow$

$3 + 64 = r^2 \Rightarrow 67 = r^2 \Rightarrow \sqrt{67} = r$

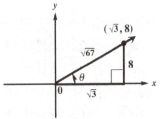

$\sin\theta = \dfrac{y}{r} = \dfrac{8}{\sqrt{67}} = \dfrac{8}{\sqrt{67}} \cdot \dfrac{\sqrt{67}}{\sqrt{67}} = \dfrac{8\sqrt{67}}{67}$

$\cos\theta = \dfrac{x}{r} = \dfrac{\sqrt{3}}{\sqrt{67}} = \dfrac{\sqrt{3}}{\sqrt{67}} \cdot \dfrac{\sqrt{67}}{\sqrt{67}}$

$\qquad = \dfrac{\sqrt{3}\sqrt{67}}{67} = \dfrac{\sqrt{201}}{67}$

$\tan\theta = \dfrac{y}{x} = \dfrac{8}{\sqrt{3}} = \dfrac{8}{\sqrt{3}} \cdot \dfrac{\sqrt{3}}{\sqrt{3}} = \dfrac{8\sqrt{3}}{3}$

$\cot\theta = \dfrac{x}{y} = \dfrac{\sqrt{3}}{8}$

$\sec\theta = \dfrac{r}{x} = \dfrac{\sqrt{67}}{\sqrt{3}} = \dfrac{\sqrt{67}}{\sqrt{3}} \cdot \dfrac{\sqrt{3}}{\sqrt{3}}$

$\qquad = \dfrac{\sqrt{67}\sqrt{3}}{3} = \dfrac{\sqrt{201}}{3}$

$\csc\theta = \dfrac{r}{y} = \dfrac{\sqrt{67}}{8}$

145. $\sin\theta = \dfrac{\sqrt{2}}{6}$, given that $\cos\theta < 0$

$\sin\theta$ is positive and $\cos\theta$ is negative when θ is in quadrant II.

$\sin\theta = \dfrac{y}{r}$ and θ in quadrant II, so let

$y = \sqrt{2}, r = 6$.

$x^2 + y^2 = r^2 \Rightarrow x^2 + \left(\sqrt{2}\right)^2 = 6^2 \Rightarrow$

$x^2 + 2 = 36 \Rightarrow x^2 = 34 \Rightarrow$

$\qquad x = \pm\sqrt{34}$

θ is in quadrant II, so $x = -\sqrt{34}$.

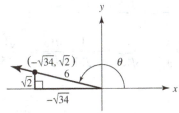

$\sin\theta = \dfrac{y}{r} = \dfrac{\sqrt{2}}{6}$; $\cos\theta = \dfrac{x}{r} = -\dfrac{\sqrt{34}}{6}$

(continued on next page)

(*continued*)

$$\tan\theta = \frac{y}{x} = -\frac{\sqrt{2}}{\sqrt{34}} = -\frac{\sqrt{2}}{\sqrt{34}}\cdot\frac{\sqrt{34}}{\sqrt{34}} = -\frac{\sqrt{68}}{34}$$

$$= -\frac{2\sqrt{17}}{34} = -\frac{\sqrt{17}}{17}$$

$$\cot\theta = \frac{x}{y} = -\frac{\sqrt{34}}{\sqrt{2}} = -\frac{\sqrt{34}}{\sqrt{2}}\cdot\frac{\sqrt{2}}{\sqrt{2}} = -\frac{\sqrt{68}}{2}$$

$$= -\frac{2\sqrt{17}}{2} = -\sqrt{17}$$

$$\sec\theta = \frac{r}{x} = -\frac{6}{\sqrt{34}} = -\frac{6}{\sqrt{34}}\cdot\frac{\sqrt{34}}{\sqrt{34}} = -\frac{6\sqrt{34}}{34}$$

$$= -\frac{3\sqrt{34}}{17}$$

$$\csc\theta = \frac{r}{y} = \frac{6}{\sqrt{2}} = \frac{6}{\sqrt{2}}\cdot\frac{\sqrt{2}}{\sqrt{2}} = \frac{6\sqrt{2}}{2} = 3\sqrt{2}$$

147. $\sec\theta = -4$, given that $\sin\theta > 0$

$\sec\theta$ is negative and $\sin\theta$ is positive when θ is in quadrant II.

$\sec\theta = \frac{r}{x}$ and θ in quadrant II, so let $x = -1, r = 4$.

$$x^2 + y^2 = r^2 \Rightarrow (-1)^2 + y^2 = 4^2 \Rightarrow$$

$$1 + y^2 = 16 \Rightarrow y^2 = 15 \Rightarrow y = \pm\sqrt{15}$$

θ is in quadrant II, so $y = \sqrt{15}$.

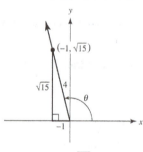

$$\sin\theta = \frac{y}{r} = \frac{\sqrt{15}}{4};\ \cos\theta = \frac{x}{r} = -\frac{1}{4}$$

$$\tan\theta = \frac{y}{x} = -\frac{\sqrt{15}}{1} = -\sqrt{15}$$

$$\cot\theta = \frac{x}{y} = -\frac{1}{\sqrt{15}} = -\frac{1}{\sqrt{15}}\cdot\frac{\sqrt{15}}{\sqrt{15}} = -\frac{\sqrt{15}}{15}$$

$$\sec\theta = \frac{r}{x} = -4$$

$$\csc\theta = \frac{r}{y} = \frac{4}{\sqrt{15}} = \frac{4}{\sqrt{15}}\cdot\frac{\sqrt{15}}{\sqrt{15}} = \frac{4\sqrt{15}}{15}$$

149. $\sin\theta = 1$

$$\sin^2\theta + \cos^2\theta = 1^2 \Rightarrow 1^2 + \cos^2\theta = 1 \Rightarrow$$

$$\cos^2\theta = 0 \Rightarrow \cos\theta = 0$$

$$\tan\theta = \frac{\sin\theta}{\cos\theta} = \frac{1}{0} \Rightarrow \tan\theta \text{ is undefined}$$

$$\cot\theta = \frac{\cos\theta}{\sin\theta} = \frac{0}{1} = 0$$

$$\sec\theta = \frac{1}{\cos\theta} = \frac{1}{0} \Rightarrow \sec\theta \text{ is undefined}$$

$$\csc\theta = \frac{1}{\sin\theta} = \frac{1}{1} = 1$$

151. $x^2 + y^2 = r^2 \Rightarrow \dfrac{x^2 + y^2}{y^2} = \dfrac{r^2}{y^2} \Rightarrow$

$$\frac{x^2}{y^2} + \frac{y^2}{y^2} = \frac{r^2}{y^2} \Rightarrow \left(\frac{x}{y}\right)^2 + 1 = \left(\frac{r}{y}\right)^2 \Rightarrow$$

$$1 + \left(\frac{x}{y}\right)^2 = \left(\frac{r}{y}\right)^2$$

Because $\cot\theta = \dfrac{x}{y}$ and $\csc\theta = \dfrac{r}{y}$, we have

$$1 + (\cot\theta)^2 = (\csc\theta)^2 \text{ or } 1 + \cot^2\theta = \csc^2\theta.$$

153. $90° < \theta < 180° \Rightarrow 180° < 2\theta < 360°$, so 2θ lies in either quadrant III or IV. Thus, $\sin 2\theta$ is negative.

155. $90° < \theta < 180° \Rightarrow 45° < \dfrac{\theta}{2} < 90°$, so $\dfrac{\theta}{2}$ lies in quadrant I. Thus, $\tan\dfrac{\theta}{2}$ is positive.

157. $-90° < \theta < 90° \Rightarrow -45° < \dfrac{\theta}{2} < 45°$, so $\dfrac{\theta}{2}$ lies in either quadrant I or quadrant IV. Thus $\cos\dfrac{\theta}{2}$ is positive.

159. $-90° < \theta < 90° \Rightarrow 90° > -\theta > -90° \Rightarrow$ $-90° < -\theta < 90°$, so $-\theta$ lies in either quadrant I or quadrant IV. Thus $\sec(-\theta)$ is positive.

Section 5.3 Trigonometric Function Values and Angle Measures

For Exercises 1–6, refer to the Function Values of Special Angles chart on page 524 of the text.

1. C; $\sin 30° = \dfrac{1}{2}$

3. B; $\tan 45° = 1$

5. E; $\csc 60° = \dfrac{1}{\sin 60°} = \dfrac{1}{\frac{\sqrt{3}}{2}} = \dfrac{2}{\sqrt{3}}$

$\qquad = \dfrac{2}{\sqrt{3}} \cdot \dfrac{\sqrt{3}}{\sqrt{3}} = \dfrac{2\sqrt{3}}{3}$

7. The value of $\sin 240°$ is <u>negative</u> because $240°$ is in quadrant <u>III</u>. The reference angle is <u>60°</u>, and the exact value of $\sin 240°$ is $-\dfrac{\sqrt{3}}{2}$.

9. The value of $\tan(-150°)$ is <u>positive</u> because $-150°$ is in quadrant <u>III</u>. The reference angle is <u>30°</u> and the exact value of $\tan(-150°)$ is $\dfrac{\sqrt{3}}{3}$.

11. $\sin A = \dfrac{\text{side opposite}}{\text{hypotenuse}} = \dfrac{21}{29}$

$\cos A = \dfrac{\text{side adjacent}}{\text{hypotenuse}} = \dfrac{20}{29}$

$\tan A = \dfrac{\text{side opposite}}{\text{side adjacent}} = \dfrac{21}{20}$

13. $\sin A = \dfrac{\text{side opposite}}{\text{hypotenuse}} = \dfrac{n}{p}$

$\cos A = \dfrac{\text{side adjacent}}{\text{hypotenuse}} = \dfrac{m}{p}$

$\tan A = \dfrac{\text{side opposite}}{\text{side adjacent}} = \dfrac{n}{m}$

15. $a = 5$, $b = 12$

$c^2 = a^2 + b^2 \Rightarrow c^2 = 5^2 + 12^2 \Rightarrow c^2 = 169 \Rightarrow$
$c = 13$

$\sin B = \dfrac{\text{side opposite}}{\text{hypotenuse}} = \dfrac{b}{c} = \dfrac{12}{13}$

$\cos B = \dfrac{\text{side adjacent}}{\text{hypotenuse}} = \dfrac{a}{c} = \dfrac{5}{13}$

$\tan B = \dfrac{\text{side opposite}}{\text{side adjacent}} = \dfrac{b}{a} = \dfrac{12}{5}$

$\cot B = \dfrac{\text{side adjacent}}{\text{side opposite}} = \dfrac{a}{b} = \dfrac{5}{12}$

$\sec B = \dfrac{\text{hypotenuse}}{\text{side adjacent}} = \dfrac{c}{a} = \dfrac{13}{5}$

$\csc B = \dfrac{\text{hypotenuse}}{\text{side opposite}} = \dfrac{c}{b} = \dfrac{13}{12}$

17. $a = 6$, $c = 7$

$c^2 = a^2 + b^2 \Rightarrow 7^2 = 6^2 + b^2 \Rightarrow$
$49 = 36 + b^2 \Rightarrow 13 = b^2 \Rightarrow \sqrt{13} = b$

$\sin B = \dfrac{\text{side opposite}}{\text{hypotenuse}} = \dfrac{b}{c} = \dfrac{\sqrt{13}}{7}$

$\cos B = \dfrac{\text{side adjacent}}{\text{hypotenuse}} = \dfrac{a}{c} = \dfrac{6}{7}$

$\tan B = \dfrac{\text{side opposite}}{\text{side adjacent}} = \dfrac{b}{a} = \dfrac{\sqrt{13}}{6}$

$\cot B = \dfrac{\text{side adjacent}}{\text{side opposite}} = \dfrac{a}{b} = \dfrac{6}{\sqrt{13}}$

$\qquad = \dfrac{6}{\sqrt{13}} \cdot \dfrac{\sqrt{13}}{\sqrt{13}} = \dfrac{6\sqrt{13}}{13}$

$\sec B = \dfrac{\text{hypotenuse}}{\text{side adjacent}} = \dfrac{c}{a} = \dfrac{7}{6}$

$\csc B = \dfrac{\text{hypotenuse}}{\text{side opposite}} = \dfrac{c}{b} = \dfrac{7}{\sqrt{13}}$

$\qquad = \dfrac{7}{\sqrt{13}} \cdot \dfrac{\sqrt{13}}{\sqrt{13}} = \dfrac{7\sqrt{13}}{13}$

19. $a = 3$, $c = 10$

$c^2 = a^2 + b^2 \Rightarrow 10^2 = 3^2 + b^2 \Rightarrow$
$100 = 9 + b^2 \Rightarrow 91 = b^2 \Rightarrow \sqrt{91} = b$

$\sin B = \dfrac{\text{side opposite}}{\text{hypotenuse}} = \dfrac{b}{c} = \dfrac{\sqrt{91}}{10}$

$\cos B = \dfrac{\text{side adjacent}}{\text{hypotenuse}} = \dfrac{a}{c} = \dfrac{3}{10}$

$\tan B = \dfrac{\text{side opposite}}{\text{side adjacent}} = \dfrac{b}{a} = \dfrac{\sqrt{91}}{3}$

$\cot B = \dfrac{\text{side adjacent}}{\text{side opposite}} = \dfrac{a}{b} = \dfrac{3}{\sqrt{91}}$

$\qquad = \dfrac{3}{\sqrt{91}} \cdot \dfrac{\sqrt{91}}{\sqrt{91}} = \dfrac{3\sqrt{91}}{91}$

$\sec B = \dfrac{\text{hypotenuse}}{\text{side adjacent}} = \dfrac{c}{a} = \dfrac{10}{3}$

$\csc B = \dfrac{\text{hypotenuse}}{\text{side opposite}} = \dfrac{c}{b} = \dfrac{10}{\sqrt{91}}$

$\qquad = \dfrac{10}{\sqrt{91}} \cdot \dfrac{\sqrt{91}}{\sqrt{91}} = \dfrac{10\sqrt{91}}{91}$

21. $a = 1, c = 2$

$$c^2 = a^2 + b^2 \Rightarrow 2^2 = 1^2 + b^2 \Rightarrow$$
$$4 = 1 + b^2 \Rightarrow 3 = b^2 \Rightarrow \sqrt{3} = b$$

$$\sin B = \frac{\text{side opposite}}{\text{hypotenuse}} = \frac{b}{c} = \frac{\sqrt{3}}{2}$$

$$\cos B = \frac{\text{side adjacent}}{\text{hypotenuse}} = \frac{a}{c} = \frac{1}{2}$$

$$\tan B = \frac{\text{side opposite}}{\text{side adjacent}} = \frac{b}{a} = \frac{\sqrt{3}}{1} = \sqrt{3}$$

$$\cot B = \frac{\text{side adjacent}}{\text{side opposite}} = \frac{a}{b} = \frac{1}{\sqrt{3}}$$
$$= \frac{1}{\sqrt{3}} \cdot \frac{\sqrt{3}}{\sqrt{3}} = \frac{\sqrt{3}}{3}$$

$$\sec B = \frac{\text{hypotenuse}}{\text{side adjacent}} = \frac{c}{a} = \frac{2}{1} = 2$$

$$\csc B = \frac{\text{hypotenuse}}{\text{side opposite}} = \frac{c}{b} = \frac{2}{\sqrt{3}}$$
$$= \frac{2}{\sqrt{3}} \cdot \frac{\sqrt{3}}{\sqrt{3}} = \frac{2\sqrt{3}}{3}$$

23. $\cos 30° = \sin (90° - 30°) = \sin 60°$

25. $\csc 60° = \sec (90° - 60°) = \sec 30°$

27. $\sec 39° = \csc (90° - 39°) = \csc 51°$

29. $\sin 38.7° = \cos (90° - 38.7°) = \cos 51.3°$

Use the following figures for exercises 31−45.

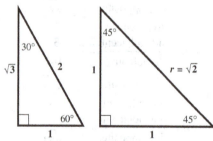

31. $\tan 30° = \dfrac{\text{side opposite}}{\text{side adjacent}} = \dfrac{1}{\sqrt{3}}$
$$= \frac{1}{\sqrt{3}} \cdot \frac{\sqrt{3}}{\sqrt{3}} = \frac{\sqrt{3}}{3}$$

33. $\sin 30° = \dfrac{\text{side opposite}}{\text{hypotenuse}} = \dfrac{1}{2}$

35. $\sec 30° = \dfrac{\text{hypotenuse}}{\text{side adjacent}} = \dfrac{2}{\sqrt{3}}$
$$= \frac{2}{\sqrt{3}} \cdot \frac{\sqrt{3}}{\sqrt{3}} = \frac{2\sqrt{3}}{3}$$

37. $\csc 45° = \dfrac{\text{hypotenuse}}{\text{side opposite}} = \dfrac{\sqrt{2}}{1} = \sqrt{2}$

39. $\cos 45° = \dfrac{\text{side adjacent}}{\text{hypotenuse}} = \dfrac{1}{\sqrt{2}}$
$$= \frac{1}{\sqrt{2}} \cdot \frac{\sqrt{2}}{\sqrt{2}} = \frac{\sqrt{2}}{2}$$

41. $\tan 45° = \dfrac{\text{side opposite}}{\text{side adjacent}} = \dfrac{1}{1} = 1$

43. $\sin 60° = \dfrac{\text{side opposite}}{\text{hypotenuse}} = \dfrac{\sqrt{3}}{2}$

45. $\tan 60° = \dfrac{\text{side opposite}}{\text{side adjacent}} = \dfrac{\sqrt{3}}{1} = \sqrt{3}$

47.

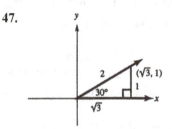

The line passes through $(0, 0)$ and $\left(\sqrt{3}, 1\right)$.

The slope is change in y over the change in x.

Thus, $m = \dfrac{1}{\sqrt{3}} = \dfrac{1}{\sqrt{3}} \cdot \dfrac{\sqrt{3}}{\sqrt{3}} = \dfrac{\sqrt{3}}{3}$ and the

equation of the line is $y = \dfrac{\sqrt{3}}{3} x$.

49. One point on the line $y = \sqrt{3}x$ is the origin $(0, 0)$. Let (x, y) be any other point on this line. Then, by the definition of slope,

$$m = \frac{y - 0}{x - 0} = \frac{y}{x} = \sqrt{3}, \text{ but also, by the}$$

definition of tangent, $\tan \theta = \sqrt{3}$. Because $\tan 60° = \sqrt{3}$, the line $y = \sqrt{3}x$ makes a $60°$ angle with the positive x-axis (See exercise 48).

51. Apply the relationships between the lengths of the sides of a $30° - 60°$ right triangle first to the triangle on the left to find the values of y and x, and then to the triangle on the right to find the values of z and w. In the $30° - 60°$ right triangle, the side opposite the $30°$ angle is $\frac{1}{2}$ the length of the hypotenuse. The longer leg is $\sqrt{3}$ times the shorter leg.

(*continued on next page*)

(continued)

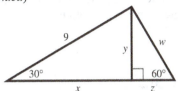

Thus, we have

$$y = \frac{1}{2}(9) = \frac{9}{2} \text{ and } x = y\sqrt{3} = \frac{9\sqrt{3}}{2}$$

$$y = z\sqrt{3}, \text{ so } z = \frac{y}{\sqrt{3}} = \frac{\frac{9}{2}}{\sqrt{3}} = \frac{9\sqrt{3}}{6} = \frac{3\sqrt{3}}{2},$$

and $w = 2z$, so $w = 2\left(\dfrac{3\sqrt{3}}{2}\right) = 3\sqrt{3}$

53. Apply the relationships between the lengths of the sides of a $45° - 45°$ right triangle to the triangle on the left to find the values of p and r. In the $45° - 45°$ right triangle, the sides opposite the $45°$ angles measure the same. The hypotenuse is $\sqrt{2}$ times the measure of a leg.

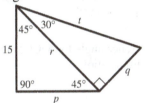

Thus, we have $p = 15$ and $r = p\sqrt{2} = 15\sqrt{2}$

Apply the relationships between the lengths of the sides of a $30° - 60°$ right triangle next to the triangle on the right to find the values of q and t. In the $30° - 60°$ right triangle, the side opposite the $60°$ angle is $\sqrt{3}$ times as long as the side opposite to the $30°$ angle. The length of the hypotenuse is 2 times as long as the shorter leg (opposite the $30°$ angle). Thus, we have $r = q\sqrt{3} \Rightarrow$

$$q = \frac{r}{\sqrt{3}} = \frac{15\sqrt{2}}{\sqrt{3}} = \frac{15\sqrt{2}}{\sqrt{3}} \cdot \frac{\sqrt{3}}{\sqrt{3}} = 5\sqrt{6} \text{ and}$$

$$t = 2q = 2\left(5\sqrt{6}\right) = 10\sqrt{6}$$

55. Because $A = \dfrac{1}{2}bh$, we have

$$A = \frac{1}{2} \cdot s \cdot s = \frac{1}{2}s^2 \text{ or } A = \frac{s^2}{2}.$$

57. C; $180° - 98° = 82°$
($98°$ is in quadrant II)

59. A; $-135° + 360° = 225°$ and
$225° - 180° = 45°$
($225°$ is in quadrant III)

61. D; $750° - 2 \cdot 360° = 30°$
($30°$ is in quadrant I)

	θ	$\sin\theta$	$\cos\theta$	$\tan\theta$	$\cot\theta$	$\sec\theta$	$\csc\theta$
63.	$30°$	$\dfrac{1}{2}$	$\dfrac{\sqrt{3}}{2}$	$\dfrac{\sqrt{3}}{3}$	$\sqrt{3}$	$\dfrac{2\sqrt{3}}{3}$	2
65.	$60°$	$\dfrac{\sqrt{3}}{2}$	$\dfrac{1}{2}$	$\sqrt{3}$	$\dfrac{\sqrt{3}}{3}$	2	$\dfrac{2\sqrt{3}}{3}$
67.	$135°$	$\dfrac{\sqrt{2}}{2}$	$-\dfrac{\sqrt{2}}{2}$	$\begin{aligned}\tan 135° \\ = -\tan 45° \\ = -1\end{aligned}$	$\begin{aligned}\cot 135° \\ = -\cot 45° \\ = -1\end{aligned}$	$-\sqrt{2}$	$\sqrt{2}$
69.	$210°$	$-\dfrac{1}{2}$	$\begin{aligned}\cos 210° \\ = -\cos 30° \\ = -\dfrac{\sqrt{3}}{2}\end{aligned}$	$\dfrac{\sqrt{3}}{3}$	$\sqrt{3}$	$\begin{aligned}\sec 210° \\ = -\sec 30° \\ = -\dfrac{2\sqrt{3}}{3}\end{aligned}$	-2

71. To find the reference angle for $300°$, sketch this angle in standard position.

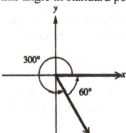

The reference angle is $360° - 300° = 60°$.
Because $300°$ lies in quadrant IV, the sine, tangent, cotangent, and cosecant are negative.

$$\sin 300° = -\sin 60° = -\frac{\sqrt{3}}{2}$$

$$\cos 300° = \cos 60° = \frac{1}{2}$$

$$\tan 300° = -\tan 60° = -\sqrt{3}$$

$$\cot 300° = -\cot 60° = -\frac{\sqrt{3}}{3}$$

$$\sec 300° = \sec 60° = 2$$

$$\csc 300° = -\csc 60° = -\frac{2\sqrt{3}}{3}$$

73. To find the reference angle for $405°$, sketch this angle in standard position.

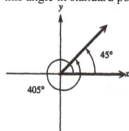

The reference angle for $405°$ is
$405° - 360° = 45°$. Because $405°$ lies in quadrant I, the values of all of its trigonometric functions will be positive, so these values will be identical to the trigonometric function values for $45°$ See the Function Values of Special Angles table on page 524 in your text.)

$$\sin 405° = \sin 45° = \frac{\sqrt{2}}{2}$$

$$\cos 405° = \cos 45° = \frac{\sqrt{2}}{2}$$

$$\tan 405° = \tan 45° = 1$$

$$\cot 405° = \cot 45° = 1$$

$$\sec 405° = \sec 45° = \sqrt{2}$$

$$\csc 405° = \csc 45° = \sqrt{2}$$

75. To find the reference angle for $480°$, sketch this angle in standard position.

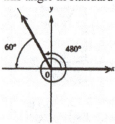

$480°$ is coterminal with $480° - 360° = 120°$.
The reference angle is $180° - 120° = 60°$.
Because $480°$ lies in quadrant II, the cosine, tangent, cotangent, and secant are negative.

$$\sin\left(480°\right) = \sin 60° = \frac{\sqrt{3}}{2}$$

$$\cos\left(480°\right) = -\cos 60° = -\frac{1}{2}$$

$$\tan\left(480°\right) = -\tan 60° = -\sqrt{3}$$

$$\cot\left(480°\right) = -\cot 60° = -\frac{\sqrt{3}}{3}$$

$$\sec\left(80°\right) = -\sec 60° = -2$$

$$\csc\left(480°\right) = \csc 60° = \frac{2\sqrt{3}}{3}$$

77. To find the reference angle for $570°$ sketch this angle in standard position.

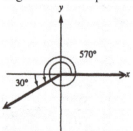

$570°$ is coterminal with $570° - 360° = 210°$.
The reference angle is $210° - 180° = 30°$.
Because $570°$ lies in quadrant III, the sine, cosine, secant, and cosecant are negative.

$$\sin 570° = -\sin 30° = -\frac{1}{2}$$

$$\cos 570° = -\cos 30° = -\frac{\sqrt{3}}{2}$$

$$\tan 570° = \tan 30° = \frac{\sqrt{3}}{3}$$

$$\cot 570° = \cot 30° = \sqrt{3}$$

$$\sec 570° = -\sec 30° = -\frac{2\sqrt{3}}{3}$$

$$\csc 570° = -\csc 30° = -2$$

79. 1305° is coterminal with
$1305° - 3 \cdot 360° = 1305° = 1080° = 225°$. The
reference angle is $225° - 180° = 45°$. Because
1305° lies in quadrant III, the sine, cosine, and
secant and cosecant are negative.

$$\sin 1305° = -\sin 45° = -\frac{\sqrt{2}}{2}$$

$$\cos 1305° = -\cos 45° = -\frac{\sqrt{2}}{2}$$

$$\tan 1305° = \tan 45° = 1$$

$$\cot 1305° = \cot 45° = 1$$

$$\sec 1305° = -\sec 45° = -\sqrt{2}$$

$$\csc 1305° = -\csc 45° = -\sqrt{2}$$

81. To find the reference angle for –300°, sketch
this angle in standard position.

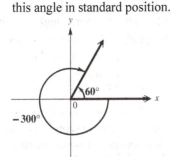

The reference angle for –300° is –
300°+ 360°= 60°. Because –300° lies in
quadrant I, the values of all of its
trigonometric functions will be positive, so
these values will be identical to the
trigonometric function values for 60°. (See the
Function Values of Special Angles table on
page 524 in your text.)

$$\sin\left(-300°\right) = \sin 60° = \frac{\sqrt{3}}{2}$$

$$\cos\left(-300°\right) = \cos 60° = \frac{1}{2}$$

$$\tan\left(-300°\right) = \tan 60° = \sqrt{3}$$

$$\cot\left(-300°\right) = \cot 60° = \frac{\sqrt{3}}{3}$$

$$\sec\left(-300°\right) = \sec 60° = 2$$

$$\csc\left(-300°\right) = \csc 60° = \frac{2\sqrt{3}}{3}$$

83. –510° is coterminal with
$-510° + 2 \cdot 360° = -510° + 720° = 210°$. The
reference angle is $210° - 180° = 30°$. Because
–510° lies in quadrant III, the sine, cosine, and
secant and cosecant are negative.

$$\sin\left(-510°\right) = -\sin 30° = -\frac{1}{2}$$

$$\cos\left(-510°\right) = -\cos 30° = -\frac{\sqrt{3}}{2}$$

$$\tan\left(-510°\right) = \tan 30° = \frac{\sqrt{3}}{3}$$

$$\cot\left(-510°\right) = \cot 30° = \sqrt{3}$$

$$\sec\left(-510°\right) = -\sec 30° = -\frac{2\sqrt{3}}{3}$$

$$\csc\left(-510°\right) = -\csc 30° = -2$$

85. –1290° is coterminal with
$-1290° + 4 \cdot 360° = -1290° + 1440° = 150°$.
The reference angle is $180° - 150° = 30°$.
Because –1290° lies in quadrant II, the cosine,
tangent, cotangent, and secant are negative.

$$\sin 2670° = \sin 30° = \frac{1}{2}$$

$$\cos 2670° = -\cos 30° = -\frac{\sqrt{3}}{2}$$

$$\tan 2670° = -\tan 30° = -\frac{\sqrt{3}}{3}$$

$$\cot 2670° = -\cot 30° = -\sqrt{3}$$

$$\sec 2670° = -\sec 30° = -\frac{2\sqrt{3}}{3}$$

$$\csc 2670° = \csc 30° = 2$$

87. –1860° is coterminal with
$-1860° + 6 \cdot 360° = -1860° + 2160° = 300°$.
The reference angle is $360° - 300° = 60°$.
Because –1860° lies in quadrant IV, the sine,
tangent, cotangent, and cosecant are negative.

$$\sin\left(-1860°\right) = -\sin 60° = -\frac{\sqrt{3}}{2}$$

$$\cos\left(-1860°\right) = \cos 60° = \frac{1}{2}$$

$$\tan\left(-1860°\right) = -\tan 60° = -\sqrt{3}$$

$$\cot\left(-1860°\right) = -\cot 60° = -\frac{\sqrt{3}}{3}$$

$$\sec\left(-1860°\right) = \sec 60° = 2$$

$$\csc\left(-1860°\right) = -\csc 60° = -\frac{2\sqrt{3}}{3}$$

89. Because 1305° is coterminal with an angle of
$1305° - 3 \cdot 360° = 1305° - 1080° = 225°$, it lies
in quadrant III. Its reference angle is
$225° - 180° = 45°$. Because the sine is
negative in quadrant III, we have

$$\sin 1305° = -\sin 45° = -\frac{\sqrt{2}}{2}.$$

91. Because $-510°$ is coterminal with an angle of $-510° + 2 \cdot 360° = -510° + 720° = 210°,$ it lies in quadrant III. Its reference angle is $210° - 180° = 30°.$ The cosine is negative in quadrant III, so

$$\cos\left(-510°\right) = -\cos 30° = -\frac{\sqrt{3}}{2}.$$

93. Because $-855°$ is coterminal with $-855° + 3 \cdot 360° = -855° + 1080° = 225°,$ it lies in quadrant III. Its reference angle is $225° - 180° = 45°.$ The cosecant is negative in quadrant III, so

$$\csc\left(-855°\right) = -\csc 45° = -\sqrt{2}.$$

95. Because $3015°$ is coterminal with $3015° - 8 \cdot 360° = 3015° - 2880° = 135°,$ it lies in quadrant II. Its reference angle is $180° - 135° = 45°.$ The tangent is negative in quadrant II, so $\tan 3015° = -\tan 45° = -1.$

97. $\sin \theta = \dfrac{1}{2}$

Because $\sin \theta$ is positive, θ must lie in quadrants I or II. Because one angle, namely $30°,$ lies in quadrant I, that angle is also the reference angle, $\theta'.$ The angle in quadrant II will be $180° - \theta' = 180° - 30° = 150°.$

99. $\tan \theta = -\sqrt{3}$

Because $\tan \theta$ is negative, θ must lie in quadrants II or IV. The absolute value of $\tan \theta$ is $\sqrt{3},$ so the reference angle, θ' must be $60°.$ The quadrant II angle θ equals $180° - \theta' = 180° - 60° = 120°,$ and the quadrant IV angle θ equals $360° - \theta' = 360° - 60° = 300°.$

101. $\cos \theta = \dfrac{\sqrt{2}}{2}$

Because $\cos \theta$ is positive, θ must lie in quadrants I or IV. One angle, namely $45°,$ lies in quadrant I, so that angle is also the reference angle, $\theta'.$ The angle in quadrant IV will be $360° - \theta' = 360° - 45° = 315°.$

103. $\csc \theta = -2$

Because $\csc \theta$ is negative, θ must lie in quadrants III or IV. The absolute value of $\csc \theta$ is 2, so the reference angle, $\theta',$ is $30°.$

The angle in quadrant III will be $180° + \theta' = 180° + 30° = 210°,$ and the quadrant IV angle is $360° - \theta' = 360° - 30° = 330°.$

105. $\tan \theta = \dfrac{\sqrt{3}}{3}$

Because $\tan \theta$ is positive, θ must lie in quadrants I or III. One angle, namely $30°,$ lies in quadrant I, so that angle is also the reference angle, $\theta'.$ The angle in quadrant III will be $180° + \theta' = 180° + 30° = 210°.$

107. $\csc \theta = -\sqrt{2}$

Because $\csc \theta$ is negative, θ must lie in quadrants III or IV. The absolute value of $\csc \theta$ is $\sqrt{2},$ so the reference angle, θ' must be $45°.$ The quadrant III angle θ equals $180° + \theta' = 180° + 45° = 225°.$ and the quadrant IV angle θ equals $360° - \theta' = 360° - 45° = 315°.$

For Exercises 109–133, be sure your calculator is in degree mode. If your calculator accepts angles in degrees, minutes, and seconds, it is not necessary to change angles to decimal degrees. Keystroke sequences may vary on the type and/or model of calculator being used. Screens shown will be from a TI-84 Plus C calculator. To obtain the degree (°) and (′) symbols, go to the ANGLE menu (2nd APPS).

For Exercises 109–133, we include TI-84 screens only for those exercises involving cotangent, secant, and cosecant.

109. $\sin 38°42' \approx 0.625243$

111. $\sec 13°15' \approx 1.027349$

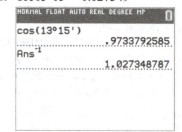

113. $\cot 183°48' \approx 15.055723$

115. $\sin(-312°12') \approx 0.740805$

117. $\csc(-317°36') \approx 1.483014$

119. $\dfrac{1}{\cot 23.4°} = \tan 23.4° \approx 0.432739$

121. $\dfrac{\cos 77°}{\sin 77°} = \cot 77° \approx 0.230868$

123. $\tan\theta = 1.4739716$
$\theta = \tan^{-1}(1.4739716) \approx 55.845496°$

125. $\sin\theta = 0.27843196$
$\theta = \sin^{-1}(0.27843196) \approx 16.166641°$

127. $\cot\theta = 1.2575516$
$\theta = \cot^{-1}(1.2575516) = \tan^{-1}\left(\dfrac{1}{1.2575516}\right)$
$\approx 38.491580°$

129. $\sec\theta = 2.7496222$
$\theta = \sec^{-1}(2.7496222) = \cos^{-1}\left(\dfrac{1}{2.7496222}\right)$
$\approx 68.673241°$

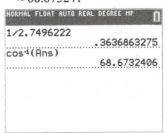

131. $\cos\theta = 0.70058013$
$\theta = \cos^{-1}(0.70058013) \approx 45.526434°$

133. $\csc\theta = 4.7216543 \Rightarrow$
$\theta = \csc^{-1}(4.7216543) = \sin^{-1}\left(\dfrac{1}{4.7216543}\right)$
$\approx 12.227282°$

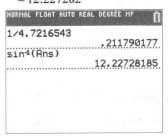

135. $F = W\sin\theta$
$F = 2100\sin 1.8° \approx 65.96 \approx 70$ lb

137. $F = W\sin\theta$
$-130 = 2600\sin\theta \Rightarrow \dfrac{-130}{2600} = \sin\theta \Rightarrow$
$-0.05 = \sin\theta \Rightarrow \theta = \sin^{-1}(-0.05) \approx -2.9°$

139. $F = W\sin\theta$
$120 = W\sin(2.7°) \Rightarrow \dfrac{120}{\sin(2.7°)} = W \Rightarrow$
$W \approx 2547 \approx 2500$ lb

141. $F = W\sin\theta$
$F = 2200\sin 2° \approx 76.77889275$ lb
$F = 2000\sin 2.2° \approx 76.77561818$ lb
The 2200-lb car on a 2° uphill grade has the greater grade resistance.

143. 45 mph = 66 ft/sec, $V = 66$, $\theta = 3°$, $g = 32.2$,
$f = 0.14$
$$R = \dfrac{V^2}{g(f + \tan\theta)} = \dfrac{66^2}{32.2(0.14 + \tan 3°)}$$
≈ 703 ft

145. Intuitively, increasing θ would make it easier to negotiate the curve at a higher speed much like is done at a race track. Mathematically, a larger value of θ (acute) will lead to a larger value for $\tan\theta$. If $\tan\theta$ increases, then the ratio determining R will *decrease*. Thus, the radius can be smaller and the curve sharper if θ is increased.

$$R = \frac{V^2}{g(f+\tan\theta)} = \frac{66^2}{32.2(0.14+\tan 4°)}$$
$$\approx 644 \text{ ft}$$

$$R = \frac{V^2}{g(f+\tan\theta)} \approx \frac{102.67^2}{32.2(0.14+\tan 4°)}$$
$$\approx 1559 \text{ ft}$$

As predicted, both values are less.

147. For Auto A, calculate $70 \cdot \cos 10° \approx 68.94$.
Auto A's reading is approximately 69 mph.
For Auto B, calculate $70 \cdot \cos 20° \approx 65.78$.
Auto B's reading is approximately 66 mph.

Chapter 5 Quiz
(Sections 5.1−5.3)

1. 19°

   (a) complement: $90° - 19° = 71°$

   (b) supplement: $180° - 19° = 161°$

3. The two angles form a right angle.
$$(5x-1)+2x = 90 \Rightarrow 7x-1 = 90 \Rightarrow$$
$$7x = 91 \Rightarrow x = 13$$
The measures of the two angles are
$$(5x-1)° = [5(13)-1] = 64° \text{ and}$$
$$2x = 2(13) = 26°.$$

5. **(a)** 410° is coterminal with
     $410° - 360° = 50°$.

   (b) −60° is coterminal with
     $-60° + 360° = 300°$.

   (c) 890° is coterminal with
     $890° - 2(360°) = 890° - 720° = 170°$.

   (d) 57° is coterminal with $57° + 360° = 417°$.

7.

$x = -24, y = 7$, and
$$r = \sqrt{x^2+y^2} = \sqrt{(-24)^2+7^2}$$
$$= \sqrt{576+49} = \sqrt{625} = 25$$

$$\sin\theta = \frac{y}{r} = \frac{7}{25} = \frac{7}{25}$$
$$\cos\theta = \frac{x}{r} = \frac{-24}{25} = -\frac{24}{25}$$
$$\tan\theta = \frac{y}{x} = \frac{7}{-24} = -\frac{7}{24}$$
$$\cot\theta = \frac{x}{y} = \frac{-24}{7} = -\frac{24}{7}$$
$$\sec\theta = \frac{r}{x} = \frac{25}{-24} = -\frac{25}{24}$$
$$\csc\theta = \frac{r}{y} = \frac{25}{7} = \frac{25}{7}$$

9.

θ	$\sin\theta$	$\cos\theta$	$\tan\theta$	$\cot\theta$	$\sec\theta$	$\csc\theta$
30°	$\frac{1}{2}$	$\frac{\sqrt{3}}{2}$	$\frac{\sqrt{3}}{3}$	$\sqrt{3}$	$\frac{2\sqrt{3}}{3}$	2
45°	$\frac{\sqrt{2}}{2}$	$\frac{\sqrt{2}}{2}$	1	1	$\sqrt{2}$	$\sqrt{2}$
60°	$\frac{\sqrt{3}}{2}$	$\frac{1}{2}$	$\sqrt{3}$	$\frac{\sqrt{3}}{3}$	2	$\frac{2\sqrt{3}}{3}$

11. $\sin 30° = \frac{w}{36} \Rightarrow w = 36\sin 30° = 36 \cdot \frac{1}{2} = 18$

$\cos 30° = \frac{x}{36} \Rightarrow x = 36\cos 30° = 36 \cdot \frac{\sqrt{3}}{2} = 18\sqrt{3}$

$\tan 45° = \frac{w}{y} \Rightarrow 1 = \frac{18}{y} \Rightarrow y = 18$

$\sin 45° = \frac{w}{z} \Rightarrow \frac{\sqrt{2}}{2} = \frac{18}{z} \Rightarrow z = \frac{36}{\sqrt{2}} = 18\sqrt{2}$

13. −150° is coterminal with $360° - 150° = 210°$. This lies in quadrant III, so the reference angle is $210° - 180° = 30°$. In quadrant III, the tangent and cotangent functions are positive, while the remaining trigonometric functions are negative.

$$\sin(-150°) = -\sin 30° = -\frac{1}{2}$$
$$\cos(-150°) = -\cos 30° = -\frac{\sqrt{3}}{2}$$
$$\tan(-150°) = \tan 30° = \frac{\sqrt{3}}{3}$$
$$\cot(-150°) = \cot 30° = \sqrt{3}$$
$$\sec(-150°) = -\sec 30° = -\frac{2\sqrt{3}}{3}$$
$$\csc(-150°) = -\csc 30° = -2$$

15. $\sin \theta = \dfrac{\sqrt{3}}{2}$

Because $\sin \theta$ is positive, θ must lie in quadrants I or II, and the reference angle, θ', is 60°. The angle in quadrant I is 60°, while the angle in quadrant II is $180° - \theta' = 180° - 60° = 120°$.

17. $\sin 42°18' \approx 0.673013$

19. $\tan \theta = 2.6743210 \Rightarrow \theta \approx 69.497888°$

Section 5.4   Solutions and Applications of Right Triangles

1. B   **3.** A   **5.** C

7. C   **9.** A   **11.** E

13. 23.825 to 23.835

15. $A = 36°20', \ c = 964$ m

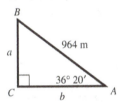

$A + B = 90° \Rightarrow B = 90° - A \Rightarrow$
$\quad B = 90° - 36°20'$
$\qquad = 89°60' - 36°20' = 53°40'$

$\sin A = \dfrac{a}{c} \Rightarrow \sin 36°20' = \dfrac{a}{964} \Rightarrow$
$\quad a = 964 \sin 36°20' \approx 571$ m (rounded to three significant digits)

$\cos A = \dfrac{b}{c} \Rightarrow \cos 36°20' = \dfrac{b}{964} \Rightarrow$
$\quad b = 964 \cos 36°20' \approx 777$ m   (rounded to three significant digits)

17. $N = 51.2°, \ m = 124$ m

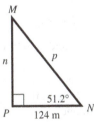

$M + N = 90° \Rightarrow M = 90° - N \Rightarrow$
$\quad M = 90° - 51.2° = 38.8°$

$\tan N = \dfrac{n}{m} \Rightarrow \tan 51.2° = \dfrac{n}{124} \Rightarrow$
$\quad n = 124 \tan 51.2° \approx 154$ m   (rounded to three significant digits)

$\cos N = \dfrac{m}{p} \Rightarrow \cos 51.2° = \dfrac{124}{p} \Rightarrow$
$\quad p = \dfrac{124}{\cos 51.2°} \approx 198$ m   (rounded to three significant digits)

19. $B = 42.0892°, b = 56.851$

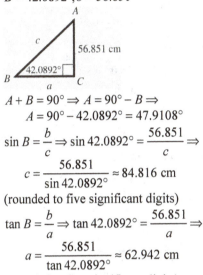

$A + B = 90° \Rightarrow A = 90° - B \Rightarrow$
$\quad A = 90° - 42.0892° = 47.9108°$

$\sin B = \dfrac{b}{c} \Rightarrow \sin 42.0892° = \dfrac{56.851}{c} \Rightarrow$
$\quad c = \dfrac{56.851}{\sin 42.0892°} \approx 84.816$ cm
(rounded to five significant digits)

$\tan B = \dfrac{b}{a} \Rightarrow \tan 42.0892° = \dfrac{56.851}{a} \Rightarrow$
$\quad a = \dfrac{56.851}{\tan 42.0892°} \approx 62.942$ cm
(rounded to five significant digits)

21. $a = 12.5, b = 16.2$

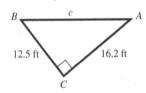

Using the Pythagorean theorem, we have
$a^2 + b^2 = c^2 \Rightarrow 12.5^2 + 16.2^2 = c^2 \Rightarrow$
$418.69 = c^2 \Rightarrow c \approx 20.5$ ft (rounded to three significant digits)

$\tan A = \dfrac{a}{b} \Rightarrow \tan A = \dfrac{12.5}{16.2} \Rightarrow$
$\quad A = \tan^{-1} \dfrac{12.5}{16.2} \approx 37.6540°$
$\qquad \approx 37° + (0.6540 \cdot 60)' \approx 37°39' \approx 37°40'$
(rounded to three significant digits)

$\tan B = \dfrac{b}{a} \Rightarrow \tan B = \dfrac{16.2}{12.5} \Rightarrow$
$\quad B = \tan^{-1} \dfrac{16.2}{12.5} \approx 52.3460°$
$\qquad \approx 52° + (0.3460 \cdot 60)' \approx 52°21' \approx 52°20'$
(rounded to three significant digits)

23. $A = 28.0°$, $c = 17.4$ ft

$A + B = 90°$
$\quad B = 90° - A$
$\quad B = 90° - 28.0° = 62.0°$

$\sin A = \dfrac{a}{c} \Rightarrow \sin 28.0° = \dfrac{a}{17.4} \Rightarrow$
$\quad a = 17.4 \sin 28.0° \approx 8.17$ ft   (rounded to three significant digits)

$\cos A = \dfrac{b}{c} \Rightarrow \cos 28.00° = \dfrac{b}{17.4} \Rightarrow$
$\quad b = 17.4 \cos 28.00° \approx 15.4$ ft   (rounded to three significant digits)

25. Solve the right triangle with $B = 73.0°$, $b = 128$ in. and $C = 90°$

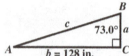

$A + B = 90° \Rightarrow A = 90° - B \Rightarrow$
$\quad A = 90° - 73.0° = 17.0°$

$\tan B° = \dfrac{b}{a} \Rightarrow \tan 73.0° = \dfrac{128}{a} \Rightarrow$
$\quad a = \dfrac{128}{\tan 73.0°} \Rightarrow a = 39.1$ in (rounded to three significant digits)

$\sin B° = \dfrac{b}{c} \Rightarrow \sin 73.0° = \dfrac{128}{c} \Rightarrow$
$\quad c = \dfrac{128}{\sin 73.0°} \Rightarrow c = 134$ in (rounded to three significant digits)

27. $A = 61.0°$, $b = 39.2$ cm
$A + B = 90° \Rightarrow B = 90° - A \Rightarrow$
$\quad B = 90° - 61.0° = 29.0°$

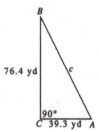

$\tan A = \dfrac{a}{b} \Rightarrow \tan 61.0° = \dfrac{a}{39.2} \Rightarrow$
$\quad a = 39.2 \tan 61.0 \approx 70.7$ cm
(rounded to three significant digits)

$\cos A = \dfrac{b}{c} \Rightarrow \cos 61.0° = \dfrac{39.2}{c} \Rightarrow$
$\quad c = \dfrac{39.2}{\cos 61.0°} \approx 80.9$ cm
(rounded to three significant digits)

29. $a = 13$ m, $c = 22$m

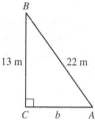

$c^2 = a^2 + b^2 \Rightarrow 22^2 = 13^2 + b^2 \Rightarrow$
$484 = 169 + b^2 \Rightarrow 315 = b^2 \Rightarrow b \approx 18$ m
(rounded to two significant digits)

We will determine the measurements of both A and B by using the sides of the right triangle. In practice, once you find one of the measurements, subtract it from $90°$ to find the other.

$\sin A = \dfrac{a}{c} \Rightarrow \sin A = \dfrac{13}{22} \Rightarrow$
$\quad A \approx \sin^{-1}\left(\dfrac{13}{22}\right) \approx 36.2215° \approx 36°$ (rounded to two significant digits)

$\cos B = \dfrac{b}{c} \Rightarrow \cos B = \dfrac{13}{22} \Rightarrow$
$\quad B \approx \cos^{-1}\left(\dfrac{13}{22}\right) \approx 53.7784° \approx 54°$

(rounded to two significant digits)

31. $a = 76.4$ yd, $b = 39.3$ yd

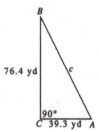

$c^2 = a^2 + b^2 \Rightarrow c = \sqrt{a^2 + b^2}$
$\quad = \sqrt{(76.4)^2 + (39.3)^2} = \sqrt{5836.96 + 1544.49}$
$\quad = \sqrt{7381.45} \approx 85.9$ yd (rounded to three significant digits)

We will determine the measurements of both A and B by using the sides of the right triangle. In practice, once you find one of the measurements, subtract it from $90°$ to find the other.

(continued on next page)

(continued)

$$\tan A = \frac{a}{b} \Rightarrow \tan A = \frac{76.4}{39.3} \Rightarrow$$

$$A \approx \tan^{-1}\left(\frac{76.4}{39.3}\right) \approx 62.7788°$$

$$\approx 62° + (0.7788 \cdot 60)' \approx 62°47' \approx 62°50'$$

(rounded to three significant digits)

$$\tan B = \frac{b}{a} \Rightarrow \tan B = \frac{39.3}{76.4} \Rightarrow$$

$$B \approx \tan^{-1}\left(\frac{39.3}{76.4}\right) \approx 27.2212°$$

$$\approx 27° + (0.2212 \cdot 60)' \approx 27°13' \approx 27°10'$$

(rounded to three significant digits)

33. $a = 18.9$ cm, $c = 46.3$ cm

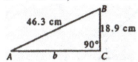

$$c^2 = a^2 + b^2 \Rightarrow 46.3^2 = 18.9^2 + b^2 \Rightarrow$$

$$2143.69 = 357.21 + b^2 \Rightarrow 1786.48 = b^2 \Rightarrow$$

$$b \approx 42.3 \text{ cm (rounded to three}$$

significant digits)

$$\sin A = \frac{a}{c} \Rightarrow \sin A = \frac{18.9}{46.3} \Rightarrow$$

$$A \approx \sin^{-1}\left(\frac{18.9}{46.3}\right) \approx 24.09227°$$

$$\approx 24° + (0.09227 \cdot 60)' \approx 24°06' \approx 24°10'$$

(rounded to three significant digits)

$$\cos B = \frac{a}{c} \Rightarrow \cos B = \frac{18.9}{46.3} \Rightarrow$$

$$B \approx \cos^{-1}\left(\frac{18.9}{46.3}\right) \approx 65.9077°$$

$$\approx 65° + (0.9077 \cdot 60)' \approx 65°54' \approx 65°50'$$

(rounded to three significant digits)

35. $A = 53°24'$, $c = 387.1$ ft

$$A + B = 90°$$
$$B = 90° - A$$
$$B = 90° - 53°24'$$
$$= 89°60' - 53°24'$$
$$= 36°36'$$

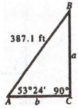

$$\sin A = \frac{a}{c} \Rightarrow \sin 53°24' = \frac{a}{387.1} \Rightarrow$$
$$a = 387.1\sin 53°24' \approx 310.8 \text{ ft (rounded}$$
to four significant digits)

$$\cos A = \frac{b}{c} \Rightarrow \cos 53°24' = \frac{b}{387.1} \Rightarrow$$
$$b = 387.1\cos 53°24' \approx 230.8 \text{ ft (rounded}$$
to four significant digits)

37. $B = 39°09'$, $c = 0.6231$ m

$$A + B = 90°$$
$$B = 90° - A$$
$$B = 90° - 39°09'$$
$$= 89°60' - 39°09'$$
$$= 50°51'$$

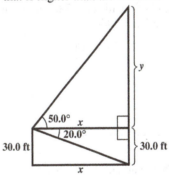

$$\sin B = \frac{b}{c} \Rightarrow \sin 39°09' = \frac{b}{0.6231} \Rightarrow$$
$$b = 0.6231\sin 39°09' \approx 0.3934 \text{ m (rounded}$$
to four significant digits)

$$\cos B = \frac{a}{c} \Rightarrow \cos 39°09' = \frac{a}{0.6231} \Rightarrow$$
$$a = 0.6231\cos 39°09' \approx 0.4832 \text{ m (rounded}$$
to four significant digits)

39. $$\sin 43°50' = \frac{d}{13.5}$$
$$d = 13.5\sin 43°50' \approx 9.3496000$$
The ladder goes up the wall 9.35 m. (rounded
to three significant digits)

41. Let x represent the horizontal distance between
the two buildings and y represent the height of
the portion of the building across the street
that is higher than the window.

$$\tan 20.0° = \frac{30.0}{x} \Rightarrow x = \frac{30.3}{\tan 20.0°} \approx 82.4$$

$$\tan 50.0° = \frac{y}{x} \Rightarrow$$

$$y = x\tan 50.0° = \left(\frac{30.0}{\tan 20.0°}\right)\tan 50.0°$$

$$\text{height} = y + 30.0 = \left(\frac{30.0}{\tan 20.0°}\right)\tan 50.0° + 30.0$$

$$\approx 128.2295$$

The height of the building across the street is
about 128 ft (rounded to three significant
digits)

43. The altitude of an isosceles triangle bisects the base as well as the angle opposite the base. The two right triangles formed have interior angles which have the same measure. The lengths of the corresponding sides also have the same measure. The altitude bisects the base, so each leg (base) of the right triangles is $\frac{42.36}{2} = 21.18$ in.

Let x = the length of each of the two equal sides of the isosceles triangle.

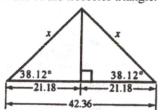

$$\cos 38.12° = \frac{21.18}{x} \Rightarrow x \cos 38.12° = 21.18 \Rightarrow$$
$$x = \frac{21.18}{\cos 38.12°} \approx 26.921918$$

The length of each of the two equal sides of the triangle is 26.92 in. (rounded to four significant digits)

45. Let h represent the height of the tower. In triangle ABC we have
$$\tan 34.6° = \frac{h}{40.6}$$
$$h = 40.6 \tan 34.6° \approx 28.0081$$
The height of the tower is 28.0 m. (rounded to three significant digits)

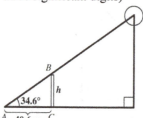

47. Let x = the length of the shadow.
$$\tan 23.4° = \frac{5.75}{x}$$
$$x \tan 23.4° = 5.75$$
$$x = \frac{5.75}{\tan 23.4°} \approx 13.2875$$

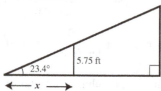

The length of the shadow is 13.3 ft. (rounded to three significant digits)

49. Let θ = the angle of depression.

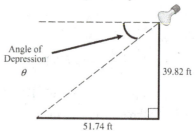

$$\tan \theta = \frac{39.82}{51.74} \Rightarrow \theta = \tan^{-1}\left(\frac{39.82}{51.74}\right)$$
$$\theta \approx 37.58° \approx 37°35'$$

51. Let θ = the angle of elevation of the sun.

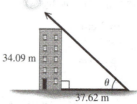

$$\tan \theta = \frac{34.09}{37.62} \Rightarrow \theta = \tan^{-1}\left(\frac{34.09}{37.62}\right)$$
$$\theta \approx 42.18°$$

53. In order to find the angle of elevation, θ, we need to first find the length of the diagonal of the square base. The diagonal forms two isosceles right triangles. Each angle formed by a side of the square and the diagonal measures 45°.

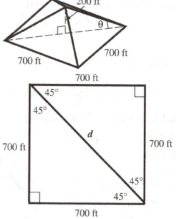

By the Pythagorean theorem,
$$700^2 + 700^2 = d^2 \Rightarrow 2 \cdot 700^2 = d^2 \Rightarrow$$
$$d = \sqrt{2 \cdot 700^2} \Rightarrow d = 700\sqrt{2}$$

Thus, length of the diagonal is $700\sqrt{2}$ ft. To to find the angle, θ, we consider the following isosceles triangle.

(continued on next page)

(continued)

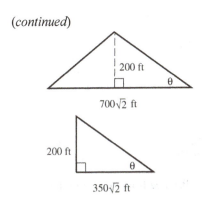

The height of the pyramid bisects the base of this triangle and forms two right triangles. We can use one of these triangles to find the angle of elevation, θ.

$$\tan \theta = \frac{200}{350\sqrt{2}} \Rightarrow$$

$$\theta \approx \tan^{-1}\left(\frac{200}{350\sqrt{2}}\right) \approx 22.0017$$

Rounding this figure to two significant digits, we have $\theta \approx 22°$.

55. **(a)** Let x = the height of the peak above 14,545 ft.

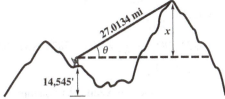

The diagonal of the right triangle formed is in miles, so we must first convert this measurement to feet. Because there are 5280 ft in one mile, we have the length of the diagonal is $27.0134(5280) =$ 142,630.752. Find the value of x by

solving $\sin 5.82° = \dfrac{x}{142,630.752}.$

$x = 142,630.752 \sin 5.82°$
$\quad \approx 14,463.2674$

Thus, the value of x rounded to five significant digits is 14,463 ft. Thus, the total height is about
$14,545 + 14,463 = 29,008 \approx 29,000$ ft.

(b) The curvature of the earth would make the peak appear shorter than it actually is. Initially the surveyors did not think Mt. Everest was the tallest peak in the Himalayas. It did not look like the tallest peak because it was farther away than the other large peaks.

57. $(-4, 0)$

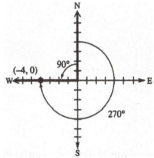

The bearing of the airplane measured in a clockwise direction from due north is 270°. The bearing can also be expressed as N 90° W, or S 90° W.

59. $(0, 4)$

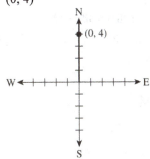

The bearing of the airplane measured in a clockwise direction from due north is 0°. The bearing can also be expressed as N 0° E or N 0° W.

61. $(-5, 5)$

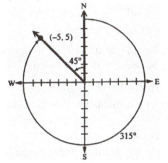

The bearing of the airplane measured in a clockwise direction from due north is 315°. The bearing can also be expressed as N 45° W.

63. $(2, -2)$

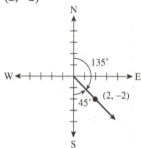

The bearing of the airplane measured in a clockwise direction from due north is 135°. The bearing can also be expressed as S 45° E.

65. Let x = the distance the plane is from its starting point. In the figure, the measure of angle ACB is
$$38° + (180° - 128°) = 38° + 52° = 90°.$$
Therefore, triangle ACB is a right triangle.

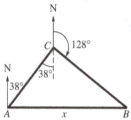

Because $d = rt$, the distance traveled in 1.5 hr is $(1.5 \text{ hr})(110 \text{ mph}) = 165 \text{ mi}$. The distance traveled in 1.3 hr is
$(1.3 \text{ hr})(110 \text{ mph}) = 143 \text{ mi}$.
Using the Pythagorean theorem, we have
$$x^2 = 165^2 + 143^2 \Rightarrow x^2 = 27,225 + 20,449 \Rightarrow$$
$$x^2 = 47,674 \Rightarrow x \approx 218.3438$$
The plane is 220 mi from its starting point. (rounded to two significant digits)

67. Let x = distance the ships are apart.
In the figure, the measure of angle CAB is $130° - 40° = 90°$. Therefore, triangle CAB is a right triangle.
Because $d = rt$, the distance traveled by the first ship in 1.5 hr is $(1.5 \text{ hr})(18 \text{ knots}) = 27$ nautical mi and the second ship is $(1.5\text{hr})(26 \text{ knots}) = 39$ nautical mi. (See figure in next column.) Applying the Pythagorean theorem, we have
$$x^2 = 27^2 + 39^2 \Rightarrow x^2 = 729 + 1521 \Rightarrow$$
$$x^2 = 2250 \Rightarrow x = \sqrt{2250} \approx 47.4342$$
The ships are 47 nautical mi apart (rounded to 2 significant digits).

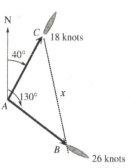

69. Let b = the distance from dock A to the coral reef C.

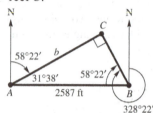

In the figure, the measure of angle CAB is $90° - 58°22' = 31°38'$, and the measure of angle CBA is $328°22' - 270° = 58°22'$. Because $31°38' + 58°22' = 90°$, ABC is a right triangle.
$$\cos A = \frac{b}{2587}$$
$$\cos 31°38' = \frac{b}{2587} \Rightarrow b = 2587\cos 31°38'$$
$$b \approx 2203 \text{ ft}$$

71. Let x = distance between the two ships.

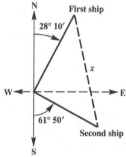

The angle between the bearings of the ships is $180° - (28°10' + 61°50') = 90°$. The triangle formed is a right triangle. The distance traveled at 24.0 mph is
$(4 \text{ hr}) (24.0 \text{ mph}) = 96 \text{ mi}$. The distance traveled at 28.0 mph is
$(4 \text{ hr})(28.0 \text{ mph}) = 112 \text{ mi}$.
Applying the Pythagorean theorem we have
$$x^2 = 96^2 + 112^2 \Rightarrow x^2 = 9216 + 12,544 \Rightarrow$$
$$x^2 = 21,760 \Rightarrow x = \sqrt{21,760} \approx 147.5127$$
The ships are 148 mi apart. (rounded to three significant digits)

73. Let b = the distance from A to C and let c = the distance from A to B.

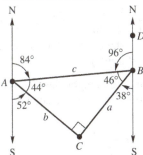

Because the bearing from A to B is N 84° E, the measure of angle ABD is $180° - 84° = 96°$. The bearing from B to C is 38°, so the measure of angle $ABC = 180° - (96° + 38°) = 46°$. The bearing of A to C is 52°, so the measure of angle $BAC = 180° - (52° + 84°) = 44°$. The measure of angle C is $180° - (44° + 46°) = 90°$, so triangle ABC is a right triangle. The distance from A to B, labeled c, is $2.4(250) = 600$ miles.

$$\sin 46° = \frac{b}{c} = \frac{b}{600}$$
$$b = 600 \sin 46° \approx 430 \text{ mi}$$

75. Draw triangle WDG with W representing Winston-Salem, D representing Danville, and G representing Goldsboro. Name any point X on the line due south from D.

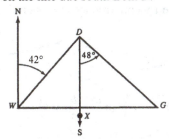

The bearing from W to D is 42° (equivalent to N 42° E), so angle WDX measures 42°. Because angle XDG measures 48°, the measure of angle D is $42° + 48° = 90°$. Thus, triangle WDG is a right triangle. Using $d = rt$ and the Pythagorean theorem, we have

$$WG = \sqrt{(WD)^2 + (DG)^2}$$
$$= \sqrt{[65(1.1)]^2 + [65(1.8)]^2}$$
$$WG = \sqrt{71.5^2 + 117^2} = \sqrt{5112.25 + 13,689}$$
$$= \sqrt{18,801.25} \approx 137$$

The distance from Winston-Salem to Goldsboro is approximately 140 mi. (rounded to two significant digits)

77. Let x = the distance from the closer point on the ground to the base of height h of the pyramid.

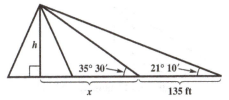

In the larger right triangle, we have

$$\tan 21°10' = \frac{h}{135 + x} \Rightarrow h = (135 + x)\tan 21°10'$$

In the smaller right triangle, we have

$$\tan 35°30' = \frac{h}{x} \Rightarrow h = x \tan 35°30'.$$

Substitute for h in this equation, and solve for x to obtain the following.

$$(135 + x)\tan 21°10' = x \tan 35°30'$$
$$135 \tan 21°10' + x \tan 21°10' = x \tan 35°30'$$
$$135 \tan 21°10' = x \tan 35°30' - x \tan 21°10'$$
$$135 \tan 21°10' = x(\tan 35°30' - \tan 21°10')$$
$$\frac{135 \tan 21°10'}{\tan 35°30' - \tan 21°10'} = x$$

Substitute for x in the equation for the smaller triangle.

$$h = \frac{135 \tan 21°10'}{\tan 35°30' - \tan 21°10'} \tan 35°30'$$
$$\approx 114.3427$$

The height of the pyramid is 114 ft. (rounded to three significant digits)

79. Let x = the height of the antenna; h = the height of the house.

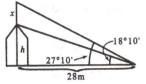

In the smaller right triangle, we have

$$\tan 18°10' = \frac{h}{28} \Rightarrow h = 28 \tan 18°10'.$$

In the larger right triangle, we have

$$\tan 27°10' = \frac{x + h}{28} \Rightarrow x + h = 28 \tan 27°10' \Rightarrow$$
$$x = 28 \tan 27°10' - h$$
$$x = 28 \tan 27°10' - 28 \tan 18°10'$$
$$\approx 5.1816$$

The height of the antenna is 5.18 m. (rounded to three significant digits)

81. Algebraic solution:
Let x = the side adjacent to 49.2° in the smaller triangle.

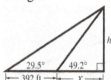

In the larger right triangle, we have

$$\tan 29.5° = \frac{h}{392 + x} \Rightarrow h = (392 + x)\tan 29.5°.$$

In the smaller right triangle, we have

$$\tan 49.2° = \frac{h}{x} \Rightarrow h = x\tan 49.2°.$$

Substituting, we have

$$x\tan 49.2° = (392 + x)\tan 29.5°$$
$$x\tan 49.2° = 392\tan 29.5°$$
$$+ x\tan 29.5°$$
$$x\tan 49.2° - x\tan 29.5° = 392\tan 29.5°$$
$$x(\tan 49.2° - \tan 29.5°) = 392\tan 29.5°$$
$$x = \frac{392\tan 29.5°}{\tan 49.2° - \tan 29.5°}$$

Now substitute this expression for x in the equation for the smaller triangle to obtain
$h = x\tan 49.2°$

$$h = \frac{392\tan 29.5°}{\tan 49.2° - \tan 29.5°} \cdot \tan 49.2°$$
$$\approx 433.4762 \approx 433 \text{ ft (rounded to three}$$
significant digits.
Graphing calculator solution:
Graph $y_1 = (\tan 29.5°)x$ and

$y_2 = (\tan 29.5°)(x - 392)$ in the window
$[0, 1000] \times [0, 500]$. Then find the intersection.

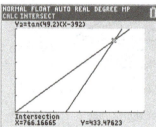

The height of the triangle is 433 ft (rounded to three significant digits.

83. Let x = the minimum distance that a plant needing full sun can be placed from the fence.

$$\tan 23°20' = \frac{4.65}{x} \Rightarrow x\tan 23°20' = 4.65 \Rightarrow$$
$$x = \frac{4.65}{\tan 23°20'} \approx 10.7799$$
The minimum distance is 10.8 ft. (rounded to three significant digits)

85. Let h = the minimum height above the surface of Earth so a pilot at A can see an object on the horizon at C.

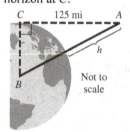

Using the Pythagorean theorem, we have
$$(4.00 \times 10^3 + h)^2 = (4.00 \times 10^3)^2 + 125^2$$
$$(4000 + h)^2 = 4000^2 + 125^2$$
$$(4000 + h)^2 = 16,000,000 + 15,625$$
$$(4000 + h)^2 = 16,015,625$$
$$4000 + h = \sqrt{16,015,625}$$
$$h = \sqrt{16,015,625} - 4000$$
$$\approx 4001.9526 - 4000 = 1.9526$$
The minimum height above the surface of Earth would be 1.95 mi. (rounded to 3 significant digits)

87. (a)

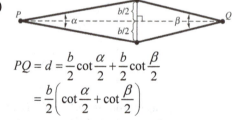

$$PQ = d = \frac{b}{2}\cot\frac{\alpha}{2} + \frac{b}{2}\cot\frac{\beta}{2}$$
$$= \frac{b}{2}\left(\cot\frac{\alpha}{2} + \cot\frac{\beta}{2}\right)$$

(b) Using the result of part (a), let
$\alpha = 37'48'', \beta = 42'03'',$ and $b = 2.000$

$$d = \frac{b}{2}\left(\cot\frac{\alpha}{2} + \cot\frac{\beta}{2}\right) \Rightarrow$$
$$d = \frac{2.000}{2}\left(\cot\frac{37'48''}{2} + \cot\frac{42'03''}{2}\right)$$
$$= \cot 0.315° + \cot 0.3504166667°$$
$$\approx 345.3951$$
The distance between the two points P and Q is about 345.4 cm.

89. (a) If $\theta = 37°$, then $\dfrac{\theta}{2} = \dfrac{37°}{2} = 18.5°$.

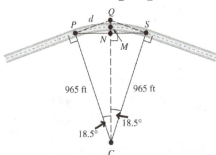

To find the distance between P and Q, d, we first note that angle QPC is a right angle. Hence, triangle QPC is a right triangle and we can solve

$$\tan 18.5° = \frac{d}{965}$$
$$d = 965 \tan 18.5° \approx 322.8845$$

The distance between P and Q, is 320 ft. (rounded to two significant digits)

(b) We are dealing with a circle, so the distance between M and C is R. If we let x be the distance from N to M, then the distance from C to N will be $R - x$.

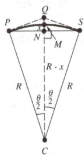

Triangle CNP is a right triangle, so we can set up the following equation.

$$\cos\frac{\theta}{2} = \frac{R-x}{R} \Rightarrow R\cos\frac{\theta}{2} = R - x \Rightarrow$$
$$x = R - R\cos\frac{\theta}{2} \Rightarrow x = R\left(1 - \cos\frac{\theta}{2}\right)$$

Chapter 5   Review Exercises

1. The complement of $35°$ is $90° - 35° = 55°$.
The supplement of $35°$ is $180° - 35° = 145°$.

3. $-174°$. is coterminal with
$-174° + 360° = 186°$

5. 650 rotations per min $= \dfrac{650}{60}$ rotations per sec

$= \dfrac{65}{6}$ rotations per sec

$= 26$ rotations per 2.4 sec

$= 26(360°)$ per 2.4 sec $= 9360°$ per 2.4 sec

A point of the edge of the propeller will move $9360°$ in 2.4 sec.

7. $119°8'3'' = 119° + \dfrac{8}{60}° + \dfrac{3}{3600}°$
$\approx 119° + 0.1333° + 0.0008°$
$\approx 119.134°$

9. $275.1005 = 275° + 0.1005(60') = 275° + 6.03'$
$= 275°06' + 0.03'$
$= 275°06' + 0.03(60'')$
$= 275°6'1.8'' \approx 275°06'02''$

11. $x = 1, \ y = -\sqrt{3}$ and

$$r = \sqrt{x^2 + y^2} = \sqrt{1^2 + \left(-\sqrt{3}\right)^2}$$
$$= \sqrt{1+3} = \sqrt{4} = 2$$

$$\sin\theta = \frac{y}{r} = \frac{-\sqrt{3}}{2} = -\frac{\sqrt{3}}{2}$$

$$\cos\theta = \frac{x}{r} = \frac{1}{2}$$

$$\tan\theta = \frac{y}{x} = \frac{-\sqrt{3}}{1} = -\sqrt{3}$$

$$\cot\theta = \frac{x}{y} = \frac{1}{-\sqrt{3}} = -\frac{1}{\sqrt{3}} \cdot \frac{\sqrt{3}}{\sqrt{3}} = -\frac{\sqrt{3}}{3}$$

$$\sec\theta = \frac{r}{x} = \frac{2}{1} = 2$$

$$\csc\theta = \frac{r}{y} = \frac{2}{-\sqrt{3}} = -\frac{2}{\sqrt{3}} \cdot \frac{\sqrt{3}}{\sqrt{3}} = -\frac{2\sqrt{3}}{3}$$

13. $x = -2, y = 0$, and

$$r = \sqrt{x^2 + y^2} = \sqrt{(-2)^2 + (0)^2} = \sqrt{4} = 2$$

$$\sin 180° = \frac{y}{r} = \frac{0}{2} = 0; \ \cos 180° = \frac{x}{r} = \frac{-2}{2} = -1$$

$$\tan 180° = \frac{y}{x} = \frac{0}{-2} = 0$$

$$\cot 180° = \frac{x}{y} = \frac{-2}{0}, \text{ undefined}$$

$$\sec 180° = \frac{r}{x} = \frac{2}{-2} = -1$$

$$\csc 180° = \frac{r}{y} = \frac{2}{0}, \text{ undefined}$$

15. $x = 3$, $y = -4$ and $r = \sqrt{3^2 + (-4)^2} = \sqrt{25} = 5$

$$\sin\theta = \frac{y}{r} = \frac{-4}{5} = -\frac{4}{5}; \ \cos\theta = \frac{x}{r} = \frac{3}{5}$$

$$\tan\theta = \frac{y}{x} = \frac{-4}{3} = -\frac{4}{3}$$

$$\cot\theta = \frac{x}{y} = \frac{3}{-4} = -\frac{3}{4}$$

$$\sec\theta = \frac{r}{x} = \frac{5}{3}; \ \csc\theta = \frac{r}{y} = \frac{5}{-4} = -\frac{5}{4}$$

17. $x = -8$, $y = 15$, and

$$r = \sqrt{(-8)^2 + 15^2} = \sqrt{289} = 17$$

$$\sin\theta = \frac{y}{r} = \frac{15}{17}; \ \cos\theta = \frac{x}{r} = \frac{-8}{17} = -\frac{8}{17}$$

$$\tan\theta = \frac{y}{x} = \frac{15}{-8} = -\frac{15}{8}$$

$$\cot\theta = \frac{x}{y} = \frac{-8}{15} = -\frac{8}{15}$$

$$\sec\theta = \frac{r}{x} = \frac{17}{-8} = -\frac{17}{8}$$

$$\csc\theta = \frac{r}{y} = \frac{17}{15}$$

19. $x = -2\sqrt{2}$, $y = 2\sqrt{2}$, and

$$r = \sqrt{\left(-2\sqrt{2}\right)^2 + \left(2\sqrt{2}\right)^2} = \sqrt{8+8} = \sqrt{16} = 4$$

$$\sin\theta = \frac{y}{r} = \frac{2\sqrt{2}}{4} = \frac{\sqrt{2}}{2}$$

$$\cos\theta = \frac{x}{r} = \frac{-2\sqrt{2}}{4} = -\frac{\sqrt{2}}{2}$$

$$\tan\theta = \frac{y}{x} = \frac{2\sqrt{2}}{-2\sqrt{2}} = -1$$

$$\cot\theta = \frac{x}{y} = \frac{-2\sqrt{2}}{2\sqrt{2}} = -1$$

$$\sec\theta = \frac{r}{x} = \frac{4}{-2\sqrt{2}} = -\frac{2}{\sqrt{2}} = -\frac{2}{\sqrt{2}} \cdot \frac{\sqrt{2}}{\sqrt{2}} = -\sqrt{2}$$

$$\csc\theta = \frac{r}{y} = \frac{4}{2\sqrt{2}} = \frac{2}{\sqrt{2}} = \frac{2}{\sqrt{2}} \cdot \frac{\sqrt{2}}{\sqrt{2}} = \sqrt{2}$$

21. The terminal side of the angle is defined by $5x - 3y = 0$, $x \geq 0$, so a point on this terminal side is $(3, 5)$.

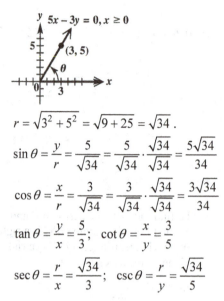

$$r = \sqrt{3^2 + 5^2} = \sqrt{9 + 25} = \sqrt{34}.$$

$$\sin\theta = \frac{y}{r} = \frac{5}{\sqrt{34}} = \frac{5}{\sqrt{34}} \cdot \frac{\sqrt{34}}{\sqrt{34}} = \frac{5\sqrt{34}}{34}$$

$$\cos\theta = \frac{x}{r} = \frac{3}{\sqrt{34}} = \frac{3}{\sqrt{34}} \cdot \frac{\sqrt{34}}{\sqrt{34}} = \frac{3\sqrt{34}}{34}$$

$$\tan\theta = \frac{y}{x} = \frac{5}{3}; \ \cot\theta = \frac{x}{y} = \frac{3}{5}$$

$$\sec\theta = \frac{r}{x} = \frac{\sqrt{34}}{3}; \ \csc\theta = \frac{r}{y} = \frac{\sqrt{34}}{5}$$

23. $\sin 180° = 0$
$\cos 180° = -1$
$\tan 180° = 0$
$\cot 180°$ is undefined
$\sec 180° = -1$
$\csc 180°$ is undefined.

25. $\cos\theta = -\frac{5}{8}$ and θ is in quadrant III

$\cos\theta = \frac{x}{r} = \frac{-5}{8}$ and θ in quadrant III, so let
$x = -5$, $r = 8$.
$x^2 + y^2 = r^2 \Rightarrow (-5)^2 + y^2 = 8^2 \Rightarrow$
$25 + y^2 = 64 \Rightarrow y^2 = 39 \Rightarrow y = \pm\sqrt{39}$
θ is in quadrant III, so $y = -\sqrt{39}$.

$$\sin\theta = \frac{y}{r} = \frac{-\sqrt{39}}{8} = -\frac{\sqrt{39}}{8}$$

$$\cos\theta = \frac{x}{r} = \frac{-5}{8} = -\frac{5}{8}$$

$$\tan\theta = \frac{y}{x} = \frac{-\sqrt{39}}{-5} = \frac{\sqrt{39}}{5}$$

$$\cot\theta = \frac{x}{y} = \frac{-5}{-\sqrt{39}} = \frac{5}{\sqrt{39}} \cdot \frac{\sqrt{39}}{\sqrt{39}} = \frac{5\sqrt{39}}{39}$$

$$\sec\theta = \frac{r}{x} = \frac{8}{-5} = -\frac{8}{5}$$

$$\csc\theta = \frac{r}{y} = \frac{8}{-\sqrt{39}} = -\frac{8}{\sqrt{39}}$$

$$= -\frac{8}{\sqrt{39}} \cdot \frac{\sqrt{39}}{\sqrt{39}} = -\frac{8\sqrt{39}}{39}$$

27. $\sec\theta = -\sqrt{5}$ and θ is in quadrant II

θ in quadrant II $\Rightarrow x < 0, y > 0$, so

$$\sec\theta = -\sqrt{5} = \frac{r}{x} = \frac{\sqrt{5}}{-1} \Rightarrow x = -1, r = \sqrt{5}$$

$$x^2 + y^2 = r^2 \Rightarrow (-1)^2 + y^2 = \left(\sqrt{5}\right)^2 \Rightarrow$$

$$1 + y^2 = 5 \Rightarrow y^2 = 4 \Rightarrow y = 2$$

$$\sin\theta = \frac{y}{r} = \frac{2}{\sqrt{5}} = \frac{2}{\sqrt{5}} \cdot \frac{\sqrt{5}}{\sqrt{5}} = \frac{2\sqrt{5}}{5}$$

$$\cos\theta = \frac{x}{r} = \frac{-1}{\sqrt{5}} = -\frac{1}{\sqrt{5}} \cdot \frac{\sqrt{5}}{\sqrt{5}} = -\frac{\sqrt{5}}{5}$$

$$\tan\theta = \frac{y}{x} = \frac{2}{-1} = -2; \ \cot\theta = \frac{x}{y} = \frac{-1}{2} = -\frac{1}{2}$$

$$\sec\theta = \frac{r}{x} = \frac{\sqrt{5}}{-1} = -\sqrt{5}; \ \csc\theta = \frac{r}{y} = \frac{\sqrt{5}}{2}$$

29. $\sec\theta = \frac{5}{4}$ and θ is in quadrant IV

θ in quadrant IV $\Rightarrow x > 0, y < 0$, so

$$\sec\theta = \frac{r}{x} = \frac{5}{4} \Rightarrow x = 4, r = 5$$

$$x^2 + y^2 = r^2 \Rightarrow (4)^2 + y^2 = 5^2 \Rightarrow$$

$$16 + y^2 = 25 \Rightarrow y^2 = 9 \Rightarrow y = -3$$

$$\sin\theta = \frac{y}{r} = \frac{-3}{5} = -\frac{3}{5}; \ \cos\theta = \frac{x}{r} = \frac{4}{5}$$

$$\tan\theta = \frac{y}{x} = \frac{-3}{4} = -\frac{3}{4}$$

$$\cot\theta = \frac{x}{y} = \frac{4}{-3} = -\frac{4}{3}$$

$$\sec\theta = \frac{r}{x} = \frac{5}{4}; \ \csc\theta = \frac{r}{y} = \frac{5}{-3} = -\frac{5}{3}$$

31. $\sin A = \dfrac{\text{side opposite}}{\text{hypotenuse}} = \dfrac{60}{61}$

$\cos A = \dfrac{\text{side adjacent}}{\text{hypotenuse}} = \dfrac{11}{61}$

$\tan A = \dfrac{\text{side opposite}}{\text{side adjacent}} = \dfrac{60}{11}$

$\cot A = \dfrac{\text{side adjacent}}{\text{side opposite}} = \dfrac{11}{60}$

$\sec A = \dfrac{\text{hypotenuse}}{\text{side adjacent}} = \dfrac{61}{11}$

$\csc A = \dfrac{\text{hypotenuse}}{\text{side opposite}} = \dfrac{61}{60}$

33. $1020°$ is coterminal with $1020° - 2 \cdot 360° = 300°$. The reference angle is $360° - 300° = 60°$. Because $1020°$ lies in quadrant IV, the sine, tangent, cotangent, and cosecant are negative.

$$\sin 1020° = -\sin 60° = -\frac{\sqrt{3}}{2}$$

$$\cos 1020° = \cos 60° = \frac{1}{2}$$

$$\tan 1020° = -\tan 60° = -\sqrt{3}$$

$$\cot 1020° = -\cot 60° = -\frac{\sqrt{3}}{3}$$

$$\sec 1020° = \sec 60° = 2$$

$$\csc 1020° = -\csc 60° = -\frac{2\sqrt{3}}{3}$$

35. $-1470°$ is coterminal with $-1470° + 5 \cdot 360° = 330°$. This angle lies in quadrant IV. The reference angle is $360° - 330° = 30°$. Because $-1470°$ is in quadrant IV, the sine, tangent, cotangent, and cosecant are negative.

$$\sin(-1470°) = -\sin 30° = -\frac{1}{2}$$

$$\cos(-1470°) = \cos 30° = \frac{\sqrt{3}}{2}$$

$$\tan(-1470°) = -\tan 30° = -\frac{\sqrt{3}}{3}$$

$$\cot(-1470°) = -\cot 30° = -\sqrt{3}$$

$$\sec(-1470°) = \sec 30° = \frac{2\sqrt{3}}{3}$$

$$\csc(-1470°) = -\csc 30° = -2$$

37. $\cos\theta = -\frac{1}{2}$

Because $\cos\theta$ is negative, θ must lie in quadrants II or III. The absolute value of $\cos\theta$ is $\frac{1}{2}$, so the reference angle, θ' must be $60°$. The quadrant II angle θ equals $180° - \theta' = 180° - 60° = 120°$, and the quadrant III angle θ equals $180° + \theta' = 180° + 60° = 240°$.

39. $\sec\theta = -\dfrac{2\sqrt{3}}{3}$

Because $\sec\theta$ is negative, θ must lie in quadrants II or III. The absolute value of $\sec\theta$ is $\dfrac{2\sqrt{3}}{3}$, so the reference angle, θ' must be 30°. The quadrant II angle θ equals $180° - \theta' = 180° - 30° = 150°,$ and the quadrant III angle θ equals $180° + \theta' = 180° + 30° = 210°.$

41. $(-3, -3)$

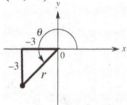

The distance from the origin is r:

$r = \sqrt{x^2 + y^2} \Rightarrow r = \sqrt{(-3)^2 + (-3)^2} \Rightarrow$
$r = \sqrt{9+9} \Rightarrow r = \sqrt{18} \Rightarrow r = 3\sqrt{2}$

$\sin\theta = \dfrac{y}{r} = -\dfrac{3}{3\sqrt{2}} = -\dfrac{1}{\sqrt{2}} \cdot \dfrac{\sqrt{2}}{\sqrt{2}} = -\dfrac{\sqrt{2}}{2}$

$\cos\theta = \dfrac{x}{r} = -\dfrac{3}{3\sqrt{2}} = -\dfrac{1}{\sqrt{2}} \cdot \dfrac{\sqrt{2}}{\sqrt{2}} = -\dfrac{\sqrt{2}}{2}$

$\tan\theta = \dfrac{y}{x} = \dfrac{-3}{-3} = 1$

For the exercises in this section, be sure your calculator is in degree mode.

43. $\sec 222°30' = \dfrac{1}{\cos 222°30'} \approx -1.356342$

45. $\csc 78°21' = \dfrac{1}{\sin 78°21'} \approx 1.021034$

47. $\tan 11.7689° \approx 0.208344$

49. $\sin\theta = 0.8254121$
$\theta = \sin^{-1}(0.8254121) \approx 55.673870°$

51. $\cos\theta = 0.97540415$
$\theta = \cos^{-1}(0.97540415) \approx 12.733938°$

53. $\tan\theta = 1.9633124$
$\theta = \tan^{-1}(1.9633124) \approx 63.008286°$

55. $\sin\theta = 0.73135370$
$\theta = \sin^{-1}(0.73135370) \approx 47°$

The value of $\sin\theta$ is positive in quadrants I and II, so the two angles in $[0°, 360°)$ are 47° and $180° - 47° = 133°$.

57. No, this will result in an angle having tangent equal to 25. The function $\tan^{-1}$ is not the reciprocal of the tangent (the cotangent), but is the *inverse tangent* function. To find cot 25°, the student must find the reciprocal of tan 25°.

$\cot 25° = \dfrac{1}{\tan 25°} \ne \tan^{-1} 25°.$

59. $A = 58°\ 30', \ c = 748$

$A + B = 90° \Rightarrow B = 90° - A \Rightarrow$
$\quad B = 90° - 58°30' = 89°60' - 58°30'$
$\quad\quad = 31°30'$

$\sin A = \dfrac{a}{c} \Rightarrow \sin 58°30' = \dfrac{a}{748} \Rightarrow$
$\quad a = 748\sin 58°30' \approx 638$ (rounded to three significant digits)

$\cos A = \dfrac{b}{c} \Rightarrow \cos 58°30' = \dfrac{b}{748} \Rightarrow$
$\quad b = 748\cos 58°30' \approx 391$ (rounded to three significant digits)

61. $A = 39.72°, \ b = 38.97$ m

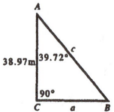

$A + B = 90° \Rightarrow B = 90° - A \Rightarrow$
$\quad B = 90° - 39.72° = 50.28°$

$\tan A = \dfrac{a}{b} \Rightarrow \tan 39.72° = \dfrac{a}{38.97} \Rightarrow$
$\quad a = 38.97\tan 39.72° \approx 32.38$ m (rounded to four significant digits)

$\cos A = \dfrac{b}{c} \Rightarrow \cos 39.72° = \dfrac{38.97}{c} \Rightarrow$
$c\cos 39.72° = 38.97 \Rightarrow$
$\quad c = \dfrac{38.97}{\cos 39.72°} \approx 50.66$ m

(rounded to five significant digits)

63. Let x = height of the tower.

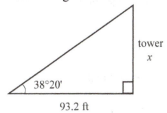

$$\tan 38°20' = \frac{x}{93.2}$$
$$x = 93.2 \tan 38°20'$$
$$x \approx 73.6930$$

The height of the tower is 73.7 ft. (rounded to three significant digits)

65. Let x = length of the diagonal

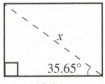

$$\cos 35.65° = \frac{15.24}{x}$$
$$x = \frac{15.24}{\cos 35.65°} \approx 18.7548$$

The length of the diagonal is 18.75 cm (rounded to three significant digits).

67. Draw triangle ABC and extend the north-south lines to a point X south of A and S to a point Y, north of C.

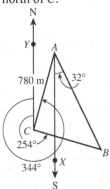

Angle $ACB = 344° - 254° = 90°$, so ABC is a right triangle.
Angle $BAX = 32°$ because it is an alternate interior angle to $32°$.
Angle $YCA = 360° - 344° = 16°$
Angle $XAC = 16°$ because it is an alternate interior angle to angle YCA.
Angle $BAC = 32° + 16° = 48°$.

In triangle ABC,
$$\cos A = \frac{AC}{AB} \Rightarrow \cos 48° = \frac{780}{AB} \Rightarrow$$
$$AB \cos 48° = 780 \Rightarrow AB = \frac{780}{\cos 48°} \approx 1165.6917$$

The distance from A to B is 1200 m. (rounded to two significant digits)

69. Suppose A is the car heading south at 55 mph, B is the car heading west, and point C is the intersection from which they start. After two hours, using $d = rt$, $AC = 55(2) = 110$. Angle ACB is a right angle, so triangle ACB is a right triangle. The bearing of A from B is 324°, so angle $CAB = 360° - 324° = 36°$.

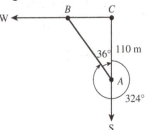

$$\cos \angle CAB = \frac{AC}{AB} \Rightarrow \cos 36° = \frac{110}{AB} \Rightarrow$$
$$AB = \frac{110}{\cos 36°} \approx 135.9675$$

The distance from A to B is about 140 mi (rounded to two significant digits).

Chapter 5   Chapter Test

1. 67°
complement: $90° - 67° = 23°$
supplement: $180° - 67° = 113°$

2. The two angles are supplements, so their sum is 180°.
$$(7x + 19) + (2x - 1) = 180 \Rightarrow 9x + 18 = 180 \Rightarrow$$
$$9x = 162 \Rightarrow x = 18$$
$$7(18) + 19 = 145; 2(18) - 1 = 35$$
The measures of the angles are 145° and 35°.

3. The two angles are complements, so their sum is 90°.
$$(-8x + 30) + (-3x + 5) = 90$$
$$-11x + 35 = 90$$
$$-11x = 55 \Rightarrow x = -5$$
$$-8(-5) + 30 = 70; -3(-5) + 5 = 20$$
The angles measure 20° and 70°.

4. $74°18'36'' = 74° + \frac{18}{60}° + \frac{36}{3600}°$
$\approx 74° + 0.3° + 0.01° = 74.31°$

5. $45.2025° = 45° + 0.2025°$
$$= 45° + 0.2025(60')$$
$$= 45° + 12.15'$$
$$= 45° + 12' + 0.15'$$
$$= 45° + 12' + 0.15(60'')$$
$$= 45° + 12' + 09'' = 45°12'09''$$

6. (a) 390° is coterminal with
$390° - 360° = 30°.$

(b) −80° is coterminal with
$-80° + 360° = 280°.$

(c) 810° is coterminal with
$810° - 2(360°) = 810° - 720° = 90°.$

7. $\dfrac{450(360°)}{1 \text{ min}} = \dfrac{450(360°)}{60 \text{ sec}} = \dfrac{450(6°)}{\text{sec}}$
$$= 2700°/\text{sec}$$

A point on the tire rotates 2700° in one second.

8.

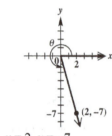

$x = 2, \; y = -7$
$$r = \sqrt{x^2 + y^2} = \sqrt{2^2 + (-7)^2} = \sqrt{4 + 49} = \sqrt{53}$$

$$\sin\theta = \frac{y}{r} = \frac{-7}{\sqrt{53}} = -\frac{7}{\sqrt{53}} \cdot \frac{\sqrt{53}}{\sqrt{53}} = -\frac{7\sqrt{53}}{53}$$

$$\cos\theta = \frac{x}{r} = \frac{2}{\sqrt{53}} = \frac{2}{\sqrt{53}} \cdot \frac{\sqrt{53}}{\sqrt{53}} = \frac{2\sqrt{53}}{53}$$

$$\tan\theta = \frac{y}{x} = \frac{-7}{2} = -\frac{7}{2}; \quad \cot\theta = \frac{x}{y} = \frac{2}{-7} = -\frac{2}{7}$$

$$\sec\theta = \frac{r}{x} = \frac{\sqrt{53}}{2}; \quad \csc\theta = \frac{r}{y} = \frac{\sqrt{53}}{-7} = -\frac{\sqrt{53}}{7}$$

9.

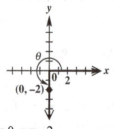

$x = 0, \; y = -2$
$$r = \sqrt{x^2 + y^2} = \sqrt{0^2 + (-2)^2} = \sqrt{0 + 4} = \sqrt{4} = 2$$

$$\sin\theta = \frac{y}{r} = \frac{-2}{2} = -1; \quad \cos\theta = \frac{x}{r} = \frac{0}{2} = 0$$

$$\tan\theta = \frac{y}{x} = \frac{-2}{0} \text{ undefined}$$

$$\cot\theta = \frac{x}{y} = \frac{0}{-2} = 0$$

$$\sec\theta = \frac{r}{x} = \frac{2}{0} \text{ undefined}$$

$$\csc\theta = \frac{r}{y} = \frac{2}{-2} = -1$$

10. Because $x \le 0$, the graph of the line
$3x - 4y = 0$ is shown to the left of the y-axis.

A point on this graph is $(-4, -3)$. The corresponding value of r is
$$r = \sqrt{(-4)^2 + (-3)^2} = \sqrt{16 + 9} = \sqrt{25} = 5.$$

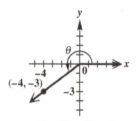

$3x - 4y = 0, x \; 0$

$x = -4, \; y = -3$

$$\sin\theta = \frac{y}{r} = -\frac{3}{5}; \quad \cos\theta = \frac{x}{r} = -\frac{4}{5}$$

$$\tan\theta = \frac{y}{x} = \frac{3}{4}; \quad \cot\theta = \frac{x}{y} = \frac{4}{3}$$

$$\sec\theta = \frac{r}{x} = -\frac{5}{4}; \quad \csc\theta = \frac{r}{y} = -\frac{5}{3}$$

11.

θ	90°	−360°	630°
$\sin\theta$	1	0	−1
$\cos\theta$	0	1	0
$\tan\theta$	undefined	0	undefined
$\cot\theta$	0	undefined	0
$\sec\theta$	undefined	1	undefined
$\csc\theta$	1	undefined	−1

12. If the terminal side of a quadrantal angle lies on the negative part of the x-axis, any point on the terminal side would have the form $(k, 0)$, where k is any real number < 0.

$$\sin\theta = \frac{y}{r} = \frac{0}{r} = 0; \quad \cos\theta = \frac{x}{r} = \frac{k}{r}$$

$$\tan\theta = \frac{y}{x} = \frac{0}{k} = 0; \quad \cot\theta = \frac{x}{y} = \frac{k}{0} \text{ undefined}$$

(continued on next page)

(continued)

$$\sec\theta = \frac{r}{x} = \frac{r}{k}; \quad \csc\theta = \frac{r}{y} = \frac{r}{0} \text{ undefined}$$

Thus, the cotangent and the cosecant are undefined.

13. (a) $\cos\theta > 0$ in quadrants I and IV, while $\tan\theta > 0$ in quadrants I and III. So, both conditions are met only in quadrant I.

(b) $\sin\theta < 0$ in quadrants III and IV. $\csc\theta$ is the reciprocal of $\sin\theta < 0$, so $\csc\theta < 0$ also in quadrants III and IV. Thus, both conditions are met in quadrants III and IV.

(c) $\cot\theta > 0$ in quadrants I and III, while $\cos < 0$ in quadrants II and III. Both conditions are met only in quadrant III.

14. $\sin\theta = \dfrac{3}{7}$ with θ in quadrant II

θ in quadrant II $\Rightarrow x < 0, y > 0$

$$\sin\theta = \frac{y}{r} = \frac{3}{7} \Rightarrow y = 3, r = 7$$

$$r^2 = x^2 + y^2 \Rightarrow 7^2 = x^2 + 3^2 \Rightarrow 49 = x^2 + 9 \Rightarrow$$
$$x^2 = 40 \Rightarrow x = -\sqrt{40} = -2\sqrt{10}$$

$$\cos\theta = \frac{x}{r} = \frac{-2\sqrt{10}}{7} = -\frac{2\sqrt{10}}{7}$$

$$\tan\theta = \frac{y}{x} = \frac{3}{-2\sqrt{10}} = -\frac{3}{2\sqrt{10}}\cdot\frac{\sqrt{10}}{\sqrt{10}} = -\frac{3\sqrt{10}}{20}$$

$$\cot\theta = \frac{x}{y} = \frac{-2\sqrt{10}}{3}$$

$$\sec\theta = \frac{r}{x} = \frac{7}{-2\sqrt{10}} = -\frac{7}{2\sqrt{10}}\cdot\frac{\sqrt{10}}{\sqrt{10}} = -\frac{7\sqrt{10}}{20}$$

$$\csc\theta = \frac{r}{y} = \frac{7}{3}$$

15. $\sin A = \dfrac{\text{side opposite}}{\text{hypotenuse}} = \dfrac{12}{13}$

$\cos A = \dfrac{\text{side adjacent}}{\text{hypotenuse}} = \dfrac{5}{13}$

$\tan A = \dfrac{\text{side opposite}}{\text{side adjacent}} = \dfrac{12}{5}$

$\cot A = \dfrac{\text{side adjacent}}{\text{side opposite}} = \dfrac{5}{12}$

$\sec A = \dfrac{\text{hypotenuse}}{\text{side adjacent}} = \dfrac{13}{5}$

$\csc A = \dfrac{\text{hypotenuse}}{\text{side opposite}} = \dfrac{13}{12}$

16. Apply the relationships between the lengths of the sides of a $30° - 60°$ right triangle first to the triangle on the right to find the values of y and w. In the $30° - 60°$ right triangle, the side opposite the $60°$ angle is $\sqrt{3}$ times as long as the side opposite to the $30°$ angle. The length of the hypotenuse is 2 times as long as the shorter leg (opposite the $30°$ angle).

Thus, we have $y = 4\sqrt{3}$ and $w = 2(4) = 8$.

Apply the relationships between the lengths of the sides of a $45° - 45°$ right triangle next to the triangle on the left to find the values of x and z. In the $45° - 45°$ right triangle, the sides opposite the $45°$ angles measure the same. The hypotenuse is $\sqrt{2}$ times the measure of a leg. Thus, we have $x = 4$ and $z = 4\sqrt{2}$

17. A $240°$ angle lies in quadrant III, so the reference angle is $240° - 180° = 60°$. Because $240°$ is in quadrant III, the sine, cosine, secant, and cosecant are negative.

$$\sin 240° = -\sin 60° = -\frac{\sqrt{3}}{2}$$

$$\cos 240° = -\cos 60° = -\frac{1}{2}$$

$$\tan 240° = \tan 60° = \sqrt{3}$$

$$\cot 240° = \cot 60° = \frac{1}{\sqrt{3}} = \frac{\sqrt{3}}{3}$$

$$\sec 240° = -\sec 60° = -2$$

$$\csc 240° = -\csc 60° = -\frac{2}{\sqrt{3}} = -\frac{2\sqrt{3}}{3}$$

18. $-135°$ is coterminal with $-135° + 360° = 225°$. This angle lies in quadrant III. The reference angle is $225° - 180° = 45°$. Because $-135°$ is in quadrant III, the sine, cosine, secant, and cosecant are negative.

$$\sin(-135°) = -\sin 45° = -\frac{\sqrt{2}}{2}$$

$$\cos(-35°) = -\cos 45° = -\frac{\sqrt{2}}{2}$$

$$\tan(-135°) = \tan 45° = 1$$

$$\cot(-135°) = \cot 45° = 1$$

(continued on next page)

(*continued*)

$$\sec(-135°) = -\sec 45° = -\sqrt{2}$$
$$\csc(-135°) = -\csc 45° = -\sqrt{2}$$

19. 990° is coterminal with $990° - 2 \cdot 360° = 270°$, which is the reference angle.

$\sin 990° = \sin 270° = -1$
$\cos 990° = \cos 270° = 0$
$\tan 990° = \tan 270°$ undefined
$\cot 990° = \cot 270° = 0$
$\sec 990° = \sec 270°$ undefined
$\csc 990° = \csc 270° = -1$

20. $\cos\theta = -\dfrac{\sqrt{2}}{2}$

Because $\cos\theta$ is negative, θ must lie in quadrant II or quadrant III. The absolute value of $\cos\theta$ is $\dfrac{\sqrt{2}}{2}$, so $\theta' = 45°$. The quadrant II angle θ equals $180° - \theta' = 180° - 45° = 135°$, and the quadrant III angle θ equals $180° + \theta' = 180° + 45° = 225°$.

21. $\csc\theta = -\dfrac{2\sqrt{3}}{3}$

Because $\csc\theta$ is negative, θ must lie in quadrant III or quadrant IV. The absolute value of $\csc\theta$ is $\dfrac{2\sqrt{3}}{3}$, so $\theta' = 60°$. The quadrant III angle θ equals $180° + \theta' = 180° + 60° = 240°$, and the quadrant IV angle θ equals $360° - \theta' = 360° - 60° = 300°$.

22. $\tan\theta = 1 \Rightarrow \theta = 45°$ or $\theta = 225°$

23. $\tan\theta = 1.6778490$

$\cot\theta = \dfrac{1}{\tan\theta} = (\tan\theta)^{-1}$, so we can use division or the inverse key (multiplicative inverse).

24. **(a)** $\sin 78°21' \approx 0.979399$

(b) $\tan 117.689° \approx -1.905608$

(c) $\sec 58.9041° = \dfrac{1}{\cos 58.9041°} \approx 1.936213$

25. $\sin\theta = 0.27843196$

$\theta = \sin^{-1}(0.27843196) \approx 16.166641°$

26. $A = 58°30', a = 748$

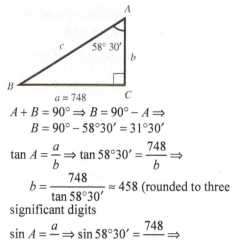

$A + B = 90° \Rightarrow B = 90° - A \Rightarrow$
$\quad B = 90° - 58°30' = 31°30'$

$\tan A = \dfrac{a}{b} \Rightarrow \tan 58°30' = \dfrac{748}{b} \Rightarrow$

$b = \dfrac{748}{\tan 58°30'} \approx 458$ (rounded to three significant digits

$\sin A = \dfrac{a}{c} \Rightarrow \sin 58°30' = \dfrac{748}{c} \Rightarrow$

$c = \dfrac{748}{\sin 58°30'} \approx 877$ (rounded to three significant digits

27. Let θ = the measure of the angle that the guy wire makes with the ground.

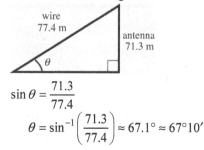

$\sin\theta = \dfrac{71.3}{77.4}$

$\theta = \sin^{-1}\left(\dfrac{71.3}{77.4}\right) \approx 67.1° \approx 67°10'$

28. Let x = the height of the flagpole.

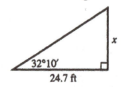

$\tan 32°10' = \dfrac{x}{24.7}$

$x = 24.7\tan 32°10' \approx 15.5344$

The flagpole is approximately 15.5 ft high. (rounded to three significant digits)

29. Let h = the height of the top of mountain above the cabin. Then $2000 + h$ = the height of the mountain.

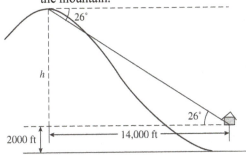

$$\tan 26° = \frac{h}{14,000} \Rightarrow h \approx 6800 \text{ (rounded to two)}$$

significant digits). Thus, the height of the mountain is about $6800 + 2000 = 8800$ ft.

30. Let x = distance the ships are apart.
In the figure, the measure of angle CAB is $122° - 32° = 90°$. Therefore, triangle CAB is a right triangle.

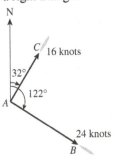

Because $d = rt$, the distance traveled by the first ship in 2.5 hr is
(2.5 hr)(16 knots) = 40 nautical mi and the second ship is
(2.5hr)(24 knots) = 60 nautical mi.
Applying the Pythagorean theorem, we have
$x^2 = 40^2 + 60^2 \Rightarrow x^2 = 1600 + 3600 \Rightarrow$
$x^2 = 5200 \Rightarrow x = \sqrt{5200} \approx 72.111$
The ships are 72 nautical mi apart (rounded to 2 significant digits).

31. Draw triangle ACB and extend north-south lines from points A and C. Angle ACD is $62°$ (alternate interior angles of parallel lines cut by a transversal have the same measure), so Angle ACB is $62° + 28° = 90°$.

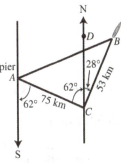

Angle ACB is a right angle, so use the Pythagorean theorem to find the distance from A to B.
$(AB)^2 = 75^2 + 53^2 \Rightarrow (AB)^2 = 5625 + 2809 \Rightarrow$
$(AB)^2 = 8434 \Rightarrow AB = \sqrt{8434} \approx 91.8368$
It is 92 km from the pier to the boat, rounded to two significant digits.

32. Let x = the side adjacent to $52.5°$ in the smaller triangle.

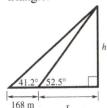

In the larger triangle, we have
$$\tan 41.2° = \frac{h}{168 + x} \Rightarrow h = (168 + x)\tan 41.2°.$$
In the smaller triangle, we have
$$\tan 52.5° = \frac{h}{x} \Rightarrow h = x \tan 52.5°.$$
Substitute for h in this equation to solve for x.
$$(168 + x)\tan 41.2° = x \tan 52.5°$$
$$168 \tan 41.2° + x \tan 41.2° = x \tan 52.5°$$
$$168 \tan 41.2° = x \tan 52.5° - x \tan 41.2°$$
$$168 \tan 41.2° = x (\tan 52.5° - \tan 41.2°)$$
$$\frac{168 \tan 41.2°}{\tan 52.5° - \tan 41.2°} = x$$
Substituting for x in the equation for the smaller triangle gives
$$h = x \tan 52.5°$$
$$h = \frac{168 \tan 41.2° \tan 52.5°}{\tan 52.5° - \tan 41.2°} \approx 448.0432$$
The height of the triangle is approximately 448 m (rounded to three significant digits).

Chapter 6

The Circular Functions and Their Graphs

Section 6.1   Radian Measure

1. An angle with its vertex at the center of a circle that intercepts an arc on the circle equal in length to the <u>radius</u> of the circle has measure 1 radian.

3. Multiply a degree measure by $\frac{\pi}{180}$ radian and simplify to convert to radians.

5. $r = 4$, $\theta = \frac{\pi}{2}$

$$s = r\theta = 4\left(\frac{\pi}{2}\right) = 2\pi$$

7. $s = 20$, $r = 10$

$$s = r\theta \Rightarrow \theta = \frac{s}{r} = \frac{20}{10} = 2$$

9. $\mathcal{A} = 8$ sq units, $r = 4$

$$\mathcal{A} = \frac{1}{2}r^2\theta \Rightarrow 8 = \frac{1}{2}(4)^2\theta \Rightarrow 8 = \frac{1}{2}(16)\theta \Rightarrow$$
$$8 = 8\theta \Rightarrow \theta = 1 \text{ radian}$$

11. $60° = 60\left(\frac{\pi}{180} \text{ radian}\right) = \frac{\pi}{3}$ radians

13. $90° = 90\left(\frac{\pi}{180} \text{ radian}\right) = \frac{\pi}{2}$ radians

15. $150° = 150\left(\frac{\pi}{180} \text{ radian}\right) = \frac{5\pi}{6}$ radians

17. $-300° = -300\left(\frac{\pi}{180} \text{ radian}\right) = -\frac{5\pi}{3}$ radians

19. $450° = 450\left(\frac{\pi}{180} \text{ radian}\right) = \frac{5\pi}{2}$ radians

21. $1800° = 1800°\left(\frac{\pi}{180°} \text{ radian}\right) = 10\pi$ radians

23. $0° = 0°\left(\frac{\pi}{180°} \text{ radian}\right) = 0$ radians

25. $-900° = -900°\left(\frac{\pi}{180°} \text{ radian}\right)$
$$= -5\pi \text{ radians}$$

27. $\frac{\pi}{3} = \frac{\pi}{3}\left(\frac{180°}{\pi}\right) = 60°$

29. $\frac{7\pi}{4} = \frac{7\pi}{4}\left(\frac{180°}{\pi}\right) = 315°$

31. $\frac{11\pi}{6} = \frac{11\pi}{6}\left(\frac{180°}{\pi}\right) = 330°$

33. $-\frac{\pi}{6} = -\frac{\pi}{6}\left(\frac{180°}{\pi}\right) = -30°$

35. $\frac{7\pi}{10} = \frac{7\pi}{10}\left(\frac{180°}{\pi}\right) = 126°$

37. $-\frac{4\pi}{15} = -\frac{4\pi}{15}\left(\frac{180°}{\pi}\right) = -48°$

39. $\frac{17\pi}{20} = \frac{17\pi}{20}\left(\frac{180°}{\pi}\right) = 153°$

41. $-5\pi = -5\pi\left(\frac{180°}{\pi}\right) = -900°$

43. $39° = 39\left(\frac{\pi}{180} \text{ radian}\right) \approx 0.681$ radian

45. $42.5° = 42.5\left(\frac{\pi}{180} \text{ radian}\right) \approx 0.742$ radian

47. $139°10' = \left(139 + \frac{10}{60}\right)°$
$$\approx 139.1666667\left(\frac{\pi}{180} \text{ radian}\right)$$
$$\approx 2.429 \text{ radians}$$

49. $64.29° = 64.29\left(\frac{\pi}{180} \text{ radian}\right) \approx 1.122$ radians

51. $56°25' = \left(56 + \frac{25}{60}\right)°$
$$\approx 56.41666667\left(\frac{\pi}{180} \text{ radian}\right)$$
$$\approx 0.985 \text{ radian}$$

53. $-47.69° = -47.69\left(\frac{\pi}{180} \text{ radian}\right)$
$$\approx -0.832 \text{ radian}$$

55. $2 \text{ radians} = 2\left(\dfrac{180°}{\pi}\right) \approx 114.591559°$

$\qquad = 114° + (0.591559 \cdot 60)'$

$\qquad \approx 114°35'$

57. $1.74 \text{ radians} = 1.74\left(\dfrac{180°}{\pi}\right) \approx 99.69465635°$

$\qquad = 99° + (0.69465635 \cdot 60)'$

$\qquad \approx 99°42'$

59. $0.3417 \text{ radian} = .3417\left(\dfrac{180°}{\pi}\right) \approx 19.57796786°$

$\qquad = 19° + (0.57796786 \cdot 60)'$

$\qquad = 19°35'$

61. $-5.01095 \text{ radian} = -5.01095\left(\dfrac{180°}{\pi}\right)$

$\qquad \approx -287.1062864°$

$\qquad = -\left[287° + (0.1062864 \cdot 60)'\right]$

$\qquad \approx -287°06'$

63. Without the degree symbol on the 30, it is assumed that 30 is measured in radians. Thus, the approximate value of sin 30 is -0.98803, not $\frac{1}{2}$.

65. Begin the calculation with the blank next to 30°, and then proceed counterclockwise from there.

$30° = 30\left(\dfrac{\pi}{180} \text{ radian}\right) = \dfrac{\pi}{6} \text{ radian}$

$\dfrac{\pi}{4} \text{ radians} = \dfrac{\pi}{4}\left(\dfrac{180°}{\pi}\right) = 45°$

$60° = 60\left(\dfrac{\pi}{180} \text{ radian}\right) = \dfrac{\pi}{3} \text{ radians}$

$\dfrac{2\pi}{3} \text{ radians} = \dfrac{2\pi}{3}\left(\dfrac{180°}{\pi}\right) = 120°$

$\dfrac{3\pi}{4} \text{ radians} = \dfrac{3\pi}{4}\left(\dfrac{180°}{\pi}\right) = 135°$

$150° = 150\left(\dfrac{\pi}{180} \text{ radian}\right) = \dfrac{5\pi}{6} \text{ radians}$

$180° = 180\left(\dfrac{\pi}{180} \text{ radian}\right) = \pi \text{ radians}$

$210° = 210\left(\dfrac{\pi}{180} \text{ radian}\right) = \dfrac{7\pi}{6} \text{ radians}$

$225° = 225\left(\dfrac{\pi}{180} \text{ radian}\right) = \dfrac{5\pi}{4} \text{ radians}$

$\dfrac{4\pi}{3} \text{ radians} = \dfrac{4\pi}{3}\left(\dfrac{180°}{\pi}\right) = 240°$

$\dfrac{5\pi}{3} \text{ radians} = \dfrac{5\pi}{3}\left(\dfrac{180°}{\pi}\right) = 300°$

$315° = 315\left(\dfrac{\pi}{180} \text{ radian}\right) = \dfrac{7\pi}{4} \text{ radians}$

$330° = 330\left(\dfrac{\pi}{180} \text{ radian}\right) = \dfrac{11\pi}{6} \text{ radians}$

67. $r = 12.3 \text{ cm}, \theta = \dfrac{2\pi}{3} \text{ radians}$

$s = r\theta = 12.3\left(\dfrac{2\pi}{3}\right) = 8.2\pi \approx 25.8 \text{ cm}$

69. $r = 1.38 \text{ ft}, \theta = \dfrac{5\pi}{6} \text{ radians}$

$s = r\theta = 1.38\left(\dfrac{5\pi}{6}\right)$

$\qquad = 1.15\pi \text{ ft} \approx 3.61 \text{ ft}$ (rounded to three significant digits)

71. $r = 4.82 \text{ m}, \theta = 60°$

Converting θ to radians, we have

$\theta = 60° = 60\left(\dfrac{\pi}{180} \text{ radian}\right) = \dfrac{\pi}{3} \text{ radians}.$

Thus, the arc is

$s = r\theta = 4.82\left(\dfrac{\pi}{3}\right) = \dfrac{4.82\pi}{3} \approx 5.05 \text{ m}.$

(rounded to three significant digits)

73. $r = 15.1 \text{ in.}, \theta = 210°$

Converting θ to radians, we have

$\theta = 210° = 210\left(\dfrac{\pi}{180} \text{ radian}\right) = \dfrac{7\pi}{6} \text{ radians}.$

Thus, the arc is

$s = r\theta = 15.1\left(\dfrac{7\pi}{6}\right) = \dfrac{105.7\pi}{6} \approx 55.3 \text{ in.}$

(rounded to three significant digits)

For Exercises 75–80, note that 6400 has two significant digits and the angles are given to the nearest degree, so we can have only two significant digits in the answers.

75. 9° N, 40° N

$\theta = 40° - 9° = 31° = 31\left(\dfrac{\pi}{180} \text{ radian}\right)$

$\qquad = \dfrac{31\pi}{180} \text{ radian}$

$s = r\theta = 6400\left(\dfrac{31\pi}{180}\right) \approx 3500 \text{ km}$

77. 41° N, 12° S
12° S = −12° N

$$\theta = 41° - (-12°) = 53° = 53\left(\frac{\pi}{180} \text{ radian}\right)$$

$$= \frac{53\pi}{180} \text{ radian}$$

$$s = r\theta = 6400\left(\frac{53\pi}{180}\right) \approx 5900 \text{ km}$$

79. $r = 6400$ km, $s = 1200$ km

$$s = r\theta \Rightarrow 1200 = 6400\theta \Rightarrow \theta = \frac{1200}{6400} = \frac{3}{16}$$

Converting $\frac{3}{16}$ radian to degrees, we have

$$\theta = \frac{3}{16}\left(\frac{180°}{\pi}\right) \approx 11°. \text{ The north-south}$$

distance between the two cities is 11°.
Let x = the latitude of Madison.
$x - 33° = 11° \Rightarrow x = 44°$ N
The latitude of Madison is 44° N.

81. The arc length on the smaller gear is

$$s = r\theta = 3.7\left(300 \cdot \frac{\pi}{180}\right) = 3.7\left(\frac{5\pi}{3}\right)$$

$$= \frac{18.5\pi}{3} \text{ cm}$$

An arc with this length on the larger gear
corresponds to an angle measure θ where

$$s = r\theta \Rightarrow \frac{18.5\pi}{3} = 7.1\theta \Rightarrow \frac{18.5\pi}{21.3} = \theta \Rightarrow$$

$$\theta = \frac{18.5\pi}{21.3} \cdot \frac{180}{\pi} \approx 156°$$

The larger gear will rotate through
approximately 156°.

83. A rotation of $\theta = 60.0\left(\frac{\pi}{180} \text{ radian}\right) = \frac{\pi}{3}$

radians on the smaller wheel moves through
an arc length of

$$s = r\theta = 5.23\left(\frac{\pi}{3}\right) = \frac{5.23\pi}{3} \text{ cm (holding on}$$

to more digits for the intermediate steps).
Both wheels move together, so the larger

wheel also moves $\frac{5.23\pi}{3}$ cm, which rotates it

through an angle θ, where

$$\frac{5.23\pi}{3} = 8.16\theta$$

$$\theta = \frac{5.23\pi}{24.48} \text{ radian} = \frac{5.23\pi}{24.48}\left(\frac{180°}{\pi}\right) \approx 38.5°$$

The larger wheel rotates through 38.5°.

85. The arc length s represents the distance
traveled by a point on the rim of a wheel. The
two wheels rotate together, so s will be the
same for both wheels. For the smaller wheel,

$$\theta = 80° = 80\left(\frac{\pi}{180}\right) = \frac{4\pi}{9} \text{ radians and}$$

$$s = r\theta = 11.7\left(\frac{4\pi}{9}\right) = 5.2\pi \text{ cm.}$$

For the larger wheel,

$$\theta = 50° = 50\left(\frac{\pi}{180} \text{ radian}\right) = \frac{5\pi}{18} \text{ radian.}$$

Thus, we can solve

$$s = r\theta \Rightarrow 5.2\pi = r\left(\frac{5\pi}{18}\right) \Rightarrow$$

$$r = 5.2\pi \cdot \frac{18}{5\pi} = 18.72$$

The radius of the larger wheel is 18.7 cm.
(rounded to 3 significant digits)

87. (a) The number of inches lifted is the arc
length in a circle with $r = 9.27$ in. and
$\theta = 71°50'$.

$$71°50' = \left(71 + \frac{50}{60}\right)\left(\frac{\pi}{180°}\right)$$

$$s = r\theta = 9.27\left(71 + \frac{50}{60}\right)\left(\frac{\pi}{180°}\right) \approx 11.6221$$

The weight will rise 11.6 in. (rounded to
three significant digits)

(b) When the weight is raised 6 in., we have
$$s = r\theta \Rightarrow 6 = 9.27\theta \Rightarrow$$

$$\theta = \frac{6}{9.27} \text{ radian} = \frac{6}{9.27}\left(\frac{180°}{\pi}\right)$$

$$\approx 37.0846° = 37° + 0.0846(60')$$

$$\approx 37°05'$$

The pulley must be rotated through
37°05′.

89. A rotation of $\theta = 180\left(\frac{\pi}{180} \text{ radian}\right) = \pi$ radians.

The chain moves a distance equal to half the
arc length of the larger gear. So, for the large
gear and pedal, $s = r\theta \Rightarrow 4.72\pi$. Thus, the
chain moves 4.72π in. The small gear rotates
through an angle as follows.

$$\theta = \frac{s}{r} \Rightarrow \theta = \frac{4.72\pi}{1.38} \approx 3.42\pi$$

θ for the wheel and θ for the small gear are
the same, or 3.42π. So, for the wheel, we have
$$s = r\theta \Rightarrow r = 13.6(3.42\pi) \approx 146.12$$

The bicycle will move 146 in. (rounded to
three significant digits)

91. Because $\dfrac{30}{60} = \dfrac{1}{2}$ rotation, we have

$\theta = \dfrac{1}{2}(2\pi) = \pi.$ Thus, $s = r\theta \Rightarrow s = 3\pi$ in.

93. Because $\theta = 4.5(2\pi) = 9\pi,$ we have

$s = r\theta \Rightarrow s = 3(9\pi) = 27\pi$ in.

95. Let t = the length of the train.
t is approximately the arc length subtended by $3°\,20'$. First convert $\theta = 3°20'$ to radians.

$\theta = 3°20' = \left(3 + \tfrac{20}{60}\right)° = 3\tfrac{1}{3}°$

$= \left(3\tfrac{1}{3}\right)\left(\dfrac{\pi}{180}\text{ radian}\right) = \left(\dfrac{10}{3}\right)\left(\dfrac{\pi}{180}\text{ radian}\right)$

$= \dfrac{\pi}{54}\text{ radian}$

The length of the train is

$t = r\theta \Rightarrow t = 3.5\left(\dfrac{\pi}{54}\right) \approx 0.20$ km long.

(rounded to two significant digits)

97. Let r = the distance of the boat.
The height of the mast, 32.0 ft, is approximately the arc length subtended by $2°\,11'$. First convert $\theta = 2°11'$ to radians.

$\theta = 2°11' = \left(2 + \tfrac{11}{60}\right)° = 2\tfrac{11}{60}°$

$= \left(2\tfrac{11}{60}\right)\left(\dfrac{\pi}{180}\text{ radian}\right) = \left(\dfrac{131}{60}\right)\left(\dfrac{\pi}{180}\text{ radian}\right)$

$= \dfrac{131\pi}{10800}\text{ radian}$

We must now find the radius, r.

$s = r\theta \Rightarrow r = \dfrac{s}{\theta} \Rightarrow$

$r = \dfrac{32}{\frac{131\pi}{10800}} = 32 \cdot \dfrac{10800}{131\pi} \approx 839.7549$

The boat is about 840 ft away. (rounded to two significant digits)

In Exercises 99–105, we will be rounding to the nearest tenth.

99. $r = 29.2$ m, $\theta = \dfrac{5\pi}{6}$ radians $0.517°$.

The area of the sector is 1116.1 m². (1120 m² rounded to three significant digits)

101. $r = 30.0$ ft, $\theta = \dfrac{\pi}{2}$ radians

$\mathcal{A} = \dfrac{1}{2}r^2\theta = \dfrac{1}{2}(30.0)^2\left(\dfrac{\pi}{2}\right) = \dfrac{1}{2}(900)\left(\dfrac{\pi}{2}\right)$

$= 225\pi \approx 706.8583$

The area of the sector is 706.9 ft². (707 ft² rounded to three significant digits)

103. $r = 12.7$ cm, $\theta = 81°$

The formula $\mathcal{A} = \dfrac{1}{2}r^2\theta$ requires that θ be measured in radians. Converting $81°$ to radians, we have

$\theta = 81\left(\dfrac{\pi}{180}\text{ radian}\right) = \dfrac{9\pi}{20}$ radians.

$\mathcal{A} = \dfrac{1}{2}(12.7)^2\left(\dfrac{9\pi}{20}\right) = \dfrac{1}{2}(161.29)\left(\dfrac{9\pi}{20}\right)$

≈ 114.0092

The area of the sector is 114.0 cm². (114 cm² rounded to three significant digits)

105. $r = 40.0$ mi, $\theta = 135°$

The formula $\mathcal{A} = \dfrac{1}{2}r^2\theta$ requires that θ be measured in radians. Converting $135°$ to radians, we have

$\theta = 135\left(\dfrac{\pi}{180}\text{ radian}\right) = \dfrac{3\pi}{4}$ radians.

$\mathcal{A} = \dfrac{1}{2}(40.0)^2\left(\dfrac{3\pi}{4}\right) = \dfrac{1}{2}(1600)\left(\dfrac{3\pi}{4}\right)$

$= 600\pi \approx 1884.9556$

The area of the sector is 1885.0 mi². (1880 mi² rounded to three significant digits)

107. $\mathcal{A} = 16$ in.^{2}, $r = 3.0$ in.

$\mathcal{A} = \dfrac{1}{2}r^2\theta \Rightarrow 16 = \dfrac{1}{2}(3)^2\,\theta \Rightarrow 16 = \dfrac{9}{2}\theta \Rightarrow$

$\theta = 16 \cdot \dfrac{2}{9} = \dfrac{32}{9} \approx 3.6$ radians

(rounded to two significant digits)

109. The formula $\mathcal{A} = \dfrac{1}{2}r^2\theta$ requires that θ be measured in radians. Converting $40.0°$ to radians, we have

$\theta = 40.0\left(\dfrac{\pi}{180}\text{ radian}\right) = \dfrac{2\pi}{9}$ radians.

$\mathcal{A} = \dfrac{1}{2}r^2\theta = \dfrac{1}{2}(152)^2\left(\dfrac{2\pi}{9}\right) = \dfrac{1}{2}(23,104)\left(\dfrac{2\pi}{9}\right)$

$= \dfrac{23,104\pi}{9} \approx 8060$ yd^2

(rounded to three significant digits)

111. $\mathcal{A} = 50$ in.^{2}, $r = 5$ in.

First find θ.

$$\mathcal{A} = \frac{1}{2}r^2\theta \Rightarrow 50 = \frac{1}{2}(5)^2\theta \Rightarrow 50 = \frac{25}{2}\theta \Rightarrow$$

$\theta = 4$ radians

To find the arc length, apply the formula $s = r\theta$.

$s = 5 \cdot 4 = 20$ in.

113. **(a)** The central angle in degrees measures

$\dfrac{360°}{27} = 13\frac{1}{3}°$. Converting to radians, we have the following.

$$13\tfrac{1}{3}° = \left(13\tfrac{1}{3}\right)\left(\frac{\pi}{180} \text{ radian}\right)$$

$$= \left(\frac{40}{3}\right)\left(\frac{\pi}{180} \text{ radian}\right) = \frac{2\pi}{27} \text{ radian}$$

(b) $C = 2\pi r$, and $r = 76$ ft, so

$C = 2\pi(76) = 152\pi \approx 477.5221$. The circumference is about 478 ft.

(c) $r = 76$ ft and $\theta = \dfrac{2\pi}{27}$, so

$$s = r\theta = 76\left(\frac{2\pi}{27}\right) = \frac{152\pi}{27} \approx 17.6860.$$

Thus, the length of the arc is 17.7 ft. Note: If this measurement is approximated to be $\dfrac{160}{9}$, then the approximated value would be 17.8 ft, rounded to three significant digits.

(d) Area of sector with $r = 76$ ft and $\theta = \dfrac{2\pi}{27}$ is as follows.

$$\mathcal{A} = \frac{1}{2}r^2\theta \Rightarrow$$

$$\mathcal{A} = \frac{1}{2}\left(76^2\right)\frac{2\pi}{27} = \frac{1}{2}(5776)\frac{2\pi}{27} = \frac{5776\pi}{27}$$

$$\approx 672.0681 \approx 672 \text{ ft}^2$$

115. (a)

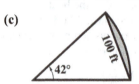

The triangle formed by the central angle and the chord is isosceles. Therefore, the bisector of the central angle is also the perpendicular bisector of the chord.

$$\sin 21° = \frac{50}{r} \Rightarrow r = \frac{50}{\sin 21°} \approx 140 \text{ ft}$$

(b) $r = \dfrac{50}{\sin 21°}$; $\theta = 42°$

Converting θ to radians, we have

$42\left(\dfrac{\pi}{180} \text{ radians}\right) = \dfrac{7\pi}{30}$ radians. Solving for the arc length, we have

$$s = r\theta \Rightarrow$$

$$s = \frac{50}{\sin 21°} \cdot \frac{7\pi}{30} = \frac{35\pi}{3\sin 21°} \approx 102 \text{ ft}$$

(c)

The area of the portion of the circle can be found by subtracting the area of the triangle from the area of the sector. From the figure in part (a), we have

$$\tan 21° = \frac{50}{h} \quad \text{so} \quad h = \frac{50}{\tan 21°}.$$

$$\mathcal{A}_{\text{sector}} = \frac{1}{2}r^2\theta \Rightarrow$$

$$\mathcal{A}_{\text{sector}} = \frac{1}{2}\left(\frac{50}{\sin 21°}\right)^2\left(\frac{7\pi}{30}\right) \approx 7135 \text{ ft}^2$$

and

$$\mathcal{A}_{\text{triangle}} = \frac{1}{2}bh \Rightarrow$$

$$\mathcal{A}_{\text{triangle}} = \frac{1}{2}(100)\left(\frac{50}{\tan 21°}\right) \approx 6513 \text{ ft}^2$$

The area bounded by the arc and the chord is $7135 - 6513 = 622 \text{ ft}^2$.

117. Use the Pythagorean theorem to find the hypotenuse of the triangle, which is also the radius of the sector of the circle.

$r^2 = 30^2 + 40^2 \Rightarrow r^2 = 900 + 1600 \Rightarrow$
$r^2 = 2500 \Rightarrow r = 50$

The total area of the lot is the sum of the areas of the triangle and the sector.

Converting $\theta = 60°$ to radians, we have

$60\left(\dfrac{\pi}{180} \text{ radian}\right) = \dfrac{\pi}{3}$ radians.

$\mathcal{A}_{\text{triangle}} = \dfrac{1}{2}bh = \dfrac{1}{2}(30)(40) = 600 \text{ yd}^2$

$\mathcal{A}_{\text{sector}} = \dfrac{1}{2}r^2\theta = \dfrac{1}{2}(50)^2\left(\dfrac{\pi}{3}\right)$
$= \dfrac{1}{2}(2500)\left(\dfrac{\pi}{3}\right) = \dfrac{1250\pi}{3} \text{ yd}^2$

Total area

$\mathcal{A}_{\text{triangle}} + \mathcal{A}_{\text{sector}} = 600 + \dfrac{1250\pi}{3} \approx 1908.9969$

or 1900 yd^2, rounded to two significant digits.

Section 6.2   The Unit Circle and Circular Functions

1.

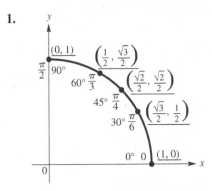

3. $\sin\dfrac{\pi}{4} = \dfrac{\sqrt{2}}{2}$

5. $\tan\dfrac{\pi}{4} = 1$

7. The measure of how fast the position of point P is changing is the <u>linear speed (or linear velocity)</u>.

9. If the angular speed of point P is 1 radian per sec, then P will move around the entire unit circle in $\underline{2\pi}$ sec.

11. An angular speed of 1 revolution per min on the unit circle is equivalent to an angular speed, ω, of $\underline{2\pi}$ radians per min.

13. An angle of $s = \dfrac{\pi}{2}$ radians intersects the unit circle at the point $(0,1)$.

(a) $\sin s = y = 1$

(b) $\cos s = x = 0$

(c) $\tan s = \dfrac{y}{x} = \dfrac{1}{0}$; undefined

15. An angle of $s = 2\pi$ radians intersects the unit circle at the point $(1,0)$.

(a) $\sin s = y = 0$ (b) $\cos s = x = 1$

(c) $\tan s = \dfrac{y}{x} = \dfrac{0}{1} = 0$

17. An angle of $s = -\pi$ radians intersects the unit circle at the point $(-1,0)$.

(a) $\sin s = y = 0$ (b) $\cos s = x = -1$

(c) $\tan s = \dfrac{y}{x} = \dfrac{0}{-1} = 0$

For Exercises 19–33, use the following copy of Figure 13 on page 579 of the text.

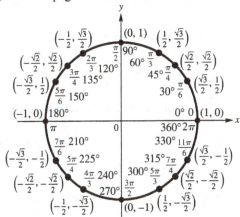

19. $\dfrac{7\pi}{6}$ is in quadrant III, so the reference angle is $\dfrac{7\pi}{6} - \pi = \dfrac{7\pi}{6} - \dfrac{6\pi}{6} = \dfrac{\pi}{6}$. In quadrant III, the sine is negative. Thus, $\sin\dfrac{7\pi}{6} = -\sin\dfrac{\pi}{6} = -\dfrac{1}{2}$. Converting $\dfrac{7\pi}{6}$ to degrees, we have $\dfrac{7\pi}{6} = \dfrac{7}{6}(180°) = 210°$.

The reference angle is $210° - 180° = 30°$.

Thus, $\sin\dfrac{7\pi}{6} = \sin 210° = -\sin 30° = -\dfrac{1}{2}$.

21. $\dfrac{3\pi}{4}$ is in quadrant II, so the reference angle is

$\pi - \dfrac{3\pi}{4} = \dfrac{4\pi}{4} - \dfrac{3\pi}{4} = \dfrac{\pi}{4}$. In quadrant II, the tangent is negative. Thus,

$\tan \dfrac{3\pi}{4} = -\tan \dfrac{\pi}{4} = -1$. Converting $\dfrac{3\pi}{4}$ to

degrees, we have $\dfrac{3\pi}{4} = \dfrac{3}{4}(180°) = 135°$. The reference angle is $180° - 135° = 45°$. Thus,

$\tan \dfrac{3\pi}{4} = \tan 135° = -\tan 45° = -1$.

23. $\dfrac{11\pi}{6}$ is in quadrant IV, so the reference angle

is $2\pi - \dfrac{11\pi}{6} = \dfrac{12\pi}{6} - \dfrac{11\pi}{6} = \dfrac{\pi}{6}$. In quadrant

IV, the cosecant is negative. Thus,

$\csc \dfrac{11\pi}{6} = -\csc \dfrac{\pi}{6} = -2$. Converting $\dfrac{11\pi}{6}$ to

degrees, we have $\dfrac{11\pi}{6} = \dfrac{11}{6}(180°) = 330°$.

The reference angle is $360° - 330° = 30°$.

Thus, $\csc \dfrac{11\pi}{6} = \csc 330° = -\csc 30° = -2$.

25. $-\dfrac{4\pi}{3}$ is coterminal with

$-\dfrac{4\pi}{3} + 2\pi = -\dfrac{4\pi}{3} + \dfrac{6\pi}{3} = \dfrac{2\pi}{3}$.

$\dfrac{2\pi}{3}$ is in quadrant II, so the reference angle is

$\pi - \dfrac{2\pi}{3} = \dfrac{3\pi}{3} - \dfrac{2\pi}{3} = \dfrac{\pi}{3}$. In quadrant II, the

cosine is negative. Thus,

$\cos\left(-\dfrac{4\pi}{3}\right) = \cos \dfrac{2\pi}{3} = -\cos \dfrac{\pi}{3} = -\dfrac{1}{2}$.

Converting $\dfrac{2\pi}{3}$ to degrees, we have

$\dfrac{2\pi}{3} = \dfrac{2}{3}(180°) = 120°$. The reference angle is

$180° - 120° = 60°$.

Thus,

$\cos\left(-\dfrac{4\pi}{3}\right) = \cos \dfrac{2\pi}{3} = \cos 120°$

$= -\cos 60° = -\dfrac{1}{2}$

27. $\dfrac{7\pi}{4}$ is in quadrant IV, so the reference angle

is $2\pi - \dfrac{7\pi}{4} = \dfrac{8\pi}{4} - \dfrac{7\pi}{4} = \dfrac{\pi}{4}$. In quadrant IV,

the cosine is positive. Thus,

$\cos \dfrac{7\pi}{4} = \cos \dfrac{\pi}{4} = \dfrac{\sqrt{2}}{2}$. Converting $\dfrac{7\pi}{4}$ to

degrees, we have $\dfrac{7\pi}{4} = \dfrac{7}{4}(180°) = 315°$. The

reference angle is $360° - 315° = 45°$. Thus,

$\cos \dfrac{7\pi}{4} = \cos 315° = \cos 45° = \dfrac{\sqrt{2}}{2}$.

29. $-\dfrac{4\pi}{3}$ is coterminal with

$-\dfrac{4\pi}{3} + 2\pi = -\dfrac{4\pi}{3} + \dfrac{6\pi}{3} = \dfrac{2\pi}{3}$.

$\dfrac{2\pi}{3}$ is in quadrant II, so the reference angle is

$\pi - \dfrac{2\pi}{3} = \dfrac{3\pi}{3} - \dfrac{2\pi}{3} = \dfrac{\pi}{3}$. In quadrant II, the

sine is positive. Thus,

$\sin\left(-\dfrac{4\pi}{3}\right) = \sin \dfrac{2\pi}{3} = \sin \dfrac{\pi}{3} = \dfrac{\sqrt{3}}{2}$.

Converting $\dfrac{2\pi}{3}$ radians to degrees, we have

$\dfrac{2\pi}{3} = \dfrac{2}{3}(180°) = 120°$. The reference angle is

$180° - 120° = 60°$. Thus,

$\sin\left(-\dfrac{4\pi}{3}\right) = \sin \dfrac{2\pi}{3} = \sin 120° = \sin 60° = \dfrac{\sqrt{3}}{2}$

31. $\dfrac{23\pi}{6}$ is coterminal with

$\dfrac{23\pi}{6} - 2\pi = \dfrac{23\pi}{6} - \dfrac{12\pi}{6} = \dfrac{11\pi}{6}$.

$\dfrac{11\pi}{6}$ is in quadrant IV, so the reference angle

is $2\pi - \dfrac{11\pi}{6} = \dfrac{12\pi}{6} - \dfrac{11\pi}{6} = \dfrac{\pi}{6}$.

In quadrant IV, the secant is positive. Thus,

$\sec \dfrac{23\pi}{6} = \sec \dfrac{11\pi}{6} = \sec \dfrac{\pi}{6} = \dfrac{2\sqrt{3}}{3}$.

Converting $\dfrac{11\pi}{6}$ radians to degrees, we have

$\dfrac{11\pi}{6} = \dfrac{11}{6}(180°) = 330°$.

(continued on next page)

(*continued*)

The reference angle is $360° - 330° = 30°$. Thus,

$$\sec \frac{23\pi}{6} = \sec \frac{11\pi}{6} = \sec 330° = \sec 30° = \frac{2\sqrt{3}}{3}.$$

33. $\dfrac{5\pi}{6}$ is in quadrant II, so the reference angle is

$$\pi - \frac{5\pi}{6} = \frac{6\pi}{6} - \frac{5\pi}{6} = \frac{\pi}{6}.$$ In quadrant II, the

tangent is negative. Thus,

$$\tan \frac{5\pi}{6} = -\tan \frac{\pi}{6} = -\frac{\sqrt{3}}{3}.$$

Converting $\dfrac{5\pi}{6}$ radians to degrees, we have

$$\frac{5\pi}{6} = \frac{5}{6}(180°) = 150°.$$ The reference angle is

$180° - 150° = 30°.$ Thus,

$$\tan \frac{5\pi}{6} = \tan 150° = -\tan 30° = -\frac{\sqrt{3}}{3}.$$

For Exercises 35–46, 63–68, and 79–88, your calculator must be set in radian mode. Keystroke sequences may vary based on the type and/or model of calculator being used. As in Example 3, we will set the calculator to show four decimal digits.

35. $\sin 0.6109 \approx 0.5736$

37. $\cos(-1.1519) \approx 0.4068$

39. $\tan 4.0203 \approx 1.2065$

41. $\csc(-9.4946) \approx 14.3338$

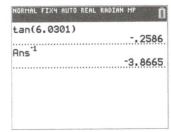

43. $\sec 2.8440 \approx -1.0460$

45. $\cot 6.0301 \approx -3.8665$

47. $\cos 0.8 \approx 0.7$

49. $\sin 2 \approx 0.9$

51. $\sin 3.8 \approx -0.6$

53. $\cos\theta = -0.65 \Rightarrow x = -0.65 \Rightarrow \theta \approx 2.3$ radians
or $\theta \approx 4$ radians

55. $\sin\theta = -0.7 \Rightarrow y = 0.7 \Rightarrow \theta \approx 0.8$ radian or
$\theta \approx 2.4$ radians

57. $\cos 2$

$\dfrac{\pi}{2} \approx 1.57$ and $\pi \approx 3.14$, so $\dfrac{\pi}{2} < 2 < \pi$. Thus,

an angle of 2 radians is in quadrant II. (The figure for Exercises 35–44 also shows that 2 radians is in quadrant II.) Because values of the cosine function are negative in quadrant II, $\cos 2$ is negative.

59. $\sin 5$

$\dfrac{3\pi}{2} \approx 4.71$ and $2\pi \approx 6.28$, so $\dfrac{3\pi}{2} < 5 < 2\pi.$

Thus, an angle of 5 radians is in quadrant IV. (The figure for Exercises 35–44 also shows that 5 radians is in quadrant IV.) Because values of the sine function are negative in quadrant IV, $\sin 5$ is negative.

61. $\tan 6.29$
$2\pi \approx 6.28$ and

$$2\pi + \frac{\pi}{2} = \frac{4\pi}{2} + \frac{\pi}{2} = \frac{5\pi}{2} \approx 7.85, \text{ so}$$

$$2\pi < 6.29 < \frac{5\pi}{2}.$$

(*continued on next page*)

(continued)

Notice that 2π is coterminal with 0 and $\dfrac{5\pi}{2}$ is coterminal with $\dfrac{\pi}{2}$. Thus, an angle of 6.29 radians is in quadrant I. Because values of the tangent function are positive in quadrant I, tan 6.29 is positive.

63. $\tan s = 0.2126 \Rightarrow s \approx 0.2095$

65. $\sin s = 0.9918 \Rightarrow s \approx 1.4426$

67. $\sec s = 1.0806 \Rightarrow s \approx 0.3887$

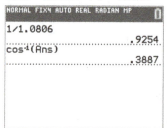

69. $\left[\dfrac{\pi}{2}, \pi\right]$; $\sin s = \dfrac{1}{2}$

Recall that $\sin\dfrac{\pi}{6} = \dfrac{1}{2}$ and in quadrant II, $\sin s$ is positive. Therefore,

$$\sin\left(\pi - \dfrac{\pi}{6}\right) = \sin\dfrac{5\pi}{6} = \dfrac{1}{2}, \text{ so } s = \dfrac{5\pi}{6}.$$

71. $\left[\pi, \dfrac{3\pi}{2}\right]$; $\tan s = \sqrt{3}$

Recall that $\tan\dfrac{\pi}{3} = \sqrt{3}$ and in quadrant III, $\tan s$ is positive. Therefore,

$$\tan\left(\pi + \dfrac{\pi}{3}\right) = \tan\dfrac{4\pi}{3} = \sqrt{3}, \text{ so } s = \dfrac{4\pi}{3}.$$

73. $\left[\dfrac{3\pi}{2}, 2\pi\right]$; $\tan s = -1$

Recall that $\tan\dfrac{\pi}{4} = 1$ and in quadrant IV, $\tan s$ is negative. Therefore,

$$\tan\left(2\pi - \dfrac{\pi}{4}\right) = \tan\dfrac{7\pi}{4} = -1, \text{ so } s = \dfrac{7\pi}{4}.$$

75. $[0, 2\pi)$; $\sin s = -\dfrac{\sqrt{3}}{2}$

Recall that $\sin\dfrac{\pi}{3} = \dfrac{\sqrt{3}}{2}$, and that $\sin s$ is negative in quadrants III and IV. Thus, the angles we are seeking have reference angle $\dfrac{\pi}{3}$ and are located in quadrants III and IV. In quadrant III, $s = \pi + \dfrac{\pi}{3} = \dfrac{4\pi}{3}$. In quadrant IV, $s = 2\pi - \dfrac{\pi}{3} = \dfrac{5\pi}{3}$. Thus, the angles are $\dfrac{4\pi}{3}$ and $\dfrac{5\pi}{3}$.

77. $[0, 2\pi)$; $\cos^2 s = \dfrac{1}{2} \Rightarrow \cos s = \pm\sqrt{\dfrac{1}{2}} = \pm\dfrac{\sqrt{2}}{2}$

Recall that $\cos\dfrac{\pi}{4} = \dfrac{\sqrt{2}}{2}$, and that $\cos s$ is positive in quadrants I and IV, and negative in quadrants II and III. Thus, the angles we are seeking have reference angle $\dfrac{\pi}{4}$. In quadrant II, $s = \pi - \dfrac{\pi}{4} = \dfrac{3\pi}{4}$. In quadrant III, $s = \pi + \dfrac{\pi}{4} = \dfrac{5\pi}{4}$. In quadrant IV, $s = 2\pi - \dfrac{\pi}{4} = \dfrac{7\pi}{4}$. Thus, the angles are $\dfrac{\pi}{4}$, $\dfrac{3\pi}{4}$, $\dfrac{5\pi}{4}$, and $\dfrac{7\pi}{4}$.

79. $[-2\pi, \pi)$; $3\tan^2 s = 1 \Rightarrow \tan^2 s = \dfrac{1}{3} \Rightarrow$

$$\tan s = \pm\sqrt{\dfrac{1}{3}} = \pm\dfrac{\sqrt{3}}{3}$$

We will split the interval into $[-2\pi, 0)$ and $[0, \pi)$. First we will find the angles in the interval $[0, \pi)$. Recall that $\tan\dfrac{\pi}{6} = \dfrac{\sqrt{3}}{3}$, and that $\tan s$ is positive in quadrants I and III, and negative in quadrants II and IV. In quadrant II, $s = \pi - \dfrac{\pi}{6} = \dfrac{5\pi}{6}$. In quadrant III, $s = \pi + \dfrac{\pi}{6} = \dfrac{7\pi}{6}$.

(continued on next page)

(*continued*)

In quadrant IV, $s = 2\pi - \dfrac{\pi}{6} = \dfrac{11\pi}{6}$. To find

the angles in the interval $[-2\pi, 0)$, recall that

moving around the unit circle $\dfrac{\pi}{6}$ in the

positive direction yields the same ending point

as moving $\dfrac{11\pi}{6}$ units in the negative

direction. So $-\dfrac{11\pi}{6}$ is one of the angles.

Moving $\dfrac{5\pi}{6}$ units in the positive direction is

the same as moving $\dfrac{7\pi}{6}$ units in the negative

direction, so $-\dfrac{7\pi}{6}$ is another angle. Now we

must find the negative angles in quadrants III

and IV. Moving $\dfrac{7\pi}{6}$ units in the positive

direction is that same as moving $\dfrac{5\pi}{6}$ units in

the negative direction, so $-\dfrac{5\pi}{6}$ is another

angle. Finally, moving $\dfrac{11\pi}{6}$ units in the

positive direction is the same as moving $\dfrac{\pi}{6}$

units in the negative direction. Thus, the

angles are $-\dfrac{11\pi}{6}, -\dfrac{7\pi}{6},$

$-\dfrac{5\pi}{6}, -\dfrac{\pi}{6}, \dfrac{\pi}{6},$ and $\dfrac{5\pi}{6}$.

81. Refer to figures 18 and 19 on page 584 in the text.

(a) $OQ = \cos 60° = \dfrac{1}{2}$

(b) $PQ = \sin 60° = \dfrac{\sqrt{3}}{2}$

(c) $VR = \tan 60° = \sqrt{3}$

(d) $OV = \sec 60° = \dfrac{1}{\cos 60°} = \dfrac{1}{\frac{1}{2}} = 2$

(e) $OU = \csc 60° = \dfrac{1}{\sin 60°} = \dfrac{1}{\frac{\sqrt{3}}{2}} = \dfrac{2}{\sqrt{3}} = \dfrac{2\sqrt{3}}{3}$

(f) $US = \cot 60° = \dfrac{1}{\tan 60°} = \dfrac{1}{\sqrt{3}} = \dfrac{\sqrt{3}}{3}$

83. $r = 20$ cm, $\omega = \dfrac{\pi}{12}$ radian per sec, $t = 6$ sec

(a) $\omega = \dfrac{\theta}{t} \Rightarrow \dfrac{\pi}{12} = \dfrac{\theta}{6} \Rightarrow \theta = \dfrac{\pi}{2}$ radians

(b) $s = r\theta \Rightarrow s = 20 \cdot \dfrac{\pi}{2} = 10\pi$ cm

(c) $v = \dfrac{r\theta}{t} \Rightarrow v = \dfrac{20 \cdot \frac{\pi}{2}}{6} = \dfrac{10\pi}{6} = \dfrac{5\pi}{3}$ cm per sec

85. $r = 8$ in., $\omega = \dfrac{\pi}{3}$ radian per min, $t = 9$ min

(a) $\omega = \dfrac{\theta}{t} \Rightarrow \dfrac{\pi}{3} = \dfrac{\theta}{9} \Rightarrow \theta = \dfrac{9\pi}{3} = 3\pi$ radians

(b) $s = r\theta \Rightarrow s = 8 \cdot 3\pi = 24\pi$ in.

(c) $v = \dfrac{r\theta}{t} \Rightarrow v = \dfrac{8 \cdot 3\pi}{9} = \dfrac{8\pi}{3}$ in. per min

87. $\omega = \dfrac{2\pi}{3}$ radians per sec, $t = 3$ sec

$\omega = \dfrac{\theta}{t} \Rightarrow \dfrac{2\pi}{3} = \dfrac{\theta}{3} \Rightarrow \theta = 2\pi$ radians

89. $\omega = 0.91$ radian per min, $t = 8.1$ min

$\omega = \dfrac{\theta}{t} \Rightarrow 0.91 = \dfrac{\theta}{8.1} \Rightarrow$

$\theta = (0.91)(8.1) \approx 7.4$ radians

91. $\theta = \dfrac{3\pi}{4}$ radians, $t = 8$ sec

$\omega = \dfrac{\theta}{t} \Rightarrow \theta = \dfrac{\frac{3\pi}{4}}{8} = \dfrac{3\pi}{4} \cdot \dfrac{1}{8} = \dfrac{3\pi}{32}$ radian per sec

93. $\theta = 3.871$ radians, $t = 21.47$ sec

$\omega = \dfrac{\theta}{t}$

$\omega = \dfrac{3.871}{21.47} \approx 0.1803$ radian per sec

95. $\theta = \dfrac{2\pi}{9}$ radian, $\omega = \dfrac{5\pi}{27}$ radian per min

$\omega = \dfrac{\theta}{t} \Rightarrow \dfrac{5\pi}{27} = \dfrac{\frac{2\pi}{9}}{t} \Rightarrow \dfrac{5\pi}{27} = \dfrac{2\pi}{9t} \Rightarrow$

$45\pi t = 54\pi \Rightarrow t = \dfrac{54\pi}{45\pi} = \dfrac{6}{5}$ min

97. $r = 12$ m, $\omega = \dfrac{2\pi}{3}$ radians per sec

$$v = r\omega \Rightarrow v = 12\left(\dfrac{2\pi}{3}\right) = 8\pi \text{ m per sec}$$

99. $v = 9$ m per sec, $r = 5$ m

$$v = r\omega \Rightarrow 9 = 5\omega \Rightarrow \omega = \dfrac{9}{5} \text{ radians per sec}$$

101. $v = 12$ m per sec, $\omega = \dfrac{3\pi}{2}$ radians per sec

$$v = r\omega \Rightarrow 12 = \dfrac{3\pi}{2}r \Rightarrow$$

$$r = 12 \cdot \dfrac{2}{3\pi} = \dfrac{8}{\pi} \text{ m}$$

103. $r = 6$ cm, $\omega = \dfrac{\pi}{3}$ radians per sec, $t = 9$ sec

$$s = r\omega t \Rightarrow s = 6\left(\dfrac{\pi}{3}\right)(9) = 18\pi \text{ cm}$$

105. $s = 6\pi$ cm, $r = 2$ cm, $\omega = \dfrac{\pi}{4}$ radian per sec

$$s = r\omega t \Rightarrow 6\pi = 2\left(\dfrac{\pi}{4}\right)t \Rightarrow 6\pi = \left(\dfrac{\pi}{2}\right)t \Rightarrow$$

$$t = 6\pi\left(\dfrac{2}{\pi}\right) = 12 \text{ sec}$$

107. $s = \dfrac{3\pi}{4}$ km, $r = 2$ km, $t = 4$ sec

$$s = r\omega t \Rightarrow \dfrac{3\pi}{4} = 2\omega \cdot 4 \Rightarrow$$

$$\dfrac{3\pi}{4} = 8\omega \Rightarrow \omega = \dfrac{3\pi}{4} \cdot \dfrac{1}{8} = \dfrac{3\pi}{32} \text{ radian per sec}$$

109. The hour hand of a clock moves through an angle of 2π radians (one complete revolution) in 12 hours, so

$$\omega = \dfrac{\theta}{t} = \dfrac{2\pi}{12} = \dfrac{\pi}{6} \text{ radian per hr.}$$

111. The minute hand makes one revolution per hour. Each revolution is 2π radians, so we have $\omega = 2\pi(1) = 2\pi$ radians per hr . There are 60 minutes in 1 hour, so $\omega = \dfrac{2\pi}{60} = \dfrac{\pi}{30}$ radian per min.

113. The minute hand of a clock moves through an angle of 2π radians in 60 min, and at the tip of the minute hand, $r = 7$ cm, so we have

$$v = \dfrac{r\theta}{t} \Rightarrow v = \dfrac{7(2\pi)}{60} = \dfrac{7\pi}{30} \text{ cm per min}$$

115. The flywheel making 42 rotations per min turns through an angle $42(2\pi) = 84\pi$ radians in 1 minute with $r = 2$ m. So,

$$v = \dfrac{r\theta}{t} \Rightarrow v = \dfrac{2(84\pi)}{1} = 168\pi \text{ m per min}$$

117. At 500 rotations per min, the propeller turns through an angle of $\theta = 500(2\pi) = 1000\pi$ radians in 1 min with $r = \dfrac{3}{2} = 1.5$ m, we have

$$v = \dfrac{r\theta}{t} \Rightarrow v = \dfrac{1.5(1000\pi)}{1} = 1500\pi \text{ m per min.}$$

119. At 215 revolutions per minute, the bicycle tire is moving $215(2\pi) = 430\pi$ radians per min. This is the angular velocity ω. The linear velocity of the bicycle is

$$v = r\omega = 13.0(430\pi) = 5590\pi \text{ in. per min.}$$

Convert this to miles per hour:

$$v = \dfrac{5590\pi \text{ in.}}{\text{min}} \cdot \dfrac{60 \text{ min}}{\text{hr}} \cdot \dfrac{1 \text{ ft}}{12 \text{ in.}} \cdot \dfrac{1 \text{ mi}}{5280 \text{ ft}}$$

$$\approx 16.6 \text{ mph}$$

121. (a) $\theta = \dfrac{1}{365}(2\pi) = \dfrac{2\pi}{365}$ radian

(b) $\omega = \dfrac{2\pi}{365}$ radian per day

$$= \dfrac{2\pi}{365} \cdot \dfrac{1}{24} \text{ radian per hr}$$

$$= \dfrac{\pi}{4380} \text{ radian per hr}$$

(c) $v = r\omega$

$$v = (93,000,000)\left(\dfrac{\pi}{4380}\right) \approx 67,000 \text{ mph}$$

123. (a) Because $s = 56$ cm of belt go around in $t = 18$ sec, the linear velocity is

$$v = \dfrac{s}{t} \Rightarrow v = \dfrac{56}{18} = \dfrac{28}{9} \approx 3.1 \text{ cm per sec}$$

(b) Because the 56-cm belt goes around in 18 sec, we have

$$v = r\omega$$

$$\dfrac{56}{18} = (12.96)\omega \Rightarrow \dfrac{28}{9} = (12.96)\omega$$

$$\omega = \dfrac{\frac{28}{9}}{12.96} \approx 0.24 \text{ radian per sec}$$

125. $\omega = (152)(2\pi) = 304\pi$ radians per min

$= \dfrac{304\pi}{60}$ radians per sec

$= \dfrac{76\pi}{15}$ radians per sec

$v = r\omega \Rightarrow 59.4 = r\left(\dfrac{76\pi}{15}\right) \Rightarrow$

$r = 59.4\left(\dfrac{15}{76\pi}\right) \approx 3.73$ cm

127. In one minute, the propeller makes 5000 revolutions. Each revolution is 2π radians, so we have $5000(2\pi) = 10{,}000\pi$ radians per min. There are 60 sec in a minute, so

$\omega = \dfrac{10{,}000\pi}{60} = \dfrac{500\pi}{3} \approx 523.6$ radians per sec

Section 6.3 Graphs of the Sine and Cosine Functions

1. The amplitude of the graphs of the sine and cosine function is <u>1</u>, and the period of each is <u>2π</u>.

3. The graph of the sine function crosses the x-axis for all numbers of the form <u>$n\pi$</u>, where n is an integer.

5. The least positive number x for which $\cos x = 0$ is $\frac{\pi}{2}$.

7. $y = -\sin x$
The graph is a sinusoidal curve with amplitude 1 and period 2π. Bcause $a = -1$, the graph is a reflection of $y = \sin x$ in the x-axis. This matches with graph E.

9. $y = \sin 2x$
The graph is a sinusoidal curve with amplitude 1 and period π. Because $\sin(2 \cdot 0) = \sin 0 = 0$, the point $(0, 0)$ is on the graph. This matches with graph B.

11. $y = 2 \sin x$
The graph is a sinusoidal curve with amplitude 2 and period 2π. Because $2\sin 0 = 2 \cdot 0 = 0$ and $2\sin \pi = 2 \cdot 1 = 2$, the points $(0, 0)$ and $(\pi, 2)$, are on the graph. This matches with graph F.

13. $y = 2 \cos x$
Amplitude: $|2| = 2$

x	0	$\dfrac{\pi}{2}$	π	$\dfrac{3\pi}{2}$	2π
$\cos x$	1	0	-1	0	1
$2\cos x$	2	0	-2	0	2

This table gives five values for graphing one period of the function. Repeat this cycle for the interval $[-2\pi, 0]$.

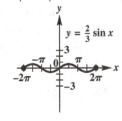

15. $y = \dfrac{2}{3}\sin x$

Amplitude: $\left|\dfrac{2}{3}\right| = \dfrac{2}{3}$

x	0	$\dfrac{\pi}{2}$	π	$\dfrac{3\pi}{2}$	2π
$\sin x$	0	1	0	-1	0
$\dfrac{2}{3}\sin x$	0	$\dfrac{2}{3} \approx 0.7$	0	$-\dfrac{2}{3} \approx -0.7$	0

This table gives five values for graphing one period of

$y = \dfrac{2}{3}\sin x$. Repeat this cycle for the interval $[-2\pi, 0]$.

17. $y = -\cos x$
Amplitude: $|-1| = 1$

x	0	$\dfrac{\pi}{2}$	π	$\dfrac{3\pi}{2}$	2π
$\cos x$	1	0	-1	0	1
$-\cos x$	-1	0	1	0	-1

This table gives five values for graphing one period of $y = -\cos x$. Repeat this cycle for the interval $[-2\pi, 0]$.

19. $y = -2 \sin x$

Amplitude: $|-2| = 2$

x	0	$\dfrac{\pi}{2}$	π	$\dfrac{3\pi}{2}$	2π
$\sin x$	0	1	0	-1	0
$-2\sin x$	0	-2	0	2	0

This table gives five values for graphing one period of $y = -2 \sin x$. Repeat this cycle for the interval $[-2\pi, 0]$.

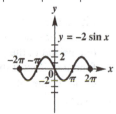

21. $y = \sin(-x)$

Amplitude: 1

x	0	$\dfrac{\pi}{2}$	π	$\dfrac{3\pi}{2}$	2π
$-x$	0	$-\dfrac{\pi}{2}$	$-\pi$	$-\dfrac{3\pi}{2}$	-2π
$\sin(-x)$	0	-1	0	1	0

This table gives five values for graphing one period of $y = \sin(-x)$. Repeat this cycle for the interval $[-2\pi, 0]$.

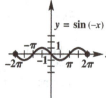

23. $y = \sin \dfrac{1}{2} x$

Period: $\dfrac{2\pi}{\frac{1}{2}} = 4\pi$ and amplitude: $|1| = 1$

Divide the interval $[0, 4\pi]$ into four equal parts to get x-values that will yield minimum and maximum points and x-intercepts. Then make a table. Repeat this cycle for the interval $[-4\pi, 0]$.

x	0	π	2π	3π	4π
$\dfrac{1}{2}x$	0	$\dfrac{\pi}{2}$	π	$\dfrac{3\pi}{2}$	2π
$\sin \dfrac{1}{2}x$	0	1	0	-1	0

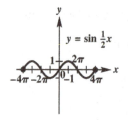

25. $y = \cos \dfrac{3}{4} x$

Period: $\dfrac{2\pi}{\frac{3}{4}} = 2\pi \cdot \dfrac{4}{3} = \dfrac{8\pi}{3}$ and amplitude: $|1| = 1$

Divide the interval $\left[0, \dfrac{8\pi}{3}\right]$ into four equal parts to get the x-values that will yield minimum and maximum points and x-intercepts. Then make a table. Repeat this cycle for the interval $\left[-\dfrac{8\pi}{3}, 0\right]$.

x	0	$\dfrac{2\pi}{3}$	$\dfrac{4\pi}{3}$	2π	$\dfrac{8\pi}{3}$
$\dfrac{3}{4}x$	0	$\dfrac{\pi}{2}$	π	$\dfrac{3\pi}{2}$	2π
$\cos \dfrac{3}{4}x$	1	0	-1	0	1

27. $y = \sin 3x$

Period: $\dfrac{2\pi}{3}$ and amplitude: $|1| = 1$

Divide the interval $\left[0, \dfrac{2\pi}{3}\right]$ into four equal parts to get x-values that will yield minimum and maximum points and x-intercepts. Then make a table. Repeat this cycle for the interval $\left[-\dfrac{2\pi}{3}, 0\right]$.

(continued on next page)

(*continued*)

x	0	$\dfrac{\pi}{6}$	$\dfrac{\pi}{3}$	$\dfrac{\pi}{2}$	$\dfrac{2\pi}{3}$
$3x$	0	$\dfrac{\pi}{2}$	π	$\dfrac{3\pi}{2}$	2π
$\sin 3x$	0	1	0	-1	0

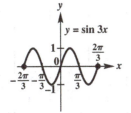

29. $y = 2\sin\dfrac{1}{4}x$

Period: $\dfrac{2\pi}{\frac{1}{4}} = 2\pi \cdot \dfrac{4}{1} = 8\pi$ and amplitude:

$|2| = 2$

Divide the interval $[0, 8\pi]$ into four equal parts to get the *x*-values that will yield minimum and maximum points and *x*-intercepts. Then make a table. Repeat this cycle for the interval $[-8\pi, 0]$.

x	0	2π	4π	6π	8π
$\dfrac{1}{4}x$	0	$\dfrac{\pi}{2}$	π	$\dfrac{3\pi}{2}$	2π
$\sin\dfrac{1}{4}x$	0	1	0	-1	0
$2\sin\dfrac{1}{4}x$	0	2	0	-2	0

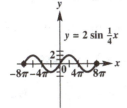

31. $y = -2\cos 3x$

Period: $\dfrac{2\pi}{3}$ and amplitude: $|-2| = 2$

Divide the interval $\left[0, \dfrac{2\pi}{3}\right]$ into four equal

parts to get the *x*-values that will yield minimum and maximum points and *x*-intercepts. Then make a table. Repeat this

cycle for the interval $\left[-\dfrac{2\pi}{3}, 0\right]$.

x	0	$\dfrac{\pi}{6}$	$\dfrac{\pi}{3}$	$\dfrac{\pi}{2}$	$\dfrac{2\pi}{3}$
$3x$	0	$\dfrac{\pi}{2}$	π	$\dfrac{3\pi}{2}$	2π
$\cos 3x$	1	0	-1	0	1
$-2\cos 3x$	-2	0	2	0	-2

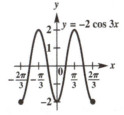

33. $y = \cos \pi x$

Period: $\dfrac{2\pi}{\pi} = 2$ and amplitude: $|1| = 1$

Divide the interval [0, 2] into four equal parts to get the *x*-values that will yield minimum and maximum points and *x*-intercepts. Then make a table. Repeat this cycle for the interval [−2, 0].

x	0	$\dfrac{1}{2}$	1	$\dfrac{3}{2}$	2
πx	0	$\dfrac{\pi}{2}$	π	$\dfrac{3\pi}{2}$	2π
$\cos \pi x$	1	0	-1	0	1

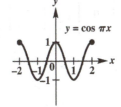

35. $y = -2 \sin 2\pi x$

Period: $\dfrac{2\pi}{2\pi} = 1$ and amplitude: $|-2| = 2$

Divide the interval [0, 1] into four equal parts to get the x-values that will yield minimum and maximum points and x-intercepts. Then make a table. Repeat this cycle for the interval [−1, 0].

x	0	$\frac{1}{4}$	$\frac{1}{2}$	$\frac{3}{4}$	1
$2\pi x$	0	$\frac{\pi}{2}$	π	$\frac{3\pi}{2}$	2π
$\sin 2\pi x$	0	1	0	−1	0
$-2\sin \pi x$	0	−2	0	2	0

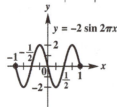

37. $y = \dfrac{1}{2}\cos \dfrac{\pi}{2}x$

Period: $\dfrac{2\pi}{\frac{\pi}{2}} = 2\pi \cdot \dfrac{2}{\pi} = 4$ and amplitude:

$\left|\dfrac{1}{2}\right| = \dfrac{1}{2}$

Divide the interval [0, 4] into four equal parts to get the x-values that will yield minimum and maximum points and x-intercepts. Then make a table. Repeat this cycle for the interval [−4, 0].

x	0	1	2	3	4
$\frac{\pi}{2}x$	0	$\frac{\pi}{2}$	π	$\frac{3\pi}{2}$	2π
$\cos\frac{\pi}{2}x$	1	0	−1	0	1
$\frac{1}{2}\cos\frac{\pi}{2}x$	$\frac{1}{2}$	0	$-\frac{1}{2}$	0	$\frac{1}{2}$

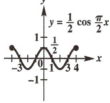

39. $y = \pi \sin \pi x$

Period: $\dfrac{2\pi}{\pi} = 2$ and amplitude: $|\pi| = \pi$

Divide the interval [0, 2] into four equal parts to get the x-values that will yield minimum and maximum points and x-intercepts. Then make a table. Repeat this cycle for the interval [−2, 0].

x	0	$\frac{1}{2}$	1	$\frac{3}{2}$	2
πx	0	$\frac{\pi}{2}$	π	$\frac{3\pi}{2}$	2π
$\sin \pi x$	0	1	0	−1	0
$\pi \sin \pi x$	0	π	0	$-\pi$	0

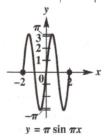

$y = \pi \sin \pi x$

41. The amplitude is $\dfrac{1}{2}[2 - (-2)] = \dfrac{1}{2}(4) = 2$, so $a = 2$. One complete cycle of the graph is achieved in π units, so the period

$\pi = \dfrac{2\pi}{b} \Rightarrow b = \dfrac{2\pi}{\pi} = 2$. Comparing the given graph with the general sine and cosine curves, we see that this graph is a cosine curve. Substituting $a = 2$ and $b = 2$, the function is $y = 2\cos 2x$. Verify by confirming minimum and maximum points and x-intercepts from the graph:

$(0, 2) \Rightarrow 2 = 2\cos(2 \cdot 0) = 2\cos 0 = 2 \cdot 1 = 2$

$\left(\dfrac{\pi}{4}, 0\right) \Rightarrow 0 = 2\cos\left(2 \cdot \dfrac{\pi}{4}\right) = 2\cos\dfrac{\pi}{2} = 2 \cdot 0 = 0$

$\left(\dfrac{\pi}{2}, -2\right) \Rightarrow -2 = 2\cos\left(2 \cdot \dfrac{\pi}{2}\right) = 2\cos\pi$

$\qquad = 2(-1) = -2$

$\left(\dfrac{3\pi}{4}, 0\right) \Rightarrow 0 = 2\cos\left(2 \cdot \dfrac{3\pi}{4}\right)$

$\qquad = 2\cos\dfrac{3\pi}{2} = 2 \cdot 0 = 0$

$(\pi, 2) \Rightarrow 2 = 2\cos(2\pi) = 2(1) = 2$

43. The amplitude is $\frac{1}{2}\big[3-(-3)\big]=\frac{1}{2}(6)=3$, so

$a = 3$. One-half of a cycle of the graph is achieved in 2π units, so the period is

$2 \cdot 2\pi = 4\pi$ and $4\pi = \dfrac{2\pi}{b} \Rightarrow b = \dfrac{2\pi}{4\pi} = \dfrac{1}{2}$.

Comparing the given graph with the general sine and cosine curves, we see that this graph is the reflection of the cosine curve in the x-axis. Thus, $a = -3$. Substituting $a = -3$ and $b = \frac{1}{2}$, the function is $y = -3\cos\frac{1}{2}x$. Verify by confirming minimum and maximum points and x–intercepts from the graph:

$(0,-3) \Rightarrow -3 = -3\cos\left(\frac{1}{2}\cdot 0\right) = -3\cos 0$
$\qquad = -3\cdot 1 = -3$

$(\pi,0) \Rightarrow 0 = -3\cos\left(\frac{1}{2}\cdot\pi\right) = -3\cos\frac{\pi}{2}$
$\qquad = -3(0) = 0$

$(2\pi,3) \Rightarrow 3 = -3\cos\left(\frac{1}{2}\cdot 2\pi\right) = -3\cos\pi$
$\qquad = -3(-1) = 3$

45. The amplitude is $\frac{1}{2}\big[3-(-3)\big]=\frac{1}{2}(6)=3$, so

$a = 3$. One complete cycle of the graph is

achieved in $\dfrac{\pi}{2}$ units, so the period

$\dfrac{\pi}{2} = \dfrac{2\pi}{b} \Rightarrow b = 2\pi\cdot\dfrac{2}{\pi} = 4$. Comparing the

given graph with the general sine and cosine curves, we see that this graph is a sine curve. Substituting $a = 3$ and $b = 4$, the function is $y = 3\sin 4x$. Verify by confirming minimum and maximum points and x–intercepts from the graph:

$(0,0) \Rightarrow 0 = 3\sin(4\cdot 0) = 3\sin 0 = 3\cdot 0 = 0$

$\left(\dfrac{\pi}{8},3\right) \Rightarrow 3 = 3\sin\left(4\cdot\dfrac{\pi}{8}\right) = 3\sin\dfrac{\pi}{2}$
$\qquad = 3(1) = 3$

$\left(\dfrac{\pi}{4},0\right) \Rightarrow 0 = 3\sin\left(4\cdot\dfrac{\pi}{4}\right) = 3\sin\pi$
$\qquad = 3(0) = 0$

$\left(\dfrac{3\pi}{8},-3\right) \Rightarrow -3 = 3\sin\left(4\cdot\dfrac{3\pi}{8}\right)$
$\qquad\qquad = 3\sin\dfrac{3\pi}{2} = 3(-1) = -3$

$\left(\dfrac{\pi}{2},0\right) \Rightarrow 0 = 3\sin\left(4\cdot\dfrac{\pi}{2}\right) = 3\sin 2\pi = 3(0) = 0$

47. **(a)** The highest temperature is 80°F; the lowest is 50°F.

(b) The amplitude is

$\dfrac{1}{2}(80-50) = \dfrac{1}{2}(30) = 15$.

(c) The period is about 35,000 yr.

(d) The trend of the temperature now is downward.

49. The graph repeats each day, so the period is 24 hours.

51. On January 20, low tide was at 6 P.M., with height approximately 0.2 ft.

53. On January 22, high tide was at $2 + 3{:}18 = 3{:}18$ A.M., with height $2.6 - 0.2 = 2.4$ feet.

55. **(a)** The graph has a general upward trend along with small annual oscillations.

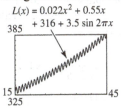
$L(x) = 0.022x^2 + 0.55x + 316 + 3.5\sin 2\pi x$

(b) The seasonal variations are caused by the term $3.5\sin 2\pi x$. The maximums will

occur when $2\pi x = \dfrac{\pi}{2} + 2n\pi$, where n is

an integer. Because x cannot be negative, n cannot be negative.
This is equivalent to

$2\pi x = \dfrac{\pi}{2} + 2n\pi,\ n = 0, 1, 2, \ldots$

$2x = \dfrac{1}{2} + 2n,\ n = 0, 1, 2, \ldots$

$x = \dfrac{1}{4} + n,\ n = 0, 1, 2, \ldots$

$x = \dfrac{4n+1}{4},\ n = 0, 1, 2, \ldots$

$x = \dfrac{1}{4}, \dfrac{5}{4}, \dfrac{9}{4}, \ldots$

Because x is in years, $x = \frac{1}{4}$ corresponds to April when the seasonal carbon dioxide levels are maximum.
The minimums will occur when

$2\pi x = \dfrac{3\pi}{2} + 2n\pi$, where n is an integer.

(continued on next page)

(continued)

Because x cannot be negative, n cannot be negative. This is equivalent to

$$2\pi x = \frac{3\pi}{2} + 2n\pi, \ n = 0, 1, 2, \ldots$$

$$2x = \frac{3}{2} + 2n, \ n = 0, 1, 2, \ldots$$

$$x = \frac{3}{4} + n, \ n = 0, 1, 2, \ldots$$

$$x = \frac{4n+3}{4}, \ n = 0, 1, 2, \ldots$$

$$x = \frac{3}{4}, \frac{7}{4}, \frac{11}{4}, \ldots$$

This is $\frac{1}{2}$ yr later, which corresponds to October.

(c) Answers will vary. Sample answer: The quadratic function provides the general increasing nature of the level, while the sine function provides the fluctuations as the years go by.

57. $T(x) = 37 + 21\sin\left[\frac{2\pi}{365}(x - 91)\right]$

(a) March 15 (day 74)

$$T(74) = 37 + 21\sin\left[\frac{2\pi}{365}(74 - 91)\right]$$

$$\approx 31°F$$

(b) April 5 (day 95)

$$T(95) = 37 + 21\sin\left[\frac{2\pi}{365}(95 - 91)\right]$$

$$\approx 38°F$$

(c) Day 200

$$T(200) = 37 + 21\sin\left[\frac{2\pi}{365}(200 - 91)\right]$$

$$\approx 57.0° \approx 57°F$$

(d) June 25 is day 176.

$$(31 + 28 + 31 + 30 + 31 + 25 = 176)$$

$$T(176) = 37 + 21\sin\left[\frac{2\pi}{365}(176 - 91)\right]$$

$$\approx 57.9° \approx 58°F$$

(e) October 1 is day 274.

$$31 + 28 + 31 + 30 + 31$$
$$+ 30 + 31 + 31 + 30 + 1 = 274$$

$$T(274) = 37 + 21\sin\left[\frac{2\pi}{365}(274 - 91)\right]$$

$$\approx 36.8° \approx 37°F$$

(f) December 31 is day 365.

$$T(365) = 37 + 21\sin\left[\frac{2\pi}{365}(365 - 91)\right]$$

$$\approx 16.0° \approx 16°F$$

59. $-1 \le y \le 1$

Amplitude: 1

Period: 8 squares $= 8(30°) = 240°$ or $\frac{4\pi}{3}$

61. No, we can't say that $\sin bx = b\sin x$. If b is not zero, then the period of $y = \sin bx$ is $\frac{2\pi}{|b|}$, and the amplitude is 1. The period of $y = b\sin x$ is 2π, and the amplitude is $|b|$.

63. X ≈ -0.4161468, Y ≈ 0.90929743. X is $\cos 2$, and Y is $\sin 2$.

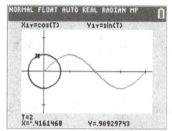

65. X $= 2$, Y ≈ -0.4161468; $\cos 2 \approx -0.4161468$

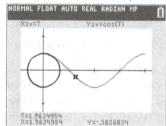

Section 6.4 Translations of the Graphs of the Sine and Cosine Functions

1. The graph of $y = \sin\left(x + \frac{\pi}{4}\right)$ is obtained by shifting the graph of $y = \sin x$ $\frac{\pi}{4}$ units <u>left</u>.

3. The graph of $y = 4\sin x$ is obtained by stretching the graph of $y = \sin x$ vertically by a factor of <u>4</u>.

5. The graph of $y = 6 + 3\sin x$ is obtained by shifting the graph of $y = 3\sin x$ <u>6</u> units <u>up</u>.

7. The graph of $y = 3 + 5\cos\left(x + \frac{\pi}{5}\right)$ is obtained

   by shifting the graph of $y = \cos x$ $\frac{\pi}{5}$ units

   horizontally to the <u>left</u>, stretching it vertically
   by a factor of <u>5</u>, and then shifting it <u>3</u> units
   vertically in the <u>up</u> direction.

9. $y = \sin\left(x - \frac{\pi}{4}\right)$ is the graph of $y = \sin x$,

   shifted to the right $\frac{\pi}{4}$ unit. This matches

   choice D.

11. $y = \cos\left(x - \frac{\pi}{4}\right)$ is the graph of $y = \cos x$,

    shifted to the right $\frac{\pi}{4}$ unit. This matches

    choice H.

13. $y = 1 + \sin x$ is the graph of $y = \sin x$,
    translated vertically 1 unit up. This matches
    choice B.

15. $y = 1 + \cos x$ is the graph of $y = \cos x$,
    translated vertically 1 unit up. This matches
    choice I.

17. The graph of $y = \sin x + 1$ is the graph of
    $y = \sin x$ translated vertically 1 unit up, while
    the graph of $y = \sin(x + 1)$ is the graph of
    $y = \sin x$ shifted horizontally 1 unit left.

19. $y = 3\sin(2x - 4) = 3\sin\left[2(x - 2)\right]$

    The amplitude $= |3| = 3$, period $= \frac{2\pi}{2} = \pi$, and
    phase shift $= 2$. This matches choice B.

21. $y = -4\sin(3x - 2) = -4\sin\left[3\left(x - \frac{2}{3}\right)\right]$

    The amplitude $= |-4| = 4$, period $= \frac{2\pi}{3}$, and

    phase shift $= \frac{2}{3}$. This matches choice C.

23. If the graph of $y = \cos x$ is translated $\frac{\pi}{2}$ units
    horizontally to the <u>right</u>, it will coincide with
    the graph of $y = \sin x$.

25. This is a sine curve that has been shifted one
    unit down, so the equation is $y = -1 + \sin x$.
    Verify by confirming minimum and maximum
    points and x–intercepts from the graph:
    $(0, -1) \Rightarrow -1 = -1 + \sin 0 = -1 + 0 = -1$
    $\left(\frac{\pi}{2}, 0\right) \Rightarrow 0 = -1 + \sin\frac{\pi}{2} = -1 + 1 = 0$
    $(\pi, -1) \Rightarrow -1 = -1 + \sin\pi = -1 + 0 = -1$

$\left(\frac{3\pi}{2}, -2\right) \Rightarrow -2 = -1 + \sin\frac{3\pi}{2} = -1 + (-1) = -2$
$(2\pi, -1) \Rightarrow -1 = 1 + \sin(2\pi) = -1 + 0 = -1$

27. The maximum is at $\left(\frac{\pi}{3}, 1\right)$, so this is a cosine

    curve that has been shifted $\frac{\pi}{3}$ units to the right.

    Thus, the equation is $y = \cos\left(x - \frac{\pi}{3}\right)$. Verify by

    confirming minimum and maximum points and
    x–intercepts from the graph.

    $\left(\frac{\pi}{3}, 1\right) \Rightarrow 1 = \cos\left(\frac{\pi}{3} - \frac{\pi}{3}\right) = \cos 0 = 1$

    $\left(\frac{4\pi}{3}, -1\right) \Rightarrow -1 = \cos\left(\frac{4\pi}{3} - \frac{\pi}{3}\right) = \cos\pi = -1$

    $\left(\frac{7\pi}{3}, 1\right) \Rightarrow 1 = \cos\left(\frac{7\pi}{3} - \frac{\pi}{3}\right) = \cos 2\pi = 1$

    $\left(\frac{10\pi}{3}, -1\right) \Rightarrow -1 = \cos\left(\frac{10\pi}{3} - \frac{\pi}{3}\right) = \cos 3\pi = -1$

    $\left(\frac{13\pi}{3}, 1\right) \Rightarrow 1 = \cos\left(\frac{13\pi}{3} - \frac{\pi}{3}\right) = \cos 4\pi = 1$

29. $y = 2\sin(x + \pi)$

    amplitude: $|2| = 2$; period: $\frac{2\pi}{1} = 2\pi$

    There is no vertical translation. The phase
    shift is π units to the left.

31. $y = -\frac{1}{4}\cos\left(\frac{1}{2}x + \frac{\pi}{2}\right) = -\frac{1}{4}\cos\left[\frac{1}{2}\left[x - (-\pi)\right]\right]$

    amplitude: $\left|-\frac{1}{4}\right| = \frac{1}{4}$; period:

    $\frac{2\pi}{\frac{1}{2}} = 2\pi \cdot \frac{2}{1} = 4\pi$

    There is no vertical translation. The phase shift
    is π units to the left.

33. $y = 3\cos\left[\frac{\pi}{2}\left(x - \frac{1}{2}\right)\right]$

    amplitude: $|3| = 3$; period: $\frac{2\pi}{\frac{\pi}{2}} = 2\pi \cdot \frac{2}{\pi} = 4$

    There is no vertical translation. The phase

    shift is $\frac{1}{2}$ unit to the right.

35. $y = 2 - \sin\left(3x - \dfrac{\pi}{5}\right) = -\sin\left[3\left(x - \dfrac{\pi}{15}\right)\right] + 2$

amplitude: $|-1| = 1$; period: $\dfrac{2\pi}{3}$

The vertical translation is 2 units up. The

phase shift is $\dfrac{\pi}{15}$ unit to the right

37. $y = \cos\left(x - \dfrac{\pi}{2}\right)$

Step 1: Find the interval whose length is $\dfrac{2\pi}{b}$.

$0 \le x - \dfrac{\pi}{2} \le 2\pi \Rightarrow 0 + \dfrac{\pi}{2} \le x \le 2\pi + \dfrac{\pi}{2} \Rightarrow$

$\dfrac{\pi}{2} \le x \le \dfrac{5\pi}{2}$

Step 2: Divide the period into four equal parts

to get the following x-values: $\dfrac{\pi}{2}$, π, $\dfrac{3\pi}{2}$, 2π,

$\dfrac{5\pi}{2}$

Step 3: Evaluate the function for each of the five x-values

x	$\dfrac{\pi}{2}$	π	$\dfrac{3\pi}{2}$	2π	$\dfrac{5\pi}{2}$
$x - \dfrac{\pi}{2}$	0	$\dfrac{\pi}{2}$	π	$\dfrac{3\pi}{2}$	2π
$\cos\left(x - \dfrac{\pi}{2}\right)$	1	0	-1	0	1

Steps 4 and 5: Plot the points found in the table and join them with a sinusoidal curve. By graphing an additional period to the right, we obtain the following graph.

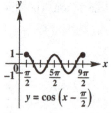

The amplitude is 1. The period is 2π. There is

no vertical translation. The phase shift is $\dfrac{\pi}{2}$

unit to the right.

39. $y = \sin\left(x + \dfrac{\pi}{4}\right)$

Step 1: Find the interval whose length is $\dfrac{2\pi}{b}$.

$0 \le x + \dfrac{\pi}{4} \le 2\pi \Rightarrow 0 - \dfrac{\pi}{4} \le x \le 2\pi - \dfrac{\pi}{4} \Rightarrow$

$-\dfrac{\pi}{4} \le x \le \dfrac{7\pi}{4}$

Step 2: Divide the period into four equal parts

to get the following x-values: $-\dfrac{\pi}{4}$, $\dfrac{\pi}{4}$,

$\dfrac{3\pi}{4}$, $\dfrac{5\pi}{4}$, $\dfrac{7\pi}{4}$

Step 3: Evaluate the function for each of the five x-values

x	$-\dfrac{\pi}{4}$	$\dfrac{\pi}{4}$	$\dfrac{3\pi}{4}$	$\dfrac{5\pi}{4}$	$\dfrac{7\pi}{4}$
$x + \dfrac{\pi}{4}$	0	$\dfrac{\pi}{2}$	π	$\dfrac{3\pi}{2}$	2π
$\sin\left(x + \dfrac{\pi}{4}\right)$	0	1	0	-1	0

Steps 4 and 5: Plot the points found in the table and join them with a sinusoidal curve. By graphing an additional period to the right, we obtain the following graph.

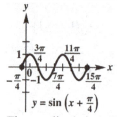

The amplitude is 1. The period is 2π. There is

no vertical translation. The phase shift is $\dfrac{\pi}{4}$

unit to the left.

41. $y = 2\cos\left(x - \dfrac{\pi}{3}\right)$

Step 1: Find the interval whose length is $\dfrac{2\pi}{b}$.

$0 \le x - \dfrac{\pi}{3} \le 2\pi \Rightarrow 0 + \dfrac{\pi}{3} \le x \le 2\pi + \dfrac{\pi}{3} \Rightarrow$

$\dfrac{\pi}{3} \le x \le \dfrac{7\pi}{3}$

(continued on next page)

(*continued*)

Step 2: Divide the period into four equal parts to get the following *x*-values $\dfrac{\pi}{3}, \dfrac{5\pi}{6}, \dfrac{4\pi}{3},$

$\dfrac{11\pi}{6}, \dfrac{7\pi}{3}$

Step 3: Evaluate the function for each of the five *x*-values.

x	$\dfrac{\pi}{3}$	$\dfrac{5\pi}{6}$	$\dfrac{4\pi}{3}$	$\dfrac{11\pi}{6}$	$\dfrac{7\pi}{3}$
$x - \dfrac{\pi}{3}$	0	$\dfrac{\pi}{2}$	π	$\dfrac{3\pi}{2}$	2π
$\cos\left(x - \dfrac{\pi}{3}\right)$	1	0	-1	0	1
$2\cos\left(x - \dfrac{\pi}{3}\right)$	2	0	-2	0	2

Steps 4 *and* 5: Plot the points found in the table and join them with a sinusoidal curve. By graphing an additional period to the right, we obtain the following graph.

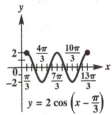

$$y = 2\cos\left(x - \dfrac{\pi}{3}\right)$$

The amplitude is 2. The period is 2π. There is no vertical translation. The phase shift is $\dfrac{\pi}{3}$ units to the right.

43. $y = \dfrac{3}{2}\sin\left[2\left(x + \dfrac{\pi}{4}\right)\right]$

Step 1: Find the interval whose length is $\dfrac{2\pi}{b}$.

$$0 \le 2\left(x + \dfrac{\pi}{4}\right) \le 2\pi \Rightarrow 0 \le x + \dfrac{\pi}{4} \le \pi \Rightarrow$$

$$-\dfrac{\pi}{4} \le x \le \dfrac{3\pi}{4}$$

Step 2: Divide the period into four equal parts to get the *x*-values: $-\dfrac{\pi}{4}, 0, \dfrac{\pi}{4}, \dfrac{\pi}{2}, \dfrac{3\pi}{4}$

Step 3: Evaluate the function for each of the five *x*-values

x	$-\dfrac{\pi}{4}$	0	$\dfrac{\pi}{4}$	$\dfrac{\pi}{2}$	$\dfrac{3\pi}{4}$
$2\left(x + \dfrac{\pi}{4}\right)$	0	$\dfrac{\pi}{2}$	π	$\dfrac{3\pi}{2}$	2π
$\sin\left[2\left(x + \dfrac{\pi}{4}\right)\right]$	0	1	0	-1	0
$\dfrac{3}{2}\sin\left[2\left(x + \dfrac{\pi}{4}\right)\right]$	0	$\dfrac{3}{2}$	0	$-\dfrac{3}{2}$	0

Steps 4 *and* 5: Plot the points found in the table and join them with a sinusoidal curve.

$$y = \dfrac{3}{2}\sin\left[2\left(x + \dfrac{\pi}{4}\right)\right]$$

The amplitude is $\dfrac{3}{2}$. The period is $\dfrac{2\pi}{2} = \pi$. There is no vertical translation. The phase shift is $\dfrac{\pi}{4}$ unit to the left.

45. $y = -4\sin\left(2x - \pi\right) = -4\sin\left[2\left(x - \dfrac{\pi}{2}\right)\right]$

Step 1: Find the interval whose length is $\dfrac{2\pi}{b}$.

$$0 \le 2\left(x - \dfrac{\pi}{2}\right) \le 2\pi \Rightarrow 0 \le x - \dfrac{\pi}{2} \le \dfrac{2\pi}{2} \Rightarrow$$

$$0 \le x - \dfrac{\pi}{2} \le \pi \Rightarrow \dfrac{\pi}{2} \le x \le \dfrac{3\pi}{2}$$

Step 2: Divide the period into four equal parts to get the following *x*-values: $\dfrac{\pi}{2}, \dfrac{3\pi}{4}, \pi,$

$\dfrac{5\pi}{4}, \dfrac{3\pi}{2}$

(*continued on next page*)

(*continued*)

Step 3: Evaluate the function for each of the five *x*-values.

x	$\dfrac{\pi}{2}$	$\dfrac{3\pi}{4}$	π	$\dfrac{5\pi}{4}$	$\dfrac{3\pi}{2}$
$2\left(x-\dfrac{\pi}{2}\right)$	0	$\dfrac{\pi}{2}$	π	$\dfrac{3\pi}{2}$	2π
$\sin\left[2\left(x-\dfrac{\pi}{2}\right)\right]$	0	1	0	−1	0
$-4\sin\left[2\left(x-\dfrac{\pi}{2}\right)\right]$	0	−4	0	4	0

Steps 4 and 5: Plot the points found in the table and join them with a sinusoidal curve.

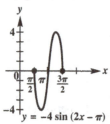

$y = -4\sin(2x - \pi)$

The amplitude is $\left|-4\right|$, which is 4. The period is $\dfrac{2\pi}{2}$, which is π. There is no vertical translation. The phase shift is $\dfrac{\pi}{2}$ units to the right

47. $y = \dfrac{1}{2}\cos\left(\dfrac{1}{2}x - \dfrac{\pi}{4}\right) = \dfrac{1}{2}\cos\left[\dfrac{1}{2}\left(x - \dfrac{\pi}{2}\right)\right]$

Step 1: Find the interval whose length is $\dfrac{2\pi}{b}$.

$0 \le \dfrac{1}{2}\left(x - \dfrac{\pi}{2}\right) \le 2\pi \Rightarrow 0 \le x - \dfrac{\pi}{2} \le 4\pi \Rightarrow$

$\dfrac{\pi}{2} \le x \le \dfrac{8\pi}{2} + \dfrac{\pi}{2} \Rightarrow \dfrac{\pi}{2} \le x \le \dfrac{9\pi}{2}$

Step 2: Divide the period into four equal parts to get the *x*-values: $\dfrac{\pi}{2}, \dfrac{3\pi}{2}, \dfrac{5\pi}{2}, \dfrac{7\pi}{2}, \dfrac{9\pi}{2}$

Step 3: Evaluate the function for each of the five *x*-values.

x	$\dfrac{\pi}{2}$	$\dfrac{3\pi}{2}$	$\dfrac{5\pi}{2}$	$\dfrac{7\pi}{2}$	$\dfrac{9\pi}{2}$
$\dfrac{1}{2}\left(x - \dfrac{\pi}{2}\right)$	0	$\dfrac{\pi}{2}$	π	$\dfrac{3\pi}{2}$	2π
$\cos\left[\dfrac{1}{2}\left(x - \dfrac{\pi}{2}\right)\right]$	1	0	−1	0	1
$\dfrac{1}{2}\cos\left[\dfrac{1}{2}\left(x - \dfrac{\pi}{2}\right)\right]$	$\dfrac{1}{2}$	0	$-\dfrac{1}{2}$	0	$\dfrac{1}{2}$

Steps 4 and 5: Plot the points found in the table and join them with a sinusoidal curve.

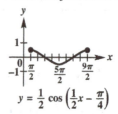

$y = \dfrac{1}{2}\cos\left(\dfrac{1}{2}x - \dfrac{\pi}{4}\right)$

The amplitude is $\dfrac{1}{2}$. The period is $\dfrac{2\pi}{\frac{1}{2}}$, which is 4π. There is no vertical translation.

The phase shift is $\dfrac{\pi}{2}$ units to the right.

49. $y = -3 + 2\sin x$

Step 1: The period is 2π.

Step 2: Divide the period into four equal parts to get the *x*-values: $0, \dfrac{\pi}{2}, \pi, \dfrac{3\pi}{2}, \pi, 2\pi$

Step 3: Evaluate the function for each of the five *x*-values:

x	0	$\dfrac{\pi}{2}$	π	$\dfrac{3\pi}{2}$	2π
$\sin x$	0	1	0	−1	0
$2\sin x$	0	2	0	−2	0
$-3 + 2\sin x$	−3	−1	−3	−5	−3

Steps 4 and 5: Plot the points found in the table and join them with a sinusoidal curve. By graphing an additional period to the left, we obtain the following graph.

$y = -3 + 2\sin x$

The amplitude is 2. The vertical translation is 3 units down. There is no phase shift.

51. $y = -1 - 5\cos 5x$

Step 1: Find the interval whose length is $\dfrac{2\pi}{b}$.

$$0 \le 5x \le 2\pi \Rightarrow 0 \le x \le \dfrac{2\pi}{5}$$

Step 2: Divide the period into four equal parts to get the *x*-values: $0, \dfrac{\pi}{10}, \dfrac{\pi}{5}, \dfrac{3\pi}{10}, \dfrac{2\pi}{5}$

Step 3: Evaluate the function for each of the five *x*-values.

x	0	$\dfrac{\pi}{10}$	$\dfrac{\pi}{5}$	$\dfrac{3\pi}{10}$	$\dfrac{2\pi}{5}$
$5x$	0	$\dfrac{\pi}{2}$	π	$\dfrac{3\pi}{2}$	2π
$\cos 5x$	1	0	-1	0	1
$-2\cos 5x$	-2	0	2	0	-2
$-1 - 2\cos 5x$	-3	-1	1	-1	-3

Steps 4 and 5: Plot the points found in the table and join them with a sinusoidal curve. By graphing an additional period to the left, we obtain the following graph.

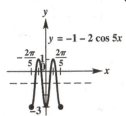

The period is $\dfrac{2\pi}{5}$. The amplitude is $\left|-2\right|$, which is 2. The vertical translation is 1 unit down. There is no phase shift.

53. $y = 1 - 2\cos\dfrac{1}{2}x$

Step 1: Find the interval whose length is $\dfrac{2\pi}{b}$.

$$0 \le \dfrac{1}{2}x \le 2\pi \Rightarrow 0 \le x \le 4\pi$$

Step 2: Divide the period into four equal parts to get the following *x*-values: $0, \pi, 2\pi, 3\pi, 4\pi$

Step 3: Evaluate the function for each of the five *x*-values.

x	0	π	2π	3π	4π
$\dfrac{1}{2}x$	0	$\dfrac{\pi}{2}$	π	$\dfrac{3\pi}{2}$	2π
$\cos\dfrac{1}{2}x$	1	0	-1	0	1
$-2\cos\dfrac{1}{2}x$	-2	0	2	0	-2
$1 - 2\cos\dfrac{1}{2}x$	-1	1	3	1	-1

Steps 4 and 5: Plot the points found in the table and join them with a sinusoidal curve. By graphing an additional period to the left, we obtain the following graph.

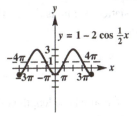

The amplitude is $\left|-2\right|$, which is 2. The period is $\dfrac{2\pi}{\frac{1}{2}}$, which is 4π. The vertical translation is 1 unit up. There is no phase shift.

55. $y = -2 + \dfrac{1}{2}\sin 3x$

Step 1: Find the interval whose length is $\dfrac{2\pi}{b}$.

$$0 \le 3x \le 2\pi \Rightarrow 0 \le x \le \dfrac{2\pi}{3}$$

Step 2: Divide the period into four equal parts to get the following *x*-values: $0, \dfrac{\pi}{6}, \dfrac{\pi}{3}, \dfrac{\pi}{2}, \dfrac{2\pi}{3}$

Step 3: Evaluate the function for each of the five *x*-values.

x	0	$\dfrac{\pi}{6}$	$\dfrac{\pi}{3}$	$\dfrac{\pi}{2}$	$\dfrac{2\pi}{3}$
$3x$	0	$\dfrac{\pi}{2}$	π	$\dfrac{3\pi}{2}$	2π
$\sin 3x$	0	1	0	-1	0
$\dfrac{1}{2}\sin 3x$	0	$\dfrac{1}{2}$	0	$-\dfrac{1}{2}$	0
$-2 + \dfrac{1}{2}\sin 3x$	-2	$-\dfrac{3}{2}$	-2	$-\dfrac{5}{2}$	-2

Steps 4 and 5: Plot the points found in the table and join them with a sinusoidal curve. By graphing an additional period to the left, we obtain the following graph.

(*continued on next page*)

(*continued*)

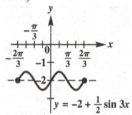

$$y = -2 + \tfrac{1}{2}\sin 3x$$

The amplitude is $\left|\dfrac{1}{2}\right| = \dfrac{1}{2}$. The period is $\dfrac{2\pi}{3}$.

The vertical translation is 2 units down. There is no phase shift.

57. $y = -3 + 2\sin\left(x + \dfrac{\pi}{2}\right)$

Step 1: Find the interval whose length is $\dfrac{2\pi}{b}$.

$$0 \le x + \frac{\pi}{2} \le 2\pi \Rightarrow 0 - \frac{\pi}{2} \le x \le 2\pi - \frac{\pi}{2} \Rightarrow$$

$$-\frac{\pi}{2} \le x \le \frac{3\pi}{2}$$

Step 2: Divide the period into four equal parts to get the following *x*-values: $-\dfrac{\pi}{2}, 0, \dfrac{\pi}{2}, \pi,$

$\dfrac{3\pi}{2}$

Step 3: Evaluate the function for each of the five *x*-values

x	$-\dfrac{\pi}{2}$	0	$\dfrac{\pi}{2}$	π	$\dfrac{3\pi}{2}$
$x + \dfrac{\pi}{2}$	0	$\dfrac{\pi}{2}$	π	$\dfrac{3\pi}{2}$	2π
$\sin\left(x + \dfrac{\pi}{2}\right)$	0	1	0	-1	0
$2\sin\left(x + \dfrac{\pi}{2}\right)$	0	2	0	-2	0
$-3 + 2\sin\left(x + \dfrac{\pi}{2}\right)$	-3	-1	-3	-5	-3

Steps 4 and 5: Plot the points found in the table and join them with a sinusoidal curve.

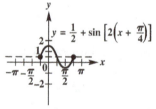

The amplitude is $\left|2\right|$, which is 2. The period is 2π. The vertical translation is 3 units down. The phase shift is $\dfrac{\pi}{2}$ units to the left.

59. $y = \dfrac{1}{2} + \sin\left[2\left(x + \dfrac{\pi}{4}\right)\right]$

Step 1: Find the interval whose length is $\dfrac{2\pi}{b}$.

$$0 \le 2\left(x + \frac{\pi}{4}\right) \le 2\pi \Rightarrow 0 \le x + \frac{\pi}{4} \le \frac{2\pi}{2} \Rightarrow$$

$$0 \le x + \frac{\pi}{4} \le \pi \Rightarrow -\frac{\pi}{4} \le x \le \frac{3\pi}{4}$$

Step 2: Divide the period into four equal parts to get the following *x*-values: $-\dfrac{\pi}{4}, 0, \dfrac{\pi}{4},$

$\dfrac{\pi}{2}, \dfrac{3\pi}{4}$

Step 3: Evaluate the function for each of the five *x*-values.

x	$-\dfrac{\pi}{4}$	0	$\dfrac{\pi}{4}$	$\dfrac{\pi}{2}$	$\dfrac{3\pi}{4}$
$2\left(x + \dfrac{\pi}{4}\right)$	0	$\dfrac{\pi}{2}$	π	$\dfrac{3\pi}{2}$	2π
$\sin\left[2\left(x + \dfrac{\pi}{4}\right)\right]$	0	1	0	-1	0
$\dfrac{1}{2} + \sin\left[2\left(x + \dfrac{\pi}{4}\right)\right]$	$\dfrac{1}{2}$	$\dfrac{3}{2}$	$\dfrac{1}{2}$	$-\dfrac{1}{2}$	$\dfrac{1}{2}$

Steps 4 and 5: Plot the points found in the table and join them with a sinusoidal curve.

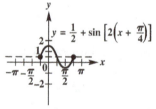

The amplitude is $\left|1\right|$, which is 1. The period is $\dfrac{2\pi}{2}$, which is π. The vertical translation is $\dfrac{1}{2}$ unit up. The phase shift is $\dfrac{\pi}{4}$ units to the left.

61. (a) Let January correspond to $x = 1$, February to $x = 2$, ..., and December of the second year to $x = 24$. Yes, the data appear to outline the graph of a translated sine graph.

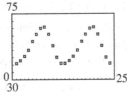

(b) The sine graph is vertically centered around the line $y = 53.5$. This line represents the average annual temperature in Seattle of 53.5°F. (This is also the actual average annual temperature.)

$y = 53.5$

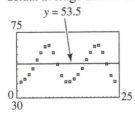

(c) The amplitude of the sine graph is 12.5 because the highest average monthly temperature is 66, the lowest average monthly temperature is 41, and

$\frac{1}{2}(66 - 41) = \frac{1}{2}(25) = 12.5$. The period is 12 because the temperature cycles every twelve months. Let $b = \frac{2\pi}{12} = \frac{\pi}{6}$. One way to determine the phase shift is to use the following technique. The minimum temperature occurs in January, $x = 1$ and the hottest temperature occurs in August, $x = 8$. Then, $\frac{1}{2}(1 + 8) = 4.5$ gives a good approximation for the phase shift.

(d) Let $f(x) = a \sin b(x - d) + c$. The amplitude is 12.5, so let $a = 12.5$. The period is equal to 1 yr or 12 mo, so $b = \frac{\pi}{6}$. The average of the maximum and minimum temperatures is

$\frac{1}{2}(66 + 41) = \frac{1}{2}(107) = 53.5$. Thus,

$f(x) = 12.5 \sin\left[\frac{\pi}{6}(x - 4.5)\right] + 53.5$

(e) Plotting the data with

$f(x) = 12.5 \sin\left[\frac{\pi}{6}(x - 4.5)\right] + 53.5$ on

the same coordinate axes gives a good fit.

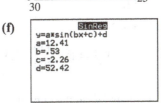

(f)

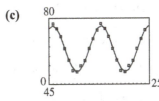

TI-84 Plus fixed to the nearest hundredth From the sine regression we have

$y \approx 12.41 \sin(0.53x - 2.26) + 52.42$

63. (a) See the graph in part (c)

(b)

$y = 12.28 \sin(0.52x + 1.06) + 63.96$

(c)

65. (a) See the graph in part (c)

(b)

```
NORMAL FLOAT AUTO REAL RADIAN MP
          SinReg
y=a*sin(bx+c)+d
a=.4944052504
b=.2144321616
c=.4093258099
d=.5245755071
```

$y = 0.49 \sin(0.21x + 0.41) + 0.52$

(c)

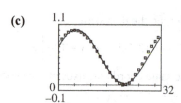

Chapter 6 Quiz
(Section 6.1–6.4)

1. $225° = 225\left(\dfrac{\pi}{180}\text{ radian}\right) = \dfrac{5\pi}{4}$ radians

3. $r = 300$, $s = 450$

$s = r\theta \Rightarrow 450 = 300\theta \Rightarrow \theta = \dfrac{450}{300} = 1.5$

5. $\dfrac{7\pi}{4}$ is in quadrant IV, so the reference angle

is $2\pi - \dfrac{7\pi}{4} = \dfrac{8\pi}{4} - \dfrac{7\pi}{4} = \dfrac{\pi}{4}$. In quadrant IV, the cosine is positive. Thus,

$\cos\dfrac{7\pi}{4} = \cos\dfrac{\pi}{4} = \dfrac{\sqrt{2}}{2}$.

Converting $\dfrac{7\pi}{4}$ to degrees, we have

$\dfrac{7\pi}{4} = \dfrac{7}{4}(180°) = 315°$. The reference angle is

$180° - 315° = 45°$. Thus,

$\cos\dfrac{7\pi}{4} = \cos 315° = \cos 45° = \dfrac{\sqrt{2}}{2}$.

7. 3π is coterminal with $3\pi - 2\pi = \pi$. So
$\tan 3\pi = \tan \pi = 0$
Converting 3π to degrees, we have

$3\pi\left(\dfrac{180}{\pi}\right) = 540°$. The reference angle is

$540° - 360° = 180°$. Thus
$\tan 3\pi = \tan 540° = \tan 180° = 0$.

9. $y = 3 - 4\sin\left(2x + \dfrac{\pi}{2}\right)$

$= 3 - 4\sin\left[2\left(x + \dfrac{\pi}{4}\right)\right] = -4\sin\left[2\left(x + \dfrac{\pi}{4}\right)\right] + 3$

Amplitude: 4; period: $\dfrac{2\pi}{2} = \pi$

Vertical translation: 3 units up

Phase shift: $\dfrac{\pi}{4}$ unit to the left

11. $y = -\dfrac{1}{2}\cos 2x$

Period: $\dfrac{2\pi}{2} = \pi$ and amplitude: $\left|-\dfrac{1}{2}\right| = \dfrac{1}{2}$

Divide the interval $[0, \pi]$ into four equal parts to get the x-values that will yield minimum and maximum points and x-intercepts. Make a table. Repeat this cycle for the interval $[-\pi, 0]$.

x	0	$\dfrac{\pi}{4}$	$\dfrac{\pi}{2}$	$\dfrac{3\pi}{4}$	π
$2x$	0	$\dfrac{\pi}{2}$	π	$\dfrac{3\pi}{2}$	2π
$\cos 2x$	1	0	-1	0	1
$-\dfrac{1}{2}\cos 2x$	$-\dfrac{1}{2}$	0	$\dfrac{1}{2}$	0	$-\dfrac{1}{2}$

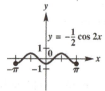

13. $y = 2 + \sin(2x - \pi) = 2 + \sin\left[2\left(x - \dfrac{\pi}{2}\right)\right]$

Step 1: Find the interval whose length is $\dfrac{2\pi}{b}$.

$0 \le 2\left(x - \dfrac{\pi}{2}\right) \le 2\pi \Rightarrow 0 \le x - \dfrac{\pi}{2} \le \dfrac{2\pi}{2} \Rightarrow$

$0 \le x - \dfrac{\pi}{2} \le \pi \Rightarrow \dfrac{\pi}{2} \le x \le \dfrac{3\pi}{2}$

Step 2: Divide the period into four equal parts

to get the following x-values: $\dfrac{\pi}{2}$, $\dfrac{3\pi}{4}$, π,

$\dfrac{5\pi}{4}$, $\dfrac{3\pi}{2}$

Step 3: Evaluate the function for each of the five x-values.

x	$\dfrac{\pi}{2}$	$\dfrac{3\pi}{4}$	π	$\dfrac{5\pi}{4}$	$\dfrac{3\pi}{2}$
$2\left(x - \dfrac{\pi}{2}\right)$	0	$\dfrac{\pi}{2}$	π	$\dfrac{3\pi}{2}$	2π
$\sin\left[2\left(x - \dfrac{\pi}{2}\right)\right]$	0	1	0	-1	0
$2 + \sin\left[2\left(x - \dfrac{\pi}{2}\right)\right]$	2	3	2	1	2

(continued on next page)

(*continued*)

Steps 4 and 5: Plot the points found in the table and join them with a sinusoidal curve.

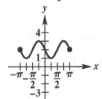

$y = 2 + \sin(2x - \pi)$

The amplitude is $|1|$, which is 1. The period is

$\dfrac{2\pi}{2}$, which is π.

15. The amplitude is $\dfrac{1}{2}[1-(-1)] = \dfrac{1}{2}(2) = 1$, so

$a = 1$. One complete cycle of the graph is achieved in π units, so the period

$\pi = \dfrac{2\pi}{b} \Rightarrow b = \dfrac{2\pi}{\pi} = 2$. Comparing the given

graph with the general sine and cosine curves, we see that this graph is a cosine curve. Substituting $a = 1$ and $b = 2$, the function is $y = \cos 2x$. Verify by confirming minimum and maximum points and x-intercepts from the graph:

$(0,1) \Rightarrow 1 = \cos(2 \cdot 0) = \cos 0 = 1$

$\left(\dfrac{\pi}{4}, 0\right) \Rightarrow 0 = \cos\left(2 \cdot \dfrac{\pi}{4}\right) = \cos\dfrac{\pi}{2} = 0$

$\left(\dfrac{\pi}{2}, -1\right) \Rightarrow -1 = \cos\left(2 \cdot \dfrac{\pi}{2}\right) = \cos\pi = -1$

$\left(\dfrac{3\pi}{4}, 0\right) \Rightarrow 0 = \cos\left(2 \cdot \dfrac{3\pi}{4}\right) = \cos\dfrac{3\pi}{2} = 0$

$(\pi, 1) \Rightarrow 1 = \cos(2\pi) = 1$

Section 6.5 Graphs of the Tangent and Cotangent Functions

1. The least positive value x for which $\tan x = 0$ is <u>π</u>.

3. Between any two successive vertical asymptotes, the graph of $y = \tan x$ <u>increases</u>.

5. The negative value k with the greatest value for which $x = k$ is a vertical asymptote of the graph of $y = \tan x$ is $-\dfrac{\pi}{2}$.

7. $y = -\tan x$
The graph is the reflection of the graph of $y = \tan x$ about the x-axis. This matches with graph C.

9. $y = \tan\left(x - \dfrac{\pi}{4}\right)$

The graph is the graph of $y = \tan x$ shifted $\dfrac{\pi}{4}$ units to the right. This matches with graph B.

11. $y = \cot\left(x + \dfrac{\pi}{4}\right)$

The graph is the graph of $y = \cot x$ shifted $\dfrac{\pi}{4}$ units to the left. This matches with graph F.

13. $y = \tan 4x$
Step 1: Find the period and locate the vertical

asymptotes. The period of tangent is $\dfrac{\pi}{b}$, so

the period for this function is $\dfrac{\pi}{4}$. Tangent has

asymptotes of the form $bx = -\dfrac{\pi}{2}$ and $bx = \dfrac{\pi}{2}$.

Therefore, the asymptotes for $y = \tan 4x$ are

$4x = -\dfrac{\pi}{2} \Rightarrow x = -\dfrac{\pi}{8}$ and $4x = \dfrac{\pi}{2} \Rightarrow x = \dfrac{\pi}{8}$.

Step 2: Sketch the two vertical asymptotes found in Step 1.
Step 3: Divide the interval into four equal

parts: $-\dfrac{\pi}{8}, -\dfrac{\pi}{16}, 0, \dfrac{\pi}{16}, \dfrac{\pi}{8}$

Step 4: Finding the first-quarter point, midpoint, and third-quarter point, we have

$\left(-\dfrac{\pi}{16}, -1\right), (0, 0), \left(\dfrac{\pi}{16}, 1\right)$

Step 5: Join the points with a smooth curve.

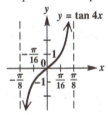

15. $y = 2\tan x$
Step 1: Find the period and locate the vertical

asymptotes. The period of tangent is $\dfrac{\pi}{b}$, so

the period for this function is π . Tangent has

asymptotes of the form $bx = -\dfrac{\pi}{2}$ and $bx = \dfrac{\pi}{2}$.

Therefore, the asymptotes for $y = 2\tan x$ are

$x = -\dfrac{\pi}{2}$ and $x = \dfrac{\pi}{2}$.

(*continued on next page*)

(continued)

Step 2: Sketch the two vertical asymptotes found in Step 1.

Step 3: Divide the interval into four equal parts: $-\dfrac{\pi}{2}, -\dfrac{\pi}{4}, 0, \dfrac{\pi}{4}, \dfrac{\pi}{2}$.

Step 4: Finding the first-quarter point, midpoint, and third-quarter point, we have

$$\left(-\dfrac{\pi}{4}, -2\right), \; (0,0), \; \left(\dfrac{\pi}{4}, 2\right)$$

Step 5: Join the points with a smooth curve. The graph is "stretched" because $a = 2$ and $|2| > 1$.

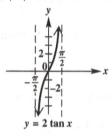

$y = 2 \tan x$

17. $y = 2 \tan \dfrac{1}{4} x$

Step 1: Find the period and locate the vertical asymptotes. The period of tangent is $\dfrac{\pi}{b}$, so the period for this function is 4π. Tangent has asymptotes of the form $bx = -\dfrac{\pi}{2}$ and $bx = \dfrac{\pi}{2}$.

Therefore, the asymptotes for $y = 2 \tan \dfrac{1}{4} x$ are

$$\dfrac{1}{4}x = -\dfrac{\pi}{2} \Rightarrow x = -2\pi \text{ and } \dfrac{1}{4}x = \dfrac{\pi}{2} \Rightarrow x = 2\pi$$

Step 2: Sketch the two vertical asymptotes found in Step 1.

Step 3: Divide the interval into four equal parts: $-2\pi, -\pi, 0, \pi, 2\pi$

Step 4: Finding the first-quarter point, midpoint, and third-quarter point, we have:

$(-\pi, -2), \; (0, 0), \; (\pi, 2)$

Step 5: Join the points with a smooth curve.

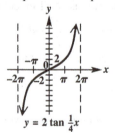

$y = 2 \tan \frac{1}{4} x$

19. $y = \cot 3x$

Step 1: Find the period and locate the vertical asymptotes. The period of cotangent is $\dfrac{\pi}{b}$, so the period for this function is $\dfrac{\pi}{3}$. Cotangent has asymptotes of the form $bx = 0$ and $bx = \pi$. The asymptotes for $y = \cot 3x$ are

$$3x = 0 \Rightarrow x = 0 \text{ and } 3x = \pi \Rightarrow x = \dfrac{\pi}{3}$$

Step 2: Sketch the two vertical asymptotes found in Step 1.

Step 3: Divide the interval into four equal parts: $0, \dfrac{\pi}{12}, \dfrac{\pi}{6}, \dfrac{\pi}{4}, \dfrac{\pi}{3}$

Step 4: Finding the first-quarter point, midpoint, and third-quarter point, we have

$$\left(\dfrac{\pi}{12}, 1\right), \; \left(\dfrac{\pi}{6}, 0\right), \; \left(\dfrac{\pi}{4}, -1\right)$$

Step 5: Join the points with a smooth curve.

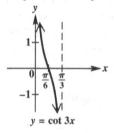

$y = \cot 3x$

21. $y = -2 \tan \dfrac{1}{4} x$

Step 1: Find the period and locate the vertical asymptotes. The period of tangent is $\dfrac{\pi}{b}$, so the period for this function is 4π . Tangent has asymptotes of the form $bx = -\dfrac{\pi}{2}$ and $bx = \dfrac{\pi}{2}$.

Therefore, the asymptotes for $y = -2 \tan \dfrac{1}{4} x$ are

$$\dfrac{1}{4}x = -\dfrac{\pi}{2} \Rightarrow x = -2\pi \text{ and } \dfrac{1}{4}x = \dfrac{\pi}{2} \Rightarrow x = 2\pi$$

Step 2: Sketch the two vertical asymptotes found in Step 1.

Step 3: Divide the interval into four equal parts: $-2\pi, -\pi, 0, \pi, 2\pi$

Step 4: Finding the first-quarter point, midpoint, and third-quarter point, we have

$(-\pi, 2), \; (0,0), \; (\pi, -2)$

(continued on next page)

(*continued*)

> *Step 5*: Join the points with a smooth curve.

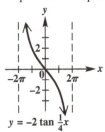

$$y = -2 \tan \tfrac{1}{4}x$$

23. $y = \dfrac{1}{2}\cot 4x$

Step 1: Find the period and locate the vertical asymptotes. The period of cotangent is $\dfrac{\pi}{b}$, so the period for this function is $\dfrac{\pi}{4}$. Cotangent has asymptotes of the form $bx = 0$ and $bx = \pi$.

The asymptotes for $y = \dfrac{1}{2}\cot 4x$ are

$$4x = 0 \Rightarrow x = 0 \text{ and } 4x = \pi \Rightarrow x = \dfrac{\pi}{4}$$

Step 2: Sketch the two vertical asymptotes found in Step 1.

Step 3: Divide the interval into four equal parts: $0, \dfrac{\pi}{16}, \dfrac{\pi}{8}, \dfrac{3\pi}{16}, \dfrac{\pi}{4}$

Step 4: Finding the first-quarter point, midpoint, and third-quarter point, we have

$$\left(\dfrac{\pi}{16}, \dfrac{1}{2}\right), \left(\dfrac{\pi}{8}, 0\right), \left(\dfrac{3\pi}{16}, -\dfrac{1}{2}\right)$$

Step 5: Join the points with a smooth curve.

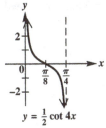

$$y = \tfrac{1}{2}\cot 4x$$

25. $y = \tan(2x - \pi) = \tan\left[2\left(x - \dfrac{\pi}{2}\right)\right]$

Period: $\dfrac{\pi}{b} = \dfrac{\pi}{2}$

Vertical translation: none

Phase shift (horizontal translation): $\dfrac{\pi}{2}$ units

to the right

Because the function is to be graphed over a two-period interval, locate three adjacent vertical asymptotes. Asymptotes of the graph $y = \tan x$ occur at $-\dfrac{\pi}{2}, \dfrac{\pi}{2}$, and $\dfrac{3\pi}{2}$, so use the following equations to locate asymptotes:

$$2\left(x - \dfrac{\pi}{2}\right) = -\dfrac{\pi}{2}, \ 2\left(x - \dfrac{\pi}{2}\right) = \dfrac{\pi}{2}, \text{ and}$$

$$2\left(x - \dfrac{\pi}{2}\right) = \dfrac{3\pi}{2}$$

Solve each of these equations:

$$2\left(x - \dfrac{\pi}{2}\right) = -\dfrac{\pi}{2} \Rightarrow x - \dfrac{\pi}{2} = -\dfrac{\pi}{4} \Rightarrow x = \dfrac{\pi}{4}$$

$$2\left(x - \dfrac{\pi}{2}\right) = \dfrac{\pi}{2} \Rightarrow x - \dfrac{\pi}{2} = \dfrac{\pi}{4} \Rightarrow x = \dfrac{3\pi}{4}$$

$$2\left(x - \dfrac{\pi}{2}\right) = \dfrac{3\pi}{2} \Rightarrow x - \dfrac{\pi}{2} = \dfrac{3\pi}{4} \Rightarrow x = \dfrac{5\pi}{4}$$

Divide the interval $\left(\dfrac{\pi}{4}, \dfrac{3\pi}{4}\right)$ into four equal parts to obtain the following key x-values:

first-quarter value: $\dfrac{3\pi}{8}$; middle value: $\dfrac{\pi}{2}$;

third-quarter value: $\dfrac{5\pi}{8}$

Evaluating the given function at these three key x-values gives the points

$$\left(\dfrac{3\pi}{8}, -1\right), \left(\dfrac{\pi}{2}, 0\right), \left(\dfrac{5\pi}{8}, 1\right)$$

Connect these points with a smooth curve and continue to graph to approach the asymptote $x = \dfrac{\pi}{4}$ and $x = \dfrac{3\pi}{4}$ to complete one period of the graph. Sketch the identical curve between the asymptotes $x = \dfrac{3\pi}{4}$ and $x = \dfrac{5\pi}{4}$ to complete a second period of the graph.

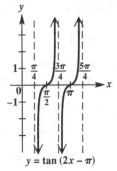

$$y = \tan(2x - \pi)$$

27. $y = \cot\left(3x + \dfrac{\pi}{4}\right) = \cot\left[3\left(x + \dfrac{\pi}{12}\right)\right]$

Period: $\dfrac{\pi}{b} = \dfrac{\pi}{3}$

Vertical translation: none

Phase shift (horizontal translation): $\dfrac{\pi}{12}$ unit to the left

Because the function is to be graphed over a two-period interval, locate three adjacent vertical asymptotes. Asymptotes of the graph $y = \cot x$ occur at multiples of π, so use the following equations to locate asymptotes:

$$3\left(x + \dfrac{\pi}{12}\right) = 0, \; 3\left(x + \dfrac{\pi}{12}\right) = \pi, \text{ and}$$

$$3\left(x + \dfrac{\pi}{12}\right) = 2\pi$$

Solve each of these equations:

$$3\left(x + \dfrac{\pi}{12}\right) = 0 \Rightarrow x + \dfrac{\pi}{12} = 0 \Rightarrow x = -\dfrac{\pi}{12}$$

$$3\left(x + \dfrac{\pi}{12}\right) = \pi \Rightarrow x + \dfrac{\pi}{12} = \dfrac{\pi}{3} \Rightarrow$$

$$x = \dfrac{\pi}{3} - \dfrac{\pi}{12} = \dfrac{\pi}{4}$$

$$3\left(x + \dfrac{\pi}{12}\right) = 2\pi \Rightarrow x + \dfrac{\pi}{12} = \dfrac{2\pi}{3} \Rightarrow$$

$$x = \dfrac{2\pi}{3} - \dfrac{\pi}{12} \Rightarrow x = \dfrac{7\pi}{12}$$

Divide the interval $\left(\dfrac{\pi}{4}, \dfrac{7\pi}{12}\right)$ into four equal parts to obtain the following key x-values:

first-quarter value: $\dfrac{\pi}{3}$; middle value: $\dfrac{5\pi}{12}$;

third-quarter value: $\dfrac{\pi}{2}$

Evaluating the given function at these three key x-values gives the points.

$$\left(\dfrac{\pi}{3}, 1\right), \; \left(\dfrac{5\pi}{12}, 0\right), \; \left(\dfrac{\pi}{2}, -1\right)$$

Connect these points with a smooth curve and continue to graph to approach the asymptote

$x = \dfrac{\pi}{4}$ and $x = \dfrac{7\pi}{12}$ to complete one period of

the graph. Sketch the identical curve between

the asymptotes $x = -\dfrac{\pi}{12}$ and $x = \dfrac{\pi}{4}$ to

complete a second period of the graph.

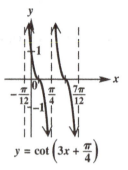

$y = \cot\left(3x + \dfrac{\pi}{4}\right)$

29. $y = 1 + \tan x$
This is the graph of $y = \tan x$ translated vertically 1 unit up.

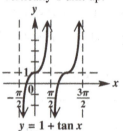

$y = 1 + \tan x$

31. $y = 1 - \cot x$
This is the graph of $y = \cot x$ reflected about the x-axis and then translated vertically 1 unit up.

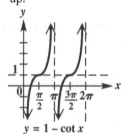

$y = 1 - \cot x$

33. $y = -1 + 2\tan x$
This is the graph of $y = 2\tan x$ translated vertically 1 unit down.

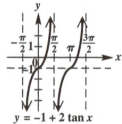

$y = -1 + 2\tan x$

35. $y = -1 + \dfrac{1}{2}\cot(2x - 3\pi)$

$\qquad = -1 + \dfrac{1}{2}\cot\left[2\left(x - \dfrac{3\pi}{2}\right)\right]$

Period: $\dfrac{\pi}{b} = \dfrac{\pi}{2}$.

Vertical translation: 1 unit down

Phase shift (horizontal translation): $\dfrac{3\pi}{2}$ units

to the right

Because the function is to be graphed over a two-period interval, locate three adjacent vertical asymptotes. Asymptotes of the graph $y = \cot x$ occur at multiples of π, use the following equations to locate asymptotes:

$2\left(x - \dfrac{3\pi}{2}\right) = -2\pi,\ 2\left(x - \dfrac{3\pi}{2}\right) = -\pi,$ and

$2\left(x - \dfrac{3\pi}{2}\right) = 0$

Solve each of these equations:

$2\left(x - \dfrac{3\pi}{2}\right) = -2\pi \Rightarrow x - \dfrac{3\pi}{2} = -\pi \Rightarrow$

$x = -\pi + \dfrac{3\pi}{2} = \dfrac{\pi}{2}$

$2\left(x - \dfrac{3\pi}{2}\right) = -\pi \Rightarrow x - \dfrac{3\pi}{2} = -\dfrac{\pi}{2} \Rightarrow$

$x = -\dfrac{\pi}{2} + \dfrac{3\pi}{2} \Rightarrow x = \dfrac{2\pi}{2} = \pi$

$2\left(x - \dfrac{3\pi}{2}\right) = 0 \Rightarrow x - \dfrac{3\pi}{2} = 0 \Rightarrow x = \dfrac{3\pi}{2}$

Divide the interval $\left(\dfrac{\pi}{2}, \pi\right)$ into four equal

parts to obtain the following key x-values:

first-quarter value: $\dfrac{5\pi}{8}$; middle value: $\dfrac{3\pi}{4}$;

third-quarter value: $\dfrac{7\pi}{8}$

Evaluating the given function at these three key x-values gives the points.

$\left(\dfrac{5\pi}{8}, -\dfrac{1}{2}\right),\ \left(\dfrac{3\pi}{4}, -1\right),\ \left(\dfrac{7\pi}{8}, -\dfrac{3}{2}\right)$

Connect these points with a smooth curve and continue to graph to approach the asymptote

$x = \dfrac{\pi}{2}$ and $x = \pi$ to complete one period of

the graph.

Sketch the identical curve between the

asymptotes $x = \pi$ and $x = \dfrac{3\pi}{2}$ to complete a

second period of the graph.

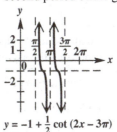

$y = -1 + \tfrac{1}{2}\cot(2x - 3\pi)$

37. $y = 1 - 2\cot\left[2\left(x + \dfrac{\pi}{2}\right)\right]$

Period: $\dfrac{\pi}{b} = \dfrac{\pi}{2}$

Vertical translation: 1 unit up

Phase shift (horizontal translation): $\dfrac{\pi}{2}$ unit to

the left

Because the function is to be graphed over a two-period interval, locate three adjacent vertical asymptotes. Asymptotes of the graph $y = \cot x$ occur at multiples of π, so use the following equations to locate asymptotes.

$2\left(x + \dfrac{\pi}{2}\right) = 0,\ 2\left(x + \dfrac{\pi}{2}\right) = \pi,$ and

$2\left(x + \dfrac{\pi}{2}\right) = 2\pi$

Solve each equation:

$2\left(x + \dfrac{\pi}{2}\right) = 0 \Rightarrow x + \dfrac{\pi}{2} = 0 \Rightarrow x = 0 - \dfrac{\pi}{2} = -\dfrac{\pi}{2}$

$2\left(x + \dfrac{\pi}{2}\right) = \pi \Rightarrow x + \dfrac{\pi}{2} = \dfrac{\pi}{2} \Rightarrow$

$\qquad x = \dfrac{\pi}{2} - \dfrac{\pi}{2} \Rightarrow x = 0$

$2\left(x + \dfrac{\pi}{2}\right) = 2\pi \Rightarrow x + \dfrac{\pi}{2} = \pi \Rightarrow$

$\qquad x = \pi - \dfrac{\pi}{2} = \dfrac{\pi}{2}$

Divide the interval $\left(0, \dfrac{\pi}{2}\right)$ into four equal

parts to obtain the following key x-values:

first-quarter value: $\dfrac{\pi}{8}$; middle value: $\dfrac{\pi}{4}$;

third-quarter value: $\dfrac{3\pi}{8}$

(*continued on next page*)

(*continued*)

Evaluating the given function at these three key *x*-values gives the points

$$\left(\frac{\pi}{8}, -1\right), \left(\frac{\pi}{4}, 1\right), \left(\frac{3\pi}{8}, 3\right)$$

Connect these points with a smooth curve and continue to graph to approach the asymptote $x = 0$ and $x = \frac{\pi}{2}$ to complete one period of the graph. Sketch the identical curve between the asymptotes $x = -\frac{\pi}{2}$ and $x = 0$ to complete a second period of the graph.

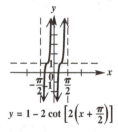

$$y = 1 - 2\cot\left[2\left(x + \frac{\pi}{2}\right)\right]$$

39. The asymptotes are at $-\frac{\pi}{2}$, $\frac{\pi}{2}$, and $\frac{3\pi}{2}$, so this is a tangent function of the form $y = a\tan x$. The graph passes through the point $\left(\frac{\pi}{4}, -2\right)$. Substituting these values into the generic equation gives

$$y = a\tan x \Rightarrow -2 = a\tan\frac{\pi}{4} \Rightarrow -2 = a \cdot 1 \Rightarrow$$
$$-2 = a$$

Thus, the equation of the graph is $y = -2\tan x$.

41. The asymptotes occur at 0, $\frac{\pi}{3}$, and $\frac{2\pi}{3}$, so this is a cotangent function of the form $y = a\cot bx$. The period of the function is $\frac{\pi}{3}$, so we have $\frac{\pi}{b} = \frac{\pi}{3} \Rightarrow b = 3$. The graph passes through the point $\left(\frac{\pi}{12}, 1\right)$. Substituting these values into the generic equation gives

$$y = a\cot bx \Rightarrow 1 = a\cot\left(3 \cdot \frac{\pi}{12}\right) \Rightarrow$$
$$1 = a\cot\frac{\pi}{4} \Rightarrow 1 = a \cdot 1 \Rightarrow 1 = a$$

Thus, the equation of the graph is $y = \cot 3x$.

43. Since the asymptotes occur at $-\pi$ and π, this is a tangent function with period 2π instead of π. Therefore, the coefficient of *x* is $\frac{1}{2}$. The graph is vertically translated 1 unit up compared to the graph of $y = \tan\frac{1}{2}x$, so an equation for this graph is $y = 1 + \tan\frac{1}{2}x$.

45. True; $\frac{\pi}{2}$ is the smallest positive value where

$$\cos\frac{\pi}{2} = 0. \text{ Since } \tan\frac{\pi}{2} = \frac{\sin\frac{\pi}{2}}{\cos\frac{\pi}{2}}, \frac{\pi}{2} \text{ is the}$$

smallest positive value where the tangent function is undefined. Thus, $k = \frac{\pi}{2}$ is the smallest positive value for which $x = k$ is an asymptote for the tangent function.

47. False; $\tan(-x) = \frac{\sin(-x)}{\cos(-x)} = \frac{-\sin x}{\cos x} = -\tan x$

(because $\sin x$ is odd and $\cos x$ is even) for all *x* in the domain. Moreover, if $\tan(-x) = \tan x$, then the graph would be symmetric about the *y*-axis, which it is not.

49. The function $\tan x$ has a period of π, so it repeats four times over the interval $(-2\pi, 2\pi]$. Since its range is $(-\infty, \infty)$, $\tan x = c$ has four solutions for every value of *c*.

51. $\tan(-x) = \frac{\sin(-x)}{\cos(-x)} = \frac{-\sin x}{\cos x} = -\tan x$,

$$\left\{ x \,\middle|\, x \neq (2n+1)\frac{\pi}{4}, \text{ where } n \text{ is any integer} \right\}.$$

53. $d = 4\tan\left[2\pi(0)\right] = 4\tan 0 \approx 4(0) = 0$ m

55. $d = 4\tan\left[2\pi(1.2)\right] = 4\tan(2.4\pi)$
$\approx 4(3.0777) \approx 12.3$ m

57. The least positive number for which $y = \cot x$ is undefined is $x = \pi$.

59. The vertical asymptotes in general occur at $x = \frac{5\pi}{4} + n\pi$, where *n* is an integer.

61. $0.32175055 + \pi \approx 3.4633432$

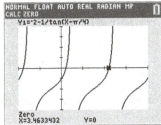

The next positive x-intercept is (3.46, 0).

Section 6.6   Graphs of the Secant and Cosecant Functions

1. A

3. D

5. C

7. $y = -\csc x$

The graph is the reflection of the graph of $y = \csc x$ about the x-axis. This matches with graph B.

9. $y = \sec\left(x - \dfrac{\pi}{2}\right)$

The graph is the graph of $y = \sec x$ shifted $\dfrac{\pi}{2}$ units to the right. This matches with graph D.

11. $y = 3\sec\dfrac{1}{4}x$

Step 1: Graph the corresponding reciprocal function $y = 3\cos\dfrac{1}{4}x$. The period is

$\dfrac{2\pi}{\frac{1}{4}} = 2\pi \cdot \dfrac{4}{1} = 8\pi$ and its amplitude is $|3| = 3$.

One period is in the interval $0 \le x \le 8\pi$. Dividing the interval into four equal parts gives us the following key points: (0, 1), $(2\pi, 0)$, $(4\pi, -1)$, $(6\pi, 0)$, $(8\pi, 1)$

Step 2: The vertical asymptotes of $y = \sec\dfrac{1}{4}x$

are at the x-intercepts of $y = \cos\dfrac{1}{4}x$, which

are $x = 2\pi$ and $x = 6\pi$. Continuing this pattern to the left, we also have a vertical asymptote of $x = -2\pi$.

Step 3: Sketch the graph.

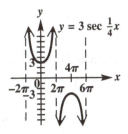

13. $y = -\dfrac{1}{2}\csc\left(x + \dfrac{\pi}{2}\right)$

Step 1: Graph the corresponding reciprocal

function $y = -\dfrac{1}{2}\sin\left(x + \dfrac{\pi}{2}\right)$. The period is

2π and its amplitude is $\left|-\dfrac{1}{2}\right| = \dfrac{1}{2}$. One period

is in the interval $-\dfrac{\pi}{2} \le x \le \dfrac{3\pi}{2}$.

Dividing the interval into four equal parts

gives us the following key points: $\left(-\dfrac{\pi}{2}, 0\right)$,

$\left(0, -\dfrac{1}{2}\right)$, $\left(\dfrac{\pi}{2}, 0\right)$, $\left(\pi, \dfrac{1}{2}\right)$, $\left(\dfrac{3\pi}{2}, 0\right)$

Step 2: The vertical asymptotes of

$y = -\dfrac{1}{2}\csc\left(x + \dfrac{\pi}{2}\right)$ are at the x-intercepts of

$y = -\dfrac{1}{2}\sin\left(x + \dfrac{\pi}{2}\right)$, which are $x = -\dfrac{\pi}{2}$,

$x = \dfrac{\pi}{2}$, and $x = \dfrac{3\pi}{2}$.

Step 3: Sketch the graph.

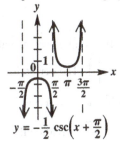

15. $y = \csc\left(x - \dfrac{\pi}{4}\right)$

Step 1: Graph the corresponding reciprocal

function $y = \sin\left(x - \dfrac{\pi}{4}\right)$ The period is 2π

and its amplitude is $|1| = 1$. One period is in

the interval $\dfrac{\pi}{4} \le x \le \dfrac{9\pi}{4}$. Dividing the

interval into four equal parts gives us the

following key points: $\left(\dfrac{\pi}{4}, 0\right)$, $\left(\dfrac{3\pi}{4}, 1\right)$,

$\left(\dfrac{5\pi}{4}, 0\right)$, $\left(\dfrac{7\pi}{4}, -1\right)$, $\left(\dfrac{9\pi}{4}, 0\right)$

Step 2: The vertical asymptotes of

$y = \csc\left(x - \dfrac{\pi}{4}\right)$ are at the x-intercepts of

$y = \sin\left(x - \dfrac{\pi}{4}\right)$, which are $x = \dfrac{\pi}{4}$, $x = \dfrac{5\pi}{4}$,

and $x = \dfrac{9\pi}{4}$.

Step 3: Sketch the graph.

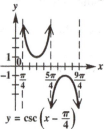

17. $y = \sec\left(x + \dfrac{\pi}{4}\right)$

Step 1: Graph the corresponding reciprocal

function $y = \cos\left(x + \dfrac{\pi}{4}\right)$. The period is 2π

and its amplitude is $|1| = 1$. One period is in

the interval $-\dfrac{\pi}{4} \le x \le \dfrac{7\pi}{4}$. Dividing the

interval into four equal parts gives us the

following key points: $\left(-\dfrac{\pi}{4}, 1\right)$, $\left(\dfrac{\pi}{4}, 0\right)$,

$\left(\dfrac{3\pi}{4}, -1\right)$, $\left(\dfrac{5\pi}{4}, 0\right)$, $\left(\dfrac{7\pi}{4}, 1\right)$

Step 2: The vertical asymptotes of

$y = \sec\left(x + \dfrac{\pi}{4}\right)$ are at the x-intercepts of

$y = \cos\left(x + \dfrac{\pi}{4}\right)$, which are $x = \dfrac{\pi}{4}$ and

$x = \dfrac{5\pi}{4}$. Continuing this pattern to the right,

we also have a vertical asymptote of $x = \dfrac{9\pi}{4}$.

Step 3: Sketch the graph.

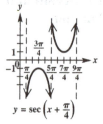

19. $y = \csc\left(\dfrac{1}{2}x - \dfrac{\pi}{4}\right) = \csc\left[\dfrac{1}{2}\left(x - \dfrac{\pi}{2}\right)\right]$

Step 1: Graph the corresponding reciprocal

function $y = \sin\left[\dfrac{1}{2}\left(x - \dfrac{\pi}{2}\right)\right]$. The period is

$\dfrac{2\pi}{\frac{1}{2}} = 2\pi \cdot \dfrac{2}{1} = 4\pi$ and its amplitude is $|1| = 1$.

One period is in the interval $\dfrac{\pi}{2} \le x \le \dfrac{9\pi}{2}$.

Dividing the interval into four equal parts

gives us the following key points: $\left(\dfrac{\pi}{2}, 0\right)$,

$\left(\dfrac{3\pi}{2}, 1\right)$, $\left(\dfrac{5\pi}{2}, 0\right)$, $\left(\dfrac{7\pi}{2}, -1\right)$, $\left(\dfrac{9\pi}{2}, 0\right)$

Step 2: The vertical asymptotes of

$y = \csc\left[\dfrac{1}{2}\left(x - \dfrac{\pi}{2}\right)\right]$ are at the x-intercepts of

$y = \sin\left[\dfrac{1}{2}\left(x - \dfrac{\pi}{2}\right)\right]$, which are $x = \dfrac{\pi}{2}$,

$x = \dfrac{5\pi}{2}$, and $x = \dfrac{9\pi}{2}$.

Step 3: Sketch the graph.

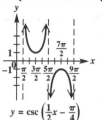

21. $y = 2 + 3\sec(2x - \pi) = 2 + 3\sec\left[2\left(x - \dfrac{\pi}{2}\right)\right]$

Step 1: Graph the corresponding reciprocal

function $y = 2 + 3\cos\left[2\left(x - \dfrac{\pi}{2}\right)\right]$. The

period is π and its amplitude is $|3| = 3$. One

period is in the interval $\dfrac{\pi}{2} \le x \le \dfrac{3\pi}{2}$.

Dividing the interval into four equal parts
gives us the following key points:

$\left(\dfrac{\pi}{2}, 5\right), \left(\dfrac{3\pi}{4}, 2\right), (\pi, -1), \left(\dfrac{5\pi}{4}, 2\right), \left(\dfrac{3\pi}{2}, 5\right)$

Step 2: The vertical asymptotes of

$y = 2 + 3\sec\left[2\left(x - \dfrac{\pi}{2}\right)\right]$ are at the x-

intercepts of $y = 3\cos 2\left(x - \dfrac{\pi}{2}\right)$, which are

$x = \dfrac{3\pi}{4}$ and $x = \dfrac{5\pi}{4}$. Continuing this pattern

to the left, we also have a vertical asymptote

of $x = \dfrac{\pi}{4}$.

Step 3: Sketch the graph.

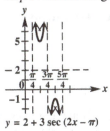

$y = 2 + 3\sec(2x - \pi)$

23. $y = 1 - \dfrac{1}{2}\csc\left(x - \dfrac{3\pi}{4}\right)$

Step 1: Graph the corresponding reciprocal

function $y = 1 - \dfrac{1}{2}\sin\left(x - \dfrac{3\pi}{4}\right)$. The period is

2π and its amplitude is $\dfrac{1}{2}$.

One period is in the interval $\dfrac{3\pi}{4} \le x \le \dfrac{11\pi}{4}$.

Dividing the interval into four equal parts
gives us the following key points:

$\left(\dfrac{3\pi}{4}, 1\right), \left(\dfrac{5\pi}{4}, \dfrac{1}{2}\right),$

$\left(\dfrac{7\pi}{4}, 1\right), \left(\dfrac{9\pi}{4}, \dfrac{3}{2}\right), \left(\dfrac{11\pi}{4}, 1\right)$

Step 2: The vertical asymptotes of

$y = 1 - \dfrac{1}{2}\csc\left(x - \dfrac{3\pi}{4}\right)$ are at the x-intercepts

of $y = -\dfrac{1}{2}\sin\left(x - \dfrac{3\pi}{4}\right)$, which are $x = \dfrac{3\pi}{4}$,

$x = \dfrac{7\pi}{4}$, and $x = \dfrac{11\pi}{4}$.

Step 3: Sketch the graph.

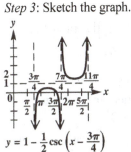

$y = 1 - \dfrac{1}{2}\csc\left(x - \dfrac{3\pi}{4}\right)$

For exercises 25–29, other answers are possible.

25. Since the graph crosses the y-axis at $(0, 1)$, this
is a secant graph with $a = 1$. The period is

$\left|-\dfrac{\pi}{4} - \dfrac{\pi}{4}\right| = \left|-\dfrac{\pi}{2}\right| = \dfrac{\pi}{2}$. Thus,

$b = \dfrac{2\pi}{\dfrac{\pi}{2}} \Rightarrow b = 4$. The equation of the graph is

$y = \sec 4x$.

27. This is the graph of $y = \csc x$ translated two
units down. Thus, the equation of the graph is
$y = -2 + \csc x$.

29. This is the graph of $y = \sec x$, reflected across
the x-axis and translated one unit down. Thus,
the equation of the graph is $y = -1 - \sec x$.

31. True. Because $\tan x = \dfrac{\sin x}{\cos x}$ and

$\sec x = \dfrac{1}{\cos x}$, the tangent and secant

functions will be undefined at the same values.

33. True. $\sec(-x) = \dfrac{1}{\cos(-x)} = \dfrac{1}{\cos(x)} = \sec(x)$

(because $\cos x$ is even) for all x in the domain.
Moreover, if $\sec(-x) = \sec x$, then the graph
would be symmetric about the y-axis, which it
is.

35. None; $|\cos x| \le 1$ for all x, so

$\dfrac{1}{|\cos x|} \ge 1$ and $|\sec x| \ge 1$. Because $|\sec x| \ge 1$,

$\sec x$ has no values in the interval $(-1, 1)$.

37. $\sec(-x) = \dfrac{1}{\cos(-x)} = \dfrac{1}{\cos(x)} = \sec(x)$,

$\left\{ x \,\middle|\, x \neq (2n+1)\dfrac{\pi}{2}, \text{ where } n \text{ is any integer} \right\}$.

39. $t = 0$

$a = 4\left|\sec 0\right| = 4\left|1\right| = 4(1) = 4$ m

41. $t = 1.24$

$a = 4\left|\sec\left[2\pi(1.24)\right]\right| \approx 4\left|15.9260\right|$

$\quad = 4(15.9260) \approx 63.7$ m

43. $y_1 = \sin x; \; y_2 = \sin 2x; \; y_3 = y_1 + y_2$

Graph the functions in the window $[-2\pi, 2\pi] \times [-3, 3]$.

y_1 is in black, y_2 is in dark grey, and y_3 is in light grey.

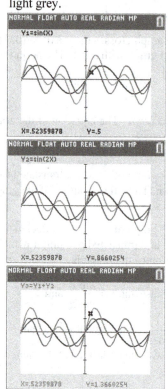

$Y_1\left(\dfrac{\pi}{6}\right) + Y_2\left(\dfrac{\pi}{6}\right) \approx 0.5 + 0.8660254$

$\quad = 1.3660254$

$\quad = Y_3\left(\dfrac{\pi}{6}\right) = (Y_1 + Y_2)\left(\dfrac{\pi}{6}\right)$

45. $y_1 = \tan x; \; y_2 = \sec x; \; y_3 = y_1 + y_2$

Graph the functions in the window $[-2\pi, 2\pi] \times [-3, 3]$.

y_1 is in black, y_2 is in dark grey, and y_3 is in light grey.

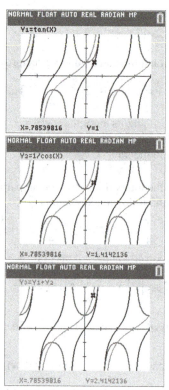

$Y_1\left(\dfrac{\pi}{4}\right) + Y_2\left(\dfrac{\pi}{4}\right) \approx 1 + 1.4142136 = 2.4142136$

$\quad = Y_3\left(\dfrac{\pi}{4}\right) = (Y_1 + Y_2)\left(\dfrac{\pi}{4}\right)$

Summary Exercises on Graphing Circular Functions

1. $y = 2\sin \pi x$

Period: $\dfrac{2\pi}{\pi} = 2$ and amplitude: $|2| = 2$

Divide the interval $[0, 2]$ into four equal parts to get the x-values that will yield minimum and maximum points and x-intercepts. Then make a table.

x	0	$\dfrac{1}{2}$	1	$\dfrac{3}{2}$	2
πx	0	$\dfrac{\pi}{2}$	π	$\dfrac{3\pi}{2}$	2π
$\sin \pi x$	0	1	0	-1	0
$2\sin \pi x$	0	2	0	-2	0

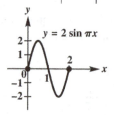

3. $y = -2 + \dfrac{1}{2}\cos\dfrac{\pi}{4}x$

Step 1: Find the interval whose length is $\dfrac{2\pi}{b}$.

$$0 \le \dfrac{\pi}{4}x \le 2\pi \Rightarrow 0 \le x \le 8$$

Step 2: Divide the period into four equal parts to get the following *x*-values: 0, 2, 4, 6, 8

Step 3: Evaluate the function for each of the five *x*-values.

x	0	2	4	6	8
$\dfrac{\pi}{4}x$	0	$\dfrac{\pi}{2}$	π	$\dfrac{3\pi}{2}$	2π
$\cos\dfrac{\pi}{4}x$	1	0	-1	0	1
$\dfrac{1}{2}\cos\dfrac{\pi}{4}x$	$\dfrac{1}{2}$	0	$-\dfrac{1}{2}$	0	$\dfrac{1}{2}$
$-2 + \dfrac{1}{2}\cos\dfrac{\pi}{4}x$	$-\dfrac{3}{2}$	-2	$-\dfrac{5}{2}$	-2	$-\dfrac{3}{2}$

Steps 4 and 5: Plot the points found in the table and join them with a sinusoidal curve. By graphing an additional period to the left, we obtain the following graph.

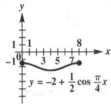

The amplitude is $\left|\dfrac{1}{2}\right| = \dfrac{1}{2}$. The period is 8. The vertical translation is 2 units down. There is no phase shift.

5. $y = -4\csc\dfrac{1}{2}x$

Step 1: Graph the corresponding reciprocal function $y = -4\sin\dfrac{1}{2}x$ The period is

$$\dfrac{2\pi}{\frac{1}{2}} = 2\pi \cdot \dfrac{2}{1} = 4\pi \text{ and its amplitude is}$$

$$|-4| = 4.$$

One period is in the interval $0 \le x \le 4\pi$. Dividing the interval into four equal parts gives us the following key points: $(0, 0)$, $(\pi, -4)$, $(2\pi, 0)$, $(3\pi, 4)$, $(4\pi, 0)$

Step 2: The vertical asymptotes of

$y = -4\csc\dfrac{1}{2}x$ are at the *x*-intercepts of

$y = -4\sin\dfrac{1}{2}x$, which are $x = 0$, $x = 2\pi$, and

$x = 4\pi$.

Step 3: Sketch the graph.

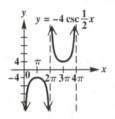

7. $y = -5\sin\dfrac{x}{3}$

Period: $\dfrac{2\pi}{\frac{1}{3}} = 2\pi \cdot \dfrac{3}{1} = 6\pi$ and amplitude:

$$|-5| = 5$$

Divide the interval $[0, 6\pi]$ into four equal parts to get the *x*-values that will yield minimum and maximum points and *x*-intercepts. Then make a table. Repeat this cycle for the interval $[-6\pi, 0]$.

x	0	$\dfrac{3\pi}{2}$	3π	$\dfrac{9\pi}{2}$	6π
$\dfrac{x}{3}$	0	$\dfrac{\pi}{2}$	π	$\dfrac{3\pi}{2}$	2π
$\sin\dfrac{x}{3}$	0	1	0	-1	0
$-5\sin\dfrac{x}{3}$	0	-5	0	5	0

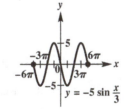

9. $y = 3 - 4\sin\left(\dfrac{5}{2}x + \pi\right) = 3 - 4\sin\left[\dfrac{5}{2}\left(x + \dfrac{2\pi}{5}\right)\right]$

Step 1: Find the interval whose length is $\dfrac{2\pi}{b}$.

$0 \le \dfrac{5}{2}\left(x + \dfrac{2\pi}{5}\right) \le 2\pi \Rightarrow$

$0 \le x + \dfrac{2\pi}{5} \le \dfrac{4\pi}{5} \Rightarrow -\dfrac{2\pi}{5} \le x \le \dfrac{2\pi}{5}$

Step 2: Divide the period into four equal parts to get the following x-values: $-\dfrac{2\pi}{5}, -\dfrac{\pi}{5}, 0,$

$\dfrac{\pi}{5}, \dfrac{2\pi}{5}$

Step 3: Evaluate the function for each of the five x-values

x	$-\dfrac{2\pi}{5}$	$-\dfrac{\pi}{5}$	0	$\dfrac{\pi}{5}$	$\dfrac{2\pi}{5}$
$\dfrac{5}{2}\left(x + \dfrac{2\pi}{5}\right)$	0	$\dfrac{\pi}{2}$	π	$\dfrac{3\pi}{2}$	2π
$\sin\left[\dfrac{5}{2}\left(x + \dfrac{2\pi}{5}\right)\right]$	0	1	0	-1	0
$-4\sin\left[\dfrac{5}{2}\left(x + \dfrac{2\pi}{5}\right)\right]$	0	-4	0	4	0
$3 - 4\sin\left[\dfrac{5}{2}\left(x + \dfrac{2\pi}{5}\right)\right]$	3	-1	3	7	3

Steps 4 and 5: Plot the points found in the table and join them with a sinusoidal curve. By graphing an additional period to the right, we obtain the following graph.

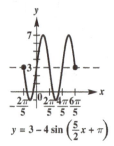

$y = 3 - 4\sin\left(\dfrac{5}{2}x + \pi\right)$

The amplitude is $|-4|$, which is 4. The period is $\dfrac{2\pi}{\frac{5}{2}}$, which is $\dfrac{4\pi}{5}$. The vertical translation is 3 units up. The phase shift is $\dfrac{2\pi}{5}$ units to the left.

Section 6.7 Harmonic Motion

1. The amplitude is 5.

3. The frequency is $\dfrac{2}{2\pi} = \dfrac{1}{\pi}$ oscillation per second.

5. $s\left(\dfrac{\pi}{2}\right) = 5\cos\left(2 \cdot \dfrac{\pi}{2}\right) = 5(-1) = -5$

7. **(a)** The weight is pulled down 4 units, so $s(0) = -4$. Thus, we have

$s(0) = -4 = a\cos[\omega(0)] \Rightarrow$
$-4 = a\cos 0 \Rightarrow -4 = a(1) \Rightarrow a = -4$

The time it takes to complete one oscillation is 3 sec, so $P = 3$ sec.

$P = 3\text{ sec} \Rightarrow 3 = \dfrac{2\pi}{\omega} \Rightarrow 3 = \dfrac{2\pi}{\omega} \Rightarrow$

$3\omega = 2\pi \Rightarrow \omega = \dfrac{2\pi}{3}$

Therefore, $s(t) = -4\cos\dfrac{2\pi}{3}t$.

(b) $s(1) = -4\cos\left[\dfrac{2\pi}{3}(1.25)\right]$

$= -4\cos\left[\dfrac{2\pi}{3}\left(\dfrac{5}{4}\right)\right] = -4\cos\dfrac{5\pi}{6}$

$= -4\left(-\dfrac{\sqrt{3}}{2}\right) = 2\sqrt{3} \approx 3.46$ units

(c) The frequency is the reciprocal of the period, or $\dfrac{1}{3}$ oscillation per second.

9. $E = 5\cos 120\pi t$

(a) Amplitude: $|5| = 5$ and period:

$\dfrac{2\pi}{120\pi} = \dfrac{1}{60}$ sec

(b) The period is $\dfrac{1}{60}$, so one oscillation is completed in $\dfrac{1}{60}$ sec. Therefore, the frequency is 60 oscillations per second.

(c) $t = 0$, $E = 5\cos 120\pi(0) = 5\cos 0 = 5(1) = 5$
$t = 0.03$,
$E = 5\cos 120\pi(0.03) = 5\cos 3.6\pi \approx 1.545$
$t = 0.06$,
$E = 5\cos 120\pi(0.06) = 5\cos 7.2\pi \approx -4.045$

(continued on next page)

(*continued*)

$$t = 0.09,$$
$$E = 5\cos 120\pi (0.09)$$
$$= 5\cos 10.8\pi \approx -4.045$$
$$t = 0.12,$$
$$E = 5\cos 120\pi (0.12) = 5\cos 14.4\pi \approx 1.545$$

(d)

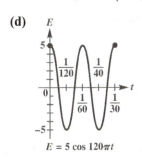

$$E = 5\cos 120\pi t$$

11. (a) $a = 2, \omega = 2$

$$s(t) = a\sin\omega t \Rightarrow s(t) = 2\sin 2t$$

amplitude $= |a| = |2| = 2$; period

$$= \frac{2\pi}{\omega} = \frac{2\pi}{2} = \pi \text{ ; frequency } = \frac{\omega}{2\pi} = \frac{1}{\pi}$$

rotation per second

(b) $a = 2, \omega = 4$

$$s(t) = a\sin\omega t \Rightarrow s(t) = 2\sin 4t$$

amplitude $= |a| = |2| = 2$; period

$$= \frac{2\pi}{\omega} = \frac{2\pi}{4} = \frac{\pi}{2}; \text{ frequency}$$

$$= \frac{\omega}{2\pi} = \frac{4}{2\pi} = \frac{2}{\pi} \text{ rotation per second}$$

13. $P = 2\pi\sqrt{\dfrac{L}{32}}; L = \dfrac{1}{2}$ ft

The period is

$$P = 2\pi\sqrt{\frac{L}{32}} \Rightarrow P = 2\pi\sqrt{\frac{\frac{1}{2}}{32}} = 2\pi\sqrt{\frac{1}{64}}$$

$$= 2\pi \cdot \frac{1}{8} = \frac{\pi}{4} \text{ sec}$$

The frequency is the reciprocal of the period,

or $\dfrac{4}{\pi}$ oscillations per second.

15. $s(t) = a\sin\sqrt{\dfrac{k}{m}}t; k = 4; P = 1$ sec

A period of 1 sec is produced when $\dfrac{2\pi}{\sqrt{\dfrac{k}{m}}} = 1$

Because $k = 4$, we can solve

$$\frac{2\pi}{\sqrt{\dfrac{k}{m}}} = 1 \Rightarrow \frac{2\pi}{\sqrt{\dfrac{4}{m}}} = 1 \Rightarrow 2\pi = \sqrt{\frac{4}{m}} \Rightarrow$$

$$4\pi^2 = \frac{4}{m} \Rightarrow 4\pi^2 m = 4 \Rightarrow m = \frac{4}{4\pi^2} = \frac{1}{\pi^2}$$

17. $s(t) = -5\cos 4\pi t$, $a = |-5| = 5$, $\omega = 4\pi$

(a) maximum height = amplitude

$$= a = |-5| = 5 \text{ in.}$$

(b) frequency

$$= \frac{\omega}{2\pi} = \frac{4\pi}{2\pi} = 2 \text{ cycles per sec; period}$$

$$= \frac{2\pi}{\omega} = \frac{1}{2} \text{ sec}$$

(c) $s(t) = -5\cos 4\pi t = 5 \Rightarrow \cos 4\pi t = -1 \Rightarrow$

$$4\pi t = \pi \Rightarrow t = \frac{1}{4}$$

The weight first reaches its maximum

height after $\dfrac{1}{4}$ sec.

(d) $s(1.3) = -5\cos\left[4\pi(1.3)\right]$
$$= -5\cos 5.2\pi \approx 4,$$

After 1.3 sec, the weight is about 4.0 in. above the equilibrium position.

19. $a = -3$

(a) We will use a model of the form $s(t) = a\cos\omega t$ with $a = -3$.

$$s(0) = -3\cos\left[\omega(0)\right]$$
$$= -3\cos 0 = -3(1) = -3$$

Using a cosine function rather than a sine function will avoid the need for a phase

shift. The frequency $= \dfrac{6}{\pi}$ cycles per sec,

so, by definition,

$$\frac{\omega}{2\pi} = \frac{6}{\pi} \Rightarrow \omega\pi = 12\pi \Rightarrow \omega = 12.$$

Therefore, a model for the position of the weight at time t seconds is $s(t) = -3\cos 12t.$

(b) The period is the reciprocal of the

frequency, or $\dfrac{\pi}{6}$ sec.

21. $s(0) = 2$ in.; $P = 0.5$ sec

   (a) Given $s(t) = a\cos\omega t$, the period is $\dfrac{2\pi}{\omega}$

      and the amplitude is $|a|$.

$$P = 0.5 \text{ sec} \Rightarrow 0.5 = \frac{2\pi}{\omega} \Rightarrow \frac{1}{2} = \frac{2\pi}{\omega} \Rightarrow$$
$$\omega = 4\pi$$
$$s(0) = 2 = a\cos\left[\omega(0)\right] \Rightarrow$$
$$2 = a\cos 0 \Rightarrow 2 = a(1) \Rightarrow a = 2$$

      Thus, $s(t) = 2\cos 4\pi t$.

   (b) $s(1) = 2\cos\left[4\pi(1)\right] = 2\cos 4\pi$
$$= 2(1) = 2$$

      The weight is neither moving upward nor downward. At $t = 1$, the motion of the weight is changing from up to down.

23. $s(0) = -3$ in.; $P = 0.8$ sec

   (a) Given $s(t) = a\cos\omega t$, the period is $\dfrac{2\pi}{\omega}$

      and the amplitude is $|a|$.

$$P = 0.8 \text{ sec} \Rightarrow 0.8 = \frac{2\pi}{\omega} \Rightarrow \frac{4}{5} = \frac{2\pi}{\omega} \Rightarrow$$
$$4\omega = 10\pi \Rightarrow \omega = \frac{10\pi}{4} = 2.5\pi$$
$$s(0) = -3 = a\cos\left[\omega(0)\right] \Rightarrow$$
$$-3 = a\cos 0 \Rightarrow -3 = a(1) \Rightarrow a = -3$$

      Thus, $s(t) = -3\cos 2.5\pi t$.

   (b) $s(1) = -3\cos\left[2.5\pi(1)\right] = -3\cos\dfrac{5\pi}{2}$
$$= -3(0) = 0$$

      The weight is moving upward.

25. The frequency is $\dfrac{\omega}{2\pi}$, so

$$27.5 = \frac{\omega}{2\pi} \Rightarrow \omega = 55\pi. \text{ Because } s(0) = 0.21,$$
$$0.21 = a\cos\left[\omega(0)\right] \Rightarrow 0.21 = a\cos 0 \Rightarrow$$
$$0.21 = a(1) \Rightarrow a = 0.21.$$
Thus, $s(t) = 0.21\cos 55\pi t$.

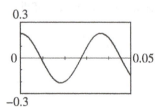

27. The frequency is $\dfrac{\omega}{2\pi}$, so

$$55 = \frac{\omega}{2\pi} \Rightarrow \omega = 110\pi. \text{ Since } s(0) = 0.14,$$
$$0.14 = a\cos\left[\omega(0)\right] \Rightarrow 0.14 = a\cos 0 \Rightarrow$$
$$0.14 = a(1) \Rightarrow a = 0.14. \text{ Thus,}$$
$$s(t) = 0.14\cos 110\pi t.$$

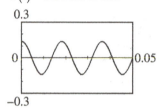

29. The weight was pulled down 11 in. from the equilibrium position before it was released.

31.

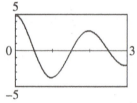

The graph intersects the horizontal axis at $t = 1, 3, 5, 7, 9, 11$.

33. (a)

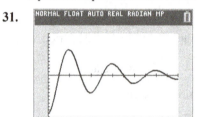

   (b) The upper envelope of the graph of $y_3 = 5e^{-0.3x}\cos(\pi x)$ is $y_1 = 5e^{-0.3x}$.

   (c) $5e^{-0.3x} = 5e^{-0.3x}\cos(\pi x)$ when $\cos(\pi x) = 1$, or when $x = 0$ and $x = 2$. This can be verified by examining the graphs.

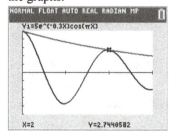

Chapter 6   Review Exercises

1. A central angle of a circle that intercepts an arc of length 2 times the radius of the circle has a measure of 2 radians.

3. To find a coterminal angle, add or subtract multiples of 2π. Three of the many possible answers are $1+2\pi$, $1+4\pi$, and $1+6\pi$.

5. $45° = 45\left(\dfrac{\pi}{180}\text{ radian}\right) = \dfrac{\pi}{4}\text{ radian}$

7. $175° = 175\left(\dfrac{\pi}{180}\text{ radian}\right) = \dfrac{35\pi}{36}\text{ radians}$

9. $800° = 800\left(\dfrac{\pi}{180}\text{ radian}\right) = \dfrac{40\pi}{9}\text{ radians}$

11. $\dfrac{5\pi}{4} = \dfrac{5\pi}{4}\left(\dfrac{180°}{\pi}\right) = 225°$

13. $\dfrac{8\pi}{3} = \dfrac{8\pi}{3}\left(\dfrac{180°}{\pi}\right) = 480°$

15. $-\dfrac{11\pi}{18} = -\dfrac{11\pi}{18}\left(\dfrac{180°}{\pi}\right) = -110°$

17. $\dfrac{15}{60} = \dfrac{1}{4}$ rotation, so we have
$\theta = \dfrac{1}{4}(2\pi) = \dfrac{\pi}{2}$. Thus,
$s = r\theta \Rightarrow s = 2\left(\dfrac{\pi}{2}\right) = \pi$ in.

19. $\theta = 3(2\pi) = 6\pi$, so we have
$s = r\theta \Rightarrow s = 2(6\pi) = 12\pi$ in.

21. $r = 15.2$ cm, $\theta = \dfrac{3\pi}{4}$
$s = r\theta \Rightarrow s = 15.2\left(\dfrac{3\pi}{4}\right) = 11.4\pi \approx 35.8$ cm

23. $s = 7.683$, $r = 8.973$ cm
$s = r\theta \Rightarrow 7.683 = 8.973\theta \Rightarrow$
$\theta = \dfrac{7.683}{8.973} \approx 0.8562$ radian
$0.8592 \text{ radian} = 0.8592\left(\dfrac{180°}{\pi}\right) \approx 49.06°$

25. $r = 38.0$ m, $\theta = 21°40'$
First convert $\theta = 21°40'$ to radians:
$\theta = 21°40' = \left(21 + \dfrac{40}{60}\right)\left(\dfrac{\pi}{180}\right)$
$= \dfrac{65}{3}\left(\dfrac{\pi}{180}\right) = \dfrac{13\pi}{108}$
$\mathcal{A} = \dfrac{1}{2}r^2\theta$
$\mathcal{A} = \dfrac{1}{2}(38.0)^2\left(\dfrac{13\pi}{108}\right) \approx 273$ m^2

27. The cities are at 28°N and 12°S.
12°S $= -12°$N, so
$\theta = 28° - (-12°) = 40° = 40\left(\dfrac{\pi}{180}\right) = \dfrac{2\pi}{9}$
radians
$s = r\theta = 6400\left(\dfrac{2\pi}{9}\right) \approx 4500$ km (rounded to
two significant digits)

29. $r = 2$, $s = 1.5$
$s = r\theta \Rightarrow 1.5 = 2\theta \Rightarrow \theta = \dfrac{1.5}{2} = \dfrac{3}{4}$ radian
$\mathcal{A} = \dfrac{1}{2}r^2\theta \Rightarrow$
$\mathcal{A} = \dfrac{1}{2}(2)^2\left(\dfrac{3}{4}\right) = \dfrac{1}{2}(4)\left(\dfrac{3}{4}\right) = \dfrac{3}{2} = 1.5$ sq units

31. $\tan\dfrac{\pi}{3}$
Converting $\dfrac{\pi}{3}$ radians to degrees, we have
$\dfrac{\pi}{3} = \dfrac{1}{3}(180°) = 60° \Rightarrow \tan\dfrac{\pi}{3} = \tan 60° = \sqrt{3}$

33. $\sin\left(-\dfrac{5\pi}{6}\right)$
$-\dfrac{5\pi}{6}$ is coterminal with $-\dfrac{5\pi}{6} + 2\pi = \dfrac{7\pi}{6}$.
$\dfrac{7\pi}{6}$ is in quadrant III, so the reference angle
is $\dfrac{7\pi}{6} - \pi = \dfrac{\pi}{6}$. In quadrant III, the sine is
negative.

(*continued on next page*)

(*continued*)

Thus, $\sin\left(-\dfrac{5\pi}{6}\right) = \sin\dfrac{7\pi}{6} = -\sin\dfrac{\pi}{6} = -\dfrac{1}{2}$

Converting $\dfrac{5\pi}{6}$ to degrees, we have

$\dfrac{7\pi}{6}\left(\dfrac{180°}{\pi}\right) = 210°$. The reference angle is

$210° - 180° = 30°$. Thus,

$\sin\left(-\dfrac{5\pi}{6}\right) = \sin\dfrac{7\pi}{6} = \sin 210°$

$= -\sin 30° = -\dfrac{1}{2}$

35. $\csc\left(-\dfrac{11\pi}{6}\right)$

$-\dfrac{11\pi}{6}$ is coterminal with

$-\dfrac{11\pi}{6} + 2\pi = -\dfrac{11\pi}{6} + \dfrac{12\pi}{6} = \dfrac{\pi}{6}$.

$\dfrac{\pi}{6}$ is in quadrant I, so

$\csc\left(-\dfrac{11\pi}{6}\right) = \csc\dfrac{\pi}{6} = 2.$

Converting $\dfrac{\pi}{6}$ to degrees, we have

$\dfrac{\pi}{6} = \dfrac{1}{6}(180°) = 30°$. Thus,

$\csc\left(-\dfrac{11\pi}{6}\right) = \csc\dfrac{\pi}{6} = \csc 30° = 2.$

37. $\sin 1.0472 \approx 0.8660$

39. $\cos(-0.2443) \approx 0.9703$

41. $\sec 7.3159 \approx 1.9513$

43. $\cos s = 0.9250 \Rightarrow s \approx 0.3898$

45. $\sin s = 0.4924 \Rightarrow s \approx 0.5148$

47. $\cot s = 0.5022 \Rightarrow s \approx 1.1054$

49. $\left[0, \dfrac{\pi}{2}\right]$, $\cos s = \dfrac{\sqrt{2}}{2}$

$\cos s = \dfrac{\sqrt{2}}{2}$, so the reference angle for s must

be $\dfrac{\pi}{4}$ because $\cos\dfrac{\pi}{4} = \dfrac{\sqrt{2}}{2}$. For s to be in the

interval $\left[0, \dfrac{\pi}{2}\right]$, s must be the reference angle.

Therefore, $s = \dfrac{\pi}{4}$.

51. $\left[\pi, \dfrac{3\pi}{2}\right]$, $\sec s = -\dfrac{2\sqrt{3}}{3}$

$\sec s = -\dfrac{2\sqrt{3}}{3}$, so the reference angle for s

must be $\dfrac{\pi}{6}$ because $\sec\dfrac{\pi}{6} = \dfrac{2\sqrt{3}}{3}$. For s to be

in the interval $\left[\pi, \dfrac{3\pi}{2}\right]$, we must add the

reference angle to π. Therefore,

$s = \pi + \dfrac{\pi}{6} = \dfrac{7\pi}{6}$.

53. $r = 15$ cm, $\omega = \dfrac{2\pi}{3}$ radians per sec, $t = 30$ sec

(a) $\omega = \dfrac{\theta}{t} \Rightarrow \dfrac{2\pi}{3} = \dfrac{\theta}{30} \Rightarrow$

$\theta = \dfrac{2\pi}{3}(30) = 20\pi$ radians

(b) $s = r\theta \Rightarrow s = 15 \cdot 20\pi = 300\pi$ cm

(c) $v = r\omega \Rightarrow v = 15\left(\dfrac{2\pi}{3}\right) = 10\pi$ cm per sec

55. The flywheel is rotating 90 times per sec or
$90(2\pi) = 180\pi$ radians per sec. Because
$r = 7$ cm, we have
$v = r\omega \Rightarrow v = 7(180\pi) = 1260\pi$ cm per sec

57. In the diagram, $a = 1.4$ in. and $b = 0.2$ in.

$$r = \frac{a^2 + b^2}{2b} = \frac{1.4^2 + 0.2^2}{2(0.2)}$$

$$= \frac{1.96 + 0.04}{0.4} = \frac{2.00}{0.4} = 5$$

Thus, the radius is 5 inches.

59. B; The amplitude is $|4| = 4$ and period is

$$\frac{2\pi}{2} = \pi.$$

61. $y = 2 \sin x$

Amplitude: 2
Period: 2π
Vertical translation: none
Phase shift: none

63. $y = -\frac{1}{2} \cos 3x$

Amplitude: $\left|-\frac{1}{2}\right| = \frac{1}{2}$

Period: $\frac{2\pi}{3}$

Vertical translation: none
Phase shift: none

65. $y = 1 + 2 \sin \frac{1}{4} x$

Amplitude: $|2| = 2$

Period: $\frac{2\pi}{\frac{1}{4}} = 8\pi$

Vertical translation: up 1 unit
Phase shift: none

67. $y = 3 \cos\left(x + \frac{\pi}{2}\right) = 3 \cos\left[x - \left(-\frac{\pi}{2}\right)\right]$

Amplitude: $|3| = 3$
Period: 2π
Vertical translation: none

Phase shift: $\frac{\pi}{2}$ units to the left

69. $y = \frac{1}{2} \csc\left(2x - \frac{\pi}{4}\right) = \frac{1}{2} \csc\left[2\left(x - \frac{\pi}{8}\right)\right]$

Amplitude: not applicable

Period: $\frac{2\pi}{2} = \pi$

Vertical translation: none

Phase shift: $\frac{\pi}{8}$ unit to the right

71. $y = \frac{1}{3} \tan\left(3x - \frac{\pi}{3}\right) = \frac{1}{3} \tan\left[3\left(x - \frac{\pi}{9}\right)\right]$

Amplitude: not applicable

Period: $\frac{\pi}{3}$

Vertical translation: none

Phase shift: $\frac{\pi}{9}$ unit to the right

73. The tangent function has a period of π and x-intercepts at integral multiples of π.

75. The cosine function has a period of 2π and has the value 0 when $x = \frac{\pi}{2}$.

77. The cotangent function has a period of π and decreases on the interval $(0, \pi)$.

79. $y = 3 \sin x$

Period: 2π and amplitude: $|3| = 3$

Divide the interval $[0, 2\pi]$ into four equal parts to get x-values that will yield minimum and maximum points and x-intercepts. Then make a table.

x	0	$\frac{\pi}{2}$	π	$\frac{3\pi}{2}$	2π
$\sin x$	0	1	0	-1	0
$3 \sin x$	0	3	0	-3	0

This table gives five values for graphing one period of $y = 3 \sin x$.

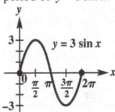

81. $y = -\tan x$

This is a reflection of the graph of $y = \tan x$ over the x-axis. The period is π and vertical

asymptotes are $x = -\frac{\pi}{2}$ and $x = \frac{\pi}{2}$.

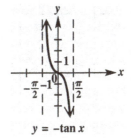

$y = -\tan x$

83. $y = 2 + \cot x$

This is the graph of $y = \cot x$ translated up 2 units. The period is π and the vertical asymptotes are $x = 0$ and $x = \pi$.

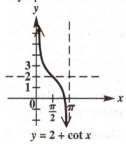

$y = 2 + \cot x$

85. $y = \sin 2x$

Period: $\dfrac{2\pi}{2} = \pi$ and amplitude: $|1| = 1$

Divide the interval $[0, \pi]$ into four equal parts to get the x-values that will yield minimum and maximum points and x-intercepts. Then make a table.

x	0	$\dfrac{\pi}{4}$	$\dfrac{\pi}{2}$	$\dfrac{3\pi}{4}$	π
$2x$	0	$\dfrac{\pi}{2}$	π	$\dfrac{3\pi}{2}$	2π
$\sin 2x$	0	1	0	-1	0

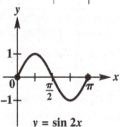

$y = \sin 2x$

87. $y = 3 \cos 2x$

Period: $\dfrac{2\pi}{2} = \pi$ and amplitude: $|3| = 3$

Divide the interval $[0, \pi]$ into four equal parts to get the x-values that will yield minimum and maximum points and x-intercepts. Then make a table.

x	0	$\dfrac{\pi}{4}$	$\dfrac{\pi}{2}$	$\dfrac{3\pi}{4}$	π
$2x$	0	$\dfrac{\pi}{2}$	π	$\dfrac{3\pi}{2}$	2π
$\cos 2x$	1	0	-1	0	1
$3\cos 2x$	3	0	-3	0	3

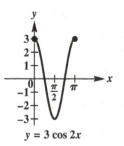

$y = 3 \cos 2x$

89. $y = \cos\left(x - \dfrac{\pi}{4}\right)$

Step 1: Find the interval whose length is $\dfrac{2\pi}{b}$.

$$0 \le x - \frac{\pi}{4} \le 2\pi \Rightarrow 0 + \frac{\pi}{4} \le x \le 2\pi + \frac{\pi}{4} \Rightarrow$$
$$\frac{\pi}{4} \le x \le \frac{9\pi}{4}$$

Step 2: Divide the period into four equal parts to get the following x-values: $\dfrac{\pi}{4}, \pi, \dfrac{3\pi}{2}, 2\pi,$ $\dfrac{9\pi}{4}$

Step 3: Evaluate the function for each of the five x-values.

x	$\dfrac{\pi}{4}$	$\dfrac{3\pi}{4}$	$\dfrac{5\pi}{4}$	$\dfrac{7\pi}{4}$	$\dfrac{9\pi}{4}$
$x - \dfrac{\pi}{4}$	0	$\dfrac{\pi}{2}$	π	$\dfrac{3\pi}{2}$	2π
$\cos\left(x - \dfrac{\pi}{4}\right)$	1	0	-1	0	1

Steps 4 and 5: Plot the points found in the table and join them with a sinusoidal curve.

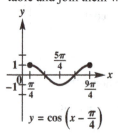

$y = \cos\left(x - \dfrac{\pi}{4}\right)$

The amplitude is 1. The period is 2π. There is no vertical translation. The phase shift is $\dfrac{\pi}{4}$ unit to the right.

91. $y = \sec\left(2x + \dfrac{\pi}{3}\right) = \sec\left[2\left(x + \dfrac{\pi}{6}\right)\right]$

Step 1: Graph the corresponding reciprocal

function $y = \cos\left[2\left(x + \dfrac{\pi}{6}\right)\right]$.

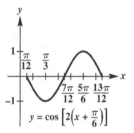

$y = \cos\left[2\left(x + \dfrac{\pi}{6}\right)\right]$

The period is $\dfrac{2\pi}{2} = \pi$, and its amplitude is

$|1| = 1$. One period is in the interval

$\dfrac{\pi}{12} \le x \le \dfrac{13\pi}{12}$.

Dividing the interval into four equal parts

gives the key points $\left(\dfrac{\pi}{12}, 0\right), \left(\dfrac{\pi}{3}, -1\right)$,

$\left(\dfrac{7\pi}{12}, 0\right), \left(\dfrac{5\pi}{6}, 1\right)$, and $\left(\dfrac{13\pi}{12}, 0\right)$.

Step 2: The vertical asymptotes of

$y = \sec 2\left(x + \dfrac{\pi}{6}\right)$ are at the x-intercepts of

$y = \cos 2\left(x + \dfrac{\pi}{6}\right)$, which are $x = \dfrac{\pi}{12}$,

$x = \dfrac{7\pi}{12}$, and $x = \dfrac{13\pi}{12}$.

Step 3: Sketch the graph.

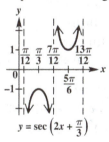

$y = \sec\left(2x + \dfrac{\pi}{3}\right)$

93. $y = 1 + 2\cos 3x$

Step 1: Find the interval whose length is $\dfrac{2\pi}{b}$.

$0 \le 3x \le 2\pi \Rightarrow 0 \le x \le \dfrac{2\pi}{3}$

Step 2: Divide the period into four equal parts
to get the following x-values:

$0, \dfrac{\pi}{6}, \dfrac{\pi}{3}, \dfrac{\pi}{2}, \dfrac{2\pi}{3}$

Step 3: Evaluate the function for each of the
five x-values.

x	0	$\dfrac{\pi}{6}$	$\dfrac{\pi}{3}$	$\dfrac{\pi}{2}$	$\dfrac{2\pi}{3}$
$3x$	0	$\dfrac{\pi}{2}$	π	$\dfrac{3\pi}{2}$	2π
$\cos 3x$	1	0	-1	0	1
$2\cos 3x$	2	0	-2	0	2
$1 + 2\cos 3x$	3	1	-1	1	3

Steps 4 and 5: Plot the points found in the
table and join them with a sinusoidal curve.

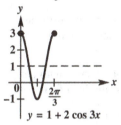

$y = 1 + 2\cos 3x$

The period is $\dfrac{2\pi}{3}$. The amplitude is $|2|$,

which is 2. The vertical translation is 1 unit
up. There is no phase shift.

95. $y = 2\sin \pi x$

Period: $\dfrac{2\pi}{\pi} = 2$ and amplitude: $|2| = 2$

Divide the interval [0, 2] into four equal parts
to get the x-values that will yield minimum
and maximum points and x-intercepts. Then
make a table.

x	0	$\dfrac{1}{2}$	1	$\dfrac{3}{2}$	2
πx	0	$\dfrac{\pi}{2}$	π	$\dfrac{3\pi}{2}$	2π
$\sin \pi x$	0	1	0	-1	0
$2\sin \pi x$	0	2	0	-2	0

Steps 4 and 5: Plot the points found in the
table and join them with a sinusoidal curve.

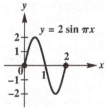

$y = 2\sin \pi x$

97. (a) See the graph in part (c).

(b)

$$y = 8.02\sin(0.52x + 0.84) + 59.83$$

(c)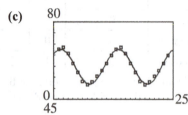

99. This is the graph of $y = \sin x$ reflected across the x-axis and translated 1 unit up. The equation is $y = -\sin x + 1$.

101. This is the graph of $y = \tan x$ with period 2π and stretched vertically by a factor of 2. The equation is $y = 2\tan\dfrac{1}{2}x$.

103. (a) The shorter leg of the right triangle has length $h_2 - h_1$. Thus, we have

$$\cot\theta = \frac{d}{h_2 - h_1} \Rightarrow d = (h_2 - h_1)\cot\theta$$

(b) When $h_2 = 55$ and $h_1 = 5$,

$$d = (55 - 5)\cot\theta = 50\cot\theta.$$

The period is π, but the graph wanted is d for $0 < \theta < \dfrac{\pi}{2}$. The asymptote is the line $\theta = 0$. Also, when

$$\theta = \frac{\pi}{4}, d = 50\cot\frac{\pi}{4} = 50(1) = 50.$$

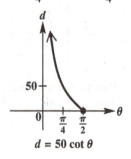

$$d = 50\cot\theta$$

105. $t = 60 - 30\cos\dfrac{x\pi}{6}$

(a) For January, $x = 0$. Thus,

$$t = 60 - 30\cos\frac{0\cdot\pi}{6} = 60 - 30\cos 0$$
$$= 60 - 30(1) = 60 - 30 = 30°F$$

(b) For April, $x = 3$. Thus,

$$t = 60 - 30\cos\frac{3\pi}{6} = 60 - 30\cos\frac{\pi}{2}$$
$$= 60 - 30(0) = 60 - 0 = 60°F$$

(c) For May, $x = 4$. Thus,

$$t = 60 - 30\cos\frac{4\pi}{6} = 60 - 30\cos\frac{2\pi}{3}$$
$$= 60 - 30\left(-\frac{1}{2}\right) = 60 + 15 = 75°F$$

(d) For June, $x = 5$. Thus,

$$t = 60 - 30\cos\frac{5\pi}{6} = 60 - 30\left(-\frac{\sqrt{3}}{2}\right)$$
$$= 60 + 15\sqrt{3} \approx 86°F$$

(e) For August, $x = 7$. Thus,

$$t = 60 - 30\cos\frac{7\pi}{6} = 60 - 30\left(-\frac{\sqrt{3}}{2}\right)$$
$$= 60 + 15\sqrt{3} \approx 86°F$$

(f) For October, $x = 9$. Thus,

$$t = 60 - 30\cos\frac{9\pi}{6} = 60 - 30\cos\frac{3\pi}{2}$$
$$= 60 - 30(0) = 60 - 0 = 60°F$$

107. $P(x) = 7(1 - \cos 2\pi x)(x + 10) + 100e^{0.2x}$

(a) January 1, base year $x = 0$
$$P(0) = 7(1 - \cos 0)(10) + 100e^0$$
$$= 7(1 - 1)(10) + 100(1)$$
$$= 7(0)(10) + 100 = 0 + 100 = 100$$

(b) July 1, base year $x = 0.5$
$$P(.5) = 7(1 - \cos\pi)(0.5 + 10) + 100e^{0.2(0.5)}$$
$$= 7[1 - (-1)](10.5) + 100e^{0.1}$$
$$= 7(2)(10.5) + 100e^{0.1}$$
$$= 147 + 100e^{0.1} \approx 258$$

(c) January 1, following year $x = 1$
$$P(1) = 7(1 - \cos 2\pi)(1 + 10) + 100e^{0.2}$$
$$= 7(1 - 1)(1 + 10) + 100e^{0.2}$$
$$= 7(0)(11) + 100e^2 = 0 + 100e^{0.2}$$
$$= 100e^{0.2} \approx 122$$

(d) July 1, following year $x = 1.5$

$$P(1.5) = 7(1 - \cos 3\pi)(1.5 + 10) + 100e^{0.2(1.5)}$$
$$= 7[1 - (-1)](11.5) + 100e^{0.3}$$
$$= 7(2)(11.5) + 100e^{0.3}$$
$$= 161 + +100e^{0.3} \approx 296$$

109. $s(t) = 4 \sin \pi t$

$a = 4, \omega = \pi$

amplitude $= |a| = 4$; period $= \dfrac{2\pi}{\omega} = \dfrac{2\pi}{\pi} = 2$

frequency $= \dfrac{\omega}{2\pi} = \dfrac{\pi}{2\pi} = \dfrac{1}{2}$ cycle per sec

111. The frequency is the number of cycles in one unit of time.

$$s(1.5) = 4 \sin 1.5\pi = 4 \sin \frac{3\pi}{2} = 4(-1) = -4$$

$$s(2) = 4 \sin 2\pi = 4(0) = 0$$

$$s(3.25) = 4 \sin 3.25\pi = 4 \sin \frac{13\pi}{4} = 4 \sin \frac{5\pi}{4}$$

$$= 4\left(-\frac{\sqrt{2}}{2}\right) = -2\sqrt{2}$$

Chapter 6   Test

1. $120° = 120\left(\dfrac{\pi}{180} \text{ radian}\right) = \dfrac{2\pi}{3}$ radians

2. $-45° = -45\left(\dfrac{\pi}{180} \text{ radian}\right) = -\dfrac{\pi}{4}$ radian

3. $5° = 5\left(\dfrac{\pi}{180} \text{ radian}\right) = \dfrac{\pi}{36} \approx 0.087$ radian

4. $\dfrac{3\pi}{4} = \dfrac{3\pi}{4}\left(\dfrac{180°}{\pi}\right) = 135°$

5. $-\dfrac{7\pi}{6} = -\dfrac{7\pi}{6}\left(\dfrac{180°}{\pi}\right) = -210°$

6. $4 = 4\left(\dfrac{180°}{\pi}\right) \approx 229.18°$

$= 229° + 0.18(60') = 229°11'$

7. $r = 150$ cm, $s = 200$ cm

(a) $s = r\theta \Rightarrow 200 = 150\theta \Rightarrow \theta = \dfrac{200}{150} = \dfrac{4}{3}$

(b) $\mathcal{A} = \dfrac{1}{2}r^2\theta$

$$\mathcal{A} = \frac{1}{2}(150)^2\left(\frac{4}{3}\right) = \frac{1}{2}(22,500)\left(\frac{4}{3}\right)$$
$$= 15,000 \text{ cm}^2$$

8. $r = \dfrac{1}{2}$ in., $s = 1$ in.

$$s = r\theta \Rightarrow 1 = \frac{1}{2}\theta \Rightarrow \theta = 2 \text{ radians}$$

For Exercises 9−14, refer to Figure 13 on page 579 of the text.

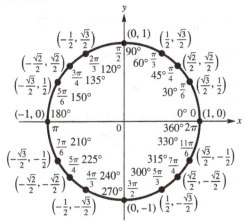

9. $\sin \dfrac{3\pi}{4}$

$\dfrac{3\pi}{4}$ is in quadrant II, so the reference angle is

$\pi - \dfrac{3\pi}{4} = \dfrac{4\pi}{4} - \dfrac{3\pi}{4} = \dfrac{\pi}{4}$. In quadrant II, the

sine is positive. Thus, $\sin \dfrac{3\pi}{4} = \sin \dfrac{\pi}{4} = \dfrac{\sqrt{2}}{2}$.

Converting $\dfrac{3\pi}{4}$ to degrees, we have

$\dfrac{3\pi}{4} = \dfrac{3}{4}(180°) = 135°$. The reference angle is

$180° - 135° = 45°$. Thus,

$$\sin \frac{3\pi}{4} = \sin 135° = \sin 45° = \frac{\sqrt{2}}{2}.$$

10. $\cos\left(-\dfrac{7\pi}{6}\right)$

$-\dfrac{7\pi}{6}$ is coterminal with

$-\dfrac{7\pi}{6}+2\pi=-\dfrac{7\pi}{6}+\dfrac{12\pi}{6}=\dfrac{5\pi}{6}.$

$\dfrac{5\pi}{6}$ is in quadrant II, so the reference angle is

$\pi-\dfrac{5\pi}{6}=\dfrac{6\pi}{6}-\dfrac{5\pi}{6}=\dfrac{\pi}{6}.$ In quadrant II, the cosine is negative. Thus,

$\cos\left(-\dfrac{7\pi}{6}\right)=\cos\dfrac{5\pi}{6}=-\cos\dfrac{\pi}{6}=-\dfrac{\sqrt{3}}{2}.$

Converting $\dfrac{5\pi}{6}$ to degrees, we have

$\dfrac{5\pi}{6}=\dfrac{5}{6}(180°)=150°.$

The reference angle is $180°-150°=30°.$

Thus, $\cos\left(-\dfrac{7\pi}{6}\right)=\cos\dfrac{5\pi}{6}=\cos150°$

$=-\cos30°=-\dfrac{\sqrt{3}}{2}$

11. $\tan\dfrac{3\pi}{2}=\tan270°$ is undefined.

12. $\sec\dfrac{8\pi}{3}$

$\dfrac{8\pi}{3}$ is coterminal with $\dfrac{8\pi}{3}-2\pi=\dfrac{2\pi}{3}.$

$\dfrac{2\pi}{3}$ is in quadrant II, so the reference angles

is $\pi-\dfrac{2\pi}{3}=\dfrac{\pi}{3}.$ In quadrant II, the secant is negative. Thus,

$\sec\dfrac{8\pi}{3}=\sec\dfrac{2\pi}{3}=-\sec\dfrac{\pi}{3}=-2.$

Converting $\dfrac{2\pi}{3}$ to degrees, we have

$\dfrac{2\pi}{3}=\dfrac{2\pi}{3}\cdot\dfrac{180°}{\pi}=120°.$ The reference angle

is $180°-120°=60°.$ Thus,

$\sec\dfrac{8\pi}{3}=\sec\dfrac{2\pi}{3}=\sec120°=-\sec60°=-2.$

13. $\tan\pi=\tan180°=0$

14. $\cos\dfrac{3\pi}{2}=\cos270°=0$

15. $s=\dfrac{7\pi}{6}$

$\dfrac{7\pi}{6}$ is in quadrant III, so the reference angle

is $\dfrac{7\pi}{6}-\pi=\dfrac{\pi}{6}.$ In quadrant III, the sine and cosine are negative.

$\sin\dfrac{7\pi}{6}=-\sin\dfrac{\pi}{6}=-\dfrac{1}{2}$

$\cos\dfrac{7\pi}{6}=-\cos\dfrac{\pi}{6}=-\dfrac{\sqrt{3}}{2}$

$\tan\dfrac{7\pi}{6}=\tan\dfrac{\pi}{6}=\dfrac{\sqrt{3}}{3}$

$\csc\dfrac{7\pi}{6}=\dfrac{1}{\sin\frac{7\pi}{6}}=\dfrac{1}{-\frac{1}{2}}=-2$

$\sec\dfrac{7\pi}{6}=\dfrac{1}{\cos\frac{7\pi}{6}}=\dfrac{1}{-\frac{\sqrt{3}}{2}}=-\dfrac{2}{\sqrt{3}}=-\dfrac{2\sqrt{3}}{3}$

$\cot\dfrac{7\pi}{6}=\dfrac{1}{\tan\frac{7\pi}{6}}=\dfrac{1}{\frac{\sqrt{3}}{3}}=\dfrac{3}{\sqrt{3}}=\dfrac{3\sqrt{3}}{3}=\sqrt{3}$

16. **(a)** $\sin s=0.8258\Rightarrow s\approx0.9716$

(b) $\cos\dfrac{\pi}{3}=\dfrac{1}{2}$ and $0\le\dfrac{\pi}{3}\le\dfrac{\pi}{2},$ so $s=\dfrac{\pi}{3}.$

17. **(a)** The speed of ray OP is $\omega=\dfrac{\pi}{12}$ radian per sec. $\omega=\dfrac{\theta}{t},$ then in 8 sec,

$\omega=\dfrac{\theta}{t}\Rightarrow\dfrac{\pi}{12}=\dfrac{\theta}{8}\Rightarrow\theta=\dfrac{8\pi}{12}=\dfrac{2\pi}{3}$ radians

(b) From part (a), P generates an angle of $\dfrac{2\pi}{3}$ radians in 8 sec. The distance traveled by P along the circle is

$s=r\theta\Rightarrow s=60\left(\dfrac{2\pi}{3}\right)=40\pi$ cm

(c) $v=\dfrac{s}{t}\Rightarrow\dfrac{40\pi}{8}=5\pi$ cm per sec.

18. (a)

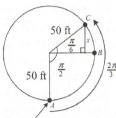

person loads here

Suppose the person takes a seat at point A. The person travels $\frac{\pi}{2}$ radians, the person is 50 ft above the ground. The person travels $\frac{\pi}{6}$ more radians, so let x be the additional vertical distance traveled.

$$\sin\frac{\pi}{6} = \frac{x}{50} \Rightarrow x = 50\sin\frac{\pi}{6} = 50\left(\frac{1}{2}\right) = 25$$

Thus, the person traveled an additional 25 ft above the ground, for a total of 75 ft above the ground.

(b) $t = 30$ sec and $\theta = \frac{2\pi}{3}$ radians, so

$$\omega = \frac{\theta}{t} \Rightarrow \omega = \frac{\frac{2\pi}{3}}{30} = \frac{2\pi}{3} \cdot \frac{1}{30} = \frac{\pi}{45} \text{ radian}$$

per sec. Thus, the Ferris wheel turned at the rate of $\frac{\pi}{45}$ radian per second.

19. (a) $y = \sec x$ **(b)** $y = \sin x$

(c) $y = \cos x$ **(d)** $y = \tan x$

(e) $y = \csc x$ **(e)** $y = \cot x$

20. (a) This is a cosine curve with period 4π, so

$$4\pi = \frac{2\pi}{b} \Rightarrow b = \frac{2\pi}{4\pi} = \frac{1}{2}.$$ The graph has

been shifted 1 unit up, so an equation is $y = 1 + \cos\frac{1}{2}x$.

(b) This is a cotangent curve that has been reflected across the y-axis. Since the graph passes through $\left(\frac{\pi}{4}, -\frac{1}{2}\right)$, the graph has been compressed vertically by a factor of $\frac{1}{2}$. Thus, an equation of the graph is $y = -\frac{1}{2}\cot x$.

21. (a) The domain of the cosine function is $(-\infty, \infty)$.

(b) The range of the sine function is $[-1, 1]$.

(c) The least positive value for which the tangent function is undefined is $\frac{\pi}{2}$.

(d) The range of the secant function is $(-\infty, -1] \cup [1, \infty)$.

22. $y = 3 - 6\sin\left(2x + \frac{\pi}{2}\right) = 3 - 6\sin\left[2\left(x + \frac{\pi}{4}\right)\right]$

$$= 3 - 6\sin\left[2\left[x - \left(-\frac{\pi}{4}\right)\right]\right]$$

(a) The period is $\frac{2\pi}{2} = \pi$.

(b) The amplitude is 6.

(c) The range is $[-3, 9]$.

(d) The y-intercept occurs when $x = 0$.

$$-6\sin\left(2\cdot 0 + \frac{\pi}{2}\right) + 3 = -6\sin\left(0 + \frac{\pi}{2}\right) + 3$$

$$= -6\sin\left(\frac{\pi}{2}\right) + 3$$

$$= -6(1) + 3 = -3$$

The y-intercept is (0, 3).

(e) The phase shift is $\frac{\pi}{4}$ unit to the left

$$\left(\text{that is, } -\frac{\pi}{4}\right)$$

23. $y = \sin(2x + \pi) = \sin\left[2\left(x + \frac{\pi}{2}\right)\right]$

$$= \sin\left[2\left(x - \left(-\frac{\pi}{2}\right)\right)\right]$$

Step 1: Find the interval whose length is $\frac{2\pi}{b}$.

$$0 \le 2\left(x + \frac{\pi}{2}\right) \le 2\pi \Rightarrow 0 \le x + \frac{\pi}{2} \le \frac{2\pi}{2} \Rightarrow$$

$$0 \le x + \frac{\pi}{2} \le \pi \Rightarrow -\frac{\pi}{2} \le x \le \frac{\pi}{2}$$

Step 2: Divide the period into four equal parts to get the x-values: $-\frac{\pi}{2}, -\frac{\pi}{4}, 0, \frac{\pi}{4}, \frac{\pi}{2}$

Step 3: Evaluate the function for each of the five x-values

(continued on next page)

(continued)

x	$-\dfrac{\pi}{2}$	$-\dfrac{\pi}{4}$	0	$\dfrac{\pi}{4}$	$\dfrac{\pi}{2}$
$x+\dfrac{\pi}{2}$	0	$\dfrac{\pi}{4}$	$\dfrac{\pi}{2}$	$\dfrac{3\pi}{4}$	π
$2\left(x+\dfrac{\pi}{2}\right)$	0	$\dfrac{\pi}{2}$	π	$\dfrac{3\pi}{2}$	2π
$\sin\left[2\left(x+\dfrac{\pi}{2}\right)\right]$	0	1	0	-1	0

Steps 4 and 5: Plot the points found in the table and join them with a sinusoidal curve.

$$y = \sin(2x + \pi)$$

The period is π. There is no vertical translation. The phase shift is $\dfrac{\pi}{2}$ units to the right.

24. $y = 2 + \cos x$

This is the graph of $y = \cos x$ translated vertically 2 units up.

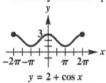

$$y = 2 + \cos x$$

25. $y = -1 + 2\sin(x + \pi)$

Step 1: Find the interval whose length is $\dfrac{2\pi}{b}$.

$0 \le x + \pi \le 2\pi \Rightarrow -\pi \le x \le \pi$

Step 2: Divide the period into four equal parts to get the x-values: $-\pi, -\dfrac{\pi}{2}, 0, \dfrac{\pi}{2}, \pi$

Step 3: Evaluate the function for each of the five x-values

x	$-\pi$	$-\dfrac{\pi}{2}$	0	$\dfrac{\pi}{2}$	π
$x+\pi$	0	$\dfrac{\pi}{2}$	π	$\dfrac{3\pi}{2}$	2π
$\sin(x+\pi)$	0	1	0	-1	0
$2\sin(x+\pi)$	0	2	0	-2	0
$-1+2\sin(x+\pi)$	-1	1	-1	-3	-1

Steps 4 and 5: Plot the points found in the table and join them with a sinusoidal curve. Repeat this cycle for the interval $[-\pi, 0]$.

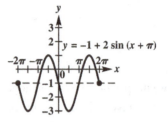

$$y = -1 + 2\sin(x + \pi)$$

The amplitude is $|2|$, which is 2. The period is 2π. The vertical translation is 1 unit down. The phase shift is π units to the left.

26. $y = \tan\left(x - \dfrac{\pi}{2}\right)$

Period: π
Vertical translation: none

Phase shift (horizontal translation): $\dfrac{\pi}{2}$ units to the right

Because the function is to be graphed over a two-period interval, locate three adjacent vertical asymptotes. Asymptotes of the graph $y = \tan x$ occur at $-\dfrac{\pi}{2}$, and $\dfrac{\pi}{2}$, so use the following equations to locate asymptotes:

$x - \dfrac{\pi}{2} = -\dfrac{\pi}{2} \Rightarrow x = 0$ and

$x - \dfrac{\pi}{2} = \dfrac{\pi}{2} \Rightarrow x = \pi$.

Divide the interval $(0, \pi)$ into four equal parts to obtain the key x-values: first-quarter value: $\dfrac{\pi}{4}$; middle value: $\dfrac{\pi}{2}$; third-quarter value: $\dfrac{3\pi}{4}$

Evaluating the given function at these three key x-values gives the points: $\left(\dfrac{\pi}{4}, -1\right)$, $\left(\dfrac{\pi}{2}, 0\right)$, $\left(\dfrac{3\pi}{4}, 1\right)$. Connect these points with a smooth curve and continue to graph to approach the asymptote $x = 0$ and $x = \pi$ to complete one period of the graph. Repeat this cycle for the interval $[-\pi, 0]$.

(continued on next page)

(*continued*)

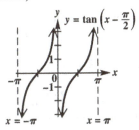

$$y = \tan\left(x - \frac{\pi}{2}\right)$$

27. $y = -2 - \cot\left(x - \frac{\pi}{2}\right)$

Period: $\dfrac{\pi}{b} = \dfrac{\pi}{1} = \pi$

Vertical translation: 2 units down

Phase shift (horizontal translation): $\dfrac{\pi}{2}$ units to the right

Because the function is to be graphed over a two-period interval, locate three adjacent vertical asymptotes. Asymptotes of the graph $y = \cot x$ occur at multiples of π, so use the following equations to locate asymptotes:

$x - \dfrac{\pi}{2} = -\pi,\ x - \dfrac{\pi}{2} = 0,$ and $x - \dfrac{\pi}{2} = \pi$.

Solve each of these equations:

$x - \dfrac{\pi}{2} = -\pi \Rightarrow x = -\dfrac{\pi}{2}$

$x - \dfrac{\pi}{2} = 0 \Rightarrow x = \dfrac{\pi}{2}$

$x - \dfrac{\pi}{2} = \pi \Rightarrow x = \dfrac{3\pi}{2}$

Divide the interval $\left(-\dfrac{\pi}{2}, \dfrac{\pi}{2}\right)$ into four equal parts to obtain the following key x-values.

first-quarter value: $-\dfrac{\pi}{4}$ middle value: 0 ;

third-quarter value: $\dfrac{\pi}{4}$

Evaluating the given function at these three key x-values gives the points: $\left(-\dfrac{\pi}{4}, -3\right)$,

$(0, -2),\ \left(\dfrac{\pi}{4}, -1\right)$

Connect these points with a smooth curve and continue to graph to approach the asymptote

$x = -\dfrac{\pi}{2}$ and $x = \dfrac{\pi}{2}$ to complete one period of

the graph. Sketch the identical curve between

the asymptotes $x = \dfrac{\pi}{2}$ and $x = \dfrac{3\pi}{2}$ to

complete a second period of the graph.

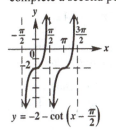

$$y = -2 - \cot\left(x - \frac{\pi}{2}\right)$$

28. $y = -\csc 2x$

Step 1: Graph the corresponding reciprocal

function $y = -\sin 2x$ The period is $\dfrac{2\pi}{2} = \pi$

and its amplitude is $|-1| = 1$. One period is in

the interval $0 \le x \le \pi$. Dividing the interval into four equal parts gives us the key points:

$(0, 0),\ \left(\dfrac{\pi}{2}, -1\right),\ \left(\dfrac{\pi}{2}, 0\right),\ \left(\dfrac{3\pi}{4}, 1\right),\ (\pi, 0)$

Step 2: The vertical asymptotes of
$y = -\csc 2x$ are at the x-intercepts of

$y = -\sin 2x$, which are $x = 0, x = \dfrac{\pi}{2}$, and

$x = \pi$. Continuing this pattern to the right, we

also have a vertical asymptotes of $x = \dfrac{3\pi}{2}$

and $x = 2\pi$.

Step 3: Sketch the graph.

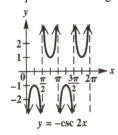

$$y = -\csc 2x$$

29. (a)

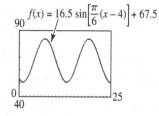

$f(x) = 16.5 \sin\left[\frac{\pi}{6}(x-4)\right] + 67.5$

(b) Amplitude: 16.5

Period: $\dfrac{2\pi}{\frac{\pi}{6}} = 2\pi \cdot \dfrac{6}{\pi} = 12$;

Phase shift: 4 units to the right
Vertical translation: 67.5 units up

(c) For the month of December, $x = 12$.

$$f(12) = 16.5 \sin\left[\frac{\pi}{6}(12-4)\right] + 67.5$$

$$= 16.5 \sin\left(\frac{4\pi}{3}\right) + 67.5$$

$$= 16.5\left(-\frac{\sqrt{3}}{2}\right) + 67.5 \approx 53°F$$

(d) Examining the graph shows that a minimum of 51 occurs at $x = 13 = 12 + 1$ implies a minimum average monthly temperature of 51°F in January.
A maximum of 84 occurs at $x = 7$ and $x = 19 = 12 + 7$ a maximum average monthly temperature of 84°F in July.

(e) The average annual temperature is about 67.5°F. This is the vertical translation.

30. $s(t) = -4\cos 8\pi t$, $a = |-4| = 4$, $\omega = 8\pi$

(a) maximum height = amplitude
$$= a = |-4| = 4 \text{ in.}$$

(b) $s(t) = -4\cos 8\pi t = 4 \Rightarrow \cos 8\pi t = -1 \Rightarrow$

$$8\pi t = \pi \Rightarrow t = \frac{1}{8}$$

The weight first reaches its maximum height after $\dfrac{1}{8}$ sec.

(c) frequency $= \dfrac{\omega}{2\pi} = \dfrac{8\pi}{2\pi} = 4$ cycles per sec;

period $= \dfrac{2\pi}{\omega} = \dfrac{2\pi}{8\pi} = \dfrac{1}{4}$ sec

Chapter 7

Trigonometric Identities and Equations

Section 7.1   Fundamental Identities

1. B; $\dfrac{\cos x}{\sin x} = \cot x$

3. E; $\cos(-x) = \cos x$

5. A; $1 = \sin^2 x + \cos^2 x$

7. By a negative angle identity, $\cos(-\theta) = \cos\theta$. Thus, if $\cos\theta = -0.65$, then $\cos(-\theta) = \underline{-0.65}$.

9. By a quotient identity, $\tan\theta = \dfrac{\sin\theta}{\cos\theta}$. By a negative angle identity $\sin(-\theta) = -\sin\theta$ and $\cos(-\theta) = \cos\theta$. Thus, if $\cos x = 0.8$ and $\sin x = 0.6$, then
$$\tan(-x) = \frac{\sin(-x)}{\cos(-x)} = \frac{-\sin x}{\cos x} = \frac{-0.6}{0.8} = -0.75.$$

11. $\cos\theta = \dfrac{3}{4}$, θ is in quadrant I.

    An identity that relates sine and cosine is $\sin^2\theta + \cos^2\theta = 1$.
    $$\sin^2\theta + \cos^2\theta = 1 \Rightarrow \sin^2\theta + \left(\frac{3}{4}\right)^2 = 1 \Rightarrow$$
    $$\sin^2\theta = 1 - \frac{9}{16} = \frac{7}{16} \Rightarrow \sin\theta = \pm\frac{\sqrt{7}}{4}$$
    θ is in quadrant I, so $\sin\theta = \dfrac{\sqrt{7}}{4}$.

13. $\cot\theta = -\dfrac{1}{5}$, θ in quadrant IV

    Use the identity $1 + \cot^2\theta = \csc^2\theta$ because $\sin\theta = \dfrac{1}{\csc\theta}$.
    $$1 + \cot^2\theta = \csc^2\theta \Rightarrow 1 + \left(-\frac{1}{5}\right)^2 = \csc^2\theta \Rightarrow$$
    $$1 + \frac{1}{25} = \csc^2\theta \Rightarrow \frac{26}{25} = \csc^2\theta \Rightarrow \csc\theta = \pm\frac{\sqrt{26}}{5}$$

θ is in quadrant IV, so $\csc\theta < 0$, so $\csc\theta = -\dfrac{\sqrt{26}}{5}$. Thus,
$$\sin\theta = \frac{1}{\csc\theta} = -\frac{5}{\sqrt{26}} = -\frac{5}{\sqrt{26}} \cdot \frac{\sqrt{26}}{\sqrt{26}} = -\frac{5\sqrt{26}}{26}$$

15. $\cos(-\theta) = \dfrac{\sqrt{5}}{5}$, $\tan\theta < 0$

    Because $\cos(-\theta) = \dfrac{\sqrt{5}}{5}$, we have $\cos\theta = \dfrac{\sqrt{5}}{5}$ by a negative angle identity. An identity that relates sine and cosine is $\sin^2\theta + \cos^2\theta = 1$.
    $$\sin^2\theta + \cos^2\theta = 1 \Rightarrow \sin^2\theta + \left(\frac{\sqrt{5}}{5}\right)^2 = 1 \Rightarrow$$
    $$\sin^2\theta + \frac{5}{25} = 1 \Rightarrow \sin^2\theta = 1 - \frac{5}{25} = 1 - \frac{1}{5} = \frac{4}{5} \Rightarrow$$
    $$\sin\theta = \pm\frac{2}{\sqrt{5}} = \pm\frac{2}{\sqrt{5}} \cdot \frac{\sqrt{5}}{\sqrt{5}} = \pm\frac{2\sqrt{5}}{5}$$
    Because $\tan\theta < 0$ and $\cos\theta > 0$, θ is in quadrant IV and $\sin\theta < 0$. Thus,
    $$\sin\theta = -\frac{2\sqrt{5}}{5}.$$

17. $\tan\theta = -\dfrac{\sqrt{6}}{2}$, $\cos\theta > 0$
    $$\tan^2\theta + 1 = \sec^2\theta \Rightarrow \left(-\frac{\sqrt{6}}{2}\right)^2 + 1 = \sec^2\theta \Rightarrow$$
    $$\frac{6}{4} + 1 = \frac{10}{4} = \frac{5}{2} = \sec^2\theta \Rightarrow$$
    $$\sec\theta = \pm\sqrt{\frac{5}{2}} = \pm\frac{\sqrt{5}}{\sqrt{2}} \cdot \frac{\sqrt{2}}{\sqrt{2}} = \pm\frac{\sqrt{10}}{2}$$
    Because $\cos\theta = \dfrac{1}{\sec\theta}$ and $\cos\theta > 0$,
    $$\cos\theta = \frac{1}{\sqrt{10}} = \frac{2}{\sqrt{10}} = \frac{2}{\sqrt{10}} \cdot \frac{\sqrt{10}}{\sqrt{10}}$$
    $$= \frac{2\sqrt{10}}{10} = \frac{\sqrt{10}}{5}.$$

    (continued on next page)

(*continued*)

Now, use the identity $\sin^2\theta + \cos^2\theta = 1$:

$$\sin^2\theta + \left(\frac{\sqrt{10}}{5}\right)^2 = 1 \Rightarrow \sin^2\theta + \frac{10}{25} = 1 \Rightarrow$$

$$\sin^2\theta = 1 - \frac{10}{25} = 1 - \frac{2}{5} = \frac{3}{5} \Rightarrow$$

$$\sin\theta = \pm\sqrt{\frac{3}{5}} = \pm\frac{\sqrt{3}}{\sqrt{5}} \cdot \frac{\sqrt{5}}{\sqrt{5}} = \pm\frac{\sqrt{15}}{5}$$

Because $\tan\theta < 0$ and $\cos\theta > 0$, θ is in quadrant IV and $\sin\theta < 0$. Thus,

$$\sin\theta = -\frac{\sqrt{15}}{5}.$$

19. $\sec\theta = \frac{11}{4}$, $\cot\theta < 0$

Because $\cos\theta = \frac{1}{\sec\theta}$, $\cos\theta = \frac{1}{\frac{11}{4}} = \frac{4}{11}$. Use

the identity $\sin^2\theta + \cos^2\theta = 1$ to obtain

$$\sin^2\theta + \cos^2\theta = 1 \Rightarrow \sin^2\theta + \left(\frac{4}{11}\right)^2 = 1 \Rightarrow$$

$$\sin^2\theta + \frac{16}{121} = 1 \Rightarrow \sin^2\theta = 1 - \frac{16}{121} \Rightarrow$$

$$\sin^2\theta = \frac{105}{121} \Rightarrow \sin\theta = \pm\frac{\sqrt{105}}{11}$$

Because $\cot\theta < 0$ and $\sec\theta > 0$, θ is in quadrant IV and $\sin\theta < 0$. Thus,

$$\sin\theta = -\frac{\sqrt{105}}{11}.$$

21. $\csc\theta = -\frac{9}{4}$

$$\sin\theta = \frac{1}{\csc\theta}, \text{ so } \sin\theta = \frac{1}{-\frac{9}{4}} = -\frac{4}{9}.$$

23. Because $\sin\theta = \frac{1}{\csc\theta}$, the sign of $\sin\theta$ will

be the same as $\csc\theta$.

25. $f(x) = \frac{\sin x}{x}$

$$f(-x) = \frac{\sin(-x)}{-x} = \frac{-\sin x}{-x} = \frac{\sin x}{x}$$

Because $f(x) = f(-x)$, the function is even.

27. This is the graph of $f(x) = \sec x$. It is symmetric about the y-axis.

$$f(-x) = \sec(-x) = \frac{1}{\cos(-x)} = \frac{1}{\cos x}$$

$$= \sec x = f(x)$$

Because $f(x) = f(-x)$, the function is even.

29. This is the graph of $f(x) = \cot x$. It is symmetric about the origin.

$$f(-x) = \cot(-x) = \frac{\cos(-x)}{\sin(-x)} = \frac{\cos x}{-\sin x}$$

$$= -\frac{\cos x}{\sin x} = -\cot x = -f(x)$$

Because $f(x) = -f(x)$, the function is odd.

31. $\sin\theta = \frac{2}{3}$, θ in quadrant II

θ is in quadrant II, so the sine and cosecant function values are positive. The cosine, tangent, cotangent, and secant function values are negative.

$$\sin^2\theta + \cos^2\theta = 1 \Rightarrow$$

$$\cos^2\theta = 1 - \sin^2\theta = 1 - \left(\frac{2}{3}\right)^2 = 1 - \frac{4}{9} = \frac{5}{9} \Rightarrow$$

$$\cos\theta = -\frac{\sqrt{5}}{3}, \text{ since } \cos\theta < 0$$

$$\tan\theta = \frac{\sin\theta}{\cos\theta} = \frac{\frac{2}{3}}{-\frac{\sqrt{5}}{3}} = -\frac{2}{\sqrt{5}}$$

$$= -\frac{2}{\sqrt{5}} \cdot \frac{\sqrt{5}}{\sqrt{5}} = -\frac{2\sqrt{5}}{5}$$

$$\cot\theta = \frac{1}{\tan\theta} = \frac{1}{-\frac{2}{\sqrt{5}}} = -\frac{\sqrt{5}}{2}$$

$$\sec\theta = \frac{1}{\cos\theta} = \frac{1}{-\frac{\sqrt{5}}{3}} = -\frac{3}{\sqrt{5}}$$

$$= -\frac{3}{\sqrt{5}} \cdot \frac{\sqrt{5}}{\sqrt{5}} = -\frac{3\sqrt{5}}{5}$$

$$\csc\theta = \frac{1}{\sin\theta} = \frac{1}{\frac{2}{3}} = \frac{3}{2}$$

33. $\tan\theta = -\dfrac{1}{4}, \theta$ in quadrant IV

θ is in quadrant IV, so the cosine and secant function values are positive. The sine, tangent, cotangent, and cosecant function values are negative.

$$\cot\theta = \frac{1}{\tan\theta} = \frac{1}{-\frac{1}{4}} = -4$$

$$\sec^2\theta = 1 + \tan^2\theta = 1 + \left(-\frac{1}{4}\right)^2$$

$$= 1 + \frac{1}{16} = \frac{17}{16} \Rightarrow$$

$$\sec\theta = \frac{\sqrt{17}}{4}, \text{ since } \sec\theta > 0$$

$$\cos\theta = \frac{1}{\sec\theta} = \frac{1}{\frac{\sqrt{17}}{4}} = \frac{4}{\sqrt{17}}$$

$$= \frac{4}{\sqrt{17}} \cdot \frac{\sqrt{17}}{\sqrt{17}} = \frac{4\sqrt{17}}{17}$$

$$\sin^2\theta + \cos^2\theta = 1 \Rightarrow$$

$$\sin^2\theta = 1 - \cos^2\theta = 1 - \left(\frac{4}{\sqrt{17}}\right)^2$$

$$\sin^2\theta = 1 - \frac{16}{17} = \frac{1}{17} \Rightarrow$$

$$\sin\theta = -\frac{1}{\sqrt{17}} = -\frac{1}{\sqrt{17}} \cdot \frac{\sqrt{17}}{\sqrt{17}} = -\frac{\sqrt{17}}{17},$$

since $\sin\theta < 0$

$$\csc\theta = \frac{1}{\sin\theta} = \frac{1}{-\frac{1}{\sqrt{17}}} = -\sqrt{17}$$

35. $\cot\theta = \dfrac{4}{3}, \sin\theta > 0$

$\cot\theta > 0$ and $\sin\theta > 0$, so θ is in quadrant I and all the function values are positive.

$$\tan = \frac{1}{\cot\theta} = \frac{1}{\frac{4}{3}} = \frac{3}{4}$$

$$\sec^2\theta = 1 + \tan^2\theta = 1 + \left(\frac{3}{4}\right)^2 = 1 + \frac{9}{16} = \frac{25}{16} \Rightarrow$$

$$\sec\theta = \frac{5}{4}, \text{ since } \sec\theta > 0$$

$$\cos\theta = \frac{1}{\sec\theta} = \frac{1}{\frac{5}{4}} = \frac{4}{5}$$

$$\sin^2\theta = 1 - \cos^2\theta = 1 - \left(\frac{4}{5}\right)^2 = 1 - \frac{16}{25} = \frac{9}{25} \Rightarrow$$

$$\sin\theta = \frac{3}{5}, \text{ because } \sin\theta > 0$$

$$\csc\theta = \frac{1}{\sin\theta} = \frac{1}{\frac{3}{5}} = \frac{5}{3}$$

37. $\sec\theta = \dfrac{4}{3}, \sin\theta < 0$

$\sec\theta > 0$ and $\sin\theta < 0$, so θ is in quadrant IV and the cosine function value is positive. The tangent, cotangent, and cosecant function values are negative.

$$\cos\theta = \frac{1}{\sec\theta} = \frac{1}{\frac{4}{3}} = \frac{3}{4}$$

$$\sin^2\theta = 1 - \cos^2\theta = 1 - \left(\frac{3}{4}\right)^2 = 1 - \frac{9}{16} = \frac{7}{16} \Rightarrow$$

$$\sin\theta = -\frac{\sqrt{7}}{4}, \text{ since } \sin\theta < 0$$

$$\tan\theta = \frac{\sin\theta}{\cos\theta} = \frac{-\frac{\sqrt{7}}{4}}{\frac{3}{4}} = -\frac{\sqrt{7}}{3}$$

$$\cot\theta = \frac{1}{\tan\theta} = \frac{1}{-\frac{\sqrt{7}}{3}} = -\frac{3}{\sqrt{7}} \cdot \frac{\sqrt{7}}{\sqrt{7}} = -\frac{3\sqrt{7}}{7}$$

$$\csc\theta = \frac{1}{\sin\theta} = \frac{1}{-\frac{\sqrt{7}}{4}} = -\frac{4}{\sqrt{7}} \cdot \frac{\sqrt{7}}{\sqrt{7}} = -\frac{4\sqrt{7}}{7}$$

39. C

$$-\tan x\cos x = -\frac{\sin x}{\cos x} \cdot \cos x$$
$$= -\sin x = \sin(-x)$$

41. E; $\dfrac{\sec x}{\csc x} = \dfrac{\frac{1}{\cos x}}{\frac{1}{\sin x}} = \dfrac{\sin x}{\cos x} = \tan x$

43. B; $\cos^2 x = \dfrac{1}{\sec^2 x}$

45. Find $\sin\theta$ if $\cos\theta = \dfrac{x}{x+1}$.

$\sin^2\theta + \cos^2\theta = 1$ and $\cos\theta = \dfrac{x}{x+1}$, so

$$\sin^2\theta = 1 - \cos^2\theta = 1 - \left(\frac{x}{x+1}\right)^2$$

$$= 1 - \frac{x^2}{(x+1)^2} = \frac{(x+1)^2 - x^2}{(x+1)^2}$$

$$= \frac{x^2 + 2x + 1 - x^2}{(x+1)^2} = \frac{2x+1}{(x+1)^2}$$

Thus, $\sin\theta = \dfrac{\pm\sqrt{2x+1}}{x+1}$.

47. $\sin^2 x + \cos^2 x = 1 \Rightarrow \sin^2 x = 1 - \cos^2 x \Rightarrow$

$\sin x = \pm\sqrt{1 - \cos^2 x}$

49. $\tan^2 x + 1 = \sec^2 x \Rightarrow \tan^2 x = \sec^2 x - 1 \Rightarrow$

$\tan x = \pm\sqrt{\sec^2 x - 1}$

51. $\csc x = \dfrac{1}{\sin x} \Rightarrow$

$\csc x = \dfrac{1}{\pm\sqrt{1 - \cos^2 x}}$

$= \dfrac{\pm 1}{\sqrt{1 - \cos^2 x}} \cdot \dfrac{\sqrt{1 - \cos^2 x}}{\sqrt{1 - \cos^2 x}}$

$= \dfrac{\pm\sqrt{1 - \cos^2 x}}{1 - \cos^2 x}$

For exercises 53–77, there may be more than one possible answer.

53. $\cot\theta \sin\theta = \dfrac{\cos\theta}{\sin\theta} \cdot \sin\theta = \cos\theta$

55. $\sec\theta \cot\theta \sin\theta = \dfrac{1}{\cos\theta} \cdot \dfrac{\cos\theta}{\sin\theta} \cdot \dfrac{\sin\theta}{1}$

$= \dfrac{\sin\theta \cos\theta}{\cos\theta \sin\theta} = 1$

57. $\cos\theta \csc\theta = \cos\theta \cdot \dfrac{1}{\sin\theta} = \dfrac{\cos\theta}{\sin\theta} = \cot\theta$

59. $\sin^2\theta \left(\csc^2\theta - 1\right) = \sin^2\theta \left(\dfrac{1}{\sin^2\theta} - 1\right)$

$= \dfrac{\sin^2\theta}{\sin^2\theta} - \sin^2\theta$

$= 1 - \sin^2\theta = \cos^2\theta$

61. $(1 - \cos\theta)(1 + \sec\theta)$

$= 1 + \sec\theta - \cos\theta - \cos\theta\sec\theta$

$= 1 + \sec\theta - \cos\theta - \cos\theta\left(\dfrac{1}{\cos\theta}\right)$

$= 1 + \sec\theta - \cos\theta - 1 = \sec\theta - \cos\theta$

63. $\dfrac{1 + \tan(-\theta)}{\tan(-\theta)} = \dfrac{1 - \tan\theta}{-\tan\theta} = \dfrac{1}{-\tan\theta} + \dfrac{-\tan\theta}{-\tan\theta}$

$= -\cot\theta + 1$

65. $\dfrac{1 - \cos^2(-\theta)}{1 + \tan^2(-\theta)} = \dfrac{\sin^2(-\theta)}{\sec^2(-\theta)} = \dfrac{\sin^2(-\theta)}{\dfrac{1}{\cos^2(-\theta)}}$

$= \sin^2(-\theta)\cos^2(-\theta)$

$= \sin^2\theta\cos^2\theta$

67. $\sec\theta - \cos\theta = \dfrac{1}{\cos\theta} - \cos\theta = \dfrac{1}{\cos\theta} - \dfrac{\cos^2\theta}{\cos\theta}$

$= \dfrac{1 - \cos^2\theta}{\cos\theta} = \dfrac{\sin^2\theta}{\cos\theta}$

$= \dfrac{\sin\theta}{\cos\theta} \cdot \sin\theta = \tan\theta \sin\theta$

Alternatively, we have

$\dfrac{\cos^2\theta}{\sin\theta} = \cot\theta\cos\theta$

69. $(\sec\theta + \csc\theta)(\cos\theta - \sin\theta)$

$= \left(\dfrac{1}{\cos\theta} + \dfrac{1}{\sin\theta}\right)(\cos\theta - \sin\theta)$

$= \dfrac{1}{\cos\theta}(\cos\theta) - \dfrac{1}{\cos\theta}(\sin\theta)$

$+ \dfrac{1}{\sin\theta}(\cos\theta) - \dfrac{1}{\sin\theta}(\sin\theta)$

$= 1 - \dfrac{\sin\theta}{\cos\theta} + \dfrac{\cos\theta}{\sin\theta} - 1 = -\tan\theta + \cot\theta$

$= \cot\theta - \tan\theta$

71. $\sin\theta(\csc\theta - \sin\theta) = \sin\theta\csc\theta - \sin^2\theta$

$= \sin\theta \cdot \dfrac{1}{\sin\theta} - \sin^2\theta$

$= 1 - \sin^2\theta = \cos^2\theta$

73. $\dfrac{1 + \tan^2\theta}{1 + \cot^2\theta} = \dfrac{\sec^2\theta}{\csc^2\theta} = \dfrac{\dfrac{1}{\cos^2\theta}}{\dfrac{1}{\sin^2\theta}}$

$= \dfrac{1}{\cos^2\theta} \cdot \dfrac{\sin^2\theta}{1} = \dfrac{\sin^2\theta}{\cos^2\theta} = \tan^2\theta$

75. $\dfrac{\csc\theta}{\cot(-\theta)} = \dfrac{\csc\theta}{-\cot\theta} = \dfrac{\dfrac{1}{\sin\theta}}{-\dfrac{\cos\theta}{\sin\theta}} = -\dfrac{1}{\cos\theta}$

$= -\sec\theta$

77. $\sin^2(-\theta) + \tan^2(-\theta) + \cos^2(-\theta)$

$= \left[\sin^2(-\theta) + \cos^2(-\theta)\right] + \tan^2(-\theta)$

$= 1 + \tan^2(-\theta) = 1 + (-\tan\theta)^2$

$= 1 + \tan^2\theta = \sec^2\theta$

79. $\cos x = \dfrac{1}{5}$, so x is in quadrant I or quadrant IV.

$$\sin x = \pm\sqrt{1-\cos^2 x} = \pm\sqrt{1-\left(\dfrac{1}{5}\right)^2} = \pm\sqrt{\dfrac{24}{25}}$$

$$= \pm\dfrac{\sqrt{24}}{5} = \pm\dfrac{2\sqrt{6}}{5}$$

$$\tan x = \dfrac{\sin x}{\cos x} = \dfrac{\pm\frac{2\sqrt{6}}{5}}{\frac{1}{5}} = \pm 2\sqrt{6}$$

$$\sec x = \dfrac{1}{\cos x} = \dfrac{1}{\frac{1}{5}} = 5$$

Quadrant I:

$$\dfrac{\sec x - \tan x}{\sin x} = \dfrac{5-2\sqrt{6}}{\frac{2\sqrt{6}}{5}} = \dfrac{25-10\sqrt{6}}{2\sqrt{6}}$$

$$= \dfrac{25-10\sqrt{6}}{2\sqrt{6}} \cdot \dfrac{\sqrt{6}}{\sqrt{6}} = \dfrac{25\sqrt{6}-60}{12}$$

Quadrant IV:

$$\dfrac{\sec x - \tan x}{\sin x} = \dfrac{5-\left(-2\sqrt{6}\right)}{-\frac{2\sqrt{5}}{5}} = \dfrac{25+10\sqrt{6}}{-2\sqrt{6}}$$

$$= \dfrac{25+10\sqrt{6}}{-2\sqrt{6}} \cdot \dfrac{-\sqrt{6}}{-\sqrt{6}} = \dfrac{-25\sqrt{6}-60}{12}$$

In Exercises 81–83, the functions are graphed in the window $\left[-2\pi, -2\pi\right] \times \left[-4, 4\right]$.

81. The equation $\cos 2x = 1-2\sin^2 x$ is an identity. $y_1 = \cos 2x$, $y_2 = 1-2\sin^2 x$

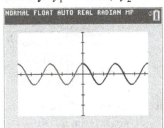

83. The equation $\sin x = \sqrt{1-\cos^2 x}$ is not an identity. $y_1 = \sin x$, $y_2 = \sqrt{1-\cos^2 x}$

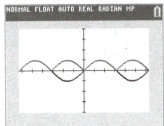

85. $y = \sin(-2x) \Rightarrow y = -\sin(2x)$

87. $y = \cos(-4x) \Rightarrow y = \cos(4x)$

89. **(a)** $y = \sin(-4x) \Rightarrow y = -\sin(4x)$

   (b) $y = \cos(-2x) \Rightarrow y = \cos(2x)$

   (c) $y = -5\sin(-3x) \Rightarrow y = -5\left[-\sin(3x)\right] \Rightarrow$
   $y = 5\sin(3x)$

Section 7.2 Verifying Trigonometric Identities

1. B   **3.** A

5. $\sin^2 \theta + \cos^2 \theta = \underline{1}$

7. $\sin(-\theta) = \underline{-\sin\theta}$

9. $\tan\theta = \dfrac{1}{\cot\theta} = \dfrac{\sin\theta}{\cos\theta}$

11. $\cot\theta + \dfrac{1}{\cot\theta} = \cot\theta + \tan\theta$

$$= \dfrac{\cos\theta}{\sin\theta} + \dfrac{\sin\theta}{\cos\theta}$$

$$= \dfrac{\cos^2\theta + \sin^2\theta}{\sin\theta\cos\theta}$$

$$= \dfrac{1}{\sin\theta\cos\theta} \text{ or } \csc\theta\sec\theta$$

13. $\tan s(\cot x + \csc x) = \dfrac{\sin x}{\cos x}\left(\dfrac{\cos x}{\sin x} + \dfrac{1}{\sin x}\right)$

$$= 1 + \dfrac{1}{\cos x}$$

$$= 1 + \sec x$$

15. $\dfrac{1}{\csc^2\theta} + \dfrac{1}{\sec^2\theta} = \sin^2\theta + \cos^2\theta = 1$

17. $(\sin\alpha - \cos\alpha)^2$

$$= \sin^2\alpha - 2\sin\alpha\cos\alpha + \cos^2\alpha$$

$$= (\sin^2\alpha + \cos^2\alpha) - 2\sin\alpha\cos\alpha$$

$$= 1 - 2\sin\alpha\cos\alpha$$

19. $(1+\sin t)^2 + \cos^2 t = 1 + 2\sin t + \sin^2 t + \cos^2 t$

$$= 1 + 2\sin t + (\sin^2 t + \cos^2 t)$$

$$= 1 + 2\sin t + 1 = 2 + 2\sin t$$

21. $\dfrac{1}{1+\cos x}-\dfrac{1}{1-\cos x}$

$=\dfrac{1-\cos x}{(1+\cos x)(1-\cos x)}-\dfrac{1+\cos x}{(1+\cos x)(1-\cos x)}$

$=\dfrac{(1-\cos x)-(1+\cos x)}{(1+\cos x)(1-\cos x)}$

$=\dfrac{1-\cos x-1-\cos x}{1-\cos^2 x}=-\dfrac{2\cos x}{\sin^2 x}$ or

$-\dfrac{2\cos x}{\sin^2 x}=-\dfrac{2\cos x}{\sin x\sin x}=-2\left(\dfrac{\cos x}{\sin x}\right)\left(\dfrac{1}{\sin x}\right)$

$=-2\cot x\csc x$

23. $\sin^2\theta-1=(\sin\theta+1)(\sin\theta-1)$

25. $(\sin x+1)^2-(\sin x-1)^2$

$=\left[(\sin x+1)+(\sin x-1)\right]$

$\qquad\cdot\left[(\sin x+1)-(\sin x-1)\right]$

$=(\sin x+1+\sin x-1)(\sin x+1-\sin x+1)$

$=(2\sin x)(2)=4\sin x$

27. $2\sin^2 x+3\sin x+1$

Let $a=\sin x$.

$2\sin^2 x+3\sin x+1=2a^2+3a+1$

$\qquad\qquad\qquad\quad=(2a+1)(a+1)$

$\qquad\qquad\qquad\quad=(2\sin x+1)(\sin x+1)$

29. $\cos^4 x+2\cos^2 x+1$

Let $\cos^2 x=a$.

$\cos^4 x+2\cos^2 x+1=a^2+2a+1$

$\qquad\qquad\qquad\quad=(a+1)^2=(\cos^2 x+1)^2$

31. $\sin^3 x-\cos^3 x$

Let $\sin x=a$ and $\cos x=b$.

$\sin^3 x-\cos^3 x$

$=a^3-b^3=(a-b)(a^2+ab+b^2)$

$=(\sin x-\cos x)(\sin^2 x+\sin x\cos x+\cos^2 x)$

$=(\sin x-\cos x)\left[(\sin^2 x+\cos^2 x)+\sin x\cos x\right]$

$=(\sin x-\cos x)(1+\sin x\cos x)$

33. $\tan\theta\cos\theta=\dfrac{\sin\theta}{\cos\theta}\cos\theta=\sin\theta$

35. $\sec r\cos r=\dfrac{1}{\cos r}\cdot\cos r=1$

37. $\dfrac{\sin\beta\tan\beta}{\cos\beta}=\tan\beta\tan\beta=\tan^2\beta$

39. $\sec^2 x-1=\dfrac{1}{\cos^2 x}-1=\dfrac{1}{\cos^2 x}-\dfrac{\cos^2 x}{\cos^2 x}$

$\qquad\qquad=\dfrac{1-\cos^2 x}{\cos^2 x}=\dfrac{\sin^2 x}{\cos^2 x}=\tan^2 x$

41. $\dfrac{\sin^2 x}{\cos^2 x}+\sin x\csc x=\tan^2 x+\sin x\cdot\dfrac{1}{\sin x}$

$\qquad\qquad\qquad\qquad=\tan^2 x+1=\sec^2 x$

43. $1-\dfrac{1}{\csc^2 x}=1-\sin^2 x=\cos^2 x$

45. Verify $\dfrac{\cot\theta}{\csc\theta}=\cos\theta$.

$\dfrac{\cot\theta}{\csc\theta}=\dfrac{\dfrac{\cos\theta}{\sin\theta}}{\dfrac{1}{\sin\theta}}=\dfrac{\cos\theta}{\sin\theta}\cdot\dfrac{\sin\theta}{1}=\cos\theta$

47. Verify $\dfrac{1-\sin^2\beta}{\cos\beta}=\cos\beta$.

$\dfrac{1-\sin^2\beta}{\cos\beta}=\dfrac{\cos^2\beta}{\cos\beta}=\cos\beta$

49. Verify $\cos^2\theta(\tan^2\theta+1)=1$.

$\cos^2\theta(\tan^2\theta+1)=\cos^2\theta\left(\dfrac{\sin^2\theta}{\cos^2\theta}+1\right)$

$\qquad\qquad=\cos^2\theta\left(\dfrac{\sin^2\theta}{\cos^2\theta}+\dfrac{\cos^2\theta}{\cos^2\theta}\right)$

$\qquad\qquad=\cos^2\theta\left(\dfrac{\sin^2\theta+\cos^2\theta}{\cos^2\theta}\right)$

$\qquad\qquad=\cos^2\theta\left(\dfrac{1}{\cos^2\theta}\right)=1$

51. Verify $\cot\theta+\tan\theta=\sec\theta\csc\theta$.

$\cot\theta+\tan\theta=\dfrac{\cos\theta}{\sin\theta}+\dfrac{\sin\theta}{\cos\theta}$

$\qquad\qquad=\dfrac{\cos^2\theta}{\sin\theta\cos\theta}+\dfrac{\sin^2\theta}{\sin\theta\cos\theta}$

$\qquad\qquad=\dfrac{\cos^2\theta+\sin^2\theta}{\cos\theta\sin\theta}=\dfrac{1}{\cos\theta\sin\theta}$

$\qquad\qquad=\dfrac{1}{\cos\theta}\cdot\dfrac{1}{\sin\theta}=\sec\theta\csc\theta$

53. Verify $\dfrac{\cos\alpha}{\sec\alpha}+\dfrac{\sin\alpha}{\csc\alpha}=\sec^2\alpha-\tan^2\alpha$.

Working with the left side, we have

$$\frac{\cos\alpha}{\sec\alpha}+\frac{\sin\alpha}{\csc\alpha}=\frac{\cos\alpha}{\dfrac{1}{\cos\alpha}}+\frac{\sin\alpha}{\dfrac{1}{\sin\alpha}}$$
$$=\cos^2\alpha+\sin^2\alpha=1$$

Working with the right side, we have

$$\sec^2\alpha-\tan^2\alpha=1.$$

$\dfrac{\cos\alpha}{\sec\alpha}+\dfrac{\sin\alpha}{\csc\alpha}=1=\sec^2\alpha-\tan^2\alpha,$ so the statement has been verified.

55. Verify $\sin^4\theta-\cos^4\theta=2\sin^2\theta-1$.

$$\sin^4\theta-\cos^4\theta$$
$$=\left(\sin^2\theta+\cos^2\theta\right)\left(\sin^2\theta-\cos^2\theta\right)$$
$$=1\cdot\left(\sin^2\theta-\cos^2\theta\right)=\sin^2\theta-\cos^2\theta$$
$$=\sin^2\theta-\left(1-\sin^2\theta\right)=2\sin^2\theta-1$$

57. Verify $\dfrac{1-\cos x}{1+\cos x}=(\cot x-\csc x)^2$.

Work with the left side.

$$\frac{1-\cos x}{1+\cos x}=\frac{(1-\cos x)(1-\cos x)}{(1+\cos x)(1-\cos x)}$$
$$=\frac{1-2\cos x+\cos^2 x}{1-\cos^2 x}$$
$$=\frac{1-2\cos x+\cos^2 x}{\sin^2 x}$$

Work with the right side.

$$\left(\cot x-\csc x\right)^2=\left(\frac{\cos x}{\sin x}-\frac{1}{\sin x}\right)^2=\left(\frac{\cos x-1}{\sin x}\right)^2$$
$$=\frac{\cos^2 x-2\cos x+1}{\sin^2 x}$$

$$\frac{1-\cos x}{1+\cos x}=\frac{\cos^2 x-2\cos x+1}{\sin^2 x}=\left(\cot x-\csc x\right)^2$$

Thus, the statement has been verified.

59. Verify $\dfrac{\cos\theta+1}{\tan^2\theta}=\dfrac{\cos\theta}{\sec\theta-1}$.

Work with the left side.

$$\frac{\cos\theta+1}{\tan^2\theta}=\frac{\cos\theta+1}{\sec^2\theta-1}=\frac{\cos\theta+1}{\dfrac{1}{\cos^2\theta}-1}$$
$$=\frac{(\cos\theta+1)\cos^2\theta}{\left(\dfrac{1}{\cos^2\theta}-1\right)\cos^2\theta}$$

$$=\frac{\cos^2\theta\left(\cos\theta+1\right)}{1-\cos^2\theta}$$
$$=\frac{\cos^2\theta\left(\cos\theta+1\right)}{(1+\cos\theta)(1-\cos\theta)}=\frac{\cos^2\theta}{1-\cos\theta}$$

Now work with the right side.

$$\frac{\cos\theta}{\sec\theta-1}=\frac{\cos\theta}{\dfrac{1}{\cos\theta}-1}=\frac{\cos\theta}{\dfrac{1}{\cos\theta}-1}\cdot\frac{\cos\theta}{\cos\theta}$$
$$=\frac{\cos^2\theta}{1-\cos\theta}$$

$$\frac{\cos\theta+1}{\tan^2\theta}=\frac{\cos^2\theta}{1-\cos\theta}=\frac{\cos\theta}{\sec\theta-1}$$

Thus, the statement has been verified.

61. Verify $\dfrac{1}{1-\sin\theta}+\dfrac{1}{1+\sin\theta}=2\sec^2\theta$.

$$\frac{1}{1-\sin\theta}+\frac{1}{1+\sin\theta}$$
$$=\frac{1+\sin\theta}{(1+\sin\theta)(1-\sin\theta)}+\frac{1-\sin\theta}{(1+\sin\theta)(1-\sin\theta)}$$
$$=\frac{(1+\sin\theta)+(1-\sin\theta)}{(1+\sin\theta)(1-\sin\theta)}=\frac{1+\sin\theta+1-\sin\theta}{(1+\sin\theta)(1-\sin\theta)}$$
$$=\frac{2}{1-\sin^2\theta}=\frac{2}{\cos^2\theta}=2\sec^2\theta$$

63. Verify $\dfrac{\cot\alpha+1}{\cot\alpha-1}=\dfrac{1+\tan\alpha}{1-\tan\alpha}$.

$$\frac{\cot\alpha+1}{\cot\alpha-1}=\frac{\dfrac{\cos\alpha}{\sin\alpha}+1}{\dfrac{\cos\alpha}{\sin\alpha}-1}=\frac{\dfrac{\cos\alpha}{\sin\alpha}+1}{\dfrac{\cos\alpha}{\sin\alpha}-1}\cdot\frac{\sin\alpha}{\sin\alpha}$$
$$=\frac{\cos\alpha+\sin\alpha}{\cos\alpha-\sin\alpha}$$
$$=\frac{\cos\alpha+\sin\alpha}{\cos\alpha-\sin\alpha}\cdot\frac{\dfrac{1}{\cos\alpha}}{\dfrac{1}{\cos\alpha}}$$
$$=\frac{\dfrac{\cos\alpha}{\cos\alpha}+\dfrac{\sin\alpha}{\cos\alpha}}{\dfrac{\cos\alpha}{\cos\alpha}-\dfrac{\sin\alpha}{\cos\alpha}}=\frac{1+\tan\alpha}{1-\tan\alpha}$$

65. Verify $\dfrac{\cos\theta}{\sin\theta\cot\theta}=1$.

$$\frac{\cos\theta}{\sin\theta\cot\theta}=\frac{\cos\theta}{\sin\theta\cdot\frac{\cos\theta}{\sin\theta}}=\frac{\cos\theta}{\cos\theta}=1$$

67. Verify $\dfrac{\sec^4\theta-\tan^4\theta}{\sec^2\theta+\tan^2\theta}=\sec^2\theta-\tan^2\theta$.

$$\frac{\sec^4\theta-\tan^4\theta}{\sec^2\theta+\tan^2\theta}$$
$$=\frac{\left(\sec^2\theta+\tan^2\theta\right)\left(\sec^2\theta-\tan^2\theta\right)}{\sec^2\theta+\tan^2\theta}$$
$$=\sec^2\theta-\tan^2\theta$$

69. Verify $\dfrac{\tan^2 t - 1}{\sec^2 t} = \dfrac{\tan t - \cot t}{\tan t + \cot t}$.

Simplify the right side

$$\dfrac{\tan t - \cot t}{\tan t + \cot t} = \dfrac{\tan t - \frac{1}{\tan t}}{\tan t + \frac{1}{\tan t}} = \dfrac{\tan t - \frac{1}{\tan t}}{\tan t + \frac{1}{\tan t}} \cdot \dfrac{\tan t}{\tan t} = \dfrac{\tan^2 t - 1}{\tan^2 t + 1} = \dfrac{\tan^2 t - 1}{\sec^2 t}$$

71. Verify $\sin^2 \alpha \sec^2 \alpha + \sin^2 \alpha \csc^2 \alpha = \sec^2 \alpha$.

$$\sin^2 \alpha \sec^2 \alpha + \sin^2 \alpha \csc^2 \alpha = \sin^2 \alpha \cdot \dfrac{1}{\cos^2 \alpha} + \sin^2 \alpha \cdot \dfrac{1}{\sin^2 \alpha} = \dfrac{\sin^2 \alpha}{\cos^2 \alpha} + 1 = \tan^2 \alpha + 1 = \sec^2 \alpha$$

73. Verify $\dfrac{\tan x}{1 + \cos x} + \dfrac{\sin x}{1 - \cos x} = \cot x + \sec x \csc x$.

$$\dfrac{\tan x}{1 + \cos x} + \dfrac{\sin x}{1 - \cos x} = \dfrac{\tan x (1 - \cos x)}{(1 + \cos x)(1 - \cos x)} + \dfrac{\sin x (1 + \cos x)}{(1 + \cos x)(1 - \cos x)}$$

$$= \dfrac{\tan x (1 - \cos x) + \sin x (1 + \cos x)}{(1 + \cos x)(1 - \cos x)} = \dfrac{\tan x - \tan x \cos x + \sin x + \sin x \cos x}{(1 + \cos x)(1 - \cos x)}$$

$$= \dfrac{\tan x - \frac{\sin x}{\cos x} \cos x + \sin x + \sin x \cos x}{(1 + \cos x)(1 - \cos x)} = \dfrac{\tan x - \sin x + \sin x + \sin x \cos x}{1 - \cos^2 x}$$

$$= \dfrac{\tan x + \sin x \cos x}{\sin^2 x} = \dfrac{\tan x}{\sin^2 x} + \dfrac{\sin x \cos x}{\sin^2 x} = \tan x \cdot \dfrac{1}{\sin^2 x} + \dfrac{\cos x}{\sin x}$$

$$= \dfrac{\sin x}{\cos x} \cdot \dfrac{1}{\sin^2 x} + \cot x = \dfrac{1}{\cos x} \cdot \dfrac{1}{\sin x} + \cot x = \sec x \csc x + \cot x$$

75. Verify $\dfrac{1 + \cos x}{1 - \cos x} - \dfrac{1 - \cos x}{1 + \cos x} = 4 \cot x \csc x$

$$\dfrac{1 + \cos x}{1 - \cos x} - \dfrac{1 - \cos x}{1 + \cos x} = \dfrac{(1 + \cos x)^2}{(1 + \cos x)(1 - \cos x)} - \dfrac{(1 - \cos x)^2}{(1 + \cos x)(1 - \cos x)}$$

$$= \dfrac{1 + 2\cos x + \cos^2 x}{(1 + \cos x)(1 - \cos x)} - \dfrac{1 - 2\cos x + \cos^2 x}{(1 + \cos x)(1 - \cos x)}$$

$$= \dfrac{1 + 2\cos x + \cos^2 x - 1 + 2\cos x - \cos^2 x}{(1 + \cos x)(1 - \cos x)}$$

$$= \dfrac{4\cos x}{1 - \cos^2 x} = \dfrac{4\cos x}{\sin^2 x} = 4 \cdot \dfrac{\cos x}{\sin x} \cdot \dfrac{1}{\sin x} = 4 \cot x \csc x$$

77. Verify $\dfrac{1 - \sin \theta}{1 + \sin \theta} = \sec^2 \theta - 2 \sec \theta \tan \theta + \tan^2 \theta$

Simplify the right side

$$\sec^2 \theta - 2 \sec \theta \tan \theta + \tan^2 \theta = \dfrac{1}{\cos^2 \theta} - 2 \cdot \dfrac{1}{\cos \theta} \cdot \dfrac{\sin \theta}{\cos \theta} + \dfrac{\sin^2 \theta}{\cos^2 \theta} = \dfrac{1 - 2\sin \theta + \sin^2 \theta}{\cos^2 \theta} = \dfrac{(1 - \sin \theta)^2}{1 - \sin^2 \theta}$$

$$= \dfrac{(1 - \sin \theta)^2}{(1 + \sin \theta)(1 - \sin \theta)} = \dfrac{1 - \sin \theta}{1 + \sin \theta}$$

79. Verify $\dfrac{-1}{\tan\alpha-\sec\alpha}+\dfrac{-1}{\tan\alpha+\sec\alpha}=2\tan\alpha.$

$$\dfrac{-1}{\tan\alpha-\sec\alpha}+\dfrac{-1}{\tan\alpha+\sec\alpha}=\dfrac{-\tan\alpha-\sec\alpha}{(\tan\alpha+\sec\alpha)(\tan\alpha-\sec\alpha)}+\dfrac{-\tan\alpha+\sec\alpha}{(\tan\alpha+\sec\alpha)(\tan\alpha-\sec\alpha)}$$

$$=\dfrac{-\tan\alpha-\sec\alpha-\tan\alpha+\sec\alpha}{(\tan\alpha+\sec\alpha)(\tan\alpha-\sec\alpha)}=\dfrac{-2\tan\alpha}{\tan^2\alpha-\sec^2\alpha}=\dfrac{-2\tan\alpha}{\tan^2\alpha-\left(\tan^2\alpha+1\right)}$$

$$=\dfrac{-2\tan\alpha}{\tan^2\alpha-\tan^2\alpha-1}=\dfrac{-2\tan\alpha}{-1}=2\tan\alpha$$

81. Verify $\left(1-\cos^2\alpha\right)\left(1+\cos^2\alpha\right)=2\sin^2\alpha-\sin^4\alpha.$

$$\left(1-\cos^2\alpha\right)\left(1+\cos^2\alpha\right)=\sin^2\alpha\left(1+\cos^2\alpha\right)=\sin^2\alpha\left(2-\sin^2\alpha\right)=2\sin^2\alpha-\sin^4\alpha$$

83. Verify $\dfrac{1-\cos x}{1+\cos x}=\csc^2 x-2\csc x\cot x+\cot^2 x$

Work with the left side:

$$\dfrac{1-\cos x}{1+\cos x}=\dfrac{1-\cos x}{1+\cos x}\cdot\dfrac{1-\cos x}{1-\cos x}=\dfrac{1-2\cos x+\cos^2 x}{1-\cos^2 x}=\dfrac{1-2\cos x+\cos^2 x}{\sin^2 x}$$

Work with the right side:

$$\csc^2 x-2\csc x\cot x+\cot^2 x=\dfrac{1}{\sin^2 x}-\dfrac{2\cos x}{\sin^2 x}+\dfrac{\cos^2 x}{\sin^2 x}=\dfrac{1-2\cos x+\cos^2 x}{\sin^2 x}$$

$$\dfrac{1-\cos x}{1+\cos x}=\dfrac{1-2\cos x+\cos^2 x}{\sin^2 x}=\csc^2 x-2\csc x\cot x+\cot^2 x,\ \text{so the statement has been verified.}$$

85. Verify $(2\sin x+\cos x)^2+(2\cos x-\sin x)^2=5$

$$(2\sin x+\cos x)^2+(2\cos x-\sin x)^2=\left(4\sin^2 x+4\sin x\cos x+\cos^2 x\right)+\left(4\cos^2 x-4\sin x\cos x+\sin^2 x\right)$$

$$=4\left(\sin^2 x+\cos^2 x\right)+\left(\cos^2 x+\sin^2 x\right)=4+1=5$$

87. Verify $\sec x-\cos x+\csc x-\sin x-\sin x\tan x=\cos x\cot x$

$$\sec x-\cos x+\csc x-\sin x-\sin x\tan x=\dfrac{1}{\cos x}-\cos x+\dfrac{1}{\sin x}-\sin x-\sin x\left(\dfrac{\sin x}{\cos x}\right)$$

$$=\left(\dfrac{1}{\cos x}-\cos x\right)+\left(\dfrac{1}{\sin x}-\sin x\right)-\dfrac{\sin^2 x}{\cos x}$$

$$=\dfrac{1-\cos^2 x}{\cos x}+\dfrac{1-\sin^2 x}{\sin x}-\dfrac{\sin^2 x}{\cos x}$$

$$=\left(\dfrac{1-\cos^2 x}{\cos x}-\dfrac{\sin^2 x}{\cos x}\right)+\dfrac{1-\sin^2 x}{\sin x}=\dfrac{1-\cos^2 x-\sin^2 x}{\cos x}+\dfrac{\cos^2 x}{\sin x}$$

$$=\dfrac{1-\left(\cos^2 x+\sin^2 x\right)}{\cos x}+\dfrac{\cos^2 x}{\sin x}=\dfrac{1-1}{\cos x}+\cos x\cdot\dfrac{\cos x}{\sin x}=\cos x\cot x$$

In Exercises 89–96, the functions are graphed in the window $[-2\pi, 2\pi] \times [-4, 4]$.

89. $(\sec\theta + \tan\theta)(1 - \sin\theta)$ appears to be equivalent to $\cos\theta$.

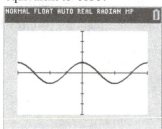

$$(\sec\theta + \tan\theta)(1 - \sin\theta)$$
$$= \left(\frac{1}{\cos\theta} + \frac{\sin\theta}{\cos\theta}\right)(1 - \sin\theta)$$
$$= \left(\frac{1 + \sin\theta}{\cos\theta}\right)(1 - \sin\theta)$$
$$= \frac{(1 + \sin\theta)(1 - \sin\theta)}{\cos\theta}$$
$$= \frac{1 - \sin^2\theta}{\cos\theta} = \frac{\cos^2\theta}{\cos\theta} = \cos\theta$$

91. $\dfrac{\cos\theta + 1}{\sin\theta + \tan\theta}$ appears to be equivalent to $\cot\theta$.

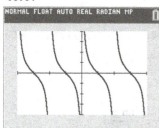

$$\frac{\cos\theta + 1}{\sin\theta + \tan\theta} = \frac{1 + \cos\theta}{\sin\theta + \dfrac{\sin\theta}{\cos\theta}} = \frac{1 + \cos\theta}{\sin\theta\left(1 + \dfrac{1}{\cos\theta}\right)}$$
$$= \frac{1 + \cos\theta}{\sin\theta\left(1 + \dfrac{1}{\cos\theta}\right)} \cdot \frac{\cos\theta}{\cos\theta}$$
$$= \frac{(1 + \cos\theta)\cos\theta}{\sin\theta(\cos\theta + 1)} = \frac{\cos\theta}{\sin\theta} = \cot\theta$$

93. Is $\dfrac{2 + 5\cos x}{\sin x} = 2\csc x + 5\cot x$ an identity?

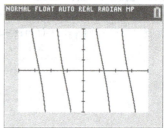

The graphs of $y_1 = \dfrac{2 + 5\cos x}{\sin x}$ and $y_2 = 2\csc x + 5\cot x$ appear to be the same.
$$\frac{2 + 5\cos x}{\sin x} = \frac{2}{\sin x} + \frac{5\cos x}{\sin x} = 2\csc x + 5\cot x$$
Thus, the given statement is an identity.

95. Is $\dfrac{\tan x - \cot x}{\tan x + \cot x} = 2\sin^2 x$ an identity?

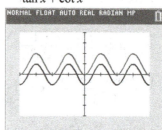

The graphs of
$$y = \frac{\tan x - \cot x}{\tan x + \cot x} \text{ and } y = 2\sin^2 x \text{ are not the}$$
same. The given statement is not an identity.

97. Show that $\sin(\csc t) = 1$ is not an identity.

We need to find only one value for which the statement is false. Let $t = 2$. Use a calculator to find that $\sin(\csc 2) \approx 0.891094$, which is not equal to 1. $\sin(\csc t) = 1$ does not hold true for *all* real numbers t. Thus, it is not an identity.

99. Show that $\csc t = \sqrt{1 + \cot^2 t}$ is not an identity.

Let $t = \frac{\pi}{4}$. We have $\csc\frac{\pi}{4} = \sqrt{2}$ and
$$\sqrt{1 + \cot^2 \frac{\pi}{4}} = \sqrt{1 + 1^2} = \sqrt{1 + 1} = \sqrt{2}.$$ But let
$t = -\frac{\pi}{4}$. We have $\csc\left(-\frac{\pi}{4}\right) = -\sqrt{2}$ and
$$\sqrt{1 + \cot^2\left(-\frac{\pi}{4}\right)} = \sqrt{1 + (-1)^2} = \sqrt{1 + 1} = \sqrt{2}.$$
$\csc t = \sqrt{1 + \cot^2 t}$ does not hold true for *all* real numbers t. Thus, it is not an identity.

101. (a) $I = k\cos^2\theta = k\left(1 - \sin^2\theta\right)$

(b) When $\theta = 0$, $\cos\theta = 1$, its maximum value. Thus, $\cos^2\theta$ will be a maximum and, as a result, I will be maximized if k is a positive constant.

103. The sum of L and C equals 3.

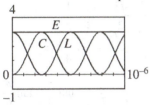

105. $E(t) = L(t) + C(t)$
$$= 3\cos^2(6,000,000t) + 3\sin^2(6,000,000t)$$
$$= 3\left[\cos^2(6,000,000t) + \sin^2(6,000,000t)\right]$$
$$= 3 \cdot 1 = 3$$

Section 7.3 Sum and Difference Identities

1. E; $\cos(x + y) = \cos x \cos y - \sin x \sin y$

3. D; $\cos\left(\dfrac{\pi}{2} - x\right) = \sin x$

5. D

$$\cos\left(x - \frac{\pi}{2}\right) = \cos\left[-\left(\frac{\pi}{2} - x\right)\right]$$
$$= \cos\left(\frac{\pi}{2} - x\right) = \sin x$$

7. D; $\sin(A + B) = \sin A \cos B + \cos A \sin B$

9. B; $\tan(A + B) = \dfrac{\tan A + \tan B}{1 - \tan A \tan B}$

11. $\cos 75° = \cos(30° + 45°)$
$$= \cos 30° \cos 45° - \sin 30° \sin 45°$$
$$= \frac{\sqrt{3}}{2} \cdot \frac{\sqrt{2}}{2} - \frac{1}{2} \cdot \frac{\sqrt{2}}{2}$$
$$= \frac{\sqrt{6}}{4} - \frac{\sqrt{2}}{4} = \frac{\sqrt{6} - \sqrt{2}}{4}$$

13. $\cos(-105°) = \cos\left[-60° + (-45°)\right]$
$$= \cos(-60°)\cos(-45°) - \sin(-60°)\sin(-45°)$$
$$= \frac{1}{2} \cdot \frac{\sqrt{2}}{2} - \left(-\frac{\sqrt{3}}{2}\right)\left(-\frac{\sqrt{2}}{2}\right)$$
$$= \frac{\sqrt{2}}{4} - \frac{\sqrt{6}}{4} = \frac{\sqrt{2} - \sqrt{6}}{4}$$

15. $\cos\left(\dfrac{7\pi}{12}\right) = \cos\left(\dfrac{4\pi}{12} + \dfrac{3\pi}{12}\right) = \cos\left(\dfrac{\pi}{3} + \dfrac{\pi}{4}\right)$
$$= \cos\frac{\pi}{3}\cos\frac{\pi}{4} - \sin\frac{\pi}{3}\sin\frac{\pi}{4}$$
$$= \frac{1}{2} \cdot \frac{\sqrt{2}}{2} - \frac{\sqrt{3}}{2} \cdot \frac{\sqrt{2}}{2}$$
$$= \frac{\sqrt{2}}{4} - \frac{\sqrt{6}}{4} = \frac{\sqrt{2} - \sqrt{6}}{4}$$

17. $\cos\left(-\dfrac{\pi}{12}\right) = \cos\left(\dfrac{2\pi}{12} - \dfrac{3\pi}{12}\right) = \cos\left(\dfrac{\pi}{6} - \dfrac{\pi}{4}\right)$
$$= \cos\frac{\pi}{6}\cos\frac{\pi}{4} + \sin\frac{\pi}{6}\sin\frac{\pi}{4}$$
$$= \frac{\sqrt{3}}{2} \cdot \frac{\sqrt{2}}{2} + \frac{1}{2} \cdot \frac{\sqrt{2}}{2}$$
$$= \frac{\sqrt{6}}{4} + \frac{\sqrt{2}}{4} = \frac{\sqrt{6} + \sqrt{2}}{4}$$

19. $\cos 40° \cos 50° - \sin 40° \sin 50°$
$$= \cos(40° + 50°) = \cos 90° = 0$$

21. $\tan 87° = \cot(90° - 87°) = \cot 3°$

23. $\cos\dfrac{\pi}{12} = \sin\left(\dfrac{\pi}{2} - \dfrac{\pi}{12}\right) = \sin\dfrac{5\pi}{12}$

25. $\csc 14°24' = \sec(90° - 14°24') = \sec 75°36'$

27. $\sin\dfrac{5\pi}{8} = \cos\left(\dfrac{\pi}{2} - \dfrac{5\pi}{8}\right) = \cos\left(\dfrac{4\pi}{8} - \dfrac{5\pi}{8}\right)$
$$= \cos\left(-\frac{\pi}{8}\right)$$

29. Because $\dfrac{\pi}{6} = \dfrac{\pi}{2} - \dfrac{\pi}{3}$, $\cot\dfrac{\pi}{3} = \tan\dfrac{\pi}{6}$.
$$\cot\frac{\pi}{3} = \underline{\tan\frac{\pi}{6}}$$

31. Because $90° - 57° = 33°$, $\sin 57° = \cos 33°$.
$$\sin 57° = \underline{\cos 33°}$$

33. Because $90° - 70° = 20°$, and $\sin x = \dfrac{1}{\csc x}$,
$$\cos 70° = \sin 20° = \frac{1}{\underline{\csc 20°}}.$$

For exercises 35–39, other answers are possible.

35. $\tan\theta = \cot(45° + 2\theta)$
Because $\tan\theta = \cot(90° - \theta)$,
$$90° - \theta = 45° + 2\theta \Rightarrow 90° = 45° + 3\theta \Rightarrow$$
$$3\theta = 45° \Rightarrow \theta = 15°$$

37. $\sec x = \csc \dfrac{2\pi}{3}$

By a cofunction identity, $\sec x = \csc\left(\dfrac{\pi}{2} - x\right)$.

Thus,

$\csc \dfrac{2\pi}{3} = \csc\left(\dfrac{\pi}{2} - x\right) \Rightarrow \dfrac{2\pi}{3} = \dfrac{\pi}{2} - x \Rightarrow$

$x = -\dfrac{\pi}{6}$

39. $\sin(3\theta - 15°) = \cos(\theta + 25°)$

Because $\sin\theta = \cos(90° - \theta)$, we have

$\sin(3\theta - 15°) = \cos\left[90° - (3\theta - 15°)\right]$
$= \cos(90° - 3\theta + 15°)$
$= \cos(105° - 3\theta)$

Solve $\cos(105° - 3\theta) = \cos(\theta + 25°)$.

$\cos(105° - 3\theta) = \cos(\theta + 25°)$
$105° - 3\theta = \theta + 25°$
$105° = 4\theta + 25°$
$4\theta = 80° \Rightarrow \theta = 20°$

41. $\sin 165° = \sin(180° - 15°)$
$= \sin 180° \cos 15° - \cos 180° \sin 15°$
$= (0)\cos 15° - (-1)\sin 15° = 0 + \sin 15°$
$= \sin 15°$

Now use a difference identity to find $\sin 15°$.
$\sin 15° = \sin(45° - 30°)$
$= \sin 45° \cos 30° - \cos 45° \sin 30°$
$= \dfrac{\sqrt{2}}{2} \cdot \dfrac{\sqrt{3}}{2} - \dfrac{\sqrt{2}}{2} \cdot \dfrac{1}{2}$
$= \dfrac{\sqrt{6}}{4} - \dfrac{\sqrt{2}}{4} = \dfrac{\sqrt{6} - \sqrt{2}}{4}$

43. $\tan 165° = \tan(180° - 15°)$
$= \dfrac{\tan 180° - \tan 15°}{1 + \tan 180° \tan 15°}$
$= \dfrac{0 - \tan 15°}{1 + 0 \cdot \tan 15°} = -\tan 15°$

Now use a difference identity to find $\tan 15°$.

$\tan 15° = \tan(45° - 30°) = \dfrac{\tan 45° - \tan 30°}{1 + \tan 45° \tan 30°}$

$= \dfrac{1 - \frac{\sqrt{3}}{3}}{1 + 1 \cdot \frac{\sqrt{3}}{3}} = \dfrac{3 - \sqrt{3}}{3 + \sqrt{3}}$

$= \dfrac{3 - \sqrt{3}}{3 + \sqrt{3}} \cdot \dfrac{3 - \sqrt{3}}{3 - \sqrt{3}} = \dfrac{9 - 3\sqrt{3} - 3\sqrt{3} + 3}{9 - 3}$

$= \dfrac{12 - 6\sqrt{3}}{6} = 2 - \sqrt{3}$

Thus,

$\tan 165° = -\tan 15° = -(2 - \sqrt{3}) = -2 + \sqrt{3}.$

45. $\sin \dfrac{5\pi}{12} = \sin\left(\dfrac{\pi}{4} + \dfrac{\pi}{6}\right)$

$= \sin \dfrac{\pi}{4} \cos \dfrac{\pi}{6} + \cos \dfrac{\pi}{4} \sin \dfrac{\pi}{6}$

$= \dfrac{\sqrt{2}}{2} \cdot \dfrac{\sqrt{3}}{2} + \dfrac{\sqrt{2}}{2} \cdot \dfrac{1}{2}$

$= \dfrac{\sqrt{6}}{4} + \dfrac{\sqrt{2}}{4} = \dfrac{\sqrt{6} + \sqrt{2}}{4}$

47. $\tan \dfrac{\pi}{12} = \tan\left(\dfrac{\pi}{4} - \dfrac{\pi}{6}\right) = \dfrac{\tan\frac{\pi}{4} - \tan\frac{\pi}{6}}{1 + \tan\frac{\pi}{4}\tan\frac{\pi}{6}}$

$= \dfrac{1 - \frac{\sqrt{3}}{3}}{1 + \frac{\sqrt{3}}{3}} = \dfrac{1 - \frac{\sqrt{3}}{3}}{1 + \frac{\sqrt{3}}{3}} \cdot \dfrac{3}{3} = \dfrac{3 - \sqrt{3}}{3 + \sqrt{3}}$

$= \dfrac{3 - \sqrt{3}}{3 + \sqrt{3}} \cdot \dfrac{3 - \sqrt{3}}{3 - \sqrt{3}} = \dfrac{\left(3 - \sqrt{3}\right)^2}{3^2 - \left(\sqrt{3}\right)^2}$

$= \dfrac{9 - 6\sqrt{3} + 3}{9 - 3} = \dfrac{12 - 6\sqrt{3}}{6} = 2 - \sqrt{3}$

49. $\sin \dfrac{7\pi}{12} = \sin\left(\dfrac{\pi}{4} + \dfrac{\pi}{3}\right)$

$= \sin \dfrac{\pi}{4} \cos \dfrac{\pi}{3} + \cos \dfrac{\pi}{4} \sin \dfrac{\pi}{3}$

$= \dfrac{\sqrt{2}}{2} \cdot \dfrac{1}{2} + \dfrac{\sqrt{2}}{2} \cdot \dfrac{\sqrt{3}}{2} = \dfrac{\sqrt{2} + \sqrt{6}}{4}$

51. $\sin\left(-\dfrac{7\pi}{12}\right) = \sin\left(-\dfrac{\pi}{3} - \dfrac{\pi}{4}\right)$

$= \sin\left(-\dfrac{\pi}{3}\right)\cos \dfrac{\pi}{4} - \cos\left(-\dfrac{\pi}{3}\right)\sin \dfrac{\pi}{4}$

$= -\sin \dfrac{\pi}{3} \cos \dfrac{\pi}{4} - \cos \dfrac{\pi}{3} \sin \dfrac{\pi}{4}$

$= -\dfrac{\sqrt{3}}{2} \cdot \dfrac{\sqrt{2}}{2} - \dfrac{1}{2} \cdot \dfrac{\sqrt{2}}{2}$

$= -\dfrac{\sqrt{6}}{4} - \dfrac{\sqrt{2}}{4} = \dfrac{-\sqrt{6} - \sqrt{2}}{4}$

53. $\sin\left(-\dfrac{5\pi}{12}\right) = \tan\left(-\dfrac{\pi}{6} - \dfrac{\pi}{4}\right)$

$= \dfrac{\tan\left(-\dfrac{\pi}{6}\right) + \tan\left(-\dfrac{\pi}{4}\right)}{1 - \tan\left(-\dfrac{\pi}{6}\right)\tan\left(-\dfrac{\pi}{4}\right)}$

$= \dfrac{-\tan\dfrac{\pi}{6} - \tan\dfrac{\pi}{4}}{1 - \tan\dfrac{\pi}{6}\tan\dfrac{\pi}{4}}$

$= \dfrac{-\dfrac{\sqrt{3}}{3} - 1}{1 - \left(\dfrac{\sqrt{3}}{3}\right)(1)} = \dfrac{-\sqrt{3} - 3}{3 - \sqrt{3}}$

$= \dfrac{-\sqrt{3} - 3}{3 - \sqrt{3}} \cdot \dfrac{3 + \sqrt{3}}{3 + \sqrt{3}}$

$= \dfrac{-3\sqrt{3} - 3 - 9 - 3\sqrt{3}}{3^2 - \left(\sqrt{3}\right)^2} = \dfrac{-12 - 6\sqrt{3}}{9 - 3}$

$= \dfrac{-12 - 6\sqrt{3}}{6} = -2 - \sqrt{3}$

55. $\tan\dfrac{11\pi}{12} = \tan\left(\pi - \dfrac{\pi}{12}\right)$

$= \dfrac{\tan\pi - \tan\dfrac{\pi}{12}}{1 + \tan\pi\tan\dfrac{\pi}{12}} = -\tan\dfrac{\pi}{12}$

Now use a difference identity to find $\tan\dfrac{\pi}{12}$.

$\tan\dfrac{\pi}{12} = \tan\left(\dfrac{\pi}{4} - \dfrac{\pi}{6}\right) = \dfrac{\tan\dfrac{\pi}{4} - \tan\dfrac{\pi}{6}}{1 + \tan\dfrac{\pi}{4}\tan\dfrac{\pi}{6}}$

$= \dfrac{1 - \dfrac{\sqrt{3}}{3}}{1 + 1 \cdot \dfrac{\sqrt{3}}{3}} = \dfrac{1 - \dfrac{\sqrt{3}}{3}}{1 + \dfrac{\sqrt{3}}{3}} = \dfrac{\dfrac{3 - \sqrt{3}}{3}}{\dfrac{3 + \sqrt{3}}{3}}$

$= \dfrac{3 - \sqrt{3}}{3 + \sqrt{3}} = \dfrac{3 - \sqrt{3}}{3 + \sqrt{3}} \cdot \dfrac{3 - \sqrt{3}}{3 - \sqrt{3}}$

$= \dfrac{9 - 6\sqrt{3} + 3}{9 - 3} = \dfrac{12 - 6\sqrt{3}}{6}$

$= \dfrac{6\left(2 - \sqrt{3}\right)}{6} = 2 - \sqrt{3}$

Thus,

$\tan\dfrac{11\pi}{12} = -\tan\dfrac{\pi}{12} = -\left(2 - \sqrt{3}\right) = -2 + \sqrt{3}.$

57. $\sin 76° \cos 31° - \cos 76° \sin 31° = \sin\left(76° - 31°\right)$

$= \sin 45° = \dfrac{\sqrt{2}}{2}$

59. $\sin\dfrac{\pi}{5}\cos\dfrac{3\pi}{10} + \cos\dfrac{\pi}{5}\sin\dfrac{3\pi}{10}$

$= \sin\left(\dfrac{\pi}{5} + \dfrac{3\pi}{10}\right) = \sin\left(\dfrac{\pi}{2}\right) = 1$

61. $\dfrac{\tan 80° + \tan 55°}{1 - \tan 80°\tan 55°} = \tan\left(80° + 55°\right)$

$= \tan 135° = -1$

63. $\cos\left(\theta - 180°\right) = \cos\theta\cos 180° + \sin\theta\sin 180°$

$= \cos\theta(-1) + \sin\theta(0)$

$= -\cos\theta + 0 = -\cos\theta$

65. $\cos\left(180° + \theta\right) = \cos 180°\cos\theta - \sin 180°\sin\theta$

$= (-1)\cos\theta - (0)\sin\theta$

$= -\cos\theta - 0 = -\cos\theta$

67. $\cos\left(60° + \theta\right) = \cos 60°\cos\theta - \sin 60°\sin\theta$

$= \dfrac{1}{2}\cos\theta - \dfrac{\sqrt{3}}{2}\sin\theta$

$= \dfrac{1}{2}\left(\cos\theta - \sqrt{3}\sin\theta\right)$

$= \dfrac{\cos\theta - \sqrt{3}\sin\theta}{2}$

69. $\cos\left(\dfrac{3\pi}{4} - x\right) = \cos\dfrac{3\pi}{4}\cos x + \sin\dfrac{3\pi}{4}\sin x$

$= \left(-\dfrac{\sqrt{2}}{2}\right)\cos x + \left(\dfrac{\sqrt{2}}{2}\right)\sin x$

$= \dfrac{\sqrt{2}}{2}\left(-\cos x + \sin x\right)$

$= \dfrac{\sqrt{2}\left(\sin x - \cos x\right)}{2}$

71. $\tan\left(\theta + 30°\right) = \dfrac{\tan\theta + \tan 30°}{1 - \tan\theta\tan 30°}$

$= \dfrac{\tan\theta + \dfrac{1}{\sqrt{3}}}{1 - \left(\dfrac{1}{\sqrt{3}}\right)\tan\theta}$

$= \dfrac{\sqrt{3}\tan\theta + 1}{\sqrt{3} - \tan\theta}$

73. $\sin\left(\dfrac{\pi}{4} + x\right) = \sin\dfrac{\pi}{4}\cos x + \cos\dfrac{\pi}{4}\sin x$

$= \dfrac{\sqrt{2}}{2}\cos x + \dfrac{\sqrt{2}}{2}\sin x$

$= \dfrac{\sqrt{2}\left(\cos x + \sin x\right)}{2}$

75. $\sin\left(270° - \theta\right) = \sin 270° \cos\theta - \cos 270° \sin\theta$
$$= (-1)(\cos\theta) - (0)(\sin\theta)$$
$$= -\cos\theta$$

77. $\tan\left(2\pi - x\right) = \dfrac{\tan 2\pi - \tan x}{1 + \tan 2\pi \tan x}$
$$= \dfrac{0 - \tan x}{1 + 0 \cdot \tan x} = -\tan x$$

79. $\sin s = \dfrac{3}{5}$ and $\sin t = -\dfrac{12}{13}$, s is in quadrant I and t is in quadrant III.

$\sin s = \dfrac{y}{r} = \dfrac{3}{5} \Rightarrow y = 3, r = 5$. Substituting into the Pythagorean theorem, we have

$x^2 + 3^2 = 5^2 \Rightarrow x^2 = 16 \Rightarrow x = 4$, because

$\cos s > 0$. Thus, $\cos s = \dfrac{x}{r} = \dfrac{4}{5}$. We will use a Pythagorean identity to find the value of $\cos t$.

$\cos t = -\sqrt{1 - \left(-\dfrac{12}{13}\right)^2} = -\sqrt{1 - \dfrac{144}{169}}$
$$= -\sqrt{\dfrac{25}{169}} = -\dfrac{5}{13}$$

$\cos\left(s + t\right) = \cos s \cos t - \sin s \sin t$
$$= \left(\dfrac{4}{5}\right)\left(-\dfrac{5}{13}\right) - \left(\dfrac{3}{5}\right)\left(-\dfrac{12}{13}\right)$$
$$= -\dfrac{20}{65} + \dfrac{36}{65} = \dfrac{16}{65}$$

$\cos\left(s - t\right) = \cos s \cos t + \sin s \sin t$
$$= \left(\dfrac{4}{5}\right)\left(-\dfrac{5}{13}\right) + \left(\dfrac{3}{5}\right)\left(-\dfrac{12}{13}\right)$$
$$= -\dfrac{20}{65} - \dfrac{36}{65} = -\dfrac{56}{65}$$

81. $\cos s = -\dfrac{1}{5}$, $\sin t = \dfrac{3}{5}$, s and t are in quadrant II.

$\cos s = \dfrac{x}{r} \Rightarrow \cos s = -\dfrac{1}{5} = \dfrac{-1}{5} \Rightarrow x = -1, r = 5$.

Substituting into the Pythagorean theorem, we have $(-1)^2 + y^2 = 5^2 \Rightarrow y^2 = 24 \Rightarrow y = \sqrt{24}$,

because $\sin x > 0$. Thus, $\sin s = \dfrac{y}{r} = \dfrac{\sqrt{24}}{5}$.

We will use a Pythagorean identity to find the value of $\cos t$.

$\cos t = -\sqrt{1 - \sin^2 t} = -\sqrt{1 - \left(\dfrac{3}{5}\right)^2}$
$$= -\sqrt{1 - \dfrac{9}{25}} = -\sqrt{\dfrac{16}{25}} = -\dfrac{4}{5}$$

$\cos(s + t) = \cos s \cos t - \sin s \sin t$
$$= \left(-\dfrac{1}{5}\right)\left(-\dfrac{4}{5}\right) - \left(\dfrac{\sqrt{24}}{5}\right)\left(\dfrac{3}{5}\right)$$
$$= \dfrac{4}{25} - \dfrac{3\sqrt{24}}{25} = \dfrac{4}{25} - \dfrac{6\sqrt{6}}{25}$$
$$= \dfrac{4 - 6\sqrt{6}}{25}$$

$\cos(s - t) = \cos s \cos t + \sin s \sin t$
$$= \left(-\dfrac{1}{5}\right)\left(-\dfrac{4}{5}\right) + \left(\dfrac{\sqrt{24}}{5}\right)\left(\dfrac{3}{5}\right)$$
$$= \dfrac{4}{25} + \dfrac{3\sqrt{24}}{25} = \dfrac{4}{25} + \dfrac{6\sqrt{6}}{25}$$
$$= \dfrac{4 + 6\sqrt{6}}{25}$$

83. $\cos s = \dfrac{3}{5}$, $\sin t = \dfrac{5}{13}$, and s and t are in quadrant I.

First find the values of $\sin s$, $\tan s$, $\cos t$, and $\tan t$. Because s and t are both in quadrant I, the values of $\sin s$ and $\cos t$, $\tan s$, and $\tan t$ will be positive.

$\sin s = \sqrt{1 - \left(\dfrac{3}{5}\right)^2} = \sqrt{1 - \dfrac{9}{25}} = \sqrt{\dfrac{16}{25}} = \dfrac{4}{5}$

$\cos t = \sqrt{1 - \left(\dfrac{5}{13}\right)^2} = \sqrt{1 - \dfrac{25}{169}} = \sqrt{\dfrac{144}{169}} = \dfrac{12}{13}$

$\tan s = \dfrac{\sin s}{\cos s} = \dfrac{\frac{4}{5}}{\frac{3}{5}} = \dfrac{4}{3}$

$\tan t = \dfrac{\sin t}{\cos t} = \dfrac{\frac{5}{13}}{\frac{12}{13}} = \dfrac{5}{12}$

(a) $\sin\left(s + t\right) = \sin s \cos t + \cos s \sin t$
$$= \left(\dfrac{4}{5}\right)\left(\dfrac{12}{13}\right) + \left(\dfrac{3}{5}\right)\left(\dfrac{5}{13}\right)$$
$$= \dfrac{48}{65} + \dfrac{15}{65} = \dfrac{63}{65}$$

(b) $\tan\left(s + t\right) = \dfrac{\tan s + \tan t}{1 - \tan s \tan t} = \dfrac{\frac{4}{3} + \frac{5}{12}}{1 - \left(\frac{4}{3}\right)\left(\frac{5}{12}\right)}$
$$= \dfrac{48 + 15}{36 - 20} = \dfrac{63}{16}$$

(c) From parts (a) and (b), $\sin(s + t) > 0$ and $\tan(s + t) > 0$. The only quadrant in which the values of both the sine and the tangent are positive is quadrant I, so $s + t$ is in quadrant I.

85. $\cos s = -\dfrac{8}{17}$ and $\cos t = -\dfrac{3}{5}$, s and t are in

quadrant III

First find the values of $\sin s$, $\sin t$, $\tan s$, and $\tan t$. Because s and t are both in quadrant III, the values of $\sin s$ and $\sin t$ will be negative, while $\tan s$ and $\tan t$ will be positive.

$$\sin s = -\sqrt{1-\cos^2 s} = -\sqrt{1-\left(-\dfrac{8}{17}\right)^2}$$

$$= -\sqrt{1-\dfrac{64}{289}} = -\sqrt{\dfrac{225}{289}} = -\dfrac{15}{17}$$

$$\sin t = -\sqrt{1-\cos^2 t} = -\sqrt{1-\left(-\dfrac{3}{5}\right)^2}$$

$$= -\sqrt{1-\dfrac{9}{25}} = -\sqrt{\dfrac{16}{25}} = -\dfrac{4}{5}$$

$$\tan s = \dfrac{\sin s}{\cos s} = \dfrac{-\frac{15}{17}}{-\frac{8}{17}} = \dfrac{15}{8}$$

$$\tan t = \dfrac{\sin t}{\cos t} = \dfrac{-\frac{4}{5}}{-\frac{3}{5}} = \dfrac{4}{3}$$

(a) $\sin(s+t) = \sin s \cos t + \cos s \sin t$

$$= \left(-\dfrac{15}{17}\right)\left(-\dfrac{3}{5}\right) + \left(-\dfrac{8}{17}\right)\left(-\dfrac{4}{5}\right)$$

$$= \dfrac{45}{85} + \dfrac{32}{85} = \dfrac{77}{85}$$

(b) $\tan(s+t) = \dfrac{\frac{15}{8}+\frac{4}{3}}{1-\left(\frac{15}{8}\right)\left(\frac{4}{3}\right)} = \dfrac{45+32}{24-60}$

$$= \dfrac{77}{-36} = -\dfrac{77}{36}$$

(c) From parts (a) and (b), $\sin(s+t) > 0$ and $\tan(s+t) < 0$. The only quadrant in which the value of the sine is positive and the value of the tangent is negative is quadrant II, so $s+t$ is in quadrant II.

87. $\sin s = \dfrac{2}{3}$ and $\sin t = -\dfrac{1}{3}$, s is in quadrant II

and t is in quadrant IV.

First find the values of $\cos s$, $\cos t$, $\tan s$, and $\tan t$. Because s is in quadrant II and t is in quadrant IV, the values of $\cos s$, $\tan s$, and $\tan t$ will be negative, while $\cos t$ will be positive.

$$\cos s = -\sqrt{1-\left(\dfrac{2}{3}\right)^2} = -\sqrt{1-\dfrac{4}{9}} = -\sqrt{\dfrac{5}{9}} = -\dfrac{\sqrt{5}}{3}$$

$$\cos t = \sqrt{1-\left(-\dfrac{1}{3}\right)^2} = \sqrt{1-\dfrac{1}{9}} = \sqrt{\dfrac{8}{9}}$$

$$= \dfrac{\sqrt{8}}{3} = \dfrac{2\sqrt{2}}{3}$$

$$\tan s = \dfrac{\sin s}{\cos s} = \dfrac{\frac{2}{3}}{-\frac{\sqrt{5}}{3}} = -\dfrac{2}{\sqrt{5}} = -\dfrac{2\sqrt{5}}{5}$$

$$\tan t = \dfrac{\sin t}{\cos t} = \dfrac{-\frac{1}{3}}{\frac{2\sqrt{2}}{3}} = -\dfrac{1}{2\sqrt{2}} = -\dfrac{\sqrt{2}}{4}$$

(a) $\sin(s+t) = \sin s \cos t + \cos s \sin t$

$$= \left(\dfrac{2}{3}\right)\left(\dfrac{2\sqrt{2}}{3}\right) + \left(-\dfrac{\sqrt{5}}{3}\right)\left(-\dfrac{1}{3}\right)$$

$$= \dfrac{4\sqrt{2}}{9} + \dfrac{\sqrt{5}}{9} = \dfrac{4\sqrt{2}+\sqrt{5}}{9}$$

(b) Different forms of $\tan(s+t)$ will be obtained depending on whether $\tan s$ and $\tan t$ are written with rationalized denominators.

$\tan(s+t)$

$$= \dfrac{-\frac{2\sqrt{5}}{5}+\left(-\frac{\sqrt{2}}{4}\right)}{1-\left(-\frac{2\sqrt{5}}{5}\right)\left(-\frac{\sqrt{2}}{4}\right)} = \dfrac{-8\sqrt{5}-5\sqrt{2}}{20-2\sqrt{10}}$$

$$= \dfrac{-8\sqrt{5}-5\sqrt{2}}{20-2\sqrt{10}} \cdot \dfrac{20+2\sqrt{10}}{20+2\sqrt{10}}$$

$$= \dfrac{-160\sqrt{5}-16\sqrt{50}-100\sqrt{2}-10\sqrt{20}}{400-40}$$

$$= \dfrac{-160\sqrt{5}-80\sqrt{2}-100\sqrt{2}-20\sqrt{5}}{360}$$

$$= \dfrac{-180\sqrt{5}-180\sqrt{2}}{360} = \dfrac{-\sqrt{5}-\sqrt{2}}{2}$$

or

$$\tan(s+t) = \dfrac{-\frac{2}{\sqrt{5}}+\left(-\frac{1}{2\sqrt{2}}\right)}{1-\left(-\frac{2}{\sqrt{5}}\right)\left(-\frac{1}{2\sqrt{2}}\right)}$$

$$= \dfrac{-4\sqrt{2}-\sqrt{5}}{2\sqrt{10}-2} = \dfrac{4\sqrt{2}+\sqrt{5}}{2-2\sqrt{10}}$$

$$= \dfrac{4\sqrt{2}+\sqrt{5}}{2-2\sqrt{10}} \cdot \dfrac{2+2\sqrt{10}}{2+2\sqrt{10}}$$

$$= \dfrac{8\sqrt{2}+8\sqrt{20}+2\sqrt{5}+2\sqrt{50}}{4-40}$$

$$= \dfrac{8\sqrt{2}+16\sqrt{5}+2\sqrt{5}+10\sqrt{2}}{-36}$$

$$= \dfrac{18\sqrt{2}+18\sqrt{5}}{-36} = \dfrac{-\sqrt{2}-\sqrt{5}}{2}$$

(c) To find the quadrant of $s + t$, notice from the preceding that $\sin(s+t) = \dfrac{4\sqrt{2}+\sqrt{5}}{9} > 0$ and

$\tan(s+t) = \dfrac{-8\sqrt{5}-5\sqrt{2}}{20-2\sqrt{10}} \approx -1.8 < 0$. The only quadrant in which the values of sine are positive and

tangent is negative is quadrant II. Therefore, $s + t$ is in quadrant II.

The graphs in exercises 89–91 are shown in the following window $[-2\pi,\, 2\pi] \times [-4,\, 4]$.

89. $\sin\left(\dfrac{\pi}{2}+\theta\right)$ appears to be equivalent to $\cos\theta$.

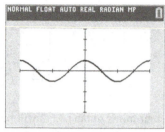

$\sin\left(\dfrac{\pi}{2}+\theta\right) = \sin\dfrac{\pi}{2}\cos\theta + \sin\theta\cos\dfrac{\pi}{2} = 1\cdot\cos\theta + \sin\theta\cdot 0 = \cos\theta + 0 = \cos\theta$

91. $\tan\left(\dfrac{\pi}{2}+\theta\right)$ appears to be equivalent to $-\cot\theta$.

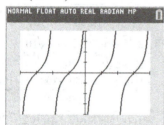

$\tan\left(\dfrac{\pi}{2}+\theta\right) = \dfrac{\sin\left(\dfrac{\pi}{2}+\theta\right)}{\cos\left(\dfrac{\pi}{2}+\theta\right)} = \dfrac{\sin\dfrac{\pi}{2}\cos\theta + \cos\dfrac{\pi}{2}\sin\theta}{\cos\dfrac{\pi}{2}\cos\theta - \sin\dfrac{\pi}{2}\sin\theta} = \dfrac{1\cdot\cos\theta + 0\cdot\sin\theta}{0\cdot\cos\theta - 1\cdot\sin\theta} = \dfrac{\cos\theta}{-\sin\theta} = -\cot\theta$

93. Verify $\sin 2x = 2\sin x\cos x$ is an identity.

$\sin 2x = \sin(x+x) = \sin x\cos x + \cos x\sin x = 2\sin x\cos x$

95. Verify $\sin\left(\dfrac{7\pi}{6}+x\right) - \cos\left(\dfrac{2\pi}{3}+x\right) = 0$ is an identity.

$\sin\left(\dfrac{7\pi}{6}+x\right) - \cos\left(\dfrac{2\pi}{3}+x\right) = \left(\sin\dfrac{7\pi}{6}\cos x + \cos\dfrac{7\pi}{6}\sin x\right) - \left(\cos\dfrac{2\pi}{3}\cos x - \sin\dfrac{2\pi}{3}\sin x\right)$

$= \left(-\dfrac{1}{2}\cos x - \dfrac{\sqrt{3}}{2}\sin x\right) - \left(-\dfrac{1}{2}\cos x - \dfrac{\sqrt{3}}{2}\sin x\right) = 0$

97. Verify $\dfrac{\cos(\alpha-\beta)}{\cos\alpha\sin\beta} = \tan\alpha + \cot\beta$ is an identity.

$\dfrac{\cos(\alpha-\beta)}{\cos\alpha\sin\beta} = \dfrac{\cos\alpha\cos\beta + \sin\alpha\sin\beta}{\cos\alpha\sin\beta} = \dfrac{\cos\alpha\cos\beta}{\cos\alpha\sin\beta} + \dfrac{\sin\alpha\sin\beta}{\cos\alpha\sin\beta} = \dfrac{\cos\beta}{\sin\beta} + \dfrac{\sin\alpha}{\cos\alpha} = \cot\beta + \tan\alpha$

99. Verify that $\dfrac{\sin(x-y)}{\sin(x+y)} = \dfrac{\tan x - \tan y}{\tan x + \tan y}$ is an identity

$$\frac{\sin(x-y)}{\sin(x+y)} = \frac{\sin x \cos y - \cos x \sin y}{\sin x \cos y + \cos x \sin y} = \frac{\dfrac{\sin x \cos y}{\cos x \cos y} - \dfrac{\cos x \sin y}{\cos x \cos y}}{\dfrac{\sin x \cos y}{\cos x \cos y} + \dfrac{\cos x \sin y}{\cos x \cos y}} = \frac{\dfrac{\sin x}{\cos x} \cdot \dfrac{\cos y}{\cos y} - \dfrac{\cos x}{\cos x} \cdot \dfrac{\sin y}{\cos y}}{\dfrac{\sin x}{\cos x} \cdot \dfrac{\cos y}{\cos y} + \dfrac{\cos x}{\cos x} \cdot \dfrac{\sin y}{\cos y}}$$

$$= \frac{\dfrac{\sin x}{\cos x} \cdot 1 - 1 \cdot \dfrac{\sin y}{\cos y}}{\dfrac{\sin x}{\cos x} \cdot 1 + 1 \cdot \dfrac{\sin y}{\cos y}} = \frac{\dfrac{\sin x}{\cos x} - \dfrac{\sin y}{\cos y}}{\dfrac{\sin x}{\cos x} + \dfrac{\sin y}{\cos y}} = \frac{\tan x - \tan y}{\tan x + \tan y}$$

101. Verify $\dfrac{\sin(s-t)}{\sin t} + \dfrac{\cos(s-t)}{\cos t} = \dfrac{\sin s}{\sin t \cos t}$ is an identity.

$$\frac{\sin(s-t)}{\sin t} + \frac{\cos(s-t)}{\cos t} = \frac{\sin s \cos t - \sin t \cos s}{\sin t} + \frac{\cos s \cos t + \sin t \sin s}{\cos t}$$

$$= \frac{\sin s \cos^2 t - \sin t \cos t \cos s}{\sin t \cos t} + \frac{\sin t \cos t \cos s + \sin^2 t \sin s}{\sin t \cos t} = \frac{\sin s \cos^2 t + \sin s \sin^2 t}{\sin t \cos t}$$

$$= \frac{\sin s \left(\cos^2 t + \sin^2 t\right)}{\sin t \cos t} = \frac{\sin s}{\sin t \cos t}$$

103. **(a)** There are 60 cycles per sec, so the number of cycles in 0.05 sec is given by
(0.05 sec)(60 cycles per sec) = 3 cycles.

(b) Because $V = 163 \sin \omega t$ and the maximum value of $\sin \omega t$ is 1, the maximum voltage is 163. Similarly, because the minimum value of $\sin \omega t$ is –1, the minimum voltage is –163.

(c) Because $V = 163 \sin \omega t$ and the minimum value of $\sin \omega t$ is –1, the minimum voltage is –163. Therefore, the voltage is not always equal to 115.

105. **(a)** $F = \dfrac{0.6W \sin(\theta + 90°)}{\sin 12°} = \dfrac{0.6(170)\sin(30+90)°}{\sin 12°} = \dfrac{102 \sin 120°}{\sin 12°} \approx 425\,\text{lb}$

(This is a good reason why people frequently have back problems.)

(b) $F = \dfrac{0.6W \sin(\theta + 90°)}{\sin 12°} = \dfrac{0.6W\left(\sin \theta \cos 90° + \sin 90° \cos \theta\right)}{\sin 12°} = \dfrac{0.6W\left(\sin \theta \cdot 0 + 1 \cdot \cos \theta\right)}{\sin 12°}$

$= \dfrac{0.6W\left(0 + \cos \theta\right)}{\sin 12°} = \dfrac{0.6}{\sin 12°} W \cos \theta \approx 2.9W \cos \theta$

(c) F will be maximum when $\cos \theta = 1$ or $\theta = 0°$. ($\theta = 0°$ corresponds to the back being horizontal which gives a maximum force on the back muscles. This agrees with intuition because stress on the back increases as one bends farther until the back is parallel with the ground.)

107. Because 90° is a quadrantal angle whose terminal side lies along the y-axis, follow the reasoning in Case 2 in the text. If θ is a small positive angle, then $90° + \theta$ lies in quadrant II and $\sin \theta$ is positive while $\cos \theta$ is negative. Thus, $\cos(90° + \theta) = -\sin \theta$.

109. Because 180° is a quadrantal angle whose terminal side lies along the x-axis, follow the reasoning in Case 1 in the text. If θ is a small positive angle, then $180° + \theta$ lies in quadrant III and $\cos \theta$ is negative. Thus, $\cos(180° + \theta) = -\cos \theta$.

111. Because $180°$ is a quadrantal angle whose terminal side lies along the x-axis, follow the reasoning in Case 1 in the text. If θ is a small positive angle, then $180° + \theta$ lies in quadrant III and $\sin \theta$ is negative. Thus,

$$\sin(180° - \theta) = -\sin \theta.$$

Chapter 7 Quiz
(Sections 7.1–7.3)

1. $\sin \theta = -\dfrac{7}{25}$, θ is in quadrant IV

In quadrant IV, the cosine and secant function values are positive. The tangent, cotangent, and cosecant function values are negative.

$$\cos \theta = \sqrt{1 - \sin^2 \theta} = \sqrt{1 - \left(-\frac{7}{25}\right)^2}$$

$$= \sqrt{1 - \frac{49}{625}} = \sqrt{\frac{576}{625}} = \frac{24}{25}$$

$$\tan \theta = \frac{\sin \theta}{\cos \theta} = \frac{-\frac{7}{25}}{\frac{24}{25}} = -\frac{7}{24}$$

$$\cot \theta = \frac{1}{\tan \theta} = \frac{1}{-\frac{7}{24}} = -\frac{24}{7}$$

$$\sec \theta = \frac{1}{\cos \theta} = \frac{1}{\frac{24}{25}} = \frac{25}{24}$$

$$\csc \theta = \frac{1}{\sin \theta} = \frac{1}{-\frac{7}{25}} = -\frac{25}{7}$$

3. $\sin\left(-\dfrac{7\pi}{12}\right) = -\sin\left(\dfrac{7\pi}{12}\right) = -\sin\left(\dfrac{\pi}{3} + \dfrac{\pi}{4}\right)$

$$= -\left(\sin\frac{\pi}{3}\cos\frac{\pi}{4} + \cos\frac{\pi}{3}\sin\frac{\pi}{4}\right)$$

$$= -\left[\frac{\sqrt{3}}{2}\left(\frac{\sqrt{2}}{2}\right) + \frac{1}{2}\left(\frac{\sqrt{2}}{2}\right)\right]$$

$$= -\left(\frac{\sqrt{6}}{4} + \frac{\sqrt{2}}{4}\right) = -\left(\frac{\sqrt{6} + \sqrt{2}}{4}\right)$$

$$= \frac{-\sqrt{6} - \sqrt{2}}{4}$$

5. $\cos A = \dfrac{3}{5}$, $\sin B = -\dfrac{5}{13}$, $0 < A < \dfrac{\pi}{2}$, and

$$\pi < B < \frac{3\pi}{2}$$

$$\cos A = \frac{3}{5} = \frac{y}{r} \Rightarrow y = 3, r = 5.$$

Substituting into the Pythagorean theorem, we have $x^2 + 3^2 = 5^2 \Rightarrow x = 4$, because $\sin A > 0$.

Thus, $\sin A = \dfrac{4}{5}$.

We will use a Pythagorean identity to find the value of $\cos B$.

$$\cos B = -\sqrt{1 - \left(-\frac{5}{13}\right)^2} = -\sqrt{1 - \frac{25}{169}}$$

$$= -\sqrt{\frac{144}{169}} = -\frac{12}{13}$$

Note that $\cos B$ is negative because B is in quadrant III.

(a) $\cos(A + B) = \cos A \cos B - \sin A \sin B$

$$= \frac{3}{5}\left(-\frac{12}{13}\right) - \frac{4}{5}\left(-\frac{5}{13}\right)$$

$$= -\frac{36}{65} + \frac{20}{65} = -\frac{16}{65}$$

(b) $\sin(A + B) = \sin A \cos B + \cos A \sin B$

$$= \frac{4}{5}\left(-\frac{12}{13}\right) + \frac{3}{5}\left(-\frac{5}{13}\right)$$

$$= -\frac{48}{65} - \frac{15}{65} = -\frac{63}{65}$$

(c) Both $\cos(A + B)$ and $\sin(A + B)$ are negative. Thus $(A + B)$ is in quadrant III.

7. Verify $\dfrac{1 + \sin \theta}{\cot^2 \theta} = \dfrac{\sin \theta}{\csc \theta - 1}$ is an identity.

Working with the right side, we have

$$\frac{\sin \theta}{\csc \theta - 1} = \frac{\sin \theta}{\csc \theta - 1} \cdot \frac{\csc \theta + 1}{\csc \theta + 1}$$

$$= \frac{\sin \theta \csc \theta + \sin \theta}{\csc^2 \theta - 1} = \frac{1 + \sin \theta}{\cot^2 \theta}$$

9. Verify $\dfrac{\sin^2 \theta - \cos^2 \theta}{\sin^4 \theta - \cos^4 \theta} = 1$ is an identity.

$$\frac{\sin^2 \theta - \cos^2 \theta}{\sin^4 \theta - \cos^4 \theta}$$

$$= \frac{\sin^2 \theta - \cos^2 \theta}{\left(\sin^2 \theta - \cos^2 \theta\right)\left(\sin^2 \theta + \cos^2 \theta\right)} = \frac{1}{1} = 1$$

Section 7.4   Double-Angle and Half-Angle Identities

1. C. $2\cos^2 15° - 1 = \cos(2 \cdot 15°) = \cos 30° = \dfrac{\sqrt{3}}{2}$

3. B.
$$2\sin 22.5° \cos 22.5° = \sin(2 \cdot 22.5°)$$
$$= \sin 45° = \dfrac{\sqrt{2}}{2}$$

5. F. $4\sin\dfrac{\pi}{3}\cos\dfrac{\pi}{3} = 2\sin\left(2 \cdot \dfrac{\pi}{3}\right) = 2\sin\dfrac{2\pi}{3}$
$$= 2 \cdot \dfrac{\sqrt{3}}{2} = \sqrt{3}$$

7. 195° lies in quadrant III, and the sine is negative in quadrant III, so choose the negative square root.

9. 225° lies in quadrant III, and the tangent is positive in quadrant III, so choose the positive square root.

11. $\sin\theta = \dfrac{2}{5}, \cos\theta < 0$

$\cos 2\theta = 1 - 2\sin^2\theta \Rightarrow$

$\cos 2\theta = 1 - 2\left(\dfrac{2}{5}\right)^2 1 - 2 \cdot \dfrac{4}{25} = 1 - \dfrac{8}{25} = \dfrac{17}{25}$

$\cos^2 2\theta + \sin^2 2\theta = 1 \Rightarrow$
$\sin^2 2\theta = 1 - \cos^2 2\theta \Rightarrow$

$\sin^2 2\theta = 1 - \left(\dfrac{17}{25}\right)^2 = 1 - \dfrac{289}{625} = \dfrac{336}{625}$

$\cos\theta < 0$, so $\sin 2\theta < 0$ because $\sin 2\theta = 2\sin\theta\cos\theta < 0$ and $\sin\theta > 0$.

$\sin 2\theta = -\sqrt{\dfrac{336}{625}} = -\dfrac{\sqrt{336}}{25} = -\dfrac{4\sqrt{21}}{25}$

13. $\tan x = 2, \cos x > 0$

$\tan 2x = \dfrac{2\tan x}{1 - \tan^2 x} = \dfrac{2(2)}{1 - 2^2} = \dfrac{4}{1 - 4} = -\dfrac{4}{3}$

Both $\tan x$ and $\cos x$ are positive, so x must be in quadrant I. Because $0° < x < 90°$, then $0° < 2x < 180°$. Thus, $2x$ must be in either quadrant I or quadrant II. However, $\tan 2x < 0$, so $2x$ is in quadrant II, and $\sec 2x$ is negative.

$\sec^2 2x = 1 + \tan^2 2x = 1 + \left(-\dfrac{4}{3}\right)^2 = 1 + \dfrac{16}{9} = \dfrac{25}{9}$

$\sec 2x = -\dfrac{5}{3} \Rightarrow \cos 2x = \dfrac{1}{\sec 2x} = -\dfrac{3}{5}$

$\cos^2 2x + \sin^2 2x = 1 \Rightarrow \sin^2 2x = 1 - \cos^2 2x \Rightarrow$

$\sin^2 2x = 1 - \left(-\dfrac{3}{5}\right)^2 \Rightarrow \sin^2 2x = 1 - \dfrac{9}{25} = \dfrac{16}{25}$

In quadrants I and II, $\sin 2x > 0$. Thus, we

have $\sin 2x = \sqrt{\dfrac{16}{25}} = \dfrac{4}{5}$.

15. $\sin\theta = -\dfrac{\sqrt{5}}{7}, \cos\theta > 0$

$\cos^2\theta = 1 - \sin^2\theta = 1 - \left(-\dfrac{\sqrt{5}}{7}\right)^2 = 1 - \dfrac{5}{49} = \dfrac{44}{49}$

$\cos\theta > 0$, so $\cos\theta = \sqrt{\dfrac{44}{49}} = \dfrac{\sqrt{44}}{7} = \dfrac{2\sqrt{11}}{7}$.

$\cos 2\theta = 1 - 2\sin^2\theta = 1 - 2\left(-\dfrac{\sqrt{5}}{7}\right)^2$

$= 1 - 2 \cdot \dfrac{5}{49} = 1 - \dfrac{10}{49} = \dfrac{39}{49}$

$\sin 2\theta = 2\sin\theta\cos\theta = 2\left(-\dfrac{\sqrt{5}}{7}\right)\left(\dfrac{2\sqrt{11}}{7}\right)$

$= -\dfrac{4\sqrt{55}}{49}$

17. $\cos 2\theta = \dfrac{3}{5}, \theta$ is in quadrant I.

$\cos 2\theta = 2\cos^2\theta - 1 \Rightarrow \dfrac{3}{5} = 2\cos^2\theta - 1 \Rightarrow$

$2\cos^2\theta = \dfrac{3}{5} + 1 = \dfrac{8}{5} \Rightarrow \cos^2\theta = \dfrac{8}{10} = \dfrac{4}{5}$

θ is in quadrant I, so $\cos\theta > 0$. Thus,

$\cos\theta = \sqrt{\dfrac{4}{5}} = \dfrac{2}{\sqrt{5}} = \dfrac{2\sqrt{5}}{5}$.

θ is in quadrant I, so $\sin\theta > 0$.

$\sin\theta = \sqrt{1 - \cos^2\theta} = \sqrt{1 - \left(\dfrac{2\sqrt{5}}{5}\right)^2} = \sqrt{1 - \dfrac{20}{25}}$

$= \sqrt{\dfrac{5}{25}} = \sqrt{\dfrac{1}{5}} = \dfrac{1}{\sqrt{5}} = \dfrac{1}{\sqrt{5}} \cdot \dfrac{\sqrt{5}}{\sqrt{5}} = \dfrac{\sqrt{5}}{5}$

19. $\cos 2\theta = -\dfrac{5}{12}, 90° < \theta < 180°$

$\cos 2\theta = 2\cos^2\theta - 1 \Rightarrow$

$2\cos^2\theta = \cos 2\theta + 1 = -\dfrac{5}{12} + 1 = \dfrac{7}{12} \Rightarrow$

$\cos^2\theta = \dfrac{7}{24}$

(*continued on next page*)

(*continued*)

θ is in quadrant II, so $\cos\theta < 0$. Thus, $\cos\theta = -\sqrt{\dfrac{7}{24}} = -\dfrac{\sqrt{7}}{\sqrt{24}} = -\dfrac{\sqrt{7}}{2\sqrt{6}} = -\dfrac{\sqrt{7}}{2\sqrt{6}} \cdot \dfrac{\sqrt{6}}{\sqrt{6}} = -\dfrac{\sqrt{42}}{12}$.

$$\sin^2\theta = 1 - \cos^2\theta = 1 - \left(-\sqrt{\dfrac{7}{24}}\right)^2 = 1 - \dfrac{7}{24} = \dfrac{17}{24}$$

θ is in quadrant II, so $\sin\theta > 0$. Thus, $\sin\theta = \sqrt{\dfrac{17}{24}} = \dfrac{\sqrt{17}}{\sqrt{24}} = \dfrac{\sqrt{17}}{2\sqrt{6}} = \dfrac{\sqrt{17}}{2\sqrt{6}} \cdot \dfrac{\sqrt{6}}{\sqrt{6}} = \dfrac{\sqrt{102}}{12}$.

21. $\cos^2 15° - \sin^2 15° = \cos\left[2(15°)\right] = \cos 30° = \dfrac{\sqrt{3}}{2}$

23. $1 - 2\sin^2 15° = \cos\left[2(15°)\right] = \cos 30° = \dfrac{\sqrt{3}}{2}$

25. $2\cos^2 67\dfrac{1}{2}° - 1 = \cos^2 67\dfrac{1}{2}° - \sin^2 67\dfrac{1}{2}° = \cos 2\left(67\dfrac{1}{2}°\right) = \cos 135° = -\dfrac{\sqrt{2}}{2}$

27. $\dfrac{\tan 51°}{1 - \tan^2 51°}$

$\dfrac{2\tan A}{1 - \tan^2 A} = \tan 2A$, so we have

$\dfrac{1}{2}\left(\dfrac{2\tan A}{1 - \tan^2 A}\right) = \dfrac{1}{2}\tan 2A \Rightarrow \dfrac{\tan A}{1 - \tan^2 A} = \dfrac{1}{2}\tan 2A \Rightarrow \dfrac{\tan 51°}{1 - \tan^2 51°} = \dfrac{1}{2}\tan\left[2(51°)\right] = \dfrac{1}{2}\tan 102°$.

29. $\dfrac{1}{4} - \dfrac{1}{2}\sin^2 47.1° = \dfrac{1}{4}\left(1 - 2\sin^2 47.1°\right) = \dfrac{1}{4}\cos\left[2(47.1°)\right] = \dfrac{1}{4}\cos 94.2°$

31. $\sin^2\dfrac{2\pi}{5} - \cos^2\dfrac{2\pi}{5} = -\left(\cos^2\dfrac{2\pi}{5} - \sin^2\dfrac{2\pi}{5}\right)$

$\cos^2 A - \sin^2 A = \cos 2A \Rightarrow -\left(\cos^2\dfrac{2\pi}{5} - \sin^2\dfrac{2\pi}{5}\right) = -\cos\left(2\cdot\dfrac{2\pi}{5}\right) = -\cos\dfrac{4\pi}{5}$

33. $\sin 4x = \sin\left[2(2x)\right] = 2\sin 2x\cos 2x = 2(2\sin x\cos x)\left(\cos^2 x - \sin^2 x\right) = 4\sin x\cos^3 x - 4\sin^3 x\cos x$

35. $\tan 3x = \tan(2x + x) = \dfrac{\tan 2x + \tan x}{1 - \tan 2x\tan x} = \dfrac{\dfrac{2\tan x}{1 - \tan^2 x} + \tan x}{1 - \dfrac{2\tan x}{1 - \tan^2 x}\cdot\tan x} = \dfrac{\dfrac{2\tan x}{1 - \tan^2 x} + \dfrac{\left(1 - \tan^2 x\right)\tan x}{1 - \tan^2 x}}{\dfrac{1 - \tan^2 x}{1 - \tan^2 x} - \dfrac{2\tan^2 x}{1 - \tan^2 x}}$

$= \dfrac{\dfrac{2\tan x}{1 - \tan^2 x} + \dfrac{\tan x - \tan^3 x}{1 - \tan^2 x}}{\dfrac{1 - \tan^2 x}{1 - \tan^2 x} - \dfrac{2\tan^2 x}{1 - \tan^2 x}} \cdot \dfrac{1 - \tan^2 x}{1 - \tan^2 x} = \dfrac{2\tan x + \tan x - \tan^3 x}{1 - \tan^2 x - 2\tan^2 x} = \dfrac{3\tan x - \tan^3 x}{1 - 3\tan^2 x}$

37. $2\sin 58°\cos 102° = 2\left(\dfrac{1}{2}\left[\sin(58° + 102°) + \sin(58° - 102°)\right]\right) = \sin 160° + \sin(-44°) = \sin 160° - \sin 44°$

39. $2\sin\dfrac{\pi}{6}\cos\dfrac{\pi}{3} = 2\cdot\dfrac{1}{2}\left[\sin\left(\dfrac{\pi}{6} + \dfrac{\pi}{3}\right) + \sin\left(\dfrac{\pi}{6} - \dfrac{\pi}{3}\right)\right] = \sin\dfrac{\pi}{2} + \sin\left(-\dfrac{\pi}{6}\right) = \sin\dfrac{\pi}{2} - \sin\dfrac{\pi}{6}$

41. $6\sin 4x\sin 5x = 6\cdot\dfrac{1}{2}\Big[\cos(4x-5x)-\cos(4x+5x)\Big] = 3\cos(-x)-3\cos 9x = 3\cos x - 3\cos 9x$

43. $\cos 4x - \cos 2x = -2\sin\left(\dfrac{4x+2x}{2}\right)\sin\left(\dfrac{4x-2x}{2}\right) = -2\sin\dfrac{6x}{2}\sin\dfrac{2x}{2} = -2\sin 3x\sin x$

45. $\sin 25° + \sin(-48°) = 2\sin\left(\dfrac{25°+(-48°)}{2}\right)\cos\left(\dfrac{25°-(-48°)}{2}\right) = 2\sin\dfrac{-23°}{2}\cos\dfrac{73°}{2}$
$$= 2\sin(-11.5°)\cos 36.5° = -2\sin 11.5°\cos 36.5°$$

47. $\cos 4x + \cos 8x = 2\cos\left(\dfrac{4x+8x}{2}\right)\cos\left(\dfrac{4x-8x}{2}\right)$
$$= 2\cos\dfrac{12x}{2}\cos\dfrac{-4x}{2}$$
$$= 2\cos 6x\cos(-2x)$$
$$= 2\cos 6x\cos 2x$$

49. $\sin 67.5° = \sin\left(\dfrac{135°}{2}\right)$

67.5° is in quadrant I, so $\sin 67.5° > 0$.

$\sin 67.5° = \sqrt{\dfrac{1-\cos 135°}{2}} = \sqrt{\dfrac{1-(-\cos 45°)}{2}}$
$$= \sqrt{\dfrac{1+\dfrac{\sqrt{2}}{2}}{2}} = \sqrt{\dfrac{1+\dfrac{\sqrt{2}}{2}}{2}\cdot\dfrac{2}{2}}$$
$$= \dfrac{\sqrt{2+\sqrt{2}}}{\sqrt{4}} = \dfrac{\sqrt{2+\sqrt{2}}}{2}$$

51. $\tan 195° = \tan\left(\dfrac{390°}{2}\right) = \dfrac{\sin 390°}{1+\cos 390°}$
$$= \dfrac{\sin 30°}{1+\cos 30°} = \dfrac{\dfrac{1}{2}}{1+\dfrac{\sqrt{3}}{2}} = \dfrac{\dfrac{1}{2}}{1+\dfrac{\sqrt{3}}{2}}\cdot\dfrac{2}{2}$$
$$= \dfrac{1}{2+\sqrt{3}} = \dfrac{1}{2+\sqrt{3}}\cdot\dfrac{2-\sqrt{3}}{2-\sqrt{3}} = \dfrac{2-\sqrt{3}}{4-3}$$
$$= \dfrac{2-\sqrt{3}}{1} = 2-\sqrt{3}$$

53. $\cos 165° = \cos\left(\dfrac{330°}{2}\right)$

165° is in quadrant II, so $\cos 165° < 0$.

$\cos 165° = -\sqrt{\dfrac{1+\cos 330°}{2}} = -\sqrt{\dfrac{1+\cos 30°}{2}}$
$$= -\sqrt{\dfrac{1+\dfrac{\sqrt{3}}{2}}{2}} = -\sqrt{\dfrac{2+\sqrt{3}}{4}}$$
$$= \dfrac{-\sqrt{2+\sqrt{3}}}{2}$$

55. Find $\cos\dfrac{x}{2}$, given $\cos x = \dfrac{1}{4}$, with
$$0 < x < \dfrac{\pi}{2}.$$
$$0 < x < \dfrac{\pi}{2} \Rightarrow 0 < \dfrac{x}{2} < \dfrac{\pi}{4}, \text{ so } \cos\dfrac{x}{2} > 0.$$
$$\cos\dfrac{x}{2} = \sqrt{\dfrac{1+\cos x}{2}} = \sqrt{\dfrac{1+\dfrac{1}{4}}{2}} = \sqrt{\dfrac{4+1}{8}} = \sqrt{\dfrac{5}{8}}$$
$$= \dfrac{\sqrt{5}}{\sqrt{8}} = \dfrac{\sqrt{5}}{2\sqrt{2}} = \dfrac{\sqrt{5}}{2\sqrt{2}}\cdot\dfrac{\sqrt{2}}{\sqrt{2}} = \dfrac{\sqrt{10}}{4}$$

57. Find $\tan\dfrac{\theta}{2}$, given $\sin\theta = \dfrac{3}{5}$, with $90° < \theta < 180°$.

To find $\tan\dfrac{\theta}{2}$, we need the values of $\sin\theta$ and $\cos\theta$. We know $\sin\theta = \dfrac{3}{5}$.

$\cos\theta = \pm\sqrt{1-\left(\dfrac{3}{5}\right)^2} = \pm\sqrt{1-\dfrac{9}{25}} = \pm\sqrt{\dfrac{16}{25}} = \pm\dfrac{4}{5}$

Because $90° < \theta < 180°$ (θ is in quadrant II), $\cos\theta < 0$. Thus, $\cos\theta = -\dfrac{4}{5}$.

$\tan\dfrac{\theta}{2} = \dfrac{\sin\theta}{1+\cos\theta} = \dfrac{\dfrac{3}{5}}{1+\left(-\dfrac{4}{5}\right)} = \dfrac{\dfrac{3}{5}}{1+\left(-\dfrac{4}{5}\right)}\cdot\dfrac{5}{5}$
$$= \dfrac{3}{5-4} = \dfrac{3}{1} = 3$$

59. Find $\sin\dfrac{x}{2}$, given $\tan x = 2$, with $0 < x < \dfrac{\pi}{2}$.

Because x is in quadrant I, $\sec x > 0$.

$\sec^2 x = \tan^2 x + 1 \Rightarrow$
$\sec^2 x = 2^2 + 1 = 4 + 1 = 5 \Rightarrow \sec x = \sqrt{5}$

$\cos x = \dfrac{1}{\sec x} = \dfrac{1}{\sqrt{5}} = \dfrac{1}{\sqrt{5}}\cdot\dfrac{\sqrt{5}}{\sqrt{5}} = \dfrac{\sqrt{5}}{5}$

(continued on next page)

(continued)

Because $0 < x < \dfrac{\pi}{2} \Rightarrow 0 < \dfrac{x}{2} < \dfrac{\pi}{4}$, $\dfrac{x}{2}$ is in

quadrant I. Thus, $\sin\dfrac{x}{2} > 0$.

$$\sin\frac{x}{2} = \sqrt{\frac{1-\cos x}{2}} = \sqrt{\frac{1-\frac{\sqrt 5}{5}}{2}} = \frac{\sqrt{50-10\sqrt 5}}{10}$$

61. Find $\tan\dfrac{\theta}{2}$, given $\tan\theta = \dfrac{\sqrt 7}{3}$, with

$180° < \theta < 270°$.

$\sec^2\theta = \tan^2\theta + 1 \Rightarrow$

$$\sec^2\theta = \left(\frac{\sqrt 7}{3}\right)^2 + 1 = \frac{7}{9} + 1 = \frac{16}{9}$$

θ is in quadrant III, so $\sec\theta < 0$ and
$\sin\theta < 0$.

$$\sec\theta = -\sqrt{\frac{16}{9}} = -\frac{4}{3} \text{ and}$$

$$\cos\theta = \frac{1}{\sec\theta} = \frac{1}{-\frac{4}{3}} = -\frac{3}{4}$$

$$\sin\theta = -\sqrt{1-\cos^2\theta}$$
$$= -\sqrt{1-\left(-\frac{3}{4}\right)^2} - \sqrt{1-\frac{9}{16}} = -\frac{\sqrt 7}{4}$$

$$\tan\frac{\theta}{2} = \frac{\sin\theta}{1+\cos\theta} = \frac{-\frac{\sqrt 7}{4}}{1+\left(-\frac{3}{4}\right)} = \frac{-\sqrt 7}{4-3} = -\sqrt 7$$

63. Find $\sin\theta$, given $\cos 2\theta = \dfrac{3}{5}$, θ is in

quadrant I. θ is in quadrant I, so $\sin\theta > 0$.

$$\sin\theta = \sqrt{\frac{1-\cos 2\theta}{2}} \Rightarrow$$

$$\sin\theta = \sqrt{\frac{1-\frac{3}{5}}{2}} = \sqrt{\frac{\frac{2}{5}}{2}} = \sqrt{\frac{2}{10}} = \sqrt{\frac{1}{5}}$$

$$= \frac{1}{\sqrt 5} \cdot \frac{\sqrt 5}{\sqrt 5} = \frac{\sqrt 5}{5}$$

65. Find $\cos x$, given $\cos 2x = -\dfrac{5}{12}, \dfrac{\pi}{2} < x < \pi$.

Because $\dfrac{\pi}{2} < x < \pi, \cos x < 0$.

$$\cos x = -\sqrt{\frac{1+\cos 2x}{2}} \Rightarrow$$

$$\cos x = -\sqrt{\frac{1+\left(-\frac{5}{12}\right)}{2}} = -\sqrt{\frac{\frac{7}{12}}{2}} = -\sqrt{\frac{7}{24}}$$

$$= -\frac{\sqrt 7}{2\sqrt 6} \cdot \frac{\sqrt 6}{\sqrt 6} = -\frac{\sqrt{42}}{12}$$

67. $\sqrt{\dfrac{1-\cos 40°}{2}} = \sin\dfrac{40°}{2} = \sin 20°$

69. $\sqrt{\dfrac{1-\cos 147°}{1+\cos 147°}} = \tan\dfrac{147°}{2} = \tan 73.5°$

71. $\dfrac{1-\cos 59.74°}{\sin 59.74°} = \tan\dfrac{59.74°}{2} = \tan 29.87°$

73. $\pm\sqrt{\dfrac{1+\cos 18x}{2}} = \cos\dfrac{18x}{2} = \cos 9x$

75. $\pm\sqrt{\dfrac{1-\cos 8\theta}{1+\cos 8\theta}} = \tan\dfrac{8\theta}{2} = \tan 4\theta$

77. $\pm\sqrt{\dfrac{1+\cos\frac{x}{4}}{2}} = \cos\dfrac{\frac{x}{4}}{2} = \cos\dfrac{x}{8}$

79. Verify $\left(\sin x + \cos x\right)^2 = \sin 2x + 1$ is an identity.

$$\left(\sin x + \cos x\right)^2 = \sin^2 x + 2\sin x\cos x + \cos^2 x$$
$$= \left(\sin^2 x + \cos^2 x\right) + 2\sin x\cos x$$
$$= 1 + \sin 2x$$

81. Verify $\left(\cos 2x + \sin 2x\right)^2 = 1 + \sin 4x$ is an
identity.

$$\left(\cos 2x + \sin 2x\right)^2$$
$$= \cos^2 2x + 2\cos 2x\sin 2x + \sin^2 2x$$
$$= \left(\cos^2 2x + \sin^2 2x\right) + 2\cos 2x\sin 2x$$
$$= 1 + \sin 4x$$

83. Verify $\tan 8\theta - \tan 8\theta\tan^2 4\theta = 2\tan 4\theta$ is an
identity.

$$\tan 8\theta - \tan 8\theta\tan^2 4\theta = \tan 8\theta\left(1-\tan^2 4\theta\right)$$
$$= \frac{2\tan 4\theta}{1-\tan^2 4\theta}\left(1-\tan^2 4\theta\right)$$
$$= 2\tan 4\theta$$

85. Verify $\cos 2\theta = \dfrac{2-\sec^2\theta}{\sec^2\theta}$ is an identity.

Working with the right side, we have

$$\dfrac{2-\sec^2\theta}{\sec^2\theta}=\dfrac{2-\frac{1}{\cos^2\theta}}{\frac{1}{\cos^2\theta}}=\dfrac{2-\frac{1}{\cos^2\theta}}{\frac{1}{\cos^2\theta}}\cdot\dfrac{\cos^2\theta}{\cos^2\theta}$$

$$=\dfrac{2\cos^2\theta-1}{1}=\cos 2\theta$$

87. Verify that $\sin 4x = 4\sin x\cos x\cos 2x$ is an identity.

$$\sin 4x = \sin 2(2x) = 2\sin 2x\cos 2x$$
$$=2(2\sin x\cos x)\cos 2x$$
$$=4\sin x\cos x\cos 2x$$

89. Verify $\dfrac{2\cos 2\theta}{\sin 2\theta}=\cot\theta-\tan\theta$ is an identity.

Work with the right side.

$$\cot\theta-\tan\theta=\dfrac{\cos\theta}{\sin\theta}-\dfrac{\sin\theta}{\cos\theta}$$

$$=\dfrac{\cos\theta}{\sin\theta}\cdot\dfrac{\cos\theta}{\cos\theta}-\dfrac{\sin\theta}{\cos\theta}\cdot\dfrac{\sin\theta}{\sin\theta}$$

$$=\dfrac{\cos^2\theta-\sin^2\theta}{\sin\theta\cos\theta}$$

$$=\dfrac{2\left(\cos^2\theta-\sin^2\theta\right)}{2\sin\theta\cos\theta}=\dfrac{2\cos 2\theta}{\sin 2\theta}$$

91. Verify $\sec^2\dfrac{x}{2}=\dfrac{2}{1+\cos x}$ is an identity.

$$\sec^2\dfrac{x}{2}=\dfrac{1}{\cos^2\frac{x}{2}}=\dfrac{1}{\left(\pm\sqrt{\frac{1+\cos x}{2}}\right)^2}=\dfrac{1}{\frac{1+\cos x}{2}}$$

$$=\dfrac{2}{1+\cos x}$$

93. Verify $\sin^2\dfrac{x}{2}=\dfrac{\tan x-\sin x}{2\tan x}$ is an identity.

Work with the left side:

$$\sin^2\dfrac{x}{2}=\left(\pm\sqrt{\dfrac{1-\cos x}{2}}\right)^2=\dfrac{1-\cos x}{2}$$

Work with the right side.

$$\dfrac{\tan x-\sin x}{2\tan x}=\dfrac{\frac{\sin x}{\cos x}-\sin x}{2\cdot\frac{\sin x}{\cos x}}$$

$$=\dfrac{\frac{\sin x}{\cos x}-\sin x}{2\cdot\frac{\sin x}{\cos x}}\cdot\dfrac{\cos x}{\cos x}$$

$$=\dfrac{\sin x-\cos x\sin x}{2\sin x}$$

$$=\dfrac{\sin x(1-\cos x)}{2\sin x}=\dfrac{1-\cos x}{2}$$

$\sin^2\dfrac{x}{2}=\dfrac{1-\cos x}{2}=\dfrac{\tan x-\sin x}{2\tan x}$, so the statement has been verified.

95. Verify $\dfrac{2}{1+\cos x}-\tan^2\dfrac{x}{2}=1$ is an identity.

$$\dfrac{2}{1+\cos x}-\tan^2\dfrac{x}{2}=\dfrac{2}{1+\cos x}-\left(\pm\sqrt{\dfrac{1-\cos x}{1+\cos x}}\right)^2$$

$$=\dfrac{2}{1+\cos x}-\dfrac{1-\cos x}{1+\cos x}$$

$$=\dfrac{2-1+\cos x}{1+\cos x}=\dfrac{1+\cos x}{1+\cos x}$$

$$=1$$

97. Verify $1-\tan^2\dfrac{\theta}{2}=\dfrac{2\cos\theta}{1+\cos\theta}$ is an identity.

$$1-\tan^2\dfrac{\theta}{2}=1-\left(\dfrac{\sin\theta}{1+\cos\theta}\right)^2=1-\dfrac{\sin^2\theta}{\left(1+\cos\theta\right)^2}$$

$$=\dfrac{\left(1+\cos\theta\right)^2-\sin^2\theta}{\left(1+\cos\theta\right)^2}$$

$$=\dfrac{1+2\cos\theta+\cos^2\theta-\sin^2\theta}{\left(1+\cos\theta\right)^2}$$

$$=\dfrac{1+2\cos\theta+\cos^2\theta-\left(1-\cos^2\theta\right)}{\left(1+\cos\theta\right)^2}$$

$$=\dfrac{1+2\cos\theta+2\cos^2\theta-1}{\left(1+\cos\theta\right)^2}$$

$$=\dfrac{2\cos^2\theta+2\cos\theta}{\left(1+\cos\theta\right)^2}$$

$$=\dfrac{2\cos\theta\left(1+\cos\theta\right)}{\left(1+\cos\theta\right)^2}=\dfrac{2\cos\theta}{1+\cos\theta}$$

Exercises 99–105 are graphed in the window $[-2\pi,\ 2\pi]$ by $[-4,\ 4]$.

99. $\cos^4 x-\sin^4 x$ appears to be equivalent to $\cos 2x$.

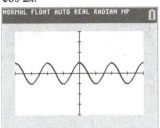

$\cos^4 x-\sin^4 x$
$$=\left(\cos^2 x+\sin^2 x\right)\left(\cos^2 x-\sin^2 x\right)$$
$$=1\cdot\cos 2x=\cos 2x$$

101. $\dfrac{2\tan x}{2-\sec^2 x}$ appears to be equivalent to $\tan 2x$.

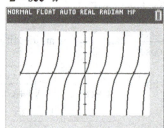

$$\frac{2\tan x}{2-\sec^2 x} = \frac{2\tan x}{1-\left(\sec^2 x-1\right)} = \frac{2\tan x}{1-\tan^2 x} = \tan 2x$$

103. $\dfrac{\sin x}{1+\cos x}$ appears to be equivalent to $\tan\dfrac{x}{2}$.

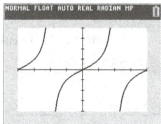

$$\frac{\sin x}{1+\cos x} = \frac{\sin 2\left(\frac{x}{2}\right)}{1+\cos 2\left(\frac{x}{2}\right)} = \frac{2\sin\left(\frac{x}{2}\right)\cos\left(\frac{x}{2}\right)}{1+\left[2\cos^2\left(\frac{x}{2}\right)-1\right]} = \frac{2\sin\left(\frac{x}{2}\right)\cos\left(\frac{x}{2}\right)}{2\cos^2\left(\frac{x}{2}\right)} = \frac{\sin\left(\frac{x}{2}\right)}{\cos\left(\frac{x}{2}\right)} = \tan\left(\frac{x}{2}\right)$$

105. $\dfrac{\tan\frac{x}{2}+\cot\frac{x}{2}}{\cot\frac{x}{2}-\tan\frac{x}{2}}$ appears to be equivalent to

$\sec x$.

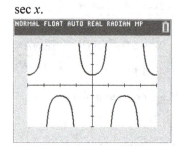

$$\frac{\tan\frac{x}{2}+\cot\frac{x}{2}}{\cot\frac{x}{2}-\tan\frac{x}{2}} = \frac{\dfrac{\sin\frac{x}{2}}{\cos\frac{x}{2}}+\dfrac{\cos\frac{x}{2}}{\sin\frac{x}{2}}}{\dfrac{\cos\frac{x}{2}}{\sin\frac{x}{2}}-\dfrac{\sin\frac{x}{2}}{\cos\frac{x}{2}}} = \frac{\dfrac{\sin\frac{x}{2}}{\cos\frac{x}{2}}+\dfrac{\cos\frac{x}{2}}{\sin\frac{x}{2}}}{\dfrac{\cos\frac{x}{2}}{\sin\frac{x}{2}}-\dfrac{\sin\frac{x}{2}}{\cos\frac{x}{2}}} \cdot \frac{\sin\frac{x}{2}\cos\frac{x}{2}}{\sin\frac{x}{2}\cos\frac{x}{2}} = \frac{\sin^2\frac{x}{2}+\cos^2\frac{x}{2}}{\cos^2\frac{x}{2}-\sin^2\frac{x}{2}} = \frac{1}{\cos 2\left(\frac{x}{2}\right)} = \frac{1}{\cos x} = \sec x$$

107. From Example 5, $W = \dfrac{\left(163\sin 120\pi t\right)^2}{15} \Rightarrow W \approx 1771.3\left(\sin 120\pi t\right)^2$. Thus, we have

$$1771.3\left(\sin 120\pi t\right)^2 = 1771.3\sin 120\pi t \cdot \sin 120\pi t$$

$$= (1771.3)\left(\frac{1}{2}\right)\left[\cos\left(120\pi t - 120\pi t\right) - \cos\left(120\pi t + 120\pi t\right)\right]$$

$$= 885.6\left(\cos 0 - \cos 240\pi t\right) = 885.6\left(1 - \cos 240\pi t\right) = -885.6\cos 240\pi t + 885.6$$

If we compare this to $W = a\cos\left(\omega t\right) + c$, then $a = -885.6$, $c = 885.6$, and $\omega = 240\pi$. The graph shows

$$y_1 = \frac{\left(163\sin 120\pi t\right)^2}{15} \text{ and } y_2 = -885.6\cos 240\pi t + 885.6$$

graphed in the window $[0, 0.05] \times [0, 2000]$.

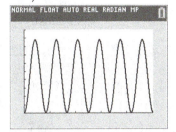

109. $\sin\dfrac{\theta}{2} = \dfrac{1}{m}, \quad m = \dfrac{5}{4}$

Because $\sin\dfrac{\theta}{2} = \dfrac{1}{\frac{5}{4}} = \dfrac{4}{5}$ and $\sin^2\dfrac{\theta}{2} = \dfrac{1-\cos\theta}{2}$, we have

$$\left(\frac{4}{5}\right)^2 = \frac{1-\cos\theta}{2} \Rightarrow \frac{16}{25} = \frac{1-\cos\theta}{2} \Rightarrow \frac{32}{25} = 1 - \cos\theta \Rightarrow \frac{7}{25} = -\cos\theta \Rightarrow \cos\theta = -\frac{7}{25}.$$

Thus, we have $\theta = \cos^{-1}\left(-\dfrac{7}{25}\right) \approx 106°$.

111. $\sin\dfrac{\theta}{2} = \dfrac{1}{m}, \quad \theta = 60°$

$$\sin\frac{60°}{2} = \frac{1}{m} \Rightarrow \sin 30° = \frac{1}{m} \Rightarrow \frac{1}{2} = \frac{1}{m} \Rightarrow m = 2$$

Summary Exercises on Verifying Trigonometric Identities

For the following exercises, other solutions are possible.

1. Verify $\tan\theta + \cot\theta = \sec\theta\csc\theta$ is an identity.

$$\tan\theta + \cot\theta = \frac{\sin\theta}{\cos\theta} + \frac{\cos\theta}{\sin\theta} = \frac{\sin^2\theta}{\cos\theta\sin\theta} + \frac{\cos^2\theta}{\cos\theta\sin\theta} = \frac{\sin^2\theta + \cos^2\theta}{\cos\theta\sin\theta} = \frac{1}{\cos\theta\sin\theta} = \frac{1}{\cos\theta}\cdot\frac{1}{\sin\theta}$$
$$= \sec\theta\csc\theta$$

3. Verify $\tan\dfrac{x}{2} = \csc x - \cot x$ is an identity.

Starting on the right side, we have $\csc x - \cot x = \dfrac{1}{\sin x} - \dfrac{\cos x}{\sin x} = \dfrac{1-\cos x}{\sin x} = \tan\dfrac{x}{2}$

5. Verify $\dfrac{\sin t}{1+\cos t} = \dfrac{1-\cos t}{\sin t}$ is an identity.

$$\frac{\sin t}{1+\cos t} = \frac{\sin t}{1+\cos t}\cdot\frac{1-\cos t}{1-\cos t} = \frac{\sin t\left(1-\cos t\right)}{1-\cos^2 t} = \frac{\sin t\left(1-\cos t\right)}{\sin^2 t} = \frac{1-\cos t}{\sin t}$$

7. Verify $\sin 2\theta = \dfrac{2\tan\theta}{1+\tan^2\theta}$ is an identity.

Starting on the right side, we have

$$\frac{2\tan\theta}{1+\tan^2\theta} = \frac{2\tan\theta}{\sec^2\theta} = \frac{2\cdot\frac{\sin\theta}{\cos\theta}}{\frac{1}{\cos^2\theta}} = 2\cdot\frac{\sin\theta}{\cos\theta}\cdot\frac{\cos^2\theta}{1} = 2\sin\theta\cos\theta = \sin 2\theta$$

9. Verify $\cot\theta - \tan\theta = \dfrac{2\cos^2\theta - 1}{\sin\theta\cos\theta}$ is an identity.

$$\cot\theta - \tan\theta = \frac{\cos\theta}{\sin\theta} - \frac{\sin\theta}{\cos\theta} = \frac{\cos^2\theta}{\sin\theta\cos\theta} - \frac{\sin^2\theta}{\sin\theta\cos\theta} = \frac{\cos^2\theta - \sin^2\theta}{\sin\theta\cos\theta} = \frac{\cos^2\theta - \left(1-\cos^2\theta\right)}{\sin\theta\cos\theta}$$

$$= \frac{\cos^2\theta - 1 + \cos^2\theta}{\sin\theta\cos\theta} = \frac{2\cos^2\theta - 1}{\sin\theta\cos\theta}$$

11. Verify $\dfrac{\sin(x+y)}{\cos(x-y)} = \dfrac{\cot x + \cot y}{1+\cot x\cot y}$ is an identity.

$$\frac{\sin(x+y)}{\cos(x-y)} = \frac{\sin x\cos y + \cos x\sin y}{\cos x\cos y + \sin x\sin y} = \frac{\sin x\cos y + \cos x\sin y}{\cos x\cos y + \sin x\sin y}\cdot\frac{\frac{1}{\cos x\cos y}}{\frac{1}{\cos x\cos y}} = \frac{\frac{\sin x\cos y}{\cos x\cos y} + \frac{\cos x\sin y}{\cos x\cos y}}{\frac{\cos x\cos y}{\cos x\cos y} + \frac{\sin x\sin y}{\cos x\cos y}}$$

$$= \frac{\frac{\sin x}{\cos x} + \frac{\sin y}{\cos y}}{1 + \frac{\sin x}{\cos x}\cdot\frac{\sin y}{\cos y}} = \frac{\cot x + \cot y}{1+\cot x\cot y}$$

13. Verify $\dfrac{\sin\theta + \tan\theta}{1+\cos\theta} = \tan\theta$ is an identity.

$$\frac{\sin\theta + \tan\theta}{1+\cos\theta} = \frac{\sin\theta + \frac{\sin\theta}{\cos\theta}}{1+\cos\theta} = \frac{\sin\theta + \frac{\sin\theta}{\cos\theta}}{1+\cos\theta}\cdot\frac{\cos\theta}{\cos\theta} = \frac{\sin\theta\cos\theta + \sin\theta}{\cos\theta(1+\cos\theta)} = \frac{\sin\theta(\cos\theta + 1)}{\cos\theta(1+\cos\theta)} = \frac{\sin\theta}{\cos\theta} = \tan\theta$$

15. Verify $\cos x = \dfrac{1-\tan^2\frac{x}{2}}{1+\tan^2\frac{x}{2}}$ is an identity.

$$\frac{1-\tan^2\frac{x}{2}}{1+\tan^2\frac{x}{2}} = \frac{1-\left(\frac{1-\cos x}{\sin x}\right)^2}{1+\left(\frac{1-\cos x}{\sin x}\right)^2} = \frac{1-\frac{(1-\cos x)^2}{\sin^2 x}}{1+\frac{(1-\cos x)^2}{\sin^2 x}} = \frac{1-\frac{(1-\cos x)^2}{\sin^2 x}}{1+\frac{(1-\cos x)^2}{\sin^2 x}}\cdot\frac{\sin^2 x}{\sin^2 x} = \frac{\sin^2 x - (1-\cos x)^2}{\sin^2 x + (1-\cos x)^2}$$

$$= \frac{\sin^2 x - \left(1-2\cos x + \cos^2 x\right)}{\sin^2 x + \left(1-2\cos x + \cos^2 x\right)} = \frac{\sin^2 x - 1 + 2\cos x - \cos^2 x}{\sin^2 x + 1 - 2\cos x + \cos^2 x} = \frac{\left(1-\cos^2 x\right) - 1 + 2\cos x - \cos^2 x}{\left(\sin^2 x + \cos^2 x\right) + 1 - 2\cos x}$$

$$= \frac{1-\cos^2 x - 1 + 2\cos x - \cos^2 x}{1+1-2\cos x} = \frac{2\cos x - 2\cos^2 x}{2-2\cos x} = \frac{2\cos x(1-\cos x)}{2(1-\cos x)} = \cos x$$

17. Verify $\dfrac{\tan^2 t + 1}{\tan t \csc^2 t} = \tan t$ is an identity.

$$\frac{\tan^2 t + 1}{\tan t \csc^2 t} = \frac{\dfrac{\sin^2 t}{\cos^2 t} + 1}{\dfrac{\sin t}{\cos t} \cdot \dfrac{1}{\sin^2 t}} = \frac{\dfrac{\sin^2 t}{\cos^2 t} + 1}{\dfrac{1}{\cos t \sin t}} = \frac{\dfrac{\sin^2 t}{\cos^2 t} + 1}{\dfrac{1}{\cos t \sin t}} \cdot \frac{\cos^2 t \sin t}{\cos^2 t \sin t} = \frac{\sin^3 t + \cos^2 t \sin t}{\cos t}$$

$$= \frac{\sin t \left(\sin^2 t + \cos^2 t \right)}{\cos t} = \frac{\sin t (1)}{\cos t} = \frac{\sin t}{\cos t} = \tan t$$

19. Verify $\tan 4\theta = \dfrac{2 \tan 2\theta}{2 - \sec^2 2\theta}$ is an identity.

$$\frac{2 \tan 2\theta}{2 - \sec^2 2\theta} = \frac{2 \cdot \dfrac{\sin 2\theta}{\cos 2\theta}}{2 - \dfrac{1}{\cos^2 2\theta}} = \frac{2 \cdot \dfrac{\sin 2\theta}{\cos 2\theta}}{2 - \dfrac{1}{\cos^2 2\theta}} \cdot \frac{\cos^2 2\theta}{\cos^2 2\theta} = \frac{2 \sin 2\theta \cos 2\theta}{2 \cos^2 2\theta - 1} = \frac{\sin \left[2(2\theta) \right]}{\cos \left[2(2\theta) \right]} = \frac{\sin 4\theta}{\cos 4\theta} = \tan 4\theta$$

21. Verify $\dfrac{\cot s - \tan s}{\cos s + \sin s} = \dfrac{\cos s - \sin s}{\sin s \cos s}$ is an identity.

$$\frac{\cot s - \tan s}{\cos s + \sin s} = \frac{\dfrac{\cos s}{\sin s} - \dfrac{\sin s}{\cos s}}{\cos s + \sin s} = \frac{\dfrac{\cos s}{\sin s} - \dfrac{\sin s}{\cos s}}{\cos s + \sin s} \cdot \frac{\sin s \cos s}{\sin s \cos s} = \frac{\cos^2 s - \sin^2 s}{\left(\cos s + \sin s \right) \sin s \cos s}$$

$$= \frac{\left(\cos s + \sin s \right)\left(\cos s - \sin s \right)}{\left(\cos s + \sin s \right) \sin s \cos s} = \frac{\cos s - \sin s}{\sin s \cos s}$$

23. Verify $\dfrac{\tan(x + y) - \tan y}{1 + \tan(x + y) \tan y} = \tan x$ is an identity.

$$\frac{\tan(x + y) - \tan y}{1 + \tan(x + y) \tan y} = \frac{\dfrac{\tan x + \tan y}{1 - \tan x \tan y} - \tan y}{1 + \dfrac{\tan x + \tan y}{1 - \tan x \tan y} \cdot \tan y} = \frac{\dfrac{\tan x + \tan y}{1 - \tan x \tan y} - \tan y}{1 + \dfrac{\tan x + \tan y}{1 - \tan x \tan y} \cdot \tan y} \cdot \frac{1 - \tan x \tan y}{1 - \tan x \tan y}$$

$$= \frac{\tan x + \tan y - \tan y \left(1 - \tan x \tan y \right)}{1 - \tan x \tan y + \left(\tan x + \tan y \right) \tan y} = \frac{\tan x + \tan x \tan^2 y}{1 - \tan x \tan y + \tan x \tan y + \tan^2 y}$$

$$= \frac{\tan x \left(1 + \tan^2 y \right)}{1 + \tan^2 y} = \tan x$$

25. Verify $\dfrac{\cos^4 x - \sin^4 x}{\cos^2 x} = 1 - \tan^2 x$ is an identity.

$$\frac{\cos^4 x - \sin^4 x}{\cos^2 x} = \frac{\left(\cos^2 x + \sin^2 x \right)\left(\cos^2 x - \sin^2 x \right)}{\cos^2 x} = \frac{(1)\left(\cos^2 x - \sin^2 x \right)}{\cos^2 x} = \frac{\cos^2 x - \sin^2 x}{\cos^2 x}$$

$$= \frac{\cos^2 x}{\cos^2 x} - \frac{\sin^2 x}{\cos^2 x} = 1 - \tan^2 x$$

Section 7.5   Inverse Circular Functions

1. For a function to have an inverse, it must be <u>one</u>-to-<u>one</u>.

3. $y = \cos^{-1} x$ means that $x = \underline{\cos y}$ for $0 \le y \le \pi$.

5. If a function f has an inverse and $f(\pi) = -1$, then $f^{-1}(-1) = \underline{\pi}$.

7. (a) $[-1, 1]$

   (b) $\left[-\dfrac{\pi}{2}, \dfrac{\pi}{2}\right]$

   (c) increasing

   (d) -2 is not in the domain.

9. (a) $(-\infty, \infty)$

   (b) $\left(-\dfrac{\pi}{2}, \dfrac{\pi}{2}\right)$

   (c) increasing

   (d) no

11. The interval must be chosen so that the function is one-to-one. The sine and cosine functions are not one-to-one on the same intervals.

13. $y = \sin^{-1} 0$

$\sin y = 0,\ -\dfrac{\pi}{2} \le y \le \dfrac{\pi}{2}$

$\sin 0 = 0,\ \text{so } y = 0.$

15. $y = \cos^{-1}(-1)$

$\cos y = -1,\ 0 \le y \le \pi$

$\cos \pi = -1,\ \text{so } y = \pi.$

17. $y = \tan^{-1} 1$

$\tan y = 1,\ -\dfrac{\pi}{2} < y < \dfrac{\pi}{2}$

$\tan \dfrac{\pi}{4} = 1,\ \text{so } y = \dfrac{\pi}{4}.$

19. $y = \arctan 0$

$\tan y = 0,\ -\dfrac{\pi}{2} < y < \dfrac{\pi}{2}$

$\tan 0 = 0,\ \text{so } y = 0.$

21. $y = \arcsin\left(-\dfrac{\sqrt{3}}{2}\right)$

$\sin y = -\dfrac{\sqrt{3}}{2},\ -\dfrac{\pi}{2} \le y \le \dfrac{\pi}{2}$

$\sin\left(-\dfrac{\pi}{3}\right) = -\dfrac{\sqrt{3}}{2},\ \text{so } y = -\dfrac{\pi}{3}.$

23. $y = \arccos\left(-\dfrac{\sqrt{3}}{2}\right)$

$\cos y = -\dfrac{\sqrt{3}}{2},\ 0 \le y \le \pi$

$\cos \dfrac{5\pi}{6} = -\dfrac{\sqrt{3}}{2},\ \text{so } y = \dfrac{5\pi}{6}.$

25. $y = \sin^{-1} \sqrt{3}$

$\sin y = \sqrt{3},\ -\dfrac{\pi}{2} \le y \le \dfrac{\pi}{2}$

$\sqrt{3} > 1,$ so there is no angle θ such that $\sin \theta = \sqrt{3}$. Thus, $\sin^{-1} \sqrt{3}$ does not exist.

27. $y = \cot^{-1}(-1)$

$\cot y = -1,\ 0 < y < \pi$

y is in quadrant II. The reference angle is $\dfrac{\pi}{4}$.

$\cot \dfrac{3\pi}{4} = 1,\ \text{so } y = \dfrac{3\pi}{4}.$

29. $y = \csc^{-1}(-2)$

$\csc y = -2,\ -\dfrac{\pi}{2} \le y \le \dfrac{\pi}{2},\ y \ne 0$

y is in quadrant IV. The reference angle is $\dfrac{\pi}{6}$.

$\csc\left(-\dfrac{\pi}{6}\right) = -2,\ \text{so } y = -\dfrac{\pi}{6}.$

31. $y = \text{arc sec}\left(\dfrac{2\sqrt{3}}{3}\right)$

$\sec y = \dfrac{2\sqrt{3}}{3},\ 0 \le y \le \pi,\ y \ne \dfrac{\pi}{2}$

$\sec \dfrac{\pi}{6} = \dfrac{2\sqrt{3}}{3},\ \text{so } y = \dfrac{\pi}{6}.$

33. $y = \sec^{-1} 1$

$\sec\ y = 1,\ 0 \le y \le \pi,\ y \ne \dfrac{\pi}{2}$

$\sec 0 = 1,\ \text{so } y = 0.$

35. $y = \csc^{-1}\left(\dfrac{\sqrt{2}}{2}\right)$

$\csc y = \dfrac{\sqrt{2}}{2},\ -\dfrac{\pi}{2} \le y \le \dfrac{\pi}{2},\ y \ne 0$

There is no angle θ such that $\csc\theta = \dfrac{\sqrt{2}}{2}$.

Thus, $\csc^{-1}\left(\dfrac{\sqrt{2}}{2}\right)$ does not exist.

37. $\theta = \arctan(-1)$

$\tan\theta = -1,\ -90° < \theta < 90°$

θ is in quadrant IV. The reference angle is $45°$. Thus, $\theta = -45°$.

39. $\theta = \arcsin\left(-\dfrac{\sqrt{3}}{2}\right)$

$\sin\theta = -\dfrac{\sqrt{3}}{2},\ -90° \le \theta \le 90°$

θ is in quadrant IV. The reference angle is $60°$. $\theta = -60°$.

41. $\theta = \arccos\left(-\dfrac{1}{2}\right)$

$\cos\theta = -\dfrac{1}{2},\ 0° \le \theta \le 180°$

θ is in quadrant II. The reference angle is $60°$. Thus, $\theta = 180° - 60° = 120°$.

43. $\theta = \cot^{-1}\left(-\dfrac{\sqrt{3}}{3}\right)$

$\cot\theta = -\dfrac{\sqrt{3}}{3},\ 0° < \theta < 180°$

θ is in quadrant II. The reference angle is $60°$. $\theta = 180° - 60° = 120°$

45. $\theta = \csc^{-1}(-2)$

$\csc\theta = -2$ and $-90° < \theta < 90°,\ \theta \ne 0°$

θ is in quadrant IV. The reference angle is $30°$. $\theta = -30°$

47. $\theta = \sin^{-1} 2$

$\sin\theta = 2,\ 0° \le \theta \le 180°$

There is no angle θ such that $\sin\theta = 2$.

For Exercises 49–57, be sure that your calculator is in degree mode.

49. $\theta = \sin^{-1}(-0.13349122) = -7.6713835°$

51. $\theta = \arccos(-0.39876459) \approx 113.500970°$

53. $\theta = \csc^{-1} 1.9422833 \approx 30.987961°$

55. $\theta = \cot^{-1}(-0.60724226)$

$= \tan^{-1}\left(-\dfrac{1}{0.60724226}\right) \approx -58.7321071°$

Note that θ is in quadrant II because $\cot^{-1}\theta$ is defined for $0° < \theta < 180°$. So,
$\theta = -58.7321071° + 180° = 121.267893°$

57. $\theta = \tan^{-1}(-7.7828641) \approx -82.67832938°$

Note that $\tan^{-1} y$ is defined for $-90° < y < 90°$, so

$\tan^{-1}(-7.7828641) \approx -82.678329°$

For Exercises 59–67, be sure that your calculator is in radian mode.

59. $y = \arcsin 0.92837781 \approx 1.1900238$

61. $y = \cos^{-1}(-0.32647891) \approx 1.9033723$

63. $y = \arctan 1.1111111 \approx 0.83798122$

65. $y = \cot^{-1}(-0.92170128)$

$= \tan^{-1}\left(\dfrac{1}{-0.92170128}\right) \approx 2.3154725$

67. $y = \sec^{-1}(-1.2871684)$

$= \cos^{-1}\left(-\dfrac{1}{1.2871684}\right) \approx 2.4605221$

69.

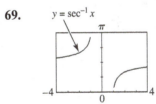

71.

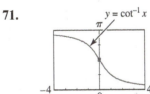

73.

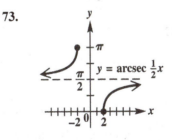

75. $\tan\left(\arccos\dfrac{3}{4}\right)$

Let $\omega = \arccos\dfrac{3}{4}$, so that $\cos\omega = \dfrac{3}{4}$.

Because arccos is defined only in quadrants I

and II, and $\dfrac{3}{4}$ is positive, ω is in quadrant I.

Sketch ω and label a triangle with the side opposite ω equal to

$$\sqrt{4^2 - 3^2} = \sqrt{16 - 9} = \sqrt{7}.$$

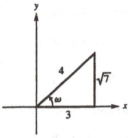

$$\tan\left(\arccos\dfrac{3}{4}\right) = \tan\omega = \dfrac{\sqrt{7}}{3}$$

77. $\cos(\tan^{-1}(-2))$

Let $\omega = \tan^{-1}(-2)$, so that $\tan\omega = -2$.

Because $\tan^{-1}$ is defined only in quadrants I and IV, and –2 is negative, ω is in quadrant IV. Sketch ω and label a triangle with the hypotenuse equal to

$$\sqrt{(-2)^2 + 1} = \sqrt{4 + 1} = \sqrt{5}.$$

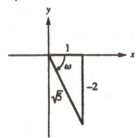

$$\cos(\tan^{-1}(-2)) = \cos\omega = \dfrac{\sqrt{5}}{5}$$

79. $\sin\left(2\tan^{-1}\dfrac{12}{5}\right)$

Let $\omega = \tan^{-1}\dfrac{12}{5}$, so that $\tan\omega = \dfrac{12}{5}$.

Because $\tan^{-1}\omega$ is defined only in quadrants

I and IV, and $\dfrac{12}{5}$ is positive, ω is in quadrant

I.

Sketch ω and label a right triangle with the hypotenuse equal to

$$\sqrt{12^2 + 5^2} = \sqrt{144 + 25} = \sqrt{169} = 13.$$

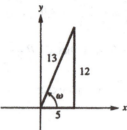

$$\sin\omega = \dfrac{12}{13}; \cos\omega = \dfrac{5}{13}$$

$$\sin\left(2\tan^{-1}\dfrac{12}{5}\right) = \sin(2\omega) = 2\sin\omega\cos\omega$$

$$= 2\left(\dfrac{12}{13}\right)\left(\dfrac{5}{13}\right) = \dfrac{120}{169}$$

81. $\cos\left(2\arctan\dfrac{4}{3}\right)$

Let $\omega = \arctan\dfrac{4}{3}$, so that $\tan\omega = \dfrac{4}{3}$.

Because arctan is defined only in quadrants I

and IV, and $\dfrac{4}{3}$ is positive, ω is in quadrant I.

Sketch ω and label a triangle with the hypotenuse equal to

$$\sqrt{4^2 + 3^2} = \sqrt{16 + 9} = \sqrt{25} = 5.$$

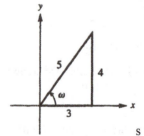

$$\cos\omega = \dfrac{3}{5}; \sin\omega = \dfrac{4}{5}$$

$$\cos\left(2\arctan\dfrac{4}{3}\right) = \cos(2\omega) = \cos^2\omega - \sin^2\omega$$

$$= \left(\dfrac{3}{5}\right)^2 - \left(\dfrac{4}{5}\right)^2$$

$$= \dfrac{9}{25} - \dfrac{16}{25} = -\dfrac{7}{25}$$

83. $\sin\left(2\cos^{-1}\dfrac{1}{5}\right)$

Let $\theta = \cos^{-1}\dfrac{1}{5}$, so that $\cos\theta = \dfrac{1}{5}$. The inverse cosine function yields values only in quadrants I and II, and because $\dfrac{1}{5}$ is positive, θ is in quadrant I. Sketch θ and label the sides of the right triangle. By the Pythagorean theorem, the length opposite to θ will be $\sqrt{5^2 - 1^2} = \sqrt{24} = 2\sqrt{6}$.

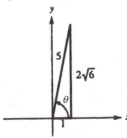

From the figure, $\sin\theta = \dfrac{2\sqrt{6}}{5}$.

Then,

$\sin\left(2\cos^{-1}\dfrac{1}{5}\right) = \sin 2\theta = 2\sin\theta\,\cos\theta$

$\qquad\qquad = 2\left(\dfrac{2\sqrt{6}}{5}\right)\left(\dfrac{1}{5}\right) = \dfrac{4\sqrt{6}}{25}$

85. $\sec(\sec^{-1} 2)$

Secant and inverse secant are inverse functions, so $\sec\left(\sec^{-1} 2\right) = 2$.

87. $\cos\left(\tan^{-1}\dfrac{5}{12} - \tan^{-1}\dfrac{3}{4}\right)$

Let $\alpha = \tan^{-1}\dfrac{5}{12}$ and $\beta = \tan^{-1}\dfrac{3}{4}$. Then $\tan\alpha = \dfrac{5}{12}$ and $\tan\beta = \dfrac{3}{4}$. Sketch angles α and β, both in quadrant I.

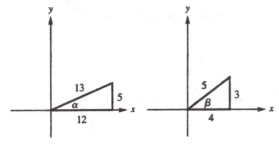

We have $\sin\alpha = \dfrac{5}{13}$, $\cos\alpha = \dfrac{12}{13}$, $\sin\beta = \dfrac{3}{5}$, and $\cos\beta = \dfrac{4}{5}$.

$\cos\left(\tan^{-1}\dfrac{5}{12} - \tan^{-1}\dfrac{3}{4}\right)$
$= \cos(\alpha - \beta) = \cos\alpha\cos\beta + \sin\alpha\sin\beta$
$= \left(\dfrac{12}{13}\right)\left(\dfrac{4}{5}\right) + \left(\dfrac{5}{13}\right)\left(\dfrac{3}{5}\right) = \dfrac{48}{65} + \dfrac{15}{65} = \dfrac{63}{65}$

89. $\sin\left(\sin^{-1}\dfrac{1}{2} + \tan^{-1}(-3)\right)$

Let $\sin^{-1}\dfrac{1}{2} = A$ and $\tan^{-1}(-3) = B$.

Then $\sin A = \dfrac{1}{2}$ and $\tan B = -3$. Sketch angle A in quadrant I and angle B in quadrant IV.

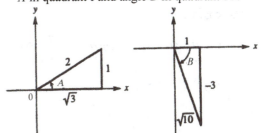

We have $\cos A = \dfrac{\sqrt{3}}{2}$, $\sin A = \dfrac{1}{2}$,

$\cos B = \dfrac{1}{\sqrt{10}} = \dfrac{\sqrt{10}}{10}$, and

$\sin B = \dfrac{-3}{\sqrt{10}} = -\dfrac{3\sqrt{10}}{10}$.

$\sin\left(\sin^{-1}\dfrac{1}{2} + \tan^{-1}(-3)\right)$
$= \sin(A + B) = \sin A\cos B + \cos A\sin B$
$= \dfrac{1}{2}\cdot\dfrac{1}{\sqrt{10}} + \dfrac{\sqrt{3}}{2}\cdot\dfrac{-3}{\sqrt{10}}$
$= \dfrac{1 - 3\sqrt{3}}{2\sqrt{10}} = \dfrac{\sqrt{10} - 3\sqrt{30}}{20}$

For Exercises 91−93, your calculator could be in either degree or radian mode.

91. $\cos(\tan^{-1} 0.5) \approx 0.894427191$

93. $\tan(\arcsin 0.12251014) \approx 0.1234399811$

95. $\sin(\arccos u)$

Let $\theta = \arccos u$, so $\cos\theta = u = \dfrac{u}{1}$. Because

$u > 0,\ 0 < \theta < \dfrac{\pi}{2}$.

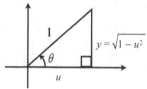

Because $y > 0$, from the Pythagorean theorem,

$y = \sqrt{1^2 - u^2} = \sqrt{1 - u^2}$.

Therefore, $\sin\theta = \dfrac{\sqrt{1 - u^2}}{1} = \sqrt{1 - u^2}$. Thus,

$\sin(\arccos u) = \sqrt{1 - u^2}$.

97. $\cos(\arcsin u)$

Let $\theta = \arcsin u$, so $\sin\theta = u = \dfrac{u}{1}$. Because

$u > 0,\ 0 < \theta < \dfrac{\pi}{2}$.

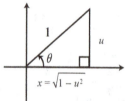

Because $x > 0$, from the Pythagorean theorem,

$x = \sqrt{1^2 - u^2} = \sqrt{1 - u^2}$.

Therefore, $\cos\theta = \dfrac{\sqrt{1 - u^2}}{1} = \sqrt{1 - u^2}$. Thus,

$\cos(\arcsin u) = \sqrt{1 - u^2}$

99. $\sin\left(2\sec^{-1}\dfrac{u}{2}\right)$

Let $\theta = \sec^{-1}\dfrac{u}{2}$, so $\sec\theta = \dfrac{u}{2}$. Because

$u > 0,\ 0 < \theta < \dfrac{\pi}{2}$.

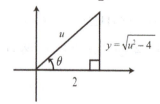

Because $y > 0$, from the Pythagorean theorem,

$y = \sqrt{u^2 - 2^2} = \sqrt{u^2 - 4}$. Now find $\sin 2\theta$.

$\sin\theta = \dfrac{\sqrt{u^2 - 4}}{u}$ and $\cos\theta = \dfrac{2}{u}$ Thus,

$\sin\left(2\sec^{-1}\dfrac{u}{2}\right) = \sin 2\theta = 2\sin\theta\cos\theta$

$= 2\left(\dfrac{\sqrt{u^2 - 4}}{u}\right)\left(\dfrac{2}{u}\right)$

$= \dfrac{4\sqrt{u^2 - 4}}{u^2}$

101. $\tan\left(\sin^{-1}\dfrac{u}{\sqrt{u^2 + 2}}\right)$

Let $\theta = \sin^{-1}\dfrac{u}{\sqrt{u^2 + 2}}$, so $\sin\theta = \dfrac{u}{\sqrt{u^2 + 2}}$.

Because $u > 0,\ 0 < \theta < \dfrac{\pi}{2}$.

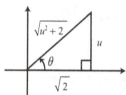

Because $x > 0$, from the Pythagorean theorem,

$x = \sqrt{\left(\sqrt{u^2 + 2}\right)^2 - u^2} = \sqrt{u^2 + 2 - u^2} = \sqrt{2}$.

Therefore, $\tan\theta = \dfrac{u}{\sqrt{2}} = \dfrac{u\sqrt{2}}{2}$. Thus,

$\tan\left(\sin^{-1}\dfrac{u}{\sqrt{u^2 + 2}}\right) = \dfrac{u\sqrt{2}}{2}$.

103. $\sec\left(\operatorname{arc\,cot}\dfrac{\sqrt{4 - u^2}}{u}\right)$

Let $\theta = \operatorname{arc\,cot}\dfrac{\sqrt{4 - u^2}}{u}$, so $\cot\theta = \dfrac{\sqrt{4 - u^2}}{u}$.

Because $u > 0,\ 0 < \theta < \dfrac{\pi}{2}$.

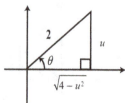

(continued on next page)

(continued)

From the Pythagorean theorem,

$$r = \sqrt{\left(\sqrt{4-u^2}\right)^2 + u^2} = \sqrt{4-u^2+u^2}$$
$$= \sqrt{4} = 2.$$

Therefore, $\sec\theta = \dfrac{2}{\sqrt{4-u^2}} = \dfrac{2\sqrt{4-u^2}}{4-u^2}.$

Thus, $\sec\left(\text{arc cot}\dfrac{\sqrt{4-u^2}}{u}\right) = \dfrac{2\sqrt{4-u^2}}{4-u^2}.$

105. From Example 8 in the text, we have

$$\theta = \arcsin\sqrt{\dfrac{v^2}{2v^2+64h}}.$$

$$\theta = \arcsin\sqrt{\dfrac{32^2}{2(32^2)+64(5.0)}} \approx 41°$$

107. $\theta = \tan^{-1}\left(\dfrac{x}{x^2+2}\right)$

(a) $x = 1,$

$$\theta = \tan^{-1}\left(\dfrac{1}{1^2+2}\right) = \tan^{-1}\left(\dfrac{1}{3}\right) \approx 18°$$

(b) $x = 2,$

$$\theta = \tan^{-1}\left(\dfrac{2}{2^2+2}\right) = \tan^{-1}\dfrac{2}{6}$$
$$= \tan^{-1}\dfrac{1}{3} \approx 18°$$

(c) $x = 3,$

$$\theta = \tan^{-1}\left(\dfrac{3}{3^2+2}\right) = \tan^{-1}\dfrac{3}{11} \approx 15°$$

(d) $\tan(\theta+\alpha) = \dfrac{1+1}{x} = \dfrac{2}{x}$ and $\tan\alpha = \dfrac{1}{x}$

$$\tan(\theta+\alpha) = \dfrac{\tan\theta+\tan\alpha}{1-\tan\theta\tan\alpha}$$
$$\dfrac{2}{x} = \dfrac{\tan\theta+\frac{1}{x}}{1-\tan\theta\left(\frac{1}{x}\right)} = \dfrac{x\tan\theta+1}{x-\tan\theta}$$
$$2(x-\tan\theta) = x(x\tan\theta+1)$$
$$2x-2\tan\theta = x^2\tan\theta+x$$
$$2x-x = x^2\tan\theta+2\tan\theta$$
$$x = \tan\theta\left(x^2+2\right)$$
$$\tan\theta = \dfrac{x}{x^2+2}$$
$$\theta = \tan^{-1}\left(\dfrac{x}{x^2+2}\right)$$

(e) If we graph $y_1 = \tan^{-1}\left(\dfrac{x}{x^2+2}\right)$ using a graphing calculator, the maximum value of the function occurs when x is 1.4142123 m. (Note: Due to the computational routine, there may be a discrepancy in the last few decimal places.)

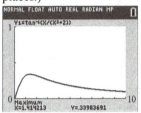

(f) $x = \sqrt{(1)(2)} = \sqrt{2}$

109. The diameter of the earth is 7927 miles at the equator, so the radius of the earth is 3963.5 miles. Then

$$\cos\theta = \dfrac{3963.5}{20,000+3963.5} = \dfrac{3963.5}{23,963.5}$$ and

$$\theta = \arccos\left(\dfrac{3963.5}{23,963.5}\right) \approx 80.48°.$$

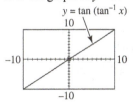

The percent of the equator that can be seen by the satellite is $\dfrac{2\theta}{360}\cdot100 = \dfrac{2(80.48)}{360} \approx 44.7\%.$

111. $f(x) = 3x-2;\ f^{-1}(x) = \dfrac{1}{3}x+\dfrac{2}{3}$

$$f\left[f^{-1}(x)\right] = f\left[\dfrac{1}{3}x+\dfrac{2}{3}\right] = 3\left(\dfrac{1}{3}x+\dfrac{2}{3}\right)-2$$
$$= x+2-2 = x$$
$$f^{-1}\left[f(x)\right] = f^{-1}[3x-2] = \dfrac{3x-2}{3}+\dfrac{2}{3}$$
$$= \dfrac{3x-2+2}{3} = x$$

In each case, the result is x.

113. It is the graph of $y = x$.

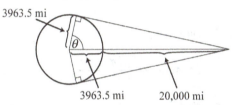

Section 7.6 Trigonometric Equations

Use the unit circle below to solve each equation.

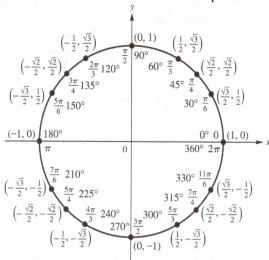

The unit circle $x^2 + y^2 = 1$

1. $\cos x = \dfrac{1}{2}$

Solution set: $\left\{\dfrac{\pi}{3}, \dfrac{5\pi}{3}\right\}$

3. $\sin x = -\dfrac{1}{2}$

Solution set: $\left\{\dfrac{7\pi}{6}, \dfrac{11\pi}{6}\right\}$

5. $\sin \theta = -1$

Solution set: $\left\{270°\right\}$

7. $\cos 2x = \dfrac{1}{2}$

$2x = \dfrac{\pi}{3}$	$2x = \dfrac{\pi}{3} + 2\pi$	$2x = \dfrac{5\pi}{3}$	$2x = \dfrac{5\pi}{3} + 2\pi$
$x = \dfrac{\pi}{6}$	$2x = \dfrac{7\pi}{3}$	$x = \dfrac{5\pi}{6}$	$2x = \dfrac{11\pi}{3}$
	$x = \dfrac{7\pi}{6}$		$x = \dfrac{11\pi}{6}$

Solution set: $\left\{\dfrac{\pi}{6}, \dfrac{5\pi}{6}, \dfrac{7\pi}{6}, \dfrac{11\pi}{6}\right\}$

9. $\sin 2x = -\dfrac{1}{2}$

$2x = \dfrac{7\pi}{6}$	$2x = \dfrac{7\pi}{6} + 2\pi$	$2x = \dfrac{11\pi}{6}$	$2x = \dfrac{11\pi}{6} + 2\pi$
$x = \dfrac{7\pi}{12}$	$2x = \dfrac{19\pi}{6}$	$x = \dfrac{11\pi}{12}$	$2x = \dfrac{23\pi}{6}$
	$x = \dfrac{19\pi}{12}$		$x = \dfrac{23\pi}{12}$

Solution set: $\left\{\dfrac{7\pi}{12}, \dfrac{11\pi}{12}, \dfrac{19\pi}{12}, \dfrac{23\pi}{12}\right\}$

11. $\sin \dfrac{\theta}{2} = -1$

$\dfrac{\theta}{2} = 270° \Rightarrow \theta = 540°$

This solution is not included in the given interval, so the solution set is $\varnothing$.

13. $-30°$ is not in the interval $\left[0°, 360°\right)$.

15. $2\cot x + 1 = -1 \Rightarrow 2\cot x = -2 \Rightarrow \cot x = -1$

Over the interval $\left[0, 2\pi\right)$, the equation $\cot x = -1$ has two solutions, the angles in quadrants II and IV that have a reference angle of $\dfrac{\pi}{4}$. These are $\dfrac{3\pi}{4}$ and $\dfrac{7\pi}{4}$.

Solution set: $\left\{\dfrac{3\pi}{4}, \dfrac{7\pi}{4}\right\}$

17. $2\sin x + 3 = 4 \Rightarrow 2\sin x = 1 \Rightarrow \sin x = \dfrac{1}{2}$

Over the interval $\left[0, 2\pi\right)$, the equation $\sin x = \dfrac{1}{2}$ has two solutions, the angles in quadrants I and II that have a reference angle of $\dfrac{\pi}{6}$. These are $\dfrac{\pi}{6}$ and $\dfrac{5\pi}{6}$.

Solution set: $\left\{\dfrac{\pi}{6}, \dfrac{5\pi}{6}\right\}$

19. $\tan^2 x + 3 = 0 \Rightarrow \tan^2 x = -3$

The square of a real number cannot be negative, so this equation has no solution. Solution set: $\varnothing$

21. $\left(\cot x - 1\right)\left(\sqrt{3}\cot x + 1\right) = 0$

$\cot x - 1 = 0 \Rightarrow \cot x = 1$ or

$\sqrt{3}\cot x + 1 = 0 \Rightarrow \sqrt{3}\cot x = -1 \Rightarrow$

$\cot x = -\dfrac{1}{\sqrt{3}} \Rightarrow \cot x = -\dfrac{\sqrt{3}}{3}$

Over the interval $[0, 2\pi)$, the equation

$\cot x = 1$ has two solutions, the angles in

quadrants I and III that have a reference angle

of $\dfrac{\pi}{4}$. These are $\dfrac{\pi}{4}$ and $\dfrac{5\pi}{4}$. In the same

interval, $\cot x = -\dfrac{\sqrt{3}}{3}$ also has two

solutions. The angles in quadrants II and IV

that have a reference angle of $\dfrac{\pi}{3}$ are $\dfrac{2\pi}{3}$ and

$\dfrac{5\pi}{3}$.

Solution set: $\left\{\dfrac{\pi}{4}, \dfrac{2\pi}{3}, \dfrac{5\pi}{4}, \dfrac{5\pi}{3}\right\}$

23. $\cos^2 x + 2\cos x + 1 = 0$

$\cos^2 x + 2\cos x + 1 = 0 \Rightarrow \left(\cos x + 1\right)^2 = 0 \Rightarrow$

$\cos x + 1 = 0 \Rightarrow \cos x = -1$

Over the interval $[0, 2\pi)$, the equation

$\cos x = -1$ has one solution. This solution is

π. Solution set: $\{\pi\}$

25. $-2\sin^2 x = 3\sin x + 1$

$2\sin^2 x + 3\sin x + 1 = 0$

$\left(2\sin x + 1\right)\left(\sin x + 1\right) = 0$

$2\sin x + 1 = 0 \Rightarrow \sin x = -\dfrac{1}{2}$ or

$\sin x + 1 = 0 \Rightarrow \sin x = -1$

Over the interval $[0, 2\pi)$, the equation

$\sin x = -\dfrac{1}{2}$ has two solutions. The angles in

quadrants III and IV that have a reference

angle of $\dfrac{\pi}{6}$ are $\dfrac{7\pi}{6}$ and $\dfrac{11\pi}{6}$.

In the same interval, $\sin x = -1$ when the

angle is $\dfrac{3\pi}{2}$. Solution set: $\left\{\dfrac{7\pi}{6}, \dfrac{3\pi}{2}, \dfrac{11\pi}{6}\right\}$

27. $\left(\cot\theta - \sqrt{3}\right)\left(2\sin\theta + \sqrt{3}\right) = 0$

$\cot\theta - \sqrt{3} = 0 \Rightarrow \cot\theta = \sqrt{3}$ or

$2\sin\theta + \sqrt{3} = 0 \Rightarrow 2\sin\theta = -\sqrt{3} \Rightarrow$

$\sin\theta = -\dfrac{\sqrt{3}}{2}$

Over the interval $[0°, 360°)$, the equation

$\cot\theta = \sqrt{3}$ has two solutions, the angles in

quadrants I and III that have a reference angle

of $30°$ These are $30°$ and $210°$. In the same

interval, the equation $\sin\theta = -\dfrac{\sqrt{3}}{2}$ has two

solutions, the angles in quadrants III and IV

that have a reference angle of $60°$. These are

$240°$ and $300°$.

Solution set: $\{30°, 210°, 240°, 300°\}$

29. $2\sin\theta - 1 = \csc\theta \Rightarrow 2\sin\theta - 1 = \dfrac{1}{\sin\theta} \Rightarrow$

$2\sin^2\theta - \sin\theta = 1 \Rightarrow$

$2\sin^2\theta - \sin\theta - 1 = 0 \Rightarrow$

$\left(2\sin\theta + 1\right)\left(\sin\theta - 1\right) = 0$

$2\sin\theta + 1 = 0 \Rightarrow \sin\theta = -\dfrac{1}{2}$ or

$\sin\theta - 1 = 0 \Rightarrow \sin\theta = 1$

Over the interval $[0°, 360°)$, the equation

$\sin\theta = -\dfrac{1}{2}$ has two solutions, the angles in

quadrants III and IV that have a reference

angle of $30°$ These are $210°$ and $330°$. In the

same interval, the only angle θ for which

$\sin\theta = 1$ is $90°$.

Solution set: $\{90°, 210°, 330°\}$

31. $\tan\theta - \cot\theta = 0$

$\tan\theta - \cot\theta = 0 \Rightarrow \tan\theta - \dfrac{1}{\tan\theta} = 0 \Rightarrow$

$\tan^2\theta - 1 = 0 \Rightarrow \tan^2\theta = 1 \Rightarrow \tan\theta = \pm 1$

Over the interval $[0°, 360°)$, the equation

$\tan\theta = 1$ has two solutions, the angles in

quadrants I and III that have a reference angle

of $45°$ These are $45°$ and $225°$.

In the same interval, the equation $\tan\theta = -1$

has two solutions, the angles in quadrants II

and IV that have a reference angle of $45°$.

These are $135°$ and $315°$.

Solution set: $\left\{45°, 135°, 225°, 315°\right\}$

33. $\csc^2 \theta - 2\cot \theta = 0$

$$\csc^2 \theta - 2\cot \theta = 0$$
$$\left(1 + \cot^2 \theta\right) - 2\cot \theta = 0$$
$$\cot^2 \theta - 2\cot \theta + 1 = 0$$
$$\left(\cot \theta - 1\right)^2 = 0$$
$$\cot \theta - 1 = 0 \Rightarrow \cot \theta = 1$$

Over the interval $[0°, 360°)$, the equation $\cot \theta = 1$ has two solutions, the angles in quadrants I and III that have a reference angle of $45°$. These are $45°$ and $225°$

Solution set: $\{45°, 225°\}$

35. $2\tan^2 \theta \sin \theta - \tan^2 \theta = 0$

$$2\tan^2 \theta \sin \theta - \tan^2 \theta = 0$$
$$\tan^2 \theta \left(2\sin \theta - 1\right) = 0$$
$$\tan^2 \theta = 0$$
$$\tan \theta = 0 \text{ or } 2\sin \theta - 1 = 0 \Rightarrow$$
$$2\sin \theta = 1 \Rightarrow \sin \theta = \frac{1}{2}$$

Over the interval $[0°, 360°)$, the equation $\tan \theta = 0$ has two solutions. These are $0°$ and $180°$. In the same interval, the equation $\sin \theta = \frac{1}{2}$ has two solutions, the angles in quadrants I and II that have a reference angle of $30°$. These are $30°$ and $150°$.

Solution set: $\{0°, 30°, 150°, 180°\}$

37. $\sec^2 \theta \tan \theta = 2\tan \theta$

$$\sec^2 \theta \tan \theta = 2\tan \theta$$
$$\sec^2 \theta \tan \theta - 2\tan \theta = 0$$
$$\tan \theta \left(\sec^2 \theta - 2\right) = 0$$
$$\tan \theta = 0 \text{ or } \sec^2 \theta - 2 = 0 \Rightarrow$$
$$\sec^2 \theta = 2 \Rightarrow \sec \theta = \pm\sqrt{2}$$

Over the interval $[0°, 360°)$, the equation $\tan \theta = 0$ has two solutions. These are $0°$ and $180°$. In the same interval, the equation $\sec \theta = \sqrt{2}$ has two solutions, the angles in quadrants I and IV that have a reference angle of $45°$ These are $45°$ and $315°$. Finally, the equation $\sec \theta = -\sqrt{2}$ has two solutions, the angles in quadrants II and III that have a reference angle of $45°$. These are $135°$ and $225°$.

Solution set:
$\{0°, 45°, 135°, 180°, 225°, 315°\}$

For Exercises 39–45, make sure your calculator is in degree mode.

39. $9\sin^2 \theta - 6\sin \theta = 1$

$9\sin^2 \theta - 6\sin \theta = 1 \Rightarrow 9\sin^2 \theta - 6\sin \theta - 1 = 0$

We use the quadratic formula with $a = 9$, $b = -6$, and $c = -1$.

$$\sin \theta = \frac{6 \pm \sqrt{36 - 4(9)(-1)}}{2(9)} = \frac{6 \pm \sqrt{36 + 36}}{18}$$
$$= \frac{6 \pm \sqrt{72}}{18} = \frac{6 \pm 6\sqrt{2}}{18} = \frac{1 \pm \sqrt{2}}{3}$$

Because $\sin \theta = \frac{1 + \sqrt{2}}{3} > 0$ (and less than 1), we will obtain two angles. One angle will be in quadrant I and the other will be in quadrant II. Using a calculator, if

$$\sin \theta = \frac{1 + \sqrt{2}}{3} \approx 0.80473787, \text{ the quadrant I}$$

angle will be approximately $53.6°$. The quadrant II angle will be approximately $180° - 53.6° = 126.4°$. Because

$$\sin \theta = \frac{1 - \sqrt{2}}{3} < 0 \text{ (and greater than } -1\text{), we}$$

will obtain two angles. One angle will be in quadrant III and the other will be in quadrant IV. Using a calculator, if

$$\sin \theta = \frac{1 - \sqrt{2}}{3} \approx -0.13807119, \text{ then}$$

$\theta \approx -7.9°$. This solution is not in the interval $[0°, 360°)$, so we must use it as a reference angle to find angles in the interval. Our reference angle will be $7.9°$. The angle in quadrant III will be approximately $180° + 7.9° = 187.9°$. The angle in quadrant IV will be approximately $360° - 7.9° = 352.1°$.

Solution set: $\{53.6°, 126.4°, 187.9°, 352.1°\}$

41. $\tan^2 \theta + 4\tan \theta + 2 = 0$

We use the quadratic formula with $a = 1$, $b = 4$, and $c = 2$.

$$\tan \theta = \frac{-4 \pm \sqrt{16 - 4(1)(2)}}{2(1)} = \frac{-4 \pm \sqrt{16 - 8}}{2}$$
$$= \frac{-4 \pm \sqrt{8}}{2} = \frac{-4 \pm 2\sqrt{2}}{2} = -2 \pm \sqrt{2}$$

Because $\tan \theta = -2 + \sqrt{2} < 0$, we will obtain two angles. One angle will be in quadrant II and the other will be in quadrant IV.

(continued on next page)

(continued)

Using a calculator, if

$\tan\theta = -2+\sqrt{2} = -0.5857864$, then

$\theta \approx -30.4°$. This solution is not in the interval $[0°, 360°)$, so we must use it as a reference angle to find angles in the interval.
Our reference angle will be 30.4°. The angle in quadrant II will be approximately $180° - 30.4° = 149.6°$. The angle in quadrant IV will be approximately $360° - 30.4° = 329.6°$. Because $\tan\theta = -2-\sqrt{2} < 0$, we will obtain two angles. One angle will be in quadrant II and the other will be in quadrant IV. Using a calculator, if $\tan\theta = -2-\sqrt{2} = -3.4142136$, then $\theta \approx -73.7°$. This solution is not in the interval $[0°, 360°)$, so we must use it as a reference angle to find angles in the interval.
Our reference angle will be 73.7°. The angle in quadrant II will be approximately $180° - 73.7° = 106.3°$ The angle in quadrant IV will be approximately $360° - 73.7° = 286.3°$.
Solution set: $\{106.3°, 149.6°, 286.3°, 329.6°\}$

43. $\sin^2\theta - 2\sin\theta + 3 = 0$

We use the quadratic formula with $a = 1$, $b = -2$, and $c = 3$.

$$\sin\theta = \frac{2 \pm \sqrt{4 - (4)(1)(3)}}{2(1)} = \frac{2 \pm \sqrt{4-12}}{2}$$

$$= \frac{2 \pm \sqrt{-8}}{2} = \frac{2 \pm 2i\sqrt{2}}{2} = 1 \pm i\sqrt{2}$$

Because $1 \pm i\sqrt{2}$ is not a real number, the equation has no real solutions.
Solution set: $\varnothing$

45. $\cot\theta + 2\csc\theta = 3$

$$\cot\theta + 2\csc\theta = 3 \Rightarrow \frac{\cos\theta}{\sin\theta} + \frac{2}{\sin\theta} = 3$$

$$\cos\theta + 2 = 3\sin\theta$$

$$(\cos\theta + 2)^2 = (3\sin\theta)^2$$

$$\cos^2\theta + 4\cos\theta + 4 = 9\sin^2\theta$$

$$\cos^2\theta + 4\cos\theta + 4 = 9(1 - \cos^2\theta)$$

$$\cos^2\theta + 4\cos\theta + 4 = 9 - 9\cos^2\theta$$

$$10\cos^2\theta + 4\cos\theta - 5 = 0$$

Now use the quadratic formula with $a = 10$, $b = 4$, and $c = -5$.

$$\cos\theta = \frac{-4 \pm \sqrt{4^2 - 4(10)(-5)}}{2(10)}$$

$$= \frac{-4 \pm \sqrt{16 + 200}}{20} = \frac{-4 \pm \sqrt{216}}{20}$$

$$= \frac{-4 \pm 6\sqrt{6}}{20} = \frac{-2 \pm 3\sqrt{6}}{10}$$

Because $\cos\theta = \dfrac{-2 + 3\sqrt{6}}{10} > 0$ (and less than 1), we will obtain two angles. One angle will be in quadrant I and the other will be in quadrant IV. Using a calculator, if

$\cos\theta = \dfrac{-2 + 3\sqrt{6}}{10} \approx 0.53484692$, the quadrant I angle will be approximately 57.7°.
The quadrant IV angle will be approximately $360° - 57.7° = 302.3°$.

Because $\cos\theta = \dfrac{-2 - 3\sqrt{6}}{10} < 0$ (and greater than −1), we will obtain two angles. One angle will be in quadrant II and the other will be in quadrant III. Using a calculator, if

$\cos\theta = \dfrac{-2 - 3\sqrt{6}}{10} \approx -0.93484692$, the quadrant II angle will be approximately 159.2°. The reference angle is $180° - 159.2° = 20.8°$. Thus, the quadrant III angle will be approximately $180° + 20.8° = 200.8°$.
The solution was found by squaring both sides of an equation, so we must check that each proposed solution is a solution of the original equation. 302.3° and 200.8° do not satisfy our original equation. Thus, they are not elements of the solution set.
Solution set: $\{57.7°, 159.2°\}$

In Exercises 47–61, if you are using a calculator, make sure it is in radian mode if you are solving for x and in degree mode if you are solving for θ.

47. $\cos\theta + 1 = 0 \Rightarrow \cos\theta = -1 \Rightarrow \theta = 180°$ in the interval $[0, 2\pi)$. The solution set is $\{180° + 360°n$, where n is any integer$\}$.

49. $3\csc x - 2\sqrt{3} = 0 \Rightarrow 3\csc x = 2\sqrt{3} \Rightarrow$

$\csc x = \dfrac{2\sqrt{3}}{3} \Rightarrow x = \dfrac{\pi}{3}, \dfrac{2\pi}{3}$ in the interval $[0, 2\pi)$. The solution set is

$\left\{\dfrac{\pi}{3} + 2n\pi, \dfrac{2\pi}{3} + 2n\pi\right.$, where n is any integer$\left.\right\}$.

51. $6\sin^2\theta + \sin\theta = 1 \Rightarrow 6\sin^2\theta + \sin\theta - 1 = 0 \Rightarrow$
$(3\sin\theta - 1)(2\sin\theta + 1) = 0 \Rightarrow$

$\sin\theta = \dfrac{1}{3} \Rightarrow \theta \approx 19.5°$ or

$\theta \approx 180° - 19.5° = 160.5°$ or

$\sin\theta = -\dfrac{1}{2} \Rightarrow \theta = 210°$ or $\theta = 330°$

The solution set is $\{19.5° + 360°n,$
$160.5° + 360°n, 210° + 360°n, 330° + 360°n,$
where n is any integer$\}$.

53. $2\cos^2 x + \cos x - 1 = 0$
$(2\cos x - 1)(\cos x + 1) = 0$

$2\cos x - 1 = 0 \Rightarrow \cos x = \dfrac{1}{2}$ or

$\cos x + 1 = 0 \Rightarrow \cos x = -1$

Over the interval $[0, 2\pi)$, the equation

$\cos x = \dfrac{1}{2}$ has two solutions. The angles in

quadrants I and IV that have a reference angle

of $\dfrac{\pi}{3}$ are $\dfrac{\pi}{3}$ and $\dfrac{5\pi}{3}$. In the same interval,

$\cos x = -1$ when the angle is π. Thus, the

solution set is $\left\{\dfrac{\pi}{3} + 2n\pi, \pi + 2n\pi,\right.$

and $\left.\dfrac{5\pi}{3} + 2n\pi, \text{ where } n \text{ is any integer}\right\}$.

55. $\sin\theta\cos\theta - \sin\theta = 0 \Rightarrow \sin\theta(\cos\theta - 1) = 0 \Rightarrow$
$\sin\theta = 0 \Rightarrow \theta = 0°$ or $\theta = 180°$ or
$\cos\theta = 1 \Rightarrow \theta = 0°$
The solution set is $\{180°n, \text{ where } n \text{ is any integer}\}$.

57. $\sin x(3\sin x - 1) = 1 \Rightarrow$
$3\sin^2 x - \sin x - 1 = 0$
Use the quadratic formula with $a = 3$, $b = -1$,
and $c = -1$.

$\sin x = \dfrac{-(-1) \pm \sqrt{(-1)^2 - 4(3)(-1)}}{2(3)} = \dfrac{1 \pm \sqrt{13}}{6}$

Because $\sin x = \dfrac{1 + \sqrt{13}}{6} > 0$ (and less than 1),

we will obtain two angles. One angle will be in
quadrant I and the other will be in quadrant II.
Using a calculator, if

$\sin x = \dfrac{1 + \sqrt{13}}{6} \approx 0.76759188$, the quadrant I

angle will be approximately 0.8751. The
quadrant II angle will be approximately
$\pi - 0.88 \approx 2.2665.$

Because $\sin x = \dfrac{1 - \sqrt{13}}{6} < 0$ (and greater than $-$

1), we will obtain two angles. One angle will
be in quadrant III and the other will be in
quadrant IV. Using a calculator, if

$\sin x = \dfrac{1 - \sqrt{13}}{6} \approx -0.43425855$, then

$x \approx -0.4492$. This solution is not in the
interval $[0, 2\pi)$, so we must use it as a
reference angle to find angles in the interval.
Our reference angle will be 0.4492. The angle
in quadrant III will be approximately
$\pi + 0.4492 \approx 3.5908$. The angle in quadrant IV
will be approximately $2\pi - 0.4492 \approx 5.8340$.
Thus, the solution set is
$\{0.8751 + 2n\pi, 2.2665 + 2n\pi, 3.5908 + 2n\pi,$
and $5.8340 + 2n\pi,$ where n is any integer$\}$.

59. $5 + 5\tan^2\theta = 6\sec\theta \Rightarrow 5(1 + \tan^2\theta) = 6\sec\theta \Rightarrow$

$5\sec^2\theta = 6\sec\theta \Rightarrow 5\sec^2\theta - 6\sec\theta = 0 \Rightarrow$
$\sec\theta(5\sec\theta - 6) = 0$

$\sec\theta = 0$ or $5\sec\theta - 6 = 0 \Rightarrow \sec\theta = \dfrac{6}{5}$

$\sec\theta = 0$ is an impossible value because the
secant function must be either ≥ 1 or ≤ -1.

Because $\sec\theta = \dfrac{6}{5} > 1$, we will obtain two angles.

One angle will be in quadrant I and the other will
be in quadrant IV. Using a calculator, if

$\sec\theta = \dfrac{6}{5} = 1.2$, the quadrant I angle will be

approximately 33.6° The quadrant IV angle will
be approximately $360° - 33.6° = 326.4°$. Thus,
the solution set is $\{33.6° + 360°n$ and
$326.4° + 360°n,$ where n is any integer$\}$.

61. $\dfrac{2\tan\theta}{3 - \tan^2\theta} = 1$
$\qquad 2\tan\theta = 3 - \tan^2\theta$
$\tan^2\theta + 2\tan\theta - 3 = 0$
$(\tan\theta - 1)(\tan\theta + 3) = 0$
$\tan\theta - 1 = 0 \Rightarrow \tan\theta = 1$ or
$\tan\theta + 3 = 0 \Rightarrow \tan\theta = -3$
Over the interval $[0°, 360°)$, the equation
$\tan\theta = 1$ has two solutions 45° and 225°.
Over the same interval, the equation
$\tan\theta = -3$ has two solutions that are
approximately $-71.6° + 180° = 108.4°$ and
$-71.6° + 360° = 288.4°$

(continued on next page)

(continued)

Thus, the solutions are $45° + 360°n$, $108.4° + 360°n$,

$225° + 360°n$ and $288.4° + 360°n$, where n is any integer. The period of the tangent function is $180°$, so the solution set can also be written as $\{45° + 180°n$ and $108.4° + 180°n$, where n is any integer$\}$.

63. The x-intercept method is shown below.

$y_1 = x^2 + \sin x - x^3 - \cos x$ is graphed in the window $[0, 2\pi] \times [-1, 1]$.

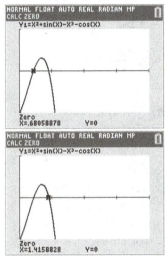

Solution set: $\{0.6806, 1.4159\}$

65. Because $2x = \dfrac{2\pi}{3}, 2\pi, \dfrac{8\pi}{3} \Rightarrow$

$x = \dfrac{2\pi}{6}, \dfrac{2\pi}{2}, \dfrac{8\pi}{6} \Rightarrow x = \dfrac{\pi}{3}, \pi, \dfrac{4\pi}{3}$, the

solution set is $\left\{ \dfrac{\pi}{3}, \pi, \dfrac{4\pi}{3} \right\}$.

67. Because $3\theta = 180°, 630°, 720°, 930° \Rightarrow$
$\theta = 60°, 210°, 240°, 310°$, the solution set is $\{60°, 210°, 240°, 310°\}$.

69. $2\cos 2x = \sqrt{3} \Rightarrow \cos 2x = \dfrac{\sqrt{3}}{2}$

Because $0 \le x < 2\pi$, $0 \le 2x < 4\pi$.
Thus,

$2x = \dfrac{\pi}{6}, \dfrac{11\pi}{6}, \dfrac{13\pi}{6}, \dfrac{23\pi}{6} \Rightarrow$

$x = \dfrac{\pi}{12}, \dfrac{11\pi}{12}, \dfrac{13\pi}{12}, \dfrac{23\pi}{12}$.

Solution set: $\dfrac{\pi}{12}, \dfrac{11\pi}{12}, \dfrac{13\pi}{12}, \dfrac{23\pi}{12}$

71. $\sin 3\theta = -1$
Because $0° \le \theta < 360°$, $0° \le 3\theta < 1080°$.
Thus, $3\theta = 270°, 630°, 990° \Rightarrow$
$x = 90°, 210°, 330°$
Solution set: $\{90°, 210°, 330°\}$

73. $3\tan 3x = \sqrt{3} \Rightarrow \tan 3x = \dfrac{\sqrt{3}}{3}$

Because $0 \le x < 2\pi$, $0 \le 3x < 6\pi$.

Thus, $3x = \dfrac{\pi}{6}, \dfrac{7\pi}{6}, \dfrac{13\pi}{6}, \dfrac{19\pi}{6}, \dfrac{25\pi}{6}, \dfrac{31\pi}{6}$

implies $x = \dfrac{\pi}{18}, \dfrac{7\pi}{18}, \dfrac{13\pi}{18}, \dfrac{19\pi}{18}, \dfrac{25\pi}{18}, \dfrac{31\pi}{18}$.

Solution set:

$\left\{ \dfrac{\pi}{18}, \dfrac{7\pi}{18}, \dfrac{13\pi}{18}, \dfrac{19\pi}{18}, \dfrac{25\pi}{18}, \dfrac{31\pi}{18} \right\}$

75. $\sqrt{2}\cos 2\theta = -1 \Rightarrow \cos 2\theta = \dfrac{-1}{\sqrt{2}} = -\dfrac{\sqrt{2}}{2}$

Because $0° \le \theta < 360°$, $0° \le 2\theta < 720°$. Thus,
$2\theta = 135°, 225°, 495°, 585° \Rightarrow$
$\theta = 67.5°, 112.5°, 247.5°, 292.5°$
Solution set: $\{67.5°, 112.5°, 247.5°, 292.5°\}$

77. $\sin \dfrac{x}{2} = \sqrt{2} - \sin \dfrac{x}{2}$

$\sin \dfrac{x}{2} = \sqrt{2} - \sin \dfrac{x}{2} \Rightarrow \sin \dfrac{x}{2} + \sin \dfrac{x}{2} = \sqrt{2} \Rightarrow$

$2\sin \dfrac{x}{2} = \sqrt{2} \Rightarrow \sin \dfrac{x}{2} = \dfrac{\sqrt{2}}{2}$

Because $0 \le x < 2\pi$, $0 \le \dfrac{x}{2} < \pi$. Thus,

$\dfrac{x}{2} = \dfrac{\pi}{4}, \dfrac{3\pi}{4} \Rightarrow x = \dfrac{\pi}{2}, \dfrac{3\pi}{2}$.

Solution set: $\left\{ \dfrac{\pi}{2}, \dfrac{3\pi}{2} \right\}$

79. $\sin x = \sin 2x$
$\sin x = \sin 2x \Rightarrow \sin x = 2\sin x \cos x \Rightarrow$
$\sin x - 2\sin x \cos x = 0 \Rightarrow \sin x(1 - 2\cos x) = 0$

Over the interval $[0, 2\pi)$, we have
$1 - 2\cos x = 0 \Rightarrow -2\cos x = -1 \Rightarrow$

$\cos x = \dfrac{1}{2} \Rightarrow x = \dfrac{\pi}{3}$ or $\dfrac{5\pi}{3}$

$\sin x = 0 \Rightarrow x = 0$ or π

Solution set: $\left\{ 0, \dfrac{\pi}{3}, \pi, \dfrac{5\pi}{3} \right\}$

Copyright © 2017 Pearson Education, Inc.

81. $8\sec^2\dfrac{x}{2} = 4 \Rightarrow \sec^2\dfrac{x}{2} = \dfrac{1}{2} \Rightarrow \sec\dfrac{x}{2} = \pm\dfrac{\sqrt{2}}{2}$

$-\dfrac{\sqrt{2}}{2}$ is not in the interval $(-\infty, -1]$ and $\dfrac{\sqrt{2}}{2}$

is not in the interval $[1, \infty)$, so this equation
has no solution. Solution set: $\varnothing$

83. $\sin\dfrac{\theta}{2} = \csc\dfrac{\theta}{2}$

$\sin\dfrac{\theta}{2} = \csc\dfrac{\theta}{2} \Rightarrow \sin\dfrac{\theta}{2} = \dfrac{1}{\sin\dfrac{\theta}{2}} \Rightarrow$

$\sin^2\dfrac{\theta}{2} = 1 \Rightarrow \sin\dfrac{\theta}{2} = \pm 1$

$0° \le \theta < 360° \Rightarrow 0 \le \dfrac{\theta}{2} < 180°.$

If $\sin\dfrac{\theta}{2} = 1, \dfrac{\theta}{2} = 90° \Rightarrow \theta = 180°.$

If $\sin\dfrac{\theta}{2} = -1, \dfrac{\theta}{2} = 270° \Rightarrow \theta = 540°,$ which is

not in the interval [0, 360°].
Solution set: $\{180°\}$

85. $\cos 2x + \cos x = 0$

We choose an identity for $\cos 2x$ that involves
only the cosine function.

$\cos 2x + \cos x = 0$

$\left(2\cos^2 x - 1\right) + \cos x = 0$

$2\cos^2 x + \cos x - 1 = 0$

$\left(2\cos x - 1\right)\left(\cos x + 1\right) = 0 \Rightarrow$

$\cos x = \dfrac{1}{2}$ or $\cos x = -1$

$\cos x = \dfrac{1}{2} \Rightarrow x = \dfrac{\pi}{3}$ or $\dfrac{5\pi}{3}$

$\cos x = -1 \Rightarrow x = \pi$

Solution set: $\left\{\dfrac{\pi}{3}, \pi, \dfrac{5\pi}{3}\right\}$

87. $\sqrt{2}\sin 3x - 1 = 0$

$\sqrt{2}\sin 3x - 1 = 0 \Rightarrow \sqrt{2}\sin 3x = 1 \Rightarrow$

$\sin 3x = \dfrac{1}{\sqrt{2}} \Rightarrow \sin 3x = \dfrac{\sqrt{2}}{2}$

In quadrant I and II, sine is positive. Thus,

$3x = \dfrac{\pi}{4} + 2n\pi, \dfrac{3\pi}{4} + 2n\pi \Rightarrow$

$x = \dfrac{\pi}{12} + \dfrac{2n\pi}{3}, \dfrac{\pi}{4} + \dfrac{2n\pi}{3}$

Solution set:

$\left\{\dfrac{\pi}{12} + \dfrac{2n\pi}{3}, \dfrac{\pi}{4} + \dfrac{2n\pi}{3},\ \text{where } n \text{ is any integer}\right\}$

89. $\cos\dfrac{\theta}{2} = 1$

$\dfrac{\theta}{2} = 0° + 360°n \Rightarrow \theta = 720°n.$

Solution set: $\{720°n, \text{ where } n \text{ is any integer}\}$

91. $2\sqrt{3}\sin\dfrac{x}{2} = 3 \Rightarrow \sin\dfrac{x}{2} = \dfrac{3}{2\sqrt{3}} \Rightarrow$

$\sin\dfrac{x}{2} = \dfrac{3\sqrt{3}}{6} \Rightarrow \sin\dfrac{x}{2} = \dfrac{\sqrt{3}}{2}$

$0 \le \theta < 360° \Rightarrow 0° \le \dfrac{\theta}{2} < 180°.$

Thus, $\dfrac{x}{2} = \dfrac{\pi}{3} + 2n\pi, \dfrac{2\pi}{3} + 2n\pi \Rightarrow$

$x = \dfrac{2\pi}{3} + 4n\pi, \dfrac{4\pi}{3} + 4n\pi$

Solution set:

$\left\{\dfrac{2\pi}{3} + 4n\pi, \dfrac{4\pi}{3} + 4n\pi,\ \text{where } n \text{ is any}\right.$

$\left.\text{integer}\right\}$

93. $2\sin\theta = 2\cos 2\theta$

$2\sin\theta = 2\cos 2\theta \Rightarrow \sin\theta = \cos 2\theta \Rightarrow$

$\sin\theta = 1 - 2\sin^2\theta \Rightarrow 2\sin^2\theta + \sin\theta - 1 = 0 \Rightarrow$

$\left(2\sin\theta - 1\right)\left(\sin\theta + 1\right) = 0 \Rightarrow$

$2\sin\theta - 1 = 0$ or $\sin\theta + 1 = 0$

Over the interval $[0°, 360°)$, we have

$2\sin\theta - 1 = 0 \Rightarrow 2\sin\theta = 1 \Rightarrow \sin\theta = \dfrac{1}{2} \Rightarrow$

$\theta = 30°$ or $150°$

$\sin\theta + 1 = 0 \Rightarrow \sin\theta = -1 \Rightarrow \theta = 270°$

Solution set: $\{30° + 360°n, 150° + 360°n,$
$270° + 360°n, \text{ where } n \text{ is any integer}\}$

95. $1 - \sin x = \cos 2x \Rightarrow 1 - \sin x = 1 - 2\sin^2 x \Rightarrow$

$2\sin^2 x - \sin x = 0 \Rightarrow \sin x\left(2\sin x - 1\right) = 0$

Over the interval $[0, 2\pi)$, we have

$\sin x = 0 \Rightarrow x = 0$ or π.

$2\sin x - 1 = 0 \Rightarrow \sin x = \dfrac{1}{2} \Rightarrow x = \dfrac{\pi}{6}$ or $\dfrac{5\pi}{6}$

Solution set:

$\left\{n\pi, \dfrac{\pi}{6} + 2n\pi, \dfrac{5\pi}{6} + 2n\pi,\ \text{where}\right.$

$\left.n \text{ is any integer}\right\}$

97. $3\csc^2\dfrac{x}{2} = 2\sec x \Rightarrow \dfrac{3}{\sin^2\frac{x}{2}} = \dfrac{2}{\cos x} \Rightarrow$

$\sin^2\dfrac{x}{2} = \dfrac{3}{2}\cos x \Rightarrow \dfrac{1-\cos x}{2} = \dfrac{3}{2}\cos x \Rightarrow$

$1-\cos x = 3\cos x \Rightarrow 1 = 4\cos x \Rightarrow \dfrac{1}{4} = \cos x$

Over the interval $[0, 2\pi)$, we have

$\cos x = \dfrac{1}{4} \Rightarrow x = 1.3181 \text{ or } x = 4.9651$

Solution set:
$\{1.3181 + 2n\pi,\ 4.9651 + 2n\pi,\text{ where } n \text{ is any}$

integer$\}$

99. $2 - \sin 2\theta = 4\sin 2\theta$

$2 - \sin 2\theta = 4\sin 2\theta \Rightarrow 2 = 5\sin 2\theta \Rightarrow$

$\sin 2\theta = \dfrac{2}{5} \Rightarrow \sin 2\theta = 0.4$

Since $0 \le \theta < 360°,\ 0° \le 2\theta < 720°$. In quadrant

I and II, sine is positive.

$\sin 2\theta = 0.4 \Rightarrow 2\theta = 23.6°, 156.4°, 383.6°, 516.4°$

Thus, $\theta = 11.8°, 78.2°, 191.8°, 258.2°$.

Solution set:
$\{11.8° + 360°n,\ 78.2° + 360°n,\ 191.8° + 360°n,$

$258.2° + 360°n,\text{where } n \text{ is any integer}\}$ or

$\{11.8° + 180°n,\ 78.2° + 180°n,$

where n is any integer$\}$

101. $2\cos^2 2\theta = 1 - \cos 2\theta$

$2\cos^2 2\theta + \cos 2\theta - 1 = 0$

$(2\cos 2\theta - 1)(\cos 2\theta + 1) = 0$

$0 \le \theta < 360° \Rightarrow 0° \le 2\theta < 720°,$ so

$2\cos 2\theta - 1 = 0 \Rightarrow 2\cos 2\theta = 1 \Rightarrow$

$\cos 2\theta = \dfrac{1}{2}.$

Thus, $2\theta = 60°, 300°, 420°, 660° \Rightarrow$

$\theta = 30°, 150°, 210°, 330°$ or

$\cos 2\theta + 1 = 0 \Rightarrow \cos 2\theta = -1$

$2\theta = 180°, 540° \Rightarrow \theta = 90°, 270°$

Solution set:
$\{30° + 360°n,\ 90° + 360°n,$

$150° + 360°n,\ 210° + 360°n,\ 270° + 360°n,$

$330° + 360°n,\text{where } n \text{ is any integer}\}$ or

$\{30° + 180°n,\ 90° + 180°n,\ 150° + 180°n,$

where n is any integer$\}$

103. The x-intercept method is shown below.

$y_1 = 2\sin 2x - x^3 + 1$ is graphed in the

window $[0, 2\pi] \times [-4, 4]$.

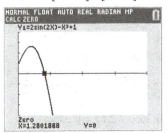

Solution set: $\{1.2802\}$

105. $P = A\sin(2\pi ft + \phi)$

(a) $0 = 0.004\sin\left[2\pi(261.63)t + \dfrac{\pi}{7}\right]$

$0 = \sin(1643.87t + 0.45)$

$1643.87t + 0.45 = n\pi,$ so $t = \dfrac{n\pi - 0.45}{1643.87},$

where n is any integer.

If $n = 0$, then $t \approx 0.000274$. If $n = 1$, then

$t \approx 0.00164$. If $n = 2$, then $t \approx 0.00355$.

If $n = 3$, then $t \approx 0.00546$. The only

solutions for t in the interval $[0, 0.005]$

are 0.00164 and 0.00355.

(b)

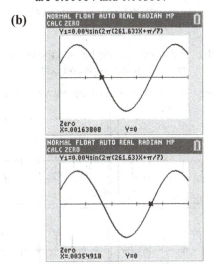

From the graphs we can estimate that

$P \le 0$ on the interval $[0.00164, 0.00355]$.

(c) $P < 0$ implies that there is a decrease in

pressure so an eardrum would be

vibrating outward.

107. $V = \cos 2\pi t, 0 \le t \le \dfrac{1}{2}$

(a) $V = 0, \cos 2\pi t = 0 \Rightarrow 2\pi t = \cos^{-1} 0 \Rightarrow$

$2\pi t = \dfrac{\pi}{2} \Rightarrow t = \dfrac{\frac{\pi}{2}}{2\pi} = \dfrac{1}{4} \sec$

(b) $V = 0.5, \cos 2\pi t = 0.5 \Rightarrow$

$2\pi t = \cos^{-1}(0.5) = \dfrac{\pi}{3} \Rightarrow t = \dfrac{\frac{\pi}{3}}{2\pi} = \dfrac{1}{6} \sec$

(c) $V = 0.25, \cos 2\pi t = 0.25 \Rightarrow$

$2\pi t = \cos^{-1}(0.25) \approx 1.3181161 \Rightarrow$

$t \approx \dfrac{1.3181161}{2\pi} \approx 0.21 \sec$

109.

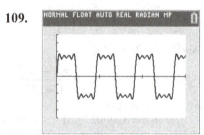

$[0, 0.03] \times [0.005, 0.005]$

$P = 0.003 \sin 220\pi t + \dfrac{0.003}{3} \sin 660\pi t$

$\qquad + \dfrac{0.003}{5} \sin 1100\pi t + \dfrac{0.003}{7} \sin 1540\pi t$

(b) The graph is periodic, and the wave has "jagged square" tops and bottoms.

(c) The eardrum is moving outward when $P < 0$.

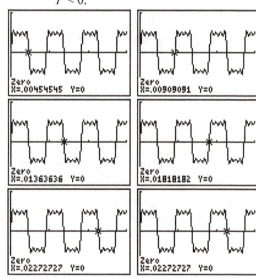

This occurs for the time intervals (0.0045, 0.0091), (0.0136, 0.0182), (0.0227, 0.0273).

111. (a) For $x = t$,

$P(t) = \dfrac{1}{2} \sin[2\pi(220)t] +$

$\qquad \dfrac{1}{3} \sin[2\pi(330)t] +$

$\qquad \dfrac{1}{4} \sin[2\pi(440)t]$

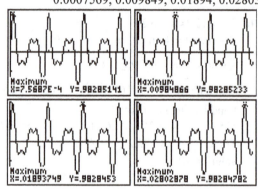

(b) Answers may vary slightly depending on the calculator settings.
0.0007569, 0.009849, 0.01894, 0.02803

(c) 110 Hz

(d) For $x = t$,

$P(t) = \sin[2\pi(110)t] +$

$\qquad \dfrac{1}{2} \sin[2\pi(220)t] +$

$\qquad \dfrac{1}{3} \sin[2\pi(330)t] +$

$\qquad \dfrac{1}{4} \sin[2\pi(440)t]$

113. (a) $14 \sin\left[\dfrac{\pi}{6}(x-4)\right] + 50 = 64$

$14 \sin\left[\dfrac{\pi}{6}(x-4)\right] = 14$

$\sin\left[\dfrac{\pi}{6}(x-4)\right] = 1$

$\dfrac{\pi}{6}(x-4) = \dfrac{\pi}{2}$

$x - 4 = 3 \Rightarrow x = 7$

The average monthly temperature is 64°F in the seventh month, July.

(b) $14\sin\left[\dfrac{\pi}{6}(x-4)\right]+50=39$

$$14\sin\left[\dfrac{\pi}{6}(x-4)\right]=-11$$

$$\sin\left[\dfrac{\pi}{6}(x-4)\right]=-\dfrac{11}{14}$$

$$\sin^{-1}\left(-\dfrac{11}{14}\right)=\dfrac{\pi}{6}(x-4)$$

$$\dfrac{6}{\pi}\left[\sin^{-1}\left(-\dfrac{11}{14}\right)\right]=x-4$$

$$\dfrac{6}{\pi}\left[\sin^{-1}\left(-\dfrac{11}{14}\right)\right]+4=x\approx 2.3 \text{ or}$$

$$\dfrac{6}{\pi}\left[\pi-\sin^{-1}\left(-\dfrac{11}{14}\right)\right]+4=x\approx 11.7$$

The average monthly temperature is 39°F in the second month, February, and in the eleventh month, November.

115. $i=I_{\max}\sin 2\pi ft$

Let $i=40,\ I_{\max}=100,\ f=60.$

$$40=100\sin\left[2\pi(60)t\right]$$

$$40=100\sin 120\pi t \Rightarrow 0.4=\sin 120\pi t$$

Using calculator,

$$120\pi t\approx 0.4115168\Rightarrow t\approx\dfrac{0.4115168}{120\pi}\Rightarrow$$

$$t\approx 0.0010916\Rightarrow t\approx 0.001 \text{ sec}$$

117. $i=I_{\max}\sin 2\pi ft$

Let $i=I_{\max},\ f=60.$

$$I_{\max}=I_{\max}\sin\left[2\pi(60)t\right]\Rightarrow 1=\sin 120\pi t\Rightarrow$$

$$120\pi t=\dfrac{\pi}{2}\Rightarrow 120t=\dfrac{1}{2}\Rightarrow t=\dfrac{1}{240}\approx 0.004 \text{ sec}$$

Chapter 7 Quiz
(Sections 7.5–7.6)

1. Domain: $[-1, 1]$; range: $\left[0,\pi\right]$

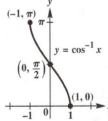

3. (a) $\theta=\arccos 0.92341853\approx 22.568922°$

(b) $\theta=\cot^{-1}\left(-1.08886767\right)\approx 137.431085°$

5. $2\sin\theta-\sqrt{3}=0\Rightarrow 2\sin\theta=\sqrt{3}\Rightarrow\sin\theta=\dfrac{\sqrt{3}}{2}$

Over the interval $[0°, 360°)$, the equation $\sin\theta=\dfrac{\sqrt{3}}{2}$ has two solutions, the angles in quadrants I and II that have a reference angle of 60°. These are 60° and 120°.
Solution set: {60°, 120°}

7. $V=\cos 2\pi t, 0\le t\le\dfrac{1}{2}$

(a) $V=1,\cos 2\pi t=1\Rightarrow 2\pi t=\cos^{-1}1\Rightarrow$

$$2\pi t=0\Rightarrow t=\dfrac{0}{2\pi}=0\sec$$

(b) $V=0.30,\cos 2\pi t=0.30$

$$2\pi t=\cos^{-1}0.30\Rightarrow$$

$$2\pi t\approx 1.266103673$$

$$t=\dfrac{1.266103673}{2\pi}=0.20\sec$$

9. $3\cot 2x-\sqrt{3}=0\Rightarrow\cot 2x=\dfrac{\sqrt{3}}{3}$

Since $0\le x<2\pi,\ 0\le 2x<4\pi.$

Thus, $2x=\dfrac{\pi}{3},\dfrac{4\pi}{3},\dfrac{7\pi}{3},\dfrac{10\pi}{3}$ implies

$$x=\dfrac{\pi}{6},\dfrac{2\pi}{3},\dfrac{7\pi}{6},\dfrac{5\pi}{3}.$$

Solution set: $\left\{\dfrac{\pi}{6},\dfrac{2\pi}{3},\dfrac{7\pi}{6},\dfrac{5\pi}{3}\right\}$

Section 7.7 Equations Involving Inverse Trigonometric Functions

1. C

3. C

5. A

7. $y=5\cos x\Rightarrow\dfrac{y}{5}=\cos x\Rightarrow x=\arccos\dfrac{y}{5},$

$0\le x\le\pi$

9. $y=3\tan 2x\Rightarrow\dfrac{y}{3}=\tan 2x\Rightarrow 2x=\arctan\dfrac{y}{3}\Rightarrow$

$$x=\dfrac{1}{2}\arctan\dfrac{y}{3},\ -\dfrac{\pi}{4}<x<\dfrac{\pi}{4}$$

11. $y=6\cos\dfrac{x}{4}\Rightarrow\dfrac{y}{6}=\cos\dfrac{x}{4}\Rightarrow\dfrac{x}{4}=\arccos\dfrac{y}{6}\Rightarrow$

$$x=4\arccos\dfrac{y}{6},\ 0\le x\le 4\pi$$

13. $y = -2\cos 5x \Rightarrow -\dfrac{y}{2} = \cos 5x \Rightarrow$

$5x = \arccos\left(-\dfrac{y}{2}\right) \Rightarrow x = \dfrac{1}{5}\arccos\left(-\dfrac{y}{2}\right),$

$0 \le x \le \dfrac{\pi}{5}$

15. $y = \sin x - 2 \Rightarrow y + 2 = \sin x \Rightarrow$

$x = \arcsin(y+2), \ -\dfrac{\pi}{2} \le x \le \dfrac{\pi}{2}$

17. $y = -4 + 2\sin x \Rightarrow y + 4 = 2\sin x \Rightarrow$

$\dfrac{y+4}{2} = \sin x \Rightarrow x = \arcsin\left(\dfrac{y+4}{2}\right),$

$-\dfrac{\pi}{2} \le x \le \dfrac{\pi}{2}$

19. $y = \dfrac{1}{2}\cot 3x \Rightarrow 2y = \cot 3x \Rightarrow$

$3x = \operatorname{arccot} 2y \Rightarrow x = \dfrac{1}{3}\operatorname{arccot} 2y, \ 0 < x < \dfrac{\pi}{3}$

21. $y = \cos(x+3) \Rightarrow x + 3 = \arccos y \Rightarrow$
$x = -3 + \arccos y, \ -3 \le x \le \pi - 3$

23. $y = \sqrt{2} + 3\sec 2x \Rightarrow y - \sqrt{2} = 3\sec 2x \Rightarrow$

$\dfrac{y-\sqrt{2}}{3} = \sec 2x \Rightarrow 2x = \sec^{-1}\left(\dfrac{y-\sqrt{2}}{3}\right) \Rightarrow$

$x = \dfrac{1}{2}\sec^{-1}\left(\dfrac{y-\sqrt{2}}{3}\right), \ 0 \le x \le \dfrac{\pi}{2}, \ x \ne \dfrac{\pi}{4}$

25. The argument of the sine function is x, not $x - 2$. To solve for x, first add 2, and then use the definition of arcsine. Another way to think about this is to think of the graph of $y = \sin x - 2$. This represents the graph of $f(x) = \sin x$, shifted 2 units down. If you think of the graph of $y = \sin(x-2)$, which represents the graph of $f(x) = \sin x$, shifted 2 units right. The graphs aren't the same, so $\sin x - 2 \ne \sin(x-2)$.

27. $-4\arcsin x = \pi \Rightarrow \arcsin x = -\dfrac{\pi}{4} \Rightarrow$

$x = \sin\left(-\dfrac{\pi}{4}\right) = -\dfrac{\sqrt{2}}{2}$

Solution set: $\left\{-\dfrac{\sqrt{2}}{2}\right\}$

29. $\dfrac{4}{3}\cos^{-1}\dfrac{x}{4} = \pi \Rightarrow \cos^{-1}\dfrac{x}{4} = \dfrac{3\pi}{4} \Rightarrow$

$\dfrac{x}{4} = \cos\dfrac{3\pi}{4} \Rightarrow \dfrac{yx}{4} = -\dfrac{\sqrt{2}}{2} \Rightarrow x = -2\sqrt{2}$

Solution set: $\left\{-2\sqrt{2}\right\}$

31. $2\arccos\left(\dfrac{x}{3} - \dfrac{\pi}{3}\right) = 2\pi \Rightarrow$

$\arccos\left(\dfrac{x}{3} - \dfrac{\pi}{3}\right) = \pi \Rightarrow \dfrac{x}{3} - \dfrac{\pi}{3} = \cos\pi$

$\dfrac{x}{3} - \dfrac{\pi}{3} = -1 \Rightarrow x - \pi = -3 \Rightarrow x = \pi - 3$

Solution set: $\{\pi - 3\}$

33. $\arcsin x = \arctan\dfrac{3}{4}$

Let $\arctan\dfrac{3}{4} = u$, so $\tan u = \dfrac{3}{4}$, u is in quadrant I. Sketch a triangle and label it. The hypotenuse is $\sqrt{3^2 + 4^2} = \sqrt{9 + 16} = \sqrt{25} = 5$.

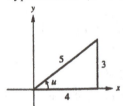

Therefore, $\sin u = \dfrac{3}{r} = \dfrac{3}{5}$. This equation becomes $\arcsin x = u$, or $x = \sin u$. Thus, $x = \dfrac{3}{5}$. Solution set: $\left\{\dfrac{3}{5}\right\}$

35. $\cos^{-1} x = \sin^{-1}\dfrac{3}{5}$

Let $\sin^{-1}\dfrac{3}{5} = u$, so $\sin u = \dfrac{3}{5}$. Sketch a triangle and label it. The hypotenuse is $\sqrt{3^2 + 4^2} = \sqrt{9 + 16} = \sqrt{25} = 5$.

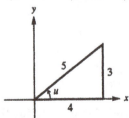

(continued on next page)

(continued)

Therefore, $\cos u = \dfrac{4}{5}$. The equation becomes

$\cos^{-1} x = u$, or $x = \cos u$. Thus, $x = \dfrac{4}{5}$.

Solution set: $\left\{\dfrac{4}{5}\right\}$

37. $\sin^{-1} x - \tan^{-1} 1 = -\dfrac{\pi}{4}$

$\sin^{-1} x - \tan^{-1} 1 = -\dfrac{\pi}{4} \Rightarrow$

$\sin^{-1} x = \tan^{-1} 1 - \dfrac{\pi}{4} \Rightarrow \sin^{-1} x = \dfrac{\pi}{4} - \dfrac{\pi}{4} \Rightarrow$

$\sin^{-1} x = 0 \Rightarrow \sin 0 = x \Rightarrow x = 0$

Solution set: $\{0\}$

39. $\arccos x + 2\arcsin \dfrac{\sqrt{3}}{2} = \pi$

$\arccos x + 2\arcsin \dfrac{\sqrt{3}}{2} = \pi \Rightarrow$

$\arccos x = \pi - 2\arcsin \dfrac{\sqrt{3}}{2} \Rightarrow$

$\arccos x = \pi - 2\left(\dfrac{\pi}{3}\right)$

$\arccos x = \pi - \dfrac{2\pi}{3} \Rightarrow \arccos x = \dfrac{\pi}{3} \Rightarrow$

$x = \cos\dfrac{\pi}{3} \Rightarrow x = \dfrac{1}{2}$

Solution set: $\left\{\dfrac{1}{2}\right\}$

41. $\arcsin 2x + \arccos x = \dfrac{\pi}{6}$

$\arcsin 2x + \arccos x = \dfrac{\pi}{6}$

$\arcsin 2x = \dfrac{\pi}{6} - \arccos x$

$2x = \sin\left(\dfrac{\pi}{6} - \arccos x\right)$

Use the identity

$\sin(A - B) = \sin A \cos B - \cos A \sin B.$

$2x = \sin\dfrac{\pi}{6}\cos(\arccos x) - \cos\dfrac{\pi}{6}\sin(\arccos x)$

Let $u = \arccos x$. Thus, $\cos u = x = \dfrac{x}{1}$.

$\sin u = \sqrt{1 - x^2}$

$2x = \sin\dfrac{\pi}{6} \cdot \cos u - \cos\dfrac{\pi}{6}\sin u \Rightarrow$

$2x = \dfrac{1}{2}x - \dfrac{\sqrt{3}}{2}\left(\sqrt{1 - x^2}\right) \Rightarrow$

$4x = x - \sqrt{3} \cdot \sqrt{1 - x^2}$

$3x = -\sqrt{3} \cdot \sqrt{1 - x^2}$

$(3x)^2 = \left(-\sqrt{3} \cdot \sqrt{1 - x^2}\right)^2 \Rightarrow 9x^2 = 3\left(1 - x^2\right)$

$9x^2 = 3 - 3x^2 \Rightarrow 12x^2 = 3$

$x^2 = \dfrac{3}{12} = \dfrac{1}{4} \Rightarrow x = \pm\dfrac{1}{2}$

Check these proposed solutions because they were found by squaring both side of an equation.

Check $x = \dfrac{1}{2}$.

$\arcsin 2x + \arccos x = \dfrac{\pi}{6}$

$\arcsin\left(2 \cdot \dfrac{1}{2}\right) + \arccos\left(\dfrac{1}{2}\right) = \dfrac{\pi}{6}$?

$\dfrac{\pi}{2} + \dfrac{\pi}{3} = \dfrac{\pi}{6}$?

$\dfrac{5\pi}{6} = \dfrac{\pi}{6}$ False

$\dfrac{1}{2}$ is not a solution.

Check $x = -\dfrac{1}{2}$.

$\arcsin 2x + \arccos x = \dfrac{\pi}{6}$

$\arcsin\left(2 \cdot -\dfrac{1}{2}\right) + \arccos\left(-\dfrac{1}{2}\right) = \dfrac{\pi}{6}$?

$-\dfrac{\pi}{2} + \dfrac{2\pi}{3} = \dfrac{\pi}{6}$?

$\dfrac{\pi}{6} = \dfrac{\pi}{6}$ True

$-\dfrac{1}{2}$ is a solution.

Solution set: $\left\{-\dfrac{1}{2}\right\}$

43. $\cos^{-1} x + \tan^{-1} x = \dfrac{\pi}{2}$

$$\cos^{-1} x + \tan^{-1} x = \frac{\pi}{2}$$

$$\cos^{-1} x = \frac{\pi}{2} - \tan^{-1} x$$

$$x = \cos\left(\frac{\pi}{2} - \tan^{-1} x\right)$$

Use the identity
$\cos(A - B) = \cos A \cos B + \sin A \sin B.$

$$x = \cos\frac{\pi}{2}\cos\left(\tan^{-1} x\right) + \sin\frac{\pi}{2}\sin\left(\tan^{-1} x\right)$$

$$x = 0\cdot\cos\left(\tan^{-1} x\right) + 1\cdot\sin\left(\tan^{-1} x\right)$$

$$x = \sin\left(\tan^{-1} x\right)$$

Let $u = \tan^{-1} x$. So, $\tan u = x.$

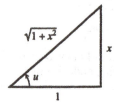

From the triangle, we find $\sin u = \dfrac{x}{\sqrt{1+x^2}}$, so

the equation $x = \sin\left(\tan^{-1} x\right)$ becomes

$x = \dfrac{x}{\sqrt{1+x^2}}$. Solve this equation.

$$x = \frac{x}{\sqrt{1+x^2}} \Rightarrow x\sqrt{1+x^2} = x$$

$$x\sqrt{1+x^2} - x = 0 \Rightarrow x\left(\sqrt{1+x^2} - 1\right) = 0$$

$x = 0$ or $\sqrt{1+x^2} - 1 = 0 \Rightarrow \sqrt{1+x^2} = 1 \Rightarrow$
$1 + x^2 = 1 \Rightarrow x^2 = 0 \Rightarrow x = 0$
Solution set: $\{0\}$

45.

$$y_1 = \sin^{-1} x - \cos^{-1} x - \frac{\pi}{6}$$

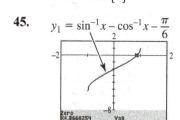

47. The x-intercept method is shown below.

$y_1 = \left(\arctan x\right)^3 - x + 2$ is graphed in the

window $[0, 6] \times [0, 6] \times [-\pi, \pi].$

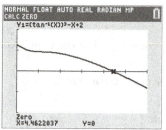

Solution set: $\{4.4622\}$

49. $A = \sqrt{\begin{array}{l}\left(A_1\cos\phi_1 + A_2\cos\phi_2\right)^2 \\ + \left(A_1\sin\phi_1 + A_2\sin\phi_2\right)^2\end{array}}$ and

$\phi = \arctan\left(\dfrac{A_1\sin\phi_1 + A_2\sin\phi_2}{A_1\cos\phi + A_2\cos\phi_2}\right)$

Make sure your calculator is in radian mode.

(a) Let $A_1 = 0.0012,\ \phi_1 = 0.052,\ A_2 = 0.004,$
and $\phi_2 = 0.61.$

$A = \sqrt{\begin{array}{l}\left(0.0012\cos.052 + 0.004\cos 0.61\right)^2 \\ + \left(0.0012\sin 0.052 + 0.004\sin 0.61\right)^2\end{array}}$
≈ 0.00506

$\phi = \arctan\left(\dfrac{0.0012\sin 0.052 + 0.004\sin 0.61}{0.0012\cos 0.052 + 0.004\cos 0.61}\right)$
≈ 0.484

If $f = 220,$ then $P = A\sin\left(2\pi ft + \phi\right)$

becomes $P = 0.00506\sin\left(440\pi t + 0.484\right).$

(b) For $x = t,$
$P(t) = 0.00506\sin(440\pi t + 0.484)$
$P_1(t) + P_2(t) = 0.0012\sin(440\pi t + 0.052)$
$\qquad\qquad\qquad + 0.004\sin(440\pi t + 0.61)$

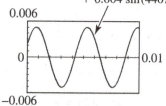

The two graphs are the same.

51. (a) $\tan\alpha = \dfrac{x}{z}$ and $\tan\beta = \dfrac{x+y}{z}$

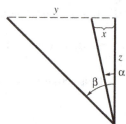

(b) Because

$$\tan\alpha = \frac{x}{z} \Rightarrow z\tan\alpha = x \Rightarrow z = \frac{x}{\tan\alpha}$$

and $\tan\beta = \frac{x+y}{z} \Rightarrow z\tan\beta = x+y \Rightarrow$

$$z = \frac{x+y}{\tan\beta}, \text{ we have } \frac{x}{\tan\alpha} = \frac{x+y}{\tan\beta}$$

(c) $(x+y)\tan\alpha = x\tan\beta \Rightarrow$

$$\tan\alpha = \frac{x\tan\beta}{x+y} \Rightarrow \alpha = \arctan\left(\frac{x\tan\beta}{x+y}\right)$$

(d) $x\tan\beta = (x+y)\tan\alpha$

$$\tan\beta = \frac{(x+y)\tan\alpha}{x}$$

$$\beta = \arctan\left(\frac{(x+y)\tan\alpha}{x}\right)$$

53. (a) $E = E_{max}\sin 2\pi ft \Rightarrow \dfrac{E}{E_{max}} = \sin 2\pi ft \Rightarrow$

$$2\pi ft = \arcsin\frac{E}{E_{max}} \Rightarrow$$

$$t = \frac{1}{2\pi f}\arcsin\frac{E}{E_{max}}$$

(b) Let $E_{max} = 12, E = 5,$ and $f = 100.$

$$t = \frac{1}{2\pi(100)}\arcsin\frac{5}{12}$$

$$= \frac{1}{200\pi}\arcsin\frac{5}{12} \approx 0.00068\,\text{sec}$$

55. $y = \dfrac{1}{3}\sin\dfrac{4\pi t}{3}$

(a) $3y = \sin\dfrac{4\pi t}{3} \Rightarrow \dfrac{4\pi t}{3} = \arcsin 3y \Rightarrow$

$$4\pi t = 3\arcsin 3y \Rightarrow t = \frac{3}{4\pi}\arcsin 3y$$

(b) If $y = 0.3$ radian,

$$t = \frac{3}{4\pi}\arcsin 0.9 \Rightarrow t \approx 0.27\,\text{sec}.$$

Chapter 7   Review Exercises

1. B   **3.** C   **5.** D

7. $\sec^2\theta - \tan^2\theta = \dfrac{1}{\cos^2\theta} - \dfrac{\sin^2\theta}{\cos^2\theta}$

$$= \frac{1-\sin^2\theta}{\cos^2\theta} = \frac{\cos^2\theta}{\cos^2\theta} = 1$$

9. $\tan^2\theta\left(1+\cot^2\theta\right) = \dfrac{\sin^2\theta}{\cos^2\theta}\left(1+\dfrac{\cos^2\theta}{\sin^2\theta}\right)$

$$= \frac{\sin^2\theta}{\cos^2\theta}\left(\frac{\sin^2\theta+\cos^2\theta}{\sin^2\theta}\right)$$

$$= \frac{\sin^2\theta}{\cos^2\theta}\left(\frac{1}{\sin^2\theta}\right) = \frac{1}{\cos^2\theta}$$

$$= \sec^2\theta$$

11. $\tan\theta - \sec\theta\csc\theta = \dfrac{\sin\theta}{\cos\theta} - \dfrac{1}{\cos\theta}\cdot\dfrac{1}{\sin\theta}$

$$= \frac{\sin\theta}{\cos\theta} - \frac{1}{\sin\theta\cos\theta}$$

$$= \frac{\sin^2\theta}{\sin\theta\cos\theta} - \frac{1}{\sin\theta\cos\theta}$$

$$= \frac{\sin^2\theta-1}{\sin\theta\cos\theta} = \frac{(1-\cos^2\theta)-1}{\sin\theta\cos\theta}$$

$$= \frac{-\cos^2\theta}{\sin\theta\cos\theta} = -\frac{\cos\theta}{\sin\theta}$$

$$= -\cot\theta$$

13. $\cos x = \dfrac{3}{5}, x$ is in quadrant IV.

$$\sin^2 x = 1-\cos^2 x = 1-\left(\frac{3}{5}\right)^2 = 1-\frac{9}{25} = \frac{16}{25}$$

Because x is in quadrant IV, $\sin x < 0.$

$$\sin x = -\sqrt{\frac{16}{25}} = -\frac{4}{5}$$

$$\tan x = \frac{\sin x}{\cos x} = \frac{-\frac{4}{5}}{\frac{3}{5}} = -\frac{4}{3}$$

$$\cot(-x) = -\cot x = \frac{1}{-\tan x} = \frac{1}{-\left(-\frac{4}{3}\right)} = \frac{3}{4}$$

15. Use the fact that $165° = 180° - 15°.$

$$\sin 165° = \sin 180°\cos 15° - \cos 180°\sin 15°$$
$$= 0\cdot\cos 15° - 1\cdot\sin 15° = \sin 15°$$
$$= \sin(45°-30°)$$
$$= \sin 45°\cos 30° - \cos 45°\sin 30°$$
$$= \frac{\sqrt{2}}{2}\cdot\frac{\sqrt{3}}{2} - \frac{\sqrt{2}}{2}\cdot\frac{1}{2} = \frac{\sqrt{6}-\sqrt{2}}{4}$$

$$\cos 165° = \cos 180°\cos 15° + \sin 180°\sin 15°$$
$$= -1\cdot\cos 15° + 0\cdot\sin 15° = -\cos 15°$$
$$= -\cos(45°-30°)$$
$$= -(\cos 45°\cos 30° + \sin 45°\sin 30°)$$
$$= -\left(\frac{\sqrt{2}}{2}\cdot\frac{\sqrt{3}}{2} + \frac{\sqrt{2}}{2}\cdot\frac{1}{2}\right)$$
$$= -\left(\frac{\sqrt{6}+\sqrt{2}}{4}\right) = \frac{-\sqrt{6}-\sqrt{2}}{4}$$

(continued on next page)

(*continued*)

$$\tan 165° = \frac{\tan 180° - \tan 15°}{1 + \tan 180° \tan 15°} = \frac{0 - \tan 15°}{1 + 0 \cdot \tan 15°}$$

$$= -\tan 15° = -\frac{\tan 45° - \tan 30°}{1 + \tan 45° \tan 30°}$$

$$= -\frac{1 - \frac{\sqrt{3}}{3}}{1 + \frac{\sqrt{3}}{3}} = -\frac{1 - \frac{\sqrt{3}}{3}}{1 + \frac{\sqrt{3}}{3}} \cdot \frac{3}{3}$$

$$= -\frac{3 - \sqrt{3}}{3 + \sqrt{3}} = -\frac{3 - \sqrt{3}}{3 + \sqrt{3}} \cdot \frac{3 - \sqrt{3}}{3 - \sqrt{3}}$$

$$= -\frac{9 - 3\sqrt{3} - 3\sqrt{3} + 3}{9 - 3} = -\frac{12 - 6\sqrt{3}}{6}$$

$$= -\left(2 - \sqrt{3}\right) = -2 + \sqrt{3}$$

$$\cot 165° = \frac{1}{\tan 165°} = \frac{1}{-2 + \sqrt{3}}$$

$$= \frac{1}{-2 + \sqrt{3}} \cdot \frac{-2 - \sqrt{3}}{-2 - \sqrt{3}} = \frac{-2 - \sqrt{3}}{4 - 3}$$

$$= -2 - \sqrt{3}$$

$$\sec 165° = \frac{1}{\cos 165°} = \frac{1}{\frac{-\sqrt{6} - \sqrt{2}}{4}} = \frac{4}{-\sqrt{6} - \sqrt{2}}$$

$$= \frac{4}{-\sqrt{6} - \sqrt{2}} \cdot \frac{-\sqrt{6} + \sqrt{2}}{-\sqrt{6} + \sqrt{2}}$$

$$= \frac{4\left(-\sqrt{6} + \sqrt{2}\right)}{6 - 2} = -\sqrt{6} + \sqrt{2}$$

$$\csc 165° = \frac{1}{\sin 165°} = \frac{1}{\frac{\sqrt{6} - \sqrt{2}}{4}} = \frac{4}{\sqrt{6} - \sqrt{2}}$$

$$= \frac{4}{\sqrt{6} - \sqrt{2}} \cdot \frac{\sqrt{6} + \sqrt{2}}{\sqrt{6} + \sqrt{2}} = \frac{4\left(\sqrt{6} + \sqrt{2}\right)}{6 - 2}$$

$$= \sqrt{6} + \sqrt{2}$$

17. I. $\cos 210° = \cos\left(150° + 60°\right)$
$$= \cos 150° \cos 60° - \sin 150° \sin 60°$$

19. H. $\tan\left(-35°\right) = \cot\left[90° - \left(-35°\right)\right] = \cot 125°$

21. G. $\cos 35° = \cos\left(-35°\right)$

23. J. $\sin 75° = \sin\left(15° + 60°\right)$
$$= \sin 15° \cos 60° + \cos 15° \sin 60°$$

25. F. $\cos 300° = \cos 2\left(150°\right)$
$$= \cos^2 150° - \sin^2 150°$$

27. Find $\sin\left(x + y\right)$, $\cos\left(x - y\right)$, and $\tan\left(x + y\right)$, given $\sin x = -\frac{3}{5}$, $\cos y = -\frac{7}{25}$, x and y are in quadrant III. Because x and y are in quadrant III, $\cos x$ and $\sin y$ are negative.

$$\cos x = -\sqrt{1 - \sin^2 x} = -\sqrt{1 - \left(-\frac{3}{5}\right)^2}$$

$$= -\sqrt{1 - \frac{9}{25}} = -\sqrt{\frac{16}{25}} = -\frac{4}{5}$$

$$\sin y = -\sqrt{1 - \cos^2 y} = -\sqrt{1 - \left(-\frac{7}{25}\right)^2}$$

$$= -\sqrt{1 - \frac{49}{625}} = -\sqrt{\frac{596}{625}} = -\frac{24}{25}$$

$$\sin\left(x + y\right) = \sin x \cos y + \cos x \sin y$$

$$= \left(-\frac{3}{5}\right)\left(-\frac{7}{25}\right) + \left(-\frac{4}{5}\right)\left(-\frac{24}{25}\right)$$

$$= \frac{21}{125} + \frac{96}{125} = \frac{117}{125}$$

$$\cos\left(x - y\right) = \cos x \cos y + \sin x \sin y$$

$$= \left(-\frac{4}{5}\right)\left(-\frac{7}{25}\right) + \left(-\frac{3}{5}\right)\left(-\frac{24}{25}\right)$$

$$= \frac{28}{125} + \frac{72}{125} = \frac{100}{125} = \frac{4}{5}$$

To find $\tan\left(x + y\right)$, first find $\cos\left(x + y\right)$.

$$\cos\left(x + y\right) = \cos x \cos y - \sin x \sin y$$

$$= \left(-\frac{4}{5}\right)\left(-\frac{7}{25}\right) - \left(-\frac{3}{5}\right)\left(-\frac{24}{25}\right)$$

$$= \frac{28}{125} - \frac{72}{125} = -\frac{44}{125}$$

$$\tan\left(x + y\right) = \frac{\sin\left(x + y\right)}{\cos\left(x + y\right)} = \frac{\frac{117}{125}}{-\frac{44}{5}} \cdot \frac{125}{125} = -\frac{117}{44}$$

Note that using the formula

$$\tan\left(x + y\right) = \frac{\tan x + \tan y}{1 - \tan x \tan y}, \text{ we have}$$

$$\tan\left(x + y\right) = \frac{\frac{3}{4} + \frac{24}{7}}{1 - \left(\frac{3}{4}\right)\left(\frac{24}{7}\right)} = \frac{21 + 96}{28 - 72} = -\frac{117}{44}$$

To find the quadrant of $x + y$, notice that $\sin(x + y) > 0$, which implies $x + y$ is in quadrant I or II. Also $\tan(x + y) < 0$, which implies that $x + y$ is in quadrant II or IV. Therefore, $x + y$ is in quadrant II.

29. Find $\sin(x+y)$, $\cos(x-y)$, and $\tan(x+y)$,

given $\sin x = -\dfrac{1}{2}$, $\cos y = -\dfrac{2}{5}$, x and y are in

quadrant III.

Because x and y are in quadrant III, $\cos x$ and $\sin y$ are negative.

$$\cos x = -\sqrt{1-\sin^2 x} = -\sqrt{1-\left(-\dfrac{1}{2}\right)^2}$$

$$= -\sqrt{1-\dfrac{1}{4}} = -\sqrt{\dfrac{3}{4}} = -\dfrac{\sqrt{3}}{2}$$

$$\sin y = -\sqrt{1-\cos^2 y} = -\sqrt{1-\left(-\dfrac{2}{5}\right)^2}$$

$$= -\sqrt{1-\dfrac{4}{25}} = -\sqrt{\dfrac{21}{25}} = -\dfrac{\sqrt{21}}{5}$$

$$\sin(x+y) = \sin x \cos y + \cos x \sin y$$

$$= \left(-\dfrac{1}{2}\right)\left(-\dfrac{2}{5}\right) + \left(-\dfrac{\sqrt{3}}{2}\right)\left(-\dfrac{\sqrt{21}}{5}\right)$$

$$= \dfrac{2}{10} + \dfrac{\sqrt{63}}{10} = \dfrac{2+3\sqrt{7}}{10}$$

$$\cos(x-y) = \cos x \cos y + \sin x \sin y$$

$$= \left(-\dfrac{\sqrt{3}}{2}\right)\left(-\dfrac{2}{5}\right) + \left(-\dfrac{1}{2}\right)\left(-\dfrac{\sqrt{21}}{5}\right)$$

$$= \dfrac{2\sqrt{3}}{10} + \dfrac{\sqrt{21}}{10} = \dfrac{2\sqrt{3}+\sqrt{21}}{10}$$

To find $\tan(x+y)$, first find $\cos(x+y)$.

$$\cos(x+y) = \cos x \cos y - \sin x \sin y$$

$$= \left(-\dfrac{2}{5}\right)\left(-\dfrac{\sqrt{3}}{2}\right) - \left(-\dfrac{1}{2}\right)\left(-\dfrac{\sqrt{21}}{5}\right)$$

$$= \dfrac{2\sqrt{3}}{10} - \dfrac{\sqrt{21}}{10} = \dfrac{2\sqrt{3}-\sqrt{21}}{10}$$

$$\tan(x+y) = \dfrac{\sin(x+y)}{\cos(x+y)} = \dfrac{\frac{2+3\sqrt{7}}{10}}{\frac{2\sqrt{3}-\sqrt{21}}{10}}$$

$$= \dfrac{2+3\sqrt{7}}{2\sqrt{3}-\sqrt{21}}$$

Using the formula $\tan(x+y) = \dfrac{\tan x + \tan y}{1 - \tan x \tan y}$,

we have

$$\tan(x+y) = \dfrac{\frac{\sqrt{3}}{3} + \frac{\sqrt{21}}{2}}{1 - \left(\frac{\sqrt{3}}{3}\right)\left(\frac{\sqrt{21}}{2}\right)} = \dfrac{2\sqrt{3}+3\sqrt{21}}{6-3\sqrt{7}}$$

$$= -\dfrac{75\sqrt{3}+24\sqrt{21}}{27} = \dfrac{-25\sqrt{3}-8\sqrt{21}}{9}$$

The two forms of $\tan(x+y)$ are equal.

To find the quadrant of $x+y$, notice that $\sin(x+y) > 0$, which implies $x+y$ is in quadrant I or II. Also $\tan(x+y) < 0$, which implies that $x+y$ is in quadrant II or IV. Therefore, $x+y$ is in quadrant II.

31. Find $\sin(x+y)$, $\cos(x-y)$, and $\tan(x+y)$,

given $\sin x = \dfrac{1}{10}$, $\cos y = \dfrac{4}{5}$, x is in

quadrant I and y is in quadrant IV.

Because x is in quadrant I, $\cos x$ is positive.

$$\cos x = \sqrt{1-\sin^2 x} = \sqrt{1-\left(\dfrac{1}{10}\right)^2}$$

$$= \sqrt{1-\dfrac{1}{100}} = \sqrt{\dfrac{99}{100}} = \dfrac{3\sqrt{11}}{10}$$

Because y is in quadrant IV, $\sin y$ is negative.

$$\sin y = -\sqrt{1-\cos^2 y} = -\sqrt{1-\left(\dfrac{4}{5}\right)^2}$$

$$= -\sqrt{1-\dfrac{16}{25}} = -\sqrt{\dfrac{9}{25}} = -\dfrac{3}{5}$$

$$\sin(x+y) = \sin x \cos y + \cos x \sin y$$

$$= \left(\dfrac{1}{10}\right)\left(\dfrac{4}{5}\right) + \left(\dfrac{3\sqrt{11}}{10}\right)\left(-\dfrac{3}{5}\right)$$

$$= \dfrac{4}{50} - \dfrac{9\sqrt{11}}{50} = \dfrac{4-9\sqrt{11}}{50}$$

$$\cos(x-y) = \cos x \cos y + \sin x \sin y$$

$$= \left(\dfrac{3\sqrt{11}}{10}\right)\left(\dfrac{4}{5}\right) + \left(\dfrac{1}{10}\right)\left(-\dfrac{3}{5}\right)$$

$$= \dfrac{12\sqrt{11}}{50} - \dfrac{3}{50} = \dfrac{12\sqrt{11}-3}{50}$$

To find $\tan(x+y)$, first find $\cos(x+y)$.

$$\cos(x+y) = \cos x \cos y - \sin x \sin y$$

$$= \left(\dfrac{3\sqrt{11}}{10}\right)\left(\dfrac{4}{5}\right) - \left(\dfrac{1}{10}\right)\left(-\dfrac{3}{5}\right)$$

$$= \dfrac{12\sqrt{11}}{50} + \dfrac{3}{50} = \dfrac{12\sqrt{11}+3}{50}$$

$$\tan(x+y) = \dfrac{\sin(x+y)}{\cos(x+y)} = \dfrac{\frac{4-9\sqrt{11}}{50}}{\frac{12\sqrt{11}+3}{50}}$$

$$= \dfrac{4-9\sqrt{11}}{12\sqrt{11}+3}$$

(continued on next page)

(*continued*)

To find $\tan(x+y)$ using the formula

$\tan(x+y) = \dfrac{\tan x + \tan y}{1 - \tan x \tan y}$, we have

$\tan x = \dfrac{\sin x}{\cos x} = \dfrac{\frac{1}{10}}{\frac{3\sqrt{11}}{10}} = \dfrac{1}{3\sqrt{11}} \cdot \dfrac{\sqrt{11}}{\sqrt{11}} = \dfrac{\sqrt{11}}{33}$

$\tan y = \dfrac{\sin y}{\cos y} = \dfrac{-\frac{3}{5}}{\frac{4}{5}} = -\dfrac{3}{4}$

$\tan(x+y) = \dfrac{\tan x + \tan y}{1 - \tan x \tan y} = \dfrac{\frac{\sqrt{11}}{33} + \left(-\frac{3}{4}\right)}{1 - \left(\frac{\sqrt{11}}{33}\right)\left(-\frac{3}{4}\right)}$

$= \dfrac{4\sqrt{11} - 99}{132 + 3\sqrt{11}} \cdot \dfrac{132 - 3\sqrt{11}}{132 - 3\sqrt{11}}$

$= \dfrac{528\sqrt{11} - 132 - 13068 + 297\sqrt{11}}{17424 - 99}$

$= \dfrac{825\sqrt{11} - 13200}{17325} = \dfrac{\sqrt{11} - 16}{21}$

The two forms of $\tan(x+y)$ are equal. To find the quadrant of $x+y$, notice that $\sin(x+y) < 0$, which implies $x+y$ is in quadrant III or IV. Also $\tan(x+y) < 0$, which implies that $x+y$ is in quadrant II or IV. Therefore, $x+y$ is in quadrant IV.

33. Find $\sin\theta$ and $\cos\theta$, given $\cos 2\theta = -\dfrac{3}{4}$, $90° < 2\theta < 180°$.
$90° < 2\theta < 180° \Rightarrow 45° < \theta < 90° \Rightarrow \theta$ is in quadrant I, so $\sin\theta$ and $\cos\theta$ are both positive.

$\cos 2\theta = 1 - 2\sin^2\theta \Rightarrow -\dfrac{3}{4} = 1 - 2\sin^2\theta \Rightarrow$

$-\dfrac{7}{4} = -2\sin^2\theta \Rightarrow \dfrac{7}{8} = \sin^2\theta \Rightarrow$

$\sin\theta = \sqrt{\dfrac{7}{8}} = \dfrac{\sqrt{7}}{2\sqrt{2}} = \dfrac{\sqrt{14}}{4}$

$\cos\theta = \sqrt{1 - \sin^2\theta} = \sqrt{1 - \dfrac{7}{8}} = \sqrt{\dfrac{1}{8}} = \dfrac{1}{\sqrt{8}}$

$= \dfrac{1}{2\sqrt{2}} = \dfrac{\sqrt{2}}{4}$

35. Find $\sin 2x$ and $\cos 2x$, given $\tan x = 3$, $\sin x < 0$.
Because $\tan x > 0$ and $\sin x < 0$, x is in quadrant III, and $2x$ is in quadrant I or II.

$\tan 2x = \dfrac{2\tan x}{1 - \tan^2 x} = \dfrac{2(3)}{1 - 3^2} = -\dfrac{6}{8} = -\dfrac{3}{4}$

Because $\tan 2x < 0$, $2x$ is in quadrant II. Thus, $\sin 2x > 0$ and $\cos 2x < 0$.

$\sec 2x = -\sqrt{1 + \tan^2 x} = -\sqrt{1 + \left(-\dfrac{3}{4}\right)^2}$

$= -\sqrt{1 + \dfrac{9}{16}} = -\sqrt{\dfrac{25}{16}} = -\dfrac{5}{4} \Rightarrow$

$\cos 2x = \dfrac{1}{-\frac{5}{4}} = -\dfrac{4}{5}$

$\sin 2x = \sqrt{1 - \cos^2(2x)} = \sqrt{1 - \left(-\dfrac{4}{5}\right)^2}$

$= \sqrt{1 - \dfrac{16}{25}} = \sqrt{\dfrac{9}{25}} = \dfrac{3}{5}$

37. Find $\cos\dfrac{\theta}{2}$, given $\cos\theta = -\dfrac{1}{2}$, $90° < \theta < 180°$.

Because $90° < \theta < 180° \Rightarrow 45° < \dfrac{\theta}{2} < 90°$, $\dfrac{\theta}{2}$ is in quadrant I and $\cos\dfrac{\theta}{2} > 0$.

$\cos\dfrac{\theta}{2} = \sqrt{\dfrac{1 + \left(-\frac{1}{2}\right)}{2}} = \sqrt{\dfrac{2 - 1}{4}} = \sqrt{\dfrac{1}{4}} = \dfrac{1}{2}$

39. Find $\tan x$, given $\tan 2x = 2$, with $\pi < x < \dfrac{3\pi}{2}$.

$\tan 2x = \dfrac{2\tan x}{1 - \tan^2 x} \Rightarrow 2 = \dfrac{2\tan x}{1 - \tan^2 x} \Rightarrow$

$2\tan x = 2\left(1 - \tan^2 x\right)$, if $\tan x \neq \pm 1$

Thus, $2\left(\tan^2 x + 2\tan x - 1\right) = 0 \Rightarrow$

$\tan^2 x + 2\tan x - 1 = 0$, so we can use the quadratic formula to solve for $\tan x$.

$x = \dfrac{-b \pm \sqrt{b^2 - 4ac}}{2a} \Rightarrow$

$\tan x = \dfrac{-1 \pm \sqrt{1^2 - 4(1)(-1)}}{2} = \dfrac{-1 \pm \sqrt{5}}{2}$

Because x is in quadrant III, $\tan x > 0$, so

$\tan x = \dfrac{-1 + \sqrt{5}}{2}$

41. Find $\tan\dfrac{x}{2}$, given $\sin x = 0.8$, with $0 < x < \dfrac{\pi}{2}$.

$\cos x = \pm\sqrt{1 - \sin^2 x} = \pm\sqrt{1 - 0.8^2} = \sqrt{1 - 0.64}$

$= \pm\sqrt{0.36} = \pm 0.6$

Because x is in quadrant I, $\cos x > 0$, so $\cos x = 0.6$

$\tan\dfrac{x}{2} = \dfrac{1 - \cos x}{\sin x} = \dfrac{1 - 0.6}{0.8} = \dfrac{0.4}{0.8} = 0.5$

Exercises 43−47 are graphed in the window $\left[-2\pi,\ 2\pi\right]$ by $[-4,\ 4]$.

43. $\dfrac{\sin 2x + \sin x}{\cos 2x - \cos x}$ appears to be equivalent to $\cot \frac{x}{2}$.

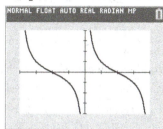

$$\frac{\sin 2x + \sin x}{\cos x - \cos 2x} = \frac{2\sin x \cos x + \sin x}{\cos x - \left(2\cos^2 x - 1\right)}$$
$$= \frac{\sin x\left(2\cos x + 1\right)}{-2\cos^2 x + \cos x + 1}$$
$$= \frac{\sin x\left(2\cos x + 1\right)}{\left(2\cos x + 1\right)\left(-\cos x + 1\right)}$$
$$= \frac{\sin x}{-\cos x + 1} = \frac{\sin x}{1 - \cos x}$$
$$= \frac{1}{\frac{1-\cos x}{\sin x}} = \frac{1}{\tan \frac{x}{2}} = \cot \frac{x}{2}$$

45. $\dfrac{\sin x}{1 - \cos x}$ appears to be equivalent to $\cot \dfrac{x}{2}$.

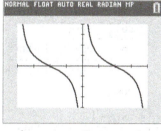

$$\frac{\sin x}{1 - \cos x} = \frac{1}{\frac{1 - \cos x}{\sin x}} = \frac{1}{\tan \frac{x}{2}} = \cot \frac{x}{2}$$

47. $\dfrac{2\left(\sin x - \sin^3 x\right)}{\cos x}$ appears to be equivalent to $\sin 2x$.

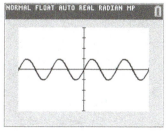

$$\frac{2\left(\sin x - \sin^3 x\right)}{\cos x} = \frac{2\sin x\left(1 - \sin^2 x\right)}{\cos x}$$
$$= \frac{2\sin x \cos^2 x}{\cos x}$$
$$= 2\sin x \cos x = \sin 2x$$

49. Verify $\sin^2 x - \sin^2 y = \cos^2 y - \cos^2 x$ is an identity.
$$\sin^2 x - \sin^2 y = \left(1 - \cos^2 x\right) - \left(1 - \cos^2 y\right)$$
$$= 1 - \cos^2 x - 1 + \cos^2 y$$
$$= \cos^2 y - \cos^2 x$$

51. Verify $\dfrac{\sin^2 x}{2 - 2\cos x} = \cos^2 \dfrac{x}{2}$ is an identity.
$$\frac{\sin^2 x}{2 - 2\cos x} = \frac{1 - \cos^2 x}{2\left(1 - \cos x\right)}$$
$$= \frac{\left(1 - \cos x\right)\left(1 + \cos x\right)}{2\left(1 - \cos x\right)}$$
$$= \frac{1 + \cos x}{2} = \cos^2 \frac{x}{2}$$

53. Verify $2\cos A - \sec A = \cos A - \dfrac{\tan A}{\csc A}$ is an identity. Work with the right side.
$$\cos A - \frac{\tan A}{\csc A} = \cos A - \frac{\frac{\sin A}{\cos A}}{\frac{1}{\sin A}} = \cos A - \frac{\frac{\sin^2 A}{\cos A}}{1}$$
$$= \frac{\cos^2 A}{\cos A} - \frac{\sin^2 A}{\cos A}$$
$$= \frac{\cos^2 A - \sin^2 A}{\cos A}$$
$$= \frac{\cos^2 A - \left(1 - \cos^2 A\right)}{\cos A}$$
$$= \frac{2\cos^2 A - 1}{\cos A} = 2\cos A - \frac{1}{\cos A}$$
$$= 2\cos A - \sec A$$

55. Verify $1 + \tan^2 \alpha = 2\tan \alpha \csc 2\alpha$ is an identity. Work with the right side.
$$2\tan \alpha \csc 2\alpha = \frac{2\tan \alpha}{\sin 2\alpha} = \frac{2 \cdot \frac{\sin \alpha}{\cos \alpha}}{2\sin \alpha \cos \alpha}$$
$$= \frac{2\sin \alpha}{2\sin \alpha \cos^2 \alpha} = \frac{1}{\cos^2 \alpha}$$
$$= \sec^2 \alpha = 1 + \tan^2 \alpha$$

57. Verify $\tan\theta \sin 2\theta = 2 - 2\cos^2\theta$ is an identity.

$$\tan\theta \sin 2\theta = \tan\theta \left(2\sin\theta\cos\theta\right)$$
$$= \frac{\sin\theta}{\cos\theta}\left(2\sin\theta\cos\theta\right) = 2\sin^2\theta$$
$$= 2\left(1 - \cos^2\theta\right) = 2 - 2\cos^2\theta$$

59. Verify $2\tan x \csc 2x - \tan^2 x = 1$ is an identity.

$$2\tan x \csc 2x - \tan^2 x$$
$$= 2\tan x \frac{1}{\sin 2x} - \tan^2 x$$
$$= 2 \cdot \frac{\sin x}{\cos x} \cdot \frac{1}{2\sin x\cos x} - \frac{\sin^2 x}{\cos^2 x}$$
$$= \frac{1}{\cos^2 x} - \frac{\sin^2 x}{\cos^2 x} = \frac{1 - \sin^2 x}{\cos^2 x}$$
$$= \frac{\cos^2 x}{\cos^2 x} = 1$$

61. Verify $\tan\theta\cos^2\theta = \dfrac{2\tan\theta\cos^2\theta - \tan\theta}{1 - \tan^2\theta}$ is an identity.

Work with the right side.

$$\frac{2\tan\theta\cos^2\theta - \tan\theta}{1 - \tan^2\theta}$$
$$= \frac{\tan\theta\left(2\cos^2\theta - 1\right)}{1 - \tan^2\theta}$$
$$= \frac{\tan\theta\left(2\cos^2\theta - 1\right)}{1 - \dfrac{\sin^2\theta}{\cos^2\theta}} \cdot \frac{\cos^2\theta}{\cos^2\theta}$$
$$= \frac{\tan\theta\cos^2\theta\left(2\cos^2\theta - 1\right)}{\cos^2\theta - \sin^2\theta}$$
$$= \frac{\tan\theta\cos^2\theta\left(2\cos^2\theta - 1\right)}{2\cos^2\theta - 1} = \tan\theta\cos^2\theta$$

63. Verify $\dfrac{\sin^2 x - \cos^2 x}{\csc x} = 2\sin^3 x - \sin x$ is an identity.

$$\frac{\sin^2 x - \cos^2 x}{\csc x} = \frac{\sin^2 x - \left(1 - \sin^2 x\right)}{\dfrac{1}{\sin x}}$$
$$= \frac{2\sin^2 x - 1}{\dfrac{1}{\sin x}} \cdot \frac{\sin x}{\sin x}$$
$$= \left(2\sin^2 x - 1\right)\sin x$$
$$= 2\sin^3 x - \sin x$$

65. Verify $\tan 4\theta = \dfrac{2\tan 2\theta}{2 - \sec^2 2\theta}$ is an identity.

$$\tan 4\theta = \tan\left[2\left(2\theta\right)\right] = \frac{2\tan 2\theta}{1 - \tan^2 2\theta}$$
$$= \frac{2\tan 2\theta}{1 - \left(\sec^2 2\theta - 1\right)} = \frac{2\tan 2\theta}{2 - \sec^2 2\theta}$$

67. Verify $\tan\left(\dfrac{x}{2} + \dfrac{\pi}{4}\right) = \sec x + \tan x$ is an identity. Working with the left side, we have

$$\tan\left(\frac{x}{2} + \frac{\pi}{4}\right) = \frac{\tan\frac{x}{2} + \tan\frac{\pi}{4}}{1 - \tan\frac{x}{2}\tan\frac{\pi}{4}} = \frac{\tan\frac{x}{2} + 1}{1 - \tan\frac{x}{2}}$$

Working with the right side, we have

$$\sec x + \tan x = \frac{1}{\cos x} + \frac{\sin x}{\cos x} = \frac{1 + \sin x}{\cos x}$$
$$= \frac{\left(\cos^2\frac{x}{2} + \sin^2\frac{x}{2}\right) + \sin\left[2\left(\frac{x}{2}\right)\right]}{\cos\left[2\left(\frac{x}{2}\right)\right]}$$
$$= \frac{\cos^2\frac{x}{2} + \sin^2\frac{x}{2} + 2\sin\frac{x}{2}\cos\frac{x}{2}}{\cos^2 x - \sin^2 x}$$
$$= \frac{\left(\cos\frac{x}{2} + \sin\frac{x}{2}\right)^2}{\left(\cos\frac{x}{2} + \sin\frac{x}{2}\right)\left(\cos\frac{x}{2} - \sin\frac{x}{2}\right)}$$
$$= \frac{\cos\frac{x}{2} + \sin\frac{x}{2}}{\cos\frac{x}{2} - \sin\frac{x}{2}} \cdot \frac{\dfrac{1}{\cos\frac{x}{2}}}{\dfrac{1}{\cos\frac{x}{2}}}$$
$$= \frac{\dfrac{\cos\frac{x}{2}}{\cos\frac{x}{2}} + \dfrac{\sin\frac{x}{2}}{\cos\frac{x}{2}}}{\dfrac{\cos\frac{x}{2}}{\cos\frac{x}{2}} - \dfrac{\sin\frac{x}{2}}{\cos\frac{x}{2}}} = \frac{1 + \tan\frac{x}{2}}{1 - \tan\frac{x}{2}}$$

$$\tan\left(\frac{x}{2} + \frac{\pi}{4}\right) = \frac{\tan\frac{x}{2} + 1}{1 - \tan\frac{x}{2}} = \sec x + \tan x, \text{ so the}$$

statement is verified.

69. Verify $-\cot\dfrac{x}{2} = \dfrac{\sin 2x + \sin x}{\cos 2x - \cos x}$ is an identity.

Work with the right side.

$$\frac{\sin 2x + \sin x}{\cos 2x - \cos x} = \frac{2\sin x \cos x + \sin x}{\left(2\cos^2 x - 1\right) - \cos x}$$

$$= \frac{\sin x\left(2\cos x + 1\right)}{2\cos^2 x - \cos x - 1}$$

$$= \frac{\sin x\left(2\cos x + 1\right)}{\left(2\cos x + 1\right)\left(\cos x - 1\right)}$$

$$= \frac{\sin x}{1 - \cos x} = -\frac{\sin x}{\cos x - 1}$$

$$= -\frac{1}{\frac{\sin x}{\cos x - 1}} = -\frac{1}{\tan\frac{x}{2}} = -\cot\frac{x}{2}$$

71.

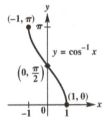

Domain: $[-1, 1]$

Range: $\left[-\dfrac{\pi}{2}, \dfrac{\pi}{2}\right]$

Domain: $[-1, 1]$

Range: $[0, \pi]$

Domain: $[-\infty, \infty]$

Range: $\left(-\dfrac{\pi}{2}, \dfrac{\pi}{2}\right)$

73. False. $\arcsin\left(-\dfrac{1}{2}\right) = -\dfrac{\pi}{6}$, not $\dfrac{11\pi}{6}$.

75. $y = \sin^{-1}\dfrac{\sqrt{2}}{2} \Rightarrow \sin y = \dfrac{\sqrt{2}}{2}$

$-\dfrac{\pi}{2} \le y \le \dfrac{\pi}{2}$, so $y = \dfrac{\pi}{4}$.

77. $y = \tan^{-1}\left(-\sqrt{3}\right) \Rightarrow \tan y = -\sqrt{3}$

$-\dfrac{\pi}{2} < y < \dfrac{\pi}{2}$, so $y = -\dfrac{\pi}{3}$.

79. $y = \cos^{-1}\left(-\dfrac{\sqrt{2}}{2}\right) \Rightarrow \cos y = -\dfrac{\sqrt{2}}{2}$

$0 \le y \le \pi$, so $y = \dfrac{3\pi}{4}$.

81. $y = \sec^{-1}\left(-2\right) \Rightarrow \sec y = -2$

$0 \le y \le \pi$ and $y \ne \dfrac{\pi}{2}$, so $y = \dfrac{2\pi}{3}$.

83. $y = \text{arccot}\left(-1\right) \Rightarrow \cot y = -1$

$0 < y < \pi$, so $y = \dfrac{3\pi}{4}$.

85. $\theta = \arcsin\left(-\dfrac{\sqrt{3}}{2}\right) \Rightarrow \sin\theta = -\dfrac{\sqrt{3}}{2}$

$-90° \le \theta \le 90°$, so $\theta = -60°$.

For Exercises 87–91, be sure that your calculator is in degree mode.

87. $\theta = \arctan 1.7804675 = 60.67924514°$

89. $\theta = \cos^{-1} 0.80396577 \approx 36.4895081°$

91. $\theta = \text{arc}\sec 3.4723155 \approx 73.26220613°$

93. $\cos\left(\arccos\left(-1\right)\right) = \cos\pi = -1$ or

$\cos\left(\arccos\left(-1\right)\right) = \cos 180° = -1$

95. $\arccos\left(\cos\dfrac{3\pi}{4}\right) = \arccos\left(-\dfrac{\sqrt{2}}{2}\right) = \dfrac{3\pi}{4}$

97. $\tan^{-1}\left(\tan\dfrac{\pi}{4}\right) = \tan^{-1}\dfrac{\sqrt{2}}{2} = \dfrac{\pi}{4}$

99. $\sin\left(\arccos\dfrac{3}{4}\right)$

Let $\omega = \arccos\dfrac{3}{4}$, so that $\cos\omega = \dfrac{3}{4}$.

Because arccos is defined only in quadrants I and II, and $\dfrac{3}{4}$ is positive, ω is in quadrant I.

Sketch ω and label a triangle with the side opposite ω equal to

$\sqrt{4^2 - 3^2} = \sqrt{16 - 9} = \sqrt{7}$. (See figure on next page.)

$\sin\left(\arccos\dfrac{3}{4}\right) = \sin\omega = \dfrac{\sqrt{7}}{4}$

(*continued on next page*)

(continued)

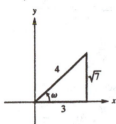

101. $\cos\left(\csc^{-1}(-2)\right)$

Let $\omega = \csc^{-1}(-2)$, so that $\csc\omega = -2$.

Because $-\dfrac{\pi}{2} \le \omega \le \dfrac{\pi}{2}$ and $\omega \ne 0$, and $\csc\omega = -2$ (negative), ω is in quadrant IV. Sketch ω and label a triangle with side adjacent to ω equal to

$$\sqrt{2^2-(-1)^2} = \sqrt{4-1} = \sqrt{3}.$$

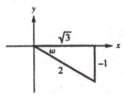

$$\cos\left(\csc^{-1}(-2)\right) = \cos\omega = \dfrac{\sqrt{3}}{2}$$

103. $\tan\left(\arcsin\dfrac{3}{5} + \arccos\dfrac{5}{7}\right)$

Let $\omega_1 = \arcsin\dfrac{3}{5}$, $\omega_2 = \arccos\dfrac{5}{7}$. Sketch angles ω_1 and ω_2 in quadrant I. The side adjacent to ω_1 is

$\sqrt{5^2-3^2} = \sqrt{25-9} = \sqrt{16} = 4$. The side opposite ω_2 is

$$\sqrt{7^2-5^2} = \sqrt{49-25} = \sqrt{24} = 2\sqrt{6}.$$

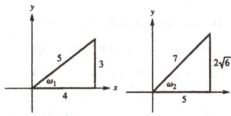

We have $\tan\omega_1 = \dfrac{3}{4}$ and $\tan\omega_2 = \dfrac{2\sqrt{6}}{5}$.

$$\tan\left(\arcsin\dfrac{3}{5} + \arccos\dfrac{5}{7}\right)$$

$$= \tan(\omega_1+\omega_2) = \dfrac{\tan\omega_1+\tan\omega_2}{1-\tan\omega_1\tan\omega_2}$$

$$= \dfrac{\dfrac{3}{4}+\dfrac{2\sqrt{6}}{5}}{1-\left(\dfrac{3}{4}\right)\left(\dfrac{2\sqrt{6}}{5}\right)} = \dfrac{\dfrac{15+8\sqrt{6}}{20}}{\dfrac{20-6\sqrt{6}}{20}}$$

$$= \dfrac{15+8\sqrt{6}}{20-6\sqrt{6}} = \dfrac{15+8\sqrt{6}}{20-6\sqrt{6}} \cdot \dfrac{20+6\sqrt{6}}{20+6\sqrt{6}}$$

$$= \dfrac{588+250\sqrt{6}}{184} = \dfrac{294+125\sqrt{6}}{92}$$

105. $\tan\left(\operatorname{arcsec}\dfrac{\sqrt{u^2+1}}{u}\right)$

Let $\theta = \operatorname{arcsec}\dfrac{\sqrt{u^2+1}}{u}$, so $\sec\theta = \dfrac{\sqrt{u^2+1}}{u}$.

If $u > 0$, $0 < \theta < \dfrac{\pi}{2}$.

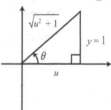

From the Pythagorean theorem,

$$y = \sqrt{\left(\sqrt{u^2+1}\right)^2 - u^2} = \sqrt{u^2+1-u^2} = \sqrt{1} = 1.$$

Therefore $\tan\theta = \dfrac{1}{u}$. Thus,

$$\tan\left(\operatorname{arcsec}\dfrac{\sqrt{u^2+1}}{u}\right) = \dfrac{1}{u}.$$

107. $2\tan x - 1 = 0$

$$2\tan x - 1 = 0 \Rightarrow 2\tan x = 1 \Rightarrow \tan x = \dfrac{1}{2}$$

Over the interval $[0, 2\pi)$, the equation

$\tan x = \dfrac{1}{2}$ has two solutions. One solution is in quadrant I and the other is in quadrant III. Using a calculator, the quadrant I solution is approximately 0.4636. The quadrant III solution would be approximately $0.4636 + \pi \approx 3.6052$.

Solution set: $\{0.4636, 3.6052\}$

109. $\tan x = \cot x$

Use the identity $\cot x = \dfrac{1}{\tan x}, \tan x \ne 0.$

$\tan x = \cot x \Rightarrow \tan x = \dfrac{1}{\tan x} \Rightarrow$

$\tan^2 x = 1 \Rightarrow \tan x = \pm 1$

Over the interval $[0, 2\pi)$, the equation

$\tan x = 1$ has two solutions. One solution is in quadrant I and the other is in quadrant III.

These solutions are $\dfrac{\pi}{4}$ and $\dfrac{5\pi}{4}$. In the same

interval, the equation $\tan x = -1$ has two solutions. One solution is in quadrant II and the other is in quadrant IV. These solutions

are $\dfrac{3\pi}{4}$ and $\dfrac{7\pi}{4}$. Solution set:

$\left\{ \dfrac{\pi}{4}, \dfrac{3\pi}{4}, \dfrac{5\pi}{4}, \dfrac{7\pi}{4} \right\}$

111. $\tan^2 2x - 1 = 0$

$\tan^2 2x - 1 = 0 \Rightarrow \tan^2 2x = 1 \Rightarrow \tan 2x = \pm 1$

$0 \le x < 2\pi \Rightarrow 0 \le 2x < 4\pi.$ Thus,

$2x = \dfrac{\pi}{4}, \dfrac{3\pi}{4}, \dfrac{5\pi}{4}, \dfrac{7\pi}{4}, \dfrac{9\pi}{4}, \dfrac{11\pi}{4}, \dfrac{13\pi}{4}, \dfrac{15\pi}{4}$

implies

$x = \dfrac{\pi}{8}, \dfrac{3\pi}{8}, \dfrac{5\pi}{8}, \dfrac{7\pi}{8}, \dfrac{9\pi}{8}, \dfrac{11\pi}{8}, \dfrac{13\pi}{8}, \dfrac{15\pi}{8}.$

Solution set:

$\left\{ \dfrac{\pi}{8}, \dfrac{3\pi}{8}, \dfrac{5\pi}{8}, \dfrac{7\pi}{8}, \dfrac{9\pi}{8}, \dfrac{11\pi}{8}, \dfrac{13\pi}{8}, \dfrac{15\pi}{8} \right\}$

113. $\cos 2x + \cos x = 0$

$\cos 2x + \cos x = 0 \Rightarrow 2\cos^2 x - 1 + \cos x = 0 \Rightarrow$

$2\cos^2 x + \cos x - 1 = 0 \Rightarrow$

$(2\cos x - 1)(\cos x + 1) = 0$

$2\cos x - 1 = 0 \Rightarrow 2\cos x = 1 \Rightarrow \cos x = \dfrac{1}{2}$ or

$\cos x + 1 = 0 \Rightarrow \cos x = -1$

Over the interval $[0, 2\pi)$, the equation

$\cos x = \dfrac{1}{2}$ has two solutions. The angles in

quadrants I and IV that have a reference angle

of $\dfrac{\pi}{3}$ are $\dfrac{\pi}{3}$ and $\dfrac{5\pi}{3}$. In the same interval,

$\cos x = -1$ when the angle is π.

Solution set:

$\left\{ \dfrac{\pi}{3} + 2n\pi, \ \pi + 2n\pi, \ \dfrac{5\pi}{3} + 2n\pi, \right.$

where n is any integer$\Big\}$

115. $\sin^2 \theta + 3\sin \theta + 2 = 0$

$\sin^2 \theta + 3\sin \theta + 2 = 0$

$(\sin \theta + 2)(\sin \theta + 1) = 0$

In the interval $[0°, 360°)$, we have

$\sin \theta + 1 = 0 \Rightarrow \sin \theta = -1 \Rightarrow \theta = 270°$ and

$\sin \theta + 2 = 0 \Rightarrow \sin \theta = -2 < -1 \Rightarrow$ no solution

Solution set: $\{270°\}$

117. $\sin 2\theta = \cos 2\theta + 1$

$\sin 2\theta = \cos 2\theta + 1$

$(\sin 2\theta)^2 = (\cos 2\theta + 1)^2$

$\sin^2 2\theta = \cos^2 2\theta + 2\cos 2\theta + 1$

$1 - \cos^2 2\theta = \cos^2 2\theta + 2\cos 2\theta + 1$

$2\cos^2 2\theta + 2\cos 2\theta = 0$

$\cos^2 2\theta + \cos 2\theta = 0$

$\cos 2\theta (\cos 2\theta + 1) = 0$

$0° \le \theta < 360° \Rightarrow 0° \le 2\theta < 720°.$

$\cos 2\theta = 0 \Rightarrow 2\theta = 90°, 270°, 450°, 630° \Rightarrow$
$\theta = 45°, 135°, 225°, 315°$

$\cos 2\theta + 1 = 0 \Rightarrow \cos 2\theta = -1$
$2\theta = 180°, 540° \Rightarrow \theta = 90°, 270°$

Possible values for θ are
$\theta = 45°, 90°, 135°, 225°, 270°, 315°.$

All proposed solutions must be checked because the solutions were found by squaring an equation. A value for θ will be a solution if $\sin 2\theta - \cos 2\theta = 1.$

$\theta = 45°, 2\theta = 90° \Rightarrow$
$\sin 90° - \cos 90° = 1 - 0 = 1$
$\theta = 90°, 2\theta = 180° \Rightarrow$
$\sin 180° - \cos 180° = 0 - (-1) = 1$
$\theta = 135°, 2\theta = 270° \Rightarrow$
$\sin 270° - \cos 270° = -1 - 0 \ne 1$
$\theta = 225°, 2\theta = 450° \Rightarrow$
$\sin 450° - \cos 450° = 1 - 0 = 1$
$\theta = 270°, 2\theta = 540° \Rightarrow$
$\sin 540° - \cos 540° = 0 - (-1) = 1$
$\theta = 315°, 2\theta = 630° \Rightarrow$
$\sin 630° - \cos 630° = -1 - 0 \ne 1$
Thus, $\theta = 45°, 90°, 225°, 270°.$
Solution set: $\{45°, 90°, 225°, 270°\}$

119. $3\cos^2\theta + 2\cos\theta - 1 = 0$

$(3\cos\theta - 1)(\cos\theta + 1) = 0$

In the interval $[0°, 360°)$, we have

$3\cos\theta - 1 = 0 \Rightarrow \cos\theta = \dfrac{1}{3} \Rightarrow$

$\theta \approx 70.5°$ and $289.5°$ (using a calculator)

$\cos\theta + 1 = 0 \Rightarrow \cos\theta = -1 \Rightarrow \theta = 180°$

Solution set: $\{70.5°, 180°, 289.5°\}$

121. $2\sqrt{3}\cos\dfrac{\theta}{2} = -3 \Rightarrow \cos\dfrac{\theta}{2} = -\dfrac{3}{2\sqrt{3}} = -\dfrac{\sqrt{3}}{2} \Rightarrow$

$\dfrac{\theta}{2} = \cos^{-1}\left(-\dfrac{\sqrt{3}}{2}\right) \Rightarrow \theta = 2\cos^{-1}\left(-\dfrac{\sqrt{3}}{2}\right) \Rightarrow$

$\theta = 2(150°) + 2(360°)n$ or

$\theta = 2(210°) + 2(360°)n \Rightarrow$

$\theta = 300° + 720°n$ or $\theta = 420° + 720°n$

Solution set: $\{300° + 720°n,\ 420° + 720°n\}$

123. $\tan\theta - \sec\theta = 1 \Rightarrow \tan\theta = 1 + \sec\theta \Rightarrow$

$\tan^2\theta = 1 + 2\sec\theta + \sec^2\theta \Rightarrow$

$\tan^2\theta = 1 + 2\sec\theta + 1 + \tan^2\theta \Rightarrow$

$-2 = 2\sec\theta \Rightarrow \sec\theta = -1 \Rightarrow \cos\theta = -1 \Rightarrow$

$\theta = 180°$

Check the proposed solution because it was obtained by squaring.

$\tan 180° - \sec 180° \overset{?}{=} 1$

$0 - (-1) \overset{?}{=} 1$

$1 = 1$ True

Solution set: $\{180° + 360°n\}$

125. $\dfrac{4}{3}\arctan\dfrac{x}{2} = \pi \Rightarrow \arctan\dfrac{x}{2} = \dfrac{3\pi}{4}$

But, by definition, the range of arctan is $\left(-\dfrac{\pi}{2}, \dfrac{\pi}{2}\right)$. So, this equation has no solution.

Solution set: $\varnothing$

127. $\arccos x + \arctan 1 = \dfrac{11\pi}{12}$

$\arccos x + \arctan 1 = \dfrac{11\pi}{12}$

$\arccos x = \dfrac{11\pi}{12} - \arctan 1$

$\arccos x = \dfrac{11\pi}{12} - \dfrac{\pi}{4} = \dfrac{8\pi}{12} = \dfrac{2\pi}{3}$

$\cos\dfrac{2\pi}{3} = x \Rightarrow x = -\dfrac{1}{2}$

Solution set: $\left\{-\dfrac{1}{2}\right\}$

129. $y = \dfrac{1}{2}\sin x \Rightarrow 2y = \sin x \Rightarrow x = \sin^{-1}(2y)$,

$-\dfrac{\pi}{2} \le x \le \dfrac{\pi}{2}$

131. $y = \dfrac{1}{2}\tan(3x + 2) \Rightarrow 2y = \tan(3x + 2) \Rightarrow$

$3x + 2 = \arctan 2y \Rightarrow 3x = \arctan 2y - 2 \Rightarrow$

$x = \left(\dfrac{1}{3}\arctan 2y\right) - \dfrac{2}{3},\ -\dfrac{2}{3} - \dfrac{\pi}{6} < x < -\dfrac{2}{3} + \dfrac{\pi}{6}$

133. (a) Let α be the angle to the left of θ.

Thus, we have

$\tan(\alpha + \theta) = \dfrac{5 + 10}{x}$

$\alpha + \theta = \tan^{-1}\left(\dfrac{15}{x}\right)$

$\theta = \tan^{-1}\left(\dfrac{15}{x}\right) - \alpha$

$\theta = \tan^{-1}\left(\dfrac{15}{x}\right) - \tan^{-1}\left(\dfrac{5}{x}\right)$

(b) The maximum occurs at approximately 8.66026 ft. There may be a discrepancy in the final digits.

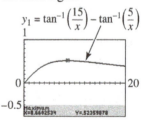

135. If $\theta_1 > 48.8°$, then $\theta_2 > 90°$ and the light beam is completely underwater. No light will emerge from the water.

Chapter 7 Test

1. $\cos\theta = \dfrac{24}{25}$, θ is in quadrant IV.

$$\sin^2\theta = 1 - \cos^2\theta = 1 - \left(\dfrac{24}{25}\right)^2 = 1 - \dfrac{576}{625} = \dfrac{49}{625}$$

θ is in quadrant IV, so $\sin\theta < 0$.

$$\sin\theta = -\sqrt{\dfrac{49}{625}} = -\dfrac{7}{25}$$

$$\tan\theta = \dfrac{\sin\theta}{\cos\theta} = \dfrac{-\frac{7}{25}}{\frac{24}{25}} = -\dfrac{7}{24}$$

$$\cot\theta = \dfrac{1}{\tan\theta} = \dfrac{1}{-\frac{7}{24}} = -\dfrac{24}{7}$$

$$\sec\theta = \dfrac{1}{\cos\theta} = \dfrac{1}{\frac{24}{25}} = \dfrac{25}{24}$$

$$\csc\theta = \dfrac{1}{\sin\theta} = \dfrac{1}{-\frac{7}{25}} = -\dfrac{25}{7}$$

2. $\sec\theta - \sin\theta\tan\theta = \dfrac{1}{\cos\theta} - \sin\theta \cdot \dfrac{\sin\theta}{\cos\theta}$

$$= \dfrac{1-\sin^2\theta}{\cos\theta} = \dfrac{\cos^2\theta}{\cos\theta} = \cos\theta$$

3. $\tan^2 x - \sec^2 x = \dfrac{\sin^2 x}{\cos^2 x} - \dfrac{1}{\cos^2 x}$

$$= \dfrac{\sin^2 x - 1}{\cos^2 x} = -\dfrac{1-\sin^2 x}{\cos^2 x}$$

$$= -\dfrac{\cos^2 x}{\cos^2 x} = -1$$

4. $\cos\dfrac{5\pi}{12} = \cos\left(\dfrac{\pi}{6} + \dfrac{\pi}{4}\right)$

$$= \cos\dfrac{\pi}{6}\cos\dfrac{\pi}{4} - \sin\dfrac{\pi}{6}\sin\dfrac{\pi}{4}$$

$$= \dfrac{\sqrt{3}}{2}\left(\dfrac{\sqrt{2}}{2}\right) - \dfrac{1}{2}\left(\dfrac{\sqrt{2}}{2}\right) = \dfrac{\sqrt{6}-\sqrt{2}}{4}$$

5. (a) $\cos(270° - x)$

$\quad = \cos 270°\cos x + \sin 270°\sin x$

$\quad = 0 \cdot \cos x + (-1)\sin x = 0 - \sin x = -\sin x$

(b) $\tan(\pi + x) = \dfrac{\tan\pi + \tan x}{1 - \tan\pi\tan x} = \tan x$

6. $\sin(-22.5°) = \pm\sqrt{\dfrac{1-\cos(-45°)}{2}} = \pm\sqrt{\dfrac{1-\frac{\sqrt{2}}{2}}{2}}$

$$= \pm\sqrt{\dfrac{2-\sqrt{2}}{4}} = \pm\dfrac{\sqrt{2-\sqrt{2}}}{2}$$

Because $-22.5°$ is in quadrant IV, $\sin(-22.5°)$

is negative. Thus, $\sin(-22.5°) = -\dfrac{\sqrt{2-\sqrt{2}}}{2}$.

7. Find $\sin(A+B)$, $\cos(A+B)$, and $\tan(A-B)$, given $\sin A = \dfrac{5}{13}$, $\cos B = -\dfrac{3}{5}$, A is in quadrant I and B is in quadrant II. Thus, $\cos A > 0$ and $\sin B > 0$.

$$\cos A = \sqrt{1-\sin^2 A} = \sqrt{1-\left(\dfrac{5}{13}\right)^2}$$

$$= \sqrt{1-\dfrac{25}{169}} = \sqrt{\dfrac{144}{169}} = \dfrac{12}{13}$$

$$\sin B = \sqrt{1-\cos^2 B} = \sqrt{1-\left(-\dfrac{3}{5}\right)^2}$$

$$= \sqrt{1-\dfrac{9}{25}} = \sqrt{\dfrac{16}{25}} = \dfrac{4}{5}$$

(a) $\sin(A+B) = \sin A\cos B + \cos A\sin B$

$$= \left(\dfrac{5}{13}\right)\left(-\dfrac{3}{5}\right) + \left(\dfrac{12}{13}\right)\left(\dfrac{4}{5}\right)$$

$$= -\dfrac{15}{65} + \dfrac{48}{65} = \dfrac{33}{65}$$

(b) $\cos(A+B) = \cos A\cos B - \sin A\sin B$

$$= \left(\dfrac{12}{13}\right)\left(-\dfrac{3}{5}\right) - \left(\dfrac{5}{13}\right)\left(\dfrac{4}{5}\right)$$

$$= -\dfrac{36}{65} - \dfrac{20}{65} = -\dfrac{56}{65}$$

(c) To use the formula

$\tan(A+B) = \dfrac{\tan A + \tan B}{1 - \tan A\tan B}$, first find $\tan A$ and $\tan B$:

$$\tan A = \dfrac{\sin A}{\cos A} = \dfrac{\frac{5}{13}}{\frac{12}{13}} = \dfrac{5}{12}$$

$$\tan B = \dfrac{\sin B}{\cos B} = \dfrac{\frac{4}{5}}{-\frac{3}{5}} = -\dfrac{4}{3}$$

$$\tan(A-B) = \dfrac{\frac{5}{12} - \left(-\frac{4}{3}\right)}{1 + \left(\frac{5}{12}\right)\left(-\frac{4}{3}\right)}$$

$$= \dfrac{15+48}{36-20} = \dfrac{63}{16}$$

(d) To find the quadrant of $A + B$, notice that $\sin(A+B) > 0$, which implies $x + y$ is in quadrant I or II. Also $\cos(A+B) < 0$, which implies that $A + B$ is in quadrant II or III. Therefore, $A + B$ is in quadrant II.

8. Given $\cos\theta = -\dfrac{3}{5}$, $90° < \theta < 180°$

θ is in quadrant II, so $\sin\theta > 0$, 2θ is in quadrant III or quadrant IV, and

$\dfrac{\pi}{4} < \dfrac{\theta}{2} < \dfrac{\pi}{2} \Rightarrow \dfrac{\theta}{2}$ is in quadrant I. Also

$\sin\theta = \sqrt{1-\cos^2\theta} = \sqrt{1-\left(-\dfrac{3}{5}\right)^2} = \sqrt{\dfrac{16}{25}} = \dfrac{4}{5}$

$\tan\theta = \dfrac{\sin\theta}{\cos\theta} = \dfrac{\frac{4}{5}}{-\frac{3}{5}} = -\dfrac{4}{3}$

(a) $\cos 2\theta = 2\cos^2\theta - 1 = 2\left(-\dfrac{3}{5}\right)^2 - 1 = -\dfrac{7}{25}$

Note that 2θ is in quadrant III because $\cos 2\theta < 0$.

(b) $\sin 2\theta = 2\sin\theta\cos\theta = 2\left(\dfrac{4}{5}\right)\left(-\dfrac{3}{5}\right) = -\dfrac{24}{25}$

(c) $\tan 2\theta = \dfrac{2\tan\theta}{1-\tan^2\theta} = \dfrac{2\left(-\frac{4}{3}\right)}{1-\left(\frac{4}{3}\right)^2} = \dfrac{-\frac{8}{3}}{1-\frac{16}{9}}$

$= \dfrac{-24}{9-16} = \dfrac{24}{7}$

(d) $\cos\frac{1}{2}\theta = \sqrt{\dfrac{1+\cos\theta}{2}} = \sqrt{\dfrac{1+\left(-\frac{3}{5}\right)}{2}}$

$= \sqrt{\dfrac{5-3}{10}} = \sqrt{\dfrac{2}{10}} = \dfrac{1}{\sqrt{5}} = \dfrac{\sqrt{5}}{5}$

(e) $\tan\frac{1}{2}\theta = \dfrac{\sin\theta}{1+\cos\theta} = \dfrac{\frac{4}{5}}{1-\frac{3}{5}} = \dfrac{4}{5-3} = 2$

9. Verify $\sec^2 B = \dfrac{1}{1-\sin^2 B}$ is an identity.

Work with the right side.

$\dfrac{1}{1-\sin^2 B} = \dfrac{1}{\cos^2 B} = \sec^2 B$

10. Verify $\cos 2A = \dfrac{\cot A - \tan A}{\csc A \sec A}$ is an identity.

Work with the right side.

$\dfrac{\cot A - \tan A}{\csc A \sec A} = \dfrac{\dfrac{\cos A}{\sin A} - \dfrac{\sin A}{\cos A}}{\left(\dfrac{1}{\sin A}\right)\left(\dfrac{1}{\cos A}\right)} \cdot \dfrac{\sin A \cos A}{\sin A \cos A}$

$= \cos^2 A - \sin^2 A = \cos 2A$

11. Verify $\tan^2 x - \sin^2 x = (\tan x \sin x)^2$ is an identity.

$\tan^2 x - \sin^2 x = \dfrac{\sin^2 x}{\cos^2 x} - \sin^2 x$

$= \dfrac{\sin^2 x - \sin^2 x \cos^2 x}{\cos^2 x}$

$= \dfrac{\sin^2 x\left(1-\cos^2 x\right)}{\cos^2 x} = \dfrac{\sin^2 x \sin^2 x}{\cos^2 x}$

$= \tan^2 x \sin^2 x = (\tan x \sin x)^2$

12. Verify $\dfrac{\tan x - \cot x}{\tan x + \cot x} = 2\sin^2 x - 1$ is an identity.

$\dfrac{\tan x - \cot x}{\tan x + \cot x} = \dfrac{\dfrac{\sin x}{\cos x} - \dfrac{\cos x}{\sin x}}{\dfrac{\sin x}{\cos x} + \dfrac{\cos x}{\sin x}}$

$= \dfrac{\dfrac{\sin x}{\cos x} - \dfrac{\cos x}{\sin x}}{\dfrac{\sin x}{\cos x} + \dfrac{\cos x}{\sin x}} \cdot \dfrac{\cos x \sin x}{\cos x \sin x}$

$= \dfrac{\sin^2 x - \cos^2 x}{\sin^2 + \cos^2 x}$

$= \sin^2 x - \cos^2 x$

$= \sin^2 x - \left(1-\sin^2 x\right)$

$= 2\sin^2 x - 1$

13. (a) $V = 163\sin\omega t$

$\sin x = \cos\left(\dfrac{\pi}{2} - x\right) \Rightarrow$

$V = 163\cos\left(\dfrac{\pi}{2} - \omega t\right).$

(b) $V = 163\sin\omega t = 163\sin 120\pi t$

$= 163\cos\left(\dfrac{\pi}{2} - 120\pi t\right) \Rightarrow$ the

maximum voltage occurs when

$\cos\left(\dfrac{\pi}{2} - 120\pi t\right) = 1$. Thus, the

maximum voltage is $V = 163$ volts.

$\cos\left(\dfrac{\pi}{2} - 120\pi t\right) = 1$ when

$\dfrac{\pi}{2} - 120\pi t = 2k\pi$, where k is any integer.

(continued on next page)

(continued)

The first maximum occurs when

$$\frac{\pi}{2} - 120\pi t = 0 \Rightarrow$$

$$\frac{\pi}{2} = 120\pi t \Rightarrow \frac{1}{120\pi} \cdot \frac{\pi}{2} = t \Rightarrow t = \frac{1}{240}$$

The maximum voltage will first occur at

$$\frac{1}{240} \sec .$$

14.

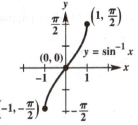

Domain: $[-1, 1]$; range: $\left[-\dfrac{\pi}{2}, \dfrac{\pi}{2}\right]$

15. **(a)** $y = \arccos\left(-\dfrac{1}{2}\right) \Rightarrow \cos y = -\dfrac{1}{2}$

$0 \le y \le \pi$, so $y = \dfrac{2\pi}{3}$.

(b) $y = \sin^{-1}\left(-\dfrac{\sqrt{3}}{2}\right) \Rightarrow \sin y = -\dfrac{\sqrt{3}}{2}$

$-\dfrac{\pi}{2} \le y \le \dfrac{\pi}{2}$, so $y = -\dfrac{\pi}{3}$.

(c) $y = \tan^{-1} 0 \Rightarrow \tan y = 0$

$-\dfrac{\pi}{2} < y < \dfrac{\pi}{2}$, so $y = 0$.

(d) $y = \operatorname{arcsec}(-2) \Rightarrow \sec y = -2$

$0 \le y \le \pi$ and $y \ne \dfrac{\pi}{2}$, so $y = \dfrac{2\pi}{3}$.

16. **(a)** $\theta = \arccos \dfrac{\sqrt{3}}{2} \Rightarrow \cos \theta = \dfrac{\sqrt{3}}{2}$

$0 \le y \le 180°$, so $y = 30°$.

(b) $\theta = \tan^{-1}(-1) \Rightarrow \tan \theta = -1$

$-90° < y < 90°$, so $y = -45°$.

(c) $\theta = \cot^{-1}(-1) \Rightarrow \cot \theta = -1$

$0° < y < 180°$, so $y = 135°$.

(d) $\theta = \csc^{-1}\left(-\dfrac{2\sqrt{3}}{3}\right) \Rightarrow \csc \theta = -\dfrac{2\sqrt{3}}{3}$

$-90 \le y \le 90°$, so $\theta = -60°$.

17. **(a)** $\sin^{-1} 0.69431882 \approx 43.97°$

(b) $\sec^{-1} 1.0840880 \approx 22.72°$

(c) $\cot^{-1}(-0.7125586) \approx 125.47°$

18. **(a)** $\cos\left(\arcsin \dfrac{2}{3}\right)$

Let $\arcsin \dfrac{2}{3} = u$, so that $\sin u = \dfrac{2}{3}$.
Because arcsine is defined only in
quadrants I and IV, and $\dfrac{2}{3}$ is positive, u
is in quadrant I. Sketch u and label a
triangle with the side adjacent u equal to
$\sqrt{3^2 - 2^2} = \sqrt{9 - 4} = \sqrt{5}$.

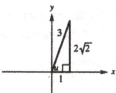

$$\cos\left(\arcsin \frac{2}{3}\right) = \cos u = \frac{\sqrt{5}}{3}$$

(b) $\sin\left(2\cos^{-1} \dfrac{1}{3}\right)$

Let $\theta = \cos^{-1} \dfrac{1}{3}$, so that $\cos \theta = \dfrac{1}{3}$.
Because arccosine is defined only in
quadrants I and II, and $\dfrac{1}{3}$ is positive, θ
is in quadrant I. Sketch θ and label a
triangle with the side opposite to θ equal
to $\theta = \sqrt{3^2 - (-1)^2} = \sqrt{9 - 1} = \sqrt{8} = 2\sqrt{2}$.

Thus, $\sin \theta = \dfrac{2\sqrt{2}}{3}$ and

$$\sin\left(2\cos^{-1} \frac{1}{3}\right) = \sin 2\theta .$$

$$\sin\left(2\cos^{-1} \frac{1}{3}\right) = \sin 2\theta = 2\sin \theta \cos \theta$$

$$= 2\left(\frac{2\sqrt{2}}{3}\right)\left(\frac{1}{3}\right) = \frac{4\sqrt{2}}{9}$$

19. $\tan\left(\arcsin u\right)$

Let $\theta = \arcsin u$, so $\sin\theta = u = \dfrac{u}{1}$. If $u > 0$,

$0 < \theta < \dfrac{\pi}{2}$.

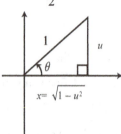

$x = \sqrt{1 - u^2}$

From the Pythagorean theorem,

$x = \sqrt{1^2 - u^2} = \sqrt{1 - u^2}$. Therefore,

$\tan\theta = \dfrac{u}{\sqrt{1 - u^2}} = \dfrac{u\sqrt{1 - u^2}}{1 - u^2}$. Thus,

$\tan\left(\arcsin u\right) = \dfrac{u\sqrt{1 - u^2}}{1 - u^2}$.

20. $-3\sec\theta + 2\sqrt{3} = 0 \Rightarrow \sec\theta = \dfrac{2\sqrt{3}}{3}$

Over the interval $[0, 360°)$, the equation

$\sec x = \dfrac{2\sqrt{3}}{3}$ has two solutions. One solution

is in quadrant I and the other is in quadrant IV. These solutions are 30° and 330°.
Solution set: {30°, 330°}

21. $\sin^2\theta = \cos^2\theta + 1$

$\sin^2\theta = 1 - \sin^2\theta + 1$

$2\sin^2\theta = 2 \Rightarrow \sin^2\theta = 1 \Rightarrow \sin\theta = \pm 1$

Over the interval $[0, 360°)$, the equation
$\sin\theta = 1$ has one solution, 90°. Over the
interval $[0, 2\pi)$, the equation $\sin\theta = -1$ has
one solution, 270°. Solution set: {90°, 270°}

22. $\csc^2\theta - 2\cot\theta = 4$

$1 + \cot^2\theta - 2\cot\theta = 4$

$\cot^2\theta - 2\cot\theta - 3 = 0$

$\left(\cot\theta - 3\right)\left(\cot\theta + 1\right) = 0 \Rightarrow$

$\cot\theta = 3$ or $\cot\theta = -1$

Over the interval $[0, 360°)$, the equation
$\cot\theta = 3$ has two solutions, 18.4° and 198.4°
(found using a calculator.) Over the interval
$[0, 2\pi)$, the equation $\cot\theta = -1$ has two

solutions, 135° and 315°.
Solution set: {18.4°, 135°, 198.4°, 315°}

23. $\cos x = \cos 2x$

$\cos x = 2\cos^2 x - 1$

$2\cos^2 x - \cos x - 1 = 0$

$\left(2\cos x + 1\right)\left(\cos x - 1\right) = 0 \Rightarrow$

$\cos x = -\dfrac{1}{2}$ or $\cos x = 1$

Over the interval $[0, 2\pi)$, the equation

$\cos x = -\dfrac{1}{2}$ has two solutions, $\dfrac{2\pi}{3}$ and $\dfrac{4\pi}{3}$.

Over the interval $[0, 2\pi)$, the equation
$\cos x = 1$ has one solution, 0.

Solution set: $\left\{0, \dfrac{2\pi}{3}, \dfrac{4\pi}{3}\right\}$

24. $\sqrt{2}\cos 3x - 1 = 0 \Rightarrow \cos 3x = \dfrac{1}{\sqrt{2}} = \dfrac{\sqrt{2}}{2}$

$0 \le x < 2\pi \Rightarrow 0 \le 3x < 6\pi$. Thus

$3x = \dfrac{\pi}{4}, \dfrac{7\pi}{4}, \dfrac{9\pi}{4}, \dfrac{15\pi}{4}, \dfrac{17\pi}{4}, \dfrac{23\pi}{4} \Rightarrow$

$x = \dfrac{\pi}{12}, \dfrac{7\pi}{12}, \dfrac{3\pi}{4}, \dfrac{5\pi}{4}, \dfrac{17\pi}{12}, \dfrac{23\pi}{12}$

Solution set: $\left\{\dfrac{\pi}{12}, \dfrac{7\pi}{12}, \dfrac{3\pi}{4}, \dfrac{5\pi}{4}, \dfrac{17\pi}{12}, \dfrac{23\pi}{12}\right\}$

25. $\sin x \cos x = \dfrac{1}{3} \Rightarrow 2\sin x \cos x = \dfrac{2}{3} \Rightarrow$

$\sin 2x = \dfrac{2}{3}$

$0 \le x < 2\pi \Rightarrow 0 \le 2x < 4\pi$. Use a calculator to
find $2x$:
$2x \approx 0.72672, 2.4118, 7.0129, 8.69505 \Rightarrow$
$x \approx 0.3649, 1.2059, 3.5065, 4.3475$
Solution set: {0.3649, 1.2059, 3.5065, 4.3475}

26. $\sin^2\theta = -\cos 2\theta$

$\sin^2\theta = -\left(1 - 2\sin^2\theta\right)$

$\sin^2\theta = 1 \Rightarrow \sin\theta = \pm 1$

Over the interval $[0, 360°)$, the equation
$\sin\theta = 1$ has one solution, 90°. Over the
interval $[0, 360°)$, the equation $\sin\theta = -1$ has
one solution, 270°. Because $270° = 90° + 180°$,
the solution set is {90° + 180°n, where n is any
integer}.

27. $2\sqrt{3}\sin\dfrac{x}{2}=3 \Rightarrow \sin\dfrac{x}{2}=\dfrac{3}{2\sqrt{3}} \Rightarrow$

$\dfrac{x}{2}=\sin^{-1}\dfrac{3}{2\sqrt{3}} \Rightarrow \dfrac{x}{2}=\sin^{-1}\dfrac{\sqrt{3}}{2}$

Since $0 \le x < 2\pi \Rightarrow 0° \le \dfrac{x}{2} < \pi$, we have

$\dfrac{x}{2}=\dfrac{\pi}{3},\dfrac{2\pi}{3} \Rightarrow x=\dfrac{2\pi}{3},\dfrac{4\pi}{3}$

Solution set:

$\left\{\dfrac{2\pi}{3}+2\pi n,\ \dfrac{4\pi}{3}+2\pi n,\right.$

where n is any integer$\Big\}$

28. (a) $y=\cos 3x \Rightarrow 3x=\arccos y \Rightarrow$

$x=\dfrac{1}{3}\arccos y,\ 0 \le x \le \dfrac{\pi}{3}$

(b) $y=4+3\cot x \Rightarrow y-4=3\cot x \Rightarrow$

$\dfrac{y-4}{3}=\cot x \Rightarrow x=\cot^{-1}\left(\dfrac{y-4}{3}\right),$

$0 < x < \pi$

29. (a) $\arcsin x = \arctan\dfrac{4}{3}$

Let $\omega = \arctan\dfrac{4}{3}$. Then $\tan\omega=\dfrac{4}{3}$.

Sketch ω in quadrant I and label a
triangle with the hypotenuse equal to

$\sqrt{4^2+3^2}=\sqrt{16+9}=\sqrt{25}=5$.

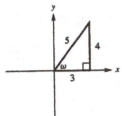

Thus, we have

$\arcsin x = \arctan\dfrac{4}{3} \Rightarrow \arcsin x = \omega \Rightarrow$

$x=\sin\omega=\dfrac{4}{5}$

Solution set: $\left\{\dfrac{4}{5}\right\}$

(b) $\operatorname{arccot} x + 2\arcsin\dfrac{\sqrt{3}}{2}=\pi \Rightarrow$

$\operatorname{arccot} x + 2\left(\dfrac{\pi}{3}\right)=\pi \Rightarrow \operatorname{arccot} x=\dfrac{\pi}{3} \Rightarrow$

$x=\cot\dfrac{\pi}{3}=\dfrac{1}{\tan\dfrac{\pi}{3}}=\dfrac{1}{\sqrt{3}}=\dfrac{\sqrt{3}}{3}$

Solution set: $\left\{\dfrac{\sqrt{3}}{3}\right\}$

30.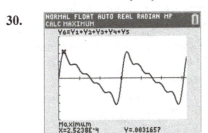

P first reaches its maximum at
approximately 2.5×10^{-4}. The maximum is
approximately 0.003166.

Chapter 8

Applications of Trigonometry

Section 8.1   The Law of Sines

1. Yes, the law of sines may be used.

3. No, there is insufficient information to use the law of sines.

5. A. These values lead to two possible measures for angle B.

$$\frac{\sin A}{a} = \frac{\sin B}{b} \Rightarrow \frac{\sin 50°}{19} = \frac{\sin B}{21} \Rightarrow$$

$$\sin B = \frac{0.7660 \cdot 21}{19} \approx 0.8467 \Rightarrow$$

$B \approx 57.9°$ or $B \approx 122.1°$

Choices B, C, and D all lead to unique triangles.

7. The vertical distance from the point $(3, 4)$ to the x-axis is 4.

   (a) If L is more than 4, two triangles can be drawn. But h must be less than 5 for both triangles to be on the positive x-axis. So, $4 < L < 5$.

   (b) If $L = 4$, then exactly one triangle is possible. If $L > 5$, then only one triangle is possible on the positive x-axis.

   (c) If $L < 4$, then no triangle is possible, because the side of length L would not reach the x-axis.

9. $a = 50$, $b = 26$, $A = 95°$

$$\frac{\sin A}{a} = \frac{\sin B}{b} \Rightarrow \frac{\sin 95°}{50} = \frac{\sin B}{26} \Rightarrow$$

$$\sin B = \frac{26 \sin 95°}{50} \approx 0.51802124 \Rightarrow B \approx 31.2°$$

Another possible value for B is $180° - 21.2° = 148.8°$, but this is too large to be in a triangle that also has a 95° angle. Therefore, only one triangle is possible.

11. $c = 50$, $b = 61$, $C = 58°$

$$\frac{\sin C}{c} = \frac{\sin B}{b} \Rightarrow \frac{\sin 58°}{50} = \frac{\sin B}{61} \Rightarrow$$

$$\sin B = \frac{61 \sin 58°}{50} \approx 1.03461868$$

Because $\sin B > 1$ is impossible, no triangle is possible for the given parts.

13. The measure of angle C is
$180° - (60° + 75°) = 180° - 135° = 45°$.

$$\frac{a}{\sin A} = \frac{c}{\sin C} \Rightarrow \frac{a}{\sin 60°} = \frac{\sqrt{2}}{\sin 45°} \Rightarrow$$

$$a = \frac{\sqrt{2} \sin 60°}{\sin 45°} = \frac{\sqrt{2} \cdot \frac{\sqrt{3}}{2}}{\frac{\sqrt{2}}{2}} = \sqrt{2} \cdot \frac{\sqrt{3}}{2} \cdot \frac{2}{\sqrt{2}}$$

$$= \sqrt{3}$$

15. $A = 37°$, $B = 48°$, $c = 18$ m

$C = 180° - A - B \Rightarrow$
$C = 180° - 37° - 48° = 95°$

$$\frac{b}{\sin B} = \frac{c}{\sin C} \Rightarrow \frac{b}{\sin 48°} = \frac{18}{\sin 95°} \Rightarrow$$

$$b = \frac{18 \sin 48°}{\sin 95°} \approx 13 \text{ m}$$

$$\frac{a}{\sin A} = \frac{c}{\sin C} \Rightarrow \frac{a}{\sin 37°} = \frac{18}{\sin 95°} \Rightarrow$$

$$a = \frac{18 \sin 37°}{\sin 95°} \approx 11 \text{ m}$$

17. $A = 27.2°$, $C = 115.5°$, $c = 76.0$ ft

$B = 180° - A - C \Rightarrow$
$B = 180° - 27.2° - 115.5° = 37.3°$

$$\frac{a}{\sin A} = \frac{c}{\sin C} \Rightarrow \frac{a}{\sin 27.2°} = \frac{76.0}{\sin 115.5°} \Rightarrow$$

$$a = \frac{76.0 \sin 27.2°}{\sin 115.5°} \approx 38.5 \text{ ft}$$

$$\frac{b}{\sin B} = \frac{c}{\sin C} \Rightarrow \frac{b}{\sin 37.3°} = \frac{76.0}{\sin 115.5°} \Rightarrow$$

$$b = \frac{76.0 \sin 37.3°}{\sin 115.5°} \approx 51.0 \text{ ft}$$

19. $A = 68.41°$, $B = 54.23°$, $a = 12.75$ ft

$C = 180° - A - B \Rightarrow$
$C = 180° - 68.41° - 54.23° = 57.36°$

$$\frac{a}{\sin A} = \frac{b}{\sin B} \Rightarrow \frac{12.75}{\sin 68.41°} = \frac{b}{\sin 54.23°} \Rightarrow$$

$$b = \frac{12.75 \sin 54.23°}{\sin 68.41°} \approx 11.13 \text{ ft}$$

$$\frac{a}{\sin A} = \frac{c}{\sin C} \Rightarrow \frac{12.75}{\sin 68.41°} = \frac{c}{\sin 57.36°} \Rightarrow$$

$$c = \frac{12.75 \sin 57.36°}{\sin 68.41°} \approx 11.55 \text{ ft}$$

21. $A = 87.2°, b = 75.9 \text{ yd}, C = 74.3°$

$B = 180° - A - C \Rightarrow$
$B = 180° - 87.2° - 74.3° = 18.5°$

$\dfrac{a}{\sin A} = \dfrac{b}{\sin B} \Rightarrow \dfrac{a}{\sin 87.2°} = \dfrac{75.9}{\sin 18.5°} \Rightarrow$

$a = \dfrac{75.9 \sin 87.2°}{\sin 18.5°} \approx 239 \text{ yd}$

$\dfrac{b}{\sin B} = \dfrac{c}{\sin C} \Rightarrow \dfrac{75.9}{\sin 18.5°} = \dfrac{c}{\sin 74.3°} \Rightarrow$

$c = \dfrac{75.9 \sin 74.3°}{\sin 18.5°} \approx 230 \text{ yd}$

23. $B = 20°50', AC = 132 \text{ ft}, C = 103°10'$

$A = 180° - B - C$
$A = 180° - 20°50' - 103°10' \Rightarrow A = 56°00'$

$\dfrac{AC}{\sin B} = \dfrac{AB}{\sin C} \Rightarrow \dfrac{132}{\sin 20°50'} = \dfrac{AB}{\sin 103°10'} \Rightarrow$

$AB = \dfrac{132 \sin 103°10'}{\sin 20°50'} \approx 361 \text{ ft}$

$\dfrac{BC}{\sin A} = \dfrac{AC}{\sin B} \Rightarrow \dfrac{BC}{\sin 56°00'} = \dfrac{132}{\sin 20°50'} \Rightarrow$

$BC = \dfrac{132 \sin 56°00'}{\sin 20°50'} \approx 308 \text{ ft}$

25. $A = 39.70°, C = 30.35°, b = 39.74 \text{ m}$

$B = 180° - A - C \Rightarrow$
$B = 180° - 39.70° - 30.35° \Rightarrow$
$B = 109.95° \approx 110.0° \text{ (rounded)}$

$\dfrac{a}{\sin A} = \dfrac{b}{\sin B} \Rightarrow \dfrac{a}{\sin 39.70°} = \dfrac{39.74}{\sin 109.95°} \Rightarrow$

$a = \dfrac{39.74 \sin 39.70°}{\sin 109.95°} \approx 27.01 \text{ m}$

$\dfrac{b}{\sin B} = \dfrac{c}{\sin C} \Rightarrow \dfrac{39.74}{\sin 109.95°} = \dfrac{c}{\sin 30.35°} \Rightarrow$

$c = \dfrac{39.74 \sin 30.35°}{\sin 109.95°} \approx 21.36 \text{ m}$

27. $B = 42.88°, C = 102.40°, b = 3974 \text{ ft}$

$A = 180° - B - C \Rightarrow$
$A = 180° - 42.88° - 102.40° = 34.72°$

$\dfrac{a}{\sin A} = \dfrac{b}{\sin B} \Rightarrow \dfrac{a}{\sin 34.72°} = \dfrac{3974}{\sin 42.88°} \Rightarrow$

$a = \dfrac{3974 \sin 34.72°}{\sin 42.88°} \approx 3326 \text{ ft}$

$\dfrac{b}{\sin B} = \dfrac{c}{\sin C} \Rightarrow \dfrac{3974}{\sin 42.88°} = \dfrac{c}{\sin 102.40°} \Rightarrow$

$c = \dfrac{3974 \sin 102.40°}{\sin 42.88°} \approx 5704 \text{ ft}$

29. $A = 39°54', a = 268.7 \text{ m}, B = 42°32'$

$C = 180° - A - B \Rightarrow$
$C = 180° - 39°54' - 42°32' = 97°34'$

$\dfrac{a}{\sin A} = \dfrac{b}{\sin B} \Rightarrow \dfrac{268.7}{\sin 39°54'} = \dfrac{b}{\sin 42°32'} \Rightarrow$

$b = \dfrac{268.7 \sin 42°32'}{\sin 39°54'} \approx 283.2 \text{ m}$

$\dfrac{a}{\sin A} = \dfrac{c}{\sin C} \Rightarrow \dfrac{268.7}{\sin 39°54'} = \dfrac{c}{\sin 97°54'} \Rightarrow$

$c = \dfrac{268.7 \sin 97°54'}{\sin 39°54'} \approx 415.2 \text{ m}$

31. $a = \sqrt{6}, b = 2, A = 60°$

$\dfrac{\sin B}{b} = \dfrac{\sin A}{a} \Rightarrow \dfrac{\sin B}{2} = \dfrac{\sin 60°}{\sqrt{6}} \Rightarrow$

$\sin B = \dfrac{2 \sin 60°}{\sqrt{6}} = \dfrac{2 \cdot \dfrac{\sqrt{3}}{2}}{\sqrt{6}} = \dfrac{\sqrt{3}}{\sqrt{6}} = \sqrt{\dfrac{3}{6}}$

$= \sqrt{\dfrac{1}{2}} = \dfrac{1}{\sqrt{2}} \cdot \dfrac{\sqrt{2}}{2} = \dfrac{\sqrt{2}}{2}$

$\sin B = \dfrac{\sqrt{2}}{2} \Rightarrow B = 45°$

There is another angle between $0°$ and $180°$ whose sine is $\dfrac{\sqrt{2}}{2}$: $180° - 45° = 135°$.
However, this is too large because $A = 60°$ and $60° + 135° = 195° > 180°$, so there is only one solution, $B = 45°$.

33. $A = 29.7°, b = 41.5 \text{ ft}, a = 27.2 \text{ ft}$

$\dfrac{\sin B}{b} = \dfrac{\sin A}{a} \Rightarrow \dfrac{\sin B}{41.5} = \dfrac{\sin 29.7°}{27.2} \Rightarrow$

$\sin B = \dfrac{41.5 \sin 29.7°}{27.2} \approx 0.75593878$

There are two angles B between $0°$ and $180°$ that satisfy the condition. Because $\sin B \approx 0.75593878$, to the nearest tenth value of B is $B_1 = 49.1°$. Supplementary angles have the same sine value, so another possible value of B is $B_2 = 180° - 49.1° = 130.9°$. This is a valid angle measure for this triangle because
$A + B_2 = 29.7° + 130.9° = 160.6° < 180°$.
Solving separately for angles C_1 and C_2 we have the following.

$C_1 = 180° - A - B_1$
$\quad = 180° - 29.7° - 49.1° = 101.2°$
$C_2 = 180° - A - B_2$
$\quad = 180° - 29.7° - 130.9° = 19.4°$

35. $C = 41°20'$, $b = 25.9$ m, $c = 38.4$ m

$$\frac{\sin B}{b} = \frac{\sin C}{c} \Rightarrow \frac{\sin B}{25.9} = \frac{\sin 41°20'}{38.4} \Rightarrow$$

$$\sin B = \frac{25.9 \sin 41°20'}{38.4} \approx 0.44545209$$

There are two angles B between $0°$ and $180°$ that satisfy the condition. Because $\sin B \approx 0.44545209$, $B_1 \approx 26.5° = 26°30'$. Supplementary angles have the same sine value, so another possible value of B is $B_2 = 180° - 26°30' = 153°30'$. This is not a valid angle measure for this triangle because $C + B_2 = 41°20' + 153°30' = 194°30' > 180°$.

Thus, $A = 180° - 26°30' - 41°20' = 112°10'$.

37. $B = 74.3°$, $a = 859$ m, $b = 783$ m

$$\frac{\sin A}{a} = \frac{\sin B}{b} \Rightarrow \frac{\sin A}{859} = \frac{\sin 74.3°}{783} \Rightarrow$$

$$\sin A = \frac{859 \sin 74.3°}{783} \approx 1.0561331$$

Because $\sin A > 1$ is impossible, no such triangle exists.

39. $A = 142.13°$, $b = 5.432$ ft, $a = 7.297$ ft

$$\frac{\sin B}{b} = \frac{\sin A}{a} \Rightarrow \sin B = \frac{b \sin A}{a} \Rightarrow$$

$$\sin B = \frac{5.432 \sin 142.13°}{7.297} \approx 0.45697580 \Rightarrow$$

$$B \approx 27.19°$$

Because angle A is obtuse, angle B must be acute, so this is the only possible value for B and there is one triangle with the given measurements.

$$C = 180° - A - B$$
$$= 180° - 142.13° - 27.19° = 10.68°$$

Thus, $B \approx 27.19°$ and $C \approx 10.68°$.

41. $A = 42.5°$, $a = 15.6$ ft, $b = 8.14$ ft

$$\frac{\sin B}{b} = \frac{\sin A}{a} \Rightarrow \frac{\sin B}{8.14} = \frac{\sin 42.5°}{15.6} \Rightarrow$$

$$\sin B = \frac{8.14 \sin 42.5°}{15.6} \approx 0.35251951$$

There are two angles B between $0°$ and $180°$ that satisfy the condition. Because $\sin B \approx 0.35251951$, to the nearest tenth value of B is $B_1 = 20.6°$. Supplementary angles have the same sine value, so another possible value of B is $B_2 = 180° - 20.6° = 159.4°$. This is not a valid angle measure for this triangle because $A + B_2 = 42.5° + 159.4° = 201.9° > 180°$.

Thus, $C = 180° - 42.5° - 20.6° = 116.9°$.

Solving for c, we have the following.

$$\frac{c}{\sin C} = \frac{a}{\sin A} \Rightarrow \frac{c}{\sin 116.9°} = \frac{15.6}{\sin 42.5°} \Rightarrow$$

$$c = \frac{15.6 \sin 116.9°}{\sin 42.5°} \approx 20.6 \text{ ft}$$

43. $B = 72.2°$, $b = 78.3$ m, $c = 145$ m

$$\frac{\sin C}{c} = \frac{\sin B}{b} \Rightarrow \frac{\sin C}{145} = \frac{\sin 72.2°}{78.3} \Rightarrow$$

$$\sin C = \frac{145 \sin 72.2°}{78.3} \approx 1.7632026$$

Because $\sin C > 1$ is impossible, no such triangle exists.

45. $A = 38°40'$, $a = 9.72$ m, $b = 11.8$ m

$$\frac{\sin B}{b} = \frac{\sin A}{a} \Rightarrow \frac{\sin B}{11.8} = \frac{\sin 38°40'}{9.72} \Rightarrow$$

$$\sin B = \frac{11.8 \sin 38°40'}{9.72} \approx 0.75848811$$

There are two angles B between $0°$ and $180°$ that satisfy the condition. Because $\sin B \approx 0.75848811$, to the nearest tenth value of B is $B_1 = 49.3° \approx 49°20'$. Supplementary angles have the same sine value, so another possible value of B is

$$B_2 = 180° - 49°20'$$
$$= 179°60' - 49°20' = 130°40'$$

This is a valid angle measure for this triangle because

$$A + B_2 = 38°40' + 130°40' = 169°20' < 180°.$$

Solving separately for triangles AB_1C_1 and AB_2C_2 we have the following.

AB_1C_1:

$$C_1 = 180° - A - B_1 = 180° - 38°40' - 49°20'$$
$$= 180° - 88°00' = 92°00'$$

$$\frac{c_1}{\sin C_1} = \frac{a}{\sin A} \Rightarrow \frac{c_1}{\sin 92°00'} = \frac{9.72}{\sin 38°40'} \Rightarrow$$

$$c_1 = \frac{9.72 \sin 92°00'}{\sin 38°40'} \approx 15.5 \text{ m}$$

AB_2C_2:

$$C_2 = 180° - A - B_2$$
$$= 180° - 38°40' - 130°40' = 10°40'$$

$$\frac{c_2}{\sin C_2} = \frac{a}{\sin A} \Rightarrow \frac{c_2}{\sin 10°40'} = \frac{9.72}{\sin 38°40'} \Rightarrow$$

$$c_2 = \frac{9.72 \sin 10°40'}{\sin 38°40'} \approx 2.88 \text{ m}$$

47. $A = 96.80°$, $b = 3.589$ ft, $a = 5.818$ ft

$$\frac{\sin B}{b} = \frac{\sin A}{a} \Rightarrow \frac{\sin B}{3.589} = \frac{\sin 96.80°}{5.818} \Rightarrow$$

$$\sin B = \frac{3.589 \sin 96.80°}{5.818} \approx 0.61253922$$

There are two angles B between $0°$ and $180°$ that satisfy the condition. Because $\sin B \approx 0.61253922$, $B_1 \approx 37.77°$.

Supplementary angles have the same sine value, so another possible value of B is $B_2 = 180° - 37.77° = 142.23°$. This is not a valid angle measure for this triangle because $A + B_2 = 96.80° + 142.23° = 239.03° > 180°$.

Thus $C = 180° - 96.80° - 37.77° = 45.43°$. Solving for c, we have the following.

$$\frac{c}{\sin C} = \frac{a}{\sin A} \Rightarrow \frac{c}{\sin 45.43°} = \frac{5.8518}{\sin 96.80°} \Rightarrow$$

$$c = \frac{5.8518 \sin 45.43°}{\sin 96.80°} \approx 4.174 \text{ ft}$$

49. $B = 39.68°$, $a = 29.81$ m, $b = 23.76$ m

$$\frac{\sin A}{a} = \frac{\sin B}{b} \Rightarrow \frac{\sin A}{29.81} = \frac{\sin 39.68°}{23.76} \Rightarrow$$

$$\sin A = \frac{29.81 \sin 39.68°}{23.76} \approx 0.80108002$$

There are two angles A between $0°$ and $180°$ that satisfy the condition.

Because $\sin A \approx 0.80108002$, to the nearest hundredth, the value of A is $A_1 = 53.23°$.

Supplementary angles have the same sine value, so another possible value of A is $A_2 = 180° - 53.23° = 126.77°$. This is a valid angle measure for this triangle because $B + A_2 = 39.68° + 126.77° = 166.45° < 180°$.

Solving separately for triangles $A_1 BC_1$ and $A_2 BC_2$ we have the following.

$A_1 BC_1$:

$$C_1 = 180° - A_1 - B$$
$$= 180° - 53.23° - 39.68° = 87.09°$$

$$\frac{c_1}{\sin C_1} = \frac{b}{\sin B} \Rightarrow \frac{c_1}{\sin 87.09°} = \frac{23.76}{\sin 39.68°} \Rightarrow$$

$$c_1 = \frac{23.76 \sin 87.09°}{\sin 39.68°} \approx 37.16 \text{ m}$$

$A_2 BC_2$:

$$C_2 = 180° - A_2 - B$$
$$= 180° - 126.77° - 39.68° = 13.55°$$

$$\frac{c_2}{\sin C_2} = \frac{b}{\sin B} \Rightarrow \frac{c_2}{\sin 13.55°} = \frac{23.76}{\sin 39.68°} \Rightarrow$$

$$c_2 = \frac{23.76 \sin 13.55°}{\sin 39.68°} \approx 8.719 \text{ m}$$

51. $a = \sqrt{5}$, $c = 2\sqrt{5}$, $A = 30°$

$$\frac{\sin C}{c} = \frac{\sin A}{a} \Rightarrow \frac{\sin C}{2\sqrt{5}} = \frac{\sin 30°}{\sqrt{5}} \Rightarrow$$

$$\sin C = \frac{2\sqrt{5} \sin 30°}{\sqrt{5}} = \frac{2\sqrt{5} \cdot \frac{1}{2}}{\sqrt{5}} = \frac{\sqrt{5}}{\sqrt{5}} = 1 \Rightarrow$$

$$C = 90°$$

This is a right triangle.

53. Answers will vary. Sample answer: Angle A is an obtuse angle, and because there can be only one obtuse angle in a triangle, the longest side, a, will be opposite this angle. However, we are given that $b > a$, so no triangle exists.

55. $B = 112°10'$; $C = 15°20'$; $BC = 354$ m

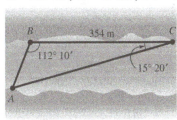

$$A = 180° - B - C$$
$$A = 180° - 112°10' - 15°20'$$
$$= 179°60' - 127°30' = 52°30'$$

$$\frac{BC}{\sin A} = \frac{AB}{\sin C}$$

$$\frac{354}{\sin 52°30'} = \frac{AB}{\sin 15°20'}$$

$$AB = \frac{354 \sin 15°20'}{\sin 52°30'} \approx 118 \text{ m}$$

57. Let $d =$ the distance the ship traveled between the two observations; $L =$ the location of the lighthouse.

$$L = 180° - 38.8° - 44.2° = 97.0°$$

$$\frac{d}{\sin 97°} = \frac{12.5}{\sin 44.2°}$$

$$d = \frac{12.5 \sin 97°}{\sin 44.2°} \approx 17.8 \text{ km}$$

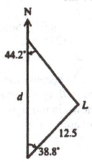

59. Let x = the distance to the lighthouse at bearing N 37° E; y = the distance to the lighthouse at bearing N 25° E.
$$\theta = 180° - 37° = 143°$$
$$\alpha = 180° - \theta - 25° = 180° - 143° - 25° = 12°$$

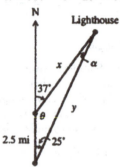

$$\frac{2.5}{\sin \alpha} = \frac{x}{\sin 25°} \Rightarrow x = \frac{2.5 \sin 25°}{\sin 12°} \approx 5.1 \text{ mi}$$

$$\frac{2.5}{\sin \alpha} = \frac{y}{\sin \theta} \Rightarrow \frac{2.5}{\sin 12°} = \frac{y}{\sin 143°} \Rightarrow$$
$$y = \frac{2.5 \sin 143°}{\sin 12°} \approx 7.2 \text{ mi}$$

61. Let A = the location of the balloon; B = the location of the farther town; C = the location of the closer town. Angle $ABC = 31°$ and angle $ACB = 35°$ because the angles of depression are alternate interior angles with the angles of the triangle.

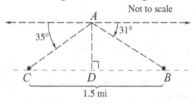

Angle $BAC = 180° - 31° - 35° = 114°$
$$\frac{1.5}{\sin BAC} = \frac{AB}{\sin ACB} \Rightarrow \frac{1.5}{\sin 114°} = \frac{AB}{\sin 35°} \Rightarrow$$
$$AB = \frac{1.5 \sin 35°}{\sin 114°} \approx 0.94178636$$
$$\sin ABC = \frac{AD}{AB} \Rightarrow \sin 31° = \frac{AD}{0.94178636} \Rightarrow$$
$$AD = 0.94178636 \cdot \sin 31° \approx 0.49$$
The balloon is 0.49 mi above the ground.

63. We cannot find θ directly because the length of the side opposite angle θ is not given. Redraw the triangle shown in the figure to the right and label the third angle as α.

$$\frac{\sin \alpha}{1.6 + 2.7} = \frac{\sin 38°}{1.6 + 3.6}$$
$$\frac{\sin \alpha}{4.3} = \frac{\sin 38°}{5.2}$$
$$\sin \alpha = \frac{4.3 \sin 38°}{5.2} \approx 0.50910468$$
$$\alpha \approx \sin^{-1}(0.50910468) \approx 31°$$

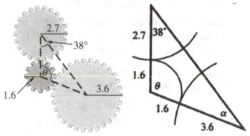

Thus,
$$\theta = 180° - 38° - \alpha \approx 180° - 38° - 31° = 111°$$

65. $Y = 43°30', Z = 95°30', XY = 960 \text{ m}$
We are looking for XZ.
$$\frac{XZ}{\sin Y} = \frac{XY}{\sin Z} \Rightarrow \frac{XZ}{\sin 43°30'} = \frac{960}{\sin 95°30'} \Rightarrow$$
$$XZ = \frac{960 \sin 43°30'}{\sin 95°30'} \approx 664 \text{ m}$$

67. The height of the building is CD.

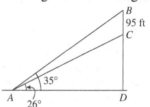

In right triangle ABD, we have $m\angle B = 90° - 35° = 55°$. In triangle ABC, we have $m\angle CAB = 35° - 26° = 9°$. This gives
$$\frac{BC}{\sin \angle CAB} = \frac{AC}{\sin \angle B} \Rightarrow \frac{95}{\sin 9°} = \frac{AC}{\sin 55°} \Rightarrow$$
$$AC = \frac{95 \sin 55°}{\sin 9°} \approx 497.5$$
In triangle ACD,
$$\sin \angle CAD = \frac{CD}{AC} \Rightarrow \sin 26° = \frac{CD}{497.5} \Rightarrow$$
$$CD = 497.5 \sin 26° \approx 218 \text{ ft}$$

69. Angle C is equal to the difference between the angles of elevation.
$C = B - A = 52.7430° - 52.6997° = 0.0433°$
The distance BC to the moon can be determined using the law of sines.
$$\frac{BC}{\sin A} = \frac{AB}{\sin C}$$
$$\frac{BC}{\sin 52.6997°} = \frac{398}{\sin 0.0433°}$$
$$BC = \frac{398 \sin 52.6997°}{\sin 0.0433°}$$
$$BC \approx 418,930 \text{ km}$$

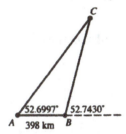

To find AC, we have
$$\frac{AC}{\sin B} = \frac{AB}{\sin C}$$
$$\frac{AC}{\sin (180° - 52.7430°)} = \frac{398}{\sin 0.0433°}$$
$$\frac{AC}{\sin 127.2570°} = \frac{398}{\sin 0.0433°}$$
$$AC = \frac{398 \sin 127.2570°}{\sin 0.0433°}$$
$$\approx 419,171 \text{ km}$$
In either case the distance is approximately 419,000 km compared to the actual value of 406,000 km.

71. To find the area of the triangle, use $\mathscr{A} = \frac{1}{2}bh$, with $b = 1$ and $h = \sqrt{3}$. $\mathscr{A} = \frac{1}{2}(1)(\sqrt{3}) = \frac{\sqrt{3}}{2}$

Now use $\mathscr{A} = \frac{1}{2}ab\sin C$, with $a = \sqrt{3}$, $b = 1$, and $C = 90°$.
$$\mathscr{A} = \frac{1}{2}(\sqrt{3})(1)\sin 90° = \frac{1}{2}(\sqrt{3})(1)(1) = \frac{\sqrt{3}}{2}$$

73. To find the area of the triangle, use $\mathscr{A} = \frac{1}{2}bh$, with $b = 1$ and $h = \sqrt{2}$. $\mathscr{A} = \frac{1}{2}(1)(\sqrt{2}) = \frac{\sqrt{2}}{2}$

Now use $\mathscr{A} = \frac{1}{2}ab\sin C$, with $a = 2$, $b = 1$, and $C = 45°$.

$$\mathscr{A} = \frac{1}{2}(2)(1)\sin 45° = \frac{1}{2}(2)(1)\left(\frac{\sqrt{2}}{2}\right) = \frac{\sqrt{2}}{2}$$

75. $A = 42.5°$, $b = 13.6$ m, $c = 10.1$ m
Angle A is included between sides b and c. Thus, we have
$$\mathscr{A} = \frac{1}{2}bc\sin A = \frac{1}{2}(13.6)(10.1)\sin 42.5°$$
$$\approx 46.4 \text{ m}^2$$

77. $B = 124.5°$, $a = 30.4$ cm, $c = 28.4$ cm
Angle B is included between sides a and c. Thus, we have
$$\mathscr{A} = \frac{1}{2}ac\sin B = \frac{1}{2}(30.4)(28.4)\sin 124.5°$$
$$\approx 356 \text{ cm}^2$$

79. $A = 56.80°$, $b = 32.67$ in., $c = 52.89$ in.
Angle A is included between sides b and c. Thus, we have
$$\mathscr{A} = \frac{1}{2}bc\sin A = \frac{1}{2}(32.67)(52.89)\sin 56.80°$$
$$\approx 722.9 \text{ in.}^2$$

81. $A = 30.50°$, $b = 13.00$ cm, $C = 112.60°$
In order to use the area formula, we need to find either a or c.
$B = 180° - A - C \Rightarrow$
$B = 180° - 30.50° - 112.60° = 36.90°$
Finding a:
$$\frac{a}{\sin A} = \frac{b}{\sin B} \Rightarrow \frac{a}{\sin 30.5°} = \frac{13.00}{\sin 36.90°} \Rightarrow$$
$$a = \frac{13.00 \sin 30.5°}{\sin 36.90°} \approx 10.9890 \text{ cm}$$
$$\mathscr{A} = \frac{1}{2}ab\sin C$$
$$= \frac{1}{2}(10.9890)(13.00)\sin 112.60°$$
$$\approx 65.94 \text{ cm}^2$$
Finding c:
$$\frac{b}{\sin B} = \frac{c}{\sin C} \Rightarrow \frac{13.00}{\sin 36.9°} = \frac{c}{\sin 112.6°} \Rightarrow$$
$$c = \frac{13.00 \sin 112.6°}{\sin 36.9°} \approx 19.9889 \text{ cm}$$
$$\mathscr{A} = \frac{1}{2}bc\sin A = \frac{1}{2}(19.9889)(13.00)\sin 30.5°$$
$$\approx 65.94 \text{ cm}^2$$

83. $\mathscr{A} = \frac{1}{2}ab\sin C$
$$= \frac{1}{2}(16.1)(15.2)\sin 125° \approx 100 \text{ m}^2$$

85. $\dfrac{a}{\sin A}=\dfrac{b}{\sin B}=\dfrac{c}{\sin C}=2r$ and $r=\dfrac{1}{2}$

(because the diameter is 1), so we have

$\dfrac{a}{\sin A}=\dfrac{b}{\sin B}=\dfrac{c}{\sin C}=2\left(\dfrac{1}{2}\right)=1$

Then, $a=\sin A$, $b=\sin B$, and $c=\sin C$.

87. Triangles ACD and BCD are right triangles, so

we have $\tan\alpha=\dfrac{x}{d+BC}$ and $\tan\beta=\dfrac{x}{BC}$.

$\tan\beta=\dfrac{x}{BC}\Rightarrow BC=\dfrac{x}{\tan\beta}$, so we can

substitute into $\tan\alpha=\dfrac{x}{d+BC}$ and solve for

x.

$\tan\alpha=\dfrac{x}{d+BC}\Rightarrow\tan\alpha=\dfrac{x}{d+\dfrac{x}{\tan\beta}}\Rightarrow$

$\dfrac{\sin\alpha}{\cos\alpha}=\dfrac{x}{d+\dfrac{x\cos\beta}{\sin\beta}}\cdot\dfrac{\sin\beta}{\sin\beta}$

$\dfrac{\sin\alpha}{\cos\alpha}=\dfrac{x\sin\beta}{d\sin\beta+x\cos\beta}$

$\sin\alpha\left(d\sin\beta+x\cos\beta\right)=x\sin\beta\left(\cos\alpha\right)$

$d\sin\alpha\sin\beta+x\sin\alpha\cos\beta=x\cos\alpha\sin\beta$

$d\sin\alpha\sin\beta=x\cos\alpha\sin\beta-x\sin\alpha\cos\beta$

$\qquad\qquad=-x\left(\sin\alpha\cos\beta-\cos\alpha\sin\beta\right)$

$\qquad\qquad=-x\sin\left(a-\beta\right)=x\sin\left[-\left(\alpha-\beta\right)\right]$

$d\sin\alpha\sin\beta=x\sin\left(\beta-\alpha\right)$

$\dfrac{d\sin\alpha\sin\beta}{\sin\left(\beta-\alpha\right)}=x$

89. We can use the area formula $\mathcal{A}=\dfrac{1}{2}rR\sin B$

for this triangle. By the law of sines, we have

$\dfrac{r}{\sin A}=\dfrac{R}{\sin C}\Rightarrow r=\dfrac{R\sin A}{\sin C}$

$\sin C=\sin\left[180°-\left(A+B\right)\right]=\sin\left(A+B\right)$, so

we have $r=\dfrac{R\sin A}{\sin C}\Rightarrow r=\dfrac{R\sin A}{\sin\left(A+B\right)}$

By substituting into the area formula, we have

$\mathcal{A}=\dfrac{1}{2}rR\sin B\Rightarrow$

$\mathcal{A}=\dfrac{1}{2}\left[\dfrac{R\sin A}{\sin\left(A+B\right)}\right]R\sin B\Rightarrow$

$\mathcal{A}=\dfrac{1}{2}\cdot\dfrac{\sin A\sin B}{\sin\left(A+B\right)}R^2$

There are a total of 10 stars, so the total area covered by the stars is

$\mathcal{A}=10\left[\dfrac{1}{2}\cdot\dfrac{\sin A\sin B}{\sin\left(A+B\right)}R^2\right]=\left[5\dfrac{\sin A\sin B}{\sin\left(A+B\right)}\right]R^2$

91. (a) $11.4\text{ in.}\cdot\dfrac{10}{13}\text{ in.}\approx8.77\text{ in.}^2$

(b) $\mathcal{A}=50\left[5\dfrac{\sin 18°\sin 36°}{\sin\left(18°+36°\right)}\right]\cdot 0.308^2$

$\approx5.32\text{ in.}^2$

Section 8.2 The Law of Cosines

1. a, b, and C

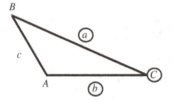

(a) SAS

(b) law of cosines

3. a, b, and A

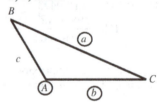

(a) SSA

(b) law of sines

5. A, B, and c

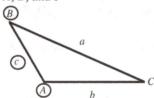

(a) ASA

(b) law of sines

7. a, b, and c

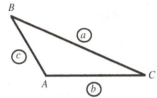

(a) SSS

(b) law of cosines

9. $a^2 = 1^2 + \left(4\sqrt{2}\right)^2 - 2(1)\left(4\sqrt{2}\right)\cos 45°$

$\quad = 1 + 32 - 8\sqrt{2}\left(\dfrac{\sqrt{2}}{2}\right) = 33 - 8 = 25$

$\quad a = \sqrt{25} = 5$

11. $\cos\theta = \dfrac{b^2 + c^2 - a^2}{2bc} = \dfrac{3^2 + 5^2 - 7^2}{2(3)(5)}$

$\quad = \dfrac{9 + 25 - 49}{30} = -\dfrac{1}{2} \Rightarrow \theta = 120°$

13. $A = 121°$, $b = 5$, $c = 3$

Start by finding a with the law of cosines.

$a^2 = b^2 + c^2 - 2bc\cos A \Rightarrow$

$a^2 = 5^2 + 3^2 - 2(5)(3)\cos 121° \approx 49.5 \Rightarrow$

$\quad a \approx 7.04 \approx 7.0$

Of the remaining angles B and C, C must be smaller because it is opposite the shorter of the two sides b and c. Therefore, C cannot be obtuse.

$\dfrac{\sin A}{a} = \dfrac{\sin C}{C} \Rightarrow \dfrac{\sin 121°}{7.04} = \dfrac{\sin C}{3} \Rightarrow$

$\sin C = \dfrac{3\sin 121°}{7.04} \approx 0.36527016 \Rightarrow C \approx 21.4°$

Thus, $B = 180° - 121° - 21.4° = 37.6°$.

15. $a = 12$, $b = 10$, $c = 10$

We can use the law of cosines to solve for any angle of the triangle. Because b and c have the same measure, this is an isosceles triangle and angles B and C also have the same measure. If we solve for B, we obtain

$b^2 = a^2 + c^2 - 2ac\cos B$

$\cos B = \dfrac{a^2 + c^2 - b^2}{2ac}$

$\cos B = \dfrac{12^2 + 10^2 - 10^2}{2(12)(10)} = \dfrac{144}{240} = \dfrac{3}{5} \Rightarrow$

$\quad B \approx 53.13° \approx 53.1°$

Therefore, $C = B \approx 53.13° \approx 53.1°$ and

$A = 180° - 2(53.13°) = 37.74° \approx 37.7°$.

If we solve for A directly, however, we obtain

$a^2 = b^2 + c^2 - 2bc\cos A$

$\cos A = \dfrac{b^2 + c^2 - a^2}{2bc}$

$\cos A = \dfrac{10^2 + 10^2 - 12^2}{2(10)(10)} = \dfrac{56}{200} = \dfrac{7}{5}$

$\quad A \approx 73.7°$

The angles may not sum to 180° due to rounding.

17. $B = 55°$, $a = 90$, $c = 100$

Start by finding b with the law of cosines.

$b^2 = a^2 + c^2 - 2ac\cos B$

$b^2 = 90^2 + 100^2 - 2(90)(100)\cos 55° \approx 7775.6$

$\quad b \approx 88.18$ (rounded to 88.2)

Of the remaining angles A and C, A must be smaller because it is opposite the shorter of the two sides a and c. Therefore, A cannot be obtuse.

$\dfrac{\sin A}{a} = \dfrac{\sin B}{b} \Rightarrow \dfrac{\sin A}{90} = \dfrac{\sin 55°}{88.18} \Rightarrow$

$\sin A = \dfrac{90\sin 55°}{88.18} \approx 0.83605902 \Rightarrow A \approx 56.7°$

Thus, $C = 180° - 55° - 56.7° = 68.3°$.

19. $A = 41.4°$, $b = 2.78$ yd, $c = 3.92$ yd

First find a.

$a^2 = b^2 + c^2 - 2bc\cos A$

$a^2 = 2.78^2 + 3.92^2 - 2(2.78)(3.92)\cos 41.4°$

$\quad \approx 6.7460 \Rightarrow a \approx 2.60$ yd

Find B next, because angle B is smaller than angle C (because $b < c$), and thus angle B must be acute.

$\dfrac{\sin B}{b} = \dfrac{\sin A}{a} \Rightarrow \dfrac{\sin B}{2.78} = \dfrac{\sin 41.4°}{2.597} \Rightarrow$

$\sin B = \dfrac{2.78\sin 41.4°}{2.597} \approx 0.707091182 \Rightarrow$

$\quad B \approx 45.1°$

Finally, $C = 180° - 41.4° - 45.1° = 93.5°$.

21. $C = 45.6°$, $b = 8.94$ m, $a = 7.23$ m

First find c.

$c^2 = a^2 + b^2 - 2ab\cos C \Rightarrow$

$c^2 = 7.23^2 + 8.94^2 - 2(7.23)(8.94)\cos 45.6°$

$\quad \approx 41.7493 \Rightarrow c \approx 6.46$ m

Find A next, because angle A is smaller than angle B (because $a < b$), and thus angle A must be acute.

$\dfrac{\sin A}{a} = \dfrac{\sin C}{c} \Rightarrow \dfrac{\sin A}{7.23} = \dfrac{\sin 45.6°}{6.461} \Rightarrow$

$\sin A = \dfrac{7.23\sin 45.6°}{6.461} \approx 0.79951052 \Rightarrow$

$\quad A \approx 53.1°$

Finally, $B = 180° - 53.1° - 45.6° = 81.3°$.

23. $a = 9.3$ cm, $b = 5.7$ cm, $c = 8.2$ cm
We can use the law of cosines to solve for any of angle of the triangle. We solve for A, the largest angle. We will know that A is obtuse if $\cos A < 0$.

$$a^2 = b^2 + c^2 - 2bc\cos A \Rightarrow$$

$$\cos A = \frac{5.7^2 + 8.2^2 - 9.3^2}{2(5.7)(8.2)} \approx 0.14163457 \Rightarrow$$

$$A \approx 82°$$

Find B next, because angle B is smaller than angle C (because $b < c$), and thus angle B must be acute.

$$\frac{\sin B}{b} = \frac{\sin A}{a} \Rightarrow \frac{\sin B}{5.7} = \frac{\sin 82°}{9.3} \Rightarrow$$

$$\sin B = \frac{5.7\sin 82°}{9.3} \approx 0.60693849 \Rightarrow B \approx 37°$$

Thus, $C = 180° - 82° - 37° = 61°$.

25. $a = 42.9$ m, $b = 37.6$ m, $c = 62.7$ m
Angle C is the largest, so find it first.

$$c^2 = a^2 + b^2 - 2ab\cos C \Rightarrow$$

$$\cos C = \frac{42.9^2 + 37.6^2 - 62.7^2}{2(42.9)(37.6)}$$

$$\cos C \approx -0.20988940 \Rightarrow C \approx 102.1° \approx 102°10'$$

Find B next, because angle B is smaller than angle A (because $b < a$), and thus angle B must be acute.

$$\frac{\sin B}{b} = \frac{\sin C}{c} \Rightarrow \frac{\sin B}{37.6} = \frac{\sin 102.1°}{62.7} \Rightarrow$$

$$\sin B = \frac{37.6\sin 102.1°}{62.7} \approx 0.58635805 \Rightarrow$$

$$B \approx 35.9° \approx 35°50'$$

Thus,
$$A = 180° - 35°50' - 102°10'$$
$$= 180° - 138° = 42°00'$$

27. $a = 965$ ft, $b = 876$ ft, $c = 1240$ ft
Angle C is the largest, so find it first.

$$c^2 = a^2 + b^2 - 2ab\cos C \Rightarrow$$

$$\cos C = \frac{965^2 + 876^2 - 1240^2}{2(965)(876)} \approx 0.09522855 \Rightarrow$$

$$C \approx 84.5° \text{ or } 84°30'$$

Find B next, because angle B is smaller than angle A (because $b < a$), and thus angle B must be acute.

$$\frac{\sin B}{b} = \frac{\sin C}{c} \Rightarrow \frac{\sin B}{876} = \frac{\sin 84°30'}{1240} \Rightarrow$$

$$\sin B = \frac{876\sin 84°30'}{1240} \approx 0.70319925 \Rightarrow$$

$$B \approx 44.7° \text{ or } 44°40'$$

Thus,
$$A = 180° - 44°40' - 84°30'$$
$$= 179°60' - 129°10' = 50°50'$$

29. $A = 80°40'$ $b = 143$ cm, $c = 89.6$ cm
First find a.

$$a^2 = b^2 + c^2 - 2bc\cos A \Rightarrow$$

$$a^2 = 143^2 + 89.6^2 - 2(143)(89.6)\cos 80°40'$$

$$\approx 24{,}321.25 \Rightarrow a \approx 156 \text{ cm}$$

Find C next, because angle C is smaller than angle B (because $c < a$), and thus angle C must be acute.

$$\frac{\sin C}{c} = \frac{\sin A}{a} \Rightarrow \frac{\sin C}{89.6} = \frac{\sin 80°40'}{156.0} \Rightarrow$$

$$\sin C = \frac{89.6\sin 80°40'}{156.0} \approx 0.56675534 \Rightarrow$$

$$C \approx 34.5° = 34°30'$$

Finally,
$$B = 180° - 80°40' - 34°30'$$
$$= 179°60' - 115°10' = 64°50'$$

31. $B = 74.8°$, $a = 8.92$ in., $c = 6.43$ in.
First find b.

$$b^2 = a^2 + c^2 - 2ac\cos B \Rightarrow$$

$$b^2 = 8.92^2 + 6.43^2$$
$$\qquad - 2(8.92)(6.43)\cos 74.8°$$

$$\approx 90.8353 \Rightarrow b \approx 9.53 \text{ in.}$$

Find C next, because angle C is smaller than angle A (because $c < a$), and thus angle C must be acute.

$$\frac{\sin C}{c} = \frac{\sin B}{b} \Rightarrow \frac{\sin C}{6.43} = \frac{\sin 74.8°}{9.53} \Rightarrow$$

$$\sin C = \frac{6.43\sin 74.8°}{9.53} \approx 0.6540561811 \Rightarrow$$

$$C \approx 40.6°$$

Thus, $A = 180° - 74.8° - 40.6° = 64.6°$.

33. $A = 112.8°$, $b = 6.28$ m, $c = 12.2$ m
First find a.

$$a^2 = b^2 + c^2 - 2bc\cos A \Rightarrow$$

$$a^2 = 6.28^2 + 12.2^2 - 2(6.28)(12.2)\cos 112.8°$$

$$\approx 247.658 \Rightarrow a \approx 15.7 \text{ m}$$

Find B next, because angle B is smaller than angle C (because $b < c$), and thus angle B must be acute.

$$\frac{\sin B}{b} = \frac{\sin A}{a} \Rightarrow \frac{\sin B}{6.28} = \frac{\sin 112.8°}{15.74} \Rightarrow$$

$$\sin B = \frac{6.28\sin 112.8°}{15.74} \approx 0.36780817 \Rightarrow$$

$$B \approx 21.6°$$

Finally, $C = 180° - 112.8° - 21.6° = 45.6°$.

35. $a = 3.0$ ft, $b = 5.0$ ft, $c = 6.0$ ft

Angle C is the largest, so find it first.

$$c^2 = a^2 + b^2 - 2ab\cos C \Rightarrow$$

$$\cos C = \frac{3.0^2 + 5.0^2 - 6.0^2}{2(3.0)(5.0)} = -\frac{2}{30} = -\frac{1}{15}$$

$$\approx -0.06666667 \Rightarrow C \approx 94°$$

Find A next, because angle A is smaller than angle B (because $a < b$), and thus angle A must be acute.

$$\frac{\sin A}{a} = \frac{\sin C}{c} \Rightarrow \frac{\sin A}{3} = \frac{\sin 94°}{6} \Rightarrow$$

$$\sin A = \frac{3\sin 94°}{6} \approx 0.49878203 \Rightarrow A \approx 30°$$

Thus, $B = 180° - 30° - 94° = 56°$.

37. There are three ways to apply the law of cosines when $a = 3$, $b = 4$, and $c = 10$.

Solving for A:

$$a^2 = b^2 + c^2 - 2bc\cos A \Rightarrow$$

$$\cos A = \frac{4^2 + 10^2 - 3^2}{2(4)(10)} = \frac{107}{80} = 1.3375$$

Solving for B:

$$b^2 = a^2 + c^2 - 2ac\cos B \Rightarrow$$

$$\cos B = \frac{3^2 + 10^2 - 4^2}{2(3)(10)} = \frac{93}{60} = \frac{31}{20} = 1.55$$

Solving for C:

$$c^2 = a^2 + b^2 - 2ab\cos C \Rightarrow$$

$$\cos C = \frac{3^2 + 4^2 - 10^2}{2(3)(4)} = \frac{-75}{24} = -\frac{25}{8} = -3.125$$

The cosine of any angle of a triangle must be between -1 and 1, so a triangle cannot have sides 3, 4, and 10.

39. A and B are on opposite sides of False River. We must find AB, or c, in the following triangle.

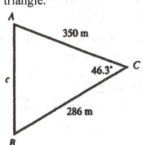

$$c^2 = a^2 + b^2 - 2ab\cos C$$

$$c^2 = 286^2 + 350^2 - 2(286)(350)\cos 46.3°$$

$$c^2 \approx 65{,}981.3 \Rightarrow c \approx 257$$

The length of AB is 257 m.

41. Using the law of cosines we can solve for the measure of angle A.

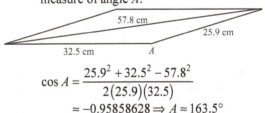

$$\cos A = \frac{25.9^2 + 32.5^2 - 57.8^2}{2(25.9)(32.5)}$$

$$\approx -0.95858628 \Rightarrow A \approx 163.5°$$

43. Find AC, or b, in the following triangle.

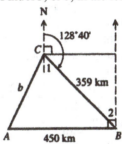

$m\angle 1 = 180° - 128°40' = 51°20'$

Angles 1 and 2 are alternate interior angles formed when parallel lines (the north lines) are cut by a transversal, line BC, so

$m\angle 2 = m\angle 1 = 51°20'$.

$m\angle ABC = 90° - m\angle 2 = 90° - 51°20' = 38°40'$

$$b^2 = a^2 + c^2 - 2ac\cos B \Rightarrow$$

$$b^2 = 359^2 + 450^2 - 2(359)(450)\cos 38°40'$$

$$\approx 79{,}106 \Rightarrow b \approx 281 \text{ km}$$

C is about 281 km from A.

45. Let $x =$ the distance between the ends of the two equal sides.

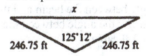

Use the law of cosines to find x.

$$x^2 = 246.75^2 + 246.75^2$$

$$\qquad - 2(246.75)(246.75)\cos 125°12'$$

$$\approx 191{,}963.937 \Rightarrow x \approx 438.14$$

The distance between the ends of the two equal sides is 438.14 feet.

47. Sketch a triangle showing the situation as follows.

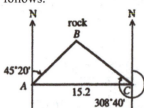

$m\angle A = 90° - 45°20' = 44°40'$

$m\angle C = 308°40' - 270° = 38°40'$

$m\angle B = 180° - A - C$
$\qquad = 180° - 44°40' - 38°40' = 96° \; 40'$

We have only one side of a triangle, so use the law of sines to find $BC = a$.

$$\frac{a}{\sin A} = \frac{b}{\sin B} \Rightarrow \frac{a}{\sin 44°\;40'} = \frac{15.2}{\sin 96°\;40'} \Rightarrow$$

$$a = \frac{15.2 \sin 44°\;40'}{\sin 96°\;40'} \approx 10.8$$

The distance between the ship and the rock is about 10.8 miles.

49. Use the law of cosines to find the angle, θ.

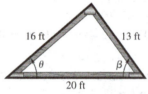

$$\cos\theta = \frac{20^2 + 16^2 - 13^2}{2(20)(16)} = \frac{487}{640}$$
$$\approx 0.76093750 \Rightarrow \theta \approx 40°$$

51. Let A = the angle between the beam and the 45-ft cable. Let B = the angle between the beam and the 60-ft cable.

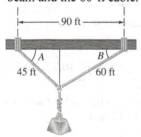

$$\cos A = \frac{45^2 + 90^2 - 60^2}{2(45)(90)} = \frac{6525}{8100} = \frac{29}{36}$$
$$\approx 0.80555556 \Rightarrow A \approx 36°$$

$$\cos B = \frac{90^2 + 60^2 - 45^2}{2(90)(60)} = \frac{9675}{10,800} = \frac{43}{48}$$
$$\approx 0.89583333 \Rightarrow B \approx 26°$$

53. Let A = home plate; B = first base; C = second base; D = third base; P = pitcher's rubber. Draw AC through P, draw PB and PD.

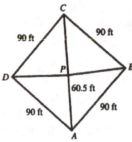

In triangle ABC, $m\angle B = 90°$, and $m\angle A = m\angle C = 45°$.

$AC = \sqrt{90^2 + 90^2} = \sqrt{2 \cdot 90^2} = 90\sqrt{2}$ and

$PC = 90\sqrt{2} - 60.5 \approx 66.8$ ft

In triangle APB, $m\angle A = 45°$.

$PB^2 = AP^2 + AB^2 - 2(AP)(AB)\cos A$

$PB^2 = 60.5^2 + 90^2 - 2(60.5)(90)\cos 45°$

$PB^2 \approx 4059.86 \Rightarrow PB \approx 63.7$ ft

Triangles APB and APD are congruent, so $PB = PD = 63.7$ ft.

The distance to second base is 66.8 ft and the distance to both first and third base is 63.7 ft.

55. Find the distance of the ship from point A.

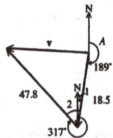

$m\angle 1 = 189° - 180° = 9°$

$m\angle 2 = 360° - 317° = 43°$

$m\angle 1 + m\angle 2 = 9° + 43° = 52°$

Use the law of cosines to find v.

$v^2 = 47.8^2 + 18.5^2 - 2(47.8)(18.5)\cos 52°$
$\qquad \approx 1538.23 \Rightarrow v \approx 39.2$ km

57. Let c = the length of the property line that cannot be directly measured.

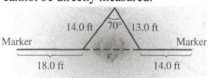

Not to scale

Using the law of cosines, we have

$c^2 = 14.0^2 + 13.0^2 - 2(14.0)(13.0)\cos 70°$
$\qquad \approx 240.5 \Rightarrow c \approx 15.5$ ft

(rounded to three significant digits)

The length of the property line is approximately $18.0 + 15.5 + 14.0 = 47.5$ feet

59. Let c = the length of the tunnel.

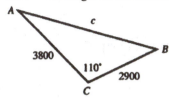

Use the law of cosines to find c.

$c^2 = 3800^2 + 2900^2 - 2(3800)(2900)\cos 110°$

$\approx 30,388,124 \Rightarrow c \approx 5512.5$

The tunnel is 5500 meters long. (rounded to two significant digits)

61. Let a be the length of the segment from $(0, 0)$ to $(6, 8)$. Use the distance formula.

$a = \sqrt{(6-0)^2 + (8-0)^2} = \sqrt{6^2 + 8^2}$

$= \sqrt{36 + 64} = \sqrt{100} = 10$

Let b be the length of the segment from $(0, 0)$ to $(4, 3)$.

$b = \sqrt{(4-0)^2 + (3-0)^2} = \sqrt{4^2 + 3^2}$

$= \sqrt{16 + 9} = \sqrt{25} = 5$

Let c be the length of the segment from $(4, 3)$ to $(6, 8)$.

$c = \sqrt{(6-4)^2 + (8-3)^2} = \sqrt{2^2 + 5^2}$

$= \sqrt{4 + 25} = \sqrt{29}$

$\cos\theta = \dfrac{a^2 + b^2 - c^2}{2ab} \Rightarrow$

$\cos\theta = \dfrac{10^2 + 5^2 - \left(\sqrt{29}\right)^2}{2(10)(5)} = \dfrac{100 + 25 - 29}{100}$

$= 0.96 \Rightarrow \theta \approx 16.26°$

63. $\mathcal{A} = \dfrac{1}{2}bh \Rightarrow$

$\mathcal{A} = \dfrac{1}{2}(16)(3\sqrt{3}) = 24\sqrt{3} \approx 41.57.$

To use Heron's Formula, first find the semiperimeter,

$s = \dfrac{1}{2}(a+b+c) = \dfrac{1}{2}(6+14+16) = \dfrac{1}{2} \cdot 36 = 18.$

Now find the area of the triangle.

$\mathcal{A} = \sqrt{s(s-a)(s-b)(s-c)}$

$= \sqrt{18(18-6)(18-14)(18-16)}$

$= \sqrt{18(12)(4)(2)} = \sqrt{1728} = 24\sqrt{3} \approx 41.57$

Both formulas give the same area.

65. $a = 12$ m, $b = 16$ m, $c = 25$ m

$s = \dfrac{1}{2}(a+b+c) = \dfrac{1}{2}(12+16+25)$

$= \dfrac{1}{2} \cdot 53 = 26.5$

$\mathcal{A} = \sqrt{s(s-a)(s-b)(s-c)}$

$= \sqrt{26.5(26.5-12)(26.5-16)(26.5-25)}$

$= \sqrt{26.5(14.5)(10.5)(1.5)} \approx 78 \text{ m}^2$

(rounded to two significant digits)

67. $a = 154$ cm, $b = 179$ cm, $c = 183$ cm

$s = \dfrac{1}{2}(a+b+c) = \dfrac{1}{2}(154+179+183)$

$= \dfrac{1}{2} \cdot 516 = 258$

$\mathcal{A} = \sqrt{s(s-a)(s-b)(s-c)}$

$= \sqrt{258(258-154)(258-179)(258-183)}$

$= \sqrt{258(104)(79)(75)} \approx 12,600 \text{ cm}^2$

(rounded to three significant digits)

69. $a = 76.3$ ft, $b = 109$ ft, $c = 98.8$ ft

$s = \dfrac{1}{2}(a+b+c) = \dfrac{1}{2}(76.3+109+98.8)$

$= \dfrac{1}{2} \cdot 284.1 = 142.05$

$\mathcal{A} = \sqrt{s(s-a)(s-b)(s-c)}$

$= \sqrt{\begin{array}{c}142.05(142.05-76.3)(142.05-109)\cdot \\ (142.05-98.8)\end{array}}$

$= \sqrt{142.05(65.75)(33.05)(43.25)} \approx 3650 \text{ ft}^2$

(rounded to three significant digits)

71. Perimeter: $9 + 10 + 17 = 36$ feet, so the semi-perimeter is $\dfrac{1}{2} \cdot 36 = 18$ feet.

Use Heron's Formula to find the area.

$\mathcal{A} = \sqrt{s(s-a)(s-b)(s-c)}$

$= \sqrt{18(18-9)(18-10)(18-17)}$

$= \sqrt{18(9)(8)(1)} = \sqrt{1296} = 36 \text{ ft}$

The perimeter and area both equal 36, so the triangle is a *perfect triangle*.

73. Find the area of the Bermuda Triangle using Heron's Formula.

$$s = \frac{1}{2}(a+b+c) = \frac{1}{2}(850+925+1300) = \frac{1}{2} \cdot 3075 = 1537.5$$

$$\mathcal{A} = \sqrt{s(s-a)(s-b)(s-c)} = \sqrt{1537.5(1537.5-850) \cdot (1537.5-925)(1537.5-1300)}$$

$$= \sqrt{1537.5(687.5)(612.5)(237.5)} \approx 392{,}128.82$$

The area of the Bermuda Triangle is about 390,000 mi^2.

75. (a) Using the law of sines, we have

$$\frac{\sin C}{c} = \frac{\sin A}{a} \Rightarrow \frac{\sin C}{15} = \frac{\sin 60°}{13} \Rightarrow \sin C = \frac{15\sin 60°}{13} = \frac{15}{13} \cdot \frac{\sqrt{3}}{2} \approx 0.99926008$$

There are two angles C between $0°$ and $180°$ that satisfy the condition. Because $\sin C \approx 0.99926008$, to the nearest tenth value of C is $C_1 = 87.8°$. Supplementary angles have the same sine value, so another possible value of C is $B_2 = 180° - 87.8° = 92.2°$.

(b) By the law of cosines, we have

$$\cos C = \frac{a^2 + b^2 - c^2}{2ab} = \frac{13^2 + 7^2 - 15^2}{2(13)(7)} = \frac{-7}{182} = -\frac{1}{26} \approx -0.03846154 \Rightarrow C \approx 92.2°$$

(c) With the law of cosines, we are required to find the inverse cosine of a negative number; therefore; we know angle C is greater than $90°$.

77.

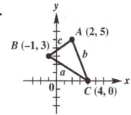

$$a = \sqrt{(-1-4)^2 + (3-0)^2} = \sqrt{25+9} = \sqrt{34}; \quad b = \sqrt{(2-4)^2 + (5-0)^2} = \sqrt{4+25} = \sqrt{29}$$

$$c = \sqrt{(-1-2)^2 + (3-5)^2} = \sqrt{9+4} = \sqrt{13}$$

79. First find the semiperimeter. $s = \frac{1}{2}(a+b+c) = \frac{1}{2}\left(\sqrt{34} + \sqrt{29} + \sqrt{13}\right)$

Using Heron's formula, we have

$$\mathcal{A} = \sqrt{s(s-a)(s-b)(s-c)}$$

$$= \sqrt{\frac{1}{2}\left(\sqrt{34} + \sqrt{29} + \sqrt{13}\right)\left(\frac{1}{2}\left(\sqrt{34} + \sqrt{29} + \sqrt{13}\right) - \sqrt{34}\right)\left(\frac{1}{2}\left(\sqrt{34} + \sqrt{29} + \sqrt{13}\right) - \sqrt{29}\right)\left(\frac{1}{2}\left(\sqrt{34} + \sqrt{29} + \sqrt{13}\right) - \sqrt{13}\right)}$$

$$= 9.5 \text{ sq units (found using a calculator)}$$

Chapter 8   Quiz
(Sections 8.1−8.2)

1. Using the law of sines, we have

$$\frac{\sin B}{b} = \frac{\sin C}{c} \Rightarrow \frac{\sin 30.6°}{7.42} = \frac{\sin C}{4.54} \Rightarrow$$

$$\sin C = \frac{4.54 \sin 30.6°}{7.42} \approx 0.311462 \Rightarrow$$

$$C \approx 18.1°$$

$$A = 180° - B - C$$

$$= 180° - 30.6° - 18.1° = 131.3° \approx 131°$$

(rounded to three significant digits)

3. Using the law of cosines, we have

$$c^2 = a^2 + b^2 - 2ab \cos C$$

$$21.2^2 = 28.4^2 + 16.9^2 - 2 \cdot 28.4 \cdot 16.9 \cos C$$

$$-642.73 = -959.92 \cos C$$

$$\frac{642.73}{959.92} = \cos C \Rightarrow$$

$$C \approx 48.0° \text{ (rounded to three}$$
significant digits)

5. First find the semiperimeter:

$$s = \frac{1}{2}(19.5 + 21.0 + 22.5) = 31.5$$

Using Heron's formula, we have

$$\mathcal{A} = \sqrt{s(s-a)(s-b)(s-c)}$$

$$= \sqrt{31.5(31.5 - 19.5)(31.5 - 21.0)(31.5 - 22.5)}$$

$$= \sqrt{31.5(12)(10.5)(9)}$$

$$= \sqrt{35,721} = 189 \text{ km}^2$$

7. $\angle C = 180° - 111° - 41° = 28°$

Using the law of sines, we have

$$\frac{a}{\sin A} = \frac{c}{\sin C} \Rightarrow \frac{a}{\sin 111°} = \frac{326}{\sin 28°} \Rightarrow$$

$$a = \frac{326 \sin 111°}{\sin 28°} \approx 648$$

$$\frac{b}{\sin B} = \frac{c}{\sin C} \Rightarrow \frac{b}{\sin 41°} = \frac{326}{\sin 28°} \Rightarrow$$

$$b = \frac{326 \sin 41°}{\sin 28°} \approx 456$$

Note that both a and b have been rounded to three significant digits.

9. $AB = 22.47928$ mi, $AC = 28.14276$ mi, $A = 58.56989°$

This is SAS, so use the law of cosines.

$$BC^2 = AC^2 + AB^2 - 2(AC)(AB)\cos A$$

$$BC^2 = 28.14276^2 + 22.47928^2$$
$$\quad - 2(28.14276)(22.47928)\cos 58.56989°$$

$$BC^2 \approx 637.55393$$

$$BC \approx 25.24983$$

BC is approximately 25.24983 mi.
(rounded to seven significant digits)

Section 8.3 Geometrically Defined Vectors and Applications

1. Equal vectors have the same magnitude and direction. Equal vectors are **m** and **p**; **n** and **r**.

3. One vector is a positive scalar multiple of another if the two vectors point in the same direction; they may have different magnitudes.
   m = 1**p**; **m** = 2**t**; **n** = 1**r**; **p** = 2**t** or

   $$\mathbf{p} = 1\mathbf{m};\ \mathbf{t} = \frac{1}{2}\mathbf{m};\ \mathbf{r} = 1\mathbf{n};\ \mathbf{t} = \frac{1}{2}\mathbf{p}$$

5.

7.

9.

11.

13.

15.

17. $\mathbf{a} + (\mathbf{b} + \mathbf{c}) = (\mathbf{a} + \mathbf{b}) + \mathbf{c}$

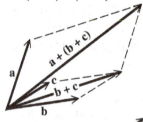

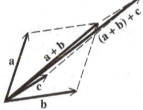

Yes, vector addition is associative.

19. $|\mathbf{u}| = 12, |\mathbf{v}| = 20, \theta = 27°$

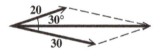

21. $|\mathbf{u}| = 20, |\mathbf{v}| = 30, \theta = 30°$

23. $\alpha = 180° - 40° = 140°$

$|\mathbf{v}|^2 = 40^2 + 60^2 - 2(40)(60)\cos 140°$

$|\mathbf{v}|^2 \approx 8877.0133 \Rightarrow |\mathbf{v}| \approx 94.2 \text{ lb}$

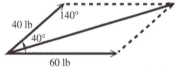

25. $\alpha = 180° - 110° = 70°$

$|\mathbf{v}|^2 = 15^2 + 25^2 - 2(15)(25)\cos 70°$

$|\mathbf{v}|^2 \approx 593.48489 \Rightarrow |\mathbf{v}| \approx 24.4 \text{ lb}$

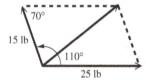

27.   Forces of 250 newtons and 450 newtons, forming an angle of 85°

$\alpha = 180° - 85° = 95°$

$|\mathbf{v}|^2 = 250^2 + 450^2 - 2(250)(450)\cos 95°$

$|\mathbf{v}|^2 \approx 284,610.04 \Rightarrow |\mathbf{v}| \approx 533.5$

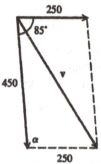

The magnitude of the resulting force is 530 newtons. (rounded to two significant digits)

29.   Forces of 116 lb and 139 lb, forming an angle of 140° 50′

$\alpha = 180° - 140°50′$
$\quad = 179°60′ - 140°50′ = 39°10′$

$|\mathbf{v}|^2 = 139^2 + 116^2 - 2(139)(116)\cos 39°10′$

$|\mathbf{v}|^2 \approx 7774.7359 \Rightarrow |\mathbf{v}| \approx 88.174$

The magnitude of the resulting force is 88.2 lb. (rounded to three significant digits)

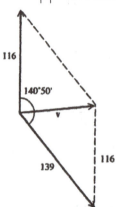

31.   Find the direction and magnitude of the equilibrant.
$A = 180° - 28.2° = 151.8°,$ so we can use the law of cosines to find the magnitude of the resultant, $\mathbf{v}$.

$|\mathbf{v}|^2 = 1240^2 + 1480^2 - 2(1240)(1480)\cos 151.8°$
$\quad \approx 6962736.2 \Rightarrow |\mathbf{v}| \approx 2639 \text{ lb}$

(will be rounded as 2640)
Use the law of sines to find α.

$\dfrac{\sin \alpha}{1240} = \dfrac{\sin 151.8°}{2639}$

$\sin \alpha = \dfrac{1240 \sin 151.8°}{2639} \approx 0.22203977$

$\alpha \approx 12.8°$

Thus, we have 2640 lb at an angle of $\theta = 180° - 12.8° = 167.2°$ with the 1480-lb force.

33. Let α = the angle between the forces.
To find α, use the law of cosines to find θ.

$$786^2 = 692^2 + 423^2 - 2(692)(423)\cos\theta$$

$$\cos\theta = \frac{692^2 + 423^2 - 786^2}{2(692)(423)} \approx 0.06832049$$

$$\theta \approx 86.1°$$

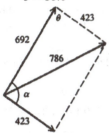

Thus, $\alpha = 180° - 86.1° = 93.9°$.

35. Use the parallelogram rule. In the figure, **x** represents the second force and **v** is the resultant.

$\alpha = 180° - 78°50' = 101°10'$ and
$\beta = 78°50' - 41°10' = 37°40'$

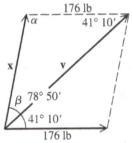

Using the law of sines, we have

$$\frac{|\mathbf{x}|}{\sin 41°10'} = \frac{176}{\sin 37°40'} \Rightarrow$$

$$|\mathbf{x}| = \frac{176\sin 41°10'}{\sin 37°40'} \approx 190$$

$$\frac{|\mathbf{v}|}{\sin\alpha} = \frac{176}{\sin 37°40'} \Rightarrow$$

$$|\mathbf{v}| = \frac{176\sin 101°10'}{\sin 37°40'} \approx 283$$

Thus, the magnitude of the second force is about 190 lb and the magnitude of the resultant is about 283 lb.

37. Let θ = the angle that the hill makes with the horizontal.
The 80-lb downward force has a 25-lb component parallel to the hill. The two right triangles are similar and have congruent angles.

$$\sin\theta = \frac{25}{80} = \frac{5}{16} = 0.3125 \Rightarrow \theta \approx 18°$$

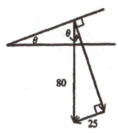

39. Find the force needed to hold a 60-ton monolith along the causeway.
The force needed to pull 60 tons is equal to the magnitude of **x**, the component parallel to the causeway. The length of the causeway is not relevant in this problem.

$$\sin 2.3° = \frac{|\mathbf{x}|}{60} \Rightarrow |\mathbf{x}| = 60\sin 2.3° \approx 2.4 \text{ tons}$$

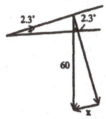

The force needed is 2.4 tons.

41. As in Example 4 on page 776 of the text, angle B equals angle θ, and here the magnitude of vector **BA** represents the weight of the stump grinder. The vector **AC** equals vector **BE**, which represents the force required to hold the stump grinder on the incline. Thus, we have

$$\sin B = \frac{18.0}{60.0} = \frac{3.0}{10.0} = 0.3 \Rightarrow B \approx 17.5°$$

43. Let **r** = the vertical component of the person exerting a 114-lb force;
s = the vertical component of the person exerting a 150-lb force.
The weight of the box is the sum of the magnitudes of the vertical components of the two vectors representing the forces exerted by the two people.

$$|\mathbf{r}| = 114\sin 54.9° \approx 93.27 \text{ and}$$

$$|\mathbf{s}| = 150\sin 62.4° \approx 132.93$$

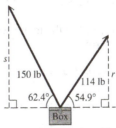

Thus, the weight of box is

$$|\mathbf{r}| + |\mathbf{s}| \approx 93.27 + 132.93 = 226.2 \approx 226 \text{ lb}.$$

45. Refer to the diagram. In order for the ship to turn due east, the ship must turn the measure of angle CAB, which is $90° - 34° = 56°$. Angle DAB is therefore $180° - 56° = 124°$.

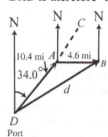

Port

Using the law of cosines, we can solve for the distance the ship is from port as follows.

$$d^2 = 10.4^2 + 4.6^2 - 2(10.4)(4.6)\cos 124°$$
$$\approx 182.824 \Rightarrow d \approx 13.52$$

Thus, the distance the ship is from port is 13.5 mi. (rounded to three significant digits) To find the bearing, we first seek the measure of angle ADB, which we will refer to as angle D. Using the law of cosines we have

$$\cos D = \frac{13.52^2 + 10.4^2 - 4.6^2}{2(13.52)(10.4)} \approx 0.95937073 \Rightarrow$$

$$D \approx 16.4°$$

Thus, the bearing is $34.0° + D = 34.0° + 16.4° = 50.4°$.

47. Find the distance of the ship from point A.
Angle 1 $= 189° - 180° = 9°$
Angle 2 $= 360° - 317° = 43°$
Angle 1 + Angle 2 $= 9° + 43° = 52°$

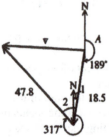

Use the law of cosines to find $|\mathbf{v}|$.

$$|\mathbf{v}|^2 = 47.8^2 + 18.5^2 - 2(47.8)(18.5)\cos 52°$$
$$\approx 1538.23 \Rightarrow |\mathbf{v}| \approx 39.2 \text{ km}$$

49. Let $x =$ be the actual speed of the motorboat; $y =$ the speed of the current.

$$\sin 10° = \frac{y}{20.0} \Rightarrow y = 20.0\sin 10° \approx 3.5$$

$$\cos 10° = \frac{x}{20.0} \Rightarrow x = 20.0\cos 10° \approx 19.7$$

The speed of the current is 3.5 mph and the actual speed of the motorboat is 19.7 mph.

51. Let $\mathbf{v} =$ the ground speed vector. Find the bearing and ground speed of the plane.
Angle $A = 233° - 114° = 119°$
Use the law of cosines to find $|\mathbf{v}|$.

$$|\mathbf{v}|^2 = 39^2 + 450^2 - 2(39)(450)\cos 119°$$
$$|\mathbf{v}|^2 \approx 221,037.82$$
$$|\mathbf{v}| \approx 470.1$$

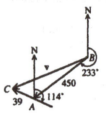

The ground speed is 470 mph. (rounded to two significant digits)
Use the law of sines to find angle B.

$$\frac{\sin B}{39} = \frac{\sin 119°}{470.1} \Rightarrow$$

$$\sin B = \frac{39\sin 119°}{470.1} \approx 0.07255939 \Rightarrow B \approx 4°$$

Thus, the bearing is
$B + 233° = 4° + 233° = 237°$.

53. Let $|\mathbf{x}| =$ the airspeed and $|\mathbf{d}| =$ the ground speed.

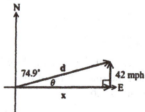

$\theta = 90° - 74.9° = 15.1°$

$$\frac{|\mathbf{x}|}{42.0} = \cot 15.1° \Rightarrow$$

$$|\mathbf{x}| = 42.0\cot 15.1° = \frac{42.0}{\tan 15.1°} \approx 156 \text{ mph}$$

$$\frac{|\mathbf{d}|}{42} = \csc 15.1° \Rightarrow$$

$$|\mathbf{d}| = 42.0\csc 15.1° = \frac{42.0}{\sin 15.1°} \approx 161 \text{ mph}$$

55. Let **c** = the ground speed vector.

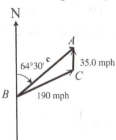

By alternate interior angles, angle $A = 64°30'$.
Use the law of sines to find B.

$$\frac{\sin B}{35.0} = \frac{\sin A}{190.0} \Rightarrow$$

$$\sin B = \frac{35.0 \sin 64°30'}{190.0} \approx 0.16626571 \Rightarrow$$

$$B \approx 9.57° \approx 9°30'$$

Thus, the bearing is
$64°30' + B = 64°30' + 9°30' = 74°00'$.
Because $C = 180° - A - B$
$= 180° - 64.50° - 9.57° = 105.93°$, we use the
law of sines to find the ground speed.

$$\frac{|\mathbf{c}|}{\sin C} = \frac{35.0}{\sin B} \Rightarrow$$

$$|\mathbf{c}| = \frac{35.0 \sin 105.93°}{\sin 9.57°} \approx 202 \text{ mph}$$

The bearing is $74°00'$; the ground speed is
202 mph.

57. Let **v** = the airspeed vector

The ground speed is $\dfrac{400 \text{ mi}}{2.5 \text{ hr}} = 160$ mph.

angle $BAC = 328° - 180° = 148°$
Using the law of cosines to find $|\mathbf{v}|$, we have

$$|\mathbf{v}|^2 = 11^2 + 160^2 - 2(11)(160)\cos 148°$$

$$|\mathbf{v}|^2 \approx 28,706.1 \Rightarrow |\mathbf{v}| \approx 169.4$$

The airspeed must be 170 mph. (rounded to
two significant digits)

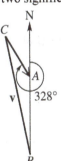

Use the law of sines to find B.

$$\frac{\sin B}{11} = \frac{\sin 148°}{169.4} \Rightarrow \sin B = \frac{11 \sin 148°}{169.4} \Rightarrow$$

$$\sin B \approx 0.03441034 \Rightarrow B \approx 2.0°$$

The bearing must be approximately
$360° - 2.0° = 358°$.

59. Find the ground speed and resulting bearing.
Angle $A = 245° - 174° = 71°$
Use the law of cosines to find $|\mathbf{v}|$.

$$|\mathbf{v}|^2 = 30^2 + 240^2 - 2(30)(240)\cos 71°$$

$$|\mathbf{v}|^2 \approx 53,811.8 \Rightarrow |\mathbf{v}| \approx 232.1$$

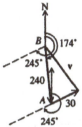

The ground speed is 230 km per hr. (rounded
to two significant digits)
Use the law of sines to find angle B.

$$\frac{\sin B}{30} = \frac{\sin 71°}{230} \Rightarrow$$

$$\sin B = \frac{30 \sin 71°}{230} \approx 0.12332851 \Rightarrow B \approx 7°$$

Thus, the bearing is
$174° - b = 174° - 7° = 167°$.

Section 8.4   Algebraically Defined
Vectors and the Dot
Product

1. The magnitude of vector **u** is <u>2</u>.

3. The horizontal component, a, of vector **v** is
$\dfrac{\sqrt{2}}{2}$.

5. The sum of the vectors $\mathbf{u} = \langle -3, 5 \rangle$ and
$\mathbf{v} = \langle 7, 4 \rangle$ is $\langle 4, 9 \rangle$.

7. The formula for the dot product of the two
vectors $\mathbf{u} = \langle a, b \rangle$ and $\mathbf{v} = \langle c, d \rangle$ is
$\mathbf{u} \cdot \mathbf{v} = \underline{ac + bd}$.

9. Magnitude:

$$\sqrt{15^2 + (-8)^2} = \sqrt{225 + 64} = \sqrt{289} = 17$$

Angle:

$$\tan\theta' = \frac{b}{a} \Rightarrow \tan\theta' = \frac{-8}{15} \Rightarrow$$

$$\theta' = \tan^{-1}\left(-\frac{8}{15}\right) \approx -28.1° \Rightarrow$$

$$\theta = -28.1° + 360° = 331.9°$$

(θ lies in quadrant IV)

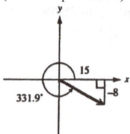

11. Magnitude:

$$\sqrt{(-4)^2 + \left(4\sqrt{3}\right)^2} = \sqrt{16 + 48} = \sqrt{64} = 8$$

Angle:

$$\tan\theta' = \frac{b}{a} \Rightarrow \tan\theta' = \frac{4\sqrt{3}}{-4} \Rightarrow$$

$$\theta' = \tan^{-1}\left(-\sqrt{3}\right) = -60° \Rightarrow$$

$$\theta = -60° + 180° = 120°$$

(θ lies in quadrant II)

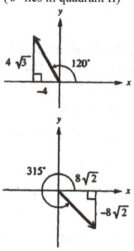

In Exercises 13–17, **x** is the horizontal component of **v**, and **y** is the vertical component of **v**. Thus, $|\mathbf{x}|$ is the magnitude of **x** and $|\mathbf{y}|$ is the magnitude of **y**.

13. $\theta = 20°$, $|\mathbf{v}| = 50$

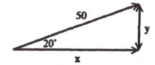

$$x = 50\cos 20° \approx 47$$
$$y = 50\sin 20° \approx 17$$

15. $\theta = 35°50'$, $|\mathbf{v}| = 47.8$

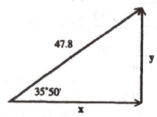

$$x = 47.8\cos 35°50' \approx 38.8$$
$$y = 47.8\sin 35°50' \approx 28.0$$

17. $\theta = 128.5°$, $|\mathbf{v}| = 198$

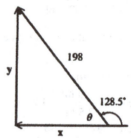

$$x = 198\cos 128.5° \approx -123$$
$$y = 198\sin 128.5° \approx 155$$

19. $\mathbf{u} = \langle a, b \rangle = \langle 5\cos(30°), 5\sin(30°) \rangle$

$$= \left\langle \frac{5\sqrt{3}}{2}, \frac{5}{2} \right\rangle$$

21. $\mathbf{v} = \langle a, b \rangle = \langle 4\cos(410°), 4\sin(140°) \rangle$

$$\approx \langle -3.0642, 2.5712 \rangle$$

23. $\mathbf{v} = \langle a, b \rangle = \langle 5\cos(-35°), 5\sin(-35°) \rangle$

$$\approx \langle 4.0958, -2.8679 \rangle$$

25. From the figure, $\mathbf{u} = \langle -8, 8 \rangle$ and $\mathbf{v} = \langle 4, 8 \rangle$.

(a) $\mathbf{u} + \mathbf{v} = \langle -8, 8 \rangle + \langle 4, 8 \rangle$

$$= \langle -8 + 4, 8 + 8 \rangle = \langle -4, 16 \rangle$$

(b) $\mathbf{u} - \mathbf{v} = \langle -8,8 \rangle - \langle 4,8 \rangle$
$= \langle -8-4, 8-8 \rangle = \langle -12,0 \rangle$

(c) $-\mathbf{u} = -\langle -8,8 \rangle = \langle 8,-8 \rangle$

27. From the figure, $\mathbf{u} = \langle 4,8 \rangle$ and $\mathbf{v} = \langle 4,-8 \rangle$.

(a) $\mathbf{u} + \mathbf{v} = \langle 4,8 \rangle + \langle 4,-8 \rangle$
$= \langle 4+4, 8-8 \rangle = \langle 8,0 \rangle$

(b) $\mathbf{u} - \mathbf{v} = \langle 4,8 \rangle - \langle 4,-8 \rangle$
$= \langle 4-4, 8-(-8) \rangle = \langle 0,16 \rangle$

(c) $-\mathbf{u} = -\langle 4,8 \rangle = \langle -4,-8 \rangle$

29. From the figure, $\mathbf{u} = \langle -8,4 \rangle$ and $\mathbf{v} = \langle 8,8 \rangle$.

(a) $\mathbf{u} + \mathbf{v} = \langle -8,4 \rangle + \langle 8,8 \rangle$
$= \langle -8+8, 4+8 \rangle = \langle 0,12 \rangle$

(b) $\mathbf{u} - \mathbf{v} = \langle -8,4 \rangle - \langle 8,8 \rangle$
$= \langle -8-8, 4-8 \rangle = \langle -16,-4 \rangle$

(c) $-\mathbf{u} = -\langle -8,4 \rangle = \langle 8,-4 \rangle$

31. (a) $2\mathbf{u} = 2(2\mathbf{i}) = 4\mathbf{i}$

(b) $2\mathbf{u} + 3\mathbf{v} = 2(2\mathbf{i}) + 3(\mathbf{i} + \mathbf{j})$
$= 4\mathbf{i} + 3\mathbf{i} + 3\mathbf{j} = 7\mathbf{i} + 3\mathbf{j}$

(c) $\mathbf{v} - 3\mathbf{u} = \mathbf{i} + \mathbf{j} - 3(2\mathbf{i}) = \mathbf{i} + \mathbf{j} - 6\mathbf{i} = -5\mathbf{i} + \mathbf{j}$

33. (a) $2\mathbf{u} = 2\langle -1,2 \rangle = \langle -2,4 \rangle$

(b) $2\mathbf{u} + 3\mathbf{v} = 2\langle -1,2 \rangle + 3\langle 3,0 \rangle$
$= \langle -2,4 \rangle + \langle 9,0 \rangle$
$= \langle -2+9, 4+0 \rangle = \langle 7,4 \rangle$

(c) $\mathbf{v} - 3\mathbf{u} = \langle 3,0 \rangle - 3\langle -1,2 \rangle = \langle 3,0 \rangle - \langle -3,6 \rangle$
$= \langle 3-(-3), 0-6 \rangle = \langle 6,-6 \rangle$

For Exercises 35–41, $\mathbf{u} = \langle -2, 5 \rangle$ and $\mathbf{v} = \langle 4, 3 \rangle$.

35. $\mathbf{u} - \mathbf{v} = \langle -2, 5 \rangle - \langle 4, 3 \rangle$
$= \langle -2-4, 5-3 \rangle = \langle -6, 2 \rangle$

37. $-4\mathbf{u} = -4\langle -2,5 \rangle = \langle -4(-2), -4(5) \rangle = \langle 8,-20 \rangle$

39. $3\mathbf{u} - 6\mathbf{v} = 3\langle -2,5 \rangle - 6\langle 4,3 \rangle$
$= \langle -6,15 \rangle - \langle 24,18 \rangle$
$= \langle -6-24, 15-18 \rangle = \langle -30,-3 \rangle$

41. $\mathbf{u} + \mathbf{v} - 3\mathbf{u} = \langle -2,5 \rangle + \langle 4,3 \rangle - 3\langle -2,5 \rangle$
$= \langle -2,5 \rangle + \langle 4,3 \rangle - \langle 3(-2), 3(5) \rangle$
$= \langle -2,5 \rangle + \langle 4,3 \rangle - \langle -6,15 \rangle$
$= \langle -2+4, 5+3 \rangle - \langle -6,15 \rangle$
$= \langle 2,8 \rangle - \langle -6,15 \rangle$
$= \langle 2-(-6), 8-15 \rangle = \langle 8,-7 \rangle$

43. $\langle -5, 8 \rangle = -5\mathbf{i} + 8\mathbf{j}$

45. $\langle 2, 0 \rangle = 2\mathbf{i} + 0\mathbf{j} = 2\mathbf{i}$

47. $\langle 6, -1 \rangle \cdot \langle 2, 5 \rangle = 6(2) + (-1)(5) = 12 - 5 = 7$

49. $\langle 5, 2 \rangle \cdot \langle -4, 10 \rangle = 5(-4) + 2(10) = -20 + 20 = 0$

51. $4\mathbf{i} = \langle 4,0 \rangle; 5\mathbf{i} - 9\mathbf{j} = \langle 5,-9 \rangle$
$\langle 4,0 \rangle \cdot \langle 5,-9 \rangle = 4(5) + 0(-9) = 20 - 0 = 20$

53. $\langle 2, 1 \rangle \cdot \langle -3, 1 \rangle$

$\cos\theta = \dfrac{\langle 2, 1 \rangle \cdot \langle -3, 1 \rangle}{\sqrt{2^2 + 1^2} \cdot \sqrt{(-3)^2 + 1^2}} = \dfrac{-6+1}{\sqrt{5} \cdot \sqrt{10}}$

$= \dfrac{-5}{5\sqrt{2}} = \dfrac{-1}{\sqrt{2}} = -\dfrac{\sqrt{2}}{2} \Rightarrow \theta = 135°$

55. $\langle 1, 2 \rangle \cdot \langle -6, 3 \rangle$

$\cos\theta = \dfrac{\langle 1, 2 \rangle \cdot \langle -6, 3 \rangle}{\sqrt{1^2 + 2^2} \cdot \sqrt{(-6)^2 + 3^2}} = \dfrac{-6+6}{\sqrt{5}\sqrt{45}}$

$= \dfrac{0}{15} = 0 \Rightarrow \theta = 90°$

57. First write the given vectors in component form: $3\mathbf{i} + 4\mathbf{j} = \langle 3,4 \rangle$ and $\mathbf{j} = \langle 0,1 \rangle$

$\cos\theta = \dfrac{\langle 3, 4 \rangle \cdot \langle 0, 1 \rangle}{|\langle 3, 4 \rangle||\langle 0, 1 \rangle|} = \dfrac{\langle 3, 4 \rangle \cdot \langle 0, 1 \rangle}{\sqrt{3^2 + 4^2} \cdot \sqrt{0^2 + 1^2}}$

$= \dfrac{0+4}{\sqrt{25} \cdot \sqrt{1}} = \dfrac{4}{5 \cdot 1} = \dfrac{4}{5} = 0.8 \Rightarrow$

$\theta = \cos^{-1} 0.8 \approx 36.87°$

For Exercises 59–61, $\mathbf{u} = \langle -2,1 \rangle$, $\mathbf{v} = \langle 3,4 \rangle$, and $\mathbf{w} = \langle -5, 12 \rangle$.

59. $(3\mathbf{u}) \cdot \mathbf{v} = (3\langle -2, 1 \rangle) \cdot \langle 3, 4 \rangle$
$= \langle -6, 3 \rangle \cdot \langle 3, 4 \rangle = -18 + 12 = -6$

61. $\mathbf{u} \cdot \mathbf{v} - \mathbf{u} \cdot \mathbf{w} = \langle -2, 1 \rangle \cdot \langle 3,4 \rangle - \langle -2,1 \rangle \cdot \langle -5, 12 \rangle$
$= (-6+4) - (10+12)$
$= -2 - 22 = -24$

63. $\langle 1,2 \rangle \cdot \langle -6,3 \rangle = -6+6 = 0,$ so the vectors are orthogonal.

65. $\langle 1,0 \rangle \cdot \langle \sqrt{2},0 \rangle = \sqrt{2}+0 = \sqrt{2} \neq 0,$ so the vectors are not orthogonal

67. $\sqrt{5}\mathbf{i}-2\mathbf{j} = \langle \sqrt{5},-2 \rangle; -5\mathbf{i}+2\sqrt{5}\mathbf{j} = \langle -5,2\sqrt{5} \rangle$

$\langle \sqrt{5},-2 \rangle \cdot \langle -5,2\sqrt{5} \rangle = -5\sqrt{5}-4\sqrt{5}$

$= -9\sqrt{5} \neq 0,$ so the vectors are not orthogonal.

69. $\mathbf{R} = \mathbf{i}-2\mathbf{j}$ and $\mathbf{A} = 0.5\mathbf{i}+\mathbf{j}$

(a) Write the given vector in component form. $\mathbf{R} = \mathbf{i}-2\mathbf{j} = \langle 1,-2 \rangle$ and

$\mathbf{A} = 0.5\mathbf{i}+\mathbf{j} = \langle 0.5,1 \rangle$

$|\mathbf{R}| = \sqrt{1^2+(-2)^2} = \sqrt{1+4} = \sqrt{5} \approx 2.2$

and $|\mathbf{A}| = \sqrt{0.5^2+1^2} = \sqrt{0.25+1} \approx 1.1$

About 2.2 in. of rain fell. The area of the opening of the rain gauge is about 1.1 in.^{2}.

(b) $V = |\mathbf{R} \cdot \mathbf{A}| = |\langle 1,-2 \rangle \cdot \langle 0.5,1 \rangle|$

$= |0.5+(-2)| = |-1.5| = 1.5$

The volume of rain was 1.5 in.^{3}.

71. Draw a line parallel to the x-axis and the vector $\mathbf{u}+\mathbf{v}$ (shown as a dashed line) Because $\theta_1 = 110°,$ its supplementary angle is $70°$. Further, because $\theta_2 = 260°,$ the angle α is $260°-180° = 80°$. Then the angle CBA becomes $180-(80+70) = 180-150 = 30°$.

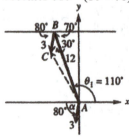

Using the law of cosines, the magnitude of $\mathbf{u}+\mathbf{v}$ is found as follows:

$|\mathbf{u}+\mathbf{v}|^2 = a^2+c^2-2ac\cos B$

$|\mathbf{u}+\mathbf{v}|^2 = 3^2+12^2-2(3)(12)\cos 30°$

$= 9+144-72 \cdot \frac{\sqrt{3}}{2} = 153-36\sqrt{3}$

≈ 90.646171

Thus, $|\mathbf{u}+\mathbf{v}| \approx 9.5208.$

Using the law of sines, we have

$\dfrac{\sin A}{a} = \dfrac{\sin B}{b} \Rightarrow \dfrac{\sin A}{3} = \dfrac{\sin 30°}{9.5208} \Rightarrow$

$\sin A = \dfrac{3\sin 30°}{9.5208} = \dfrac{3 \cdot \frac{1}{2}}{9.5208} \approx 0.15754979 \Rightarrow$

$A \approx 9.0647°$

The direction angle of $\mathbf{u}+\mathbf{v}$ is $110°+9.0647° = 119.0647°.$

73. $c = 3\cos 260° \approx -0.52094453$ and $d = 3\sin 260° \approx -2.95442326,$ so $\langle c,d \rangle \approx \langle -0.5209,-2.9544 \rangle.$

75. Magnitude:

$\sqrt{(-4.62518625)^2+8.32188819^2} \approx 9.5208$

Angle:

$\tan \theta' = \dfrac{8.32188819}{-4.625186258} \Rightarrow \theta' \approx -60.9353° \Rightarrow$

$\theta = -60.9353°+180° = 119.0647°$

(θ lies in quadrant II)

Summary Exercises on Applications of Trigonometry and Vectors

1. Consider the diagram below.

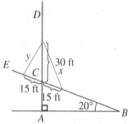

If we extend the flagpole, a right triangle CAB is formed. Thus, the measure of angle BCA is $90°-20° = 70°$. Angles DCB and BCA are supplementary, so Sthe measure of angle DCB is $180°-70° = 110°$. We can now use the law of cosines to find the measure of the support wire on the right, x.

$x^2 = 30^2+15^2-2(30)(15)\cos 110°$

$\approx 1432.818 \Rightarrow x \approx 37.85 \approx 38$ ft

Now, to find the length of the support wire on the left, we have different ways to find it. One way would be to use the approximation for x and use the law of cosines. To avoid using the approximate value, we will find y with the same method as for x. Angles DCB and DCE are supplementary, so the measure of angle DCE is $180°-70° = 110°$.

(continued on next page)

(*continued*)

We can now use the law of cosines to find the measure of the support wire on the left, y.

$$y^2 = 30^2 + 15^2 - 2(30)(15)\cos 70°$$
$$\approx 817.182 \Rightarrow x \approx 28.59 \approx 29 \text{ ft}$$

The lengths of the two wires are about 29 ft and 38 ft.

3. Let c be the distance between the two lighthouses. Refer to the figure on the next page.

   Angles DAC and CAB form a line, so angle CAB is the supplement of angle DAC. Thus, the measure of angle CAB is the following.
   $$180° - 129°43' = 179°60' - 129°43' = 50°17'$$
   The angles of a triangle must add up to $180°$, so the measure of angle ACB is
   $$180° - 39°43' - 50°17' = 180° - 90° = 90°$$

   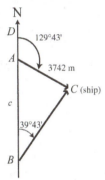

   Thus, we have the following.
   $$\cos 50°17' = \frac{3742}{c} \Rightarrow c = \frac{3742}{\cos 50°17'} \approx 5856$$
   The two lighthouses are 5856 m apart.

5. Let **x** be the horizontal force.
   $$\tan 40° = \frac{|\mathbf{x}|}{50} \Rightarrow |\mathbf{x}| = 50\tan 40° \approx 42 \text{ lb}$$

7. $\mathbf{v} = 6\mathbf{i} + 8\mathbf{j} = \langle 6, 8 \rangle$

   (a) The speed of the wind is
   $$|\mathbf{v}| = \sqrt{6^2 + 8^2} = \sqrt{36 + 64}$$
   $$= \sqrt{100} = 10 \text{ mph}$$

   (b) $3\mathbf{v} = \langle 3 \cdot 6, 3 \cdot 8 \rangle = \langle 18, 24 \rangle = 18\mathbf{i} + 24\mathbf{j}$;
   $$|3\mathbf{v}| = \sqrt{18^2 + 24^2} = \sqrt{324 + 576} = 30$$
   This represents a 30 mph wind in the direction of **v**.

   (c) **u** represents a southeast wind of
   $$|\mathbf{u}| = \sqrt{(-8)^2 + 8^2} = \sqrt{64 + 64}$$
   $$= \sqrt{128} = 8\sqrt{2} \approx 11.3 \text{ mph}$$

9.

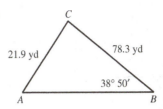

Using the law of sines, we have
$$\frac{\sin A}{BC} = \frac{\sin B}{AC} \Rightarrow \frac{\sin A}{78.3} = \frac{\sin 38°50'}{21.9} \Rightarrow$$
$$\sin A = \frac{78.3\sin 38°50'}{21.9} \approx 2.24$$

Because $-1 \le \sin A \le 1$, the triangle cannot exist.

Section 8.5 Trigonometric (Polar) Form of Complex Numbers; Products and Quotients

1. (a) 2

   (b) $2(\cos 0° + i\sin 0°)$

3. (a) $2i$

   (b) $2(\cos 90° + i\sin 90°)$

5. (a) $2 + 2i$

   (b) $2\sqrt{2}(\cos 45° + i\sin 45°)$

7. When multiplying two complex numbers in trigonometric form, we <u>multiply</u> their absolute values and <u>add</u> their arguments.

9. $\left[5(\cos 150° + i\sin 150°)\right] \cdot$
   $$\left[2(\cos 30° + i\sin 30°)\right]$$
   $$= \underline{10}\left[\cos \underline{180°} + i\sin \underline{180°}\right] = \underline{-10} + \underline{0}i$$

11. $\text{cis}\,(-1000°) \cdot \text{cis}\,1000° = \text{cis}\,\underline{0°} = \underline{1} + \underline{0}i$

13.

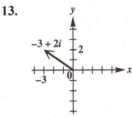

15.

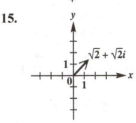

17.

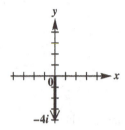

19.

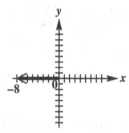

21. $(4-3i)+(-1+2i)=3-i$

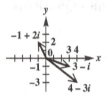

23. $(5-6i)+(-5+3i)=-3i$

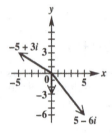

25. $-3+3i$

27. $(-5-8i)-1=-6-8i$

29. $(7+6i)+3i=7+9i$

31. $\left(\dfrac{1}{2}+\dfrac{2}{3}i\right)+\left(\dfrac{2}{3}+\dfrac{1}{2}i\right)=\dfrac{7}{6}+\dfrac{7}{6}i$

33. $2(\cos 45°+i\sin 45°)=2\left(\dfrac{\sqrt{2}}{2}+i\dfrac{\sqrt{2}}{2}\right)$
$$=\sqrt{2}+i\sqrt{2}$$

35. $10(\cos 90°+i\sin 90°)=10(0+i)$
$$=0+10i=10i$$

37. $4(\cos 240°+i\sin 240°)=4\left(-\dfrac{1}{2}-i\dfrac{\sqrt{3}}{2}\right)$
$$=-2-2i\sqrt{3}$$

39. $3\operatorname{cis}150°=3(\cos 150°+i\sin 150°)$
$$=3\left(-\dfrac{\sqrt{3}}{2}+\dfrac{1}{2}i\right)=-\dfrac{3\sqrt{3}}{2}+\dfrac{3}{2}i$$

41. $5\operatorname{cis}300°=5(\cos 300°+i\sin 300°)$
$$=5\left[\dfrac{1}{2}+\left(-\dfrac{\sqrt{3}}{2}\right)i\right]=\dfrac{5}{2}-\dfrac{5\sqrt{3}}{2}i$$

43. $\sqrt{2}\operatorname{cis}225°=\sqrt{2}(\cos 225°+i\sin 225°)$
$$=\sqrt{2}\left[-\dfrac{\sqrt{2}}{2}+\left(-\dfrac{\sqrt{2}}{2}i\right)\right]$$
$$=-1-i$$

45. $4(\cos(-30°)+i\sin(-30°))$
$$=4(\cos 30°-i\sin 30°)=4\left(\dfrac{\sqrt{3}}{2}-\dfrac{1}{2}i\right)$$
$$=2\sqrt{3}-2i$$

47. $-3-3i\sqrt{3}$

Sketch a graph of $-3-3i\sqrt{3}$ in the complex plane.

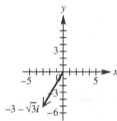

$x=-3$ and $y=-3\sqrt{3}$,
$$r=\sqrt{(-3)^2+(-3\sqrt{3})^2}=\sqrt{9+27}=\sqrt{36}=6$$

and $\tan\theta=\dfrac{-3\sqrt{3}}{-3}=\sqrt{3}$. Thus, the reference angle for θ is 60°. The graph shows that θ is in quadrant III, so $\theta=180°+60°=240°$. Therefore,
$$-3-3i\sqrt{3}=6(\cos 240°+i\sin 240°)$$

49. $\sqrt{3} - i$

Sketch a graph of $\sqrt{3} - i$ in the complex plane.

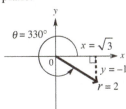

$x = \sqrt{3}$ and $y = -1$, so

$$r = \sqrt{\left(\sqrt{3}\right)^2 + \left(-1\right)^2} = \sqrt{3+1} = \sqrt{4} = 2.$$

$\tan \theta = \dfrac{-1}{\sqrt{3}} = -\dfrac{\sqrt{3}}{3}$, so the reference angle for θ is 30°. The graph shows that θ is in quadrant IV, so $\theta = 360° - 30° = 330°$.

Therefore, $\sqrt{3} - i = 2\left(\cos 330° + i \sin 330°\right)$.

51. $-5 - 5i$

Sketch a graph of $-5 - 5i$ in the complex plane.

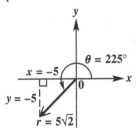

$x = -5$ and $y = -5$, so

$$r = \sqrt{\left(-5\right)^2 + \left(-5\right)^2} = \sqrt{25+25} = \sqrt{50} = 5\sqrt{2}.$$

$\tan \theta = \dfrac{y}{x} = \dfrac{-5}{-5} = 1$, so the reference angle for θ is 45°. The graph shows that θ is in quadrant III, so $\theta = 180° + 45° = 225°$.

Therefore,

$$-5 - 5i = 5\sqrt{2}\left(\cos 225° + i \sin 225°\right).$$

53. $2 + 2i$

Sketch a graph of $2 + 2i$ in the complex plane.

$x = 2, y = 2$, so

$$r = \sqrt{2^2 + 2^2} = \sqrt{4+4} = \sqrt{8} = 2\sqrt{2} \text{ and}$$

$\tan \theta = \dfrac{2}{2} = 1$. Thus, the reference angle for θ is 45°. The graph shows that θ is in quadrant I, so $\theta = 45°$. Therefore,

$$2 + 2i = 2\sqrt{2}\left(\cos 45° + i \sin 45°\right).$$

55. $5i = 0 + 5i$

$0 + 5i$ is on the positive y-axis, so $\theta = 90°$ and $x = 0, y = 5 \Rightarrow r = \sqrt{0^2 + 5^2} = 5$

Thus, $5i = 5\left(\cos 90° + i \sin 90°\right)$.

57. $-4 = -4 + 0i$

$-4 + 0i$ is on the negative x-axis, so $\theta = 180°$ and $x = -4, y = 0$. $r = \sqrt{\left(-4\right)^2 + 0^2} = \sqrt{16} = 4$

Thus, $-4 = 4\left(\cos 180° + i \sin 180°\right)$.

59. $2 + 3i$

$x = 2, y = 3 \Rightarrow r = \sqrt{2^2 + 3^2} = \sqrt{4+9} = \sqrt{13}$

$\tan \theta = \dfrac{3}{2}$

$2 + 3i$ is in quadrant I, so $\theta = 56.31°$.

$2 + 3i = \sqrt{13}\left(\cos 56.31° + i \sin 56.31°\right)$

61. $3\left(\cos 250° + i \sin 250°\right) = -1.0261 - 2.8191i$

63. $12i = 0 + 12i$

$x = 0, y = 12 \Rightarrow r = \sqrt{0^2 + 12^2} = 12$

$0 + 12i$ is on the positive y-axis, so $0 = 90°$.

$12i = 12(\cos 90° + i \sin 90°)$

65. $3 + 5i$

$x = 3, y = 5 \Rightarrow r = \sqrt{3^2 + 5^2} = \sqrt{9+25} = \sqrt{34}$

$\tan \theta = \dfrac{5}{3} \Rightarrow \theta = \tan^{-1}\left(\dfrac{5}{3}\right)$

$3 + 5i$ is in quadrant I, so $\theta = 59.04°$.

$3 + 5i = \sqrt{34}\left(\cos 59.04° + i \sin 59.04°\right)$

67. The modulus represents the magnitude of the vector in the complex plane, so $z = 1$ represents a circle of radius one centered at the origin.

69. The real part of $z = x + yi$ is 1, the graph of $1 + yi$ is the vertical line $x = 1$.

71. $z = -0.2i$

$z^2 - 1 = (-0.2i)^2 - 1 = 0.04i^2 - 1 = 0.04(-1) - 1 = -0.04 - 1 = -1.04$

The modulus is 1.04.

$(z^2 - 1)^2 - 1 = (-1.04)^2 - 1 = 0.0816$

The modulus is 0.0816.

$\left[\left(z^2 - 1\right)^2 - 1\right]^2 - 1 = (0.0816)^2 - 1 = -0.99334144$

The modulus is 0.99334144. The moduli do not exceed 2. Therefore, z is in the Julia set.

73. $\left[3(\cos 60° + i\sin 60°)\right]\left[2(\cos 90° + i\sin 90°)\right] = 3\cdot 2\left[\cos(60° + 90°) + i\sin(60° + 90°)\right]$

$$= 6(\cos 150° + i\sin 150°) = 6\left(-\frac{\sqrt{3}}{2} + \frac{1}{2}i\right) = -3\sqrt{3} + 3i$$

75. $\left[4(\cos 60° + i\sin 60°)\right]\cdot\left[6(\cos 330° + i\sin 330°)\right] = 4\cdot 6\left[(\cos(60° + 330°) + i\sin(60° + 330°)\right]$

$$= 24(\cos 390° + i\sin 390°) = 24(\cos 30° + i\sin 30°)$$

$$= 24\left(\frac{\sqrt{3}}{2} + i\frac{1}{2}\right) = 12\sqrt{3} + 12i$$

77. $\left[2(\cos 135° + i\sin 135°)\right]\cdot\left[2(\cos 225° + i\sin 225°)\right] = 2\cdot 2\left[\cos(135° + 225°) + i\sin(135° + 225°)\right]$

$$= 4(\cos 360° + i\sin 360°) = 4(1 - 0i) = 4$$

79. $\left[\sqrt{3}\operatorname{cis} 45°\right]\left[\sqrt{3}\operatorname{cis} 225°\right] = \sqrt{3}\cdot\sqrt{3}\left[\operatorname{cis}(45° + 225°)\right]$

$$= 3\operatorname{cis} 270° = 3(\cos 270° + i\sin 270°) = 3(0 - i) = 0 - 3i \text{ or } -3i$$

81. $[5\operatorname{cis} 90°][3\operatorname{cis} 45°] = 5\cdot 3\left[\operatorname{cis}(90° + 45°)\right] = 15\operatorname{cis} 135°$

$$= 15(\cos 135° + i\sin 135°) = 15\left(-\frac{\sqrt{2}}{2} + \frac{\sqrt{2}}{2}i\right) = -\frac{15\sqrt{2}}{2} + \frac{15\sqrt{2}}{2}i$$

83. $\dfrac{4(\cos 150° + i\sin 150°)}{2(\cos 120° + i\sin 120°)} = \dfrac{4}{2}\left[\cos(150° - 120°) + i\sin(150° - 120°)\right] = 2(\cos 30° + i\sin 30°) = 2\left(\frac{\sqrt{3}}{2} + \frac{1}{2}i\right)$

$$= \sqrt{3} + i$$

85. $\dfrac{10(\cos 50° + i\sin 50°)}{5(\cos 230° + i\sin 230°)} = \dfrac{10}{5}\left[\cos(50° - 230°) + i\sin(50° - 230°)\right]$

$$= 2(\cos(-180°) + i\sin(-180°)) = 2(-1 - 0\cdot i) = -2 + 0i \text{ or } -2$$

87. $\dfrac{3\operatorname{cis} 305°}{9\operatorname{cis} 65°} = \dfrac{1}{3}\operatorname{cis}(305° - 65°) = \dfrac{1}{3}(\operatorname{cis} 240°) = \dfrac{1}{3}(\cos 240° + i\sin 240°) = \dfrac{1}{3}\left(-\frac{1}{2} - \frac{\sqrt{3}}{2}i\right) = -\frac{1}{6} - \frac{\sqrt{3}}{6}i$

89. $\dfrac{8}{\sqrt{3}+i}$

numerator: $8 = 8 + 0i$ and $r = \sqrt{8^2 + 0^2} = 8$

$\theta = 0°$ because $\cos 0° = 1$ and $\sin 0° = 0$, so $8 = 8 \operatorname{cis} 0°$.

denominator: $\sqrt{3} + i$ and

$r = \sqrt{\left(\sqrt{3}\right)^2 + 1^2} = \sqrt{3+1} = \sqrt{4} = 2$

$\tan \theta = \dfrac{1}{\sqrt{3}} = \dfrac{\sqrt{3}}{3}$

Because x and y are both positive, θ is in quadrant I, so $\theta = 30°$.

Thus $\sqrt{3} + i = 2 \operatorname{cis} 30°$.

$\dfrac{8}{\sqrt{3}+i} = \dfrac{8 \operatorname{cis} 0°}{2 \operatorname{cis} 30°} = \dfrac{8}{2} \operatorname{cis}\left(0 - 30°\right)$

$= 4\left[\cos\left(-30°\right) + i \sin\left(-30°\right)\right]$

$= 4\left(\dfrac{\sqrt{3}}{2} - \dfrac{1}{2} i\right) = 2\sqrt{3} - 2i$

91. $\dfrac{-i}{1+i}$

numerator: $-i = 0 - i$ and

$r = \sqrt{0^2 + (-1)^2} = \sqrt{0+1} = \sqrt{1} = 1$

$\theta = 270°$ because $\cos 270° = 0$ and $\sin 270° = -1$, so $-i = 1 \operatorname{cis} 270°$.

denominator: $1 + i$

$r = \sqrt{1^2 + 1^2} = \sqrt{1+1} = \sqrt{2}$ and

$\tan \theta = \dfrac{y}{x} = \dfrac{1}{1} = 1$

Because x and y are both positive, θ is in quadrant I, so $\theta = 45°$. Thus,

$1 + i = \sqrt{2} \operatorname{cis} 45°$

$\dfrac{-i}{1+i} = \dfrac{\operatorname{cis} 270°}{\sqrt{2} \operatorname{cis} 45°} = \dfrac{1}{\sqrt{2}} \operatorname{cis}\left(270° - 45°\right)$

$= \dfrac{\sqrt{2}}{2} \operatorname{cis} 225°$

$= \dfrac{\sqrt{2}}{2}\left(\cos 225° + i \sin 225°\right)$

$= \dfrac{\sqrt{2}}{2}\left(-\dfrac{\sqrt{2}}{2} - i \cdot \dfrac{\sqrt{2}}{2}\right) = -\dfrac{1}{2} - \dfrac{1}{2} i$

93. $\dfrac{2\sqrt{6} - 2i\sqrt{2}}{\sqrt{2} - i\sqrt{6}}$

numerator: $2\sqrt{6} - 2i\sqrt{2}$ and

$r = \sqrt{\left(2\sqrt{6}\right)^2 + \left(-2\sqrt{2}\right)^2} = \sqrt{24+8}$

$= \sqrt{32} = 4\sqrt{2}$

$\tan \theta = \dfrac{-2\sqrt{2}}{2\sqrt{6}} = -\dfrac{1}{\sqrt{3}} = -\dfrac{\sqrt{3}}{3}$

Because x is positive and y is negative, θ is in quadrant IV, so $\theta = -30°$. Thus,

$2\sqrt{6} - 2i\sqrt{2} = 4\sqrt{2} \operatorname{cis}\left(-30°\right)$.

denominator: $\sqrt{2} - i\sqrt{6}$ and

$r = \sqrt{\left(\sqrt{2}\right)^2 + \left(-\sqrt{6}\right)^2} = \sqrt{2+6} = \sqrt{8} = 2\sqrt{2}$

$\tan \theta = \dfrac{-\sqrt{6}}{\sqrt{2}} = -\sqrt{3}$

Because x is positive and y is negative, θ is in quadrant IV, so $\theta = -60°$. Thus,

$\sqrt{2} - i\sqrt{6} = 2\sqrt{2} \operatorname{cis}\left(-30°\right)$

$\dfrac{2\sqrt{6} - 2i\sqrt{2}}{\sqrt{2} - i\sqrt{6}} = \dfrac{4\sqrt{2} \operatorname{cis}\left(-30°\right)}{2\sqrt{2} \operatorname{cis}\left(-60°\right)}$

$= \dfrac{4\sqrt{2}}{2\sqrt{2}} \operatorname{cis}\left[-30° - \left(-60°\right)\right]$

$= 2 \operatorname{cis} 30° = 2\left(\cos 30° + i \sin 30°\right)$

$= 2\left(\dfrac{\sqrt{3}}{2} + i\dfrac{1}{2}\right) = \sqrt{3} + i$

95. $\left[2.5\left(\cos 35° + i \sin 35°\right)\right]\left[3.0\left(\cos 50° + i \sin 50°\right)\right]$

$= 2.5 \cdot 3.0\left[\cos\left(35° + 50°\right) + i \sin\left(35° + 50°\right)\right]$

$= 7.5\left(\cos 85° + i \sin 85°\right) \approx 0.6537 + 7.4715i$

97. $\left(12 \operatorname{cis} 18.5°\right)\left(3 \operatorname{cis} 12.5°\right)$

$= 12 \cdot 3 \operatorname{cis}\left(18.5° + 12.5°\right)$

$= 36 \operatorname{cis} 31° = 36\left(\cos 31° + i \sin 31°\right)$

$\approx 30.8580 + 18.5414i$

99. $\dfrac{45\left(\cos \dfrac{2\pi}{3} + i \sin \dfrac{2\pi}{3}\right)}{22.5\left(\cos \dfrac{3\pi}{5} + i \sin \dfrac{3\pi}{5}\right)}$

$= \dfrac{45}{22.5}\left[\cos\left(\dfrac{2\pi}{3} - \dfrac{3\pi}{5}\right) + i \sin\left(\dfrac{2\pi}{3} - \dfrac{3\pi}{5}\right)\right]$

$= 2\left(\cos \dfrac{\pi}{15} + i \sin \dfrac{\pi}{15}\right) \approx 1.9563 + 0.4158i$

101. $\left[2\operatorname{cis}\dfrac{5\pi}{9}\right]^2 = \left[2\operatorname{cis}\dfrac{5\pi}{9}\right]\left[2\operatorname{cis}\dfrac{5\pi}{9}\right]$

$= 2\cdot 2\operatorname{cis}\left(\dfrac{5\pi}{9}+\dfrac{5\pi}{9}\right)$

$= 4\left(\cos\dfrac{10\pi}{9}+i\sin\dfrac{10\pi}{9}\right)$

$\approx -3.7588 - 1.3681i$

103. To square a complex number in trigonometric form, square its absolute value and double its argument.

105. $z = r(\cos\theta + i\sin\theta)$

Because $1 = 1 + 0i = 1(\cos 0° + i\sin 0°)$,

$\dfrac{1}{z} = \dfrac{1(\cos 0° + i\sin 0°)}{r(\cos\theta + i\sin\theta)}$

$= \dfrac{1}{r}\left[\cos(0° - \theta) + i\sin(0° - \theta)\right]$

$= \dfrac{1}{r}\left[\cos(-\theta) + i\sin(-\theta)\right]$

$= \dfrac{1}{r}\left[\cos\theta - i\sin\theta\right]$

107. $E = 8(\cos 20° + i\sin 20°), R = 6, X_L = 3,$

$I = \dfrac{E}{Z}, Z = R + X_L i$

Write $Z = 6 + 3i$ in trigonometric form.

$x = 6,$ and $y = 3 \Rightarrow r = \sqrt{6^2 + 3^2} = \sqrt{36 + 9}$

$= \sqrt{45} = 3\sqrt{5}$.

$\tan\theta = \dfrac{3}{6} = \dfrac{1}{2},$ so $\theta \approx 26.6°.$ Thus,

$Z = 3\sqrt{5}\operatorname{cis} 26.6°.$

$I = \dfrac{8\operatorname{cis} 20°}{3\sqrt{5}\operatorname{cis} 26.6°} = \dfrac{8}{3\sqrt{5}}\operatorname{cis}(20° - 26.6°)$

$= \dfrac{8\sqrt{5}}{15}\operatorname{cis}(-6.6°)$

$= \dfrac{8\sqrt{5}}{15}\left[\cos(-6.6°) + i\sin(-6.6°)\right]$

$\approx 1.18 - 0.14i$

109. $Z_1 = 50 + 25i$ and $Z_2 = 60 + 20i,$ so

$\dfrac{1}{Z_1} = \dfrac{1}{50+25i}\cdot\dfrac{50-25i}{50-25i} = \dfrac{50-25i}{50^2 - 25^2 i^2}$

$= \dfrac{50-25i}{2500 - 625(-1)} = \dfrac{50-25i}{2500 + 625}$

$= \dfrac{50-25i}{3125} = \dfrac{2}{125} - \dfrac{1}{125}i$

$\dfrac{1}{Z_2} = \dfrac{1}{60+20i}\cdot\dfrac{60-20i}{60-20i} = \dfrac{60-20i}{60^2 - 20^2 i^2}$

$= \dfrac{60-20i}{3600 - 400(-1)} = \dfrac{60-20i}{3600 + 400}$

$= \dfrac{60-20i}{4000} = \dfrac{3}{200} - \dfrac{1}{200}i$

$\dfrac{1}{Z_1} + \dfrac{1}{Z_2} = \left(\dfrac{2}{125} - \dfrac{1}{125}i\right) + \left(\dfrac{3}{200} - \dfrac{1}{200}i\right)$

$= \left(\dfrac{2}{125} + \dfrac{3}{200}\right) - \left(\dfrac{1}{125} + \dfrac{1}{200}\right)i$

$= \left(\dfrac{16}{1000} + \dfrac{15}{1000}\right) - \left(\dfrac{8}{1000} + \dfrac{5}{1000}\right)i$

$= \dfrac{31}{1000} - \dfrac{13}{1000}i$

$Z = \dfrac{1}{\dfrac{1}{Z_1} + \dfrac{1}{Z_2}} = \dfrac{1}{\dfrac{31}{1000} - \dfrac{13}{1000}i}$

$= \dfrac{1000}{31 - 13i}\cdot\dfrac{31 + 13i}{31 + 13i} = \dfrac{1000(31 + 13i)}{31^2 - 13^2 i^2}$

$= \dfrac{31,000 + 13,000i}{961 - 169(-1)} = \dfrac{31,000 + 13,000i}{961 + 169}$

$= \dfrac{31,000 + 13,000i}{1130} = \dfrac{3100}{113} + \dfrac{1300}{113}i$

$\approx 27.43 + 11.50i$

In Exercises 111–117, $w = -1 + i$ and $z = -1 - i.$

111. $w\cdot z = (-1+i)(-1-i)$

$= -1(-1) + (-1)(-i) + i(-1) + i(-i)$

$= 1 + i - i - i^2 = 1 - (-1) = 2$

113. $w\cdot z = \left(\sqrt{2}\operatorname{cis} 135°\right)\left(\sqrt{2}\operatorname{cis} 225°\right)$

$= \sqrt{2}\cdot\sqrt{2}\left[\operatorname{cis}(135° + 225°)\right]$

$= 2\operatorname{cis} 360° = 2\operatorname{cis} 0°$

115. $\dfrac{w}{z} = \dfrac{-1+i}{-1-i} = \dfrac{-1+i}{-1-i}\cdot\dfrac{-1+i}{-1+i} = \dfrac{1 - i - i + i^2}{1 - i^2}$

$= \dfrac{1 - 2i + (-1)}{1 - (-1)} = \dfrac{-2i}{2} = -i$

117. $\operatorname{cis}(-90°) = \cos(-90°) + i\sin(-90°)$

$= 0 + i(-1) = 0 - i = -i$

Section 8.6 DeMoivre's Theorem; Powers and Roots of Complex Numbers

1. Given that $z = 3(\cos 30° + i\sin 30°)$, it
   follows that
   $$z^3 = \underline{27}(\cos\underline{90°} + i\sin\underline{90°})$$
   $$= \underline{27}(\underline{0} + i\cdot\underline{1})$$
   $$= \underline{0} + \underline{27}i$$
   $$= \underline{27}i$$

3. $[\cos 6° + i\sin 6°]^{30} = \cos\underline{180°} + i\sin\underline{180°}$
   $$= \underline{-1} + \underline{0}i$$

5. Two. 1 and -1

7. $[3(\cos 30° + i\sin 30°)]^3$
   $$= 3^3[\cos(3\cdot 30°) + i\sin(3\cdot 30°)]$$
   $$= 27(\cos 90° + i\sin 90°)$$
   $$= 27(0 + 1\cdot i) = 0 + 27i \text{ or } 27i$$

9. $(\cos 45° + i\sin 45°)^8$
   $$= [\cos(8\cdot 45°) + i\sin(8\cdot 45°)]$$
   $$= \cos 360° + i\sin 360° = 1 + 0\cdot i \text{ or } 1$$

11. $[3\operatorname{cis} 100°]^3$
    $$= 3^3\operatorname{cis}(3\cdot 100°)$$
    $$= 27\operatorname{cis} 300° = 27(\cos 300° + i\sin 300°)$$
    $$= 27\left(\frac{1}{2} - \frac{\sqrt{3}}{2}i\right) = \frac{27}{2} - \frac{27\sqrt{3}}{2}i$$

13. $(\sqrt{3} + i)^5$

    First write $\sqrt{3} + i$ in trigonometric form.
    $$r = \sqrt{(\sqrt{3})^2 + 1^2} = \sqrt{3+1} = \sqrt{4} = 2 \text{ and}$$
    $$\tan\theta = \frac{1}{\sqrt{3}} = \frac{\sqrt{3}}{3}$$
    Because x and y are both positive, θ is in
    quadrant I, so $\theta = 30°$.
    $$\sqrt{3} + i = 2(\cos 30° + i\sin 30°)$$
    $$(\sqrt{3}+i)^5 = [2(\cos 30° + i\sin 30°)]^5$$
    $$= 2^5[\cos(5\cdot 30°) + i\sin(5\cdot 30°)]$$
    $$= 32(\cos 150° + i\sin 150°)$$
    $$= 32\left(-\frac{\sqrt{3}}{2} + i\frac{1}{2}\right) = -16\sqrt{3} + 16i$$

15. $(2\sqrt{2} - 2i\sqrt{2})^6$

    First write $2\sqrt{2} - 2i\sqrt{2}$ in trigonometric form.
    $$r = \sqrt{(2\sqrt{2})^2 + (-2\sqrt{2})^2} = \sqrt{8+8} = \sqrt{16} = 4$$
    and $\tan\theta = \dfrac{-2\sqrt{2}}{2\sqrt{2}} = -1$

    Because x is positive and y is negative, θ is
    in quadrant IV, so $\theta = 315°$.
    $$2\sqrt{2} - 2i\sqrt{2} = 4(\cos 315° + i\sin 315°)^6$$
    $$(2\sqrt{2} - 2i\sqrt{2})^6$$
    $$= [4(\cos 315° + i\sin 315°)]^6$$
    $$= 4^6[\cos(6\cdot 315°) + i\sin(6\cdot 315°)]$$
    $$= 4096[\cos 1890° + i\sin 1890°]$$
    $$= 4096(\cos 90° + i\sin 90°)$$
    $$= 4096(0 + 1\cdot i) = 0 + 4096i \text{ or } 4096i$$

17. $(-2 - 2i)^5$

    First write $-2 - 2i$ in trigonometric form.
    $$r = \sqrt{(-2)^2 + (-2)^2} = \sqrt{4+4} = \sqrt{8} = 2\sqrt{2} \text{ and}$$
    $$\tan\theta = \frac{-2}{-2} = 1$$
    Because x and y are both negative, θ is in
    quadrant III, so $\theta = 225°$.
    $$-2 - 2i = 2\sqrt{2}(\cos 225° + i\sin 225°)$$
    $$(-2 - 2i)^5$$
    $$= [2\sqrt{2}(\cos 225° + i\sin 225°)]^5$$
    $$= (2\sqrt{2})^5[\cos(5\cdot 225°) + i\sin(5\cdot 225°)]$$
    $$= 32\sqrt{32}(\cos 1125° + i\sin 1125°)$$
    $$= 128\sqrt{2}(\cos 45° + i\sin 45°)$$
    $$= 128\sqrt{2}\left(\frac{\sqrt{2}}{2} + \frac{\sqrt{2}}{2}i\right) = 128 + 128i$$

19. (a) $\cos 0° + i \sin 0° = 1(\cos 0° + i \sin 0°)$

We have $r = 1$ and $\theta = 0°$. Because

$r^3 (\cos 3\alpha + i \sin 3\alpha) = 1(\cos 0° + i \sin 0°),$

then we have $r^3 = 1 \Rightarrow r = 1$ and

$3\alpha = 0° + 360° \cdot k \Rightarrow \alpha = \dfrac{0° + 360° \cdot k}{3}$

$= 0° + 120° \cdot k = 120° \cdot k,$ k any integer.

If $k = 0$, then $\alpha = 0°$.

If $k = 1$, then $\alpha = 120°$.

If $k = 2$, then $\alpha = 240°$.

So, the cube roots are

$\cos 0° + i \sin 0°,$ $\cos 120° + i \sin 120°,$

and $\cos 240° + i \sin 240°.$

(b)

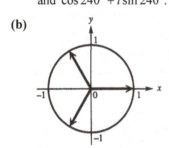

21. (a) Find the cube roots of 8 cis 60°

We have $r = 8$ and $\theta = 60°$.

Because $r^3 (\cos 3\alpha + i \sin 3\alpha)$

$= 8(\cos 60° + i \sin 60°),$ we have

$r^3 = 8 \Rightarrow r = 2$ and

$3\alpha = 60° + 360° \cdot k \Rightarrow$

$\alpha = \dfrac{60° + 360° \cdot k}{3} = 20° + 120° \cdot k,$ k any

integer.

If $k = 0$, then $\alpha = 20° + 0° = 20°$.

If $k = 1$, then $\alpha = 20° + 120° = 140°$.

If $k = 2$, then $\alpha = 20° + 240° = 260°$.

So, the cube roots are

$2 \text{ cis } 20°, 2 \text{ cis } 140°,$ and $2 \text{ cis } 260°.$

(b)

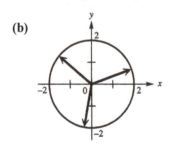

23. (a) Find the cube roots of

$-8i = 8(\cos 270° + i \sin 270°)$

We have $r = 8$ and $\theta = 270°$.

Because $r^3 (\cos 3\alpha + i \sin 3\alpha)$

$= 8(\cos 270° + i \sin 270°),$ then we have

$r^3 = 8 \Rightarrow r = 2$ and

$3\alpha = 270° + 360° \cdot k \Rightarrow$

$\alpha = \dfrac{270° + 360° \cdot k}{3} = 90° + 120° \cdot k,$ k

any integer.

If $k = 0$, then $\alpha = 90° + 0° = 90°$.

If $k = 1$, then $\alpha = 90° + 120° = 210°$.

If $k = 2$, then $\alpha = 90° + 240° = 330°$.

So, the cube roots are

$2(\cos 90° + i \sin 90°),$

$2(\cos 210° + i \sin 210°),$ and

$2(\cos 330° + i \sin 330°).$

(b)

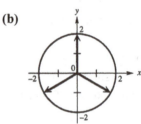

25. (a) Find the cube roots of

$-64 = 64(\cos 180° + i \sin 180°)$

We have $r = 64$ and $\theta = 180°$.

Because $r^3 (\cos 3\alpha + i \sin 3\alpha)$

$= 64(\cos 180° + i \sin 180°),$ then we have

$r^3 = 64 \Rightarrow r = 4$ and

$3\alpha = 180° + 360° \cdot k \Rightarrow$

$\alpha = \dfrac{180° + 360° \cdot k}{3} = 60° + 120° \cdot k,$ k any

integer.

If $k = 0$, then $\alpha = 60° + 0° = 60°$.

If $k = 1$, then $\alpha = 60° + 120° = 180°$.

If $k = 2$, then $\alpha = 60° + 240° = 300°$.

So, the cube roots are

$4(\cos 60° + i \sin 60°),$

$4(\cos 180° + i \sin 180°),$ and

$4(\cos 300° + i \sin 300°).$

(b)

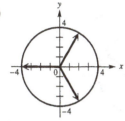

27. (a) Find the cube roots of $1 + i\sqrt{3}$.
We have
$$r = \sqrt{1^2 + \left(\sqrt{3}\right)^2} = \sqrt{1+3} = \sqrt{4} = 2 \text{ and}$$

$\tan\theta = \dfrac{\sqrt{3}}{1} = \sqrt{3}$. Because θ is in

quadrant I, $\theta = 60°$. Thus,
$$1 + i\sqrt{3} = 2\left(\frac{1}{2} + i\frac{\sqrt{3}}{2}\right)$$
$$= 2\left(\cos 60° + i\sin 60°\right).$$

Because $r^3\left(\cos 3\alpha + i\sin 3\alpha\right)$
$= 2\left(\cos 60° + i\sin 60°\right)$, we have

$r^3 = 2 \Rightarrow r = \sqrt[3]{2}$ and
$3\alpha = 60° + 360° \cdot k \Rightarrow$
$$\alpha = \frac{60° + 360° \cdot k}{3} = 20° + 120° \cdot k, \ k \text{ any}$$

integer.
If $k = 0$, then $\alpha = 20° + 0° = 20°$.
If $k = 1$, then $\alpha = 20° + 120° = 140°$.
If $k = 2$, then $\alpha = 20° + 240° = 260°$.
So, the cube roots are
$\sqrt[3]{2}\left(\cos 20° + i\sin 20°\right)$,
$\sqrt[3]{2}\left(\cos 140° + i\sin 140°\right)$, and
$\sqrt[3]{2}\left(\cos 260° + i\sin 260°\right)$.

(b)

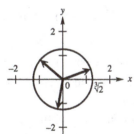

29. (a) Find the cube roots of $-2\sqrt{3} + 2i$.

We have $r = \sqrt{\left(-2\sqrt{3}\right)^2 + 2^2} = \sqrt{12 + 4}$

$= \sqrt{16} = 4$ and $\tan\theta = \dfrac{2}{-2\sqrt{3}} = -\dfrac{\sqrt{3}}{3}$.

θ is in quadrant II, so $\theta = 150°$. Thus,
$$-2\sqrt{3} + 2i = 4\left(-\frac{\sqrt{3}}{2} + \frac{1}{2}i\right)$$
$$= 4\left(\cos 150° + i\sin 150°\right).$$

Because $r^3\left(\cos 3\alpha + i\sin 3\alpha\right)$
$= 4\left(\cos 150° + i\sin 150°\right)$, we have
$r^3 = 4 \Rightarrow r = \sqrt[3]{4}$ and
$3\alpha = 150° + 360° \cdot k \Rightarrow$
$$\alpha = \frac{150° + 360° \cdot k}{3} = 50° + 120° \cdot k, \ k \text{ any}$$

integer.
If $k = 0$, then $\alpha = 50° + 0° = 50°$.
If $k = 1$, then $\alpha = 50° + 120° = 170°$.
If $k = 2$, then $\alpha = 50° + 240° = 290°$.

So, the cube roots are
$\sqrt[3]{4}\left(\cos 50° + i\sin 50°\right)$,
$\sqrt[3]{4}\left(\cos 170° + i\sin 170°\right)$, and
$\sqrt[3]{4}\left(\cos 290° + i\sin 290°\right)$.

(b)

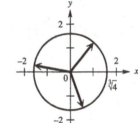

31. Find all the second (or square) roots of
$1 = 1\left(\cos 0° + i\sin 0°\right)$.

Because $r^2\left(\cos 2\alpha + i\sin 2\alpha\right)$
$= 1\left(\cos 0° + i\sin 0°\right)$, we have
$2\alpha = 0° + 360° \cdot k \Rightarrow$
$$\alpha = \frac{0° + 360° \cdot k}{2} = 0° + 180° \cdot k = 180° \cdot k, \ k$$

any integer. If $k = 0$, then $\alpha = 0°$.
If $k = 1$, then $\alpha = 180°$. So, the second roots
of 1 are
$\cos 0° + i\sin 0°$, and $\cos 180° + i\sin 180°$ (or 1
and -1)

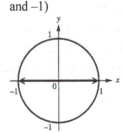

33. Find all the sixth roots of
$$1 = 1(\cos 0° + i \sin 0°).$$

Because $r^6 (\cos 6\alpha + i \sin 6\alpha)$

$= 1(\cos 0° + i \sin 0°),$ we have $r^6 = 1 \Rightarrow r = 1$

and $6\alpha = 0° + 360° \cdot k \Rightarrow$

$\alpha = \dfrac{0° + 360° \cdot k}{6} = 0° + 60° \cdot k = 60° \cdot k,\ k$ any

integer. If $k = 0$, then $\alpha = 0°$.

If $k = 1$, then $\alpha = 60°$.

If $k = 2$, then $\alpha = 120°$.

If $k = 3$, then $\alpha = 180°$.

If $k = 4$, then $\alpha = 240°$.

If $k = 5$, then $\alpha = 300°$.

So, the sixth roots of 1 are

$\cos 0° + i \sin 0°,\ \cos 60° + i \sin 60°,$

$\cos 120° + i \sin 120°,\ \cos 180° + i \sin 180°,$

$\cos 240° + i \sin 240°,$ and $\cos 300° + i \sin 300°.$

$\left(\text{or } 1,\ \dfrac{1}{2} + \dfrac{\sqrt{3}}{2} i,\ -\dfrac{1}{2} + \dfrac{\sqrt{3}}{2} i,\ -1,\right.$

$\left. -\dfrac{1}{2} - \dfrac{\sqrt{3}}{2} i,\ \text{and } \dfrac{1}{2} - \dfrac{\sqrt{3}}{2} i \right)$

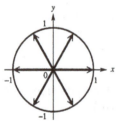

35. Find all the third (cube) roots of
$$i = 1(\cos 90° + i \sin 90°).$$

Because

$r^3 (\cos 3\alpha + i \sin 3\alpha) = 1(\cos 90° + i \sin 90°),$

we have $r^3 = 1 \Rightarrow r = 1$ and

$3\alpha = 90° + 360° \cdot k \Rightarrow$

$\alpha = \dfrac{90° + 360° \cdot k}{3} = 30° + 120° \cdot k,\ k$ any

integer. If $k = 0$, then $\alpha = 30° + 0° = 30°$.

If $k = 1$, then $\alpha = 30° + 120° = 150°$.

If $k = 2$, then $\alpha = 30° + 240° = 270°$. So, the

third roots of i are $\cos 30° + i \sin 30°,$

$\cos 150° + i \sin 150°,$ and $\cos 270° + i \sin 270°.$

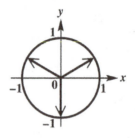

37. $x^3 - 1 = 0 \Rightarrow x^3 = 1$

We have $r = 1$ and $\theta = 0°$.

$x^3 = 1 = 1 + 0i = 1(\cos 0° + i \sin 0°)$

Because $r^3 (\cos 3\alpha + i \sin 3\alpha)$

$= 1(\cos 0° + i \sin 0°),$ we have $r^3 = 1 \Rightarrow r = 1$

and $3\alpha = 0° + 360° \cdot k \Rightarrow$

$\alpha = \dfrac{0° + 360° \cdot k}{3} = 0° + 120° \cdot k = 120° \cdot k,\ k$

any integer. If $k = 0$, then $\alpha = 0°$.

If $k = 1$, then $\alpha = 120°$.

If $k = 2$, then $\alpha = 240°$.

Solution set:

$\{\cos 0° + i \sin 0°,\ \cos 120° + i \sin 120°,$

$\quad \cos 240° + i \sin 240°\}$

or $\left\{ 1,\ -\dfrac{1}{2} + \dfrac{\sqrt{3}}{2} i,\ -\dfrac{1}{2} - \dfrac{\sqrt{3}}{2} i \right\}$

39. $x^3 + i = 0 \Rightarrow x^3 = -i$

We have $r = 1$ and $\theta = 270°$.

$x^3 = -i = 0 - i = 1(\cos 270° + i \sin 270°)$

Because $r^3 (\cos 3\alpha + i \sin 3\alpha)$

$= 1(\cos 270° + i \sin 270°),$ we have

$r^3 = 1 \Rightarrow r = 1$ and $3\alpha = 270° + 360° \cdot k \Rightarrow$

$\alpha = \dfrac{270° + 360° \cdot k}{3} = 90° + 120° \cdot k,\ k$ any

integer. If $k = 0$, then $\alpha = 90° + 0° = 90°$.

If $k = 1$, then $\alpha = 90° + 120° = 210°$.

If $k = 2$, then $\alpha = 90° + 240° = 330°$.

Solution set:

$\{\cos 90° + i \sin 90°,\ \cos 210° + i \sin 210°,$

$\quad \cos 330° + i \sin 330°\}$ or

$\left\{ i,\ -\dfrac{\sqrt{3}}{2} - \dfrac{1}{2} i,\ \dfrac{\sqrt{3}}{2} - \dfrac{1}{2} i \right\}$

41. $x^3 - 8 = 0 \Rightarrow x^3 = 8$

We have $r = 8$ and $\theta = 0°$.

$x^3 = 8 = 8 + 0i = 8(\cos 0° + i \sin 0°)$

Because $r^3(\cos 3\alpha + i \sin 3\alpha)$

$= 8(\cos 0° + i \sin 0°)$, we have $r^3 = 8 \Rightarrow r = 2$

and $3\alpha = 0° + 360° \cdot k \Rightarrow$

$\alpha = \dfrac{0° + 360° \cdot k}{3} = 0° + 120° \cdot k = 120° \cdot k,\ k$

any integer. If $k = 0$, then $\alpha = 0°$.

If $k = 1$, then $\alpha = 120°$. If $k = 2$, then $\alpha = 240°$.

Solution set:

$\{2(\cos 0° + i \sin 0°),\ 2(\cos 120° + i \sin 120°),$

$2(\cos 240° + i \sin 240°)\}$ or

$\left\{2,\ -1 + \sqrt{3}i,\ -1 - \sqrt{3}i\right\}$

43. $x^4 + 1 = 0 \Rightarrow x^4 = -1\rangle$

We have $r = 1$ and $\theta = 180°$.

$x^4 = -1 = -1 + 0i = 1(\cos 180° + i \sin 180°)$

Because $r^4(\cos 4\alpha + i \sin 4\alpha)$

$= 1(\cos 180° + i \sin 180°)$, we have

$r^4 = 1 \Rightarrow r = 1$ and $4\alpha = 180° + 360° \cdot k \Rightarrow$

$\alpha = \dfrac{180° + 360° \cdot k}{4} = 45° + 90° \cdot k,\ k$ any

integer. If $k = 0$, then $\alpha = 45° + 0° = 45°$.

If $k = 1$, then $\alpha = 45° + 90° = 135°$.

If $k = 2$, then $\alpha = 45° + 180° = 225°$.

If $k = 3$, then $\alpha = 45° + 270° = 315°$.

Solution set:

$\{\cos 45° + i \sin 45°,\ \cos 135° + i \sin 135°,$

$\cos 225° + i \sin 225°,\ \cos 315° + i \sin 315°\}$ or

$\left\{\dfrac{\sqrt{2}}{2} + \dfrac{\sqrt{2}}{2}i,\ -\dfrac{\sqrt{2}}{2} + \dfrac{\sqrt{2}}{2}i,\ -\dfrac{\sqrt{2}}{2} - \dfrac{\sqrt{2}}{2}i,\right.$

$\left.\dfrac{\sqrt{2}}{2} - \dfrac{\sqrt{2}}{2}i\right\}$

45. $x^4 - i = 0 \Rightarrow x^4 = i$

We have $r = 1$ and $\theta = 90°$.

$x^4 = i = 0 + i = 1(\cos 90° + i \sin 90°)$

Because $r^4(\cos 4\alpha + i \sin 4\alpha)$

$= 1(\cos 90° + i \sin 90°)$, we have

$r^4 = 1 \Rightarrow r = 1$ and $4\alpha = 90° + 360° \cdot k \Rightarrow$

$\alpha = \dfrac{90° + 360° \cdot k}{4} = 22.5° + 90° \cdot k,\ k$ any

integer. If $k = 0$, then $\alpha = 22.5° + 0° = 22.5°$.

If $k = 1$, then $\alpha = 22.5° + 90° = 112.5°$.

If $k = 2$, then $\alpha = 22.5° + 180° = 202.5°$.

If $k = 3$, then $\alpha = 22.5° + 270° = 292.5°$.

Solution set:

$\{\cos 22.5° + i \sin 22.5°,\ \cos 112.5° + i \sin 112.5°,$

$\cos 202.5° + i \sin 202.5°,$

$\cos 292.5° + i \sin 292.5°\}$

47. $x^3 - (4 + 4i\sqrt{3}) = 0 \Rightarrow x^3 = 4 + 4i\sqrt{3}$

We have

$r = \sqrt{4^2 + (4\sqrt{3})^2} = \sqrt{16 + 48} = \sqrt{64} = 8$ and

$\tan \theta = \dfrac{4\sqrt{3}}{4} = \sqrt{3}$. Because θ is in quadrant I,

$\theta = 60°$.

$x^3 = 4 + 4i\sqrt{3} = 8\left(\dfrac{1}{2} + i\dfrac{\sqrt{3}}{2}\right)$

$= 8(\cos 60° + i \sin 60°)$

Because $r^3(\cos 3\alpha + i \sin 3\alpha)$

$= 8(\cos 60° + i \sin 60°)$, we have

$r^3 = 8 \Rightarrow r = 2$ and $3\alpha = 60° + 360° \cdot k \Rightarrow$

$\alpha = \dfrac{60° + 360° \cdot k}{3} = 20° + 120° \cdot k,\ k$ any

integer. If $k = 0$, then $\alpha = 20° + 0° = 20°$.

If $k = 1$, then $\alpha = 20° + 120° = 140°$.

If $k = 2$, then $\alpha = 20° + 240° = 260°$.

Solution set:

$\{2(\cos 20° + i \sin 20°),\ 2(\cos 140° + i \sin 140°),$

$2(\cos 260° + i \sin 260°)\}$

49. $x^3 - 1 = 0 \Rightarrow (x-1)(x^2 + x + 1) = 0$

Setting each factor equal to zero, we have
$x - 1 = 0 \Rightarrow x = 1$ and
$x^2 + x + 1 = 0 \Rightarrow$

$$x = \frac{-1 \pm \sqrt{1^2 - 4 \cdot 1 \cdot 1}}{2 \cdot 1} = \frac{-1 \pm \sqrt{-3}}{2}$$

$$= \frac{-1 \pm i\sqrt{3}}{2} = -\frac{1}{2} \pm \frac{\sqrt{3}}{2}i.$$

Thus, $x = 1, -\frac{1}{2} + \frac{\sqrt{3}}{2}i, -\frac{1}{2} - \frac{\sqrt{3}}{2}i.$ We see
that the solutions are the same as Exercise 37.

51. (a) If $z = 0 + 0i$, then $z = 0$, $0^2 + 0 = 0$,
$0^2 + 0 = 0$, and so on. The calculations
repeat as 0, 0, 0, . . ., and will never
exceed a modulus of 2. The point (0, 0) is
part of the Mandelbrot set. The pixel at
the origin should be turned on.

(b) If $z = 1 - 1i$, then $(1 - i)^2 + (1 - i) = 1 - 3i$.
The modulus of $1 - 3i$ is

$$\sqrt{1^2 + (-3)^2} = \sqrt{1 + 9} = \sqrt{10}, \text{ which is}$$

greater than 2. Therefore, $1 - 1i$ is not
part of the Mandelbrot set, and the pixel
at $(1, -1)$ should be left off.

(c) If $z = -0.5i$, then

$(-0.5i)^2 - 0.5i = -0.25 - 0.5i;$

$(-0.25 - 0.5i)^2 + (-0.25 - 0.5i)$
$\quad = -0.4375 - 0.25i;$

$(-0.4375 - 0.25i)^2 + (-0.4375 - 0.25i)$
$\quad = -0.308593 - 0.03125i;$

$(-0.308593 - 0.03125i)^2$
$\qquad + (-0.308593 - 0.03125i)$
$\quad = -0.214339 - 0.0119629i;$

$(-0.214339 - 0.0119629i)^2$
$\qquad + (-0.214339 - 0.0119629i)$
$\quad = -0.16854 - 0.00683466i$

This sequence appears to be approaching
the origin, and no number has a modulus
greater than 2. Thus, $-0.5i$ is part of the
Mandelbrot set, and the pixel at $(0, -0.5i)$
should be turned on.

53. Using the trace function, we find that the other
four fifth roots of 1 are:
$0.30901699 + 0.95105652i,$
$-0.809017 + 0.58778525i,$

$-0.809017 - 0.5877853i,$
$0.30901699 - 0.9510565i.$

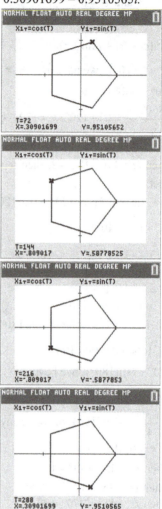

55. $x^2 - 3 + 2i = 0 \Rightarrow x^2 = 3 - 2i$

$r = \sqrt{3^2 + (-2)^2} = \sqrt{9 + 4} = \sqrt{13} \Rightarrow$

$r^{1/n} = r^{1/2} = \left(\sqrt{13}\right)^{1/2} \approx 1.89883$ and

$\tan \theta = -\frac{2}{3}$. Because θ is in quadrant IV,

$\theta \approx 326.3099°$ and

$\alpha = \dfrac{326.3099° + 360° \cdot k}{2} \approx 163.155° + 180° \cdot k,$

where k is an integer.

$x \approx 1.89833(\cos 163.155° + i \sin 163.155°),$
$\quad 1.89833(\cos 343.155° + i \sin 343.155°)$

Solution set:
$\{-1.8174 + 0.5503i, 1.8174 - 0.5503i\}$

57. $x^5 + 2 + 3i = 0 \Rightarrow x^5 = -2 - 3i$

$r = \sqrt{(-2)^2 + (-3)^2} = \sqrt{4 + 9} = \sqrt{13} \Rightarrow$

$r^{1/n} = r^{1/5} = \left(\sqrt{13}\right)^{1/5} = 13^{1/10} \approx 1.2924$

and $\tan\theta = \dfrac{-3}{-2} = 1.5$

Because θ is in quadrant III, $\theta \approx 236.310°$ and

$\alpha = \dfrac{236.310° + 360° \cdot k}{5} = 47.262° + 72° \cdot k,$

where k is an integer.

$x \approx 1.29239\left(\cos 47.262° + i\sin 47.262°\right),$
$\qquad 1.29239\left(\cos 119.262° + i\sin 119.2622°\right),$
$\qquad 1.29239\left(\cos 191.262° + i\sin 191.2622°\right),$
$\qquad 1.29239\left(\cos 263.262° + i\sin 263.2622°\right),$
$\qquad 1.29239\left(\cos 335.262° + i\sin 335.2622°\right)$

Solution set:
$\{0.8771 + 0.9492i, -0.6317 + 1.1275i,$
$-1.2675 - 0.2524i, -0.1516 - 1.2835i,$
$1.1738 - 0.54083i\}$

59. De Moivre's theorem states that

$\left(\cos\theta + i\sin\theta\right)^2 = 1^2\left(\cos 2\theta + i\sin 2\theta\right)$
$\qquad\qquad\qquad = \underline{\cos 2\theta + i\sin 2\theta}$

61. Two complex numbers $a + bi$ and $c + di$ are equal only if $a = c$ and $b = d$. Thus, $a = c$ implies $\cos^2\theta - \sin^2\theta = \cos 2\theta$.

Chapter 8 Quiz
(Sections 8.3–8.6)

1. $\mathbf{a} = \langle -1, 4 \rangle,\ \mathbf{b} = \langle 5, 2 \rangle$

  (a) $3\mathbf{a} = 3\langle -1, 4 \rangle = \langle -3, 12 \rangle$

  (b) $4\mathbf{a} - 2\mathbf{b} = 4\langle -1, 4 \rangle - 2\langle 5, 2 \rangle$
$\qquad\qquad = \langle -4, 16 \rangle - \langle 10, 4 \rangle = \langle -14, 12 \rangle$

  (c) $|\mathbf{a}| = \sqrt{(-1)^2 + 4^2} = \sqrt{17}$

  (d) $\mathbf{a} \cdot \mathbf{b} = (-1)(5) + (4)(2) = 3$

  (e) $\cos\theta = \dfrac{\mathbf{a} \cdot \mathbf{b}}{|\mathbf{a}||\mathbf{b}|} = \dfrac{(-1)(5) + (4)(2)}{\sqrt{(-1)^2 + 4^2} \cdot \sqrt{5^2 + 2^2}}$

$\qquad\quad = \dfrac{3}{\sqrt{17} \cdot \sqrt{29}} \approx .1351132047 \Rightarrow$

$\qquad \theta \approx 82.23°$

3. $w = 3 + 5i,\ z = -4 + i$

$w + z = (3 + 5i) + (-4 + i) = -1 + 6i$

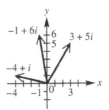

5. (a) Sketch a graph of $-4i$ in the complex plane.

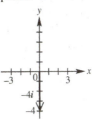

Because $-4i = 0 - 4i$, we have $x = 0$ and $y = -4$, so $r = \sqrt{0^2 + (-4)^2} = 4$. We cannot find θ by using $\tan\theta = \dfrac{y}{x}$ because $x = 0$. From the graph, we see that $-4i$ is on the negative y-axis, so $\theta = 270°$. Thus,

$-4i = 4\left(\cos 270° + i\sin 270°\right)$

  (b) $1 - i\sqrt{3}$

Sketch a graph of $1 - i\sqrt{3}$ in the complex plane.

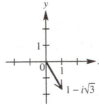

Because $x = 1$ and $y = -\sqrt{3}$,

$r = \sqrt{1^2 + \left(-\sqrt{3}\right)^2} = \sqrt{1 + 3} = \sqrt{4} = 2$ and

$\tan\theta = \dfrac{-\sqrt{3}}{1} = -\sqrt{3}$. The graph shows that θ is in quadrant IV, so $\theta = 300°$. Therefore,

$1 - i\sqrt{3} = 2\left(\cos 300° + i\sin 300°\right)$

(c) $-3-i$

Sketch a graph of $-3-i$ in the complex plane.

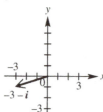

Using a calculator, we find that the reference angle is 18.4°. The graph shows that θ is in quadrant III, so $\theta = 180° + 18.4° = 198.4°$. Therefore,

$-3-i = \sqrt{10}\left(\cos 198.4° + i \sin 198.4°\right).$

7. $w = 12\left(\cos 80° + i \sin 80°\right),$

$z = 3\left(\cos 50° + i \sin 50°\right)$

(a) $wz = 12\left(\cos 80° + i \sin 80°\right)$

$\qquad\qquad \cdot 3\left(\cos 50° + i \sin 50°\right)$

$= 12 \cdot 3\begin{bmatrix} \cos\left(80° + 50°\right) \\ + i \sin\left(80° + 50°\right) \end{bmatrix}$

$= 36\left(\cos 130° + i \sin 130°\right)$

(b) $\dfrac{w}{z} = \dfrac{12\left(\cos 80° + i \sin 80°\right)}{3\left(\cos 50° + i \sin 50°\right)}$

$= 4\left[\cos\left(80° - 50°\right) + i \sin\left(80° - 50°\right)\right]$

$= 4\left(\cos 30° + i \sin 30°\right)$

$= 4\left(\dfrac{\sqrt{3}}{2} + \dfrac{1}{2}i\right) = 2\sqrt{3} + 2i$

(c) $z^3 = \left[3\left(\cos 50° + i \sin 50°\right)\right]^3$

$= 3^3\left[\cos\left(3 \cdot 50°\right) + i \sin\left(3 \cdot 50°\right)\right]$

$= 27\left(\cos 150° + i \sin 150°\right)$

$= 27\left(-\dfrac{\sqrt{3}}{2} + \dfrac{1}{2}i\right) = -\dfrac{27\sqrt{3}}{2} + \dfrac{27}{2}i$

Section 8.7 Polar Equations and Graphs

1. For the polar equation $r = 3\cos\theta$, if $\theta = 60°$, then $r = \frac{3}{2}$.

3. For the polar equation $r^2 = 4\sin 2\theta$, if $\theta = 15°$, then $r = \pm\sqrt{2}$.

5. II (because $r > 0$ and $90° < \theta < 180°$)

7. IV (because $r > 0$ and $-90° < \theta < 0°$)

9. positive x-axis

11. negative y-axis

For Exercises 13(b)–24(b), answers may vary.

13. **(a)**

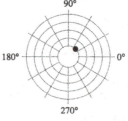

(b) Two other pairs of polar coordinates for $(1, 45°)$ are $(1, 405°)$ and $(-1, 225°)$.

(c) $x = r\cos\theta \Rightarrow x = 1 \cdot \cos 45° = \dfrac{\sqrt{2}}{2}$ and

$y = r\sin\theta \Rightarrow y = 1 \cdot \sin 45° = \dfrac{\sqrt{2}}{2}$, so the

point is $\left(\dfrac{\sqrt{2}}{2}, \dfrac{\sqrt{2}}{2}\right).$

15. **(a)**

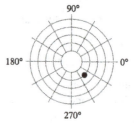

(b) Two other pairs of polar coordinates for $(-2, 135°)$ are $(-2, 495°)$ and $(2, 315°)$.

(c) $x = r\cos\theta \Rightarrow x = (-2)\cos 135° = \sqrt{2}$ and

$y = r\sin\theta \Rightarrow y = (-2)\sin 135° = \sqrt{2}$, so

the point is $\left(\sqrt{2}, -\sqrt{2}\right).$

17. **(a)**

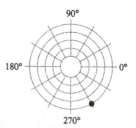

(b) Two other pairs of polar coordinates for $(5, -60°)$ are $(5, 300°)$ and $(-5, 120°)$.

(c) $x = r\cos\theta \Rightarrow x = 5\cos(-60°) = \dfrac{5}{2}$ and

$y = r\sin\theta \Rightarrow y = 5\sin(-60°) = -\dfrac{5\sqrt{3}}{2}$,

so the point is $\left(\dfrac{5}{2}, -\dfrac{5\sqrt{3}}{2}\right)$.

19. (a)

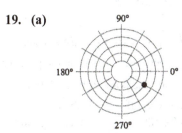

(b) Two other pairs of polar coordinates for $(-3, -210°)$ are $(-3, 150°)$ and $(3, -30°)$.

(c) $x = r\cos\theta \Rightarrow x = (-3)\cos(-210°) = \dfrac{3\sqrt{3}}{2}$

and

$y = r\sin\theta \Rightarrow y = (-3)\sin(-210°) = -\dfrac{3}{2}$,

so the point is $\left(\dfrac{3\sqrt{3}}{2}, -\dfrac{3}{2}\right)$.

21. (a)

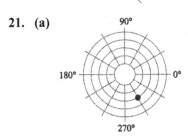

(b) Two other pairs of polar coordinates for $\left(3, \dfrac{5\pi}{3}\right)$ are $\left(3, \dfrac{11\pi}{3}\right)$ and $\left(-3, \dfrac{2\pi}{3}\right)$.

(c) $x = r\cos\theta \Rightarrow x = 3\cos\dfrac{5\pi}{3} = \dfrac{3}{2}$ and

$y = r\sin\theta \Rightarrow y = 3\sin\dfrac{5\pi}{3} = -\dfrac{3\sqrt{3}}{2}$, so

the point is $\left(\dfrac{3}{2}, -\dfrac{3\sqrt{3}}{2}\right)$.

23. (a)

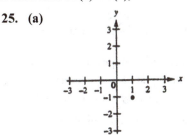

(b) Two other pairs of polar coordinates for $\left(-2, \dfrac{\pi}{3}\right)$ are $\left(-2, \dfrac{7\pi}{3}\right)$ and $\left(2, \dfrac{4\pi}{3}\right)$.

(c) $x = r\cos\theta \Rightarrow x = -2\cos\dfrac{\pi}{3} = -1$ and

$y = r\sin\theta \Rightarrow y = -2\sin\dfrac{\pi}{3} = -\sqrt{3}$, so the

point is $\left(-1, -\sqrt{3}\right)$.

For Exercises 25(b)–35(b), answers may vary.

25. (a)

(b) $r = \sqrt{1^2 + (-1)^2} = \sqrt{1+1} = \sqrt{2}$ and

$\theta = \tan^{-1}\left(\dfrac{-1}{1}\right) = \tan^{-1}(-1) = -45°$,

because θ is in quadrant IV.
$360° - 45° = 315°$, so one possibility is
$\left(\sqrt{2}, 315°\right)$. Alternatively, if $r = -\sqrt{2}$,
then $\theta = 315° - 180° = 135°$. Thus, a
second possibility is $\left(-\sqrt{2}, 135°\right)$.

27. (a)

(b) $r = \sqrt{0^2 + 3^2} = \sqrt{0+9} = \sqrt{9} = 3$ and $\theta = 90°$, because $(0,3)$ is on the positive y-axis. So, one possibility is $(3, 90°)$. Alternatively, if $r = -3$, then $\theta = 90° + 180° = 270°$. Thus, a second possibility is $(-3, 270°)$.

29. (a)

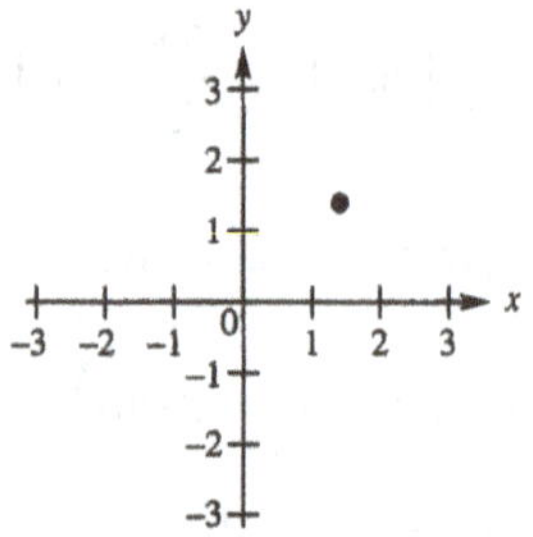

(b) $r = \sqrt{\left(\sqrt{2}\right)^2 + \left(\sqrt{2}\right)^2} = \sqrt{2+2} = \sqrt{4} = 2$

and $\theta = \tan^{-1}\left(\dfrac{\sqrt{2}}{\sqrt{2}}\right) = \tan^{-1} 1 = 45°$,

because θ is in quadrant I. So, one possibility is $(2, 45°)$. Alternatively, if $r = -2$, then $\theta = 45° + 180° = 225°$. Thus, a second possibility is $(-2, 225°)$.

31. (a)

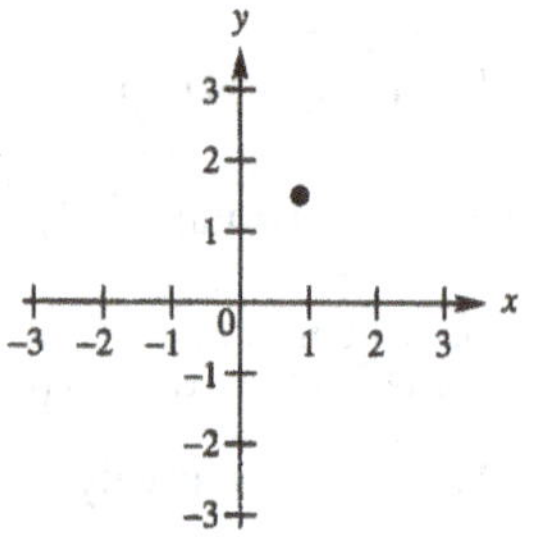

(b) $r = \sqrt{\left(\dfrac{\sqrt{3}}{2}\right)^2 + \left(\dfrac{3}{2}\right)^2} = \sqrt{\dfrac{3}{4} + \dfrac{9}{4}}$

$= \sqrt{\dfrac{12}{4}} = \sqrt{3}$ and $\theta = \arctan\left(\dfrac{3}{2} \cdot \dfrac{2}{\sqrt{3}}\right)$

$= \tan^{-1}\left(\sqrt{3}\right) = 60°$, because θ is in quadrant I. So, one possibility is $\left(\sqrt{3}, 60°\right)$. Alternatively, if $r = -\sqrt{3}$, then $\theta = 60° + 180° = 240°$. Thus, a second possibility is $\left(-\sqrt{3}, 240°\right)$.

33. (a)

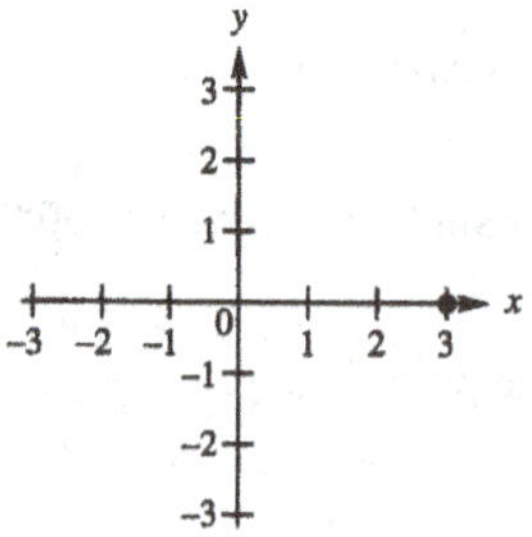

(b) $r = \sqrt{3^2 + 0^2} = \sqrt{9+0} = \sqrt{9} = 3$ and $\theta = 0°$, because $(3,0)$ is on the positive x-axis. So, one possibility is $(3, 0°)$. Alternatively, if $r = -3$, then $\theta = 0° + 180° = 180°$. Thus, a second possibility is $(-3, 180°)$.

35. (a)

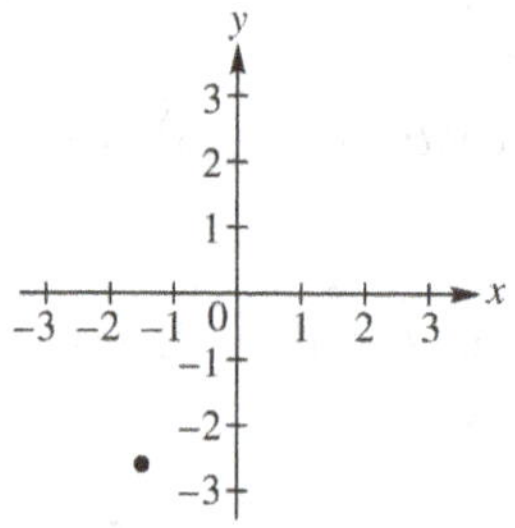

(b) $r = \sqrt{\left(-\dfrac{3}{2}\right)^2 + \left(-\dfrac{3\sqrt{3}}{2}\right)^2} = \sqrt{\dfrac{9}{4} + \dfrac{27}{4}}$

$= \sqrt{\dfrac{36}{4}} = 3$

and $\theta = \tan^{-1}\left(\dfrac{-\frac{3\sqrt{3}}{2}}{-\frac{3}{2}}\right) = \tan^{-1}\sqrt{3}.$

Because θ is in quadrant III, we have $\theta = 240°$. So, one possibility is $(3, 240°)$. Alternatively, if $r = -3$, then $\theta = 240° - 180° = 60°$. Thus, a second possibility is $(-3, 60°)$.

37. $x - y = 4$

Using the general form for the polar equation of a line, $r = \dfrac{c}{a\cos\theta + b\sin\theta}$, with $a = 1, b = -1$, and $c = 4$, the polar equation is

$r = \dfrac{4}{\cos\theta - \sin\theta}.$

(continued on next page)

(*continued*)

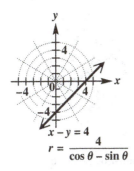

$x - y = 4$

$r = \dfrac{4}{\cos \theta - \sin \theta}$

39. $x^2 + y^2 = 16 \Rightarrow r^2 = 16 \Rightarrow r = \pm 4$

The equation of the circle in polar form is $r = 4$ or $r = -4$.

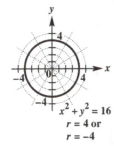

$x^2 + y^2 = 16$

$r = 4$ or

$r = -4$

41. $2x + y = 5$

Using the general form for the polar equation

of a line, $r = \dfrac{c}{a\cos\theta + b\sin\theta}$, with

$a = 2, b = 1$, and $c = 5$, the polar equation is

$r = \dfrac{5}{2\cos\theta + \sin\theta}$.

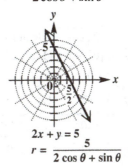

$2x + y = 5$

$r = \dfrac{5}{2\cos\theta + \sin\theta}$

43. C. $r = 3$ represents the set of all points 3 units from the pole.

45. A. $r = \cos 2\theta$ is a rose curve with $2 \cdot 2 = 4$ petals.

47. $r = 2 + 2\cos\theta$ (cardioid)

θ	0°	30°	60°	90°	120°	150°
$\cos\theta$	1	0.9	0.5	0	−0.5	−0.9
$r = 2 + 2\cos\theta$	4	3.7	3	2	1	0.3

θ	180°	210°	240°	270°	300°	330°
$\cos\theta$	−1	−0.9	−0.5	0	0.5	0.9
$r = 2 + 2\cos\theta$	0	0.3	1	2	3	3.7

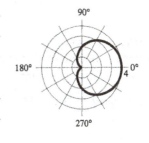

49. $r = 3 + \cos\theta$ (limaçon)

θ	0°	30°	60°	90°	120°	150°
$r = 3 + \cos\theta$	4	3.9	3.5	3	2.5	2.1

θ	180°	210°	240°	270°	300°	330°
$r = 3 + \cos\theta$	2	2.1	2.5	3	3.5	3.9

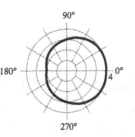

51. $r = 4\cos 2\theta$ (four-leaved rose)

θ	0°	30°	45°	60°	90°	120°	135°	150°
$r = 4\cos 2\theta$	4	2	0	−2	−4	−2	0	2

θ	180°	210°	225°	240°	270°	300°	315°	330°
$r = 4\cos 2\theta$	4	2	0	−2	−4	−2	0	2

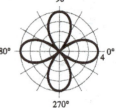

53. $r^2 = 4\cos 2\theta \Rightarrow r = \pm 2\sqrt{\cos 2\theta}$ (lemniscate)

Graph only exists for $[0°, 45°]$, $[135°, 225°]$, and $[315°, 360°]$ because $\cos 2\theta$ must be positive.

θ	0°	30°	45°	135°	150°
$r = \pm 2\sqrt{\cos 2\theta}$	±2	±1.4	0	0	±1.4

θ	180°	210°	225°	315°	330°
$r = \pm 2\sqrt{\cos 2\theta}$	±2	±1.4	0	0	±1.4

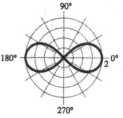

55. $r = 4 - 4\cos\theta$ (cardioid)

θ	0°	30°	60°	90°	120°	150°
$r = 4 - 4\cos\theta$	0	0.5	2	4	6	7.5
θ	180°	210°	240°	270°	300°	330°
$r = 4 - 4\cos\theta$	8	7.5	6	4	2	0.5

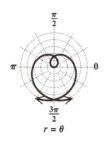

57. $r = 2\sin\theta\tan\theta$ (cissoid)

r is undefined at $\theta = 90°$ and $\theta = 270°$.

θ	0°	30°	45°	60°	90°	120°	135°	150°	180°
$r = 2\sin\theta\tan\theta$	0	0.6	1.4	3	undefined	−3	−1.4	−0.6	0

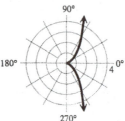

Notice that for $[180°, 360°)$, the graph retraces the path traced for $[0°, 180°)$.

59. Graph $r = \theta$

θ	−360°	−270°	−180°	−90°	0°	90°	180°	270°	360°
θ (radians)	−6.3	−4.7	−3.1	−1.6	0	1.6	3.1	4.7	6.3
$r = \theta$	−6.3	−4.7	−3.1	−1.6	0	1.6	3.1	4.7	6.3

61. $r = 2 \sin \theta$

Multiply both sides by r to obtain

$r^2 = 2r \sin \theta$.

$r^2 = x^2 + y^2$ and $y = r \sin \theta \Rightarrow x^2 + y^2 = 2y$.

Complete the square on y to obtain

$x^2 + y^2 - 2y + 1 = 1 \Rightarrow x^2 + (y-1)^2 = 1$.

The graph is a circle with center at (0, 1) and radius 1.

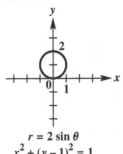

$r = 2 \sin \theta$
$x^2 + (y-1)^2 = 1$

63. $r = \dfrac{2}{1 - \cos \theta}$

Multiply both sides by $1 - \cos \theta$ to obtain

$r - r \cos \theta = 2$.

Substitute $r = \sqrt{x^2 + y^2}$ to obtain

$\sqrt{x^2 + y^2} - x = 2 \Rightarrow \sqrt{x^2 + y^2} = 2 + x \Rightarrow$

$x^2 + y^2 = (2 + x)^2 \Rightarrow x^2 + y^2 = 4 + 4x + x^2 \Rightarrow$

$y^2 = 4(1 + x)$

The graph is a parabola with vertex at $(-1, 0)$ and axis $y = 0$.

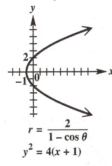

$r = \dfrac{2}{1 - \cos \theta}$
$y^2 = 4(x + 1)$

65. $r + 2 \cos \theta = -2 \sin \theta$

$r + 2 \cos \theta = -2 \sin \theta \Rightarrow$

$r^2 = -2r \sin \theta - 2r \cos \theta \Rightarrow x^2 + y^2 = -2y - 2x$

$x^2 + 2x + y^2 + 2y = 0 \Rightarrow$

$x^2 + 2x + 1 + y^2 + 2y + 1 = 2 \Rightarrow$

$(x + 1)^2 + (y + 1)^2 = 2$

The graph is a circle with center $(-1, -1)$ and radius $\sqrt{2}$.

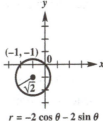

$(-1, -1)$

$r = -2 \cos \theta - 2 \sin \theta$
$(x + 1)^2 + (y + 1)^2 = 2$

67. $r = 2 \sec \theta$

$r = 2 \sec \theta \Rightarrow r = \dfrac{2}{\cos \theta} \Rightarrow r \cos \theta = 2 \Rightarrow x = 2$

The graph is a vertical line, intercepting the x-axis at 2.

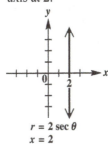

$r = 2 \sec \theta$
$x = 2$

69. $r = \dfrac{2}{\cos \theta + \sin \theta}$

Using the general form for the polar equation

of a line, $r = \dfrac{c}{a \cos \theta + b \sin \theta}$, with

$a = 1$, $b = 1$, and $c = 2$, we have $x + y = 2$.

The graph is a line with intercepts (0, 2) and (2, 0).

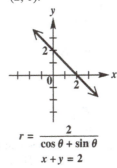

$r = \dfrac{2}{\cos \theta + \sin \theta}$
$x + y = 2$

71. In rectangular coordinates, the line passes through $(1,0)$ and $(0,2)$. So

$$m = \frac{2-0}{0-1} = \frac{2}{-1} = -2 \text{ and}$$

$$(y-0) = -2(x-1) \Rightarrow y = -2x+2 \Rightarrow$$

$2x + y = 2$. Converting to polar form

$$r = \frac{c}{a\cos\theta + b\sin\theta}, \text{ we have:}$$

$$r = \frac{2}{2\cos\theta + \sin\theta}.$$

73. (a) $(r, -\theta)$

(b) $(r, \pi-\theta)$ or $(-r, -\theta)$

(c) $(r, \pi+\theta)$ or $(-r, \theta)$

75.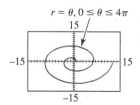
$r = \theta, 0 \le \theta \le 4\pi$

77.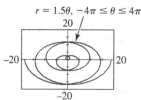
$r = 1.5\theta, -4\pi \le \theta \le 4\pi$

79. $r = 4\sin\theta$, $r = 1+2\sin\theta$, $0 \le \theta < 2\pi$

$4\sin\theta = 1+2\sin\theta \Rightarrow 2\sin\theta = 1 \Rightarrow$

$$\sin\theta = \frac{1}{2} \Rightarrow \theta = \frac{\pi}{6} \text{ or } \frac{5\pi}{6}$$

The points of intersection are

$$\left(4\sin\frac{\pi}{6}, \frac{\pi}{6}\right) = \left(2, \frac{\pi}{6}\right) \text{ and}$$

$$\left(4\sin\frac{5\pi}{6}, \frac{5\pi}{6}\right) = \left(2, \frac{5\pi}{6}\right).$$

Using a graphing calculator, we see that the pole, $(0, 0)$ is also an intersection. However, it is not reached at the same value of theta for each equation, which is why it does not appear as a solution of the equation $4\sin\theta = 1+2\sin\theta$.

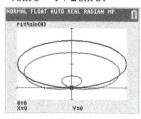

81. $r = 2+\sin\theta$, $r = 2+\cos\theta$, $0 \le \theta < 2\pi$

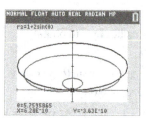

$2+\sin\theta = 2+\cos\theta \Rightarrow \sin\theta = \cos\theta \Rightarrow$

$$\theta = \frac{\pi}{4} \text{ or } \frac{5\pi}{4}$$

$$r = 2+\sin\frac{\pi}{4} = 2+\frac{\sqrt{2}}{2} = \frac{4+\sqrt{2}}{2} \text{ and}$$

$$r = 2+\sin\frac{5\pi}{4} = 2-\frac{\sqrt{2}}{2} = \frac{4-\sqrt{2}}{2}$$

The points of intersection are

$$\left(\frac{4+\sqrt{2}}{2}, \frac{\pi}{4}\right) \text{ and } \left(\frac{4-\sqrt{2}}{2}, \frac{5\pi}{4}\right).$$

83. (a) Plot the following polar equations on the same polar axis in radian mode:

Mercury: $r = \dfrac{0.39(1-0.206^2)}{1+0.206\cos\theta}$

Venus: $r = \dfrac{0.78(1-0.007^2)}{1+0.007\cos\theta}$

Earth: $r = \dfrac{1(1-0.017^2)}{1+0.017\cos\theta}$

Mars: $r = \dfrac{1.52(1-0.093^2)}{1+0.093\cos\theta}$

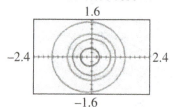

(b) Plot the following polar equations on the same polar axis:

Earth: $r = \dfrac{1(1-0.017^2)}{1+0.017\cos\theta}$

Jupiter: $r = \dfrac{5.2(1-0.048^2)}{1+0.048\cos\theta}$

Uranus: $r = \dfrac{19.2(1-0.047^2)}{1+0.047\cos\theta}$

Pluto: $r = \dfrac{39.4(1-0.249^2)}{1+0.249\cos\theta}$

(continued on next page)

(continued)

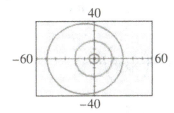

(c) We must determine if the orbit of Pluto is always outside the orbits of the other satellites. Neptune is closest to Pluto, so plot the orbits of Neptune and Pluto on the same polar axes.

Neptune: $r = \dfrac{30.1(1 - 0.009^2)}{1 + 0.009 \cos \theta}$

Pluto: $r = \dfrac{39.4(1 - 0.249^2)}{1 + 0.249 \cos \theta}$

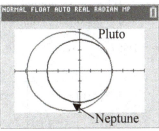

$[-60, 60] \times [-60, 60]$

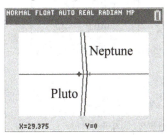

The graph shows that their orbits are very close near the polar axis. Use ZOOM or change your window to see that the orbit of Pluto does indeed pass inside the orbit of Neptune. Therefore, there are times when Neptune, not Pluto, is the farthest planet from the sun. (However, Pluto's average distance from the sun is considerably greater than Neptune's average distance.)

85. $r \sin \theta = k$

87. $r = \dfrac{k}{\sin \theta} \Rightarrow r = k \csc \theta$

89. $r \cos \theta = k$

91. $r = \dfrac{k}{\cos \theta} \Rightarrow r = k \sec \theta$

Section 8.8 Parametric Equations, Graphs, and Applications

1. For the plane curve defined by $x = t^2 + 1$, $y = 2t + 3$, for t in $[-4, 4]$, the ordered pair that corresponds to $t = -3$ is (10, -3).

3. For the plane curve defined by $x = \cos t$, $y = 2 \sin t$, for t in $[0, 2\pi]$, the ordered pair that corresponds to $t = \frac{\pi}{3}$ is $\left(\frac{1}{2}, \sqrt{3} \right)$.

5. C. At $t = 2$, $x = 3(2) + 6 = 12$ and $y = -2(2) + 4 = 0$.

7. A. At $t = 5$, $x = 5$ and $y = 5^2 = 25$.

9. (a) $x = t + 2$, $y = t^2$, for t in $[-1, 1]$

t	$x = t + 2$	$y = t^2$
-1	$-1 + 2 = 1$	$(-1)^2 = 1$
0	$0 + 2 = 2$	$0^2 = 0$
1	$1 + 2 = 3$	$1^2 = 1$

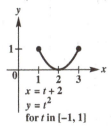

for t in [-1, 1]

(b) $x - 2 = t$, therefore $y = (x - 2)^2$ or $y = x^2 - 4x + 4$. Because t is in $[-1, 1]$, x is in $[-1 + 2, 1 + 2]$ or $[1, 3]$.

11. (a) $x = \sqrt{t}$, $y = 3t - 4$, for t in $[0, 4]$.

t	$x = \sqrt{t}$	$y = 3t - 4$
0	$\sqrt{0} = 0$	$3(0) - 4 = -4$
1	$\sqrt{1} = 1$	$3(1) - 4 = -1$
2	$\sqrt{2} = 1.4$	$3(2) - 4 = 2$
3	$\sqrt{3} = 1.7$	$3(3) - 4 = 5$
4	$\sqrt{4} = 2$	$3(4) - 4 = 8$

(continued on next page)

(continued)

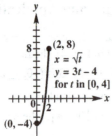

(b) $x = \sqrt{t}$, $y = 3t - 4$

$x = \sqrt{t} \Rightarrow x^2 = t$, so $y = 3x^2 - 4$.
Because t is in $[0, 4]$, x is in
$[\sqrt{0}, \sqrt{4}]$ or $[0, 2]$.

13. (a) $x = t^3 + 1$, $y = t^3 - 1$, for t in $(-\infty, \infty)$

t	$x = t^3 + 1$	$y = t^3 - 1$
-2	$(-2)^3 + 1 = -7$	$(-2)^3 - 1 = -9$
-1	$(-1)^3 + 1 = 0$	$(-1)^3 - 1 = -2$
0	$0^3 + 1 = 1$	$0^3 - 1 = -1$
1	$1^3 + 1 = 2$	$1^3 - 1 = 0$
2	$2^3 + 1 = 9$	$2^3 - 1 = 7$
3	$3^3 + 1 = 28$	$3^3 - 1 = 26$

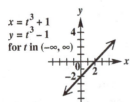

(b) $x = t^3 + 1$, so $x - 1 = t^3$.
$y = t^3 - 1$, so $y = (x - 1) - 1 = x - 2$.
Because t is in $(-\infty, \infty)$, x is in $(-\infty, \infty)$.

15. (a) $x = 2 \sin t$, $y = 2 \cos t$, for t in $[0, 2\pi]$

t	$x = 2 \sin t$	$y = 2 \cos t$
0	$2 \sin 0 = 0$	$2 \cos \theta = 2$
$\frac{\pi}{6}$	$2 \sin \frac{\pi}{6} = 1$	$2 \cos \frac{\pi}{6} = \sqrt{3}$
$\frac{\pi}{4}$	$2 \sin \frac{\pi}{4} = \sqrt{2}$	$2 \cos \frac{\pi}{4} = \sqrt{2}$
$\frac{\pi}{3}$	$2 \sin \frac{\pi}{3} = \sqrt{3}$	$2 \cos \frac{\pi}{3} = 1$
$\frac{\pi}{2}$	$2 \sin \frac{\pi}{2} = 2$	$2 \cos \frac{\pi}{2} = 0$

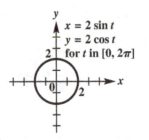

(b) $x = 2 \sin t$ and $y = 2 \cos t$, so

$\dfrac{x}{2} = \sin t$ and $\dfrac{y}{2} = \cos t$.

$\sin^2 t + \cos^2 t = 1 \Rightarrow$

$\left(\dfrac{x}{2}\right)^2 + \left(\dfrac{y}{2}\right)^2 = 1 \Rightarrow \dfrac{x^2}{4} + \dfrac{y^2}{4} = 1 \Rightarrow$

$x^2 + y^2 = 4$. Because t is in $[0, 2\pi]$, x is
in $[-2, 2]$ because the graph is a circle,
centered at the origin, with radius 2.

17. (a) $x = 3 \tan t$, $y = 2 \sec t$, for t in $\left(-\frac{\pi}{2}, \frac{\pi}{2}\right)$

t	$x = 3 \tan t$	$y = 2 \sec t$
$-\frac{\pi}{3}$	$3 \tan \left(-\frac{\pi}{3}\right) = -3\sqrt{3}$	$2 \sec \left(-\frac{\pi}{3}\right) = 4$
$-\frac{\pi}{6}$	$3 \tan \left(-\frac{\pi}{6}\right) = -\sqrt{3}$	$2 \sec \left(-\frac{\pi}{6}\right) = \frac{4\sqrt{3}}{3}$
0	$3 \tan 0 = 0$	$2 \sec 0 = 2$
$\frac{\pi}{6}$	$3 \tan \frac{\pi}{6} = \sqrt{3}$	$2 \sec \frac{\pi}{6} = \frac{4\sqrt{3}}{3}$
$\frac{\pi}{3}$	$3 \tan \frac{\pi}{3} = 3\sqrt{3}$	$2 \sec \frac{\pi}{3} = 4$

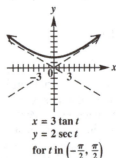

(b) $\dfrac{x}{3} = \tan t$, $\dfrac{y}{2} = \sec t$, and

$1 + \tan^2 t = \left(\dfrac{y}{2}\right)^2 = \sec^2 t$, so

$1 + \left(\dfrac{x}{3}\right)^2 = \left(\dfrac{y}{2}\right)^2 \Rightarrow 1 + \dfrac{x^2}{9} = \dfrac{y^2}{4} \Rightarrow$

$y^2 = 4\left(1 + \dfrac{x^2}{9}\right) \Rightarrow y = 2\sqrt{1 + \dfrac{x^2}{9}}$

This graph is the top half of a hyperbola,
so x is in $(-\infty, \infty)$.

19. (a) $x = \sin t,\ y = \csc t$ for t in $(0, \pi)$

Because t is in $(0, \pi)$ and $x = \sin t$, x is in $(0, 1]$.

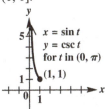

$x = \sin t$
$y = \csc t$
for t in $(0, \pi)$
$(1, 1)$

(b) Because $x = \sin t$ and $y = \csc t = \dfrac{1}{\sin t}$,

we have $y = \dfrac{1}{x}$, where x is in $(0, 1]$.

21. (a) $x = t,\ y = \sqrt{t^2 + 2}$, for t in $(-\infty, \infty)$

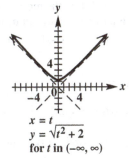

$x = t$
$y = \sqrt{t^2 + 2}$
for t in $(-\infty, \infty)$

(b) $x = t$ and $y = \sqrt{t^2 + 2}$, so $y = \sqrt{x^2 + 2}$.
Because t is in $(-\infty\ \infty)$ and $x = t$, x is in $(-\infty, \infty)$.

23. (a) $x = 2 + \sin t,\ y = 1 + \cos t$, for t in $\left[0, 2\pi\right]$

This is a circle centered at $(2, 1)$ with radius 1 and t is in $[0, 2\pi]$, so x is in $[1, 3]$.

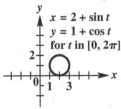

$x = 2 + \sin t$
$y = 1 + \cos t$
for t in $[0, 2\pi]$

(b) $x = 2 + \sin t \Rightarrow x - 2 = \sin t$
$y = 1 + \cos t \Rightarrow y - 1 = \cos t$
$\sin^2 t + \cos^2 t = 1$, so
$(x - 2)^2 + (y - 1)^2 = 1$, for x in $[1, 3]$.

25. (a) $x = t + 2,\ y = \dfrac{1}{t + 2}$, for $t \neq 2$

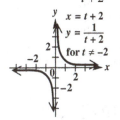

$x = t + 2$
$y = \dfrac{1}{t + 2}$
for $t \neq -2$

(b) $x = t + 2$ and $y = \dfrac{1}{t + 2} \Rightarrow y = \dfrac{1}{x}$.

Because $t \neq -2, x \neq -2 + 2, x \neq 0$.
Therefore, x is in $(-\infty, 0) \cup (0, \infty)$.

27. (a) $x = t + 2,\ y = t - 4$, for t in $\left(-\infty, \infty\right)$

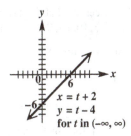

$x = t + 2$
$y = t - 4$
for t in $(-\infty, \infty)$

(b) $x = t + 2 \Rightarrow t = x - 2$.
$y = t - 4 \Rightarrow y = (x - 2) - 4 = x - 6$.
Because t is in $(-\infty, \infty)$, x is in $(-\infty, \infty)$.

29. $x = 3\cos t,\ y = 3\sin t$

$x = 3\cos t \Rightarrow \cos t = \dfrac{x}{3}$,

$y = 3\sin t \Rightarrow \sin t = \dfrac{y}{3}$, and

$\sin^2 t + \cos^2 t = 1$. Thus,

$\left(\dfrac{y}{3}\right)^2 + \left(\dfrac{x}{3}\right)^2 = 1 \Rightarrow \dfrac{y^2}{9} + \dfrac{x^2}{9} = 1 \Rightarrow$

$x^2 + y^2 = 9$.

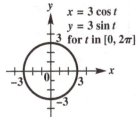

$x = 3\cos t$
$y = 3\sin t$
for t in $[0, 2\pi]$

This is a circle centered at the origin with radius 3.

31. $x = 3\sin t,\ y = 2\cos t$

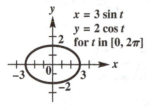

$x = 3\sin t \Rightarrow \sin t = \dfrac{x}{3},$

$y = 2\cos t \Rightarrow \cos t = \dfrac{y}{2},$ and

$\sin^2 t + \cos^2 t = 1.$

Thus, $\left(\dfrac{x}{3}\right)^2 + \left(\dfrac{y}{2}\right)^2 = 1 \Rightarrow \dfrac{x^2}{9} + \dfrac{y^2}{4} = 1.$

This is an ellipse centered at the origin with axes endpoints $(-3, 0)$, $(3, 0)$, $(0, -2)$, $(0, 2)$.

In Exercises 33–35, answers may vary.

33. $y = (x + 3)^2 - 1$

$x = t,\ y = (t + 3)^2 - 1$ for t in $(-\infty, \infty)$;

$x = t - 3,\ y = t^2 - 1$ for t in $(-\infty, \infty)$

35. $y = x^2 - 2x + 3 = (x - 1)^2 + 2$

$x = t,\ y = (t - 1)^2 + 2 = t^2 - 2t + 3$ for t in

$(-\infty, \infty)$; $x = t + 1,\ y = t^2 + 2$ for t in $(-\infty, \infty)$

37. $x = 2t - 2\sin t,\ y = 2 - 2\cos t$, for t in $[0, 4\pi]$

t	0	$\dfrac{\pi}{2}$	π	$\dfrac{3\pi}{2}$
$x = 2t - 2\sin t$	0	1.14	2π	11.4
$y = 2 - 2\cos t$	0	2	4	2

t	2π	3π	4π
$x = 2t - 2\sin t$	4π	6π	8π
$y = 2 - 2\cos t$	0	4	0

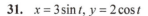

39. $x = 2\cos t,\ y = 3\sin 2t$

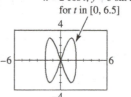

41. $x = 3\sin 4t,\ y = 3\cos 3t$

For Exercises 43–45, recall that the motion of a projectile (neglecting air resistance) can be modeled by: $x = (v_0 \cos\theta)t,\ y = (v_0 \sin\theta)t - 16t^2$ for t in $[0, k]$.

43. (a) $x = (v\cos\theta)t \Rightarrow$

$x = (48\cos 60°)t = 48\left(\dfrac{1}{2}\right)t = 24t$

$y = (v\sin\theta)t - 16t^2 = (48\sin 60°)t - 16t^2$

$= 48\cdot\dfrac{\sqrt{3}}{2}t - 16t^2 = -16t^2 + 24\sqrt{3}t$

(b) $t = \dfrac{x}{24}$, so $y = -16\left(\dfrac{x}{24}\right)^2 + 24\sqrt{3}\left(\dfrac{x}{24}\right)$

$= -\dfrac{x^2}{36} + \sqrt{3}x$

(c) $y = -16t^2 + 24\sqrt{3}t$

When the rocket is no longer in flight, $y = 0$. Solve $0 = -16t^2 + 24\sqrt{3}t \Rightarrow$

$0 = t\left(-16t + 24\sqrt{3}\right).$

$t = 0$ or $-16t + 24\sqrt{3} = 0 \Rightarrow$

$-16t = -24\sqrt{3} \Rightarrow t = \dfrac{24\sqrt{3}}{16} \Rightarrow$

$t = \dfrac{3\sqrt{3}}{2} \approx 2.6$

The flight time is about 2.6 seconds. The horizontal distance at $t = \dfrac{3\sqrt{3}}{2}$ is

$x = 24t = 24\left(\dfrac{3\sqrt{3}}{2}\right) \approx 62$ ft

45. (a) $x = (v\cos\theta)t \Rightarrow x = (88\cos 20°)t$

$y = (v\sin\theta)t - 16t^2 + 2 \Rightarrow$

$y = (88\sin 20°)t - 16t^2 + 2$

(b) $t = \dfrac{x}{88\cos 20°}$, so

$y = 88\sin 20°\left(\dfrac{x}{88\cos 20°}\right)$

$\qquad - 16\left(\dfrac{x}{88\cos 20°}\right)^2 + 2$

$= (\tan 20°)x - \dfrac{x^2}{484\cos^2 20°} + 2$

(c) Solving $0 = -16t^2 + (88\sin 20°)t + 2$ by the quadratic formula, we have

$t = \dfrac{-88\sin 20° \pm \sqrt{(88\sin 20°)^2 - 4(-16)(2)}}{(2)(-16)}$

$= \dfrac{-30.098 \pm \sqrt{905.8759 + 128}}{-32} \Rightarrow$

$t \approx -0.064$ or 1.9

Discard $t = -0.064$ because it is an unacceptable answer.

At $t \approx 1.9$ sec, $x = (88\cos 20°)t \approx 161$ ft .

The softball traveled 1.9 sec and 161 feet.

47. (a) $x = (v\cos\theta)t \Rightarrow$

$x = (128\cos 60°)t = 128\left(\dfrac{1}{2}\right)t = 64t$

$y = (v\sin\theta)t - 16t^2 + 8 \Rightarrow$

$y = (128\sin 60°)t - 16t^2 + 8$

$= 128\left(\dfrac{\sqrt{3}}{2}\right)t - 16t^2 + 8$

$= 64\sqrt{3}t - 16t^2 + 8$

$t = \dfrac{x}{64}$, so

$y = 64\sqrt{3}\left(\dfrac{x}{64}\right) - 16\left(\dfrac{x}{64}\right)^2 + 8 \Rightarrow$

$y = -\dfrac{1}{256}x^2 + \sqrt{3}x + 8.$

This is a parabolic path.

(b) Solving $0 = -16t^2 + 64\sqrt{3}t + 8$ by the quadratic formula, we have

$t = \dfrac{-64\sqrt{3} \pm \sqrt{\left(64\sqrt{3}\right)^2 - 4(-16)(8)}}{2(-16)}$

$= \dfrac{-64\sqrt{3} \pm \sqrt{12,800}}{-32}$

$= \dfrac{-64\sqrt{3} \pm 80\sqrt{2}}{-32} \Rightarrow t \approx -0.07, \ 7.0$

Discard $t = -0.07$ because it gives an unacceptable answer. At $t \approx 7.0$ sec, $x = 64t = 448$ ft. The rocket traveled approximately 7 sec and 448 feet.

49. (a) $x = (v\cos\theta)t \Rightarrow$

$x = (64\cos 60°)t = 64\left(\dfrac{1}{2}\right)t = 32t$

$y = (v\sin\theta)t - 16t^2 + 3 \Rightarrow$

$y = (64\sin 60°)t - 16t^2 + 3$

$= 64\left(\dfrac{\sqrt{3}}{2}\right)t - 16t^2 + 3$

$= 32\sqrt{3}t - 16t^2 + 3$

(b) Solving $0 = -16t^2 + 32\sqrt{3}t + 3$ by the quadratic formula, we have

$t = \dfrac{-32\sqrt{3} \pm \sqrt{\left(32\sqrt{3}\right)^2 - 4(-16)(3)}}{2(-16)}$

$= \dfrac{-32\sqrt{3} \pm \sqrt{3264}}{-32} = \dfrac{-32\sqrt{3} \pm 8\sqrt{51}}{-32} \Rightarrow$

$t \approx -0.05, \ 3.52$

Discard $t = -0.05$ because it gives an unacceptable answer. At $t \approx 3.52$ sec, $x = 32t \approx 112.6$ ft. The ball traveled approximately 112.6 feet.

(c) To find the maximum height, find the vertex of $y = -16t^2 + 32\sqrt{3}t + 3$.

$y = -16t^2 + 32\sqrt{3}t + 3$

$= -16\left(t^2 - 2\sqrt{3}t\right) + 3$

$= -16\left(t^2 - 2\sqrt{3}t + 3 - 3\right) + 3$

$y = -16\left(t - \sqrt{3}\right)^2 + 48 + 3$

$= -16\left(t - \sqrt{3}\right)^2 + 51$

The maximum height of 51 ft is reached at $\sqrt{3} \approx 1.73$ sec. because $x = 32t$, the ball has traveled horizontally $32\sqrt{3} \approx 55.4$ ft.

(d) To determine if the ball would clear a 5-ft-high fence that is 100 ft from the batter, we need to first determine at what time is the ball 100 ft from the batter. Because $x = 32t$, the time the ball is 100 ft from the batter is $t = \dfrac{100}{32} = 3.125$ sec.

We next need to determine how high off the ground the ball is at this time. Because $y = 32\sqrt{3}t - 16t^2 + 3$, the height of the ball is

$$y = 32\sqrt{3}(3.125) - 16(3.125)^2 + 3$$

≈ 20.0 ft .This height exceeds 5 ft, so the ball will clear the fence.

For Exercises 51−53, other answers are possible.

51. $y = a(x - h)^2 + k$

To find one parametric representation, let $x = t$. We therefore have, $y = a(t - h)^2 + k$. For another representation, let $x = t + h$. We therefore have $y = a(t + h - h)^2 + k = at^2 + k.$

53. $\dfrac{x^2}{a^2} + \dfrac{y^2}{b^2} = 1$

To find a parametric representation, let $x = a \sin t$. We therefore have

$$\frac{(a\sin t)^2}{a^2} + \frac{y^2}{b^2} = 1 \Rightarrow \frac{a^2 \sin^2 t}{a^2} + \frac{y^2}{b^2} = 1 \Rightarrow$$

$$\sin^2 t + \frac{y^2}{b^2} = 1 \Rightarrow \frac{y^2}{b^2} = 1 - \sin^2 t$$

$$y^2 = b^2\left(1 - \sin^2 t\right) \Rightarrow y^2 = b^2 \cos^2 t \Rightarrow$$

$$y = b \cos t$$

55. To show that $r\theta = a$ is given parametrically by $x = \dfrac{a\cos\theta}{\theta}$, $y = \dfrac{a\sin\theta}{\theta}$, for θ in $(-\infty, 0) \cup (0, \infty)$, we must show that the parametric equations yield $r\theta = a$, where $r^2 = x^2 + y^2$.

$$r^2 = x^2 + y^2 \Rightarrow$$

$$r^2 = \left(\frac{a\cos\theta}{\theta}\right)^2 + \left(\frac{a\sin\theta}{\theta}\right)^2 \Rightarrow$$

$$r^2 = \frac{a^2 \cos^2\theta}{\theta^2} + \frac{a^2 \sin^2\theta}{\theta^2} \Rightarrow$$

$$r^2 = \frac{a^2}{\theta^2}\cos^2\theta + \frac{a^2}{\theta^2}\sin^2\theta \Rightarrow$$

$$r^2 = \frac{a^2}{\theta^2}\left(\cos^2\theta + \sin^2\theta\right) \Rightarrow r^2 = \frac{a^2}{\theta^2} \Rightarrow$$

$$r = \pm\frac{a}{\theta} \text{ or just } r = \frac{a}{\theta}$$

This implies that the parametric equations satisfy $r\theta = a$.

Chapter 8　　Review Exercises

1. Find b, given $C = 74.2°$, $c = 96.3$ m, $B = 39.5°$.
Use the law of sines to find b.

$$\frac{b}{\sin B} = \frac{c}{\sin C} \Rightarrow \frac{b}{\sin 39.5°} = \frac{96.3}{\sin 74.2°} \Rightarrow$$

$$b = \frac{96.3 \sin 39.5°}{\sin 74.2°} \approx 63.7 \text{ m}$$

3. Find B, given $C = 51.3°$, $c = 68.3$ m, $b = 58.2$ m.
Use the law of sines to find B.

$$\frac{\sin B}{b} = \frac{\sin C}{c} \Rightarrow \frac{\sin B}{58.2} = \frac{\sin 51.3°}{68.3} \Rightarrow$$

$$\sin B = \frac{58.2 \sin 51.3°}{68.3} \approx 0.66502269$$

There are two angles B between $0°$ and $180°$ that satisfy the condition. Because $\sin B \approx 0.66502269$, to the nearest tenth, th value of B is $B_1 = 41.7°$.

Supplementary angles have the same sine value, so another possible value of B is $B_2 = 180° - 41.7° = 138.3°$. This is not a valid angle measure for this triangle because $C + B_2 = 51.3° + 138.3° = 189.6° > 180°$. Thus, $B = 41.7°$.

5. Find A, given $B = 39°50'$, $b = 268$ m, $a = 340$ m.
Use the law of sines to find A.

$$\frac{\sin A}{a} = \frac{\sin B}{b} \Rightarrow \frac{\sin A}{340} = \frac{\sin 39°50'}{268} \Rightarrow$$

$$\sin A = \frac{340 \sin 39°50'}{268} \approx 0.81264638$$

There are two angles A between $0°$ and $180°$ that satisfy the condition. Because $\sin A \approx 0.81264638$, to the nearest tenth value of A is $A_1 = 54.4° \approx 54°20'$. Supplementary angles have the same sine value, so another possible value of A is $A_2 = 180° - 54°20' = 179°60' - 54°20' = 125°40'$. This is a valid angle measure for this triangle because $B + A_2 = 39°50' + 125°40' = 165°30' < 180°$.

$A = 54°20'$ or $A = 125°40'$

7. No. If two angles of a triangle are given, then the third angle is known because the sum of the measures of the three angles is 180°. One side is given, so there will only be one triangle that will satisfy the conditions.

9. $a = 10$, $B = 30°$

   (a) The value of b that forms a right triangle would yield exactly one value for A. That is, $b = 10 \sin 30° = 5$. Also, any value of b greater than or equal to 10 would yield a unique value for A.

   (b) Any value of b between 5 and 10, would yield two possible values for A.

   (c) If b is less than 5, then no value for A is possible.

11. Find A, given $a = 86.14$ in., $b = 253.2$ in., $c = 241.9$ in.
    Use the law of cosines to find A.
    $$a^2 = b^2 + c^2 - 2bc \cos A \Rightarrow$$
    $$\cos A = \frac{b^2 + c^2 - a^2}{2bc}$$
    $$= \frac{253.2^2 + 241.9^2 - 86.14^2}{2(253.2)(241.9)}$$
    $$\approx 0.94046923$$
    Thus, $A \approx 19.87°$ or $19°52'$.

13. Find a, given $A = 51°20'$, $c = 68.3$ m, $b = 58.2$ m.
    Use the law of cosines to find a.
    $$a^2 = b^2 + c^2 - 2bc \cos A \Rightarrow$$
    $$a^2 = 58.2^2 + 68.3^2 - 2(58.2)(68.3)\cos 51°20'$$
    $$\approx 3084.99 \Rightarrow a \approx 55.5 \text{ m}$$

15. Find a, given $A = 60°$, $b = 5.0$ cm, $c = 21$ cm.
    Use the law of cosines to find a.
    $$a^2 = b^2 + c^2 - 2bc \cos A \Rightarrow$$
    $$a^2 = 5.0^2 + 21^2 - 2(5.0)(21)\cos 60° = 361 \Rightarrow$$
    $$a = 19 \text{ cm}$$

17. Solve the triangle, given $A = 25.2°$, $a = 6.92$ yd, $b = 4.82$ yd.
    $$\frac{\sin B}{b} = \frac{\sin A}{a} \Rightarrow \sin B = \frac{b \sin A}{a} \Rightarrow$$
    $$\sin B = \frac{4.82 \sin 25.2°}{6.92} \approx 0.29656881$$
    There are two angles B between 0° and 180° that satisfy the condition. because $\sin B \approx 0.29656881$, to the nearest tenth value of B is $B_1 = 17.3°$. Supplementary angles have the same sine value, so another possible value of B is $B_2 = 180° - 17.3° = 162.7°$.

This is not a valid angle measure for this triangle because
$$A + B_2 = 25.2° + 162.7° = 187.9° > 180°.$$
$$C = 180° - A - B \Rightarrow$$
$$C = 180° - 25.2° - 17.3° \Rightarrow C = 137.5°$$
Use the law of sines to find c.
$$\frac{c}{\sin C} = \frac{a}{\sin A} \Rightarrow \frac{c}{\sin 137.5°} = \frac{6.92}{\sin 25.2°} \Rightarrow$$
$$c = \frac{6.92 \sin 137.5°}{\sin 25.2°} \approx 11.0 \text{ yd}$$

19. Solve the triangle, given $a = 27.6$ cm, $b = 19.8$ cm, $C = 42° 30'$.
    This is a SAS case, so using the law of cosines.
    $$c^2 = a^2 + b^2 - 2ab \cos C \Rightarrow$$
    $$c^2 = 27.6^2 + 19.8^2 - 2(27.6)(19.8)\cos 42° 30'$$
    $$\approx 347.985 \Rightarrow c \approx 18.65 \text{ cm}$$
    (rounded to 18.7)
    Of the remaining angles A and B, B must be smaller because it is opposite the shorter of the two sides a and b. Therefore, B cannot be obtuse.
    $$\frac{\sin B}{b} = \frac{\sin C}{c} \Rightarrow \frac{\sin B}{19.8} = \frac{\sin 42°30'}{18.65} \Rightarrow$$
    $$\sin B = \frac{19.8 \sin 42°30'}{18.65} \approx 0.717124859 \Rightarrow$$
    $$B \approx 45.8° \approx 45°50'$$
    Thus,
    $$A = 180° - B - C = 180° - 45° 50' - 42° 30'$$
    $$= 179°60' - 88°20' = 91° 40'$$

21. Given $b = 840.6$ m, $c = 715.9$ m, $A = 149.3°$, find the area.
    Angle A is included between sides b and c. Thus, we have
    $$\mathcal{A} = \frac{1}{2}bc \sin A = \frac{1}{2}(840.6)(715.9)\sin 149.3°$$
    $$\approx 153,600 \text{ m}^2$$
    (rounded to four significant digits)

23. Given $a = 0.913$ km, $b = 0.816$ km, $c = 0.582$ km, find the area.
    Use Heron's formula to find the area.
    $$s = \frac{1}{2}(a+b+c) = \frac{1}{2}(0.913 + 0.816 + 0.582)$$
    $$= \frac{1}{2} \cdot 2.311 = 1.1555$$

(continued on next page)

(*continued*)

$$\mathcal{A} = \sqrt{s(s-a)(s-b)(s-c)}$$
$$= \sqrt{\begin{array}{c}1.1555(1.1555-0.913)\cdot\\(1.1555-0.816)(1.1555-0.582)\end{array}}$$
$$= \sqrt{1.1555(0.2425)(0.3395)(0.5735)}$$
$$\approx 0.234 \text{ km}^2$$

(rounded to three significant digits)

25. $B = 58.4°$ and $C = 27.9°$, so
$A = 180° - B - C = 180° - 58.4° - 27.9° = 93.7°$.
Using the law of sines, we have
$$\frac{AB}{\sin C} = \frac{125}{\sin A} \Rightarrow \frac{AB}{\sin 27.9°} = \frac{125}{\sin 93.7°} \Rightarrow$$
$$AB = \frac{125\sin 27.9°}{\sin 93.7°} \approx 58.61$$
The canyon is 58.6 feet across. (rounded to three significant digits)

27. Let $AC = x = $ the height of the tree.

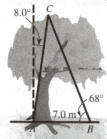

Angle $A = 90° - 8.0° = 82°$
Angle $C = 180° - B - A = 30°$
Use the law of sines to find $AC = b$.
$$\frac{b}{\sin B} = \frac{c}{\sin C}$$
$$\frac{b}{\sin 68°} = \frac{7.0}{\sin 30°}$$
$$b = \frac{7.0\sin 68°}{\sin 30°}$$
$$b \approx 12.98$$
The tree is 13 meters tall (rounded to two significant digits).

29. Let $h = $ the height of tree.
$$\theta = 27.2° - 14.3° = 12.9°$$
$$\alpha = 90° - 27.2° = 62.8°$$
$$\frac{h}{\sin\theta} = \frac{212}{\sin\alpha}$$
$$\frac{h}{\sin 12.9°} = \frac{212}{\sin 62.8°}$$
$$h = \frac{212\sin 12.9°}{\sin 62.8°} \approx 53.21$$

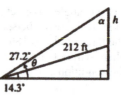

The height of the tree is 53.2 ft. (rounded to three significant digits)

31. Let $x = $ the distance between the boats.
In 3 hours the first boat travels $3(36.2) = 108.6$ km and the second travels $3(45.6) = 136.8$ km.
Use the law of cosines to find x.
$$x^2 = 108.6^2 + 136.8^2$$
$$\qquad - 2(108.6)(136.8)\cos 54°10'$$
$$\approx 13,113.359 \Rightarrow x \approx 115 \text{ km}$$
They are 115 km apart.

33. Use the distance formula to find the distances between the points.
Distance between $(-8, 6)$ and $(0, 0)$:
$$\sqrt{(-8-0)^2 + (6-0)^2} = \sqrt{(-8)^2 + 6^2}$$
$$= \sqrt{64+36} = \sqrt{100} = 10$$
Distance between $(-8, 6)$ and $(3, 4)$:
$$\sqrt{(-8-3)^2 + (6-4)^2} = \sqrt{(-11)^2 + 2^2}$$
$$= \sqrt{121+4} = \sqrt{125}$$
$$= 5\sqrt{5} \approx 11.18$$
Distance between $(3, 4)$ and $(0, 0)$:
$$\sqrt{(3-0)^2 + (4-0)^2} = \sqrt{3^2 + 4^2}$$
$$= \sqrt{9+16} = \sqrt{25} = 5$$
$$s \approx \frac{1}{2}(10 + 11.18 + 5) = \frac{1}{2}\cdot 26.18 = 13.09$$
$$\mathcal{A} = \sqrt{s(s-a)(s-b)(s-c)}$$
$$= \sqrt{\begin{array}{c}13.09(13.09-10)\cdot\\(13.09-11.18)(13.09-5)\end{array}}$$
$$= \sqrt{13.09(3.09)(1.91)(8.09)}$$
$$\approx 25 \text{ sq units (rounded to two significant digits)}$$

35. $\mathbf{a} - \mathbf{b}$

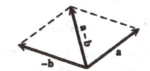

37. $\alpha = 180° - 52° = 128°$

$|\mathbf{v}|^2 = 100^2 + 130^2 - 2(100)(130)\cos 128°$

$|\mathbf{v}|^2 \approx 42907.2 \Rightarrow |\mathbf{v}| \approx 207$ lb

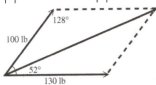

39. $|\mathbf{v}| = 964, \theta = 154°20'$

horizontal:

$x = |\mathbf{v}|\cos\theta = 964\cos 154° \, 20' \approx -869$

vertical: $y = |\mathbf{v}|\sin\theta = 964\sin 154° \, 20' \approx 418$

41. $\mathbf{u} = \langle -9, 12 \rangle$

magnitude:

$|\mathbf{u}| = \sqrt{(-9)^2 + 12^2} = \sqrt{81 + 144} = \sqrt{225} = 15$

Angle:

$\tan\theta' = \dfrac{b}{a} \Rightarrow \tan\theta' = \dfrac{12}{-9} \Rightarrow$

$\theta' = \tan^{-1}\left(-\dfrac{4}{3}\right) \approx -53.1° \Rightarrow$

$\theta = -53.1° + 180° = 126.9°$

(θ lies in quadrant II)

43. $\mathbf{v} = 2\mathbf{i} - \mathbf{j}, \quad \mathbf{u} = -3\mathbf{i} + 2\mathbf{j}$

First write the given vectors in component form.

$\mathbf{v} = 2\mathbf{i} - \mathbf{j} = \langle 2, -1 \rangle$ and $\mathbf{u} = -3\mathbf{i} + 2\mathbf{j} = \langle -3, 2 \rangle$

(a) $2\mathbf{v} + \mathbf{u} = 2\langle 2, -1 \rangle + \langle -3, 2 \rangle$

$= \langle 2 \cdot 2, 2(-1) \rangle + \langle -3, 2 \rangle$

$= \langle 4, -2 \rangle + \langle -3, 2 \rangle$

$= \langle 4 + (-3), -2 + 2 \rangle = \langle 1, 0 \rangle = \mathbf{i}$

(b) $2\mathbf{v} = 2\langle 2, -1 \rangle = \langle 2 \cdot 2, 2(-1) \rangle$

$= \langle 4, -2 \rangle = 4\mathbf{i} - 2\mathbf{j}$

(c) $\mathbf{v} - 3\mathbf{u} = \langle 2, -1 \rangle - 3\langle -3, 2 \rangle$

$= \langle 2, -1 \rangle - \langle 3(-3), 3 \cdot 2 \rangle$

$= \langle 2, -1 \rangle - \langle -9, 6 \rangle$

$= \langle 2 - (-9), -1 - 6 \rangle$

$= \langle 11, -7 \rangle = 11\mathbf{i} - 7\mathbf{j}$

45. $\mathbf{u} = \langle 5, -3 \rangle, \mathbf{v} = \langle 3, 5 \rangle$

$\mathbf{u} \cdot \mathbf{v} = \langle 5, -3 \rangle \cdot \langle 3, 5 \rangle = 5(3) + (-3) \cdot 5$

$= 15 - 15 = 0$

$\cos\theta = \dfrac{\mathbf{u} \cdot \mathbf{v}}{|\mathbf{u}||\mathbf{v}|} \Rightarrow \cos\theta = \dfrac{0}{|\langle 5, -3 \rangle||\langle 3, 5 \rangle|} = 0$

So, $\theta = \cos^{-1} 0 = 90°$

The vectors are orthogonal.

47. Let $|\mathbf{x}|$ be the resultant force.

$\theta = 180° - 15° - 10° = 155°$

$|\mathbf{x}|^2 = 12^2 + 18^2 - 2(12)(18)\cos 155°$

$|\mathbf{x}|^2 \approx 859.5 \Rightarrow |\mathbf{x}| \approx 29$

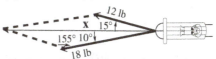

The magnitude of the resultant force on Jessie and the sled is 29 lb.

49. Let $\mathbf{v}$ = the ground speed vector.

$\alpha = 212° - 180° = 32°$ and $\beta = 50°$ because they are alternate interior angles. Angle opposite to 520 is $\alpha + \beta = 82°$.

Using the law of sines, we have

$\dfrac{\sin\theta}{37} = \dfrac{\sin 82°}{520} \Rightarrow \sin\theta = \dfrac{37\sin 82°}{520} \Rightarrow$

$\sin\theta \approx 0.07046138 \Rightarrow \theta \approx 4°$

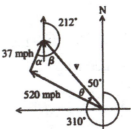

Thus, the bearing is $360° - 50° - \theta = 306°$.

The angle opposite $\mathbf{v}$ is $180° - 82° - 4° = 94°$.

Using the laws of sines, we have

$\dfrac{|\mathbf{v}|}{\sin 94°} = \dfrac{520}{\sin 82°} \Rightarrow$

$|\mathbf{v}| = \dfrac{520\sin 94°}{\sin 82°} \approx 524$ mph

The pilot should fly on a bearing of 306°. Her actual speed is 524 mph.

51. Refer to Example 3 in section 8.3. The magnitude of vector **AC** gives the magnitude of the required force.

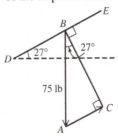

$\sin 27° = \dfrac{|\mathbf{AC}|}{75} \Rightarrow |\mathbf{AC}| = 75\sin 27° \approx 34$ lb

53. $\left[5\left(\cos 90° + i \sin 90°\right)\right]$
$\qquad \cdot \left[6\left(\cos 180° + i \sin 180°\right)\right]$
$= 5 \cdot 6 \left[\cos\left(90° + 180°\right) + i \sin\left(90° + 180°\right)\right]$
$= 30 \left(\cos 270° + i \sin 270°\right)$
$= 30 \left(0 - i\right)$
$= 0 - 30i \ \text{ or } \ -30i$

55. $\dfrac{2\left(\cos 60° + i \sin 60°\right)}{8\left(\cos 300° + \sin 300°\right)}$

$= \dfrac{2}{8}\left[\cos\left(60° - 300°\right) + i \sin\left(60° - 300°\right)\right]$

$= \dfrac{1}{4}\left[\cos\left(-240°\right) + i \sin\left(-240°\right)\right]$

$= \dfrac{1}{4}\left[\cos\left(240°\right) - i \sin\left(240°\right)\right]$

$= \dfrac{1}{4}\left[-\cos 60° + i \sin 60°\right]$

$= \dfrac{1}{4}\left(-\dfrac{1}{2} + \dfrac{\sqrt{3}}{2}\right) = -\dfrac{1}{8} + \dfrac{\sqrt{3}}{8}i$

57. $\left(\sqrt{3} + i\right)^3$

$r = \sqrt{\left(\sqrt{3}\right)^2 + 1^2} = \sqrt{3 + 1} = \sqrt{4} = 2$ and

because θ is in quadrant I,

$\tan \theta = \dfrac{1}{\sqrt{3}} = \dfrac{\sqrt{3}}{3} \Rightarrow \theta = 30°.$

$\left(\sqrt{3} + i\right)^3 = \left[2\left(\cos 30° + i \sin 30°\right)\right]^3$

$= 2^3 \left[\cos\left(3 \cdot 30°\right) + i \sin\left(3 \cdot 30°\right)\right]$

$= 8\left[\cos 90° + i \sin 90°\right]$

$= 8\left(0 + i\right) = 0 + 8i = 8i$

59. $\left(\cos 100° + i \sin 100°\right)^6$

$= \cos\left(6 \cdot 100°\right) + i \sin\left(6 \cdot 100°\right)$

$= \cos 600° + i \sin 600° = \cos 240° + i \sin 240°$

$= -\cos 60° - i \sin 60° = -\dfrac{1}{2} - \dfrac{\sqrt{3}}{2}i$

61.

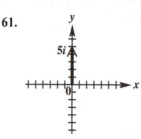

63.

65. $-2 + 2i$

$r = \sqrt{\left(-2\right)^2 + 2^2} = \sqrt{4 + 4} = \sqrt{8} = 2\sqrt{2}$

Because θ is in quadrant II,

$\tan \theta = \dfrac{2}{-2} = -1 \Rightarrow \theta = 135°.$ Thus,

$-2 + 2i = 2\sqrt{2}\left(\cos 135° + i \sin 135°\right).$

67. $2\left(\cos 225° + i \sin 225°\right)$

$= 2\left(-\cos 45° - i \sin 45°\right)$

$= 2\left(-\dfrac{\sqrt{2}}{2} - \dfrac{i\sqrt{2}}{2}\right) = -\sqrt{2} - i\sqrt{2}$

69. $1 - i$

$r = \sqrt{1^2 + \left(-1\right)^2} = \sqrt{1 + 1} = \sqrt{2}$ and

$\tan \theta = \dfrac{-1}{1} = -1 \Rightarrow \theta = 315°,$ because θ is in

quadrant IV. Thus,

$1 - i = \sqrt{2}\left(\cos 315° + i \sin 315°\right).$

71. $-4i$

Because $r = 4$ and the point $(0, -4)$ intersects
the negative y-axis, $\theta = 270°$ and
$-4i = 4(\cos 270° + i \sin 270°).$

73. $z = x + yi$

Because the imaginary part of z is the negative
of the real part of z, we are saying $y = -x$. This
is a line.

75. Convert $1 - i$ to polar form

$r = \sqrt{1^2 + \left(-1\right)^2} = \sqrt{1 + 1} = \sqrt{2}$ and

$\tan \theta = \dfrac{-1}{1} = -1 \Rightarrow \theta = 315°,$ because θ is in

quadrant IV. Thus,

$1 - i = \sqrt{2}\left(\cos 315° + i \sin 315°\right).$

$r^3\left(\cos 3\alpha + i \sin 3\alpha\right)$

$= \sqrt{2}\left(\cos 315° + i \sin 315°\right),$ so we have

$r^3 = \sqrt{2} \Rightarrow r = \sqrt[6]{2}$ and

$3\alpha = 315° + 360° \cdot k \Rightarrow$

$\alpha = \dfrac{315° + 360° \cdot k}{3} = 105° + 120° \cdot k, \ k$ any

integer.

(continued on next page)

(*continued*)

If $k = 0$, then $\alpha = 105° + 0° = 105°$.

If $k = 1$, then $\alpha = 105° + 120° = 225°$.

If $k = 2$, then $\alpha = 105° + 240° = 345°$.

So, the cube roots of $1 - i$ are

$\sqrt[6]{2}\left(\cos 105° + i \sin 105°\right)$,

$\sqrt[6]{2}\left(\cos 225° + i \sin 225°\right)$, and

$\sqrt[6]{2}\left(\cos 345° + i \sin 345°\right)$.

77. The number –64 has no real sixth roots because a real number raised to the sixth power will never be negative.

79. $x^4 + 16 = 0 \Rightarrow x^4 = -16$

We have, $r = 16$ and $\theta = 180°$.

$x^4 = -16 = -16 + 0i = 16\left(\cos 180° + i \sin 180°\right)$

$r^4\left(\cos 4\alpha + i \sin 4\alpha\right)$

$= 16\left(\cos 180° + i \sin 180°\right)$, so we have

$r^4 = 16 \Rightarrow r = 2$ and $4\alpha = 180° + 360° \cdot k \Rightarrow$

$\alpha = \dfrac{180° + 360° \cdot k}{4} = 45° + 90° \cdot k,\ k$ any

integer.

If $k = 0$, then $\alpha = 45° + 0° = 45°$.

If $k = 1$, then $\alpha = 45° + 90° = 135°$.

If $k = 2$, then $\alpha = 45° + 180° = 225°$.

If $k = 3$, then $\alpha = 45° + 270° = 315°$.

Solution set:

$\{2\left(\cos 45° + i \sin 45°\right),\ 2\left(\cos 135° + i \sin 135°\right),$

$2\left(\cos 225° + i \sin 225°\right),$

$2\left(\cos 315° + i \sin 315°\right)\}$

81. $x^2 + i = 0 \Rightarrow x^2 = -i$

We have $r = 1$ and $\theta = 270°$.

$x^2 = -i = 0 - i = 1\left(\cos 270° + i \sin 270°\right)$

$r^2\left(\cos 2\alpha + i \sin 2\alpha\right)$

$= 1\left(\cos 270° + i \sin 270°\right)$, so we have

$r^2 = 1 \Rightarrow r = 1$ and $2\alpha = 270° + 360° \cdot k \Rightarrow$

$\alpha = \dfrac{270° + 360° \cdot k}{2} = 135° + 180° \cdot k,\ k$ any

integer. If $k = 0$, then $\alpha = 135° + 0° = 135°$.

If $k = 1$, then $\alpha = 135° + 180° = 315°$.

Solution set: $\{\cos 135° + i \sin 135°,$

$\cos 315° + i \sin 315°\}$

83. $\left(-1, \sqrt{3}\right)$

$r = \sqrt{\left(-1\right)^2 + \left(\sqrt{3}\right)^2} = \sqrt{1 + 3} = \sqrt{4} = 2$ and

$\theta = \tan^{-1}\left(-\dfrac{\sqrt{3}}{1}\right) = \tan^{-1}\left(-\sqrt{3}\right) = 120°$,

because θ is in quadrant II. Thus, the polar coordinates are $(2, 120°)$.

85. $r = 4 \cos\theta$ is a circle.

θ	0°	30°	45°
$r = 4\cos\theta$	4	3.5	2.8
θ	60°	90°	120°
$r = 4\cos\theta$	2	0	–2
θ	135°	150°	180°
$r = 4\cos\theta$	–2.8	–3.5	–4

$r = 4 \cos\theta$

Graph is retraced in the interval $(180°, 360°)$.

87. $r = 2 \sin 4\theta$ is an eight-leaved rose.

θ	0°	7.5°	15°	22.5°
$r = 2\sin 4\theta$	0	1	$\sqrt{3}$	2
θ	30°	37.5°	45°	52.5°
$r = 2\sin 4\theta$	$\sqrt{3}$	1	0	–1
θ	60°	67.5°	75°	82.5°
$r = 2\sin 4\theta$	$-\sqrt{3}$	–2	$-\sqrt{3}$	–1
θ	90°	52.5°		
$r = 2\sin 4\theta$	0	–1		

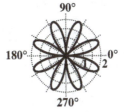

$r = 2 \sin 4\theta$

The graph continues to form eight petals for the interval $[0°, 360°]$.

89. $r = \dfrac{3}{1+\cos\theta}$

$r = \dfrac{3}{1+\cos\theta} \Rightarrow r(1+\cos\theta) = 3 \Rightarrow$

$r + r\cos\theta = 3 \Rightarrow \sqrt{x^2+y^2} + x = 3 \Rightarrow$

$\sqrt{x^2+y^2} = 3 - x$

$x^2 + y^2 = (3-x)^2 \Rightarrow x^2 + y^2 = 9 - 6x + x^2 \Rightarrow$

$y^2 = 9 - 6x \Rightarrow y^2 + 6x - 9 = 0 \Rightarrow y^2 = -6x + 9$

$y^2 = -6\left(x - \dfrac{3}{2}\right)$ or $y^2 + 6x - 9 = 0$

91. $r = 2 \Rightarrow \sqrt{x^2+y^2} = 2 \Rightarrow x^2 + y^2 = 4$

93. $y = x^2 \Rightarrow r\sin\theta = r^2\cos^2\theta \Rightarrow$

$\sin\theta = r\cos^2\theta \Rightarrow r = \dfrac{\sin\theta}{\cos^2\theta} \Rightarrow$

$r = \dfrac{\sin\theta}{\cos\theta} \cdot \dfrac{1}{\cos\theta} = \tan\theta\sec\theta$

$r = \tan\theta\sec\theta$ or $r = \dfrac{\tan\theta}{\cos\theta}$

95. $x = 2$

$x = r\cos\theta \Rightarrow r\cos\theta = 2 \Rightarrow$

$r = \dfrac{2}{\cos\theta}$ or $r = 2\sec\theta$

97. $x + 2y = 4$

$x = r\cos\theta$ and $y = r\sin\theta$, so we have

$r\cos\theta + 2r\sin\theta = 4 \Rightarrow$

$r(\cos\theta + 2\sin\theta) = 4 \Rightarrow r = \dfrac{4}{\cos\theta + 2\sin\theta}$

99. $x = t + \cos t,\ y = \sin t$ for t in $[0, 2\pi]$

t	0	$\frac{\pi}{6}$	$\frac{\pi}{3}$	$\frac{\pi}{2}$	$\frac{3\pi}{4}$	π
$x = t + \cos t$	1	$\frac{\pi}{6} + \frac{\sqrt{3}}{2} \approx 1.4$	$\frac{\pi}{3} + \frac{1}{2} \approx 1.5$	$\frac{\pi}{2} \approx 1.6$	$\frac{3\pi}{4} - \frac{\sqrt{2}}{2} \approx 1.6$	$\pi - 1 \approx 2.1$
$y = \sin t$	0	$\frac{1}{2} = 0.5$	$\frac{\sqrt{3}}{2} \approx 1.7$	1	$\frac{\sqrt{2}}{2} \approx 0.7$	0

t	$\frac{7\pi}{6}$	$\frac{5\pi}{4}$	$\frac{4\pi}{3}$	$\frac{3\pi}{2}$	$\frac{7\pi}{4}$	2π
$x = t + \cos t$	$\frac{7\pi}{6} - \frac{\sqrt{3}}{2} \approx 2.8$	$\frac{5\pi}{4} - \frac{\sqrt{2}}{2} \approx 3.2$	$\frac{4\pi}{3} - \frac{1}{2} \approx 3.7$	$\frac{3\pi}{2} \approx 4.7$	$\frac{7\pi}{4} + \frac{\sqrt{2}}{2} \approx 6.2$	$2\pi + 1 \approx 7.3$
$y = \sin t$	$-\frac{1}{2} = -0.5$	$-\frac{\sqrt{2}}{2} \approx -0.7$	$-\frac{\sqrt{3}}{2} \approx -0.9$	-1	$-\frac{\sqrt{2}}{2} \approx -0.7$	0

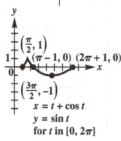

$\left(\frac{\pi}{2}, 1\right)$
$(\pi - 1, 0)\ (2\pi + 1, 0)$
$\left(\frac{3\pi}{2}, -1\right)$
$x = t + \cos t$
$y = \sin t$
for t in $[0, 2\pi]$

101. $x = \sqrt{t-1},\ y = \sqrt{t}$, for t in $[1, \infty)$

$x = \sqrt{t-1} \Rightarrow x^2 = t - 1 \Rightarrow t = x^2 + 1$, so substitute $x^2 + 1$ for t in the equation for y to obtain $y = \sqrt{x^2 + 1}$.

Because t is in $[1, \infty)$, x is in $[\sqrt{1-1}, \infty)$ or $[0, \infty)$.

103. $x = 5 \tan t$, $y = 3 \sec t$, for t in $\left(-\dfrac{\pi}{2}, \dfrac{\pi}{2}\right)$

$\dfrac{x}{5} = \tan t$, $\dfrac{y}{3} = \sec t$, and $1 + \tan^2 t = \sec^2 t$, so

we have $1 + \left(\dfrac{x}{5}\right)^2 = \left(\dfrac{y}{3}\right)^2 \Rightarrow 1 + \dfrac{x^2}{25} = \dfrac{y^2}{9} \Rightarrow$

$9\left(1 + \dfrac{x^2}{25}\right) = y^2 \Rightarrow y = \sqrt{9\left(1 + \dfrac{x^2}{25}\right)} \Rightarrow$

$y = 3\sqrt{1 + \dfrac{x^2}{25}}$.

y is positive because $y = 3\sec t > 0$ for t in

$\left(-\dfrac{\pi}{2}, \dfrac{\pi}{2}\right)$. Because t is in $\left(-\dfrac{\pi}{2}, \dfrac{\pi}{2}\right)$ and

$x = 5\tan t$ is undefined at $-\dfrac{\pi}{2}$ and $\dfrac{\pi}{2}$, x is in

$(-\infty, \infty)$.

105. $x = \cos 2t$, $y = \sin t$ for t in $(-\pi, \pi)$

$\cos 2t = \cos^2 t - \sin^2 t$ (double angle formula)

$\cos^2 t + \sin^2 t = 1$, so we have

$\cos^2 t + \sin^2 t = 1 \Rightarrow$

$\left(\cos^2 t - \sin^2 t\right) + 2\sin^2 t = 1 \Rightarrow$

$x + 2y^2 = 1 \Rightarrow 2y^2 = -x + 1 \Rightarrow 2y^2 = -(x - 1)$

$y^2 = -\dfrac{1}{2}(x - 1)$ or $2y^2 + x - 1 = 0$

$x \geq 0 + 5 = 5$ t is in $(-\pi, \pi)$ and $\cos 2t$ is in

$[-1, 1]$, x is in $[-1, 1]$.

107. (a) $x = (v\cos\theta)t \Rightarrow x = (118\cos 27°)t$ and

$y = (v\sin\theta)t - 16t^2 + h \Rightarrow$

$y = (118\sin 27°)t - 16t^2 + 3.2$

(b) $t = \dfrac{x}{118\cos 27°}$, so we have

$y = 118\sin 27° \cdot \dfrac{x}{118\cos 27°}$

$\quad - 16\left(\dfrac{x}{118\cos 27°}\right)^2 + 3.2$

$\quad = 3.2 - \dfrac{4}{3481\cos^2 27°}x^2 + (\tan 27°)x$

(c) Solving $0 = -16t^2 + (118\sin 27°)t + 3.2$ by

the quadratic formula, we have

$t = \dfrac{-118\sin 27° \pm \sqrt{(118\sin 27°)^2 - 4(-16)(3.2)}}{2(-16)} \Rightarrow$

$t \approx -0.06, 3.406$

Discard $t = -0.06$ sec because it is an
unacceptable answer. At $t = 3.4$ sec, the
baseball traveled

$x = (118\cos 27°)(3.406) \approx 358$ ft.

Chapter 8   Test

1. Find C, given $A = 25.2°$, $a = 6.92$ yd,
$b = 4.82$ yd.
Use the law of sines to first find the measure of
angle B.

$\dfrac{\sin 25.2°}{6.92} = \dfrac{\sin B}{4.82} \Rightarrow \sin B = \dfrac{4.82\sin 25.2°}{6.92} \Rightarrow$

$B = \sin^{-1}\left(\dfrac{4.82\sin 25.2°}{6.92}\right) \approx 17.3°$

Use the fact that the angles of a triangle sum to
180° to find the measure of angle C.
$C = 180° - A - B = 180° - 25.2° - 17.3° = 137.5°$

2. Find c, given $C = 118°$, $b = 131$ km,
$a = 75.0$ km.
Using the law of cosines to find the length
of c.

$c^2 = a^2 + b^2 - 2ab\cos C \Rightarrow c^2$

$\quad = 75.0^2 + 131^2 - 2(75.0)(131)\cos 118°$

$\quad \approx 32011.12 \Rightarrow c \approx 178.9$ km

c is approximately 179 km. (rounded to two
significant digits)

3. Find B, given $a = 17.3$ ft, $b = 22.6$ ft,
$c = 29.8$ ft.
Using the law of cosines, find the measure of
angle B.

$b^2 = a^2 + c^2 - 2ac\cos B \Rightarrow$

$\cos B = \dfrac{a^2 + c^2 - b^2}{2ac} = \dfrac{17.3^2 + 29.8^2 - 22.6^2}{2(17.3)(29.8)}$

$\quad \approx 0.65617605 \Rightarrow B \approx 49.0°$

B is approximately 49.0°.

4. $a = 14$, $b = 30$, $c = 40$
We can use Heron's formula to find the area.

$s = \tfrac{1}{2}(a + b + c) = \tfrac{1}{2}(14 + 30 + 40) = 42$

$\mathcal{A} = \sqrt{s(s - a)(s - b)(s - c)}$

$\quad = \sqrt{42(42 - 14)(42 - 30)(42 - 40)}$

$\quad = \sqrt{42 \cdot 28 \cdot 12 \cdot 2} = \sqrt{28,224} = 168$ sq units

5. This is SAS, so we can use the formula
$\mathcal{A} = \tfrac{1}{2}zy\sin X$.

$\mathcal{A} = \dfrac{1}{2} \cdot 6 \cdot 12\sin 30° = \dfrac{1}{2} \cdot 6 \cdot 12 \cdot \dfrac{1}{2} = 18$ sq units

6. B is obtuse, so b must be the longest side of the triangle.

  (a) $b > 10$

  (b) none

  (c) $b \le 10$

7. $A = 60°, b = 30$ m, $c = 45$ m
This is SAS, so use the law of cosines to find
a: $a^2 = b^2 + c^2 - 2bc \cos A \Rightarrow$

$$a^2 = 30^2 + 45^2 - 2 \cdot 30 \cdot 45 \cos 60° = 1575 \Rightarrow$$

$$a = 15\sqrt{7} \approx 40 \text{ m}$$

Now use the law of sines to find B:

$$\frac{\sin B}{b} = \frac{\sin A}{a} \Rightarrow \frac{\sin B}{30} = \frac{\sin 60°}{15\sqrt{7}} \Rightarrow$$

$$\sin B = \frac{30 \sin 60°}{15\sqrt{7}} \Rightarrow B \approx 41°$$

$$C = 180° - A - B = 180° - 60° - 41° = 79°$$

8. $b = 1075$ in., $c = 785$ in., $C = 38°30'$
We can use the law of sines.

$$\frac{\sin B}{b} = \frac{\sin C}{c} \Rightarrow \frac{\sin B}{1075} = \frac{\sin 38°30'}{785} \Rightarrow$$

$$\sin B = \frac{1075 \sin 38°30'}{785} \Rightarrow$$

$$B_1 \approx 58.5° = 58°30' \text{ or}$$

$$B_2 = 180° - 58°30' = 121°30'$$

Solving separately for triangles
$A_1 B_1 C$ and $A_2 B_2 C,$ we have the following.

$A_1 B_1 C$:

$$A_1 = 180° - B_1 - C = 180° - 58°30' - 38°30'$$
$$= 83°00'$$

$$\frac{a_1}{\sin A_1} = \frac{b}{\sin B_1} \Rightarrow \frac{a_1}{\sin 83°} = \frac{1075}{\sin 58°30'} \Rightarrow$$

$$a_1 = \frac{1075 \sin 83°}{\sin 58°30'} \approx 1251 \approx 1250 \text{ in. (rounded}$$
to three significant digits)

$A_2 B_2 C$:

$$A_2 = 180° - B_2 - C = 180° - 121°30' - 38°30'$$
$$= 20°00'$$

$$\frac{a_2}{\sin A_2} = \frac{b}{\sin B_2} \Rightarrow \frac{a_2}{\sin 20°} = \frac{1075}{\sin 121°30'} \Rightarrow$$

$$a_2 = \frac{1075 \sin 20°}{\sin 121°30'} \approx 431 \text{ in. (rounded to}$$
three significant digits)

9. magnitude:

$$|\mathbf{v}| = \sqrt{(-6)^2 + 8^2} = \sqrt{36 + 64} = \sqrt{100} = 10$$

angle:

$$\tan \theta' = \frac{y}{x} \Rightarrow$$

$$\tan \theta' = \frac{8}{-6} = -\frac{4}{3} \approx -1.33333333 \Rightarrow$$

$$\theta' \approx -53.1° \Rightarrow \theta = -53.1° + 180° = 126.9°$$

(θ lies in quadrant II)
The magnitude $|\mathbf{v}|$ is 10 and $\theta = 126.9°$.

10. $\mathbf{u} = \langle -1, 3 \rangle, \mathbf{v} = \langle 2, -6 \rangle$

  (a) $\mathbf{u} + \mathbf{v} = \langle -1, 3 \rangle + \langle 2, -6 \rangle$
  $$= \langle -1 + 2, 3 + (-6) \rangle = \langle 1, -3 \rangle$$

  (b) $-3\mathbf{v} = -3 \langle 2, -6 \rangle = \langle -3 \cdot 2, -3(-6) \rangle$
  $$= \langle -6, 18 \rangle$$

  (c) $\mathbf{u} \cdot \mathbf{v} = \langle -1, 3 \rangle \cdot \langle 2, -6 \rangle = -1(2) + 3(-6)$
  $$= -2 - 18 = -20$$

  (d) $|\mathbf{u}| = \sqrt{(-1)^2 + 3^2} = \sqrt{1 + 9} = \sqrt{10}$

11. $\mathbf{u} = \langle 4, 3 \rangle, \mathbf{v} = \langle 1, 5 \rangle$

$$\mathbf{u} \cdot \mathbf{v} = \langle 4, 3 \rangle \cdot \langle 1, 5 \rangle = 4(1) + 3(5) = 19$$

$$|\mathbf{u}| = \sqrt{4^2 + 3^2} = \sqrt{25} = 5$$

$$|\mathbf{v}| = \sqrt{1^2 + 5^2} = \sqrt{26}$$

$$\cos \theta = \frac{\mathbf{u} \cdot \mathbf{v}}{|\mathbf{u}||\mathbf{v}|} \Rightarrow \cos \theta = \frac{19}{5\sqrt{26}} \Rightarrow$$

$$\theta = \cos^{-1} \left(\frac{19}{5\sqrt{26}} \right) \approx 41.8°$$

12. Given $A = 24° \ 50', B = 47° \ 20'$ and
$AB = 8.4$ mi, first find the measure of angle C.
$C = 180° - 47° \ 20' - 24° \ 50'$
$\quad = 179°60' - 72° \ 10' = 107° \ 50'$

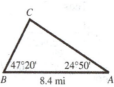

Use this information and the law of sines to
find AC.

$$\frac{AC}{\sin 47° \ 20'} = \frac{8.4}{\sin 107° \ 50'} \Rightarrow$$

$$AC = \frac{8.4 \sin 47° \ 20'}{\sin 107°50'} \approx 6.49 \text{ mi}$$

(continued on next page)

(*continued*)

Drop a perpendicular line from C to segment AB.

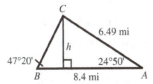

Thus, $\sin 24°50' = \dfrac{h}{6.49} \Rightarrow$

$h \approx 6.49 \sin 24°50' \approx 2.7$ mi.

The balloon is 2.7 miles off the ground.

13. horizontal:

$x = |\mathbf{v}| \cos \theta = 569 \cos 127.5° \approx -346$ and

vertical: $y = |\mathbf{v}| \sin \theta = 569 \sin 127.5° \approx 451$

The vector is $\langle -346, 451 \rangle$.

14.

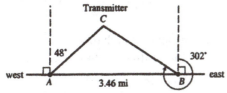

The bearing is $48°$ from A, so angle A in ABC must be $90° - 48° = 42°$. The bearing is $302°$ from B, so angle B in ABC must be $302° - 270° = 32°$. The angles of a triangle sum to $180°$, so

$C = 180° - A - B = 180° - 42° - 32° = 106°$.

Using the law of sines, we have

$\dfrac{b}{\sin B} = \dfrac{c}{\sin C} \Rightarrow \dfrac{b}{\sin 32°} = \dfrac{3.46}{\sin 106°} \Rightarrow$

$b = \dfrac{3.46 \sin 32°}{\sin 106°} \approx 1.91$ mi

The distance from A to the transmitter is 1.91 miles. (rounded to two significant digits)

15.

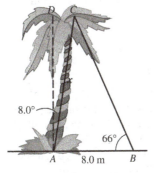

$m\angle DAC = 8.0°$, so

$m\angle CAB = 90° - 8.0° = 82.0°$. $m\angle B = 66°$, so

$m\angle C = 180° - 82° - 66° = 32°$. Now use the law of sines to find AC:

$\dfrac{AC}{\sin B} = \dfrac{AB}{\sin C} \Rightarrow \dfrac{AC}{\sin 66°} = \dfrac{8.0}{\sin 32°} \Rightarrow$

$AC = \dfrac{8.0 \sin 66°}{\sin 32°} \approx 13.8 \approx 14$ m

16. **AX** is the airspeed vector. The plane is flying 630 miles due north in 3 hours, so the ground speed is 210 mph. The measure of angle ACX is $180° - 42° = 138°$.

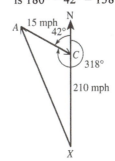

$|\overrightarrow{\mathbf{AX}}|^2 = 15^2 + 210^2 - 2(15)(210)\cos 138°$

$= 49006.8124 \Rightarrow$

$|\overrightarrow{\mathbf{AX}}| \approx 221.3748 \approx 220$ mph (rounded to two

significant digits)

Using the law of sines to find the measure of angle X, we have

$\dfrac{\sin X}{15} = \dfrac{\sin 138°}{220} \Rightarrow \sin X = \dfrac{15 \sin 138°}{220} \Rightarrow$

$X = \sin^{-1}\left(\dfrac{15 \sin 138°}{220}\right) \approx 2.6°$

The plane's bearing is $360° - 2.6° = 357.4° \approx 357°$.

17. $|\overrightarrow{\mathbf{AC}}| = 16.0$ lb; $|\overrightarrow{\mathbf{BA}}| = 50.0$ lb

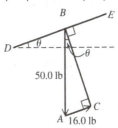

$\sin \theta = \dfrac{16.0}{50.0} \Rightarrow \theta = \sin^{-1}\left(\dfrac{16.0}{50.0}\right) \approx 18.7°$

18. Forces of 135 newtons and 260 newtons, forming an angle of $115°$

$180° - 115° = 65°$

$|\mathbf{u}|^2 = 135^2 + 260^2 - 2(135)(260)\cos 65°$

$|\mathbf{u}|^2 \approx 56157.2 \Rightarrow |\mathbf{u}| \approx 237$ newtons

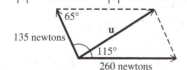

19. $w = 2 - 4i, z = 5 + i$

$w + z = (2 - 4i) + (5 + i) = 7 - 3i$

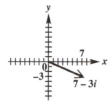

20. (a) $3i$

$r = \sqrt{0^2 + 3^2} = \sqrt{0 + 9} = \sqrt{9} = 3$

The point $(0, 3)$ is on the positive y-axis, so, $\theta = 90°$. Thus,

$3i = 3(\cos 90° + i \sin 90°)$.

(b) $1 + 2i$

$r = \sqrt{1^2 + 2^2} = \sqrt{1 + 4} = \sqrt{5}$

Because θ is in quadrant I,

$\theta = \tan^{-1}\left(\dfrac{2}{1}\right) = \tan^{-1} 2 \approx 63.43°$. Thus,

$1 + 2i = \sqrt{5}(\cos 63.43° + i \sin 63.43°)$.

(c) $-1 - \sqrt{3}i$

$r = \sqrt{(-1)^2 + (-\sqrt{3})^2} = \sqrt{1 + 3} = \sqrt{4} = 2$

Because θ is in quadrant III,

$\theta = \tan^{-1}\left(\dfrac{-\sqrt{3}}{-1}\right) = \tan^{-1}\sqrt{3} = 240°$.

Thus, $-1 - \sqrt{3}i = 2(\cos 240° + i \sin 240°)$

21. (a) $3(\cos 30° + i \sin 30°) = 3\left(\dfrac{\sqrt{3}}{2} + \dfrac{1}{2}i\right)$

$= \dfrac{3\sqrt{3}}{2} + \dfrac{3}{2}i$

(b) $4 \operatorname{cis} 40° = 3.06 + 2.57i$

(c) $3(\cos 90° + i \sin 90°) = 3(0 + 1 \cdot i)$

$= 0 + 3i = 3i$

22. $w = 8(\cos 40° + i \sin 40°)$,

$z = 2(\cos 10° + i \sin 10°)$

(a) wz

$= 8 \cdot 2\left[\cos(40° + 10°) + i \sin(40° + 10°)\right]$

$= 16(\cos 50° + i \sin 50°)$

(b) $\dfrac{w}{z} = \dfrac{8}{2}\left[\cos(40° - 10°) + i \sin(40° - 10°)\right]$

$= 4(\cos 30° + i \sin 30°) = 4\left(\dfrac{\sqrt{3}}{2} + \dfrac{1}{2}i\right)$

$= 2\sqrt{3} + 2i$

(c) $z^3 = \left[2(\cos 10° + i \sin 10°)\right]^3$

$= 2^3(\cos 3 \cdot 10° + i \sin 3 \cdot 10°)$

$= 8(\cos 30° + i \sin 30°)$

$= 8\left(\dfrac{\sqrt{3}}{2} + \dfrac{1}{2}i\right) = 4\sqrt{3} + 4i$

23. Find all the fourth roots of

$-16i = 16(\cos 270° + i \sin 270°)$.

$r^4(\cos 4\alpha + i \sin 4\alpha)$

$= 16(\cos 270° + i \sin 270°)$, so we have

$r^4 = 16 \Rightarrow r = 2$ and $4\alpha = 270° + 360° \cdot k \Rightarrow$

$\alpha = \dfrac{270° + 360° \cdot k}{4} = 67.5° + 90° \cdot k$, k any

integer. If $k = 0$, then $\alpha = 67.5°$.

If $k = 1$, then $\alpha = 157.5°$.

If $k = 2$, then $\alpha = 247.5°$. If $k = 3$, then $\alpha = 337.5°$.

The fourth roots of $-16i$ are

$2(\cos 67.5° + \sin 67.5)$,

$2(\cos 157.5° + i \sin 157.5°)$,

$2(\cos 247.5° + i \sin 247.5°)$, and

$2(\cos 337.5° + i \sin 337.5°)$.

24. Answers may vary.

(a) $(0, 5)$

$r = \sqrt{0^2 + 5^2} = \sqrt{0 + 25} = \sqrt{25} = 5$

The point $(0, 5)$ is on the positive y-axis. Thus, $\theta = 90°$. One possibility is $(5, 90°)$. Alternatively, if $\theta = 90° - 360° = -270°$, a second possibility is $(5, -270°)$.

(b) $(-2, -2)$

$$r = \sqrt{(-2)^2 + (-2)^2} = \sqrt{4+4} = \sqrt{8} = 2\sqrt{2}$$

Because θ is in quadrant III, $\theta = \tan^{-1}\left(\dfrac{-2}{-2}\right) = \tan^{-1} 1 = 225°$. One possibility is $\left(2\sqrt{2}, 225°\right)$.

Alternatively, if $\theta = 225° - 360° = -135°$, a second possibility is $\left(2\sqrt{2}, -135°\right)$.

25. (a) $(3, 315°)$

$$x = r\cos\theta \Rightarrow x = 3\cos 315° = 3 \cdot \frac{\sqrt{2}}{2} = \frac{3\sqrt{2}}{2} \text{ and } y = r\sin\theta \Rightarrow y = 3\sin 315° = 3\left(-\frac{\sqrt{2}}{2}\right) = \frac{-3\sqrt{2}}{2}.$$

The rectangular coordinates are $\left(\dfrac{3\sqrt{2}}{2}, \dfrac{-3\sqrt{2}}{2}\right)$.

(b) $(-4, 90°)$

$$x = r\cos\theta \Rightarrow x = -4\cos 90° = 0 \text{ and } y = r\sin\theta \Rightarrow y = -4\sin 90° = -4$$

The rectangular coordinates are $(0, -4)$.

26. $r = 1 - \cos\theta$ is a cardioid.

θ	0°	30°	45°	60°	90°	135°
$r = 1 - \cos\theta$	0	0.1	0.3	0.5	1	1.7

θ	180°	225°	270°	315°	360°
$r = 1 - \cos\theta$	2	1.7	1	0.3	0

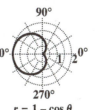

$r = 1 - \cos\theta$

27. $r = 3\cos 3\theta$ is a three-leaved rose.

θ	0°	30°	45°	60°	90°	120°	135°	150°	180°
$r = 3\cos 3\theta$	3	0	−2.1	−3	0	3	2.1	0	−3

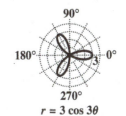

$r = 3\cos 3\theta$

Graph is retraced in the interval $(180°, 360°)$.

28. (a) $r = \dfrac{4}{2\sin\theta - \cos\theta} = \dfrac{4}{-1 \cdot \cos\theta + 2\sin\theta}$, so we can use the general form for the polar equation of a line,

$r = \dfrac{c}{a\cos\theta + b\sin\theta}$, with $a = -1$, $b = 2$, and $c = 4$, we have $-x + 2y = 4$ or $x - 2y = -4$. The graph is a line with intercepts $(-4, 0)$ and $(0, 2)$.

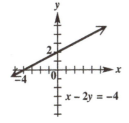

$x - 2y = -4$

(b) $r = 6$ represents the equation of a circle centered at the origin with radius 6, namely $x^2 + y^2 = 36$.

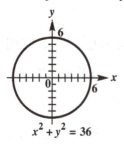

$x^2 + y^2 = 36$

29. $x = 4t - 3$, $y = t^2$ for t in $[-3, 4]$

t	x	y
-3	-15	9
-1	-7	1
0	-3	0
1	1	1
2	5	4
4	13	16

$x = 4t - 3 \Rightarrow t = \dfrac{x+3}{4}$ and $y = t^2$, so we have $y = \left(\dfrac{x+3}{4}\right)^2 = \dfrac{1}{4}(x+3)^2$, where x is in $[-15, 13]$

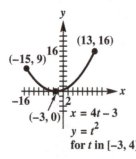

$(-15, 9)$ $(13, 16)$ $(-3, 0)$ $x = 4t - 3$ $y = t^2$ for t in $[-3, 4]$

30. $x = 2\cos 2t$, $y = 2\sin 2t$ for t in $[0, 2\pi]$

t	0	$\frac{\pi}{8}$	$\frac{\pi}{4}$	$\frac{3\pi}{8}$	$\frac{\pi}{2}$	$\frac{5\pi}{8}$	$\frac{3\pi}{4}$	π	$\frac{5\pi}{4}$	$\frac{3\pi}{2}$	$\frac{7\pi}{4}$	2π
x	2	$\sqrt{2}$	0	$-\sqrt{2}$	-2	$-\sqrt{2}$	0	2	0	-2	0	2
y	0	$\sqrt{2}$	2	$\sqrt{2}$	0	$-\sqrt{2}$	-2	0	2	0	-2	0

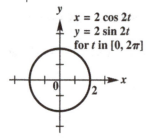

$x = 2\cos 2t$ $y = 2\sin 2t$ for t in $[0, 2\pi]$

$x = 2\cos 2t \Rightarrow \cos 2t = \dfrac{x}{2}$, $y = 2\sin 2t \Rightarrow \sin 2t = \dfrac{y}{2}$, and $\cos^2(2t) + \sin^2(2t) = 1$, so we have

$\left(\dfrac{x}{2}\right)^2 + \left(\dfrac{y}{2}\right)^2 = 1 \Rightarrow \dfrac{x^2}{4} + \dfrac{y^2}{4} = 1 \Rightarrow x^2 + y^2 = 4$, where x is in $[-1, 1]$.

Chapter 9

SYSTEMS AND MATRICES

Section 9.1   Systems of Linear Equations

1. The solution set of the following system is
    $\left\{(1, \underline{4})\right\}$.
    $$-2x + 5y = 18$$
    $$x + y = 5$$
    Find the value for y by substituting 1 for x in either equation and then solving for y.

3. One way of solving the following system by elimination is to multiply equation (2) by the integer $\underline{-11}$ to eliminate the y-terms by direct addition.
    $$14x + 11y = 80 \quad (1)$$
    $$2x + y = 19 \quad (2)$$

5. If a system of linear equation in two variables has two graphs that coincide, there are <u>infinitely many</u> solutions to the system.

7. $4x + 3y = -13 \quad (1)$
    $-x + y = 5 \quad (2)$
    Solve equation (2) for y.
    $-x + y = 5 \quad (2)$
    $y = x + 5 \quad (3)$
    Replace y with $x + 5$ in equation (1), and solve for x.
    $$4x + 3y = -13 \Rightarrow 4x + 3(x+5) = -13 \Rightarrow$$
    $$4x + 3x + 15 = -13 \Rightarrow 7x + 15 = -13 \Rightarrow$$
    $$7x = -28 \Rightarrow x = -4$$
    Replace x with -4 in equation (3) to obtain
    $y = -4 + 5 = 1$.
    Check:

$4x + 3y = -13$ (1)	$-x + y = 5$ (2)
$4(-4) + 3(1) = -13$?	$-(-4) + 1 = 5$?
$-16 + 3 = -13$	$4 + 1 = 5$
$-13 = -13$ True	$5 = 5$ True

    Solution set: $\left\{(-4, 1)\right\}$

9. $x - 5y = 8 \quad (1)$
    $x = 6y \quad (2)$
    Replace x with $6y$ in equation (1), and solve
    for y: $x - 5y = 8 \Rightarrow 6y - 5y = 8 \Rightarrow y = 8$
    Replace y with 8 in equation (2) to obtain
    $x = 6(8) = 48$.
    Check:

$x - 5y = 8$ (1)	$x = 6y$ (2)
$48 - 5(8) = 8$?	$48 = 6(8)$?
$48 - 40 = 8$	$48 = 48$ True
$8 = 8$ True	

    Solution set: $\left\{(48, 8)\right\}$

11. $8x - 10y = -22 \quad (1)$
    $3x + y = 6 \quad (2)$
    Solve equation (2) for y.
    $3x + y = 6 \quad (2)$
    $y = -3x + 6 \quad (3)$
    Replace y with $-3x + 6$ in equation (1), and solve for x.
    $$8x - 10y = -22$$
    $$8x - 10(-3x + 6) = -22$$
    $$8x + 30x - 60 = -22$$
    $$38x - 60 = -22$$
    $$38x = 38$$
    $$x = 1$$
    Replace x with 1 in equation (3) to obtain
    $y = -3(1) + 6 = 3$.
    Check:

$8x - 10y = -22$ (1)	$3x + y = 6$ (2)
$8(1) - 10(3) = -22$?	$3(1) + 3 = 6$?
$8 - 30 = -22$	$3 + 3 = 6$
$-22 = -22$ True	$6 = 6$ True

    Solution set: $\left\{(1, 3)\right\}$

13. $7x - y = -10$ (1)
$3y - x = 10$ (2)

Solve equation (1) for y.
$7x - y = -10$ (1)
$y = 7x + 10$ (3)

Replace y with $7x + 10$ in equation (2), and solve for x.
$$3y - x = 10 \Rightarrow 3(7x + 10) - x = 10 \Rightarrow$$
$$21x + 30 - x = 10 \Rightarrow 20x + 30 = 10 \Rightarrow$$
$$20x = -20 \Rightarrow x = -1$$

Replace x with -1 in equation (3) to obtain
$y = 7(-1) + 10 = 3$.
Check:

$$
\begin{array}{l|l}
7x - y = -10 \;\; (1) & 3y - x = 10 \;\; (2) \\
7(-1) - 3 = -10 \;\; ? & 3(3) - (-1) = 10 \;\; ? \\
-7 - 3 = -10 & 9 + 1 = 10 \\
-10 = -10 \;\text{True} & 10 = 10 \;\text{True}
\end{array}
$$

Solution set: $\{(-1, 3)\}$

15. $-2x = 6y + 18$ (1)
$-29 = 5y - 3x$ (2)

Solve equation (1) for x.
$-2x = 6y + 18$ (1)
$x = -3y - 9$ (3)

Replace x with $-3y - 9$ in equation (2), and solve for y.
$$-29 = 5y - 3(-3y - 9)$$
$$-29 = 5y + 9y + 27 \Rightarrow -29 = 14y + 27 \Rightarrow$$
$$-56 = 14y \Rightarrow -4 = y$$

Replace y with -4 in equation (3) to obtain
$x = -3(-4) - 9 = 12 - 9 = 3$.

Check: $-2x = 6y + 18$ (1)
$$-2(3) = 6(-4) + 18 \;\; ?$$
$$-6 = -24 + 18 \Rightarrow -6 = -6 \;\text{True}$$
$$-29 = 5y - 3x \;\;\;\; (2)$$
$$-29 = 5(-4) - 3(3) \;\; ?$$
$$-29 = -20 - 9 \Rightarrow -29 = -29 \;\text{True}$$

Solution set: $\{(3, -4)\}$

17. $3y = 5x + 6$ (1)
$x + y = 2$ (2)

Solve equation (2) for y. (You could solve equation (2) for x just as easily.)
$x + y = 2$ (2)
$y = 2 - x$ (3)

Replace y with $2 - x$ in equation (1), and solve for x.
$$3y = 5x + 6 \Rightarrow 3(2 - x) = 5x + 6 \Rightarrow$$
$$6 - 3x = 5x + 6 \Rightarrow 6 = 8x + 6 \Rightarrow$$
$$0 = 8x \Rightarrow 0 = x$$

Replace x with 0 in equation (3) to obtain
$y = 2 - 0 = 2$.

$$
\begin{array}{l|l}
Check: \;\; 3y = 5x + 6 \;\; (1) & x + y = 2 \;\; (2) \\
\quad\quad\quad 3(2) = 5(0) + 6 \;\; ? & 0 + 2 = 2 \;\; ? \\
\quad\quad\quad\quad\; 6 = 0 + 6 & \quad\; 2 = 2 \;\text{True} \\
\quad\quad\quad\quad\; 6 = 6 \quad\quad \text{True} &
\end{array}
$$

Solution set: $\{(0, 2)\}$

19. $3x - y = -4$ (1)
$x + 3y = 12$ (2)

Multiply equation (2) by -3 and add the result to equation (1).
$$
\begin{array}{rcl}
3x - y &=& -4 \\
-3x - 9y &=& -36 \\
\hline
-10y &=& -40 \Rightarrow y = 4
\end{array}
$$

Substitute 4 for y in equation (2) and solve for x: $x + 3(4) = 12 \Rightarrow x + 12 = 12 \Rightarrow x = 0$

Check:

$$
\begin{array}{l|l}
3x - y = -4 \;\; (1) & x + 3y = 12 \;\; (2) \\
3(0) - 4 = -4 \;\; ? & 0 + 3(4) = 12 \;\; ? \\
\quad 0 - 4 = -4 & \quad 0 + 12 = 12 \\
\quad\quad -4 = -4 \;\text{True} & \quad\quad 12 = 12 \;\text{True}
\end{array}
$$

Solution set: $\{(0, 4)\}$

21. $2x - 3y = -7$ (1)
$5x + 4y = 17$ (2)

Multiply equation (1) by 4 and equation (2) by 3 and then add the resulting equations.
$$
\begin{array}{rcl}
8x - 12y &=& -28 \\
15x + 12y &=& 51 \\
\hline
23x &=& 23 \Rightarrow x = 1
\end{array}
$$

Substitute 1 for x in equation (2) and solve for y.
$$5(1) + 4y = 17 \Rightarrow 5 + 4y = 17 \Rightarrow$$
$$4y = 12 \Rightarrow y = 3$$

Check:

$$
\begin{array}{l|l}
2x - 3y = -7 \;\; (1) & 5x + 4y = 17 \;\; (2) \\
2(1) - 3(3) = -7 \;\; ? & 5(1) + 4(3) = 17 \;\; ? \\
\quad\quad 2 - 9 = -7 & \quad\quad 5 + 12 = 17 \\
\quad\quad\quad -7 = -7 \;\text{True} & \quad\quad\quad 17 = 17 \;\text{True}
\end{array}
$$

Solution set: $\{(1, 3)\}$

23. $5x + 7y = 6$ (1)
$10x - 3y = 46$ (2)

Multiply equation (1) by –2 and add to equation (2).

$-10x - 14y = -12$
$\underline{10x - \ \ 3y = \ \ 46}$
$\qquad -17y = \ \ 34 \Rightarrow y = -2$

Substitute –2 for y in equation (2) and solve for x.

$10x - 3(-2) = 46 \Rightarrow 10x + 6 = 46$
$\qquad 10x = 40 \Rightarrow x = 4$

Check: $5x + 7y = 6$ (1)
$\qquad 5(4) + 7(-2) = 6$?
$\qquad\qquad 20 - 14 = 6$
$\qquad\qquad\qquad 6 = 6$ True

$\qquad 10x - 3y = 46$ (2)
$\qquad 10(4) - 3(-2) = 46$?
$\qquad\qquad 40 + 6 = 46$
$\qquad\qquad 46 = 46$ True

Solution set: $\{(4, -2)\}$

25. $6x + 7y + 2 = 0$ (1)
$7x - 6y - 26 = 0$ (2)

Multiply equation (1) by 6 and equation (2) by 7 and then add the resulting equations.
$36x + 42y + 12 = 0$
$\underline{49x - 42y - 182 = 0}$
$\overline{85x \qquad -170 = 0} \Rightarrow 85x = 170 \Rightarrow x = 2$

Substitute 2 for x in equation (1).

$6(2) + 7y + 2 = 0$
$\quad 12 + 7y + 2 = 0$
$\qquad\quad 7y + 14 = 0$
$\qquad\qquad 7y = -14 \Rightarrow y = -2$

Check: $6x + 7y + 2 = 0$ (1)
$\qquad 6(2) + 7(-2) + 2 = 0$?
$\qquad\qquad 12 - 14 + 2 = 0 \Rightarrow 0 = 0$ True

$\qquad 7x - 6y - 26 = 0$ (2)
$\qquad 7(2) - 6(-2) - 26 = 0$?
$\qquad\qquad 14 + 12 - 26 = 0 \Rightarrow 0 = 0$ True

Solution set: $\{(2, -2)\}$

27. $\frac{x}{2} + \frac{y}{3} = 4$ (1)
$\frac{3x}{2} + \frac{3y}{2} = 15$ (2)

To clear denominators, multiply equation (1) by 6 and equation (2) by 2.

$6\left(\frac{x}{2} + \frac{y}{3}\right) = 6(4)$

$2\left(\frac{3x}{2} + \frac{3y}{2}\right) = 2(15)$

This gives the system
$3x + 2y = 24$ (3)
$3x + 3y = 30$ (4).

Multiply equation (3) by –1 and add to equation (4).
$-3x - 2y = -24$
$\underline{3x + 3y = \ \ 30}$
$\qquad\quad y = \ \ 6$

Substitute 6 for y in equation (1).
$\frac{x}{2} + \frac{6}{3} = 4 \Rightarrow \frac{x}{2} + 2 = 4 \Rightarrow \frac{x}{2} = 2 \Rightarrow x = 4$

Check: $\frac{x}{2} + \frac{y}{3} = 4$ (1) $\Big|$ $\frac{3x}{2} + \frac{3y}{2} = 15$ (2)
$\quad \frac{4}{2} + \frac{6}{3} = 4$? $\Big|$ $\frac{3(4)}{2} + \frac{3(6)}{2} = 15$?
$\qquad 2 + 2 = 4$ $\Big|$ $\qquad 6 + 9 = 15$
$\qquad\qquad 4 = 4$ True$\Big|$ $\qquad 15 = 15$ True

Solution set: $\{(4, 6)\}$

29. $\frac{2x-1}{3} + \frac{y+2}{4} = 4$ (1)
$\frac{x+3}{2} - \frac{x-y}{3} = 3$ (2)

Multiply equation (1) by 12 and equation (2) by 6 to clear denominators. Also, remove parentheses and combine like terms.

$12\left(\frac{2x-1}{3} + \frac{y+2}{4}\right) = 12(4)$
$4(2x - 1) + 3(y + 2) = 48$
$\quad 8x - 4 + 3y + 6 = 48 \Rightarrow 8x + 3y = 46$ (3)
$\qquad 6\left(\frac{x+3}{2} - \frac{x-y}{3}\right) = 6(3)$
$3(x + 3) - 2(x - y) = 18$
$\quad 3x + 9 - 2x + 2y = 18 \Rightarrow x + 2y = 9$ (4)

Multiply equation (4) by –8 and then add the result to equation (3).
$8x + \ \ 3y = \ \ 46$
$\underline{-8x - 16y = -72}$
$\qquad -13y = -26 \Rightarrow y = 2$

Substitute 2 for y into equation (4) and solve for x: $x + 2(2) = 9 \Rightarrow x + 4 = 9 \Rightarrow x = 5$

Check:

$\frac{2x-1}{3} + \frac{y+2}{4} = 4$ (1) $\Big|$ $\frac{x+3}{2} - \frac{x-y}{3} = 3$ (2)
$\frac{2(5)-1}{3} + \frac{2+2}{4} = 4$? $\Big|$ $\frac{5+3}{2} - \frac{5-2}{3} = 3$?
$\quad \frac{10-1}{3} + \frac{4}{4} = 4$ $\Big|$ $\qquad \frac{8}{2} - \frac{3}{3} = 3$
$\qquad \frac{9}{3} + 1 = 4$ $\Big|$ $\qquad 4 - 1 = 3$
$\qquad 3 + 1 = 4$ $\Big|$ $\qquad 3 = 3$ True
$\qquad 4 = 4$ True$\Big|$

Solution set: $\{(5, 2)\}$

31.
$$9x - 5y = 1 \quad (1)$$
$$-18x + 10y = 1 \quad (2)$$

Multiply equation (1) by 2 and add the result to equation (2).
$$18x - 10y = 2$$
$$\underline{-18x + 10y = 1}$$
$$0 = 3$$

This is a false statement. The solution set is $\varnothing$, and the system is inconsistent.

33.
$$4x - y = 9 \quad (1)$$
$$-8x + 2y = -18 \quad (2)$$

Multiply equation (1) by 2 and add the result to equation (2).
$$8x - 2y = 18$$
$$\underline{-8x + 2y = -18}$$
$$0 = 0$$

This is a true statement. There are infinitely many solutions. We will now express the solution set with y as the arbitrary variable. Solve equation (1) for x.
$$4x - y = 9 \Rightarrow 4x = y + 9 \Rightarrow x = \tfrac{y+9}{4}$$

Solution set: $\left\{\left(\tfrac{y+9}{4}, y\right)\right\}$

35.
$$5x - 5y - 3 = 0 \quad (1)$$
$$x - y - 12 = 0 \quad (2)$$

Multiply equation (2) by –5 and then add the resulting equations.
$$5x - 5y - 3 = 0$$
$$\underline{-5x + 5y + 60 = 0}$$
$$57 = 0$$

This is a false statement. The solution set is $\varnothing$, and the system is inconsistent.

37.
$$7x + 2y = 6 \quad (1)$$
$$14x + 4y = 12 \quad (2)$$

Multiply equation (1) by –2 and add the result to equation (2).
$$-14x - 4y = -12$$
$$\underline{14x + 4y = 12}$$
$$0 = 0$$

This is a true statement. There are infinitely many solutions. We will express the solution set with y as the arbitrary variable. Solve equation (1) for x.
$$7x + 2y = 6 \Rightarrow 7x = 6 - 2y \Rightarrow x = \tfrac{6-2y}{7}$$

Solution set: $\left\{\left(\tfrac{6-2y}{7}, y\right)\right\}$

39.
$$2x - 6y = 0 \quad (1)$$
$$-7x + 21y = 10 \quad (2)$$

Multiply equation (1) by $\tfrac{7}{2}$ and then add the result to equation (1).
$$7x - 21y = 0$$
$$\underline{-7x + 21y = 10}$$
$$0 = 10$$

This is a false statement. The solution set is $\varnothing$, and the system is inconsistent.

41. Because $y = ax + b$ and the line passes through (2, 0) and (0, 3), we have the equations $0 = a(2) + b$ and $3 = a(0) + b$
These becomes the following system.
$$2a + b = 0 \quad (1)$$
$$b = 3 \quad (2)$$
Substitute $b = 3$ into equation (1) to solve for a: $2a + 3 = 0 \Rightarrow 2a = -3 \Rightarrow a = -\tfrac{3}{2}$. Thus, the equation is $y = -\tfrac{3}{2}x + 3 \Rightarrow 3x + 2y = 6$.
The other line goes through the points (0, 1) and (–3, 0), so we have the equations $1 = a(0) + b$ and $0 = a(-3) + b$. These become the following system:
$$b = 1 \quad (3)$$
$$-3a + b = 0 \quad (4)$$
Substitute $b = 1$ into equation (4), then solve for a:
$$-3a + 1 = 0 \Rightarrow a = \tfrac{1}{3}$$
Thus, the equation of this line is
$$y = \tfrac{1}{3}x + 1 \Rightarrow 3y = x + 3 \Rightarrow x - 3y = -3$$
The equations of the two lines are $3x + 2y = 6$ and $x - 3y = -3$.

43. Solve each equation for y.
$$\tfrac{11}{3}x + y = 0.5 \Rightarrow y = 0.5 - \tfrac{11}{3}x \text{ and}$$
$$0.6x - y = 3 \Rightarrow -y = 3 - 0.6x \Rightarrow y = 0.6x - 3.$$
Then graph each equation in the window $[-10, 10] \times [-10, 10]$ and find the intersection.

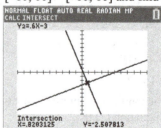

Solution set: $\{(0.820, -2.508)\}$

45. Solve each equation for y.

$$\sqrt{7}x + \sqrt{2}y = 3 \Rightarrow \sqrt{2}y = 3 - \sqrt{7}x \Rightarrow$$

$$y = \frac{3 - \sqrt{7}x}{\sqrt{2}} \text{ and } \sqrt{6}x - y = \sqrt{3} \Rightarrow$$

$$y = \sqrt{6}x - \sqrt{3}.$$

Then graph each equation in the window $[-10, 10] \times [-10, 10]$ and find the intersection.

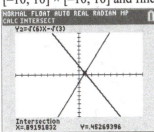

Solution set: $\{0.892, 0.453)\}$

47. $x + y + z = 2$ (1)
$2x + y - z = 5$ (2)
$x - y + z = -2$ (3)

Eliminate z by adding equations (1) and (2) to get $3x + 2y = 7$ (4).

Eliminate z by adding equations (2) and (3) to get $3x = 3 \Rightarrow x = 1$ (5).

Using $x = 1$, find y from equation (4) by substitution.

$$3(1) + 2y = 7 \Rightarrow 3 + 2y = 7 \Rightarrow 2y = 4 \Rightarrow y = 2$$

Substitute 1 for x and 2 for y in equation (1) to find z.

$$1 + 2 + z = 2 \Rightarrow 3 + z = 2 \Rightarrow z = -1$$

Verify that the ordered triple $(1, 2, -1)$ satisfies all three equations.

Check:

$x + y + z = 2$ (1)	$2x + y - z = 5$ (2)
$1 + 2 + (-1) = 2$?	$2(1) + 2 - (-1) = 5$?
$2 = 2$ True	$2 + 2 + 1 = 5$
	$5 = 5$ True

$$x - y + z = -2 \ (3)$$
$$1 - 2 + (-1) = -2 \ ?$$
$$-1 - 1 = -2$$
$$-2 = -2 \text{ True}$$

Solution set: $\{(1, 2, -1)\}$

49. $x + 3y + 4z = 14$ (1)
$2x - 3y + 2z = 10$ (2)
$3x - y + z = 9$ (3)

Eliminate y by adding equations (1) and (2) to get $3x + 6z = 24$ (4).

Multiply equation (3) by 3 and add the result to equation (1).

$$\begin{array}{r} x + 3y + 4z = 14 \\ 9x - 3y + 3z = 27 \\ \hline 10x + 7z = 41 \ (5) \end{array}$$

Multiply equation (4) by 10 and equation (5) by -3 and add in order to eliminate y.

$$\begin{array}{r} 30x + 60z = 240 \\ -30x - 21z = -123 \\ \hline 39z = 117 \Rightarrow z = 3 \end{array}$$

Using $z = 3$, find x from equation (5) by substitution.

$$10x + 7(3) = 41 \Rightarrow 10x + 21 = 41 \Rightarrow$$
$$10x = 20 \Rightarrow x = 2$$

Substitute 2 for x and 3 for z in equation (1) to find y.

$$2 + 3y + 4(3) = 14 \Rightarrow 3y = 0 \Rightarrow y = 0$$

Verify that the ordered triple $(2, 0, 3)$ satisfies all three equations.

Check:

$$\begin{array}{c} x + 3y + 4z = 14 \ (1) \\ 2 + 3(0) + 4(3) = 14 \ \ ? \\ 2 + 0 + 12 = 14 \\ 14 = 14 \text{ True} \end{array}$$

$$\begin{array}{c} 2x - 3y + 2z = 10 \ \ (2) \\ 2(2) - 3(0) + 2(3) = 10 \ \ ? \\ 4 - 0 + 6 = 10 \\ 10 = 10 \text{ True} \end{array}$$

$$\begin{array}{c} 3x - y + z = 9 \ (3) \\ 3(2) - 0 + 3 = 9 \ \ ? \\ 6 - 0 + 3 = 9 \\ 9 = 9 \text{ True} \end{array}$$

Solution set: $\{(2, 0, 3)\}$

51. $x + 4y - z = 6$ (1)
$2x - y + z = 3$ (2)
$3x + 2y + 3z = 16$ (3)

Eliminate z by adding equations (1) and (2) to get $3x + 3y = 9$ or $x + y = 3$ (4).

Multiply equation (1) by 3 and add the result to equation (3).

$$\begin{array}{r} 3x + 12y - 3z = 18 \\ 3x + 2y + 3z = 16 \\ \hline 6x + 14y = 34 \ (5) \end{array}$$

Multiply equation (4) by -6 and then add the result to equation (5) in order to eliminate x.

$$\begin{array}{r} -6x - 6y = -18 \\ 6x + 14y = 34 \\ \hline 8y = 16 \Rightarrow y = 2 \end{array}$$

(continued on next page)

(continued)

Using $y = 2$ find x from equation (4) by substitution: $x + 2 = 3 \Rightarrow x = 1$

Substitute 1 for x and 2 for y in equation (1) to find z.

$$1 + 4(2) - z = 6 \Rightarrow 1 + 8 - z = 6 \Rightarrow$$
$$9 - z = 6 \Rightarrow -z = -3 \Rightarrow z = 3$$

Verify that the ordered triple $(1, 2, 3)$ satisfies all three equations.

Check:

$x + 4y - z = 6$ (1)	$2x - y + z = 3$ (2)
$1 + 4(2) - 3 = 6$?	$2(1) - 2 + 3 = 3$?
$1 + 8 - 3 = 6$	$2 - 2 + 3 = 3$
$6 = 6$ True	$3 = 3$ True

$$3x + 2y + 3z = 16 \quad (3)$$
$$3(1) + 2(2) + 3(3) = 16 \quad ?$$
$$3 + 4 + 9 = 16$$
$$16 = 16 \text{ True}$$

Solution set: $\{(1, 2, 3)\}$

53. $\quad x - 3y - 2z = -3 \quad (1)$
$\quad 3x + 2y - z = 12 \quad (2)$
$\quad -x - y + 4z = 3 \quad (3)$

Eliminate x by adding equations (1) and (3) to get $-4y + 2z = 0$ (4).

Eliminate x by multiplying equation (3) by 3 and add to equation (2).

$$-3x - 3y + 12z = 9$$
$$\underline{3x + 2y - \quad z = 12}$$
$$-y + 11z = 21 \quad (5)$$

Multiply equation (5) by -4 and add to equation (4) in order to eliminate y.

$$-4y + 2z = 0$$
$$\underline{4y - 44z = -84}$$
$$-42z = -84 \Rightarrow z = 2$$

Using $z = 2$, find y from equation (4) by substitution.

$$-4y + 2(2) = 0 \Rightarrow -4y + 4 = 0 \Rightarrow$$
$$-4y = -4 \Rightarrow y = 1$$

Substitute 1 for y and 2 for z in equation (1) to find x.

$$x - 3(1) - 2(2) = -3 \Rightarrow x - 3 - 4 = -3 \Rightarrow$$
$$x - 7 = -3 \Rightarrow x = 4$$

Verify that the ordered triple $(4, 1, 2)$ satisfies all three equations.

Check:
$$x - 3y - 2z = -3 \quad (1)$$
$$4 - 3(1) - 2(2) = -3 \quad ?$$
$$4 - 3 - 4 = -3$$
$$-3 = -3 \text{ True}$$

$$3x + 2y - z = 12 \quad (2)$$
$$3(4) + 2(1) - 2 = 12 \quad ?$$
$$12 + 2 - 2 = 12$$
$$12 = 12 \text{ True}$$

$$-x - y + 4z = 3 \quad (3)$$
$$-4 - 1 + 4(2) = 3 \quad ?$$
$$-4 - 1 + 8 = 3$$
$$3 = 3 \text{ True}$$

Solution set: $\{(4, 1, 2)\}$

55. $\quad 2x + 6y - z = 6 \quad (1)$
$\quad 4x - 3y + 5z = -5 \quad (2)$
$\quad 6x + 9y - 2z = 11 \quad (3)$

Eliminate y by multiplying equation (2) by 2 and add to equation (1).

$$2x + 6y - \quad z = \quad 6$$
$$\underline{8x - 6y + 10z = -10}$$
$$10x \qquad + 9z = -4 \quad (4)$$

Eliminate y by multiplying equation (2) by 3 and add to equation (3).

$$6x + 9y - \quad 2z = \quad 11$$
$$\underline{12x - 9y + 15z = -15}$$
$$18x \qquad + 13z = -4 \quad (5)$$

Multiply equation (4) by 9 and equation (5) by -5 and add in order to eliminate x.

$$90x + 81z = -36$$
$$\underline{-90x - 65z = 20}$$
$$16z = -16 \Rightarrow z = -1$$

Using $z = -1$, find x from equation (4) by substitution.

$$10x + 9(-1) = -4 \Rightarrow 10x - 9 = -4$$
$$10x = 5 \Rightarrow x = \tfrac{1}{2}$$

Substitute $\tfrac{1}{2}$ for x and -1 for z in equation (1) to find y.

$$2\left(\tfrac{1}{2}\right) + 6y - (-1) = 6 \Rightarrow 1 + 6y + 1 = 6 \Rightarrow$$
$$6y + 2 = 6 \Rightarrow 6y = 4 \Rightarrow y = \tfrac{2}{3}$$

Verify that the ordered triple $\left(\tfrac{1}{2}, \tfrac{2}{3}, -1\right)$ satisfies all three equations.

(continued on next page)

(*continued*)

Check:

$$2x + 6y - z = 6 \quad (1)$$
$$2\left(\tfrac{1}{2}\right) + 6\left(\tfrac{2}{3}\right) - (-1) = 6 \quad ?$$
$$1 + 4 + 1 = 6$$
$$6 = 6 \text{ True}$$

$$4x - 3y + 5z = -5 \quad (2)$$
$$4\left(\tfrac{1}{2}\right) - 3\left(\tfrac{2}{3}\right) + 5(-1) = -5 \quad ?$$
$$2 - 2 - 5 = -5$$
$$-5 = -5 \text{ True}$$

$$6x + 9y - 2z = 11 \quad (3)$$
$$6\left(\tfrac{1}{2}\right) + 9\left(\tfrac{2}{3}\right) - 2(-1) = 11 \quad ?$$
$$3 + 6 + 2 = 11$$
$$11 = 11 \text{ True}$$

Solution set: $\left\{ \left(\tfrac{1}{2}, \tfrac{2}{3}, -1 \right) \right\}$.

57.
$$2x - 3y + 2z \;\; - 3 = 0 \;\; (1)$$
$$4x + 8y + \;\; z \;\; - 2 = 0 \;\; (2)$$
$$-x - 7y + 3z - 14 = 0 \;\; (3)$$

Eliminate x by multiplying equation (3) by 2 and add to equation (1).

$$\begin{aligned} 2x - \;\; 3y + 2z - \;\; 3 &= 0 \\ -2x - 14y + 6z - 28 &= 0 \\ \hline -17y + 8z - 31 &= 0 \;\; (4) \end{aligned}$$

Eliminate x by multiplying equation (3) by 4 and add to equation (2).

$$\begin{aligned} 4x + \;\; 8y + \;\; z - \;\; 2 &= 0 \\ -4x - 28y + 12z - 56 &= 0 \\ \hline -20y + 13z - 58 &= 0 \;\; (5) \end{aligned}$$

Multiply equation (5) by 17, equation (4) by $-20,$ and add in order to eliminate y.

$$\begin{aligned} 340y - 160z + 620 &= 0 \\ -340y + 221z - 986 &= 0 \\ \hline 61z - 366 &= 0 \Rightarrow z = 6 \end{aligned}$$

Using $z = 6,$ find y from equation (5) by substitution.

$$-20y + 13(6) - 58 = 0 \Rightarrow -20y + 78 - 58 = 0$$
$$-20y + 20 = 0 \Rightarrow -20y = -20 \Rightarrow y = 1$$

Substitute 1 for y and 6 for z in equation (1) to find x.

$$2x - 3(1) + 2(6) - 3 = 0 \Rightarrow 2x - 3 + 12 - 3 = 0 \Rightarrow$$
$$2x + 6 = 0 \Rightarrow 2x = -6 \Rightarrow x = -3$$

Verify that the ordered triple $(-3, 1, 6)$ satisfies all three equations.

Check:
$$2x - 3y + 2z - 3 = 0 \quad (1)$$
$$2(-3) - 3(1) + 2(6) - 3 = 0 \quad ?$$
$$-6 - 3 + 12 - 3 = 0$$
$$0 = 0 \text{ True}$$

$$4x + 8y + z - 2 = 0 \quad (2)$$
$$4(-3) + 8(1) + 6 - 2 = 0$$
$$-12 + 8 + 6 - 2 = 0 \quad ?$$
$$0 = 0 \text{ True}$$

$$-x - 7y + 3z - 14 = 0 \quad (3)$$
$$-(-3) - 7(1) + 3(6) - 14 = 0 \quad ?$$
$$3 - 7 + 18 - 14 = 0$$
$$0 = 0 \text{ True}$$

Solution set: $\{(-3, 1, 6)\}$

59.
$$x - 2y + 3z = 6 \quad (1)$$
$$2x - y + 2z = 5 \quad (2)$$

Multiply equation (2) by -2 and add to equation (1) in order to eliminate y.

$$\begin{aligned} x - 2y + 3z &= \;\; 6 \\ -4x + 2y - 4z &= -10 \\ \hline -3x \;\;\;\;\;\;\;\; - z &= -4 \;\; (3) \end{aligned}$$

Solve equation (3) for x.

$$-3x - z = -4 \Rightarrow -3x = z - 4 \Rightarrow x = \frac{-z + 4}{3}$$

Express y in terms of z by solving equation (2) for y and substituting $\frac{-z+4}{3}$ for x.

$$2x - y + 2z = 5 \Rightarrow -y = -2x - 2z + 5 \Rightarrow$$
$$y = 2x + 2z - 5 \Rightarrow$$

$$y = 2\left(\frac{-z + 4}{3} \right) + 2z - 5 = \frac{-2z + 8}{3} + 2z - 5$$
$$= \frac{-2z + 8}{3} + \frac{6z}{3} - \frac{15}{3} = \frac{4z - 7}{3}$$

With z arbitrary, the solution set is of the form

$$\left\{ \left(\frac{-z + 4}{3}, \frac{4z - 7}{3}, z \right) \right\}.$$

61.
$$5x - 4y + z = \;\; 9 \quad (1)$$
$$y + z = 15 \quad (2)$$

Solve equation (2) for y in terms of z.

$$y + z = 15 \Rightarrow y = -z + 15$$

Substitute $-z + 15$ for y in equation (1) and then solve for x in terms of z.

$$5x - 4y + z = 9 \Rightarrow 5x - 4(-z + 15) + z = 9 \Rightarrow$$
$$5x + 4z - 60 + z = 9 \Rightarrow 5x + 5z = 69 \Rightarrow$$
$$5x = -5z + 69 \Rightarrow x = \frac{-5z + 69}{5}$$

With z arbitrary, the solution set is of the form

$$\left\{ \left(\frac{-5z + 69}{5}, -z + 15, z \right) \right\}.$$

63. $3x + 4y - z = 13$ (1)

$x + y + 2z = 15$ (2)

Multiply equation (2) by -4 and add to equation (1) in order to eliminate y.

$$\begin{array}{r} 3x + 4y - z = 13 \\ -4x - 4y - 8z = -60 \\ \hline -x \qquad - 9z = -47 \end{array} \text{ (3)}$$

Solve equation (3) for x in terms of z.

$-x - 9z = -47 \Rightarrow -x = 9z - 47 \Rightarrow$

$x = -9z + 47$

Express y in terms of z by solving equation (2) for y and substituting $-9z + 47$ for x.

$x + y + 2z = 15 \Rightarrow y = -x - 2z + 15 \Rightarrow$

$y = -(-9z + 47) - 2z + 15$

$\quad = 9z - 47 - 2z + 15 = 7z - 32$

With z arbitrary, the solution set is of the form

$\{(-9z + 47, \ 7z - 32, \ z)\}$.

65. $3x + 5y - z = -2$ (1)

$4x - y + 2z = 1$ (2)

$-6x - 10y + 2z = 0$ (3)

Multiply equation (1) by 2 and add the result to equation (3).

$$\begin{array}{r} 6x + 10y - 2z = -4 \\ -6x - 10y + 2z = 0 \\ \hline 0 = -4 \end{array}$$

We obtain a false statement. The solution set is $\varnothing$, and the system is inconsistent.

67. $5x - 4y + z = 0$ (1)

$x + y = 0$ (2)

$-10x + 8y - 2z = 0$ (3)

Multiply equation (2) by 4 and add the result to equation (1) in order to eliminate y.

$$\begin{array}{r} 5x - 4y + z = 0 \\ 4x + 4y = 0 \\ \hline 9x \qquad + z = 0 \end{array} \text{ (4)}$$

Multiply equation (2) by -8 and add the result to equation (3) in order to eliminate y.

$$\begin{array}{r} -8x - 8y = 0 \\ -10x + 8y - 2z = 0 \\ \hline -18x \qquad - 2z = 0 \end{array} \text{ (5)}$$

Multiply equation (4) by 2 and add the result to equation (5).

$$\begin{array}{r} 18x + 2z = 0 \\ -18x - 2z = 0 \\ \hline 0 = 0 \end{array}$$

This is a true statement and the system has infinitely many solutions.

Solve equation (4) for x.

$9x + z = 0 \Rightarrow 9x = -z \Rightarrow x = -\frac{z}{9}$

Express y in terms of z by solving equation (1) for y and substituting $-\frac{z}{9}$ for x.

$5x - 4y + z = 0 \Rightarrow -4y = -5x - z \Rightarrow$

$y = \dfrac{-5x - z}{-4} = \dfrac{5x + z}{4} = \dfrac{5\left(-\frac{z}{9}\right) + z}{4}$

$\quad = \dfrac{-5z + 9z}{36} = \dfrac{4z}{36} = \dfrac{z}{9}$

With z arbitrary, the solution set is of the form

$\left\{\left(-\dfrac{z}{9}, \ \dfrac{z}{9}, \ z\right)\right\}$.

69. $\dfrac{2}{x} + \dfrac{1}{y} = \dfrac{3}{2}$ (1)

$\dfrac{3}{x} - \dfrac{1}{y} = 1$ (2)

Let $\frac{1}{x} = t$ and $\frac{1}{y} = u$. With these substitutions, the system becomes

$2t + u = \frac{3}{2}$ (3)

$3t - u = 1$ (4)

Add these equations, eliminate u, and solve for t.

$5t = \frac{5}{2} \Rightarrow t = \frac{1}{2}$

Substitute $\frac{1}{2}$ for t in equation (3) and solve for u.

$2\left(\frac{1}{2}\right) + u = \frac{3}{2} \Rightarrow 1 + u = \frac{3}{2} \Rightarrow u = \frac{1}{2}$

Now find the values of x and y, the variables in the original system. So, $\frac{1}{x} = t$, $tx = 1$, and $x = \frac{1}{t}$. Likewise $y = \frac{1}{u}$.

$x = \dfrac{1}{t} = \dfrac{1}{\frac{1}{2}} = 2$ and $y = \dfrac{1}{u} = \dfrac{1}{\frac{1}{2}} = 2$

Solution set: $\{(2, 2)\}$

71. $\dfrac{2}{x} + \dfrac{1}{y} = 11$ (1)

$\dfrac{3}{x} - \dfrac{5}{y} = 10$ (2)

Let $\frac{1}{x} = t$ and $\frac{1}{y} = u$. With these substitutions, the system becomes

$2t + u = 11$ (3)

$3t - 5u = 10$ (4)

Multiply equation (3) by 5 and add to equation (4), eliminating u, and solve for t.

$$\begin{array}{r} 10t + 5u = 55 \\ 3t - 5u = 10 \\ \hline 13t \qquad = 65 \Rightarrow t = 5 \end{array}$$

(continued on next page)

(*continued*)

Substitute 5 for t in equation (3) and solve for u: $2(5) + u = 11 \Rightarrow 10 + u = 11 \Rightarrow u = 1$

Now find the values of x and y, the variables in the original system.

$$x = \frac{1}{t} = \frac{1}{5} \text{ and } y = \frac{1}{u} = \frac{1}{1} = 1$$

Solution set: $\left\{\left(\frac{1}{5}, 1\right)\right\}$

73. $\dfrac{2}{x} + \dfrac{3}{y} - \dfrac{2}{z} = -1$ (1)

$\dfrac{8}{x} - \dfrac{12}{y} + \dfrac{5}{z} = 5$ (2)

$\dfrac{6}{x} + \dfrac{3}{y} - \dfrac{1}{z} = 1$ (3)

Let $\frac{1}{x} = t$, $\frac{1}{y} = u$, and $\frac{1}{z} = v$. With these substitutions, the system becomes

$2t + 3u - 2v = -1$ (4)
$8t - 12u + 5v = 5$ (5)
$6t + 3u - v = 1$ (6)

Eliminate v by multiplying equation (6) by 5 and add to equation (5).

$$
\begin{array}{r}
8t - 12u + 5v = 5 \\
30t + 15u - 5v = 5 \\
\hline
38t + 3u \phantom{{}- 5v} = 10 \quad (7)
\end{array}
$$

Eliminate v by multiplying equation (6) by -2 and add to equation (4).

$$
\begin{array}{r}
2t + 3u - 2v = -1 \\
-12t - 6u + 2v = -2 \\
\hline
-10t - 3u \phantom{{}+ 2v} = -3 \quad (8)
\end{array}
$$

Add equations (7) and (8) in order to eliminate u.

$$
\begin{array}{r}
38t + 3u = 10 \\
-10t - 3u = -3 \\
\hline
28t \phantom{{}- 3u} = 7 \Rightarrow t = \frac{1}{4}
\end{array}
$$

Using $t = \frac{1}{4}$, find u from equation (7) by substitution.

$$38\left(\tfrac{1}{4}\right) + 3u = 10 \Rightarrow \tfrac{19}{2} + 3u = 10 \Rightarrow$$
$$3u = \tfrac{20}{2} - \tfrac{19}{2} = \tfrac{1}{2} \Rightarrow u = \tfrac{1}{6}$$

Substitute $\frac{1}{4}$ for t and $\frac{1}{6}$ for u in equation (4) to find v.

$$2\left(\tfrac{1}{4}\right) + 3\left(\tfrac{1}{6}\right) - 2v = -1 \Rightarrow \tfrac{1}{2} + \tfrac{1}{2} - 2v = -1 \Rightarrow$$
$$1 - 2v = -1 \Rightarrow -2v = -2 \Rightarrow v = 1$$

Now find the values of x, y, and z, the variables in the original system.

$$x = \frac{1}{t} = \frac{1}{\frac{1}{4}} = 4, \quad y = \frac{1}{u} = \frac{1}{\frac{1}{6}} = 6, \text{ and}$$

$$z = \frac{1}{v} = \frac{1}{1} = 1$$

Solution set: $\{(4, 6, 1)\}$

75. $x - 2y = 3$ (1)
$-2x + 4y = k$ (2)

Multiply equation (1) by 2 and add the result to equation (2).

$$
\begin{array}{r}
2x - 4y = 6 \\
-2x + 4y = k \\
\hline
0 = 6 + k
\end{array}
$$

The system will have no solution when $6 + k \neq 0 \Rightarrow k \neq -6$.

The system will have infinitely many solutions when $6 + k = 0 \Rightarrow k = -6$.

77. Because $y = ax + b$ and the line passes through $(-2, 1)$ and $(-1, -2)$, we have the equations $1 = a(-2) + b$ and $-2 = a(-1) + b$.

This becomes the following system.

$-2a + b = 1$ (1)
$-a + b = -2$ (2)

Multiply equation (1) by -1 and add the result to equation (2) in order to eliminate b.

$$
\begin{array}{r}
2a - b = -1 \\
-a + b = -2 \\
\hline
a \phantom{{}- b} = -3
\end{array}
$$

Substitute this value into equation (1).

$$-2(-3) + b = 1 \Rightarrow 6 + b = 1 \Rightarrow b = -5$$

An equation is $y = -3x - 5$.

79. Because $y = ax^2 + bx + c$ and the parabola passes through the points $(2, 3)$, $(-1, 0)$, and $(-2, 2)$, we have the following equations.

$$3 = a(2)^2 + b(2) + c \Rightarrow 4a + 2b + c = 3 \ (1)$$
$$0 = a(-1)^2 + b(-1) + c \Rightarrow a - b + c = 0 \ (2)$$
$$2 = a(-2)^2 + b(-2) + c \Rightarrow 4a - 2b + c = 2 \ (3)$$

Multiply equation (2) by -1 and add the result to equation (1) in order to eliminate c.

$$
\begin{array}{r}
4a + 2b + c = 3 \\
-a + b - c = 0 \\
\hline
3a + 3b \phantom{{}+ c} = 3 \quad (4)
\end{array}
$$

(*continued on next page*)

(*continued*)

Multiply equation (2) by −1 and add the result to equation (3) in order to eliminate c.

$$-a + b - c = 0$$
$$\underline{4a - 2b + c = 2}$$
$$3a - b \quad = 2 \quad (5)$$

Multiply equation (4) by −1 and then add the result to equation (5) and solve for b.

$$-3a - 3b = -3$$
$$\underline{3a - b = 2}$$
$$-4b = -1 \Rightarrow b = \tfrac{1}{4}$$

Substitute this value into equation (5).

$$3a - \tfrac{1}{4} = 2 \Rightarrow 3a = \tfrac{8}{4} + \tfrac{1}{4} = \tfrac{9}{4} \Rightarrow a = \tfrac{3}{4}$$

Substitute $a = \tfrac{3}{4}$ and $b = \tfrac{1}{4}$ into equation (1) in order to solve for c.

$$4\left(\tfrac{3}{4}\right) + 2\left(\tfrac{1}{4}\right) + c = 3$$
$$3 + \tfrac{1}{2} + c = 3 \Rightarrow c = -\tfrac{1}{2}$$

An equation of the parabola is
$$y = \tfrac{3}{4}x^2 + \tfrac{1}{4}x - \tfrac{1}{2}.$$

81. Because $y = ax + b$ and the line passes through the points $(2, 5)$ and $(-1, -4)$, we have the equations $5 = a(2) + b$ and $-4 = a(-1) + b$.

This becomes the following system.
$$2a + b = 5 \quad (1)$$
$$-a + b = -4 \quad (2)$$

Multiply equation (2) by −1 and add the result to equation (1) in order to eliminate b.

$$2a + b = 5$$
$$\underline{a - b = 4}$$
$$3a \quad = 9 \Rightarrow a = 3$$

Substitute this value into equation (1).
$$2(3) + b = 5 \Rightarrow 6 + b = 5 \Rightarrow b = -1$$

An equation is $y = 3x - 1$.

83. Because $y = ax^2 + bx + c$ and the parabola passes through the points $(-2, -3.75)$, $(4, -3.75)$, and $(-1, -1.25)$, we have the following equations.

$$-3.75 = a(-2)^2 + b(-2) + c$$
$$4a - 2b + c = -3.75 \quad (1)$$
$$-3.75 = a(4)^2 + b(4) + c$$
$$16a + 4b + c = -3.75 \quad (2)$$
$$-1.25 = a(-1)^2 + b(-1) + c$$
$$a - b + c = -1.25 \quad (3)$$

Multiply equation (2) by −1 and add the result to equation (1) in order to eliminate c.

$$4a - 2b + c = -3.75$$
$$\underline{-16a - 4b - c = 3.75}$$
$$-12a - 6b \quad = 0 \text{ or } 2a + b = 0 \quad (4)$$

Multiply equation (2) by −1 and add the result to equation (3) in order to eliminate c.

$$-16a - 4b - c = 3.75$$
$$\underline{a - b + c = -1.25}$$
$$-15a - 5b \quad = 2.50 \text{ or } -3a - b = 0.5 \ (5)$$

Add equations (4) and (5) in order to solve for b.

$$2a + b = 0$$
$$\underline{-3a - b = 0.5}$$
$$-a \quad = 0.5 \Rightarrow a = -0.5$$

Substitute $a = -0.5$ into equation (4) in order to solve for b.

$$2(-0.5) + b = 0 \Rightarrow -1 + b = 0 \Rightarrow b = 1$$

Substitute $a = -0.5$ and $b = 1$ into equation (3) in order to solve for c.

$$-.5 - 1 + c = -1.25$$
$$-1.5 + c = -1.25 \Rightarrow c = 0.25$$

An equation of the parabola is
$$y = -0.5x^2 + x + 0.25 \text{ or } y = -\tfrac{1}{2}x^2 + x + \tfrac{1}{4}.$$

85. Because $x^2 + y^2 + ax + by + c = 0$ and the circle passes through the points $(-1, 3)$, $(6, 2)$, and $(-2, -4)$, we have the following equations.

$$(-1)^2 + 3^2 + a(-1) + b(3) + c = 0$$
$$-a + 3b + c = -10 \quad (1)$$
$$6^2 + 2^2 + a(6) + b(2) + c = 0 \Rightarrow$$
$$6a + 2b + c = -40 \quad (2)$$
$$(-2)^2 + (-4)^2 + a(-2) + b(-4) + c = 0$$
$$-2a - 4b + c = -20 \quad (3)$$

Multiply equation (1) by −2 and add the result to equation (3) in order to eliminate a.

$$2a - 6b - 2c = 20$$
$$\underline{-2a - 4b + c = -20}$$
$$-10b - c = 0 \ (4)$$

Multiply equation (1) by 6 and add the result to equation (2) in order to eliminate a.

$$-6a + 18b + 6c = -60$$
$$\underline{6a + 2b + c = -40}$$
$$20b + 7c = -100 \ (5)$$

Multiply equation (4) by 7 and add the result to equation (5) in order to eliminate c.

$$-70b - 7c = 0$$
$$\underline{20b + 7c = -100}$$
$$-50b = -100 \Rightarrow b = 2$$

(*continued on next page*)

(*continued*)

We substitute this value into equation (4) in order to solve for c.

$-10(2) - c = 0 \Rightarrow -20 - c = 0 \Rightarrow -20 = c$

Substitute $b = 2$ and $c = -20$ into equation (1).

$-a + 3(2) - 20 = -10 \Rightarrow -a + 6 - 20 = -10 \Rightarrow$
$-a - 14 = -10 \Rightarrow -a = 4 \Rightarrow a = -4$

An equation of the circle is

$x^2 + y^2 - 4x + 2y - 20 = 0.$

87. Because $x^2 + y^2 + ax + by + c = 0$ and the circle passes through the points $(2, 1)$, $(-1, 0)$, and $(3, 3)$, we have the following equations.

$$2^2 + 1^2 + a(2) + b(1) + c = 0$$
$$2a + b + c = -5 \quad (1)$$

$$(-1)^2 + 0^2 + a(-1) + b(0) + c = 0$$
$$-a + c = -1 \quad (2)$$

$$3^2 + 3^2 + a(3) + b(3) + c = 0$$
$$3a + 3b + c = -18 \quad (3)$$

Multiply equation (1) by -3 and add the result to equation (3) in order to eliminate b.

$$-6a - 3b - 3c = 15$$
$$\underline{3a + 3b \quad + c = -18}$$
$$-3a \quad - 2c = -3 \quad (4)$$

Multiply equation (2) by 2 and add the result to equation (4) in order to eliminate c.

$$-2a + 2c = -2$$
$$\underline{-3a - 2c = -3}$$
$$-5a \quad = -5 \Rightarrow a = 1$$

We substitute this value into equation (2) in order to solve for c.

$-1 + c = -1 \Rightarrow c = 0$

Substitute $a = 1$ and $c = 0$ into equation (1).

$2(1) + b + 0 = -5 \Rightarrow 2 + b = -5 \Rightarrow b = -7$

An equation of the circle is

$x^2 + y^2 + x - 7y = 0$.

89. Because $x^2 + y^2 + ax + by + c = 0$ and the circle passes through points $(0, 3)$, $(4, 2)$, and $(-5, -2)$, we have the following equations.

$$0^2 + 3^2 + a(0) + b(3) + c = 0$$
$$3b + c = -9 \quad (1)$$

$$4^2 + 2^2 + a(4) + b(2) + c = 0$$
$$4a + 2b + c = -20 \quad (2)$$

$$(-5)^2 + (-2)^2 + a(-5) + b(-2) + c = 0$$
$$-5a - 2b + c = -29 \quad (3)$$

Multiply equation (2) by 5 and multiply equation (3) by 4, and then add the results in order to eliminate a.

$$20a + 10b + 5c = -100$$
$$\underline{-20a - 8b + 4c = -116}$$
$$2b + 9c = -216 \quad (4)$$

Multiply equation (1) by -9 and add the result to equation (4) in order to eliminate c.

$$-27b - 9c = \quad 81$$
$$\underline{2b + 9c = -216}$$
$$-25b \quad = -135 \Rightarrow b = \tfrac{27}{5}$$

Substitute $b = \tfrac{27}{5}$ into equation (1) in order to solve for c.

$$3\left(\tfrac{27}{5}\right) + c = -9 \Rightarrow \tfrac{81}{5} + c = -9 \Rightarrow$$
$$c = -\tfrac{81}{5} - \tfrac{45}{5} = -\tfrac{126}{5}$$

Substitute $b = \tfrac{27}{5}$ and $c = -\tfrac{126}{5}$ into equation (2).

$$4a + 2\left(\tfrac{27}{5}\right) + \left(-\tfrac{126}{5}\right) = -20 \Rightarrow$$
$$4a + \tfrac{54}{5} - \tfrac{126}{5} = -20$$
$$4a - \tfrac{72}{5} = -20 \Rightarrow 4a = \tfrac{72}{5} - \tfrac{100}{5} \Rightarrow$$
$$4a = -\tfrac{28}{5} \Rightarrow a = -\tfrac{7}{5}$$

An equation of the circle is

$x^2 + y^2 - \tfrac{7}{5}x + \tfrac{27}{5}y - \tfrac{126}{5} = 0.$

91. (a) Because $C = at^2 + bt + c$ and we have the ordered pairs $(0, 317)$, $(20, 339)$, and $(53, 396)$, we have the following equations.

$$317 = a(0)^2 + b(0) + c$$
$$c = 317 \quad (1)$$

$$339 = a(20)^2 + b(20) + c$$
$$400a + 20b + c = 339 \quad (2)$$

$$396 = a(53)^2 + b(53) + c$$
$$2809a + 53b + c = 396 \quad (3)$$

Substitute this 317 for c into equations (2) and (3) to obtain the following system.

$$400a + 20b + 317 = 339$$
$$400a + 20b = 22 \quad (4)$$
$$2809a + 53b + 317 = 396$$
$$2809a + 53b = 79 \quad (5)$$

Multiply equation (4) by -53, multiply equation (5) by 20, then add the results in order to eliminate b.

$$-21200a - 1060b = -1166$$
$$\underline{56180a + 1060b = \quad 1580}$$
$$34980a \quad = \quad 414$$
$$a = \frac{414}{34980} = \frac{69}{5830} \approx 0.0118$$

(*continued on next page*)

(continued)

Substitute $a = \frac{69}{5830}$ into equation (4) in order to solve for b.

$$400\left(\frac{69}{5830}\right) + 20b = 22$$

$$20b = 22 - 400\left(\frac{69}{5830}\right)$$

$$b = \frac{22 - 400\left(\frac{69}{5830}\right)}{20}$$

$$b = \frac{5033}{5830} \approx 0.8633$$

The constants are

$a = \frac{69}{5830} \approx 0.0118$, $b = \frac{5033}{5830} \approx 0.8633$, and $c = 317$.

The equation is $C = \frac{69}{5830}t^2 + \frac{5033}{5830}t + 317$

or $C = 0.0118t^2 + 0.8633t + 317$.

(b) Because $t = 0$ corresponds to 1960, the amount of carbon dioxide will be double its 1960 level when

$$2 \cdot 317 = \frac{69}{5830}t^2 + \frac{5033}{5830}t + 317$$

We will solve this equation graphically by graphing $y_1 = \frac{69}{5830}t^2 + \frac{5033}{5830}t + 317$

and $y_2 = 2 \cdot 317$ in the window $[0, 200] \times [100, 900]$ and then finding the intersection.

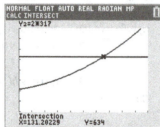

According to the model, the amount of carbon dioxide in the atmosphere will double 131 years after 1960 or in 2091.

93. The population of the Jacksonville metropolitan area was greater than that of the New Orleans metropolitan area from 2006−2013.

95. (2005.2, 1.26)

97. If equation of the form $y = f(t)$ were determined that modeled either of the two graphs, then the variable t would represent year and y would represent population (in millions).

99. Let x = one number; let y = the other number
We have the following system of equations.
$x + y = 47$ (1)
$x - y = 1$ (2)
Add the two equations to eliminate y.
$2x = 48 \Rightarrow x = 24$. Substitute this value into equation (1) and solve for y:
$24 + y = 47 \Rightarrow y = 23$.
The two numbers are 23 and 24.

101. Let x = the FCI for football; let y = the FCI for baseball. We have the following system of equations.
$$\frac{x + y}{2} = 345.53 \Rightarrow x + y = 691.06 \quad (1)$$
$$x - y = 266.13 \quad (2)$$
Add the two equations to eliminate y.
$2x = 957.19 \Rightarrow x = 478.60$
Substitute this value into equation (1) and solve for y:
$478.60 + y = 691.06 \Rightarrow y = 212.47$
Verify that (266.13, 212.47) satisfies both equations in the original system.
The FCI for football was \$478.60, and the FCI for baseball was \$212.47.

103. Let x = the number of \$3.00 gallons;
y = the number of \$4.50 gallons;
z = the number of \$9.00 gallons.
We have the following equations.
$$x + y + z = 300 \quad (1)$$
$$2x = y \Rightarrow$$
$$2x - y = 0 \quad (2)$$
$$3.00x + 4.50y + 9.00z = 6.00(300) \Rightarrow$$
$$3x + 4.5y + 9z = 1800 \quad (3)$$
Multiply equation (1) by −9 and add to equation (3) in order to eliminate z.
$$-9x - 9y - 9z = -2700$$
$$\underline{3x + 4.5y + 9z = 1800}$$
$$-6x - 4.5y = -900 \ (4)$$
Multiply equation (2) by 3 and add to equation (4) in order to eliminate x.
$$6x - 3y = 0$$
$$\underline{-6x - 4.5y = -900}$$
$$-7.5y = -900 \Rightarrow y = 120$$
Substitute this value into equation (2) to solve for x: $2x - 120 = 0 \Rightarrow 2x = 120 \Rightarrow x = 60$
Substitute $x = 60$ and $y = 120$ into equation (1) and solve for z.
$60 + 120 + z = 300 \Rightarrow 180 + z = 300 \Rightarrow z = 120$
She should use 60 gal of the \$3.00 water, 120 gal of the \$4.50 water, and 120 gal of the \$9.00 water.

105. Let x = the length of the shortest side;
y = the length of the medium side;
z = the length of the longest side.
We have the following equations.

$$x + y + z = 59 \quad (1)$$
$$z = y + 11 \Rightarrow \quad -y + z = 11 \quad (2)$$
$$y = x + 3 \Rightarrow -x + y \quad = 3 \quad (3)$$

Add equations (1) and (3) together in order to eliminate x.

$$\begin{array}{r} x + y + z = 59 \\ -x + y \quad = 3 \\ \hline 2y + z = 62 \ (4) \end{array}$$

Multiply equation (2) by 2 and then add the result to equation (4) to solve for z.

$$\begin{array}{r} -2y + 2z = 22 \\ 2y + z = 62 \\ \hline 3z = 84 \Rightarrow z = 28 \end{array}$$

Substitute this value into equation (2) in order to solve for y.

$$-y + 28 = 11 \Rightarrow -y = -17 \Rightarrow y = 17$$

Substitute $y = 17$ into equation (3) in order to solve for x: $-x + 17 = 3 \Rightarrow -x = -14 \Rightarrow x = 14$

The lengths of the sides of the triangle are 14 inches, 17 inches, and 28 inches.
Note: $14 + 17 + 28 = 59$

107. Let x = the amount invested in real estate (at 3%); y = the amount invested in a money market account (at 2.5%); z = the amount invested in CDs (at 1.5%).
Completing the table we have the following.

	Amount Invested	Rate (as a decimal)	Annual Interest
Real Estate	x	0.03	0.03x
Money Market	y	0.025	0.025y
CDs	z	0.015	0.015z

We have the following equations.

$$x + y + z = 200,000 \quad (1)$$
$$z = x + y - 80,000 \Rightarrow$$
$$x + y - z = 80,000 \quad (2)$$
$$0.03x + 0.025y + 0.015z = 4900 \Rightarrow$$
$$30x + 25y + 15z = 4,900,000 \quad (3)$$

Add equations (1) and (2) in order to eliminate z.

$$\begin{array}{r} x + y + z = 200,000 \\ x + y - z = 80,000 \\ \hline 2x + 2y = 280,000 \text{ or } x + y = 140,000 \ (4) \end{array}$$

Multiply equation (2) by 15 and add the result to equation (3) in order to eliminate z.

$$\begin{array}{r} 15x + 15y - 15z = 1,200,000 \\ 30x + 25y + 15z = 4,900,000 \\ \hline 45x + 40y = 6,100,000 \text{ or} \end{array}$$
$$9x + 8y = 1,220,000 \quad (5)$$

Multiply equation (4) by -8 and add the result to equation (5).

$$\begin{array}{r} -8x - 8y = -1,120,000 \\ 9x + 8y = 1,220,000 \\ \hline x = 100,000 \end{array}$$

Substitute this value into equation (4) in order to solve for y.

$$100,000 + y = 140,000 \Rightarrow y = 40,000$$

Substitute $x = 100,000$ and $y = 40,000$ into equation (1) in order to solve for z.

$$100,000 + 40,000 + z = 200,000$$
$$140,000 + z = 200,000 \Rightarrow z = 60,000$$

The amounts invested were $100,000 at 3% (real estate), $40,000 at 2.5% (money market), and $60,000 at 1.5% (CDs).

109. $25x + 40y + 20z = 2200 \quad (4)$
$4x + 2y + 3z = 280 \quad (5)$
$3x + 2y + z = 180 \quad (6)$

Eliminate z by multiplying equation (6) by -3 and add to equation (5).

$$\begin{array}{r} -9x - 6y - 3z = -540 \\ 4x + 2y + 3z = 280 \\ \hline -5x - 4y = -260 \ (7) \end{array}$$

Eliminate z by multiplying equation (6) by -20 and add to equation (4).

$$\begin{array}{r} -60x - 40y - 20z = -3600 \\ 25x + 40y + 20z = 2200 \\ \hline -35x = -1400 \Rightarrow x = 40 \end{array}$$

Substitute this value in equation (7) in order to solve for y.

$$-5(40) - 4y = -260 \Rightarrow -200 - 4y = -260 \Rightarrow$$
$$-4y = -60 \Rightarrow y = 15$$

Substitute $x = 40$ and $y = 15$ into equation (6) to solve for z.

$$3(40) + 2(15) + z = 180 \Rightarrow$$
$$120 + 30 + z = 180 \Rightarrow$$
$$150 + z = 180 \Rightarrow z = 30$$

Solution set: $\{(40, 15, 30)\}$

111. Let x = the number of pounds of Arabian Mocha Sanani; y = the number of pounds of Organic Shade-Grown Mexico; z = the number of pounds of Guatemala Antigua. Completing the table we have the following.

	Number of Pounds	Cost per Pound	Total Cost
Arabian Mocha	x	15.99	$15.99x$
Organic Mexico	y	12.99	$12.99y$
Guatemala Antigua	z	10.19	$10.19z$
Total	50	12.37	$50(12.37) =$ 618.50

We have the following equations.

$$x + y + z = 50 \quad (1)$$
$$z = 2x \quad (2)$$
$$15.99x + 12.99y + 10.19z = 618.50 \quad (3)$$

Substitute $z = 2x$ into equations (1) and (3):

$$x + y + 2x = 50 \Rightarrow$$
$$3x + y = 50 \quad (4)$$
$$15.99x + 12.99y + 10.19(2x) = 618.50 \Rightarrow$$
$$36.37x + 12.99y = 618.50 \quad (5)$$

Solve equation (4) for y, then substitute that value into equation (5) and solve for x:
$$3x + y = 50 \Rightarrow y = -3x + 50$$

$$36.37x + 12.99(-3x + 50) = 618.50 \Rightarrow$$
$$36.37x - 38.97x + 649.50 = 618.50 \Rightarrow$$
$$-2.6x = -31 \Rightarrow$$
$$x \approx 11.9231 \approx 11.92$$

Substitute $x = 11.9231$ into equation (2) to solve for z: $z \approx 2(11.9231) \approx 23.85$

Substitute the values for x and z into equation (1) to solve for y:
$$11.92 + y + 23.85 = 50 \Rightarrow y \approx 14.23$$

11.92 pounds of Arabian Mocha Sanani, 14.23 pounds of Organic Shade-Grown Mexico, and 23.85 pounds of Guatemala Antigua are needed. (Answers are approximations.)

113. $p = 16 - \frac{5}{4}q$

   (a) $p = 16 - \frac{5}{4}(0) = 16 - 0 = 16$
     The price is \$16.

   (b) $p = 16 - \frac{5}{4}(4) = 16 - 5 = 16 - 5 = 11$
     The price is \$11.

   (c) $p = 16 - \frac{5}{4}(8) = 16 - 10 = 6$
     The price is \$6.

115. See answer to Exercise 117.

117.

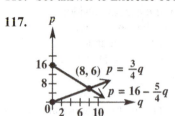

Section 9.2 Matrix Solution of Linear Systems

1. $\begin{bmatrix} -2 & 5 & 8 & 0 \\ 1 & 13 & -6 & 9 \end{bmatrix}$

This matrix has 2 rows and 4 columns. It has dimension is 2×4.

3. $-3x + 5y = 2$
    $6x + 2y = 7$

The augmented matrix is $\begin{bmatrix} -3 & 5 & | & 2 \\ 6 & 2 & | & 7 \end{bmatrix}$.

5. $3x + 2y \quad\quad = 5$
   $-9x + \quad\quad 6z = 1$
        $-8y + z = 4$

The augmented matrix is $\begin{bmatrix} 3 & 2 & 0 & | & 5 \\ -9 & 0 & 6 & | & 1 \\ 0 & -8 & 1 & | & 4 \end{bmatrix}$.

7. $\begin{bmatrix} 1 & 4 \\ 4 & 7 \end{bmatrix}$; -4 times row 1 added to row 2

$$\begin{bmatrix} 1 & 4 \\ 4 + (-4)(1) & 7 + (-4)(4) \end{bmatrix} = \begin{bmatrix} 1 & 4 \\ 0 & -9 \end{bmatrix}$$

9. $\begin{bmatrix} 1 & 5 & 6 \\ -2 & 3 & -1 \\ 4 & 7 & 0 \end{bmatrix}$; 2 times row 1 added to row 2

$$\begin{bmatrix} 1 & 5 & 6 \\ -2+2(1) & 3+2(5) & -1+2(6) \\ 4 & 7 & 0 \end{bmatrix} = \begin{bmatrix} 1 & 5 & 6 \\ 0 & 13 & 11 \\ 4 & 7 & 0 \end{bmatrix}$$

11. $2x + 3y = 11$
    $x + 2y = 8$

The augmented matrix is $\begin{bmatrix} 2 & 3 & | & 11 \\ 1 & 2 & | & 8 \end{bmatrix}$. The size is 2×3.

13. $2x + y + z - 3 = 0$
  $3x - 4y + 2z + 7 = 0$
  $x + y + z - 2 = 0$

Each equation of a linear system must have the constant term isolated on one side of the equal sign. Rewriting this system, we have

$2x + y + z = 3$
$3x - 4y + 2z = -7$
  $x + y + z = 2$

The augmented matrix is $\begin{bmatrix} 2 & 1 & 1 & 3 \\ 3 & -4 & 2 & -7 \\ 1 & 1 & 1 & 2 \end{bmatrix}$.

The size is 3×4.

15. $\begin{bmatrix} 3 & 2 & 1 & 1 \\ 0 & 2 & 4 & 22 \\ -1 & -2 & 3 & 15 \end{bmatrix}$ is associated with the

system $\begin{array}{l} 3x + 2y + z = 1 \\ 2y + 4z = 22 \\ -x - 2y + 3z = 15. \end{array}$

17. $\begin{bmatrix} 1 & 0 & 0 & 2 \\ 0 & 1 & 0 & 3 \\ 0 & 0 & 1 & -2 \end{bmatrix}$ is associated with the system

$x = 2$
$y = 3$
$z = -2.$

19. $\begin{bmatrix} 1 & 1 & 0 & 3 \\ 0 & 2 & 1 & -4 \\ 1 & 0 & -1 & 5 \end{bmatrix}$ is associated with the

system $\begin{array}{l} x + y = 3 \\ 2y + z = -4 \\ x - z = 5. \end{array}$

21. $x + y = 5$
  $x - y = -1$

This system has the augmented matrix
$\begin{bmatrix} 1 & 1 & 5 \\ 1 & -1 & -1 \end{bmatrix}$.

$\begin{bmatrix} 1 & 1 & 5 \\ 0 & -2 & -6 \end{bmatrix} \begin{array}{l} \\ -1R1+R2 \end{array} \Rightarrow$

$\begin{bmatrix} 1 & 1 & 5 \\ 0 & 1 & 3 \end{bmatrix} -\frac{1}{2}R2 \Rightarrow \begin{bmatrix} 1 & 0 & 2 \\ 0 & 1 & 3 \end{bmatrix} -1R2+R1$

Solution set: $\{(2,3)\}$

23. $3x + 2y = -9$
  $2x - 5y = -6$

This system has the augmented matrix
$\begin{bmatrix} 3 & 2 & -9 \\ 2 & -5 & -6 \end{bmatrix}$.

$\begin{bmatrix} 1 & 7 & -3 \\ 2 & -5 & -6 \end{bmatrix} \begin{array}{l} -1R2+R1 \\ \\ \end{array} \Rightarrow$

$\begin{bmatrix} 1 & 7 & -3 \\ 0 & -19 & 0 \end{bmatrix} \begin{array}{l} \\ -2R1+R2 \end{array} \Rightarrow$

$\begin{bmatrix} 1 & 7 & -3 \\ 0 & 1 & 0 \end{bmatrix} -\frac{1}{19}R2 \Rightarrow$

$\begin{bmatrix} 1 & 0 & -3 \\ 0 & 1 & 0 \end{bmatrix} -7R2+R1$

Solution set: $\{(-3,0)\}$

25. $6x - 3y - 4 = 0$
  $3x + 6y - 7 = 0$

Rewrite the system as $\begin{array}{l} 6x - 3y = 4 \\ 3x + 6y = 7 \end{array}$

The augmented matrix is $\begin{bmatrix} 6 & -3 & 4 \\ 3 & 6 & 7 \end{bmatrix}$.

$\begin{bmatrix} 1 & -\frac{1}{2} & \frac{2}{3} \\ 3 & 6 & 7 \end{bmatrix} \frac{1}{6}R1 \Rightarrow \begin{bmatrix} 1 & -\frac{1}{2} & \frac{2}{3} \\ 0 & \frac{15}{2} & 5 \end{bmatrix} -3R1+R2 \Rightarrow$

$\begin{bmatrix} 1 & -\frac{1}{2} & \frac{2}{3} \\ 0 & 1 & \frac{2}{3} \end{bmatrix} \frac{2}{15}R2 \Rightarrow \begin{bmatrix} 1 & 0 & 1 \\ 0 & 1 & \frac{2}{3} \end{bmatrix} \frac{1}{2}R2+R1$

Solution set: $\left\{\left(1, \frac{2}{3}\right)\right\}$

27. $2x - y = 6$
  $4x - 2y = 0$

This system has the augmented matrix
$\begin{bmatrix} 2 & -1 & 6 \\ 4 & -2 & 0 \end{bmatrix}$.

$\begin{bmatrix} 1 & -\frac{1}{2} & 3 \\ 4 & -2 & 0 \end{bmatrix} \frac{1}{2}R1 \Rightarrow \begin{bmatrix} 1 & -\frac{1}{2} & 3 \\ 0 & 0 & -12 \end{bmatrix} -4R1+R2$

The second row of the augmented matrix corresponds to the equation $0x + 0y = -12$, which has no solution. Thus, the solution set is $\varnothing$.

29. $\frac{3}{8}x - \frac{1}{2}y = \frac{7}{8}$
$-6x + 8y = -14$

This system has the augmented matrix $\begin{bmatrix} \frac{3}{8} & -\frac{1}{2} & \Big| & \frac{7}{8} \\ -6 & 8 & \Big| & -14 \end{bmatrix}$

$\begin{bmatrix} 3 & -4 & \Big| & 7 \\ -6 & 8 & \Big| & -14 \end{bmatrix} \begin{matrix} 8R1 \\ \Rightarrow \end{matrix} \begin{bmatrix} 1 & -\frac{4}{3} & \Big| & \frac{7}{3} \\ -6 & 8 & \Big| & -14 \end{bmatrix} \begin{matrix} \frac{1}{3}R1 \\ \Rightarrow \end{matrix} \begin{bmatrix} 1 & -\frac{4}{3} & \Big| & \frac{7}{3} \\ 0 & 0 & \Big| & 0 \end{bmatrix} 6R1 + R2$

It is impossible to go further. The equation that corresponds to the first row in the final matrix is

$x - \frac{4}{3}y = \frac{7}{3} \Rightarrow x = \frac{4}{3}y + \frac{7}{3} = \frac{4y+7}{3}$

Solution set: $\left\{ \frac{4y+7}{3}, \ y \right\}$

31. $x + y - 5z = -18$
$3x - 3y + z = 6$
$x + 3y - 2z = -13$

This system has the augmented matrix $\begin{bmatrix} 1 & 1 & -5 & \Big| & -18 \\ 3 & -3 & 1 & \Big| & 6 \\ 1 & 3 & -2 & \Big| & -13 \end{bmatrix}$.

$\begin{bmatrix} 1 & 1 & -5 & \Big| & -18 \\ 0 & -6 & 16 & \Big| & 60 \\ 1 & 3 & -2 & \Big| & -13 \end{bmatrix} \begin{matrix} \\ -3R1+R2 \\ \\ \end{matrix} \Rightarrow \begin{bmatrix} 1 & 1 & -5 & \Big| & -18 \\ 0 & -6 & 16 & \Big| & 60 \\ 0 & 2 & 3 & \Big| & 5 \end{bmatrix} \begin{matrix} \\ \\ -1R1+R3 \end{matrix} \Rightarrow \begin{bmatrix} 1 & 1 & -5 & \Big| & -18 \\ 0 & 1 & -\frac{8}{3} & \Big| & -10 \\ 0 & 2 & 3 & \Big| & 5 \end{bmatrix} \begin{matrix} \\ -\frac{1}{6}R2 \\ \\ \end{matrix}$

$\begin{bmatrix} 1 & 0 & -\frac{7}{3} & \Big| & -8 \\ 0 & 1 & -\frac{8}{3} & \Big| & -10 \\ 0 & 2 & 3 & \Big| & 5 \end{bmatrix} \begin{matrix} -1R2+R1 \\ \\ \\ \end{matrix} \Rightarrow \begin{bmatrix} 1 & 0 & -\frac{7}{3} & \Big| & -8 \\ 0 & 1 & -\frac{8}{3} & \Big| & -10 \\ 0 & 0 & \frac{25}{3} & \Big| & 25 \end{bmatrix} \begin{matrix} \\ \\ -2R2+R3 \end{matrix} \Rightarrow \begin{bmatrix} 1 & 0 & -\frac{7}{3} & \Big| & -8 \\ 0 & 1 & -\frac{8}{3} & \Big| & -10 \\ 0 & 0 & 1 & \Big| & 3 \end{bmatrix} \begin{matrix} \\ \\ \frac{3}{25}R3 \end{matrix}$

$\begin{bmatrix} 1 & 0 & 0 & \Big| & -1 \\ 0 & 1 & -\frac{8}{3} & \Big| & -10 \\ 0 & 0 & 1 & \Big| & 3 \end{bmatrix} \begin{matrix} \frac{7}{3}R3+R1 \\ \\ \\ \end{matrix} \Rightarrow \begin{bmatrix} 1 & 0 & 0 & \Big| & -1 \\ 0 & 1 & 0 & \Big| & -2 \\ 0 & 0 & 1 & \Big| & 3 \end{bmatrix} \begin{matrix} \\ \frac{8}{3}R3+R2 \\ \\ \end{matrix}$

Solution set: $\left\{ (-1, -2, 3) \right\}$

33. $x + y - z = 6$
$2x - y + z = -9$
$x - 2y + 3z = 1$

This system has the augmented matrix $\begin{bmatrix} 1 & 1 & -1 & \Big| & 6 \\ 2 & -1 & 1 & \Big| & -9 \\ 1 & -2 & 3 & \Big| & 1 \end{bmatrix}$.

$\begin{bmatrix} 1 & 1 & -1 & \Big| & 6 \\ 0 & -3 & 3 & \Big| & -21 \\ 1 & -2 & 3 & \Big| & 1 \end{bmatrix} \begin{matrix} \\ -2R1+R2 \\ \\ \end{matrix} \Rightarrow \begin{bmatrix} 1 & 1 & -1 & \Big| & 6 \\ 0 & -3 & 3 & \Big| & -21 \\ 0 & -3 & 4 & \Big| & -5 \end{bmatrix} \begin{matrix} \\ \\ -1R1+R3 \end{matrix} \Rightarrow \begin{bmatrix} 1 & 1 & -1 & \Big| & 6 \\ 0 & 1 & -1 & \Big| & 7 \\ 0 & -3 & 4 & \Big| & -5 \end{bmatrix} \begin{matrix} \\ -\frac{1}{3}R2 \\ \\ \end{matrix}$

$\begin{bmatrix} 1 & 0 & 0 & \Big| & -1 \\ 0 & 1 & -1 & \Big| & 7 \\ 0 & -3 & 4 & \Big| & -5 \end{bmatrix} \begin{matrix} -1R2+R1 \\ \\ \\ \end{matrix} \Rightarrow \begin{bmatrix} 1 & 0 & 0 & \Big| & -1 \\ 0 & 1 & -1 & \Big| & 7 \\ 0 & 0 & 1 & \Big| & 16 \end{bmatrix} \begin{matrix} \\ \\ 3R2+R3 \end{matrix} \Rightarrow \begin{bmatrix} 1 & 0 & 0 & \Big| & -1 \\ 0 & 1 & 0 & \Big| & 23 \\ 0 & 0 & 1 & \Big| & 16 \end{bmatrix} \begin{matrix} \\ R3+R2 \\ \\ \end{matrix}$

Solution set: $\left\{ (-1, 23, 16) \right\}$

35. $x - z = -3$
$y + z = 9$
$x + z = 7$

This system has the augmented matrix $\begin{bmatrix} 1 & 0 & -1 & | & -3 \\ 0 & 1 & 1 & | & 9 \\ 1 & 0 & 1 & | & 7 \end{bmatrix}$.

$\begin{bmatrix} 1 & 0 & -1 & | & -3 \\ 0 & 1 & 1 & | & 9 \\ 0 & 0 & 2 & | & 10 \end{bmatrix} \begin{matrix} \\ \\ -1R1+R3 \end{matrix} \Rightarrow \begin{bmatrix} 1 & 0 & -1 & | & -3 \\ 0 & 1 & 1 & | & 9 \\ 0 & 0 & 1 & | & 5 \end{bmatrix} \begin{matrix} \\ \\ \frac{1}{2}R3 \end{matrix} \Rightarrow \begin{bmatrix} 1 & 0 & 0 & | & 2 \\ 0 & 1 & 1 & | & 9 \\ 0 & 0 & 1 & | & 5 \end{bmatrix} \begin{matrix} R3+R1 \\ \\ \end{matrix} \Rightarrow \begin{bmatrix} 1 & 0 & 0 & | & 2 \\ 0 & 1 & 0 & | & 4 \\ 0 & 0 & 1 & | & 5 \end{bmatrix} \begin{matrix} \\ -1R3+R2 \\ \end{matrix}$

Solution set: $\{(2,4,5)\}$

37. $y = -2x - 2z + 1$
$x = -2y - z + 2$
$z = x - y$

Rewrite the system as $\begin{aligned} 2x + y + 2z &= 1 \\ x + 2y + z &= 2 \\ x - y - z &= 0 \end{aligned}$. The augmented matrix is $\begin{bmatrix} 2 & 1 & 2 & | & 1 \\ 1 & 2 & 1 & | & 2 \\ 1 & -1 & -1 & | & 0 \end{bmatrix}$.

$\begin{bmatrix} 1 & 2 & 1 & | & 2 \\ 2 & 1 & 2 & | & 1 \\ 1 & -1 & -1 & | & 0 \end{bmatrix} \begin{matrix} R1 \leftrightarrow R2 \\ \\ \end{matrix} \Rightarrow \begin{bmatrix} 1 & 2 & 1 & | & 2 \\ 0 & -3 & 0 & | & -3 \\ 1 & -1 & -1 & | & 0 \end{bmatrix} \begin{matrix} \\ -2R1+R2 \\ \end{matrix} \Rightarrow \begin{bmatrix} 1 & 2 & 1 & | & 2 \\ 0 & -3 & 0 & | & -3 \\ 0 & -3 & -2 & | & -2 \end{bmatrix} \begin{matrix} \\ \\ -1R1+R3 \end{matrix}$

$\begin{bmatrix} 1 & 2 & 1 & | & 2 \\ 0 & 1 & 0 & | & 1 \\ 0 & -3 & -2 & | & -2 \end{bmatrix} \begin{matrix} \\ -\frac{1}{3}R2 \\ \end{matrix} \Rightarrow \begin{bmatrix} 1 & 0 & 1 & | & 0 \\ 0 & 1 & 0 & | & 1 \\ 0 & -3 & -2 & | & -2 \end{bmatrix} \begin{matrix} -2R2+R1 \\ \\ \end{matrix} \Rightarrow \begin{bmatrix} 1 & 0 & 1 & | & 0 \\ 0 & 1 & 0 & | & 1 \\ 0 & 0 & -2 & | & 1 \end{bmatrix} \begin{matrix} \\ \\ 3R2+R3 \end{matrix}$

$\begin{bmatrix} 1 & 0 & 1 & | & 0 \\ 0 & 1 & 0 & | & 1 \\ 0 & 0 & 1 & | & -\frac{1}{2} \end{bmatrix} \begin{matrix} \\ \\ -\frac{1}{2}R3 \end{matrix} \Rightarrow \begin{bmatrix} 1 & 0 & 0 & | & \frac{1}{2} \\ 0 & 1 & 0 & | & 1 \\ 0 & 0 & 1 & | & -\frac{1}{2} \end{bmatrix} \begin{matrix} -1R3+R1 \\ \\ \end{matrix}$

Solution set: $\left\{ \left(\frac{1}{2}, 1, -\frac{1}{2} \right) \right\}$

39. $2x - y + 3z = 0$
$x + 2y - z = 5$
$2y + z = 1$

This system has the augmented matrix $\begin{bmatrix} 2 & -1 & 3 & | & 0 \\ 1 & 2 & -1 & | & 5 \\ 0 & 2 & 1 & | & 1 \end{bmatrix}$.

$\begin{bmatrix} 1 & 2 & -1 & | & 5 \\ 2 & -1 & 3 & | & 0 \\ 0 & 2 & 1 & | & 1 \end{bmatrix} \begin{matrix} R1 \leftrightarrow R2 \\ \\ \end{matrix} \Rightarrow \begin{bmatrix} 1 & 2 & -1 & | & 5 \\ 0 & -5 & 5 & | & -10 \\ 0 & 2 & 1 & | & 1 \end{bmatrix} \begin{matrix} \\ -2R1+R2 \\ \end{matrix} \Rightarrow \begin{bmatrix} 1 & 2 & -1 & | & 5 \\ 0 & 1 & -1 & | & 2 \\ 0 & 2 & 1 & | & 1 \end{bmatrix} \begin{matrix} \\ -\frac{1}{5}R2 \\ \end{matrix}$

$\begin{bmatrix} 1 & 0 & 1 & | & 1 \\ 0 & 1 & -1 & | & 2 \\ 0 & 2 & 1 & | & 1 \end{bmatrix} \begin{matrix} -2R2+R1 \\ \\ \end{matrix} \Rightarrow \begin{bmatrix} 1 & 0 & 1 & | & 1 \\ 0 & 1 & -1 & | & 2 \\ 0 & 0 & 3 & | & -3 \end{bmatrix} \begin{matrix} \\ \\ -2R2+R3 \end{matrix} \Rightarrow \begin{bmatrix} 1 & 0 & 1 & | & 1 \\ 0 & 1 & -1 & | & 2 \\ 0 & 0 & 1 & | & -1 \end{bmatrix} \begin{matrix} \\ \\ \frac{1}{3}R3 \end{matrix}$

$\begin{bmatrix} 1 & 0 & 0 & | & 2 \\ 0 & 1 & -1 & | & 2 \\ 0 & 0 & 1 & | & -1 \end{bmatrix} \begin{matrix} -1R3+R1 \\ \\ \end{matrix} \Rightarrow \begin{bmatrix} 1 & 0 & 0 & | & 2 \\ 0 & 1 & 0 & | & 1 \\ 0 & 0 & 1 & | & -1 \end{bmatrix} \begin{matrix} \\ R3+R2 \\ \end{matrix}$

Solution set: $\{(2,1,-1)\}$

41. $3x + 5y - z + 2 = 0$
$4x - y + 2z - 1 = 0$
$-6x - 10y + 2z = 0$

$$\begin{aligned} 3x + 5y - z &= -2 \\ \text{Rewrite the system as } \quad 4x - y + 2z &= 1 \\ -6x - 10y + 2z &= 0. \end{aligned}$$

The augmented matrix is $\begin{bmatrix} 3 & 5 & -1 & | & -2 \\ 4 & -1 & 2 & | & 1 \\ -6 & -10 & 2 & | & 0 \end{bmatrix}$.

$\begin{bmatrix} 1 & \frac{5}{3} & -\frac{1}{3} & | & -\frac{2}{3} \\ 4 & -1 & 2 & | & 1 \\ -6 & -10 & 2 & | & 0 \end{bmatrix} \begin{matrix} \frac{1}{3}R1 \\ \\ \\ \end{matrix} \Rightarrow \begin{bmatrix} 1 & \frac{5}{3} & -\frac{1}{3} & | & -\frac{2}{3} \\ 0 & -\frac{23}{3} & \frac{10}{3} & | & \frac{11}{3} \\ -6 & -10 & 2 & | & 0 \end{bmatrix} \begin{matrix} \\ -4R1 + R2 \\ \\ \end{matrix} \Rightarrow \begin{bmatrix} 1 & \frac{5}{3} & -\frac{1}{3} & | & -\frac{2}{3} \\ 0 & -\frac{23}{3} & \frac{10}{3} & | & \frac{11}{3} \\ 0 & 0 & 0 & | & -4 \end{bmatrix} \begin{matrix} \\ \\ 6R1 + R3 \end{matrix}$

The last row indicates that there is no solution. The solution set is $\varnothing$.

43. $x - 8y + z = 4$
$3x - y + 2z = -1$

This system has the augmented matrix $\begin{bmatrix} 1 & -8 & 1 & | & 4 \\ 3 & -1 & 2 & | & -1 \end{bmatrix}$.

$\begin{bmatrix} 1 & -8 & 1 & | & 4 \\ 0 & 23 & -1 & | & -13 \end{bmatrix} \begin{matrix} \\ -3R1 + R2 \end{matrix} \Rightarrow \begin{bmatrix} 1 & -8 & 1 & | & 4 \\ 0 & 1 & -\frac{1}{23} & | & -\frac{13}{23} \end{bmatrix} \begin{matrix} \\ \frac{1}{23}R2 \end{matrix} \Rightarrow \begin{bmatrix} 1 & 0 & \frac{15}{23} & | & -\frac{12}{23} \\ 0 & 1 & -\frac{1}{23} & | & -\frac{13}{23} \end{bmatrix} \begin{matrix} 8R2 + R1 \\ \\ \end{matrix}$

The equations that correspond to the final matrix are $x + \frac{15}{23}z = -\frac{12}{23}$ and $y - \frac{1}{23}z = -\frac{13}{23}$.

This system has infinitely many solutions. We will express the solution set with z as the arbitrary variable.

Therefore, $\begin{aligned} x &= -\frac{15}{23}z - \frac{12}{23} = -\frac{15z - 12}{23} \text{ and} \\ y &= \frac{1}{23}z - \frac{13}{23} = \frac{z - 13}{23}. \end{aligned}$

Solution set: $\left\{ \left(-\frac{15z - 12}{23}, \frac{z - 13}{23}, z \right) \right\}$

45. $x - y + 2z + w = 4$
$y + z = 3$
$z - w = 2$
$x - y = 0$

This system has the augmented matrix $\begin{bmatrix} 1 & -1 & 2 & 1 & | & 4 \\ 0 & 1 & 1 & 0 & | & 3 \\ 0 & 0 & 1 & -1 & | & 2 \\ 1 & -1 & 0 & 0 & | & 0 \end{bmatrix}$

$\begin{bmatrix} 1 & -1 & 2 & 1 & | & 4 \\ 0 & 1 & 1 & 0 & | & 3 \\ 0 & 0 & 1 & -1 & | & 2 \\ 0 & 0 & -2 & -1 & | & -4 \end{bmatrix} \begin{matrix} \\ \\ \\ -1R1 + R4 \end{matrix} \Rightarrow \begin{bmatrix} 1 & 0 & 3 & 1 & | & 7 \\ 0 & 1 & 1 & 0 & | & 3 \\ 0 & 0 & 1 & -1 & | & 2 \\ 0 & 0 & -2 & -1 & | & -4 \end{bmatrix} \begin{matrix} R2 + R1 \\ \\ \\ \end{matrix} \Rightarrow$

$\begin{bmatrix} 1 & 0 & 0 & 4 & | & 1 \\ 0 & 1 & 1 & 0 & | & 3 \\ 0 & 0 & 1 & -1 & | & 2 \\ 0 & 0 & -2 & -1 & | & -4 \end{bmatrix} \begin{matrix} -3R3 + R1 \\ \\ \\ \end{matrix} \Rightarrow \begin{bmatrix} 1 & 0 & 0 & 4 & | & 1 \\ 0 & 1 & 0 & 1 & | & 1 \\ 0 & 0 & 1 & -1 & | & 2 \\ 0 & 0 & -2 & -1 & | & -4 \end{bmatrix} \begin{matrix} \\ -1R3 + R2 \\ \\ \end{matrix} \Rightarrow$

(continued on next page)

(continued)

$$\begin{bmatrix} 1 & 0 & 0 & 4 & | & 1 \\ 0 & 1 & 0 & 1 & | & 1 \\ 0 & 0 & 1 & -1 & | & 2 \\ 0 & 0 & 0 & -3 & | & 0 \end{bmatrix} \begin{matrix} \\ \\ \\ 2R3+R4 \end{matrix} \Rightarrow \begin{bmatrix} 1 & 0 & 0 & 4 & | & 1 \\ 0 & 1 & 0 & 1 & | & 1 \\ 0 & 0 & 1 & -1 & | & 2 \\ 0 & 0 & 0 & 1 & | & 0 \end{bmatrix} \begin{matrix} \\ \\ \\ -\frac{1}{3}R4 \end{matrix} \Rightarrow$$

$$\begin{bmatrix} 1 & 0 & 0 & 0 & | & 1 \\ 0 & 1 & 0 & 1 & | & 1 \\ 0 & 0 & 1 & -1 & | & 2 \\ 0 & 0 & 0 & 1 & | & 0 \end{bmatrix} \begin{matrix} -4R4+R1 \\ \\ \\ \end{matrix} \Rightarrow \begin{bmatrix} 1 & 0 & 0 & 0 & | & 1 \\ 0 & 1 & 0 & 0 & | & 1 \\ 0 & 0 & 1 & -1 & | & 2 \\ 0 & 0 & 0 & 1 & | & 0 \end{bmatrix} \begin{matrix} \\ -1R4+R2 \\ \\ \end{matrix} \Rightarrow$$

$$\begin{bmatrix} 1 & 0 & 0 & 0 & | & 1 \\ 0 & 1 & 0 & 0 & | & 1 \\ 0 & 0 & 1 & 0 & | & 2 \\ 0 & 0 & 0 & 1 & | & 0 \end{bmatrix} \begin{matrix} \\ \\ R4+R3 \\ \end{matrix}$$

Solution set: $\{(1,1,2,0)\}$

47. $\begin{aligned} x + 3y - 2z - w &= 9 \\ 4x + y + z + 2w &= 2 \\ -3x - y + z - w &= -5 \\ x - y - 3z - 2w &= 2 \end{aligned}$

This system has the augmented matrix $\begin{bmatrix} 1 & 3 & -2 & -1 & | & 9 \\ 4 & 1 & 1 & 2 & | & 2 \\ -3 & -1 & 1 & -1 & | & -5 \\ 1 & -1 & -3 & -2 & | & 2 \end{bmatrix}.$

$$\begin{bmatrix} 1 & 3 & -2 & -1 & | & 9 \\ 0 & -11 & 9 & 6 & | & -34 \\ -3 & -1 & 1 & -1 & | & -5 \\ 1 & -1 & -3 & -2 & | & 2 \end{bmatrix} \begin{matrix} \\ -4R1+R2 \\ \\ \end{matrix} \Rightarrow \begin{bmatrix} 1 & 3 & -2 & -1 & | & 9 \\ 0 & -11 & 9 & 6 & | & -34 \\ 0 & 8 & -5 & -4 & | & 22 \\ 1 & -1 & -3 & -2 & | & 2 \end{bmatrix} \begin{matrix} \\ \\ 3R1+R3 \\ \end{matrix} \Rightarrow$$

$$\begin{bmatrix} 1 & 3 & -2 & -1 & | & 9 \\ 0 & -11 & 9 & 6 & | & -34 \\ 0 & 8 & -5 & -4 & | & 22 \\ 0 & -4 & -1 & -1 & | & -7 \end{bmatrix} \begin{matrix} \\ \\ \\ -1R1+R4 \end{matrix} \Rightarrow \begin{bmatrix} 1 & 3 & -2 & -1 & | & 9 \\ 0 & 1 & -\frac{9}{11} & -\frac{6}{11} & | & \frac{34}{11} \\ 0 & 8 & -5 & -4 & | & 22 \\ 0 & -4 & -1 & -1 & | & -7 \end{bmatrix} \begin{matrix} \\ -\frac{1}{11}R2 \\ \\ \end{matrix} \Rightarrow$$

$$\begin{bmatrix} 1 & 0 & \frac{5}{11} & \frac{7}{11} & | & -\frac{3}{11} \\ 0 & 1 & -\frac{9}{11} & -\frac{6}{11} & | & \frac{34}{11} \\ 0 & 8 & -5 & -4 & | & 22 \\ 0 & -4 & -1 & -1 & | & -7 \end{bmatrix} \begin{matrix} -3R2+R1 \\ \\ \\ \end{matrix} \Rightarrow \begin{bmatrix} 1 & 0 & \frac{5}{11} & \frac{7}{11} & | & -\frac{3}{11} \\ 0 & 1 & -\frac{9}{11} & -\frac{6}{11} & | & \frac{34}{11} \\ 0 & 0 & \frac{17}{11} & \frac{4}{11} & | & -\frac{30}{11} \\ 0 & -4 & -1 & -1 & | & -7 \end{bmatrix} \begin{matrix} \\ \\ -8R2+R3 \\ \end{matrix} \Rightarrow$$

$$\begin{bmatrix} 1 & 0 & \frac{5}{11} & \frac{7}{11} & | & -\frac{3}{11} \\ 0 & 1 & -\frac{9}{11} & -\frac{6}{11} & | & \frac{34}{11} \\ 0 & 0 & \frac{17}{11} & \frac{4}{11} & | & -\frac{30}{11} \\ 0 & 0 & -\frac{47}{11} & -\frac{35}{11} & | & \frac{59}{11} \end{bmatrix} \begin{matrix} \\ \\ \\ 4R2+R4 \end{matrix} \Rightarrow \begin{bmatrix} 1 & 0 & \frac{5}{11} & \frac{7}{11} & | & -\frac{3}{11} \\ 0 & 1 & -\frac{9}{11} & -\frac{6}{11} & | & \frac{34}{11} \\ 0 & 0 & 1 & \frac{4}{17} & | & -\frac{30}{17} \\ 0 & 0 & -\frac{47}{11} & -\frac{35}{11} & | & \frac{59}{11} \end{bmatrix} \begin{matrix} \\ \\ \frac{11}{17}R3 \\ \end{matrix} \Rightarrow$$

(continued on next page)

(*continued*)

$$\begin{bmatrix} 1 & 0 & 0 & \frac{9}{17} & \frac{9}{17} \\ 0 & 1 & -\frac{9}{11} & -\frac{6}{11} & \frac{34}{11} \\ 0 & 0 & 1 & \frac{4}{17} & -\frac{30}{17} \\ 0 & 0 & -\frac{47}{11} & -\frac{35}{11} & \frac{59}{11} \end{bmatrix} \begin{matrix} \\ \\ \\ \end{matrix} -\frac{5}{11}R3 + R1 \Rightarrow \begin{bmatrix} 1 & 0 & 0 & \frac{9}{17} & \frac{9}{17} \\ 0 & 1 & 0 & -\frac{6}{17} & \frac{28}{71} \\ 0 & 0 & 1 & \frac{4}{17} & -\frac{30}{17} \\ 0 & 0 & -\frac{47}{11} & -\frac{35}{11} & \frac{59}{11} \end{bmatrix} \frac{9}{11}R3 + R2 \Rightarrow$$

$$\begin{bmatrix} 1 & 0 & 0 & \frac{9}{17} & \frac{9}{17} \\ 0 & 1 & 0 & -\frac{6}{17} & \frac{28}{17} \\ 0 & 0 & 1 & \frac{4}{17} & -\frac{30}{17} \\ 0 & 0 & 0 & -\frac{37}{17} & -\frac{37}{17} \end{bmatrix} \begin{matrix} \\ \\ \\ \frac{47}{11}R3 + R4 \end{matrix} \Rightarrow \begin{bmatrix} 1 & 0 & 0 & \frac{9}{17} & \frac{9}{17} \\ 0 & 1 & 0 & -\frac{6}{17} & \frac{28}{17} \\ 0 & 0 & 1 & \frac{4}{17} & -\frac{30}{17} \\ 0 & 0 & 0 & 1 & 1 \end{bmatrix} -\frac{17}{37}R4 \Rightarrow$$

$$\begin{bmatrix} 1 & 0 & 0 & 0 & 0 \\ 0 & 1 & 0 & -\frac{6}{17} & \frac{28}{17} \\ 0 & 0 & 1 & \frac{4}{17} & -\frac{30}{17} \\ 0 & 0 & 0 & 1 & 1 \end{bmatrix} \begin{matrix} -\frac{9}{17}R4 + R1 \\ \\ \\ \end{matrix} \Rightarrow \begin{bmatrix} 1 & 0 & 0 & 0 & 0 \\ 0 & 1 & 0 & 0 & 2 \\ 0 & 0 & 1 & \frac{4}{17} & -\frac{30}{17} \\ 0 & 0 & 0 & 1 & 1 \end{bmatrix} \frac{6}{17}R4 + R2 \Rightarrow$$

$$\begin{bmatrix} 1 & 0 & 0 & 0 & 0 \\ 0 & 1 & 0 & 0 & 2 \\ 0 & 0 & 1 & 0 & -2 \\ 0 & 0 & 0 & 1 & 1 \end{bmatrix} -\frac{4}{17}R4 + R3$$

Solution set: $\{(0, 2, -2, 1)\}$

49. $0.3x + 2.7y - \sqrt{2}z = 3$
$\sqrt{7}x - 20y + 12z = -2$
$4x + \sqrt{3}y - 1.2z = \frac{3}{4}$

This system has the augmented matrix

$$\begin{bmatrix} 0.3 & 2.7 & -\sqrt{2} & 3 \\ \sqrt{7} & -20 & 12 & -2 \\ 4 & \sqrt{3} & -1.2 & \frac{3}{4} \end{bmatrix}.$$

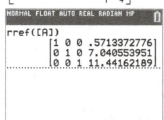

Using a graphing calculator. we obtain the approximate solution set.
Solution set: $\{(0.571, 7.041, 11.442)\}$.

51. $2x + 3y = 5 \Rightarrow 3y = 5 - 2x \Rightarrow y = \frac{5-2x}{3}$
$-3x + 5y = 22 \Rightarrow 5y = 3x + 22 \Rightarrow y = \frac{3x+22}{5}$
$2x + y = -1 \Rightarrow y = -2x - 1$

Graph the three equations in the window $[-3, -1] \times [2, 4]$.

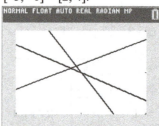

There are no solutions because the three lines do not intersect at one point.

53. $\dfrac{1}{(x-1)(x+1)} = \dfrac{A}{x-1} + \dfrac{B}{x+1}$

If we add the rational expression on the right, we get the following.

$\dfrac{1}{(x-1)(x+1)} = \dfrac{A(x+1)}{(x-1)(x+1)} + \dfrac{B(x-1)}{(x-1)(x+1)}$

$\dfrac{1}{(x-1)(x+1)} = \dfrac{A(x+1)+B(x-1)}{(x-1)(x+1)}$

Because the denominators are equal, the numerators must be equal. Thus we have the equation $1 = A(x+1) + B(x-1)$.

$1 = A(x+1) + B(x-1) \Rightarrow$
$1 = Ax + A + Bx - B$
$1 = (A-B) + (Ax + Bx) \Rightarrow$
$1 = (A-B) + (A+B)x$

Equating the coefficients of like powers of x gives the following system of equations.

$A + B = 0$
$A - B = 1$

Solve this system by the Gauss-Jordan method.

$\begin{bmatrix} 1 & 1 & | & 0 \\ 1 & -1 & | & 1 \end{bmatrix} \Rightarrow \begin{bmatrix} 1 & 1 & | & 0 \\ 0 & -2 & | & 1 \end{bmatrix} \begin{matrix} \\ -R1+R2 \end{matrix} \Rightarrow$

$\begin{bmatrix} 1 & 1 & | & 0 \\ 0 & 1 & | & -\frac{1}{2} \end{bmatrix} \begin{matrix} \\ -\frac{1}{2}R2 \end{matrix} \Rightarrow$

$\begin{bmatrix} 1 & 0 & | & \frac{1}{2} \\ 0 & 1 & | & -\frac{1}{2} \end{bmatrix} \begin{matrix} -1R2+R1 \\ \\ \end{matrix}$

Thus, $A = \frac{1}{2}$ and $B = -\frac{1}{2}$.

55. $\dfrac{x}{(x-a)(x+a)} = \dfrac{A}{x-a} + \dfrac{B}{x+a}$

If we add the rational expression on the right, we obtain the following.

$\dfrac{x}{(x-a)(x+a)} = \dfrac{A(x+a)}{(x-a)(x+a)} + \dfrac{B(x-a)}{(x-a)(x+a)}$

$\dfrac{x}{(x-a)(x+a)} = \dfrac{A(x+a)+B(x-a)}{(x-a)(x+a)}$

The denominators are equal, so the numerators must be equal. Thus we have the equation $x = A(x+a) + B(x-a)$.

$x = A(x+a) + B(x-a) \Rightarrow$
$x = Ax + Aa + Bx - Ba \Rightarrow$
$x = (A+B)x + (A-B)a$

Equating the coefficients of like powers of x gives the following system of equations.

$A + B = 1$
$A - B = 0$

Use the Gauss-Jordan method.

$\begin{bmatrix} 1 & 1 & | & 1 \\ 1 & -1 & | & 0 \end{bmatrix} \Rightarrow \begin{bmatrix} 1 & 1 & | & 1 \\ 0 & -2 & | & -1 \end{bmatrix} \begin{matrix} \\ -R1+R2 \end{matrix} \Rightarrow$

$\begin{bmatrix} 1 & 1 & | & 1 \\ 0 & 1 & | & \frac{1}{2} \end{bmatrix} \begin{matrix} \\ -\frac{1}{2}R2 \end{matrix} \Rightarrow \begin{bmatrix} 1 & 0 & | & \frac{1}{2} \\ 0 & 1 & | & \frac{1}{2} \end{bmatrix} \begin{matrix} -1R2+R1 \\ \\ \end{matrix}$

Thus, $A = \frac{1}{2}$ and $B = \frac{1}{2}$.

57. Let $x =$ the daily wage for a day laborer
$y =$ the daily wage for a concrete finisher
From the information, we can write the system
$7x + 2y = 1384$
$x + 5y = 952$

Use the Gauss-Jordan method:

$\begin{bmatrix} 7 & 2 & | & 1384 \\ 1 & 5 & | & 952 \end{bmatrix} \Rightarrow$

$\begin{bmatrix} 1 & 5 & | & 952 \\ 7 & 2 & | & 1384 \end{bmatrix} \begin{matrix} R1 \leftrightarrow R2 \\ \\ \end{matrix} \Rightarrow$

$\begin{bmatrix} 1 & 5 & | & 952 \\ 0 & -33 & | & -5280 \end{bmatrix} \begin{matrix} \\ -7R1+R2 \end{matrix} \Rightarrow$

$\begin{bmatrix} 1 & 5 & | & 952 \\ 0 & 1 & | & 160 \end{bmatrix} \begin{matrix} \\ -\frac{1}{33}R2 \end{matrix} \Rightarrow$

$\begin{bmatrix} 1 & 0 & | & 152 \\ 0 & 1 & | & 160 \end{bmatrix} \begin{matrix} -5R2+R1 \\ \\ \end{matrix}$

From the final matrix, $x = 152$ and $y = 160$, so the day laborers earn \$152 per day and the concreter finishers earn \$160 per day.

59. Let $x =$ the first number; $y =$ the second number; $z =$ the third number.
From the information, we can write the system

$x + y + z = 20$
$x = 3(y-z) \Rightarrow x = 3y - 3z \Rightarrow x - 3y + 3z = 0$
$y = 2 + 2z \Rightarrow y - 2z = 2$

Solve this system by the Gauss-Jordan method:

$\begin{bmatrix} 1 & 1 & 1 & | & 20 \\ 1 & -3 & 3 & | & 0 \\ 0 & 1 & -2 & | & 2 \end{bmatrix} \Rightarrow$

$\begin{bmatrix} 1 & 1 & 1 & | & 20 \\ 0 & 1 & -\frac{1}{2} & | & 5 \\ 0 & 1 & -2 & | & 2 \end{bmatrix} \begin{matrix} \\ \frac{1}{4}(R1-R2) \\ \\ \end{matrix} \Rightarrow$

$\begin{bmatrix} 1 & 1 & 1 & | & 20 \\ 0 & 1 & -\frac{1}{2} & | & 5 \\ 0 & 0 & 1 & | & 2 \end{bmatrix} \begin{matrix} \\ \\ \frac{2}{3}(R2-R3) \end{matrix} \Rightarrow$

$\begin{bmatrix} 1 & 1 & 1 & | & 20 \\ 0 & 1 & 0 & | & 6 \\ 0 & 0 & 1 & | & 2 \end{bmatrix} \begin{matrix} \\ R2+\frac{1}{2}R3 \\ \\ \end{matrix} \Rightarrow$

(continued on next page)

(*continued*)

$$\begin{bmatrix} 1 & 0 & 1 & | & 14 \\ 0 & 1 & 0 & | & 6 \\ 0 & 0 & 1 & | & 2 \end{bmatrix} \begin{matrix} R1-R2 \\ \\ \end{matrix} \Rightarrow$$

$$\begin{bmatrix} 1 & 0 & 0 & | & 12 \\ 0 & 1 & 0 & | & 6 \\ 0 & 0 & 1 & | & 2 \end{bmatrix} \begin{matrix} R1-R3 \\ \\ \end{matrix}$$

From the final matrix, $x = 12$, $y = 6$, and $z = 2$.
Thus, the numbers are 12, 6, and 2.

61. Let $x =$ number of cubic centimeters of the 2% solution; $y =$ number of cubic centimeters of the 7% solution.
From the information we can write the system

$$\begin{aligned} x + \quad y &= 40 & \Rightarrow & \quad x + \quad y &= 40 \\ 0.02x + 0.07y &= 0.032(40) & \Rightarrow & \quad 0.02x + 0.07y &= 1.28 \end{aligned}$$

Solve this system by the Gauss-Jordan method.

$$\begin{bmatrix} 1 & 1 & | & 40 \\ 0.02 & 0.07 & | & 1.28 \end{bmatrix} \Rightarrow$$

$$\begin{bmatrix} 1 & 1 & | & 40 \\ 0 & 0.05 & | & 0.48 \end{bmatrix} -0.02R1+R2$$

$$\begin{bmatrix} 1 & 1 & | & 40 \\ 0 & 1 & | & 9.6 \end{bmatrix} \tfrac{1}{0.05}R2 \Rightarrow \begin{bmatrix} 1 & 0 & | & 30.4 \\ 0 & 1 & | & 9.6 \end{bmatrix} -1R2+R1$$

From the final matrix, $x = 30.4$ and $y = 9.6$,

so the chemist should mix 30.4 cm^3 of the 2%

solution with 9.6 cm^3 of the 7% solution.

63. Let $x =$ the amount deposited at 1.5%; $y =$ the amount deposited at 2.2%; $z =$ the amount deposited at 2.4%.
Completing the table we have the following.

Amount Invested	Rate	Annual Interest
x	0.015	$0.015x$
y	0.022	$0.022y$
z	0.024	$0.024z$
		535

From the information given in the exercise, we can write the system.

$$\begin{cases} x + y + z = 25,000 \\ \qquad\quad y = z \\ 0.015x + 0.022y + 0.024z = 535 \end{cases} \Rightarrow$$

$$\begin{aligned} x + \quad y + \quad z &= 25,000 \\ y - \quad z &= 0 \\ 15x + 22y + 24z &= 535,000 \end{aligned}$$

This system has the augmented matrix

$$\begin{bmatrix} 1 & 1 & 1 & | & 25,000 \\ 0 & 1 & -1 & | & 0 \\ 15 & 22 & 24 & | & 535,000 \end{bmatrix}.$$

Solve by the Gauss-Jordan method.

$$\begin{bmatrix} 1 & 1 & 1 & | & 25,000 \\ 0 & 1 & -1 & | & 0 \\ 0 & 7 & 9 & | & 160,000 \end{bmatrix} -15R1+R3$$

$$\begin{bmatrix} 1 & 1 & 1 & | & 25,000 \\ 0 & 1 & -1 & | & 0 \\ 0 & 0 & 16 & | & 160,000 \end{bmatrix} -7R2+R3$$

$$\begin{bmatrix} 1 & 1 & 1 & | & 25,000 \\ 0 & 1 & -1 & | & 0 \\ 0 & 0 & 1 & | & 10,000 \end{bmatrix} \tfrac{1}{16}R3$$

$$\begin{bmatrix} 1 & 1 & 1 & | & 25,000 \\ 0 & 1 & 0 & | & 10,000 \\ 0 & 0 & 1 & | & 10,000 \end{bmatrix} R2+R3$$

$$\begin{bmatrix} 1 & 1 & 0 & | & 15,000 \\ 0 & 1 & 0 & | & 10,000 \\ 0 & 0 & 1 & | & 10,000 \end{bmatrix} -R3+R1$$

$$\begin{bmatrix} 1 & 0 & 0 & | & 5,000 \\ 0 & 1 & 0 & | & 10,000 \\ 0 & 0 & 1 & | & 10,000 \end{bmatrix} -R2+R1$$

From the final matrix, we have $x = 5,000$, $y = 10,000$, and $z = 10,000$. The investor deposited $5,000 at 1.5%, $10,000 at 2.2%, and $10,000 at 2.4%.

65. Let $x =$ number of grams of food A; $y =$ number of grams of food B; $z =$ number of grams of food C.
Completing the table we have the following.

Food Group	A	B	C	Total
Grams/Meal	x	y	z	400

From the information given in the exercise, we can write the system.

$$\begin{aligned} x + \quad y + z &= 400 \\ x = \tfrac{1}{3}y \Rightarrow 3x - \quad y \quad &= 0 \\ x + z = 2y \Rightarrow x - 2y + z &= 0 \end{aligned}$$

This system has the augmented matrix

$$\begin{bmatrix} 1 & 1 & 1 & | & 400 \\ 3 & -1 & 0 & | & 0 \\ 1 & -2 & 1 & | & 0 \end{bmatrix}$$

Solve by the Gauss-Jordan method.

$$\begin{bmatrix} 1 & 1 & 1 & | & 400 \\ 0 & -4 & -3 & | & -1200 \\ 1 & -2 & 1 & | & 0 \end{bmatrix} -3R1+R2 \Rightarrow$$

$$\begin{bmatrix} 1 & 1 & 1 & | & 400 \\ 0 & -4 & -3 & | & -1200 \\ 0 & -3 & 0 & | & -400 \end{bmatrix} -1R1+R3 \Rightarrow$$

(*continued on next page*)

(*continued*)

$$\begin{bmatrix} 1 & 1 & 1 & | & 400 \\ 0 & 1 & \frac{3}{4} & | & 300 \\ 0 & -3 & 0 & | & -400 \end{bmatrix} \begin{matrix} \\ -\frac{1}{4}R2 \\ \\ \end{matrix} \Rightarrow$$

$$\begin{bmatrix} 1 & 0 & \frac{1}{4} & | & 100 \\ 0 & 1 & \frac{3}{4} & | & 300 \\ 0 & -3 & 0 & | & -400 \end{bmatrix} \begin{matrix} -1R2+R1 \\ \\ \\ \end{matrix} \Rightarrow$$

$$\begin{bmatrix} 1 & 0 & \frac{1}{4} & | & 100 \\ 0 & 1 & \frac{3}{4} & | & 300 \\ 0 & 0 & \frac{9}{4} & | & 500 \end{bmatrix} \begin{matrix} \\ \\ 3R2+R3 \end{matrix} \Rightarrow$$

$$\begin{bmatrix} 1 & 0 & \frac{1}{4} & | & 100 \\ 0 & 1 & \frac{3}{4} & | & 300 \\ 0 & 0 & 1 & | & \frac{2000}{9} \end{bmatrix} \begin{matrix} \\ \\ \frac{4}{9}R3 \end{matrix}$$

$$\begin{bmatrix} 1 & 0 & 0 & | & \frac{400}{9} \\ 0 & 1 & \frac{3}{4} & | & 300 \\ 0 & 0 & 1 & | & \frac{2000}{9} \end{bmatrix} \begin{matrix} -\frac{1}{4}R3+R1 \\ \\ \\ \end{matrix} \Rightarrow$$

$$\begin{bmatrix} 1 & 0 & 0 & | & \frac{400}{9} \\ 0 & 1 & 0 & | & \frac{400}{3} \\ 0 & 0 & 1 & | & \frac{2000}{9} \end{bmatrix} \begin{matrix} \\ -\frac{3}{4}R3+R2 \\ \\ \end{matrix}$$

From the final matrix, we have

$x = \frac{400}{9} \approx 44.4$, $y = \frac{400}{3} \approx 133.3$, and

$z = \frac{2000}{9} \approx 222.2$.

The diet should include 44.4 g of food A, 133.3 g of food B, and 222.2 g of food C.

67. (a) For the 65 or older group:
With $x = 0$ representing 2015 and $x = 35$ representing 2050, we have information that correlates to the points $(0, 0.148)$ and $(35, 0.209)$.

$$m = \frac{0.209 - 0.148}{35 - 0} = \frac{0.061}{35} \approx 0.0017$$

We have the point $(0, 0.148)$, the equation in slope-intercept form is
$y = 0.0017x + 0.148$.

For the 25–34 group:
With $x = 0$ representing 2015 and $x = 35$ representing 2050, we have information that correlates to the points $(0, 0.2)$ and $(35, 0.103)$.

$$m = \frac{0.103 - 0.2}{35 - 0} = \frac{-0.097}{35} = -0.0028$$

We have the point $(0, 0.2)$, so the equation in slope-intercept form is
$y = -0.0028x + 0.2$.

(b) Solve the following system.

$\begin{cases} y = 0.0017x + 0.148 \\ y = -0.0028x + 0.2 \end{cases} \Rightarrow$

$\begin{cases} -0.0017x + y = 0.148 \\ 0.0028x + y = 0.2 \end{cases} \Rightarrow$

$-17x + 10000y = 1480$
$28x + 10000y = 2000$

This system has the augmented matrix

$$\begin{bmatrix} -17 & 10000 & | & 1480 \\ 28 & 10000 & | & 2000 \end{bmatrix}$$

Solve by the Gauss-Jordan method.

$$\begin{bmatrix} 45 & 0 & | & 520 \\ 28 & 10000 & | & 2000 \end{bmatrix} \begin{matrix} -R1+R2 \\ \\ \end{matrix}$$

$$\begin{bmatrix} 1 & 0 & | & \frac{104}{9} \\ 28 & 10000 & | & 2000 \end{bmatrix} \begin{matrix} \frac{1}{45}R1 \\ \\ \end{matrix}$$

$$\begin{bmatrix} 1 & 0 & | & \frac{104}{9} \\ 0 & 10000 & | & \frac{15088}{9} \end{bmatrix} \begin{matrix} \\ -28R1+R2 \end{matrix}$$

$$\begin{bmatrix} 1 & 0 & | & \frac{104}{9} \\ 0 & 1 & | & \frac{15088}{90000} \end{bmatrix} \begin{matrix} \\ \frac{1}{10000}R1 \end{matrix}$$

The solution set is

$\left\{ \left(\frac{104}{9}, \frac{15088}{90000} \right) \right\} \approx \left\{ (11.5556, 0.1676) \right\}$.

$x \approx 11.5556$ represents the year 2026 and $y \approx 0.1676 = 16.76\%$. The two groups will have the same percentage of the population during 2026, about 16.8%.

(c) Answers will vary. Sample answer: The *percentage* of people in the U.S. population aged 25–39 is decreasing, but not necessarily the *number* of people in this age group.

69. (a) A height of 6′11″ is 83″.
If $W = 7.46H - 374$, then
$W = 7.46(83) - 374 = 245.18$.
Using the first equation, the predicted weight is approximately 245 pounds.
If $W = 7.93H - 405$, then
$W = 7.93(83) - 405 = 253.19$
Using the second equation, the predicted weight is approximately 253 pounds.

(b) For the first model $W = 7.46H - 374$, a 1-inch increase in height results in a 7.46-pound increase in weight.
For the second model $W = 7.93H - 405$, a 1-inch increase in height results in a 7.93-pound increase in weight. In each case, the change is given by the slope of the line that is the graph of the given equation.

(c) $W - 7.46H = -374$
$W - 7.93H = -405$

This system has the augmented matrix

$$\begin{bmatrix} 1 & -7.46 & | & -374 \\ 1 & -7.93 & | & -405 \end{bmatrix}.$$

Solve this system by the Gauss-Jordan method.

$$\begin{bmatrix} 1 & -7.46 & | & -374 \\ 0 & -0.47 & | & -31 \end{bmatrix} -1R1 + R2 \Rightarrow$$

$$\begin{bmatrix} 1 & -7.46 & | & -374 \\ 0 & 1 & | & 65.957 \end{bmatrix} -\frac{1}{0.47}R2 \Rightarrow$$

$$\begin{bmatrix} 1 & 0 & | & 118.043 \\ 0 & 1 & | & 65.957 \end{bmatrix} 7.46R2 + R1$$

From the last matrix, we have $W \approx 118$ and $H \approx 66$. The two models agree at a height of 66 inches and a weight of 118 pounds.

71. $T(n) = \frac{2}{3}n^3 + \frac{3}{2}n^2 - \frac{7}{6}n$

Continuing in this manner we have the following.

n	T
3	28
6	191
10	805
29	17,487
100	681,550
200	5,393,100
400	42,906,200
1000	668,165,500
5000	8.3×10^{10}
10,000	6.7×10^{11}
100,000	6.7×10^{14}

73. If the number of variables doubles, the number of operations increases by a factor of 8. 100 variables require 681,550 operations and 200 variables requires 5,393,100 operations.

The ratio $\frac{5,393,100}{681,550} \approx 7.91 \approx 8$.

75. $F = a + bA + cP + dW$

Substituting the values, we have the following system of equations.

$a + 871b + 11.5c + 3d = 239$
$a + 847b + 12.2c + 2d = 234$
$a + 685b + 10.6c + 5d = 192$
$a + 969b + 14.2c + 1d = 343.$

77. Using these values,
$F = -715.457 + 0.348A + 48.659P$
$\qquad\qquad + 30.720W$

Section 9.3 Determinant Solution of Linear Systems

1. $\begin{vmatrix} 4 & 0 \\ -2 & 0 \end{vmatrix} = 4 \cdot 0 - (-2)(0) = 0$

3. $\begin{vmatrix} x & 4 \\ 3 & x \end{vmatrix} = x^2 - 12$

5. $\begin{vmatrix} x & 0 \\ 0 & x \end{vmatrix} = 9 \Rightarrow x^2 = 9 \Rightarrow x = \pm 3$

7. $\begin{vmatrix} -5 & 9 \\ 4 & -1 \end{vmatrix} = -5(-1) - 4 \cdot 9 = 5 - 36 = -31$

```
[A]
        [[-5  9 ]
         [4  -1]]
det([A])
               -31
```

9. $\begin{vmatrix} -1 & -2 \\ 5 & 3 \end{vmatrix} = -1 \cdot 3 - 5(-2) = -3 - (-10) = 7$

```
[A]
        [[-1  -2]
         [5   3]]
det([A])
                 7
```

11. $\begin{vmatrix} 9 & 3 \\ -3 & -1 \end{vmatrix} = 9(-1) - (-3) \cdot 3 = -9 - (-9) = 0$

```
[A]
        [[9   3 ]
         [-3  -1]]
det([A])
                 0
```

13. $\begin{vmatrix} 3 & 4 \\ 5 & -2 \end{vmatrix} = 3(-2) - 5 \cdot 4 = -6 - 20 = -26$

```
[A]
        [[3  4 ]
         [5  -2]]
det([A])
               -26
```

15. $\begin{vmatrix} -7 & 0 \\ 3 & 0 \end{vmatrix} = -7 \cdot 0 - 3 \cdot 0 = 0 - 0 = 0$

```
[A]
        [[-7 0]
         [3  0]]
det([A])
                0
```

17. $\begin{vmatrix} -2 & 0 & 1 \\ 1 & 2 & 0 \\ 4 & 2 & 1 \end{vmatrix}$

To find the cofactor of 1, we have
$i = 2, j = 1,$

$M_{21} = \begin{vmatrix} 0 & 1 \\ 2 & 1 \end{vmatrix} = 0 \cdot 1 - 2 \cdot 1 = 0 - 2 = -2$

Thus, the cofactor is

$(-1)^{2+1}(-2) = (-1)^3(-2) = (-1)(-2) = 2.$

To find the cofactor of 2, we have
$i = 2, j = 2,$

$M_{22} = \begin{vmatrix} -2 & 1 \\ 4 & 1 \end{vmatrix} = -2 \cdot 1 - 4 \cdot 1 = -2 - 4 = -6$

Thus, the cofactor is

$(-1)^{2+2}(-6) = (-1)^4(-6) = 1(-6) = -6.$

To find the cofactor of 0, we have
$i = 2, j = 3,$

$M_{23} = \begin{vmatrix} -2 & 0 \\ 4 & 2 \end{vmatrix} = -2 \cdot 2 - 4 \cdot 0 = -4 - 0 = -4$

Thus, the cofactor is

$(-1)^{2+3}(-4) = (-1)^5(-4) = (-1)(-4) = 4.$

19. $\begin{vmatrix} 1 & 2 & -1 \\ 2 & 3 & -2 \\ -1 & 4 & 1 \end{vmatrix}$

To find the cofactor of 2, we have $i = 2, j = 1,$

$M_{21} = \begin{vmatrix} 2 & -1 \\ 4 & 1 \end{vmatrix} = 2 \cdot 1 - 4(-1) = 2 - (-4) = 6$

Thus, the cofactor is

$(-1)^{2+1}(6) = (-1)^3(6) = (-1)(6) = -6.$

To find the cofactor of 3, we have $i = 2, j = 2,$

$M_{22} = \begin{vmatrix} 1 & -1 \\ -1 & 1 \end{vmatrix} = 1 \cdot 1 - (-1)(-1) = 1 - 1 = 0$

Thus, the cofactor is

$(-1)^{2+2}(0) = (-1)^4(0) = 1(0) = 0.$

To find the cofactor of -2, we have
$i = 2, j = 3,$

$M_{23} = \begin{vmatrix} 1 & 2 \\ -1 & 4 \end{vmatrix} = 1 \cdot 4 - (-1) \cdot 2 = 4 - (-2) = 6$

Thus, the cofactor is

$(-1)^{2+3}(6) = (-1)^5(6) = (-1)(6) = -6.$

For Exercises 21–37, an answer can be arrived at by expanding on any row or column. In the solutions, we will expand on a row or column that allows a minimum number of calculations. Any row or column containing zero will reduce the number of calculations. You may want to refer to the Determinant Theorems on page 880 of your text.

21. $\begin{vmatrix} 4 & -7 & 8 \\ 2 & 1 & 3 \\ -6 & 3 & 0 \end{vmatrix}$

If we expand by the third row, we will need to find $a_{31} \cdot A_{31} + a_{32} \cdot A_{32} + a_{33} \cdot A_{33}$. However, we do not need to calculate A_{33} because $a_{33} = 0.$

$A_{31} = (-1)^{3+1} \begin{vmatrix} -7 & 8 \\ 1 & 3 \end{vmatrix} = (-1)^4(-7 \cdot 3 - 1 \cdot 8)$

$\qquad = 1(-21 - 8) = 1(-29) = -29$

$A_{32} = (-1)^{3+2} \begin{vmatrix} 4 & 8 \\ 2 & 3 \end{vmatrix} = (-1)^5(4 \cdot 3 - 2 \cdot 8)$

$\qquad = -1(12 - 16) = -1(-4) = 4$

$a_{31} \cdot A_{31} + a_{32} \cdot A_{32} + a_{33} \cdot A_{33}$
$\quad = a_{31} \cdot A_{31} + a_{32} \cdot A_{32} + 0 \cdot A_{33}$
$\quad = -6(-29) + 3(4) + 0 = 174 + 12 = 186$

23. $\begin{vmatrix} 1 & 2 & 0 \\ -1 & 2 & -1 \\ 0 & 1 & 4 \end{vmatrix}$

If we expand by the third row, we will need to find $a_{31} \cdot A_{31} + a_{32} \cdot A_{32} + a_{33} \cdot A_{33}$. However, we do not need to calculate A_{31} because $a_{31} = 0.$

$A_{32} = (-1)^{3+2} \begin{vmatrix} 1 & 0 \\ -1 & -1 \end{vmatrix}$

$\qquad = (-1)^5 [1(-1) - (-1) \cdot 0]$

$\qquad = -1(-1 - 0) = -1(-1) = 1$

(continued on next page)

(*continued*)

$$A_{33} = (-1)^{3+3} \begin{vmatrix} 1 & 2 \\ -1 & 2 \end{vmatrix}$$

$$= (-1)^6 \left[1 \cdot 2 - (-1) \cdot 2 \right]$$

$$= 1 \left[2 - (-2) \right] = 1(4) = 4$$

$$a_{31} \cdot A_{31} + a_{32} \cdot A_{32} + a_{33} \cdot A_{33}$$
$$= 0 \cdot A_{31} + a_{32} \cdot A_{32} + a_{33} \cdot A_{33}$$
$$= 0 + 1(1) + 4(4) = 1 + 16 = 17$$

For Exercises 25–33, use the sign checkerboard shown on page 879 of the text for the calculations.

25. $\begin{vmatrix} 10 & 2 & 1 \\ -1 & 4 & 3 \\ -3 & 8 & 10 \end{vmatrix}$

If we expand by the first row, we will need to find $a_{11} \cdot M_{11} - a_{12} \cdot M_{12} + a_{13} \cdot M_{13}$.

$$M_{11} = \begin{vmatrix} 4 & 3 \\ 8 & 10 \end{vmatrix} = 4 \cdot 10 - 8 \cdot 3 = 40 - 24 = 16,$$

$$M_{12} = \begin{vmatrix} -1 & 3 \\ -3 & 10 \end{vmatrix} = -1 \cdot 10 - (-3) \cdot 3$$
$$= -10 - (-9) = -1, \text{ and}$$

$$M_{13} = \begin{vmatrix} -1 & 4 \\ -3 & 8 \end{vmatrix} = -1 \cdot 8 - (-3) \cdot 4 = -8 - (-12) = 4$$

$$a_{11} \cdot M_{11} - a_{12} \cdot M_{12} + a_{13} \cdot M_{13}$$
$$= 10 \cdot 16 - 2(-1) + 1 \cdot 4$$
$$= 160 - (-2) + 4 = 166$$

27. $\begin{vmatrix} 1 & -2 & 3 \\ 0 & 0 & 0 \\ 1 & 10 & -12 \end{vmatrix}$

If we expand by the second row, we will need to find $-a_{21} \cdot M_{21} + a_{22} \cdot M_{22} - a_{23} \cdot M_{23}$. Because $a_{21} = a_{22} = a_{23} = 0,$ the result will be zero.

29. $\begin{vmatrix} 3 & 3 & -1 \\ 2 & 6 & 0 \\ -6 & -6 & 2 \end{vmatrix}$

If we expand by the second row, we will need to find $-a_{21} \cdot M_{21} + a_{22} \cdot M_{22} - a_{23} \cdot M_{23}$. However, we do not need to calculate $M_{23},$ because $a_{23} = 0.$

$$M_{21} = \begin{vmatrix} 3 & -1 \\ -6 & 2 \end{vmatrix} = 3 \cdot 2 - (-6)(-1) = 6 - 6 = 0$$

and

$$M_{22} = \begin{vmatrix} 3 & -1 \\ -6 & 2 \end{vmatrix} = 3 \cdot 2 - (-6)(-1) = 6 - 6 = 0$$

$$a_{21} \cdot M_{21} - a_{22} \cdot M_{22} + a_{23} \cdot M_{23}$$
$$= a_{21} \cdot M_{21} - a_{22} \cdot M_{22} + 0 \cdot M_{23}$$
$$= 2 \cdot 0 - 6 \cdot 0 + 0 = 0 - 0 = 0$$

31. $\begin{vmatrix} 1 & 0 & 0 \\ 0 & 1 & 0 \\ 0 & 0 & 1 \end{vmatrix}$

If we expand by the first row, we will need to find $a_{11} \cdot M_{11} - a_{12} \cdot M_{12} + a_{13} \cdot M_{13}$. However, we do not need to calculate M_{12} or M_{13} because $a_{12} = a_{13} = 0.$

$$M_{11} = \begin{vmatrix} 1 & 0 \\ 0 & 1 \end{vmatrix} = 1 \cdot 1 - 0 \cdot 0 = 1 - 0 = 1$$

$$a_{11} \cdot M_{11} - a_{12} \cdot M_{12} + a_{13} \cdot M_{13}$$
$$= a_{11} \cdot M_{11} - 0 \cdot M_{12} + 0 \cdot M_{13}$$
$$= 1 \cdot 1 - 0 + 0 = 1$$

33. $\begin{vmatrix} -2 & 0 & 1 \\ 0 & 1 & 0 \\ 0 & 0 & -1 \end{vmatrix}$

If we expand by the first column, we will need to find $a_{11} \cdot M_{11} - a_{21} \cdot M_{21} + a_{31} \cdot M_{31}$. However, we do not need to calculate M_{21} or M_{31} because $a_{21} = a_{31} = 0.$

$$M_{11} = \begin{vmatrix} 1 & 0 \\ 0 & -1 \end{vmatrix} = 1(-1) - 0 \cdot 0 = -1 - 0 = -1$$

$$a_{11} \cdot M_{11} - a_{21} \cdot M_{21} + a_{31} \cdot M_{31}$$
$$= a_{11} \cdot M_{11} - 0 \cdot M_{21} + 0 \cdot M_{31}$$
$$= -2(-1) - 0 + 0 = 2$$

35. $\begin{vmatrix} \sqrt{2} & 4 & 0 \\ 1 & -\sqrt{5} & 7 \\ -5 & \sqrt{5} & 1 \end{vmatrix}$

If we expand by the third column, we will need to find $a_{13} \cdot M_{13} - a_{23} \cdot M_{23} + a_{33} \cdot M_{33}$. However, we do not need to calculate M_{13} because $a_{13} = 0.$

$$M_{23} = \begin{vmatrix} \sqrt{2} & 4 \\ -5 & \sqrt{5} \end{vmatrix} = \sqrt{2} \cdot \sqrt{5} - (4)(-5)$$
$$= \sqrt{10} + 20$$

(*continued on next page*)

(*continued*)

$$M_{33} = \begin{vmatrix} \sqrt{2} & 4 \\ 1 & -\sqrt{5} \end{vmatrix} = \sqrt{2}\left(-\sqrt{5}\right) - (1)(4)$$

$$= -\sqrt{10} - 4$$

$$a_{13} \cdot M_{13} - a_{23} \cdot M_{23} + a_{33} \cdot M_{33}.$$

$$= 0 \cdot M_{13} - 7\left(\sqrt{10} + 20\right) + 1\left(-\sqrt{10} - 4\right)$$

$$= -7\sqrt{10} - 140 - \sqrt{10} - 4$$

$$= -144 - 8\sqrt{10}$$

37. $\begin{vmatrix} 0.4 & -0.8 & 0.6 \\ 0.3 & 0.9 & 0.7 \\ 3.1 & 4.1 & -2.8 \end{vmatrix}$

If we expand by the first column, we will need to find $a_{11} \cdot M_{11} - a_{21} \cdot M_{21} + a_{31} \cdot M_{31}.$

$$M_{11} = \begin{vmatrix} 0.9 & 0.7 \\ 4.1 & -2.8 \end{vmatrix}$$

$$= 0.9(-2.8) - 4.1(0.7) = -5.39$$

$$M_{21} = \begin{vmatrix} -0.8 & 0.6 \\ 4.1 & -2.8 \end{vmatrix}$$

$$= -0.8(-2.8) - 4.1(0.6) = -0.22$$

$$M_{31} = \begin{vmatrix} -0.8 & 0.6 \\ 0.9 & 0.7 \end{vmatrix}$$

$$= -0.8(0.7) - 0.9(0.6) = -1.1$$

$$a_{11} \cdot M_{11} - a_{21} \cdot M_{21} + a_{31} \cdot M_{31}$$
$$= 0.4(-5.39) - 0.3(-0.22) + 3.1(-1.1)$$
$$= -5.5$$

For exercises 39−61, refer to the determinant theorems on page 880 in the text. Also, let

$$|A| = \begin{vmatrix} 1 & 2 & 3 \\ 4 & 5 & 6 \\ 7 & 9 & 10 \end{vmatrix}.$$

39. $\begin{vmatrix} 4 & 5 & 6 \\ 1 & 2 & 3 \\ 7 & 9 & 10 \end{vmatrix}$

Interchange rows 1 and 2.

$$\begin{vmatrix} 4 & 5 & 6 \\ 1 & 2 & 3 \\ 7 & 9 & 10 \end{vmatrix} = -\begin{vmatrix} 1 & 2 & 3 \\ 4 & 5 & 6 \\ 7 & 9 & 10 \end{vmatrix} = -3$$

41. If each element of a row si multiplied by a real number k, then the determinant of the new matrix equals the determinant of the original matrix.

$$\begin{vmatrix} 5 & 10 & 15 \\ 4 & 5 & 6 \\ 7 & 9 & 10 \end{vmatrix} = \begin{vmatrix} 5\cdot1 & 5\cdot2 & 5\cdot3 \\ 4 & 5 & 6 \\ 7 & 9 & 10 \end{vmatrix} = 5\cdot3 = 15$$

43. Adding a row to another row does not change the value of the determinant.

$$\begin{vmatrix} 1 & 2 & 3 \\ 4 & 5 & 6 \\ 8 & 11 & 13 \end{vmatrix} = \begin{vmatrix} 1 & 2 & 3 \\ 4 & 5 & 6 \\ 7+1 & 9+2 & 10+3 \end{vmatrix} = 3$$

45. $\begin{vmatrix} 1 & 0 & 0 \\ 1 & 0 & 1 \\ 3 & 0 & 0 \end{vmatrix}$

By Theorem 1, the determinant is 0 because every entry in column 2 is 0.

47. $\begin{vmatrix} 6 & 8 & -12 \\ -1 & 0 & 2 \\ 4 & 0 & -8 \end{vmatrix}$

Given $\begin{bmatrix} 6 & 8 & -12 \\ -1 & 0 & 2 \\ 4 & 0 & -8 \end{bmatrix}$, add 2 times column 1

to column 3 to obtain $\begin{bmatrix} 6 & 8 & 0 \\ -1 & 0 & 0 \\ 4 & 0 & 0 \end{bmatrix}.$

By Theorem 6 the following statement is true.

$$\begin{vmatrix} 6 & 8 & -12 \\ -1 & 0 & 2 \\ 4 & 0 & -8 \end{vmatrix} = \begin{vmatrix} 6 & 8 & 0 \\ -1 & 0 & 0 \\ 4 & 0 & 0 \end{vmatrix}$$

By Theorem 1, the determinant is 0 because every entry in column 3 is 0.

49. $\begin{vmatrix} -4 & 1 & 4 \\ 2 & 0 & 1 \\ 0 & 2 & 4 \end{vmatrix}$

Because

$$\begin{bmatrix} -4 & 1 & 4 \\ 2 & 0 & 1 \\ 0 & 2 & 4 \end{bmatrix} \Rightarrow \begin{bmatrix} 0 & 1 & 6 \\ 2 & 0 & 1 \\ 0 & 2 & 4 \end{bmatrix} \begin{matrix} 2R2+R1 \\ \\ \end{matrix}, \text{ by}$$

Theorem 6 we have the following.

$$\begin{vmatrix} -4 & 1 & 4 \\ 2 & 0 & 1 \\ 0 & 2 & 4 \end{vmatrix} = \begin{vmatrix} 0 & 1 & 6 \\ 2 & 0 & 1 \\ 0 & 2 & 4 \end{vmatrix}$$

(*continued on next page*)

(*continued*)

Expanding by the first column, we have the following.

$$\begin{vmatrix} 0 & 1 & 6 \\ 2 & 0 & 1 \\ 0 & 2 & 4 \end{vmatrix} = 0\begin{vmatrix} 0 & 1 \\ 2 & 4 \end{vmatrix} - 2\begin{vmatrix} 1 & 6 \\ 2 & 4 \end{vmatrix} + 0\begin{vmatrix} 1 & 6 \\ 0 & 1 \end{vmatrix}$$

$$= 0 - 2(4-12) + 0 = -2(-8) = 16$$

51. $\begin{vmatrix} 0 & 1 & -3 \\ 7 & 5 & 2 \\ 1 & -2 & 6 \end{vmatrix}$

Add 2 times row 1 to row 3 to obtain

$$\begin{vmatrix} 0 & 1 & -3 \\ 7 & 5 & 2 \\ 1 & 0 & 0 \end{vmatrix}.$$

By Theorem 6 the following statement is true.

$$\begin{vmatrix} 0 & 1 & -3 \\ 7 & 5 & 2 \\ 1 & -2 & 6 \end{vmatrix} = \begin{vmatrix} 0 & 1 & -3 \\ 7 & 5 & 2 \\ 1 & 0 & 0 \end{vmatrix}$$

Expanding by the third row, we have the following.

$$\begin{vmatrix} 0 & 1 & -3 \\ 7 & 5 & 2 \\ 1 & 0 & 0 \end{vmatrix} = 1\begin{vmatrix} 1 & -3 \\ 5 & 2 \end{vmatrix} - 0\begin{vmatrix} 0 & -3 \\ 7 & 2 \end{vmatrix} + 0\begin{vmatrix} 0 & 1 \\ 7 & 5 \end{vmatrix}$$

$$= 1(2+15) - 0 + 0 = 1(17) = 17$$

53. $\begin{vmatrix} 1 & 6 & 7 \\ 0 & 6 & 7 \\ 0 & 0 & 9 \end{vmatrix}$

Add −1 times row 2 to row 1 to obtain

$$\begin{vmatrix} 1 & 0 & 0 \\ 0 & 6 & 7 \\ 0 & 0 & 9 \end{vmatrix}.$$

By Theorem 6 the following statement is true.

$$\begin{vmatrix} 1 & 6 & 7 \\ 0 & 6 & 7 \\ 0 & 0 & 9 \end{vmatrix} = \begin{vmatrix} 1 & 0 & 0 \\ 0 & 6 & 7 \\ 0 & 0 & 9 \end{vmatrix}$$

Expanding by the third row, we have the following.

$$\begin{vmatrix} 1 & 0 & 0 \\ 0 & 6 & 7 \\ 0 & 0 & 9 \end{vmatrix} = 0\begin{vmatrix} 0 & 0 \\ 6 & 7 \end{vmatrix} - 0\begin{vmatrix} 1 & 0 \\ 0 & 7 \end{vmatrix} + 9\begin{vmatrix} 1 & 0 \\ 0 & 6 \end{vmatrix}$$

$$= 0 - 0 + 9(6) = 54$$

55. $\begin{vmatrix} 2 & -1 & 3 \\ 6 & 4 & 10 \\ 4 & 5 & 7 \end{vmatrix}$

Add 4 times row 1 to row 2 to obtain

$$\begin{vmatrix} 2 & -1 & 3 \\ 14 & 0 & 22 \\ 4 & 5 & 7 \end{vmatrix}$$

Now add 5 times row 1 to row 3 to obtain

$$\begin{vmatrix} 2 & -1 & 3 \\ 14 & 0 & 22 \\ 14 & 0 & 22 \end{vmatrix}$$

By Theorem 6 the following statement is true.

$$\begin{vmatrix} 2 & -1 & 3 \\ 6 & 4 & 10 \\ 4 & 5 & 7 \end{vmatrix} = \begin{vmatrix} 2 & -1 & 3 \\ 14 & 0 & 22 \\ 4 & 5 & 7 \end{vmatrix} = \begin{vmatrix} 2 & -1 & 3 \\ 14 & 0 & 22 \\ 14 & 0 & 22 \end{vmatrix}$$

Theorem 5 states that if any two rows of a matrix are identical, then the determinant of the matrix is 0. Thus,

$$\begin{vmatrix} 2 & -1 & 3 \\ 6 & 4 & 10 \\ 4 & 5 & 7 \end{vmatrix} = \begin{vmatrix} 2 & -1 & 3 \\ 14 & 0 & 22 \\ 4 & 5 & 7 \end{vmatrix} = \begin{vmatrix} 2 & -1 & 3 \\ 14 & 0 & 22 \\ 14 & 0 & 22 \end{vmatrix} = 0$$

57. $\begin{vmatrix} -1 & 0 & 2 & 3 \\ 5 & 4 & -3 & 7 \\ 8 & 2 & 9 & -5 \\ 4 & 4 & -1 & 10 \end{vmatrix}$

Add 5 times row 1 to row 2 to obtain

$$\begin{vmatrix} -1 & 0 & 2 & 3 \\ 0 & 4 & 7 & 22 \\ 8 & 2 & 9 & -5 \\ 4 & 4 & -1 & 10 \end{vmatrix}$$

Add 4 times row 1 to row 4 to obtain

$$\begin{vmatrix} -1 & 0 & 2 & 3 \\ 0 & 4 & 7 & 22 \\ 8 & 2 & 9 & -5 \\ 0 & 4 & 7 & 22 \end{vmatrix}$$

Theorem 5 states that if any two rows of a matrix are identical, then the determinant of the matrix is 0. Thus,

$$\begin{vmatrix} -1 & 0 & 2 & 3 \\ 5 & 4 & -3 & 7 \\ 8 & 2 & 9 & -5 \\ 4 & 4 & -1 & 10 \end{vmatrix} = 0$$

For exercises 59–61, as in previous exercises, we can arrive at a solution by expanding on any row or column. We will also use determinant theorems to reduce the number of calculations.

59.
$$\begin{vmatrix} 4 & 0 & 0 & 2 \\ -1 & 0 & 3 & 0 \\ 2 & 4 & 0 & 1 \\ 0 & 0 & 1 & 2 \end{vmatrix}$$

Expanding by the second column, we have

$$\begin{vmatrix} 4 & 0 & 0 & 2 \\ -1 & 0 & 3 & 0 \\ 2 & 4 & 0 & 1 \\ 0 & 0 & 1 & 2 \end{vmatrix} = (-1)^{3+2} \cdot 4 \begin{vmatrix} 4 & 0 & 2 \\ -1 & 3 & 0 \\ 0 & 1 & 2 \end{vmatrix}$$

$$= -4 \begin{vmatrix} 4 & 0 & 2 \\ -1 & 3 & 0 \\ 0 & 1 & 2 \end{vmatrix}$$

Using the definition of the determinant in the text, we have the following.

$$-4 \begin{vmatrix} 4 & 0 & 2 \\ -1 & 3 & 0 \\ 0 & 1 & 2 \end{vmatrix}$$

$$= -4 \left(\begin{bmatrix} 4(3)(2) + 0(0)(0) + 2(-1)(1) \end{bmatrix} \\ - \begin{bmatrix} 0(3)(2) + 1(0)(4) + 2(-1)(0) \end{bmatrix} \right)$$

$$= -4 \big[(24 + 0 - 2) - (0 + 0 + 0) \big]$$

$$= -4(22 - 0) = -4(22) = -88$$

61.
$$\begin{vmatrix} 3 & -6 & 5 & -1 \\ 0 & 2 & -1 & 3 \\ -6 & 4 & 2 & 0 \\ -7 & 3 & 1 & 1 \end{vmatrix}$$

Add 3 times row 1 to row 2 and add row 1 to row 4 to obtain

$$\begin{vmatrix} 3 & -6 & 5 & -1 \\ 0 & 2 & -1 & 3 \\ -6 & 4 & 2 & 0 \\ -7 & 3 & 1 & 1 \end{vmatrix} = \begin{vmatrix} 3 & -6 & 5 & -1 \\ 9 & -16 & 14 & 0 \\ -6 & 4 & 2 & 0 \\ -4 & -3 & 6 & 0 \end{vmatrix}.$$

Now expand by the fourth column.

$$\begin{vmatrix} 3 & -6 & 5 & -1 \\ 9 & -16 & 14 & 0 \\ -6 & 4 & 2 & 0 \\ -4 & -3 & 6 & 0 \end{vmatrix}$$

$$= (-1)^{1+4} \cdot (-1) \begin{vmatrix} 9 & -16 & 14 \\ -6 & 4 & 2 \\ -4 & -3 & 6 \end{vmatrix}$$

$$= \begin{vmatrix} 9 & -16 & 14 \\ -6 & 4 & 2 \\ -4 & -3 & 6 \end{vmatrix}$$

Adding −7 times row 2 to row 1 and −3 times row 2 added to row 3 we have the following.

$$\begin{vmatrix} 9 & -16 & 14 \\ -6 & 4 & 2 \\ -4 & -3 & 6 \end{vmatrix} = \begin{vmatrix} 51 & -44 & 0 \\ -6 & 4 & 2 \\ 14 & -15 & 0 \end{vmatrix}$$

Expanding by the third column, we have

$$-2 \begin{vmatrix} 51 & -44 \\ 14 & -15 \end{vmatrix} = -2(-765 + 616)$$

$$= -2(-149) = 298$$

63. $x + y = 4$
$2x - y = 2$

$$D = \begin{vmatrix} 1 & 1 \\ 2 & -1 \end{vmatrix} = 1(-1) - 2(1) = -1 - 2 = -3$$

$$D_x = \begin{vmatrix} 4 & 1 \\ 2 & -1 \end{vmatrix} = 4(-1) - 2(1) = -4 - 2 = -6$$

$$D_y = \begin{vmatrix} 1 & 4 \\ 2 & 2 \end{vmatrix} = 1(2) - 2(4) = 2 - 8 = -6 \Rightarrow$$

$x = \frac{D_x}{D} = \frac{-6}{-3} = 2$ and $y = \frac{D_y}{D} = \frac{-6}{-3} = 2$.
Solution set: $\{(2, 2)\}$

65. $4x + 3y = -7$
$2x + 3y = -11$

$$D = \begin{vmatrix} 4 & 3 \\ 2 & 3 \end{vmatrix} = 4(3) - 2(3) = 12 - 6 = 6,$$

$$D_x = \begin{vmatrix} -7 & 3 \\ -11 & 3 \end{vmatrix} = -7(3) - (-11)(3)$$

$$= -21 + 33 = 12,$$

$$D_y = \begin{vmatrix} 4 & -7 \\ 2 & -11 \end{vmatrix} = 4(-11) - 2(-7)$$

$$= -44 + 14 = -30 \Rightarrow$$

$x = \frac{D_x}{D} = \frac{12}{6} = 2$ and $y = \frac{D_y}{D} = \frac{-30}{6} = -5$.
Solution set: $\{(2, -5)\}$

67. $5x + 4y = 10$
$3x - 7y = 6$

$$D = \begin{vmatrix} 5 & 4 \\ 3 & -7 \end{vmatrix} = 5(-7) - 3(4) = -35 - 12 = -47,$$

$$D_x = \begin{vmatrix} 10 & 4 \\ 6 & -7 \end{vmatrix} = 10(-7) - 6(4) = -70 - 24 = -94,$$

$$D_y = \begin{vmatrix} 5 & 10 \\ 3 & 6 \end{vmatrix} = 5(6) - 3(10) = 30 - 30 = 0 \Rightarrow$$

$x = \frac{D_x}{D} = \frac{-94}{-47} = 2$ and $y = \frac{D_y}{D} = \frac{0}{-47} = 0.$

Solution set: $\{(2, 0)\}$

69. $1.5x + 3y = 5$ (1)
$2x + 4y = 3$ (2)

$$D = \begin{vmatrix} 1.5 & 3 \\ 2 & 4 \end{vmatrix} = (1.5)(4) - 2(3) = 6 - 6 = 0$$

Because $D = 0$, Cramer's rule does not apply. To determine whether the system is inconsistent or has infinitely many solutions, use the elimination method.

$6x + 12y = 20$ Multiply equation (1) by 4.
$\underline{-6x - 12y = -9}$ Multiply equation (2) by -3.
$ 0 = 11$ False

The system is inconsistent.
Solution set: $\varnothing$

71. $3x + 2y = 4$ (1)
$6x + 4y = 8$ (2)

$$D = \begin{vmatrix} 3 & 2 \\ 6 & 4 \end{vmatrix} = 3(4) - 6(2) = 12 - 12 = 0$$

Because $D = 0$, Cramer's rule does not apply. To determine whether the system is inconsistent or has infinitely many solutions, use the elimination method.

$-6x - 4y = -8$ Multiply equation (1) by -2.
$\underline{6x + 4y = 8}$
$ 0 = 0$ True

This shows that equations (1) and (2) are dependent. To write the solution set with y as the arbitrary variable, solve equation (1) for x in terms of y.

$3x + 2y = 4 \Rightarrow 3x = 4 - 2y \Rightarrow x = \frac{4-2y}{3}$

Solution set: $\left\{ \left(\frac{4-2y}{3}, y \right) \right\}$

73. $\frac{1}{2}x + \frac{1}{3}y = 2$
$\frac{3}{2}x - \frac{1}{2}y = -12$

$$D = \begin{vmatrix} \frac{1}{2} & \frac{1}{3} \\ \frac{3}{2} & -\frac{1}{2} \end{vmatrix} = \frac{1}{2}\left(-\frac{1}{2}\right) - \frac{3}{2}\left(\frac{1}{3}\right)$$

$$= -\frac{1}{4} - \frac{1}{2} = -\frac{1}{4} - \frac{2}{4} = -\frac{3}{4}$$

$$D_x = \begin{vmatrix} 2 & \frac{1}{3} \\ -12 & -\frac{1}{2} \end{vmatrix} = 2\left(-\frac{1}{2}\right) - (-12)\left(\frac{1}{3}\right)$$ and

$$= -1 + 4 = 3$$

$$D_y = \begin{vmatrix} \frac{1}{2} & 2 \\ \frac{3}{2} & -12 \end{vmatrix} = \frac{1}{2}(-12) - \frac{3}{2}(2) = -6 - 3 = -9$$

Thus, we have $x = \frac{D_x}{D} = \frac{3}{-\frac{3}{4}} = 3\left(-\frac{4}{3}\right) = -4$ and

$y = \frac{D_y}{D} = \frac{-9}{-\frac{3}{4}} = -9\left(-\frac{4}{3}\right) = 12$

Solution set: $\left\{ (-4, 12) \right\}$

In Exercises 75–85, we will be using the Determinant Theorems on page 880 of the text to reduce the number of calculations necessary in simplifying determinants.

75. $2x - y + 4z = -2$
$3x + 2y - z = -3$
$x + 4y + 2z = 17$

Adding 2 times row 1 to row 2 and 4 times row 1 to row 3 we have

$$D = \begin{vmatrix} 2 & -1 & 4 \\ 3 & 2 & -1 \\ 1 & 4 & 2 \end{vmatrix} = \begin{vmatrix} 2 & -1 & 4 \\ 7 & 0 & 7 \\ 9 & 0 & 18 \end{vmatrix}.$$

Expanding by column two, we have

$$D = -(-1)\begin{vmatrix} 7 & 7 \\ 9 & 18 \end{vmatrix} = 126 - 63 = 63.$$

Adding 2 times row 1 to row 2 and 4 times row 1 to row 3, we have

$$D_x = \begin{vmatrix} -2 & -1 & 4 \\ -3 & 2 & -1 \\ 17 & 4 & 2 \end{vmatrix} = \begin{vmatrix} -2 & -1 & 4 \\ -7 & 0 & 7 \\ 9 & 0 & 18 \end{vmatrix}.$$

Expanding about column two, we have

$$D_x = -(-1)\begin{vmatrix} -7 & 7 \\ 9 & 18 \end{vmatrix} = -126 - 63 = -189.$$

Adding column 2 to column 1, we have

$$D_y = \begin{vmatrix} 2 & -2 & 4 \\ 3 & -3 & -1 \\ 1 & 17 & 2 \end{vmatrix} = \begin{vmatrix} 0 & -2 & 4 \\ 0 & -3 & -1 \\ 18 & 17 & 2 \end{vmatrix}.$$

(continued on next page)

(*continued*)

Expanding by column one, we have

$$D_y = 18\begin{vmatrix} -2 & 4 \\ -3 & -1 \end{vmatrix} = 18(2+12) = 18(14) = 252.$$

Adding column 3 to column 1, we have

$$D_z = \begin{vmatrix} 2 & -1 & -2 \\ 3 & 2 & -3 \\ 1 & 4 & 17 \end{vmatrix} = \begin{vmatrix} 0 & -1 & -2 \\ 0 & 2 & -3 \\ 18 & 4 & 17 \end{vmatrix}.$$

Expanding by column one we have

$$D_z = 18\begin{vmatrix} -1 & -2 \\ 2 & -3 \end{vmatrix} = 18(3+4) = 18(7) = 126.$$

Thus, we have $x = \frac{D_x}{D} = \frac{-189}{63} = -3$,

$y = \frac{D_y}{D} = \frac{252}{63} = 4$, and $z = \frac{D_z}{D} = \frac{126}{63} = 2$.

Solution set: $\{(-3, 4, 2)\}$

77. $x + 2y + 3z = 4$ (1)
$4x + 3y + 2z = 1$ (2)
$-x - 2y - 3z = 0$ (3)

Adding row 1 to row 3, we have

$$D = \begin{vmatrix} 1 & 2 & 3 \\ 4 & 3 & 2 \\ -1 & -2 & -3 \end{vmatrix} = \begin{vmatrix} 1 & 2 & 3 \\ 4 & 3 & 2 \\ 0 & 0 & 0 \end{vmatrix}.$$

Because we have a row of zeros, $D = 0$ and we cannot use Cramer's rule.
Using the elimination method, we can add equations (1) and (3).

$x + 2y + 3z = 4$
$\underline{-x - 2y - 3z = 0}$
$\qquad 0 = 4$ False

The system is inconsistent.
Solution set: $\varnothing$

79. $-2x - 2y + 3z = 4$ (1)
$5x + 7y - z = 2$ (2)
$2x + 2y - 3z = -4$ (3)

Adding row 1 to row 3, we have

$$D = \begin{vmatrix} -2 & -2 & 3 \\ 5 & 7 & -1 \\ 2 & 2 & -3 \end{vmatrix} = \begin{vmatrix} -2 & -2 & 3 \\ 5 & 7 & -1 \\ 0 & 0 & 0 \end{vmatrix}.$$

Because we have a row of zeros, $D = 0$ and we cannot use Cramer's rule. Add equations (1) and (3) using the elimination method.

$-2x - 2y + 3z = 4$
$\underline{2x + 2y - 3z = -4}$
$\qquad 0 = 0$ True

This system will have infinitely many solutions.
Solve the system made up of equations (2) and (3) in terms of the arbitrary variable z. To eliminate x, multiply equation (2) by -2 and equation (3) by 5 and add the results
$-10x - 14y + 2z = -4$
$\underline{10x + 10y - 15z = -20}$
$\qquad -4y - 13z = -24$

Solve for y in terms of z.
$-4y - 13z = -24 \Rightarrow -4y = -24 + 13z \Rightarrow$
$y = \frac{-24 + 13z}{-4} = \frac{24 - 13z}{4}$

Express x also in terms of z by solving equation (3) for x and substituting $\frac{24-13z}{4}$ for y.
$2x + 2y - 3z = -4 \Rightarrow 2x = -2y + 3z - 4 \Rightarrow$
$x = \frac{-2y + 3z - 4}{2}$

$x = \frac{-2\left(\frac{24-13z}{4}\right) + 3z - 4}{2} = \frac{\frac{-24+13z}{2} + 3z - 4}{2}$

$= \frac{-24 + 13z + 6z - 8}{4} = \frac{-32 + 19z}{4}$

Solution set (with z arbitrary):

$$\left\{\left(\frac{-32+19z}{4}, \frac{24-13z}{4}, z\right)\right\}$$

81. $4x - 3y + z + 1 = 0 \qquad 4x - 3y + z = -1$
$5x + 7y + 2z - 2 = 0 \Rightarrow 5x + 7y + 2z = -2$
$3x - 5y - z - 1 = 0 \qquad 3x - 5y - z = 1$

Adding row 3 to row 1 and 2 times row 3 to row 2, we have

$$D = \begin{vmatrix} 4 & -3 & 1 \\ 5 & 7 & 2 \\ 3 & -5 & -1 \end{vmatrix} = \begin{vmatrix} 7 & -8 & 0 \\ 11 & -3 & 0 \\ 3 & -5 & -1 \end{vmatrix}.$$

Expanding by column three, we have

$$D = -1\begin{vmatrix} 7 & -8 \\ 11 & -3 \end{vmatrix} = -1(-21+88) = -1(67) = -67$$

Adding column 1 to column 3 we have

$$D_x = \begin{vmatrix} -1 & -3 & 1 \\ -2 & 7 & 2 \\ 1 & -5 & -1 \end{vmatrix} = \begin{vmatrix} -1 & 3 & 0 \\ -2 & 7 & 0 \\ 1 & -5 & 0 \end{vmatrix}.$$

Because we have a column of zeros, $D_x = 0$.

(continued on next page)

(*continued*)

Adding column 2 to column 3, we have

$$D_y = \begin{vmatrix} 4 & -1 & 1 \\ 5 & -2 & 2 \\ 3 & 1 & -1 \end{vmatrix} = \begin{vmatrix} 4 & -1 & 0 \\ 5 & -2 & 0 \\ 3 & 1 & 0 \end{vmatrix}.$$

Because we have a column of zeros, $D_y = 0$.

Adding row 3 to row 1 and 2 times row 3 to row 2, we have

$$D_z = \begin{vmatrix} 4 & -3 & -1 \\ 5 & 7 & -2 \\ 3 & -5 & 1 \end{vmatrix} = \begin{vmatrix} 7 & -8 & 0 \\ 11 & -3 & 0 \\ 3 & -5 & 1 \end{vmatrix}.$$

Expanding by the third column, we have

$$D_z = 1 \begin{vmatrix} 7 & -8 \\ 11 & -3 \end{vmatrix} = -21 + 88 = 67.$$

Thus, we have $x = \frac{D_x}{D} = \frac{0}{-67} = 0$,

$y = \frac{D_y}{D} = \frac{0}{-67} = 0$, and $z = \frac{D_z}{D} = \frac{67}{-67} = -1$.

Solution set: $\{(0,0,-1)\}$

83. $5x - y = -4$
$\quad 3x + 2z = 4$
$\quad 4y + 3z = 22$

Adding 4 times row 1 to row 3, we have

$$D = \begin{vmatrix} 5 & -1 & 0 \\ 3 & 0 & 2 \\ 0 & 4 & 3 \end{vmatrix} = \begin{vmatrix} 5 & -1 & 0 \\ 3 & 0 & 2 \\ 20 & 0 & 3 \end{vmatrix}.$$

Expanding by column two, we have

$$D = -(-1) \begin{vmatrix} 3 & 2 \\ 20 & 3 \end{vmatrix} = 9 - 40 = -31.$$

Adding 4 times row 1 to row 3, we have

$$D_x = \begin{vmatrix} -4 & -1 & 0 \\ 4 & 0 & 2 \\ 22 & 4 & 3 \end{vmatrix} = \begin{vmatrix} -4 & -1 & 0 \\ 4 & 0 & 2 \\ 6 & 0 & 3 \end{vmatrix}.$$

Expand by column two, we have

$$D_x = -(-1) \begin{vmatrix} 4 & 2 \\ 6 & 3 \end{vmatrix} = 12 - 12 = 0.$$

Adding column 2 to column 1, we have

$$D_y = \begin{vmatrix} 5 & -4 & 0 \\ 3 & 4 & 2 \\ 0 & 22 & 3 \end{vmatrix} = \begin{vmatrix} 1 & -4 & 0 \\ 7 & 4 & 2 \\ 22 & 22 & 3 \end{vmatrix}.$$

Adding 4 times column 1 to column 2 gives

$$D_y = \begin{vmatrix} 1 & 0 & 0 \\ 7 & 32 & 2 \\ 22 & 110 & 3 \end{vmatrix}.$$

Expanding by row one, we have

$$D_y = 1 \begin{vmatrix} 32 & 2 \\ 110 & 3 \end{vmatrix} = 96 - 220 = -124.$$

Adding 4 times row 1 to row 3, we have

$$D_z = \begin{vmatrix} 5 & -1 & -4 \\ 3 & 0 & 4 \\ 0 & 4 & 22 \end{vmatrix} = \begin{vmatrix} 5 & -1 & -4 \\ 3 & 0 & 4 \\ 20 & 0 & 6 \end{vmatrix}.$$

Expanding by column two, we have

$$D_z = -(-1) \begin{vmatrix} 3 & 4 \\ 20 & 6 \end{vmatrix} = 18 - 80 = -62.$$

Thus, we have

$$x = \frac{D_x}{D} = \frac{0}{-31} = 0, \quad y = \frac{D_y}{D} = \frac{-124}{-31} = 4, \text{ and}$$

$$z = \frac{D_z}{D} = \frac{-62}{-31} = 2.$$

Solution set: $\{(0,4,2)\}$

85. $x + 2y = 10$
$\quad 3x + 4z = 7$
$\quad -y - z = 1$

Adding -3 times row 1 to row 2, we have

$$D = \begin{vmatrix} 1 & 2 & 0 \\ 3 & 0 & 4 \\ 0 & -1 & -1 \end{vmatrix} = \begin{vmatrix} 1 & 2 & 0 \\ 0 & -6 & 4 \\ 0 & -1 & -1 \end{vmatrix}.$$

Expanding by column one, we have

$$D = 1 \begin{vmatrix} -6 & 4 \\ -1 & -1 \end{vmatrix} = 6 + 4 = 10.$$

Adding column 1 to column 2 and column 1 to column 3, we have

$$D_x = \begin{vmatrix} 10 & 2 & 0 \\ 7 & 0 & 4 \\ 1 & -1 & -1 \end{vmatrix} = \begin{vmatrix} 10 & 12 & 10 \\ 7 & 7 & 11 \\ 1 & 0 & 0 \end{vmatrix}.$$

Expanding by row three, we have

$$D_x = 1 \begin{vmatrix} 12 & 10 \\ 7 & 11 \end{vmatrix} = 132 - 70 = 62.$$

Adding column 2 to column 3, we have

$$D_y = \begin{vmatrix} 1 & 10 & 0 \\ 3 & 7 & 4 \\ 0 & 1 & -1 \end{vmatrix} = \begin{vmatrix} 1 & 10 & 10 \\ 3 & 7 & 11 \\ 0 & 1 & 0 \end{vmatrix}.$$

Expanding by row three, we have

$$D_y = -1 \begin{vmatrix} 1 & 10 \\ 3 & 11 \end{vmatrix} = -1(11 - 30) = -1(-19) = 19.$$

Adding column 3 to column 2, we have

$$D_z = \begin{vmatrix} 1 & 2 & 10 \\ 3 & 0 & 7 \\ 0 & -1 & 1 \end{vmatrix} = \begin{vmatrix} 1 & 12 & 10 \\ 3 & 7 & 7 \\ 0 & 0 & 1 \end{vmatrix}.$$

(*continued on next page*)

(*continued*)

Expanding by row three, we have

$$D_z = 1 \begin{vmatrix} 1 & 12 \\ 3 & 7 \end{vmatrix} = 7 - 36 = -29.$$

Thus, we have $x = \frac{D_x}{D} = \frac{62}{10} = \frac{31}{5}$, $y = \frac{D_y}{D} = \frac{19}{10}$,

and $z = \frac{D_z}{D} = \frac{-29}{10} = -\frac{29}{10}$.

Solution set: $\left\{ \left(\frac{31}{5}, \frac{19}{10}, -\frac{29}{10} \right) \right\}$

87. $\frac{\sqrt{3}}{2}(W_1 + W_2) = 100$

$\quad\quad W_1 - W_2 = 0$

Using the distributive property, we have the following system.

$\frac{\sqrt{3}}{2}W_1 + \frac{\sqrt{3}}{2}W_2 = 100$

$\quad\quad W_1 - W_2 = 0$

Using Cramer's rule, we have

$$D = \begin{vmatrix} \frac{\sqrt{3}}{2} & \frac{\sqrt{3}}{2} \\ 1 & -1 \end{vmatrix} = -\frac{\sqrt{3}}{2} - \frac{\sqrt{3}}{2} = -\sqrt{3},$$

$$D_{W_1} = \begin{vmatrix} 100 & \frac{\sqrt{3}}{2} \\ 0 & -1 \end{vmatrix} = -100 - 0 = -100, \text{ and}$$

$$D_{W_2} = \begin{vmatrix} \frac{\sqrt{3}}{2} & 100 \\ 1 & 0 \end{vmatrix} = 0 - 100 = -100.$$

This yields the following solution.

$W_1 = \frac{D_{W_1}}{D} = \frac{-100}{-\sqrt{3}} = \frac{100}{\sqrt{3}} = \frac{100\sqrt{3}}{3} \approx 58$ and

$W_2 = \frac{D_{W_2}}{D} = \frac{-100}{-\sqrt{3}} \approx 58$

Both W_1 and W_2 are approximately 58 lb.

89. $P(0,0), Q(0,2), R(1,4)$

Find $D = \frac{1}{2} \begin{vmatrix} x_1 & y_1 & 1 \\ x_2 & y_2 & 1 \\ x_3 & y_3 & 1 \end{vmatrix}$, where

$P = (x_1, y_1) = (0,0), Q(x_2, y_2) = (0,2)$, and

$R = (x_3, y_3) = (1,4)$

Expanding by the first row, we have the following

$$D = \frac{1}{2} \begin{vmatrix} 0 & 0 & 1 \\ 0 & 2 & 1 \\ 1 & 4 & 1 \end{vmatrix} = \frac{1}{2} \left[0 \begin{vmatrix} 2 & 1 \\ 4 & 1 \end{vmatrix} - 0 \begin{vmatrix} 0 & 1 \\ 1 & 1 \end{vmatrix} + 1 \begin{vmatrix} 0 & 2 \\ 1 & 4 \end{vmatrix} \right]$$

$$= \frac{1}{2} \left[0(2-4) - 0(0-1) + 1(0-2) \right]$$

$$= \frac{1}{2} \left[0(-2) - 0(-1) + 1(-2) \right] = \frac{1}{2}(-2) = -1$$

Area of triangle $= |D| = |-1| = 1 \text{ unit}^2.$

91. $P(2,5), Q(-1,3), R(4,0)$

Find $D = \frac{1}{2} \begin{vmatrix} x_1 & y_1 & 1 \\ x_2 & y_2 & 1 \\ x_3 & y_3 & 1 \end{vmatrix}$, where

$P = (x_1, y_1) = (2,5), Q(x_2, y_2) = (-1,3)$, and

$R = (x_3, y_3) = (4,0)$.

Expanding by the third row, we have the following.

$$D = \frac{1}{2} \begin{vmatrix} 2 & 5 & 1 \\ -1 & 3 & 1 \\ 4 & 0 & 1 \end{vmatrix}$$

$$= \frac{1}{2} \left[4 \begin{vmatrix} 5 & 1 \\ 3 & 1 \end{vmatrix} - 0 \begin{vmatrix} 2 & 1 \\ -1 & 1 \end{vmatrix} + 1 \begin{vmatrix} 2 & 5 \\ -1 & 3 \end{vmatrix} \right]$$

$$= \frac{1}{2} \left[4(5-3) - 0(2+1) + 1(6+5) \right]$$

$$= \frac{1}{2} \left[4(2) - 0(3) + 1(11) \right] = \frac{1}{2}(8 - 0 + 11)$$

$$= \frac{1}{2}(19) = \frac{19}{2} = 9.5$$

Area of triangle $= |D| = |9.5| = 9.5 \text{ units}^2.$

93. $(101.3, 52.7), (117.2, 253.9), (313.1, 301.6)$

Label the points as follows.

$P = (x_1, y_1) = (101.3, 52.7),$

$Q(x_2, y_2) = (117.2, 253.9),$ and

$R = (x_3, y_3) = (313.1, 301.6)$

Because

$$D = \frac{1}{2} \begin{vmatrix} x_1 & y_1 & 1 \\ x_2 & y_2 & 1 \\ x_3 & y_3 & 1 \end{vmatrix} = \frac{1}{2} \begin{vmatrix} 101.3 & 52.7 & 1 \\ 117.2 & 253.9 & 1 \\ 313.1 & 301.6 & 1 \end{vmatrix},$$

we will enter the 3×3 matrix as

$\begin{bmatrix} 101.3 & 52.7 & 1 \\ 117.2 & 253.9 & 1 \\ 313.1 & 301.6 & 1 \end{bmatrix}$ and perform the

calculations as shown below.

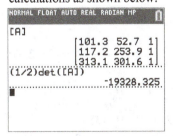

The area of triangular lot is $|-19,328.325| \text{ ft}^2$

or approximately $19,328.3 \text{ ft}^2.$

95. To solve the equation $\begin{vmatrix} 5 & x \\ -3 & 2 \end{vmatrix} = 6$, we need to

solve $5 \cdot 2 - (-3) \cdot x = 6$.

$$5 \cdot 2 - (-3) \cdot x = 6 \Rightarrow 10 + 3x = 6 \Rightarrow$$
$$3x = -4 \Rightarrow x = -\tfrac{4}{3}$$

Verifying $\begin{vmatrix} 5 & -\tfrac{4}{3} \\ -3 & 2 \end{vmatrix} = 6$, we have

$$5 \cdot 2 - (-3)\left(-\tfrac{4}{3}\right) = 10 - 4 = 6.$$

Solution set: $\left\{-\tfrac{4}{3}\right\}$

97. To solve the equation $\begin{vmatrix} x & 3 \\ x & x \end{vmatrix} = 4$, we need to

solve

$$x \cdot x - 3x = 4 \Rightarrow x^2 - 3x = 4 \Rightarrow$$
$$x^2 - 3x - 4 = 0 \Rightarrow (x+1)(x-4) = 0$$
$$x + 1 = 0 \Rightarrow x = -1 \text{ or } x - 4 = 0 \Rightarrow x = 4$$

Verify $x = -1$.

$$\begin{vmatrix} -1 & 3 \\ -1 & -1 \end{vmatrix} = -1(-1) - (-1) \cdot 3 = 1 - (-3) = 4$$

Verify $x = 4$.

$$\begin{vmatrix} 4 & 3 \\ 4 & 4 \end{vmatrix} = 4 \cdot 4 - 4 \cdot 3 = 16 - 12 = 4$$

Solution set: $\{-1, 4\}$

99. To solve the equation $\begin{vmatrix} -2 & 0 & 1 \\ -1 & 3 & x \\ 5 & -2 & 0 \end{vmatrix} = 3$, expand

by the first row. In order to do this, we will
need to find $a_{11} \cdot M_{11} - a_{12} \cdot M_{12} + a_{13} \cdot M_{13}$.
However, we do not need to calculate M_{12},
because $a_{12} = 0$.

$$M_{11} = \begin{vmatrix} 3 & x \\ -2 & 0 \end{vmatrix} = 3 \cdot 0 - (-2) \cdot x = 0 - (-2x) = 2x$$

$$M_{13} = \begin{vmatrix} -1 & 3 \\ 5 & -2 \end{vmatrix} = (-1)(-2) - 5 \cdot 3 = 2 - 15 = -13$$

$$a_{11} \cdot M_{11} - a_{12} \cdot M_{12} + a_{13} \cdot M_{13}$$
$$= a_{11} \cdot M_{11} - 0 \cdot M_{12} + a_{13} \cdot M_{13}$$
$$= -2(2x) - 0 + 1(-13)$$
$$= -4x + (-13) = -4x - 13$$

Set this equal to 3 and solve to get
$$-4x - 13 = 3 \Rightarrow -4x = 16 \Rightarrow x = -4.$$

Verify $\begin{vmatrix} -2 & 0 & 1 \\ -1 & 3 & -4 \\ 5 & -2 & 0 \end{vmatrix} = 3$.

Because

$$M_{11} = \begin{vmatrix} 3 & -4 \\ -2 & 0 \end{vmatrix} = 3 \cdot 0 - (-2)(-4) = 0 - 8 = -8$$

and $M_{13} = \begin{vmatrix} -1 & 3 \\ 5 & -2 \end{vmatrix} = -13$, we have

$$a_{11} \cdot M_{11} - a_{12} \cdot M_{12} + a_{13} \cdot M_{13}$$
$$= a_{11} \cdot M_{11} - 0 \cdot M_{12} + a_{13} \cdot M_{13}$$
$$= -2(-8) - 0 + 1(-13) = 16 + (-13) = 3$$

Solution set: $\{-4\}$

101. To solve the equation $\begin{vmatrix} 5 & 3x & -3 \\ 0 & 2 & -1 \\ 4 & -1 & x \end{vmatrix} = -7$,

expand by the second row. We will need to
find $-a_{21} \cdot M_{21} + a_{22} \cdot M_{22} - a_{23} \cdot M_{23}$.
However, we do not need to calculate M_{21},
because $a_{21} = 0$.

$$M_{22} = \begin{vmatrix} 5 & -3 \\ 4 & x \end{vmatrix} = 5x - 4(-3) = 5x - (-12) = 5x + 12$$

and $M_{23} = \begin{vmatrix} 5 & 3x \\ 4 & -1 \end{vmatrix} = 5(-1) - 4 \cdot 3x = -5 - 12x$

$$-a_{21} \cdot M_{21} + a_{22} \cdot M_{22} - a_{23} \cdot M_{23}$$
$$= -0 \cdot M_{21} + a_{22} \cdot M_{22} - a_{23} \cdot M_{23}$$
$$= 0 + 2(5x + 12) - (-1)(-5 - 12x)$$
$$= 2(5x + 12) + (-5 - 12x)$$
$$= 10x + 24 - 5 - 12x = 19 - 2x$$

Set this equal to -7 and solve to get
$$19 - 2x = -7 \Rightarrow -2x = -26 \Rightarrow x = 13.$$

Verify $\begin{vmatrix} 5 & 3(13) & -3 \\ 0 & 2 & -1 \\ 4 & -1 & 13 \end{vmatrix} = \begin{vmatrix} 5 & 39 & -3 \\ 0 & 2 & -1 \\ 4 & -1 & 13 \end{vmatrix} = -7$.

$$M_{22} = \begin{vmatrix} 5 & -3 \\ 4 & 13 \end{vmatrix} = 5 \cdot 13 - 4(-3) = 65 - (-12) = 77$$

and

$$M_{23} = \begin{vmatrix} 5 & 39 \\ 4 & -1 \end{vmatrix} = 5(-1) - 4 \cdot 39 = -5 - 156 = -161,$$

so, we have the following.
$$-a_{21} \cdot M_{21} + a_{22} \cdot M_{22} - a_{23} \cdot M_{23}$$
$$= -0 \cdot M_{21} + a_{22} \cdot M_{22} - a_{23} \cdot M_{23}$$
$$= 0 + 2 \cdot 77 - (-1)(-161) = 154 - 161 = -7$$

Solution set: $\{13\}$

103. To solve the equation $\begin{vmatrix} x & 0 & -1 \\ 2 & -3 & x \\ x & 0 & 7 \end{vmatrix} = 12,$ expand

by the second column. We will need to find $-a_{12} \cdot M_{12} + a_{22} \cdot M_{22} - a_{32} \cdot M_{32}.$ However, we do not need to calculate M_{12} or M_{32}

because $a_{12} = 0$ and $a_{32} = 0.$

$$M_{22} = \begin{vmatrix} x & -1 \\ x & 7 \end{vmatrix} = 7x + x = 8x$$

$-a_{12} \cdot M_{12} + a_{22} \cdot M_{22} - a_{32} \cdot M_{32}.$
$\quad = 0 \cdot M_{12} + (-3)8x + 0 \cdot M_{32} = -24x$

Set this equal to 12 and solve to get

$-24x = 12 \Rightarrow x = -\frac{1}{2}$

Verify $\begin{vmatrix} -\frac{1}{2} & 0 & -1 \\ 2 & -3 & -\frac{1}{2} \\ -\frac{1}{2} & 0 & 7 \end{vmatrix} = 12.$

$$M_{22} = \begin{vmatrix} -\frac{1}{2} & -1 \\ -\frac{1}{2} & 7 \end{vmatrix} = -\frac{1}{2} \cdot 7 - \left(-\frac{1}{2}\right)(-1)$$

$$= -\frac{7}{2} - \frac{1}{2} = -\frac{8}{2} = -4$$

$-a_{12} \cdot M_{12} + a_{22} \cdot M_{22} - a_{32} \cdot M_{32}.$
$\quad = 0 \cdot M_{12} + (-3)(-4) + 0 \cdot M_{32} = 12$

Solution set: $\left\{ -\frac{1}{2} \right\}$

105. $bx + y = a^2$
$ax + y = b^2$
Using Cramer's rule, we have

$$D = \begin{vmatrix} b & 1 \\ a & 1 \end{vmatrix} = b - a, \ D_x = \begin{vmatrix} a^2 & 1 \\ b^2 & 1 \end{vmatrix} = a^2 - b^2, \text{ and}$$

$$D_y = \begin{vmatrix} b & a^2 \\ a & b^2 \end{vmatrix} = b^3 - a^3.$$

$$x = \frac{D_x}{D} = \frac{a^2 - b^2}{b - a} = \frac{(a+b)(a-b)}{b-a}$$

$$= \frac{(a+b)(a-b)}{-(a-b)} = -(a+b) = -a - b$$

$$y = \frac{D_y}{D} = \frac{b^3 - a^3}{b - a} = \frac{(b-a)(b^2 + ab + a^2)}{b-a}$$

$$= b^2 + ab + a^2$$

Solution set: $\left\{ \left(-a - b, \ a^2 + ab + b^2\right) \right\}$

107. $b^2 x + a^2 y = b^2$
$ax + by = a$
Using Cramer's rule, we have

$$D = \begin{vmatrix} b^2 & a^2 \\ a & b \end{vmatrix} = b^3 - a^3,$$

$$D_x = \begin{vmatrix} b^2 & a^2 \\ a & b \end{vmatrix} = b^3 - a^3, \text{ and}$$

$$D_y = \begin{vmatrix} b^2 & b^2 \\ a & a \end{vmatrix} = ab^2 - ab^2 = 0.$$

$$x = \frac{D_x}{D} = \frac{b^3 - a^3}{b^3 - a^3} = 1$$

$$y = \frac{D_y}{D} = \frac{0}{b^3 - a^3} = 0$$

Note: In order for $D \neq 0,$ we also assumed $a \neq b.$

Solution set: $\{(1, 0)\}$

109. $ax + by = c$
$dx + ey = f$
Because $a, b, c, d, e,$ and f are consecutive integers, we can rewrite any integer in terms of any other.
Using Cramer's rule, we have

$$D = \begin{vmatrix} a & b \\ d & e \end{vmatrix} = \begin{vmatrix} a & a+1 \\ a+3 & a+4 \end{vmatrix}$$

$$= a^2 + 4a - \left(a^2 + 4a + 3\right) = -3$$

$$D_x = \begin{vmatrix} c & b \\ f & e \end{vmatrix} = \begin{vmatrix} b+1 & b \\ b+4 & b+3 \end{vmatrix}$$

$$= \left(b^2 + 4b + 3\right) - \left(b^2 + 4b\right) = 3$$

$$D_y = \begin{vmatrix} a & c \\ d & f \end{vmatrix} = \begin{vmatrix} a & a+2 \\ a+3 & a+5 \end{vmatrix}$$

$$= \left(a^2 + 5a\right) - \left(a^2 + 5a + 6\right) = -6$$

$$x = \frac{D_x}{D} = \frac{3}{-3} = -1$$

$$y = \frac{D_y}{D} = \frac{-6}{-3} = 2$$

Solution set: $\{(-1, 2)\}$

111. $\begin{vmatrix} a_{11} & a_{12} & a_{13} \\ a_{21} & a_{22} & a_{23} \\ a_{31} & a_{32} & a_{33} \end{vmatrix} \begin{matrix} a_{11} & a_{12} \\ a_{21} & a_{22} \\ a_{31} & a_{32} \end{matrix}$

$d_1 = a_{11}a_{22}a_{33}; d_2 = a_{12}a_{23}a_{31}; d_3 = a_{13}a_{21}a_{32}$
$d_4 = a_{13}a_{22}a_{31}; d_5 = a_{11}a_{23}a_{32}; d_6 = a_{12}a_{21}a_{33}$
$(d_1 + d_2 + d_3) - (d_4 + d_5 + d_6)$

$= (a_{11}a_{22}a_{33} + a_{12}a_{23}a_{31} + a_{13}a_{21}a_{32})$
$\quad - (a_{13}a_{22}a_{31} + a_{11}a_{23}a_{32} + a_{12}a_{21}a_{33})$

$= (a_{11}a_{22}a_{33} + a_{12}a_{23}a_{31} + a_{13}a_{21}a_{32})$
$\quad - (a_{31}a_{22}a_{13} + a_{32}a_{23}a_{11} + a_{33}a_{21}a_{12})$

113. $\begin{vmatrix} 1 & 3 & 2 \\ 0 & 2 & 6 \\ 7 & 1 & 5 \end{vmatrix}$

If we expand by the first column, we will need to find $a_{11} \cdot M_{11} - a_{21} \cdot M_{21} + a_{31} \cdot M_{31}$.

$(-1)^{1+1} \cdot 1 \cdot \begin{vmatrix} 2 & 6 \\ 1 & 5 \end{vmatrix} - (-1)^{1+2} \cdot 0 \cdot \begin{vmatrix} 3 & 2 \\ 1 & 5 \end{vmatrix}$

$\quad + (-1)^{1+3} \cdot 7 \cdot \begin{vmatrix} 3 & 2 \\ 2 & 6 \end{vmatrix}$

$= 1(10 - 6) - 0 + 7(18 - 4) = 4 + 98 = 102$

Both methods give the same determinant.

Section 9.4   Partial Fractions

1. Multiply each side of the equation by the LCD, $3x(2x + 1)$, to clear the fractions.

3. Multiply each side of the equation by the LCD, $(x + 4)(3x^2 + 1)$, to clear the fractions.

5. Multiply each side of the equation by the LCD, $x(2x^2 + 1)^2$, to clear the fractions.

7. $\dfrac{5}{3x(2x + 1)} = \dfrac{A}{3x} + \dfrac{B}{2x + 1}$

Multiply both sides by $3x(2x + 1)$ to get

$5 = A(2x + 1) + B(3x)$. (1)

First substitute 0 for x to get
$5 = A(2 \cdot 0 + 1) + B(3 \cdot 0) \Rightarrow A = 5$.

Replace A with 5 in equation (1) and substitute $-\frac{1}{2}$ for x to get the following.

$5 = 5\left[2\left(-\frac{1}{2}\right) + 1\right] + B\left[3\left(-\frac{1}{2}\right)\right]$

$= 5(-1 + 1) - \frac{3}{2}B = 5(0) - \frac{3}{2}B \Rightarrow$

$5 = -\frac{3}{2}B \Rightarrow -\frac{10}{3} = B$

Thus,

$\dfrac{5}{3x(2x + 1)} = \dfrac{5}{3x} + \dfrac{-\frac{10}{3}}{2x + 1} = \dfrac{5}{3x} + \dfrac{-10}{3(2x + 1)}$

9. $\dfrac{4x + 2}{(x + 2)(2x - 1)} = \dfrac{A}{x + 2} + \dfrac{B}{2x - 1}$

Multiply both sides by $(x + 2)(2x - 1)$ to get

$4x + 2 = A(2x - 1) + B(x + 2)$. (1)

First substitute -2 for x to get the following.

$4(-2) + 2 = A\left[2(-2) - 1\right] + B(-2 + 2)$

$-8 + 2 = A(-4 - 1) + B(0) = -6 = -5A$

$A = \frac{6}{5}$

Replace A with $\frac{6}{5}$ in equation (1) and substitute $\frac{1}{2}$ for x to get the following.

$4 \cdot \frac{1}{2} + 2 = \frac{6}{5}\left(2 \cdot \frac{1}{2} - 1\right) + B\left(\frac{1}{2} + 2\right)$

$2 + 2 = \frac{6}{5}(1 - 1) + B \cdot \frac{5}{2} \Rightarrow 4 = \frac{6}{5}(0) + \frac{5}{2}B \Rightarrow$

$4 = \frac{5}{2}B \Rightarrow B = \frac{8}{5}$

Thus,

$\dfrac{4x + 2}{(x + 2)(2x - 1)} = \dfrac{\frac{6}{5}}{(x + 2)} + \dfrac{\frac{8}{5}}{(2x - 1)}$

$\qquad\qquad = \dfrac{6}{5(x + 2)} + \dfrac{8}{5(2x - 1)}$

11. $\dfrac{x}{x^2 + 4x - 5} = \dfrac{x}{(x + 5)(x - 1)} = \dfrac{A}{x + 5} + \dfrac{B}{x - 1}$

Multiply both sides by $(x + 5)(x - 1)$ to get

$x = A(x - 1) + B(x + 5)$. (1)

First substitute -5 for x to get

$-5 = A(-5 - 1) + B(-5 + 5) \Rightarrow$

$-5 = A(-6) + B(0) \Rightarrow -5 = -6A \Rightarrow A = \frac{5}{6}$

Replace A with $\frac{5}{6}$ in equation (1) and substitute 1 for x to get

$1 = \frac{5}{6}(1 - 1) + B(1 + 5) \Rightarrow 1 = \frac{5}{6}(0) + B(6) \Rightarrow$

$1 = 6B \Rightarrow B = \frac{1}{6}$

Thus,

$\dfrac{x}{(x + 5)(x - 1)} = \dfrac{\frac{5}{6}}{x + 5} + \dfrac{\frac{1}{6}}{x - 1}$

$\qquad\qquad = \dfrac{5}{6(x + 5)} + \dfrac{1}{6(x - 1)}$

13. $\dfrac{4}{x(1-x)} = \dfrac{A}{x} + \dfrac{B}{1-x}$

Multiply both sides by $x(1-x)$ to get

$4 = A(1-x) + Bx.$ (1)

First substitute 0 for x to get the following.

$4 = A(1-0) + B(0) \Rightarrow 4 = A + 0 \Rightarrow A = 4$

Replace A with 4 in equation (1) and substitute 1 for x to get the following.

$4 = 4(1-1) + B(1) \Rightarrow 4 = 0 + B \Rightarrow B = 4$

Thus, $\dfrac{4}{x(1-x)} = \dfrac{4}{x} + \dfrac{4}{1-x}$.

15. $\dfrac{4x^2 - x - 15}{x(x+1)(x-1)} = \dfrac{A}{x} + \dfrac{B}{x+1} + \dfrac{C}{x-1}$

Multiply both sides by $x(x+1)(x-1)$ to get

$4x^2 - x - 15$
$\quad = A(x+1)(x-1) + Bx(x-1) + Cx(x+1)$ (1)

First substitute 0 for x to get the following.

$4(0)^2 - 0 - 15$
$\quad = A(0+1)(0-1) + B(0)(0-1) + C(0)(0+1)$

$0 - 0 - 15 = -A + 0 + 0 \Rightarrow -15 = -A \Rightarrow A = 15$

Replace A with 15 in equation (1) and substitute -1 for x to get the following.

$4(-1)^2 - (-1) - 15$
$\quad = 15(-1+1)(-1-1) + B(-1)(-1-1)$
$\qquad\qquad + C(-1)(-1+1)$

$4 + 1 - 15$
$\quad = 15(0)(-2) + B(-1)(-2) + C(-1)(0)$

$-10 = 0 + 2B + 0 \Rightarrow -10 = 2B \Rightarrow B = -5$

$4x^2 - x - 15$
$\quad = 15(x+1)(x-1) - 5x(x-1) + Cx(x+1)$ (2)

Now substitute 1 for x in equation (2) to get the following.

$4(1)^2 - 1 - 15 = 15(1+1)(1-1) - 5(1)(1-1)$
$\qquad\qquad\qquad + C(1)(1+1)$

$4 - 1 - 15 = 15(2)(0) - 5(1)(0) + C(1)(2)$

$-12 = 0 - 0 + 2C$

$-12 = 2C \Rightarrow C = -6$

Thus,

$\dfrac{4x^2 - x - 15}{x(x+1)(x-1)} = \dfrac{15}{x} + \dfrac{-5}{x+1} + \dfrac{-6}{x-1}$.

17. $\dfrac{2x+1}{(x+2)^3} = \dfrac{A}{x+2} + \dfrac{B}{(x+2)^2} + \dfrac{C}{(x+2)^3}$

Multiply both sides by $(x+2)^3$ to get

$2x+1 = A(x+2)^2 + B(x+2) + C.$ (1)

First substitute -2 for x to get the following.

$2(-2)+1 = A(-2+2)^2 + B(-2+2) + C$

$-4+1 = 0 + 0 + C \Rightarrow C = -3$

Replace C with -3 in equation (1) and substitute 0 (arbitrary choice) for x to get the following.

$2(0)+1 = A(0+2)^2 + B(0+2) - 3$

$1 = 4A + 2B - 3 \Rightarrow 4A + 2B = 4$

$2A + B = 2$ (3)

Replace C with -3 in equation (1) and substitute -1 (arbitrary choice) for x to get the following.

$2(-1)+1 = A(-1+2)^2 + B(-1+2) - 3$

$-1 = A + B - 3 \Rightarrow A + B = 2$ (4)

Solve the system of equations by multiplying equation (4) by -1 and adding to equation 3.

$\begin{array}{r} 2A + B = 2 \\ \underline{-A - B = -2} \\ A = 0 \end{array}$

Substituting 0 for A in equation (4), we obtain $0 + B = 2 \Rightarrow B = 2$.

Thus,

$\dfrac{2x+1}{(x+2)^3} = \dfrac{0}{x+2} + \dfrac{2}{(x+2)^2} + \dfrac{-3}{(x+2)^3}$

$\qquad\qquad = \dfrac{2}{(x+2)^2} + \dfrac{-3}{(x+2)^3}$

19. $\dfrac{x^2}{x^2 + 2x + 1}$

This is not a proper fraction; the numerator has degree greater than or equal to that of the denominator. Divide the numerator by the denominator.

$$\begin{array}{r} 1 \\ x^2 + 2x + 1 \overline{\smash{\big)}\, x^2 } \\ \underline{x^2 + 2x + 1} \\ -2x - 1 \end{array}$$

Find the partial fraction decomposition for

$\dfrac{-2x-1}{x^2 + 2x + 1} = \dfrac{-2x-1}{(x+1)^2}$.

$\dfrac{-2x-1}{(x+1)^2} = \dfrac{A}{x+1} + \dfrac{B}{(x+1)^2}$

(continued on next page)

(continued)

Find the partial fraction decomposition for

$$\frac{-2x-1}{x^2+2x+1} = \frac{-2x-1}{(x+1)^2}.$$

$$\frac{-2x-1}{(x+1)^2} = \frac{A}{x+1} + \frac{B}{(x+1)^2}$$

Multiply both sides by $(x+1)^2$ to get

$$-2x-1 = A(x+1) + B. \quad (1)$$

First substitute -1 for x to get the following.

$$-2(-1)-1 = A(-1+1) + B$$
$$2-1 = 0 + B \Rightarrow 1 = B$$

Replace B with 1 in equation (1) and substitute 2 (arbitrary choice) for x to get the following.

$$-2(2)-1 = A(2+1)+1 \Rightarrow -5 = 3A+1$$
$$-6 = 3A \Rightarrow A = -2$$

Thus, $\dfrac{x^2}{x^2+2x+1} = 1 + \dfrac{-2}{x+1} + \dfrac{1}{(x+1)^2}.$

21. $\dfrac{x^3+4}{9x^3-4x}$

Find the quotient because the degrees of the numerator and denominator are the same.

$$
\begin{array}{r}
\frac{1}{9} \\
9x^3-4x \overline{\smash{\big)}\, x^3 + 0x^2 + 0x + 4} \\
\underline{x^3 \qquad\quad -\frac{4}{9}x} \\
\frac{4}{9}x + 4
\end{array}
$$

Find the partial fraction decomposition for

$$\frac{\frac{4}{9}x+4}{9x^3-4x}.$$

$$\frac{\frac{4}{9}x+4}{9x^3-4x} = \frac{\frac{4}{9}x+4}{x(9x^2-4)} = \frac{\frac{4}{9}x+4}{x(3x+2)(3x-2)}$$

$$= \frac{A}{x} + \frac{B}{3x+2} + \frac{C}{3x-2}$$

Multiply both sides of

$$\frac{\frac{4}{9}x+4}{9x^3-4x} = \frac{A}{x} + \frac{B}{3x+2} + \frac{C}{3x-2} \text{ by}$$

$x(3x+2)(3x-2)$ to get the following.

$$\frac{4}{9}x+4 = A(3x+2)(3x-2) + Bx(3x-2)$$
$$+ Cx(3x+2) \quad (1)$$

First substitute 0 for x to get the following.

$$\frac{4}{9}(0) + 4$$
$$= A[3(0)+2][3(0)-2] + B(0)[3(0)-2]$$
$$+ C(0)[3(0)+2]$$

$$0+4 = A(2)(-2) + 0 + 0 \Rightarrow 4 = -4A \Rightarrow A = -1$$

Replace A with -1 in equation (1) and substitute $-\frac{2}{3}$ for x to get the following.

$$\frac{4}{9}\left(-\frac{2}{3}\right)+4 = -\left[3\left(-\frac{2}{3}\right)+2\right]\left[3\left(-\frac{2}{3}\right)-2\right]$$
$$+ B\left(-\frac{2}{3}\right)\left[3\left(-\frac{2}{3}\right)-2\right]$$
$$+ C\left(-\frac{2}{3}\right)\left[3\left(-\frac{2}{3}\right)+2\right]$$

$$-\frac{8}{27}+4 = 0 + B\left(-\frac{2}{3}\right)(-4) + 0$$

$$\frac{100}{27} = \frac{8}{3}B \Rightarrow B = \frac{25}{18}$$

$$\frac{4}{9}x+4 = -(3x+2)(3x-2)$$
$$+ \frac{25}{18}x(3x-2) + Cx(3x+2) \quad (2)$$

Substitute $\frac{2}{3}$ in equation (2) for x to get the following.

$$\frac{4}{9}\left(\frac{2}{3}\right)+4 = -\left[3\left(\frac{2}{3}\right)+2\right]\left[3\left(\frac{2}{3}\right)-2\right]$$
$$+ \frac{25}{18}\left(\frac{2}{3}\right)\left[3\left(\frac{2}{3}\right)-2\right] + C\left(\frac{2}{3}\right)\left[3\left(\frac{2}{3}\right)+2\right]$$

$$\frac{8}{27}+4 = 0 + 0 + C\left(\frac{2}{3}\right)(4)$$

$$\frac{116}{27} = \frac{8}{3}C \Rightarrow C = \frac{29}{18}$$

Thus,

$$\frac{x^3+4}{9x^3-4x} = \frac{1}{9} + \frac{-1}{x} + \frac{25}{18(3x+2)} + \frac{29}{18(3x-2)}.$$

23. $\dfrac{-3}{x^2(x^2+5)} = \dfrac{A}{x} + \dfrac{B}{x^2} + \dfrac{Cx+D}{x^2+5}$

Multiply both sides by $x^2(x^2+5)$ to get the following.

$$-3 = Ax(x^2+5) + B(x^2+5) + (Cx+D)x^2$$

Distributing and combining like terms on the right side of the equation, we have the following.

$$-3 = Ax(x^2+5) + B(x^2+5) + (Cx+D)x^2$$
$$-3 = Ax^3 + 5Ax + Bx^2 + 5B + Cx^3 + Dx^2$$
$$-3 = (A+C)x^3 + (B+D)x^2 + (5A)x + 5B$$

Equate the coefficients of like powers of x on the two sides of the equation.

For the x^3-term, we have $0 = A+C$.

For the x^2-term, we have $0 = B+D$.

For the x-term, we have $0 = 5A \Rightarrow A = 0$.

(continued on next page)

(*continued*)

For the constant term, we have

$-3 = 5B \Rightarrow B = -\frac{3}{5}.$

Because $A = 0$ and $0 = A + C,$ we have $C = 0.$

Because $B = -\frac{3}{5}$ and $0 = B + D,$ we have $D = \frac{3}{5}.$

Thus, $\dfrac{-3}{x^2\left(x^2 + 5\right)} = \dfrac{-3}{5x^2} + \dfrac{3}{5\left(x^2 + 5\right)}.$

25. $\dfrac{3x - 2}{(x + 4)\left(3x^2 + 1\right)} = \dfrac{A}{x + 4} + \dfrac{Bx + C}{3x^2 + 1}$

Multiply both sides by $(x + 4)\left(3x^2 + 1\right)$ to get the following.

$3x - 2 = A\left(3x^2 + 1\right) + (Bx + C)(x + 4)$ (1)

First substitute -4 for x to get the following.

$3(-4) - 2 = A\left[3(-4)^2 + 1\right]$
$\qquad\qquad + \left[B(-4) + C\right](-4 + 4)$
$-12 - 2 = A(48 + 1) + 0$
$\qquad -14 = 49A \Rightarrow -\frac{14}{49} = A \Rightarrow A = -\frac{2}{7}$

Replace A with $-\frac{2}{7}$ in equation (1) and substitute 0 for x to get the following.

$3(0) - 2 = -\frac{2}{7}\left[3(0)^2 + 1\right] + \left[B(0) + C\right](0 + 4)$
$\qquad -2 = -\frac{2}{7} + 4C \Rightarrow -\frac{12}{7} = 4C \Rightarrow C = -\frac{3}{7}$

$3x - 2 = -\frac{2}{7}\left(3x^2 + 1\right) + \left(Bx - \frac{3}{7}\right)(x + 4)$ (2)

Substitute 1 (arbitrary value) in equation (2) for x to get the following.

$3(1) - 2 = -\frac{2}{7}\left[3(1)^2 + 1\right] + \left[B(1) - \frac{3}{7}\right](1 + 4)$
$\qquad 3 - 2 = -\frac{2}{7}(4) + \left[B - \frac{3}{7}\right](5)$
$\qquad\quad 1 = -\frac{8}{7} + 5B - \frac{15}{7}$
$\qquad\quad 1 = 5B - \frac{23}{7} \Rightarrow \frac{30}{7} = 5B \Rightarrow B = \frac{6}{7}$

Thus,

$\dfrac{3x - 2}{(x + 4)\left(3x^2 + 1\right)} = \dfrac{-\frac{2}{7}}{x + 4} + \dfrac{\frac{6}{7}x + \left(-\frac{3}{7}\right)}{3x^2 + 1}$

$\qquad\qquad = \dfrac{-2}{7(x + 4)} + \dfrac{6x - 3}{7\left(3x^2 + 1\right)}$

27. $\dfrac{1}{x(2x + 1)\left(3x^2 + 4\right)} = \dfrac{A}{x} + \dfrac{B}{2x + 1} + \dfrac{Cx + D}{3x^2 + 4}$

Multiply both sides by $x(2x + 1)\left(3x^2 + 4\right)$ to get the following.

$1 = A(2x + 1)\left(3x^2 + 4\right) + Bx\left(3x^2 + 4\right)$
$\qquad\qquad + (Cx + D)(x)(2x + 1)$

Expanding and combining like terms on the right side of the equation, we have the following.

$1 = A(2x + 1)\left(3x^2 + 4\right) + Bx\left(3x^2 + 4\right)$
$\qquad\qquad + (Cx + D)(x)(2x + 1)$
$1 = A\left(6x^3 + 3x^2 + 8x + 4\right) + B\left(3x^3 + 4x\right)$
$\qquad\qquad + C\left(2x^3 + x^2\right)$
$\qquad\qquad + D\left(2x^2 + x\right)$
$1 = 6Ax^3 + 3Ax^2 + 8Ax + 4A + 3Bx^3$
$\qquad + 4Bx + 2Cx^3 + Cx^2 + 2Dx^2 + Dx$
$1 = (6A + 3B + 2C)x^3 + (3A + C + 2D)x^2$
$\qquad\qquad + (8A + 4B + D)x + 4A$

Equate the coefficients of like powers of x on the two sides of the equation.

For the x^3-term, we have $0 = 6A + 3B + 2C.$

For the x^2-term, we have $0 = 3A + C + 2D.$

For the x-term, we have $0 = 8A + 4B + D.$

For the constant term, we have $1 = 4A.$

If we use the Gauss-Jordan method, we begin with the following augmented matrix.

$\begin{bmatrix} 6 & 3 & 2 & 0 & | & 0 \\ 3 & 0 & 1 & 2 & | & 0 \\ 8 & 4 & 0 & 1 & | & 0 \\ 4 & 0 & 0 & 0 & | & 1 \end{bmatrix}$

$\begin{bmatrix} 6 & 3 & 2 & 0 & | & 0 \\ 3 & 0 & 1 & 2 & | & 0 \\ 8 & 4 & 0 & 1 & | & 0 \\ 1 & 0 & 0 & 0 & | & \frac{1}{4} \end{bmatrix} \frac{1}{4}R4 \quad\Rightarrow$

$\begin{bmatrix} 1 & 0 & 0 & 0 & | & \frac{1}{4} \\ 3 & 0 & 1 & 2 & | & 0 \\ 8 & 4 & 0 & 1 & | & 0 \\ 6 & 3 & 2 & 0 & | & 0 \end{bmatrix} R1 \leftrightarrow R4 \quad\Rightarrow$

$\begin{bmatrix} 1 & 0 & 0 & 0 & | & \frac{1}{4} \\ 0 & 0 & 1 & 2 & | & -\frac{3}{4} \\ 0 & 1 & 0 & \frac{1}{4} & | & -\frac{1}{2} \\ 0 & 3 & 2 & 0 & | & -\frac{3}{2} \end{bmatrix} \frac{1}{4}R3 \quad\Rightarrow$

(continued on next page)

(*continued*)

$$\begin{bmatrix} 1 & 0 & 0 & 0 & \Big| & \frac{1}{4} \\ 0 & 0 & 1 & 2 & \Big| & -\frac{3}{4} \\ 0 & 4 & 0 & 1 & \Big| & -2 \\ 0 & 3 & 2 & 0 & \Big| & -\frac{3}{2} \end{bmatrix} \begin{matrix} \\ -3R1 + R2 \\ -8R1 + R3 \\ -6R1 + R4 \end{matrix} \Rightarrow$$

$$\begin{bmatrix} 1 & 0 & 0 & 0 & \Big| & \frac{1}{4} \\ 0 & 1 & 0 & \frac{1}{4} & \Big| & -\frac{1}{2} \\ 0 & 0 & 1 & 2 & \Big| & -\frac{3}{4} \\ 0 & 3 & 2 & 0 & \Big| & -\frac{3}{2} \end{bmatrix} \begin{matrix} \\ R2 \leftrightarrow R3 \\ \\ \end{matrix} \Rightarrow$$

$$\begin{bmatrix} 1 & 0 & 0 & 0 & \Big| & \frac{1}{4} \\ 0 & 1 & 0 & \frac{1}{4} & \Big| & -\frac{1}{2} \\ 0 & 0 & 1 & 2 & \Big| & -\frac{3}{4} \\ 0 & 0 & 2 & -\frac{3}{4} & \Big| & 0 \end{bmatrix} \begin{matrix} \\ \\ \\ -3R2 + R4 \end{matrix} \Rightarrow$$

$$\begin{bmatrix} 1 & 0 & 0 & 0 & \Big| & \frac{1}{4} \\ 0 & 1 & 0 & \frac{1}{4} & \Big| & -\frac{1}{2} \\ 0 & 0 & 1 & 2 & \Big| & -\frac{3}{4} \\ 0 & 0 & 0 & -\frac{19}{4} & \Big| & \frac{3}{2} \end{bmatrix} \begin{matrix} \\ \\ \\ -2R3 + R4 \end{matrix} \Rightarrow$$

$$\begin{bmatrix} 1 & 0 & 0 & 0 & \Big| & \frac{1}{4} \\ 0 & 1 & 0 & \frac{1}{4} & \Big| & -\frac{1}{2} \\ 0 & 0 & 1 & 2 & \Big| & -\frac{3}{4} \\ 0 & 0 & 0 & 1 & \Big| & -\frac{6}{19} \end{bmatrix} \begin{matrix} \\ \\ \\ -\frac{4}{19}R4 \end{matrix} \Rightarrow$$

$$\begin{bmatrix} 1 & 0 & 0 & 0 & \Big| & \frac{1}{4} \\ 0 & 1 & 0 & 0 & \Big| & -\frac{8}{19} \\ 0 & 0 & 1 & 0 & \Big| & -\frac{9}{76} \\ 0 & 0 & 0 & 1 & \Big| & -\frac{6}{19} \end{bmatrix} \begin{matrix} \\ -\frac{1}{4}R4 + R2 \\ -2R4 + R3 \\ \end{matrix}$$

Thus,

$$\frac{1}{x(2x+1)(3x^2+4)}$$

$$= \frac{\frac{1}{4}}{x} + \frac{-\frac{8}{19}}{2x+1} + \frac{-\frac{9}{76}x + \left(-\frac{6}{19}\right)}{3x^2+4}$$

$$= \frac{\frac{1}{4}}{x} + \frac{-\frac{8}{19}}{2x+1} + \frac{-\frac{9}{76}x - \frac{24}{76}}{3x^2+4}$$

$$= \frac{1}{4x} + \frac{-8}{19(2x+1)} + \frac{-9x-24}{76(3x^2+4)}$$

29. $\dfrac{2x^5 + 3x^4 - 3x^3 - 2x^2 + x}{2x^2 + 5x + 2}$

The degree of the numerator is greater than the degree of the denominator, so first find the quotient.

$$\begin{array}{r} x^3 - x^2 \\ 2x^2+5x+2 \overline{\big) 2x^5 + 3x^4 - 3x^3 - 2x^2 + x} \\ \underline{2x^5 + 5x^4 + 2x^3 } \\ -2x^4 - 5x^3 - 2x^2 + x \\ \underline{-2x^4 - 5x^3 - 2x^2 } \\ x \end{array}$$

Find the partial fraction decomposition for

$$\frac{x}{2x^2 + 5x + 2} = \frac{x}{(2x+1)(x+2)}.$$

$$\frac{x}{(2x+1)(x+2)} = \frac{A}{2x+1} + \frac{B}{x+2}$$

Multiply both sides by $(2x+1)(x+2)$ to get

$$x = A(x+2) + B(2x+1). \ (1)$$

First substitute $-\frac{1}{2}$ for x to get the following.

$$-\tfrac{1}{2} = A\left(-\tfrac{1}{2}+2\right) + B\left[2\left(-\tfrac{1}{2}\right)+1\right]$$

$$-\tfrac{1}{2} = A\left(\tfrac{3}{2}\right) + B(0) \Rightarrow -\tfrac{1}{2} = \tfrac{3}{2}A \Rightarrow A = -\tfrac{1}{3}$$

Replace A with $-\frac{1}{3}$ in equation (1) and substitute -2 for x to get the following.

$$-2 = -\tfrac{1}{3}(-2+2) + B\left[2(-2)+1\right]$$

$$-2 = -\tfrac{1}{3}(0) + B(-3) \Rightarrow -2 = -3B \Rightarrow B = \tfrac{2}{3}$$

Thus, we have

$$\frac{2x^5 + 3x^4 - 3x^3 - 2x^2 + x}{2x^2 + 5x + 2}$$

$$= x^3 - x^2 + \frac{-1}{3(2x+1)} + \frac{2}{3(x+2)}$$

31. $\dfrac{3x-1}{x\left(2x^2+1\right)^2} = \dfrac{A}{x} + \dfrac{Bx+C}{2x^2+1} + \dfrac{Dx+E}{\left(2x^2+1\right)^2}$

Multiply both sides by $x\left(2x^2+1\right)^2$ to get the following.

$3x-1 = A\left(2x^2+1\right)^2 + (Bx+C)(x)\left(2x^2+1\right)$
$\qquad + (Dx+E)x$

Expanding and combining like terms on the right side of the equation, we have the following.

$3x-1 = A\left(2x^2+1\right)^2 + (Bx+C)(x)\left(2x^2+1\right)$
$\qquad\qquad + (Dx+E)x$
$\qquad = A\left(4x^4+4x^2+1\right) + B\left(2x^4+x^2\right)$
$\qquad\qquad + C\left(2x^3+x\right)$
$\qquad\qquad + Dx^2 + Ex$
$\qquad = 4Ax^4 + 4Ax^2 + A + 2Bx^4 + Bx^2$
$\qquad\qquad + 2Cx^3 + Cx + Dx^2 + Ex$
$3x-1 = (4A+2B)x^4 + 2Cx^3 + (4A+B+D)x^2$
$\qquad\qquad + (C+E)x + A$

Equate the coefficients of like powers of x on the two sides of the equation.

For the x^4-term, we have $0 = 4A+2B$.

For the x^3-term, we have $0 = 2C \Rightarrow C = 0$.

For the x^2-term, we have $0 = 4A+B+D$.

For the x-term, we have $3 = C+E$.

For the constant term, we have $-1 = A$.

Because $A = -1$ and $0 = 4A+2B$, we have $B = 2$. Because $A = -1$, $B = 2$ and $0 = 4A+B+D$, we have $D = 2$. Because $C = 0$ and $3 = C+E$, we have $E = 3$.

Thus,

$\dfrac{3x-1}{x\left(2x^2+1\right)^2} = \dfrac{-1}{x} + \dfrac{2x+0}{2x^2+1} + \dfrac{2x+3}{\left(2x^2+1\right)^2}$

$\qquad\qquad = \dfrac{-1}{x} + \dfrac{2x}{2x^2+1} + \dfrac{2x+3}{\left(2x^2+1\right)^2}$

33. $\dfrac{-x^4-8x^2+3x-10}{(x+2)\left(x^2+4\right)^2}$

$\qquad = \dfrac{A}{x+2} + \dfrac{Bx+C}{x^2+4} + \dfrac{Dx+E}{\left(x^2+4\right)^2}$

Multiply both sides by $(x+2)\left(x^2+4\right)^2$ to get the following.

$-x^4-8x^2+3x-10$
$\quad = A\left(x^2+4\right)^2 + (Bx+C)(x+2)\left(x^2+4\right)$
$\qquad + (Dx+E)(x+2)$ (1)

Substitute then expand and combine like terms on the right side of the equation.

Substitute -2 in equation (1) for x to get the following.

$-(-2)^4 - 8(-2)^2 + 3(-2) - 10$
$\quad = A\left[(-2)^2+4\right]^2$
$\qquad + \left[B(-2)+C\right](-2+2)\left[(-2)^2+4\right]$
$\qquad + \left[D(-2)+E\right](-2+2)$
$-16-32+(-6)-10 = A(4+4)^2 + 0 + 0$
$\qquad\qquad -64 = 64A \Rightarrow A = -1$

Substituting -1 for A in equation (1) and expanding and combining like terms, we have the following.

$-x^4-8x^2+3x-10$
$\quad = -1\left(x^2+4\right)^2 + (Bx+C)(x+2)\left(x^2+4\right)$
$\qquad + (Dx+E)(x+2)$
$\quad = -1\left(x^4+8x^2+16\right)$
$\qquad + (Bx+C)\left(x^3+2x^2+4x+8\right)$
$\qquad + D\left(x^2+2x\right) + E(x+2)$
$\quad = -x^4-8x^2-16 + Bx^4 + 2Bx^3 + 4Bx^2 + 8Bx$
$\qquad + Cx^3 + 2Cx^2 + 4Cx + 8C + Dx^2 + 2Dx$
$\qquad + Ex + 2E$
$\quad = (-1+B)x^4 + (2B+C)x^3$
$\qquad\qquad + (-8+4B+2C+D)x^2$
$\qquad\qquad + (8B+4C+2D+E)x$
$\qquad\qquad + (-16+8C+2E)$

Equate the coefficients of like powers of x on the two sides of the equation.

For the x^4-term, we have
$-1 = -1+B \Rightarrow B = 0$.

For the x^3-term, we have $0 = 2B+C$.

For the x^2-term, we have
$-8 = -8+4B+2C+D \Rightarrow 0 = 4B+2C+D$.

For the x-term, we have
$3 = 8B+4C+2D+E$.

For the constant term, we have
$-10 = -16+8C+2E \Rightarrow 6 = 8C+2E$.

Because $B = 0$ and $0 = 2B+C$, we have $C = 0$. Because $B = 0$, $C = 0$, and $0 = 4B+2C+D$, we have $D = 0$. Because $C = 0$ and $6 = 8C+2E$, we have $E = 3$.

(continued)

(*continued*)

Thus,

$$\frac{-x^4 - 8x^2 + 3x - 10}{(x+2)(x^2+4)^2} = \frac{-1}{x+2} + \frac{0 \cdot x + 0}{x^2+4} + \frac{0 \cdot x + 3}{(x^2+4)^2} = \frac{-1}{x+2} + \frac{3}{(x^2+4)^2}$$

35. $\dfrac{5x^5 + 10x^4 - 15x^3 + 4x^2 + 13x - 9}{x^3 + 2x^2 - 3x}$

Because the degree of the numerator is higher than the degree of the denominator, first find the quotient.

$$\begin{array}{r} 5x^2 \\ x^3 + 2x^2 - 3x \overline{)5x^5 + 10x^4 - 15x^3 + 4x^2 + 13x - 9} \\ \underline{5x^5 + 10x^4 - 15x^3} \\ 4x^2 + 13x - 9 \end{array}$$

Find the partial fraction decomposition of $\dfrac{4x^2 + 13x - 9}{x^3 + 2x^2 - 3x}$.

$$\frac{4x^2 + 13x - 9}{x^3 + 2x^2 - 3x} = \frac{4x^2 + 13x - 9}{x(x^2 + 2x - 3)} = \frac{4x^2 + 13x - 9}{x(x+3)(x-1)} = \frac{A}{x} + \frac{B}{x+3} + \frac{C}{x-1}$$

Multiply by $x(x+3)(x-1)$ to obtain $4x^2 + 13x - 9 = A(x+3)(x-1) + Bx(x-1) + Cx(x+3)$ (1)

Substitute 0 for x in equation (1) to get the following.

$$4(0)^2 + 13(0) - 9 = A(0+3)(0-1) + B(0)(0-1) + C(0)(0+3) \Rightarrow -9 = -3A \Rightarrow A = 3$$

Replace A with 3 in equation (1) and substitute -3 for x to get the following.

$$4(-3)^2 + 13(-3) - 9 = 3(-3+3)(-3-1) + B(-3)(-3-1) + C(-3)(-3+3) \Rightarrow -12 = 12B \Rightarrow B = -1$$

$$4x^2 + 13x - 9 = 3(x+3)(x-1) - x(x-1) + Cx(x+3) \quad (2)$$

Now substitute 1 (arbitrary choice) for x in equation (2) to get the following.

$$4(1)^2 + 13(1) - 9 = 3(1+3)(1-1) - 1(1-1) + C(1)(1+3) \Rightarrow 8 = 4C \Rightarrow C = 2$$

Thus,

$$\frac{5x^5 + 10x^4 - 15x^3 + 4x^2 + 13x - 9}{x^3 + 2x^2 - 3x} = 5x^2 + \frac{3}{x} + \frac{-1}{x+3} + \frac{2}{x-1}$$

Note that the fractions in exercises 37–39 are proper fractions.

37. $\dfrac{x^2}{x^4 - 1} = \dfrac{x^2}{(x^2+1)(x^2-1)} = \dfrac{x^2}{(x^2+1)(x+1)(x-1)} = \dfrac{A}{x+1} + \dfrac{B}{x-1} + \dfrac{C}{x^2+1}$

Multiply both sides by $(x^2+1)(x+1)(x-1)$ to get the following.

$$x^2 = A(x^2+1)(x-1) + B(x^2+1)(x+1) + C(x+1)(x-1) \quad (1)$$

Substitute, then expand and combine like terms on the right side of the equation.
Substitute -1 in equation (1) for x to obtain the following.

$$(-1)^2 = A\left[(-1)^2 + 1\right](-1-1) + B\left[(-1)^2 + 1\right](-1+1) + \left[C(-1) + D\right](-1+1)(-1-1)$$

$$1 = A(1+1)(-1-1) + 0 + 0 \Rightarrow 1 = -4A \Rightarrow A = -\tfrac{1}{4}$$

Substituting $-\tfrac{1}{4}$ for A and 1 for x in equation (1), we have:

$$1^2 = -\tfrac{1}{4}(1^2 + 1)(1-1) + B(1^2+1)(1+1) + C(1+1)(1-1)$$

$$1 = 0 + 0 + 4B \Rightarrow 1 = 4B \Rightarrow B = \tfrac{1}{4}$$

(*continued on next page*)

(continued)

$$x^2 = -\tfrac{1}{4}\left(x^2+1\right)(x-1)+\tfrac{1}{4}\left(x^2+1\right)(x+1)+C(x+1)(x-1) \quad (2)$$

Expanding and combining like terms, we have the following.

$$x^2 = -\tfrac{1}{4}\left(x^2+1\right)(x-1)+\tfrac{1}{4}\left(x^2+1\right)(x+1)+C(x+1)(x-1)$$

$$= -\tfrac{1}{4}\left(x^3-x^2+x-1\right)+\tfrac{1}{4}\left(x^3+x^2+x+1\right)+C\left(x^2-1\right)=Cx^2+\tfrac{1}{2}x^2-C+\tfrac{1}{2}=\left(C+\tfrac{1}{2}\right)x^2-C+\tfrac{1}{2}$$

Equate the coefficients of like powers of x on the two sides of the equation.

For the x^2-term, we have $1=C+\tfrac{1}{2}\Rightarrow C=\tfrac{1}{2}$.

For the constant term, we have $0=-C+\tfrac{1}{2}\Rightarrow C=\tfrac{1}{2}$ (as above)

Thus, $\dfrac{x^2}{x^4-1}=\dfrac{-\tfrac{1}{4}}{x+1}+\dfrac{\tfrac{1}{4}}{x-1}+\dfrac{\tfrac{1}{2}}{x^2+1}=\dfrac{-1}{4(x+1)}+\dfrac{1}{4(x-1)}+\dfrac{1}{2\left(x^2+1\right)}$

39. $\dfrac{4x^2-3x-4}{x^3+x^2-2x}=\dfrac{4x^2-3x-4}{x\left(x^2+x-2\right)}=\dfrac{4x^2-3x-4}{x(x-1)(x+2)}=\dfrac{A}{x}+\dfrac{B}{x-1}+\dfrac{C}{x+2}$

Multiply both sides by $x(x-1)(x+2)$ to get the following.

$$4x^2-3x-4=A(x-1)(x+2)+Bx(x+2)+Cx(x-1) \quad (1)$$

Substitute, then expand and combine like terms on the right side of the equation.
Substitute 1 in equation (1) for x to get the following.

$$4(1)^2-3(1)-4=A(1-1)(1+2)+B(1)(1+2)+C(1)(1-1)\Rightarrow -3=0+3B+0\Rightarrow B=-1$$

Now substitute -1 for B and -2 for x in equation (1).

$$4(-2)^2-3(-2)-4=A(-2-1)(-2+2)+(-1)(-2)(-2+2)+C(-2)(-2-1)\Rightarrow 18=0+0+6C\Rightarrow C=3$$

Substitute the values for B and C into equation (1), then expand and combine like terms.

$$4x^2-3x-4=A(x-1)(x+2)-1x(x+2)+3x(x-1)=A\left(x^2+x-2\right)-x^2-2x+3x^2-3x$$

$$=Ax^2+Ax-2A-x^2-2x+3x^2-3x=Ax^2-x^2+3x^2+Ax-2x-3x-2A$$

$$=(A+2)x^2+(A-5)x-2A$$

Equate the coefficients of like powers of x on the two sides of the equation.

For the x^2-term, we have $4=A+2\Rightarrow 2=A$.

For the x-term, we have $-3=A-5\Rightarrow 2=A$ (as above).

For the constant term, we have $-4=-2A\Rightarrow 2=A$ (again).

Thus, $\dfrac{4x^2-3x-4}{x^3+x^2-2x}=\dfrac{2}{x}+\dfrac{-1}{x-1}+\dfrac{3}{x+2}$

Chapter 9 Quiz
(Sections 9.1–9.4)

1. $2x + y = -4$ (1)
$-x + 2y = 2$ (2)

Solve equation (1) for y:

$2x + y = -4 \Rightarrow y = -2x - 4$ (3)

Replace y in equation (2) with $-2x - 4$ and solve for x:

$-x + 2y = 2 \Rightarrow -x + 2(-2x - 4) = 2$
$-5x - 8 = 2 \Rightarrow x = -2$

Substitute -2 for x in equation (3) and solve for y: $y = -2(-2) - 4 = 0$

Verify that the ordered pair $(-2, 0)$ satisfies both equations.
Check:

$2x + y = -4$ (1)	$-x + 2y = 2$ (2)
$2(-2) + 0 = -4$?	$-(-2) + 2(0) = 2$?
$-4 = -4$ True	$2 = 2$ True

Solution set: $\{(-2, 0)\}$

3. $x - y = 6$ (1)
$x - y = 4$ (2)

Multiply equation (2) by -1, then add the result to equation (1):

$x - y = 6$
$\underline{-x + y = -4}$
$0 = 2$

This is a false statement. The system is inconsistent and the solution set is $\varnothing$.

5. $3x + 5y = -5$ (1)
$-2x + 3y = 16$ (2)

This system has the augmented matrix

$\begin{bmatrix} 3 & 5 & | & -5 \\ -2 & 3 & | & 16 \end{bmatrix}$.

$\begin{bmatrix} 1 & 8 & | & 11 \\ -2 & 3 & | & 16 \end{bmatrix} \begin{matrix} R1 + R2 \\ \end{matrix} \Rightarrow$

$\begin{bmatrix} 1 & 8 & | & 11 \\ 0 & 19 & | & 38 \end{bmatrix} \begin{matrix} \\ 2R1 + R2 \end{matrix} \Rightarrow$

$\begin{bmatrix} 1 & 8 & | & 11 \\ 0 & 1 & | & 2 \end{bmatrix} \frac{1}{19} R2 \Rightarrow$

$\begin{bmatrix} 1 & 0 & | & -5 \\ 0 & 1 & | & 2 \end{bmatrix} R1 - 8R2$

Verify that the ordered pair $(-5, 2)$ satisfies both equations.
Check: $3x + 5y = -5$ (1)
$3(-5) + 5(2) = -5$?
$-15 + 10 = -5$
$-5 = -5$ True

$-2x + 3y = 16$ (2)
$-2(-5) + 3(2) = 16$?
$10 + 6 = 16$
$16 = 16$ True

Solution set: $\{(-5, 2)\}$

7. $x + y + z = 1$ (1)
$-x + y + z = 5$ (2)
$y + 2z = 5$ (3)

Eliminate x by adding equations (1) and (2) to get $2y + 2z = 6 \Rightarrow y + z = 3$ (4).

Now solve the system consisting of equations (3) and (4) by subtracting equation (4) from equation (3).

$y + 2z = 5$ (3)
$\underline{y + z = 3}$ (4)
$z = 2$

Substitute the value for z into equation (4) to solve for y: $y + 2 = 3 \Rightarrow y = 1$

Now, substitute the values for y and z into equation (1) to solve for x:
$x + 1 + 2 = 1 \Rightarrow x = -2$

Verify that the ordered triple $(-2, 1, 2)$ satisfies all three equations.
Check:

$x + y + z = 1$ (1)	$-x + y + z = 5$ (2)
$-2 + 1 + 2 = 1$?	$-(-2) + 1 + 2 = 5$?
$1 = 1$ True	$5 = 5$ True

$y + 2z = 5$ (3)
$1 + 2(2) = 5$?
$5 = 5$ True

Solution set: $\{(-2, 1, 2)\}$

9. $7x + y - z = 4$ (1)
$2x - 3y + z = 2$ (2)
$-6x + 9y - 3z = -6$ (3)

$D = \begin{vmatrix} 7 & 1 & -1 \\ 2 & -3 & 1 \\ -6 & 9 & -3 \end{vmatrix}$

Adding 3 times row 2 to row 3, we have

$D = \begin{vmatrix} 7 & 1 & -1 \\ 2 & -3 & 1 \\ 0 & 0 & 0 \end{vmatrix}$

Because we have a row of zeros, $D = 0$ and we cannot use Cramer's rule.

(continued on next page)

(*continued*)

Using the elimination method, we can add 3 times equation (2) to equation (3).

$$\underline{\begin{array}{r} 6x - 9y + 3z = 6 \qquad (2) \\ -6x + 9y - 3z = -6 \qquad (3) \end{array}}$$
$$\qquad\qquad 0 = 0 \qquad \text{True}$$

This system will have infinitely many solutions.

Solve the system made up of equations (1) and (3) in terms of the arbitrary variable y. To eliminate z, multiply equation (1) by -3 and add the result to equation (3):

$$\underline{\begin{array}{r} -21x - 3y + 3z = -12 \\ -6x + 9y - 3z = -6 \end{array}}$$
$$-27x + 6y = -18 \Rightarrow -27x = -6y - 18 \Rightarrow$$
$$x = \frac{6y + 18}{27} = \frac{2y + 6}{9}$$

Now, express z also in terms of y by solving equation (1) for y and substituting $\frac{2y+6}{9}$ for x in the result.

$$7x + y - z = 4 \Rightarrow z = 7x + y - 4 \Rightarrow$$
$$z = 7\left(\frac{2y+6}{9}\right) + y - 4 = \frac{14y + 42}{9} + y - 4$$
$$= \frac{23y + 6}{9}$$

Solution set: $\left\{ \left(\dfrac{2y+6}{9}, y, \dfrac{23y+6}{9} \right) \right\}$

11. Let $x =$ the amount invested at 2%; $y =$ the amount invested at 3%; $z =$ the amount invested at 4%

The information in the problem gives the system:

$$\begin{cases} x + y + z = 5000 \\ z = x + y \Rightarrow \\ 0.02x + 0.03y + 0.04z = 165 \end{cases}$$

$$\begin{cases} x + y + z = 5000 & (1) \\ -x - y + z = 0 & (2) \\ 0.02x + 0.03y + 0.04z = 165 & (3) \end{cases}$$

This system has the augmented matrix

$$\begin{bmatrix} 1 & 1 & 1 & 5000 \\ -1 & -1 & 1 & 0 \\ 0.02 & 0.03 & 0.04 & 165 \end{bmatrix}.$$

$$\begin{bmatrix} 1 & 1 & 1 & 5000 \\ 0 & 0 & 1 & 2500 \\ 0.02 & 0.03 & 0.04 & 165 \end{bmatrix} \tfrac{1}{2}(R1 + R2) \Rightarrow$$

$$\begin{bmatrix} 1 & 1 & 1 & 5000 \\ 0.02 & 0.03 & 0.04 & 165 \\ 0 & 0 & 1 & 2500 \end{bmatrix} R2 \leftrightarrow R3 \Rightarrow$$

$$\begin{bmatrix} 1 & 1 & 1 & 5000 \\ 2 & 3 & 4 & 16{,}500 \\ 0 & 0 & 1 & 2500 \end{bmatrix} 100R2 \qquad \Rightarrow$$

$$\begin{bmatrix} 1 & 1 & 1 & 5000 \\ 0 & 1 & 2 & 6500 \\ 0 & 0 & 1 & 2500 \end{bmatrix} -2R1 + R2 \qquad \Rightarrow$$

$$\begin{bmatrix} 1 & 1 & 1 & 5000 \\ 0 & 1 & 0 & 1500 \\ 0 & 0 & 1 & 2500 \end{bmatrix} R2 - 2R3 \qquad \Rightarrow$$

$$\begin{bmatrix} 1 & 0 & 1 & 3500 \\ 0 & 1 & 0 & 1500 \\ 0 & 0 & 1 & 2500 \end{bmatrix} R1 - R2 \qquad \Rightarrow$$

$$\begin{bmatrix} 1 & 0 & 0 & 1000 \\ 0 & 1 & 0 & 1500 \\ 0 & 0 & 1 & 2500 \end{bmatrix} R1 - R3 \qquad \Rightarrow$$

$1000 was invested at 2%, $1500 was invested at 3%, and $2500 was invested at 4%.

13. $\begin{vmatrix} -1 & 2 & 4 \\ -3 & -2 & -3 \\ 2 & -1 & 5 \end{vmatrix}$

To expand by the first row, we will need to find $a_{11} \cdot M_{11} - a_{12} \cdot M_{12} + a_{13} \cdot M_{13}$.

$$M_{11} = \begin{vmatrix} -2 & -3 \\ -1 & 5 \end{vmatrix} = (-2)(5) - (-1)(-3) = -13$$

$$M_{12} = \begin{vmatrix} -3 & -3 \\ 2 & 5 \end{vmatrix} = (-3)(5) - (2)(-3) = -9$$

$$M_{13} = \begin{vmatrix} -3 & -2 \\ 2 & -1 \end{vmatrix} = (-3)(-1) - (2)(-2) = 7$$

$$\begin{aligned} a_{11} \cdot M_{11} &- a_{12} \cdot M_{12} + a_{13} \cdot M_{13} \\ &= (-1)(-13) - (2)(-9) + (4)(7) = 59 \end{aligned}$$

15. $\dfrac{2x^2 - 15x - 32}{(x-1)(x^2 + 6x + 8)} = \dfrac{2x^2 - 15x - 32}{(x-1)(x+2)(x+4)}$

$$= \frac{A}{x-1} + \frac{B}{x+2} + \frac{C}{x+4}$$

Multiply both sides by $(x-1)(x+2)(x+4)$:

$$\begin{aligned} 2x^2 &- 15x - 32 \\ &= A(x+2)(x+4) + B(x-1)(x+4) \\ &\qquad + C(x-1)(x+2) \end{aligned}$$

(*continued on next page*)

(*continued*)

Expanding and combining like terms on the right side of the equation, we have

$$2x^2 - 15x - 32$$
$$= A(x+2)(x+4) + B(x-1)(x+4)$$
$$\qquad\qquad + C(x-1)(x+2)$$
$$= Ax^2 + 6Ax + 8A + Bx^2 + 3Bx - 4B$$
$$\qquad\qquad + Cx^2 + Cx - 2C$$
$$= x^2(A+B+C) + x(6A+3B+C)$$
$$\qquad\qquad + (8A - 4B - 2C)$$

Equate the coefficients of like powers of x on the two sides of the equation.

For the x^2-term, $2 = A + B + C$
For the x-term, $-15 = 6A + 3B + C$
For the constant term, $-32 = 8A - 4B - 2C$
Using the Gauss-Jordan method, we have

$$\begin{bmatrix} 1 & 1 & 1 & | & 2 \\ 6 & 3 & 1 & | & -15 \\ 8 & -4 & -2 & | & -32 \end{bmatrix} \Rightarrow$$

$$\begin{bmatrix} 1 & 1 & 1 & | & 2 \\ 0 & -3 & -5 & | & -27 \\ 0 & -12 & -10 & | & -48 \end{bmatrix} \begin{matrix} \\ -6R1 + R2 \\ -8R1 + R2 \end{matrix} \Rightarrow$$

$$\begin{bmatrix} 1 & 1 & 1 & | & 2 \\ 0 & -3 & -5 & | & -27 \\ 0 & 0 & 1 & | & 6 \end{bmatrix} \frac{1}{10}(-4R2 + R3) \Rightarrow$$

$$\begin{bmatrix} 1 & 1 & 1 & | & 2 \\ 0 & 1 & 0 & | & -1 \\ 0 & 0 & 1 & | & 6 \end{bmatrix} -\frac{1}{3}(5R3 + R2) \Rightarrow$$

$$\begin{bmatrix} 1 & 0 & 1 & | & 3 \\ 0 & 1 & 0 & | & -1 \\ 0 & 0 & 1 & | & 6 \end{bmatrix} R1 - R2 \Rightarrow$$

$$\begin{bmatrix} 1 & 0 & 0 & | & -3 \\ 0 & 1 & 0 & | & -1 \\ 0 & 0 & 1 & | & 6 \end{bmatrix} R1 - R3$$

$A = -3$, $B = -1$, $C = 6$.
Thus

$$\frac{2x^2 - 15x - 32}{(x-1)(x^2 + 6x + 8)} = \frac{-3}{x-1} + \frac{-1}{x+2} + \frac{6}{x+4}.$$

Section 9.5 Nonlinear Systems of Equations

1. The following nonlinear system

$$x + y = 7$$
$$x^2 + y^2 = 25$$

has two solutions, one of which is (3, 4̲).

3. The following nonlinear system

$$2x + y = 1$$
$$x^2 + y^2 = 10$$

has two solutions, one of which is (−1̲, 3).

5. When equation (2) is solved for y, we have $x = y$. Substituting x for y into equation (1) gives $x^2 + x = 2$.

7. The system is

$$x^2 = y - 1$$
$$y = 3x + 5$$

The proposed solution set is $\{(-1, 2), (4, 17)\}$.

Check $(-1, 2)$.

$x^2 = y - 1$	$y = 3x + 5$
$(-1)^2 = 2 - 1$?	$2 = 3(-1) + 5$?
$1 = 1$	$2 = -3 + 5$
	$2 = 2$

Check $(4, 17)$.

$x^2 = y - 1$	$y = 3x + 5$
$4^2 = 17 - 1$?	$17 = 3(4) + 5$?
$16 = 16$	$17 = 12 + 5$
	$17 = 17$

Both solutions are valid.

9.
$$x^2 + y^2 = 5$$
$$-3x + 4y = 2$$

The proposed solution set is
$$\left\{(-2, -1), \left(\tfrac{38}{25}, \tfrac{41}{25}\right)\right\}.$$

Check $(-2, -1)$.

$x^2 + y^2 = 5$	$-3x + 4y = 2$
$(-2)^2 + (-1)^2 = 5$?	$-3(-2) + 4(-1) = 2$?
$4 + 1 = 5$	$6 + (-4) = 2$
$5 = 5$	$2 = 2$

Check $\left(\tfrac{38}{25}, \tfrac{41}{25}\right)$.

$x^2 + y^2 = 5$	$-3x + 4y = 2$
$\left(\tfrac{38}{25}\right)^2 + \left(\tfrac{41}{25}\right)^2 = 5$?	$-3\left(\tfrac{38}{25}\right) + 4\left(\tfrac{41}{25}\right) = 2$?
$\tfrac{1444}{625} + \tfrac{1681}{625} = 5$	$-\tfrac{114}{25} + \tfrac{164}{25} = 2$
$\tfrac{3125}{625} = 5$	$\tfrac{50}{25} = 2$
$5 = 5$	$2 = 2$

Both solutions are valid.

11. $y = 3x^2$
   $x^2 + y^2 = 10$

The proposed solution set is $\{(-1, 3), (1, 3)\}$.

Check $(-1, 3)$.

$y = 3x^2$	$x^2 + y^2 = 10$
$3 = 3(-1)^2$?	$(-1)^2 + 3^2 = 10$?
$3 = 3(1)$	$1 + 9 = 10$
$3 = 3$	$10 = 10$

Check $(1, 3)$.

$y = 3x^2$	$x^2 + y^2 = 10$
$3 = 3(1)^2$?	$(1)^2 + 3^2 = 10$?
$3 = 3(1)$	$1 + 9 = 10$
$3 = 3$	$10 = 10$

Both solutions are valid.

13. The system
   $x^2 - y = 4$
   $x + y = -2$

cannot have more than two solutions because a parabola and a line cannot intersect in more than two points.

In the solutions to Exercises 15–19, we include both the algebraic solution and graphing calculator solution as in Example 1. For Exercises 21–49, we include just the algebraic solution. Also, in the solutions to Exercises 21–49, we omit the checking of solutions. Recall, though, when checking elements of the solution set, substitute the ordered pair(s) into both equations of the system.

15. $x^2 - y = 0$ (1)
   $x + y = 2$ (2)

Algebraic Solution:
Solving equation (2) for y, we have $y = 2 - x$.

Substitute this result into equation (1).
$$2 - x = x^2 \Rightarrow x^2 + x - 2 = 0$$
$$(x + 2)(x - 1) = 0 \Rightarrow x = -2 \text{ or } x = 1$$

If $x = -2$, then $y = 2 - (-2) = 4$. If $x = 1$, then $y = 2 - 1 = 1$.

Graphing Calculator Solution
$x^2 - y = 0 \Rightarrow y = x^2$ and $x + y = 2 \Rightarrow y = 2 - x$

Graph the equations in the window $[-10, 10] \times [-10, 10]$.

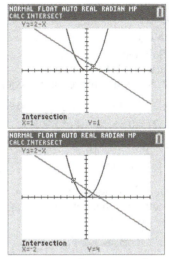

Solution set: $\{(1,1), (-2,4)\}$

17. $y = x^2 - 2x + 1$ (1)
   $x - 3y = -1$ (2)

Algebraic Solution:
Solving equation (2) for y, we have $y = \frac{x+1}{3}$.

Substitute this result into equation (1).
$$\frac{x+1}{3} = x^2 - 2x + 1 \Rightarrow x + 1 = 3x^2 - 6x + 3 \Rightarrow$$
$$3x^2 - 7x + 2 = 0 \Rightarrow (3x - 1)(x - 2) = 0 \Rightarrow$$
$$x = \tfrac{1}{3} \text{ or } x = 2$$

If $x = 2$, then $y = \frac{2+1}{3} = 1$. If $x = \frac{1}{3}$, then
$$y = \frac{\frac{1}{3}+1}{3} = \frac{1+3}{9} = \frac{4}{9}.$$

Graphing Calculator Solution
Graph the equations in the window $[-5, 5] \times [-5, 5]$.

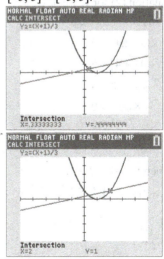

Solution set: $\left\{(2,1), \left(\tfrac{1}{3}, \tfrac{4}{9}\right)\right\}$

19. $y = x^2 + 4x$ (1)
$2x - y = -8$ (2)
Algebraic Solution:
Solving equation (2) for y, we have
$y = 2x + 8$. Substitute this result into equation

(1): $2x + 8 = x^2 + 4x \Rightarrow x^2 + 2x - 8 = 0 \Rightarrow$
$(x + 4)(x - 2) = 0 \Rightarrow x = -4$ or $x = 2$
If $x = -4$, then $y = 2(-4) + 8 = 0$. If $x = 2$,

then $y = 2(2) + 8 = 12$.

Graphing Calculator Solution
Graph the equations in the window
$[-10, 10] \times [-10, 20]$.

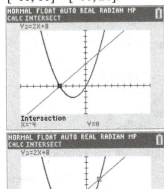

Solution set: $\{(2,12), (-4,0)\}$

21. $3x^2 + 2y^2 = 5$ (1)
$x - y = -2$ (2)
Solving equation (2) for y, we have $y = x + 2$.
Substitute this result into equation (1).
$$3x^2 + 2(x + 2)^2 = 5$$
$$3x^2 + 2(x^2 + 4x + 4) = 5$$
$$3x^2 + 2x^2 + 8x + 8 = 5$$
$$5x^2 + 8x + 3 = 0$$
$$(5x + 3)(x + 1) = 0 \Rightarrow x = -\tfrac{3}{5} \text{ or } x = -1$$

If $x = -\frac{3}{5}$, then $y = -\frac{3}{5} + 2 = -\frac{3}{5} + \frac{10}{5} = \frac{7}{5}$.
If $x = -1$, then $y = -1 + 2 = 1$.
Solution set: $\left\{\left(-\frac{3}{5}, \frac{7}{5}\right), (-1, 1)\right\}$

23. $x^2 + y^2 = 8$ (1)
$x^2 - y^2 = 0$ (2)
Using the elimination method, we add
equations (1) and (2).

$x^2 + y^2 = 8$
$\underline{x^2 - y^2 = 0}$
$2x^2 \phantom{{}+y^2} = 8 \Rightarrow x^2 = 4 \Rightarrow x = \pm 2$
Find y by substituting back into equation (2).
If $x = 2$, then $2^2 - y^2 = 0 \Rightarrow 4 - y^2 = 0 \Rightarrow$
$y^2 = 4 \Rightarrow y = \pm 2$. If $x = -2$, then
$(-2)^2 - y^2 = 0 \Rightarrow 4 - y^2 = 0 \Rightarrow y^2 = 4 \Rightarrow$
$y = \pm 2$.
Solution set: $\{(2,2), (2,-2), (-2,2), (-2,-2)\}$

25. $5x^2 - y^2 = 0$ (1)
$3x^2 + 4y^2 = 0$ (2)
Using the elimination method, we multiply
equation (1) by 4 and add to equation (2).
$20x^2 - 4y^2 = 0$
$\underline{3x^2 + 4y^2 = 0}$
$23x^2 \phantom{{}+4y^2} = 0 \Rightarrow x^2 = 0 \Rightarrow x = 0$
Find y by substituting back into equation (1).
If $x = 0$, then
$5(0)^2 - y^2 = 0 \Rightarrow 0 - y^2 = 0 \Rightarrow y^2 = 0 \Rightarrow y = 0$.
Solution set: $\{(0,0)\}$

27. $3x^2 + y^2 = 3$ (1)
$4x^2 + 5y^2 = 26$ (2)
Using the elimination method, we multiply
equation (1) by -5 and add to equation (2).
$-15x^2 - 5y^2 = -15$
$\underline{4x^2 + 5y^2 = 26}$
$-11x^2 \phantom{{}+5y^2} = 11 \Rightarrow x^2 = -1 \Rightarrow x = \pm i$
Find y by substituting back into equation (1).
If $x = i$, then $3(i)^2 + y^2 = 3 \Rightarrow$
$3(-1) + y^2 = 3 \Rightarrow -3 + y^2 = 3 \Rightarrow y^2 = 6 \Rightarrow$
$y = \pm\sqrt{6}$. If $x = -i$, then $3(-i)^2 + y^2 = 3 \Rightarrow$
$3(-1) + y^2 = 3 \Rightarrow -3 + y^2 = 3 \Rightarrow y^2 = 6 \Rightarrow$
$y = \pm\sqrt{6}$.
Solution set:
$\left\{\left(i, \sqrt{6}\right), \left(-i, \sqrt{6}\right), \left(i, -\sqrt{6}\right), \left(-i, -\sqrt{6}\right)\right\}$

29. $2x^2 + 3y^2 = 5$ (1)

$3x^2 - 4y^2 = -1$ (2)

Using the elimination method, we multiply equation (1) by 4 and equation (2) by 3 and then add the resulting equations.

$8x^2 + 12y^2 = 20$

$9x^2 - 12y^2 = -3$

$\overline{17x^2 \qquad = 17} \Rightarrow x^2 = 1 \Rightarrow x = \pm 1$

Find y by substituting back into equation (1).

If $x = 1$, then $2(1)^2 + 3y^2 = 5 \Rightarrow$

$2 + 3y^2 = 5 \Rightarrow 3y^2 = 3 \Rightarrow y^2 = 1 \Rightarrow y = \pm 1$.

If $x = -1$, then $2(-1)^2 + 3y^2 = 5 \Rightarrow$

$2 + 3y^2 = 5 \Rightarrow 3y^2 = 3 \Rightarrow y^2 = 1 \Rightarrow y = \pm 1$.

Solution set: $\left\{ (1, -1), (-1, 1), (1, 1), (-1, -1) \right\}$

31. $2x^2 + 2y^2 = 20$ (1)

$4x^2 + 4y^2 = 30$ (2)

Using the elimination method, we multiply equation (1) by -2 and add to equation (2).

$-4x^2 - 4y^2 = -40$

$4x^2 + 4y^2 = 30$ |

$\overline{\qquad\qquad 0 = -10}$

This is a false statement.

Solution set: $\varnothing$

33. $2x^2 - 3y^2 = 12$ (1)

$6x^2 + 5y^2 = 36$ (2)

Using the elimination method, we multiply equation (1) by -3 and add to equation (2).

$-6x^2 + 9y^2 = -36$

$6x^2 + 5y^2 = 36$

$\overline{14y^2 = \quad 0} \Rightarrow y^2 = 0 \Rightarrow y = 0$

Find x by substituting back into equation (2).

If $y = 0$, then $6x^2 + 5(0)^2 = 36 \Rightarrow$

$6x^2 + 0 = 36 \Rightarrow x^2 = 6 \Rightarrow x = \pm\sqrt{6}$.

Solution set: $\left\{ (-\sqrt{6}, 0), (\sqrt{6}, 0) \right\}$

35. $xy = -15$ (1)

$4x + 3y = 3$ (2)

Solving equation (1) for y, we have $y = -\frac{15}{x}$.

Substitute this result into equation (2).

$4x + 3\left(-\frac{15}{x} \right) = 3 \Rightarrow 4x^2 - 45 = 3x \Rightarrow$

$4x^2 - 3x - 45 = 0 \Rightarrow (x + 3)(4x - 15) = 0 \Rightarrow$

$x = -3$ or $x = \frac{15}{4}$

If $x = -3$, then $y = -\frac{15}{-3} = 5$. If $x = \frac{15}{4}$, then

$y = -\frac{15}{\frac{15}{4}} = -4$.

Solution set: $\left\{ (-3, 5), \left(\frac{15}{4}, -4 \right) \right\}$

37. $2xy + 1 = 0$ (1)

$x + 16y = 2$ (2)

Solving equation (2) for x, we have $x = 2 - 16y$.

Substitute this result into equation (1).

$2(-16y + 2)y + 1 = 0 \Rightarrow -32y^2 + 4y + 1 = 0 \Rightarrow$

$\qquad 32y^2 - 4y - 1 = 0 \Rightarrow (8y + 1)(4y - 1) = 0 \Rightarrow$

$y = -\frac{1}{8}$ or $y = \frac{1}{4}$

If $y = -\frac{1}{8}$, then $x = 2 - 16\left(-\frac{1}{8} \right) = 2 + 2 = 4$. If

$y = \frac{1}{4}$, then $x = 2 - 16\left(\frac{1}{4} \right) = 2 - 4 = -2$.

Solution set: $\left\{ \left(4, -\frac{1}{8} \right), \left(-2, \frac{1}{4} \right) \right\}$

39. $3x^2 - y^2 = 11$ (1)

$xy = 12$ (2)

Solving equation (2) for x, we have $x = \frac{12}{y}$.

Substitute this result into equation (1).

$3\left(\frac{12}{y} \right)^2 - y^2 = 11 \Rightarrow \frac{432}{y^2} - y^2 = 11 \Rightarrow$

$432 - y^4 = 11y^2 \Rightarrow y^4 + 11y^2 - 432 = 0 \Rightarrow$

$\left(y^2 + 27 \right)\left(y^2 - 16 \right) = 0$

$y^2 + 27 = 0 \Rightarrow y = \pm\sqrt{-27} = \pm 3i\sqrt{3}$ or

$y^2 - 16 = 0 \Rightarrow y^2 = \sqrt{16} \Rightarrow y = \pm 4$

If $y = 3i\sqrt{3}$, then $x = \frac{12}{3i\sqrt{3}} = \frac{4}{i\sqrt{3}} = -\frac{4\sqrt{3}}{3}i$.

If $y = -3i\sqrt{3}$, then $x = \frac{12}{-3i\sqrt{3}} = -\frac{4}{i\sqrt{3}} = \frac{4\sqrt{3}}{3}i$.

If $y = -4$, then $x = \frac{12}{-4} = -3$. If $y = 4$, then

$x = \frac{12}{4} = 3$.

Solution set:

$\left\{ \left(-\frac{4\sqrt{3}}{3}i, 3i\sqrt{3} \right), \left(\frac{4\sqrt{3}}{3}i, -3i\sqrt{3} \right), (3, 4), \right.$

$\left. (-3, -4) \right\}$

41. $x^2 - xy + y^2 = 5$ (1)

$2x^2 + xy - y^2 = 10$ (2)

Using the elimination method, we add equations (1) and (2).

$$x^2 - xy + y^2 = 5$$
$$\underline{2x^2 + xy - y^2 = 10}$$
$$3x^2 \qquad = 15 \Rightarrow x^2 = 5 \Rightarrow x = \pm\sqrt{5}$$

Find y by substituting back into equation (1).

If $x = \sqrt{5}$, then

$$\left(\sqrt{5}\right)^2 - \sqrt{5}y + y^2 = 5 \Rightarrow 5 - \sqrt{5}y + y^2 = 5 \Rightarrow$$
$$y^2 - \sqrt{5}y = 0 \Rightarrow y\left(y - \sqrt{5}\right) = 0$$

Thus, we have $y = 0$ or $y = \sqrt{5}$.

If $x = -\sqrt{5}$, then

$$\left(-\sqrt{5}\right)^2 - \left(-\sqrt{5}\right)y + y^2 = 5$$
$$5 + \sqrt{5}y + y^2 = 5 \Rightarrow y^2 + \sqrt{5}y = 0 \Rightarrow$$
$$y\left(y + \sqrt{5}\right) = 0 \Rightarrow y = 0 \text{ or } y = -\sqrt{5}$$

Thus, we have $y = 0$ or $y = -\sqrt{5}$.

Solution set:

$$\left\{\left(\sqrt{5}, 0\right), \left(-\sqrt{5}, 0\right), \left(\sqrt{5}, \sqrt{5}\right), \left(-\sqrt{5}, -\sqrt{5}\right)\right\}$$

43. $x^2 + 2xy - y^2 = 14$ (1)

$x^2 - y^2 = -16$ (2)

This system can be solved using a combination of the elimination and substitution methods.

Using the elimination method, multiply equation (2) by -1 and add to equation (1).

$$x^2 + 2xy - y^2 = 14$$
$$\underline{-x^2 + \qquad y^2 = 16}$$
$$2xy \qquad = 30 \Rightarrow xy = 15 \, (3)$$

Solve equation (3) for y.

$$xy = 15 \Rightarrow y = \frac{15}{x} \ (4)$$

Find x by substituting equation (4) into equation (2).

$$x^2 - \left(\frac{15}{x}\right)^2 = -16 \Rightarrow x^2 - \frac{225}{x^2} = -16 \Rightarrow$$
$$x^4 - 225 = -16x^2 \Rightarrow$$
$$x^4 + 16x^2 - 225 = 0 \Rightarrow \left(x^2 - 9\right)\left(x^2 + 25\right) = 0$$

$$x^2 - 9 = 0 \Rightarrow x^2 = 9 \Rightarrow x = \pm 3$$
$$x^2 + 25 = 0 \Rightarrow x^2 = -25 \Rightarrow x = \pm 5i$$

If $x = 3$, then $y = \frac{15}{3} = 5$. If $x = -3$, then

$$y = \frac{15}{-3} = -5.$$

If $x = 5i$, then $y = \frac{15}{5i} = \frac{3}{i} = -3i$. If $x = -5i$,

then $y = \frac{15}{-5i} = -\frac{3}{i} = 3i$.

Solution set:

$$\left\{(3, 5), (-3, -5), (5i, -3i), (-5i, 3i)\right\}$$

45. $x^2 + y^2 = 25$ (1)

$|x| - y = 5$ (2)

Solve equation (2) for x, then substitute this value into equation (1) and solve for y.

$|x| = y + 5 \Rightarrow x = y + 5$ or $x = -(y + 5)$

If $x = y + 5$, then

$$(y + 5)^2 + y^2 = 25 \Rightarrow$$
$$y^2 + 10y + 25 + y^2 = 25 \Rightarrow 2y^2 + 10y = 0 \Rightarrow$$
$$2y(y + 5) = 0 \Rightarrow 2y = 0 \text{ or } y + 5 = 0 \Rightarrow$$
$$y = 0 \text{ or } y = -5$$

If $y = 0$, then $x = 0 + 5 = 5$.

If $y = -5$, then $x = -5 + 5 = 0$.

If $x = -(y + 5)$, then

$$\left[-(y + 5)\right]^2 + y^2 = 25 \Rightarrow$$
$$y^2 + 10y + 25 + y^2 = 25 \Rightarrow 2y^2 + 10y = 0 \Rightarrow$$
$$2y(y + 5) = 0 \Rightarrow 2y = 0 \text{ or } y + 5 = 0 \Rightarrow$$
$$y = 0 \text{ or } y = -5$$

If $y = 0$, then $x = -(0 + 5) = -5$.

If $y = -5$, then $x = -(-5 + 5) = 0$.

Solution set: $\{(5, 0), (-5, 0), (0, -5)\}$

47. $x = |y|$ (1)

$x^2 + y^2 = 18$ (2)

If $x = |y|$, then $x^2 = y^2$ because $|y|^2 = y^2$.

Substitute $x^2 = y^2$ into equation (2).

$$y^2 + y^2 = 18 \Rightarrow 2y^2 = 18 \Rightarrow y^2 = 9 \Rightarrow y = \pm 3$$

If $y = 3$, then $x = |y| = |3| = 3$. If $y = -3$, then

$x = |y| = |-3| = 3$.

Solution set: $\{(3, 3), (3, 3)\}$

49. $2x^2 - y^2 = 4$ (1)

$|x| = |y|$ (2)

Because $|x|^2 = x^2$ and $|y|^2 = y^2$, we substitute

$y^2 = x^2$ into equation (1) and solve for x.

$$2x^2 - x^2 = 4 \Rightarrow x^2 = 4 \Rightarrow x = \pm 2$$

By substituting these values back into equation (2) we have the following.

If $x = 2$, then $|2| = |y| \Rightarrow 2 = |y| \Rightarrow y = \pm 2$. If

$x = -2$, then $|-2| = |y| \Rightarrow 2 = |y| \Rightarrow y = \pm 2$.

Solution set: $\left\{(2, 2), (-2, -2), (2, -2), (-2, 2)\right\}$

51. $y = \log(x + 5)$

$y = x^2$

Graph the equations in the window
$[-5, 5] \times [-5, 5]$.

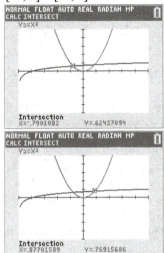

Solution set: $\{(-0.79, 0.62), (0.88, 0.77)\}$

53. $y = e^{x+1}$

$2x + y = 3 \Rightarrow y = 3 - 2x$

Graph the equations in the window
$[-5, 5] \times [-5, 5]$.

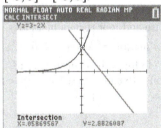

Solution set: $\{(0.06, 2.88)\}$

55. Let x and y represent the numbers.
We obtain the following system.

$x + y = -17$ (1)

$xy = 42$ (2)

Solving equation (1) for y we have
$y = -17 - x$. (3). Substituting this into
equation (2) we have $x(-17 - x) = 42$.

Solving this equation for x we have the
following.

$$x(-17 - x) = 42 \Rightarrow -17x - x^2 = 42 \Rightarrow$$
$$0 = x^2 + 17x + 42$$
$$(x + 3)(x + 14) = 0 \Rightarrow x = -3 \text{ or } x = -14$$

Using equation (3), if $x = -3$ then
$-3 + y = -17 \Rightarrow y = -14$.

If $x = -14$, then $-14 + y = -17 \Rightarrow y = -3$.

The two numbers are -14 and -3.

57. Let x and y represent the numbers.
We obtain the following system.

$x^2 + y^2 = 100$ (1)

$x^2 - y^2 = 28$ (2)

Adding equations (1) and (2) in order to
eliminate y^2, we have $2x^2 = 128$.

Solving this equation for x we have

$2x^2 = 128 \Rightarrow x^2 = 64 \Rightarrow x = \pm 8$

Using equation (1), if $x = 8$ then

$8^2 + y^2 = 100 \Rightarrow 64 + y^2 = 100 \Rightarrow$
$$y^2 = 36 \Rightarrow y = \pm 6$$

If $x = -8$ then $(-8)^2 + y^2 = 100 \Rightarrow$

$64 + y^2 = 100 \Rightarrow y^2 = 36 \Rightarrow y = \pm 6$

The two numbers are 8 and 6, or 8 and -6, or
-8 and 6, or -8 and -6.

59. Let x and y represent the numbers.
We obtain the following system.

$\dfrac{x}{y} = \dfrac{9}{2}$ (1)

$xy = 162$ (2)

Solve equation (1) for x: $x = \frac{9}{2}y$

Substitute $\frac{9}{2}y$ for x in equation (2) and solve
for y.

$$\left(\tfrac{9}{2}y\right)y = 162 \Rightarrow \tfrac{9}{2}y^2 = 162 \Rightarrow$$
$$\tfrac{2}{9}\left(\tfrac{9}{2}y^2\right) = \tfrac{2}{9}(162) \Rightarrow y^2 = 36 \Rightarrow y = \pm 6$$

If $y = 6$, $x = \frac{9}{2}(6) = 27$.

If $y = -6$, $x = \frac{9}{2}(-6) = -27$.

The two numbers are either 27 and 6, or
-27 and -6.

61. Let $x =$ the length of the second side and
$y =$ the length of the third side.
We obtain the following system.

$x^2 + y^2 = 13^2 = 169$ (1)

$x = y + 7$ (2)

Substitute $y + 7$ for x in equation (1) and solve
for y.

$$(y + 7)^2 + y^2 = 169$$
$$y^2 + 14y + 49 + y^2 = 169$$
$$2y^2 + 14y - 120 = 0$$
$$y^2 + 7y - 60 = 0$$
$$(y + 12)(y - 5) = 0 \Rightarrow y = -12 \text{ or } y = 5$$

(continued on next page)

(*continued*)

Disregard the negative solution because a length cannot be negative. If $y = 5$, then
$x = 5 + 7 = 12$.
The lengths of the two shorter sides are 5 m and 12 m.

63. If the system formed by the following equations has a solution, the line and the circle intersect.

$3x - 2y = 9$ (1)

$x^2 + y^2 = 25$ (2)

Solving equation (1) for x we have

$$3x = 9 + 2y \Rightarrow x = \frac{9 + 2y}{3}. \ (3)$$

Substituting into equation (3) into equation (2) we have $\left(\dfrac{9 + 2y}{3}\right)^2 + y^2 = 25.$

Solving this equation for y we have the following.

$$\left(\frac{9 + 2y}{3}\right)^2 + y^2 = 25 \Rightarrow$$

$$\frac{(9 + 2y)^2}{9} + y^2 = 25$$

$$(9 + 2y)^2 + 9y^2 = 225$$

$$81 + 36y + 4y^2 + 9y^2 = 225$$

$$13y^2 + 36y + 81 = 225$$

$$13y^2 + 36y - 144 = 0$$

Using the quadratic formula where $a = 13$, $b = 36$, and $c = -144$, we have the following.

$$y = \frac{-36 \pm \sqrt{(36)^2 - 4(13)(-144)}}{2(13)}$$

$$= \frac{-36 \pm \sqrt{1296 + 7488}}{26} = \frac{-36 \pm \sqrt{8784}}{26}$$

$$y = \frac{-36 - \sqrt{8784}}{26} \approx -4.989 \text{ and}$$

$$y = \frac{-36 + \sqrt{8784}}{26} \approx 2.220$$

If $y = \frac{-36 + \sqrt{8784}}{26}$, then

$$x = \frac{9 + 2\left(\frac{-36 + \sqrt{8784}}{26}\right)}{3} = \frac{234 + 2\left(-36 + \sqrt{8784}\right)}{78}$$

$$= \frac{162 + 2\sqrt{8784}}{78} \approx 4.480$$

If $y = \frac{-36 - \sqrt{8784}}{26}$, then

$$x = \frac{9 + 2\left(\frac{-36 - \sqrt{8784}}{26}\right)}{3} = \frac{234 + 2\left(-36 - \sqrt{8784}\right)}{78}$$

$$= \frac{162 - 2\sqrt{8784}}{78} \approx -0.326$$

Thus, the circle and line do intersect, in fact twice, at approximately (4.48, 2.22) and (−0.326, −4.99).

65. We must first find the solution to the following system.

$y = x^2$ (1)

$x^2 + y^2 = 90$ (2)

Substitute y for x^2 in equation (2) and solve for y.

$$y + y^2 = 90 \Rightarrow y^2 + y - 90 = 0$$

$$(y + 10)(y - 9) = 0 \Rightarrow y = -10 \text{ or } y = 9$$

If $y = -10$, then $-10 = x^2 \Rightarrow x = \pm i\sqrt{10}$. We are seeking points of intersection, so we reject these nonreal solutions. If $y = 9$, then $9 = x^2 \Rightarrow x = \pm 3$. The two points of intersection of the two graphs are (3, 9) and (−3, 9). The slope of the line through these two points is $m = \frac{9-9}{-3-3} = 0$, so this is the horizontal line with equation $y = 9$.

67. Let x represent the length and width of the square base, and let y represent the height. Using the formula for the volume of a box, $V = LWH$, we have the equation

$x^2 y = 360$ (1). The surface area consists of a square base and four rectangular sides, so we have the equation $x^2 + 4xy = 276$ (2). Solve equation (1) for y, then substitute this expression into equation (2) and solve for x.

$$x^2 y = 360 \Rightarrow y = \frac{360}{x^2}$$

$$x^2 + 4xy = 276 \Rightarrow x^2 + 4x\left(\frac{360}{x^2}\right) = 276 \Rightarrow$$

$$x^2 + \frac{1440}{x} = 276 \Rightarrow x^3 + 1440 = 276x \Rightarrow$$

$$x^3 - 276x + 1440 = 0$$

We are looking for the dimensions of the box, so we are restricted to positive solutions. Also, the solutions should be relatively small. By the rational zeros theorem, factors of 1440 are the only possible rational solutions. Using synthetic division, we see that 6 is a solution.

$$\begin{array}{r|rrrr} 6) & 1 & 0 & -276 & 1440 \\ & & 6 & 36 & -1440 \\ \hline & 1 & 6 & -240 & 0 \end{array}$$

Therefore, one possible solution is $x = 6$ and $6^2 y = 360 \Rightarrow 36y = 360 \Rightarrow y = 10$.

(*continued on next page*)

(continued)

To find any other possible solutions, solve $x^2 + 6x - 240 = 0$, the new quotient polynomial.

Using the quadratic formula, we have

$$x = \frac{-6 \pm \sqrt{6^2 - 4(1)(-240)}}{2(1)} = \frac{-6 \pm \sqrt{996}}{2}$$

$$\approx 12.780 \text{ or } -18.780$$

We disregard the negative solution.

If $x = 12.780$, then $12.780^2\, y = 360 \Rightarrow$

$163.3284 y = 360 \Rightarrow y \approx 2.204$.

The solutions are:
length = width = 6 ft; height = 10 ft or
length = width $\approx$ 12.780 ft; height $\approx$ 2.204 ft

69. The system is $\begin{array}{l} x^2 y = 75 \\ x^2 + 4xy = 85 \end{array}$

The first possible solution is $x = 5$ and $y = 3$.

$\begin{array}{ll} x^2 y = 75 & x^2 + 4xy = 85 \\ 5^2 \cdot 3 = 75\ ? & 5^2 + 4 \cdot 5 \cdot 3 = 85\ ? \\ 25 \cdot 3 = 75 & 25 + 60 = 85 \\ \quad 75 = 75 \text{ True} & \quad 85 = 85 \text{ True} \end{array}$

Thus, the solution length = width = 5 in. and height = 3 in. is valid.

Ideally we should check the exact values in the system in order to determine if the approximate dimensions are valid.

From Example 6 we have

$$x = \frac{-5 + \sqrt{5^2 - 4(1)(-60)}}{2(1)} = \frac{-5 + \sqrt{25 + 240}}{2} = \frac{-5 + \sqrt{265}}{2}.$$

The corresponding y-value is

$$y = \frac{75}{x^2} = \frac{75}{\left(\frac{-5+\sqrt{265}}{2}\right)^2} = \frac{75}{\frac{\left(-5+\sqrt{265}\right)^2}{4}} = \frac{300}{\left(-5+\sqrt{265}\right)^2}.$$

The second possible solution is $x = \frac{-5+\sqrt{265}}{2}$

and $y = \frac{300}{\left(-5+\sqrt{265}\right)^2}$.

$$x^2 y = 75$$

$$\left(\frac{-5+\sqrt{265}}{2}\right)^2 \cdot \frac{300}{\left(-5+\sqrt{265}\right)^2} = 75\ ?$$

$$\frac{\left(-5+\sqrt{265}\right)^2}{4} \cdot \frac{300}{\left(-5+\sqrt{265}\right)^2} = 75$$

$$75 = 75 \text{ True}$$

$$x^2 + 4xy = 85$$

$$\frac{\left(-5+\sqrt{265}\right)^2}{4} + \frac{600}{-5+\sqrt{265}} = 85\ ?$$

$$\frac{\left(-5+\sqrt{265}\right)^3 + 2400}{4\left(-5+\sqrt{265}\right)} = 85$$

$$\frac{\left(-5+\sqrt{265}\right)^2 \left(-5+\sqrt{265}\right) + 2400}{4\left(-5+\sqrt{265}\right)} = 85$$

$$\frac{\left(25 - 10\sqrt{265} + 265\right)\left(-5+\sqrt{265}\right) + 2400}{4\left(-5+\sqrt{265}\right)} = 85$$

$$\frac{\left(290 - 10\sqrt{265}\right)\left(-5+\sqrt{265}\right) + 2400}{4\left(-5+\sqrt{265}\right)} = 85$$

$$\frac{\left(-1450 + 340\sqrt{265} - 2650\right) + 2400}{4\left(-5+\sqrt{265}\right)} = 85$$

$$\frac{-4100 + 340\sqrt{265} + 2400}{4\left(-5+\sqrt{265}\right)} = 85$$

$$\frac{-1700 + 340\sqrt{265}}{4\left(-5+\sqrt{265}\right)} = 85$$

$$\frac{340\left(-5+\sqrt{265}\right)}{4\left(-5+\sqrt{265}\right)} = 85$$

$$85 = 85 \text{ True}$$

71. supply: $p = \sqrt{0.1q + 9} - 2$
demand: $p = \sqrt{25 - 0.1q}$

(a) Equilibrium occurs when supply equals demand, so solve the system formed by the supply and demand equations. This system can be solved by substitution.

Substitute $\sqrt{0.1q + 9} - 2$ for p in the demand equation and solve the resulting equation for q.

$$\sqrt{0.1q + 9} - 2 = \sqrt{25 - 0.1q}$$

$$\left(\sqrt{0.1q + 9} - 2\right)^2 = \left(\sqrt{25 - 0.1q}\right)^2$$

$$0.1q + 9 - 4\sqrt{0.1q + 9} + 4 = 25 - 0.1q$$

$$0.2q - 12 = 4\sqrt{0.1q + 9}$$

$$\left(0.2q - 12\right)^2 = \left(4\sqrt{0.1q + 9}\right)^2$$

$$0.04q^2 - 4.8q + 144 = 16\left(0.1q + 9\right)$$

$$0.04q^2 - 4.8q + 144 = 1.6q + 144$$

$$0.04q^2 - 6.4q = 0$$

$$0.04q\left(q - 160\right) = 0 \Rightarrow$$

$$q = 0 \text{ or } q = 160$$

Disregard an equilibrium demand of 0. The equilibrium demand is 160 units.

(b) Substitute 160 for q in either equation and solve for p.

$$p = \sqrt{0.1\left(160\right) + 9} - 2 = \sqrt{16 + 9} - 2$$

$$= \sqrt{25} - 2 = 5 - 2 = 3$$

The equilibrium price is $3.

73. (a) Revenue for public colleges from state sources is decreasing, and revenue from tuition is increasing.

(b) The revenue from state sources was increasing from 2007–2008.

(c) The sources of revenue were equal in 2011, when they each contributed about 24%.

75. Shift the graph of $y = |x|$ one unit to the right to obtain the graph of $y = |x - 1|$.

77. If $x - 1 \geq 0 \ (x \geq 1)$, $|x - 1| = x - 1$. If $x - 1 < 0 \ (x < 1)$, $|x - 1| = -(x - 1) = 1 - x$.

Therefore, $y = \begin{cases} x - 1 & \text{if } x \geq 1 \\ 1 - x & \text{if } x < 1. \end{cases}$

79. $x^2 - 4 = x - 1$ if $x \geq 1$

$x^2 - x - 3 = 0$

Solve by the quadratic formula where $a = 1, b = -1,$ and $c = -3$.

$x = \dfrac{-(-1) \pm \sqrt{(-1)^2 - 4(1)(-3)}}{2(1)} = \dfrac{1 \pm \sqrt{1 + 12}}{2}$

$= \dfrac{1 \pm \sqrt{13}}{2}$

If $\frac{1 + \sqrt{13}}{2} \approx 2.3 \geq 1$, but $\frac{1 - \sqrt{13}}{2} \approx -1.3 < 1$.

Therefore, $x = \frac{1 + \sqrt{13}}{2}$.

$x^2 - 4 = 1 - x$ if $x < 1$

$x^2 + x - 5 = 0$

Solve by the quadratic formula where $a = 1, b = 1,$ and $c = -5$.

$x = \dfrac{-1 \pm \sqrt{1^2 - 4(1)(-5)}}{2(1)} = \dfrac{-1 \pm \sqrt{1 + 20}}{2}$

$= \dfrac{-1 \pm \sqrt{21}}{2}$

If $\frac{-1 - \sqrt{21}}{2} \approx -2.8 < 1$, but $\frac{-1 + \sqrt{21}}{2} \approx 1.8 > 1$,

Therefore, $x = \frac{-1 - \sqrt{21}}{2}$.

Summary Exercises on Systems of Equations

Different methods of solving equations have been introduced. In the solutions to these exercises, we will present the solution using one method (or combination of methods). In general, nonlinear systems cannot use the matrix, Gauss-Jordan, or Cramer's rule methods. Only the methods of substitution or elimination should be considered for such systems.

1. $2x + 5y = 4$
$3x - 2y = -13$

$D = \begin{vmatrix} 2 & 5 \\ 3 & -2 \end{vmatrix} = 2(-2) - 3(5) = -4 - 15 = -19,$

$D_x = \begin{vmatrix} 4 & 5 \\ -13 & -2 \end{vmatrix} = 4(-2) - (-13)(5) = 57$

$D_y = \begin{vmatrix} 2 & 4 \\ 3 & -13 \end{vmatrix} = 2(-13) - 3(4) = -38 \Rightarrow$

$x = \frac{D_x}{D} = \frac{57}{-19} = -3$ and $y = \frac{D_y}{D} = \frac{-38}{-19} = 2.$

Solution set: $\{(-3, 2)\}$

3. $2x^2 + y^2 = 5$ (1)
$3x^2 + 2y^2 = 10$ (2)

Using the elimination method, we add -2 times equation (1) to equation (2).

$\begin{array}{r} -4x^2 - 2y^2 = -10 \\ 3x^2 + 2y^2 = 10 \\ \hline -x^2 \qquad\quad = 0 \end{array} \Rightarrow x^2 = 0 \Rightarrow x = 0$

If $x = 0$, then $2(0)^2 + y^2 = 5 \Rightarrow$

$0 + y^2 = 5 \Rightarrow y^2 = 5 \Rightarrow y = \pm\sqrt{5}.$

Solution set: $\{(0, \sqrt{5}), (0, -\sqrt{5})\}$

5. $6x - y = 5$ (1)
$xy = 4$ (2)

Solving equation (2) for y, we have $y = \frac{4}{x}$.

Substitute this result into equation (1).

$6x - \dfrac{4}{x} = 5 \Rightarrow 6x^2 - 4 = 5x \Rightarrow$

$6x^2 - 5x - 4 = 0 \Rightarrow (2x + 1)(3x - 4) = 0 \Rightarrow$

$x = -\frac{1}{2}$ or $x = \frac{4}{3}$

If $x = \frac{4}{3}$, then $y = \frac{4}{\frac{4}{3}} = 3$. If $x = -\frac{1}{2}$, then

$y = \frac{4}{-\frac{1}{2}} = -8.$

Solution set: $\left\{ \left(\frac{4}{3}, 3\right), \left(-\frac{1}{2}, -8\right) \right\}$

7. $x + 2y + z = 5$ (1)

  $y + 3z = 9$ (2)

This system has infinitely many solutions. We will express the solution set with z as the arbitrary variable. Solving equation (2) for y, we have $y = 9 - 3z$. Substituting y (in terms of z) into equation (1) and solving for x we have the following.

$x + 2(9 - 3z) + z = 5 \Rightarrow x + 18 - 6z + z = 5 \Rightarrow$

  $x + 18 - 5z = 5 \Rightarrow x = -13 + 5z$

Solution set: $\{(-13 + 5z, 9 - 3z, z)\}$

9. $3x + 6y - 9z = 1$

  $2x + 4y - 6z = 1$

  $3x + 4y + 5z = 0$

This system has the augmented matrix

$$\begin{bmatrix} 3 & 6 & -9 & | & 1 \\ 2 & 4 & -6 & | & 1 \\ 3 & 4 & 5 & | & 0 \end{bmatrix}.$$

$$\begin{bmatrix} 2 & 4 & -6 & | & 1 \\ 3 & 6 & -9 & | & 1 \\ 3 & 4 & 5 & | & 0 \end{bmatrix} \begin{matrix} R1 \leftrightarrow R2 \\ \\ \\ \end{matrix} \Rightarrow$$

$$\begin{bmatrix} 1 & 2 & -3 & | & \frac{1}{2} \\ 3 & 6 & -9 & | & 1 \\ 3 & 4 & 5 & | & 0 \end{bmatrix} \begin{matrix} \frac{1}{2}R1 \\ \\ \\ \end{matrix} \Rightarrow$$

$$\begin{bmatrix} 1 & 2 & -3 & | & \frac{1}{2} \\ 0 & 0 & 0 & | & -\frac{1}{2} \\ 0 & -2 & 14 & | & -\frac{3}{2} \end{bmatrix} \begin{matrix} \\ -3R1 + R2 \\ -3R1 + R3 \end{matrix}$$

The second row of the augmented matrix corresponds to the statement $0 = -\frac{1}{2}$. This is a false statement. Thus the solution set is $\varnothing$.

11. $x^2 + y^2 = 4$ (1)

  $y = x + 6$ (2)

Substitute equation (1) into equation (2).

$$x^2 + (x + 6)^2 = 4$$

$$x^2 + x^2 + 12x + 36 = 4$$

$$2x^2 + 12x + 36 = 4$$

$$2x^2 + 12x + 32 = 0 \Rightarrow x^2 + 6x + 16 = 0$$

Using the quadratic formula where $a = 1, b = 6,$ and $c = 16,$ we have the following.

$$x = \frac{-6 \pm \sqrt{6^2 - 4(1)(16)}}{2(1)} = \frac{-6 \pm \sqrt{36 - 64}}{2}$$

$$= \frac{-6 \pm \sqrt{-28}}{2} = \frac{-6 \pm 2i\sqrt{7}}{2} = -3 \pm i\sqrt{7}$$

If $x = -3 + i\sqrt{7}$, then

$y = -3 + i\sqrt{7} + 6 = 3 + i\sqrt{7}$. If $x = -3 - i\sqrt{7}$,

then $y = -3 - i\sqrt{7} + 6 = 3 - i\sqrt{7}$.

Solution set:

$$\left\{ \left(-3 + i\sqrt{7}, 3 + i\sqrt{7}\right), \left(-3 - i\sqrt{7}, 3 - i\sqrt{7}\right) \right\}$$

13. $y + 1 = x^2 + 2x$ (1)

  $y + 2x = 4$ (2)

Solving equation (2) for y, we have $y = 4 - 2x$. Substitute this result into equation (1).

$$4 - 2x + 1 = x^2 + 2x \Rightarrow 5 - 2x = x^2 + 2x \Rightarrow$$

$$0 = x^2 + 4x - 5 \Rightarrow (x + 5)(x - 1) = 0 \Rightarrow$$

  $x = -5$ or $x = 1$

If $x = -5$, then $y = 4 - 2(-5) = 4 + 10 = 14$. If

$x = 1$, then $y = 4 - 2(1) = 4 - 2 = 2$.

Solution set: $\{(1, 2), (-5, 14)\}$

15. $2x + 3y + 4z = 3$

  $-4x + 2y - 6z = 2$

    $4x + 3z = 0$

This system has the augmented matrix

$$\begin{bmatrix} 2 & 3 & 4 & | & 3 \\ -4 & 2 & -6 & | & 2 \\ 4 & 0 & 3 & | & 0 \end{bmatrix}.$$

$$\begin{bmatrix} 1 & \frac{3}{2} & 2 & | & \frac{3}{2} \\ -4 & 2 & -6 & | & 2 \\ 4 & 0 & 3 & | & 0 \end{bmatrix} \begin{matrix} \frac{1}{2}R1 \\ \\ \\ \end{matrix} \Rightarrow$$

$$\begin{bmatrix} 1 & \frac{3}{2} & 2 & | & \frac{3}{2} \\ 0 & 8 & 2 & | & 8 \\ 0 & -6 & -5 & | & -6 \end{bmatrix} \begin{matrix} \\ 4R1 + R2 \\ -4R1 + R3 \end{matrix} \Rightarrow$$

$$\begin{bmatrix} 1 & \frac{3}{2} & 2 & | & \frac{3}{2} \\ 0 & 1 & \frac{1}{4} & | & 1 \\ 0 & -6 & -5 & | & -6 \end{bmatrix} \begin{matrix} \\ \frac{1}{8}R2 \\ \\ \end{matrix} \Rightarrow$$

$$\begin{bmatrix} 1 & 0 & \frac{13}{8} & | & 0 \\ 0 & 1 & \frac{1}{4} & | & 1 \\ 0 & 0 & -\frac{7}{2} & | & 0 \end{bmatrix} \begin{matrix} -\frac{3}{2}R2 + R1 \\ \\ 6R2 + R3 \end{matrix} \Rightarrow$$

$$\begin{bmatrix} 1 & 0 & \frac{13}{8} & | & 0 \\ 0 & 1 & \frac{1}{4} & | & 1 \\ 0 & 0 & 1 & | & 0 \end{bmatrix} \begin{matrix} \\ \\ -\frac{2}{7}R3 \end{matrix} \Rightarrow$$

$$\begin{bmatrix} 1 & 0 & 0 & | & 0 \\ 0 & 1 & 0 & | & 1 \\ 0 & 0 & 1 & | & 0 \end{bmatrix} \begin{matrix} -\frac{13}{8}R3 + R1 \\ -\frac{1}{4}R3 + R2 \\ \end{matrix}$$

Solution set: $\{(0, 1, 0)\}$

17. $-5x + 2y + z = 5$ (1)
$-3x - 2y - z = 3$ (2)
$-x + 6y \quad = 1$ (3)

Add equations (1) and (2) in order to eliminate z to obtain $-8x = 8$ or $x = -1$.

Substitute $x = -1$ into equation (3) to solve for y.

$-(-1) + 6y = 1 \Rightarrow 1 + 6y = 1 \Rightarrow 6y = 0 \Rightarrow y = 0$

Substitute $x = -1$ and $y = 0$ into equation (1) to solve for z.

$-5(-1) + 2(0) + z = 5 \Rightarrow 5 + 0 + z = 5 \Rightarrow$
$5 + z = 5 \Rightarrow z = 0$

Solution set: $\{(-1, 0, 0)\}$

19. $2x^2 + y^2 = 9$ (1)
$3x - 2y = -6$ (2)

Solving equation (2) for x, we have

$3x = 2y - 6 \Rightarrow x = \dfrac{2y - 6}{3}$.

Substitute this result into equation (1).

$$2\left(\frac{2y - 6}{3}\right)^2 + y^2 = 9$$

$$2 \cdot \frac{(2y - 6)^2}{9} + y^2 = 9$$

$$2 \cdot (2y - 6)^2 + 9y^2 = 81$$

$$2\left(4y^2 - 24y + 36\right) + 9y^2 = 81$$

$$8y^2 - 48y + 72 + 9y^2 = 81$$

$$17y^2 - 48y + 72 = 81$$

$$17y^2 - 48y - 9 = 0$$

$$(y - 3)(17y + 3) = 0 \Rightarrow y - 3 \text{ or } y = -\tfrac{3}{17}$$

If $y = 3$, then $x = \frac{2(3) - 6}{3} = \frac{6 - 6}{3} = \frac{0}{3} = 0$.

If $y = -\frac{3}{17}$, then $x = \frac{2\left(-\frac{3}{17}\right) - 6}{3} = \frac{2(-3) - 102}{51}$

$= \frac{-6 - 102}{51} = \frac{-108}{51} = -\frac{36}{17}$.

Solution set: $\left\{(0, 3), \left(-\frac{36}{17}, -\frac{3}{17}\right)\right\}$

21. $x + y - z = 0$ (1)
$2y - z = 1$ (2)
$2x + 3y - 4z = -4$ (3)

Add -2 times equation (1) to equation (3) in order to eliminate x.

$-2x - 2y + 2z = 0$
$\underline{2x + 3y - 4z = -4}$
$\quad y - 2z = -4$ (4)

Add -2 times equation (4) to equation (2) in order to eliminate y and solve for z.

$2y - z = 1$
$\underline{-2y + 4z = 8}$
$\quad 3z = 9 \Rightarrow z = 3$

Substitute 3 for z in equation (4) to obtain
$y - 2(3) = -4 \Rightarrow y - 6 = -4 \Rightarrow y = 2$.

Substitute 3 for z and 2 for y in equation (1) to obtain $x + 2 - 3 = 0 \Rightarrow x - 1 = 0 \Rightarrow x = 1$.

Solution set: $\{(1, 2, 3)\}$

23. $2x - 3y \quad = -2$ (1)
$x + y - 4z = -16$ (2)
$3x - 2y + z = 7$ (3)

Add 4 times equation (3) to equation (2) in order to eliminate z.

$x + y - 4z = -16$
$\underline{12x - 8y + 4z = 28}$
$13x - 7y \quad = 12$ (4)

Add 3 times equation (4) to -7 times equation (1) in order to eliminate y and solve for x.

$-14x + 21y = 14$
$\underline{39x - 21y = 36}$
$25x \quad = 50 \Rightarrow x = 2$

Substitute 2 for x in equation (4) to obtain
$13(2) - 7y = 12 \Rightarrow 26 - 7y = 12 \Rightarrow$
$-7y = -14 \Rightarrow y = 2$

Substitute 2 for x and 2 for y in equation (2) to obtain the following.
$2 + 2 - 4z = -16 \Rightarrow 4 - 4z = -16 \Rightarrow$
$-4z = -20 \Rightarrow z = 5$

Solution set: $\{(2, 2, 5)\}$

25. $x^2 + 3y^2 = 28$ (1)
$y - x = -2$ (2)

Solving equation (2) for y, we have $y = x - 2$. Substitute this result into equation (1) and solve for y

$$x^2 + 3(x - 2)^2 = 28$$

$$x^2 + 3\left(x^2 - 4x + 4\right) = 28$$

$$x^2 + 3x^2 - 12x + 12 = 28$$

$$4x^2 - 12x + 12 = 28$$

$$4x^2 - 12x - 16 = 0$$

$$x^2 - 3x - 4 = 0$$

$$(x + 1)(x - 4) = 0 \Rightarrow x = -1 \text{ or } x = 4$$

If $x = -1$, then $y = -1 - 2 = -3$. If $x = 4$, then $y = 4 - 2 = 2$.

Solution set: $\{(4, 2), (-1, -3)\}$

27. $2x^2 + 3y^2 = 20$ (1)

$x^2 + 4y^2 = 5$ (2)

Using the elimination method, we add -2 times equation (2) to equation (1).

$2x^2 + 3y^2 = 20$

$\underline{-2x^2 - 8y^2 = -10}$

$-5y^2 = 10 \Rightarrow y^2 = -2 \Rightarrow y = \pm i\sqrt{2}$

Find x by substituting back into equation (2).

If $y = i\sqrt{2}$, then $x^2 + 4\left(i\sqrt{2}\right)^2 = 5 \Rightarrow$

$x^2 + 4(-1)(2) = 5 \Rightarrow x^2 - 8 = 5 \Rightarrow x^2 = 13 \Rightarrow$

$x = \pm\sqrt{13}.$

If $y = -i\sqrt{2}$, then

$x^2 + 4\left(-i\sqrt{2}\right)^2 = 5 \Rightarrow x^2 + 4(-1)(2) = 5 \Rightarrow$

$x^2 - 8 = 5 \Rightarrow x^2 = 13 \Rightarrow x = \pm\sqrt{13}$

Solution set: $\left\{\left(\sqrt{13}, i\sqrt{2}\right), \left(-\sqrt{13}, i\sqrt{2}\right),\right.$

$\left. \left(\sqrt{13}, -i\sqrt{2}\right), \left(-\sqrt{13}, -i\sqrt{2}\right)\right\}$

29. $x + 2z = 9$

$y + z = 1$

$3x - 2y = 9$

$D = \begin{vmatrix} 1 & 0 & 2 \\ 0 & 1 & 1 \\ 3 & -2 & 0 \end{vmatrix}$

Adding -3 times row 1 to row 3, we have

$D = \begin{vmatrix} 1 & 0 & 2 \\ 0 & 1 & 1 \\ 3 & -2 & 0 \end{vmatrix} = \begin{vmatrix} 1 & 0 & 2 \\ 0 & 1 & 1 \\ 0 & -2 & -6 \end{vmatrix}.$

Expanding by column one, we have

$D = 1 \cdot \begin{vmatrix} 1 & 1 \\ -2 & -6 \end{vmatrix} = -6 + 2 = -4.$

$D_x = \begin{vmatrix} 9 & 0 & 2 \\ 1 & 1 & 1 \\ 9 & -2 & 0 \end{vmatrix}$

Adding 2 times row 2 to row 3, we have

$D_x = \begin{vmatrix} 9 & 0 & 2 \\ 1 & 1 & 1 \\ 9 & -2 & 0 \end{vmatrix} = \begin{vmatrix} 9 & 0 & 2 \\ 1 & 1 & 1 \\ 11 & 0 & 2 \end{vmatrix}.$

Expanding by column two, we have

$D_x = 1 \cdot \begin{vmatrix} 9 & 2 \\ 11 & 2 \end{vmatrix} = 18 - 22 = -4.$

$D_y = \begin{vmatrix} 1 & 9 & 2 \\ 0 & 1 & 1 \\ 3 & 9 & 0 \end{vmatrix}$

Adding -3 times row 1 to row 3 we have

$D_y = \begin{vmatrix} 1 & 9 & 2 \\ 0 & 1 & 1 \\ 3 & 9 & 0 \end{vmatrix} = \begin{vmatrix} 1 & 9 & 2 \\ 0 & 1 & 1 \\ 0 & -18 & -6 \end{vmatrix}.$

Expanding by column one, we have

$D_y = 1 \cdot \begin{vmatrix} 1 & 1 \\ -18 & -6 \end{vmatrix} = -6 + 18 = 12.$

$D_z = \begin{vmatrix} 1 & 0 & 9 \\ 0 & 1 & 1 \\ 3 & -2 & 9 \end{vmatrix}$

Adding -3 times row 1 to row 3, we have

$D_z = \begin{vmatrix} 1 & 0 & 9 \\ 0 & 1 & 1 \\ 3 & -2 & 9 \end{vmatrix} = \begin{vmatrix} 1 & 0 & 9 \\ 0 & 1 & 1 \\ 0 & -2 & -18 \end{vmatrix}.$

Expanding by column one, we have

$D_z = 1 \cdot \begin{vmatrix} 1 & 1 \\ -2 & -18 \end{vmatrix} = -18 + 2 = -16.$

Thus, we have

$x = \frac{D_x}{D} = \frac{-4}{-4} = 1, \ y = \frac{D_y}{D} = \frac{12}{-4} = -3,$

and $z = \frac{D_z}{D} = \frac{-16}{-4} = 4.$

Solution set: $\{(1, -3, 4)\}$

31. $-x + y = -1$

$x + z = 4$

$6x - 3y + 2z = 10$

This system has the augmented matrix

$\begin{bmatrix} -1 & 1 & 0 & | & -1 \\ 1 & 0 & 1 & | & 4 \\ 6 & -3 & 2 & | & 10 \end{bmatrix}.$

$\begin{bmatrix} 1 & -1 & 0 & | & 1 \\ 1 & 0 & 1 & | & 4 \\ 6 & -3 & 2 & | & 10 \end{bmatrix} \begin{matrix} -1R1 \\ \\ \end{matrix} \Rightarrow$

$\begin{bmatrix} 1 & -1 & 0 & | & 1 \\ 0 & 1 & 1 & | & 3 \\ 0 & 3 & 2 & | & 4 \end{bmatrix} \begin{matrix} \\ -1R1+R2 \\ -6R1+R3 \end{matrix} \Rightarrow$

$\begin{bmatrix} 1 & 0 & 1 & | & 4 \\ 0 & 1 & 1 & | & 3 \\ 0 & 0 & -1 & | & -5 \end{bmatrix} \begin{matrix} R2+R1 \\ \\ -3R2+R3 \end{matrix} \Rightarrow$

(continued on next page)

(*continued*)

$$\begin{bmatrix} 1 & 0 & 1 & | & 4 \\ 0 & 1 & 1 & | & 3 \\ 0 & 0 & 1 & | & 5 \end{bmatrix} {\scriptstyle -1R3} \quad \Rightarrow$$

$$\begin{bmatrix} 1 & 0 & 0 & | & -1 \\ 0 & 1 & 0 & | & -2 \\ 0 & 0 & 1 & | & 5 \end{bmatrix} {\scriptstyle -1R3+R1 \atop -1R3+R2}$$

Solution set: $\{(-1, -2, 5)\}$

33. $xy = -3$ (1)
$x + y = -2$ (2)

Solving equation (1) for y, we have $y = -\dfrac{3}{x}$.

Substitute this result into equation (2).

$$x + \left(-\frac{3}{x}\right) = -2 \Rightarrow x^2 - 3 = -2x \Rightarrow$$

$$x^2 + 2x - 3 = 0 \Rightarrow (x+3)(x-1) = 0 \Rightarrow$$
$$x = -3 \text{ or } x = 1$$

If $x = -3$, then $y = -\frac{3}{-3} = 1$. If $x = 1$, then

$y = -\frac{3}{1} = -3$.

Solution set: $\{(1, -3), (-3, 1)\}$

35. $y = x^2 + 6x + 9$ (1)
$x + y = 3$ (2)

Solve equation (2) for y to obtain
$x + y = 3 \Rightarrow y = 3 - x.$ (3)

Substitute equation (3) into equation (1) and solve for x.

$$3 - x = x^2 + 6x + 9$$
$$0 = x^2 + 7x + 6$$
$$(x+6)(x+1) = 0 \Rightarrow x = -6 \text{ or } x = -1$$

If $x = -6$, then $y = 3 - (-6) = 9$. If $x = -1$,

then $y = 3 - (-1) = 4$.

Solution set: $\{(-6, 9), (-1, 4)\}$

Section 9.6 Systems of Inequalities and Linear Programming

1. The answer is C because this represents the region where $x \geq 5$ and (at the same time) $y \leq -3$.

3. The answer is B because this represents the region where $x > 5$ and (at the same time) $y < -3$.

5. The test point (0, 0) <u>does not</u> satisfy the inequality $-3x - 4y \geq 12$.

7. The coordinates of the point of intersection are (2, –5).

9. The graph of $4x - 7y < 28$ has a <u>dashed</u> boundary line.

11. $x + 2y \leq 6$

The boundary is the line $x + 2y = 6$, which can be graphed using the x-intercept 6 and y-intercept 3. The boundary is included in the graph, so draw a solid line. Solving for y, we have the following.

$$x + 2y \leq 6 \Rightarrow 2y \leq -x + 6 \Rightarrow y \leq -\tfrac{1}{2}x + 3$$

Because y is *less than* or equal to $-\tfrac{1}{2}x + 3$, the graph of the solution set is the line and the half-plane *below* the boundary. We also can use (0, 0) as a test point. Because $0 + 2(0) \leq 6 \Rightarrow 0 \leq 6$ is a true statement, shade line and the side of the graph containing the test point (0, 0) (the half-plane below the boundary).

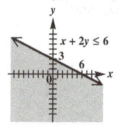

13. $2x + 3y \geq 4$

The boundary is the line $2x + 3y = 4$, which can be graphed using the x-intercept 2 and y-intercept $\frac{4}{3}$. The boundary is included in the graph, so draw a solid line. Solving for y, we have the following.

$$2x + 3y \geq 4 \Rightarrow 3y \geq -2x + 4 \Rightarrow y \geq -\tfrac{2}{3}x + \tfrac{4}{3}$$

Because y is *greater than* or equal to $-\tfrac{2}{3}x + \tfrac{4}{3}$, the graph of the solution set is the line and the half-plane *above* the boundary. We also can use (0, 0) as a test point. Because $2(0) + 3(0) \geq 4 \Rightarrow 0 \geq 4$ is a false statement, shade the line and the side of the graph not containing the test point (0,0) (the half-plane above the boundary).

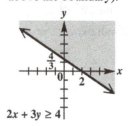

15. $3x - 5y > 6$

The boundary is the line $3x - 5y = 6$, which can be graphed using the x-intercept 2 and y-intercept $-\frac{6}{5}$. The boundary is not included in the graph, so draw a dashed line. Solving for y, we have the following.

$3x - 5y > 6 \Rightarrow -5y > -3x + 6 \Rightarrow y < \frac{3}{5}x - \frac{6}{5}$

Because y is *less than* $\frac{3}{5}x - \frac{6}{5}$, the graph of the solution set is the half-plane *below* the boundary. We also can use $(0, 0)$ as a test point. Because $3(0) - 5(0) > 6 \Rightarrow 0 > 6$ is a false statement, shade the side of the graph not containing the test point $(0, 0)$ (the half-plane below the boundary).

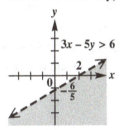

17. $5x < 4y - 2$

The boundary is the line $5x = 4y - 2$, which can be graphed using the x-intercept $-\frac{2}{5}$ and y-intercept $\frac{1}{2}$. The boundary is not included in the graph, so draw a dashed line. Solving for y, we have the following.

$5x < 4y - 2 \Rightarrow -4y < -5x - 2 \Rightarrow y > \frac{5}{4}x + \frac{1}{2}$

Because y is greater than $\frac{5}{4}x + \frac{1}{2}$, the graph of the solution set is the half-plane *above* the boundary. We also can use $(0, 0)$ as a test point. Because $5(0) < 4(0) - 2 \Rightarrow 0 < -2$ is a false statement, shade the side of the graph not containing the test point $(0, 0)$ (the half-plane above the boundary).

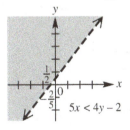

19. $x \le 3$

The boundary is the vertical line $x = 3$, which intersects the x-axis at 3. The boundary is included in the graph, so draw a solid line. Because $x \le 3$, it can easily be determined that we should shade to the left of the boundary. We also can use $(0, 0)$ as a test point. Because $0 \le 3$ is a true statement, shade the line and the side of the graph containing the test point $(0, 0)$ (the half-plane to the left of the boundary).

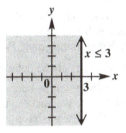

21. $y < 3x^2 + 2$

The boundary is the parabola $y = 3x^2 + 2$, which opens upwards. It has vertex $(0, 2)$, y-intercept 2, and no x-intercepts. The inequality symbol is $<$, so draw a dashed curve. Because y is *less than* $3x^2 + 2$, the graph of the solution set is the half-plane *below* the boundary. We also can use $(0, 0)$ as a test point. Because $0 < 3(0)^2 + 2 \Rightarrow 0 < 2$ is a true statement, shade the region of the graph containing the test point $(0, 0)$.

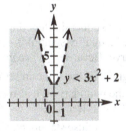

23. $y > (x - 1)^2 + 2$

The boundary is the parabola $y = (x - 1)^2 + 2$, which opens upwards. It has vertex $(1, 2)$, y-intercept 3, and no x-intercepts. The inequality symbol is $>$, so draw a dashed curve. Because y is *greater than* $(x - 1)^2 + 2$, the graph of the solution set is the half-plane *above* the boundary. We also can use $(0, 0)$ as a test point. Because $0 > (0 - 1)^2 + 2 \Rightarrow 0 > 3$ is a false statement, shade the region of the graph not containing the test point $(0, 0)$.

(continued on next page)

(*continued*)

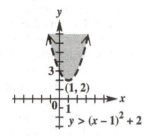

$y > (x - 1)^2 + 2$

25. $x^2 + (y + 3)^2 \leq 16$

The boundary is a circle with center $(0, -3)$ and radius 4. Draw a solid circle to show that the boundary is included in the graph. Because $x^2 + (y + 3)^2 \leq 16$, it can easily be determined that points in the interior or on the boundary would satisfy this relation. We also can use $(0, 0)$ as a test point. Because we have

$0^2 + (0 + 3)^2 \leq 16 \Rightarrow 9 \leq 16$ is a true

statement, shade the region of the graph containing the test point $(0, 0)$.

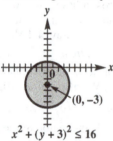

$x^2 + (y + 3)^2 \leq 16$

27. $y > 2^x + 1$

The boundary is an exponential function with y-intercept 2. The inequality symbol is $>$, so draw a dashed curve. Because y is *greater than* $2^x + 1$, the graph of the solution set is the half-plane *above* the boundary. We also can use $(0, 0)$ as a test point. Because

$0 > 2^0 + 1 \Rightarrow 0 > 2$ is a false statement, shade the region of the graph not containing the test point $(0, 0)$.

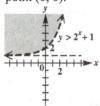

29. $Ax + By \geq C, B > 0$

Solving for y we have
$Ax + By \geq C \Rightarrow By \geq -Ax + C \Rightarrow$
$y \geq -\frac{A}{B}x + \frac{C}{B}$
Because $B > 0$, the inequality symbol was not reversed when both sides are divided by B.
Because $y \geq -\frac{A}{B}x + \frac{C}{B}$, you would shade above the line.

31. B. The graph of $(x - 5)^2 + (y - 2)^2 = 4$ is a circle with center $(5, 2)$ and radius $r = \sqrt{4} = 2$. The graph of $(x - 5)^2 + (y - 2)^2 < 4$ is the region in the interior of this circle.

33. C. The graph of $y \leq 3x - 6$ is the region below the line with slope 3 and y-intercept -6.

35. A. The graph of $y \leq -3x - 6$ is the region below the line with slope -3 and y-intercept -6.

37. $x + y \geq 0$
$2x - y \geq 3$

Graph $x + y = 0$ as a solid line through the origin with a slope of -1. Shade the region above this line. Graph $2x - y = 3$ as a solid line with x-intercept $\frac{3}{2}$ and y-intercept -3. Shade the region below this line.
To find where the boundaries of the two lines intersect, solve the system
$x + y = 0$ (1)
$2x - y = 3$ (2)
by adding the two equations together to obtain $3x = 3 \Rightarrow x = 1$. Substituting this value into equation (1), we obtain $1 + y = 0 \Rightarrow y = -1$.
Thus, the boundaries intersect at $(1, -1)$. The solution set is the common region, which is shaded in the final graph.

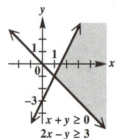

$x + y \geq 0$
$2x - y \geq 3$

For Exercises 39–69, we omit the steps to find the points in which the boundaries intersect.

39. $2x + y > 2$
   $x - 3y < 6$

Graph $2x + y = 2$ as a dashed line with y-intercept 2 and x-intercept 1. Shade the region above this line. Graph $x - 3y = 6$ as a dashed line with y-intercept -2 and x-intercept 6. Shade the region above this line. The solution set is the common region, which is shaded in the final graph. The lines intersect at $\left(\frac{12}{7}, -\frac{10}{7}\right)$.

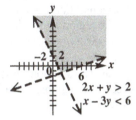

41. $3x + 5y \leq 15$
   $x - 3y \geq 9$

Graph $3x + 5y = 15$ as a solid line with y-intercept 3 and x-intercept 5. Shade the region below this line. Graph $x - 3y = 9$ as a solid line with y-intercept -3 and x-intercept 9. Shade the region below this line. The solution set is the common region, which is shaded in the final graph. The lines intersect at $\left(\frac{45}{7}, -\frac{6}{7}\right)$.

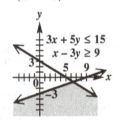

43. $4x - 3y \leq 12$
   $y \leq x^2$

Graph $4x - 3y = 12$ as a solid line with y-intercept -4 and x-intercept 3. Shade the region above this line. Graph the solid parabola $y = x^2$. Shade the region outside of this parabola. The solution set is the intersection of these two regions, which is shaded in the final graph. The line and the parabola do not intersect.

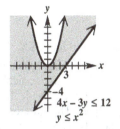

$4x - 3y \leq 12$
$y \leq x^2$

45. $x + 2y \leq 4$
   $y \geq x^2 - 1$

Graph $x + 2y = 4$ as a solid line with y-intercept 2 and x-intercept 4. Shade the region below the line. Graph the solid parabola $y = x^2 - 1$. Shade the region inside of the parabola. The solution set is the common region, which is shaded in the final graph. The parabola and the line intersect at $\left(\frac{3}{2}, \frac{5}{4}\right)$ and $(-2, 3)$.

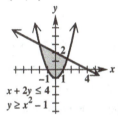

$x + 2y \leq 4$
$y \geq x^2 - 1$

47. $y \leq (x + 2)^2$
   $y \geq -2x^2$

Graph $y = (x + 2)^2$ as a solid parabola opening up with a vertex at $(-2, 0)$. Shade the region below the parabola. Graph $y = -2x^2$ as a solid parabola opening down with a vertex at the origin. Shade the region above the parabola. The solution set is the intersection of these two regions, which is shaded in the final graph. The parabolas do not intersect.

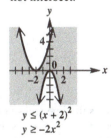

$y \leq (x + 2)^2$
$y \geq -2x^2$

49. $x + y \le 36$
$\quad -4 \le x \le 4$

Graph $x + y = 36$ as a solid line with y-intercept 36 and x-intercept 36. Shade the region below this line. Graph the vertical lines $x = -4$ and $x = 4$ as solid lines. Shade the region between these lines. The solution set is the intersection of these two regions, which is shaded in the final graph. The lines intersect at $(-4, 40)$ and $(4, 32)$.

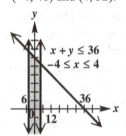

51. $y \ge x^2 + 4x + 4$
$\quad y < -x^2$

Graph the solid parabola $y = x^2 + 4x + 4$ or $y = (x + 2)^2$, which has vertex $(-2, 0)$ and opens upward. Shade the region inside of this parabola. Graph the dashed parabola $y = -x^2$, which has vertex $(0, 0)$ and opens downward. Shade the region inside this parabola.

These two regions have no points in common, so the system has no solution. The parabolas do not intersect.

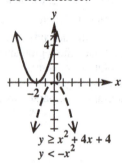

53. $3x - 2y \ge 6$ (1)
$\quad x + y \le -5$ (2)
$\quad\quad y \le 4$ (3)

Graph $3x - 2y = 6$ as a solid line and shade the region below it. Graph $x + y = -5$ as a solid line and shade the region below it. Graph $y = 4$ as a solid horizontal line and shade the region below it. The solution set is the intersection of these three regions, which is shaded in the final graph. Lines (1) and (2) intersect at $\left(-\frac{4}{5}, -\frac{21}{5}\right)$.

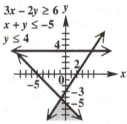

55. $-2 < x < 2$
$\quad y > 1$
$\quad x - y > 0$

Graph the vertical lines $x = -2$ and $x = 2$ as a dashed line. Shade the region between the two lines. Graph the horizontal line $y = 1$ as a dashed line. Shade the region above the line. Graph the line $x - y = 0$ as a dashed line through the origin with a slope of 1. Shade the region below this line. The solution set is the intersection of these three regions, which is shaded in the final graph.

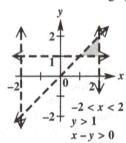

57. $x \le 4$
$\quad x \ge 0$
$\quad y \ge 0$
$\quad x + 2y \ge 2$

Graph $x = 4$ as a solid vertical line. Shade the region to the left of this line. Graph $x = 0$ as a solid vertical line. (This is the y-axis.) Shade the region to the right of this line. Graph $y = 0$ as a solid horizontal line. (This is the x-axis.) Shade the region above the line. Graph $x + 2y = 2$ as a solid line with x-intercept 2 and y-intercept 1. Shade the region above the line. The solution set is the intersection of these four regions, which is shaded in the final graph.

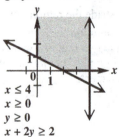

59. $2x + 3y \leq 12$
$2x + 3y > -6$
$3x + \ y < 4$
$\quad\quad x \geq 0$
$\quad\quad\quad y \geq 0$

Graph $2x + 3y = 12$ as a solid line and shade the region below it. Graph $2x + 3y = 6$ as a dashed line and shade the region above it. Graph $3x + y = 4$ as a dashed line and shade the region below it. $x = 0$ is the y-axis. Shade the region to the right of it. $y = 0$ is the x-axis. Shade the region above it. The solution set is the intersection of these five regions, which is shaded in the final graph. The open circles at $(0, 4)$ and $\left(\frac{4}{3}, 0\right)$ indicate that those points are not included in the solution (due to the fact that the boundary line on which they lie, $3x + y = 4$, is not included).

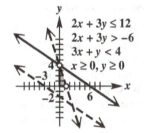

61. $y \leq \left(\frac{1}{2}\right)^x$
$y \geq 4$

Graph $y = \left(\frac{1}{2}\right)^x$ using a solid curve passing through the points $(-2, 4)$, $(-1, 2)$, $(0, 1)$, $\left(1, \frac{1}{2}\right)$, and $\left(2, \frac{1}{4}\right)$. Shade the region below this curve. Graph the solid horizontal line $y = 4$ and shade the region above it. The solution set consists of the intersection of these two regions, which is shaded in the final graph. The curves intersect at $(-2, 4)$.

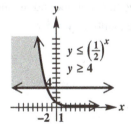

63. $y \leq \log x$
$y \geq |x - 2|$

Graph $y = \log x$ using a solid curve because $y \leq \log x$ is a nonstrict inequality. (Recall that "$\log x$" means $\log_{10} x$.) This graph contains the points $(0.1, -1)$, $(1, 0)$ and $(10, 1)$. Use a calculator to approximate other points on the graph, such as $(2, 0.30)$ and $(4, 0.60)$. Because the symbol is $\leq$, shade the region *below* the curve. Now graph $y = |x - 2|$. Make this boundary solid because $y \geq |x - 2|$ is also a nonstrict inequality. This graph can be obtained by translating the graph of $y = |x|$ to the right 2 units. It contains points $(0, 2)$, $(2, 0)$, and $(4, 2)$. Because the symbol is $\geq$, shade the region *above* the absolute value graph. The solution set is the intersection of the two regions, which is shaded in the final graph.

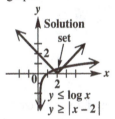

65. $y > x^3 + 1$
$y \geq -1$

Graph $y > x^3 + 1$ as a dashed curve, which is the graph of $y = x^3$ translated up 1 unit. Shade the region above the curve. Graph $y = -1$ as a solid horizontal line through $(0, -1)$. Shade the area above the line. The solution set is the common region, which is shaded in the final graph.

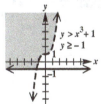

67. $3x + 2y \geq 6$

Solving the inequality for y we have

$3x + 2y \geq 6 \Rightarrow 2y \geq -3x + 6 \Rightarrow y \geq -\frac{3}{2}x + 3.$

Enter $y_1 = (-3/2)x + 3$ and use a graphing calculator to shade the region above the line. Notice the icon to the left of y_1. The graphing window is $[-10, 10] \times [-10, 10]$.

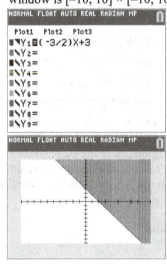

69. $x + y \geq 2$
$x + y \leq 6$

Solving each inequality for y, we have
$y = -x + 2$ and $y = -x + 6$. Enter

$y_1 = -x + 2$ and $y_2 = -x + 6$ and use the graphing calculator to shade the region above the graph of $y_1 = -x + 2$ and below the graph of $y_2 = -x + 6$. The solution set of the system is the overlapping region. Notice the icons to the left of y_1 and y_2. The graphing window is
$[-10, 10] \times [-10, 10]$.

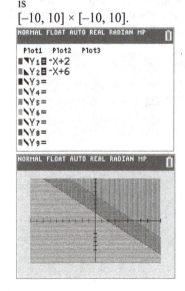

71. The upper line passes through $(0, 2)$ and $(4, 0)$, so $m = \frac{0-2}{4-0} = -\frac{1}{2}$, and the equation is

$y = -\frac{1}{2}x + 2 \Rightarrow x + 2y = 4$. The line is solid, so it is included in the inequality. The lower line passes through $(0, -3)$ and $(4, 0)$, so

$m = \frac{0-(-3)}{4-0} = \frac{3}{4}$ and the equation is

$y = \frac{3}{4}x - 3 \Rightarrow 3x - 4y = 12$. The line is solid, so it is included in the inequality. The shaded region contains the point $(0, 0)$, so test $(0, 0)$ in each equation to determine the direction of the inequalities:

$0 + 2(0) = 0 < 4 \Rightarrow x + 2y \leq 4$,

$3(0) + 4(0) = 0 < 12 \Rightarrow 3x - 4y \leq 12$

The system is $\quad x + 2y \leq 4 \quad .$
$\qquad\qquad\quad 3x - 4y \leq 12$

73. The circle has center $(0, 0)$ and radius 4, so its equation is $x^2 + y^2 = 16$. The curve is solid, so it is included in the inequality. The horizontal line passes through $(0, 2)$, so its equation is $y = 2$. It is solid, so it is included in the inequality. The shaded region includes the point $(-1, 3)$, so test $(-1, 3)$ in each equation to determine the direction of the inequalities:

$(-1)^2 + 3^2 = 10 < 16 \Rightarrow x^2 + y^2 \leq 16$,

$3 > 2 \Rightarrow y > 2$. The system is $x^2 + y^2 \leq 16$.
$\qquad\qquad\qquad\qquad\qquad\qquad\qquad y \geq 2$

75. The graph is in the first quadrant, so $x > 0$ and $y > 0$. A circle with radius 2 centered at the origin has the equation $x^2 + y^2 = 4$. The region includes and is inside the circle, so the inequality is $x^2 + y^2 \leq 4$. A line that passes through $(0, -1)$ and $(2, 2)$ has slope

$m = \frac{2-(-1)}{2-0} = \frac{3}{2}$ and equation $y = \frac{3}{2}x - 1$. The region is above the line, but does not include the line, so the inequality is $y > \frac{3}{2}x - 1$. The system is

$\qquad\quad x > 0 \qquad .$
$\qquad\quad y > 0$
$\quad x^2 + y^2 \leq 4$
$\qquad\quad y > \frac{3}{2}x - 1$

77.

Point	Value of $3x + 5y$	
$(1,1)$	$3(1)+5(1)=8$	← Minimum
$(2,7)$	$3(2)+5(7)=41$	
$(5,10)$	$3(5)+5(10)=65$	← Maximum
$(6,3)$	$3(6)+5(3)=33$	

The maximum value is 65 at (5, 10).
The minimum value is 8 at (1, 1).

79.

Point	Value of $3x + 5y$	
$(1,0)$	$3(1)+5(0)=3$	← Minimum
$(1,10)$	$3(1)+5(10)=53$	
$(7,9)$	$3(7)+5(9)=66$	← Maximum
$(7,6)$	$3(7)+5(6)=51$	

The maximum value is 66 at (7, 9).
The minimum value is 3 at (1, 0).

81.

Point	Value of $10y$	
$(1,0)$	$10(0)=0$	← Minimum
$(1,10)$	$10(10)=100$	← Maximum
$(7,9)$	$10(9)=90$	
$(7,6)$	$10(6)=60$	

The maximum value is 100 at $(1,10)$.

The minimum value is 0 at $(1,0)$.

83. Let x = the number of Brand X pills;
y = the number of Brand Y pills.
The following table is helpful in organizing the information.

	Number of Brand X pills (x)	Number of Brand Y pills (y)	Restrictions
Vitamin A	3000	1000	At least 6000
Vitamin C	45	50	At least 195
Vitamin D	75	200	At least 600

We have
$$3000x + 1000y \geq 6000$$
$$45x + 50y \geq 195$$
$$75x + 200y \geq 600$$
$$x \geq 0, y \geq 0.$$

Graph $3000x + 1000y = 6000$ as a solid line with x-intercept 2 and y-intercept 6. Shade the region above the line. Graph $45x + 50y = 195$ as a solid line with x-intercept $4.\overline{3}$ and y-intercept 3.9. Shade the region above the line. Graph $75x + 200y = 600$ as a solid line with x-intercept 8 and y-intercept 3. Shade the region above the line. Graph $x = 0$ (the y-axis) as a solid line and shade the region to the right of it. Graph $y = 0$ (the x-axis) as a solid line and shade the region above it. The region of feasible solutions is the intersection of these five regions.

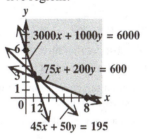

85. Let x = number of cartons of food;
y = number of cartons of clothing.
The following table is helpful in organizing the information.

	Number of cartons of food (x)	Number of cartons of clothing (y)	Restrictions
Weight	40	10	Cannot Exceed 16,000
Volume	20	30	No more than 18,000

We have
$$40x + 10y \leq 16,000$$
$$20x + 30y \leq 18,000$$
$$x \geq 0, y \geq 0.$$

Maximize objective function, number of people = $10x + 8y$.

Find the region of feasible solutions by graphing the system of inequalities that is made up of the constraints. To graph $40x + 10y \leq 16,000$, draw the line with x-intercept 400 and y-intercept 1600 as a solid line. Because the test point (0, 0) satisfies this inequality, shade the region *below* the line. To graph $20x + 30y \leq 18,000$, draw the line with x-intercept 900 and y-intercept 600 as a solid line. Because the test point (0, 0) satisfies this inequality, shade the region *below* the line.

(continued on next page)

(continued)

The constraints $x \geq 0$ and $y \geq 0$ restrict the graph to the first quadrant. The graph of the feasible region is the intersection of the regions that are the graphs of the individual constraints.

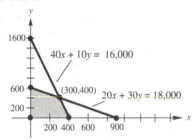

From the graph, observe that three are the vertices are $(0,0), (0,600),$ and $(400,0).$ The fourth vertex is the intersection point of the lines $40x + 10y = 16,000$ and $20x + 30y = 18,000.$ To find this point, solve the system
$40x + 10y = 16,000$
$20x + 30y = 18,000.$

The first equation can be written as
$4x + y = 1600 \Rightarrow y = 1600 - 4x.$ Substituting this equation into $20x + 30y = 18,000,$ we have

$$20x + 30(1600 - 4x) = 18,000$$
$$20x + 48,000 - 120x = 18,000$$
$$48,000 - 100x = 18,000$$
$$-100x = -30,000$$
$$x = 300$$

Substituting $x = 300$ into $y = 1600 - 4x,$ we have $y = 1600 - 4(300) = 1600 - 1200 = 400.$

Thus, the fourth vertex is (300, 400). Next, evaluate the objective function at each vertex.

Point	Number of people $= 10x + 8y$	
$(0,0)$	$10(0) + 8(0) = 0$	
$(0,600)$	$10(0) + 8(600) = 4800$	
$(300,400)$	$10(300) + 8(400) = 6200$	← Maximum
$(400,0)$	$10(400) + 8(0) = 4000$	

The maximum value of $10x + 8y$ occurs at $(300, 400),$ so they should send 300 cartons of food and 400 cartons of clothes to maximize the number of people helped. The maximum number of people helped is 6200.

87. Let $x =$ number of cabinet A;
$y =$ number of cabinet B.
The cost constraint is $10x + 20y \leq 140.$ The space constraint is $6x + 8y \leq 72.$ The number of cabinets cannot be negative, so we also have $x \geq 0$ and $y \geq 0.$ We want to maximize the objective function, storage capacity $= 8x + 12y.$ Find the region of feasible solutions by graphing the system of inequalities that is made up of the constraints. To graph $10x + 20y \leq 140,$ draw the line with x-intercept 14 and y-intercept 7 as a solid line. Because the test point (0, 0) satisfies this inequality, shade the region below the line. To graph $6x + 8y \leq 72,$ draw the line with x-intercept 12 and y-intercept 9 as a solid line. Because the test point (0, 0) satisfies this inequality, shade the region below the line. The constraints $x \geq 0$ and $y \geq 0$ restrict the graph to the first quadrant. The graph of the feasible region is the intersection of the regions that are the graphs of the individual constraints.

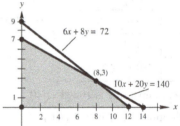

From the graph, observe that three vertices are (0, 0), (0, 7), and (12, 0). The fourth vertex is the intersection point of the lines $10x + 20y = 140$ and $6x + 8y = 72.$ To find this point, solve the system
$10x + 20y = 140$
$6x + 8y = 72.$

The first equation can be written as
$x + 2y = 14 \Rightarrow x = 14 - 2y.$ Substituting this equation into $6x + 8y = 72,$ we have
$6(14 - 2y) + 8y = 72 \Rightarrow 84 - 12y + 8y = 72 \Rightarrow$
$84 - 4y = 72 \Rightarrow -4y = -12 \Rightarrow y = 3$

Substituting $y = 3$ into $x = 14 - 2y,$ we have $x = 14 - 2(3) = 8.$ Thus, the fourth vertex is (8, 3). Now evaluate the objective function at each vertex.

(continued on next page)

(*continued*)

Point	Storage Capacity $= 8x + 12y$	
$(0,0)$	$8(0)+12(0)=0$	
$(0,7)$	$8(0)+12(7)=84$	
$(8,3)$	$8(8)+12(3)=100$	$\leftarrow$ Maximum
$(12,0)$	$8(12)+12(0)=96$	

The maximum value of $8x + 12y$ occurs at $(8, 3)$, so the office manager should buy 8 of cabinet A and 3 of cabinet B for a total storage capacity of 100 ft^3.

89. Let x = number of servings of product A; y = number of servings of product B. The Supplement I constraint is $3x + 2y \geq 15$. The Supplement II constraint is $2x + 4y \geq 15$.

The numbers of servings cannot be negative, so we also have $x \geq 0$ and $y \geq 0$. We want to minimize the objective function, cost = $0.25x + 0.40y$. Find the region of feasible solutions by graphing the system of inequalities that is made up of the constraints. To graph $3x + 2y \geq 15$, draw the line with x-intercept 5 and y-intercept $\frac{15}{2} = 7\frac{1}{2}$ as a solid line. Because the test point $(0, 0)$ does not satisfy this inequality, shade the region *above* the line. To graph $2x + 4y \geq 15$, draw the line with x-intercept $\frac{15}{2} = 7\frac{1}{2}$ and y-intercept $\frac{15}{4} = 3\frac{3}{4}$ as a solid line. Because the test point $(0, 0)$ does not satisfy this inequality, shade the region *above* the line.

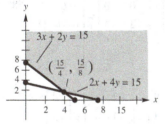

The constraints $x \geq 0$ and $y \geq 0$ restrict the graph to the first quadrant. The graph of the feasible region is the intersection of the regions that are the graphs of the individual constraints.

From the graph, observe that two vertices are $\left(0, \frac{15}{2}\right)$ and $\left(\frac{15}{2}, 0\right)$. The third vertex is the intersection point of the lines $3x + 2y = 15$ and $2x + 4y = 15$. To find this point, solve the system
$$3x + 2y = 15$$
$$2x + 4y = 15.$$
Multiply the first equation by –2 and add it to the second equation.
$$-6x - 4y = -30$$
$$\underline{2x + 4y = 15}$$
$$-4x = -15 \Rightarrow x = \frac{15}{4}$$
Substituting this equation into $2x + 4y = 15$, we have $2\left(\frac{15}{4}\right) + 4y = 15 \Rightarrow \frac{15}{2} + 4y = 15 \Rightarrow$ $15 + 8y = 30 \Rightarrow 8y = 15 \Rightarrow y = \frac{15}{8}$.

Thus, the third vertex is $\left(\frac{15}{4}, \frac{15}{8}\right)$. Next, evaluate the objective function at each vertex.

Point	Cost = $0.25x + 0.40y$	
$\left(0, \frac{15}{2}\right)$	$0.25(0)+0.40\left(\frac{15}{2}\right)=3.00$	
$\left(\frac{15}{4}, \frac{15}{8}\right)$	$0.25\left(\frac{15}{4}\right)+0.40\left(\frac{15}{8}\right)=1.69$	$\leftarrow$ Minimum
$\left(\frac{15}{2}, 0\right)$	$0.25\left(\frac{15}{2}\right)+0.40(0)=1.88$	

The minimum cost is \$1.69 for $\frac{15}{4} = 3\frac{3}{4}$ servings of A and $\frac{15}{8} = 1\frac{7}{8}$ servings of B.

Section 9.7   Properties of Matrices

1. For the following statement to be true, the value of x must be $\underline{2}$ and the value of y must be $\underline{0}$.
$$\begin{bmatrix} 3 & -6 \\ 5 & 1 \end{bmatrix} = \begin{bmatrix} x+1 & -6 \\ 5 & y+1 \end{bmatrix}$$

3. For the following difference to be true, we must have $w = \underline{6}$, $x = \underline{-1}$, $y = \underline{-1}$, and $z = \underline{12}$.
$$\begin{bmatrix} 7 & 2 \\ 5 & 12 \end{bmatrix} - \begin{bmatrix} 1 & 3 \\ 6 & 0 \end{bmatrix} = \begin{bmatrix} w & x \\ y & z \end{bmatrix}$$

5. If the dimension of matrix A is 3×2 and the dimension of matrix B is 2×6, then the dimension of AB is $\underline{3 \times 6}$.

7. $\begin{bmatrix} -4 & 8 \\ 2 & 3 \end{bmatrix}$

This matrix has 2 rows and 2 columns, so it is a 2×2 square matrix.

9. $\begin{bmatrix} -6 & 8 & 0 & 0 \\ 4 & 1 & 9 & 2 \\ 3 & -5 & 7 & 1 \end{bmatrix}$

This matrix has 3 rows and 4 columns, so it is a 3×4 matrix.

11. $\begin{bmatrix} 2 \\ 4 \end{bmatrix}$

This matrix has 2 rows and 1 column, so it is a 2×1 column matrix.

13. $\begin{bmatrix} -3 & a \\ b & 5 \end{bmatrix} = \begin{bmatrix} c & 0 \\ 4 & d \end{bmatrix}$

Because corresponding elements are equal, we have the following.
$a = 0, b = 4, c = -3, d = 5$

15. $\begin{bmatrix} x+2 & y-6 \\ z-3 & w+5 \end{bmatrix} = \begin{bmatrix} -2 & 8 \\ 0 & 3 \end{bmatrix}$

Because corresponding elements are equal, we have the following.
$x + 2 = -2 \Rightarrow x = -4$
$y - 6 = 8 \Rightarrow y = 14$
$z - 3 = 0 \Rightarrow z = 3$
$w + 5 = 3 \Rightarrow w = -2$

17. Because $\begin{bmatrix} x & y & z \end{bmatrix}$ is a 1×3 matrix and $\begin{bmatrix} 21 & 6 \end{bmatrix}$ is a 1×2 matrix, the statement cannot be true, hence we cannot find values of $x, y,$ and z.

19. $\begin{bmatrix} 0 & 5 & x \\ -1 & 3 & y+2 \\ 4 & 1 & z \end{bmatrix} = \begin{bmatrix} 0 & w & 6 \\ -1 & 3 & 0 \\ 4 & 1 & 8 \end{bmatrix}$

Because corresponding elements are equal, we have the following.
$x = 6$
$y + 2 = 0 \Rightarrow y = -2$
$z = 8$
$w = 5$

21. $\begin{bmatrix} -7+z & 4r & 8s \\ 6p & 2 & 5 \end{bmatrix} + \begin{bmatrix} -9 & 8r & 3 \\ 2 & 5 & 4 \end{bmatrix}$
$= \begin{bmatrix} 2 & 36 & 27 \\ 20 & 7 & 12a \end{bmatrix} \Rightarrow$
$\begin{bmatrix} -16+z & 12r & 8s+3 \\ 6p+2 & 7 & 9 \end{bmatrix} = \begin{bmatrix} 2 & 36 & 27 \\ 20 & 7 & 12a \end{bmatrix}$

Because corresponding elements are equal, we have
$-16 + z = 2 \Rightarrow z = 18, 12r = 36 \Rightarrow r = 3,$
$8s + 3 = 27 \Rightarrow s = 3, 6p + 2 = 20 \Rightarrow p = 3,$
$9 = 12a \Rightarrow a = \frac{3}{4}$

Thus, $z = 18, r = 3, s = 3, p = 3,$ and $a = \frac{3}{4}$.

23. Answers will vary. Sample answer: Be sure that the two matrices have the same dimension. The sum will have this dimension as well. To find the elements of the sum, add the corresponding elements of the two matrices.

25. $\begin{bmatrix} -4 & 3 \\ 12 & -6 \end{bmatrix} + \begin{bmatrix} 2 & -8 \\ 5 & 10 \end{bmatrix} = \begin{bmatrix} -4+2 & 3+(-8) \\ 12+5 & -6+10 \end{bmatrix}$
$= \begin{bmatrix} -2 & -5 \\ 17 & 4 \end{bmatrix}$

27. $\begin{bmatrix} 6 & -9 & 2 \\ 4 & 1 & 3 \end{bmatrix} + \begin{bmatrix} -8 & 2 & 5 \\ 6 & -3 & 4 \end{bmatrix}$
$= \begin{bmatrix} 6+(-8) & -9+2 & 2+5 \\ 4+6 & 1+(-3) & 3+4 \end{bmatrix}$
$= \begin{bmatrix} -2 & -7 & 7 \\ 10 & -2 & 7 \end{bmatrix}$

29. Because $\begin{bmatrix} 2 & 4 & 6 \end{bmatrix}$ is a 1×3 matrix and $\begin{bmatrix} -2 \\ -4 \\ -6 \end{bmatrix}$ is a 3×1 matrix, $\begin{bmatrix} 2 & 4 & 6 \end{bmatrix} + \begin{bmatrix} -2 \\ -4 \\ -6 \end{bmatrix}$ cannot be added.

31. $\begin{bmatrix} -6 & 8 \\ 0 & 0 \end{bmatrix} - \begin{bmatrix} 0 & 0 \\ -4 & -2 \end{bmatrix} = \begin{bmatrix} -6-0 & 8-0 \\ 0-(-4) & 0-(-2) \end{bmatrix}$
$= \begin{bmatrix} -6 & 8 \\ 4 & 2 \end{bmatrix}$

33. $\begin{bmatrix} 12 \\ -1 \\ 3 \end{bmatrix} - \begin{bmatrix} 8 \\ 4 \\ -1 \end{bmatrix} = \begin{bmatrix} 12-8 \\ -1-4 \\ 3-(-1) \end{bmatrix} = \begin{bmatrix} 4 \\ -5 \\ 4 \end{bmatrix}$

35. Because $\begin{bmatrix} -4 & 3 \end{bmatrix}$ is a 1×2 matrix and $\begin{bmatrix} 5 & 8 & 2 \end{bmatrix}$ is a 1×3 matrix, $\begin{bmatrix} -4 & 3 \end{bmatrix} - \begin{bmatrix} 5 & 8 & 2 \end{bmatrix}$ cannot be subtracted.

37. $\begin{bmatrix} \sqrt{3} & -4 \\ 2 & -\sqrt{5} \\ -8 & \sqrt{8} \end{bmatrix} - \begin{bmatrix} 2\sqrt{3} & 9 \\ -2 & \sqrt{5} \\ -7 & 3\sqrt{2} \end{bmatrix}$

$= \begin{bmatrix} \sqrt{3} - 2\sqrt{3} & -4-9 \\ 2-(-2) & -\sqrt{5} - \sqrt{5} \\ -8-(-7) & \sqrt{8} - 3\sqrt{2} \end{bmatrix} = \begin{bmatrix} -\sqrt{3} & -13 \\ 4 & -2\sqrt{5} \\ -1 & -\sqrt{2} \end{bmatrix}$

39. $\begin{bmatrix} 3x+y & -2y \\ x+2y & 3y \end{bmatrix} + \begin{bmatrix} 2x & 3y \\ 5x & x \end{bmatrix}$

$= \begin{bmatrix} (3x+y)+2x & -2y+3y \\ (x+2y)+5x & 3y+x \end{bmatrix}$

$= \begin{bmatrix} 5x+y & y \\ 6x+2y & x+3y \end{bmatrix}$

In Exercises 41–47, $A = \begin{bmatrix} -2 & 4 \\ 0 & 3 \end{bmatrix}$ and $B = \begin{bmatrix} -6 & 2 \\ 4 & 0 \end{bmatrix}$.

41. $2A = 2\begin{bmatrix} -2 & 4 \\ 0 & 3 \end{bmatrix} = \begin{bmatrix} 2(-2) & 2(4) \\ 2(0) & 2(3) \end{bmatrix} = \begin{bmatrix} -4 & 8 \\ 0 & 6 \end{bmatrix}$

43. $\frac{3}{2}B = \frac{3}{2}\begin{bmatrix} -6 & 2 \\ 4 & 0 \end{bmatrix} = \begin{bmatrix} \frac{3}{2}(-6) & \frac{3}{2}(2) \\ \frac{3}{2}(4) & \frac{3}{2}(0) \end{bmatrix} = \begin{bmatrix} -9 & 3 \\ 6 & 0 \end{bmatrix}$

45. $2A - 3B = 2\begin{bmatrix} -2 & 4 \\ 0 & 3 \end{bmatrix} - 3\begin{bmatrix} -6 & 2 \\ 4 & 0 \end{bmatrix}$

$= \begin{bmatrix} -4 & 8 \\ 0 & 6 \end{bmatrix} - \begin{bmatrix} -18 & 6 \\ 12 & 0 \end{bmatrix}$

$= \begin{bmatrix} -4-(-18) & 8-6 \\ 0-12 & 6-0 \end{bmatrix} = \begin{bmatrix} 14 & 2 \\ -12 & 6 \end{bmatrix}$

47. $-A + \frac{1}{2}B = -\begin{bmatrix} -2 & 4 \\ 0 & 3 \end{bmatrix} + \frac{1}{2}\begin{bmatrix} -6 & 2 \\ 4 & 0 \end{bmatrix}$

$= \begin{bmatrix} 2 & -4 \\ 0 & -3 \end{bmatrix} + \begin{bmatrix} -3 & 1 \\ 2 & 0 \end{bmatrix}$

$= \begin{bmatrix} 2+(-3) & -4+1 \\ 0+2 & -3+0 \end{bmatrix} = \begin{bmatrix} -1 & -3 \\ 2 & -3 \end{bmatrix}$

49. AB can be calculated and the result will be a 2×5 matrix.

Matrix A Matrix B
2×3 3×5
matches
size of AB
2×5

51. BA cannot be calculated.

Matrix B Matrix A
3×5 2×3
different

53. BC can be calculated and the result will be a 3×2 matrix.

Matrix B Matrix C
3×5 5×2
matches
size of CA
3×2

55. A 2×2 matrix multiplied by a 2×1 matrix results in a 2×1 matrix.

$\begin{bmatrix} 1 & 2 \\ 3 & 4 \end{bmatrix}\begin{bmatrix} -1 \\ 7 \end{bmatrix} = \begin{bmatrix} 1(-1)+2(7) \\ 3(-1)+4(7) \end{bmatrix}$

$= \begin{bmatrix} -1+14 \\ -3+28 \end{bmatrix} = \begin{bmatrix} 13 \\ 25 \end{bmatrix}$

57. A 2×3 matrix multiplied by a 3×1 matrix results in a 2×1 matrix.

$\begin{bmatrix} 3 & -4 & 1 \\ 5 & 0 & 2 \end{bmatrix}\begin{bmatrix} -1 \\ 4 \\ 2 \end{bmatrix} = \begin{bmatrix} 3(-1)+(-4)(4)+1(2) \\ 5(-1)+0(4)+2(2) \end{bmatrix}$

$= \begin{bmatrix} -3+(-16)+2 \\ -5+0+4 \end{bmatrix} = \begin{bmatrix} -17 \\ -1 \end{bmatrix}$

59. A 2×3 matrix multiplied by a 3×2 matrix results in a 2×2 matrix.

$$\begin{bmatrix} \sqrt{2} & \sqrt{2} & -\sqrt{18} \\ \sqrt{3} & \sqrt{27} & 0 \end{bmatrix} \begin{bmatrix} 8 & -10 \\ 9 & 12 \\ 0 & 2 \end{bmatrix} = \begin{bmatrix} \sqrt{2} & \sqrt{2} & -3\sqrt{2} \\ \sqrt{3} & 3\sqrt{3} & 0 \end{bmatrix} \begin{bmatrix} 8 & -10 \\ 9 & 12 \\ 0 & 2 \end{bmatrix}$$

$$= \begin{bmatrix} 8\sqrt{2} + 9\sqrt{2} & -10\sqrt{2} + 12\sqrt{2} \\ \quad -3\sqrt{2}(0) & -3\sqrt{2}(2) \\ 8\sqrt{3} + 3\sqrt{3}(9) & -10\sqrt{3} + 3\sqrt{3}(12) \\ \quad + 0(0) & + 0(2) \end{bmatrix} = \begin{bmatrix} 17\sqrt{2} & -4\sqrt{2} \\ 35\sqrt{3} & 26\sqrt{3} \end{bmatrix}$$

61. A 2×2 matrix multiplied by a 2×2 matrix results in a 2×2 matrix.

$$\begin{bmatrix} \sqrt{3} & 1 \\ 2\sqrt{5} & 3\sqrt{2} \end{bmatrix} \begin{bmatrix} \sqrt{3} & -\sqrt{6} \\ 4\sqrt{3} & 0 \end{bmatrix} = \begin{bmatrix} \sqrt{3}(\sqrt{3}) + 1(4\sqrt{3}) & \sqrt{3}(-\sqrt{6}) + 1(0) \\ 2\sqrt{5}(\sqrt{3}) + 3\sqrt{2}(4\sqrt{3}) & 2\sqrt{5}(-\sqrt{6}) + 3\sqrt{2}(0) \end{bmatrix}$$

$$= \begin{bmatrix} 3 + 4\sqrt{3} & -\sqrt{18} \\ 2\sqrt{15} + 12\sqrt{6} & -2\sqrt{30} \end{bmatrix} = \begin{bmatrix} 3 + 4\sqrt{3} & -3\sqrt{2} \\ 2\sqrt{15} + 12\sqrt{6} & -2\sqrt{30} \end{bmatrix}$$

63. $\begin{bmatrix} -3 & 0 & 2 & 1 \\ 4 & 0 & 2 & 6 \end{bmatrix} \begin{bmatrix} -4 & 2 \\ 0 & 1 \end{bmatrix}$

A 2×4 matrix cannot be multiplied by a 2×2 matrix because the number of columns of the first matrix (four) is not equal to the number of rows of the second matrix (two).

65. A 1×3 matrix multiplied by a 3×3 matrix results in a 1×3 matrix.

$$\begin{bmatrix} -2 & 4 & 1 \end{bmatrix} \begin{bmatrix} 3 & -2 & 4 \\ 2 & 1 & 0 \\ 0 & -1 & 4 \end{bmatrix} = \begin{bmatrix} -2(3) + 4(2) + 1(0) & -2(-2) + 4(1) + 1(-1) & -2(4) + 4(0) + 1(4) \end{bmatrix}$$

$$= \begin{bmatrix} -6 + 8 + 0 & 4 + 4 + (-1) & -8 + 0 + 4 \end{bmatrix} = \begin{bmatrix} 2 & 7 & -4 \end{bmatrix}$$

67. A 3×3 matrix multiplied by a 3×3 matrix results in a 3×3 matrix.

$$\begin{bmatrix} -2 & -3 & -4 \\ 2 & -1 & 0 \\ 4 & -2 & 3 \end{bmatrix} \begin{bmatrix} 0 & 1 & 4 \\ 1 & 2 & -1 \\ 3 & 2 & -2 \end{bmatrix}$$

$$= \begin{bmatrix} -2(0) + (-3)(1) + (-4)(3) & -2(1) + (-3)(2) + (-4)(2) & -2(4) + (-3)(-1) + (-4)(-2) \\ 2(0) + (-1)(1) + 0(3) & 2(1) + (-1)(2) + 0(2) & 2(4) + (-1)(-1) + 0(-2) \\ 4(0) + (-2)(1) + 3(3) & 4(1) + (-2)(2) + 3(2) & 4(4) + (-2)(-1) + 3(-2) \end{bmatrix}$$

$$= \begin{bmatrix} 0 + (-3) + (-12) & -2 + (-6) + (-8) & -8 + 3 + 8 \\ 0 + (-1) + 0 & 2 + (-2) + 0 & 8 + 1 + 0 \\ 0 + (-2) + 9 & 4 + (-4) + 6 & 16 + 2 + (-6) \end{bmatrix} = \begin{bmatrix} -15 & -16 & 3 \\ -1 & 0 & 9 \\ 7 & 6 & 12 \end{bmatrix}$$

In Exercises 69–75, $A = \begin{bmatrix} 4 & -2 \\ 3 & 1 \end{bmatrix}$, $B = \begin{bmatrix} 5 & 1 \\ 0 & -2 \\ 3 & 7 \end{bmatrix}$ and $C = \begin{bmatrix} -5 & 4 & 1 \\ 0 & 3 & 6 \end{bmatrix}$.

69. A 3×2 matrix multiplied by a 2×2 matrix results in a 3×2 matrix.

$$BA = \begin{bmatrix} 5 & 1 \\ 0 & -2 \\ 3 & 7 \end{bmatrix} \begin{bmatrix} 4 & -2 \\ 3 & 1 \end{bmatrix} = \begin{bmatrix} 5(4) + 1(3) & 5(-2) + 1(1) \\ 0(4) + (-2)(3) & 0(-2) + (-2)(1) \\ 3(4) + 7(3) & 3(-2) + 7(1) \end{bmatrix} = \begin{bmatrix} 20 + 3 & -10 + 1 \\ 0 + (-6) & 0 + (-2) \\ 12 + 21 & -6 + 7 \end{bmatrix} = \begin{bmatrix} 23 & -9 \\ -6 & -2 \\ 33 & 1 \end{bmatrix}$$

71. A 3×2 matrix multiplied by a 2×3 matrix results in a 3×3 matrix.

$$BC = \begin{bmatrix} 5 & 1 \\ 0 & -2 \\ 3 & 7 \end{bmatrix} \begin{bmatrix} -5 & 4 & 1 \\ 0 & 3 & 6 \end{bmatrix} = \begin{bmatrix} 5(-5)+1(0) & 5(4)+1(3) & 5(1)+1(6) \\ 0(-5)+(-2)(0) & 0(4)+(-2)(3) & 0(1)+(-2)(6) \\ 3(-5)+7(0) & 3(4)+7(3) & 3(1)+7(6) \end{bmatrix}$$

$$= \begin{bmatrix} -25+0 & 20+3 & 5+6 \\ 0+0 & 0+(-6) & 0+(-12) \\ -15+0 & 12+21 & 3+42 \end{bmatrix} = \begin{bmatrix} -25 & 23 & 11 \\ 0 & -6 & -12 \\ -15 & 33 & 45 \end{bmatrix}$$

73. $AB = \begin{bmatrix} 4 & -2 \\ 3 & 1 \end{bmatrix} \begin{bmatrix} 5 & 1 \\ 0 & -2 \\ 3 & 7 \end{bmatrix}$ is the product of a 2×2 matrix multiplied by a 3×2 matrix, which is not possible.

75. Because $A^2 = AA$, we are finding the product of two 2×2 matrices, which results in a 2×2 matrix.

$$A^2 = \begin{bmatrix} 4 & -2 \\ 3 & 1 \end{bmatrix} \begin{bmatrix} 4 & -2 \\ 3 & 1 \end{bmatrix} = \begin{bmatrix} 4(4)+(-2)(3) & 4(-2)+(-2)(1) \\ 3(4)+1(3) & 3(-2)+1(1) \end{bmatrix} = \begin{bmatrix} 16+(-6) & -8+(-2) \\ 12+3 & -6+1 \end{bmatrix} = \begin{bmatrix} 10 & -10 \\ 15 & -5 \end{bmatrix}$$

77. The answers to 69 and 73 are not equal, so $BA \neq AB$.
The answers to 71 and 72 are not equal, so $BC \neq CB$.
The answers to 70 and 74 are not equal, so $AC \neq CA$.

79. $A = \begin{bmatrix} 3 & 4 \\ -2 & 1 \end{bmatrix}$ and $B = \begin{bmatrix} 6 & 0 \\ 5 & -2 \end{bmatrix}$

(a) $AB = \begin{bmatrix} 3 & 4 \\ -2 & 1 \end{bmatrix} \begin{bmatrix} 6 & 0 \\ 5 & -2 \end{bmatrix} = \begin{bmatrix} 3(6)+4(5) & 3(0)+4(-2) \\ -2(6)+1(5) & -2(0)+1(-2) \end{bmatrix} = \begin{bmatrix} 18+20 & 0+(-8) \\ -12+5 & 0+(-2) \end{bmatrix} = \begin{bmatrix} 38 & -8 \\ -7 & -2 \end{bmatrix}$

(b) $BA = \begin{bmatrix} 6 & 0 \\ 5 & -2 \end{bmatrix} \begin{bmatrix} 3 & 4 \\ -2 & 1 \end{bmatrix} = \begin{bmatrix} 6(3)+0(-2) & 6(4)+0(1) \\ 5(3)+(-2)(-2) & 5(4)+(-2)(1) \end{bmatrix} = \begin{bmatrix} 18+0 & 24+0 \\ 15+4 & 20+(-2) \end{bmatrix} = \begin{bmatrix} 18 & 24 \\ 19 & 18 \end{bmatrix}$

Note: $AB \neq BA$

81. $A = \begin{bmatrix} 0 & 1 & -1 \\ 0 & 1 & 0 \\ 0 & 0 & 1 \end{bmatrix}$ and $B = \begin{bmatrix} 1 & 0 & 0 \\ 0 & 1 & 0 \\ 0 & 0 & 1 \end{bmatrix}$

(a) $AB = \begin{bmatrix} 0 & 1 & -1 \\ 0 & 1 & 0 \\ 0 & 0 & 1 \end{bmatrix} \begin{bmatrix} 1 & 0 & 0 \\ 0 & 1 & 0 \\ 0 & 0 & 1 \end{bmatrix} = \begin{bmatrix} 0(1)+1(0)+(-1)(0) & 0(0)+1(1)+(-1)(0) & 0(0)+1(0)+(-1)(1) \\ 0(1)+1(0)+0(0) & 0(0)+1(1)+0(0) & 0(0)+1(0)+0(1) \\ 0(1)+0(0)+1(0) & 0(0)+0(1)+1(0) & 0(0)+0(0)+1(1) \end{bmatrix}$

$$= \begin{bmatrix} 0+0+0 & 0+1+0 & 0+0+(-1) \\ 0+0+0 & 0+1+0 & 0+0+0 \\ 0+0+0 & 0+0+0 & 0+0+1 \end{bmatrix} = \begin{bmatrix} 0 & 1 & -1 \\ 0 & 1 & 0 \\ 0 & 0 & 1 \end{bmatrix}$$

(b) $BA = \begin{bmatrix} 1 & 0 & 0 \\ 0 & 1 & 0 \\ 0 & 0 & 1 \end{bmatrix} \begin{bmatrix} 0 & 1 & -1 \\ 0 & 1 & 0 \\ 0 & 0 & 1 \end{bmatrix} = \begin{bmatrix} 1(0)+0(0)+0(0) & 1(1)+0(1)+0(0) & 1(-1)+0(0)+0(1) \\ 0(0)+1(0)+0(0) & 0(1)+1(1)+0(0) & 0(-1)+1(0)+0(1) \\ 0(0)+0(0)+1(0) & 0(1)+0(1)+1(0) & 0(-1)+0(0)+1(1) \end{bmatrix}$

$$= \begin{bmatrix} 0+0+0 & 1+0+0 & -1+0+0 \\ 0+0+0 & 0+1+0 & 0+0+0 \\ 0+0+0 & 0+0+0 & 0+0+1 \end{bmatrix} = \begin{bmatrix} 0 & 1 & -1 \\ 0 & 1 & 0 \\ 0 & 0 & 1 \end{bmatrix}$$

Note: $AB = BA = A$

83. In Exercise 81, $AB = A$ and $BA = A$. For this pair of matrices, B acts in the same way for multiplication as the number $\underline{1}$ acts for multiplication of real numbers.

85. (a) The sales figure information may be written as the 3×3 matrix.

$$
\begin{array}{c}
\\
\text{Location I} \\
\text{Location II} \\
\text{Location III}
\end{array}
\begin{array}{ccc}
\text{nonfat} & \text{regular} & \text{super creamy} \\
\end{array}
\begin{bmatrix}
50 & 100 & 30 \\
10 & 90 & 50 \\
60 & 120 & 40
\end{bmatrix}
$$

(b) The income per gallon information may be written as the 3×1 matrix

$$
\begin{array}{c}
\text{nonfat} \\
\text{regular} \\
\text{super creamy}
\end{array}
\begin{bmatrix}
12 \\
10 \\
15
\end{bmatrix}.
$$

Note: If the matrix in part (a) had been written with its rows and columns interchanged, then this income per gallon information would be written instead as a 1×3 matrix.

(c)
$$
\begin{bmatrix}
50 & 100 & 30 \\
10 & 90 & 50 \\
60 & 120 & 40
\end{bmatrix}
\begin{bmatrix}
12 \\
10 \\
15
\end{bmatrix}
=
\begin{bmatrix}
2050 \\
1770 \\
2520
\end{bmatrix}
$$

Note: This result may be written as a 1×3 matrix instead.

(d) $2050 + 1770 + 2520 = 6340$; The total daily income from the three locations is $6340.

87. Answers will vary depending on when numbers are rounded.

(a) $j_1 = 690, s_p = 210, a_1 = 2100$

1st year:
$$j_2 = 0.33a_1 = 0.33(2100) = 693$$
$$s_2 = 0.18j_1 = 0.18(690)$$
$$\approx 124.2 \approx 124$$
$$a_2 = 0.71s_1 + 0.94a_1$$
$$= 0.71(210) + 0.94(2100)$$
$$\approx 2123.1 \approx 2123$$
$$j_2 + s_2 + a_2 = 693 + 124 + 2123 = 2940$$

2nd year:
$$j_3 = 0.33a_2 = 0.33(2123) \approx 700.6$$
$$s_3 = 0.18j_2 = 0.18(693) \approx 124.7$$

$$a_3 = 0.71s_2 + 0.94a_2$$
$$= 0.71(124) + 0.94(2123)$$
$$\approx 2083.7$$
$$j_3 + s_3 + a_3 = 700.6 + 124.7 + 2083.7$$
$$= 2909$$

3rd year:
$$j_4 = 0.33a_3 = 0.33(2084) \approx 687$$
$$s_4 = 0.18j_3 = 0.18(700) = 126$$
$$a_4 = 0.72s_3 + 0.94a_3$$
$$= 0.71(125) + 0.94(2084)$$
$$= 2048$$
$$j_3 + s_3 + a_3 = 687 + 126 + 2048 = 2861$$

4th year:
$$j_5 = 0.33a_4 = 0.33(2048) \approx 676$$
$$s_5 = 0.18j_4 = 0.18(687) \approx 124$$
$$a_5 = 0.72s_4 + 0.94a_4$$
$$= 0.71(126) + 0.94(2048)$$
$$= 2014$$
$$j_5 + s_5 + a_5 = 686 + 124 + 2048 = 2814$$

5th year:
$$j_6 = 0.33a_5 = 0.33(2014) \approx 664$$
$$s_6 = 0.18j_5 = 0.18(676) \approx 122$$
$$a_6 = 0.72s_5 + 0.94a_5$$
$$= 0.71(124) + 0.94(2014)$$
$$\approx 1981$$
$$j_6 + s_6 + a_6 = 664 + 122 + 1981 = 2767$$

(b) The northern spotted owl will become extinct.

(c) $j_1 = 690, s_1 = 210, a_1 = 2100$

1st year:
$$j_2 = 0.33a_1 = 0.33(2100) = 693$$
$$s_2 = 0.3j_1 = 0.3(690) = 207$$
$$a_2 = 0.71s_1 + 0.94a_1$$
$$= 0.71(210) + 0.94(2100)$$
$$\approx 2123$$
$$j_2 + s_2 + a_2 = 693 + 207 + 2123 = 3023$$

2nd year:
$$j_3 = 0.33a_2 = 0.33(2123) \approx 701$$
$$s_3 = 0.3j_2 = 0.3(693) \approx 208$$
$$a_3 = 0.71s_2 + 0.94a_2$$
$$= 0.71(207) + 0.94(2123)$$
$$\approx 2143$$
$$j_3 + s_3 + a_3 = 701 + 208 + 2143 = 3052$$

(continued on next page)

3rd year: $j_4 = 0.33a_3 = 0.33(2143) \approx 707$

$s_4 = 0.3 j_3 = 0.3(701) = 210$

$a_4 = 0.71s_3 + 0.94a_3 = 0.71(208) + 0.94(2143) \approx 2162$

$j_4 + s_4 + a_4 = 707 + 210 + 2162 = 3079$

4th year: $j_5 = 0.33a_4 = 0.33(2162) \approx 714$

$s_5 = 0.3 j_4 = 0.3(707) \approx 212$

$a_5 = 0.72s_4 + 0.94a_4 = 0.71(210) + 0.94(2162) \approx 2181$

$j_5 + s_5 + a_5 = 714 + 212 + 2181 = 3107$

5th year: $j_6 = 0.71s_5 = 0.33a_5 = 0.33(2181) \approx 720$

$s_6 = 0.3 j_5 = 0.3(714) \approx 214$

$a_6 = 0.71s_5 + 0.94a_5 = 0.71(212) + 0.94(2181) = 2201$

$j_6 + s_6 + a_6 = 720 + 214 + 2201 = 3135$

89. Expanding $\begin{vmatrix} -x & 0 & 0.33 \\ 0.18 & -x & 0 \\ 0 & 0.71 & 0.94-x \end{vmatrix}$ by row one, we have the following.

$$-x\begin{vmatrix} -x & 0 \\ 0.71 & 0.94-x \end{vmatrix} - 0 + 0.33\begin{vmatrix} 0.18 & -x \\ 0 & 0.71 \end{vmatrix} = -x\left[-x(0.94-x) - 0.71(0)\right] + 0.33\left[(0.18)(0.71) - 0(-x)\right]$$

$$= -x\left(-0.94x + x^2 - 0\right) + 0.33(0.1278 - 0) = 0.94x^2 - x^3 + 0.042174$$

Thus, the polynomial is $-x^3 + 0.94x^2 + 0.042174$. Evaluating this polynomial with $x = 0.98359$, we have

$-(0.98359)^3 + 0.94(0.98359)^2 + 0.042174 \approx 0.0000029$. Thus 0.98359 is an approximate zero.

In Exercises 91–97, $A = \begin{bmatrix} a_{11} & a_{12} \\ a_{21} & a_{22} \end{bmatrix}$, $B = \begin{bmatrix} b_{11} & b_{12} \\ b_{21} & b_{22} \end{bmatrix}$, and $C = \begin{bmatrix} c_{11} & c_{12} \\ c_{21} & c_{22} \end{bmatrix}$.

91. $A + B = B + A$

$$A + B = \begin{bmatrix} a_{11} & a_{12} \\ a_{21} & a_{22} \end{bmatrix} + \begin{bmatrix} b_{11} & b_{12} \\ b_{21} & b_{22} \end{bmatrix} = \begin{bmatrix} a_{11}+b_{11} & a_{12}+b_{12} \\ a_{21}+b_{21} & a_{22}+b_{22} \end{bmatrix} = \begin{bmatrix} b_{11}+a_{11} & b_{12}+a_{12} \\ b_{21}+a_{21} & b_{22}+a_{22} \end{bmatrix} = B + A$$

93. $(AB)C = \left(\begin{bmatrix} a_{11} & a_{12} \\ a_{21} & a_{22} \end{bmatrix}\begin{bmatrix} b_{11} & b_{12} \\ b_{21} & b_{22} \end{bmatrix}\right)C = \begin{bmatrix} a_{11}b_{11}+a_{12}b_{21} & a_{11}b_{12}+a_{12}b_{22} \\ a_{21}b_{11}+a_{22}b_{21} & a_{21}b_{12}+a_{22}b_{22} \end{bmatrix}\begin{bmatrix} c_{11} & c_{12} \\ c_{21} & c_{22} \end{bmatrix}$

$$= \begin{bmatrix} (a_{11}b_{11}+a_{12}b_{21})c_{11} + (a_{11}b_{12}+a_{12}b_{22})c_{21} & (a_{11}b_{11}+a_{12}b_{21})c_{12} + (a_{11}b_{12}+a_{12}b_{22})c_{22} \\ (a_{21}b_{11}+a_{22}b_{21})c_{11} + (a_{21}b_{12}+a_{22}b_{22})c_{21} & (a_{21}b_{11}+a_{22}b_{21})c_{12} + (a_{21}b_{12}+a_{22}b_{22})c_{22} \end{bmatrix}$$

$$= \begin{bmatrix} a_{11}b_{11}c_{11}+a_{12}b_{21}c_{11}+a_{11}b_{12}c_{21}+a_{12}b_{22}c_{21} & a_{11}b_{11}c_{12}+a_{12}b_{21}c_{12}+a_{11}b_{12}c_{22}+a_{12}b_{22}c_{22} \\ a_{21}b_{11}c_{11}+a_{22}b_{21}c_{11}+a_{21}b_{12}c_{21}+a_{22}b_{22}c_{21} & a_{21}b_{11}c_{12}+a_{22}b_{21}c_{12}+a_{21}b_{12}c_{22}+a_{22}b_{22}c_{22} \end{bmatrix}$$

$A(BC) = A\left(\begin{bmatrix} b_{11} & b_{12} \\ b_{21} & b_{22} \end{bmatrix}\begin{bmatrix} c_{11} & c_{12} \\ c_{21} & c_{22} \end{bmatrix}\right) = \begin{bmatrix} a_{11} & a_{12} \\ a_{21} & a_{22} \end{bmatrix}\begin{bmatrix} b_{11}c_{11}+b_{12}c_{21} & b_{11}c_{12}+b_{12}c_{22} \\ b_{21}c_{11}+b_{22}c_{21} & b_{21}c_{12}+b_{22}c_{22} \end{bmatrix}$

$$= \begin{bmatrix} a_{11}(b_{11}c_{11}+b_{12}c_{21})+a_{12}(b_{21}c_{11}+b_{22}c_{21}) & a_{11}(b_{11}c_{12}+b_{12}c_{22})+a_{12}(b_{21}c_{12}+b_{22}c_{22}) \\ a_{21}(b_{11}c_{11}+b_{12}c_{21})+a_{22}(b_{21}c_{11}+b_{22}c_{21}) & a_{21}(b_{11}c_{12}+b_{12}c_{22})+a_{22}(b_{21}c_{12}+b_{22}c_{22}) \end{bmatrix}$$

$$= \begin{bmatrix} a_{11}b_{11}c_{11}+a_{11}b_{12}c_{21}+a_{12}b_{21}c_{11}+a_{12}b_{22}c_{21} & a_{11}b_{11}c_{12}+a_{11}b_{12}c_{22}+a_{12}b_{21}c_{12}+a_{12}b_{22}c_{22} \\ a_{21}b_{11}c_{11}+a_{21}b_{12}c_{21}+a_{22}b_{21}c_{11}+a_{22}b_{22}c_{21} & a_{21}b_{11}c_{12}+a_{21}b_{12}c_{22}+a_{22}b_{21}c_{12}+a_{22}b_{22}c_{22} \end{bmatrix}$$

$$= \begin{bmatrix} a_{11}b_{11}c_{11}+a_{12}b_{21}c_{11}+a_{11}b_{12}c_{21}+a_{12}b_{22}c_{21} & a_{11}b_{11}c_{12}+a_{12}b_{21}c_{12}+a_{11}b_{12}c_{22}+a_{12}b_{22}c_{22} \\ a_{21}b_{11}c_{11}+a_{22}b_{21}c_{11}+a_{21}b_{12}c_{21}+a_{22}b_{22}c_{21} & a_{21}b_{11}c_{12}+a_{22}b_{21}c_{12}+a_{21}b_{12}c_{22}+a_{22}b_{22}c_{22} \end{bmatrix}$$

The final matrix for $(AB)C$ and $A(BC)$ are the same, so we have obtained the desired results.

95. $c(A+B) = cA + cB$, for any real number c.

$$c(A+B) = c\left(\begin{bmatrix} a_{11} & a_{12} \\ a_{21} & a_{22} \end{bmatrix} + \begin{bmatrix} b_{11} & b_{12} \\ b_{21} & b_{22} \end{bmatrix}\right) = c\begin{bmatrix} a_{11}+b_{11} & a_{12}+b_{12} \\ a_{21}+b_{21} & a_{22}+b_{22} \end{bmatrix} = \begin{bmatrix} c(a_{11}+b_{11}) & c(a_{12}+b_{12}) \\ c(a_{21}+b_{21}) & c(a_{22}+b_{22}) \end{bmatrix}$$

$$= \begin{bmatrix} c \cdot a_{11}+c \cdot b_{11} & c \cdot a_{12}+c \cdot b_{12} \\ c \cdot a_{21}+c \cdot b_{21} & c \cdot a_{22}+c \cdot b_{22} \end{bmatrix} = \begin{bmatrix} c \cdot a_{11} & c \cdot a_{12} \\ c \cdot a_{21} & c \cdot a_{22} \end{bmatrix} + \begin{bmatrix} c \cdot b_{11} & c \cdot b_{12} \\ c \cdot b_{21} & c \cdot b_{22} \end{bmatrix}$$

$$= c\begin{bmatrix} a_{11} & a_{12} \\ a_{21} & a_{22} \end{bmatrix} + c\begin{bmatrix} b_{11} & b_{12} \\ b_{21} & b_{22} \end{bmatrix} = cA + cB$$

97. $c(A)d = (cd)A$, for any real numbers c and d.

$$c(A)d = \left(c\begin{bmatrix} a_{11} & a_{12} \\ a_{21} & a_{22} \end{bmatrix}\right)d = \begin{bmatrix} c \cdot a_{11} & c \cdot a_{12} \\ c \cdot a_{21} & c \cdot a_{22} \end{bmatrix}d = \begin{bmatrix} c \cdot a_{11} \cdot d & c \cdot a_{12} \cdot d \\ c \cdot a_{21} \cdot d & c \cdot a_{22} \cdot d \end{bmatrix} = \begin{bmatrix} c \cdot d \cdot a_{11} & c \cdot d \cdot a_{12} \\ c \cdot d \cdot a_{21} & c \cdot d \cdot a_{22} \end{bmatrix}$$

$$= \begin{bmatrix} (cd) \cdot a_{11} & (cd) \cdot a_{12} \\ (cd) \cdot a_{21} & (cd) \cdot a_{22} \end{bmatrix} = (cd)\begin{bmatrix} a_{11} & a_{12} \\ a_{21} & a_{22} \end{bmatrix} = (cd)A$$

Section 9.8   Matrix Inverses

1. The product of $\begin{bmatrix} 6 & 4 \\ -1 & 8 \end{bmatrix}$ and I_2 (in either order) is $\begin{bmatrix} 6 & 4 \\ -1 & 8 \end{bmatrix}$.

3. Because the matrices are inverses, their product is $\begin{bmatrix} 1 & 0 \\ 0 & 1 \end{bmatrix}$, or I_2.

5. The coefficient matrix of the system $\begin{array}{l} 3x-6y = 8 \\ -x+3y = 4 \end{array}$ is $\begin{bmatrix} 3 & -6 \\ -1 & 3 \end{bmatrix}$.

7. $A = \begin{bmatrix} -2 & 4 & 0 \\ 3 & 5 & 9 \\ 0 & 8 & -6 \end{bmatrix}$ and $I_3 = \begin{bmatrix} 1 & 0 & 0 \\ 0 & 1 & 0 \\ 0 & 0 & 1 \end{bmatrix}$

$$I_3 A = \begin{bmatrix} 1 & 0 & 0 \\ 0 & 1 & 0 \\ 0 & 0 & 1 \end{bmatrix}\begin{bmatrix} -2 & 4 & 0 \\ 3 & 5 & 9 \\ 0 & 8 & -6 \end{bmatrix} = \begin{bmatrix} 1(-2)+0(3)+0(0) & 1(4)+0(5)+0(8) & 1(0)+0(9)+0(-6) \\ 0(-2)+1(3)+0(0) & 0(4)+1(5)+0(8) & 0(0)+1(9)+0(-6) \\ 0(-2)+0(3)+1(0) & 0(4)+0(5)+1(8) & 0(0)+0(9)+1(-6) \end{bmatrix}$$

$$= \begin{bmatrix} -2+0+0 & 4+0+0 & 0+0+0 \\ 0+3+0 & 0+5+0 & 0+9+0 \\ 0+0+0 & 0+0+8 & 0+0+(-6) \end{bmatrix} = \begin{bmatrix} -2 & 4 & 0 \\ 3 & 5 & 9 \\ 0 & 8 & -6 \end{bmatrix}$$

9. $\begin{bmatrix} 5 & 7 \\ 2 & 3 \end{bmatrix} \begin{bmatrix} 3 & -7 \\ -2 & 5 \end{bmatrix} = \begin{bmatrix} 5(3)+7(-2) & 5(-7)+7(5) \\ 2(3)+3(-2) & 2(-7)+3(5) \end{bmatrix} = \begin{bmatrix} 15+(-14) & -35+35 \\ 6+(-6) & (-14)+15 \end{bmatrix} = \begin{bmatrix} 1 & 0 \\ 0 & 1 \end{bmatrix}$

$\begin{bmatrix} 3 & -7 \\ -2 & 5 \end{bmatrix} \begin{bmatrix} 5 & 7 \\ 2 & 3 \end{bmatrix} = \begin{bmatrix} 3(5)+(-7)(2) & 3(7)+(-7)(3) \\ (-2)(5)+5(2) & (-2)(7)+5(3) \end{bmatrix} = \begin{bmatrix} 15+(-14) & 21+(-21) \\ (-10)+10 & (-14)+15 \end{bmatrix} = \begin{bmatrix} 1 & 0 \\ 0 & 1 \end{bmatrix}$

The products obtained by multiplying the matrices in either order are both the 2×2 identity matrix, so the given matrices are inverses of each other.

11. $\begin{bmatrix} -1 & 2 \\ 3 & -5 \end{bmatrix} \begin{bmatrix} -5 & -2 \\ -3 & -1 \end{bmatrix} = \begin{bmatrix} -1(-5)+2(-3) & -1(-2)+2(-1) \\ 3(-5)+(-5)(-3) & 3(-2)+(-5)(-1) \end{bmatrix} = \begin{bmatrix} 5+(-6) & 2+(-2) \\ -15+15 & -6+5 \end{bmatrix} = \begin{bmatrix} -1 & 0 \\ 0 & -1 \end{bmatrix}$

This product is not the 2×2 identity matrix, so the given matrices are not inverses of each other.

13. $\begin{bmatrix} 0 & 1 & 0 \\ 0 & 0 & -2 \\ 1 & -1 & 0 \end{bmatrix} \begin{bmatrix} 1 & 0 & 1 \\ 1 & 0 & 0 \\ 0 & -1 & 0 \end{bmatrix} = \begin{bmatrix} 0(1)+1(1)+0(0) & 0(0)+1(0)+0(-1) & 0(1)+1(0)+0(0) \\ 0(1)+0(1)+(-2)(0) & 0(0)+0(0)+(-2)(-1) & 0(1)+0(0)+(-2)(0) \\ 1(1)+(-1)(1)+0(0) & 1(0)+(-1)(0)+0(-1) & 1(1)+(-1)(0)+0(0) \end{bmatrix}$

$= \begin{bmatrix} 0+1+0 & 0+0+0 & 0+0+0 \\ 0+0+0 & 0+0+2 & 0+0+0 \\ 1+(-1)+0 & 0+0+0 & 1+0+0 \end{bmatrix} = \begin{bmatrix} 1 & 0 & 0 \\ 0 & 2 & 0 \\ 0 & 0 & 1 \end{bmatrix}$

This product is not the 3×3 identity matrix, so the given matrices are not inverses of each other.

15. $\begin{bmatrix} 1 & 0 & 0 \\ 0 & -1 & 0 \\ 1 & 0 & 1 \end{bmatrix} \begin{bmatrix} 1 & 0 & 0 \\ 0 & -1 & 0 \\ -1 & 0 & 1 \end{bmatrix}$

$= \begin{bmatrix} 1(1)+0(0)+0(-1) & 1(0)+0(-1)+0(0) & 1(0)+0(0)+0(1) \\ 0(1)-1(0)+0(-1) & 0(0)-1(-1)+0(0) & 0(0)-1(0)+0(1) \\ 1(1)+0(0)+1(-1) & 1(0)+0(-1)+1(0) & 1(0)+0(0)+1(1) \end{bmatrix} = \begin{bmatrix} 1 & 0 & 0 \\ 0 & 1 & 0 \\ 0 & 0 & 1 \end{bmatrix} = I_3$

$\begin{bmatrix} 1 & 0 & 0 \\ 0 & -1 & 0 \\ -1 & 0 & 1 \end{bmatrix} \begin{bmatrix} 1 & 0 & 0 \\ 0 & -1 & 0 \\ 1 & 0 & 1 \end{bmatrix}$

$= \begin{bmatrix} 1(1)+0(0)+0(1) & 1(0)+0(-1)+0(0) & 1(0)+0(0)+0(1) \\ 0(1)-1(0)+0(1) & 0(0)-1(-1)+0(0) & 0(0)-1(0)+0(1) \\ -1(1)+0(0)+1(1) & -1(0)+0(-1)+1(0) & -1(0)+0(0)+1(1) \end{bmatrix} = \begin{bmatrix} 1 & 0 & 0 \\ 0 & 1 & 0 \\ 0 & 0 & 1 \end{bmatrix} = I_3$

The given matrices are inverses of each other.

17. Find the inverse of $A = \begin{bmatrix} -1 & 2 \\ -2 & -1 \end{bmatrix}$, if it exists. Because $[A \mid I_2] = \begin{bmatrix} -1 & 2 & 1 & 0 \\ -2 & -1 & 0 & 1 \end{bmatrix}$, we have

$\begin{bmatrix} 1 & -2 & -1 & 0 \\ -2 & -1 & 0 & 1 \end{bmatrix} \begin{matrix} -1R1 \\ \\ \end{matrix} \Rightarrow \begin{bmatrix} 1 & -2 & -1 & 0 \\ 0 & -5 & -2 & 1 \end{bmatrix} \begin{matrix} \\ 2R1+R2 \end{matrix} \Rightarrow$

$\begin{bmatrix} 1 & -2 & -1 & 0 \\ 0 & 1 & \frac{2}{5} & -\frac{1}{5} \end{bmatrix} \begin{matrix} \\ -\frac{1}{5}R2 \end{matrix} \Rightarrow \begin{bmatrix} 1 & 0 & -\frac{1}{5} & -\frac{2}{5} \\ 0 & 1 & \frac{2}{5} & -\frac{1}{5} \end{bmatrix} 2R2+R1$

Thus, $A^{-1} = \begin{bmatrix} -\frac{1}{5} & -\frac{2}{5} \\ \frac{2}{5} & -\frac{1}{5} \end{bmatrix}$.

19. Find the inverse of $A = \begin{bmatrix} -1 & -2 \\ 3 & 4 \end{bmatrix}$, if it exists. Because $\left[A \mid I_2\right] = \begin{bmatrix} -1 & -2 & 1 & 0 \\ 3 & 4 & 0 & 1 \end{bmatrix}$, we have

$$\begin{bmatrix} -1 & -2 & 1 & 0 \\ 0 & -2 & 3 & 1 \end{bmatrix} 3R1+R2 \Rightarrow \begin{bmatrix} 1 & 2 & -1 & 0 \\ 0 & -2 & 3 & 1 \end{bmatrix} \begin{matrix} -1R1 \\ {} \end{matrix} \Rightarrow \begin{bmatrix} 1 & 0 & 2 & 1 \\ 0 & -2 & 3 & 1 \end{bmatrix} R2+R1 \Rightarrow \begin{bmatrix} 1 & 0 & 2 & 1 \\ 0 & 1 & -\frac{3}{2} & -\frac{1}{2} \end{bmatrix} -\tfrac{1}{2}R2$$

Thus, $A^{-1} = \begin{bmatrix} 2 & 1 \\ -\frac{3}{2} & -\frac{1}{2} \end{bmatrix}$.

21. Find the inverse of $\begin{bmatrix} 5 & 10 \\ -3 & -6 \end{bmatrix}$, if it exists. Because $\left[A \mid I_2\right] = \begin{bmatrix} 5 & 10 & 1 & 0 \\ -3 & -6 & 0 & 1 \end{bmatrix}$, we have

$$\begin{bmatrix} 1 & 2 & \frac{1}{5} & 0 \\ -3 & -6 & 0 & 1 \end{bmatrix} \tfrac{1}{5}R1 \Rightarrow \begin{bmatrix} 1 & 2 & \frac{1}{5} & 0 \\ 0 & 0 & \frac{3}{5} & 1 \end{bmatrix} 3R1+R2$$

At this point, the matrix should be changed so that the second-row, second-column element will be 1. That element is now 0, so the desired transformation cannot be completed. Therefore, the inverse of the given matrix does not exist.

23. Find the inverse of $A = \begin{bmatrix} 1 & 0 & 1 \\ 0 & -1 & 0 \\ 2 & 1 & 1 \end{bmatrix}$, if it exists. Because $\left[A \mid I_3\right] = \begin{bmatrix} 1 & 0 & 1 & 1 & 0 & 0 \\ 0 & -1 & 0 & 0 & 1 & 0 \\ 2 & 1 & 1 & 0 & 0 & 1 \end{bmatrix}$ we have

$$\begin{bmatrix} 1 & 0 & 1 & 1 & 0 & 0 \\ 0 & -1 & 0 & 0 & 1 & 0 \\ 0 & 1 & -1 & -2 & 0 & 1 \end{bmatrix} -2R1+R3 \Rightarrow \begin{bmatrix} 1 & 0 & 1 & 1 & 0 & 0 \\ 0 & 1 & 0 & 0 & -1 & 0 \\ 0 & 0 & -1 & -2 & 1 & 1 \end{bmatrix} \begin{matrix} -1R2 \\ R2+R3 \end{matrix}$$

$$\begin{bmatrix} 1 & 0 & 1 & 1 & 0 & 0 \\ 0 & 1 & 0 & 0 & -1 & 0 \\ 0 & 0 & 1 & 2 & -1 & -1 \end{bmatrix} -1R3 \Rightarrow \begin{bmatrix} 1 & 0 & 0 & -1 & 1 & 1 \\ 0 & 1 & 0 & 0 & -1 & 0 \\ 0 & 0 & 1 & 2 & -1 & -1 \end{bmatrix} -1R3+R1 \,.$$

Thus, $A^{-1} = \begin{bmatrix} -1 & 1 & 1 \\ 0 & -1 & 0 \\ 2 & -1 & -1 \end{bmatrix}$.

25. Find the inverse of $A = \begin{bmatrix} 2 & 3 & 3 \\ 1 & 4 & 3 \\ 1 & 3 & 4 \end{bmatrix}$, if it exists. Because $\left[A \mid I_3\right] = \begin{bmatrix} 2 & 3 & 3 & 1 & 0 & 0 \\ 1 & 4 & 3 & 0 & 1 & 0 \\ 1 & 3 & 4 & 0 & 0 & 1 \end{bmatrix}$, we have

$$\begin{bmatrix} 2 & 3 & 3 & 1 & 0 & 0 \\ 1 & 4 & 3 & 0 & 1 & 0 \\ 0 & -1 & 1 & 0 & -1 & 1 \end{bmatrix} -R2+R3 \Rightarrow \begin{bmatrix} 1 & -1 & 0 & 1 & -1 & 0 \\ 1 & 4 & 3 & 0 & 1 & 0 \\ 0 & -1 & 1 & 0 & -1 & 1 \end{bmatrix} -R2+R1 \Rightarrow \begin{bmatrix} 1 & -1 & 0 & 1 & -1 & 0 \\ 1 & 7 & 0 & 0 & 4 & -3 \\ 0 & -1 & 1 & 0 & -1 & 1 \end{bmatrix} -3R3+R2 \Rightarrow$$

$$\begin{bmatrix} 8 & 0 & 0 & 7 & -3 & -3 \\ 1 & 7 & 0 & 0 & 4 & -3 \\ 0 & -1 & 1 & 0 & -1 & 1 \end{bmatrix} 7R1+R2 \Rightarrow \begin{bmatrix} 1 & 0 & 0 & \frac{7}{8} & -\frac{3}{8} & -\frac{3}{8} \\ 1 & 7 & 0 & 0 & 4 & -3 \\ 0 & -1 & 1 & 0 & -1 & 1 \end{bmatrix} \tfrac{1}{8}R1 \Rightarrow \begin{bmatrix} 1 & 0 & 0 & \frac{7}{8} & -\frac{3}{8} & -\frac{3}{8} \\ 0 & 7 & 0 & -\frac{7}{8} & \frac{35}{8} & -\frac{21}{8} \\ 0 & -1 & 1 & 0 & -1 & 1 \end{bmatrix} -R1+R2 \Rightarrow$$

$$\begin{bmatrix} 1 & 0 & 0 & \frac{7}{8} & -\frac{3}{8} & -\frac{3}{8} \\ 0 & 1 & 0 & -\frac{1}{8} & \frac{5}{8} & -\frac{3}{8} \\ 0 & -1 & 1 & 0 & -1 & 1 \end{bmatrix} \tfrac{1}{7}R2 \Rightarrow \begin{bmatrix} 1 & 0 & 0 & \frac{7}{8} & -\frac{3}{8} & -\frac{3}{8} \\ 0 & 1 & 0 & -\frac{1}{8} & \frac{5}{8} & -\frac{3}{8} \\ 0 & 0 & 1 & -\frac{1}{8} & -\frac{3}{8} & \frac{5}{8} \end{bmatrix} R2+R3 \Rightarrow$$

Thus, $A^{-1} = \begin{bmatrix} \frac{7}{8} & -\frac{3}{8} & -\frac{3}{8} \\ -\frac{1}{8} & \frac{5}{8} & -\frac{3}{8} \\ -\frac{1}{8} & -\frac{3}{8} & \frac{5}{8} \end{bmatrix}$.

27. Find the inverse of $A = \begin{bmatrix} 2 & 2 & -4 \\ 2 & 6 & 0 \\ -3 & -3 & 5 \end{bmatrix}$, if it exists. Because $\begin{bmatrix} A \mid I_3 \end{bmatrix} = \begin{bmatrix} 2 & 2 & -4 & 1 & 0 & 0 \\ 2 & 6 & 0 & 0 & 1 & 0 \\ -3 & -3 & 5 & 0 & 0 & 1 \end{bmatrix}$,

$$\begin{bmatrix} 1 & 1 & -2 & \tfrac{1}{2} & 0 & 0 \\ 2 & 6 & 0 & 0 & 1 & 0 \\ -3 & -3 & 5 & 0 & 0 & 1 \end{bmatrix} \begin{matrix} \tfrac{1}{2}R1 \\ \\ \\ \end{matrix} \Rightarrow \begin{bmatrix} 1 & 1 & -2 & \tfrac{1}{2} & 0 & 0 \\ 0 & 4 & 4 & -1 & 1 & 0 \\ 0 & 0 & -1 & \tfrac{3}{2} & 0 & 1 \end{bmatrix} \begin{matrix} \\ -2R1+R2 \\ 3R1+R3 \end{matrix}$$

$$\begin{bmatrix} 1 & 1 & -2 & \tfrac{1}{2} & 0 & 0 \\ 0 & 1 & 1 & -\tfrac{1}{4} & \tfrac{1}{4} & 0 \\ 0 & 0 & -1 & \tfrac{3}{2} & 0 & 1 \end{bmatrix} \begin{matrix} \\ \tfrac{1}{4}R2 \\ \\ \end{matrix} \Rightarrow \begin{bmatrix} 1 & 0 & -3 & \tfrac{3}{4} & -\tfrac{1}{4} & 0 \\ 0 & 1 & 1 & -\tfrac{1}{4} & \tfrac{1}{4} & 0 \\ 0 & 0 & -1 & \tfrac{3}{2} & 0 & 1 \end{bmatrix} \begin{matrix} -1R2+R1 \\ \\ \\ \end{matrix}$$

$$\begin{bmatrix} 1 & 0 & -3 & \tfrac{3}{4} & -\tfrac{1}{4} & 0 \\ 0 & 1 & 1 & -\tfrac{1}{4} & \tfrac{1}{4} & 0 \\ 0 & 0 & 1 & -\tfrac{3}{2} & 0 & -1 \end{bmatrix} \begin{matrix} \\ \\ -1R3 \end{matrix} \Rightarrow \begin{bmatrix} 1 & 0 & 0 & -\tfrac{15}{4} & -\tfrac{1}{4} & -3 \\ 0 & 1 & 0 & \tfrac{5}{4} & \tfrac{1}{4} & 1 \\ 0 & 0 & 1 & -\tfrac{3}{2} & 0 & -1 \end{bmatrix} \begin{matrix} 3R3+R1 \\ -1R3+R2. \\ \end{matrix}$$

Thus, $A^{-1} = \begin{bmatrix} -\tfrac{15}{4} & -\tfrac{1}{4} & -3 \\ \tfrac{5}{4} & \tfrac{1}{4} & 1 \\ -\tfrac{3}{2} & 0 & -1 \end{bmatrix}$.

29. Find the inverse of $A = \begin{bmatrix} 1 & 1 & 0 & 2 \\ 2 & -1 & 1 & -1 \\ 3 & 3 & 2 & -2 \\ 1 & 2 & 1 & 0 \end{bmatrix}$, if it exists.

$$\begin{bmatrix} A \mid I_4 \end{bmatrix} = \begin{bmatrix} 1 & 1 & 0 & 2 & 1 & 0 & 0 & 0 \\ 2 & -1 & 1 & -1 & 0 & 1 & 0 & 0 \\ 3 & 3 & 2 & -2 & 0 & 0 & 1 & 0 \\ 1 & 2 & 1 & 0 & 0 & 0 & 0 & 1 \end{bmatrix} \Rightarrow \begin{bmatrix} 1 & 1 & 0 & 2 & 1 & 0 & 0 & 0 \\ 0 & -3 & 1 & -5 & -2 & 1 & 0 & 0 \\ 0 & 0 & 2 & -8 & -3 & 0 & 1 & 0 \\ 0 & 1 & 1 & -2 & -1 & 0 & 0 & 1 \end{bmatrix} \begin{matrix} \\ -2R1+R2 \\ -3R1+R3 \\ -1R1+R4 \end{matrix}$$

$$\Rightarrow \begin{bmatrix} 1 & 1 & 0 & 2 & 1 & 0 & 0 & 0 \\ 0 & 1 & -\tfrac{1}{3} & \tfrac{5}{3} & \tfrac{2}{3} & -\tfrac{1}{3} & 0 & 0 \\ 0 & 0 & 2 & -8 & -3 & 0 & 1 & 0 \\ 0 & 1 & 1 & -2 & -1 & 0 & 0 & 1 \end{bmatrix} \begin{matrix} \\ -\tfrac{1}{3}R2 \\ \\ \\ \end{matrix} \Rightarrow \begin{bmatrix} 1 & 0 & \tfrac{1}{3} & \tfrac{1}{3} & \tfrac{1}{3} & \tfrac{1}{3} & 0 & 0 \\ 0 & 1 & -\tfrac{1}{3} & \tfrac{5}{3} & \tfrac{2}{3} & -\tfrac{1}{3} & 0 & 0 \\ 0 & 0 & 2 & -8 & -3 & 0 & 1 & 0 \\ 0 & 0 & \tfrac{4}{3} & -\tfrac{11}{3} & -\tfrac{5}{3} & \tfrac{1}{3} & 0 & 1 \end{bmatrix} \begin{matrix} -1R2+R1 \\ \\ \\ -1R2+R4 \end{matrix}$$

$$\Rightarrow \begin{bmatrix} 1 & 0 & \tfrac{1}{3} & \tfrac{1}{3} & \tfrac{1}{3} & \tfrac{1}{3} & 0 & 0 \\ 0 & 1 & -\tfrac{1}{3} & \tfrac{5}{3} & \tfrac{2}{3} & -\tfrac{1}{3} & 0 & 0 \\ 0 & 0 & 1 & -4 & -\tfrac{3}{2} & 0 & \tfrac{1}{2} & 0 \\ 0 & 0 & \tfrac{4}{3} & -\tfrac{11}{3} & -\tfrac{5}{3} & \tfrac{1}{3} & 0 & 1 \end{bmatrix} \begin{matrix} \\ \\ \tfrac{1}{2}R3 \\ \\ \end{matrix} \Rightarrow \begin{bmatrix} 1 & 0 & 0 & \tfrac{5}{3} & \tfrac{5}{6} & \tfrac{1}{3} & -\tfrac{1}{6} & 0 \\ 0 & 1 & 0 & \tfrac{1}{3} & \tfrac{1}{6} & -\tfrac{1}{3} & \tfrac{1}{6} & 0 \\ 0 & 0 & 1 & -4 & -\tfrac{3}{2} & 0 & \tfrac{1}{2} & 0 \\ 0 & 0 & 0 & \tfrac{5}{3} & \tfrac{1}{3} & \tfrac{1}{3} & -\tfrac{2}{3} & 1 \end{bmatrix} \begin{matrix} -\tfrac{1}{3}R3+R1 \\ \tfrac{1}{3}R3+R2 \\ \\ -\tfrac{4}{3}R3+R4 \end{matrix}$$

$$\Rightarrow \begin{bmatrix} 1 & 0 & 0 & \tfrac{5}{3} & \tfrac{5}{6} & \tfrac{1}{3} & -\tfrac{1}{6} & 0 \\ 0 & 1 & 0 & \tfrac{1}{3} & \tfrac{1}{6} & -\tfrac{1}{3} & \tfrac{1}{6} & 0 \\ 0 & 0 & 1 & -4 & -\tfrac{3}{2} & 0 & \tfrac{1}{2} & 0 \\ 0 & 0 & 0 & 1 & \tfrac{1}{5} & \tfrac{1}{5} & -\tfrac{2}{5} & \tfrac{3}{5} \end{bmatrix} \begin{matrix} \\ \\ \\ \tfrac{3}{5}R4 \end{matrix} \Rightarrow \begin{bmatrix} 1 & 0 & 0 & 0 & \tfrac{1}{2} & 0 & \tfrac{1}{2} & -1 \\ 0 & 1 & 0 & 0 & \tfrac{1}{10} & -\tfrac{2}{5} & \tfrac{3}{10} & -\tfrac{1}{5} \\ 0 & 0 & 1 & 0 & -\tfrac{7}{10} & \tfrac{4}{5} & -\tfrac{11}{10} & \tfrac{12}{5} \\ 0 & 0 & 0 & 1 & \tfrac{1}{5} & \tfrac{1}{5} & -\tfrac{2}{5} & \tfrac{3}{5} \end{bmatrix} \begin{matrix} -\tfrac{5}{3}R4+R1 \\ -\tfrac{1}{3}R4+R2 \\ 4R4+R3 \\ \end{matrix}$$

Thus, $A^{-1} = \begin{bmatrix} \tfrac{1}{2} & 0 & \tfrac{1}{2} & -1 \\ \tfrac{1}{10} & -\tfrac{2}{5} & \tfrac{3}{10} & -\tfrac{1}{5} \\ -\tfrac{7}{10} & \tfrac{4}{5} & -\tfrac{11}{10} & \tfrac{12}{5} \\ \tfrac{1}{5} & \tfrac{1}{5} & -\tfrac{2}{5} & \tfrac{3}{5} \end{bmatrix}$.

31. Find the inverse of $A = \begin{bmatrix} 3 & 2 & 0 & -1 \\ 2 & 0 & 1 & 2 \\ 1 & 2 & -1 & 0 \\ 2 & -1 & 1 & 1 \end{bmatrix}$, if it exists.

$$[A \mid I_4] = \begin{bmatrix} 3 & 2 & 0 & -1 & 1 & 0 & 0 & 0 \\ 2 & 0 & 1 & 2 & 0 & 1 & 0 & 0 \\ 1 & 2 & -1 & 0 & 0 & 0 & 1 & 0 \\ 2 & -1 & 1 & 1 & 0 & 0 & 0 & 1 \end{bmatrix} \Rightarrow \begin{bmatrix} 5 & 1 & 1 & 0 & 1 & 0 & 0 & 1 \\ 0 & 1 & 0 & 1 & 0 & 1 & 0 & -1 \\ 1 & 2 & -1 & 0 & 0 & 0 & 1 & 0 \\ 2 & -1 & 1 & 1 & 0 & 0 & 0 & 1 \end{bmatrix} \begin{matrix} R1+R4 \\ -R4+R2 \\ \\ \end{matrix} \Rightarrow$$

$$\begin{bmatrix} 6 & 3 & 0 & 0 & 1 & 0 & 1 & 1 \\ 0 & 1 & 0 & 1 & 0 & 1 & 0 & -1 \\ 1 & 2 & -1 & 0 & 0 & 0 & 1 & 0 \\ 0 & -5 & 3 & 1 & 0 & 0 & -2 & 1 \end{bmatrix} \begin{matrix} R1+R3 \\ \\ \\ -2R3+R4 \end{matrix} \Rightarrow \begin{bmatrix} 6 & 3 & 0 & 0 & 1 & 0 & 1 & 1 \\ 0 & 1 & 0 & 1 & 0 & 1 & 0 & -1 \\ 0 & -9 & 6 & 0 & 1 & 0 & -5 & 1 \\ 0 & -5 & 3 & 1 & 0 & 0 & -2 & 1 \end{bmatrix} \begin{matrix} \\ \\ -6R3+R1 \\ \end{matrix} \Rightarrow$$

$$\begin{bmatrix} 6 & 3 & 0 & 0 & 1 & 0 & 1 & 1 \\ 0 & 1 & 0 & 1 & 0 & 1 & 0 & -1 \\ 0 & 0 & 6 & 9 & 1 & 9 & -5 & -8 \\ 0 & 0 & 3 & 6 & 0 & 5 & -2 & -4 \end{bmatrix} \begin{matrix} \\ \\ 9R2+R3 \\ 5R2+R4 \end{matrix} \Rightarrow \begin{bmatrix} 6 & 3 & 0 & 0 & 1 & 0 & 1 & 1 \\ 0 & 1 & 0 & 1 & 0 & 1 & 0 & -1 \\ 0 & 0 & 6 & 9 & 1 & 9 & -5 & -8 \\ 0 & 0 & 0 & \frac{3}{2} & -\frac{1}{2} & \frac{1}{2} & \frac{1}{2} & 0 \end{bmatrix} \begin{matrix} \\ \\ \\ -\frac{1}{2}R3+R4 \end{matrix} \Rightarrow$$

$$\begin{bmatrix} 6 & 3 & 0 & 0 & 1 & 0 & 1 & 1 \\ 0 & 1 & 0 & 1 & 0 & 1 & 0 & -1 \\ 0 & 0 & 6 & 9 & 1 & 9 & -5 & -8 \\ 0 & 0 & 0 & 1 & -\frac{1}{3} & \frac{1}{3} & \frac{1}{3} & 0 \end{bmatrix} \begin{matrix} \\ \\ \\ \frac{2}{3}R4 \end{matrix} \Rightarrow \begin{bmatrix} 6 & 3 & 0 & 0 & 1 & 0 & 1 & 1 \\ 0 & 1 & 0 & 0 & \frac{1}{3} & \frac{2}{3} & -\frac{1}{3} & -1 \\ 0 & 0 & 6 & 0 & 4 & 6 & -8 & -8 \\ 0 & 0 & 0 & 1 & -\frac{1}{3} & \frac{1}{3} & \frac{1}{3} & 0 \end{bmatrix} \begin{matrix} \\ -R4+R2 \\ -9R4+R3 \\ \end{matrix} \Rightarrow$$

$$\begin{bmatrix} 6 & 0 & 0 & 0 & 0 & -2 & 2 & 4 \\ 0 & 1 & 0 & 0 & \frac{1}{3} & \frac{2}{3} & -\frac{1}{3} & -1 \\ 0 & 0 & 1 & 0 & \frac{2}{3} & 1 & -\frac{4}{3} & -\frac{4}{3} \\ 0 & 0 & 0 & 1 & -\frac{1}{3} & \frac{1}{3} & \frac{1}{3} & 0 \end{bmatrix} \begin{matrix} -3R2+R1 \\ \\ \frac{1}{6}R3 \\ \end{matrix} \Rightarrow \begin{bmatrix} 1 & 0 & 0 & 0 & 0 & -\frac{1}{3} & \frac{1}{3} & \frac{2}{3} \\ 0 & 1 & 0 & 0 & \frac{1}{3} & \frac{2}{3} & -\frac{1}{3} & -1 \\ 0 & 0 & 1 & 0 & \frac{2}{3} & 1 & -\frac{4}{3} & -\frac{4}{3} \\ 0 & 0 & 0 & 1 & -\frac{1}{3} & \frac{1}{3} & \frac{1}{3} & 0 \end{bmatrix} \begin{matrix} \frac{1}{6}R1 \\ \\ \\ \end{matrix}$$

Thus, $A^{-1} = \begin{bmatrix} 0 & -\frac{1}{3} & \frac{1}{3} & \frac{2}{3} \\ \frac{1}{3} & \frac{2}{3} & -\frac{1}{3} & -1 \\ \frac{2}{3} & 1 & -\frac{4}{3} & -\frac{4}{3} \\ -\frac{1}{3} & \frac{1}{3} & \frac{1}{3} & 0 \end{bmatrix}.$

33. $-x + y = 1$
$2x - y = 1$

Given $A = \begin{bmatrix} -1 & 1 \\ 2 & -1 \end{bmatrix}$, $X = \begin{bmatrix} x \\ y \end{bmatrix}$, and $B = \begin{bmatrix} 1 \\ 1 \end{bmatrix}$, find A^{-1}.

Because $[A \mid I_2] = \begin{bmatrix} -1 & 1 & 1 & 0 \\ 2 & -1 & 0 & 1 \end{bmatrix}$,

$\begin{bmatrix} 1 & -1 & -1 & 0 \\ 2 & -1 & 0 & 1 \end{bmatrix} \begin{matrix} -1R1 \\ \\ \end{matrix} \Rightarrow \begin{bmatrix} 1 & -1 & -1 & 0 \\ 0 & 1 & 2 & 1 \end{bmatrix} \begin{matrix} \\ -2R1+R2 \end{matrix} \Rightarrow \begin{bmatrix} 1 & 0 & 1 & 1 \\ 0 & 1 & 2 & 1 \end{bmatrix} \begin{matrix} R2+R1 \\ \end{matrix}$

Thus, $A^{-1} = \begin{bmatrix} 1 & 1 \\ 2 & 1 \end{bmatrix}.$

$X = A^{-1}B = \begin{bmatrix} 1 & 1 \\ 2 & 1 \end{bmatrix}\begin{bmatrix} 1 \\ 1 \end{bmatrix} = \begin{bmatrix} 1(1)+1(1) \\ 2(1)+1(1) \end{bmatrix} = \begin{bmatrix} 1+1 \\ 2+1 \end{bmatrix} = \begin{bmatrix} 2 \\ 3 \end{bmatrix}$

Solution set: $\{(2,3)\}$

35. $2x - y = -8$
$3x + y = -2$

Given $A = \begin{bmatrix} 2 & -1 \\ 3 & 1 \end{bmatrix}$, $X = \begin{bmatrix} x \\ y \end{bmatrix}$, and $B = \begin{bmatrix} -8 \\ -2 \end{bmatrix}$,

find A^{-1}. Because $\begin{bmatrix} A \,|\, I_2 \end{bmatrix} = \begin{bmatrix} 2 & -1 & | & 1 & 0 \\ 3 & 1 & | & 0 & 1 \end{bmatrix}$,

$\begin{bmatrix} 5 & 0 & | & 1 & 1 \\ 3 & 1 & | & 0 & 1 \end{bmatrix} \begin{matrix} R1 + R2 \\ \\ \end{matrix} \Rightarrow$

$\begin{bmatrix} 1 & 0 & | & \frac{1}{5} & \frac{1}{5} \\ 3 & 1 & | & 0 & 1 \end{bmatrix} \begin{matrix} \frac{1}{5}R1 \\ \\ \end{matrix} \Rightarrow$

$\begin{bmatrix} 1 & 0 & | & \frac{1}{5} & \frac{1}{5} \\ 0 & 1 & | & -\frac{3}{5} & \frac{2}{5} \end{bmatrix} \begin{matrix} R1 \\ -3R1+R2 \end{matrix}$

Thus, $A^{-1} = \begin{bmatrix} \frac{1}{5} & \frac{1}{5} \\ -\frac{3}{5} & \frac{2}{5} \end{bmatrix}$.

$X = A^{-1}B = \begin{bmatrix} \frac{1}{5} & \frac{1}{5} \\ -\frac{3}{5} & \frac{2}{5} \end{bmatrix} \begin{bmatrix} -8 \\ -2 \end{bmatrix}$

$= \begin{bmatrix} \frac{1}{5}(-8) + \frac{1}{5}(-2) \\ -\frac{3}{5}(-8) + \frac{2}{5}(-2) \end{bmatrix} = \begin{bmatrix} -\frac{8}{5} + \left(-\frac{2}{5}\right) \\ \frac{24}{5} + \left(-\frac{4}{5}\right) \end{bmatrix}$

$= \begin{bmatrix} -\frac{10}{5} \\ \frac{20}{5} \end{bmatrix} = \begin{bmatrix} -2 \\ 4 \end{bmatrix}$

Solution set: $\left\{ (-2, 4) \right\}$

37. $3x + 4y = -3$
$-5x + 8y = 16$

Given $A = \begin{bmatrix} 3 & 4 \\ -5 & 8 \end{bmatrix}$, $X = \begin{bmatrix} x \\ y \end{bmatrix}$, and $B = \begin{bmatrix} -3 \\ 16 \end{bmatrix}$,

find A^{-1}. Because $\begin{bmatrix} A \,|\, I_2 \end{bmatrix} = \begin{bmatrix} 3 & 4 & | & 1 & 0 \\ -5 & 8 & | & 0 & 1 \end{bmatrix}$,

$\begin{bmatrix} 3 & 4 & | & 1 & 0 \\ 0 & 44 & | & 5 & 3 \end{bmatrix} \begin{matrix} \\ 3R2 + 5R1 \end{matrix} \Rightarrow$

$\begin{bmatrix} 3 & 4 & | & 1 & 0 \\ 0 & 1 & | & \frac{5}{44} & \frac{3}{44} \end{bmatrix} \begin{matrix} \\ \frac{1}{44}R2 \end{matrix} \Rightarrow$

$\begin{bmatrix} 3 & 0 & | & \frac{6}{11} & -\frac{3}{11} \\ 0 & 1 & | & \frac{5}{44} & \frac{3}{44} \end{bmatrix} \begin{matrix} -4R2 + R1 \\ \end{matrix} \Rightarrow$

$\begin{bmatrix} 1 & 0 & | & \frac{2}{11} & -\frac{1}{11} \\ 0 & 1 & | & \frac{5}{44} & \frac{3}{44} \end{bmatrix} \begin{matrix} \frac{1}{3}R1 \\ \end{matrix}$

Thus, $A^{-1} = \begin{bmatrix} \frac{2}{11} & -\frac{1}{11} \\ \frac{5}{44} & \frac{3}{44} \end{bmatrix}$.

$X = A^{-1}B = \begin{bmatrix} \frac{2}{11} & -\frac{1}{11} \\ \frac{5}{44} & \frac{3}{44} \end{bmatrix} \begin{bmatrix} -3 \\ 16 \end{bmatrix}$

$= \begin{bmatrix} \frac{2}{11}(-3) - \frac{1}{11}(16) \\ \frac{5}{44}(-3) + \frac{3}{44}(16) \end{bmatrix}$

$= \begin{bmatrix} -\frac{6}{11} - \frac{16}{11} \\ -\frac{15}{44} + \frac{48}{44} \end{bmatrix} = \begin{bmatrix} -\frac{22}{11} \\ \frac{33}{44} \end{bmatrix} = \begin{bmatrix} -2 \\ \frac{3}{4} \end{bmatrix}$

Solution set: $\left\{ \left(-2, \frac{3}{4}\right) \right\}$

39. $6x + 9y = 3$
$-8x + 3y = 6$

Given $A = \begin{bmatrix} 6 & 9 \\ -8 & 3 \end{bmatrix}$, $X = \begin{bmatrix} x \\ y \end{bmatrix}$, and $B = \begin{bmatrix} 3 \\ 6 \end{bmatrix}$,

find A^{-1}. Because $\begin{bmatrix} A \,|\, I_2 \end{bmatrix} = \begin{bmatrix} 6 & 9 & | & 1 & 0 \\ -8 & 3 & | & 0 & 1 \end{bmatrix}$,

$\begin{bmatrix} 30 & 0 & | & 1 & -3 \\ -8 & 3 & | & 0 & 1 \end{bmatrix} \begin{matrix} -3R2 + R1 \\ \end{matrix} \Rightarrow$

$\begin{bmatrix} 1 & 0 & | & \frac{1}{30} & -\frac{1}{10} \\ -8 & 3 & | & 0 & 1 \end{bmatrix} \begin{matrix} \frac{1}{30}R1 \\ \end{matrix} \Rightarrow$

$\begin{bmatrix} 1 & 0 & | & \frac{1}{30} & -\frac{1}{10} \\ 0 & 3 & | & \frac{4}{15} & \frac{1}{5} \end{bmatrix} \begin{matrix} \\ 8R1 + R2 \end{matrix} \Rightarrow$

$\begin{bmatrix} 1 & 0 & | & \frac{1}{30} & -\frac{1}{10} \\ 0 & 1 & | & \frac{4}{45} & \frac{1}{15} \end{bmatrix} \begin{matrix} \\ \frac{1}{3}R2 \end{matrix} \Rightarrow$

Thus $A^{-1} = \begin{bmatrix} \frac{1}{30} & -\frac{1}{10} \\ \frac{4}{45} & \frac{1}{15} \end{bmatrix}$.

$X = A^{-1}B = \begin{bmatrix} \frac{1}{30} & -\frac{1}{10} \\ \frac{4}{45} & \frac{1}{15} \end{bmatrix} \begin{bmatrix} 3 \\ 6 \end{bmatrix}$

$= \begin{bmatrix} \frac{1}{30}(3) + \left(-\frac{1}{10}\right)(6) \\ \frac{4}{45}(3) + \frac{1}{15}(6) \end{bmatrix} = \begin{bmatrix} \frac{1}{10} + \left(-\frac{6}{10}\right) \\ \frac{4}{15} + \frac{2}{5} \end{bmatrix}$

$= \begin{bmatrix} -\frac{5}{10} \\ \frac{4}{15} + \frac{6}{15} \end{bmatrix} = \begin{bmatrix} -\frac{1}{2} \\ \frac{10}{15} \end{bmatrix} = \begin{bmatrix} -\frac{1}{2} \\ \frac{2}{3} \end{bmatrix}$

Solution set: $\left\{ \left(-\frac{1}{2}, \frac{2}{3}\right) \right\}$

41. $0.2x + 0.3y = -1.9$
$0.7x - 0.2y = 4.6$

Given $A = \begin{bmatrix} 0.2 & 0.3 \\ 0.7 & -0.2 \end{bmatrix}$, $X = \begin{bmatrix} x \\ y \end{bmatrix}$, and $B = \begin{bmatrix} -1.9 \\ 4.6 \end{bmatrix}$, find A^{-1}. Because $\left[A \mid I_2 \right] = \begin{bmatrix} 0.2 & 0.3 & 1 & 0 \\ 0.7 & -0.2 & 0 & 1 \end{bmatrix}$,

$\begin{bmatrix} 1 & 1.5 & 5 & 0 \\ 0.7 & -0.2 & 0 & 1 \end{bmatrix} 5\text{R}1 \Rightarrow \begin{bmatrix} 1 & 1.5 & 5 & 0 \\ 0 & -1.25 & -3.5 & 1 \end{bmatrix} -0.7\text{R}1 + \text{R}2 \Rightarrow$

$\begin{bmatrix} 1 & 1.5 & 5 & 0 \\ 0 & 1 & 2.8 & -0.8 \end{bmatrix} \frac{1}{-1.25}\text{R}2 \Rightarrow \begin{bmatrix} 1 & 0 & 0.8 & 1.2 \\ 0 & 1 & 2.8 & -0.8 \end{bmatrix} -1.5\text{R}2 + \text{R}1$

Thus, $A^{-1} = \begin{bmatrix} 0.8 & 1.2 \\ 2.8 & -0.8 \end{bmatrix}$.

$X = A^{-1}B = \begin{bmatrix} 0.8 & 1.2 \\ 2.8 & -0.8 \end{bmatrix}\begin{bmatrix} -1.9 \\ 4.6 \end{bmatrix} = \begin{bmatrix} 0.8(-1.9) + 1.2(4.6) \\ 2.8(-1.9) + (-0.8)(4.6) \end{bmatrix} = \begin{bmatrix} -1.52 + 5.52 \\ -5.32 + (-3.68) \end{bmatrix} = \begin{bmatrix} 4 \\ -9 \end{bmatrix}$

Solution set: $\{(4, -9)\}$

43. $\frac{1}{2}x + \frac{1}{3}y = \frac{49}{18}$
$\frac{1}{2}x + 2y = \frac{4}{3}$

Given $A = \begin{bmatrix} \frac{1}{2} & \frac{1}{3} \\ \frac{1}{2} & 2 \end{bmatrix}$, $X = \begin{bmatrix} x \\ y \end{bmatrix}$, and $B = \begin{bmatrix} \frac{49}{18} \\ \frac{4}{3} \end{bmatrix}$, find A^{-1}. Because $\left[A \mid I_2 \right] = \begin{bmatrix} \frac{1}{2} & \frac{1}{3} & 1 & 0 \\ \frac{1}{2} & 2 & 0 & 1 \end{bmatrix}$,

$\begin{bmatrix} \frac{1}{2} & \frac{1}{3} & 1 & 0 \\ 0 & \frac{5}{3} & -1 & 1 \end{bmatrix} -\text{R}1 + \text{R}2 \Rightarrow \begin{bmatrix} 1 & \frac{2}{3} & 2 & 0 \\ 0 & 1 & -\frac{3}{5} & \frac{3}{5} \end{bmatrix} \begin{matrix} 2\text{R}1 \\ \frac{3}{5}\text{R}2 \end{matrix} \Rightarrow \begin{bmatrix} 1 & 0 & \frac{12}{5} & -\frac{2}{5} \\ 0 & 1 & -\frac{3}{5} & \frac{3}{5} \end{bmatrix} -\frac{2}{3}\text{R}2 + \text{R}1$

Thus $A^{-1} = \begin{bmatrix} \frac{12}{5} & -\frac{2}{5} \\ -\frac{3}{5} & \frac{3}{5} \end{bmatrix}$.

$X = A^{-1}B = \begin{bmatrix} \frac{12}{5} & -\frac{2}{5} \\ -\frac{3}{5} & \frac{3}{5} \end{bmatrix}\begin{bmatrix} \frac{49}{18} \\ \frac{4}{3} \end{bmatrix} = \begin{bmatrix} \frac{12}{5}\left(\frac{49}{18}\right) - \frac{2}{5}\left(\frac{4}{3}\right) \\ -\frac{3}{5}\left(\frac{49}{18}\right) + \frac{3}{5}\left(\frac{4}{3}\right) \end{bmatrix} = \begin{bmatrix} \frac{98}{15} - \frac{8}{15} \\ -\frac{49}{30} + \frac{4}{5} \end{bmatrix} = \begin{bmatrix} \frac{90}{15} \\ -\frac{5}{6} \end{bmatrix} = \begin{bmatrix} 6 \\ -\frac{5}{6} \end{bmatrix}$

Solution set: $\left\{ \left(6, -\frac{5}{6}\right) \right\}$

45. In order to solve $\begin{matrix} 7x - 2y = 3 \\ 14x - 4y = 1 \end{matrix}$ by the matrix inverse method, we must first find A^{-1} given that

$A = \begin{bmatrix} 7 & -2 \\ 14 & -4 \end{bmatrix}$. Because $\left[A \mid I_2 \right] = \begin{bmatrix} 7 & -2 & 1 & 0 \\ 14 & -4 & 0 & 1 \end{bmatrix}$, we have the following.

$\begin{bmatrix} 1 & -\frac{2}{7} & \frac{1}{7} & 0 \\ 14 & -4 & 0 & 1 \end{bmatrix} \frac{1}{7}\text{R}1 \Rightarrow$

$\begin{bmatrix} 1 & -\frac{2}{7} & \frac{1}{7} & 0 \\ 0 & 0 & -2 & 1 \end{bmatrix} -14\text{R}1 + \text{R}2$

It is not possible to get a 1 in the second entry of the second column, so the inverse of A does not exist, hence the matrix inverse method cannot be used to solve the system.

47. $x + y + z = 6$
$2x + 3y - z = 7$
$3x - y - z = 6$

Given $A = \begin{bmatrix} 1 & 1 & 1 \\ 2 & 3 & -1 \\ 3 & -1 & -1 \end{bmatrix}$, $X = \begin{bmatrix} x \\ y \\ z \end{bmatrix}$, and $B = \begin{bmatrix} 6 \\ 7 \\ 6 \end{bmatrix}$, find A^{-1}. Because $\begin{bmatrix} A \mid I_3 \end{bmatrix} = \begin{bmatrix} 1 & 1 & 1 & 1 & 0 & 0 \\ 2 & 3 & -1 & 0 & 1 & 0 \\ 3 & -1 & -1 & 0 & 0 & 1 \end{bmatrix}$,

$\begin{bmatrix} 1 & 1 & 1 & 1 & 0 & 0 \\ 0 & 1 & -3 & -2 & 1 & 0 \\ 0 & -4 & -4 & -3 & 0 & 1 \end{bmatrix} \begin{matrix} \\ -2R1 + R2 \\ -3R1 + R3 \end{matrix} \Rightarrow \begin{bmatrix} 1 & 0 & 4 & 3 & -1 & 0 \\ 0 & 1 & -3 & -2 & 1 & 0 \\ 0 & 0 & -16 & -11 & 4 & 1 \end{bmatrix} \begin{matrix} -R2 + R1 \\ \\ 4R2 + R3 \end{matrix} \Rightarrow$

$\begin{bmatrix} 1 & 0 & 4 & 3 & -1 & 0 \\ 0 & 1 & -3 & -2 & 1 & 0 \\ 0 & 0 & 1 & \frac{11}{16} & -\frac{1}{4} & -\frac{1}{16} \end{bmatrix} \begin{matrix} \\ \\ -\frac{1}{16}R3 \end{matrix} \Rightarrow \begin{bmatrix} 1 & 0 & 0 & \frac{1}{4} & 0 & \frac{1}{4} \\ 0 & 1 & 0 & \frac{1}{16} & \frac{1}{4} & -\frac{3}{16} \\ 0 & 0 & 1 & \frac{11}{16} & -\frac{1}{4} & -\frac{1}{16} \end{bmatrix} \begin{matrix} -4R3 + R1 \\ 3R3 + R2. \\ \end{matrix}$

Thus, $A^{-1} = \begin{bmatrix} \frac{1}{4} & 0 & \frac{1}{4} \\ \frac{1}{16} & \frac{1}{4} & -\frac{3}{16} \\ \frac{11}{16} & -\frac{1}{4} & -\frac{1}{16} \end{bmatrix}$.

$X = A^{-1}B = \begin{bmatrix} \frac{1}{4} & 0 & \frac{1}{4} \\ \frac{1}{16} & \frac{1}{4} & -\frac{3}{16} \\ \frac{11}{16} & -\frac{1}{4} & -\frac{1}{16} \end{bmatrix} \begin{bmatrix} 6 \\ 7 \\ 6 \end{bmatrix} = \begin{bmatrix} \frac{1}{4}(6) + 0(7) + \frac{1}{4}(6) \\ \frac{1}{16}(6) + \frac{1}{4}(7) + \left(-\frac{3}{16}\right)(6) \\ \frac{11}{16}(6) + \left(-\frac{1}{4}\right)(7) + \left(-\frac{1}{16}\right)(6) \end{bmatrix} = \begin{bmatrix} \frac{3}{2} + 0 + \frac{3}{2} \\ \frac{3}{8} + \frac{7}{4} + \left(-\frac{9}{8}\right) \\ \frac{33}{8} + \left(-\frac{7}{4}\right) + \left(-\frac{3}{8}\right) \end{bmatrix} = \begin{bmatrix} \frac{6}{2} \\ \frac{7}{4} + \left(-\frac{6}{8}\right) \\ \frac{30}{8} + \left(-\frac{7}{4}\right) \end{bmatrix}$

$= \begin{bmatrix} 3 \\ \frac{7}{4} + \left(-\frac{3}{4}\right) \\ \frac{15}{4} + \left(-\frac{7}{4}\right) \end{bmatrix} = \begin{bmatrix} 3 \\ \frac{4}{4} \\ \frac{8}{4} \end{bmatrix} = \begin{bmatrix} 3 \\ 1 \\ 2 \end{bmatrix}$

Solution set: $\{(3, 1, 2)\}$

49. $2x + 3y + 3z = 1$
$x + 4y + 3z = 0$
$x + 3y + 4z = -1$

We have $A = \begin{bmatrix} 2 & 3 & 3 \\ 1 & 4 & 3 \\ 1 & 3 & 4 \end{bmatrix}$, $X = \begin{bmatrix} x \\ y \\ z \end{bmatrix}$, $B = \begin{bmatrix} 1 \\ 0 \\ -1 \end{bmatrix}$, and $A^{-1} = \begin{bmatrix} \frac{7}{8} & -\frac{3}{8} & -\frac{3}{8} \\ -\frac{1}{8} & \frac{5}{8} & -\frac{3}{8} \\ -\frac{1}{8} & -\frac{3}{8} & \frac{5}{8} \end{bmatrix}$. (from Exercise 25).

$X = A^{-1}B = \begin{bmatrix} \frac{7}{8} & -\frac{3}{8} & -\frac{3}{8} \\ -\frac{1}{8} & \frac{5}{8} & -\frac{3}{8} \\ -\frac{1}{8} & -\frac{3}{8} & \frac{5}{8} \end{bmatrix} \cdot \begin{bmatrix} 1 \\ 0 \\ -1 \end{bmatrix} = \begin{bmatrix} \frac{7}{8}(1) - \frac{3}{8}(0) - \frac{3}{8}(-1) \\ -\frac{1}{8}(1) + \frac{5}{8}(0) - \frac{3}{8}(-1) \\ -\frac{1}{8}(1) - \frac{3}{8}(0) + \frac{5}{8}(-1) \end{bmatrix} = \begin{bmatrix} \frac{7}{8} + 0 + \frac{3}{8} \\ -\frac{1}{8} + 0 + \frac{3}{8} \\ -\frac{1}{8} + 0 - \frac{5}{8} \end{bmatrix} = \begin{bmatrix} \frac{10}{8} \\ \frac{2}{8} \\ -\frac{6}{8} \end{bmatrix} = \begin{bmatrix} \frac{5}{4} \\ \frac{1}{4} \\ -\frac{3}{4} \end{bmatrix}$

Solution set: $\left\{ \left(\frac{5}{4}, \frac{1}{4}, -\frac{3}{4} \right) \right\}$

51. $2x + 2y - 4z = 12$
$2x + 6y = 16$
$-3x - 3y + 5z = -20$

We have $A = \begin{bmatrix} 2 & 2 & -4 \\ 2 & 6 & 0 \\ -3 & -3 & 5 \end{bmatrix}$, $X = \begin{bmatrix} x \\ y \\ z \end{bmatrix}$, $B = \begin{bmatrix} 12 \\ 16 \\ -20 \end{bmatrix}$, and $A^{-1} = \begin{bmatrix} -\frac{15}{4} & -\frac{1}{4} & -3 \\ \frac{5}{4} & \frac{1}{4} & 1 \\ -\frac{3}{2} & 0 & -1 \end{bmatrix}$ (from Exercise 27).

$X = A^{-1}B = \begin{bmatrix} -\frac{15}{4} & -\frac{1}{4} & -3 \\ \frac{5}{4} & \frac{1}{4} & 1 \\ -\frac{3}{2} & 0 & -1 \end{bmatrix}\begin{bmatrix} 12 \\ 16 \\ -20 \end{bmatrix} = \begin{bmatrix} -\frac{15}{4}(12) + \left(-\frac{1}{4}\right)(16) + (-3)(-20) \\ \frac{5}{4}(12) + \frac{1}{4}(16) + 1(-20) \\ -\frac{3}{2}(12) + 0(16) + (-1)(-20) \end{bmatrix} = \begin{bmatrix} -45 + (-4) + 60 \\ 15 + 4 + (-20) \\ -18 + 0 + 20 \end{bmatrix} = \begin{bmatrix} 11 \\ -1 \\ 2 \end{bmatrix}$

Solution set: $\{(11, -1, 2)\}$

53. $x + y + 2w = 3$
$2x - y + z - w = 3$
$3x + 3y + 2z - 2w = 5$
$x + 2y + z = 3$

We have $A = \begin{bmatrix} 1 & 1 & 0 & 2 \\ 2 & -1 & 1 & -1 \\ 3 & 3 & 2 & -2 \\ 1 & 2 & 1 & 0 \end{bmatrix}$, $X = \begin{bmatrix} x \\ y \\ z \\ w \end{bmatrix}$, $B = \begin{bmatrix} 3 \\ 3 \\ 5 \\ 3 \end{bmatrix}$, and $A^{-1} = \begin{bmatrix} \frac{1}{2} & 0 & \frac{1}{2} & -1 \\ \frac{1}{10} & -\frac{2}{5} & \frac{3}{10} & -\frac{1}{5} \\ -\frac{7}{10} & \frac{4}{5} & -\frac{11}{10} & \frac{12}{5} \\ \frac{1}{5} & \frac{1}{5} & -\frac{2}{5} & \frac{3}{5} \end{bmatrix}$ (from Exercise 29).

$X = A^{-1}B = \begin{bmatrix} \frac{1}{2} & 0 & \frac{1}{2} & -1 \\ \frac{1}{10} & -\frac{2}{5} & \frac{3}{10} & -\frac{1}{5} \\ -\frac{7}{10} & \frac{4}{5} & -\frac{11}{10} & \frac{12}{5} \\ \frac{1}{5} & \frac{1}{5} & -\frac{2}{5} & \frac{3}{5} \end{bmatrix}\begin{bmatrix} 3 \\ 3 \\ 5 \\ 3 \end{bmatrix} = \begin{bmatrix} \frac{1}{2}(3) + 0(3) + \frac{1}{2}(5) + (-1)(3) \\ \frac{1}{10}(3) + \left(-\frac{2}{5}\right)(3) + \frac{3}{10}(5) + \left(-\frac{1}{5}\right)(3) \\ -\frac{7}{10}(3) + \frac{4}{5}(3) + \left(-\frac{11}{10}\right)(5) + \frac{12}{5}(3) \\ \frac{1}{5}(3) + \frac{1}{5}(3) + \left(-\frac{2}{5}\right)(5) + \frac{3}{5}(3) \end{bmatrix}$

$= \begin{bmatrix} \frac{3}{2} + 0 + \frac{5}{2} + (-3) \\ \frac{3}{10} + \left(-\frac{6}{5}\right) + \frac{3}{2} + \left(-\frac{3}{5}\right) \\ -\frac{21}{10} + \frac{12}{5} + \left(-\frac{11}{2}\right) + \frac{36}{5} \\ \frac{3}{5} + \frac{3}{5} + (-2) + \frac{9}{5} \end{bmatrix} = \begin{bmatrix} \frac{8}{2} + (-3) \\ \frac{3}{10} + \left(-\frac{9}{5}\right) + \frac{3}{2} \\ -\frac{21}{10} + \left(-\frac{11}{2}\right) + \frac{48}{5} \\ (-2) + \frac{15}{5} \end{bmatrix} = \begin{bmatrix} 4 + (-3) \\ \frac{3}{10} + \left(-\frac{18}{10}\right) + \frac{15}{10} \\ -\frac{21}{10} + \left(-\frac{55}{10}\right) + \frac{96}{10} \\ (-2) + 3 \end{bmatrix} = \begin{bmatrix} 1 \\ \frac{0}{10} \\ \frac{20}{10} \\ 1 \end{bmatrix} = \begin{bmatrix} 1 \\ 0 \\ 2 \\ 1 \end{bmatrix}$

Solution set: $\{(1, 0, 2, 1)\}$

55. **(a)** $602.7 = a + 5.543b + 37.14c$
$656.7 = a + 6.933b + 41.30c$
$778.5 = a + 7.638b + 45.62c$

(b) $A = \begin{bmatrix} 1 & 5.543 & 37.14 \\ 1 & 6.933 & 41.30 \\ 1 & 7.638 & 45.62 \end{bmatrix}$, $X = \begin{bmatrix} a \\ b \\ c \end{bmatrix}$

and $B = \begin{bmatrix} 602.7 \\ 656.7 \\ 778.5 \end{bmatrix}$

Using a graphing calculator with matrix capabilities, we obtain the following.

```
NORMAL FLOAT AUTO REAL RADIAN MP
[A]⁻¹[B]
            [-490.547375]
            [    -89    ]
            [  42.71875 ]
```

$X = A^{-1}B$ or $\begin{bmatrix} a \\ b \\ c \end{bmatrix} = \begin{bmatrix} -490.547375 \\ -89 \\ 42.71875 \end{bmatrix}$.

Thus, $a \approx -490.5$, $b = -89$, $c \approx 42.72$.

(c) $S = -490.5 - 89A + 42.72B$

(d) If $A = 7.752$ and $B = 47.38$, the predicted value of S is given by the following.

$S = -490.5 - 89(7.752) + 42.72(47.38) \approx 843.6$

The predicted value is approximately 843.6.

(e) If $A = 8.9$ and $B = 66.25$, the predicted value of S is given by the following.

$S = -490.5 - 89(8.9) + 42.72(66.25) \approx 1547.6$

The predicted value is approximately 1547.6.

Using only three consecutive years to forecast six years into the future, it is probably not very accurate. Predictions made beyond the scope of the data are valid only if current trends continue.

57. Answers will vary.

59. Given $A = \begin{bmatrix} \frac{2}{3} & 0.7 \\ 22 & \sqrt{3} \end{bmatrix}$, we obtain $A^{-1} \approx \begin{bmatrix} -0.1215875322 & 0.0491390161 \\ 1.544369078 & -0.046799063 \end{bmatrix}$ using a graphing calculator.

61. Given $A = \begin{bmatrix} \frac{1}{2} & \frac{1}{4} & \frac{1}{3} \\ 0 & \frac{1}{4} & \frac{1}{3} \\ \frac{1}{2} & \frac{1}{2} & \frac{1}{3} \end{bmatrix}$, we obtain $A^{-1} = \begin{bmatrix} 2 & -2 & 0 \\ -4 & 0 & 4 \\ 3 & 3 & -3 \end{bmatrix}$ using a graphing calculator.

63. $2.1x + y = \sqrt{5}$

$\sqrt{2}x - 2y = 5$

Given $A = \begin{bmatrix} 2.1 & 1 \\ \sqrt{2} & -2 \end{bmatrix}$, $X = \begin{bmatrix} x \\ y \end{bmatrix}$, and $B = \begin{bmatrix} \sqrt{5} \\ 5 \end{bmatrix}$, using a graphing calculator, we have

$X = A^{-1}B = \begin{bmatrix} 1.68717058 \\ -1.306990242 \end{bmatrix}$. Thus, the solution set is $\{(1.68717058, -1.306990242)\}$.

65. $(\log 2)x + (\ln 3)y + (\ln 4)z = 1$

$(\ln 3)x + (\log 2)y + (\ln 8)z = 5$

$(\log 12)x + (\ln 4)y + (\ln 8)z = 9$

Given $A = \begin{bmatrix} \log 2 & \ln 3 & \ln 4 \\ \ln 3 & \log 2 & \ln 8 \\ \log 12 & \ln 4 & \ln 8 \end{bmatrix}$, $X = \begin{bmatrix} x \\ y \\ z \end{bmatrix}$, and $B = \begin{bmatrix} 1 \\ 5 \\ 9 \end{bmatrix}$, using a graphing calculator, we have

$X = A^{-1}B = \begin{bmatrix} 13.58736702 \\ 3.929011993 \\ -5.342780076 \end{bmatrix}$.

Solution set: $\{(13.58736702, 3.929011993, -5.342780076)\}$

67. $A = \begin{bmatrix} a & b \\ c & d \end{bmatrix}$ and $O = \begin{bmatrix} 0 & 0 \\ 0 & 0 \end{bmatrix}$

$A \cdot O = \begin{bmatrix} a & b \\ c & d \end{bmatrix}\begin{bmatrix} 0 & 0 \\ 0 & 0 \end{bmatrix} = \begin{bmatrix} a(0)+b(0) & a(0)+b(0) \\ c(0)+d(0) & c(0)+d(0) \end{bmatrix} = \begin{bmatrix} 0+0 & 0+0 \\ 0+0 & 0+0 \end{bmatrix} = \begin{bmatrix} 0 & 0 \\ 0 & 0 \end{bmatrix} = O$

$O \cdot A = \begin{bmatrix} 0 & 0 \\ 0 & 0 \end{bmatrix}\begin{bmatrix} a & b \\ c & d \end{bmatrix} = \begin{bmatrix} 0(a)+0(c) & 0(b)+0(d) \\ 0(a)+0(c) & 0(b)+0(d) \end{bmatrix} = \begin{bmatrix} 0+0 & 0+0 \\ 0+0 & 0+0 \end{bmatrix} = \begin{bmatrix} 0 & 0 \\ 0 & 0 \end{bmatrix} = O$

Thus, $A \cdot O = O \cdot A = O$.

69. Suppose A is an invertible square matrix, and suppose that matrices B and C are both inverses of A. We must show that $B = C$ in order to prove that any square matrix has no more than one inverse, or in other words, that the inverse for a matrix is unique. Because B is an inverse of A, we know that $AB = I$. Multiply both sides by C to obtain $C(AB) = CI = C$ (1). Using the associative law of matrix multiplication, we have $C(AB) = CA(B) = C = IB = B$ (2). Combining (1) and (2) gives $C = C(AB) = B$, or $C = B$. Thus, any square matrix has no more than one inverse.

71. Suppose A and B are matrices, where A^{-1}, B^{-1}, and AB all exist. Show that $(AB)^{-1} = B^{-1}A^{-1}$. Because A^{-1}, B^{-1}, and AB all exist, we can assume that A and B are $n \times n$ (square) matrices.

$$(AB)(AB)^{-1} = I_n \Rightarrow A^{-1}\left[(AB)(AB)^{-1}\right] = A^{-1}I_n \Rightarrow \left[A^{-1}(AB)\right](AB)^{-1} = A^{-1} \Rightarrow$$

$$\left[(A^{-1}A)B\right](AB)^{-1} = A^{-1} \Rightarrow (I_nB)(AB)^{-1} = A^{-1} \Rightarrow B(AB)^{-1} = A^{-1} \Rightarrow B^{-1}\left[B(AB)^{-1}\right] = B^{-1}A^{-1} \Rightarrow$$

$$(B^{-1}B)(AB)^{-1} = B^{-1}A^{-1} \Rightarrow I_n(AB)^{-1} = B^{-1}A^{-1} \Rightarrow (AB)^{-1} = B^{-1}A^{-1}$$

73. $A = \begin{bmatrix} 1 & 0 & 0 \\ 0 & 0 & -1 \\ 0 & 1 & -1 \end{bmatrix}$

$$A^2 = AA = \begin{bmatrix} 1 & 0 & 0 \\ 0 & 0 & -1 \\ 0 & 1 & -1 \end{bmatrix}\begin{bmatrix} 1 & 0 & 0 \\ 0 & 0 & -1 \\ 0 & 1 & -1 \end{bmatrix}$$

$$= \begin{bmatrix} 1(1)+0(0)+0(0) & 1(0)+0(0)+0(1) & 1(0)+0(-1)+0(-1) \\ 0(1)+0(0)+(-1)(0) & 0(0)+0(0)+(-1)(1) & 0(0)+0(-1)+(-1)(-1) \\ 0(1)+1(0)+(-1)(0) & 0(0)+1(0)+(-1)(1) & 0(0)+1(-1)+(-1)(-1) \end{bmatrix}$$

$$= \begin{bmatrix} 1+0+0 & 0+0+0 & 0+0+0 \\ 0+0+0 & 0+0+(-1) & 0+0+1 \\ 0+0+0 & 0+0+(-1) & 0+(-1)+1 \end{bmatrix} = \begin{bmatrix} 1 & 0 & 0 \\ 0 & -1 & 1 \\ 0 & -1 & 0 \end{bmatrix}$$

$$A^3 = AA^2 = \begin{bmatrix} 1 & 0 & 0 \\ 0 & 0 & -1 \\ 0 & 1 & -1 \end{bmatrix}\begin{bmatrix} 1 & 0 & 0 \\ 0 & -1 & 1 \\ 0 & -1 & 0 \end{bmatrix}$$

$$= \begin{bmatrix} 1(1)+0(0)+0(0) & 1(0)+0(-1)+0(-1) & 1(0)+0(1)+0(0) \\ 0(1)+0(0)+(-1)(0) & 0(0)+0(-1)+(-1)(-1) & 0(0)+0(1)+(-1)(0) \\ 0(1)+1(0)+(-1)(0) & 0(0)+1(-1)+(-1)(-1) & 0(0)+1(1)+(-1)(0) \end{bmatrix}$$

$$= \begin{bmatrix} 1+0+0 & 0+0+0 & 0+0+0 \\ 0+0+0 & 0+0+1 & 0+0+0 \\ 0+0+0 & 0+(-1)+1 & 0+1+0 \end{bmatrix} = \begin{bmatrix} 1 & 0 & 0 \\ 0 & 1 & 0 \\ 0 & 0 & 1 \end{bmatrix}$$

Because $AA^2 = I$, $A^2 = A^{-1}$. Therefore, $A^{-1} = \begin{bmatrix} 1 & 0 & 0 \\ 0 & -1 & 1 \\ 0 & -1 & 0 \end{bmatrix}$.

74. Find the inverses of I_n, $-A$, and kA.

I_n is its own inverse because $I_n \cdot I_n = I_n$.

The inverse of $-A$ is $-(A^{-1})$ because $(-A)\left[-(A^{-1})\right] = (-1)(-1)(A \cdot A^{-1}) = I_n$.

The inverse of kA (k a scalar) is $\frac{1}{k}A^{-1}$ because $(kA)\left(\frac{1}{k}A^{-1}\right) = \left(k \cdot \frac{1}{k}\right)(A \cdot A^{-1}) = I_n$.

Chapter 9   Review Exercises

1. $2x + 6y = 6$ (1)
$5x + 9y = 9$ (2)
Solving equation (1) for x, we have
$2x + 6y = 6 \Rightarrow 2x = 6 - 6y \Rightarrow x = 3 - 3y$ (3).
Substituting equation (3) into equation (2) and solving for y, we have the following.
$5(3 - 3y) + 9y = 9 \Rightarrow 15 - 15y + 9y = 9 \Rightarrow$
$\quad 15 - 6y = 9 \Rightarrow -6y = -6 \Rightarrow y = 1$
Substituting 1 for y in equation (3), we have
$x = 3 - 3(1) = 3 - 3 = 0.$
Solution set: $\{(0,1)\}$

3. $x + 5y = 9$ (1)
$2x + 10y = 18$ (2)
Multiply equation (1) by -2 and add the resulting equation to equation (2).
$\quad -2x - 10y = -18$
$\quad \underline{2x + 10y = \ 18}$
$\qquad\qquad 0 = 0$
This is a true statement and implies that the system has infinitely many solutions. Solving equation (1) for x we have $x = -5y + 9$. Given y as an arbitrary value, the solution set is
$\{(-5y + 9, y)\}.$

5. $y = -x + 3$ (1)
$2x + 2y = 1$ (2)
Substituting equation (1) into equation (2), we have $2x + 2(-x + 3) = 1$. Solving for x we
have $2x + 2(-x + 3) = 1 \Rightarrow$
$2x + (-2x) + 6 = 1 \Rightarrow 6 = 1$. This is a false statement and the solution is inconsistent. Thus, the solution set is $\varnothing$.

7. $3x - 2y = 0$ (1)
$9x + 8y = 7$ (2)
Multiply equation (1) by -3 and add the result to equation (2).
$\quad -9x + 6y = 0$
$\quad \underline{9x + 8y = 7}$
$\qquad 14y = 7 \Rightarrow y = \frac{1}{2}$
Substituting $\frac{1}{2}$ for y in equation (1) and solving for x, we have the following.
$3x - 2\left(\frac{1}{2}\right) = 0 \Rightarrow 3x - 1 = 0 \Rightarrow$
$3x = 1 \Rightarrow x = \frac{1}{3}$
Solution set: $\left\{\left(\frac{1}{3}, \frac{1}{2}\right)\right\}$

9. $2x - 5y + 3z = -1$ (1)
$x + 4y - 2z = 9$ (2)
$-x + 2y + 4z = 5$ (3)
First, we eliminate x. Multiply equation (2) by -2 and add the result to equation (1).
$\quad 2x - 5y + 3z = -1$
$\quad \underline{-2x - 8y + 4z = -18}$
$\qquad -13y + 7z = -19$ (4)
Next, add equations (2) and (3) to obtain
$6y + 2z = 14$ (5). Now, we solve the system
$\quad -13y + 7z = -19$ (4)
$\quad\ \ 6y + 2z = \ \ 14$ (5)
Multiply equation (4) by 2, multiply equation (5) by -7, and add the resulting equations.
$\quad -26y + 14z = -38$
$\quad \underline{-42y - 14z = -98}$
$\quad -68y \qquad\ = -136 \Rightarrow y = 2$
Substituting this value into equation (4), we have $-13(2) + 7z = -19 \Rightarrow$
$-26 + 7z = -19 \Rightarrow 7z = 7 \Rightarrow z = 1.$
Substituting 2 for y and 1 for z in equation (2), we have $x + 4(2) - 2(1) = 9 \Rightarrow$
$x + 8 - 2 = 9 \Rightarrow x = 3.$
Solution set: $\{(3, 2, 1)\}$

11. $5x - y = 26$ (1)
$4y + 3z = -4$ (2)
$3x + 3z = 15$ (3)
Multiply equation (2) by -1 and add the result to equation (3) in order to eliminate z.
$\quad -4y - 3z = 4$
$\quad \underline{3x \qquad + 3z = 15}$
$\quad 3x - 4y \qquad = 19$ (4)
Multiply equation (1) by -4 and add the result to equation (4).
$\quad -20x + 4y = -104$
$\quad \underline{3x - 4y = \ \ 19}$
$\quad -17x \qquad = -85 \Rightarrow x = 5$
Substitute 5 for x in equation (1) and solve for y to obtain
$5(5) - y = 26 \Rightarrow 25 - y = 26 \Rightarrow y = -1.$
Substitute 5 for x equation (3) and solve for z:
$3(5) + 3z = 15 \Rightarrow 15 + 3z = 15 \Rightarrow$
$3z = 0 \Rightarrow z = 0$
Solution set: $\{(5, -1, 0)\}$

13. Answers will vary. One possible answer is
$x + y = 2$
$x + y = 3.$

15. Let x = amount of rice; y = amount of soybeans.

We get the following system of equations. The first equation relates protein and the second relates calories.

$15x + 22.5y = 9.5$ (1)
$810x + 270y = 324$ (2)

Multiply equation (1) by -12 and add the result to equation (2).

$-180x - 270y = -114$
$\underline{810x + 270y = 324}$
$630x = 210 \Rightarrow x = \frac{1}{3}$

Substitute $\frac{1}{3}$ for x in equation (1) and solve for y.

$15\left(\frac{1}{3}\right) + 22.5y = 9.5 \Rightarrow 5 + 22.5y = 9.5 \Rightarrow$
$22.5y = 4.5 \Rightarrow y = 0.20 = \frac{1}{5}$

$\frac{1}{3}$ cup of rice and $\frac{1}{5}$ cup of soybeans should be used.

17. Let x = the number of blankets, y = the number of rugs, z = the number of skirts.

The following table is helpful in organizing the information.

	Number of blankets (x)	Number of rugs (y)	Number of skirts (z)	Available
Spinning yarn	24	30	12	306
Dying	4	5	3	59
Weaving	15	18	9	201

We have the following system of equations.

$24x + 30y + 12z = 306 \Rightarrow 4x + 5y + 2z = 51$ (1)
$4x + 5y + 3z = 59$ (2)
$15x + 18y + 9z = 201 \Rightarrow 5x + 6y + 3z = 67$ (3)

Multiply equation (1) by -1 and add the result to equation (2).

$-4x - 5y - 2z = -51$
$\underline{4x + 5y + 3z = 59}$
$z = 8$

Substitute 8 for z in equations (1) and (3) and simplify.

$4x + 5y + 2(8) = 51 \Rightarrow 4x + 5y + 16 = 51 \Rightarrow$
$4x + 5y = 35$ (4)
$5x + 6y + 3(8) = 67 \Rightarrow 5x + 6y + 24 = 67 \Rightarrow$
$5x + 6y = 43$ (5)

Multiply equation (4) by 5 and equation (5) by -4 and add the results.

$20x + 25y = 175$
$\underline{-20x - 24y = -172}$
$y = 3$

Substituting 3 for y in equation (4) and solving for x, we obtain $4x + 5(3) = 35 \Rightarrow$
$4x + 15 = 35 \Rightarrow 4x = 20 \Rightarrow x = 5$

5 blankets, 3 rugs, and 8 skirts can be made.

19. $p = \frac{3}{2}q$ (1)
$p = 81 - \frac{3}{4}q$ (2)

(a)

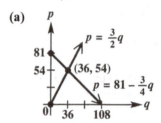

(b) To find the equilibrium demand, replace p with $\frac{3}{2}q$ in equation (2), and solve for q. The value of q will give the equilibrium demand.

$\frac{3}{2}q = 81 - \frac{3}{4}q \Rightarrow 4\left(\frac{3}{2}q\right) = 4\left(81 - \frac{3}{4}q\right) \Rightarrow$
$6q = 324 - 3q \Rightarrow 9q = 324 \Rightarrow q = 36$

The equilibrium demand is 36.

(c) To find p, substitute $q = 36$ into equation (1) to obtain $p = \frac{3}{2}(36) = 54$.

The equilibrium price is $54.

21. Because $y = ax^2 + bx + c$ and the points $(1, -2.3)$, $(2, -1.3)$, and $(3, 4.5)$ are on the parabola, we have the following system of equations.

$-2.3 = a(1)^2 + b(1) + c \Rightarrow a + b + c = -2.3$ (1)
$-1.3 = a(2)^2 + b(2) + c \Rightarrow 4a + 2b + c = -1.3$ (2)
$4.5 = a(3)^2 + b(3) + c \Rightarrow 9a + 3b + c = 4.5$ (3)

First, eliminate c by multiplying equation (1) by -1 and adding the result to equation (2).

$-a - b - c = 2.3$
$\underline{4a + 2b + c = -1.3}$
$3a + b = 1$ (4)

Next, multiply equation (2) by -1 and add to equation (3).

$-4a - 2b - c = 1.3$
$\underline{9a + 3b + c = 4.5}$
$5a + b = 5.8$ (5)

(continued on next page)

(*continued*)

Next, eliminate b by multiplying equation (4) by -1 and adding the result to equation (5).

$$
\begin{aligned}
-3a - b &= -1 \\
\underline{5a + b} &= \underline{5.8} \\
2a \quad\; &= 4.8 \Rightarrow a = 2.4
\end{aligned}
$$

Substitute this value into equation (4) to obtain

$$3(2.4) + b = 1 \Rightarrow 7.2 + b = 1 \Rightarrow b = -6.2.$$

Substitute 2.4 for a and -6.2 for b into equation (1) in order to solve for c.

$$
\begin{aligned}
(2.4) + (-6.2) + c &= -2.3 \\
-3.8 + c &= -2.3 \Rightarrow c = 1.5
\end{aligned}
$$

The equation of the parabola is

$$y_1 = 2.4x^2 - 6.2x + 1.5 \text{ or}$$

$$y_1 = \tfrac{12}{5}x^2 - \tfrac{31}{5}x + \tfrac{3}{2}.$$

23. $\quad 3x - 4y + z = 2 \;\; (1)$
$\qquad\quad 2x + y = 1 \;\; (2)$

Solving equation (2) for y, we obtain $y = 1 - 2x$. Substitute $1 - 2x$ for y in equation (1) and solve for z.

$$3x - 4(1 - 2x) + z = 2 \Rightarrow 3x - 4 + 8x + z = 2 \Rightarrow$$
$$11x + z = 6 \Rightarrow z = 6 - 11x$$

Solution set: $\{(x,\, 1 - 2x,\, 6 - 11x)\}$

25. Writing

$$
\begin{aligned}
5x + 2y &= -10 \\
3x - 5y &= -6
\end{aligned}
$$

as an augmented matrix, we have

$$\left[\begin{array}{rr|r} 5 & 2 & -10 \\ 3 & -5 & -6 \end{array}\right].$$

$$\left[\begin{array}{rr|r} 1 & \frac{2}{5} & -2 \\ 3 & -5 & -6 \end{array}\right]\begin{array}{l}\frac{1}{5}\text{R1}\end{array} \Rightarrow$$

$$\left[\begin{array}{rr|r} 1 & \frac{2}{5} & -2 \\ 0 & -\frac{31}{5} & 0 \end{array}\right]{-3\text{R1}+\text{R2}} \Rightarrow$$

$$\left[\begin{array}{rr|r} 1 & \frac{2}{5} & -2 \\ 0 & 1 & 0 \end{array}\right]{-\frac{5}{31}\text{R2}} \Rightarrow$$

$$\left[\begin{array}{rr|r} 1 & 0 & -2 \\ 0 & 1 & 0 \end{array}\right]{-\frac{2}{5}\text{R2}+\text{R1}}$$

Solution set: $\{(-2,\, 0)\}$

27. Writing

$$
\begin{aligned}
3x + y &= -7 \\
x - y &= -5
\end{aligned}
$$

as an augmented matrix, we have

$$\left[\begin{array}{rr|r} 3 & 1 & -7 \\ 1 & -1 & -5 \end{array}\right].$$

$$\left[\begin{array}{rr|r} 1 & -1 & -5 \\ 3 & 1 & -7 \end{array}\right]\text{R1} \leftrightarrow \text{R2} \quad\Rightarrow$$

$$\left[\begin{array}{rr|r} 1 & -1 & -5 \\ 0 & 4 & 8 \end{array}\right]{-3\text{R1}+\text{R2}} \quad\Rightarrow$$

$$\left[\begin{array}{rr|r} 1 & -1 & -5 \\ 0 & 1 & 2 \end{array}\right]\tfrac{1}{4}\text{R2} \quad\Rightarrow$$

$$\left[\begin{array}{rr|r} 1 & 0 & -3 \\ 0 & 1 & 2 \end{array}\right]\text{R2}+\text{R1}$$

Solution set: $\{(-3,\, 2)\}$

29. Writing

$$
\begin{aligned}
x - z &= -3 \\
y + z &= 6 \\
2x - 3z &= -9
\end{aligned}
$$

as an augmented matrix, we have

$$\left[\begin{array}{rrr|r} 1 & 0 & -1 & -3 \\ 0 & 1 & 1 & 6 \\ 2 & 0 & -3 & -9 \end{array}\right].$$

$$\left[\begin{array}{rrr|r} 1 & 0 & -1 & -3 \\ 0 & 1 & 1 & 6 \\ 0 & 0 & -1 & -3 \end{array}\right]{-2\text{R1}+\text{R3}} \Rightarrow$$

$$\left[\begin{array}{rrr|r} 1 & 0 & -1 & -3 \\ 0 & 1 & 1 & 6 \\ 0 & 0 & 1 & 3 \end{array}\right]{-1\text{R3}} \Rightarrow$$

$$\left[\begin{array}{rrr|r} 1 & 0 & -1 & -3 \\ 0 & 1 & 0 & 3 \\ 0 & 0 & 1 & 3 \end{array}\right]{-1\text{R3}+\text{R2}} \Rightarrow$$

$$\left[\begin{array}{rrr|r} 1 & 0 & 0 & 0 \\ 0 & 1 & 0 & 3 \\ 0 & 0 & 1 & 3 \end{array}\right]\text{R3}+\text{R1}$$

Solution set: $\{(0,\, 3,\, 3)\}$

31. Let x = number of pounds of \$4.60 tea;
y = the number of pounds of \$5.75 tea;
z = the number of pounds of \$6.50 tea.
From the information in the exercise, we obtain the system

$$x + y + z = 20$$
$$4.6x + 5.75y + 6.5z = 20(5.25).$$
$$x = y + z$$

Rewriting the system, we have

$$x + y + z = 20$$
$$4.6x + 5.75y + 6.5z = 105.$$
$$x - y - z = 0$$

With an augmented matrix of

$$\begin{bmatrix} 1 & 1 & 1 & | & 20 \\ 4.6 & 5.75 & 6.5 & | & 105 \\ 1 & -1 & -1 & | & 0 \end{bmatrix},$$

we solve by the Gauss-Jordan method.

$$\begin{bmatrix} 1 & 1 & 1 & | & 20 \\ 0 & 1.15 & 1.9 & | & 13 \\ 0 & -2 & -2 & | & -20 \end{bmatrix} \begin{array}{l} -4.6R1 + R2 \Rightarrow \\ -1R1 + R3 \end{array}$$

$$\begin{bmatrix} 1 & 1 & 1 & | & 20 \\ 0 & 1 & \frac{38}{23} & | & \frac{260}{23} \\ 0 & -2 & -2 & | & -20 \end{bmatrix} \begin{array}{l} \frac{20}{23}R2 \Rightarrow \end{array}$$

$$\begin{bmatrix} 1 & 0 & -\frac{15}{23} & | & \frac{200}{23} \\ 0 & 1 & \frac{38}{23} & | & \frac{260}{23} \\ 0 & 0 & \frac{30}{23} & | & \frac{60}{23} \end{bmatrix} \begin{array}{l} -1R2 + R1 \\ \\ 2R2 + R3 \end{array} \Rightarrow$$

$$\begin{bmatrix} 1 & 0 & -\frac{15}{23} & | & \frac{200}{23} \\ 0 & 1 & \frac{38}{23} & | & \frac{260}{23} \\ 0 & 0 & 1 & | & 2 \end{bmatrix} \frac{23}{30}R3 \Rightarrow$$

$$\begin{bmatrix} 1 & 0 & 0 & | & 10 \\ 0 & 1 & 0 & | & 8 \\ 0 & 0 & 1 & | & 2 \end{bmatrix} \begin{array}{l} \frac{15}{23}R3 + R1 \\ -\frac{38}{23}R3 + R2 \end{array}$$

From the final matrix, $x = 10$, $y = 8$, and $z = 2$. Therefore, 10 lb of \$4.60 tea, 8 lb of \$5.75 tea, and 2 lb of \$6.50 tea should be used.

33. $y = 3.79x + 128$ (1)
$y = 8.89x + 80.2$ (2)

The given system can be solved by substitution. Substitute equation (1) into equation (2) to obtain
$3.79x + 128 = 8.89x + 80.2$. Solving this equation, we have $47.8 = 5.1x \Rightarrow x \approx 9.4$.
Because $x = 0$ corresponds 1975, $x = 9.4$ implies that during 1984 males and females earned the same number of master's degrees.

The number of masters degrees earned by males in this year is
$y = 3.79 \cdot 9.4 + 128 \approx 163.6$ thousand. The masters degrees earned by females in this year (which is the same as the males) is
$y = 8.89 \cdot 9.4 + 80.2 \approx 163.7$ thousand. The total number of master's degrees was therefore approximately 327 thousand. (Note that the difference in the tenths place is due to rounding when x was computed.)

35. $\begin{vmatrix} -1 & 8 \\ 2 & 9 \end{vmatrix} = -1(9) - 2(8) = -9 - 16 = -25$

37. $\begin{vmatrix} x & 4x \\ 2x & 8x \end{vmatrix} = x(8x) - 2x(4x) = 8x^2 - 8x^2 = 0$

39. Expanding

$$\begin{vmatrix} -1 & 2 & 3 \\ 4 & 0 & 3 \\ 5 & -1 & 2 \end{vmatrix}$$

by the second column, we have

$$-2\begin{vmatrix} 4 & 3 \\ 5 & 2 \end{vmatrix} + 0\begin{vmatrix} -1 & 3 \\ 5 & 2 \end{vmatrix} - (-1)\begin{vmatrix} -1 & 3 \\ 4 & 3 \end{vmatrix}$$
$$= -2[4(2) - 5(3)] + 0 + [(-1)(3) - (4)(3)]$$
$$= -2(8 - 15) + (-3 - 12) = -2(-7) + (-15)$$
$$= 14 - 15 = -1$$

41. $3x + 7y = 2$
$5x - y = -22$

$$D = \begin{vmatrix} 3 & 7 \\ 5 & -1 \end{vmatrix} = 3(-1) - 5(7) = -3 - 35 = -38$$

$$D_x = \begin{vmatrix} 2 & 7 \\ -22 & -1 \end{vmatrix} = 2(-1) - (-22)(7)$$
$$= -2 - (-154) = 152$$

$$D_y = \begin{vmatrix} 3 & 2 \\ 5 & -22 \end{vmatrix} = 3(-22) - 5(2)$$
$$= -66 - 10 = -76$$

$x = \frac{D_x}{D} = \frac{152}{-38} = -4$ and $y = \frac{D_y}{D} = \frac{-76}{-38} = 2$.

Solution set: $\{(-4, 2)\}$

43. Given $\begin{array}{l} 6x + y = -3 \quad (1) \\ 12x + 2y = 1 \quad (2) \end{array}$, we have

$$D = \begin{vmatrix} 6 & 1 \\ 12 & 2 \end{vmatrix} = 6(2) - 12(1) = 12 - 12 = 0.$$

Because $D = 0$, Cramer's rule does not apply. To determine whether the system is inconsistent or has infinitely many solutions, use the elimination method.

$$\begin{array}{l} -12x - 2y = 6 \quad \text{Multiply equation (1) by } -2. \\ \underline{12x + 2y = 1} \\ 0 = 7 \quad \text{False} \end{array}$$

The system is inconsistent. Thus the solution set is $\varnothing$.

45. $\begin{array}{l} x + y = -1 \\ 2y + z = 5 \\ 3x - 2z = -28 \end{array}$

Adding -3 times row 1 to row 3, we have

$$D = \begin{vmatrix} 1 & 1 & 0 \\ 0 & 2 & 1 \\ 3 & 0 & -2 \end{vmatrix} = \begin{vmatrix} 1 & 1 & 0 \\ 0 & 2 & 1 \\ 0 & -3 & -2 \end{vmatrix}$$

Expanding by column one, we have

$$D = \begin{vmatrix} 2 & 1 \\ -3 & -2 \end{vmatrix} = -4 - (-3) = -1$$

Adding -2 times row 1 to row 2, we have

$$D_x = \begin{vmatrix} -1 & 1 & 0 \\ 5 & 2 & 1 \\ -28 & 0 & -2 \end{vmatrix} = \begin{vmatrix} -1 & 1 & 0 \\ 7 & 0 & 1 \\ -28 & 0 & -2 \end{vmatrix}$$

Expanding about column two, we have

$$D_x = -(1)\begin{vmatrix} 7 & 1 \\ -28 & -2 \end{vmatrix} = -[-14 - (-28)] = -14$$

Adding 2 times row 2 to row 3, we have

$$D_y = \begin{vmatrix} 1 & -1 & 0 \\ 0 & 5 & 1 \\ 3 & -28 & -2 \end{vmatrix} = \begin{vmatrix} 1 & -1 & 0 \\ 0 & 5 & 1 \\ 3 & -18 & 0 \end{vmatrix}$$

Expanding by column three, we have

$$D_y = -(1)\begin{vmatrix} 1 & -1 \\ 3 & -18 \end{vmatrix}$$
$$= -1[-18 - (-3)] = -(-15) = 15$$

Adding -3 times row 1 to row 3, we have

$$D_z = \begin{vmatrix} 1 & 1 & -1 \\ 0 & 2 & 5 \\ 3 & 0 & -28 \end{vmatrix} = \begin{vmatrix} 1 & 1 & -1 \\ 0 & 2 & 5 \\ 0 & -3 & -25 \end{vmatrix}$$

Expanding by column one, we have

$$D_z = 1\begin{vmatrix} 2 & 5 \\ -3 & -25 \end{vmatrix} = -50 - (-15) = -35.$$

Thus, we have

$$x = \frac{D_x}{D} = \frac{-14}{-1} = 14, \quad y = \frac{D_y}{D} = \frac{15}{-1} = -15,$$
$$z = \frac{D_z}{D} = \frac{-35}{-1} = 35$$

Solution set: $\{(14, -15, 35)\}$

47. $\begin{vmatrix} 3x & 7 \\ -x & 4 \end{vmatrix} = 8 \Rightarrow 12x - (-7x) = 8 \Rightarrow 19x = 8 \Rightarrow$

$x = \frac{8}{19}$

Solution set: $\left\{\frac{8}{19}\right\}$

49. $\dfrac{2}{3x^2 - 5x + 2} = \dfrac{2}{(x-1)(3x-2)} = \dfrac{A}{x-1} + \dfrac{B}{3x-2}$

Multiply both sides by $(x-1)(3x-2)$ to get

$$2 = A(3x - 2) + B(x - 1). \quad (1)$$

First substitute 1 for x to get

$$2 = A[3(1) - 2] + B(1 - 1)$$
$$2 = A(1) + B(0) \Rightarrow A = 2$$

Replace A with 2 in equation (1) and substitute $\frac{2}{3}$ for x to get

$$2 = 2\left[3\left(\tfrac{2}{3}\right) - 2\right] + B\left(\tfrac{2}{3} - 1\right)$$
$$2 = 2(0) + B\left(-\tfrac{1}{3}\right) \Rightarrow 2 = -\tfrac{1}{3}B \Rightarrow B = -6$$

Thus, we have

$$\frac{2}{3x^2 - 5x + 2} = \frac{2}{x-1} + \frac{-6}{3x-2} \text{ or } \frac{2}{x-1} - \frac{6}{3x-2}$$

51. $\dfrac{5 - 2x}{(x^2 + 2)(x - 1)} = \dfrac{A}{x-1} + \dfrac{Bx + C}{x^2 + 2}$

Multiply both sides by $(x^2 + 2)(x - 1)$ to get the following.

$$5 - 2x = A(x^2 + 2) + (Bx + C)(x - 1) \quad (1)$$

First substitute 1 for x to get the following.

$$5 - 2(1) = A(1^2 + 2) + [B(1) + C](1 - 1)$$
$$5 - 2 = A(1 + 2) + 0 \Rightarrow 3 = 3A \Rightarrow A = 1$$

Replace A with 1 in equation (1) and substitute 0 for x to get the following.

$$5 - 2(0) = (0^2 + 2) + [B(0) + C](0 - 1)$$
$$5 - 0 = (0 + 2) + C(-1) \Rightarrow 5 = 2 - C \Rightarrow$$
$$3 = -C \Rightarrow C = -3$$
$$5 - 2x = (x^2 + 2) + (Bx - 3)(x - 1) \quad (2)$$

Substitute 2 (arbitrary value) in equation (2) for x to get the following.

(continued on next page)

(*continued*)

$$5 - 2(2) = (2^2 + 2) + [B(2) - 3](2 - 1)$$
$$5 - 4 = (4 + 2) + (2B - 3)(1)$$
$$1 = 6 + 2B - 3 \Rightarrow 1 = 3 + 2B$$
$$-2 = 2B \Rightarrow B = -1$$

Thus, we have

$$\frac{5 - 2x}{(x^2 + 2)(x - 1)} = \frac{1}{x - 1} + \frac{(-1)x + (-3)}{x^2 + 2} \text{ or }$$

$$\frac{1}{x - 1} - \frac{x + 3}{x^2 + 2}$$

53. $y = 2x + 10$ (1)

$x^2 + y = 13$ (2)

Substituting equation (1) into equation (2), we have $x^2 + (2x + 10) = 13$. Solving for x we get the following.

$$x^2 + (2x + 10) = 13 \Rightarrow x^2 + 2x - 3 = 0 \Rightarrow$$
$$(x + 3)(x - 1) = 0 \Rightarrow x = -3 \text{ or } x = 1$$

For each value of x, use equation (1) to find the corresponding value of y.

If $x = -3$, $y = 2(-3) + 10 = -6 + 10 = 4$.

If $x = 1$, $y = 2(1) + 10 = 2 + 10 = 12$.

Solution set: $\{(-3, 4), (1, 12)\}$

55. $x^2 + y^2 = 17$ (1)

$2x^2 - y^2 = 31$ (2)

Add equations (1) and (2) to obtain

$3x^2 = 48 \Rightarrow x^2 = 16 \Rightarrow x = \pm 4$.

For each value of x, use equation (1) to find the corresponding value of y.

If $x = -4$, $(-4)^2 + y^2 = 17 \Rightarrow 16 + y^2 = 17 \Rightarrow$

$y^2 = 1 \Rightarrow y = \pm 1$. If $x = 4$, $4^2 + y^2 = 17 \Rightarrow$

$16 + y^2 = 17 \Rightarrow y^2 = 1 \Rightarrow y = \pm 1$.

Solution set: $\{(-4, 1), (-4, -1), (4, -1), (4, 1)\}$

57. $xy = -10$ (1)

$x + 2y = 1$ (2)

Solving equation (2) for x gives

$x = 1 - 2y$ (3). Substituting this expression into equation (1), we obtain $(1 - 2y)y = -10$.

Solving for y we have

$$(1 - 2y)y = -10 \Rightarrow y - 2y^2 = -10$$
$$0 = 2y^2 - y - 10$$
$$(y + 2)(2y - 5) = 0 \Rightarrow y = -2 \text{ or } y = \tfrac{5}{2}$$

For each value of y, use equation (3) to find the corresponding value of x.

If $y = -2$, $x = 1 - 2(-2) = 1 - (-4) = 5$.

If $y = \tfrac{5}{2}$, $x = 1 - 2(\tfrac{5}{2}) = 1 - 5 = -4$.

Solution set: $\{(5, -2), (-4, \tfrac{5}{2})\}$

59. $x^2 + 2xy + y^2 = 4$ (1)

$x - 3y = -2$ (2)

Solve equation (2) for x: $x = 3y - 2$ (3).

Substituting this expression into equation (1), we obtain $(3y - 2)^2 + 2(3y - 2)y + y^2 = 4$.

Solving for x we have the following.

$$(3y - 2)^2 + 2(3y - 2)y + y^2 = 4$$
$$9y^2 - 12y + 4 + 6y^2 - 4y + y^2 = 4$$
$$16y^2 - 16y + 4 = 4$$
$$16y^2 - 16y = 0$$
$$y^2 - y = 0$$
$$y(y - 1) = 0$$
$$y = 0 \text{ or } y = 1$$

For each value of y, use equation (3) to find the corresponding value of x.

If $y = 0$, $x = 3(0) - 2 = 0 - 2 = -2$.

If $y = 1$, $x = 3(1) - 2 = 3 - 2 = 1$.

Solution set: $\{(-2, 0), (1, 1)\}$

61. $2x^2 - 3y^2 = 18$ (1)

$2x^2 - 2y^2 = 14$ (2)

Multiply equation (2) by -1, then add the result to equation (1) and then solve for y.

$$2x^2 - 3y^2 = 18$$
$$\underline{-2x^2 + 2y^2 = -14}$$
$$-y^2 = 4 \Rightarrow y^2 = -4 \Rightarrow y = \pm 2i$$

If $y = -2i$, then, using equation (1), we have

$2x^2 - 3(-2i)^2 = 18 \Rightarrow 2x^2 - 3(-4) = 18 \Rightarrow$

$2x^2 + 12 = 18 \Rightarrow 2x^2 = 6 \Rightarrow x^2 = 3 \Rightarrow$

$x = \pm\sqrt{3}$

If $y = 2i$, then, using equation (1), we have

$2x^2 - 3(2i)^2 = 18 \Rightarrow 2x^2 - 3(-4) = 18 \Rightarrow$

$2x^2 + 12 = 18 \Rightarrow 2x^2 = 6 \Rightarrow x^2 = 3 \Rightarrow$

$x = \pm\sqrt{3}$

Solution set: $\left\{ \left(-\sqrt{3}, -2i\right), \left(-\sqrt{3}, 2i\right), \left(\sqrt{3}, -2i\right), \left(\sqrt{3}, 2i\right) \right\}$

63. $x^2 + y^2 = 144$ (1)
  $x + 2y = 8$ (2)

Solving equation (2) for x, we have $x = 8 - 2y$ (3). Substitute $-2y + 8$ for x in equation (1) to obtain

$$(8 - 2y)^2 + y^2 = 144$$
$$64 - 32y + 4y^2 + y^2 = 144$$
$$5y^2 - 32y - 80 = 0$$

We solve this quadratic equation for x by using the quadratic formula, with $a = 5$, $b = -32$, and $c = -80$.

$$y = \frac{-(-32) \pm \sqrt{(-32)^2 - 4(5)(-80)}}{2(5)}$$

$$= \frac{32 \pm \sqrt{1,024 + 1,600}}{10} = \frac{32 + \sqrt{2,624}}{10}$$

$$= \frac{32 \pm 8\sqrt{41}}{10} = \frac{16 \pm 4\sqrt{41}}{5}$$

For each value of y, use equation (3) to find the corresponding value of x.

If $y = \frac{16 + 4\sqrt{41}}{5}$,

$$x = 8 - 2\left(\frac{16 + 4\sqrt{41}}{5}\right) = \frac{40}{5} + \frac{-32 - 8\sqrt{41}}{5} = \frac{8 - 8\sqrt{41}}{5}$$

If $y = \frac{16 - 4\sqrt{41}}{5}$,

$$x = 8 - 2\left(\frac{16 - 4\sqrt{41}}{5}\right) = \frac{40}{5} + \frac{-32 + 8\sqrt{41}}{5} = \frac{8 + 8\sqrt{41}}{5}$$

Yes, the circle and the line have two points in common,

$$\left(\frac{8 - 8\sqrt{41}}{5}, \frac{16 + 4\sqrt{41}}{5}\right) \text{ and } \left(\frac{8 + 8\sqrt{41}}{5}, \frac{16 - 4\sqrt{41}}{5}\right).$$

65. $x + y \le 6$
  $2x - y \ge 3$

Graph the solid line $x + y = 6$, which has x-intercept 6 and y-intercept 6. Because $x + y \le 6 \Rightarrow y \le -x + 6$, shade the region below this line. Graph the solid line $2x - y = 3$, which has x-intercept $\frac{3}{2}$ and y-intercept -3. Because $2x - y \ge 3 \Rightarrow -y \ge -2x + 3 \Rightarrow y \le 2x - 3$, shade the region below this line. The solution set is the intersection of these two regions, which is shaded in the final graph.

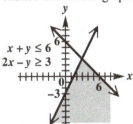

67. Maximize the objective function $2x + 4y$ subject to the restraints
  $3x + 2y \le 12$
  $5x + y \ge 5$
  $x \ge 0,\ y \ge 0$

To graph $3x + 2y \le 12$, draw the line with x-intercept 4 and y-intercept 6 as a solid line. Because the test point $(0, 0)$ satisfies this inequality, shade the region *below* the line. To graph $5x + y \ge 5$, draw the line with x-intercept 1 and y-intercept 5 as a solid line. Because the test point $(0, 0)$ does not satisfy this inequality, shade the region *above* the line. The constraints $x \ge 0$ and $y \ge 0$ restrict the graph to the first quadrant. The graph of the feasible region is the intersection of the regions that are the graphs of the individual constraints. The four vertices are $(1, 0)$, $(4, 0)$, $(0, 5)$ and $(0, 6)$.

Point	Value of $2x + 4y$	
$(1, 0)$	$2(1) + 4(0) = 2$	
$(4, 0)$	$2(4) + 4(0) = 8$	
$(0, 5)$	$2(0) + 4(5) = 20$	
$(0, 6)$	$2(0) + 4(6) = 24$	$\leftarrow$ Maximum

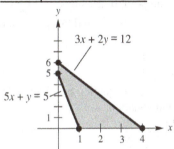

The maximum value is 24, which occurs at $(0, 6)$.

69. Let x = number of servings of food A; y = number of servings of food B. The protein constraint is $2x + 6y \ge 30$. The fat constraint is $4x + 2y \ge 20$. Also, $y \ge 2$ and because the numbers of servings cannot be negative, we also have $x \ge 0$. We want to minimize the objective function, cost $= 0.18x + 0.12y$. Find the region of feasible solutions by graphing the system of inequalities that is made up of the constraints. To graph $2x + 6y \ge 30$, draw the line with x-intercept 15 and y-intercept 5 as a solid line.
  (continued on next page)

(continued)

Because the test point (0, 0) does not satisfy this inequality, shade the region above the line. To graph $4x + 2y \geq 20$, draw the line with x-intercept 5 and y-intercept 10 as a solid line. Because the test point (0, 0) does not satisfy this inequality, shade the region above the line. To graph $y \geq 2$, draw the horizontal line with y-intercept 2 as a solid line and shade the region above the line. The graph of the feasible region is the intersection of the regions that are the graphs of the individual constraints.

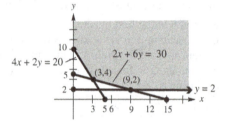

From the graph, observe that one vertex is (0, 10). The second occurs when $y = 2$ and
$2x + 6y = 30 \Rightarrow 2x + 6(2) = 30 \Rightarrow$
$2x = 18 \Rightarrow x = 9$.
Thus, (9, 2) is a vertex. The third vertex is the intersection point of the lines $2x + 6y = 30$ and $4x + 2y = 20$. To find this point, solve the system
$2x + 6y = 30$
$4x + 2y = 20$.
Multiply the first equation by –2 and add it to the second equation.
$-4x - 12y = -60$
$\underline{4x + 2y = 20}$
$-10y = -40 \Rightarrow y = 4$
Substituting 4 for y into $2x + 6y = 30$, we have the following.
$2x + 6(4) = 30 \Rightarrow 2x + 24 = 30 \Rightarrow$
$2x = 6 \Rightarrow x = 3$
Thus, the third vertex is (3, 4). Next, evaluate the objective function at each vertex.

Point	Cost $= 0.18x + 0.12y$	
$(0, 10)$	$0.18(0) + 0.12(10) = 1.20$	
$(3, 4)$	$0.18(3) + 0.12(4) = 1.02$	← Minimum
$(9, 2)$	$0.18(9) + 0.12(2) = 1.86$	

The minimum cost of \$1.02 per serving will be produced by 3 units of food A and 4 units of food B.

71. $\begin{bmatrix} 5 & x+2 \\ -6y & z \end{bmatrix} = \begin{bmatrix} a & 3x-1 \\ 5y & 9 \end{bmatrix}$

$a = 5; \quad x + 2 = 3x - 1 \Rightarrow 3 = 2x \Rightarrow x = \frac{3}{2};$
$-6y = 5y \Rightarrow 0 = 11y \Rightarrow y = 0; \quad z = 9$

Thus, $a = 5$, $x = \frac{3}{2}$, $y = 0$, and $z = 9$.

73. $\begin{bmatrix} 3 \\ 2 \\ 5 \end{bmatrix} - \begin{bmatrix} 8 \\ -4 \\ 6 \end{bmatrix} + \begin{bmatrix} 1 \\ 0 \\ 2 \end{bmatrix} = \begin{bmatrix} 3-8 \\ 2-(-4) \\ 5-6 \end{bmatrix} + \begin{bmatrix} 1 \\ 0 \\ 2 \end{bmatrix}$

$= \begin{bmatrix} -5 \\ 6 \\ -1 \end{bmatrix} + \begin{bmatrix} 1 \\ 0 \\ 2 \end{bmatrix} = \begin{bmatrix} -5+1 \\ 6+0 \\ -1+2 \end{bmatrix}$

$= \begin{bmatrix} -4 \\ 6 \\ 1 \end{bmatrix}$

75. $\begin{bmatrix} 2 & 5 & 8 \\ 1 & 9 & 2 \end{bmatrix} - \begin{bmatrix} 3 & 4 \\ 7 & 1 \end{bmatrix}$

This operation is not possible because one matrix is 2×3 and the other 2×2. Matrices of different sizes cannot be added or subtracted.

77. $\begin{bmatrix} -1 & 0 \\ 2 & 5 \end{bmatrix}\begin{bmatrix} -3 & 4 \\ 2 & 8 \end{bmatrix} = \begin{bmatrix} -1(-3)+0(2) & -1(4)+0(8) \\ 2(-3)+5(2) & 2(4)+5(8) \end{bmatrix} = \begin{bmatrix} 3+0 & -4+0 \\ -6+10 & 8+40 \end{bmatrix} = \begin{bmatrix} 3 & -4 \\ 4 & 48 \end{bmatrix}$

79. The product of a 2×3 matrix and a 3×2 matrix is a 2×2 matrix.

$\begin{bmatrix} 3 & 2 & -1 \\ 4 & 0 & 6 \end{bmatrix}\begin{bmatrix} -2 & 0 \\ 0 & 2 \\ 3 & 1 \end{bmatrix} = \begin{bmatrix} 3(-2)+2(0)+(-1)(3) & 3(0)+2(2)+(-1)(1) \\ 4(-2)+0(0)+6(3) & 4(0)+0(2)+6(1) \end{bmatrix} = \begin{bmatrix} -6+0-3 & 0+4-1 \\ -8+0+18 & 0+0+6 \end{bmatrix} = \begin{bmatrix} -9 & 3 \\ 10 & 6 \end{bmatrix}$

81. The product of a 3×3 matrix and a 3×3 matrix is a 3×3 matrix.

$\begin{bmatrix} -2 & 5 & 5 \\ 0 & 1 & 4 \\ 3 & -4 & -1 \end{bmatrix}\begin{bmatrix} 1 & 0 & -1 \\ -1 & 0 & 0 \\ 1 & 1 & -1 \end{bmatrix}$

$= \begin{bmatrix} -2(1)+5(-1)+5(1) & -2(0)+5(0)+5(1) & -2(-1)+5(0)+5(-1) \\ 0(1)+1(-1)+4(1) & 0(0)+1(0)+4(1) & 0(-1)+1(0)+4(-1) \\ 3(1)+(-4)(-1)+(-1)(1) & 3(0)+(-4)(0)+(-1)(1) & 3(-1)+(-4)(0)+(-1)(-1) \end{bmatrix}$

$= \begin{bmatrix} -2+(-5)+5 & 0+0+5 & 2+0+(-5) \\ 0+(-1)+4 & 0+0+4 & 0+0+(-4) \\ 3+4+(-1) & 0+0+(-1) & -3+0+1 \end{bmatrix} = \begin{bmatrix} -2 & 5 & -3 \\ 3 & 4 & -4 \\ 6 & -1 & -2 \end{bmatrix}$

83. Find the inverse of $A = \begin{bmatrix} 2 & 1 \\ 5 & 3 \end{bmatrix}$, if it exists. We have $[A|I_2] = \begin{bmatrix} 2 & 1 & | & 1 & 0 \\ 5 & 3 & | & 0 & 1 \end{bmatrix}$, which yields

$\begin{bmatrix} 1 & \frac{1}{2} & | & \frac{1}{2} & 0 \\ 5 & 3 & | & 0 & 1 \end{bmatrix} \begin{matrix} \frac{1}{2}R1 \\ \Rightarrow \end{matrix} \begin{bmatrix} 1 & \frac{1}{2} & | & \frac{1}{2} & 0 \\ 0 & \frac{1}{2} & | & -\frac{5}{2} & 1 \end{bmatrix} \begin{matrix} \\ -5R1+R2 \end{matrix} \Rightarrow \begin{bmatrix} 1 & \frac{1}{2} & | & \frac{1}{2} & 0 \\ 0 & 1 & | & -5 & 2 \end{bmatrix} 2R2 \Rightarrow \begin{bmatrix} 1 & 0 & | & 3 & -1 \\ 0 & 1 & | & -5 & 2 \end{bmatrix} \begin{matrix} -\frac{1}{2}R2+R1 \\ \\ \end{matrix}$

Thus, $A^{-1} = \begin{bmatrix} 3 & -1 \\ -5 & 2 \end{bmatrix}$.

85. Find the inverse of $A = \begin{bmatrix} 2 & -1 & 0 \\ 1 & 0 & 1 \\ 1 & -2 & 0 \end{bmatrix}$ if it exists. We have $[A|I_3] = \begin{bmatrix} 2 & -1 & 0 & | & 1 & 0 & 0 \\ 1 & 0 & 1 & | & 0 & 1 & 0 \\ 1 & -2 & 0 & | & 0 & 0 & 1 \end{bmatrix}$.

$\begin{bmatrix} 1 & 0 & 1 & | & 0 & 1 & 0 \\ 2 & -1 & 0 & | & 1 & 0 & 0 \\ 1 & -2 & 0 & | & 0 & 0 & 1 \end{bmatrix} \begin{matrix} R1 \leftrightarrow R2 \\ \\ \end{matrix} \Rightarrow \begin{bmatrix} 1 & 0 & 1 & | & 0 & 1 & 0 \\ 0 & -1 & -2 & | & 1 & -2 & 0 \\ 0 & -2 & -1 & | & 0 & -1 & 1 \end{bmatrix} \begin{matrix} \\ -2R1+R2 \\ -1R1+R3 \end{matrix} \Rightarrow$

$\begin{bmatrix} 1 & 0 & 1 & | & 0 & 1 & 0 \\ 0 & 1 & 2 & | & -1 & 2 & 0 \\ 0 & -2 & -1 & | & 0 & -1 & 1 \end{bmatrix} \begin{matrix} \\ -1R2 \\ \\ \end{matrix} \Rightarrow \begin{bmatrix} 1 & 0 & 1 & | & 0 & 1 & 0 \\ 0 & 1 & 2 & | & -1 & 2 & 0 \\ 0 & 0 & 3 & | & -2 & 3 & 1 \end{bmatrix} 2R2+R3 \Rightarrow$

$\begin{bmatrix} 1 & 0 & 1 & | & 0 & 1 & 0 \\ 0 & 1 & 2 & | & -1 & 2 & 0 \\ 0 & 0 & 1 & | & -\frac{2}{3} & 1 & \frac{1}{3} \end{bmatrix} \begin{matrix} \\ \\ \frac{1}{3}R3 \end{matrix} \Rightarrow \begin{bmatrix} 1 & 0 & 0 & | & \frac{2}{3} & 0 & -\frac{1}{3} \\ 0 & 1 & 0 & | & \frac{1}{3} & 0 & -\frac{2}{3} \\ 0 & 0 & 1 & | & -\frac{2}{3} & 1 & \frac{1}{3} \end{bmatrix} \begin{matrix} -1R3+R1 \\ -2R3+R2 \\ \\ \end{matrix}$

Thus, $A^{-1} = \begin{bmatrix} \frac{2}{3} & 0 & -\frac{1}{3} \\ \frac{1}{3} & 0 & -\frac{2}{3} \\ -\frac{2}{3} & 1 & \frac{1}{3} \end{bmatrix}$

87. $5x - 4y = 1 \atop x + 4y = 3$ $\Rightarrow A = \begin{bmatrix} 5 & -4 \\ 1 & 4 \end{bmatrix}$, $X = \begin{bmatrix} x \\ y \end{bmatrix}$, and $B = \begin{bmatrix} 1 \\ 3 \end{bmatrix}$. We have

$$\left[A \mid I_2\right] = \begin{bmatrix} 5 & -4 & \vline & 1 & 0 \\ 1 & 4 & \vline & 0 & 1 \end{bmatrix} \Rightarrow \begin{bmatrix} 6 & 0 & \vline & 1 & 1 \\ 1 & 4 & \vline & 0 & 1 \end{bmatrix} {\small R1+R2} \Rightarrow \begin{bmatrix} 1 & 0 & \vline & \frac{1}{6} & \frac{1}{6} \\ 1 & 4 & \vline & 0 & 1 \end{bmatrix} {\small \frac{1}{6}R1} \Rightarrow$$

$$\begin{bmatrix} 1 & 0 & \vline & \frac{1}{6} & \frac{1}{6} \\ 0 & 4 & \vline & -\frac{1}{6} & \frac{5}{6} \end{bmatrix} {\small -R1+R2} \Rightarrow \begin{bmatrix} 1 & 0 & \vline & \frac{1}{6} & \frac{1}{6} \\ 0 & 1 & \vline & -\frac{1}{24} & \frac{5}{24} \end{bmatrix} {\small \frac{1}{4}R2}$$

Thus, $A^{-1} = \begin{bmatrix} \frac{1}{6} & \frac{1}{6} \\ -\frac{1}{24} & \frac{5}{24} \end{bmatrix}$.

$$X = A^{-1}B = \begin{bmatrix} \frac{1}{6} & \frac{1}{6} \\ -\frac{1}{24} & \frac{5}{24} \end{bmatrix} \begin{bmatrix} 1 \\ 3 \end{bmatrix} = \begin{bmatrix} \frac{1}{6}(1) + \frac{1}{6}(3) \\ -\frac{1}{24}(1) + \frac{5}{24}(3) \end{bmatrix} = \begin{bmatrix} \frac{4}{6} \\ \frac{14}{24} \end{bmatrix} = \begin{bmatrix} \frac{2}{3} \\ \frac{7}{12} \end{bmatrix}$$

Solution set: $\left\{ \left(\frac{2}{3}, \frac{7}{12} \right) \right\}$

89. Given $\begin{aligned} 3x + 2y + \ z &= -5 \\ x - \ y + 3z &= -5 \\ 2x + 3y + \ z &= 0 \end{aligned}$ we have $A = \begin{bmatrix} 3 & 2 & 1 \\ 1 & -1 & 3 \\ 2 & 3 & 1 \end{bmatrix}$, $X = \begin{bmatrix} x \\ y \\ z \end{bmatrix}$, and $B = \begin{bmatrix} -5 \\ -5 \\ 0 \end{bmatrix}$.

$$\left[A \mid I_3\right] = \begin{bmatrix} 3 & 2 & 1 & \vline & 1 & 0 & 0 \\ 1 & -1 & 3 & \vline & 0 & 1 & 0 \\ 2 & 3 & 1 & \vline & 0 & 0 & 1 \end{bmatrix} \Rightarrow \begin{bmatrix} 1 & -1 & 3 & \vline & 0 & 1 & 0 \\ 3 & 2 & 1 & \vline & 1 & 0 & 0 \\ 2 & 3 & 1 & \vline & 0 & 0 & 1 \end{bmatrix} {\small R1 \leftrightarrow R2} \Rightarrow$$

$$\begin{bmatrix} 1 & -1 & 3 & \vline & 0 & 1 & 0 \\ 0 & 5 & -8 & \vline & 1 & -3 & 0 \\ 0 & 5 & -5 & \vline & 0 & -2 & 1 \end{bmatrix} {\small \begin{matrix} -3R1+R2 \\ -2R1+R3 \end{matrix}} \Rightarrow \begin{bmatrix} 1 & -1 & 3 & \vline & 0 & 1 & 0 \\ 0 & 1 & -\frac{8}{5} & \vline & \frac{1}{5} & -\frac{3}{5} & 0 \\ 0 & 5 & -5 & \vline & 0 & -2 & 1 \end{bmatrix} {\small \frac{1}{5}R2} \Rightarrow$$

$$\begin{bmatrix} 1 & 0 & \frac{7}{5} & \vline & \frac{1}{5} & \frac{2}{5} & 0 \\ 0 & 1 & -\frac{8}{5} & \vline & \frac{1}{5} & -\frac{3}{5} & 0 \\ 0 & 0 & 3 & \vline & -1 & 1 & 1 \end{bmatrix} {\small \begin{matrix} R2+R1 \\ \ \\ -5R2+R3 \end{matrix}} \Rightarrow \begin{bmatrix} 1 & 0 & \frac{7}{5} & \vline & \frac{1}{5} & \frac{2}{5} & 0 \\ 0 & 1 & -\frac{8}{5} & \vline & \frac{1}{5} & -\frac{3}{5} & 0 \\ 0 & 0 & 1 & \vline & -\frac{1}{3} & \frac{1}{3} & \frac{1}{3} \end{bmatrix} {\small \frac{1}{3}R3} \Rightarrow$$

$$\begin{bmatrix} 1 & 0 & 0 & \vline & \frac{2}{3} & -\frac{1}{15} & -\frac{7}{15} \\ 0 & 1 & 0 & \vline & -\frac{1}{3} & -\frac{1}{15} & \frac{8}{15} \\ 0 & 0 & 1 & \vline & -\frac{1}{3} & \frac{1}{3} & \frac{1}{3} \end{bmatrix} {\small \begin{matrix} -\frac{7}{5}R3+R1 \\ \frac{8}{5}R3+R2 \end{matrix}}$$

Thus, $A^{-1} = \begin{bmatrix} \frac{2}{3} & -\frac{1}{15} & -\frac{7}{15} \\ -\frac{1}{3} & -\frac{1}{15} & \frac{8}{15} \\ -\frac{1}{3} & \frac{1}{3} & \frac{1}{3} \end{bmatrix}$

Finally, we need to find $X = A^{-1}B$.

$$X = A^{-1}B = \begin{bmatrix} \frac{2}{3} & -\frac{1}{15} & -\frac{7}{15} \\ -\frac{1}{3} & -\frac{1}{15} & \frac{8}{15} \\ -\frac{1}{3} & \frac{1}{3} & \frac{1}{3} \end{bmatrix} \begin{bmatrix} -5 \\ -5 \\ 0 \end{bmatrix} = \begin{bmatrix} \frac{2}{3}(-5) + \left(-\frac{1}{15}\right)(-5) + \left(-\frac{7}{15}\right)(0) \\ -\frac{1}{3}(-5) + \left(-\frac{1}{15}\right)(-5) + \frac{8}{15}(0) \\ -\frac{1}{3}(-5) + \frac{1}{3}(-5) + \frac{1}{3}(0) \end{bmatrix} = \begin{bmatrix} -\frac{10}{3} + \frac{1}{3} + 0 \\ \frac{5}{3} + \frac{1}{3} + 0 \\ \frac{5}{3} + \left(-\frac{5}{3}\right) + 0 \end{bmatrix} = \begin{bmatrix} -\frac{9}{3} \\ \frac{6}{3} \\ 0 \end{bmatrix} = \begin{bmatrix} -3 \\ 2 \\ 0 \end{bmatrix}$$

Solution set: $\{(-3, 2, 0)\}$

Chapter 9   Test

1. $3x - y = 9$ (1)

$x + 2y = 10$ (2)

Solving equation (2) for x, we obtain
$x = 10 - 2y$ (3). Substituting this result into
equation (1) and solving for y, we obtain
$3(10 - 2y) - y = 9 \Rightarrow 30 - 6y - y = 9 \Rightarrow$
$-7y = -21 \Rightarrow y = 3$

Substituting $y = 3$ back into equation (3) in order
to find x, we obtain $x = 10 - 2(3) = 10 - 6 = 4$.

Solution set: $\{(4, 3)\}$

2. $6x + 9y = -21$ (1)

$4x + 6y = -14$ (2)

Solving equation (1) for x, we obtain

$6x = -9y - 21 \Rightarrow x = \dfrac{-9y - 21}{6} \Rightarrow$

$x = \dfrac{-3y - 7}{2}$ (3).

Substitute this result into equation (2) and solve
for y.

$4\left(\dfrac{-3y - 7}{2}\right) + 6y = -14$

$2(-3y - 7) + 6y = -14$

$-6y - 14 + 6y = -14 \Rightarrow -14 = -14$

This true statement implies there are infinitely
many solutions. We can represent the solution
set with y as the arbitrary variable.

Solution set: $\left\{\left(\dfrac{-3y - 7}{2}, y\right)\right\}$

3. $x - 2y = 4$ (1)

$-2x + 4y = 6$ (2)

Multiply equation (1) by 2 and add the result to
equation (2).

$2x - 4y = 8$

$\underline{-2x + 4y = 6}$

$0 = 14$

The system is inconsistent. The solution set is
$\varnothing$.

4. $\frac{1}{4}x - \frac{1}{3}y = -\frac{5}{12}$ (1)

$\frac{1}{10}x + \frac{1}{5}y = \frac{1}{2}$ (2)

To eliminate fractions, multiply equation (1) by
12 and equation (2) by 10.

$3x - 4y = -5$ (3)

$x + 2y = 5$ (4)

Multiply equation (4) by 2 and add the result ;to
equation (3).

$3x - 4y = -5$

$\underline{2x + 4y = 10}$

$5x \qquad = 5 \Rightarrow x = 1$

Substituting $x = 1$ in equation (4) to find y, we
obtain $1 + 2y = 5 \Rightarrow 2y = 4 \Rightarrow y = 2$.

Solution set: $\{(1, 2)\}$

5. $2x + y + z = 3$ (1)

$x + 2y - z = 3$ (2)

$3x - y + z = 5$ (3)

Eliminate z first by adding equations (1) and (2)
to obtain $3x + 3y = 6$ (4). Add equations (2) and
(3) to eliminate z and obtain $4x + y = 8$ (5).
Multiply equation (5) by -3 and add the results
to equation (4).

$3x + 3y = 6$

$\underline{-12x - 3y = -24}$

$-9x \qquad = -18 \Rightarrow x = 2$

Substituting $x = 2$ in equation (5) to find y, we
have $4(2) + y = 8 \Rightarrow 8 + y = 8 \Rightarrow y = 0$.

Substituting $x = 2$ and $y = 0$ in equation (1) to
find z, we have $2(2) + 0 + z = 3 \Rightarrow z = -1$.

Solution set: $\{(2, 0, -1)\}$

6. Writing $\begin{array}{l} 3x - 2y = 13 \\ 4x - y = 19 \end{array}$ as an augmented matrix,

we have $\begin{bmatrix} 3 & -2 & | & 13 \\ 4 & -1 & | & 19 \end{bmatrix}$.

$\begin{bmatrix} 1 & -\frac{2}{3} & | & \frac{13}{3} \\ 4 & -1 & | & 19 \end{bmatrix} \begin{array}{l} \frac{1}{3}R1 \end{array} \Rightarrow$

$\begin{bmatrix} 1 & -\frac{2}{3} & | & \frac{13}{3} \\ 0 & \frac{5}{3} & | & \frac{5}{3} \end{bmatrix} \begin{array}{l} \\ -4R1 + R2 \end{array} \Rightarrow$

$\begin{bmatrix} 1 & -\frac{2}{3} & | & \frac{13}{3} \\ 0 & 1 & | & 1 \end{bmatrix} \begin{array}{l} \\ \frac{3}{5}R2 \end{array} \Rightarrow \begin{bmatrix} 1 & 0 & | & 5 \\ 0 & 1 & | & 1 \end{bmatrix} \begin{array}{l} \frac{2}{3}R2 + R1 \end{array}$

Solution set $\{(5, 1)\}$

7. Writing
$$3x - 4y + 2z = 15$$
$$2x - y + z = 13$$
$$x + 2y - z = 5$$
as an augmented matrix, we have $\begin{bmatrix} 3 & -4 & 2 & | & 15 \\ 2 & -1 & 1 & | & 13 \\ 1 & 2 & -1 & | & 5 \end{bmatrix}$.

$\begin{bmatrix} 1 & 2 & -1 & | & 5 \\ 2 & -1 & 1 & | & 13 \\ 3 & -4 & 2 & | & 15 \end{bmatrix}$ R1 $\leftrightarrow$ R3 $\Rightarrow$

$\begin{bmatrix} 1 & 2 & -1 & | & 5 \\ 0 & -5 & 3 & | & 3 \\ 0 & -10 & 5 & | & 0 \end{bmatrix}$ $\begin{matrix} \\ -2R1 + R2 \\ -3R1 + R3 \end{matrix}$ $\Rightarrow$

$\begin{bmatrix} 1 & 2 & -1 & | & 5 \\ 0 & 1 & -\frac{3}{5} & | & -\frac{3}{5} \\ 0 & -10 & 5 & | & 0 \end{bmatrix}$ $-\frac{1}{5}$R2 $\Rightarrow$

$\begin{bmatrix} 1 & 0 & \frac{1}{5} & | & \frac{31}{5} \\ 0 & 1 & -\frac{3}{5} & | & -\frac{3}{5} \\ 0 & 0 & -1 & | & -6 \end{bmatrix}$ $\begin{matrix} -2R2 + R1 \\ \\ 10R2 + R3 \end{matrix}$ $\Rightarrow$

$\begin{bmatrix} 1 & 0 & \frac{1}{5} & | & \frac{31}{5} \\ 0 & 1 & -\frac{3}{5} & | & -\frac{3}{5} \\ 0 & 0 & 1 & | & 6 \end{bmatrix}$ -1R3 $\Rightarrow$

$\begin{bmatrix} 1 & 0 & 0 & | & 5 \\ 0 & 1 & 0 & | & 3 \\ 0 & 0 & 1 & | & 6 \end{bmatrix}$ $\begin{matrix} -\frac{1}{5}R3 + R1 \\ \frac{3}{5}R3 + R2 \\ \end{matrix}$

Solution set: $\{(5, 3, 6)\}$

8. Because $y = ax^2 + bx + c$, and the points $(1, 5)$, $(2, 3)$, and $(4, 11)$ are on the graph, we have the following equations which are then simplified.
$$5 = a(1)^2 + b(1) + c \Rightarrow \quad a + b + c = 5 \quad (1)$$
$$3 = a(2)^2 + b(2) + c \Rightarrow \quad 4a + 2b + c = 3 \quad (2)$$
$$11 = a(4)^2 + b(4) + c \Rightarrow 16a + 4b + c = 11 \quad (3)$$

First, eliminate c by adding -1 times equation (1) to equation (2).
$$\begin{array}{r} -a - b - c = -5 \\ 4a + 2b + c = 3 \\ \hline 3a + b \quad = -2 \ (4) \end{array}$$

Eliminating c again by adding -1 times equation (1) to equation (3), we have the following.
$$\begin{array}{r} -a - b - c = -5 \\ 16a + 4b + c = 11 \\ \hline 15a + 3b \quad = 6 \ (5) \end{array}$$

Next, eliminate b by adding -3 times equation (4) to equation (5).

$$\begin{array}{r} -9a - 3b = 6 \\ 15a + 3b = 6 \\ \hline 6a \qquad = 12 \Rightarrow a = 2 \end{array}$$

Substituting this value into equation (4), we obtain $3(2) + b = -2 \Rightarrow 6 + b = -2 \Rightarrow b = -8$.

Substituting 2 for a and -8 for b into equation (1), we have
$$2 + (-8) + c = 5 \Rightarrow -6 + c = 5 \Rightarrow c = 11.$$

The equation of the parabola is
$$y = 2x^2 - 8x + 11.$$

9. Let x = number of units from Toronto;
y = number of units from Montreal;
z = number of units from Ottawa.
The information in the problem gives the system
$$x + y + z = 100$$
$$80x + 50y + 65z = 5990$$
$$x = z$$
Multiply the first equation by -50 and add to the second equation.
$$\begin{array}{r} -50x - 50y - 50z = -5000 \\ 80x + 50y + 65z = 5990 \\ \hline 30x \qquad + 15z = 990 \end{array}$$

Substitute x for z in this equation to obtain
$$30x + 15x = 990 \Rightarrow 45x = 990 \Rightarrow x = 22.$$
If $x = 22$, then $z = 22$. Substitute 22 for x and for z in the first equation and solve for y:
$$22 + y + 22 = 100 \Rightarrow y = 56$$

22 units from Toronto, 56 units from Montreal, and 22 units from Ottawa were ordered.

10. $\begin{vmatrix} 6 & 8 \\ 2 & -7 \end{vmatrix} = 6(-7) - 2(8) = -58$

11. $\begin{vmatrix} 2 & 0 & 8 \\ -1 & 7 & 9 \\ 12 & 5 & -3 \end{vmatrix}$

This determinant may be evaluated by expanding about any row or any column. If we expand by the first row, we have the following.

$\begin{vmatrix} 2 & 0 & 8 \\ -1 & 7 & 9 \\ 12 & 5 & -3 \end{vmatrix} = 2\begin{vmatrix} 7 & 9 \\ 5 & -3 \end{vmatrix} - 0\begin{vmatrix} -1 & 9 \\ 12 & -3 \end{vmatrix} + 8\begin{vmatrix} -1 & 7 \\ 12 & 5 \end{vmatrix}$
$$= 2[7(-3) - 5(9)] - 0$$
$$+ 8[(-1)(5) - 12(7)]$$
$$= 2(-21 - 45) + 8(-5 - 84)$$
$$= 2(-66) + 8(-89) = -844$$

12. $2x - 3y = -33$
$4x + 5y = 11$

$$D = \begin{vmatrix} 2 & -3 \\ 4 & 5 \end{vmatrix} = 2(5) - 4(-3) = 10 - (-12) = 22$$

$$D_x = \begin{vmatrix} -33 & -3 \\ 11 & 5 \end{vmatrix} = -33(5) - 11(-3)$$
$$= -165 - (-33) = -132$$

$$D_y = \begin{vmatrix} 2 & -33 \\ 4 & 11 \end{vmatrix} = 2(11) - 4(-33)$$
$$= 22 - (-132) = 154$$

$$x = \frac{D_x}{D} = \frac{-132}{22} = -6; \; y = \frac{D_y}{D} = \frac{154}{22} = 7$$

Solution set: $\left\{(-6, 7)\right\}$

13. $x + y - z = -4$
$2x - 3y - z = 5$
$x + 2y + 2z = 3$

Adding column 3 to columns 1 and 2, we have

$$D = \begin{vmatrix} 1 & 1 & -1 \\ 2 & -3 & -1 \\ 1 & 2 & 2 \end{vmatrix} = \begin{vmatrix} 0 & 0 & -1 \\ 1 & -4 & -1 \\ 3 & 4 & 2 \end{vmatrix}.$$

Expanding by row one, we have

$$D = -1 \begin{vmatrix} 1 & -4 \\ 3 & 4 \end{vmatrix} = -[4 - (-12)] = -16.$$

Adding -1 times row 1 to row 2 and 2 times row 1 to row 3, we have the following.

$$D_x = \begin{vmatrix} -4 & 1 & -1 \\ 5 & -3 & -1 \\ 3 & 2 & 2 \end{vmatrix} = \begin{vmatrix} -4 & 1 & -1 \\ 9 & -4 & 0 \\ -5 & 4 & 0 \end{vmatrix}$$

Expanding by column three, we have

$$D_x = -1 \begin{vmatrix} 9 & -4 \\ -5 & 4 \end{vmatrix} = -(36 - 20) = -16.$$

Adding -1 times row 1 to row 2 and 2 times row 1 to row 3, we have the following.

$$D_y = \begin{vmatrix} 1 & -4 & -1 \\ 2 & 5 & -1 \\ 1 & 3 & 2 \end{vmatrix} = \begin{vmatrix} 1 & -4 & -1 \\ 1 & 9 & 0 \\ 3 & -5 & 0 \end{vmatrix}.$$

Expanding by column three, we have

$$D_y = -1 \begin{vmatrix} 1 & 9 \\ 3 & -5 \end{vmatrix} = -1(-5 - 27) = -(-32) = 32.$$

Adding -2 times row 1 to row 2 and -1 times row 1 to row 3, we have

$$D_z = \begin{vmatrix} 1 & 1 & -4 \\ 2 & -3 & 5 \\ 1 & 2 & 3 \end{vmatrix} = \begin{vmatrix} 1 & 1 & -4 \\ 0 & -5 & 13 \\ 0 & 1 & 7 \end{vmatrix}.$$

Expanding by column one, we have

$$D_z = 1 \begin{vmatrix} -5 & 13 \\ 1 & 7 \end{vmatrix} = 1(-35 - 13) = -48.$$

Thus we have

$$x = \frac{D_x}{D} = \frac{-16}{-16} = 1, \; y = \frac{D_y}{D} = \frac{32}{-16} = -2,$$

and $z = \frac{D_z}{D} = \frac{-48}{-16} = 3.$

Solution set: $\{(1, -2, 3)\}$

14. $\dfrac{9x + 19}{x^2 + 2x - 3} = \dfrac{9x + 19}{(x+3)(x-1)} = \dfrac{A}{x+3} + \dfrac{B}{x-1}$

Multiply both sides by $(x+3)(x-1)$ to get

$$9x + 19 = A(x-1) + B(x+3). \quad (1)$$

Substitute -3 for x in equation (1), to get

$$9(-3) + 19 = A(-3-1) + B(-3+3) \Rightarrow$$
$$-8 = -4A \Rightarrow A = 2$$

Now substitute 2 for A and 1 for x in equation (1) to obtain

$$9(1) + 19 = 2(1-1) + B(1+3) \Rightarrow 28 = 4B \Rightarrow$$
$$B = 7$$

Thus,

$$\frac{9x+19}{x^2+2x-3} = \frac{9x+19}{(x+3)(x-1)} = \frac{2}{x+3} + \frac{7}{x-1}$$

15. $\dfrac{x+2}{x^3 + 2x^2 + x} = \dfrac{x+2}{x(x^2 + 2x + 1)} = \dfrac{x+2}{x(x+1)^2}$

$$= \frac{A}{x} + \frac{B}{x+1} + \frac{C}{(x+1)^2}$$

Multiply both sides by $x(x+1)^2$ to get

$$x + 2 = A(x+1)^2 + Bx(x+1) + Cx. \quad (1)$$

Substitute 0 for x in equation (1), to get

$$0 + 2 = A(0+1)^2 + B(0)(0+1) + C(0) \Rightarrow$$
$$A = 2$$

Substitute 2 for A and -1 for x in equation (1), we to obtain

$$-1 + 2 = 2(-1+1)^2 + B(-1)(-1+1) + C(-1)$$
$$1 = -C \Rightarrow C = -1$$

We now have

$$x + 2 = 2(x+1)^2 + Bx(x+1) + (-1)x \quad (2).$$

(continued on next page)

(*continued*)

Now substitute 1 for x (arbitrary choice) in equation (2) to obtain the following.

$$1+2 = 2(1+1)^2 + B(1)(1+1) + (-1)(1)$$
$$3 = 2(4) + 2B - 1$$
$$3 = 8 + 2B - 1 \Rightarrow 3 = 7 + 2B$$
$$-4 = 2B \Rightarrow B = -2$$

Thus,

$$\frac{x+2}{x^3+2x^2+x} = \frac{x+2}{x(x^2+2x+1)} = \frac{x+2}{x(x+1)^2}$$
$$= \frac{2}{x} + \frac{-2}{x+1} + \frac{-1}{(x+1)^2}.$$

16. $2x^2 + y^2 = 6$ (1)

$x^2 - 4y^2 = -15$ (2)

Multiply equation (1) by 4 and add to equation (2), then solve for x.

$$8x^2 + 4y^2 = 24$$
$$\underline{x^2 - 4y^2 = -15}$$
$$9x^2 = 9 \Rightarrow x^2 = 1 \Rightarrow x = \pm 1$$

Substitute these values into equation (1) and solve for y.

If $x = 1$, then $2(1)^2 + y^2 = 6 \Rightarrow 2 + y^2 = 6 \Rightarrow$

$y^2 = 4 \Rightarrow y = \pm 2.$

If $x = -1$, then $2(-1)^2 + y^2 = 6 \Rightarrow$

$2 + y^2 = 6 \Rightarrow y^2 = 4 \Rightarrow y = \pm 2.$

Solution set: $\{(1,2),(-1,2),(1,-2),(-1,-2)\}$

17. $x^2 + y^2 = 25$ (1)

$x + y = 7$ (2)

Solving equation (2) for x, we have $x = 7 - y$ (3). Substituting this result into equation (1), we have

$$(7-y)^2 + y^2 = 25$$
$$49 - 14y + y^2 + y^2 = 25$$
$$2y^2 - 14y + 24 = 0$$
$$y^2 - 7y + 12 = 0$$
$$(y-3)(y-4) = 0 \Rightarrow y = 3 \text{ or } y = 4$$

Substitute these values into equation (3) and solve for x. If $y = 3$, $x = 7 - 3 = 4$.

If $y = 4$, $x = 7 - 4 = 3$.

Solution set: $\{(3, 4), (4, 3)\}$

18. Let x and y be the numbers.

$x + y = -1$ (1)

$x^2 + y^2 = 61$ (2)

Solving equation (1) for y, we have $y = -x - 1$ (3). Substituting this result into equation (2), we have

$$x^2 + (-x-1)^2 = 61 \Rightarrow x^2 + x^2 + 2x + 1 = 61$$
$$2x^2 + 2x - 60 = 0 \Rightarrow x^2 + x - 30 = 0$$
$$(x+6)(x-5) = 0 \Rightarrow x = -6 \text{ or } x = 5$$

Substitute these values in equation (3) to find the corresponding values of y.

If $x = -6$, $y = -(-6) - 1 = 6 - 1 = 5$.

If $x = 5$, $y = -5 - 1 = -6$.

The same pair of numbers results from both cases. The numbers are 5 and –6.

19. $x - 3y \geq 6$

$y^2 \leq 16 - x^2$

Graph $x - 3y = 6$ as a solid line with x-intercept 6 and y-intercept of –2. Because the test point $(0,0)$ does not satisfies this inequality, shade the region *below* the line. Graph $y^2 = 16 - x^2$ or $x^2 + y^2 = 16$ as a solid circle with a center at the origin and radius 4. Shade the region, which is the interior of the circle. The solution set is the intersection of these two regions, which is the region shaded in the final graph.

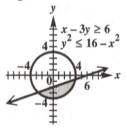

20. Find $x \geq 0$ and $y \geq 0$ such that

$$x + 2y \leq 24$$
$$3x + 4y \leq 60$$

and $2x + 3y$ is maximized.

Find the region of feasible solutions by graphing the system of inequalities that is made up of the constraints. To graph $x + 2y \leq 24$, draw the line with x-intercept 24 and y-intercept 12 as a solid line. Because the test point (0, 0) satisfies this inequality, shade the region *below* the line. To graph $3x + 4y \leq 60$, draw the line with x-intercept 20 and y-intercept 15 as a solid line. Because the test point (0, 0) satisfies this inequality, shade the region *below* the line.

(*continued on next page*)

(*continued*)

The constraints $x \geq 0$ and $y \geq 0$ restrict the graph to the first quadrant. The graph of the feasible region is the intersection of the regions that are the graphs of the individual constraints.

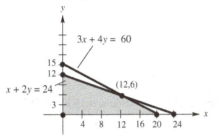

From the graph, observe that three vertices are $(0, 0)$, $(0, 12)$, and $(20, 0)$. A fourth vertex is the intersection point of the lines $x + y = 24$ and $3x + 4y = 60$. To find this point, solve the system

$$x + 2y = 24 \quad (1)$$
$$3x + 4y = 60 \quad (2)$$

To eliminate x, multiply equation (1) by -3 and add the result to equation (2).

$$-3x - 6y = -72$$
$$\underline{3x + 4y = 60}$$
$$-2y = -12 \Rightarrow y = 6$$

Substituting this value into equation (1), we obtain $x + 2(6) = 24 \Rightarrow x + 12 = 24 \Rightarrow x = 12$.

The fourth vertex is $(12, 6)$. Next, evaluate the objective function at each vertex.

Point	Profit $= 2x + 3y$.
$(0, 0)$	$2(0) + 3(0) = 0$
$(0, 12)$	$2(0) + 3(12) = 36$
$(20, 0)$	$2(20) + 3(0) = 40$
$(12, 6)$	$2(12) + 3(6) = 42$ ← Maximum

The maximum value is 42 at $(12, 6)$.

21. Let x = the number of VIP rings; y = the number of SST rings.
The constraints are
$$x + y \leq 24$$
$$3x + 2y \leq 60$$
$$x \geq 0, \ y \geq 0.$$

Maximize profit function, profit $= 30x + 40y$. Find the region of feasible solutions by graphing the system of inequalities that is made up of the constraints. To graph $x + y \leq 24$, draw the line with x-intercept 24 and y-intercept 24 as a solid line. Because the test point $(0, 0)$ satisfies this inequality, shade the region *below* the line. To graph $3x + 2y \leq 60$, draw the line with x-intercept 20 and y-intercept 30 as a solid line.

Because the test point $(0, 0)$ satisfies this inequality, shade the region *below* the line. The constraints $x \geq 0$ and $y \geq 0$ restrict the graph to the first quadrant. The graph of the feasible region is the intersection of the regions that are the graphs of the individual constraints.

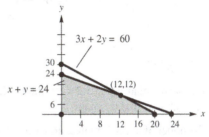

From the graph, observe that three vertices are $(0, 0)$, $(20, 0)$, and $(0, 24)$. A fourth vertex is the intersection point of the lines $x + y = 24$ and $3x + 2y = 60$. To find this point, solve the system

$$x + y = 24$$
$$3x + 2y = 60.$$

The first equation can be written as $x = 24 - y$. Substituting this equation into $3x + 2y = 60$, we have the following.

$$3(24 - y) + y = 60 \Rightarrow 72 - 3y + 2y = 60 \Rightarrow$$
$$72 - y = 60 \Rightarrow -y = -12 \Rightarrow y = 12$$

Substituting $y = 12$ into $x = 24 - y$, we have $x = 24 - 12 = 12$. Thus, the four vertex is $(12, 12)$. Next, evaluate the objective function at each vertex.

Point	Profit $= 30x + 40y$.
$(0, 0)$	$30(0) + 40(0) = 0$
$(20, 0)$	$30(20) + 40(0) = 600$
$(0, 24)$	$30(0) + 40(24) = 960$ ← Maximum
$(12, 12)$	$30(12) + 40(12) = 840$

0 VIP rings and 24 SST rings should be made daily for a daily profit of $960.

22. $\begin{bmatrix} 5 & x+6 \\ 0 & 4 \end{bmatrix} = \begin{bmatrix} y-2 & 4-x \\ 0 & w+7 \end{bmatrix}$

All corresponding elements, position by position, of the two matrices must be equal.

$x + 6 = 4 - x \Rightarrow 2x = -2 \Rightarrow x = -1;$
$5 = y - 2 \Rightarrow y = 7; 4 = w + 7 \Rightarrow w = -3$
Thus, $x = -1$, $y = 7$, and $w = -3$.

23. $3\begin{bmatrix} 2 & 3 \\ 1 & -4 \\ 5 & 9 \end{bmatrix} - \begin{bmatrix} -2 & 6 \\ 3 & -1 \\ 0 & 8 \end{bmatrix} = \begin{bmatrix} 3(2) & 3(3) \\ 3(1) & 3(-4) \\ 3(5) & 3(9) \end{bmatrix} - \begin{bmatrix} -2 & 6 \\ 3 & -1 \\ 0 & 8 \end{bmatrix} = \begin{bmatrix} 6 & 9 \\ 3 & -12 \\ 15 & 27 \end{bmatrix} + \begin{bmatrix} 2 & -6 \\ -3 & 1 \\ 0 & -8 \end{bmatrix}$

$= \begin{bmatrix} 6+2 & 9+(-6) \\ 3+(-3) & -12+1 \\ 15+0 & 27+(-8) \end{bmatrix} = \begin{bmatrix} 8 & 3 \\ 0 & -11 \\ 15 & 19 \end{bmatrix}$

24. $\begin{bmatrix} 1 \\ 2 \end{bmatrix} + \begin{bmatrix} 4 \\ -6 \end{bmatrix} + \begin{bmatrix} 2 & 8 \\ -7 & 5 \end{bmatrix}$

The first two matrices are 2×1 and the third is 2×2. Matrices must be the same size to be added, so it is not possible to find this sum.

25. The product of a 2×3 matrix and a 3×2 matrix is a 2×2 matrix.

$\begin{bmatrix} 2 & 1 & -3 \\ 4 & 0 & 5 \end{bmatrix} \begin{bmatrix} 1 & 3 \\ 2 & 4 \\ 3 & -2 \end{bmatrix} = \begin{bmatrix} 2(1)+1(2)+(-3)(3) & 2(3)+1(4)+(-3)(-2) \\ 4(1)+0(2)+5(3) & 4(3)+0(4)+5(-2) \end{bmatrix}$

$= \begin{bmatrix} 2+2+(-9) & 6+4+6 \\ 4+0+15 & 12+0+(-10) \end{bmatrix} = \begin{bmatrix} -5 & 16 \\ 19 & 2 \end{bmatrix}$

26. $\begin{bmatrix} 2 & -4 \\ 3 & 5 \end{bmatrix} \begin{bmatrix} 4 \\ 2 \\ 7 \end{bmatrix}$

The first matrix is 2×3 and the second is 3×1. The first matrix has two columns and the second has three rows, so it is not possible to find this product.

27. There are associative, distributive, and identity properties that apply to multiplication of matrices, but matrix multiplication is not commutative. The correct choice is A.

28. Find the inverse of $A = \begin{bmatrix} -8 & 5 \\ 3 & -2 \end{bmatrix}$, if it exists. The augmented matrix is $\begin{bmatrix} A | I_2 \end{bmatrix} = \begin{bmatrix} -8 & 5 & | & 1 & 0 \\ 3 & -2 & | & 0 & 1 \end{bmatrix}$.

$\begin{bmatrix} 1 & -\frac{5}{8} & | & -\frac{1}{8} & 0 \\ 3 & -2 & | & 0 & 1 \end{bmatrix} -\frac{1}{8}R1 \Rightarrow \begin{bmatrix} 1 & -\frac{5}{8} & | & -\frac{1}{8} & 0 \\ 0 & -\frac{1}{8} & | & \frac{3}{8} & 1 \end{bmatrix} -3R1+R2 \Rightarrow \begin{bmatrix} 1 & -\frac{5}{8} & | & -\frac{1}{8} & 0 \\ 0 & 1 & | & -3 & -8 \end{bmatrix} -8R2 \Rightarrow$

$\begin{bmatrix} 1 & 0 & | & -2 & -5 \\ 0 & 1 & | & -3 & -8 \end{bmatrix} \frac{5}{8}R2+R1$

Thus, $A^{-1} = \begin{bmatrix} -2 & -5 \\ -3 & -8 \end{bmatrix}$.

29. Find the inverse of $A = \begin{bmatrix} 4 & 12 \\ 2 & 6 \end{bmatrix}$, if it exists. The augmented matrix is $\begin{bmatrix} A | I_2 \end{bmatrix} = \begin{bmatrix} 4 & 12 & | & 1 & 0 \\ 2 & 6 & | & 0 & 1 \end{bmatrix}$.

$\begin{bmatrix} 1 & 3 & | & \frac{1}{4} & 0 \\ 2 & 6 & | & 0 & 1 \end{bmatrix} \frac{1}{4}R1 \Rightarrow \begin{bmatrix} 1 & 3 & | & \frac{1}{4} & 0 \\ 0 & 0 & | & -\frac{1}{2} & 1 \end{bmatrix} -2R1+R2$

The second row, second column element is now 0, so the desired transformation cannot be completed. Therefore, the inverse of the given matrix does not exist.

30. Find the inverse of $A = \begin{bmatrix} 1 & 3 & 4 \\ 2 & 7 & 8 \\ -2 & -5 & -7 \end{bmatrix}$, if it exists. Performing row operations on the augmented matrix, we

have the following.

$$\begin{bmatrix} A \mid I_3 \end{bmatrix} = \begin{bmatrix} 1 & 3 & 4 & 1 & 0 & 0 \\ 2 & 7 & 8 & 0 & 1 & 0 \\ -2 & -5 & -7 & 0 & 0 & 1 \end{bmatrix} \Rightarrow \begin{bmatrix} 1 & 3 & 4 & 1 & 0 & 0 \\ 0 & 1 & 0 & -2 & 1 & 0 \\ 0 & 1 & 1 & 2 & 0 & 1 \end{bmatrix} \begin{matrix} \\ -2R1+R2 \\ 2R1+R3 \end{matrix} \Rightarrow$$

$$\begin{bmatrix} 1 & 0 & 4 & 7 & -3 & 0 \\ 0 & 1 & 0 & -2 & 1 & 0 \\ 0 & 0 & 1 & 4 & -1 & 1 \end{bmatrix} \begin{matrix} -3R2+R1 \\ \\ -1R2+R3 \end{matrix} \Rightarrow \begin{bmatrix} 1 & 0 & 0 & -9 & 1 & -4 \\ 0 & 1 & 0 & -2 & 1 & 0 \\ 0 & 0 & 1 & 4 & -1 & 1 \end{bmatrix} \begin{matrix} -4R3+R1 \\ \\ \end{matrix}$$

Thus,

$$A^{-1} = \begin{bmatrix} -9 & 1 & -4 \\ -2 & 1 & 0 \\ 4 & -1 & 1 \end{bmatrix}.$$

31. The system $\begin{matrix} 2x + y = -6 \\ 3x - y = -29 \end{matrix}$ yields the matrix equation $AX = B$ where $A = \begin{bmatrix} 2 & 1 \\ 3 & -1 \end{bmatrix}$, $X = \begin{bmatrix} x \\ y \end{bmatrix}$, and $B = \begin{bmatrix} -6 \\ -29 \end{bmatrix}$.

Find A^{-1}. The augmented matrix is $\begin{bmatrix} A \mid I_2 \end{bmatrix} = \begin{bmatrix} 2 & 1 & 1 & 0 \\ 3 & -1 & 0 & 1 \end{bmatrix}$.

$$\begin{bmatrix} 1 & \frac{1}{2} & \frac{1}{2} & 0 \\ 3 & -1 & 0 & 1 \end{bmatrix} \begin{matrix} \frac{1}{2}R1 \\ \\ \end{matrix} \Rightarrow \begin{bmatrix} 1 & \frac{1}{2} & \frac{1}{2} & 0 \\ 0 & -\frac{5}{2} & -\frac{3}{2} & 1 \end{bmatrix} \begin{matrix} \\ -3R1+R2 \end{matrix} \Rightarrow$$

$$\begin{bmatrix} 1 & 0 & \frac{1}{5} & \frac{1}{5} \\ 0 & -\frac{5}{2} & -\frac{3}{2} & 1 \end{bmatrix} \begin{matrix} \frac{1}{5}R2+R1 \\ \\ \end{matrix} \Rightarrow \begin{bmatrix} 1 & 0 & \frac{1}{5} & \frac{1}{5} \\ 0 & 1 & \frac{3}{5} & -\frac{2}{5} \end{bmatrix} \begin{matrix} \\ -\frac{2}{5}R2 \end{matrix}$$

Thus,

$$A^{-1} = \begin{bmatrix} \frac{1}{5} & \frac{1}{5} \\ \frac{3}{5} & -\frac{2}{5} \end{bmatrix}.$$

Because $X = A^{-1}B$, we have

$$A^{-1}B = \begin{bmatrix} \frac{1}{5} & \frac{1}{5} \\ \frac{3}{5} & -\frac{2}{5} \end{bmatrix} \begin{bmatrix} -6 \\ -29 \end{bmatrix} = \begin{bmatrix} \frac{1}{5}(-6) + \frac{1}{5}(-29) \\ \frac{3}{5}(-6) + \left(-\frac{2}{5}\right)(-29) \end{bmatrix} = \begin{bmatrix} -\frac{6}{5} + \left(-\frac{29}{5}\right) \\ -\frac{18}{5} + \frac{58}{5} \end{bmatrix} = \begin{bmatrix} -\frac{35}{5} \\ \frac{40}{5} \end{bmatrix} = \begin{bmatrix} -7 \\ 8 \end{bmatrix}$$

Solution set: $\{(-7, 8)\}$

32. The system $\begin{matrix} x + y = 5 \\ y - 2z = 23 \\ x + 3z = -27 \end{matrix}$ yields the matrix equation $AX = B$ where $A = \begin{bmatrix} 1 & 1 & 0 \\ 0 & 1 & -2 \\ 1 & 0 & 3 \end{bmatrix}$,

$X = \begin{bmatrix} x \\ y \\ z \end{bmatrix}$, and $B = \begin{bmatrix} 5 \\ 23 \\ -27 \end{bmatrix}$.

Find A^{-1}. The augmented matrix is $\begin{bmatrix} A \mid I_3 \end{bmatrix} = \begin{bmatrix} 1 & 1 & 0 & 1 & 0 & 0 \\ 0 & 1 & -2 & 0 & 1 & 0 \\ 1 & 0 & 3 & 0 & 0 & 1 \end{bmatrix}$.

(continued on next page)

(*continued*)

$$\begin{bmatrix} 1 & 1 & 0 & | & 1 & 0 & 0 \\ 0 & 1 & -2 & | & 0 & 1 & 0 \\ 0 & -1 & 3 & | & -1 & 0 & 1 \end{bmatrix}\begin{matrix} \\ \\ -R1+R3 \end{matrix} \Rightarrow \begin{bmatrix} 1 & 0 & 2 & | & 1 & -1 & 0 \\ 0 & 1 & -2 & | & 0 & 1 & 0 \\ 0 & 0 & 1 & | & -1 & 1 & 1 \end{bmatrix}\begin{matrix} -1R2+R1 \\ \\ R2+R3 \end{matrix} \Rightarrow$$

$$\begin{bmatrix} 1 & 0 & 0 & | & 3 & -3 & -2 \\ 0 & 1 & 0 & | & -2 & 3 & 2 \\ 0 & 0 & 1 & | & -1 & 1 & 1 \end{bmatrix}\begin{matrix} -2R3+R1 \\ 2R3+R2 \\ \end{matrix}$$

Thus, $A^{-1} = \begin{bmatrix} 3 & -3 & -2 \\ -2 & 3 & 2 \\ -1 & 1 & 1 \end{bmatrix}$.

Because $X = A^{-1}B$, we have

$$A^{-1}B = \begin{bmatrix} 3 & -3 & -2 \\ -2 & 3 & 2 \\ -1 & 1 & 1 \end{bmatrix}\begin{bmatrix} 5 \\ 23 \\ -27 \end{bmatrix} = \begin{bmatrix} 3(5)+(-3)(23)+(-2)(-27) \\ -2(5)+3(23)+2(-27) \\ -1(5)+1(23)+1(-27) \end{bmatrix} = \begin{bmatrix} 15+(-69)+54 \\ -10+(69)+(-54) \\ -5+23+(-27) \end{bmatrix} = \begin{bmatrix} 0 \\ 5 \\ -9 \end{bmatrix} = X$$

Solution set: $\{(0, 5, -9)\}$

Chapter 10

ANALYTIC GEOMETRY

Section 10.1   Parabolas

1. **(a)** The relation $y - 2 = (x + 4)^2$ has vertex $(-4, 2)$ and opens up, so the correct choice is D.

   (b) The relation $y - 4 = (x + 2)^2$ has vertex $(-2, 4)$ and opens up, so the correct choice is B.

   (c) The relation $y - 2 = -(x + 4)^2$ has vertex $(-4, 2)$ and opens down, so the correct choice is C.

   (d) The relation $y - 4 = -(x + 2)^2$ has vertex $(-2, 4)$ and opens down, so the correct choice is A.

   (e) The relation $x - 2 = (y + 4)^2$ has vertex $(2, -4)$ and opens to the right, so the correct choice is F.

   (f) The relation $x - 4 = (y + 2)^2$ has vertex $(4, -2)$ and opens to the right, so the correct choice is H.

   (g) The relation $x - 2 = -(y + 4)^2$ has vertex $(2, -4)$ and opens to the left, so the correct choice is E.

   (h) The relation $x - 4 = -(y + 2)^2$ has vertex $(4, -2)$ and opens to the left, so the correct choice is G.

3. Vertex: $(2, -1)$; focus: $(2, 1)$; directrix: $y = -3$; axis of symmetry: $x = 2$; domain: $(-\infty, \infty)$, range: $[-1, \infty)$

5. $x + 4 = y^2 \Rightarrow x = y^2 - 4 \Rightarrow x - 0 = y^2 - 4$
   The vertex is $(0, -4)$. The graph opens to the right and has the same shape as $x = y^2$. It is a translation 4 units to the left of the graph of $x = y^2$. The domain is $[-4, \infty)$. The range is $(-\infty, \infty)$. The graph is symmetric about its axis, the horizontal line $y = 0$ (the x-axis).

Use the vertex and axis and plot a few additional points.

x	y
-4	0
-3	± 1
0	± 2

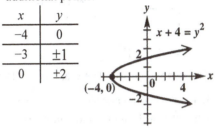

7. $x = (y - 3)^2 \Rightarrow x - 0 = (y - 3)^2$
   The vertex is $(0, 3)$. The graph opens to the right and has the same shape as $x = y^2$. It is a translation 3 units up of the graph of $x = y^2$. The domain is $[0, \infty)$. The range is $(-\infty, \infty)$. The graph is symmetric about its axis, the horizontal line $y = 3$. Use the vertex and axis and plot a few additional points.

x	y
4	1
1	2
0	3
1	4
4	5

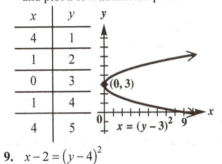

9. $x - 2 = (y - 4)^2$
   The vertex is $(2, 4)$. The graph opens to the right and has the same shape as $x = y^2$. It is a translation 2 units to the right and 4 units up of the graph of $x = y^2$. The domain is $[2, \infty)$. The range is $(-\infty, \infty)$. The graph is symmetric about its axis, the horizontal line $y = 4$. Use the vertex and axis and plot a few additional points.

(continued on next page)

(*continued*)

x	y
6	6
3	5
2	4
3	3
6	2

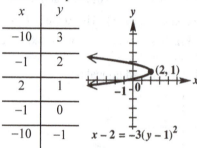

$x - 2 = (y - 4)^2$

11. $x - 2 = -3(y - 1)^2$

The vertex is $(2,1)$. The graph opens to the left and has the same shape as $x = -3y^2$. It is a translation 1 unit up and 2 units to the right of the graph of $x = -3y^2$. The domain is $(-\infty, 2]$. The range is $(-\infty, \infty)$. The graph is symmetric about its axis, the horizontal line $y = 1$. Use the vertex and axis and plot a few additional points.

x	y
-10	3
-1	2
2	1
-1	0
-10	-1

$(2, 1)$

$x - 2 = -3(y - 1)^2$

13. $-\frac{1}{2}x = (y + 3)^2 \Rightarrow x = -2(y + 3)^2 \Rightarrow$

$x - 0 = -2\left[y - (-3)\right]^2$

The vertex is $(0, -3)$. The graph opens to the left and has the same shape as $x = -2y^2$. It is a translation 3 units down of the graph of $x = -2y^2$. The domain is $(-\infty, 0]$. The range is $(-\infty, \infty)$. The graph is symmetric about its axis, the horizontal line $y = -3$. Use the vertex and axis and plot a few additional points.

x	y
-8	-1
-2	-2
0	-3
-2	-4
-8	-5

$-\frac{1}{2}x = (y + 3)^2$

$(0, -3)$

15. $x = y^2 + 4y + 2$

Complete the square on y to find the vertex and the axis.

$x = y^2 + 4y + 2 \Rightarrow x = \left(y^2 + 4y \quad \right) + 2$

$x = \left(y^2 + 4y + 4 - 4\right) + 2$

$x = \left(y^2 + 4y + 4\right) - 4 + 2$

$x = (y + 2)^2 - 2 \Rightarrow x - (-2) = \left[y - (-2)\right]^2$

The vertex is $(-2, -2)$. The graph opens to the right and has the same shape as $x = y^2$, translated 2 units to the left and 2 units down. The domain is $[-2, \infty)$. The range is $(-\infty, \infty)$. The graph is symmetric about its axis, the horizontal line $y = -2$. Use the vertex and axis and plot a few additional points.

x	y
2	0
-1	-1
-2	-2
-1	-3
2	-4

$(-2, -2)$

$x = y^2 + 4y + 2$

17. $x = -4y^2 - 4y + 3$

Complete the square on y to find the vertex and the axis.

$x = -4y^2 - 4y + 3 \Rightarrow x = -4\left(y^2 + y \quad \right) + 3$

$x = -4\left(y^2 + y + \frac{1}{4} - \frac{1}{4}\right) + 3$

$x = -4\left(y^2 + y + \frac{1}{4}\right) - 4\left(-\frac{1}{4}\right) + 3$

$x = -4\left(y^2 + y + \frac{1}{4}\right) + 1 + 3$

$x = -4\left(y + \frac{1}{2}\right)^2 + 4 \Rightarrow x - 4 = -4\left[y - \left(-\frac{1}{2}\right)\right]^2$

The vertex is $\left(4, -\frac{1}{2}\right)$. The graph opens to the left and has the same shape as $x = -4y^2$, translated 4 units to the right and $\frac{1}{2}$ unit down. The domain is $(-\infty, 4]$. The range is $(-\infty, \infty)$. The graph is symmetric about its axis, the horizontal line $y = -\frac{1}{2}$. Use the vertex and axis and plot a few additional points.

(*continued on next page*)

(*continued*)

x	y
-12	$1\frac{1}{2}$
0	$\frac{1}{2}$
4	$-\frac{1}{2}$
0	$-1\frac{1}{2}$
-12	$-2\frac{1}{2}$

$\left(4, -\frac{1}{2}\right)$

$x = -4y^2 - 4y + 3$

19. $2x - y^2 + 4y - 6 = 0$

Solve for x, then complete the square on y to find the vertex and the axis.

$2x - y^2 + 4y - 6 = 0 \Rightarrow 2x = y^2 - 4y + 6 \Rightarrow$
$x = \frac{1}{2}y^2 - 2y + 3$

$x = \frac{1}{2}\left(y^2 - 4y \quad\right) + 3$

$x = \frac{1}{2}\left(y^2 - 4y + 4 - 4\right) + 3$

$x = \frac{1}{2}\left(y^2 - 4y + 4\right) + \frac{1}{2}(-4) + 3$

$x = \frac{1}{2}\left(y^2 - 4y + 4\right) - 2 + 3$

$x = \frac{1}{2}(y - 2)^2 + 1 \Rightarrow x - 1 = \frac{1}{2}(y - 2)^2$

The vertex is $(1, 2)$. The graph opens to the right and has the same shape as $x = \frac{1}{2}y^2$, translated 1 unit to the right and 2 units up. The domain is $[1, \infty)$. The range is $(-\infty, \infty)$.

The graph is symmetric about its axis, the horizontal line $y = 2$. Use the vertex and axis and plot a few additional points.

x	y
3	4
1.5	3
1	2
1.5	1
3	0

$(1, 2)$

$2x - y^2 + 4y - 6 = 0$

21. $-x = 3y^2 + 6y + 2 \Rightarrow x = -3y^2 - 6y - 2$

Complete the square on y to find the vertex and the axis.

$x = -3y^2 - 6y - 2$

$x = -3\left(y^2 + 2y \quad\right) - 2$

$x = -3\left(y^2 + 2y + 1 - 1\right) - 2$

$x = -3\left(y^2 + 2y + 1\right) - 3(-1) - 2$

$x = -3\left(y^2 + 2y + 1\right) + 3 - 2$

$x = -3(y + 1)^2 + 1$

The vertex is $(1, -1)$ The graph opens to the left and has the same shape as $x = -3y^2$, translated 1 unit to the right and 1 unit down. The domain is $(-\infty, 1]$. The range is $(-\infty, \infty)$.

The graph is symmetric about its axis, the horizontal line $y = -1$. Use the vertex and axis and plot a few additional points.

x	y
-11	-3
-2	-2
1	-1
-2	0
-11	1

$(1, -1)$

$-x = 3y^2 + 6y + 2$

23. The equation $x^2 = 24y$ has the form $x^2 = 4py$, so $4p = 24$, from which $p = 6$. Because the x-term is squared, the parabola is vertical, with focus $(0, p) = (0, 6)$ and directrix, $y = -p$, is $y = -6$. The vertex is $(0, 0)$ and the axis of the parabola is the y-axis.

25. The equation $y = -4x^2 \Rightarrow -\frac{1}{4}y = x^2$ has the form $x^2 = 4py$, so $4p = -\frac{1}{4}$, from which $p = -\frac{1}{16}$. The x-term is squared, so the parabola is vertical, with focus $(0, p) = \left(0, -\frac{1}{16}\right)$ and directrix, $y = -p$, is $y = \frac{1}{16}$. The vertex is $(0, 0)$, and the axis of the parabola is the y-axis.

27. The equation $y^2 = -4x$ has the form $y^2 = 4px$, so $4p = -4$, from which $p = -1$. Because the y-term is squared, the parabola is horizontal, with focus $(p, 0) = (-1, 0)$ and directrix, $x = -p$, is $x = 1$. The vertex is $(0, 0)$, and the axis of the parabola is the x-axis.

29. The equation $x = -32y^2 \Rightarrow -\frac{1}{32}x = y^2$ has the form $y^2 = 4px$, so $4p = -\frac{1}{32}$, from which $p = -\frac{1}{128}$. Because the y-term is squared, the parabola is horizontal, with focus $(p, 0) = \left(-\frac{1}{128}, 0\right)$ and directrix, $x = -p$, is $x = \frac{1}{128}$. The vertex is $(0, 0)$, and the axis of the parabola is the x-axis.

31. $(y-3)^2 = 12(x-1)$ has the form

$(y-k)^2 = 4p(x-h)$, with

$h = 1$, $k = 3$, and $4p = 12$, so $p = 3$. The graph of the given equation is a parabola with vertical axis. The vertex (h, k) is $(1, 3)$. Because this parabola has a horizontal axis and $p > 0$, the parabola opens right, so the focus is the distance $p = 3$ units to the right of the vertex. Thus the focus is $(4, 3)$ The directrix is the vertical line $p = 3$ units to the left of the vertex, so the directrix is the line $x = -2$. The axis is the horizontal line through the vertex, so the equation of the axis is $y = 3$.

33. $(x-7)^2 = 16(y+5)$ can be written as

$(x-7)^2 = 16[y-(-5)]$. Thus, the parabola

has the form $(x-h)^2 = 4p(y-k)$, with $h =$ 7, $k = -5$, and $4p = 16$, so $p = 4$. The graph of the given equation is a parabola with vertical axis. The vertex (h, k) is $(7, -5)$. Because this parabola has a vertical axis and $p > 0$, the parabola opens up, so the focus is the distance $p = 4$ units above the vertex. Thus, the focus is $(7, -1)$. The directrix is the horizontal line $p = 4$ units below the vertex, so the directrix is the line $y = -9$. The axis is the vertical line through the vertex, so the equation of the axis is $x = 7$.

35. A parabola with focus $(5,0)$ and vertex at the origin is a horizontal parabola. The equation has the form $y^2 = 4px$. Because $p = 5$ is positive, it opens to the right. Substituting 5 for p, we find that an equation for this parabola is $y^2 = 4(5)x \Rightarrow y^2 = 20x$.

37. A parabola with directrix $y = -\frac{1}{4}$ and vertex at the origin is a vertical parabola with focus $\left(0, \frac{1}{4}\right)$. The equation has the form $x^2 = 4py$.

Because $p = \frac{1}{4}$ is positive, it opens up.

Substituting $\frac{1}{4}$ for p, we find that an equation for this parabola is $x^2 = 4\left(\frac{1}{4}\right)y \Rightarrow x^2 = y$.

39. A parabola passing through the point $\left(\sqrt{3},3\right)$, opening up, and vertex at the origin has an equation of the form $x^2 = 4py$. Use this equation with the coordinates of the point $\left(\sqrt{3},3\right)$ to find the value of p.

$x^2 = 4py \Rightarrow \left(\sqrt{3}\right)^2 = 4p \cdot 3 \Rightarrow$

$3 = 12p \Rightarrow \frac{1}{4} = p$

Thus, an equation of the parabola is

$x^2 = 4\left(\frac{1}{4}\right)y \Rightarrow x^2 = y$.

41. A parabola through the point $(3, 2)$, symmetric with respect to the x-axis, and vertex at the origin has a horizontal axis (x-axis) and the equation is of the form $y^2 = 4px$. Use this equation with the coordinates of the point $(3, 2)$ to find the value of p.

$y^2 = 4px \Rightarrow 2^2 = 4p \cdot 3 \Rightarrow 4 = 12p \Rightarrow p = \frac{1}{3}$

Thus, an equation for the parabola is

$y^2 = 4\left(\frac{1}{3}\right)x \Rightarrow y^2 = \frac{4}{3}x$.

43. The vertex is $(4, 3)$ and the focus is $(4, 5)$. The focus is above the vertex, so the axis is vertical and the parabola opens upward. The distance between the vertex and the focus is $5 - 3 = 2$. The parabola opens upward, so choose $p = 2$. The equation will have the form $(x-h)^2 = 4p(y-k)$. Substitute $p = 2$, $h = 4$, and $k = 3$ to find the required equation.

$(x-h)^2 = 4p(y-k)$

$(x-4)^2 = 4(2)(y-3) \Rightarrow (x-4)^2 = 8(y-3)$

45. The vertex is $(-5, 6)$ and the directrix is $x = -12$. The directrix is 7 units to the left of the vertex, so the focus is 7 units to the right of the vertex, at $(2, 6)$ and $p = 7$. The focus is to the right of the vertex, so the axis is horizontal and the parabola opens to the right. The equation will have the form $(y-k)^2 = 4p(x-h)$. Substitute $p = 7$, $h = -5$, and $k = 6$.

$(y-k)^2 = 4p(x-h)$

$(y-6)^2 = 4(7)[x-(-5)]$

$(y-6)^2 = 28(x+5)$

47. Complete the square on y.

$$x = 3y^2 + 6y - 4 \Rightarrow 3\left(y^2 + 2y \quad \right) - 4 = x \Rightarrow$$

$$3\left(y^2 + 2y + 1 - 1\right) - 4 = x \Rightarrow$$

$$3\left(y^2 + 2y + 1\right) + 3(-1) - 4 = x \Rightarrow$$

$$3(y+1)^2 - 7 = x \Rightarrow 3(y+1)^2 = x + 7 \Rightarrow$$

$$(y+1)^2 = \frac{x+7}{3} \Rightarrow y + 1 = \pm\sqrt{\frac{x+7}{3}} \Rightarrow$$

$$y = -1 \pm \sqrt{\frac{x+7}{3}}$$

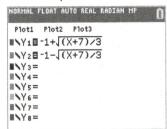

The graphing window is $[-10, 2] \times [-4, 4]$.

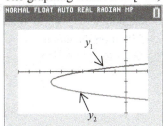

49. Solving for y we have

$$-(y+1)^2 = x + 2 \Rightarrow (y+1)^2 = -x - 2 \Rightarrow$$

$$y + 1 = \pm\sqrt{-x-2} \Rightarrow y = -1 \pm \sqrt{-x-2}$$

The graphing window is $[-10, 2] \times [-4, 4]$.

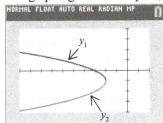

51. (a) Sketch a cross-section of the dish. Place this parabola on a coordinate system with the vertex at the origin. (This solution will show that the focus lies outside of the dish.)

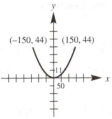

The parabola has vertex $(0, 0)$ and a vertical axis (the y-axis), so it has an equation of the form $x^2 = 4py$.

Substitute $x = 150$ and $y = 44$ to find the value of p.

$$x^2 = 4py \Rightarrow 150^2 = 4p \cdot 44 \Rightarrow$$

$$22{,}500 = 176p \Rightarrow p = \frac{22{,}500}{176} = \frac{5625}{44}$$

The required equation is

$$x^2 = 4\left(\frac{5625}{44}\right)y \Rightarrow x^2 = \frac{5625}{11}y$$

$$y = \frac{11}{5625}x^2$$

(b) Because $p = \frac{5625}{44} \approx 127.8,$ the receiver should be placed approximately 127.8 ft from the vertex.

53. Place the parabola that represents the arch on a coordinate system with the center of the bottom of the arch at the origin. Then the vertex will be $(0, 12)$ and the points $(-6, 0)$ and $(6, 0)$ will also be on the parabola. Because the axis of the parabola is the y-axis and the vertex is $(0, 12)$, the equation will have the form $x^2 = 4p(y - 12)$. Use the coordinates of the point $(6, 0)$ to find the value of p.

$$x^2 = 4p(y - 12) \Rightarrow 6^2 = 4p(0 - 12) \Rightarrow$$

$$36 = -48p \Rightarrow p = \frac{36}{-48} \Rightarrow p = -\frac{3}{4}$$

Thus, the equation of the parabola is

$$x^2 = 4\left(-\frac{3}{4}\right)(y - 12) \text{ or } x^2 = -3(y - 12).$$

Now find the x-coordinate of point whose y-coordinate is 9 and whose x-coordinate is positive: $x^2 = -3(y - 12) \Rightarrow$

$$x^2 = -3(9 - 12) \Rightarrow x^2 = 9 \Rightarrow x = \sqrt{9} = 3$$

From the symmetry of the parabola, we see that the width of the arch 9 ft up is $2(3 \text{ ft}) = 6$ ft.

55. (a) Locate the cannon at the origin. With $v = 252.982$, the equation becomes

$$y = x - \frac{32}{v^2}x^2 \Rightarrow y = x - \frac{32}{252.982^2}x^2 \Rightarrow$$

$$y \approx x - \frac{1}{2000}x^2$$

Complete the square on x.

$$y = x - \frac{1}{2000}x^2$$

$$y = -\frac{1}{2000}\left(x^2 - 2000x \right)$$

$$y = -\frac{1}{2000}\left(\begin{array}{l} x^2 - 2000x + 1,000,000 \\ -1,000,000 \end{array}\right)$$

$$y = -\frac{1}{2000}\left(x^2 - 2000x + 1,000,000\right)$$
$$+ \left(-\frac{1}{2000}\right)\left(-1,000,000\right)$$

$$y = -\frac{1}{2000}\left(x - 1000\right)^2 + 500$$

$$y - 500 = -\frac{1}{2000}\left(x - 1000\right)^2$$

Thus, the vertex of the parabola is located at $(1000, 500)$. Because of symmetry, the shell then travels an additional 1000 feet for a maximum distance of 2000 feet.

(b) The envelope parabola has x-intercepts located at $(-2000, 0)$ and $(2000, 0)$. The vertex is easily found to be located at $(0, 1000)$. Because the axis of the parabola is the y-axis and it opens down, the equation is of the form

$x^2 = 4p(y - 1000)$. Use the coordinates of the point $(2000, 0)$ to find the value of p. $2000^2 = 4p(0 - 1000) \Rightarrow$

$$4,000,000 = -4000p \Rightarrow p = -1000$$

The equation of the parabola is

$$x^2 = 4(-1000)(y - 1000)$$

$$x^2 = -4000(y - 1000)$$

$$y - 1000 = -0.00025x^2$$

$$y = 1000 - 0.00025x^2$$

(c) Using the equation of the envelope parabola in part (b), we calculate the maximum possible height of a shell when x is 1500 feet.

$$y = 1000 - 0.00025x^2$$

$$y = 1000 - 0.00025(1500)^2$$

$$y = 1000 - 0.00025(2,250,000)$$

$$y = 1000 - 562.5 \Rightarrow y = 437.5$$

If the helicopter flies at a height of 450 feet, a shell fired by the cannon would never reach the helicopter.

57. (a) $y = \sqrt{3}x - \frac{g}{3872}x^2$

For the moon, $g = 5.2$, so the equation is $y = \sqrt{3}x - \frac{5.2}{3872}x^2 = \sqrt{3}x - \frac{1.3}{968}x^2$.

For Mars, $g = 12.6$ (see exercise 56), so the equation is

$$y = \sqrt{3}x - \frac{12.6}{3872}x^2 = \sqrt{3}x - \frac{6.3}{1936}x^2.$$

Graph both equations on the same screen.

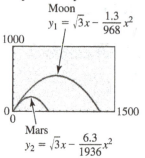

Moon
$$y_1 = \sqrt{3}x - \frac{1.3}{968}x^2$$

Mars
$$y_2 = \sqrt{3}x - \frac{6.3}{1936}x^2$$

(b) Using the "maximum" option in the CALC menu, we see that the ball reaches a maximum height of $y \approx 230$ ft on Mars and $y \approx 558$ ft on the moon.

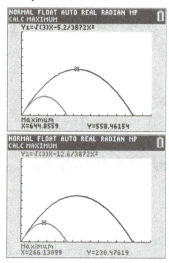

59. The equation is of the form $x = ay^2 + by + c$.

Substituting $x = -5, y = 1$, we get

$$-5 = a(1)^2 + b(1) + c \Rightarrow -5 = a + b + c \quad (1)$$

Substituting $x = -14, y = -2$, we get

$$-14 = a(-2)^2 + b(-2) + c \Rightarrow$$

$$-14 = 4a - 2b + c \quad (2)$$

Substituting $x = -10, y = 2$, we obtain

$$-10 = a(2)^2 + b(2) + c \Rightarrow$$

$$-10 = 4a + 2b + c. \quad (3)$$

61. Because $a = -2 < 0$, the parabola opens to the left.

Section 10.2 Ellipses

1. **(a)** $36x^2 + 9y^2 = 324 \Rightarrow \dfrac{x^2}{9} + \dfrac{y^2}{36} = 1 \Rightarrow$

    $\dfrac{x^2}{3^2} + \dfrac{y^2}{6^2} = 1$

    This ellipse has endpoints of the minor axis of $(\pm 3, 0)$, which are also x-intercepts. It also has vertices of $(0, \pm 6)$, which are also y-intercepts. The correct choice is A.

    (b) $9x^2 + 36y^2 = 324 \Rightarrow \dfrac{x^2}{36} + \dfrac{y^2}{9} = 1 \Rightarrow$

    $\dfrac{x^2}{6^2} + \dfrac{y^2}{3^2} = 1$

    This ellipse has vertices of $(\pm 6, 0)$, which are also x-intercepts. It also has endpoints of the minor axis of $(0, \pm 3)$, which are also y-intercepts. The correct choice is C.

    (c) $\dfrac{x^2}{25} + \dfrac{y^2}{16} = 1 \Rightarrow \dfrac{x^2}{5^2} + \dfrac{y^2}{4^2} = 1$

    This ellipse has vertices of $(\pm 5, 0)$, which are also x-intercepts. It also has endpoints of the minor axis of $(0, \pm 4)$, which are also y-intercepts. The correct choice is D.

    (d) $\dfrac{x^2}{16} + \dfrac{y^2}{25} = 1 \Rightarrow \dfrac{x^2}{4^2} + \dfrac{y^2}{5^2} = 1$

    This ellipse has endpoints of the minor axis of $(\pm 4, 0)$, which are also x-intercepts. It also has vertices of $(0, \pm 5)$, which are also y-intercepts. The correct choice is B.

For exercises 3–5, find the c, the distance from the center to a focus, using the formula $c^2 = a^2 - b^2$.

3. Domain: [–4, 4]; range: [–3, 3]; center: (0, 0); vertices: (–4, 0), (4, 0); foci: $\left(-\sqrt{7}, 0\right), \left(\sqrt{7}, 0\right)$

5. Domain: [–3, 7]; range: [–1, 3]; center: (2, 1); vertices: (–3, 1), (7, 1); foci: $\left(2 - \sqrt{21}, 1\right), \left(2 + \sqrt{21}, 1\right)$

7. $\dfrac{x^2}{25} + \dfrac{y^2}{9} = 1$

    The graph is an ellipse with center (0, 0). Rewriting the given equation, we have

    $\dfrac{x^2}{5^2} + \dfrac{y^2}{3^2} = 1$. Because 5 > 3, we have a = 5 and b = 3, and the major axis is horizontal. Thus, the vertices are (–5, 0) and (5, 0). The endpoints of the minor axis are (0, –3) and (0, 3). The domain is [–5, 5]. The range is [–3, 3].

    To find the foci we need to find c such that $c^2 = a^2 - b^2$.

    $c^2 = a^2 - b^2 \Rightarrow c^2 = 25 - 9 \Rightarrow c^2 = 16 \Rightarrow c = 4$

    The major axis lies on the x-axis, so the foci are (–4, 0) and (4, 0).

    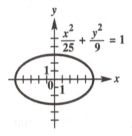

9. $\dfrac{x^2}{9} + y^2 = 1$

    Rewriting the equation, we have $\dfrac{x^2}{3^2} + \dfrac{y^2}{1^2} = 1$.

    The center is (0, 0). Because 3 > 1, we have a = 3 and b = 1, and the major axis is horizontal. The vertices are (–3, 0) and (3, 0). The endpoints of the minor axis are (0, –1) and (0, 1). The domain is [–3, 3]. The range is [–1, 1]. To find the foci, we need to find c such that $c^2 = a^2 - b^2$.

    $c^2 = a^2 - b^2 \Rightarrow c^2 = 9 - 1 \Rightarrow c^2 = 8 \Rightarrow$
    $c = \sqrt{8} = 2\sqrt{2}$

    Because the major axis lies on the x-axis, the foci are $(-2\sqrt{2}, 0)$ and $(2\sqrt{2}, 0)$.

    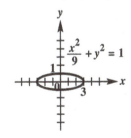

11. $9x^2 + y^2 = 81$

Rewriting the equation, we have

$\frac{x^2}{9} + \frac{y^2}{81} = 1$ or $\frac{x^2}{3^2} + \frac{y^2}{9^2} = 1$. The center is

$(0, 0)$. Because $9 > 3$, we have $a = 9$ and $b = 3$, and the major axis is vertical. The vertices are $(0, -9)$ and $(0, 9)$. The endpoints of the minor axis are $(-3, 0)$ and $(3, 0)$. The domain is $[-3, 3]$. The range is $[-9, 9]$. To find the foci, we need to find c such that $c^2 = a^2 - b^2$.

$c^2 = a^2 - b^2 \Rightarrow c^2 = 81 - 9 \Rightarrow c^2 = 72 \Rightarrow$
$c = \sqrt{72} = 6\sqrt{2}$

Because the major axis lies on the y-axis, the foci are $\left(0, -6\sqrt{2}\right)$ and $\left(0, 6\sqrt{2}\right)$.

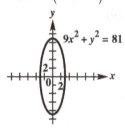

13. $4x^2 = 100 - 25y^2 \Rightarrow 4x^2 + 25y^2 = 100$

Rewriting the equation, we have $\frac{x^2}{25} + \frac{y^2}{4} = 1$

or $\frac{x^2}{5^2} + \frac{y^2}{2^2} = 1$. The center is $(0, 0)$. Because $5 > 2$, we have $a = 5$ and $b = 2$, and the major axis is horizontal. The vertices are $(-5, 0)$ and $(5, 0)$. The endpoints of the minor axis are $(0, -2)$ and $(0, 2)$. The domain is $[-5, 5]$. The range is $[-2, 2]$. To find the foci, we need to find c such that $c^2 = a^2 - b^2$.

$c^2 = a^2 - b^2 \Rightarrow c^2 = 25 - 4 \Rightarrow$
$c^2 = 21 \Rightarrow c = \sqrt{21}$

Because the major axis lies on the x-axis, the foci are $\left(-\sqrt{21}, 0\right)$ and $\left(\sqrt{21}, 0\right)$.

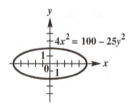

15. $\frac{(x-2)^2}{25} + \frac{(y-1)^2}{4} = 1$

Rewriting the equation, we have

$\frac{(x-2)^2}{5^2} + \frac{(y-1)^2}{2^2} = 1$. The center is $(2, 1)$.

Because $a = 5$ is associated with x^2, the major axis of the ellipse is horizontal. The vertices are on a horizontal line through $(2, 1)$, while the endpoints of the minor axis are on the vertical line through $(2, 1)$. The vertices are 5 units to the left and right of the center at $(-3, 1)$ and $(7, 1)$. The endpoints of the minor axis are 2 units below and 2 units above the center at $(2, -1)$ and $(2, 3)$. The domain is $[-3, 7]$. The range is $[-1, 3]$. To find the foci, we need to find c such that $c^2 = a^2 - b^2$.

$c^2 = a^2 - b^2 \Rightarrow c^2 = 25 - 4 \Rightarrow c^2 = 21 \Rightarrow$
$c = \sqrt{21}$

Because the major axis lies on $y = 1$, the foci are $\left(2 - \sqrt{21}, 1\right)$ and $\left(2 + \sqrt{21}, 1\right)$.

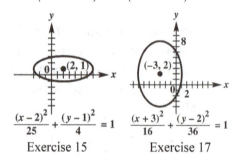

$\frac{(x-2)^2}{25} + \frac{(y-1)^2}{4} = 1$ $\frac{(x+3)^2}{16} + \frac{(y-2)^2}{36} = 1$

Exercise 15   Exercise 17

17. $\frac{(x+3)^2}{16} + \frac{(y-2)^2}{36} = 1$

Rewriting the equation, we have

$\frac{(x+3)^2}{4^2} + \frac{(y-2)^2}{6^2} = 1$. The center is $(-3, 2)$. We

have $a = 6$ and $b = 4$. Because $a = 6$ is associated with y^2, the major axis of the ellipse is vertical.

The vertices are on the vertical line through $(-3, 2)$, and the endpoints of the minor axis are on the horizontal line through $(-3, 2)$. The vertices are 6 units below and 6 units above the center at $(-3, -4)$ and $(-3, 8)$. The endpoints of the minor axis are 4 units to the left and 4 units to the right of the center at $(-7, 2)$ and $(1, 2)$. The domain is $[-7, 1]$. The range is $[-4, 8]$. To find the foci, we need to find c such that $c^2 = a^2 - b^2$.

$c^2 = a^2 - b^2 \Rightarrow c^2 = 36 - 16 \Rightarrow c^2 = 20 \Rightarrow$
$c = \sqrt{20} = 2\sqrt{5}$

Because the major axis lies on $x = -3$, the foci are $\left(-3, 2 - 2\sqrt{5}\right)$ and $\left(-3, 2 + 2\sqrt{5}\right)$.

19. x-intercepts $(\pm 5, 0)$; y-intercepts $(0, \pm 4)$

From the given information, $a = 5$ and $b = 4$. Because $5 > 4$, the x-intercepts represent vertices. The equation has the form $\frac{x^2}{a^2} + \frac{y^2}{b^2} = 1$. Because $a^2 = 25$ and $b^2 = 16$, the equation of the ellipse is $\frac{x^2}{25} + \frac{y^2}{16} = 1$.

21. Major axis with length 6, foci at $(0, 2)$, $(0, -2)$.

The length of the major axis is $2a$. Thus, we have $2a = 6 \Rightarrow a = 3$. From the foci, we have $c = 2$. Solving for b^2 we have
$$c^2 = a^2 - b^2 \Rightarrow 4 = 9 - b^2 \Rightarrow b^2 = 5.$$
The foci are on the y-axis and the ellipse is centered at the origin, so the equation has the form $\frac{x^2}{b^2} + \frac{y^2}{a^2} = 1$. Because $a = 3 \Rightarrow a^2 = 9$ and and $b^2 = 5$, the equation of the ellipse is $\frac{x^2}{5} + \frac{y^2}{9} = 1$.

23. Center at $(3, 1)$; minor axis vertical, with length 8; $c = 3$.

The center is $(3, 1)$ and the minor axis is vertical, so the equation has the form $\frac{(x-3)^2}{a^2} + \frac{(y-1)^2}{b^2} = 1$. The length of the minor axis is $2b$, so $b = 4$. Solving for a^2, we have
$$c^2 = a^2 - b^2 \Rightarrow 9 = a^2 - 16 \Rightarrow a^2 = 25.$$
Because $b = 4 \Rightarrow b^2 = 16$ and $a^2 = 25$, the equation of the ellipse is $\frac{(x-3)^2}{25} + \frac{(y-1)^2}{16} = 1$.

25. Foci at $(0, 4)$, $(0, -4)$; sum of distances from foci to point on ellipse is 10.

The center is halfway between the foci, so the center is $(0, 0)$. The distance between the foci is $4 - (-4) = 8$, so $2c = 8$ and thus $c = 4$. The sum of the distances from the foci to any point on the ellipse is 10, so $2a = 10 \Rightarrow a = 5$. The foci lie on the y-axis, so the equation is of the form $\frac{x^2}{b^2} + \frac{y^2}{a^2} = 1$. Solving for b^2, we have
$$c^2 = a^2 - b^2 \Rightarrow 4^2 = 5^2 - b^2 \Rightarrow$$
$$16 = 25 - b^2 \Rightarrow b^2 = 9.$$
Because $a = 5 \Rightarrow a^2 = 25$ and $b^2 = 9$, the equation of the ellipse is $\frac{x^2}{9} + \frac{y^2}{25} = 1$.

27. Foci at $(0, -3)$, $(0, 3)$; $(8, 3)$ on the ellipse.

The distance between the foci is $3 - (-3) = 6$, so $2c = 6$ and thus $c = 3$. The center is the midpoint between the foci, so the center is $(0, 0)$. The foci lie on the y-axis and the major axis is vertical, so the equation is of the form $\frac{x^2}{b^2} + \frac{y^2}{a^2} = 1$. Now recall from Figure 15 on page 970 of the text, $d(P, F) + d(P, F') = 2a$. If we let $P(8, 3)$, $F'(0, -3)$, and $F(0, 3)$ represent the point on the ellipse and the foci, respectively, we have the following.
$$d(P, F) + d(P, F')$$
$$= \sqrt{(8-0)^2 + (3+3)^2} + \sqrt{(8-0)^2 + (3-3)^2}$$
$$= 10 + 8 = 18 = 2a \Rightarrow a = 9$$
Solving for b^2, we have
$$c^2 = a^2 - b^2 \Rightarrow 9 = 81 - b^2 \Rightarrow b^2 = 72.$$
Because $a = 9 \Rightarrow a^2 = 81$ and $b^2 = 72$, the equation of the ellipse is $\frac{x^2}{72} + \frac{y^2}{81} = 1$.

29. $e = \frac{4}{5}$; vertices at $(-5, 0)$, $(5, 0)$.

From the vertices, we have $a = 5$ and $a^2 = 25$. The vertices lie on the x-axis, so the equation is of the form $\frac{x^2}{a^2} + \frac{y^2}{b^2} = 1$. Thus, the equation has the form $\frac{x^2}{25} + \frac{y^2}{b^2} = 1$. Use the eccentricity and a to find c.
$$e = \frac{c}{a} \Rightarrow \frac{4}{5} = \frac{c}{5} \Rightarrow 4 = c$$
Now solving for b^2, we have
$$c^2 = a^2 - b^2 \Rightarrow 4^2 = 5^2 - b^2 \Rightarrow$$
$$16 = 25 - b^2 \Rightarrow b^2 = 9 \Rightarrow b = 3$$
The equation is therefore $\frac{x^2}{25} + \frac{y^2}{9} = 1$.

31. $e = \frac{3}{4}$; foci at $(0, -2)$, $(0, 2)$.

The foci are on the y-axis, so the equation has the form $\frac{x^2}{b^2} + \frac{y^2}{a^2} = 1$. From the foci, we have $c = 2$. Use the eccentricity and c to find a.
$$e = \frac{c}{a} \Rightarrow \frac{3}{4} = \frac{2}{a} \Rightarrow 3a = 8 \Rightarrow a = \frac{8}{3}$$
Solving for b^2, we have
$$c^2 = a^2 - b^2 \Rightarrow 2^2 = \left(\frac{8}{3}\right)^2 - b^2 \Rightarrow$$
$$4 = \frac{64}{9} - b^2 \Rightarrow b^2 = \frac{64}{9} - \frac{36}{9} = \frac{28}{9}$$
Because $a = \frac{8}{3} \Rightarrow a^2 = \frac{64}{9}$ and $b^2 = \frac{28}{9}$, the equation is $\frac{x^2}{\frac{28}{9}} + \frac{y^2}{\frac{64}{9}} = 1$ or $\frac{9x^2}{28} + \frac{9y^2}{64} = 1$.

33. $\dfrac{y}{2} = \sqrt{1 - \dfrac{x^2}{25}}$

Square both sides to get

$\dfrac{y^2}{4} = 1 - \dfrac{x^2}{25} \Rightarrow \dfrac{x^2}{25} + \dfrac{y^2}{4} = 1 \Rightarrow \dfrac{x^2}{5^2} + \dfrac{y^2}{2^2} = 1,$

which is the equation of an ellipse centered at the origin with x-intercepts ± 5 (vertices) and y-intercepts ± 2 (endpoints of minor axis).

Because $\sqrt{1 - \dfrac{x^2}{25}} \geq 0,$ the only possible values

of y are those making $\dfrac{y}{2} \geq 0 \Rightarrow y \geq 0.$ The domain is $[-5, 5]$. The range is $[0, 2]$. The graph of the original equation is the upper half of the ellipse. By applying the vertical line test, we see that this is the graph of a function.

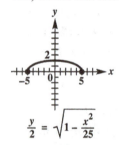

$$\dfrac{y}{2} = \sqrt{1 - \dfrac{x^2}{25}}$$

35. $x = -\sqrt{1 - \dfrac{y^2}{64}}$

Square both sides to get

$x^2 = 1 - \dfrac{y^2}{64} \Rightarrow \dfrac{x^2}{1} + \dfrac{y^2}{64} = 1 \Rightarrow \dfrac{x^2}{1^2} + \dfrac{y^2}{8^2} = 1,$ the

equation of an ellipse centered at the origin with x-intercepts ± 1 (endpoints of minor axis) and y-intercepts ± 8 (vertices). Because

$-\sqrt{1 - \dfrac{y^2}{64}} \leq 0,$ we must have $x \leq 0,$ so the

graph of the original equation is the left half of the ellipse. The domain is $[-1, 0]$. The range is $[-8, 8]$. The vertical line test shows that this is not the graph of a function.

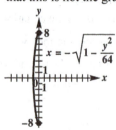

$$x = -\sqrt{1 - \dfrac{y^2}{64}}$$

37. Solve for y in the equation of the ellipse.

$\dfrac{x^2}{16} + \dfrac{y^2}{4} = 1 \Rightarrow \dfrac{y^2}{4} = 1 - \dfrac{x^2}{16} \Rightarrow$

$y^2 = 4\left(1 - \dfrac{x^2}{16}\right) \Rightarrow y = \pm\sqrt{4\left(1 - \dfrac{x^2}{16}\right)} \Rightarrow$

$y = \pm 2\sqrt{1 - \dfrac{x^2}{16}}$

The graphing window is $[-6.6, 6.6] \times [-4.1, 4.1]$.

$y_1 = 2\sqrt{1 - \dfrac{x^2}{16}}$

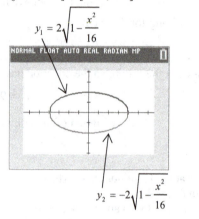

$y_2 = -2\sqrt{1 - \dfrac{x^2}{16}}$

39. Solve for y in the equation of the ellipse.

$\dfrac{(x-3)^2}{25} + \dfrac{y^2}{9} = 1 \Rightarrow \dfrac{y^2}{9} = 1 - \dfrac{(x-3)^2}{25} \Rightarrow$

$y^2 = 9\left(1 - \dfrac{(x-3)^2}{25}\right) \Rightarrow$

$y = \pm\sqrt{9\left(1 - \dfrac{(x-3)^2}{25}\right)} = \pm 3\sqrt{1 - \dfrac{(x-3)^2}{25}}$

The graphing window is $[-9.9, 9.9] \times [-8.2, 8.2]$.

$y_1 = 3\sqrt{1 - \dfrac{(x-3)^2}{25}}$

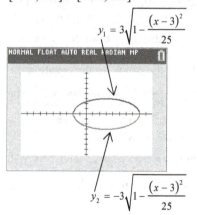

$y_2 = -3\sqrt{1 - \dfrac{(x-3)^2}{25}}$

41. $\dfrac{x^2}{3} + \dfrac{y^2}{4} = 1$

Because $4 > 3$, $a^2 = 4$, which gives $a = 2$.

Also $c = \sqrt{a^2 - b^2} = \sqrt{4 - 3} = \sqrt{1} = 1$. Thus,

$e = \dfrac{c}{a} = \dfrac{1}{2}$.

43. $4x^2 + 7y^2 = 28 \Rightarrow \dfrac{x^2}{7} + \dfrac{y^2}{4} = 1$

Because $7 > 4$, $a^2 = 7$, which gives $a = \sqrt{7}$.

Also $c = \sqrt{a^2 - b^2} = \sqrt{7 - 4} = \sqrt{3}$. Thus,

$e = \dfrac{c}{a} = \dfrac{\sqrt{7}}{\sqrt{3}} = \dfrac{\sqrt{21}}{3} \approx 0.65$.

45. Answers will vary. Sample answer: Fixing a string to two points (foci) and sweeping a curve as shown will sketch an ellipse because the sum of the distances from the two fixed points to the pencile's tip remains constant.

47. Place the half-ellipse that represents the overpass on a coordinate system with the center of the bottom of the overpass at the origin. If the complete ellipse were drawn, the center of the ellipse would be (0, 0). Then the half-ellipse will include the points (0, 15), (−10, 0), and (10, 0). The equation is of the form

$\dfrac{x^2}{b^2} + \dfrac{y^2}{a^2} = 1$. Thus, for the complete ellipse, we

have $a = 15$ and $b = 10$. Thus, we have

$\dfrac{x^2}{10^2} + \dfrac{y^2}{15^2} = 1 \Rightarrow \dfrac{x^2}{100} + \dfrac{y^2}{225} = 1$.

To find the equation of the half-ellipse, solve this equation for y and use the positive square root because the overpass is represented by the upper half of the ellipse.

$\dfrac{x^2}{100} + \dfrac{y^2}{225} = 1 \Rightarrow \dfrac{y^2}{225} = 1 - \dfrac{x^2}{100} \Rightarrow$

$y^2 = 225\left(1 - \dfrac{x^2}{100}\right) \Rightarrow y = \sqrt{225\left(1 - \dfrac{x^2}{100}\right)} \Rightarrow$

$y = 15\sqrt{1 - \dfrac{x^2}{100}}$

Find the y-coordinate of the point whose x-coordinate is $\frac{1}{2}(12) = 6$.

$y = 15\sqrt{1 - \dfrac{x^2}{100}} \Rightarrow y = 15\sqrt{1 - \dfrac{6^2}{100}} \Rightarrow$

$y = 15\sqrt{1 - \dfrac{36}{100}} = 15\sqrt{\dfrac{64}{100}} = 15\left(\dfrac{4}{5}\right) = 12$

The tallest truck that can pass under the overpass is 12 feet tall.

49. This problem is similar to Example 6 in the text. The greatest distance between the comet and the sun is $a + c = 3281$ million miles. If

$a + c = 3281 \Rightarrow c = 3281 - a$ and $e = \dfrac{c}{a}$, we

have $e = \dfrac{3281 - a}{a}$. Because $e = 0.9673$, we

have

$0.9673 = \dfrac{3281 - a}{a} \Rightarrow 0.9673a = 3281 - a \Rightarrow$

$1.9673a = 3281 \Rightarrow a = \dfrac{3281}{1.9673} \approx 1668$.

We then have $c = 3281 - 1668 = 1613$, which implies $a - c = 1668 - 1613 = 55$. Thus, the shortest distance between Halley's Comet and the sun is about 55 million miles.

51. (a) Use the given values of e and a to find the value of c for each planet. Then use the values of a and c to find the value of b. Neptune:

$e = \dfrac{c}{a} \Rightarrow c = ea \Rightarrow c = (0.009)(30.1) \Rightarrow$

$c = 0.2709$

$b^2 = a^2 - c^2 \Rightarrow b^2 = (30.1)^2 - (0.2709)^2 \Rightarrow$

$b^2 \approx 905.9366 \Rightarrow b \approx 30.1$

Because $c = 0.2709$, the graph should be translated 0.2709 units to the right so that the sun will be located at the origin. It's essentially circular with equation

$\dfrac{(x - 0.2709)^2}{30.1^2} + \dfrac{y^2}{30.1^2} = 1$.

Pluto:

$e = \dfrac{c}{a} \Rightarrow c = ea \Rightarrow c = (0.249)(39.4) \Rightarrow$

$c = 9.8106$

$b^2 = a^2 - c^2 \Rightarrow b^2 = (39.4)^2 - (9.8106)^2 \Rightarrow$

$b^2 \approx 1456.1121 \Rightarrow b \approx 38.2$

As with Neptune, we translate the graph by c units to the right so that the sun will be located at the origin. The equation is

$\dfrac{(x - 9.8106)^2}{39.4^2} + \dfrac{y^2}{38.2^2} = 1$.

(b) In order to graph these equations on a graphing calculator, we must solve each equation for y. Each equation will be broken down into two functions, so we will need to graph four functions.

Neptune:

$$\frac{(x-0.2709)^2}{30.1^2} + \frac{y^2}{30.1^2} = 1 \Rightarrow$$

$$(x-0.2709)^2 + y^2 = 30.1^2 \Rightarrow$$

$$y = \pm\sqrt{30.1^2 - (x-0.2709)^2}$$

Pluto:

$$\frac{(x-9.8106)^2}{39.4^2} + \frac{y^2}{38.2^2} = 1 \Rightarrow$$

$$\frac{y^2}{38.2^2} = 1 - \frac{(x-9.8106)^2}{39.4^2} \Rightarrow$$

$$y^2 = 38.2^2\left(1 - \frac{(x-9.8106)^2}{39.4^2}\right) \Rightarrow$$

$$y = \pm\sqrt{38.2^2\left(1 - \frac{(x-9.8106)^2}{39.4^2}\right)}$$

$$= \pm 38.2\sqrt{1 - \frac{(x-9.8106)^2}{39.4^2}}$$

Graph the following four functions on the same screen.

$$y_1 = \sqrt{30.1^2 - (x-0.2709)^2}$$

$$y_2 = -\sqrt{30.1^2 - (x-0.2709)^2} = -Y_1$$

$$y_3 = 38.2\sqrt{1 - \frac{(x-9.8106)^2}{39.4^2}}$$

$$y_4 = -38.2\sqrt{1 - \frac{(x-9.8106)^2}{39.4^2}} = -Y_3$$

53. The stone and the wave source must be placed at the foci, (c, 0) and (–c, 0). Here $a^2 = 36$ and $b^2 = 9$, so $c^2 = a^2 - b^2 \Rightarrow c^2 = 36 - 9 \Rightarrow c^2 = 27 \Rightarrow c = \sqrt{27} = 3\sqrt{3}$. Thus, the kidney stone and source of the beam must be placed $3\sqrt{3}$ units from the center.

Chapter 10   Quiz
(Sections 10.1–10.2)

1. (a) B; $x+3 = 4(y-1)^2$

This is the equation of a parabola with vertex (−3, 1). The parabola opens right.

(b) A; $(x+3)^2 + (y-1)^2 = 81$

This is the equation of a circle with center (−3, 1).

(c) E

$$25(x-2)^2 + (y-1)^2 = 100 \Rightarrow$$

$$\frac{(x-2)^2}{4} + \frac{(y-1)^2}{100} = 1$$

This is the equation of an ellipse with center (2, 1) and major axis vertical.

(d) C; $\dfrac{(x-2)^2}{16} + \dfrac{(y-1)^2}{9} = 1$

This is the equation of an ellipse with center (2, 1) and major axis horizontal.

(e) D

$$-2(x+3)^2 + 1 = y \Rightarrow -2(x+3)^2 = y - 1$$

This is the equation of a parabola with vertex (−3, 1). The parabola opens down.

3. A parabola passing through $\left(\sqrt{10}, -5\right)$, opening downward, and vertex at the origin has an equation of the form $x^2 = 4py$. Use this equation with the coordinates of the point $\left(\sqrt{10}, -5\right)$ to find the value of p.

$$x^2 = 4py \Rightarrow \left(\sqrt{10}\right)^2 = 4p(-5) \Rightarrow$$

$$10 = -20p \Rightarrow p = -\tfrac{1}{2}$$

Thus, an equation of the parabola is

$$x^2 = 4\left(-\tfrac{1}{2}\right)y = -2y \text{ or } y = -\tfrac{1}{2}x^2.$$

5. Foci at $(-3, 3)$, $(-3, 11)$, major axis with length 10. The length of the major axis is $2a$. Thus, we have $2a = 10 \Rightarrow a = 5$. From the foci, we have $c = \frac{11-3}{2} = 4$. Solving for b^2 we have $c^2 = a^2 - b^2 \Rightarrow 16 = 25 - b^2 \Rightarrow b^2 = 9$. The center is the midpoint between the foci. Thus, the center is located at
$$\left(\frac{(-3)+(-3)}{2}, \frac{3+11}{2}\right) = \left(\frac{-6}{2}, \frac{14}{2}\right) = (-3, 7).$$ The foci lie on the vertical line $x = -3$, so the major axis is vertical, so the equation is of the form
$$\frac{(x+3)^2}{b^2} + \frac{(y-7)^2}{a^2} = 1.$$
Because $a = 5 \Rightarrow a^2 = 25$ and and $b^2 = 9$, the equation of the ellipse is
$$\frac{(x+3)^2}{9} + \frac{(y-7)^2}{25} = 1.$$

7. $4x^2 + 9y^2 = 36$

   Rewriting the equation, we have $\frac{4x^2}{36} + \frac{9y^2}{36} = 1$ or $\frac{x^2}{9} + \frac{y^2}{4} = 1 \Rightarrow \frac{x^2}{3^2} + \frac{y^2}{2^2} = 1$. This is an ellipse with center $(0, 0)$. Because $3 > 2$, we have $a = 3$ and $b = 2$, and the major axis is horizontal. The vertices are $(-3, 0)$ and $(3, 0)$. The endpoints of the minor axis are $(0, -2)$ and $(0, 2)$. To find the foci, we need to find c such that $c^2 = a^2 - b^2$.
   $$c^2 = a^2 - b^2 \Rightarrow c^2 = 9 - 4 \Rightarrow c^2 = 5 \Rightarrow c = \sqrt{5}$$
   The major axis lies on the x-axis, so the foci are $\left(-\sqrt{5}, 0\right)$ and $\left(\sqrt{5}, 0\right)$.

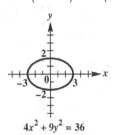

$4x^2 + 9y^2 = 36$

9. $\dfrac{(x+3)^2}{25} + \dfrac{(y+2)^2}{36} = 1$
   Rewriting the equation, we have
   $\frac{(x+3)^2}{5^2} + \frac{(y+2)^2}{6^2} = 1$. This is an ellipse with center $(-3, -2)$. We have $a = 6$ and $b = 5$. Because $a = 6$ is associated with y^2, the major axis of the ellipse is vertical.

The vertices are on the vertical line through $(-3, -2)$, and the endpoints of the minor axis are on the horizontal line through $(-3, -2)$. The vertices are 6 units below and 6 units above the center at $(-3, 4)$ and $(-3, -8)$. The endpoints of the minor axis are 5 units to the left and 5 units to the right of the center at $(-8, -2)$ and $(2, -2)$. To find the foci, we need to find c such that $c^2 = a^2 - b^2$.
$$c^2 = a^2 - b^2 \Rightarrow c^2 = 36 - 25 \Rightarrow c^2 = 11 \Rightarrow$$
$$c = \sqrt{11}$$
The major axis lies on $x = -3$, so the foci are $\left(-3, -2+\sqrt{11}\right)$ and $\left(-3, -2-\sqrt{11}\right)$.

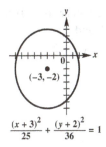

$\dfrac{(x+3)^2}{25} + \dfrac{(y+2)^2}{36} = 1$

Section 10.3 Hyperbolas

1. A; $\dfrac{(x-1)^2}{49} - \dfrac{(y-2)^2}{64} = 1$
   This is the equation of a hyperbola with center $(1, 2)$ and a horizontal transverse axis.

3. D; $\dfrac{(y-1)^2}{9} - \dfrac{(x-2)^2}{25} = 1$
   This is the equation of a hyperbola with center $(2, 1)$ and a vertical transverse axis.

5. C; $\dfrac{x^2}{25} + \dfrac{y^2}{9} = 1$
   This is an equation of an ellipse with x-intercepts $(\pm 5, 0)$ and y-intercepts $(0, \pm 3)$.

7. D; $\dfrac{x^2}{9} - \dfrac{y^2}{25} = 1$
   This is the graph of a hyperbola with a horizontal transverse axis and x-intercepts $(\pm 3, 0)$ (no y-intercepts).

9. $\dfrac{x^2}{16} - \dfrac{y^2}{9} = 1$,

This equation may be written as $\dfrac{x^2}{4^2} - \dfrac{y^2}{3^2} = 1$,

which has the form $\dfrac{x^2}{a^2} - \dfrac{y^2}{b^2} = 1$. The hyperbola is centered at (0, 0) with branches opening to the left and right. The graph has vertices and x-intercepts at (−4, 0) and (4, 0). There are no y-intercepts. The domain is $(-\infty, -4] \cup [4, \infty)$. The range is $(-\infty, \infty)$. The foci are on the x-axis. Because $c^2 = a^2 + b^2$, we have $c^2 = 16 + 9 \Rightarrow$ $c^2 = 25 \Rightarrow c = 5$. The foci are (−5, 0) and (5, 0). Because $a = 4$ and $b = 3$, the asymptotes are $y = \pm \dfrac{b}{a}x = \pm \dfrac{3}{4}x$.

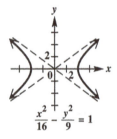

11. $\dfrac{y^2}{25} - \dfrac{x^2}{49} = 1$

This equation may be written as $\dfrac{y^2}{5^2} - \dfrac{x^2}{7^2} = 1$,

which has the form $\dfrac{y^2}{a^2} - \dfrac{x^2}{b^2} = 1$. The hyperbola is centered at (0, 0) with branches opening upward and downward. The graph has vertices and y-intercepts at (0, −5) and (0, 5). There are no x-intercepts. The domain is $(-\infty, \infty)$. The range is $(-\infty, -5] \cup [5, \infty)$. The foci are on the y-axis. Because $c^2 = a^2 + b^2$, we have $c^2 = 25 + 49 \Rightarrow$ $c^2 = 74 \Rightarrow c = \sqrt{74}$. The foci are $\left(0, -\sqrt{74}\right)$ and $\left(0, \sqrt{74}\right)$. Because $a = 5$ and $b = 7$, the asymptotes are $y = \pm \dfrac{a}{b}x = \pm \dfrac{5}{7}x$.

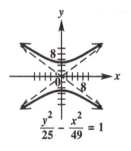

13. $x^2 - y^2 = 9$

This equation may be written as $\dfrac{x^2}{9} - \dfrac{y^2}{9} = 1 \Rightarrow \dfrac{x^2}{3^2} - \dfrac{y^2}{3^2} = 1$, which has the form $\dfrac{x^2}{a^2} - \dfrac{y^2}{b^2} = 1$. The hyperbola is centered at (0, 0) with branches opening to the left and right. The graph has vertices and x-intercepts at (−3, 0) and (3, 0). There are no y-intercepts. The domain is $(-\infty, -3] \cup [3, \infty)$. The range is $(-\infty, \infty)$. The foci are on the x-axis. Because $c^2 = a^2 + b^2$, we have $c^2 = 9 + 9 \Rightarrow$ $c^2 = 18 \Rightarrow c = \sqrt{18} = 3\sqrt{2}$. The foci are $\left(-3\sqrt{2}, 0\right)$ and $\left(3\sqrt{2}, 0\right)$. Because $a = 3$ and $b = 3$, the asymptotes are $y = \pm \dfrac{b}{a}x = \pm \dfrac{3}{3}x = \pm x$

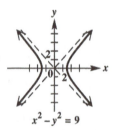

15. $9x^2 - 25y^2 = 225$

This equation may be written as $\dfrac{x^2}{25} - \dfrac{y^2}{9} = 1 \Rightarrow \dfrac{x^2}{5^2} - \dfrac{y^2}{3^2} = 1$, which has the form $\dfrac{x^2}{a^2} - \dfrac{y^2}{b^2} = 1$. The hyperbola is centered at (0, 0) with branches opening to the left and right. The graph has vertices and x-intercepts at (−5, 0) and (5, 0). There are no y-intercepts. The domain is $(-\infty, -5] \cup [5, \infty)$. The range is $(-\infty, \infty)$. The foci are on the x-axis. Because $c^2 = a^2 + b^2$, we have $c^2 = 25 + 9 \Rightarrow c^2 = 34 \Rightarrow c = \sqrt{34}$. The foci are $\left(-\sqrt{34}, 0\right)$ and $\left(\sqrt{34}, 0\right)$. Because $a = 5$ and $b = 3$, the asymptotes are $y = \pm \dfrac{b}{a}x = \pm \dfrac{3}{5}x$.

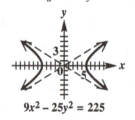

17. $4y^2 - 25x^2 = 100$

This equation may be written as

$\frac{y^2}{25} - \frac{x^2}{4} = 1 \Rightarrow \frac{y^2}{5^2} - \frac{x^2}{2^2} = 1$, which has the form

$\frac{y^2}{a^2} - \frac{x^2}{b^2} = 1$. The hyperbola is centered at

(0, 0) with branches opening upward and downward. The graph has vertices and y-intercepts at $(0, -5)$ and $(0, 5)$. There are no x-intercepts. The domain is $(-\infty, \infty)$. The range is $(-\infty, -5] \cup [5, \infty)$. The foci are on the y-axis. Because $c^2 = a^2 + b^2$, we have $c^2 = 25 + 4 \Rightarrow c^2 = 29 \Rightarrow c = \sqrt{29}$. The foci are $\left(0, -\sqrt{29}\right)$ and $\left(0, \sqrt{29}\right)$.

Because $a = 5$ and $b = 2$, the asymptotes are

$y = \pm \frac{a}{b}x = \pm \frac{5}{2}x$.

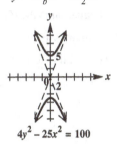

$4y^2 - 25x^2 = 100$

19. $9x^2 - 4y^2 = 1$

This equation may be written as

$\frac{x^2}{\frac{1}{9}} - \frac{y^2}{\frac{1}{4}} = 1 \Rightarrow \frac{x^2}{\left(\frac{1}{3}\right)^2} - \frac{y^2}{\left(\frac{1}{2}\right)^2} = 1$, which has the

form $\frac{x^2}{a^2} - \frac{y^2}{b^2} = 1$. The hyperbola is centered at (0, 0) with branches opening to the left and right. The graph has vertices and x-intercepts at $\left(-\frac{1}{3}, 0\right)$ and $\left(\frac{1}{3}, 0\right)$. There are no y-intercepts. The domain is $\left(-\infty, -\frac{1}{3}\right] \cup \left[\frac{1}{3}, \infty\right)$. The range is $(-\infty, \infty)$. The foci are on the x-axis. Because $c^2 = a^2 + b^2$, we have $c^2 = \frac{1}{9} + \frac{1}{4} \Rightarrow c^2 = \frac{4+9}{36} \Rightarrow c^2 = \frac{13}{36} \Rightarrow$ $c = \sqrt{\frac{13}{36}} = \frac{\sqrt{13}}{6}$.

The foci are $\left(-\frac{\sqrt{13}}{6}, 0\right)$ and $\left(\frac{\sqrt{13}}{6}, 0\right)$. Because

$a = \frac{1}{3}$ and $b = \frac{1}{2}$, the asymptotes are

$y = \pm \frac{b}{a}x = \pm \frac{\frac{1}{2}}{\frac{1}{3}}x = \pm \frac{3}{2}x$.

$9x^2 - 4y^2 = 1$

21. $\frac{(y-7)^2}{36} - \frac{(x-4)^2}{64} = 1$

Because this equation can be written as

$\frac{(y-7)^2}{6^2} - \frac{(x-4)^2}{8^2} = 1$, it has the form

$\frac{(y-k)^2}{a^2} - \frac{(x-h)^2}{b^2} = 1$ where $h = 4$, $k = 7$, $a = 6$, and $b = 8$. The center is (4, 7). The vertices are 6 units above and below the center (4, 7). These points are (4, 1) and (4, 13). The domain is $(-\infty, \infty)$. The range is $(-\infty, 1] \cup [13, \infty)$. Because $c^2 = a^2 + b^2 \Rightarrow$ $c^2 = 36 + 64 \Rightarrow c^2 = 100 \Rightarrow c = 10$, the foci are 10 units below and above the center (4, 7). Thus, the foci are (4, –3) and (4, 17). Because $h = 4$, $k = 7$, $a = 6$, $b = 8$, and $y = k \pm \frac{a}{b}(x - h)$, the asymptotes are

$y = 7 \pm \frac{6}{8}(x - 4) \Rightarrow y = 7 \pm \frac{3}{4}(x - 4)$

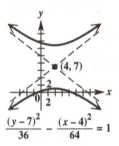

$\frac{(y-7)^2}{36} - \frac{(x-4)^2}{64} = 1$

23. $\frac{(x+3)^2}{16} - \frac{(y-2)^2}{9} = 1$,

Because this equation can be written as

$\frac{[x-(-3)]^2}{4^2} - \frac{(y-2)^2}{3^2} = 1$, it has the form

$\frac{(x-h)^2}{a^2} - \frac{(y-k)^2}{b^2} = 1$ where $h = -3$, $k = 2$, $a = 4$, and $b = 3$. The center is (–3, 2). The vertices are 4 units left and right of the center (–3, 2). These points are (–7, 2) and (1, 2) The domain is $(-\infty, -7] \cup [1, \infty)$. The range is $(-\infty, \infty)$.

(continued on next page)

(continued)

Because $c^2 = a^2 + b^2 \Rightarrow c^2 = 16 + 9 \Rightarrow$
$c^2 = 25 \Rightarrow c = 5$, the foci are 5 units left and
right of the center $(-3, 2)$. Thus, the foci are
$(-8, 2)$ and $(2, 2)$. Because $h = -3$, $k = 2$,
$a = 4$, $b = 3$, and $y = k \pm \frac{b}{a}(x - h)$, the

asymptotes are
$$y = 2 \pm \frac{3}{4}\left[x - (-3)\right] \Rightarrow y = 2 \pm \frac{3}{4}(x + 3).$$

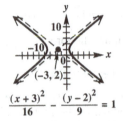

$$\frac{(x+3)^2}{16} - \frac{(y-2)^2}{9} = 1$$

25. $16(x + 5)^2 - (y - 3)^2 = 1$

This equation can be written as
$$\frac{\left[x - (-5)\right]^2}{\left(\frac{1}{4}\right)^2} - \frac{(y-3)^2}{1^2} = 1, \text{ so it has the form}$$

$$\frac{(x-h)^2}{a^2} - \frac{(y-k)^2}{b^2} = 1 \text{ where } h = -5, k = 3,$$

$a = \frac{1}{4}$, and $b = 1$. The center is $(-5, 3)$.

The vertices are $\frac{1}{4}$ unit left and right of the

center $(-5, 3)$. These points are $\left(-\frac{21}{4}, 3\right)$ and

$\left(-\frac{19}{4}, 3\right)$. The domain is

$\left(-\infty, -\frac{21}{4}\right] \cup \left[-\frac{19}{4}, \infty\right)$. The range is

$(-\infty, \infty)$. Because $c^2 = a^2 + b^2 \Rightarrow$

$c^2 = \frac{1}{16} + 1 \Rightarrow c^2 = \frac{1+16}{16} \Rightarrow c^2 = \frac{17}{16} \Rightarrow$

$c = \sqrt{\frac{17}{16}} = \frac{\sqrt{17}}{4}$, the foci are $\frac{\sqrt{17}}{4}$ units left and

right of the center $(-5, 3)$. Thus, the foci are

$\left(-5 - \frac{\sqrt{17}}{4}, 3\right)$ and $\left(-5 + \frac{\sqrt{17}}{4}, 3\right)$. Because

$h = -5$, $k = 3$, $a = \frac{1}{4}$, $b = 1$, and

$y = k \pm \frac{b}{a}(x - h)$, the asymptotes are

$$y = 3 \pm \frac{1}{\frac{1}{4}}\left[x + (-5)\right] \Rightarrow y = 3 \pm 4(x + 5)$$

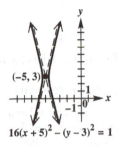

$$16(x + 5)^2 - (y - 3)^2 = 1$$

27. $\dfrac{y}{3} = \sqrt{1 + \dfrac{x^2}{16}}$

Square both sides to get $\frac{y^2}{9} = 1 + \frac{x^2}{16}$

or $\frac{y^2}{9} - \frac{x^2}{16} = 1$ or $\frac{y^2}{3^2} - \frac{x^2}{4^2} = 1$ This is the

equation of a hyperbola with center $(0, 0)$ and
vertices $(0, -3)$ and $(0, 3)$. Because $a = 3$ and

$b = 4$ and $y = \pm \frac{a}{b}x$, we have asymptotes

$y = \pm \frac{3}{4}x$. The original equation is the top half

of the hyperbola. The domain is $(-\infty, \infty)$. The

range is $[3, \infty)$. The vertical line test shows

this is a graph of a function.

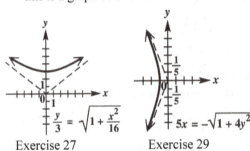

Exercise 27   Exercise 29

29. $5x = -\sqrt{1 + 4y^2}$

Square both sides to get $25x^2 = 1 + 4y^2$ or

$25x^2 - 4y^2 = 1$ or $\frac{x^2}{\left(\frac{1}{5}\right)^2} - \frac{y^2}{\left(\frac{1}{2}\right)^2} = 1$. This is the

equation of a hyperbola with center $(0, 0)$ and
vertices $\left(-\frac{1}{5}, 0\right)$ and $\left(\frac{1}{5}, 0\right)$. Because

$a = \frac{1}{5}$ and $b = \frac{1}{2}$ and $y = \pm \frac{b}{a}x$, we have

asymptotes $y = \pm \frac{\frac{1}{2}}{\frac{1}{5}}x$ or $y = \pm \frac{5}{2}x$. The

original equation is the left half of the

hyperbola. The domain is $\left(-\infty, -\frac{1}{5}\right]$. The

range is $(-\infty, \infty)$. The vertical line test shows

that this is not the graph of a function.

31. $\dfrac{x^2}{8} - \dfrac{y^2}{8} = 1$

Because $a^2 = 8$, we have $a = 2\sqrt{2}$. Also,

$c = \sqrt{a^2 + b^2} = \sqrt{8 + 8} = \sqrt{16} = 4$. Thus,

$e = \dfrac{c}{a} = \dfrac{4}{2\sqrt{2}} = \dfrac{4\sqrt{2}}{4} = \sqrt{2} \approx 1.4$.

33. $16y^2 - 8x^2 = 16 \Rightarrow \dfrac{y^2}{1} - \dfrac{x^2}{2} = 1$

Because $a^2 = 1$, we have $a = 1$. Also,

$c = \sqrt{a^2 + b^2} = \sqrt{1 + 2} = \sqrt{3}$. Thus,

$e = \dfrac{c}{a} = \dfrac{\sqrt{3}}{1} = \sqrt{3} \approx 1.7$.

35. x-intercepts $(\pm 3, 0)$; foci at $(-5, 0)$, $(5, 0)$

The center is halfway between the foci, so the center is $(0,0)$. The foci are on a horizontal transverse axis, so the equation has the form $\dfrac{x^2}{a^2} - \dfrac{y^2}{b^2} = 1$. The x-intercepts are also vertices, so $a = 3$ and thus $a^2 = 9$. With the given foci, we have $c = 5$. Because $c^2 = a^2 + b^2$, we have

$c^2 = a^2 + b^2 \Rightarrow 5^2 = 3^2 + b^2 \Rightarrow$

$25 = 9 + b^2 \Rightarrow b^2 = 16$.

Thus, the equation of the hyperbola is

$\dfrac{x^2}{9} - \dfrac{y^2}{16} = 1$.

37. Vertices at $(0, 6)$, $(0, -6)$; asymptotes

$y = \pm \frac{1}{2}x$

The center is halfway between the vertices, so the center is $(0, 0)$. Because the vertices are on a vertical transverse axis, the equation has the form $\dfrac{y^2}{a^2} - \dfrac{x^2}{b^2} = 1$ and $a = 6$, which implies $a^2 = 36$. Because $y = \pm \frac{1}{2}x = \pm \frac{a}{b}x$, we have

$\frac{a}{b} = \frac{1}{2} \Rightarrow \frac{6}{b} = \frac{1}{2} \Rightarrow b = 12 \Rightarrow b^2 = 144$. Thus,

the equation of the hyperbola is $\dfrac{y^2}{36} - \dfrac{x^2}{144} = 1$

39. Vertices at $(-3, 0)$, $(3, 0)$; passing through $(-6, -1)$

Because the center is halfway between the vertices, the center is $(0,0)$. Because the vertices are on a horizontal transverse axis, the equation has the form $\dfrac{x^2}{a^2} - \dfrac{y^2}{b^2} = 1$ and $a = 3$, which implies $a^2 = 9$. Thus, we have $\dfrac{x^2}{9} - \dfrac{y^2}{b^2} = 1$. The hyperbola goes through the point $(-6, -1)$, so substitute $x = -6$ and $y = -1$ into the equation and solve for b^2.

$\dfrac{x^2}{9} - \dfrac{y^2}{b^2} = 1 \Rightarrow \dfrac{(-6)^2}{9} - \dfrac{(-1)^2}{b^2} = 1 \Rightarrow \dfrac{36}{9} - \dfrac{1}{b^2} = 1 \Rightarrow$

$4 - \dfrac{1}{b^2} = 1 \Rightarrow -\dfrac{1}{b^2} = -3 \Rightarrow b^2 = \dfrac{1}{3}$

Thus, the equation is

$\dfrac{x^2}{9} - \dfrac{y^2}{\frac{1}{3}} = 1$ or $\dfrac{x^2}{9} - 3y^2 = 1$.

41. Foci at $\left(0, \sqrt{13}\right)$, $\left(0, -\sqrt{13}\right)$; asymptotes

$y = \pm 5x$

The center is halfway between the foci, so the center is $(0, 0)$. Because the foci are on a vertical transverse axis, the equation has the form $\dfrac{y^2}{a^2} - \dfrac{x^2}{b^2} = 1$ and $c = \sqrt{13}$, which implies $c^2 = 13$. Also, $y = \pm 5x = \pm \frac{5}{1}x = \pm \frac{a}{b}x$, so we have $\frac{a}{b} = \frac{5}{1} \Rightarrow a = 5b$ (1). Because $c^2 = a^2 + b^2$, we have $13 = a^2 + b^2$ (2). Substituting equation (1) into equation (2) and solving for b^2, we have

$13 = a^2 + b^2 \Rightarrow 13 = (5b)^2 + b^2 \Rightarrow$

$13 = 25b^2 + b^2 \Rightarrow 13 = 26b^2 \Rightarrow b^2 = \frac{1}{2}$.

Because $13 = a^2 + b^2$ and $b^2 = \frac{1}{2}$, we have

$13 = a^2 + \frac{1}{2} \Rightarrow 13 - \frac{1}{2} = a^2 \Rightarrow a^2 = \frac{26}{2} - \frac{1}{2} = \frac{25}{2}$.

Thus, the equation of the hyperbola is

$\dfrac{y^2}{\frac{25}{2}} - \dfrac{x^2}{\frac{1}{2}} = 1$ or $\dfrac{2y^2}{25} - 2x^2 = 1$.

43. Vertices at $(4, 5)$, $(4, 1)$; asymptotes

$$y = \pm 7(x - 4) + 3$$

The center is halfway between the vertices, so the center is located at

$\left(\frac{4+4}{2}, \frac{5+1}{2}\right) = \left(\frac{8}{2}, \frac{6}{2}\right) = (4, 3)$. (This could have also been determined from the equation of the asymptotes.) Because the vertices are on a vertical transverse axis, the equation has the

form $\frac{(y-k)^2}{a^2} - \frac{(x-h)^2}{b^2} = 1$. The distance between

the vertices is 4, so we have $2a = 4 \Rightarrow$ $a = 2$. The slopes of the asymptotes $\pm 7 = \pm \frac{a}{b}$.

This yields $\frac{7}{1} = \frac{a}{b} \Rightarrow 7b = a \Rightarrow b = \frac{a}{7}$. Because

$a = 2 \Rightarrow b = \frac{2}{7}$. With $a = 2 \Rightarrow a^2 = 4$,

$b = \frac{2}{7} \Rightarrow b^2 = \frac{4}{49}$, $h = 4$, and $k = 3$ we have

$\frac{(y-3)^2}{4} - \frac{(x-4)^2}{\frac{4}{49}} = 1$ or $\frac{(y-3)^2}{4} - \frac{49(x-4)^2}{4} = 1$.

45. Center at $(1, -2)$; focus at $(-2, -2)$; vertex at $(-1, -2)$

The center, focus, and vertices are on a horizontal transverse axis, so the equation has

the form $\frac{(x-h)^2}{a^2} - \frac{(y-k)^2}{b^2} = 1$. The distance

from the center to the vertex being 2 implies $a = 2$. The focus, $(-2, -2)$, is 3 units from the center, so $c = 3$. Given that $c^2 = a^2 + b^2$, $a = 2$, and $c = 3$, we have $c^2 = a^2 + b^2 \Rightarrow$ $3^2 = 2^2 + b^2 \Rightarrow 9 = 4 + b^2 \Rightarrow b^2 = 5$.

With $a = 2 \Rightarrow a^2 = 4$, $b^2 = 5$, $h = 1$, and

$k = -2$, we have $\frac{(x-1)^2}{4} - \frac{(y+2)^2}{5} = 1$.

47. Eccentricity 3; center at $(0, 0)$; vertex at $(0, 7)$

The center and vertex lie on a vertical transverse axis (the y-axis), so the equation is

of the form $\frac{y^2}{a^2} - \frac{x^2}{b^2} = 1$. The distance between

the center and a vertex is 7, so $a = 7$. Use the

eccentricity to find c: $e = \frac{c}{a} \Rightarrow 3 = \frac{c}{7} \Rightarrow c = 21$

Now find the value of b^2 given $c^2 = a^2 + b^2$, $a = 7$ and $c = 21$.

$c^2 = a^2 + b^2 \Rightarrow 21^2 = 7^2 + b^2 \Rightarrow$

$441 = 49 + b^2 \Rightarrow b^2 = 392$

Because $a = 7 \Rightarrow a^2 = 49$ and $b^2 = 392$, the

equation of the hyperbola is $\frac{y^2}{49} - \frac{x^2}{392} = 1$.

49. Foci at $(9, -1)$, $(-11, -1)$; eccentricity $\frac{25}{9}$

The foci lie on a horizontal transverse axis, so

the equation is of the form $\frac{(x-h)^2}{a^2} - \frac{(y-k)^2}{b^2} = 1$.

The center is halfway between the foci, so the center is located at

$\left(\frac{9+(-11)}{2}, \frac{(-1)+(-1)}{2}\right) = \left(\frac{-2}{2}, \frac{-2}{2}\right) = (-1, -1)$.

The distance from the center to each foci is 10, so $c = 10$. Use the eccentricity to find a.

$e = \frac{c}{a} \Rightarrow \frac{25}{9} = \frac{10}{a} \Rightarrow 25a = 90 \Rightarrow a = \frac{90}{25} = \frac{18}{5}$

Now find the value of b^2 given $c^2 = a^2 + b^2$, $a = \frac{18}{5}$, and $c = 10$.

$c^2 = a^2 + b^2 \Rightarrow 10^2 = \left(\frac{18}{5}\right)^2 + b^2 \Rightarrow$

$100 = \frac{324}{25} + b^2 \Rightarrow$

$b^2 = 100 - \frac{324}{25} = \frac{2500}{25} - \frac{324}{25} = \frac{2176}{25}$

Because

$a = \frac{18}{5} \Rightarrow a^2 = \frac{324}{25}$, $b^2 = \frac{2176}{25}$, $h = -1$, and

$k = 2$, we have $\frac{(x+1)^2}{\frac{324}{25}} - \frac{(y+1)^2}{\frac{2176}{25}} = 1$ or

$\frac{25(x+1)^2}{324} - \frac{25(y+1)^2}{2176} = 1$.

51. Solve $\frac{x^2}{4} - \frac{y^2}{16} = 1$ for y.

$\frac{x^2}{4} - \frac{y^2}{16} = 1 \Rightarrow \frac{x^2}{4} - 1 = \frac{y^2}{16} \Rightarrow$

$y^2 = 16\left(\frac{x^2}{4} - 1\right) \Rightarrow y^2 = 4x^2 - 16 \Rightarrow$

$y = \sqrt{4x^2 - 16} \Rightarrow y = \pm\sqrt{4(x^2 - 4)} \Rightarrow$

$y = \pm 2\sqrt{x^2 - 4}$

The graphing window is $[-9.9, 9.9] \times [-8, 8]$.

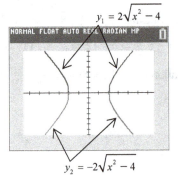

53. Solve $4y^2 - 36x^2 = 144$ for y.

$$4y^2 - 36x^2 = 144 \Rightarrow 4y^2 = 36x^2 + 144 \Rightarrow$$
$$y^2 = 9x^2 + 36 \Rightarrow y = \pm\sqrt{9x^2 + 36}$$
$$y = \pm\sqrt{9\left(x^2 + 4\right)} \Rightarrow y = \pm 3\sqrt{x^2 + 4}$$

The graphing window is $[-10, 10] \times [-15, 15]$.

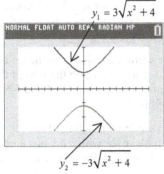

$$y_1 = 3\sqrt{x^2 + 4}$$

$$y_2 = -3\sqrt{x^2 + 4}$$

55. (a) We must determine a and b in the equation $\frac{x^2}{a^2} - \frac{y^2}{b^2} = 1$. The asymptotes are $y = x$ and $y = -x$, which have slopes of 1 and -1, respectively, so $a = b$. Look at the small right triangle that is shown in quadrant III. The line $y = x$ intersects the x-axis to form $45°$ angles. The right angle vertex of the triangle lies on the line $y = x$, so this triangle is a $45°$–$45°$–$90°$ triangle (an isosceles right triangle). Thus, both legs of the triangle have length d, and by the Pythagorean theorem, $c^2 = d^2 + d^2 \Rightarrow c^2 = 2d^2 \Rightarrow c = d\sqrt{2}$ (1). Thus, the coordinates of N are $\left(-d\sqrt{2}, 0\right)$.

Because $a = b$, we have
$c^2 = a^2 + a^2 \Rightarrow c^2 = 2a^2 \Rightarrow c = a\sqrt{2}$ (2).
From equations (1) and (2), we have
$d\sqrt{2} = a\sqrt{2} \Rightarrow d = a$. Thus,
$a = b = d = 5 \times 10^{-14}$. The equation of the trajectory of A is given by $\frac{x^2}{a^2} - \frac{y^2}{b^2} = 1$.

$$\frac{x^2}{a^2} - \frac{y^2}{b^2} = 1$$
$$\frac{x^2}{\left(5 \times 10^{-14}\right)^2} - \frac{y^2}{\left(5 \times 10^{-14}\right)^2} = 1$$
$$x^2 - y^2 = \left(5 \times 10^{-14}\right)^2$$
$$x^2 - y^2 = 25 \times 10^{-28}$$
$$x^2 - y^2 = 2.5 \times 10^{-27}$$

$$x^2 = y^2 + 2.5 \times 10^{-27}$$
$$x = \sqrt{y^2 + 2.5 \times 10^{-27}}$$

(We choose the positive square root because the trajectory occurs only where $x > 0$. This equation represents the right half of the hyperbola, as shown in the figure in the exercise.)

(b) The minimum distance between the centers of the alpha particle and the gold nucleus is $c + a = d\sqrt{2} + d$.

$$c + a = d\sqrt{2} + d$$
$$c + a = \left(5 \times 10^{-14}\right)\sqrt{2} + \left(5 \times 10^{-14}\right)$$
$$c + a = \left(\sqrt{2} + 1\right)\left(5 \times 10^{-14}\right)$$
$$c + a \approx 12.07 \times 10^{-14} \approx 1.2 \times 10^{-13} \, \text{m}$$

57. The center is the origin and the foci lie on a horizontal transverse axis, so the equation has the form $\frac{x^2}{a^2} - \frac{y^2}{b^2} = 1$. We know from page 982 of the text that $\left|d\left(P, F'\right) - d\left(P, F\right)\right| = 2a$ where F and F' are the locations of the foci, in this case $(-c, 0)$ and $(c, 0)$. Because $d = rt$, the difference between $d\left(P, F'\right)$ and $d\left(P, F\right)$ is $330t$. Thus, $2a = 330t \Rightarrow a = \frac{330t}{2}$. Because $c^2 = a^2 + b^2$, we have the following.

$$c^2 = a^2 + b^2 \Rightarrow c^2 = \left(\frac{330t}{2}\right)^2 + b^2 \Rightarrow$$
$$b^2 = c^2 - \frac{330^2 t^2}{4} \Rightarrow b^2 = \frac{4c^2 - 330^2 t^2}{4}$$

Thus, the equation of the hyperbola is

$$\frac{x^2}{\left(\frac{330t}{2}\right)^2} - \frac{y^2}{\frac{4c^2 - 330^2 t^2}{4}} = 1 \text{ or } \frac{x^2}{330^2 t^2} - \frac{y^2}{4c^2 - 330^2 t^2} = \frac{1}{4}.$$

59. (a) Rewriting the equation of the hyperbola we have the following.

$$400x^2 - 625y^2 = 250{,}000$$
$$\frac{x^2}{625} - \frac{y^2}{400} = 1 \Rightarrow \frac{x^2}{25^2} - \frac{y^2}{20^2} = 1$$

The two branches of a hyperbola are closest at the vertices and this distance is $2a$. Because $a = 25$, the buildings are 50 m apart at their closest point.

(b) First, solve for y when $x = 50$.

$$\frac{x^2}{25^2} - \frac{y^2}{20^2} = 1 \Rightarrow \frac{50^2}{25^2} - \frac{y^2}{20^2} = 1 \Rightarrow$$

$$\left(\frac{50^2}{25^2}\right)^2 - \frac{y^2}{20^2} = 1 \Rightarrow 2^2 - \frac{y^2}{20^2} = 1 \Rightarrow$$

$$4 - \frac{y^2}{20^2} = 1 \Rightarrow \frac{y^2}{20^2} = 3 \Rightarrow$$

$$y^2 = 20^2 \cdot 3 \Rightarrow y = 20\sqrt{3}$$

Thus,

$$d = 2y = 2 \cdot 20\sqrt{3} = 40\sqrt{3} \approx 69.3 \text{ m}.$$

61. Solving $\frac{x^2}{4} - y^2 = 1$ for y, we have

$\frac{x^2}{4} - y^2 = 1 \Rightarrow y = \pm\sqrt{\frac{x^2}{4} - 1}$. The positive

square root is $y = \sqrt{\frac{x^2}{4} - 1} \Rightarrow y = \sqrt{\frac{x^2}{4} - \frac{4}{4}} \Rightarrow$

$y = \sqrt{\frac{x^2 - 4}{4}} \Rightarrow y = \frac{1}{2}\sqrt{x^2 - 4}$.

63. $y = \frac{1}{2}\sqrt{x^2 - 4}$

Use the graphing window
$[-65, 65] \times [-40, 40]$.

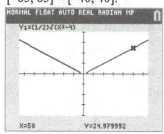

At $x = 50$, $y \approx 24.98$.

65. Because $24.98 < 25$, the graph of

$y_1 = \frac{1}{2}\sqrt{x^2 - 4}$ lies below the graph of

$y_2 = \frac{1}{2}x$ when $x = 50$. This can be verified by

zooming in at $x = 50$ several times.

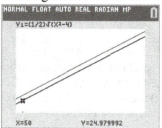

Section 10.4 Summary of the Conic Sections

1. The conic section consisting of the set of points in a plane that lie a given distance from a given point is a <u>circle</u>.

3. The conic section consisting of the set of points in the plane for which the distance from the point $(1, 3)$ is equal to a distance from the line $y = 1$ is a <u>parabola</u>.

5. The conic section consisting of the set of points in a plane for which the sum of the distances from the points $(5, 0)$ and $(-5, 0)$ is 14 is an <u>ellipse</u>.

7. The conic section consisting of the set of points in a plane for which the distance from the point $(3, 0)$ is one and one-half times the distance from the line $x = \frac{4}{3}$ is a <u>hyperbola</u>.

9. $x^2 + y^2 = 144 \Rightarrow (x - 0)^2 + (y - 0)^2 = 12^2$
   The graph of this equation is a circle with center $(0, 0)$ and radius 12. Also, note in our original equation, the x^2- and y^2-terms have the same positive coefficient.

11. $y = 2x^2 + 3x - 4 \Rightarrow y = 2\left(x^2 + \frac{3}{2}x\right) - 4$
    $y = 2\left(x^2 + \frac{3}{2}x + \frac{9}{16} - \frac{9}{16}\right) - 4$
    $y = 2\left(x + \frac{3}{4}\right)^2 + 2\left(-\frac{9}{16}\right) - 4$
    $y = 2\left(x + \frac{3}{4}\right)^2 - \frac{9}{8} - 4$
    $y = 2\left(x + \frac{3}{4}\right)^2 - \frac{41}{8} \Rightarrow y - \left(-\frac{41}{8}\right) = 2\left[x - \left(-\frac{3}{4}\right)\right]^2$
    The graph of this equation is a parabola opening upwards with a vertex of $\left(-\frac{3}{4}, -\frac{41}{8}\right)$.
    Also, note our original equation has an x^2-term, but no y^2-term.

13. $x - 1 = -3(y - 4)^2$
    The graph of this equation is a parabola opening to the left with a vertex of $(1, 4)$. Also, note when expanded, our original equation has a y^2-term, but no x^2-term.

15. $\dfrac{x^2}{49} + \dfrac{y^2}{100} = 1 \Rightarrow \dfrac{x^2}{7^2} + \dfrac{y^2}{10^2} = 1$

The graph of this equation is an ellipse centered at the origin and x-intercepts of 7 and -7, and y-intercepts of 10 and -10. Also, note in our original equation, the x^2- and y^2-terms both have different positive coefficients.

17. $\dfrac{x^2}{4} - \dfrac{y^2}{16} = 1 \Rightarrow \dfrac{x^2}{2^2} - \dfrac{y^2}{4^2} = 1$

The graph of this equation is a hyperbola centered at the origin with x-intercepts of 2 and -2, and asymptotes of $y = \pm\frac{4}{2}x = \pm 2x$.

Also, note in our original equation, the x^2- and y^2-terms have coefficient that are opposite in sign.

19. $\dfrac{x^2}{25} - \dfrac{y^2}{25} = 1 \Rightarrow \dfrac{x^2}{5^2} - \dfrac{y^2}{5^2} = 1$

The graph of this equation is a hyperbola centered at the origin with x-intercepts of 5 and -5, and asymptotes of $y = \pm\frac{5}{5}x = \pm x$.

Also, note in our original equation, the x^2- and y^2-terms have coefficients that are opposite in sign.

21. $\dfrac{x^2}{4} = 1 - \dfrac{y^2}{9} \Rightarrow \dfrac{x^2}{4} + \dfrac{y^2}{9} = 1 \Rightarrow$

$\dfrac{(x-0)^2}{2^2} + \dfrac{(y-0)^2}{3^2} = 1$

The equation is of the form $\dfrac{(x-h)^2}{b^2} + \dfrac{(y-k)^2}{a^2} = 1$ with $a = 3$, $b = 2$, $h = 0$, and $k = 0$, so the graph of the given equation is an ellipse.

23. $\dfrac{(x+3)^2}{16} + \dfrac{(y-2)^2}{16} = 1$

$(x+3)^2 + (y-2)^2 = 16$

$[x-(-3)]^2 + (y-2)^2 = 4^2$

The equation is of the form $(x-h)^2 + (y-k)^2 = r^2$ with $r = 4$, $h = -3$, and $k = 2$, so the graph of the given equation is a circle.

25. $x^2 - 6x + y = 0 \Rightarrow y = -x^2 + 6x$

$y = -\left(x^2 - 6x + 9 - 9\right)$

$y = -(x-3)^2 + 9$

$y - 9 = -(x-3)^2$

The equation is of the form $y - k = a(x-h)^2$ with $a = -1$, $h = 3$, and $k = 9$, so the graph of the given equation is a parabola.

27. $4(x-3)^2 + 3(y+4)^2 = 0$

$\dfrac{4(x-3)^2}{12} + \dfrac{3(y+4)^2}{12} = 0$

$\dfrac{(x-3)^2}{3} + \dfrac{[y-(-4)]^2}{4} = 0$

The graph is the point $(3, -4)$.

29. $x - 4y^2 - 8y = 0 \Rightarrow x = 4y^2 + 8y$

$x = 4\left(y^2 + 2y + 1 - 1\right)$

$x = 4(y+1)^2 - 4$

$x - (-4) = 4[y-(-1)]^2$

The equation is of the form $x - h = a(y-k)^2$ with $a = 4$, $h = -4$, and $k = -1$, so the graph of the given equation is a parabola.

31. $4x^2 - 24x + 5y^2 + 10y + 41 = 0$

$4\left(x^2 - 6x + 9 - 9\right) + 5\left(y^2 + 2y + 1 - 1\right) = -41$

$4(x-3)^2 - 36 + 5(y+1)^2 - 5 = -41$

$4(x-3)^2 + 5(y+1)^2 = 0$

$\dfrac{4(x-3)^2}{20} + \dfrac{5(y+1)^2}{20} = 0$

$\dfrac{(x-3)^2}{5} + \dfrac{[y-(-1)]^2}{4} = 0$

The graph is the point $(3, -1)$.

33. $\dfrac{x^2}{4} + \dfrac{y^2}{4} = -1 \Rightarrow x^2 + y^2 = -4$

A sum of squares can never be negative. This equation has no graph.

35. $\quad x^2 = 25 + y^2 \Rightarrow x^2 - y^2 = 25$

$$\frac{x^2}{25} - \frac{y^2}{25} = 1 \Rightarrow \frac{(x-0)^2}{5^2} - \frac{(y-0)^2}{5^2} = 1$$

The equation is of the form $\frac{(x-h)^2}{a^2} - \frac{(y-k)^2}{b^2} = 1$

with $a = 5$, $b = 5$, $h = 0$, and $k = 0$, so the graph of the given equation is a hyperbola with center $(0, 0)$, vertices $(-5, 0)$ and $(5, 0)$, and asymptotes $y = \pm x$.

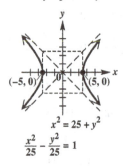

$$x^2 = 25 + y^2$$
$$\frac{x^2}{25} - \frac{y^2}{25} = 1$$

37. $\quad x^2 = 4y - 8 \Rightarrow x^2 = 4(y-2) \Rightarrow$

$$y - 2 = \tfrac{1}{4}(x-0)^2$$

The equation is of the form $y - k = a(x - h)^2$

with $a = \tfrac{1}{4}$, $h = 0$, and $k = 2$, so the graph of

the given equation is a parabola with vertex $(0, 2)$ and vertical axis $x = 0$ (the y-axis). Use the vertex and axis and plot a few additional points.

x	y
-4	6
-2	3
0	2
2	3
4	6

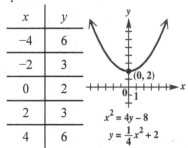

$$x^2 = 4y - 8$$
$$y = \tfrac{1}{4}x^2 + 2$$

39. $\quad y^2 - 4y = x + 4 \Rightarrow y^2 - 4y + 4 - 4 = x + 4$

$$(y-2)^2 - 4 = x + 4$$

$$x + 8 = (y-2)^2 \Rightarrow x - (-8) = (y-2)^2$$

The equation is of the form $x - h = a(y - k)^2$

with $a = 1$, $h = -8$, and $k = 2$, so the graph of the given equation is a parabola with vertex $(-8, 2)$ and horizontal axis $y = 2$. Use the vertex and axis and plot a few additional points.

x	y
0	-0.8
-4	0
-8	2
-4	4
0	4.8

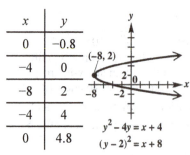

$$y^2 - 4y = x + 4$$
$$(y-2)^2 = x + 8$$

41.
$$3x^2 + 6x + 3y^2 - 12y = 12$$
$$x^2 + 2x + y^2 - 4y = 4$$
$$(x^2 + 2x + 1 - 1) + (y^2 - 4y + 4 - 4) = 4$$
$$(x+1)^2 - 1 + (y-2)^2 - 4 = 4$$
$$(x-1)^2 + (y-2)^2 = 4 + 1 + 4$$
$$(x+1)^2 + (y-2)^2 = 9$$
$$[x - (-1)]^2 + (y-2)^2 = 3^2$$

The equation is of the form

$(x-h)^2 + (y-k)^2 = r^2$ with $r = 3$, $h = -1$,

and $k = 2$, so the graph of the given equation is a circle with center $(-1, 2)$ and radius 3.

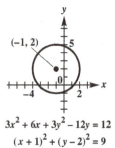

$$3x^2 + 6x + 3y^2 - 12y = 12$$
$$(x+1)^2 + (y-2)^2 = 9$$

43.
$$4x^2 - 8x + 9y^2 - 36y = -4$$
$$4(x^2 - 2x + 1 - 1) + 9(y^2 - 4y + 4 - 4) = -4$$
$$4(x-1)^2 - 4 + 9(y-2)^2 - 36 = -4$$
$$4(x-1)^2 + 9(y-2)^2 = 36$$
$$\frac{4(x-1)^2}{36} + \frac{9(y-2)^2}{36} = 1$$
$$\frac{(x-1)^2}{9} + \frac{(y-2)^2}{4} = 1$$
$$\frac{(x-1)^2}{3^2} + \frac{(y-2)^2}{2^2} = 1$$

The equation is of the form $\frac{(x-h)^2}{a^2} + \frac{(y-k)^2}{b^2} = 1$

with $a = 3$, $b = 2$, $h = 1$, and $k = 2$, so the graph of the given equation is an ellipse with center $(1, 2)$ and vertices $(-2, 2)$, $(4, 2)$, $(1, 0)$ and $(1, 4)$.

(*continued on next page*)

(*continued*)

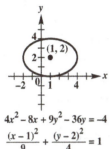

$4x^2 - 8x + 9y^2 - 36y = -4$

$\dfrac{(x-1)^2}{9} + \dfrac{(y-2)^2}{4} = 1$

45. From the graph, the coordinates of P (a point on the graph) are $(-3, 8)$, the coordinates of F (a focus) are $(3, 0)$, the equation of L (the directrix) is $x = 27$. By the distance formula, the distance from P to F is

$$\sqrt{(x_2 - x_1)^2 + (y_2 - y_1)^2}$$
$$= \sqrt{[3-(-3)]^2 + (0-8)^2} = \sqrt{6^2 + (-8)^2}$$
$$= \sqrt{36+64} = \sqrt{100} = 10$$

The distance from a point to a line is defined as the perpendicular distance, so the distance from P to L is $|27-(-3)| = 30$. Thus,

$$e = \frac{\text{Distance of } P \text{ from } F}{\text{Distance of } P \text{ from } L} = \frac{10}{30} = \frac{1}{3}.$$

47. From the graph, we see that $F = \left(\sqrt{2}, 0\right)$ and

L is the vertical line $x = -\sqrt{2}$. Choose $(0, 0)$, the vertex of the parabola, as P.

Distance of P from $F = \sqrt{2}$, and

distance of P from $L = \sqrt{2}$. Thus, we have

$$e = \frac{\text{Distance of } P \text{ from } F}{\text{Distance of } P \text{ from } L} = \frac{\sqrt{2}}{\sqrt{2}} = 1.$$

49. From the graph, we see that $P = (9, -7.5)$, $F = (9, 0)$ and L is the vertical line $x = 4$. Distance of P from $F = 7.5$, and distance of P from $L = 5$. Thus,

$$e = \frac{\text{Distance of } P \text{ from } F}{\text{Distance of } P \text{ from } L} = \frac{7.5}{5} = 1.5.$$

51. If $\dfrac{k}{\sqrt{D}} = \dfrac{2.82 \times 10^7}{\sqrt{42.5 \times 10^6}} = \dfrac{2.82 \times 10^7}{\sqrt{42.5} \times 10^3}$

$\approx 0.432568 \times 10^4 \approx 4326$ and $V = 2090$, then

we have $V < \dfrac{k}{\sqrt{D}}$. Thus, the shape of the

satellite's trajectory was elliptical.

53. Complete the square on x and on y for $Ax^2 + Cy^2 + Dx + Ey + F = 0$.

$$Ax^2 + Cy^2 + Dx + Ey + F = 0$$
$$\left(Ax^2 + Dx\right) + \left(Cy^2 + Ey\right) = -F$$
$$A\left(x^2 + \tfrac{D}{A}x\right) + C\left(y^2 + \tfrac{E}{C}y\right) = -F$$

Because $\left(\tfrac{1}{2} \cdot \tfrac{D}{A}\right)^2 = \tfrac{D^2}{4A^2}$ and $\left(\tfrac{1}{2} \cdot \tfrac{E}{C}\right)^2 = \tfrac{E^2}{4C^2}$,

we have the following.

$$A\left(x^2 + \tfrac{D}{A}x + \tfrac{D^2}{4A^2} - \tfrac{D^2}{4A^2}\right)$$
$$+ C\left(y^2 + \tfrac{E}{C}y + \tfrac{E^2}{4C^2} - \tfrac{E^2}{4C^2}\right) = -F$$

$$A\left(x + \tfrac{D}{2A}\right)^2 - \tfrac{D^2}{4A} + C\left(y + \tfrac{E}{2C}\right)^2 - \tfrac{E^2}{4C} = -F$$

$$A\left(x + \tfrac{D}{2A}\right)^2 + C\left(y + \tfrac{E}{2C}\right)^2 = \tfrac{D^2}{4A} + \tfrac{E^2}{4C} - F$$

$$A\left(x + \tfrac{D}{2A}\right)^2 + C\left(y + \tfrac{E}{2C}\right)^2 = \tfrac{CD^2}{4AC} + \tfrac{AE^2}{4AC} - \tfrac{4ACF}{4AC}$$

$$A\left(x + \tfrac{D}{2A}\right)^2 + C\left(y + \tfrac{E}{2C}\right)^2 = \tfrac{CD^2 + AE^2 - 4ACF}{4AC}$$

$$\frac{\left[x - \left(-\tfrac{D}{2A}\right)\right]^2}{\frac{CD^2 + AE^2 - 4ACF}{4A^2C}} + \frac{\left[y - \left(-\tfrac{E}{2C}\right)\right]^2}{\frac{CD^2 + AE^2 - 4ACF}{4AC^2}} = 1$$

The center of the ellipse is at $\left(-\tfrac{D}{2A}, -\tfrac{E}{2C}\right)$.

55. To enter the relation $\dfrac{x^2}{16} + \dfrac{y^2}{12} = 1$ into the

graphing calculator, we must solve for y.

$$\frac{x^2}{16} + \frac{y^2}{12} = 1 \Rightarrow \frac{y^2}{12} = 1 - \frac{x^2}{16} \Rightarrow$$

$$y^2 = 12\left(1 - \frac{x^2}{16}\right) \Rightarrow y = \pm\sqrt{12\left(1 - \frac{x^2}{16}\right)}$$

Graph $y_1 = \sqrt{12\left(1 - \dfrac{x^2}{16}\right)}$ and

$y_2 = -\sqrt{12\left(1 - \dfrac{x^2}{16}\right)}$ in the window

$[-6.6, 6.6] \times [-6.6, 6.6]$.

A sampling of points is as follows.

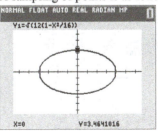

(*continued on next page*)

(continued)

The distance from $P(0, 3.464106)$ to the point $(2, 0)$ is

$$\sqrt{(2-0)^2 + (0-3.4641016)^2}$$
$$= \sqrt{4 + (-3.4641016)^2} \approx 3.999999987$$

The distance of P from the line $x = 8$ is $|0 - 8| = 8$. We have $3.999999987 \approx \frac{1}{2}(8) = 4$.

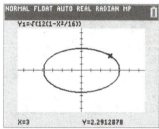

The distance from $P(3, 2.912878)$ to the point $(2, 0)$ is

$$\sqrt{(2-3)^2 + (0-2.2912878)^2}$$
$$= \sqrt{1 + (-2.2912878)^2} \approx 2.499999956$$

The distance of P from the line $x = 8$ is $|3 - 8| = 5$. We have

$$2.499999956 \approx \frac{1}{2}(5) = 2.5.$$

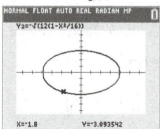

The distance from $P(-1.8, -3.093542)$ to the point $(2, 0)$ is

$$\sqrt{[2-(-1.8)]^2 + [0-(-3.093542)]^2}$$
$$= \sqrt{14.44 + (3.093542)^2} \approx 4.900000215$$

The distance of P from the line $x = 8$ is $|-1.8 - 8| = 9.8$. We have

$$4.900000215 \approx \frac{1}{2}(9.8) = 4.9.$$

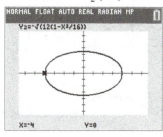

The distance from $P(-4, 0)$ to the point $(2, 0)$ is $\sqrt{[2-(-4)]^2 + (0-0)^2} = \sqrt{36} = 6$

The distance of P from the line $x = 8$ is $|-4 - 8| = 12$. We have $6 = \frac{1}{2}(12)$.

Thus, the distance of P from $(2, 0) = \frac{1}{2}$ [distance of P from the line $x = 8$].

Chapter 10   Review Exercises

1. $x = 4(y-5)^2 + 2 \Rightarrow x - 2 = 4(y-5)^2$

The vertex is $(2, 5)$. The graph opens to the right and has the same shape as $x = 4y^2$. It is a translation 5 units up and 2 units to the right of the graph of $x = 4y^2$. The domain is $[2, \infty)$. The range is $(-\infty, \infty)$. The graph is symmetric about its axis, the horizontal line $y = 5$. Use the vertex and axis and plot a few additional points.

x	y
18	7
6	6
2	5
6	4
18	3

$$x = 4(y-5)^2 + 2$$

3. $x = 5y^2 - 5y + 3$

Complete the square on y to find the vertex and the axis.

$$x = 5y^2 - 5y + 3$$
$$x = 5\left(y^2 - y \quad\right) + 3$$
$$x = 5\left(y^2 - y + \frac{1}{4} - \frac{1}{4}\right) + 3$$
$$x = 5\left(y - \frac{1}{2}\right)^2 + 5\left(-\frac{1}{4}\right) + 3$$
$$x = 5\left(y - \frac{1}{2}\right)^2 + \left(-\frac{5}{4}\right) + \frac{12}{4}$$
$$x = 5\left(y - \frac{1}{2}\right)^2 + \frac{7}{4} \Rightarrow x - \frac{7}{4} = 5\left(y - \frac{1}{2}\right)^2$$

The vertex is $\left(\frac{7}{4}, \frac{1}{2}\right)$. The graph opens to the right and has the same shape as $x = 5y^2$, translated $\frac{7}{4}$ unit to the right and $\frac{1}{2}$ unit up. The domain is $\left[\frac{7}{4}, \infty\right)$. The range is $(-\infty, \infty)$. The graph is symmetric about its axis, the horizontal line $y = \frac{1}{2}$. Use the vertex and axis and plot a few additional points.

(continued on next page)

(*continued*)

x	y
$\frac{27}{4}$	$\frac{3}{2}$
3	1
$\frac{7}{4}$	$\frac{1}{2}$
3	0
$\frac{27}{4}$	$-\frac{1}{2}$

$x = 5y^2 - 5y + 3$

5. The equation $y^2 = -\frac{2}{3}x$ has the form

   $y^2 = 4px$, so $4p = -\frac{2}{3}$, from which

   $p = -\frac{1}{6}$. Because the y-term is squared, the
   parabola is horizontal, with focus

   $(p, 0) = \left(-\frac{1}{6}, 0\right)$ and directrix, $x = -p$, is

   $x = \frac{1}{6}$. The vertex is $(0, 0)$, and the axis of
   the parabola is the x-axis. The domain is
   $(-\infty, 0]$. The range is $(-\infty, \infty)$. Use this
   information and plot a few additional points.

x	y
-6	2
$-\frac{3}{2}$	1
0	0
$-\frac{3}{2}$	-1
-6	-2
2	-2

$y^2 = -\frac{2}{3}x$

7. The equation $3x^2 = y \Rightarrow x^2 = \frac{1}{3}y$ has the

   form $x^2 = 4py$, so $4p = \frac{1}{3}$, from which

   $p = \frac{1}{12}$. Because the x-term is squared, the

   parabola is vertical, with focus $(0, p) = \left(0, \frac{1}{12}\right)$

   and directrix, $y = -p$, is $y = -\frac{1}{12}$. The vertex
   is $(0, 0)$, and the axis of the parabola is the y-
   axis. The domain is $(-\infty, \infty)$. The range is

   $[0, \infty)$. Use this information and plot a few
   additional points.

x	y
2	12
1	3
0	0
-1	3
-2	12

$3x^2 = y$

9. A parabola with focus $(4, 0)$ and vertex at the
   origin is a horizontal parabola. The equation
   has the form $y^2 = 4px$. Because $p = 4$ is
   positive, it opens to the right. Substituting 4
   for p, we find that an equation for this
   parabola is $y^2 = 4(4)x \Rightarrow y^2 = 16x$.

11. A parabola passing through $(-3, 4)$, opening up,
    and vertex at the origin has an equation of the
    form $x^2 = 4py$. Use this equation with the
    coordinates of the point $(-3, 4)$ to find the value
    of p.

    $x^2 = 4py \Rightarrow (-3)^2 = 4p \cdot 4 \Rightarrow 9 = 16p \Rightarrow \frac{9}{16} = p$

    Thus, an equation of the parabola is

    $x^2 = 4\left(\frac{9}{16}\right)y \Rightarrow x^2 = \frac{9}{4}y$.

13. $y^2 + 9x^2 = 9 \Rightarrow \dfrac{x^2}{1} + \dfrac{y^2}{9} = 1 \Rightarrow$

    $\dfrac{(x-0)^2}{1^2} + \dfrac{(y-0)^2}{3^2} = 1$

    The equation is of the form $\dfrac{(x-h)^2}{b^2} + \dfrac{(y-k)^2}{a^2} = 1$
    with $a = 3$, $b = 1$, $h = 0$, and $k = 0$, so the
    graph of the given equation is an ellipse.

15. $3y^2 - 5x^2 = 30 \Rightarrow \dfrac{y^2}{10} - \dfrac{x^2}{6} = 1 \Rightarrow$

    $\dfrac{(y-0)^2}{\left(\sqrt{10}\right)^2} - \dfrac{(x-0)^2}{\left(\sqrt{6}\right)^2} = 1$

    The equation is of the form $\dfrac{(y-k)^2}{a^2} - \dfrac{(x-h)^2}{b^2} = 1$
    with $a = \sqrt{10}$, $b = \sqrt{6}$, $h = 0$, and $k = 0$, so the
    graph of the given equation is a hyperbola.

17. $4x^2 - y = 0 \Rightarrow y = 4x^2 \Rightarrow y - 0 = 4(x - 0)^2$

    The equation is of the form $y - k = a(x - h)^2$
    with $a = 4$, $h = 0$, and $k = 0$, so the graph of
    the given equation is a parabola.

19.

$$4x^2 - 8x + 9y^2 + 36y = -4$$
$$4\left(x^2 - 2x + 1 - 1\right) + 9\left(y^2 + 4y + 4 - 4\right) = -4$$
$$4(x-1)^2 - 4 + 9(y+2)^2 - 36 = -4$$
$$4(x-1)^2 + 9(y+2)^2 = -4 + 4 + 36$$
$$4(x-1)^2 + 9(y+2)^2 = 36$$
$$\frac{(x-1)^2}{9} + \frac{(y+2)^2}{4} = 1$$
$$\frac{(x-1)^2}{3^2} + \frac{\left[y-(-2)\right]^2}{2^2} = 1$$

The equation is of the form $\frac{(x-h)^2}{a^2} + \frac{(y-k)^2}{b^2} = 1$ with $a = 3$, $b = 2$, $h = 1$, and $k = -2$, so the graph of the given equation is an ellipse.

21. $4x^2 + y^2 = 36 \Rightarrow \dfrac{x^2}{9} + \dfrac{y^2}{36} = 1 \Rightarrow$

$$\frac{(x-0)^2}{3^2} + \frac{(y-0)^2}{6^2} = 1$$

The equation is of the form $\frac{(x-h)^2}{b^2} + \frac{(y-k)^2}{a^2} = 1$ with $a = 6$, $b = 3$, $h = 0$, and $k = 0$. Thus, the graph of the given equation is an ellipse. It is centered at the origin with vertices at $(0, -6)$ and $(0, 6)$ and endpoints of the minor axis at $(-3, 0)$ and $(3, 0)$. The correct graph is F.

23.

$$(x-2)^2 + (y+3)^2 = 36$$
$$(x-2)^2 + \left[y-(-3)\right]^2 = 6^2$$

The equation is of the form
$(x-h)^2 + (y-k)^2 = r^2$ with
$r = 6$, $h = 2$, and $k = -3$, so the graph of the given equation is a circle. This circle is centered at $(2, -3)$ and has radius 6. The correct graph is A.

25.

$$(y-1)^2 - (x-2)^2 = 36$$
$$\frac{(y-1)^2}{36} - \frac{(x-2)^2}{36} = 1$$
$$\frac{(y-1)^2}{6^2} - \frac{(x-2)^2}{6^2} = 1$$

The equation is of the form $\frac{(y-k)^2}{a^2} - \frac{(x-h)^2}{b^2} = 1$ with $a = 6$, $b = 6$, $h = 2$, and $k = 1$, so the graph of the given equation is a hyperbola. This hyperbola is centered at $(2, 1)$, opening upward and downward, with vertices at $(2, -5)$ and $(2, 7)$. The correct graph is B.

27. $\dfrac{x^2}{4} + \dfrac{y^2}{9} = 1 \Rightarrow \dfrac{(x-0)^2}{2^2} + \dfrac{(y-0)^2}{3^2} = 1$

The graph is an ellipse centered at the origin with domain $[-2, 2]$, range $[-3, 3]$, and vertices at $(0, -3)$ and $(0, 3)$. The endpoints of the minor axis are $(-2, 0)$ and $(2, 0)$.

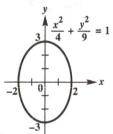

29. $\dfrac{x^2}{64} - \dfrac{y^2}{36} = 1 \Rightarrow \dfrac{(x-0)^2}{8^2} - \dfrac{(y-0)^2}{6^2} = 1$

The graph is a hyperbola centered at the origin with domain $\left(-\infty, -8\right] \cup \left[8, \infty\right)$, range $\left(-\infty, \infty\right)$, and vertices $(-8, 0)$ and $(8, 0)$. The asymptotes are the lines $y = \pm\frac{6}{8}x \Rightarrow$ $y = \pm\frac{3}{4}x$.

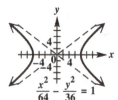

31.

$$\frac{(x+1)^2}{16} + \frac{(y-1)^2}{16} = 1$$
$$(x+1)^2 + (y-1)^2 = 16$$
$$\left[x-(-1)\right]^2 + (y-1)^2 = 4^2$$

The graph is a circle centered at $(-1, 1)$ and radius 4. Its domain is $\left[-1-4, -1+4\right] = \left[-5, 3\right]$. The range is $\left[1-4, 1+4\right] = \left[-3, 5\right]$.

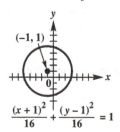

33. $4x^2 + 9y^2 = 36 \Rightarrow \dfrac{x^2}{9} + \dfrac{y^2}{4} = 1 \Rightarrow$

$\dfrac{(x-0)^2}{3^2} + \dfrac{(y-0)^2}{2^2} = 1$

The graph is an ellipse centered at the origin with domain [–3, 3], range [–2, 2], vertices at (–3, 0) and (3, 0). The endpoints of the minor axis at (0, –2) and (0, 2).

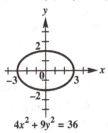

$4x^2 + 9y^2 = 36$

35. $\dfrac{(x-3)^2}{4} + (y+1)^2 = 1 \Rightarrow$

$\dfrac{(x-3)^2}{2^2} + \dfrac{\left[y-(-1)\right]^2}{1^2} = 1$

The graph is an ellipse centered at (3, –1). The vertices are (3 – 2, –1) = (1, –1) and (3 + 2, –1) = (5, –1). The endpoints of the minor axis are (3, –1 – 1) = (3, –2) and (3, –1 + 1) = (3, 0). The domain is [1, 5] and the range is [–2, 0].

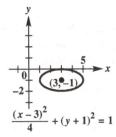

$\dfrac{(x-3)^2}{4} + (y+1)^2 = 1$

37. $\dfrac{(y+2)^2}{4} - \dfrac{(x+3)^2}{9} = 1 \Rightarrow$

$\dfrac{\left[y-(-2)\right]^2}{2^2} - \dfrac{\left[x-(-3)\right]^2}{3^2} = 1$

The graph is a hyperbola centered at $(-3,-2)$. The vertices are located at
$(-3,-2-2) = (-3,-4)$ and
$(-3,-2+2) = (-3,0)$. The domain is $(-\infty,\infty)$
and the range is $(-\infty, -4]\cup[0, \infty)$. The asymptotes are the lines
$y-(-2) = \pm\tfrac{2}{3}\left[x-(-3)\right] \Rightarrow$
$\quad y+2 = \pm\tfrac{2}{3}(x+3) \Rightarrow y = -2 \pm \tfrac{2}{3}(x+3).$

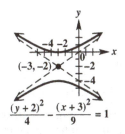

$\dfrac{(y+2)^2}{4} - \dfrac{(x+3)^2}{9} = 1$

39. $x^2 - 4x + y^2 + 6y = -12$

Complete the squares on x and y:

$x^2 - 4x + y^2 + 6y = -12$

$\left(x^2 - 4x + \quad\right) + \left(y^2 + 6y + \quad\right) = -12$

$\left(x^2 - 4x + 4\right) + \left(y^2 + 6y + 9\right) = -12 + 4 + 9$

$(x-2)^2 + (y+3)^2 = 1$

This is the graph of a circle with center (2, –3) and radius 1. The domain is [1, 3], and the range is [–4, –2].

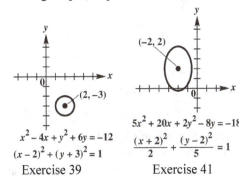

$x^2 - 4x + y^2 + 6y = -12$ $5x^2 + 20x + 2y^2 - 8y = -18$
$(x-2)^2 + (y+3)^2 = 1$ $\dfrac{(x+2)^2}{2} + \dfrac{(y-2)^2}{5} = 1$

Exercise 39   Exercise 41

41. $5x^2 + 20x + 2y^2 - 8y = -18$

Complete the squares on x and y :

$5x^2 + 20x + 2y^2 - 8y = -18$

$5\left(x^2 + 4x + \quad\right) + 2\left(y^2 - 4y + \quad\right) = -18$

$5\left(x^2 + 4x + 4\right) + 2\left(y^2 - 4y + 4\right) = -18 + 20 + 8$

$5(x+2)^2 + 2(y-2)^2 = 10$

$\dfrac{(x+2)^2}{2} + \dfrac{(y-2)^2}{5} = 1$

$a^2 = 2 \Rightarrow a = \sqrt{2}$ and $b^2 = 5 \Rightarrow b = \sqrt{5}$. The graph is an ellipse with center (–2, 2) and vertices $\left(-2, 2+\sqrt{5}\right)$ and $\left(-2, 2-\sqrt{5}\right)$. The domain is $\left[-2-\sqrt{2}, -2+\sqrt{2}\right]$, and the range is $\left[2-\sqrt{5}, 2+\sqrt{5}\right]$.

43. $\dfrac{x}{3} = -\sqrt{1 - \dfrac{y^2}{16}}$

Square both sides to get $\dfrac{x^2}{9} = 1 - \dfrac{y^2}{16}$ or

$\dfrac{x^2}{9} + \dfrac{y^2}{16} = 1$ or $\dfrac{x^2}{3^2} + \dfrac{y^2}{4^2} = 1$. This is the

equation of an ellipse with center (0, 0) and vertices (0, −4) and (0, 4). It has endpoints of the minor axis of (−3, 0) and (3, 0). The graph of the original equation is the left half of this ellipse. The domain is [−3, 0] and the range is [−4, 4]. The vertical line test shows that this relation is not a function.

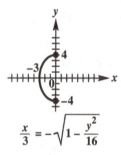

$\dfrac{x}{3} = -\sqrt{1 - \dfrac{y^2}{16}}$

45. $y = -\sqrt{1 + x^2}$

Square both sides of the equation to get

$y^2 = 1 + x^2$ or $\dfrac{y^2}{1} - \dfrac{x^2}{1} = 1$ or $\dfrac{y^2}{1^2} - \dfrac{x^2}{1^2} = 1$. This

is the equation of a hyperbola with center (0, 0) and vertices (0, −1) and (0, 1). Because $a = 1$, $b = 1$, and $y = \pm\dfrac{a}{b}x$, we have

asymptotes $y = \pm\dfrac{1}{1}x \Rightarrow y = \pm x$. The original equation is the bottom half of the hyperbola. The domain is $(-\infty, \infty)$. The range is

$(-\infty, -1]$. The vertical line test shows this is a graph of a function.

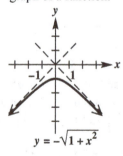

$y = -\sqrt{1 + x^2}$

47. Ellipse; vertex at (0, −4), focus at (0, −2), center at the origin
The vertex is (0, −4) and the focus is (0, −2), so the major axis is vertical and thus, the equation is of the form $\dfrac{x^2}{b^2} + \dfrac{y^2}{a^2} = 1$.

Because the ellipse is centered at the origin, we have $a = 4$ which implies $a^2 = 16$. The ellipse is centered at the origin and a focus is at (0, 2), so $c = 2$ which implies $c^2 = 4$. $c^2 = a^2 - b^2 \Rightarrow 4 = 16 - b^2 \Rightarrow b^2 = 12$. Thus, an equation of the ellipse is $\dfrac{x^2}{12} + \dfrac{y^2}{16} = 1$.

49. Hyperbola; focus at (0, 5) transverse axis with length 8, center at the origin
The focus is (0, 5) so the transverse axis is vertical and thus, the equation is of the form $\dfrac{y^2}{a^2} - \dfrac{x^2}{b^2} = 1$. Because the length of the transverse axis is 8, we have $2a = 8 \Rightarrow a = 4$. This implies that $a^2 = 16$. The hyperbola is centered at the origin and a focus is at (0, 5), so $c = 5$ which implies $c^2 = 25$. $c^2 = a^2 + b^2, \Rightarrow 25 = 16 + b^2 \Rightarrow b^2 = 9$. Thus, an equation of the hyperbola is $\dfrac{y^2}{16} - \dfrac{x^2}{9} = 1$.

51. Parabola with focus at (3, 2) and directrix $x = -3$
The directrix is $x = -3$, a vertical line, so the parabola is of the form $(y - k)^2 = 4p(x - h)$. The vertex is halfway between the focus and the directrix, so the vertex is $(0, 2) = (h, k)$. Because of the orientation of the directrix versus the focus, we know that the parabola opens to the right, so the p value must be positive. By examining the distance between the vertex and the focus or the directrix and the vertex, we know that $p = 3$. Thus, an equation of the parabola is

$(y - 2)^2 = 4 \cdot 3(x - 0)$ or $(y - 2)^2 = 12x$.

53. Ellipse with foci at (−2, 0) and (2, 0) and major axis width length 10
The foci are at (−2, 0) and (2, 0), so the major axis is horizontal and the ellipse is centered at the origin. Thus, the equation is of the form $\dfrac{x^2}{a^2} + \dfrac{y^2}{b^2} = 1$. The foci are at (−2, 0) and (2, 0), so $c = 2$, which implies $c^2 = 4$. Because the major axis is of length 10, we have $2a = 10$. This implies that $a = 5$ and $a^2 = 25$. $c^2 = a^2 - b^2 \Rightarrow 4 = 25 - b^2 \Rightarrow b^2 = 21$. Thus, an equation of the ellipse is $\dfrac{x^2}{25} + \dfrac{y^2}{21} = 1$.

55. Hyperbola with x-intercepts $(-3, 0)$ and $(3, 0)$ foci at $(-5, 0)$, $(5, 0)$

The center is halfway between the foci, so the center is at the origin. Because the foci are on a horizontal transverse axis, the equation has the form $\frac{x^2}{a^2} - \frac{y^2}{b^2} = 1$. The x-intercepts are also vertices, so $a = 3$. This implies $a^2 = 9$. The foci are $(-5, 0)$ and $(5, 0)$, so $c = 5$. This implies $c^2 = 25$.

$c^2 = a^2 + b^2 \Rightarrow 25 = 9 + b^2 \Rightarrow b^2 = 16$.

Thus, an equation of the hyperbola is

$\frac{x^2}{9} - \frac{y^2}{16} = 1$.

57. The points $F'(0,0)$ and $F(4,0)$ are the foci, so the center of the ellipse is $(2,0) = (h,k)$. The foci are on a horizontal axis, so the equation is of the form $\frac{(x-h)^2}{a^2} + \frac{(y-k)^2}{b^2} = 1$.

The distance from the center to each of the foci is 2, so $c = 2 \Rightarrow c^2 = 4$. For any point P on the ellipse, $d(P,F) + d(P,F') = 2a = 8$,

so $a = 4$. Thus, we have $a^2 = 16$.

$c^2 = a^2 - b^2 \Rightarrow 4 = 16 - b^2 \Rightarrow b^2 = 12$.

Thus, an equation of the ellipse is

$\frac{(x-2)^2}{16} + \frac{y^2}{12} = 1$.

59. The points $F'(-5,0)$ and $F(5,0)$ are the foci, so the hyperbola is centered at the origin and has a horizontal transverse axis. The equation has the form $\frac{x^2}{a^2} - \frac{y^2}{b^2} = 1$. The points $(-5, 0)$ and $(5, 0)$ are the foci, giving $c = 5$. This implies $c^2 = 25$. Also, for any point P on the hyperbola we have $|d(P,F') - d(P,F)| = 2a$. Thus, we have $2a = 8 \Rightarrow a = 4$. This implies $a^2 = 16$.

$c^2 = a^2 + b^2 \Rightarrow 25 = 16 + b^2 \Rightarrow b^2 = 9$.

Thus, an equation of the hyperbola is

$\frac{x^2}{16} - \frac{y^2}{9} = 1$.

61. The eccentricity is 0.964 and $e = \frac{c}{a}$, so

$0.964 = \frac{c}{a} \Rightarrow c = 0.964a$. The closest distance to the sun is 89 million mi, giving $a - c = 89$. Substituting $c = 0.964a$ into this equation, we have the following.

$a - c = 89 \Rightarrow a - 0.964a = 89 \Rightarrow$

$0.036a = 89 \Rightarrow a = \frac{89}{0.036} = \frac{22{,}250}{9}$

$c = 0.964a \Rightarrow c = 0.964\left(\frac{22{,}250}{9}\right) = \frac{21{,}449}{9}$.

$c^2 = a^2 - b^2 \Rightarrow \left(\frac{21{,}449}{9}\right)^2 = \left(\frac{22{,}250}{9}\right)^2 - b^2 \Rightarrow$

$\frac{460{,}059{,}601}{81} - \frac{495{,}062{,}500}{81} = -b^2 \Rightarrow b^2 = \frac{35{,}002{,}899}{81}$

$a^2 = \left(\frac{22{,}250}{9}\right)^2 = \frac{495{,}062{,}500}{81} \approx 6{,}111{,}883$ and

$b^2 = \frac{35{,}002{,}899}{81} \approx 432{,}135$, so an equation is

$\frac{x^2}{6{,}111{,}883} + \frac{y^2}{432{,}135} = 1$.

Chapter 10   Test

1. $y = -x^2 + 6x$

Completing the square on x, we have

$y = -x^2 + 6x \Rightarrow y = -\left(x^2 - 6x + 9 - 9\right) \Rightarrow$

$y = -(x-3)^2 + 9 \Rightarrow y - 9 = -(x-3)^2$

The equation has the form $x - h = a(y - k)^2$. The vertex, (h, k), is $(3, 9)$ The x-term is squared, so the axis of symmetry is the vertical line $x = h$, or $x = 3$. The domain is $(-\infty, \infty)$. The range is $(-\infty, 9]$. Using this information and plot a few additional points.

x	y
5	5
4	8
3	9
2	8
1	5

2. $x = 4y^2 + 8y$

Completing the square on y, we have

$x = 4y^2 + 8y \Rightarrow x = 4\left(y^2 + 2y + 1 - 1\right) \Rightarrow$

$x = 4(y+1)^2 - 4 \Rightarrow x - (-4) = 4\left[y - (-1)\right]^2$

The equation has the form $y - k = a(x - h)^2$. The vertex, (h, k), is $(-4, -1)$. The y-term is squared, so the axis of symmetry is the horizontal line $y = k$, or $y = -1$. The domain is $[-4, \infty)$. The range is $(-\infty, \infty)$. Using this information and plot a few additional points.

(continued on next page)

(*continued*)

x	y
12	1
0	0
–4	–1
0	–2
12	–3

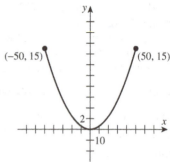

$x = 4y^2 + 8y$

3. Because $x = 8y^2 \Rightarrow y^2 = \frac{1}{8}x$, the parabola is

    of the form $y^2 = 4px$. Thus, we have

    $4p = \frac{1}{8} \Rightarrow p = \frac{1}{32}$. The y-term is squared, so

    the parabola is horizontal, with focus

    $(p, 0) = \left(\frac{1}{32}, 0\right)$ and directrix, $x = -p$, is

    $x = -\frac{1}{32}$.

4. Parabola; vertex at (2, 3) passing through
    (–18, 1) opening to the left
    The parabola opens to the left, its equation is of

    the form $(y - k)^2 = 4p(x - h)$. The vertex is

    located at (2, 3), so we have

    $(y - 3)^2 = 4p(x - 2)$. We can now substitute

    $x = -18$ and $y = 1$, and solve for p.

    $(1 - 3)^2 = 4p(-18 - 2) \Rightarrow$

    $(-2)^2 = 4p(-20) \Rightarrow 4 = -80p \Rightarrow p = -\frac{1}{20}$

    Thus, the equation is

    $(y - 3)^2 = 4\left(-\frac{1}{20}\right)(x - 2) \Rightarrow (y - 3)^2 = -\frac{1}{5}(x - 2)$

    or alternatively written as $x - 2 = -5(y - 3)^2$.

    We could have also stated that because the
    parabola opens to the left, it is of the form

    $x - h = a(y - k)^2$. With $(h, k) = (2, 3)$, we have

    $x - 2 = a(y - 3)^2$.

    Substituting $x = -18$ and $y = 1$, we have the
    following.

    $-18 - 2 = a(1 - 3)^2 \Rightarrow -20 = a(-2)^2 \Rightarrow$
    $-20 = 4a \Rightarrow a = 5$

    Thus, $x - 2 = -5(y - 3)^2$ or

    $(y - 3)^2 = -\frac{1}{5}(x - 2)$.

5. (a) Sketch a cross-section of the dish. Place
        this parabola on a coordinate system with
        the vertex at the origin. (This solution
        will show that the focus lies outside of
        the dish.)

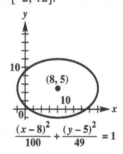

The parabola has vertex (0, 0) and a
vertical axis (the y-axis), so it has an

equation of the form $x^2 = 4py$.

Substitute $x = 50$ and $y = 15$ to find the
value of p.

$x^2 = 4py \Rightarrow 50^2 = 4p \cdot 15 \Rightarrow$

$2500 = 60p \Rightarrow p = \frac{2500}{60} = \frac{125}{3}$

The required equation is

$x^2 = 4\left(\frac{125}{3}\right)y \Rightarrow x^2 = \frac{500}{3}y$

$y = \frac{3}{500}x^2$

(b) Because $p = \frac{125}{3} \approx 41.7$, the receiver

    should be placed approximately 41.7 ft
    from the vertex.

6. $\dfrac{(x - 8)^2}{100} + \dfrac{(y - 5)^2}{49} = 1 \Rightarrow$

    $\dfrac{(x - 8)^2}{10^2} + \dfrac{(y - 5)^2}{7^2} = 1$

    The graph is an ellipse centered at (8, 5). We
    also have the vertices located at

    $(8 - 10, 5) = (-2, 5)$ and $(8 + 10, 5) = (18, 5)$.

    The endpoints of the minor axis are located at

    $(8, 5 - 7) = (8, -2)$ and $(8, 5 + 7) = (8, 12)$.

    The domain is [–2, 18] and the range is
    [–2, 12].

(8, 5)

$\dfrac{(x - 8)^2}{100} + \dfrac{(y - 5)^2}{49} = 1$

7. $16x^2 + 4y^2 = 64 \Rightarrow \dfrac{x^2}{4} + \dfrac{y^2}{16} = 1 \Rightarrow$

$\dfrac{x^2}{2^2} + \dfrac{y^2}{4^2} = 1$

The graph is an ellipse centered at the origin
The vertices are located at $(0, -4)$ and $(0, 4)$.
The endpoints of the minor axis are located at
$(-2, 0)$ and $(2, 0)$. The domain is $[-2, 2]$ and
the range is $[-4, 4]$.

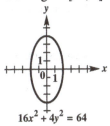

$16x^2 + 4y^2 = 64$

8. $y = -\sqrt{1 - \dfrac{x^2}{36}}$

Square both sides to get

$y = -\sqrt{1 - \dfrac{x^2}{36}} \Rightarrow y^2 = 1 - \dfrac{x^2}{36} \Rightarrow \dfrac{x^2}{36} + y^2 = 1 \Rightarrow$
$\dfrac{x^2}{6^2} + \dfrac{y^2}{1^2} = 1.$

This is the equation of an ellipse with vertices
$(-6, 0)$ and $(6, 0)$. The endpoints of the minor
axis are at $(0, -1)$ and $(0, 1)$. The graph of the
original equation is the bottom half of the
ellipse. The vertical line test shows that this
relation is a function.

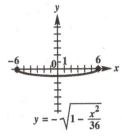

$y = -\sqrt{1 - \dfrac{x^2}{36}}$

9. Ellipse; centered at the origin, horizontal
major axis with length 6, minor axis with
length 4
The ellipse is centered at the origin and major
axis is horizontal, so the ellipse has the form
$\dfrac{x^2}{a^2} + \dfrac{y^2}{b^2} = 1$. The major axis has length 6, so
we have $2a = 6 \Rightarrow a = 3 \Rightarrow a^2 = 9$. Because
the minor axis has length 4, we have
$2b = 4 \Rightarrow b = 2 \Rightarrow b^2 = 4$. Thus, an equation
of the ellipse is $\dfrac{x^2}{9} + \dfrac{y^2}{4} = 1$.

10. Place the half-ellipse that represents the
overpass on a coordinate system with the
center of the bottom of the overpass at the
origin. If the complete ellipse were drawn, the
center of the ellipse would be $(0, 0)$. Then the
half-ellipse will include the points $(0, 12)$,
$(-20, 0)$, and $(20, 0)$. The equation is of the
form $\dfrac{x^2}{a^2} + \dfrac{y^2}{b^2} = 1$. Thus, for the complete
ellipse, we have $a = 20$ and $b = 12$. Thus, we
have $\dfrac{x^2}{20^2} + \dfrac{y^2}{12^2} = 1 \Rightarrow \dfrac{x^2}{400} + \dfrac{y^2}{144} = 1$. At a
distance of 10 ft from the center of the bottom,
$x = 10$. Find the positive y-coordinate of the
point on the ellipse with x-coordinate 10.

$\dfrac{x^2}{400} + \dfrac{y^2}{144} = 1 \Rightarrow \dfrac{10^2}{400} + \dfrac{y^2}{144} = 1 \Rightarrow$
$\dfrac{100}{400} + \dfrac{y^2}{144} = 1 \Rightarrow \dfrac{1}{4} + \dfrac{y^2}{144} = 1 \Rightarrow \dfrac{y^2}{144} = \dfrac{3}{4} \Rightarrow$
$y^2 = 108 \Rightarrow y = \sqrt{108} \approx 10.4$

The arch is approximately 10.4 ft high 10 ft
from the center of the bottom.

11. $\dfrac{x^2}{4} - \dfrac{y^2}{4} = 1 \Rightarrow \dfrac{x^2}{2^2} - \dfrac{y^2}{2^2} = 1$

The graph of the hyperbola is centered at the
origin. The x^2-term comes first, so the
branches open to the right and left, with
vertices $(-2, 0)$ and $(2, 0)$. The domain is
$(-\infty, -2] \cup [2, \infty)$. The range is $(-\infty, \infty)$. The
asymptotes are the lines $y = \pm \dfrac{2}{2} x = \pm x$.

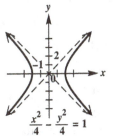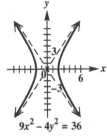

$\dfrac{x^2}{4} - \dfrac{y^2}{4} = 1$ $9x^2 - 4y^2 = 36$

Exercise 11   Exercise 12

12. $9x^2 - 4y^2 = 36 \Rightarrow \dfrac{x^2}{4} - \dfrac{y^2}{9} = 1 \Rightarrow \dfrac{x^2}{2^2} - \dfrac{y^2}{3^2} = 1$

The graph of the hyperbola is centered at the
origin. The x^2-term comes first, so the
branches open to the right and left, with
vertices $(-2, 0)$ and $(2, 0)$. The domain is
$(-\infty, -2] \cup [2, \infty)$. The range is $(-\infty, \infty)$. The
asymptotes are the lines $y = \pm \dfrac{3}{2} x$.

13. Hyperbola; y-intercepts $(0, -5)$ and $(0, 5)$; foci at $(0, -6)$ and $(0, 6)$

The center is halfway between the foci, so the center is $(0, 0)$. The equation has the form

$\frac{y^2}{a^2} - \frac{x^2}{b^2} = 1$ because the foci are on a vertical

transverse axis. The y-intercepts are also vertices, so $a = 5$, and thus $a^2 = 25$. With the given foci, we have $c = 6$. $c^2 = a^2 + b^2 \Rightarrow$

$6^2 = 5^2 + b^2 \Rightarrow 36 = 25 + b^2 \Rightarrow b^2 = 11$

Thus, an equation of the hyperbola is

$\frac{y^2}{25} - \frac{x^2}{11} = 1$.

14.
$$x^2 + 8x + y^2 - 4y + 2 = 0$$
$$\left(x^2 + 8x + 16 - 16\right) + \left(y^2 - 4y + 4 - 4\right) = -2$$
$$\left(x + 4\right)^2 - 16 + \left(y - 2\right)^2 - 4 = -2$$
$$\left(x + 4\right)^2 + \left(y - 2\right)^2 = -2 + 16 + 4$$
$$\left(x + 4\right)^2 + \left(y - 2\right)^2 = 18$$
$$\left[x - (-4)\right]^2 + \left(y - 2\right)^2 = \left(\sqrt{18}\right)^2$$

The graph of this equation is a circle with center $(-4, 2)$ and radius $\sqrt{18}$. Also note in our original equation, the x^2- and y^2- terms have the same positive coefficient.

15.
$$5x^2 + 10x - 2y^2 - 12y - 23 = 0$$
$$5\left(x^2 + 2x + 1 - 1\right) - 2\left(y^2 + 6y + 9 - 9\right) = 23$$
$$5\left(x + 1\right)^2 - 5 - 2\left(y + 3\right)^2 + 18 = 23$$
$$5\left(x + 1\right)^2 - 2\left(y + 3\right)^2 = 23 + 5 - 18$$
$$5\left(x + 1\right)^2 - 2\left(y + 3\right)^2 = 10$$
$$\frac{(x+1)^2}{2} - \frac{(y+3)^2}{5} = 1$$
$$\frac{\left[x - (-1)\right]^2}{\left(\sqrt{2}\right)^2} - \frac{\left[y - (-3)\right]^2}{\left(\sqrt{5}\right)^2} = 1$$

The graph of this equation is a hyperbola centered at $(-1, -3)$ and vertices $\left(-1 \pm \sqrt{2}, -3\right)$. The asymptotes are

$y - (-3) = \pm \frac{\sqrt{5}}{\sqrt{2}}\left[x - (-2)\right] \Rightarrow y + 3 = \pm \frac{\sqrt{10}}{2}(x + 2)$.

Also note in our original equation, the x^2- and y^2-terms have coefficients that are opposite in sign.

16. $3x^2 + 10y^2 - 30 = 0 \Rightarrow 3x^2 + 10y^2 = 30 \Rightarrow$

$\frac{x^2}{10} + \frac{y^2}{3} = 1 \Rightarrow \frac{(x-0)^2}{\left(\sqrt{10}\right)^2} + \frac{(y-0)^2}{\left(\sqrt{3}\right)^2} = 1$

The graph of this equation is an ellipse centered at the origin and x-intercepts (vertices) of $\pm\sqrt{10}$, and y-intercepts (endpoints of minor axis) of $\pm\sqrt{3}$. Also note in our original equation, the x^2- and y^2- terms both have different positive coefficients.

17. $x^2 - 4y = 0 \Rightarrow 4y = x^2 \Rightarrow y = \frac{1}{4}x^2$

This is a parabola with its vertex at the origin, opening upward. Also, note our original equation has an x^2-term, but no y^2-term.

18.
$$\left(x + 9\right)^2 + \left(y - 3\right)^2 = 0$$
$$\left[x - (-9)\right]^2 + \left(y - 3\right)^2 = 0$$

This is the equation of a "circle" centered at $(-9, 3)$ with radius 0. The graph of this equation is a point.

19.
$$x^2 + 4x + y^2 - 6y + 30 = 0$$
$$\left(x^2 + 4x + 4 - 4\right) + \left(y^2 - 6y + 9 - 9\right) = -30$$
$$\left(x^2 + 4x + 4\right) - 4 + \left(y^2 - 6y + 9\right) - 9 = -30$$
$$\left(x + 2\right)^2 + \left(y - 3\right)^2 = -30 + 4 + 9$$
$$\left[x - (-2)\right]^2 + \left(y - 3\right)^2 = -17$$

This equation has the form of the equation of a circle. However, because r^2 cannot be negative, there is no graph of this equation.

20. Solve the equation $\frac{x^2}{25} - \frac{y^2}{49} = 1$ for y.

$\frac{x^2}{25} - \frac{y^2}{49} = 1 \Rightarrow -\frac{y^2}{49} = 1 - \frac{x^2}{25} \Rightarrow \frac{y^2}{49} = \frac{x^2}{25} - 1 \Rightarrow$

$y^2 = 49\left(\frac{x^2}{25} - 1\right) \Rightarrow y = \pm\sqrt{49\left(\frac{x^2}{25} - 1\right)} \Rightarrow$

$y = \pm 7\sqrt{\frac{x^2}{25} - 1}$

Thus, the functions $y_1 = 7\sqrt{\frac{x^2}{25} - 1}$ and

$y_2 = -7\sqrt{\frac{x^2}{25} - 1}$ were used to obtain the graph.

Chapter 11

FURTHER TOPICS IN ALGEBRA

Section 11.1   Sequences and Series

1. A <u>sequence</u> is a function that computes an ordered list.

3. Some sequences are defined by a <u>recursive</u> definition, one in which each term after the first term or the first few terms is defined as an expression involving the previous term or terms.

5. $a_n = 5n + 2$

   $n = 1:\ a_1 = 5(1) + 2 = 7$
   $n = 2:\ a_2 = 5(2) + 2 = 12$
   $n = 3:\ a_3 = 5(3) + 2 = 17$
   $n = 4:\ a_4 = 5(4) + 2 = 22$
   $n = 5:\ a_5 = 5(5) + 2 = 27$

7. $\displaystyle\sum_{i=1}^{5}(5i + 2)$

   $= \left[5(1) + 2\right] + \left[5(2) + 2\right] + \left[5(3) + 2\right]$
   $\qquad + \left[5(4) + 2\right] + \left[5(5) + 2\right]$
   $= 7 + 12 + 17 + 22 + 27 = 85$

9. $a_n = 3(-3)^{n-1}$

   $n = 1:\ a_1 = 3(-3)^{1-1} = 3(-3)^0 = 3(1) = 3$
   $n = 2:\ a_2 = 3(-3)^{2-1} = 3(-3)^1 = 3(-3) = -9$
   $n = 3:\ a_3 = 3(-3)^{3-1} = 3(-3)^2 = 3(9) = 27$
   $n = 4:\ a_4 = 3(-3)^{4-1} = 3(-3)^3 = 3(-27) = -81$
   $n = 5:\ a_5 = 3(-3)^{5-1} = 3(-3)^4 = 3(81) = 243$

11. $a_n = 4n + 10$

   Replace n with 1, 2, 3, 4, and 5.
   $n = 1: a_1 = 4(1) + 10 = 14$
   $n = 2: a_2 = 4(2) + 10 = 18$
   $n = 3: a_3 = 4(3) + 10 = 22$
   $n = 4: a_4 = 4(4) + 10 = 26$
   $n = 5: a_5 = 4(5)\ + 10 = 30$
   The first five terms are 14, 18, 22, 26, and 30.

13. $a_n = \dfrac{n+5}{n+4}$

   Replace n with 1, 2, 3, 4, and 5.
   $n = 1:\ a_1 = \dfrac{1+5}{1+4} = \dfrac{6}{5}$
   $n = 2:\ a_2 = \dfrac{2+5}{2+4} = \dfrac{7}{6}$
   $n = 3:\ a_3 = \dfrac{3+5}{3+4} = \dfrac{8}{7}$
   $n = 4:\ a_4 = \dfrac{4+5}{4+4} = \dfrac{9}{8}$
   $n = 5:\ a_5 = \dfrac{5+5}{5+4} = \dfrac{10}{9}$
   The first five terms are $\dfrac{6}{5}, \dfrac{7}{6}, \dfrac{8}{7}, \dfrac{9}{8}$, and $\dfrac{10}{9}$.

15. $a_n = \left(\dfrac{1}{3}\right)^n (n-1)$

   Replace n with 1, 2, 3, 4, and 5.
   $n = 1:\ a_1 = \left(\dfrac{1}{3}\right)^1 (1-1) = \left(\dfrac{1}{3}\right)(0) = 0$
   $n = 2:\ a_2 = \left(\dfrac{1}{3}\right)^2 (2-1) = \left(\dfrac{1}{9}\right)(1) = \dfrac{1}{9}$
   $n = 3:\ a_3 = \left(\dfrac{1}{3}\right)^3 (3-1) = \left(\dfrac{1}{27}\right)(2) = \dfrac{2}{27}$
   $n = 4:\ a_4 = \left(\dfrac{1}{3}\right)^4 (4-1) = \left(\dfrac{1}{81}\right)(3) = \dfrac{1}{27}$
   $n = 5:\ a_5 = \left(\dfrac{1}{3}\right)^5 (5-1) = \left(\dfrac{1}{243}\right)(4) = \dfrac{4}{243}$
   The first five terms are $0, \dfrac{1}{9}, \dfrac{2}{27}, \dfrac{1}{27}$, and $\dfrac{4}{243}$.

17. $a_n = (-1)^n (2n)$

   Replace n with 1, 2, 3, 4, and 5.
   $n = 1:\ a_1 = (-1)^1 \left[2(1)\right] = -1(2) = -2$
   $n = 2:\ a_2 = (-1)^2 \left[2(2)\right] = 1(4) = 4$
   $n = 3:\ a_3 = (-1)^3 \left[2(3)\right] = -1(6) = -6$
   $n = 4:\ a_4 = (-1)^4 \left[2(4)\right] = 1(8) = 8$
   $n = 5:\ a_5 = (-1)^5 \left[2(5)\right] = -1(10) = -10$
   The first five terms are $-2, 4, -6, 8$, and -10.

19. $a_n = \dfrac{4n-1}{n^2+2}$

Replace n with 1, 2, 3, 4, and 5.

$n = 1: a_1 = \dfrac{4(1)-1}{(1)^2+2} = \dfrac{4-1}{1+2} = \dfrac{3}{3} = 1$

$n = 2: a_2 = \dfrac{4(2)-1}{(2)^2+2} = \dfrac{8-1}{4+2} = \dfrac{7}{6}$

$n = 3: a_3 = \dfrac{4(3)-1}{(3)^2+2} = \dfrac{12-1}{9+2} = \dfrac{11}{11} = 1$

$n = 4: a_4 = \dfrac{4(4)-1}{(4)^2+2} = \dfrac{16-1}{16+2} = \dfrac{15}{18} = \dfrac{5}{6}$

$n = 5: a_5 = \dfrac{4(5)-1}{(5)^2+2} = \dfrac{20-1}{25+2} = \dfrac{19}{27}$

The first five terms are $1, \frac{7}{6}, 1, \frac{5}{6},$ and $\frac{19}{27}$.

21. $a_n = \dfrac{n^3+8}{n+2}$

Replace n with 1, 2, 3, 4, and 5.

$n = 1: a_1 = \dfrac{1^3+8}{1+2} = \dfrac{1+8}{3} = \dfrac{9}{3} = 3$

$n = 2: a_2 = \dfrac{2^3+8}{2+2} = \dfrac{8+8}{4} = \dfrac{16}{4} = 4$

$n = 3: a_3 = \dfrac{3^3+8}{3+2} = \dfrac{27+8}{5} = \dfrac{35}{5} = 7$

$n = 4: a_4 = \dfrac{4^3+8}{4+2} = \dfrac{64+8}{6} = \dfrac{72}{6} = 12$

$n = 5: a_5 = \dfrac{5^3+8}{5+2} = \dfrac{125+8}{7} = \dfrac{133}{7} = 19$

The first five terms are 3, 4, 7, 12, and 19.

23. The sequence of the days of the week has as its domain $\{1, 2, 3, 4, 5, 6, 7\}$. Therefore, it is a finite sequence.

25. The sequence 1, 2, 3, 4, 5 has as its domain $\{1, 2, 3, 4, 5\}$. Therefore, it is a finite sequence.

27. The sequence 1, 2, 3, 4, 5, …has as its domain $\{1, 2, 3, 4, 5, …\}$. Therefore, the sequence is infinite.

29. The sequence $a_1 = 4$ and for $n \geq 2$, $a_n = 4 \cdot a_{n-1}$ has as its domain $\{2, 3, 4, …\}$. Therefore, the sequence is infinite.

31. $a_1 = -2, a_n = a_{n-1} + 3$, for $n > 1$

This is an example of a recursive definition. We know $a_1 = -2$. Because $a_n = a_{n-1} + 3$,

$n = 2: a_2 = a_1 + 3 = -2 + 3 = 1$
$n = 3: a_3 = a_2 + 3 = 1 + 3 = 4$
$n = 4: a_4 = a_3 + 3 = 4 + 3 = 7.$

The first four terms are –2, 1, 4, and 7.

33. $a_1 = 1, a_2 = 1, a_n = a_{n-1} + a_{n-2}$, for $n \geq 3$

$n = 3: a_3 = a_2 + a_1 = 1 + 1 = 2$
$n = 4: a_4 = a_3 + a_2 = 2 + 1 = 3$

The first four terms are 1, 1, 2, and 3.

35. $a_1 = 2, a_n = n \cdot a_{n-1}$, for $n > 1$
$a_2 = 2 \cdot a_1 = 2 \cdot 2 = 4$
$a_3 = 3 \cdot a_2 = 3 \cdot 4 = 12$
$a_4 = 4 \cdot a_3 = 4 \cdot 12 = 48$

The first four terms are 2, 4, 12, and 48.

37. $\displaystyle\sum_{i=1}^{5}(2i+1)$

$= \left[2(1)+1\right] + \left[2(2)+1\right] + \left[2(3)+1\right]$
$\qquad + \left[2(4)+1\right] + \left[2(5)+1\right]$
$= (2+1) + (4+1) + (6+1) + (8+1) + (10+1)$
$= 3 + 5 + 7 + 9 + 11 = 35$

39. $\displaystyle\sum_{j=1}^{4} j^{-1} = \sum_{j=1}^{4}\dfrac{1}{j} = \dfrac{1}{1} + \dfrac{1}{2} + \dfrac{1}{3} + \dfrac{1}{4}$

$= \dfrac{12}{12} + \dfrac{6}{12} + \dfrac{4}{12} + \dfrac{3}{12} = \dfrac{25}{12}$

41. $\displaystyle\sum_{i=1}^{4} i^i = 1^1 + 2^2 + 3^3 + 4^4 = 1 + 4 + 27 + 256 = 288$

43. $\displaystyle\sum_{k=1}^{6}(-1)^k \cdot k$

$= (-1)^1 \cdot 1 + (-1)^2 \cdot 2 + (-1)^3 \cdot 3 + (-1)^4 \cdot 4$
$\qquad + (-1)^5 \cdot 5 + (-1)^6 \cdot 6$
$= -1 \cdot 1 + 1 \cdot 2 + (-1) \cdot 3 + 1 \cdot 4 + (-1) \cdot 5 + 1 \cdot 6$
$= -1 + 2 - 3 + 4 - 5 + 6 = 3$

45. $\displaystyle\sum_{i=2}^{5}(6-3i)$

$= \left[6 - 3(2)\right] + \left[6 - 3(3)\right]$
$\qquad + \left[6 - 3(4)\right] + \left[6 - 3(5)\right]$
$= (6-6) + (6-9) + (6-12) + (6-15)$
$= 0 + (-3) + (-6) + (-9) = -18$

47. $\displaystyle\sum_{i=-2}^{3} 2(3)^i = 2(3)^{-2} + 2(3)^{-1} + 2(3)^0 + 2(3)^1 + 2(3)^2 + 2(3)^3$

$$= 2\left(\tfrac{1}{9}\right) + 2\left(\tfrac{1}{3}\right) + 2(1) + 2(3) + 2(9) + 2(27) = \tfrac{2}{9} + \tfrac{2}{3} + 2 + 6 + 18 + 54$$

$$= \tfrac{2}{9} + \tfrac{6}{9} + 80 = 80\tfrac{8}{9} = \tfrac{728}{9}$$

49. $\displaystyle\sum_{i=-1}^{5} \left(i^2 - 2i\right) = \left[(-1)^2 - 2(-1)\right] + \left[0^2 - 2(0)\right] + \left[1^2 - 2(1)\right] + \left[2^2 - 2(2)\right] + \left[3^2 - 2(3)\right]$

$$+ \left[4^2 - 2(4)\right] + \left[5^2 - 2(5)\right]$$

$$= \left[1 - (-2)\right] + (0 - 0) + (1 - 2) + (4 - 4) + (9 - 6) + (16 - 8) + (25 - 10)$$

$$= 3 + 0 + (-1) + 0 + 3 + 8 + 15 = 28$$

51. $\displaystyle\sum_{i=1}^{5}\left(3^i - 4\right) = \left(3^1 - 4\right) + \left(3^2 - 4\right) + \left(3^3 - 4\right) + \left(3^4 - 4\right) + \left(3^5 - 4\right)$

$$= (3 - 4) + (9 - 4) + (27 - 4) + (81 - 4) + (243 - 4) = (-1) + 5 + 23 + 77 + 239 = 343$$

53. $\displaystyle\sum_{i=1}^{3}\left(i^3 - i\right) = \left(1^3 - 1\right) + \left(2^3 - 2\right) + \left(3^3 - 3\right) = (1 - 1) + (8 - 2) + (27 - 3) = 0 + 6 + 24 = 30$

For exercises 55–57, to find the sum(feature on a TI-84, go to the LIST menu (2nd STAT), then go to MATH and choose option 5. To find the seq(feature on a TI-84, go to the LIST menu (2nd STAT), then go to OPS and choose option 5.

In Exercises 59–67, $x_1 = -2$, $x_2 = -1$, $x_3 = 0$, $x_4 = 1$, and $x_5 = 2$.

59. $\displaystyle\sum_{i=1}^{5} x_i = x_1 + x_2 + x_3 + x_4 + x_5 = -2 + (-1) + 0 + 1 + 2 = 0$

61. $\displaystyle\sum_{i=1}^{5}\left(2x_i + 3\right) = \left(2x_1 + 3\right) + \left(2x_2 + 3\right) + \left(2x_3 + 3\right) + \left(2x_4 + 3\right) + \left(2x_5 + 3\right)$

$$= \left[2(-2) + 3\right] + \left[2(-1) + 3\right] + \left[2(0) + 3\right] + \left[2(1) + 3\right] + \left[2(2) + 3\right]$$

$$= (-4 + 3) + (-2 + 3) + (0 + 3) + (2 + 3) + (4 + 3) = -1 + 1 + 3 + 5 + 7 = 15$$

63. $\displaystyle\sum_{i=1}^{3}\left(3x_i - x_i^2\right) = \left(3x_1 - x_1^2\right) + \left(3x_2 - x_2^2\right) + \left(3x_3 - x_3^2\right)$

$$= \left[3(-2) - (-2)^2\right] + \left[3(-1) - (-1)^2\right] + \left[3(0) - 0^2\right]$$

$$= (-6 - 4) + (-3 - 1) + (0 - 0) = -10 + (-4) + 0 = -14$$

65. $\displaystyle\sum_{i=2}^{5} \frac{x_i + 1}{x_i + 2} = \frac{x_2 + 1}{x_2 + 2} + \frac{x_3 + 1}{x_3 + 2} + \frac{x_4 + 1}{x_4 + 2} + \frac{x_5 + 1}{x_5 + 2} = \frac{-1 + 1}{-1 + 2} + \frac{0 + 1}{0 + 2} + \frac{1 + 1}{1 + 2} + \frac{2 + 1}{2 + 2}$

$$= \frac{0}{1} + \frac{1}{2} + \frac{2}{3} + \frac{3}{4} = 0 + \frac{6}{12} + \frac{8}{12} + \frac{9}{12} = \frac{23}{12}$$

67. $\displaystyle\sum_{i=1}^{4} \frac{x_i^3 + 1000}{x_i + 10} = \frac{x_1^3 + 1000}{x_1 + 10} + \frac{x_2^3 + 1000}{x_2 + 10} + \frac{x_3^3 + 1000}{x_3 + 10} + \frac{x_4^3 + 1000}{x_4 + 10}$

$\displaystyle = \frac{(-2)^3 + 1000}{(-2) + 10} + \frac{(-1)^3 + 1000}{(-1) + 10} + \frac{0^3 + 1000}{0 + 10} + \frac{1^3 + 1000}{1 + 10}$

$\displaystyle = \frac{-8 + 1000}{8} + \frac{-1 + 1000}{9} + \frac{1000}{10} + \frac{1 + 1000}{11} = \frac{992}{8} + \frac{999}{9} + \frac{1000}{10} + \frac{1001}{11}$

$= 124 + 111 + 100 + 91 = 426$

In Exercises 69–73, $x_1 = 0$, $x_2 = 2$, $x_3 = 4$, $x_4 = 6$, and $\Delta x = 0.5$.

69. $f(x) = 4x - 7 \Rightarrow f(x_i) = 4x_i - 7$

$\displaystyle\sum_{i=1}^{4} f(x_i)\Delta x = f(x_1)\Delta x + f(x_2)\Delta x + f(x_3)\Delta x + f(x_4)\Delta x$

$= (4x_1 - 7)\Delta x + (4x_2 - 7)\Delta x + (4x_3 - 7)\Delta x + (4x_4 - 7)\Delta x$

$= [4(0) - 7](0.5) + [4(2) - 7](0.5) + [4(4) - 7](0.5) + [4(6) - 7](0.5)$

$= (0 - 7)(0.5) + (8 - 7)(0.5) + (16 - 7)(0.5) + (24 - 7)(0.5)$

$= -7(0.5) + 1(0.5) + 9(0.5) + 17(0.5)$

$= -3.5 + 0.5 + 4.5 + 8.5 = 10$

71. $f(x) = 2x^2 \Rightarrow f(x_i) = 2x_i^2$

$\displaystyle\sum_{i=1}^{4} f(x_i)\Delta x = f(x_1)\Delta x + f(x_2)\Delta x + f(x_3)\Delta x + f(x_4)\Delta x$

$= 2(x_1^2)\Delta x + 2(x_2^2)\Delta x + 2(x_3^2)\Delta x + 2(x_4^2)\Delta x$

$= 2(0^2)(0.5) + 2(2^2)(0.5) + 2(4^2)(0.5) + 2(6^2)(0.5)$

$= 2(0)(0.5) + 2(4)(0.5) + 2(16)(0.5) + 2(36)(0.5)$

$= 0 + 4 + 16 + 36 = 56$

73. $f(x) = \dfrac{-2}{x+1} \Rightarrow f(x_i) = \dfrac{-2}{x_i + 1}$

$\displaystyle\sum_{i=1}^{4} f(x_i)\Delta x = f(x_1)\Delta x + f(x_2)\Delta x + f(x_3)\Delta x + f(x_4)\Delta x$

$= \left(\dfrac{-2}{x_1 + 1}\right)(0.5) + \left(\dfrac{-2}{x_2 + 1}\right)(0.5) + \left(\dfrac{-2}{x_3 + 1}\right)(0.5) + \left(\dfrac{-2}{x_4 + 1}\right)(0.5)$

$= \left(\dfrac{-2}{0 + 1}\right)(0.5) + \left(\dfrac{-2}{2 + 1}\right)(0.5) + \left(\dfrac{-2}{4 + 1}\right)(0.5) + \left(\dfrac{-2}{6 + 1}\right)(0.5)$

$= \left(\dfrac{-2}{1}\right)(0.5) + \left(-\dfrac{2}{3}\right)(0.5) + \left(-\dfrac{2}{5}\right)(0.5) + \left(-\dfrac{2}{7}\right)(0.5)$

$= -1 + \left(-\dfrac{1}{3}\right) + \left(-\dfrac{1}{5}\right) + \left(-\dfrac{1}{7}\right)$

$= \dfrac{-105}{105} + \left(\dfrac{-35}{105}\right) + \left(\dfrac{-21}{105}\right) + \left(\dfrac{-15}{105}\right) = -\dfrac{176}{105}$

75. $\displaystyle\sum_{i=1}^{100} 6 = 100(6) = 600$

77. $\displaystyle\sum_{i=1}^{15} i^2 = \frac{15(15+1)\left[2(15)+1\right]}{6} = \frac{15(16)(30+1)}{6} = \frac{15(16)(31)}{6} = \frac{7440}{6} = 1240$

79. $\displaystyle\sum_{i=1}^{5}(5i+3) = \sum_{i=1}^{5} 5i + \sum_{i=1}^{5} 3 = 5\sum_{i=1}^{5} i + 5(3) = 5 \cdot \frac{5(5+1)}{2} + 15 = 5 \cdot \frac{5(6)}{2} + 15 = 5 \cdot \frac{30}{2} + 15 = 75 + 15 = 90$

81. $\displaystyle\sum_{i=1}^{5}\left(4i^2 - 2i + 6\right) = \sum_{i=1}^{5} 4i^2 - \sum_{i=1}^{5} 2i + \sum_{i=1}^{5} 6 = 4\sum_{i=1}^{5} i^2 - 2\sum_{i=1}^{5} i + 5(6) = 4 \cdot \frac{5(5+1)(10+1)}{6} - 2 \cdot \frac{5(5+1)}{2} + 30$

$\qquad = 4 \cdot \frac{5(6)(11)}{6} - 2 \cdot \frac{5(6)}{2} + 30 = 4 \cdot \left[5(11)\right] - 2 \cdot \left[5(3)\right] + 30 = 220 - 30 + 30 = 220$

83. $\displaystyle\sum_{i=1}^{4}\left(3i^3 + 2i - 4\right) = \sum_{i=1}^{4} 3i^3 + \sum_{i=1}^{4} 2i - \sum_{i=1}^{4} 4 = 3\sum_{i=1}^{4} i^3 + 2\sum_{i=1}^{4} i - 4(4) = 3 \cdot \frac{4^2(4+1)^2}{4} + 2 \cdot \frac{4(4+1)}{2} - 16$

$\qquad = 3 \cdot \frac{16(5)^2}{4} + 2 \cdot \frac{4(5)}{2} - 16 = 3 \cdot \left[4(25)\right] + 2 \cdot \left[2(5)\right] - 16 = 300 + 20 - 16 = 304$

In Exercises 85−88, there are other acceptable forms of the answers.

85. $\displaystyle\frac{1}{3(1)} + \frac{1}{3(2)} + \frac{1}{3(3)} + \cdots + \frac{1}{3(9)} = \sum_{i=1}^{9} \frac{1}{3i}$

87. $\displaystyle 1 - \frac{1}{2} + \frac{1}{4} - \frac{1}{8} + \cdots - \frac{1}{128} = \frac{1}{2^0} - \frac{1}{2^1} + \frac{1}{2^2} - \frac{1}{2^3} + \cdots - \frac{1}{2^7}$

$\qquad = \left(\frac{1}{2}\right)^0 - \left(\frac{1}{2}\right)^1 + \left(\frac{1}{2}\right)^2 - \left(\frac{1}{2}\right)^3 + \cdots - \left(\frac{1}{2}\right)^7$

$\qquad = \left(\frac{1}{2}\right)^{1-1} - \left(\frac{1}{2}\right)^{2-1} + \left(\frac{1}{2}\right)^{3-1} - \left(\frac{1}{2}\right)^{4-1} + \cdots - \left(\frac{1}{2}\right)^{8-1}$

$\qquad = \left(-\frac{1}{2}\right)^{1-1} + \left(-\frac{1}{2}\right)^{2-1} + \left(-\frac{1}{2}\right)^{3-1} + \left(-\frac{1}{2}\right)^{4-1} + \cdots + \left(-\frac{1}{2}\right)^{8-1} = \sum_{k=1}^{8}\left(-\frac{1}{2}\right)^{k-1}$

For Exercises 89−93, enter the set $\{1, 2, 3, \ldots, 10\}$ into L_1. Store the given sequence in L_2. Turn the STAT PLOTS on and clear any functions you have in the $Y =$ screen.

89. $a_n = \dfrac{n+4}{2n}$

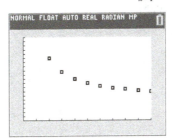

The sequence is graphed in the window $[0, 10] \times [0, 2]$.

It appears that the sequence converges to $\frac{1}{2}$.

The line $y = \frac{1}{2}$ is graphed in the same screen as the plot.

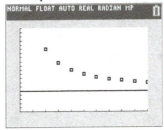

91. $a_n = 2e^n$

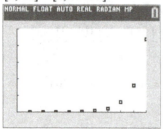

$L_2 = \{5.436563656918, 14.77811$

The sequence is graphed in the window $[0, 10] \times [0, 50000]$.

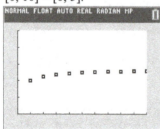

It appears that the sequence diverges.

93. $a_n = \left(1 + \dfrac{1}{n}\right)^n$

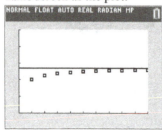

$L_2 = \{2, 2.25, 2.3703703703702,$

The sequence is graphed in the window $[0, 10] \times [0, 5]$.

It appears that the sequence converges to $e \approx 2.71828$ The line $y = e$ is graphed in the same screen as the plot.

95. (a) $a_1 = 8;\ a_n = 2.9a_{n-1} - 0.2a_{n-1}^{\,2},$ for $n > 1$

$a_1 = 8$ thousand per acre

$a_2 = 2.9a_1 - 0.2a_1^{\,2} = 2.9(8) - 0.2(8)^2$
$\quad = 23.2 - 12.8 = 10.4$ thousand per acre

$a_3 = 2.9a_2 - 0.2a_2^{\,2} = 2.9(10.4) - 0.2(10.4)^2$
$\quad = 30.16 - 21.632 = 8.528$ thousand per acre

(b) To enter the data from the recursive sequence, first select SEQ on the mode screen. Next, press $\boxed{Y=}$ to enter the sequence.

$u(n) = 2.9u(n-1) - 0.2(u(n-1))^2$

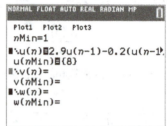

Plotting the points in the window $[0, 21] \times [0, 14]$ we have

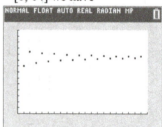

The population density oscillates above and below, and converges to, 9.5 thousand per acre. The line $y = 9.5$ is graphed in the same screen as the plot.

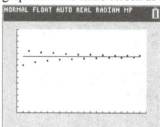

97. (a) The number of bacteria doubles every 40 minutes, so it follows that $N_{j+1} = 2N_j$ for $j \geq 1$.

(b) Two hours is 120 minutes. If $120 = 40(j-1)$, then $3 = j - 1 \Rightarrow j = 4$. $N_1 = 230,\ N_2 = 460,\ N_3 = 920,$ and $N_4 = 1840,$ so there will be 1840 bacteria after two hours.

(c) We must graph the sequence $N_{j+1} = 2N_j$ for $j = 1, 2, 3, \ldots, 7$ if $N_1 = 230$. Be sure your calculator is in SEQ mode.

Plot the sequence in the window $[0, 10] \times [0, 15000]$.

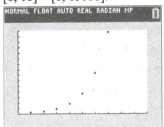

(d) Answers will vary. Sample answer: The growth is very rapid. There is a doubling of the bacteria at equal intervals, so the growth is exponential. Alternatively, as j becomes large, the values of N_j increase without bound.

99. (a) $\ln 1.02 = \ln(1 + 0.02)$

$$\approx 0.02 - \frac{0.02^2}{2} + \frac{0.02^3}{3} - \frac{0.02^4}{4}$$
$$+ \frac{0.02^5}{5} - \frac{0.02^6}{6}$$
$$\approx 0.0198026273$$

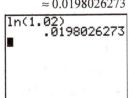

(b) $\ln 1.97 = \ln(1 - 0.03)$

$$\approx (-0.03) - \frac{(-0.03)^2}{2} + \frac{(-0.03)^3}{3}$$
$$- \frac{(-0.03)^4}{4} + \frac{(-0.03)^5}{5} - \frac{(-0.03)^6}{6}$$
$$\approx -0.0304592075$$

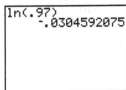

101. $e^a \approx 1 + a + \dfrac{a^2}{2!} + \dfrac{a^3}{3!} + \cdots + \dfrac{a^n}{n!}$ where

$n! = 1 \cdot 2 \cdot 3 \cdots n$

(a) Calculate

$$e^1 \approx 1 + 1 + \frac{1^2}{2!} + \frac{1^3}{3!} + \frac{1^4}{4!} + \frac{1^5}{5!} + \frac{1^6}{6!}$$
$$+ \frac{1^7}{7!} + \frac{1^8}{8!}$$

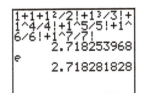

The approximation is accurate to four decimal places.

(b) Calculate

$$e^{-1} \approx 1 + (-1) + \frac{(-1)^2}{2!} + \frac{(-1)^3}{3!} + \frac{(-1)^4}{4!}$$
$$+ \frac{(-1)^5}{5!} + \frac{(-1)^6}{6!} + \frac{(-1)^7}{7!} + \frac{(-1)^8}{8!}$$

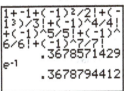

The approximation is accurate to four decimal places.

Section 11.2 Arithmetic Sequences and Series

1. In an arithmetic sequence, each term after the first differs from the preceding term by a fixed constant called the common <u>difference</u>.

3. For the arithmetic sequence having $a_1 = 10$ and $d = -2$, the term $a_3 = \underline{6}$.

5. Domain: $\{1, 2, 3, 4, 5\}$
range: $\{-4, -1, 2, 5, 8\}$

7. The common difference is 3.

9. $\sum_{i=1}^{5}(4i-1)$

This is the sum of the arithmetic sequence with $a_1 = 4(1)-1 = 3$, $d = 4$, and $n = 5$. Using the formula $S_n = \frac{n}{2}\left[2a_1 + (n-1)d\right]$, we have

$$\sum_{i=1}^{5}(4i-1) = S_5 = \frac{5}{2}\left[2(3)+4(4)\right]$$
$$= \frac{5}{2}(22) = 55$$

11. $2, 5, 8, 11, \ldots$
$d = a_2 - a_1 = 5 - 2 = 3$

13. $3, -2, -7, -12, \ldots$
$d = a_2 - a_1 = -2 - 3 = -5$

15. $x + 3y, 2x + 5y, 3x + 7y, \ldots$
$d = a_2 - a_1 = (2x+5y)-(x+3y)$
$= 2x + 5y - x - 3y = x + 2y$

17. $a_1 = 8$ and $d = 6$

Starting with $a_1 = 8$, add $d = 6$ to each term to get the next term.
$a_2 = 8 + 6 = 14;\quad a_3 = 14 + 6 = 20$
$a_4 = 20 + 6 = 26;\quad a_5 = 26 + 6 = 32$
The first five terms are 8, 14, 20, 26, and 32.

19. $a_1 = 5$, $d = -2$

$a_2 = 5 + (-2) = 3;\ a_3 = 3 + (-2) = 1$
$a_4 = 1 + (-2) = -1$
$a_5 = -1 + (-2) = -3$
The first five terms are 5, 3, 1, −1, and −3.

21. $a_1 = 10 + \sqrt{7}$, $a_2 = 10$
$d = a_2 - a_1 = 10 - \left(10 + \sqrt{7}\right) = -\sqrt{7}$
$a_3 = a_2 + d = 10 + \left(-\sqrt{7}\right) = 10 - \sqrt{7}$
$a_4 = a_3 + d = \left(10 - \sqrt{7}\right) - \sqrt{7} = 10 - 2\sqrt{7}$
$a_5 = a_4 + d = \left(10 - 2\sqrt{7}\right) - \sqrt{7} = 10 - 3\sqrt{7}$

The first five terms are $10 + \sqrt{7}$, 10, $10 - \sqrt{7}$, $10 - 2\sqrt{7}$, and $10 - 3\sqrt{7}$.

23. $5, 7, 9, \ldots$
$d = a_2 - a_1 = 7 - 5 = 2$
Because $a_n = a_1 + (n-1)d$, we have
$a_8 = 5 + (8-1)2 = 5 + 7 \cdot 2 = 5 + 14 = 19$. Also, we have
$a_n = 5 + (n-1)2 = 5 + 2n - 2 = 2n + 3$.

25. $a_1 = 5$, $a_4 = 15$

Because $a_n = a_1 + (n-1)d$, we have
$a_4 = 5 + (4-1)d = 15 \Rightarrow 5 + 3d = 15 \Rightarrow$
$3d = 10 \Rightarrow d = \frac{10}{3}$
We now have
$a_8 = 5 + (8-1)\left(\frac{10}{3}\right) = 5 + 7\left(\frac{10}{3}\right) = \frac{15}{3} + \frac{70}{3} = \frac{85}{3}$.
Also, we have
$a_n = 5 + (n-1)\left(\frac{10}{3}\right)$
$= 5 + \frac{10}{3}n - \frac{10}{3} = \frac{15}{3} + \frac{10}{3}n - \frac{10}{3} = \frac{10}{3}n + \frac{5}{3}$

27. $a_{10} = 6$, $a_{11} = 10.5$

Because a_{10} and a_{11} are successive terms in the sequence, we have
$d = a_{11} - a_{10} = 10.5 - 6 = 4.5$ Now find a_1
using the formula $a_n = a_1 + (n-1)d$.
$a_{11} = a_1 + (11-1)d \Rightarrow 10.5 = a_1 + 10(4.5) \Rightarrow$
$10.5 = a_1 + 45 \Rightarrow a_1 = -34.5$
Now find a_8 and a_n. We have
$a_8 = -34.5 + (8-1)(4.5) = -34.5 + 7(4.5)$
$= -34.5 + 31.5 = -3$

$a_n = -34.5 + (n-1)(4.5) = -34.5 + 4.5n - 4.5$
$= 4.5n - 39$

29. $a_1 = x$, $a_2 = x + 3$
$d = a_2 - a_1 = (x+3) - x = 3$
Because $a_n = a_1 + (n-1)d$, we have
$a_8 = x + (8-1)3 = x + 7 \cdot 3 = x + 21$. Also, we
have $a_n = x + (n-1)3 = x + 3n - 3$.

31. $a_4 = s + 6p, d = 2p$

First, find a_1. Because $a_n = a_1 + (n-1)d$, we
have $a_4 = a_1 + 3d \Rightarrow a_1 + 3(2p) = s + 6p \Rightarrow$
$a_1 + 6p = s + 6p \Rightarrow a_1 = s$. Thus,
$a_8 = s + 7(2p) = s + 14p$ and
$a_n = s + (n-1)(2p) = s + 2pn - 2p$

33. $a_5 = 27$, $a_{15} = 87$

Because $a_{15} = a_5 + 10d$, it follows that
$10d = a_{15} - a_5$. This implies
$10d = 87 - 27 \Rightarrow 10d = 60 \Rightarrow d = 6$.
Because $a_5 = a_1 + 4d$, we have
$27 = a_1 + 4(6) \Rightarrow 27 = a_1 + 24 \Rightarrow a_1 = 3$.

35. $a_8 = -15, \ a_{18} = -85$

Because $a_{18} = a_8 + 10d,$ it follows that

$10d = a_{18} - a_8 = -85 - (-15) = -70 \Rightarrow d = -7$

Because $a_8 = a_1 + 7d,$ we have

$-15 = a_1 + 7(-7) \Rightarrow a_1 = -15 + 49 = 34$

37. $a_7 = 23.5, \ a_{20} = 69$

Because $a_{20} = a_7 + 13d,$ it follows that

$13d = a_{20} - a_7 = 69 - 23.5 = 45.5 \Rightarrow d = 3.5$

Because $a_7 = a_1 + 6d,$ we have

$23.5 = a_1 + 6(3.5) \Rightarrow a_1 = 23.5 - 21 = 2.5$

39. The sequence is comprised of the points

$\{(1,-2),(2,-1),(3,0),(4,1),(5,2),(6,3)\}.$ If

the points were connected, they would form a

line with slope $m = \frac{y_2 - y_1}{x_2 - x_1} = \frac{-1-(-2)}{2-1} = \frac{1}{1} = 1.$

Using the point-slope form of a line, we have

$y - y_1 = m(x - x_1) \Rightarrow y - (-2) = 1(x - 1) \Rightarrow$

$y + 2 = x - 1 \Rightarrow y = x - 3$

The nth term of the sequence is determined by

$a_n = n - 3.$ The domain is $\{1, 2, 3, 4, 5, 6\}$ and

the range is $\{-2, -1, 0, 1, 2, 3\}.$

41. The sequence is comprised of the points

$\{(1,\frac{5}{2}),(2,2),(3,\frac{3}{2}),(4,1),(5,\frac{1}{2}),(6,0)\}.$ If the

points were connected, they would form a line

with slope $m = \frac{y_2 - y_1}{x_2 - x_1} = \frac{2-\frac{5}{2}}{2-1} = \frac{-\frac{1}{2}}{1} = -\frac{1}{2}.$ Using

the point-slope form of a line, we have

$y - y_1 = m(x - x_1) \Rightarrow y - \frac{5}{2} = -\frac{1}{2}(x - 1) \Rightarrow$

$y - \frac{5}{2} = -\frac{1}{2}x + \frac{1}{2} \Rightarrow y = -\frac{1}{2}x + 3$

The nth term of the sequence is determined by

$a_n = -\frac{1}{2}n + 3$ or $a_n = 3 - \frac{1}{2}n.$ The domain is

$\{1, 2, 3, 4, 5, 6\}$ and the range is

$\{0, \frac{1}{2}, 1, \frac{3}{2}, 2, \frac{5}{2}\}$ or $\{0, 0.5, 1, 1.5, 2, 2.5\}.$

43. The sequence is comprised of the points

$\{(1,10),(2,-10),(3,-30),(4,-50),(5,-70)\}.$

If the points were connected, they would form

a line with slope

$m = \frac{y_2 - y_1}{x_2 - x_1} = \frac{-10-10}{2-1} = \frac{-20}{1} = -20.$ Using the

point-slope form of a line, we have

$y - y_1 = m(x - x_1) \Rightarrow y - 10 = -20(x - 1) \Rightarrow$

$y - 10 = -20x + 20 \Rightarrow y = -20x + 30$

The nth term of the sequence is determined by

$a_n = -20n + 30$ or $a_n = 30 - 20n.$ The domain

is $\{1, 2, 3, 4, 5\}$ and the range is

$\{-70, -50, -30, -10, 10\}.$

45. 8, 11, 14, ...

Because $a_1 = 8, \ d = a_2 - a_1 = 11 - 8 = 3,$ and

$S_n = \frac{n}{2}[2a_1 + (n-1)d],$ we have

$S_{10} = \frac{10}{2}[2(8) + (10-1)(3)] = 5[16 + 9(3)]$

$= 5(16 + 27) = 5(43) = 215$

47. 5, 9, 13, ...

Because $a_1 = 5, \ d = a_2 - a_1 = 9 - 5 = 4,$ and

$S_n = \frac{n}{2}[2a_1 + (n-1)d],$ we have the

following.

$S_{10} = \frac{10}{2}[2(5) + (10-1)(4)] = 5[10 + 9(4)]$

$= 5(10 + 36) = 5(46) = 230$

49. $a_2 = 9, \ a_4 = 13$

To find $S_{10},$ we need a_1 and $d.$ Because

$a_4 = a_2 + 2d,$ we have

$13 = 9 + 2d \Rightarrow 4 = 2d \Rightarrow d = 2.$

$d = a_2 - a_1 \Rightarrow 2 = 9 - a_1 \Rightarrow a_1 = 7.$

$S_n = \frac{n}{2}[2a_1 + (n-1)d]$

$S_{10} = \frac{10}{2}[2(7) + (10-1)(2)] = 5[14 + 9(2)]$

$= 5(14 + 18) = 5(32) = 160$

51. $a_1 = 10, \ a_{10} = 5.5$

To find $S_{10},$ we need a_1 and $d.$ Because

$a_{10} = a_1 + 9d,$ we have

$5.5 = 10 + 9d \Rightarrow 9d = -4.5 \Rightarrow d = -0.5.$

$S_n = \frac{n}{2}[2a_1 + (n-1)d],$

$S_{10} = \frac{10}{2}[2(10) + (10-1)(-0.5)]$

$= 5[20 + 9(-0.5)] = 5[20 + (-4.5)]$

$= 5(15.5) = 77.5$

53. $a_1 = \pi, \ a_{10} = 10\pi$

To find $S_{10},$ we need a_1 and $d.$ Because

$a_{10} = a_1 + 9d,$ we have

$10\pi = \pi + 9d \Rightarrow 9\pi = 9d \Rightarrow d = \pi.$ Thus,

$S_n = \frac{n}{2}[2a_1 + (n-1)d]$

$S_{10} = \frac{10}{2}[2\pi + (10-1)\pi]$

$= 5[2\pi + 9\pi] = 5(11\pi) = 55\pi$

55. The first 80 positive integers form the arithmetic sequence 1, 2, 3, 4, …, 80. Thus, in the formula $S_n = \frac{n}{2}(a_1 + a_n)$, we have $n = 80$, $a_1 = 1$, and $a_{80} = 80$. Thus, the sum of the first 80 positive integers is
$$S_{80} = \frac{80}{2}(1 + 80) = 40(81) = 3240.$$

57. The positive odd integers form an arithmetic sequence 1, 3, 5, 7, … with $a_1 = 1$ and $d = 2$. Find the sum of the first 50 terms of this sequence.
$$S_n = \frac{n}{2}\left[2a_1 + (n-1)d\right]$$
$$S_{50} = \frac{50}{2}\left[2(1) + (50-1)(2)\right] = 25\left[2 + 49(2)\right]$$
$$= 25(2 + 98) = 25(100) = 2500$$

59. The positive even integers form an arithmetic sequence 2, 4, 6, 8, … with $a_1 = 2$ and $d = 2$. Find the sum of the first 60 terms of this sequence.
$$S_n = \frac{n}{2}\left[2a_1 + (n-1)d\right]$$
$$S_{60} = \frac{60}{2}\left[2(2) + (60-1)(2)\right] = 30\left[4 + 59(2)\right]$$
$$= 30(4 + 118) = 30(122) = 3660$$

61. $S_{20} = 1090$, $a_{20} = 102$

The formula for the sum is $S_n = \frac{n}{2}(a_1 + a_n)$, so $S_{20} = \frac{20}{2}(a_1 + a_{20})$. This yields
$$1090 = 10(a_1 + 102) \Rightarrow 109 = a_1 + 102 \Rightarrow a_1 = 7.$$
To solve for d, we use the fact that $a_n = a_1 + (n-1)d$.
$$a_n = a_1 + (n-1)d \Rightarrow a_{20} = a_1 + (20-1)d \Rightarrow$$
$$102 = 7 + 19d \Rightarrow 95 = 19d \Rightarrow d = 5$$
Thus, $a_1 = 7$ and $d = 5$.

63. $S_{16} = -160$, $a_{16} = -25$

The formula for the sum is $S_n = \frac{n}{2}(a_1 + a_n)$, so $S_{16} = \frac{16}{2}(a_1 + a_{16})$. This yields
$$-160 = 8\left[a_1 + (-25)\right] \Rightarrow -20 = a_1 + (-25) \Rightarrow$$
$a_1 = 5$. To solve for d, we use the fact that $a_n = a_1 + (n-1)d$.
$$a_n = a_1 + (n-1)d \Rightarrow a_{16} = a_1 + (16-1)d \Rightarrow$$
$$-25 = 5 + 15d \Rightarrow -30 = 15d \Rightarrow d = -2$$
Thus, $a_1 = 5$ and $d = -2$.

65. $S_{12} = -108$, $a_{12} = -19$

The formula for the sum is $S_n = \frac{n}{2}(a_1 + a_n)$, so $S_{12} = \frac{12}{2}(a_1 + a_{12})$. This yields
$$-108 = 6\left[a_1 + (-19)\right] \Rightarrow$$
$$-18 = a_1 + (-19) \Rightarrow a_1 = 1.$$
To solve for d, we use the fact that $a_n = a_1 + (n-1)d$.
$$a_n = a_1 + (n-1)d \Rightarrow a_{12} = a_1 + (12-1)d \Rightarrow$$
$$-19 = 1 + 11d \Rightarrow -20 = 11d \Rightarrow d = -\frac{20}{11}$$
Thus, $a_1 = 1$ and $d = -\frac{20}{11}$.

67. $\displaystyle\sum_{i=1}^{3}(i+4)$

This is a sum of three terms having a common difference of 1. Thus, in the formula $S_n = \frac{n}{2}(a_1 + a_n)$, we have $a_1 = 1 + 4 = 5$, $n = 3$, and $a_n = a_3 = 3 + 4 = 7$. The sum can, therefore, be determined to be
$$\sum_{i=1}^{3}(i+4) = S_3 = \frac{3}{2}(a_1 + a_3) = \frac{3}{2}(5 + 7) = 18$$

69. $\displaystyle\sum_{j=1}^{10}(2j+3)$

This is the sum of the arithmetic sequence with $a_1 = 2(1) + 3 = 5$, $d = 2$, and $n = 10$.

Using the formula $S_n = \frac{n}{2}\left[2a_1 + (n-1)d\right]$, we have
$$\sum_{j=1}^{10}(2j+3) = S_{10} = \frac{10}{2}\left[2(5) + 9(2)\right]$$
$$= 5(10 + 18) = 5(28) = 140$$

71. $\displaystyle\sum_{i=4}^{12}(-5 - 8i)$

This is the sum of an arithmetic sequence with $d = -8$ and $a_1 = -5 - 8(4) = -37$. If i started at 1, there would be 12 terms. Because three terms are missing, $n = 9$. Using the formula $S_n = \frac{n}{2}\left[2a_1 + (n-1)d\right]$, we have
$$\sum_{i=4}^{12}(-5 - 8i) = S_9 = \frac{9}{2}\left[2(-37) + 8(-8)\right]$$
$$= \frac{9}{2}\left[-74 + (-64)\right] = -621$$

73. $\displaystyle\sum_{i=1}^{1000} i$

This is the sum of 1000 terms having a common difference of 1. Thus, in the formula $S_n = \frac{n}{2}(a_1 + a_n)$, we have $a_1 = 1$, $n = 1000$, and $a_n = a_{1000} = 1000$. The sum can, therefore, be determined as follows.

$$\sum_{i=1}^{1000} i = S_{1000} = \frac{1000}{2}(1 + 1000) = 500,500$$

75. $\displaystyle\sum_{k=1}^{100} 2k$

This is the sum of 100 terms with $a_1 = 2$, $d = 2$ and $n = 100$. Using the formula $S_n = \frac{n}{2}\left[2a_1 + (n-1)d\right]$, we have

$$\sum_{k=1}^{100} 2k = S_{100} = \frac{100}{2}\left[2(2) + 99(2)\right]$$
$$= 50(4 + 198) = 50(202) = 10,100$$

77. $\displaystyle\sum_{j=10}^{50} 5j$

This is the sum of 41 terms with $a_1 = 50$ $d = 5$ and $n = 41$. Using the formula $S_n = \frac{n}{2}\left[2a_1 + (n-1)d\right]$, we have

$$\sum_{j=10}^{50} 5j = S_{41} = \frac{41}{2}\left[2(50) + 40(5)\right]$$
$$= \frac{41}{2}(300) = 6150$$

For Exercises 79–81, you will need the sum(and seq(features of your graphing calculator. On the TI-84, the sum(feature is located under the LIST then MATH menu. The seq(feature is under the LIST then OPS menu.

79. $a_n = 4.2n + 9.73$

Using the sequence feature of a graphing calculator, we obtain $S_{10} = 328.3$.

```
NORMAL FLOAT AUTO REAL RADIAN MP

                     seq
Expr:4.2n+9.73
Variable:n
start:1
end:10
step:1
Paste
```

```
NORMAL FLOAT AUTO REAL RADIAN MP

sum(seq(4.2n+9.73,n,1,10,▸
                        328.3
```

81. $a_n = \sqrt{8}n + \sqrt{3}$

Using the sequence feature of a graphing calculator, we obtain $S_{10} \approx 172.884$.

```
NORMAL FLOAT AUTO REAL RADIAN MP

                     seq
Expr:√(8)n+√(3)
Variable:n
start:1
end:10
step:1
Paste
```

```
NORMAL FLOAT AUTO REAL RADIAN MP

sum(seq(√8n+√3,n,1,10,1)
              172.8839999
```

83. Find the sum of all the integers from 51 to 71. This can be written as

$$\sum_{i=51}^{71} i = \sum_{i=1}^{71} i - \sum_{i=1}^{50} i = S_{71} - S_{50}.$$

$a_1 = 1$, $d = 1$, $a_{50} = 50$, and $a_{71} = 71$, so $S_{71} = \frac{71}{2}(1 + 71) = 71(36) = 2556$ and $S_{50} = \frac{50}{2}(1 + 50) = 25(51) = 1275$. Thus, the sum is $S_{71} - S_{50} = 2556 - 1275 = 1281$.

85. In every 12-hour cycle, the clock will chime $1 + 2 + 3 + \cdots + 12$ times. Because $a_1 = 1$, $n = 12$, $a_{12} = 12$, and $S_n = \frac{n}{2}(a_1 + a_n)$, we have $S_{12} = \frac{12}{2}(1 + 12) = 6(13) = 78$. There are two 12-hour cycles in 1 day, every day, so the clock will chime $2(78) = 156$ times. There are 30 days in this month, so the clock will chime $156(30) = 4680$ times.

87. In this exercise, we have $a_1 = 49,000$, $d = 580$, and $n = 5$. Using $a_n = a_1 + (n-1)d$, we need to find a_{10}.

Because
$$a_{11} = 49,000 + (11-1)580 = 49,000 + 10(580)$$
$$= 49,000 + 5800 = 54,800$$

we would expect that five years from now, the population will be 54,800.

89. The longest rung measures 28 in., and the shortest measures 18 in. The rungs are uniformly tapered, so the sum of the lengths of these 31 supports can be thought of as the sum of an arithmetic sequence with $a_1 = 28$, $a_n = a_{31} = 18$, and $n = 31$.

$$S_n = \frac{n}{2}(a_1 + a_n) \Rightarrow$$
$$S_{31} = \frac{31}{2}(a_1 + a_{31}) = \frac{31}{2}(28 + 18) = 713$$

Thus, a total of 713 in. of material would be needed.

91. Consider the arithmetic sequence $a_1, a_2, a_3, \ldots$. Then, consider the sequence $a_1, a_3, a_5, \ldots$. We have

$$a_3 - a_1 = (a_1 + 2d) - a_1 = 2d \text{ and}$$
$$a_5 - a_3 = (a_1 + 4d) - (a_1 + 2d) = 2d.$$

In general, we have the following.

$a_n - a_{n-2}$
$$= \left[a_1 + (n-1)d\right] - \left[a_1 + ((n-2)-1)d\right]$$
$$= a_1 + nd - d - \left[a_1 + (n-3)d\right]$$
$$= a_1 + nd - d - (a_1 + nd - 3d)$$
$$= a_1 + nd - d - a_1 - nd + 3d = 2d$$

Thus, $a_1, a_3, a_5, \ldots$ is an arithmetic sequence whose first term is a_1 and common difference is $2d$.

Section 11.3 Geometric Sequences and Series

1. In a geometric sequence, each term after the first is obtained by multiplying the preceding term by a fixed nonzero real number called the common <u>ratio</u>.

3. For the geometric sequence having $a_1 = 6$ and $r = 2$, the term $a_3 = \underline{24}$.

5. The sum of the first five terms of the geometric sequence 5, 10, 20, 40, ... is <u>155</u>.

7. 4, −8, 16, −32, ...
Geometric sequence; $r = -2$

9. 5, 10, 20, 35, ...
Neither arithmetic nor geometric

For Exercises 11−13, we are examining the geometric sequence 1, 2, 4, 8, ... The nth term of a geometric sequence is given by $a_n = a_1 r^{n-1}$, where a_1 is the first term and r is the common ratio. For this situation, $a_n = 1 \cdot 2^{n-1} = 2^{n-1}$.

11. day 10

(a) $a_{10} = 2^{10-1} = 2^9 = 512$ cents or $5.12

(b) $S_n = \dfrac{a_1(1-r^n)}{1-r}$ for $r \neq 1$, so we have

$$S_{10} = \frac{1(1-2^{10})}{1-2} = \frac{1-1024}{-1} = \frac{-1023}{-1} = 1023$$

cents or $10.23.

13. day 15

(a) $a_{15} = 2^{15-1} = 2^{14} = 16,384$ cents or $163.84

(b) $S_n = \dfrac{a_1(1-r^n)}{1-r}$ for $r \neq 1$, so we have

$$S_{15} = \frac{1(1-2^{15})}{1-2} = \frac{1-32,768}{-1} = \frac{-32,767}{-1} = 32,767$$

cents or $327.67.

15. $a_1 = 5$, $r = -2$
$$a_5 = a_1 r^{5-1} = 5(-2)^4 = 5(16) = 80$$
$$a_n = a_1 r^{n-1} = 5(-2)^{n-1}$$

17. $a_2 = -4$, $r = 3$
We need to determine a_1.
$$a_2 = a_1 r^{2-1} \Rightarrow -4 = a_1(3)^1 \Rightarrow$$
$$-4 = a_1(3) \Rightarrow a_1 = -\frac{4}{3}, \text{ so}$$
$$a_5 = a_1 r^{5-1} = -\frac{4}{3}(3)^4 = -\frac{4}{3}(81) = -108 \text{ and}$$
$$a_n = a_1 r^{n-1} = -\frac{4}{3}(3)^{n-1}$$

19. $a_4 = 243$, $r = -3$

We need to determine a_1.

$a_4 = a_1 r^{4-1} \Rightarrow 243 = a_1 (-3)^3 \Rightarrow$

$243 = a_1 (-27) \Rightarrow a_1 = -9$, so

$a_5 = a_1 r^{5-1} = -9(-3)^4 = -9(81) = -729$

$a_n = a_1 r^{n-1} = -9(-3)^{n-1}$

Note that

$(-9)(-3)^{n-1} = -(-3)^2 (-3)^{n-1} = (-3)^{2+n-1}$

$= -(-3)^{n+1}$, so $a_n = -(-3)^{n+1}$ is an

equivalent formula for the nth term of this sequence.

21. $-4, -12, -36, -108, \ldots$

First find r: $r = \frac{a_2}{a_1} = \frac{-12}{-4} = 3$

Also, given that $a_1 = -4$, we have

$a_5 = a_1 r^{5-1} = -4(3)^4 = -4(81) = -324$

$a_n = a_1 r^{n-1} = -4(3)^{n-1}$

23. $\frac{4}{5}, 2, 5, \frac{25}{2}, \ldots$

First find r: $r = \frac{a_2}{a_1}, = \frac{2}{\frac{4}{5}} = 2\left(\frac{5}{4}\right) = \frac{5}{2}$

Also, given that $a_1 = \frac{4}{5}$, we have

$a_5 = a_1 r^{5-1} = \frac{4}{5}\left(\frac{5}{2}\right)^4 = \frac{4}{5}\left(\frac{625}{16}\right) = \frac{125}{4}$ and

$a_n = a_1 r^{n-1} = \frac{4}{5}\left(\frac{5}{2}\right)^{n-1}$

Note that $\left(\frac{4}{5}\right)\left(\frac{5}{2}\right)^{n-1} = \frac{2^2}{5^1} \cdot \frac{5^{n-1}}{2^{n-1}} = \frac{5^{n-2}}{2^{n-3}}$, so

$a_n = \frac{5^{n-2}}{2^{n-3}}$ is an equivalent formula for the nth

term of this sequence.

25. $10, -5, \frac{5}{2}, -\frac{5}{4}, \ldots$

First find r: $r = \frac{a_2}{a_1} = \frac{-5}{10} = -\frac{1}{2}$

Also, given that $a_1 = 10$, we have

$a_5 = a_1 r^{5-1} = 10\left(-\frac{1}{2}\right)^4 = 10\left(\frac{1}{16}\right) = \frac{5}{8}$ and

$a_n = a_1 r^{n-1} = 10\left(-\frac{1}{2}\right)^{n-1}$

27. $a_2 = -6$, $a_7 = -192$

Using $a_n = a_1 r^{n-1}$, we have $a_2 = a_1 r = -6$.

This yields $a_1 = \frac{-6}{r}$. Also, $a_7 = a_1 r^6 = -192$.

Substituting $a_1 = \frac{-6}{r}$ into $a_1 r^6 = -192$ and

solving for r, we have

$a_1 r^6 = -192 \Rightarrow \left(\frac{-6}{r}\right) r^6 = -192 \Rightarrow$

$-6r^5 = -192 \Rightarrow r^5 = 32 \Rightarrow r^5 = 2^5 \Rightarrow r = 2$

$a_1 = \frac{-6}{r} \Rightarrow a_1 = \frac{-6}{2} = -3$.

29. $a_3 = 5$ and $a_8 = \frac{1}{625}$

Using $a_n = a_1 r^{n-1}$, we have $a_3 = a_1 r^2 = 5$.

This yields $a_1 = \frac{5}{r^2}$. Also, $a_8 = a_1 r^7 = \frac{1}{625}$.

Substituting $a_1 = \frac{5}{r^2}$ into $a_1 r^7 = \frac{1}{625}$ and

solving for r, we have

$a_1 r^7 = \frac{1}{625} \Rightarrow \left(\frac{5}{r^2}\right) r^7 = \frac{1}{625} \Rightarrow 5r^5 = \frac{1}{625} \Rightarrow$

$r^5 = \frac{1}{3125} \Rightarrow r^5 = \left(\frac{1}{5}\right)^5 \Rightarrow r = \frac{1}{5}$

$a_1 = \frac{5}{r^2} \Rightarrow a_1 = \frac{5}{\left(\frac{1}{5}\right)^2} = \frac{5}{\frac{1}{25}} = 5 \cdot 25 = 125$.

31. $a_3 = 50$, $a_7 = 0.005$

Using $a_n = a_1 r^{n-1}$, we have $a_3 = a_1 r^2 = 50$.

This yields $a_1 = \frac{50}{r^2}$. Also, $a_7 = a_1 r^6 = 0.005$.

Substituting $a_1 = \frac{50}{r^2}$ into $a_1 r^6 = 0.005$

equation and solving for r, we have

$a_1 r^6 = 0.005 \Rightarrow \left(\frac{50}{r^2}\right) r^6 = 0.005 \Rightarrow$

$50r^4 = 0.005 \Rightarrow r^4 = 0.0001 \Rightarrow r = \pm 0.1$

$a_1 = \frac{50}{(\pm 0.1)^2} = \frac{50}{0.01} = 5000$.

33. $2, 8, 32, 128, \ldots$

First find r: $r = \frac{a_2}{a_1} = \frac{8}{2} = 4$

Also, given that $a_1 = 2$, $n = 5$, and

$S_n = \frac{a_1(1-r^n)}{1-r}$, we have

$S_n = \frac{a_1(1-r^n)}{1-r} = \frac{2(1-4^5)}{1-4} = \frac{2(1-1024)}{-3}$

$= \frac{2(-1023)}{-3} = \frac{-2046}{-3} = 682$

35. $18, -9, \frac{9}{2}, -\frac{9}{4}, \ldots$

First find r.

$$r = \frac{a_2}{a_1} \Rightarrow r = \frac{-9}{18} = -\frac{1}{2}.$$

$a_1 = 18$, $n = 5$, so

$$S_n = \frac{a_1\left(1 - r^n\right)}{1 - r} \Rightarrow$$

$$S_5 = \frac{18\left[1 - \left(-\frac{1}{2}\right)^5\right]}{1 - \left(-\frac{1}{2}\right)} = \frac{18\left[1 - \left(-\frac{1}{32}\right)\right]}{\frac{3}{2}}$$

$$= 18\left(\frac{33}{32}\right)\left(\frac{2}{3}\right) = \frac{99}{8}$$

37. $a_1 = 8.423$, $r = 2.859$

$$S_n = \frac{a_1\left(1 - r^n\right)}{1 - r} \Rightarrow$$

$$S_5 = \frac{8.423\left[1 - \left(2.859\right)^5\right]}{1 - 2.859} \approx 860.95.$$

39. $\displaystyle\sum_{i=1}^{5}(-3)^i$

$a_1 = (-3)^1 = -3$, $r = -3$, and $n = 5$.

$$S_n = \frac{a_1\left(1 - r^n\right)}{1 - r} \Rightarrow$$

$$\sum_{i=1}^{5}(-3)^i = S_5 = \frac{-3\left(1 - (-3)^5\right)}{1 - (-3)} = \frac{-3(1 + 243)}{4}$$

$$= \frac{-3(244)}{4} = -183$$

41. $\displaystyle\sum_{j=1}^{6} 48\left(\frac{1}{2}\right)^j$

$a_1 = 48\left(\frac{1}{2}\right)^1 = 48\left(\frac{1}{2}\right) = 24$, $r = \frac{1}{2}$, and $n = 6$.

$$S_n = \frac{a_1\left(1 - r^n\right)}{1 - r} \Rightarrow$$

$$\sum_{j=1}^{6} 48\left(\frac{1}{2}\right)^j = S_6 = \frac{24\left[1 - \left(\frac{1}{2}\right)^6\right]}{1 - \frac{1}{2}} = \frac{24\left(1 - \frac{1}{64}\right)}{\frac{1}{2}}$$

$$= \frac{24\left(\frac{64}{64} - \frac{1}{64}\right)}{\frac{1}{2}} = 24\left(\frac{63}{64}\right)(2)$$

$$= 24\left(\frac{63}{32}\right) = \frac{189}{4}$$

43. $\displaystyle\sum_{k=4}^{10} 2^k$

This series is the sum of the fourth through tenth terms of a geometric sequence with $a_1 = 2^1 = 2$ and $r = 2$. To find this sum, find the difference between the sum of the first ten terms and the sum of the first three terms.

$$\sum_{k=4}^{10} 2^k = \sum_{k=1}^{10} 2^k - \sum_{k=1}^{3} 2^k = \frac{2\left(1 - 2^{10}\right)}{1 - 2} - \frac{2\left(1 - 2^3\right)}{1 - 2}$$

$$= \frac{2(1 - 1024)}{-1} - \frac{2(1 - 8)}{-1} = \frac{2(-1023)}{-1} - \frac{2(-7)}{-1}$$

$$= 2046 - 14 = 2032$$

Note: We could also consider $\displaystyle\sum_{k=4}^{10} 2^k$ as a geometric series with $a_1 = 2^4 = 16$ and $r = 2$. If the sequence started with $k = 1$, there would be 10 terms. Because it starts with 4, three of the terms are missing, so there are seven terms and $n = 7$.

Thus we have

$$\sum_{k=4}^{10} 2^k = S_7 = \frac{16\left(1 - 2^7\right)}{1 - 2} = \frac{16\left(1 - 128\right)}{-1}$$

$$= \frac{16(-127)}{-1} = 2032$$

45. The sum of an infinite geometric series exists if $|r| < 1$.

47. $12, 24, 48, 96, \ldots$

$r = \frac{a_2}{a_1} \Rightarrow r = \frac{24}{12} = 2.$

The sum of the terms of this infinite geometric sequence would not converge because $r = 2$ is not between -1 and 1.

49. $-48, -24, -12, -6, \ldots$

$r = \frac{a_2}{a_1} \Rightarrow r = \frac{-24}{-48} = \frac{1}{2}.$

Because $-1 < r < 1$, the sum converges.

51. $2 + 1 + \frac{1}{2} + \frac{1}{4} + \cdots$ is a geometric series with

$r = \frac{a_2}{a_1} = \frac{1}{2}$ and $a_1 = 2$.

$S_n = \frac{a_1\left(1 - r^n\right)}{1 - r} \Rightarrow$

$S_n = \frac{2\left[1 - \left(\frac{1}{2}\right)^n\right]}{1 - \frac{1}{2}} = \frac{2\left[1 - \left(\frac{1}{2}\right)^n\right]}{\frac{1}{2}}.$

As in Example 7, we have $\lim\limits_{n \to \infty} \left(\frac{1}{2}\right)^n = 0$. Thus,

we obtain

$\lim\limits_{n \to \infty} S_n = \lim\limits_{n \to \infty} \frac{2\left[1 - \left(\frac{1}{2}\right)^n\right]}{\frac{1}{2}} = \frac{2(1 - 0)}{\frac{1}{2}} = \frac{2}{\frac{1}{2}} = 4.$

53. $18 + 6 + 2 + \frac{2}{3} + \cdots$

For this geometric series, $a_1 = 18$ and

$r = \frac{6}{18} = \frac{1}{3}$. Because $-1 < r < 1$, this series

converges. We have

$S_\infty = \frac{a_1}{1 - r} = \frac{18}{1 - \frac{1}{3}} = \frac{18}{\frac{2}{3}} = 18 \cdot \frac{3}{2} = 27.$

55. $\frac{1}{4} - \frac{1}{6} + \frac{1}{9} - \frac{2}{27} + \cdots$

For this geometric series, $a_1 = \frac{1}{4}$ and

$r = \frac{-\frac{1}{6}}{\frac{1}{4}} = -\frac{1}{6} \cdot \frac{4}{1} = -\frac{2}{3}$. Because $-1 < r < 1$,

this series converges. We have

$S_\infty = \frac{a_1}{1 - r} = \frac{\frac{1}{4}}{1 - \left(-\frac{2}{3}\right)} = \frac{\frac{1}{4}}{\frac{5}{3}} = \frac{1}{4} \cdot \frac{3}{5} = \frac{3}{20}.$

57. $\sum\limits_{i=1}^{\infty} 3\left(\frac{1}{4}\right)^{i-1}$

For this geometric series,

$a_1 = 3\left(\frac{1}{4}\right)^{1-1} = 3\left(\frac{1}{4}\right)^0 = 3 \cdot 1 = 3$ and $r = \frac{1}{4}$.

Because $-1 < r < 1$, this series converges. We

have $S_\infty = \frac{a_1}{1 - r} = \frac{3}{1 - \frac{1}{4}} = \frac{3}{\frac{3}{4}} = 3 \cdot \frac{4}{3} = 4.$

59. $\sum\limits_{k=1}^{\infty} 3^{-k}$

For this geometric series, $a_1 = 3^{-1} = \frac{1}{3}$ and

$r = \frac{1}{3}$. Because $-1 < r < 1$ this series

converges. We have

$S_\infty = \frac{a_1}{1 - r} = \frac{\frac{1}{3}}{1 - \frac{1}{3}} = \frac{\frac{1}{3}}{\frac{2}{3}} = \frac{1}{2}.$

61. $\sum\limits_{i=1}^{\infty} \left(-\frac{2}{3}\right)\left(-\frac{1}{4}\right)^{i-1}$

For this geometric series,

$a_1 = \left(-\frac{2}{3}\right)\left(-\frac{1}{4}\right)^{1-1} = -\frac{2}{3}$ and $r = -\frac{1}{4}$.

$S_\infty = \frac{a_1}{1 - r} = \frac{-\frac{2}{3}}{1 - \left(-\frac{1}{4}\right)} = \frac{-\frac{2}{3}}{\frac{5}{4}} = -\frac{8}{15}.$

63. $\sum\limits_{i=1}^{\infty} \left(\frac{3}{7}\right)^{i}$

For this geometric series, $a_1 = \left(\frac{3}{7}\right)^1 = \frac{3}{7}$ and

$r = \frac{3}{7}.$

$S_\infty = \frac{a_1}{1 - r} = \frac{\frac{3}{7}}{1 - \frac{3}{7}} = \frac{\frac{3}{7}}{\frac{4}{7}} = \frac{3}{4}.$

For Exercises 65–67, you will need the sum(and
seq(features of your graphing calculator. On the
TI-84, the sum(feature is located under the LIST
then MATH menu. The seq(feature is under the
LIST then OPS menu.

65. $\sum\limits_{i=1}^{10} (1.4)^{i}$

Using the sequence feature of a graphing
calculator, we obtain $S_{10} \approx 97.739$.

67. $\displaystyle\sum_{j=3}^{8} 2(0.4)^j$

Using the sequence feature of a graphing calculator, the sum is approximately 0.212.

69. (a) $a_n = 1276(0.916)^n$

$a_1 = 1276(0.916)^1 = 1168.816 \approx 1169$

and $r = 0.916$

(b) $a_n = 1276(0.916)^n$

$a_{10} = 1276(0.916)^{10} \approx 531$

$a_{20} = 1276(0.916)^{20} \approx 221$

This means that a person who is 10 years from retirement should have savings of 531% of his or her annual salary; a person 20 years from retirement should have savings of 221% of his or her annual salary.

71. (a) The first term is a_1 and the common ratio is r, so $a_n = a_1 \cdot 2^{n-1}$.

(b) If $a_1 = 100$, we have $a_n = 100 \cdot 2^{n-1}$.

Because $100 = 10^2$ and $1,000,000 = 10^6$, we need to solve the equation $10^2 \cdot 2^{n-1} = 10^6$. Divide both sides by 10^2 to obtain $2^{n-1} = 10^4$. Take the common logarithm (base 10) of both sides and solve for n.

$\log 2^{n-1} = \log 10^4 \Rightarrow (n-1)\log 2 = 4 \Rightarrow$

$n-1 = \dfrac{4}{\log 2} \Rightarrow n = \dfrac{4}{\log 2} + 1 \approx 14.28$

The number of bacteria is increasing, so the first value of n where $a_n > 1,000,000$ is 15.

(c) Because a_n represents the number of bacteria after $40(n-1)$ minutes, a_{15} represents the number after $40(15-1) = 40 \cdot 14 = 560$ minutes or 9 hours, 20 minutes.

73. There are 200 insects in the first generation, so $a_1 = 200$. The r value will be 1.25. Thus,

$a_n = a_1 r^{n-1} \Rightarrow$

$a_5 = 200(1.25)^{5-1} = 200(1.25)^4 = 488.28125$.

There are approximately 488 fruit flies in the fifth generation.

75. Use the formula for the sum of the first n terms of a geometric sequence with $a_1 = 2$, $r = 2$, and $n = 5$.

$S_n = \dfrac{a_1\left(1-r^n\right)}{1-r} \Rightarrow$

$S_5 = \dfrac{2\left(1-2^5\right)}{1-2} = \dfrac{2(1-32)}{-1} = \dfrac{2(-31)}{-1} = 62$

Going back five generations, the total number of ancestors is 62. Next, use the same formula with $a_1 = 2$, $r = 2$, and $n = 10$.

$S_{10} = \dfrac{2\left(1-2^{10}\right)}{1-2} = \dfrac{2(1-1024)}{-1} = 2046$

Going back ten generations, the total number of ancestors is 2046.

77. When the midpoints of the sides of an equilateral triangle are connected, the length of a side of the new triangle is one-half the length of a side of the original triangle. Use the formula for the nth term of a geometric sequence with $a_1 = 2$, $r = \frac{1}{2}$, and $n = 8$.

$a_n = a_1 r^{n-1} \Rightarrow$

$a_8 = (2)\left(\tfrac{1}{2}\right)^{8-1} = (2)\left(\tfrac{1}{2}\right)^7 = (2)\left(\tfrac{1}{128}\right) = \tfrac{1}{64}$

The eighth triangle has sides of length $\frac{1}{64}$ m.

79. Option 1 is modeled by the arithmetic sequence $a_n = 5000 + 10,000(n-1)$ with the following sum.

$S_n = \tfrac{n}{2}\left[2a_1 + (n-1)d\right]$

$S_{30} = \tfrac{30}{2}\left[2a_1 + (30-1)d\right] = 15\left[2a_1 + 29d\right]$

$= 15\left[2(5000) + 29 \cdot 10,000\right]$

$= 15\left[10,000 + 290,000\right]$

$S_{30} = 15(300,000) = 4,500,000$

Thus, the first option will pay $4,500,000 for the month's work. Option 2 is modeled by the sequence $a_n = 0.01(2)^{n-1}$ with the following sum.

(continued on next page)

(*continued*)

$$S_n = \frac{a_1\left(1 - r^n\right)}{1 - r} \Rightarrow S_{30} = \frac{a_1\left(1 - r^{30}\right)}{1 - r} \Rightarrow$$

$$S_{30} = \frac{0.01\left(1 - 2^{30}\right)}{1 - 2}$$

$$= \frac{0.01\left(1 - 1,073,741,824\right)}{-1} \Rightarrow$$

$$S_{30} = -0.01\left(-1,073,741,823\right) = 10,737,418.23$$

The second will pay $10,737,418.23. Option 2 pays better.

In exercises 81–87, you are asked to use the future value capability of the TI-84 Plus to find the answers. We include the grapher screen along with the algebraic solution for each exercise.

81. The future value of an annuity uses the

formula $S_n = \dfrac{a_1(1 - r^n)}{1 - r}$, where

$r = 1 +$ interest rate.
The payments are $1000 for 9 yr at 3% compounded annually, so $a_1 = 1000$, $r = 1.03$, and $n = 9$.

$$S_9 = \frac{1000\left[1 - (1.03)^9\right]}{1 - 1.03} \approx 10,159.11$$

The future value is $10,159.11.

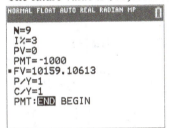

83. The future value of an annuity uses the

formula $S_n = \dfrac{a_1(1 - r^n)}{1 - r}$, where

$r = 1 +$ interest rate.
The payments are $2430 for 10 yr at 1% compounded annually, so $a_1 = 2430$, $r = 1.01$, and $n = 10$.

$$S_{10} = \frac{2430\left[1 - (1.01)^{10}\right]}{1 - 1.01} \approx 25,423.18$$

The future value is $25,423.18.

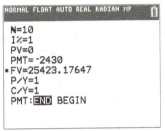

85. $R = 2430$, $i = 0.01$, and $n = 11$.
$$S_{11} = 25423.18(1.01) + 2430 \approx 28,107.41$$
The balance after 11 years is $28,107.41

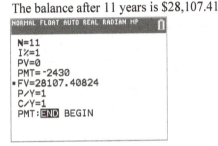

87. For deposits made annually for twenty-five

years, we need to find $S = R\left[\dfrac{(1 + i)^n - 1}{i}\right]$

where $R = 2000$, $i = 0.02$, and $n = 25$. Thus,

we have $S = 2000\left[\dfrac{(1.02)^{25} - 1}{0.02}\right] \approx 64,060.60$.

The total amount in Michael's IRA will be $64,060.60

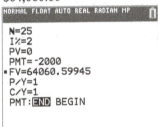

89. In the formula $S_n = \dfrac{a_1\left(1 - r^n\right)}{1 - r}$, we are

considering the sum
$a_1 + a_1 r + a_1 r^2 + \cdots + a_1 r^{n-1}$. With an annuity with deposits at the end of each year, this summation is actually reversed because the last deposit will not receive interest. Reversing the summation we have the following.
$$a_1 r^{n-1} + a_1 r^{n-2} + \cdots + a_1 r^2 + a_1 r + a_1$$
Now the first term of this series represents the first amount deposited having gone through $n - 1$ compound periods (because it is deposited at the end of the first year).

(*continued on next page*)

(continued)

The value of r is $1 + i$, where i is the interest rate in decimal form. The value of a_1 is the equal amounts of money that are deposited annually. If we call this value R, then the formula $S_n = \dfrac{a_1\left(1 - r^n\right)}{1 - r}$ becomes

$$S = \frac{R\left(1 - (1 + i)^n\right)}{1 - (1 + i)}.$$

This simplifies as follows.

$$S = \frac{R\left(1 - (1 + i)^n\right)}{1 - (1 + i)} \Rightarrow S = \frac{R\left(1 - (1 + i)^n\right)}{1 - 1 - i} \Rightarrow$$

$$S = R\frac{1 - (1 + i)^n}{-i} \Rightarrow S = R\frac{(1 + i)^n - 1}{i}$$

91. If $a_1, a_2, a_3, \ldots$ is a geometric sequence, then $a_n = a_1 r_a^{\,n-1}$ (where r_a is the common ratio for this sequence).
$a_3 = a_1 \cdot r^2, a_5 = a_3 \cdot r^2, a_7 = a_5 \cdot r^2, \ldots$ So, $a_1, a_3, a_5, \ldots$ is a geometric sequence, with common ratio r^2.

Summary Exercises on Sequences and Series

1. 2, 4, 8, 16, 32, ... is a geometric sequence with $r = 2$.
Notice $2(2) = 4$, $4(2) = 8$, $8(2) = 16$, and $16(2) = 32$.

3. $3, \frac{1}{2}, -2, -\frac{9}{2}, -7, \ldots$ is an arithmetic sequence with $d = -\frac{5}{2}$.
Notice $3 + \left(-\frac{5}{2}\right) = \frac{1}{2}$, $\frac{1}{2} + \left(-\frac{5}{2}\right) = -2$, $-2 + \left(-\frac{5}{2}\right) = -\frac{9}{2}$, and $-\frac{9}{2} + \left(-\frac{5}{2}\right) = -7$.

5. $\frac{3}{4}, 1, \frac{4}{3}, \frac{16}{9}, \frac{64}{27}, \ldots$ is a geometric sequence with $r = \frac{4}{3}$.
Notice $\frac{3}{4}\left(\frac{4}{3}\right) = 1, 1\left(\frac{4}{3}\right) = \frac{4}{3}, \frac{4}{3}\left(\frac{4}{3}\right) = \frac{16}{9}$, and $\frac{16}{9}\left(\frac{4}{3}\right) = \frac{64}{27}$.

7. $\frac{1}{2}, \frac{1}{3}, \frac{1}{4}, \frac{1}{5}, \frac{1}{6}, \ldots$ is neither an arithmetic sequence nor a geometric sequence.
Notice $\frac{1}{2} + \left(-\frac{1}{6}\right) = \frac{1}{3}$ but $\frac{1}{3} + \left(-\frac{1}{6}\right) \neq \frac{1}{4}$, so the sequence is not arithmetic. Also, $\frac{1}{2}\left(\frac{2}{3}\right) = \frac{1}{3}$ but $\frac{1}{3}\left(\frac{2}{3}\right) \neq \frac{1}{4}$, so the sequence is not geometric.

9. 1, 9, 10, 19, 29, ... is neither an arithmetic sequence nor a geometric sequence.
Notice $1 + 8 = 9$, but $9 + 8 \neq 10$, so the sequence is not arithmetic. Also, $1 \cdot 9 = 9$, but $9 \cdot 9 \neq 10$, so the sequence is not geometric.

11. The sequence 3, 6, 12, 24, 48, ... is geometric with $a_1 = 3$ and $r = 2$. Thus, we have

$$a_n = a_1 r^{n-1} \Rightarrow a_n = 3(2)^{n-1}$$

$$S_n = \frac{a_1\left(1 - r^n\right)}{1 - r} \Rightarrow$$

$$S_{10} = \frac{3\left(1 - 2^{10}\right)}{1 - 2} = \frac{3(1 - 1024)}{-1} = 3069$$

13. The sequence $4, \frac{5}{2}, 1, -\frac{1}{2}, -2, \ldots$ is arithmetic with $a_1 = 4$ and $d = -\frac{3}{2}$. Thus, we have

$$a_n = a_1 + (n-1)d \Rightarrow a_n = 4 + (n-1)\left(-\frac{3}{2}\right) \Rightarrow$$

$$a_n = 4 - \frac{3}{2}n + \frac{3}{2} \Rightarrow a_n = \frac{11}{2} - \frac{3}{2}n$$

$$S_n = \frac{n}{2}\left[2a_1 + (n-1)d\right] \Rightarrow$$

$$\sum_{i=1}^{10} \frac{11}{2} - \frac{3}{2}i = S_{10} = \frac{10}{2}\left[2(4) + (10-1)\left(-\frac{3}{2}\right)\right]$$

$$= 5\left[8 + \left(-\frac{27}{2}\right)\right] = 5\left(-\frac{11}{2}\right) = -\frac{55}{2}$$

15. The sequence 3, −6, 12, −24, 48, ... is geometric with $a_1 = 3$ and $r = -2$. Thus, we have $a_n = a_1 r^{n-1} \Rightarrow a_n = 3(-2)^{n-1}$

$$S_n = \frac{a_1\left(1 - r^n\right)}{1 - r} \Rightarrow$$

$$S_{10} = \frac{3\left[1 - (-2)^{10}\right]}{1 - (-2)} = \frac{3(1 - 1024)}{3} = -1023$$

17. $\displaystyle\sum_{i=1}^{\infty}\tfrac{1}{3}(-2)^{i-1}$ is an infinite geometric series with

$a_1 = \tfrac{1}{3}$ and $r = -2$. Because $|r| = 2 > 1$, this series diverges.

19. $\displaystyle\sum_{i=1}^{25}(4-6i)$ is an arithmetic series with $a_1 = -2$

and $d = -6$.

$S_n = \dfrac{n}{2}\left[2a_1 + (n-1)d\right] \Rightarrow$

$\displaystyle\sum_{i=1}^{25}(4-6i) = S_{25} = \dfrac{25}{2}\left[2(-2) + (25-1)(-6)\right]$

$\qquad\qquad = \dfrac{25}{2}(-148) = -1850$

21. $\displaystyle\sum_{i=1}^{\infty}4\left(-\tfrac{1}{2}\right)^{i}$ is an infinite geometric series with

$a_1 = -2$ and $r = -\tfrac{1}{2}$. Because $|r| = \tfrac{1}{2} < 1$, then the sum of this series is

$S = \dfrac{a_1}{1-r} = \dfrac{-2}{1-\left(-\tfrac{1}{2}\right)} = \dfrac{-2}{\tfrac{3}{2}} = -\dfrac{4}{3}.$

23. $\displaystyle\sum_{j=1}^{12}(2j-1)$ is an arithmetic series with $a_1 = 1$

and $d = 2$. To use the formula

$S_n = \dfrac{n}{2}(a_1 + a_n)$, we need find a_{12}.

$a_{12} = 2(12) - 1 = 24 - 1 = 23$

$\displaystyle\sum_{j=1}^{12}(2j-1) = S_{12} = \dfrac{12}{2}(1+23) = 6(24) = 144$

25. $\displaystyle\sum_{k=1}^{\infty}1.0001^{i}$ is an infinite geometric series with

$a_1 = 1.0001$ and $r = 1.0001$ Because $|r| = 1.0001 > 1$, this series diverges.

Section 11.4   The Binomial Theorem

1. Each number that is not a 1 in Pascal's triangle is the <u>sum</u> of the two numbers directly above it (one to the right and one to the left).

3. The value of 0! is <u>1</u>.

5. $_{12}C_4 = {}_{12}C_8$.

$_{12}C_4 = \dfrac{12!}{4!(12-4)!} = \dfrac{12!}{4!8!} = \dfrac{12!}{8!(12-8)!} = {}_{12}C_8$

7. In the expansion of $(x+y)^8$, the first term is $\underline{x^8}$ and the last term is $\underline{y^8}$.

9. The second term in the expansion of $(p+q)^5$ is $\underline{5p^4q}$.

11. $\dfrac{6!}{3!3!} = \dfrac{6\cdot5\cdot4\cdot3\cdot2\cdot1}{3\cdot2\cdot1\cdot3\cdot2\cdot1} = \dfrac{6\cdot5\cdot4}{3\cdot2\cdot1} = 20$

13. $\dfrac{7!}{3!4!} = \dfrac{7\cdot6\cdot5\cdot4\cdot3\cdot2\cdot1}{3\cdot2\cdot1\cdot4\cdot3\cdot2\cdot1} = \dfrac{7\cdot6\cdot5}{3\cdot2\cdot1} = 35$

15. $\dbinom{8}{5} = \dfrac{8!}{5!3!} = \dfrac{8\cdot7\cdot6\cdot5\cdot4\cdot3\cdot2\cdot1}{5\cdot4\cdot3\cdot2\cdot1\cdot3\cdot2\cdot1} = \dfrac{8\cdot7\cdot6}{3\cdot2\cdot1} = 56$

17. $\dbinom{10}{2} = \dfrac{10!}{2!8!} = \dfrac{10\cdot9\cdot8!}{2\cdot1\cdot8!} = \dfrac{10\cdot9}{2\cdot1} = 45$

19. $\dbinom{14}{14} = \dfrac{14!}{14!(14-14)!} = \dfrac{14!}{14!0!} = \dfrac{14!}{14!\cdot1} = 1$

21. $\dbinom{n}{n-1} = \dfrac{n!}{(n-1)!\left[n-(n-1)\right]!}$

$\qquad = \dfrac{n!}{(n-1)!(n-n+1)!} = \dfrac{n!}{(n-1)!1!}$

$\qquad = \dfrac{n(n-1)!}{(n-1)!} = n$

23. $_8C_3 = \dfrac{8!}{3!(8-3)!} = \dfrac{8!}{3!5!} = \dfrac{8\cdot7\cdot6}{3\cdot2\cdot1} = 56$

25. $_{100}C_{98} = \dfrac{100!}{98!(100-98)!} = \dfrac{100!}{98!2!}$

$\qquad = \dfrac{100\cdot99}{2\cdot1} = 4950$

27. $_9C_0 = \dfrac{9!}{0!(9-0)!} = \dfrac{9!}{1\cdot9!} = 1$

29. $_{12}C_1 = \dfrac{12!}{1!(12-1)!} = \dfrac{12\cdot11!}{1!11!} = \dfrac{12}{1} = 12$

31. $(x+y)^6 = x^6 + \binom{6}{1}x^5y + \binom{6}{2}x^4y^2 + \binom{6}{3}x^3y^3 + \binom{6}{4}x^2y^4 + \binom{6}{5}xy^5 + y^6$

$\quad = x^6 + \dfrac{6!}{1!5!}x^5y + \dfrac{6!}{2!4!}x^4y^2 + \dfrac{6!}{3!3!}x^3y^3 + \dfrac{6!}{4!2!}x^2y^4 + \dfrac{6!}{5!1!}xy^5 + y^6$

$\quad = x^6 + 6x^5y + \dfrac{6\cdot5}{2\cdot1}x^4y^2 + \dfrac{6\cdot5\cdot4}{3\cdot2\cdot1}x^3y^3 + \dfrac{6\cdot5}{2\cdot1}x^2y^4 + 6xy^5 + y^6$

$\quad = x^6 + 6x^5y + 15x^4y^2 + 20x^3y^3 + 15x^2y^4 + 6xy^5 + y^6$

33. $(p-q)^5 = \left[p+(-q)\right]^5$

$\quad = p^5 + \binom{5}{1}p^4(-q) + \binom{5}{2}p^3(-q)^2 + \binom{5}{3}p^2(-q)^3 + \binom{5}{4}p(-q)^4 + (-q)^5$

$\quad = p^5 + \dfrac{5!}{1!4!}p^4(-q) + \dfrac{5!}{2!3!}p^3q^2 + \dfrac{5!}{3!2!}p^2(-q)^3 + \dfrac{5!}{4!1!}pq^4 + (-q)^5$

$\quad = p^5 + 5p^4(-q) + \dfrac{5\cdot4}{2\cdot1}p^3q^2 + \dfrac{5\cdot4}{2\cdot1}p^2(-q)^3 + 5pq^4 + (-q)^5$

$\quad = p^5 - 5p^4q + 10p^3q^2 - 10p^2q^3 + 5pq^4 - q^5$

35. $(r^2+s)^5 = (r^2)^5 + \binom{5}{1}(r^2)^4 s + \binom{5}{2}(r^2)^3 s^2 + \binom{5}{3}(r^2)^2 s^3 + \binom{5}{4}(r^2)s^4 + s^5$

$\quad = r^{10} + \dfrac{5!}{1!4!}r^8s + \dfrac{5!}{2!3!}r^6s^2 + \dfrac{5!}{3!2!}r^4s^3 + \dfrac{5!}{4!1!}r^2s^4 + s^5$

$\quad = r^{10} + 5r^8s + \dfrac{5\cdot4}{2\cdot1}r^6s^2 + \dfrac{5\cdot4}{2\cdot1}r^4s^3 + 5r^2s^4 + s^5$

$\quad = r^{10} + 5r^8s + 10r^6s^2 + 10r^4s^3 + 5r^2s^4 + s^5$

37. $(p+2q)^4 = p^4 + \binom{4}{1}p^3(2q) + \binom{4}{2}p^2(2q)^2 + \binom{4}{3}p(2q)^3 + (2q)^4$

$\quad = p^4 + \dfrac{4!}{1!3!}p^3(2q) + \dfrac{4!}{2!2!}p^2(4q^2) + \dfrac{4!}{3!1!}p(8q^3) + 16q^4$

$\quad = p^4 + 4p^3(2q) + \dfrac{4\cdot3}{2\cdot1}p^2(4q^2) + 4p(8q^3) + 16q^4 = p^4 + 8p^3q + 24p^2q^2 + 32pq^3 + 16q^4$

39. $(7p-2q)^4 = (7p)^4 + \binom{4}{1}(7p)^3(-2q) + \binom{4}{2}(7p)^2(-2q)^2 + \binom{4}{3}(7p)(-2q)^3 + (-2q)^4$

$\quad = 2401p^4 + \dfrac{4!}{1!3!}(343p^3)(-2q) + \dfrac{4!}{2!2!}(49p^2)(4q^2) + \dfrac{4!}{3!1!}(7p)(-8q^3) + 16q^4$

$\quad = 2401p^4 - 4(686p^3q) + \dfrac{4\cdot3}{2\cdot1}(196p^2q^2) - 4(56pq^3) + 16q^4$

$\quad = 2401p^4 - 2744p^3q + 6(196p^2q^2) - 224pq^3 + 16q^4$

$\quad = 2401p^4 - 2744p^3q + 1176p^2q^2 - 224pq^3 + 16q^4$

41. $(3x-2y)^6 = (3x)^6 + \binom{6}{1}(3x)^5(-2y) + \binom{6}{2}(3x)^4(-2y)^2 + \binom{6}{3}(3x)^3(-2y)^3$

$$+ \binom{6}{4}(3x)^2(-2y)^4 + \binom{6}{5}(3x)(-2y)^5 + (-2y)^6$$

$$= 729x^6 + \frac{6!}{1!5!}(243x^5)(-2y) + \frac{6!}{2!4!}(81x^4)(4y^2) + \frac{6!}{3!3!}(27x^3)(-8y^3)$$

$$+ \frac{6!}{4!2!}(9x^2)(16y^4) + \frac{6!}{5!1!}(3x)(-32y^5) + 64y^6$$

$$= 729x^6 - 6(486x^5y) + \frac{6\cdot5}{2\cdot1}(324x^4y^2) - \frac{6\cdot5\cdot4}{3\cdot2\cdot1}(216x^3y^3)$$

$$+ \frac{6\cdot5}{2\cdot1}(144x^2y^4) - 6(96xy^5) + 64y^6$$

$$= 729x^6 - 2916x^5y + 15(324x^4y^2) - 20(216x^3y^3) + 15(144x^2y^4) - 576xy^5 + 64y^6$$

$$= 729x^6 - 2916x^5y + 4860x^4y^2 - 4320x^3y^3 + 2160x^2y^4 - 576xy^5 + 64y^6$$

43. $\left(\frac{m}{2}-1\right)^6 = \left(\frac{m}{2}\right)^6 + \binom{6}{1}\left(\frac{m}{2}\right)^5(-1) + \binom{6}{2}\left(\frac{m}{2}\right)^4(-1)^2 + \binom{6}{3}\left(\frac{m}{2}\right)^3(-1)^3$

$$+ \binom{6}{4}\left(\frac{m}{2}\right)^2(-1)^4 + \binom{6}{5}\left(\frac{m}{2}\right)(-1)^5 + (-1)^6$$

$$= \frac{m^6}{64} - \frac{6!}{1!5!}\left(\frac{m^5}{32}\right) + \frac{6!}{2!4!}\left(\frac{m^4}{16}\right) - \frac{6!}{3!3!}\left(\frac{m^3}{8}\right) + \frac{6!}{4!2!}\left(\frac{m^2}{4}\right) - \frac{6!}{5!1!}\left(\frac{m}{2}\right) + 1$$

$$= \frac{m^6}{64} - 6\left(\frac{m^5}{32}\right) + \frac{6\cdot5}{2\cdot1}\left(\frac{m^4}{16}\right) - \frac{6\cdot5\cdot4}{3\cdot2\cdot1}\left(\frac{m^3}{8}\right) + \frac{6\cdot5}{2\cdot1}\left(\frac{m^2}{4}\right) - 6\left(\frac{m}{2}\right) + 1$$

$$= \frac{m^6}{64} - \frac{3m^5}{16} + 15\left(\frac{m^4}{16}\right) - 20\left(\frac{m^3}{8}\right) + 15\left(\frac{m^2}{4}\right) - 3m + 1$$

$$= \frac{m^6}{64} - \frac{3m^5}{16} + \frac{15m^4}{16} - \frac{5m^3}{2} + \frac{15m^2}{4} - 3m + 1$$

45. $\left(\sqrt{2}r + \frac{1}{m}\right)^4 = (\sqrt{2}r)^4 + \binom{4}{1}(\sqrt{2}r)^3\left(\frac{1}{m}\right) + \binom{4}{2}(\sqrt{2}r)^2\left(\frac{1}{m}\right)^2 + \binom{4}{3}(\sqrt{2}r)\left(\frac{1}{m}\right)^3 + \left(\frac{1}{m}\right)^4$

$$= (\sqrt{2}r)^4 + \frac{4!}{1!3!}(\sqrt{2}r)^3\left(\frac{1}{m}\right) + \frac{4!}{2!2!}(\sqrt{2}r)^2\frac{1}{m^2} + \frac{4!}{3!1!}(\sqrt{2}r)\frac{1}{m^3} + \frac{1}{m^4}$$

$$= 4r^4 + 4(2\sqrt{2})r^3\left(\frac{1}{m}\right) + \frac{4\cdot3}{2\cdot1}(2r^2)\frac{1}{m^2} + 4(\sqrt{2}r)\frac{1}{m^3} + \frac{1}{m^4}$$

$$= 4r^4 + \frac{8\sqrt{2}r^3}{m} + (6)\frac{2r^2}{m^2} + \frac{4\sqrt{2}r}{m^3} + \frac{1}{m^4}$$

$$= 4r^4 + \frac{8\sqrt{2}r^3}{m} + \frac{12r^2}{m^2} + \frac{4\sqrt{2}r}{m^3} + \frac{1}{m^4}$$

47. $\left(\frac{1}{x^4}+x^4\right)^4 = \left(\frac{1}{x^4}\right)^4 + \binom{4}{1}\left(\frac{1}{x^4}\right)^3(x^4) + \binom{4}{2}\left(\frac{1}{x^4}\right)^2(x^4)^2 + \binom{4}{3}\left(\frac{1}{x^4}\right)(x^4)^3 + (x^4)^4$

$$= \frac{1}{x^{16}} + \frac{4!}{1!3!}\left(\frac{1}{x^{12}}\right)(x^4) + \frac{4!}{2!2!}\left(\frac{1}{x^8}\right)(x^8) + \frac{4!}{3!1!}\left(\frac{1}{x^4}\right)(x^{12}) + x^{16}$$

$$= \frac{1}{x^{16}} + 4\left(\frac{1}{x^8}\right) + 6 + 4x^8 + x^{16} = \frac{1}{x^{16}} + \frac{4}{x^8} + 6 + 4x^8 + x^{16}$$

49. $(4h - j)^8$

Using the formula $\binom{n}{k-1} x^{n-(k-1)} y^{k-1}$ with

$n = 8$, $k - 1 = 5$, and $n - (k - 1) = 3$, the sixth term of the expansion is

$$\binom{8}{5} (4h)^3 (-j)^5 = \frac{8!}{5!3!} (64h^3)(-j^5)$$
$$= \frac{8 \cdot 7 \cdot 6}{3 \cdot 2 \cdot 1} (64h^3)(-j^5)$$
$$= 56(-64h^3 j^5) = -3584 h^3 j^5$$

51. $(a^2 + b)^{22}$

Use the formula $\binom{n}{k-1} x^{n-(k-1)} y^{k-1}$ with

$n = 22$, $k = 17$, $k - 1 = 16$, and $n - (k - 1) = 6$. The seventeenth term of the expansion is

$$\binom{22}{16} (a^2)^6 (b)^{16}$$
$$= \frac{22!}{16!6!} (a^{12})(b^{16})$$
$$= \frac{22 \cdot 21 \cdot 20 \cdot 19 \cdot 18 \cdot 17}{6 \cdot 5 \cdot 4 \cdot 3 \cdot 2 \cdot 1} (a^{12} b^{16})$$
$$= 74,613 a^{12} b^{16}$$

53. $(x - y^3)^{20}$

Use the formula $\binom{n}{k-1} x^{n-(k-1)} y^{k-1}$ with

$n = 20$, $k = 15$, $k - 1 = 14$, and $n - (k - 1) = 6$. The fifteenth term of the expansion is

$$\binom{20}{14} (x)^6 (-y^3)^{14}$$
$$= \frac{20!}{14!6!} (x^6)(y^{42})$$
$$= \frac{20 \cdot 19 \cdot 18 \cdot 17 \cdot 16 \cdot 15}{6 \cdot 5 \cdot 4 \cdot 3 \cdot 2 \cdot 1} (x^6 y^{42})$$
$$= 38,760 x^6 y^{42}$$

55. $(3x^7 + 2y^3)^8$

This expansion has nine terms, so the middle term is the fifth term. Using the formula

$\binom{n}{k-1} x^{n-(k-1)} y^{k-1}$ with $n = 8$, $k = 5$,

$k - 1 = 4$, and $n - (k - 1) = 4$, the fifth term is

$$\binom{8}{4} (3x^7)^4 (2y^3)^4 = \frac{8!}{4!4!} (3^4 x^{28})(2^4 y^{12})$$
$$= \frac{8 \cdot 7 \cdot 6 \cdot 5}{4 \cdot 3 \cdot 2 \cdot 1} (81 x^{28})(16 y^{12})$$
$$= (70)(1296 x^{28} y^{12})$$
$$= 90,720 x^{28} y^{12}$$

57. If the coefficients of the fifth and eighth terms in the expansion of $(x + y)^n$ are the same, then the symmetry of the expansion can be used to determine n. There are four terms before the fifth term, so there must be four terms after the eighth term. This means that the last term of the expansion is the twelfth term. This in turn means that $n = 11$, because $(x + y)^{11}$ is the expansion that has twelve terms.

59. Using a calculator, we obtain the exact value $10! = 3,628,800$. Using Stirling's formula, we obtain

$$10! \approx \sqrt{2\pi(10)} \cdot 10^{10} \cdot e^{-10} \approx 3,598,695.619$$

61. Using a calculator, we obtain the exact value $12! = 479,001,600$. Using Stirling's formula, we obtain

$$12! \approx \sqrt{2\pi(12)} \cdot 12^{12} \cdot e^{-12} \approx 475,687,486.5$$
$$\frac{479,001,600 - 475,687,486.5}{479,001,600} \approx 0.00692, \text{ so}$$

the percent error is approximately 0.692%.

Section 11.5   Mathematical Induction

1. $S_4 : 3 + 6 + 9 + 12 = \dfrac{3(4)(4+1)}{2} \Rightarrow 30 = 30$,

which is true.

3. $S_4 : \dfrac{1}{2} + \dfrac{1}{2^2} + \dfrac{1}{2^3} + \dfrac{1}{2^4} = \dfrac{2^4 - 1}{2^4} \Rightarrow \dfrac{15}{16} = \dfrac{15}{16}$,

which is true.

5. $S_4 : 2^4 < 2(4) \Rightarrow 16 < 8$, which is false.

7. Let S_n be the statement $1+3+5+\cdots+(2n-1)=n^2$.

$S_1: 1=1^2 \Rightarrow 1=1$, which is true.

$S_2: 1+3=2^2 \Rightarrow 4=4$, which is true.

$S_3: 1+3+5=3^2 \Rightarrow 9=9$, which is true.

$S_4: 1+3+5+7=4^2 \Rightarrow 16=16$, which is true.

$S_5: 1+3+5+7+9=5^2 \Rightarrow 25=25$, which is true.

Now prove that S_n is true for every positive integer n.

(a) *Verify the statement for* $n=1$.

S_1 is the statement $1=1^2 \Rightarrow 1=1$, which is true.

(b) Write the statement for $n=k$: $S_k = 1+3+5+\cdots+(2k-1)=k^2$

(c) *Write the statement for* $n=k+1$: $S_{k+1}=1+3+5+\cdots+(2k-1)+\left[2(k+1)-1\right]=(k+1)^2$

(d) *Assume that S_k is true. Use algebra to change S_k to S_{k+1}:*

Start with $S_k: 1+3+5+\cdots+(2k-1)=k^2$ and add the $(k+1)$ st term, $\left[2(k+1)-1\right]$, to both sides of this equation.

$$1+3+5+\cdots+(2k-1)=k^2$$
$$1+3+5+\cdots+(2k-1)+\left[2(k+1)-1\right]=k^2+\left[2(k+1)-1\right]$$
$$1+3+5+\cdots+(2k-1)+\left[2(k+1)-1\right]=k^2+(2k+2-1)$$
$$1+3+5+\cdots+(2k-1)+\left[2(k+1)-1\right]=k^2+2k+1$$
$$1+3+5+\cdots+(2k-1)+\left[2(k+1)-1\right]=(k+1)^2$$

(e) *Write a conclusion based on Steps (a)–(d).*

Because S_n is true for $n=1$ and S_n is true for $n=k+1$ when it is true for $n=k$, S_n is true for every positive integer n.

9. S_n is the statement $3+6+9+\cdots+3n=\dfrac{3n(n+1)}{2}$.

(a) *Verify the statement for* $n=1$.

$3(1)=\dfrac{3(1)(1+1)}{2} \Rightarrow 3=\dfrac{6}{2} \Rightarrow 3=3$, which is true. Thus, S_n is true for $n=1$.

(b) *Write the statement for* $n=k$: $S_k=3+6+9+\cdots+3k=\dfrac{3k(k+1)}{2}$

(c) *Write the statement for* $n=k+1$: $S_{k+1}=3+6+9+\cdots+3k+3(k+1)=\dfrac{3(k+1)\left[(k+1)+1\right]}{2}$

(d) *Assume that S_k is true. Use algebra to change S_k to S_{k+1}:*

Add the $(k+1)$ st term, $3(k+1)$, to both sides of this equation.

$$3+6+9+\cdots+3k = \frac{3k(k+1)}{2} \Rightarrow 3+6+9+\cdots+3k+3(k+1) = \frac{3k(k+1)}{2}+3(k+1) \Rightarrow$$

$$3+6+9+\cdots+3k+3(k+1) = \frac{3k(k+1)+6(k+1)}{2} \Rightarrow 3+6+9+\cdots+3k+3(k+1) = \frac{3k^2+3k+6k+6}{2} \Rightarrow$$

$$3+6+9+\cdots+3k+3(k+1) = \frac{3k^2+9k+6}{2} \Rightarrow 3+6+9+\cdots+3k+3(k+1) = \frac{3(k^2+3k+2)}{2} \Rightarrow$$

$$3+6+9+\cdots+3k+3(k+1) = \frac{3(k+1)(k+2)}{2} \Rightarrow 3+6+9+\cdots+3k+3(k+1) = \frac{3(k+1)[(k+1)+1]}{2}$$

(e) *Write a conclusion based on Steps (a)–(d).*

Because S_n is true for $n=1$ and S_n is true for $n=k+1$ when it is true for $n=k$, S_n is true for every positive integer n.

11. S_n is the statement $2+4+8+\cdots+2^n = 2^{n+1}-2$.

(a) *Verify the statement for $n=1$.*

$2^1 = 2^{1+1}-2 \Rightarrow 2 = 2^2-2 \Rightarrow 2 = 4-2 \Rightarrow 2 = 2$, which is true. Thus, S_n is true for $n=1$.

(b) *Write the statement for $n=k$:* $S_k = 2+4+8+\cdots+2^k = 2^{k+1}-2$

(c) *Write the statement for $n=k+1$:* $S_{k+1} = 2+4+8+\cdots+2^k+2^{k+1} = 2^{(k+1)+1}-2$

(d) *Assume that S_k is true. Use algebra to change S_k to S_{k+1}:*

Add the $(k+1)$ st term, 2^{k+1}, to both sides of this equation.

$$2+4+8+\cdots+2^k = 2^{k+1}-2 \Rightarrow 2+4+8+\cdots+2^k+2^{k+1} = 2^{k+1}-2+2^{k+1} \Rightarrow$$

$$2+4+8+\cdots+2^k+2^{k+1} = 2\cdot2^{k+1}-2 \Rightarrow 2+4+8+\cdots+2^k+2^{k+1} = 2^1\cdot2^{k+1}-2 \Rightarrow$$

$$2+4+8+\cdots+2^k+2^{k+1} = 2^{(k+1)+1}-2$$

(e) *Write a conclusion based on Steps (a)–(d).*

Because S_n is true for $n=1$ and S_n is true for $n=k+1$ when it is true for $n=k$, S_n is true for every positive integer n.

13. S_n is the statement $1^2+2^2+3^2+\cdots+n^2 = \dfrac{n(n+1)(2n+1)}{6}$.

(a) *Verify the statement for $n=1$.*

$1^2 = \dfrac{1(1+1)(2\cdot1+1)}{6} \Rightarrow 1 = \dfrac{1(2)(2+1)}{6} \Rightarrow 1 = \dfrac{2(3)}{6} \Rightarrow 1 = 1$, which is true. Thus, S_n is true for $n=1$.

(b) *Write the statement for $n=k$:* $S_k = 1^2+2^2+3^2+\cdots+k^2 = \dfrac{k(k+1)(2k+1)}{6}$

(c) *Write the statement for $n=k+1$:*

$$S_{k+1} = 1^2+2^2+3^2+\cdots+k^2+(k+1)^2 = \frac{(k+1)[(k+1)+1][2(k+1)+1]}{6}$$

(d) *Assume that S_k is true. Use algebra to change S_k to S_{k+1}:*

Add the $(k+1)$ st term, $(k+1)^2$, to both sides of this equation.

$$1^2 + 2^2 + 3^2 + \cdots + k^2 = \frac{k(k+1)(2k+1)}{6}$$

$$1^2 + 2^2 + 3^2 + \cdots + k^2 + (k+1)^2 = \frac{k(k+1)(2k+1)}{6} + (k+1)^2$$

$$1^2 + 2^2 + 3^2 + \cdots + k^2 + (k+1)^2 = \frac{k(k+1)(2k+1) + 6(k+1)^2}{6}$$

$$1^2 + 2^2 + 3^2 + \cdots + k^2 + (k+1)^2 = \frac{(k+1)\left[k(2k+1) + 6(k+1)\right]}{6}$$

$$1^2 + 2^2 + 3^2 + \cdots + k^2 + (k+1)^2 = \frac{(k+1)\left(2k^2 + k + 6k + 6\right)}{6}$$

$$1^2 + 2^2 + 3^2 + \cdots + k^2 + (k+1)^2 = \frac{(k+1)\left(2k^2 + 7k + 6\right)}{6}$$

$$1^2 + 2^2 + 3^2 + \cdots + k^2 + (k+1)^2 = \frac{(k+1)(k+2)(2k+3)}{6}$$

$$1^2 + 2^2 + 3^2 + \cdots + k^2 + (k+1)^2 = \frac{(k+1)(k+2)(2k+2+1)}{6}$$

$$1^2 + 2^2 + 3^2 + \cdots + k^2 + (k+1)^2 = \frac{(k+1)\left[(k+1)+1\right]\left[2(k+1)+1\right]}{6}$$

(e) *Write a conclusion based on Steps (a)–(d).*
Because S_n is true for $n=1$ and S_n is true for $n=k+1$ when it is true for $n=k$, S_n is true for every positive integer n.

15. S_n is the statement $5 \cdot 6 + 5 \cdot 6^2 + 5 \cdot 6^3 + \cdots + 5 \cdot 6^n = 6\left(6^n - 1\right)$.

(a) *Verify the statement for $n=1$.*
$5 \cdot 6^1 = 6\left(6^1 - 1\right) \Rightarrow 5 \cdot 6 = 6(6-1) \Rightarrow 30 = 6(5) \Rightarrow 30 = 30$, which is true. Thus, S_n is true for $n=1$.

(b) *Write the statement for $n=k$:* $S_k = 5 \cdot 6 + 5 \cdot 6^2 + 5 \cdot 6^3 + \cdots + 5 \cdot 6^k = 6\left(6^k - 1\right)$

(c) *Write the statement for $n=k+1$:* $S_{k+1} = 5 \cdot 6 + 5 \cdot 6^2 + 5 \cdot 6^3 + \cdots + 5 \cdot 6^k + 5 \cdot 6^{k+1} = 6\left(6^{k+1} - 1\right)$

(d) *Assume that S_k is true. Use algebra to change S_k to S_{k+1}:*
Add the $(k+1)$ st term, $5 \cdot 6^{k+1}$, to both sides of this equation.

$$5 \cdot 6 + 5 \cdot 6^2 + 5 \cdot 6^3 + \cdots + 5 \cdot 6^k = 6\left(6^k - 1\right)$$

$$5 \cdot 6 + 5 \cdot 6^2 + 5 \cdot 6^3 + \cdots + 5 \cdot 6^k + 5 \cdot 6^{k+1} = 6\left(6^k - 1\right) + 5 \cdot 6^{k+1}$$

$$5 \cdot 6 + 5 \cdot 6^2 + 5 \cdot 6^3 + \cdots + 5 \cdot 6^k + 5 \cdot 6^{k+1} = 6\left[\left(6^k - 1\right) + 5 \cdot 6^k\right]$$

$$5 \cdot 6 + 5 \cdot 6^2 + 5 \cdot 6^3 + \cdots + 5 \cdot 6^k + 5 \cdot 6^{k+1} = 6\left[6 \cdot 6^k - 1\right]$$

$$5 \cdot 6 + 5 \cdot 6^2 + 5 \cdot 6^3 + \cdots + 5 \cdot 6^k + 5 \cdot 6^{k+1} = 6\left(6^{k+1} - 1\right)$$

(e) *Write a conclusion based on Steps (a)–(d).*
Because S_n is true for $n=1$ and S_n is true for $n=k+1$ when it is true for $n=k$, S_n is true for every positive integer n.

17. S_n is the statement $\dfrac{1}{1\cdot 2}+\dfrac{1}{2\cdot 3}+\dfrac{1}{3\cdot 4}+\cdots+\dfrac{1}{n(n+1)}=\dfrac{n}{n+1}$.

(a) *Verify the statement for* $n=1$.

$\dfrac{1}{(1)(1+1)}=\dfrac{1}{1+1}\Rightarrow\dfrac{1}{(1)(2)}=\dfrac{1}{2}\Rightarrow\dfrac{1}{2}=\dfrac{1}{2}$, which is true. Thus, S_n is true for $n=1$.

(b) *Write the statement for* $n=k$: $S_k=\dfrac{1}{1\cdot 2}+\dfrac{1}{2\cdot 3}+\dfrac{1}{3\cdot 4}+\cdots+\dfrac{1}{k(k+1)}=\dfrac{k}{k+1}$

(c) *Write the statement for* $n=k+1$:

$S_{k+1}=\dfrac{1}{1\cdot 2}+\dfrac{1}{2\cdot 3}+\dfrac{1}{3\cdot 4}+\cdots+\dfrac{1}{k(k+1)}+\dfrac{1}{(k+1)[(k+1)+1]}=\dfrac{k+1}{(k+1)+1}$

(d) *Assume that* S_k *is true. Use algebra to change* S_k *to* S_{k+1}:

Add the $(k+1)$ st term, $\dfrac{1}{(k+1)[(k+1)+1]}$, to both sides of this equation.

$$\dfrac{1}{1\cdot 2}+\dfrac{1}{2\cdot 3}+\dfrac{1}{3\cdot 4}+\cdots+\dfrac{1}{k(k+1)}=\dfrac{k}{k+1}$$

$$\dfrac{1}{1\cdot 2}+\dfrac{1}{2\cdot 3}+\dfrac{1}{3\cdot 4}+\cdots+\dfrac{1}{k(k+1)}+\dfrac{1}{(k+1)[(k+1)+1]}=\dfrac{k}{k+1}+\dfrac{1}{(k+1)[(k+1)+1]}$$

$$\dfrac{1}{1\cdot 2}+\dfrac{1}{2\cdot 3}+\dfrac{1}{3\cdot 4}+\cdots+\dfrac{1}{k(k+1)}+\dfrac{1}{(k+1)[(k+1)+1]}=\dfrac{k}{k+1}+\dfrac{1}{(k+1)(k+2)}$$

$$\dfrac{1}{1\cdot 2}+\dfrac{1}{2\cdot 3}+\dfrac{1}{3\cdot 4}+\cdots+\dfrac{1}{k(k+1)}+\dfrac{1}{(k+1)[(k+1)+1]}=\dfrac{k(k+2)+1}{(k+1)(k+2)}$$

$$\dfrac{1}{1\cdot 2}+\dfrac{1}{2\cdot 3}+\dfrac{1}{3\cdot 4}+\cdots+\dfrac{1}{k(k+1)}+\dfrac{1}{(k+1)[(k+1)+1]}=\dfrac{k^2+2k+1}{(k+1)(k+2)}$$

$$\dfrac{1}{1\cdot 2}+\dfrac{1}{2\cdot 3}+\dfrac{1}{3\cdot 4}+\cdots+\dfrac{1}{k(k+1)}+\dfrac{1}{(k+1)[(k+1)+1]}=\dfrac{(k+1)^2}{(k+1)(k+2)}$$

$$\dfrac{1}{1\cdot 2}+\dfrac{1}{2\cdot 3}+\dfrac{1}{3\cdot 4}+\cdots+\dfrac{1}{k(k+1)}+\dfrac{1}{(k+1)[(k+1)+1]}=\dfrac{k+1}{k+2}$$

$$\dfrac{1}{1\cdot 2}+\dfrac{1}{2\cdot 3}+\dfrac{1}{3\cdot 4}+\cdots+\dfrac{1}{k(k+1)}+\dfrac{1}{(k+1)[(k+1)+1]}=\dfrac{k+1}{(k+1)+1}$$

(e) *Write a conclusion based on Steps (a)–(d).*
Because S_n is true for $n=1$ and S_n is true for $n=k+1$ when it is true for $n=k$, S_n is true for every positive integer n.

19. S_n is the statement $\dfrac{1}{2}+\dfrac{1}{2^2}+\dfrac{1}{2^3}+\cdots+\dfrac{1}{2^n}=1-\dfrac{1}{2^n}$.

(a) *Verify the statement for* $n=1$.

$\dfrac{1}{2^1}=1-\dfrac{1}{2^1}\Rightarrow\dfrac{1}{2}=1-\dfrac{1}{2}\Rightarrow\dfrac{1}{2}=\dfrac{1}{2}$, which is true. Thus, S_n is true for $n=1$.

(b) *Write the statement for* $n=k$: $S_k=\dfrac{1}{2}+\dfrac{1}{2^2}+\dfrac{1}{2^3}+\cdots+\dfrac{1}{2^k}=1-\dfrac{1}{2^k}$

(c) *Write the statement for* $n = k+1$: $S_{k+1} = \dfrac{1}{2} + \dfrac{1}{2^2} + \dfrac{1}{2^3} + \cdots + \dfrac{1}{2^k} + \dfrac{1}{2^{k+1}} = 1 - \dfrac{1}{2^{k+1}}$

(d) *Assume that* S_k *is true. Use algebra to change* S_k *to* S_{k+1}:

Add the $(k+1)$ st term, $\dfrac{1}{2^{k+1}}$, to both sides of this equation.

$$\dfrac{1}{2} + \dfrac{1}{2^2} + \dfrac{1}{2^3} + \cdots + \dfrac{1}{2^k} = 1 - \dfrac{1}{2^k} \Rightarrow \dfrac{1}{2} + \dfrac{1}{2^2} + \dfrac{1}{2^3} + \cdots + \dfrac{1}{2^k} + \dfrac{1}{2^{k+1}} = 1 - \dfrac{1}{2^k} + \dfrac{1}{2^{k+1}} \Rightarrow$$

$$\dfrac{1}{2} + \dfrac{1}{2^2} + \dfrac{1}{2^3} + \cdots + \dfrac{1}{2^k} + \dfrac{1}{2^{k+1}} = 1 - \dfrac{1 \cdot 2}{2^k \cdot 2} + \dfrac{1}{2^{k+1}} \Rightarrow \dfrac{1}{2} + \dfrac{1}{2^2} + \dfrac{1}{2^3} + \cdots + \dfrac{1}{2^k} + \dfrac{1}{2^{k+1}} = 1 - \dfrac{2}{2^{k+1}} + \dfrac{1}{2^{k+1}} \Rightarrow$$

$$\dfrac{1}{2} + \dfrac{1}{2^2} + \dfrac{1}{2^3} + \cdots + \dfrac{1}{2^k} + \dfrac{1}{2^{k+1}} = 1 - \dfrac{1}{2^{k+1}}$$

(e) *Write a conclusion based on Steps* (a)–(d).
Because S_n is true for $n = 1$ and S_n is true for $n = k+1$ when it is true for $n = k$, S_n is true for every positive integer n.

21. S_n is the statement: If $a > 1$, then $a^n > 1$.

(a) *Verify the statement for* $n = 1$.
S_1 is the statement: If $a > 1$, then $a^1 > 1$, which is obviously true because $a = a^1$.

(b) *Write the statement for* S_k: If $a > 1$, then $a^k > 1$.

(c) *Write the statement for* S_{k+1}: If $a > 1$, then $a^{k+1} > 1$.

(d) *Assume that* S_k *is true. Use algebra to change* S_k *to* S_{k+1}:
$a^k > 1 \Rightarrow a^k \cdot a > 1 \cdot a \Rightarrow a^k \cdot a^1 > a \Rightarrow a^{k+1} > a$
Because $a > 1$ we have $a^{k+1} > a > 1 \Rightarrow a^{k+1} > 1$
Note: $a > 1 > 0$, so we did not need to be concerned about changing the direction of the inequality.

(e) *Write a conclusion based on Steps* (a)–(d).
Because S_n is true for $n = 1$ and S_n is true for $n = k+1$ when it is true for $n = k$, S_n is true for every positive integer n.

23. S_n is the statement: If $0 < a < 1$, then $a^n < a^{n-1}$.

(a) *Verify the statement for* $n = 1$.
S_1 is the statement: If $0 < a < 1$, then $a^1 < a^{1-1}$.
$a^1 < a^{1-1} \Rightarrow a < a^0 \Rightarrow a < 1$, which is true.

(b) *Write the statement for* S_k: If $0 < a < 1$, then $a^k < a^{k-1}$.

(c) *Write the statement for* S_{k+1}: If $0 < a < 1$, then $a^{k+1} < a^{(k+1)-1}$.

(d) *Assume that* S_k *is true. Use algebra to change* S_k *to* S_{k+1}:
If $0 < a < 1$, then $a^k < a^{k-1}$.
$a^k < a^{k-1} \Rightarrow a^k \cdot a^1 < a^{k-1} a^1 \Rightarrow$
$a^{k+1} < a^{(k-1)+1} \Rightarrow a^{k+1} < a^{(k+1)-1}$
Note: $0 < a < 1$, we did not need to be concerned about changing the direction of the inequality.

(e) *Write a conclusion based on Steps (a)–(d).*

Because S_n is true for $n = 1$ and S_n is true for $n = k + 1$ when it is true for $n = k$, S_n is true for every positive integer n.

25. S_n is the statement $(a^m)^n = a^{mn}$. (Assume that a and m are constant.)

(a) *Verify the statement for $n = 1$.*
S_1 is the statement
$$\left(a^m\right)^1 = a^{m \cdot 1} \Rightarrow a^m = a^m, \text{ which is true.}$$

(b) *Write the statement for S_k:*
$$\left(a^m\right)^k = a^{mk}$$

(c) *Write the statement for S_{k+1}:*
$$\left(a^m\right)^{k+1} = a^{m(k+1)}$$

(d) *Assume that S_k is true. Use algebra to change S_k to S_{k+1}:*
$$\left(a^m\right)^k = a^{mk}$$
$$\left(a^m\right)^k \cdot \left(a^m\right)^1 = a^{mk} \cdot \left(a^m\right)^1$$
$$\left(a^m\right)^{k+1} = a^{mk} \cdot a^m$$
$$\left(a^m\right)^{k+1} = a^{mk+m}$$
$$\left(a^m\right)^{k+1} = a^{m(k+1)}$$

(e) *Write a conclusion based on Steps (a)–(d).*

Because S_n is true for $n = 1$ and S_n is true for $n = k + 1$ when it is true for $n = k$, S_n is true for every positive integer n.

27. Let S_n be the statement $2^n > 2n$, if $n \geq 3$.
Step 1: Show that the statement is true when $n = 3$: S_3 is the statement $2^3 > 2 \cdot 3 \Rightarrow 8 > 6$, which is true.
Step 2: Show that S_k implies S_{k+1}, where S_k is the statement $2^k > 2k$, and S_{k+1} is the statement $2^{k+1} > 2(k+1)$.
$$2^k > 2k \Rightarrow 2^k \cdot 2 > 2k \cdot 2 \Rightarrow$$
$$2^k \cdot 2^1 > 4k \Rightarrow 2^{k+1} > 2k + 2k$$
$$k \geq 3, \text{ so } 2k > 2.$$

$$2^{k+1} > 2k + 2k > 2k + 2 \Rightarrow 2^{k+1} > 2k + 2 \Rightarrow$$
$$2^{k+1} > 2(k+1)$$
Thus, we have shown that S_k implies S_{k+1}. Therefore, the statement S_n is true for every positive integer value of n greater than or equal to 3.

29. Let S_n be the statement $2^n > n^2$, for $n \geq 5$.
Step 1: Show that the statement is true when $n = 5$: S_5 is the statement $2^5 > 5^2 \Rightarrow 32 > 25$, which is true.
Step 2: Show that S_k implies S_{k+1}, where S_k is the statement $2^k > k^2$, and S_{k+1} is the statement $2^{k+1} > (k+1)^2$.
$$2^k > k^2 \Rightarrow 2^k \cdot 2 > k^2 \cdot 2 \Rightarrow$$
$$2^k \cdot 2^1 > 2k^2 \Rightarrow 2^{k+1} > k^2 + k^2$$
Because $k > 4$, we have $2k > 8$. Thus, we have $2k > 1$. Also, because $k > 4$, we have $k^2 > 4k$.
$$2^{k+1} > k^2 + k^2 > k^2 + 4k \Rightarrow 2^{k+1} > k^2 + 4k \Rightarrow$$
$$2^{k+1} > k^2 + 2k + 2k \Rightarrow$$
$$2^{k+1} > k^2 + 2k + 2k > k^2 + 2k + 1 \Rightarrow$$
$$2^{k+1} > k^2 + 2k + 1 \Rightarrow 2^{k+1} > (k+1)^2$$
Thus, we have shown that S_k implies S_{k+1}. Therefore, the statement S_n is true for every positive integer value of n greater than or equal to 5.

31. Let S_n be the statement: $n! > 2^n$, for all n such that $n \geq 4$.
Step 1: Show that the statement is true when $n = 4$: S_4 is the statement
$$4! > 2^4 \Rightarrow 4(3)(2)(1) > (2)(2)(2)(2) \Rightarrow 24 > 16,$$
which is true.
Step 2: Show that S_k implies S_{k+1}, where S_k is the statement $k! > 2^k$, and S_{k+1} is the statement $(k+1)! > 2^{k+1}$.
$$k! > 2^k \Rightarrow k!(k+1) > 2^k(k+1) \Rightarrow$$
$$(k+1)! > 2^k(k+1)$$
Because $k + 1 > 2$, we have
$$(k+1)! > 2^k(k+1) > 2^k \cdot 2 \Rightarrow$$
$$(k+1)! > 2^k \cdot 2 \Rightarrow (k+1)! > 2^{k+1}$$
We have shown that S_k implies S_{k+1}. Therefore, the statement S_n is true for every positive integer value of n greater than or equal to 4.

33. Let S_n be the number of handshakes of the n people is $\dfrac{n^2 - n}{2}$.

Because 2 is the smallest number of people who can shake hands, we need to prove this statement for every positive integer $n \geq 2$.

Step 1: Show that the statement is true for $n = 2$. S_2 is the statement that for two people, the number of handshakes is $\dfrac{2^2 - 2}{2} = \dfrac{4 - 2}{2} = \dfrac{2}{2} = 1$, which is true.

Step 2: Show that S_k implies S_{k+1}, where S_k is the statement: k people shake hands in $\dfrac{k^2 - k}{2}$ ways and S_{k+1} is the statement: $k + 1$ people shake hands in $\dfrac{(k+1)^2 - (k+1)}{2}$ ways.

Start with S_k and assume it is a true statement: For k people, there are $\dfrac{k^2 - k}{2}$ handshakes. If one more person joins the k people, this $(k + 1)$st person will shake hands with the previous k people one time each. Thus, there will be k additional handshakes. Thus, the number of handshakes for $k + 1$ people is as follows.

$$\frac{k^2 - k}{2} + k = \frac{k^2 - k}{2} + \frac{2k}{2} = \frac{k^2 + k}{2}$$
$$= \frac{(k^2 + 2k + 1) - (k + 1)}{2}$$
$$= \frac{(k+1)^2 - (k+1)}{2}$$

This shows that if S_k is true, S_{k+1} is true.

Both steps for a proof by the generalized principle of mathematical induction have been completed, so the given statement is true for every positive integer $n \geq 2$.

35. The number of sides of the nth figure is $3 \cdot 4^{k-1}$ (from Exercise 34). To find the perimeter of each figure, multiply the number of sides by the length of each side. In each figure, the lengths of the sides are $\frac{1}{3}$ the lengths of the sides in the preceding figure. To find the perimeter of the nth figure, P_n, we try to find a pattern.

$P_1 = 3(1) = 3$

$P_2 = 3 \cdot 4\left(\frac{1}{3}\right) = 4$

$P_3 = 3 \cdot 4^2\left(\frac{1}{9}\right) = \frac{16}{3}$, and so on.

This gives a geometric sequence with $a_1 = 3$ and $r = \frac{4}{3}$.

Thus, $P_n = a_1 r^{n-1} \Rightarrow P_n = 3\left(\frac{4}{3}\right)^{n-1}$.

The result may also be written as

$P_n = \frac{3^1 \cdot 4^{n-1}}{3^{n-1}} \Rightarrow P_n = \frac{4^{n-1}}{3^{-1} \cdot 3^{n-1}} \Rightarrow P_n = \frac{4^{n-1}}{3^{n-2}}$.

37. With 1 ring, 1 move is required. With 2 rings, 3 moves are required. Note that $2^2 - 1 = 3 = 2 + 1$. With 3 rings, 7 moves are required. Note that $2^3 - 1 = 7 = 2^2 + 2 + 1$. With n rings, the number of moves required are $2^{n-1} + 2^{n-2} + \cdots + 2^1 + 1$. This is a geometric series with $a_1 = 1$ and $r = 2$. The sum is

$$\frac{a_1(1 - r^n)}{1 - r} = \frac{1(1 - 2^n)}{1 - 2} = \frac{1 - 2^n}{-1} = 2^n - 1.$$

Chapter Quiz
(Sections 11.1–11.5)

1. $a_n = -4n + 2$

$n = 1: \ a_1 = -4(1) + 2 = -2$

$n = 2: \ a_2 = -4(2) + 2 = -6$

$n = 3: \ a_3 = -4(3) + 2 = -10$

$n = 4: \ a_4 = -4(4) + 2 = -14$

$n = 5: \ a_5 = -4(5) + 2 = -18$

The first five terms are $-2, -6, -10, -14,$ and -18. There is a common difference of -4, so the sequence is arithmetic.

3. $a_1 = 5, a_2 = 3, a_n = a_{n-1} + 3a_{n-2}$, for $n \geq 3$

$a_3 = a_2 + 3a_1 = 3 + 3(5) = 18$

$a_4 = a_3 + 3a_2 = 18 + 3(3) = 27$

$a_5 = a_4 + 3a_3 = 27 + 3(18) = 81$

The first five terms are $5, 3, 18, 27,$ and 81. There is neither a common difference nor a common ratio, so the sequence is neither arithmetic nor geometric.

5. (a) $a_1 = -20, d = 14, S_n = \frac{n}{2}\left[2a_1 + (n-1)d\right]$

$S_{10} = \frac{10}{2}\left[2(-20) + (10 - 1)(14)\right]$

$\qquad = 5\left[-40 + 9(14)\right] = 5(86) = 430$

(b) $a_1 = -20, r = -\frac{1}{2}, S_n = \dfrac{a_1\left(1 - r^n\right)}{1 - r}$

$$S_{10} = \frac{-20\left[1 - \left(-\frac{1}{2}\right)^{10}\right]}{1 - \left(-\frac{1}{2}\right)} = \frac{-20\left[1 - \frac{1}{1024}\right]}{\frac{3}{2}} = \frac{-20\left(\frac{1023}{1024}\right)}{\frac{3}{2}} = -\frac{1705}{128}$$

7. $(x - 3y)^5 = \left[x + (-3y)\right]^5$

$\quad = x^5 + \binom{5}{1}x^4(-3y) + \binom{5}{2}x^3(-3y)^2 + \binom{5}{3}x^2(-3y)^3 + \binom{5}{4}x(-3y)^4 + (-3y)^5$

$\quad = x^5 + \dfrac{5!}{1!4!}x^4(-3y) + \dfrac{5!}{2!3!}x^3(9)y^2 + \dfrac{5!}{3!2!}x^2(-27)y^3 + \dfrac{5!}{4!1!}x(81)y^4 + (-243)y^5$

$\quad = x^5 + 5(-3)x^4 y + \dfrac{5 \cdot 4}{2 \cdot 1}(9)x^3 y^2 + \dfrac{5 \cdot 4}{2 \cdot 1}(-27)x^2 y^3 + 5(81)xy^4 - 243y^5$

$\quad = x^5 - 15x^4 y + 90x^3 y^2 - 270x^2 y^3 + 405xy^4 - 243y^5$

9. **(a)** $9! = 9 \cdot 8 \cdot 7 \cdot 6 \cdot 5 \cdot 4 \cdot 3 \cdot 2 \cdot 1 = 362{,}880$

(b) $\binom{10}{4} = \dfrac{10!}{4!(10-4)!} = \dfrac{10!}{4!6!} = \dfrac{10 \cdot 9 \cdot 8 \cdot 7}{4 \cdot 3 \cdot 2 \cdot 1} = 210$

Section 11.6   Basics of Counting Theory

1. From the two choices *permutation* and *combination*, a computer password is an example of a <u>permutation</u> and a hand of cards is an example of a <u>combination</u>.

3. There are $\underline{3 \cdot 2 \cdot 1 = 6}$ ways to form a three-digit number consisting of the digits 4, 5, and 9.

5. If a fair die is rolled and a fair coin is tossed, there are $\underline{6 \cdot 2 = 12}$ possible outcomes.

7. There are $3 \cdot 4 \cdot 1 \cdot 2 = 24$ possible outfits.

9. There are $2 \cdot 20 \cdot 4 = 160$ ways an attendee can choose.

11. There are $7 \cdot 6 \cdot 5 \cdot 4 \cdot 3 \cdot 2 \cdot 1 = 5040$ ways to arrange the portraits.

13. There are $16 \cdot 15 \cdot 14 = 3360$ ways to select a first-place, second-place, and third-place winner.

15. $P(12, 2) = \dfrac{12!}{(12-2)!} = \dfrac{12!}{10!}$

$\quad = \dfrac{12 \cdot 11 \cdot 10 \cdot 9 \cdot 8 \cdot 7 \cdot 6 \cdot 5 \cdot 4 \cdot 3 \cdot 2 \cdot 1}{10 \cdot 9 \cdot 8 \cdot 7 \cdot 6 \cdot 5 \cdot 4 \cdot 3 \cdot 2 \cdot 1}$

$\quad = 12 \cdot 11 = 132$

17. $P(9, 2) = \dfrac{9!}{(9-2)!} = \dfrac{9!}{7!} = \dfrac{9 \cdot 8 \cdot 7!}{7!} = 9 \cdot 8 = 72$

19. $P(5, 1) = \dfrac{5!}{(5-1)!} = \dfrac{5!}{4!} = \dfrac{5 \cdot 4!}{4!} = 5$

21. $C(4, 2) = \binom{4}{2} = \dfrac{4!}{(4-2)!2!} = \dfrac{4!}{2!2!}$

$\quad = \dfrac{4 \cdot 3 \cdot 2!}{2!2!} = \dfrac{4 \cdot 3}{2 \cdot 1} = 6$

23. $C(6, 0) = \binom{6}{0} = \dfrac{6!}{(6-0)!0!} = \dfrac{6!}{6! \cdot 1} = 1$

25. $C(12, 4) = \dfrac{12!}{(12-4)!4!} = \dfrac{12!}{8!4!}$

$\quad = \dfrac{12 \cdot 11 \cdot 10 \cdot 9 \cdot 8!}{8!4!} = \dfrac{12 \cdot 11 \cdot 10 \cdot 9}{4!}$

$\quad = \dfrac{12 \cdot 11 \cdot 10 \cdot 9}{4 \cdot 3 \cdot 2 \cdot 1} = 495$

27. $_{20}P_5 = 1{,}860{,}480$

29. $_{15}P_8 = 259{,}459{,}200$

31. $_{20}C_5 = 15{,}504$

33. $_{15}C_8 = 6435$

35. **(a)** Because the order of digits in a telephone number does matter, this involves a permutation.

(b) Because the order of digits in a Social Security number does matter, this involves a permutation.

(c) Because the order of the cards in a poker hand does not matter, this involves a combination.

(d) Because the order of members on a committee of politicians does not matter, this involves a combination.

(e) Because the order of numbers of the "combination" on a combination lock does matter, this involves a permutation.

(f) Because the order of digits and/or letters on a license plate does matter, this involves a permutation.

(g) Because the order does not matter, the lottery choice of six numbers involves a combination.

37. Using the fundamental principle of counting we have $5 \cdot 4 \cdot 2 = 40$. There are 40 different homes available if a builder offers a choice of 5 basic plans, 4 roof styles, and 2 exterior finishes.

39. (a) There are two choices for the first letter, K and W. The second letter can be any of the 26 letters of the alphabet except for the one chosen for the first letter, so there are 25 choices. The third letter can be any of the remaining 24 letters of the alphabet, so there are 24 choices. The fourth letter can be any of the remaining 23 letters of the alphabet, so there are 23 choices. Therefore, by the fundamental principle of counting, the number of possible 4-letter radio-station call letters is $2 \cdot 25 \cdot 24 \cdot 23 = 27,600$.

(b) There are two choices for the first letter. Because repeats are allowed, there are 26 choices for each of the remaining 3 letters.Therefore, by the fundamental principle of counting, the number of possible 4-letter radio-station call letters is $2 \cdot 26 \cdot 26 = 35,152$.

(c) There are two choices for the first letter. There are 24 choices for the second letter because it cannot repeat the first letter and cannot be R. There are 23 choices for the third letter because it cannot repeat any of the first two letters and cannot be R. The last letter must be R, so there is only once choice. Therefore, by the fundamental principle of counting, the number of possible 4-letter radio-station call letters is $2 \cdot 24 \cdot 23 \cdot 1 = 1104$.

41. Each ordering of the blocks is considered a different arrangement, so we have the permutation, $P(7,7)$.

$$P(7,7) = \frac{7!}{(7-7)!} = \frac{7!}{0!}$$
$$= \frac{7 \cdot 6 \cdot 5 \cdot 4 \cdot 3 \cdot 2 \cdot 1}{1} = 5040$$

There are 5040 different arrangements of the 7 blocks. Note: This problem could also be solved by using the fundamental principle of counting. There are 7 choices for the first block, 6 for the second, 5 for the third, etc. The number of arrangements would be $7 \cdot 6 \cdot 5 \cdot 4 \cdot 3 \cdot 2 \cdot 1 = 5040$.

43. (a) The first three positions could each be any one of 26 letters, and the second three positions could each be any one of 10 digits. Using the fundamental principle of counting, we have $26 \cdot 26 \cdot 26 \cdot 10 \cdot 10 \cdot 10 = 17,576,000$. Thus, 17,576,000 license plates were possible.

(b) Using the fundamental principle of counting, we have $10 \cdot 10 \cdot 10 \cdot 26 \cdot 26 \cdot 26 = 17,576,000$. 17,576,000 additional license plates were made possible by the reversal.

(c) Using the fundamental principle of counting, we have $26 \cdot 10 \cdot 10 \cdot 10 \cdot 26 \cdot 26 \cdot 26 = 456,976,000$. 456,976,000 plates were provided by prefixing the previous pattern with an additional letter.

45. Because order does matter, we have the permutation,

$$P(9,9) = \frac{9!}{(9-9)!} = \frac{9!}{0!} = \frac{9 \cdot 8 \cdot 7 \cdot 6 \cdot 5 \cdot 4 \cdot 3 \cdot 2 \cdot 1}{1}$$
$$= 362,880$$

There are 362,880 ways to seat 9 people in 9 seats in a row.

47. He has 6 choices for the first course, 5 choices for the second, and 4 choices for the third. Thus, the student has $6 \cdot 5 \cdot 4 = 120$ possible schedules according to the fundamental principle of counting.

49. The number of ways in which the 3 officers can be chosen from the 15 members is given by the following permutation.

$$P(15,3) = \frac{15!}{(15-3)!} = \frac{15!}{12!} = \frac{15 \cdot 14 \cdot 13 \cdot 12!}{12!}$$
$$= 15 \cdot 14 \cdot 13 = 2730$$

51. (a) $P(5, 5) = \dfrac{5!}{(5-5)!} = \dfrac{5!}{0!} = \dfrac{5 \cdot 4 \cdot 3 \cdot 2 \cdot 1}{1}$
$= 120$

The letters can be arranged in 120 ways.

(b) $P(3, 3) = \dfrac{3!}{(3-3)!} = \dfrac{3!}{0!} = \dfrac{3 \cdot 2 \cdot 1}{1} = 6$

The first three letters can be arranged in 6 ways.

53. We want to choose 6 group members out of 40 and the order is not important. The number of possible groups is

$C(40, 6) = \dfrac{40!}{(40-6)!6!} = \dfrac{40!}{34!6!}$

$ = \dfrac{40 \cdot 39 \cdot 38 \cdot 37 \cdot 36 \cdot 35 \cdot 34!}{34!6!}$

$ = \dfrac{40 \cdot 39 \cdot 38 \cdot 37 \cdot 36 \cdot 35}{6 \cdot 5 \cdot 4 \cdot 3 \cdot 2 \cdot 1}$

$ = 3,838,380$

55. We want to choose 4 apples from a crate of 25 and order is not important. The number of ways the apples may be sampled is

$C(25, 4) = \dfrac{25!}{(25-4)!4!} = \dfrac{25!}{21!4!}$

$ = \dfrac{25 \cdot 24 \cdot 23 \cdot 22 \cdot 21!}{21!4!}$

$ = \dfrac{25 \cdot 24 \cdot 23 \cdot 22}{4 \cdot 3 \cdot 2 \cdot 1} = 12,650$

57. (a) We want to choose 4 extras from a choice of 6 extras and order is not important. The number of different hamburgers that can be made is

$C(6,4) = \dfrac{6!}{(6-4)!4!} = \dfrac{6!}{2!4!}$

$ = \dfrac{6 \cdot 5 \cdot 4!}{2!4!} = \dfrac{6 \cdot 5}{2 \cdot 1} = 15$

(b) Because one of the four choices is fixed, we now want to choose 3 extras from a choice of 5 extras and order is not important. The number of different hamburgers that can be made is given by

$C(5,3) = \dfrac{5!}{(5-3)!3!} = \dfrac{5!}{2!3!}$

$ = \dfrac{5 \cdot 4 \cdot 3!}{2!3!} = \dfrac{5 \cdot 4}{2 \cdot 1} = 10$

59. The number of 2-marble samples that can be chosen from a total of 15 marbles is $C(15,2)$.

$C(15,2) = \dfrac{15!}{(15-2)!2!} = \dfrac{15!}{13!2!}$

$ = \dfrac{15 \cdot 14 \cdot 13!}{13!2!} = \dfrac{15 \cdot 14}{2 \cdot 1} = 105$

The number of 4-marble samples that can be chosen from a total of 15 marbles is $C(15,4)$.

$C(15,4) = \dfrac{15!}{(15-4)!4!} = \dfrac{15!}{11!4!}$

$ = \dfrac{15 \cdot 14 \cdot 13 \cdot 12 \cdot 11!}{11!4!}$

$ = \dfrac{15 \cdot 14 \cdot 13 \cdot 12}{4 \cdot 3 \cdot 2 \cdot 1} = 1365$

61. There are 5 liberal and 4 conservatives, giving a total of 9 members. Three members are chosen as delegates to a convention.

(a) There are $C(9,3)$ ways of doing this.

$C(9,3) = \dfrac{9!}{(9-3)!3!} = \dfrac{9!}{6!3!}$

$ = \dfrac{9 \cdot 8 \cdot 7 \cdot 6!}{6!3!} = \dfrac{9 \cdot 8 \cdot 7}{3 \cdot 2 \cdot 1} = 84$

84 delegations are possible.

(b) To get all liberals, we must choose 3 members from a set of 5, which can be done $C(5,3)$ ways.

$C(5,3) = \dfrac{5!}{(5-3)!3!} = \dfrac{5!}{2!3!}$

$ = \dfrac{5 \cdot 4 \cdot 3!}{2!3!} = \dfrac{5 \cdot 4}{2} = 10$

10 delegations could have all liberals.

(c) To get 2 liberals and 1 conservative involves two independent events. First select the liberals. The number of ways to do this is

$C(5,2) = \dfrac{5!}{3!2!} = \dfrac{5 \cdot 4 \cdot 3!}{3!2!} = \dfrac{5 \cdot 4}{2 \cdot 1} = 10$

Now select the conservative. The number of ways to do this is as follows.

$C(4,1) = \dfrac{4!}{(4-1)!1!} = \dfrac{4!}{3!1!} = \dfrac{4 \cdot 3!}{3!1!} = \dfrac{4}{1} = 4$

To find the number of delegations, use the fundamental principle of counting. The number of delegations with 2 liberals and 1 conservative is $10 \cdot 4 = 40$.

(d) If one particular person must be on the delegation, then there are 2 people left to choose from a set consisting of 8 members. The number of ways to do this is as follows.

$$C(8,2) = \frac{8!}{(8-2)!2!} = \frac{8!}{6!2!} = \frac{8 \cdot 7 \cdot 6!}{6!2!} = \frac{8 \cdot 7}{2 \cdot 1} = 28$$

28 delegations are possible, which includes the mayor.

63. The problem asks how many ways can Dwight arrange his schedule. Therefore, order is important, and this is a permutation problem. There are $P(8,4) = \frac{8!}{(8-4)!} = \frac{8!}{4!} = \frac{8 \cdot 7 \cdot 6 \cdot 5 \cdot 4!}{4!} = 8 \cdot 7 \cdot 6 \cdot 5 = 1680$

ways to arrange his schedule.

65. The order of the vegetables in the soup is not important, so this is a combination problem. There are

$$C(6,4) = \frac{6!}{(6-4)!4!} = \frac{6!}{2!4!} = \frac{6 \cdot 5 \cdot 4!}{2!4!} = \frac{6 \cdot 5}{2 \cdot 1} = 15 \text{ different soups she can make.}$$

67. Order is important in the seating, so this is a permutation problem. All thirteen children will have a specific location; the first twelve will sit down and thirteenth will be left standing.

$$P(13,13) = \frac{13!}{(13-13)!} = \frac{13!}{0!} = \frac{13 \cdot 12 \cdot 11 \cdot 10 \cdot 9 \cdot 8 \cdot 7 \cdot 6 \cdot 5 \cdot 4 \cdot 3 \cdot 2 \cdot 1}{1} = 6,227,020,800$$

There are 6,227,020,800 seatings possible.

69. A club has 8 women and 11 men members. There are a total of $8 + 11 = 19$ members, and 5 of them are to be chosen. Order is not important, so this is a combination problem.

(a) Choosing all women implies of the 8 women, choose 5.

$$C(8,5) = \frac{8!}{(8-5)!5!} = \frac{8!}{3!5!} = \frac{8 \cdot 7 \cdot 6 \cdot 5!}{3!5!} = \frac{8 \cdot 7 \cdot 6}{3 \cdot 2 \cdot 1} = 56$$

56 committees having 5 women can be chosen.

(b) Choose all men implies of the 11 men, choose 5.

$$C(11,5) = \frac{11!}{(11-5)!5!} = \frac{11!}{6!5!} = \frac{11 \cdot 10 \cdot 9 \cdot 8 \cdot 7 \cdot 6!}{6!5!} = \frac{11 \cdot 10 \cdot 9 \cdot 8 \cdot 7}{5 \cdot 4 \cdot 3 \cdot 2 \cdot 1} = 462$$

462 committees having 5 men can be chosen.

(c) Choosing 3 women and 2 men involves two independent events, each of which involves combinations. First, select the women. There are 8 women in the club, so the number of ways to do this is as follows.

$$C(8,3) = \frac{8!}{(8-3)!3!} = \frac{8!}{5!3!} = \frac{8 \cdot 7 \cdot 6 \cdot 5!}{5!3!} = \frac{8 \cdot 7 \cdot 6}{3 \cdot 2 \cdot 1} = 56$$

Now, select the men. There are 11 men in the club, so the number of ways to do this is as follows.

$$C(11,2) = \frac{11!}{(11-2)!2!} = \frac{11!}{9!2!} = \frac{11 \cdot 10 \cdot 9!}{9!2!} = \frac{11 \cdot 10}{2 \cdot 1} = 55$$

To find the number of committees, use the fundamental principle of counting. The number of committees with 3 women and 2 men is $56 \cdot 55 = 3080$.

(d) Choose no more than 3 men.
This means choose 0 men (and 5 women) or choose 1 man (and 4 women) or choose 2 men (and 3 women) or choose 3 men (and 2 women). For each of these choices, we use the fundamental principle of counting. Thus, the number of possible committees with no more than 3 men is
$$C(11,0) \cdot C(8,5) + C(11,1) \cdot C(8,4) + C(11,2) \cdot C(8,3) + C(11,3) \cdot C(8,2).$$

(*continued on next page*)

(continued)

$$C(11,0) \cdot C(8,5) + C(11,1) \cdot C(8,4) + C(11,2) \cdot C(8,3) + C(11,3) \cdot C(8,2)$$

$$= \frac{11!}{(11-0)!0!} \cdot \frac{8!}{(8-5)!5!} + \frac{11!}{(11-1)!1!} \cdot \frac{8!}{(8-4)!4!} + \frac{11!}{(11-2)!2!} \cdot \frac{8!}{(8-3)!3!} + \frac{11!}{(11-3)!3!} \cdot \frac{8!}{(8-2)!2!}$$

$$= \frac{11!}{11!0!} \cdot \frac{8!}{3!5!} + \frac{11!}{10!1!} \cdot \frac{8!}{4!4!} + \frac{11!}{9!2!} \cdot \frac{8!}{5!3!} + \frac{11!}{8!3!} \cdot \frac{8!}{6!2!}$$

$$= 1 \cdot \frac{8 \cdot 7 \cdot 6 \cdot 5!}{3!5!} + \frac{11 \cdot 10!}{10!1!} \cdot \frac{8 \cdot 7 \cdot 6 \cdot 5 \cdot 4!}{4!4!} + \frac{11 \cdot 10 \cdot 9!}{9!2!} \cdot \frac{8 \cdot 7 \cdot 6 \cdot 5!}{5!3!} + \frac{11 \cdot 10 \cdot 9 \cdot 8!}{8!3!} \cdot \frac{8 \cdot 7 \cdot 6!}{6!2!}$$

$$= 1 \cdot \frac{8 \cdot 7 \cdot 6}{3 \cdot 2 \cdot 1} + 11 \cdot \frac{8 \cdot 7 \cdot 6 \cdot 5}{4 \cdot 3 \cdot 2 \cdot 1} + \frac{11 \cdot 10}{2 \cdot 1} \cdot \frac{8 \cdot 7 \cdot 6}{3 \cdot 2 \cdot 1} + \frac{11 \cdot 10 \cdot 9}{3 \cdot 2 \cdot 1} \cdot \frac{8 \cdot 7}{2 \cdot 1}$$

$$= 1 \cdot 56 + 11 \cdot 70 + 55 \cdot 56 + 165 \cdot 28 = 56 + 770 + 3080 + 4620 = 8526$$

8526 committees having no more than 3 men can be chosen.

71. For the first lock, there are 10 choices for the first, second, and third digits. Thus, the number of combinations on the first lock is $10 \cdot 10 \cdot 10 = 1000$. This is the same for the second lock, that is, it also has 1000 different combinations. Each of the locks is independent from the other, so we use the fundamental principle of counting. We have a total of $1000 \cdot 1000 = 1,000,000$ combinations possible.

73. Each of the 12 switches has two possible settings. Each switch is independent from the others, so there is a total of $2 \cdot 2 \cdot 2 \cdot 2 \cdot 2 \cdot 2 \cdot 2 \cdot 2 \cdot 2 \cdot 2 \cdot 2 \cdot 2 = 2^{12} = 4096$ codes possible.

75. Because the keys are arranged in a circle, there is no "first" key. The number of distinguishable arrangements is the number of ways to arrange the other three keys in relation to any one of the keys, which is

$$P(3,3) = \frac{3!}{(3-3)!} = \frac{3!}{0!} = \frac{3!}{1} = 3 \cdot 2 \cdot 1 = 6.$$

In Exercises 77–85, use the formulas

$$P(n, r) = \frac{n!}{(n-r)!} \text{ and } C(n, r) = \frac{n!}{(n-r)!r!}, \text{ where}$$

$r \le n$.

77. Show $P(n, n-1) = P(n, n)$.

$$P(n, n-1) = \frac{n!}{[n-(n-1)]!} = \frac{n!}{(n-n+1)!}$$

$$= \frac{n!}{1!} = \frac{n!}{1} = \frac{n!}{0!} = \frac{n!}{(n-n)!} = P(n, n)$$

79. Show $P(n, 0) = 1$.

$$P(n, 0) = \frac{n!}{(n-0)!} = \frac{n!}{n!} = 1$$

81. Show $C(n, n) = 1$.

$$C(n, n) = \frac{n!}{(n-n)!n!} = \frac{n!}{0!n!} = \frac{n!}{1 \cdot n!} = \frac{n!}{n!} = 1$$

83. Show $C(0, 0) = 1$.

$$C(0, 0) = \frac{0!}{(0-0)!0!} = \frac{1!}{1!0!} = \frac{1!}{1!1} = 1$$

85. Show $C(n, n-r) = C(n, r)$.

$$C(n, n-r) = \frac{n!}{[n-(n-r)]!(n-r)!}$$

$$= \frac{n!}{(n-n+r)!(n-r)!} = \frac{n!}{r!(n-r)!}$$

$$= \frac{n!}{(n-r)!r!} = C(n, r)$$

Section 11.7   Basics of Probability

1. When a fair coin is tossed, there are <u>2</u> possible outcomes, and the probability of each outcome is $\frac{1}{2}$.

3. When two different denominations of fair coins are tossed, there are $2^2 = 4$ possible outcomes, and the probability of each outcome is $\frac{1}{4}$.

5. When a fair coin is tossed 4 times the probability of obtaining heads on all tosses is is $\left(\frac{1}{2}\right)^4 = \frac{1}{16}$.

7. Two fair coins are tossed. For each coin, either heads, H, or tails, T, lands up. The sample space is
$$S = \{(H, H), (H, T), (T, H), (T, T)\}.$$

9. Each coin can be a head or a tail and there are 3 coins, so the sample space is
$$S = \{(H,H,H),(H,H,T),(H,T,H),\\
(T,H,H),(H,T,T),(T,H,T),\\
(T,T,H),(T,T,T)\}$$

11. The sample space is
$$S = \{(1, 1), (1, 2), (1, 3), (2, 1), (2, 2),\\
(2, 3), (3, 1), (3, 2), (3, 3)\}$$

13. The sample space is
$$S = \{(H, H), (H, T), (T, H), (T, T)\},$$
so $n(S) = 4$.

   (a) The event E_1 of both coins showing the same face is $E_1 = \{(H, H), (T, T)\}$, so $n(E_1) = 2$. The probability is
$$P(E_1) = \frac{n(E_1)}{n(S)} = \frac{2}{4} = \frac{1}{2}.$$

   (b) The event E_2 where at least one head appears is
$$E_2 = \{(H, H), (H, T), (T, H)\}, \text{ so}$$
$n(E_2) = 3$. The probability is
$$P(E_2) = \frac{n(E_2)}{n(S)} = \frac{3}{4}.$$

15. The sample space is $S = \{(H, H, H), (H, H, T), (H, T, H), (H, T, T), (T, H, H), (T, H, T), (T, T, H), (T, T, T)\}$ so $n(S) = 8$.

   (a) The event E_1 of all three coins showing the same face is
$$E_1 = \{(H, H, H), (T, T, T)\}, \text{ so}$$
$n(E_1) = 2$. The probability is
$$P(E_1) = \frac{n(E_1)}{n(S)} = \frac{2}{8} = \frac{1}{4}.$$

   (b) The event E_2 where at least two coins are tails is $E_2 = \{(H, T, T), (T, H, T), (T, T, H), (T, T, T)\}$, so $n(E_2) = 4$. The probability is
$$P(E_2) = \frac{n(E_2)}{n(S)} = \frac{4}{8} = \frac{1}{2}.$$

17. The sample space is $S = \{(1, 1), (1, 2), (1, 3), (2, 1), (2, 2), (2, 3), (3, 1), (3, 2), (3, 3)\}$, so $n(S) = 9$.

   (a) The result is a repeated number" is the event $E_1 = \{(1, 1), (2, 2), (3, 3)\}$, so $n(E_1) = 3$. The probability is
$$P(E_1) = \frac{n(E_1)}{n(S)} = \frac{3}{9} = \frac{1}{3}.$$

   (b) "The second number is 1 or 3" is the event
$$E_2 = \{(1, 1),(1, 3),(2, 1),(2, 3),(3, 1),(3, 3)\},$$
so $n(E_2) = 6$. The probability is
$$P(E_2) = \frac{n(E_2)}{n(S)} = \frac{6}{9} = \frac{2}{3}.$$

   (c) "The first number is even and the second number is odd" is the event $E_3 = \{(2, 1), (2, 3)\}$. So, $n(E_3) = 2$. The probability is $P(E_3) = \frac{n(E_3)}{n(S)} = \frac{2}{9}.$

19. $P(E) = 0.857$, so $P(E') = 1 - 0.857 = 0.143$.

21. A batting average of .300 means for every 10 times at bat (the sample space), the batter g0t 3 hits. If the event E is "getting a hit," then $P(E) = 0.300$. Thus, we have $P(E') = 1 - 0.300 = 0.700$. The odds in favor of him getting a hit are $\dfrac{P(E)}{P(E')} = \dfrac{0.300}{0.700} = \dfrac{3}{7}$ or 3 to 7.

23. There are 52 cards, so $n(S) = 52$. The deck consists of thirteen spades, thirteen clubs, thirteen hearts, and thirteen diamonds. There are three face cards in each suit, or twelve in total.

   (a) E_1: The card is a spade.
   $n(E_1) = 13$. The probability is
$$P(E_1) = \tfrac{13}{52} = \tfrac{1}{4}.$$

   (b) E_2: The card is not a spade.
   The probability that a card is a spade is $\tfrac{1}{4}$, so $P(E_2) = 1 - \tfrac{1}{4} = \tfrac{3}{4}$.

   (c) E_3: The card is a spade or a heart.
   There are 13 spades and 13 hearts so $n(E_3) = 13 + 13 = 26$. The probability is
$$P(E_3) = \tfrac{26}{52} = \tfrac{1}{2}.$$

(d) E_4 : The card is a spade or a face card.

There are three face cards that are spades, so

$$P\left(E_4\right) = P\left(\text{spade}\right) + P\left(\text{face card}\right) - P\left(\text{spade and face card}\right) = \frac{13}{52} + \frac{12}{52} - \frac{3}{52} = \frac{22}{52} = \frac{11}{26}$$

(e) Using the results from (a) and (b), the odds in favor of drawing a spade are

$$\frac{P(E_1)}{P(E_2)} = \frac{\frac{1}{4}}{\frac{3}{4}} = \frac{1}{3} \text{ or 1 to 3}$$

25. The total of all foreign-born U.S. residents in 2012 is $11{,}587 + 11{,}596 + 21{,}034 + 2809 = 47{,}026$ thousand.

(a) $P\left(\text{born in Asia}\right) = \dfrac{11{,}587}{47{,}026} \approx 0.246$

(b) $P\left(\text{not born in Europe}\right) = 1 - P\left(\text{born in Europe}\right) = 1 - \dfrac{11{,}596}{47{,}026} = \dfrac{35{,}430}{47{,}026} \approx 0.753$

(c) $P\left(\text{born in Asia or Europe}\right) = \dfrac{11{,}587 + 11{,}596}{47{,}026} = \dfrac{23{,}183}{47{,}026} \approx 0.493$

(d) Let E be "selected resident was born in Latin America."

We have $P(E) = \dfrac{21{,}034}{47{,}026}$ and $P(E') = 1 - P(E) = 1 - \dfrac{21{,}034}{47{,}026} = \dfrac{25{,}992}{47{,}026}$.

The odds that a randomly selected foreign-born U.S. resident was born in Latin America are

$$\frac{P(E)}{P(E')} = \frac{\frac{21{,}034}{47{,}026}}{\frac{25{,}992}{47{,}026}} = \frac{21{,}034}{25{,}992} = \frac{10{,}517}{12{,}996} \text{ or 10,517 to 12,996.}$$

27. Each suit has thirteen cards, and the probability of choosing the correct card in that suit is $\frac{1}{13}$. Thus, we have

$$P(4 \text{ correct choices}) = \frac{1}{13} \cdot \frac{1}{13} \cdot \frac{1}{13} \cdot \frac{1}{13} = \frac{1}{28{,}561}. \text{ The probability of getting all four picks correct and}$$

winning $\$5000 = \dfrac{1}{13} \cdot \dfrac{1}{13} \cdot \dfrac{1}{13} \cdot \dfrac{1}{13} = \dfrac{1}{28{,}561} \approx 0.000035.$

29. (a) A 40-year-old man who lives 30 more years will be a 70-year old man. Let E be "selected man will live to be 70." Then $n(E) = 73{,}355$. For this situation, the sample space S is the set of all 40-year-old men, so $n(S) = 95{,}889$. Thus, the probability that a 40-year-old man will live 30 more years is

$$P(E) = \frac{n(E)}{n(S)} = \frac{73{,}355}{95{,}889} \approx 0.765.$$

(b) Using the notation and results from part (a), the probability that a 40-year-old man will not live 30 more years is $P(E') = 1 - P(E) = 1 - 0.765 = 0.235.$

(c) Use the notation and results from parts (a) and (b). In this binomial experiment, we call "a 40-year-old man survives to age 70" a success. Then $p = P\left(E\right) = 0.765$ and $1 - p = P\left(E'\right) = 0.235$.

There are 5 independent trials and we need the probability of 3 successes, so $n = 5$, $r = 3$. The probability that exactly 3 of the 40-year-old men survive to age 70 is

$$\binom{n}{r} p^r \left(1 - p\right)^{n-r} = \binom{5}{3}(0.765)^3 (0.235)^{5-3} = \frac{5!}{(5-3)!3!}(0.765)^3 (0.235)^2$$

$$= \frac{5!}{2!3!}(0.765)^3 (0.235)^2 = \frac{5 \cdot 4}{2 \cdot 1}(0.765)^3 (0.235)^2 = 10(0.765)^3 (0.235)^2 \approx 0.247$$

(d) Let F be the event "at least one man survives to age 70." The easiest way to find $P(F)$ is to first find the probability of the complement of the event, F'. F' is the event that "neither man survives to age 70."

$$P(F') = P(E') \cdot P(E') = (0.235)^2 \approx 0.055, \text{ so } P(F) = 1 - P(F') \approx 1 - 0.055 = 0.945$$

31. The amount of growth would be $11{,}400 - 10{,}000 = 1400$, so the percent growth would be $\dfrac{1400}{10{,}000} = 0.14$.

Let E be the event "worth at least \$11,400 by the end of the year." This is equivalent to "at least 14 percent growth" which is equivalent to "14 or 18 percent growth." Thus, we have
$$P(E) = P(14\% \text{ growth or } 18\% \text{ growth}) = P(14\% \text{ growth}) + P(18\% \text{ growth}) = 0.20 + 0.10 = 0.30 = 0.3.$$

33. In this binomial experiment, the outcome of interest is "has 2–7 representatives." Then $n = 10$, $r = 4$, $p = 0.44$ and $q = 1 - 0.44 = 0.56$.

$$P(\text{have 2-7 representatives}) = \binom{n}{r} p^r q^{n-r} = \binom{10}{4}(0.44)^4(0.56)^{10-4} = \frac{10!}{(10-4)!4!}(0.44)^4(0.56)^6$$
$$= \frac{10 \cdot 9 \cdot 8 \cdot 7}{4 \cdot 3 \cdot 2 \cdot 1}(0.44)^4(0.56)^6 = 210(0.44)^4(0.56)^6 \approx 0.243$$

35. In this binomial experiment, the outcome of interest is "has 8 or more representatives"
$$P(8 \text{ or more reps}) = P(8\text{-}15 \text{ reps}) + P(\text{more then } 15 \text{ reps})$$
$$= 0.28 + 0.14 = 0.42$$
Thus, $p = 0.42$ and $q = 1 - p = 0.58$. "Fewer than 2" means 0 or 1, so $P(\text{fewer than 2})$ is

$$P(0) + P(1) = \binom{10}{0}(0.42)^0(0.58)^{10} + \binom{10}{1}(0.42)^1(0.88)^9 = 1(1)(0.58)^{10} + 10(0.42)(0.58)^9 \approx 0.036$$

37. $P(\text{a student applied to fewer than 4 colleges})$
$$= P(\text{a student applied to 1 college}) + P(\text{a student applied to 2 or 3 colleges})$$
$$= 0.10 + 0.28 = 0.28$$

39. $P(\text{a student applied to more than 3 colleges})$
$$= P(\text{a student applied to 4-6 colleges}) + P(\text{a student applied to 7 or more colleges})$$
$$= 0.37 + 0.35 = 0.72$$

41. In these binomial experiments, we call "the man is color-blind" a success.
(a) For "exactly 5 are color-blind", we have $n = 53$, $r = 5$, $p = 0.042$, and $q = 1 - p = 0.958$.

$$P(\text{exactly 5 are color-blind}) = \binom{53}{5}(0.042)^5(0.958)^{53-5} = \frac{53!}{(53-5)!5!}(0.042)^5(0.958)^{48}$$
$$= \frac{53!}{48!5!}(0.042)^5(0.958)^{48} = \frac{53 \cdot 52 \cdot 51 \cdot 50 \cdot 49}{5 \cdot 4 \cdot 3 \cdot 2 \cdot 1}(0.042)^5(0.958)^{48}$$
$$= 2{,}869{,}685(0.042)^5(0.958)^{48} \approx 0.047822 \approx 0.048$$

(b) "No more than 5 are color-blind" corresponds to "0, 1, 2, 3, 4, or 5 are color-blind." Then $n = 53$, $r = 0$, 1, 2, 3, 4, or 5, $p = 0.042$, and $q = 1 - p = 0.958$.

P(no more than 5 are color-blind) $= P$(0 are color-blind) $+ P$(1 is color-blind) $+ P$(2 are color-blind)
$+ P$(3 are color-blind) $+ P$(4 are color-blind) $+ P$(5 are color-blind)

$$P = \binom{53}{0}(0.042)^0 (0.958)^{55-0} + \binom{53}{1}(0.042)^1 (0.958)^{53-1} + \binom{53}{2}(0.042)^2 (0.958)^{53-2}$$
$$+ \binom{53}{3}(0.042)^3 (0.958)^{53-3} + \binom{53}{4}(0.042)^4 (0.958)^{53-4}$$
$$+ \binom{53}{5}(0.042)^5 (0.958)^{53-5}$$

$$= \frac{53!}{(53-0)!0!}(1)(0.958)^{53} + \frac{53!}{(53-1)!1!}(0.042)^1 (0.958)^{52} + \frac{53!}{(53-2)!2!}(0.042)^2 (0.958)^{51}$$
$$+ \frac{53!}{(53-3)!3!}(0.042)^3 (0.958)^{50} + \frac{53!}{(53-4)!4!}(0.042)^4 (0.958)^{49}$$
$$+ \frac{53!}{(53-5)!5!}(0.042)^5 (0.958)^{48}$$

$$= \frac{53!}{53!0!}(0.958)^{53} + \frac{53 \cdot 52!}{52!1!}(0.042)^1 (0.958)^{52} + \frac{53 \cdot 52 \cdot 51!}{51!2!}(0.042)^2 (0.958)^{51}$$
$$+ \frac{53 \cdot 52 \cdot 51 \cdot 50!}{50!3!}(0.042)^3 (0.958)^{50} + \frac{53 \cdot 52 \cdot 51 \cdot 50 \cdot 49!}{49!4!}(0.042)^4 (0.958)^{49}$$
$$+ \frac{53 \cdot 52 \cdot 51 \cdot 50 \cdot 49 \cdot 48!}{48!5!}(0.042)^5 (0.958)^{48}$$

$$= 1(0.958)^{53} + 53(0.042)^1 (0.958)^{52} + 1378(0.042)^2 (0.958)^{51} + 23,426(0.042)^3 (0.958)^{50}$$
$$+ 292,825(0.042)^4 (0.958)^{49} + 2,869,685(0.042)^5 (0.958)^{48}$$
$$\approx 0.102890 + 0.239074 + 0.272514 + 0.203105 + 0.111305 + 0.047822 = 0.976710 \approx 0.977$$

(c) Refer to part (b): P(none are color-blind) $= \binom{53}{0}(0.042)^0 (0.958)^{53-0} \approx 0.102890 \approx 0.103$

(d) Let E be "at least 1 is color-blind." Consider the complementary event E': "0 are color-blind." Thus, we have

$$P(E) = 1 - P(E') = 1 - \binom{53}{0}(0.042)^0 (0.958)^{53-0} = 1 - \frac{53!}{(53-0)!0!}(1)(0.958)^{53}$$

$$= 1 - \frac{53!}{53!0!}(1)(0.958)^{53} = 1 - (1)(0.958)^{53} = 1 - (0.958)^{53}$$

$$= 1 - 0.102890 = 0.897110 \approx 0.897$$

43. **(a)** First compute

$$q = (1-p)^J = (1 - 0.1)^2 = 0.9^2 = 0.81.$$
$S = 4$, $k = 3$, and $q = 0.81$, so we have

$$P = \binom{S}{k} q^k (1-q)^{S-k}$$
$$= \binom{4}{3}(0.81)^3 (1 - 0.81)^{4-3}$$
$$= \frac{4!}{(4-3)!3!}(0.81)^3 (0.19)$$
$$= 4(0.81^3)(0.19) \approx 0.404$$

There is about a 40.4% chance of exactly 3 people not becoming infected.

(b) Compute

$$q = (1-p)^J = (1 - 0.5)^2 = 0.5^2 = 0.25$$
$S = 4$, $k = 3$, and $q = 0.25$, so we have

$$P = \binom{S}{k} q^k (1-q)^{S-k}$$
$$= \binom{4}{3}(0.25)^3 (1 - 0.25)^{4-3}$$
$$= \frac{4!}{(4-3)!3!}(0.25)^3 (0.75)$$
$$= 4(0.25^3)(0.75) \approx 0.047$$

There is about 4.7% chance of this occurring when the disease is highly infectious.

(c) Compute $q = (1-p)^l = (1-0.5)^1 = 0.5$.

Then, with $S = 9$, $k = 0$, and $q = 0.5$, we have

$$P = \binom{S}{k} q^k (1-q)^{S-k}$$
$$= \binom{9}{0}(0.5)^0 (1-0.5)^{9-0}$$
$$= \frac{9!}{(9-0)!0!}(1)(0.5)^9$$
$$= 1(1)(0.5^9) \approx 0.002$$

There is about a 0.2% chance of everyone becoming infected. This means that in a large family or group of people, it is highly unlikely that everyone will become sick even though the disease is highly infectious.

Chapter 11 Review Exercises

1. $a_n = \dfrac{n}{n+1}$

$a_1 = \dfrac{1}{1+1} = \dfrac{1}{2}; a_2 = \dfrac{2}{2+1} = \dfrac{2}{3}; a_3 = \dfrac{3}{3+1} = \dfrac{3}{4};$

$a_4 = \dfrac{4}{4+1} = \dfrac{4}{5}; a_5 = \dfrac{5}{5+1} = \dfrac{5}{6}$

The first five terms are $\frac{1}{2}, \frac{2}{3}, \frac{3}{4}, \frac{4}{5}$, and $\frac{5}{6}$.

This sequence does not have a common difference or a common ratio, so the sequence is neither arithmetic nor geometric.

3. $a_n = 2(n+3)$

$a_1 = 2(1+3) = 2(4) = 8;$
$a_2 = 2(2+3) = 2(5) = 10;$
$a_3 = 2(3+3) = 2(6) = 12;$
$a_4 = 2(4+3) = 2(7) = 14;$
$a_5 = 2(5+3) = 2(8) = 16$

The first five terms are 8, 10, 12, 14, and 16. There is a common difference, $d = 2$, so the sequence is arithmetic.

5. $a_1 = 5$; for $n \geq 2$, $a_n = a_{n-1} - 3$

$a_2 = a_{2-1} - 3 = a_1 - 3 = 5 - 3 = 2;$
$a_3 = a_{3-1} - 3 = a_2 - 3 = 2 - 3 = -1;$
$a_4 = a_{4-1} - 3 = a_3 - 3 = -1 - 3 = -4;$
$a_5 = a_{5-1} - 3 = a_4 - 3 = -4 - 3 = -7$

The first five terms are 5, 2, −1, −4, and −7. There is a common difference, $d = -3$, so the sequence is arithmetic.

7. $a_1 = 4$; $S_5 = 25$

Using the formula $S_n = \dfrac{n}{2}\left[2a_1 + (n-1)d\right]$ with $n = 5$, we have

$25 = \frac{5}{2}\left[2(4) + (5-1)d\right] \Rightarrow 25 = \frac{5}{2}(8+4d) \Rightarrow$
$25 = 5(4+2d) \Rightarrow 5 = 4 + 2d \Rightarrow d = \frac{1}{2}$

$a_1 = 4; a_2 = 4 + \frac{1}{2} = 4.5; a_3 = 4 + 2\left(\frac{1}{2}\right) = 5;$
$a_4 = 4 + 3\left(\frac{1}{2}\right) = 5.5; a_5 = 4 + 4\left(\frac{1}{2}\right) = 6$

The five terms are 4, 4.5, 5, 5.5, and 6.

9. Arithmetic, $a_3 = \pi$, $a_4 = 1$

$d = a_4 - a_3 = 1 - \pi$
$a_3 = a_1 + 2d \Rightarrow \pi = a_1 + 2(1 - \pi) \Rightarrow$
$\pi = a_1 + 2 - 2\pi \Rightarrow 3\pi - 2 = a_1$
$a_2 = a_1 + d \Rightarrow a_2 = (3\pi - 2) + (1 - \pi) \Rightarrow$
$a_2 = 2\pi - 1$
$a_5 = a_1 + 4d \Rightarrow a_5 = (3\pi - 2) + 4(1 - \pi) \Rightarrow$
$a_5 = 3\pi - 2 + 4 - 4\pi \Rightarrow a_5 = -\pi + 2$

The first five terms are $3\pi - 2$, $2\pi - 1$, π, 1, $-\pi + 2$.

11. Geometric, $a_1 = -5$, $a_2 = -1$

$r = \dfrac{a_2}{a_1} = \dfrac{-1}{-5} = \dfrac{1}{5}$

$a_3 = a_1 r^2 = -5\left(\frac{1}{5}\right)^2 = -5 \cdot \frac{1}{25} = -\frac{1}{5}$
$a_4 = a_1 r^3 = -5\left(\frac{1}{5}\right)^3 = -5 \cdot \frac{1}{125} = -\frac{1}{25}$
$a_5 = a_1 r^4 = -5\left(\frac{1}{5}\right)^4 = -5 \cdot \frac{1}{625} = -\frac{1}{125}$

The first five terms are -5, -1, $-\frac{1}{5}$, $-\frac{1}{25}$, and $-\frac{1}{125}$.

13. Geometric, $a_1 = -8$ and $a_7 = -\dfrac{1}{8}$

$a_n = a_1 r^{n-1}$, so $a_7 = a_1 r^{7-1} \Rightarrow$
$-\frac{1}{8} = -8r^6 \Rightarrow r^6 = \frac{1}{64} \Rightarrow r = \pm\frac{1}{2}$.

There are two geometric sequences that satisfy the given conditions.

If $r = \frac{1}{2}$, $a_4 = (-8)\left(\frac{1}{2}\right)^3 = (-8)\left(\frac{1}{8}\right) = -1$.

Also, $a_n = -8\left(\frac{1}{2}\right)^{n-1}$ or $a_n = -2^3\left(\frac{1}{2}\right)^{n-1}$

$= -\frac{2^3}{2^{n-1}} = -\frac{1}{2^{(n-1)-3}} = -\frac{1}{2^{n-4}} = -\left(\frac{1}{2}\right)^{n-4}$.

If $r = -\frac{1}{2}$, $a_4 = (-8)\left(-\frac{1}{2}\right)^3 = (-8)\left(-\frac{1}{8}\right) = 1$.

(continued on next page)

(*continued*)

Also, $a_n = -8\left(-\frac{1}{2}\right)^{n-1}$ or $a_n = -2^3\left(-\frac{1}{2}\right)^{n-1}$

$= (-2)^3\left(-\frac{1}{2}\right)^{n-1} = \dfrac{(-2)^3}{(-2)^{n-1}} = \dfrac{1}{(-2)^{(n-1)-3}}$

$= \dfrac{1}{(-2)^{n-4}} = \left(-\frac{1}{2}\right)^{n-4}$.

15. $a_1 = 6x - 9,\ a_2 = 5x + 1$

$d = a_2 - a_1 = (5x+1) - (6x-9)$
$\quad = 5x + 1 - 6x + 9 = -x + 10$

$a_n = a_1 + (n-1)d$

$a_8 = a_1 + (8-1)d \Rightarrow a_8 = a_1 + 7d \Rightarrow$

$a_8 = (6x - 9) + 7(-x + 10)$

$\quad = 6x - 9 - 7x + 70 = -x + 61$

17. $a_2 = 6,\ d = 10$

To first find a_1, we have

$a_2 = a_1 + d \Rightarrow 6 = a_1 + 10 \Rightarrow a_1 = -4$

Now, to find S_{12}, we use

$S_n = \frac{n}{2}\left[2a_1 + (n-1)d\right]$.

$S_{12} = \frac{12}{2}\left[2(-4) + (12-1)(10)\right]$

$\quad = 6\left[-8 + (11)(10)\right] = 6(-8 + 110)$

$\quad = 6(102) = 612$

19. $a_3 = 4,\ r = \dfrac{1}{5}$

$a_4 = a_3 r = 4 \cdot \frac{1}{5} = \frac{4}{5} \Rightarrow a_5 = a_4 r = \frac{4}{5} \cdot \frac{1}{5} = \frac{4}{25}$

21. $a_1 = -1,\ r = 3$. $\ S_n = \dfrac{a_1\left(1-r^n\right)}{1-r}$

$S_4 = \dfrac{a_1\left(1-r^4\right)}{1-r} = \dfrac{-1\left(1-3^4\right)}{1-3} = \dfrac{-1(1-81)}{-2}$

$\quad = \frac{1}{2}(-80) = -40$

23. $\displaystyle\sum_{i=1}^{7}(-1)^{i-1}$; $\ S_n = \dfrac{a_1\left(1-r^n\right)}{1-r}$

This is a geometric series with

$a_1 = (-1)^{1-1} = (-1)^0 = 1$ and $r = -1$.

$S_7 = \dfrac{a_1\left(1-r^7\right)}{1-r} \Rightarrow$

$S_7 = \dfrac{1\left[1-(-1)^7\right]}{1-(-1)} = \dfrac{1\left[1-(-1)\right]}{2} = \dfrac{1 \cdot 2}{2} = 1$

25. $\displaystyle\sum_{i=1}^{4}\dfrac{i+1}{i} = \dfrac{2}{1} + \dfrac{3}{2} + \dfrac{4}{3} + \dfrac{5}{4} = \dfrac{24}{12} + \dfrac{18}{12} + \dfrac{16}{12} + \dfrac{15}{12}$

$\quad = \dfrac{73}{12}$

27. Because $\displaystyle\sum_{i=1}^{n} i = \dfrac{n(n+1)}{2}$,

$\displaystyle\sum_{j=1}^{2500} j = \dfrac{2500(2500+1)}{2}$

$\quad = 1250(2501) = 3,126,250$

29. $\displaystyle\sum_{i=1}^{\infty}\left(\dfrac{4}{7}\right)^i$

This is an infinite geometric series with

$a_1 = \left(\frac{4}{7}\right)^1 = \frac{4}{7}$ and $r = \frac{4}{7}$. Because $|r| = \left|\frac{4}{7}\right| < 1$,

we can use the formula $S_\infty = \dfrac{a_1}{1-r}$. Thus, we have

$\displaystyle\sum_{i=1}^{\infty}\left(\frac{4}{7}\right)^i = S_\infty = \dfrac{a_1}{1-r} = \dfrac{\frac{4}{7}}{1-\frac{4}{7}} = \dfrac{\frac{4}{7}}{\frac{3}{7}} = \dfrac{4}{7}\cdot\dfrac{7}{3} = \dfrac{4}{3}$.

31. $\displaystyle\sum_{i=1}^{\infty} 2\left(-\dfrac{2}{3}\right)^i$

This is an infinite geometric series with

$a_1 = 2\left(-\frac{2}{3}\right)^1 = -\frac{4}{3}$ and $r = -\frac{2}{3}$. Because

$|r| = \left|-\frac{2}{3}\right| < 1$, we can use the formula

$S_\infty = \dfrac{a_1}{1-r}$. Thus, we have

$\displaystyle\sum_{i=1}^{\infty} 2\left(-\frac{2}{3}\right)^i = S_\infty = \dfrac{a_1}{1-r} = \dfrac{-\frac{4}{3}}{1-\left(-\frac{2}{3}\right)} = \dfrac{-\frac{4}{3}}{\frac{5}{3}} = -\dfrac{4}{5}$

33. $24 + 8 + \dfrac{8}{3} + \dfrac{8}{9} + \cdots$

This is an infinite geometric series with

$a_1 = 24$ and $r = \frac{8}{24} = \frac{1}{3}$. Because $|r| = \left|\frac{1}{3}\right| < 1$,

we can use the formula $S_\infty = \dfrac{a_1}{1-r}$. Thus, we

have $S_\infty = \dfrac{a_1}{1-r} = \dfrac{24}{1-\frac{1}{3}} = \dfrac{24}{\frac{2}{3}} = 24 \cdot \dfrac{3}{2} = 36$.

35. $\frac{1}{12} + \frac{1}{6} + \frac{1}{3} + \frac{2}{3} + \cdots$

This is an infinite geometric series with $a_1 = \frac{1}{12}$ and $r = \frac{\frac{1}{6}}{\frac{1}{12}} = 2$. Because $|r| > 1$, the series diverges.

37. $x_1 = 0$, $x_2 = 1$, $x_3 = 2$, $x_4 = 3$, $x_5 = 4$, and $x_6 = 5$.

$$\sum_{i=1}^{4}\left(x_i^2 - 6\right) = \left(x_1^2 - 6\right) + \left(x_2^2 - 6\right) + \left(x_3^2 - 6\right) + \left(x_4^2 - 6\right) = \left(0^2 - 6\right) + \left(1^2 - 6\right) + \left(2^2 - 6\right) + \left(3^2 - 6\right)$$

$$= (0 - 6) + (1 - 6) + (4 - 6) + (9 - 6) = -6 + (-5) + (-2) + 3 = -10$$

In Exercises 39–41, other answers may be possible.

39. $4 - 1 - 6 - \cdots - 66 = 4 + (-1) + (-6) + \cdots + (-66)$

This series is the sum of an arithmetic sequence with $a_1 = 4$ and $d = -1 - 4 = -5$. Therefore, the nth term is

$a_n = a_1 + (n-1)d = 4 + (n-1)(-5) = 4 - 5n + 5 = -5n + 9$ or, equivalently, $a_i = -5i + 9$.

To find the number of terms in the series, we realize that the last term of the series is –66. Because

$a_i = -5i + 9$ we solve $-66 = -5i + 9$ for i: $-66 = -5i + 9 \Rightarrow -75 = -5i \Rightarrow i = 15$

This indicates that the series consists of 15 terms and we have $4 - 1 - 6 - \cdots - 66 = \sum_{i=1}^{15}(-5i + 9)$.

41. $4 + 12 + 36 + \cdots + 972$

This series is the sum of a geometric sequence with $a_1 = 4$ and $r = \frac{12}{4} = 3$.

Therefore, the nth term is $a_n = a_1 r^{n-1} \Rightarrow a_n = 4(3)^{n-1}$, or, equivalently, $a_i = 4(3)^{i-1}$. To find the number of

terms in the series, we realize that the last term of the series is 972. Because $a_i = 4(3)^{i-1}$ we solve

$972 = 4(3)^{i-1}$ for i. $972 = 4(3)^{i-1} \Rightarrow 243 = 3^{i-1} \Rightarrow 3^5 = 3^{i-1} \Rightarrow 5 = i - 1 \Rightarrow i = 6$

This indicates that the series consists of 6 terms and we have $4 + 12 + 36 + \cdots + 972 = \sum_{i=1}^{6} 4(3)^{i-1}$.

43. $(x + 2y)^4 = x^4 + \binom{4}{1}x^3(2y)^1 + \binom{4}{2}x^2(2y)^2 + \binom{4}{3}x^1(2y)^3 + (2y)^4$

$$= x^4 + \frac{4!}{1!(4-1)!}x^3(2y) + \frac{4!}{(4-2)!2!}x^2(4y^2) + \frac{4!}{3!(4-3)!}x(8y^3) + 16y^4$$

$$= x^4 + \frac{4!}{1!3!}x^3(2y) + \frac{4!}{2!2!}x^2(4y^2) + \frac{4!}{3!1!}x(8y^3) + 16y^4$$

$$= x^4 + 4x^3(2y) + \frac{4 \cdot 3}{2 \cdot 1}x^2(4y^2) + 4x(8y^3) + 16y^4$$

$$= x^4 + 8x^3y + 24x^2y^2 + 32xy^3 + 16y^4$$

45. $\left(3\sqrt{x}-\dfrac{1}{\sqrt{x}}\right)^5=\left[3x^{1/2}+\left(-x^{-1/2}\right)\right]^5$

$$=\left(3x^{1/2}\right)^5+\binom{5}{1}\left(3x^{1/2}\right)^4\left(-x^{-1/2}\right)+\binom{5}{2}\left(3x^{1/2}\right)^3\left(-x^{-1/2}\right)^2$$

$$+\binom{5}{3}\left(3x^{1/2}\right)^2\left(-x^{-1/2}\right)^3+\binom{5}{4}\left(3x^{1/2}\right)\left(-x^{-1/2}\right)^4+\left(-x^{-1/2}\right)^5$$

$$=243x^{5/2}+\dfrac{5!}{1!4!}\left(81x^2\right)\left(-x^{-1/2}\right)+\dfrac{5!}{2!3!}\left(27x^{3/2}\right)^3\left(x^{-1}\right)$$

$$-\dfrac{5!}{3!2!}\left(9x\right)x^{-3/2}+\dfrac{5!}{4!1!}\left(3x^{1/2}\right)\left(x^{-2}\right)+\left(-x^{-5/2}\right)$$

$$=243x^{5/2}-5\left(81x^{2+(-1/2)}\right)+\dfrac{5\cdot4}{2\cdot1}\left(27x^{3/2+(-1)}\right)-\dfrac{5\cdot4}{2\cdot1}\left(9x^{1+(-3/2)}\right)+5\left(3x^{1/2+(-2)}\right)-x^{-5/2}$$

$$=243x^{5/2}-405x^{3/2}+270x^{1/2}-90x^{-1/2}+15x^{-3/2}-x^{-5/2}$$

47. The sixth term of the expansion of $(4x-y)^8$ is $\binom{8}{5}(4x)^3(-y)^5$.

$$\binom{8}{5}(4x)^3(-y)^5=\dfrac{8!}{5!3!}\left(64x^3\right)\left(-y^5\right)=\dfrac{8\cdot7\cdot6}{3\cdot2\cdot1}\left(-64x^3y^5\right)=56\left(-64x^3y^5\right)=-3584x^3y^5$$

49. The first four terms of $(x+2)^{12}$ are as follows.

$$(x)^{12}+\binom{12}{1}(x)^{11}(2)+\binom{12}{2}(x)^{10}(2)^2+\binom{12}{3}(x)^9(2)^3=x^{12}+\dfrac{12!}{1!11!}\left(x^{11}\right)(2)+\dfrac{12!}{2!10!}\left(x^{10}\right)(4)+\dfrac{12!}{3!9!}\left(x^9\right)(8)$$

$$=x^{12}+12\left(2x^{11}\right)+\dfrac{12\cdot11}{2\cdot1}\left(4x^{10}\right)+220\left(8x^9\right)$$

$$=x^{12}+24x^{11}+264x^{10}+1760x^9$$

51. Let S_n be the statement $1+3+5+7+\cdots+(2n-1)=n^2$.

(a) *Verify the statement for $n=1$.*

$2(1)-1=1^2\Rightarrow2-1=1\Rightarrow1=1$, which is true.

(b) *Write the statement for $n=k$:* S_k is the statement $1+3+5+7+\cdots+(2k-1)=k^2$.

(c) *Write the statement for $n=k+1$:* S_{k+1} is the statement $1+3+5+7+\cdots+\left[2(k+1)-1\right]=(k+1)^2$.

(d) *Assume that S_k is true. Use algebra to change S_k to S_{k+1}:*

Start with S_k: $1+3+5+7+\cdots+(2k-1)=k^2$. Add the $(k+1)$ st term, $\left[2(k+1)-1\right]$, to both sides of this equation.

$$1+3+5+7+\ldots+(2k-1)=k^2$$
$$1+3+5+7+\cdots+(2k-1)+\left[2(k+1)-1\right]=k^2+\left[2(k+1)-1\right]$$
$$1+3+5+7+\cdots+(2k-1)+\left[2(k+1)-1\right]=k^2+(2k+2-1)$$
$$1+3+5+7+\cdots+(2k-1)+\left[2(k+1)-1\right]=k^2+(2k+1)$$
$$1+3+5+7+\cdots+(2k-1)+\left[2(k+1)-1\right]=k^2+2k+1$$
$$1+3+5+7+\cdots+(2k-1)+\left[2(k+1)-1\right]=(k+1)^2$$

(e) *Write a conclusion based on Steps (a)–(d).*

Because S_n is true for $n=1$ and S_n is true for $n=k+1$ when it is true for $n=k$, S_n is true for every positive integer n.

53. Let S_n be the statement $2 + 2^2 + 2^3 + \cdots + 2^n = 2\left(2^n - 1\right)$.

(a) *Verify the statement for* $n = 1$.

$2^1 = 2\left(2^1 - 1\right) \Rightarrow 2 = 2(2 - 1) \Rightarrow 2 = 2(1) \Rightarrow 2 = 2$, which is true.

(b) *Write the statement for* $n = k$: S_k is the statement. $2 + 2^2 + 2^3 + \cdots + 2^k = 2\left(2^k - 1\right)$

(c) *Write the statement for* $n = k + 1$: S_{k+1} is the statement $2 + 2^2 + 2^3 + \cdots + 2^{k+1} = 2\left(2^{k+1} - 1\right)$.

(d) Assume that S_k is true. Use algebra to change S_k to S_{k+1}:

Start with S_k: $2 + 2^2 + 2^3 + \cdots + 2^k = 2\left(2^k - 1\right)$ and add the $(k+1)$ st term, 2^{k+1}, to both sides of this equation.

$$2 + 2^2 + 2^3 + \cdots + 2^k = 2\left(2^k - 1\right)$$
$$2 + 2^2 + 2^3 + \cdots + 2^k + 2^{k+1} = 2\left(2^k - 1\right) + 2^{k+1}$$
$$2 + 2^2 + 2^3 + \cdots + 2^k + 2^{k+1} = 2^{k+1} - 2 + 2^{k+1}$$
$$2 + 2^2 + 2^3 + \cdots + 2^k + 2^{k+1} = 2 \cdot 2^{k+1} - 2 \cdot 1$$
$$2 + 2^2 + 2^3 + \cdots + 2^k + 2^{k+1} = 2\left(2^{k+1} - 1\right)$$

(e) *Write a conclusion based on Steps (a)–(d).*

Because S_n is true for $n = 1$ and S_n is true for $n = k + 1$ when it is true for $n = k$, S_n is true for every positive integer n.

55. $P(9, 2) = \dfrac{9!}{(9-2)!} = \dfrac{9!}{7!} = \dfrac{9 \cdot 8 \cdot 7!}{7!} = 72$

57. $C(8, 3) = \dfrac{8!}{(8-3)!3!} = \dfrac{8!}{5!3!} = \dfrac{8 \cdot 7 \cdot 6 \cdot 5!}{3 \cdot 2 \cdot 1 \cdot 5!} = 56$

59. $C(10, 5) = \dfrac{10!}{(10-5)!5!} = \dfrac{10!}{5!5!}$

$= \dfrac{10 \cdot 9 \cdot 8 \cdot 7 \cdot 6 \cdot 5!}{5!5!}$

$= \dfrac{10 \cdot 9 \cdot 8 \cdot 7 \cdot 6}{5 \cdot 4 \cdot 3 \cdot 2 \cdot 1} = 252$

61. A person who starts working at age 18 and works until age 66 will work for 48 years. We are seeking the total amount earned during those 48 years. The first year, the person will earn \$30,000, the next year, the person will earn \$30,000 + \$268; the third year, the person will earn \$30,000 + 2(\$268), and so on. We are seeking the sum of the arithmetic sequence defined by

$a_n = 30{,}000 + (n-1)268$.

$a_1 = 30{,}000$; $a_{48} = 30{,}000 + 47(268) = 42{,}596$

$S_{48} = \dfrac{48}{2}(30{,}000 + 42{,}596) = 24(72{,}596)$

$= 1{,}742{,}304$

The person will earn \$1,742,304.

63. The person with four years of college will earn \$2,832,698 − \$1,742,304 = \$1,090,394 more than a person with no college attendance. Answers will vary.

65. Three independent events are involved. There are 5 choices of style, 3 choices of fabric, and 6 choices of color. Therefore, using the fundamental principle of counting, there are $5 \cdot 3 \cdot 6 = 90$ different couches.

67. There are nine people on the student body council.

(a) In selecting a 3-member delegation, $C(9, 3)$ choices are possible.

$C(9, 3) = \dfrac{9!}{(9-3)!3!} = \dfrac{9!}{6!3!}$

$= \dfrac{9 \cdot 8 \cdot 7 \cdot 6!}{6!3!} = \dfrac{9 \cdot 8 \cdot 7}{3 \cdot 2 \cdot 1} = 84$

84 delegations are possible.

(b) If two seniors and one junior must attend, then the two seniors can be chosen in $C(6, 2)$ ways and the junior can be chosen in $C(3, 1)$ ways.

$$C(6, 2) = \frac{6!}{(6-2)!2!} = \frac{6!}{4!2!}$$
$$= \frac{6 \cdot 5 \cdot 4!}{4!2!} = \frac{6 \cdot 5}{2 \cdot 1} = 15$$
$$C(3, 1) = \frac{3!}{(3-1)!1!} = \frac{3!}{2!1!}$$
$$= \frac{3 \cdot 2!}{2!1!} = \frac{3}{1} = 3$$

Thus, there are $15 \cdot 3 = 45$ possible delegations.

69. There are 4 spots to be filled by 26 letters and 3 spots by 10 digits. If repeats are allowed, then $26 \cdot 10 \cdot 10 \cdot 10 \cdot 26 \cdot 26 \cdot 26 = 456{,}976{,}000$ different license plates can be formed. If no repeats are allowed, then
$26 \cdot 10 \cdot 9 \cdot 8 \cdot 25 \cdot 24 \cdot 23 = 258{,}336{,}000$
different license plates can be formed.

71. (a) Let E be the event "picking a green marble."

$n(E) = 4$ and $n(S) = 4 + 5 + 6 = 15$

(total number of marbles) Thus,

$$P(E) = \frac{n(E)}{n(S)} = \frac{4}{15}$$

(b) Let E be the event "picking a black marble."

$n(E) = 5$ and $n(S) = 15$ (from part (a)).

Thus, $P(E) = \dfrac{n(E)}{n(S)} = \dfrac{5}{15} = \dfrac{1}{3}$.

The probability that the marble is not black is given by

$$P(E') = 1 - P(E) = 1 - \frac{1}{3} = \frac{2}{3}.$$

(c) Let E be the event "picking a blue marble."

$n(E) = 0$ because there are no blue

marbles and $n(S) = 15$ (from part (a)).

Thus, we have $P(E) = \dfrac{n(E)}{n(S)} = \dfrac{0}{15} = 0.$

(d) Let E be the event "picking a marble that is not white." We have

$$P(E) = \frac{n(E)}{n(S)} = \frac{9}{15} \text{ and}$$

$$P(E') = \frac{n(E')}{n(S)} = \frac{6}{15}. \text{ Thus, the odds in}$$

favor of E are

$$\frac{P(E)}{P(E')} = \frac{\frac{9}{15}}{\frac{6}{15}} = \frac{9}{15} \cdot \frac{15}{6} = \frac{3}{2} \text{ or 3 to 2.}$$

73. (a) Let B be the event "selected student earned a degree in business." Thus, we

have $P(B) = \dfrac{n(B)}{n(S)} = \dfrac{103{,}253}{754{,}299} \approx 0.137$

(b) Let H be the event "selected student earned a degree in health professions and related clinical studies" and let V be the event "selected student earned a degree in visual and performing arts." We have

$$P(H \text{ or } V) = \frac{n(H \text{ or } V)}{n(S)}$$
$$= \frac{83{,}893 + 17{,}331}{754{,}299} \approx 0.134$$

(c) Let E be the event "selected student earned a degree in education." Then E' is the event "selected student earned a degree that was not in education."

$$P(E') = \frac{n(E')}{n(S)} = \frac{754{,}299 - 178{,}062}{754{,}299}$$
$$\approx 0.764$$

75. In this binomial experiment, we call rolling a five a success. Then $n = 12$, $r = 2$, and $p = \frac{1}{6}$.

$$P(2 \text{ fives}) = \binom{n}{r} p^r (1-p)^{n-r}$$
$$= \binom{12}{2}\left(\frac{1}{6}\right)^2\left(1 - \frac{1}{6}\right)^{12-2}$$
$$= \frac{12!}{(12-2)!2!}\left(\frac{1}{6^2}\right)\left(\frac{5}{6}\right)^{10}$$
$$= \frac{12!}{10!2!}\left(\frac{1}{6^2}\right)\left(\frac{5^{10}}{6^{10}}\right)$$
$$= \frac{12 \cdot 11 \cdot 10!}{10!2!}\left(\frac{5^{10}}{6^{12}}\right)$$
$$= \frac{12 \cdot 11}{2 \cdot 1}\left(\frac{5^{10}}{6^{12}}\right) = 66\left(\frac{5^{10}}{6^{12}}\right) \approx 0.296$$

Chapter 11   Test

1. $a_n = (-1)^n (n^2 + 2)$

$n = 1: a_1 = (-1)^1 (1^2 + 2) = (-1)(1 + 2) = -3$

$n = 2: a_2 = (-1)^2 (2^2 + 2) = (1)(4 + 2) = 6$

$n = 3: a_3 = (-1)^3 (3^2 + 2) = (-1)(9 + 2) = -11$

$n = 4: a_4 = (-1)^4 (4^2 + 2) = (1)(16 + 2) = 18$

$n = 5: a_5 = (-1)^5 (5^2 + 2) = (-1)(25 + 2) = -27$

The first five terms are –3, 6, –11, 18, and –27. This sequence does not have either a common difference or a common ratio, so the sequence is neither arithmetic nor geometric.

2. $a_n = -3 \cdot \left(\dfrac{1}{2}\right)^n$

$n = 1: a_1 = -3\left(\dfrac{1}{2}\right)^1 = -3\left(\dfrac{1}{2}\right) = -\dfrac{3}{2}$

$n = 2: a_2 = -3\left(\dfrac{1}{2}\right)^2 = -3\left(\dfrac{1}{4}\right) = -\dfrac{3}{4}$

$n = 3: a_3 = -3\left(\dfrac{1}{2}\right)^3 = -3\left(\dfrac{1}{8}\right) = -\dfrac{3}{8}$

$n = 4: a_4 = -3\left(\dfrac{1}{2}\right)^4 = -3\left(\dfrac{1}{16}\right) = -\dfrac{3}{16}$

$n = 5: a_5 = -3\left(\dfrac{1}{2}\right)^5 = -3\left(\dfrac{1}{32}\right) = -\dfrac{3}{32}$

The first five terms are
$-\dfrac{3}{2}, -\dfrac{3}{4}, -\dfrac{3}{8}, -\dfrac{3}{16},$ and $-\dfrac{3}{32}$. This sequence

has a common ratio, $r = \dfrac{1}{2}$, so the sequence is geometric.

3. $a_1 = 2,\ a_2 = 3,\ a_n = a_{n-1} + 2a_{n-2},$ for $n \ge 3$

$n = 3: a_3 = a_2 + 2a_1 = 3 + 2(2) = 3 + 4 = 7$

$n = 4: a_4 = a_3 + 2a_2 = 7 + 2(3) = 7 + 6 = 13$

$n = 5: a_5 = a_4 + 2a_3 = 13 + 2(7) = 13 + 14 = 27$

The first five terms are 2, 3, 7, 13, and 27. There is no common difference or common ratio, so the sequence is neither arithmetic nor geometric.

4. $a_1 = 1$ and $a_3 = 25$

$a_n = a_1 + (n-1)d \Rightarrow a_3 = a_1 + 2d \Rightarrow$
$25 = 1 + 2d \Rightarrow 2d = 24 \Rightarrow d = 12$
$a_5 = a_1 + (5-1)d \Rightarrow$
$a_5 = a_1 + 4d = 1 + 4(12) = 1 + 48 = 49$

5. $a_1 = 81$ and $r = -\dfrac{2}{3}$

$a_n = a_1 r^{n-1} \Rightarrow a_6 = 81\left(-\dfrac{2}{3}\right)^5 = 81\left(-\dfrac{32}{243}\right) = -\dfrac{32}{3}$

6. Arithmetic, with $a_1 = -43$ and $d = 12$

$S_n = \dfrac{n}{2}\big[2a_1 + (n-1)d\big] \Rightarrow$
$S_{10} = \dfrac{10}{2}\big[2a_1 + (10-1)d\big] = 5\big[2(-43) + 9(12)\big]$
$= 5(-86 + 108)$
$= 5(22) = 110$

7. Geometric, with $a_1 = 5$ and $r = -2$

$S_n = \dfrac{a_1\left(1 - r^n\right)}{1 - r} \Rightarrow$

$S_{10} = \dfrac{5\left[1 - (-2)^{10}\right]}{1 - (-2)} = \dfrac{5(1 - 1024)}{3} = \dfrac{5(-1023)}{3}$
$= 5(-341) = -1705$

8. $\displaystyle\sum_{i=1}^{30} (5i + 2)$

This sum represents the sum of the first 30 terms of the arithmetic sequence having
$a_1 = 5 \cdot 1 + 2 = 7$ and
$a_n = a_{30} = 5 \cdot 30 + 2 = 150 + 2 = 152.$ Thus,

$\displaystyle\sum_{i=1}^{30} (5i + 2) = S_{30} = \dfrac{n}{2}\left(a_1 + a_n\right) = \dfrac{30}{2}\left(a_1 + a_{30}\right)$
$= 15(7 + 152) = 15(159) = 2385$

We could have also

$\displaystyle\sum_{i=1}^{30}(5i + 2) = \sum_{i=1}^{30} 5i + \sum_{i=1}^{30} 2 = 5\sum_{i=1}^{30} i + \sum_{i=1}^{30} 2$

$\displaystyle\sum_{i=1}^{n} c = cn$ and $\displaystyle\sum_{i=1}^{n} i = \dfrac{n(n+1)}{2},$ so

$5\displaystyle\sum_{i=1}^{30} i + \sum_{i=1}^{30} 2 = 5 \cdot \dfrac{30(30+1)}{2} + 2(30)$
$= 5 \cdot \dfrac{30(31)}{2} + 60$
$= 5(15)(31) + 60 = 2325 + 60$
$= 2385$

9. $\displaystyle\sum_{i=1}^{5}(-3\cdot 2^{i})$

This sum represents the sum of the first five terms of the geometric sequence having $a_1 = -3\cdot 2^1 = -6$ and

$r = 2$. Thus, $\displaystyle\sum_{i=1}^{5}-3\cdot 2^{i} = S_5 = \dfrac{a_1\left(1-r^n\right)}{1-r} = \dfrac{-6\left(1-2^5\right)}{1-2} = \dfrac{-6(1-32)}{-1} = 6(-31) = -186$

10. $\displaystyle\sum_{i=1}^{\infty}(2^{i})\cdot 4$

This is the sum of an infinite geometric sequence with $r = 2$. Because $|r| > 1$, the sum does not exist.

11. $\displaystyle\sum_{i=1}^{\infty}54\left(\dfrac{2}{9}\right)^{i}$

This is the sum of the infinite geometric sequence with $a_1 = 54\left(\frac{2}{9}\right)^1 = 54\left(\frac{2}{9}\right) = 12$ and $r = \frac{2}{9}$. Because $\left|\frac{2}{9}\right| < 1$,

we can use the summation formula $S_\infty = \dfrac{a_1}{1-r}$.

$\displaystyle\sum_{i=1}^{\infty}54\left(\tfrac{2}{9}\right)^{i} = S_\infty = \dfrac{a_1}{1-r} = \dfrac{12}{1-\frac{2}{9}} = \dfrac{12}{\frac{7}{9}} = 12\cdot\dfrac{9}{7} = \dfrac{108}{7}$

12. $(x+y)^6 = x^6 + \dbinom{6}{1}x^5 y + \dbinom{6}{2}x^4 y^2 + \dbinom{6}{3}x^3 y^3 + \dbinom{6}{4}x^2 y^4 + \dbinom{6}{5}xy^5 + y^6$

$\qquad = x^6 + \dfrac{6!}{1!5!}x^5 y + \dfrac{6!}{2!4!}x^4 y^2 + \dfrac{6!}{3!3!}x^3 y^3 + \dfrac{6!}{4!2!}x^2 y^4 + \dfrac{6!}{5!1!}xy^5 + y^6$

$\qquad = x^6 + 6x^5 y + \dfrac{6\cdot 5}{2\cdot 1}x^4 y^2 + \dfrac{6\cdot 5\cdot 4}{3\cdot 2\cdot 1}x^3 y^3 + \dfrac{6\cdot 5}{2\cdot 1}x^2 y^4 + 6xy^5 + y^6$

$\qquad = x^6 + 6x^5 y + 15x^4 y^2 + 20x^3 y^3 + 15x^2 y^4 + 6xy^5 + y^6$

13. $(2x-3y)^4 = (2x)^4 + \dbinom{4}{1}(2x)^3(-3y) + \dbinom{4}{2}(2x)^2(-3y)^2 + \dbinom{4}{3}(2x)(-3y)^3 + (-3y)^4$

$\qquad = 16x^4 + \dfrac{4!}{1!3!}\left(8x^3\right)(-3y) + \dfrac{4!}{2!2!}\left(4x^2\right)\left(9y^2\right) + \dfrac{4!}{1!3!}(2x)\left(-27y^3\right) + 81y^4$

$\qquad = 16x^4 - 4\left(24x^3 y\right) + \dfrac{4\cdot 3}{2\cdot 1}\left(36x^2 y^2\right) - 4\left(54xy^3\right) + 81y^4$

$\qquad = 16x^4 - 96x^3 y + 216x^2 y^2 - 216xy^3 + 81y^4$

14. To find the third term in the expansion of $(w-2y)^6$, use the formula $\dbinom{n}{k-1}x^{n-(k-1)}y^{k-1}$ with $n = 6$ and

$k = 3$. Then $k - 1 = 2$ and $n - (k-1) = 4$. Thus, the third term is as follows.

$\dbinom{n}{k-1}x^{n-(k-1)}y^{k-1} = \dbinom{6}{2}w^{6-2}(-2y)^2 = \dfrac{6!}{2!4!}w^4\left(4y^2\right) = \dfrac{6\cdot 5\cdot 4!}{2!4!}\left(4w^4 y^2\right) = \dfrac{6\cdot 5}{2\cdot 1}\left(4w^4 y^2\right) = 60w^4 y^2$

15. $8! = 8\cdot 7\cdot 6\cdot 5\cdot 4\cdot 3\cdot 2\cdot 1 = 40,320$

16. $C(10,2) = \dfrac{10!}{(10-2)!2!} = \dfrac{10!}{8!2!} = \dfrac{10\cdot 9\cdot 8!}{8!2!} = \dfrac{10\cdot 9}{2\cdot 1} = 45$

17. $C(7,3) = \dfrac{7!}{(7-3)!3!} = \dfrac{7!}{4!3!} = \dfrac{7\cdot 6\cdot 5\cdot 4!}{4!3!} = \dfrac{7\cdot 6\cdot 5}{3\cdot 2\cdot 1} = 35$

18. $P(11, 3) = \dfrac{11!}{(11-3)!} = \dfrac{11!}{8!} = \dfrac{11 \cdot 10 \cdot 9 \cdot 8!}{8!} = 11 \cdot 10 \cdot 9 = 990$

19. Let S_n be the statement $1 + 7 + 13 + \cdots + (6n - 5) = n(3n - 2)$.

(a) *Verify the statement for $n = 1$.*

$6(1) - 5 = 1(3(1) - 2) \Rightarrow 6 - 5 = 1(1) \Rightarrow 1 = 1$, which is true.

(b) *Write the statement for $n = k$:* S_k is the statement. $1 + 7 + 13 + \cdots + (6k - 5) = k(3k - 2)$

(c) *Write the statement for $n = k + 1$:* S_{k+1} is the statement

$1 + 7 + 13 + \cdots + [6(k + 1) - 5] = (k + 1)[3(k + 1) - 2]$.

(d) *Assume that S_k is true. Use algebra to change S_k to S_{k+1}:*

Start with S_k: $1 + 7 + 13 + \cdots + (6k - 5) = k(3k - 2)$ and add the $(k+1)$ st term, $[6(k + 1) - 5]$, to both sides of this equation.

$1 + 7 + 13 + \cdots + (6k - 5) = k(3k - 2)$

$1 + 7 + 13 + \cdots + (6k - 5) + [6(k + 1) - 5] = k(3k - 2) + [6(k + 1) - 5]$

$1 + 7 + 13 + \cdots + (6k - 5) + [6(k + 1) - 5] = 3k^2 - 2k + (6k + 6 - 5)$

$1 + 7 + 13 + \cdots + (6k - 5) + [6(k + 1) - 5] = 3k^2 + 4k + 1$

$1 + 7 + 13 + \cdots + (6k - 5) + [6(k + 1) - 5] = (k + 1)(3k + 1)$

$1 + 7 + 13 + \cdots + (6k - 5) + [6(k + 1) - 5] = (k + 1)[(3k + 3) - 2]$

$1 + 7 + 13 + \cdots + (6k - 5) + [6(k + 1) - 5] = (k + 1)[3(k + 1) - 2]$

(e) *Write a conclusion based on Steps (a)–(d).*

Because S_n is true for $n = 1$ and S_n is true for $n = k + 1$ when it is true for $n = k$, S_n is true for every positive integer n.

20. Using the fundamental principle of counting, we have $4 \cdot 3 \cdot 2 = 24$.

There are 24 different kinds of shoes if there are 4 styles, 3 different colors, and 2 different shades.

21. We are choosing four people from a group of ten without regard to order, so there are $C(10, 4)$ ways to do this.

$C(10, 4) = \dfrac{10!}{(10-4)!4!} = \dfrac{10!}{6!4!} = \dfrac{10 \cdot 9 \cdot 8 \cdot 7 \cdot 6!}{6!4!}$

$= \dfrac{10 \cdot 9 \cdot 8 \cdot 7}{4 \cdot 3 \cdot 2 \cdot 1} = 210$

There are 210 ways that four officers can be selected to attend a seminar.

If two women and two men must be included, we have $C(4, 2)$ ways to choose the women and $C(6, 2)$ ways to choose the men.

$C(4, 2) = \dfrac{4!}{(4-2)!2!} = \dfrac{4!}{2!2!} = \dfrac{4 \cdot 3 \cdot 2!}{2!2!}$

$= \dfrac{4 \cdot 3}{2 \cdot 1} = 6$

$C(6, 2) = \dfrac{6!}{(6-2)!2!} = \dfrac{6!}{4!2!} = \dfrac{6 \cdot 5 \cdot 4!}{4!2!} = \dfrac{6 \cdot 5}{2 \cdot 1}$

$= 15$

Altogether, then, we have $6 \cdot 15 = 90$ ways to select two women and two men.

22. The student can select the courses in $15 \cdot 14 \cdot 13 \cdot 12 = 32,760$ ways.

23. (a) $P(\text{red three}) = \dfrac{n(\text{red three cards})}{n(\text{cards in deck})}$

$= \dfrac{2}{52} = \dfrac{1}{26}$

(b) Let F be the event "draw a face card." Each suit contains 3 face cards (jack, queen, and king), so the deck contains 12 face cards. Thus $n(F) = 12$ and

$P(F) = \dfrac{12}{52} = \dfrac{3}{13}$ The probability of drawing a card that is not a face card is

$P(F') = 1 - P(F) = 1 - \dfrac{3}{13} = \dfrac{10}{13}$.

(c) Let K be the event "king is drawn" and S be the event "spade is drawn." There are 4 kings, so $n(K) = 4$ and

$P(K) = \dfrac{4}{52} = \dfrac{1}{13}$ There are 13 spades, so

$n(S) = 13$ and $P(S) = \dfrac{13}{52} = \dfrac{1}{4}$. There is one card which is both a king and a spade, so $n(K \cap S) = 1$ and

$P(K \cap S) = \dfrac{1}{52}$. Thus we have

$\begin{aligned} P(K \text{ or } S) &= P(K \cup S) \\ &= P(K) + P(S) - P(K \cap S) \\ &= \frac{1}{13} + \frac{1}{4} - \frac{1}{52} = \frac{16}{52} = \frac{4}{13} \end{aligned}$

(d) Consider the event F: "draw a face card." As shown in the solution to part (b),

$P(F) = \dfrac{3}{13}$ and $P(F') = \dfrac{10}{13}$. Thus, the odds in favor of drawing a face card are

$\dfrac{P(F)}{P(F')} = \dfrac{\frac{3}{13}}{\frac{10}{13}} = \dfrac{3}{13} \cdot \dfrac{13}{10} = \dfrac{3}{10}$ or 3 to 10

24. "At most 2" means "0 or 1 or 2."

$\begin{aligned} P(\text{at most } 2) &= P(0 \text{ or } 1 \text{ or } 2) \\ &= P(0) + P(1) + P(2) \\ &= 0.19 + 0.43 + 0.30 = 0.92 \end{aligned}$

The probability that at most 2 light bulbs are defective is 0.92.

25. In this binomial experiment, we call rolling a five a success. Then $n = 6$, $r = 2$, and $p = \dfrac{1}{6}$.

$\begin{aligned} P(2 \text{ fives}) &= \binom{n}{r} p^r (1-p)^{n-r} \\ &= \binom{6}{2} \left(\frac{1}{6}\right)^2 \left(1 - \frac{1}{6}\right)^{6-2} \\ &= \frac{6!}{(6-2)!\,2!} \left(\frac{1}{6^2}\right) \left(\frac{5}{6}\right)^4 \\ &= \frac{6!}{4!\,2!} \left(\frac{1}{6^2}\right)\left(\frac{5^4}{6^4}\right) = \frac{6 \cdot 5 \cdot 4!}{4!\,2!}\left(\frac{5^4}{6^6}\right) \\ &= \frac{6 \cdot 5}{2 \cdot 1}\left(\frac{5^4}{6^6}\right) = 15\left(\frac{5^4}{6^6}\right) \approx 0.201 \end{aligned}$

Appendices

Appendix A   Polar Form of Conic Sections

Exercises 1–9 are graphed in degree mode.

1.

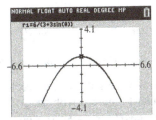

3.

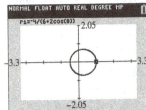

5.

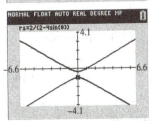

7.

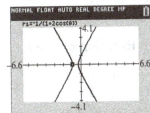

9.

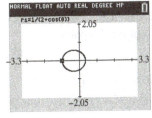

11. A parabola with vertical directrix 3 units to the right of the pole has the equation

$r = \dfrac{ep}{1 + e\cos\theta}$ where $e = 1$ and $p = 3$, so the

equation is $r = \dfrac{1 \cdot 3}{1 + 1\cos\theta} = \dfrac{3}{1 + \cos\theta}$.

13. A parabola with a horizontal directrix 5 units

below the pole has the equation $r = \dfrac{ep}{1 - e\sin\theta}$

where $e = 1$ and $p = 5$, so the equation is

$r = \dfrac{1 \cdot 5}{1 - 1\sin\theta} = \dfrac{5}{1 - \sin\theta}$.

15. A conic section with vertical directrix 5 units
to the right of the pole has a polar equation of

the form $r = \dfrac{ep}{1 + e\cos\theta}$ where $p = 5$. When

$e = \dfrac{4}{5}$, we get $r = \dfrac{\frac{4}{5} \cdot 5}{1 + \frac{4}{5}\cos\theta} \Rightarrow$

$r = \dfrac{4}{\frac{5 + 4\cos\theta}{5}} \Rightarrow r = \dfrac{20}{5 + 4\cos\theta}$. Because

$e < 1$, the graph of the equation is an ellipse.

17. A conic section with horizontal directrix 8
units below the pole has a polar equation of

the form $r = \dfrac{ep}{1 - e\sin\theta}$ where $p = 8$. When

$e = \dfrac{5}{4}$, we get $r = \dfrac{\frac{5}{4} \cdot 8}{1 - \frac{5}{4}\sin\theta} \Rightarrow$

$r = \dfrac{10}{\frac{4 - 5\sin\theta}{4}} \Rightarrow r = \dfrac{40}{4 - 5\sin\theta}$. Because $e > 1$,

the graph of the equation is a hyperbola.

19. To identify the type of conic, rewrite the
equation in standard form to find e.

$r = \dfrac{6}{3 - \cos\theta} \Rightarrow r = \dfrac{3 \cdot 2}{3\left(1 - \frac{1}{3}\cos\theta\right)} \Rightarrow$

$r = \dfrac{2}{1 - \frac{1}{3}\cos\theta}$. Thus, $e = \dfrac{1}{3}$. Because $e < 1$,

the graph of the equation is an ellipse. To
convert to rectangular form, start with the
given equation and use $r\cos\theta = x, r = \sin\theta$,
and $r^2 = x^2 + y^2$:

$$r = \dfrac{6}{3 - \cos\theta}$$
$$r(3 - \cos\theta) = 6$$
$$3r - r\cos\theta = 6$$
$$3r = r\cos\theta + 6$$
$$(3r)^2 = (r\cos\theta + 6)^2$$

(continued on next page)

(*continued*)

$$(3r)^2 = (x+6)^2$$
$$9r^2 = x^2 + 12x + 36$$
$$9(x^2 + y^2) = x^2 + 12x + 36$$
$$9x^2 + 9y^2 = x^2 + 12x + 36$$
$$8x^2 + 9y^2 - 12x - 36 = 0$$

21. To identify the type of conic, rewrite the equation in standard form to find e.

$r = \dfrac{-2}{1 + 2\cos\theta}$ is in standard form. Thus, $e = 2$. Because $e > 1$, the graph of the equation is a hyperbola. To convert to rectangular form, start with the given equation and use $r\cos\theta = x, r = \sin\theta$, and $r^2 = x^2 + y^2$:

$$r = \frac{-2}{1 + 2\cos\theta}$$
$$r(1 + 2\cos\theta) = -2$$
$$r + 2r\cos\theta = -2$$
$$r = -2r\cos\theta - 2$$
$$r^2 = (-2r\cos\theta - 2)^2$$
$$r^2 = (-2x - 2)^2$$
$$x^2 + y^2 = 4x^2 + 8x + 4$$
$$0 = 3x^2 - y^2 + 8x + 4$$

23. To identify the type of conic, rewrite the equation in standard form to find e.

$$r = \frac{-6}{4 + 2\sin\theta} \Rightarrow r = \frac{4 \cdot \frac{-6}{4}}{4\left(1 + \frac{1}{2}\sin\theta\right)} \Rightarrow$$

$r = \dfrac{\frac{-3}{2}}{1 + \frac{1}{2}\sin\theta}$. Thus, $e = \dfrac{1}{2}$. Because $e < 1$, the graph of the equation is an ellipse. To convert to rectangular form, start with the given equation and use $r\cos\theta = x, r = \sin\theta$, and $r^2 = x^2 + y^2$:

$$r = \frac{-6}{4 + 2\sin\theta}$$
$$r(4 + 2\sin\theta) = -6$$
$$4r + 2r\sin\theta = -6$$
$$4r = -2r\sin\theta - 6$$
$$(4r)^2 = (-2r\sin\theta - 6)^2$$
$$(4r)^2 = (-2y - 6)^2$$
$$16r^2 = 4y^2 + 24y + 36$$

$$16(x^2 + y^2) = 4y^2 + 24y + 36$$
$$16x^2 + 16y^2 = 4y^2 + 24y + 36$$
$$16x^2 + 12y^2 - 24y - 36 = 0$$
$$4x^2 + 3y^2 - 6y - 9 = 0$$

25. To identify the type of conic, rewrite the equation in standard form to find e.

$$r = \frac{10}{2 - 2\sin\theta} \Rightarrow r = \frac{2 \cdot 5}{2(1 - \sin\theta)} \Rightarrow$$

$r = \dfrac{5}{1 - \sin\theta}$. Thus, $e = 1$ and the graph of the equation is a parabola. To convert to rectangular form, start with the given equation and use $r\cos\theta = x, r = \sin\theta$, and $r^2 = x^2 + y^2$:

$$r = \frac{10}{2 - 2\sin\theta}$$
$$r(2 - 2\sin\theta) = 10$$
$$2r - 2r\sin\theta = 10$$
$$2r = 2r\sin\theta + 10$$
$$(2r)^2 = (2r\sin\theta + 10)^2$$
$$(2r)^2 = (2y + 10)^2$$
$$4r^2 = 4y^2 + 40y + 100$$
$$4(x^2 + y^2) = 4y^2 + 40y + 100$$
$$4x^2 + 4y^2 = 4y^2 + 40y + 100$$
$$4x^2 - 40y - 100 = 0$$
$$x^2 - 10y - 25 = 0$$

Appendix B   Rotation of Axes

1. $4x^2 + 3y^2 + 2xy - 5x = 8$

$A = 4$, $B = 2$, $C = 3$

$B^2 - 4AC = 2^2 - 4(4)(3) = 4 - 48 = -44$

Because $B^2 - 4AC < 0$, the graph will be a circle or ellipse or a point.

3. $2x^2 + 3xy - 4y^2 = 0$

$A = 2$, $B = 3$, $C = -4$

$B^2 - 4AC = 3^2 - 4(2)(-4) = 9 + 32 = 41$

Because $B^2 - 4AC > 0$, the graph will be a hyperbola or two intersecting lines.

5. $4x^2 + 4xy + y^2 + 15 = 0$

$A = 4$, $B = 4$, $C = 1$

$B^2 - 4AC = 4^2 - 4(4)(1) = 16 - 16 = 0$

Because $B^2 - 4AC = 0$, the graph will be a parabola or one line or two parallel lines.

7. The *xy*-term is removed from $2x^2 + \sqrt{3}xy + y^2 + x = 5$ by a rotation of the axes through an angle θ

satisfying $\cot 2\theta = \dfrac{A-C}{B}$, where $A = 2$, $B = \sqrt{3}$, and $C = 1$.

$\cot 2\theta = \dfrac{2-1}{\sqrt{3}} \Rightarrow \cot 2\theta = \dfrac{1}{\sqrt{3}} \Rightarrow \cot 2\theta = \dfrac{\sqrt{3}}{3} \Rightarrow 2\theta = \cot^{-1}\dfrac{\sqrt{3}}{3} \Rightarrow 2\theta = 60° \Rightarrow \theta = 30°$

9. The *xy*-term is removed from $3x^2 + \sqrt{3}xy + 4y^2 + 2x - 3y = 12$ by a rotation of the axes through an angle θ

satisfying $\cot 2\theta = \dfrac{A-C}{B}$, where $A = 3$, $B = \sqrt{3}$, , and $C = 4$.

$\cot 2\theta = \dfrac{3-4}{\sqrt{3}} \Rightarrow \cot 2\theta = \dfrac{-1}{\sqrt{3}} \Rightarrow \cot 2\theta = -\dfrac{\sqrt{3}}{3} \Rightarrow 2\theta = \cot^{-1}\left(-\dfrac{\sqrt{3}}{3}\right) \Rightarrow 2\theta = 120° \Rightarrow \theta = 60°$

11. The *xy*-term is removed from $x^2 - 4xy + 5y^2 = 18$ by a rotation of the axes through $\cot 2\theta = \dfrac{A-C}{B}$, where

$A = 1$, $B = -4$, and $C = 5$.

$\cot 2\theta = \dfrac{1-5}{-4} \Rightarrow \cot 2\theta = \dfrac{-4}{-4} \Rightarrow \cot 2\theta = 1 \Rightarrow 2\theta = \cot^{-1} 1 \Rightarrow 2\theta = 45° \Rightarrow \theta = 22.5°$

13. Because $\theta = 45°$, then $\sin\theta = \dfrac{\sqrt{2}}{2}$ and $\cos\theta = \dfrac{\sqrt{2}}{2}$, and the rotation equations are $x = \dfrac{\sqrt{2}}{2}x' - \dfrac{\sqrt{2}}{2}y'$ and

$y = \dfrac{\sqrt{2}}{2}x' + \dfrac{\sqrt{2}}{2}y'$. Substituting these values into the given equation yields

$$x^2 - xy + y^2 = 6$$

$$\left(\frac{\sqrt{2}}{2}x' - \frac{\sqrt{2}}{2}y'\right)^2 - \left(\frac{\sqrt{2}}{2}x' - \frac{\sqrt{2}}{2}y'\right)\left(\frac{\sqrt{2}}{2}x' + \frac{\sqrt{2}}{2}y'\right) + \left(\frac{\sqrt{2}}{2}x' + \frac{\sqrt{2}}{2}y'\right)^2 = 6$$

$$\left[\frac{1}{2}x'^2 - 2\left(\frac{1}{2}x'y'\right) + \frac{1}{2}y'^2\right] - \left(\frac{1}{2}x'^2 + \frac{1}{2}x'y' - \frac{1}{2}x'y' - \frac{1}{2}y'^2\right) + \left[\frac{1}{2}x'^2 + 2\left(\frac{1}{2}x'y'\right) + \frac{1}{2}y'^2\right] = 6$$

$$\left(\frac{1}{2}x'^2 - x'y' + \frac{1}{2}y'^2\right) - \left(\frac{1}{2}x'^2 - \frac{1}{2}y'^2\right) + \left(\frac{1}{2}x'^2 + x'y' + \frac{1}{2}y'^2\right) = 6$$

$$\frac{1}{2}x'^2 - x'y' + \frac{1}{2}y'^2 - \frac{1}{2}x'^2 + \frac{1}{2}y'^2 + \frac{1}{2}x'^2 + x'y' + \frac{1}{2}y'^2 = 6$$

$$\frac{1}{2}x'^2 + \frac{3}{2}y'^2 = 6$$

$$\frac{x'^2}{12} + \frac{y'^2}{4} = 1$$

The graph of the equation is an ellipse. This ellipse is centered at the origin and has x'-intercepts $\pm\sqrt{12} = \pm 2\sqrt{3} \approx \pm 3.5$ and y'-intercepts $\pm\sqrt{4} = \pm 2$.

$\frac{x'^2}{12} + \frac{y'^2}{4} = 1$

15. $x^2 - 4xy + y^2 = -5$

Because $A = 1$, $B = -4$, and $C = 1$, we have $\cot 2\theta = \dfrac{A-C}{B} = \dfrac{1-1}{-4} = 0$. Also, because $\cot 2\theta = \dfrac{\cos 2\theta}{\sin 2\theta} = 0$,

we know $\cos 2\theta = 0$. This yields $\sin \theta = \sqrt{\dfrac{1-\cos 2\theta}{2}} = \sqrt{\dfrac{1-0}{2}} = \sqrt{\dfrac{1}{2}} = \dfrac{\sqrt{2}}{2}$ and

$\cos \theta = \sqrt{\dfrac{1+\cos 2\theta}{2}} = \sqrt{\dfrac{1+0}{2}} = \sqrt{\dfrac{1}{2}} = \dfrac{\sqrt{2}}{2}$. Thus, $\theta = 45°$ and the rotation equations are $x = \dfrac{\sqrt{2}}{2}x' - \dfrac{\sqrt{2}}{2}y'$

and $y = \dfrac{\sqrt{2}}{2}x' + \dfrac{\sqrt{2}}{2}y'$. Substituting these values into the given equation yields

$$x^2 - 4xy + y^2 = -5$$

$$\left(\frac{\sqrt{2}}{2}x' - \frac{\sqrt{2}}{2}y'\right)^2 - 4\left(\frac{\sqrt{2}}{2}x' - \frac{\sqrt{2}}{2}y'\right)\left(\frac{\sqrt{2}}{2}x' + \frac{\sqrt{2}}{2}y'\right) + \left(\frac{\sqrt{2}}{2}x' + \frac{\sqrt{2}}{2}y'\right)^2 = -5$$

$$\left[\frac{1}{2}x'^2 - 2\left(\frac{1}{2}x'y'\right) + \frac{1}{2}y'^2\right] - 4\left(\frac{1}{2}x'^2 + \frac{1}{2}x'y' - \frac{1}{2}x'y' - \frac{1}{2}y'^2\right) + \left[\frac{1}{2}x'^2 + 2\left(\frac{1}{2}x'y'\right) + \frac{1}{2}y'^2\right] = -5$$

$$\left(\frac{1}{2}x'^2 - x'y' + \frac{1}{2}y'^2\right) - 4\left(\frac{1}{2}x'^2 - \frac{1}{2}y'^2\right) + \left(\frac{1}{2}x'^2 + x'y' + \frac{1}{2}y'^2\right) = -5$$

$$\frac{1}{2}x'^2 - x'y' + \frac{1}{2}y'^2 - 2x'^2 + 2y'^2 + \frac{1}{2}x'^2 + x'y' + \frac{1}{2}y'^2 = -5$$

$$-x'^2 + 3y'^2 = -5$$

$$\frac{x'^2}{5} - \frac{3y'^2}{5} = 1$$

The graph of the equation is a hyperbola. This hyperbola is centered at the origin and has vertices on the x'-axis. These vertices are $\pm\sqrt{5} \approx \pm 2.2$. The equations of the asymptotes are

$$y' = \pm\frac{\sqrt{\frac{5}{3}}}{\sqrt{5}}x' \text{ or } y' = \pm\frac{\sqrt{3}}{3}x'.$$

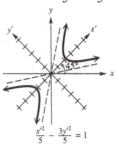

$\frac{x'^2}{5} - \frac{3y'^2}{5} = 1$

17. $7x^2 + 6\sqrt{3}xy + 13y^2 = 64$

Because $A = 7$, $B = 6\sqrt{3}$, and $C = 13$, we have $\cot 2\theta = \dfrac{A - C}{B} = \dfrac{7 - 13}{6\sqrt{3}} = -\dfrac{1}{\sqrt{3}}$. Thus, we have

$\cot 2\theta = -\dfrac{1}{\sqrt{3}} \Rightarrow \cot 2\theta = -\dfrac{\sqrt{3}}{3} \Rightarrow 2\theta = \cot^{-1}\left(-\dfrac{\sqrt{3}}{3}\right) \Rightarrow 2\theta = 120° \Rightarrow \theta = 60°$, so $\sin\theta = \sin 60° = \dfrac{\sqrt{3}}{2}$ and

$\cos\theta = \cos 60° = \dfrac{1}{2}$. This yields the rotation equations $x = \dfrac{1}{2}x' - \dfrac{\sqrt{3}}{2}y'$ and $y = \dfrac{\sqrt{3}}{2}x' + \dfrac{1}{2}y'$.

Substituting these values into the given equation yields

$$7x^2 + 6\sqrt{3}xy + 13y^2 = 64$$

$$7\left(\frac{1}{2}x' - \frac{\sqrt{3}}{2}y'\right)^2 + 6\sqrt{3}\left(\frac{1}{2}x' - \frac{\sqrt{3}}{2}y'\right)\left(\frac{\sqrt{3}}{2}x' + \frac{1}{2}y'\right) + 13\left(\frac{\sqrt{3}}{2}x' + \frac{1}{2}y'\right)^2 = 64$$

$$7\left[\frac{1}{4}x'^2 - 2\left(\frac{\sqrt{3}}{4}x'y'\right) + \frac{3}{4}y'^2\right] + 6\sqrt{3}\left(\frac{\sqrt{3}}{4}x'^2 + \frac{1}{4}x'y' - \frac{3}{4}x'y' + \frac{\sqrt{3}}{4}y'^2\right)$$

$$+ 13\left[\frac{3}{4}x'^2 + 2\left(\frac{\sqrt{3}}{4}x'y'\right) + \frac{1}{4}y'^2\right] = 64$$

$$7\left[\frac{1}{4}x'^2 - \frac{\sqrt{3}}{2}x'y' + \frac{3}{4}y'^2\right] + 6\sqrt{3}\left(\frac{\sqrt{3}}{4}x'^2 - \frac{1}{2}x'y' + \frac{\sqrt{3}}{4}y'^2\right) + 13\left[\frac{3}{4}x'^2 + \frac{\sqrt{3}}{2}x'y' + \frac{1}{4}y'^2\right] = 64$$

$$\frac{7}{4}x'^2 - \frac{7\sqrt{3}}{2}x'y' + \frac{21}{4}y'^2 + \frac{18}{4}x'^2 - \frac{6\sqrt{3}}{2}x'y' - \frac{18}{4}y'^2 + \frac{39}{4}x'^2 + \frac{13\sqrt{3}}{2}x'y' + \frac{13}{4}y'^2 = 64$$

$$\frac{64}{4}x'^2 + \frac{16}{4}y'^2 = 64$$

$$\frac{x'^2}{4} + \frac{y'^2}{16} = 1$$

The graph of the equation is an ellipse. This ellipse is centered at the origin and has x'-intercepts $\pm\sqrt{4} = \pm 2$ and y'-intercepts $\pm\sqrt{16} = \pm 4$.

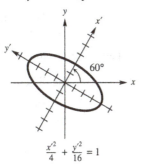

$\frac{x'^2}{4} + \frac{y'^2}{16} = 1$

19. $3x^2 - 2\sqrt{3}xy + y^2 - 2x - 2\sqrt{3}y = 0$

Because $A = 3$, $B = -2\sqrt{3}$, and $C = 1$, we have $\cot 2\theta = \dfrac{A-C}{B} = \dfrac{3-1}{-2\sqrt{3}} = -\dfrac{1}{\sqrt{3}}$. Thus, we have

$\cot 2\theta = -\dfrac{1}{\sqrt{3}} \Rightarrow \cot 2\theta = -\dfrac{\sqrt{3}}{3} \Rightarrow 2\theta = \cot^{-1}\left(-\dfrac{\sqrt{3}}{3}\right) \Rightarrow 2\theta = 120° \Rightarrow \theta = 60°$. So, $\sin\theta = \sin 60° = \dfrac{\sqrt{3}}{2}$

and $\cos\theta = \cos 60° = \dfrac{1}{2}$. This yields the rotation equations $x = \dfrac{1}{2}x' - \dfrac{\sqrt{3}}{2}y'$ and $y = \dfrac{\sqrt{3}}{2}x' + \dfrac{1}{2}y'$.

Substituting these values into the given equation yields

$$3x^2 - 2\sqrt{3}xy + y^2 - 2x - 2\sqrt{3}y = 0$$

$$3\left(\frac{1}{2}x' - \frac{\sqrt{3}}{2}y'\right)^2 - 2\sqrt{3}\left(\frac{1}{2}x' - \frac{\sqrt{3}}{2}y'\right)\left(\frac{\sqrt{3}}{2}x' + \frac{1}{2}y'\right) + \left(\frac{\sqrt{3}}{2}x' + \frac{1}{2}y'\right)^2$$

$$-2\left(\frac{1}{2}x' - \frac{\sqrt{3}}{2}y'\right) - 2\sqrt{3}\left(\frac{\sqrt{3}}{2}x' + \frac{1}{2}y'\right) = 0$$

$$3\left[\frac{1}{4}x'^2 - 2\left(\frac{\sqrt{3}}{4}x'y'\right) + \frac{3}{4}y'^2\right] - 2\sqrt{3}\left(\frac{\sqrt{3}}{4}x'^2 + \frac{1}{4}x'y' - \frac{3}{4}x'y' - \frac{\sqrt{3}}{4}y'^2\right)$$

$$+\left[\frac{3}{4}x'^2 + 2\left(\frac{\sqrt{3}}{4}x'y'\right) + \frac{1}{4}y'^2\right] - 2\left(\frac{1}{2}x' - \frac{\sqrt{3}}{2}y'\right) - 2\sqrt{3}\left(\frac{\sqrt{3}}{2}x' + \frac{1}{2}y'\right) = 0$$

$$3\left(\frac{1}{4}x'^2 - \frac{\sqrt{3}}{2}x'y' + \frac{3}{4}y'^2\right) - 2\sqrt{3}\left(\frac{\sqrt{3}}{4}x'^2 - \frac{1}{2}x'y' - \frac{\sqrt{3}}{4}y'^2\right) + \left(\frac{3}{4}x'^2 + \frac{\sqrt{3}}{2}x'y' + \frac{1}{4}y'^2\right)$$

$$-2\left(\frac{1}{2}x' - \frac{\sqrt{3}}{2}y'\right) - 2\sqrt{3}\left(\frac{\sqrt{3}}{2}x' + \frac{1}{2}y'\right) = 0$$

$$\frac{3}{4}x'^2 - \frac{3\sqrt{3}}{2}x'y' + \frac{9}{4}y'^2 - \frac{6}{4}x'^2 + \sqrt{3}x'y' + \frac{6}{4}y'^2 + \frac{3}{4}x'^2 + \frac{\sqrt{3}}{2}x'y' + \frac{1}{4}y'^2 - x' + \sqrt{3}y' - 3x' - \sqrt{3}y' = 0$$

$$\frac{16}{4}y'^2 - 4x' = 0$$

$$4y'^2 = 4x'$$

$$y'^2 = x'$$

The graph of the equation is a parabola. This parabola has its vertex on the origin and opens to the right, relative to the x'-axis. Its focus is located on the x'-axis. This point is $(x', y') = \left(\dfrac{1}{4}, 0\right)$. The directrix is the line $x' = -\dfrac{1}{4}$.

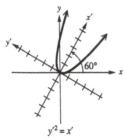

21. $x^2 + 3xy + y^2 - 5\sqrt{2}y - 15 = 0 \Rightarrow x^2 + 3xy + y^2 - 5\sqrt{2}y = 15$

Because $A = 1$, $B = 3$, and $C = 1$, we have $\cot 2\theta = \dfrac{A - C}{B} = \dfrac{1-1}{3} = 0$. Also, because $\cot 2\theta = \dfrac{\cos 2\theta}{\sin 2\theta} = 0$,

we know $\cos 2\theta = 0$. This yields $\sin\theta = \sqrt{\dfrac{1 - \cos 2\theta}{2}} = \sqrt{\dfrac{1 - 0}{2}} = \sqrt{\dfrac{1}{2}} = \dfrac{\sqrt{2}}{2}$ and

$\cos\theta = \sqrt{\dfrac{1 + \cos 2\theta}{2}} = \sqrt{\dfrac{1 + 0}{2}} = \sqrt{\dfrac{1}{2}} = \dfrac{\sqrt{2}}{2}$. Thus, $\theta = 45°$ and the rotation equations are $x = \dfrac{\sqrt{2}}{2}x' - \dfrac{\sqrt{2}}{2}y'$

and $y = \dfrac{\sqrt{2}}{2}x' + \dfrac{\sqrt{2}}{2}y'$. Substituting these values into the given equation yields

$$x^2 + 3xy + y^2 - 5\sqrt{2}y = 15$$

$$\left(\frac{\sqrt{2}}{2}x' - \frac{\sqrt{2}}{2}y'\right)^2 + 3\left(\frac{\sqrt{2}}{2}x' - \frac{\sqrt{2}}{2}y'\right)\left(\frac{\sqrt{2}}{2}x' + \frac{\sqrt{2}}{2}y'\right) + \left(\frac{\sqrt{2}}{2}x' + \frac{\sqrt{2}}{2}y'\right)^2 - 5\sqrt{2}\left(\frac{\sqrt{2}}{2}x' + \frac{\sqrt{2}}{2}y'\right) = 15$$

$$\left[\frac{1}{2}x'^2 - 2\left(\frac{1}{2}x'y'\right) + \frac{1}{2}y'^2\right] + 3\left(\frac{1}{2}x'^2 + \frac{1}{2}x'y' - \frac{1}{2}x'y' - \frac{1}{2}y'^2\right)$$

$$+ \left[\frac{1}{2}x'^2 + 2\left(\frac{1}{2}x'y'\right) + \frac{1}{2}y'^2\right] - 5\sqrt{2}\left(\frac{\sqrt{2}}{2}x' + \frac{\sqrt{2}}{2}y'\right) = 15$$

$$\left[\frac{1}{2}x'^2 - x'y' + \frac{1}{2}y'^2\right] + 3\left(\frac{1}{2}x'^2 - \frac{1}{2}y'^2\right) + \left[\frac{1}{2}x'^2 + x'y' + \frac{1}{2}y'^2\right] - 5\sqrt{2}\left(\frac{\sqrt{2}}{2}x' + \frac{\sqrt{2}}{2}y'\right) = 15$$

$$\frac{1}{2}x'^2 - x'y' + \frac{1}{2}y'^2 + \frac{3}{2}x'^2 - \frac{3}{2}y'^2 + \frac{1}{2}x'^2 + x'y' + \frac{1}{2}y'^2 - 5x' - 5y' = 15$$

$$\frac{5}{2}x'^2 - \frac{1}{2}y'^2 - 5x' - 5y' = 15$$

$$5x'^2 - y'^2 - 10x' - 10y' = 30$$

$$5x'^2 - 10x' - y'^2 - 10y' = 30$$

$5\left(x'^2 - 2x' + 1\right) - \left(y'^2 + 10y' + 25\right) = 30 + 5 - 25 \Rightarrow 5\left(x' - 1\right)^2 - \left(y' + 5\right)^2 = 10 \Rightarrow \dfrac{\left(x' - 1\right)^2}{2} - \dfrac{\left(y' + 5\right)^2}{10} = 1$

The graph of the equation is a hyperbola with its center at $x' = 1$ and $y' = -5$. By translating the axes of the $x'\,y'$ system down 5 units and right 1 unit, we get an $x''y''$-coordinate system, in which the hyperbola is centered at its origin. This hyperbola has vertices on the x'-axis. These vertices are $\pm\sqrt{2} \approx \pm1.4$. The

equations of the asymptotes are $y'' = \pm\dfrac{\sqrt{2}}{\sqrt{10}}x''$ or $y'' = \pm\dfrac{\sqrt{5}}{5}x''$.

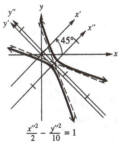

23. $4x^2 + 4xy + y^2 - 24x + 38y - 19 = 0$

Because $A = 4$, $B = 4$, and $C = 1$, we have $\cot 2\theta = \dfrac{A-C}{B} = \dfrac{4-1}{4} = \dfrac{3}{4}$.

The hypotenuse of the triangle is $\sqrt{3^2 + 4^2} = \sqrt{9+16} = \sqrt{25} = 5$. Thus,

$\cos 2\theta = \dfrac{3}{5}$.

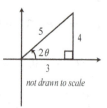

not drawn to scale

Thus, we have $\sin\theta = \sqrt{\dfrac{1-\cos 2\theta}{2}} = \sqrt{\dfrac{1-\frac{3}{5}}{2}} = \sqrt{\dfrac{2}{10}} = \dfrac{\sqrt{5}}{5}$ and $\cos\theta = \sqrt{\dfrac{1+\cos 2\theta}{2}} = \sqrt{\dfrac{1+\frac{3}{5}}{2}} = \sqrt{\dfrac{8}{10}} = \dfrac{2\sqrt{5}}{5}$.

The rotation equations are $x = \dfrac{2\sqrt{5}}{5}x' - \dfrac{\sqrt{5}}{5}y'$ and $y = \dfrac{\sqrt{5}}{5}x' + \dfrac{2\sqrt{5}}{5}y'$. Substituting these values into the given equation yields

$$4x^2 + 4xy + y^2 - 24x + 38y - 19 = 0$$

$$4\left(\dfrac{2\sqrt{5}}{5}x' - \dfrac{\sqrt{5}}{5}y'\right)^2 + 4\left(\dfrac{2\sqrt{5}}{5}x' - \dfrac{\sqrt{5}}{5}y'\right)\left(\dfrac{\sqrt{5}}{5}x' + \dfrac{2\sqrt{5}}{5}y'\right) + \left(\dfrac{\sqrt{5}}{5}x' + \dfrac{2\sqrt{5}}{5}y'\right)^2$$

$$-24\left(\dfrac{2\sqrt{5}}{5}x' - \dfrac{\sqrt{5}}{5}y'\right) + 38\left(\dfrac{\sqrt{5}}{5}x' + \dfrac{2\sqrt{5}}{5}y'\right) - 19 = 0$$

$$4\left[\dfrac{4}{5}x'^2 - 2\left(\dfrac{2}{5}x'y'\right) + \dfrac{1}{5}y'^2\right] + 4\left(\dfrac{2}{5}x'^2 + \dfrac{4}{5}x'y' - \dfrac{1}{5}x'y' - \dfrac{2}{5}y'^2\right) + \left[\dfrac{1}{5}x'^2 + 2\left(\dfrac{2}{5}x'y'\right) + \dfrac{4}{5}y'^2\right]$$

$$-24\left(\dfrac{2\sqrt{5}}{5}x' - \dfrac{\sqrt{5}}{5}y'\right) + 38\left(\dfrac{\sqrt{5}}{5}x' + \dfrac{2\sqrt{5}}{5}y'\right) - 19 = 0$$

$$4\left(\dfrac{4}{5}x'^2 - \dfrac{4}{5}x'y' + \dfrac{1}{5}y'^2\right) + 4\left(\dfrac{2}{5}x'^2 + \dfrac{3}{5}x'y' - \dfrac{2}{5}y'^2\right) + \left(\dfrac{1}{5}x'^2 + \dfrac{4}{5}x'y' + \dfrac{4}{5}y'^2\right)$$

$$-24\left(\dfrac{2\sqrt{5}}{5}x' - \dfrac{\sqrt{5}}{5}y'\right) + 38\left(\dfrac{\sqrt{5}}{5}x' + \dfrac{2\sqrt{5}}{5}y'\right) - 19 = 0$$

$$\dfrac{16}{5}x'^2 - \dfrac{16}{5}x'y' + \dfrac{4}{5}y'^2 + \dfrac{8}{5}x'^2 + \dfrac{12}{5}x'y' - \dfrac{8}{5}y'^2 + \dfrac{1}{5}x'^2 + \dfrac{4}{5}x'y' + \dfrac{4}{5}y'^2$$

$$-\dfrac{48\sqrt{5}}{5}x' + \dfrac{24\sqrt{5}}{5}y' + \dfrac{38\sqrt{5}}{5}x' + \dfrac{76\sqrt{5}}{5}y' - 19 = 0$$

$$5x'^2 - 2\sqrt{5}x' + 20\sqrt{5}y' - 19 = 0 \Rightarrow 5\left(x'^2 - \dfrac{2\sqrt{5}}{5}x' + \dfrac{1}{5}\right) + 20\sqrt{5}y' - 19 = 1$$

$$5\left(x' - \dfrac{\sqrt{5}}{5}\right)^2 = 20 - 20\sqrt{5}y' \Rightarrow \left(x' - \dfrac{\sqrt{5}}{5}\right)^2 = 4 - 4\sqrt{5}y' \Rightarrow \left(x' - \dfrac{\sqrt{5}}{5}\right)^2 = -4\sqrt{5}\left(y' - \dfrac{\sqrt{5}}{5}\right)$$

The graph of the equation is a parabola. The angle of rotation is

$\theta = \sin^{-1}\left(\dfrac{\sqrt{5}}{5}\right) \approx 26.57°$. Its vertex is located at $x' = -\dfrac{\sqrt{5}}{5}$ and

$y' = \dfrac{\sqrt{5}}{5}$ and opens in the downward direction, relative to the

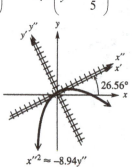

x'-axis. By translating the axes of the $x'y'$ system up $\dfrac{3}{2}$ and left

$\dfrac{\sqrt{3}}{2}$ units, we get an $x''y''$-coordinate system in which the vertex of

the parabola is at the origin.

25. $16x^2 + 24xy + 9y^2 - 130x + 90y = 0$

Because $A = 16$ $B = 24$, and $C = 9$, we have $\cot 2\theta = \dfrac{A - C}{B} = \dfrac{16 - 9}{24} = \dfrac{7}{24}$. The hypotenuse of the triangle

is $\sqrt{7^2 + 24^2} = \sqrt{49 + 576} = \sqrt{625} = 25$. Thus, $\cos 2\theta = \dfrac{7}{25}$.

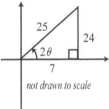

not drawn to scale

Thus, we have $\sin\theta = \sqrt{\dfrac{1 - \cos 2\theta}{2}} = \sqrt{\dfrac{1 - \frac{7}{25}}{2}} = \sqrt{\dfrac{9}{25}} = \dfrac{3}{5}$ and $\cos\theta = \sqrt{\dfrac{1 + \cos 2\theta}{2}} = \sqrt{\dfrac{1 + \frac{7}{25}}{2}} = \sqrt{\dfrac{16}{25}} = \dfrac{4}{5}$.

The rotation equations are $x = \dfrac{4}{5}x' - \dfrac{3}{5}y'$ and $y = \dfrac{3}{5}x' + \dfrac{4}{5}y'$. Substituting these values into the given

equation yields

$$16x^2 + 24xy + 9y^2 - 130x + 90y = 0$$

$$16\left(\dfrac{4}{5}x' - \dfrac{3}{5}y'\right)^2 + 24\left(\dfrac{4}{5}x' - \dfrac{3}{5}y'\right)\left(\dfrac{3}{5}x' + \dfrac{4}{5}y'\right) + 9\left(\dfrac{3}{5}x' + \dfrac{4}{5}y'\right)^2$$

$$-130\left(\dfrac{4}{5}x' - \dfrac{3}{5}y'\right) + 90\left(\dfrac{3}{5}x' + \dfrac{4}{5}y'\right) = 0$$

$$16\left[\dfrac{16}{25}x'^2 - 2\left(\dfrac{12}{25}x'y'\right) + \dfrac{9}{25}y'^2\right] + 24\left(\dfrac{12}{25}x'^2 + \dfrac{16}{25}x'y' - \dfrac{9}{25}x'y' - \dfrac{12}{25}y'^2\right)$$

$$+9\left[\dfrac{9}{25}x'^2 + 2\left(\dfrac{12}{25}x'y'\right) + \dfrac{16}{25}y'^2\right] - 130\left(\dfrac{4}{5}x' - \dfrac{3}{5}y'\right) + 90\left(\dfrac{3}{5}x' + \dfrac{4}{5}y'\right) = 0$$

$$\dfrac{256}{25}x'^2 - \dfrac{384}{25}x'y' + \dfrac{144}{25}y'^2 + \dfrac{288}{25}x'^2 + \dfrac{168}{25}x'y' - \dfrac{288}{25}y'^2 + \dfrac{81}{25}x'^2$$

$$+\dfrac{216}{25}x'y' + \dfrac{144}{25}y'^2 - 104x' + 78y' + 54x' + 72y' = 0$$

$$25x'^2 - 50x' + 150y' = 0$$

$$25\left(x'^2 - 2x'\right) = -150y'$$

$$25\left(x'^2 - 2x' + 1\right) = -150y' + 25$$

$$25\left(x' - 1\right)^2 = -150\left(y' - \dfrac{1}{6}\right)$$

$$\left(x' - 1\right)^2 = -6\left(y' - \dfrac{1}{6}\right)$$

The angle of rotation is $\theta = \sin^{-1}\left(\dfrac{3}{5}\right) \approx 36.87°$. Its vertex is located

at $x' = 1$ and $y' = \dfrac{1}{6}$ and opens in the downward direction, relative to

the x'-axis. By translating the axes of the x' y' system up $\dfrac{1}{6}$ and

right 1 unit, we get an x'' y'' –coordinate system in which the vertex

of the parabola is at the origin.

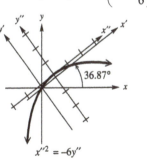

$x''^2 = -6y''$